Neuropsychiatry, Neuropsychology, and Clinical Neuroscience

Emotion, Evolution, Cognition, Language, Memory, Brain Damage, and Abnormal Behavior

Second Edition

Neuropsychiatry, Neuropsychology, and Clinical Neuroscience

Emotion, Evolution, Cognition, Language, Memory, Brain Damage, and Abnormal Behavior

Second Edition

Rhawn Joseph

Director
Brain Research Laboratory
San Jose, California
Palo Alto and Menlo Park Veterans Affairs Medical Centers
Palo Alto, California
Panel of Forensic Experts, Santa Clara County Superior Courts
California

Williams & Wilkins
A WAVERLY COMPANY

BALTIMORE • PHILADELPHIA • LONDON • PARIS • BANGKOK
BUENOS AIRES • HONG KONG • MUNICH • SYDNEY • TOKYO • WROCLAW

Managing Editors: Kathleen Courtney Millet, Leah Kiehne Hayes
Production Coordinator: Linda C. Carlson
Copy Editors: Caroline Helwick, Jack Daniel
Designer: Wilma E. Rosenberger
Illustration Planner: Raymond Lowman
Typesetter: Graphic Sciences Corporation
Manufacturing: R.R. Donnelley and Sons

Copyright © 1996 Williams & Wilkins

351 West Camden Street
Baltimore, Maryland 21201-2436 USA

Rose Tree Corporate Center
1400 North Providence Road
Building II, Suite 5025
Media, Pennsylvania 19063-2043 USA

Printed in the United States of America

First Edition, 1990 © Plenum Press

Library of Congress Cataloging-in-Publication Data

Joseph, Rhawn.
   Neuropsychiatry, neuropsychology, and clinical neuroscience / R.
Joseph. — 2nd ed.
       p.   cm.
   Rev. ed. of: Neuropsychology, neuropsychiatry, and behavioral
neurology. c1990.
   Includes bibliographical references and index.
   ISBN 0-683-04485-0
   1. Neuropsychiatry.   2. Neuropsychology.   3. Neurosciences.
I. Joseph, Rhawn.   Neuropsychology, neuropsychiatry, and behavioral
neurology.   II. Title.
   [DNLM: 1. Neuropsychology.   2. Psychophysiology.   3. Mental
Disorders—physiopathology.   4. Mental Disorders—psychology.   WL
103.5 J83n 1996]
RC341.J67 1996
616.8—dc20
DNLM/DLC
for Library of Congress   95-38650
                                                             CIP

The publishers have made every effort to trace the copyright holders for borrowed material. If they have inadvertently overlooked any, they will be pleased to make the necessary arrangements at the first opportunity.

To purchase additional copies of this book, call our customer service department at **(800) 638-0672** or fax orders to **(800) 447-8438.** For other book services, including chapter reprints and large quantity sales, ask for the Special Sales department.

Canadian customers should call **(800) 268-4178,** or fax **(905) 470-6780.** For all other calls originating outside of the United States, please call **(410) 528-4223** or fax us at **(410) 528-8550.**

Visit *Williams & Wilkins* on the Internet: http://www.wwilkins.com or contact our customer service department at **custserv@wwilkins.com.** Williams & Wilkins customer service representatives are available from 8:30 am to 6:00 pm, EST, Monday through Friday, for telephone access.

1 2 3 4 5 6 7 8 9 10   96 97 98 99 00

Preface

This book is written for clinicians, neuroscientists, practitioners of neuropsychiatry, neuropsychology, and behavioral neurology, and philosophers and neuroanatomists of the mind. It has been my intent throughout to present a multidisciplinary synthesis of facts, theories, and research findings about what is known, debated, established, and theorized regarding the functional neuroanatomy of the brain.

To produce this text I have reviewed and synthesized thousands of articles and research reports from diverse sources in neurology, neurosurgery, neuropsychiatry, neurophysiology, neuroanatomy, evolutionary anthropology, ethology, psychoanalysis, neuropsychology, aphasiology, and cognitive and developmental psychology. I have also drawn from my experience as a neuroscientist (as reflected in various published neuroanatomical, neurophysiological, and neuropsychological experiments conducted on humans, nonhuman primates, and mammals) and from thousands of hours spent directly assessing and examining thousands of brain-injured, neurosurgical, psychotic, criminally insane, and emotionally disturbed individuals. This includes patients with epilepsy, neoplasm, head injuries, strokes, and degenerative disturbances and those who have undergone lobectomies, lobotomies, corpus-callosotomies ("split-brain" neurosurgery), as well as murderers, alcoholics, drug addicts, pedophiles, rapists, and victims of crime and catastrophe. Hence, some of my research findings, theories, and clinical impressions are represented.

The second edition is comprised of 21 chapters that are organized in what I believe to be a logical sequence, although they may be read in any order. Earlier versions of Chapters 3–5, 11–14, and 19–21 can be found in my 1990 text, *Neuropsychology, Neuropsychiatry, and Behavioral Neurology*. These 10 chapters have been thoroughly updated and/or revised and expanded for the second edition.

The 11 new chapters cover issues such as neurological evolution, the role of the amygdala in Parkinson's and Alzheimer's disease and in the functioning of the basal ganglia (generally ignored in all other neurology texts and the scientific literature), sex differences in cognitive and related neurological structural and functional organization, disconnection syndromes, neural networks, neural plasticity, the neurophysiology of repression and religious experience, and the limbic system foundations of language, memory, psychosis, trauma, love, attachment, and social emotional disorders and development. The inclusion of these 11 new chapters, coupled with the emphasis on neuropsychiatric disorders, clinical neuroscience, and functional neuroanatomy, account for the new title, which differs slightly from the original 1990 edition.

Together these 21 chapters present what I believe to be a cohesive and in-depth overview of the functional and neuroanatomical organization of the brain as related to cognition, emotion, memory, evolution, language, psychosis, and sex differences in human experience.

Rhawn Joseph

Acknowledgments

I have been very fortunate to have had the opportunity to conduct a variety of studies in the neurosciences and to examine an exceedingly diverse patient population. I have also prospered from the conversation and critical feedback, support, or assistance from colleagues, friends, reviewers, editors, and numerous critics over the last 25 years. In this regard I am most grateful to Drs. Arlene Kasprisin, Oakley Ray, Roberta Gallagher, Douglas Watt, L. Gillikin, Robert Novelly, David Duvall, Jon Kass, Vivian Casagrande, Jerome Siegel, Peter Como, Jay Gonen, Peter Kaustenbaum, Nancy Forrest, J. Dalton, T. Jobe, L. Miller, Elizabeth Birecree, and J. Iwanaga; Shauna Borden, R.N.; Linda Regan; Plenum Press, Eliot Werner, and series editors Drs. A. E. Puente and C. R. Reynolds; UHS/The Chicago Medical School; the North Chicago V.A. Medical Center, Departments of Psychology, Neurology, and Speech Pathology; the Yale/V.A.M.C. Epilepsy and Seizure Unit and Yale Medical School Neurology and Neurosurgical staff; Vanderbilt University Medical School, Departments of Anatomy and Pharmacology; the University of Delaware and Institute for Neuroscience; the Journal of Clinical Psychology; the United States and California Department of Disability, Rehabilitation, and Social Services; California Home Care; Santa Cruz County Superior Courts; the Palo Alto and Menlo Park Veterans Affairs Medical Centers, Department of Speech Pathology; the Santa Clara County Public Defenders and District Attorney's Offices, and the Santa Clara County Superior Courts. I am also most appreciative of those long contemplative walks in the mountains and the woods and along the ocean and the sea with Sara, "Nietzscha," and Jesse.

Can a lizard comprehend a man?
Can a man comprehend a god?
Who dares speak for god?
Perhaps. . .
. . .even the gods have gods.

. . .and the spiraling universe swirled back and coiled 'round on planets' knees and
shooting stars to ponder its own depths in the temple of human consciousness.
. . .in the mirror of the sea of human consciousness
. . .to peer and reflect upon its own soul. . .
as mirrored in the rising tides of human consciousness. . .

Contents

Preface   v
Acknowledgments   vi
Brain Plates   P1

NEURODYNAMICS

SECTION I   *Evolution*

1   The Evolution of the Brain: *The Neuron, Nerve Net, Limbic System,*
Brainstem, Midbrain, Basal Ganglia, and Telencephalon   **3**
   Oxygen and Single-Cell Diversification   4
   SINGLE AND MULTICELLULAR LIFE: NEURONS AND MICROTUBULES   4
   Sex, DNA, and Microtubules   5
   THE NEURONAL KINGDOM OF LIFE   5
   Pheromones and Chemical Messengers   6
   Sponges and Neuronal Evolution   6
   THE NERVE NET AND THE EVOLUTION OF NEURAL GANGLIA   7
   Hydra   7
   Photosensitivity and Activation   7
   The Nerve Net   8
   Arthropod Versus Vertebrate Patterns of Neural Organization   9
   The Olfactory Limbic System (Rhinencephalon) and Memory   9
   THE BRAINSTEM AND MIDBRAIN   9
   The Dorsal Visual Thalamus   11
   JAWLESS AND CARTILAGINOUS FISH   13
   Sharks   13
   THE FOREBRAIN/TELENCEPHALON   16
   Neocortex   16
   TELEOSTS   17
   LOBED-FINNED (SARCOPTERYGIAN) FISH   17
   LUNG FISH   18
   EUSTHENOPTERONS, ICHTHYOSTEGA, AND AMPHIBIANS   19
   Social-Emotional and Motivational Motor Programs and the Striatum   21
   REPTILES   21
   Forebrain and Midbrain   21
   Cortex and Thalamus   22
   Motor Cortex and Striatum   22
   THE EVOLUTION OF AUDITORY COMMUNICATION   22
   Middle Ear/Midbrain Evolution   23
   The Evolution of Emotional-Vocal Perception and Communication   24
   THE REPTO-MAMMALS (THERAPSIDS)   24
   THE ANTERIOR CINGULATE, MATERNAL BEHAVIOR, AND THE SEPARATION CRY   25
   Limbic Language   26
   The Evolution of Social and Group Attachments   27
   THE NEOCORTEX   28
   The Evolution of the Lobes   28
   Neocortex   29
   Cortical Receiving Areas   30
   THE ASCENDANCY OF MAMMALS   30

**2   Paleo-Neurology and the Evolution of the Human Mind
and Brain:** *The Frontal and Inferior Parietal Lobes, Sex Differences,
Language, Tool Technology, and the Cro-Magnon/Neanderthal Wars*   **31**
PRIMATE EVOLUTION   31
AUSTRALOPITHECUS AND HOMO HABILIS   32
HOMO ERECTUS   34
MULTIREGIONAL EVOLUTION AND REPLACEMENT   35
   "Replacement"   35
   "Multiregional Replacement"   36
"REPLACEMENT" AND THE NEANDERTHAL FRONTAL LOBE   38
   Neanderthal Brain Development   40
   Paleoneurology and the Paleolithic Cultural Transition   42
EVOLUTION OF "MODERN" AND NEANDERTHAL HOMO SAPIENS:
   THE FRONTAL LOBES   43
   Frontal Cranial Evolution   43
THE FRONTAL LOBES   44
THE FRONTAL LOBES AND THE NEUROPSYCHOLOGY OF THE
   MIDDLE-TO-UPPER PALEOLITHIC TRANSITION   45
ARCHAIC, "EARLY MODERN," AND NEANDERTHAL MORTUARY PRACTICES
   AND SYMBOLISM   47
THE INFERIOR TEMPORAL LOBE: 49
   Visual-Shape, Face, and "Cross" Neurons   49
THE INFERIOR PARIETAL LOBULE AND TEMPORAL-SEQUENTIAL MOTOR CONTROL   49
   Handedness   50
   The Parietal Lobe   50
   Apraxia and Temporal Sequencing   52
   Tool Making and the Inferior Parietal Lobe   52
   The Inferior Parietal Lobule and the Middle-Upper Paleolithic Transition   53
   Tool Technology   54
TRANSITIONAL HUMAN FORMS   57
   An Intermediate Stage of Frontal and Inferior Parietal Development   57
SEX ROLES, TOOL MAKING, GATHERING, AND LANGUAGE   58
   Limbic Language and Temporal Sequencing   59
PRIMATES, HOMO SAPIENS, AND APHASIA   59
   Laterality, Language, and Functional Specialization in Primates   60
LANGUAGE, GATHERING, AND HUNTING SILENCE   61
HUNTING, GATHERING, LANGUAGE, AND THE MIDDLE/UPPER
   PALEOLITHIC TRANSITION   62
LANGUAGE AND ENDOCAST CAVEATS   63
CONCLUSION: THE CRO-MAGNON–NEANDERTHAL WARS   65
THE EVOLUTION OF SEX DIFFERENCES IN LANGUAGE AND COGNITION   66
   Gathering, and Why Women Like to Talk and Share Their Feelings   67
   Female Language Superiorities   67
   Overview: Sex Differences, the Parietal Lobe, Corpus Callosum, and Language   68
OVERVIEW: UPPER PALEOLITHIC HOMO SAPIENS SAPIENS AND
   SPOKEN LANGUAGE   69
LINGUISTIC COMPETITION FOR NEOCORTICAL SPACE   70

SECTION II   *The Cerebral Hemispheres*

3   The Right Cerebral Hemisphere: *Emotion, Language, Music,
Visual-Spatial Skills, Confabulation, Body Image, Dreams, and
Social-Emotional Intelligence*   **75**
LEFT HEMISPHERE OVERVIEW   75
   Broca's Aphasia   76

Wernicke's Aphasia   76
THE RIGHT CEREBRAL HEMISPHERE   78
COMPREHENSION AND EXPRESSION OF EMOTIONAL SPEECH   79
Right Hemisphere Emotional-Melodic Language Axis   80
Sex Differences in Emotional Sound Production and Perception   82
CONFABULATION   83
Gap Filling   84
MUSIC AND NONVERBAL ENVIRONMENTAL SOUNDS   84
Environmental Sounds   85
Music and Emotion   86
MUSIC, MATH, AND GEOMETRIC SPACE   87
CONSTRUCTIONAL AND SPATIAL PERCEPTUAL SKILLS   88
Sex Differences in Spatial Ability   89
Visual-Perceptual Abnormalities   89
Drawing and Constructional Deficits   90
INATTENTION AND VISUAL-SPATIAL NEGLECT   92
The Right Frontal Lobe: Arousal and Neglect   93
DISTURBANCES OF BODY IMAGE   94
Pain and Hysteria   96
FACIAL-EMOTIONAL RECOGNITION AND PROSOPAGNOSIA   97
Delusions and Facial Recognition   98
"POSITIVE" EMOTIONS AND THE LEFT HALF OF THE BRAIN   99
DISTURBANCES OF EMOTION AND PERSONALITY OVERVIEW   99
Mania and Emotional Incontinence   100
CONSCIOUSNESS, AWARENESS, MEMORY, AND DREAMING   101
Right Hemisphere Mental Functioning   101
Right Brain Perversity   102
Lateralized Goals and Attitudes   105
LATERALIZED MEMORY FUNCTIONING   105
Unilateral Memory Storage   106
DREAMING AND HEMISPHERIC OSCILLATION   108
Day Dreams, Night Dreams, and Hemispheric Oscillation   108
Hallucinations   109
Imagery   110
Dream Stimulants   110
Dreams and Emotional Trauma   112
LONG, LOST CHILDHOOD MEMORIES   113
Emotion and Right Brain Functioning in Children   113
Functional Commissurotomies and Limited Interhemispheric Transfer   114
Split-Brain Functioning in Children: The Ontology of Emotional Conflict   115
OVERVIEW AND CONCLUDING COMMENTS   116

4   The Left Cerebral Hemisphere: *Language, Aphasia, Apraxia,*
Alexia Agraphia, Psychosis, the Evolution of Reading and Writing,
and the Origin of Thought   **118**
Overview   118
Left Hemisphere Dominance for Language   118
Language and Motor Control   120
Evolution of Handedness, Language, and Left Hemisphere Specialization   121
Limbic Language and the Inferior Parietal Lobule   123
Multimodal Properties   124
THE ORGANIZATION OF LINGUISTIC THOUGHT ASSIMILATION AND
ASSOCIATION WITHIN THE INFERIOR PARIETAL LOBE   124
Stimulus Anchors and the Train of Thought   125
Confabulation and Gap Filling   126
THE ORIGIN OF EGOCENTRIC AND LINGUISTIC THOUGHT   127

Thought	127
The Purpose of Verbal Thought	127
THE DEVELOPMENT OF LANGUAGE AND THOUGHT	128
Three Linguistic Stages	128
Limbic and Brainstem Cognitions	128
Early and Late Babbling and Probable Speech	129
Late Babbling and Temporal-Sequential Speech	130
Egocentric Speech	130
The Internalization of Egocentric Speech	131
Self-Explanation and Interhemispheric Communication	132
DISORDERS OF LANGUAGE AND THOUGHT	133
The Language Axis	133
Conduction Aphasia	133
Broca's Aphasia	134
Anomia or Word Finding Difficulty	134
Expressive Aphasia	135
Depression and Broca's Aphasia	137
Apathetic States	137
Psychosis and Blunted (Negative) Schizophrenia	137
LEFT TEMPORAL LOBE: SOUND PERCEPTION, LANGUAGE, AND APHASIA	138
Pure Word Deafness	138
Wernicke's Aphasia	139
Receptive Aphasia	139
Anosognosia	140
Aphasia and Emotion	140
"Schizophrenia"	141
Global Aphasia	141
Isolation of the Speech Area (or Transcortical Aphasia)	141
LANGUAGE AND TEMPORAL-SEQUENTIAL MOTOR CONTROL	143
NAMING, KNOWING, COUNTING, FINGER RECOGNITION, AND HAND CONTROL	143
AGNOSIA, APRAXIA, ACALCULIA, AND ORIENTATION IN SPACE	144
Finger Agnosia	144
Visual Agnosia	145
Simultanagnosia	146
Right–Left Disorientation	146
Acalculia	147
Apraxia	147
Pantomime Recognition	148
ALEXIA AND AGRAPHIA	148
The Evolution of Reading and Writing	148
The Evolution of Visual Symbols and Written Signs	150
ALEXIA	154
Reading Abnormalities	154
Pure Word Blindness: Alexia Without Agraphia	154
Alexia for Sentences	155
Verbal Alexia	155
Literal/Frontal Alexia	155
Spatial Alexia	155
AGRAPHIA	156
Exner's Writing Area	156
Pure Agraphia	157
Alexic Agraphia	157
Apraxis Agraphia	157
Spatial Agraphia	158

Aphasia and Agraphia   158
Sex Differences   158

SECTION III   *The Limbic System*

5   The Limbic System: *Sex, Emotion, Emotional Intelligence,*
Pheromones, Sexual Differentiation, Attention, Memory, the
Pleasure Principle, and Primary Process   **161**
The Rational and Irrational Mind   161
AFFECTIVE ORIGINS: OLFACTION AND PHEROMONAL COMMUNICATION   162
Olfaction and Memory   163
PHEROMONES   164
Olfactory Blends: A Symphony of Smells   165
Pheromonal-Olfactory Evolution/Involution   165
Sex and Pheromones   166
Somesthesis and Emotion   168
HYPOTHALAMUS   169
Sexual Dimorphism in the Hypothalamus   169
Lateral and Ventromedial Hypothalamic Nuclei   171
Hunger and Thirst   171
Pleasure and Reward   171
Aversion   172
Hypothalamic Damage and Emotional Incontinence: Laughter and Rage   172
Uncontrolled Laughter   173
Hypothalamic Rage   173
Circadian Rhythm Generation and Seasonal Affective Disorder   174
The Hypothalamus-Pituitary-Adrenal Axis   175
Psychic Manifestations of Hypothalamic Activity: The Id   177
The Pleasure Principle   178
AMYGDALA   178
Medial and Lateral Amygdaloid Nuclei   179
Lateral Amygdala   181
Attention   181
Fear, Rage, and Aggression   182
Docility and Amygdaloid Destruction   183
Social-Emotional Agnosia   184
Emotional Language and the Amygdala   185
Emotion and Temporal Lobe Seizures   185
Crying Seizures   187
Sexual Seizures   187
Frontal Lobe Sexual Seizures   187
The Amygdala, the Anterior Commissure, Sexuality, and Emotion   187
Male Aggression is a Source of Male Pleasure   189
The Limbic System and Testosterone   190
Sexual Orientation and Heterosexual Desire   191
Overview: The Amygdala   192
HIPPOCAMPUS   193
Hippocampal Arousal, Attention, and Inhibitory Influences   194
Arousal   194
Attention and Inhibition   197
Learning and Memory: The Hippocampus   197
Hippocampal and Amygdaloid Interactions: Memory   198
THE PRIMARY PROCESS   200
Dreams, Hallucinations, and The Amygdala and Pleasure   200
Amygdala and Hippocampal Interactions During Infancy   200

The Primary Process   203
Primary Imagery   204

6   The Hippocampus, Amygdala, Memory, and Amnesia:
Synaptic Potentiation and Cognitive and Emotional Neural Networks   **206**
AMNESIA   206
    Anesthesia and Unconscious Learning   208
    Unconscious Knowledge: Verbal and Source Amnesia   208
ANTEROGRADE AND RETROGRADE AMNESIA   209
    Post-Traumatic/Anterograde Amnesia   210
    Retrograde Amnesia   210
NEURAL NETWORKS   212
NEURAL CIRCUITS AND LONG-TERM POTENTIATION   213
    Parallel, Sequential, and Isolated Neural Networks   214
    The Amygdala and Hippocampus: Monitoring, Inhibition, Activation   215
SHORT-TERM AND LONG-TERM MEMORY: THE ANTERIOR AND
    POSTERIOR HIPPOCAMPUS   217
    Short-Term Versus Long-Term Memory Loss, Retrieval, and
        Hippocampal Damage   218
    Bilateral Hippocampal Destruction and Amnesia   218
    Learning and Memory in the Absence of Hippocampal Participation   219
THE FRONTAL LOBES, GLOBAL AMNESIA, AND THE DORSAL MEDIAL
    AND ANTERIOR THALAMUS   220
    The Dorsal Medial Thalamus   221
    The Dorsal Medial Thalamus, Frontal Lobes, Korsakoff's Syndrome,
        Search and Retrieval   222
THE HIPPOCAMPUS, DORSAL MEDIAL THALAMUS, AND NEOCORTEX   224
    Hippocampal Arousal and Memory Loss   225
THE AMYGDALA AND EMOTIONAL MEMORY   226
    Fear, Anxiety, Startle, and Traumatic Stress   227
    Fear, Anticipation, and Repression   228
    Fear and Hippocampal Deactivation   229
    Memory Loss and Amygdala Dysfunction   231
    The Amygdala, Hippocampus, and Dorsal Medial Nucleus   232
    Overview: The Amygdala, Hippocampus, and Memory   233

7   Limbic Language, Social-Emotional Intelligence, Development,
and Attachment: *Amygdala, Septal Nuclei, Cingulate Gyrus, Maternal*
Language, Sex Differences, Emotional Deprivation, Contact Comfort,
and Limbic Love   **235**
    Infant Vocalizations and the Innate Languages of the Limbic System   235
    Languages of the Limbic System   237
    Hierarchical Vocal Organization   237
PERIAQUEDUCTAL GRAY AND VOCALIZATION   239
AMYGDALA SOCIAL-EMOTIONAL LIMBIC LANGUAGE AND BEHAVIOR   239
THE CINGULATE GYRUS   242
    The Posterior Cingulate   242
    The Anterior Cingulate Gyrus   243
    The Cingulate Gyrus and Emotional Free Will   244
    The Cingulate and the Evolution of Maternal Language   246
MATERNAL BEHAVIOR AND THE EVOLUTION OF INFANT SEPARATION CRIES   246
    Evolution of Maternal-Infantile Vocalizations   246
    Mother-Infant Vocalization   247
    Female Superiorities in Limbic Language   248
    Maternal Behavior, Attachment, and the Female Limbic System   249

The Male Limbic System and Infant Care   251
EMOTIONAL PROSODY AND HEMISPHERIC LANGUAGE SYSTEMS   251
TEMPORAL SEQUENCING, GRAMMAR, AND THE LEFT HEMISPHERE   253
    Evolution of Broca's Area, Superior Temporal and Inferior Parietal Lobes   253
    Disconnection: Expressive Aphasia Versus Mutism   253
    The Amygdala, Inferior Parietal Lobe, and Neocortical-Limbic Language Yoke   254
    Temporal Sequencing   255
    Overview   256
EMOTIONAL INTELLIGENCE: LIMBIC LOVE AND SOCIAL-EMOTIONAL
    ATTACHMENT   257
    Septal Nuclei   257
    Aversion and Internal Inhibition   258
    Septal Social Functioning   259
    Socialization and Contact Comfort   259
    Attachment and Amygdaloid-Septal Developmental Interactions   260
    Attachment   261
    Contact Comfort   261
    Social-Emotional Deprivation in Humans   262
    Deprivation and Amygdala, Septal, and Cingulate Functioning   264
    Temporary Separation and Progressive Deterioration   265
    Deprivation and Abuse   266
IMPLICATIONS: LOVE, HATE, AND RELATIONSHIPS   266

8 The Limbic System and the Soul: *Evolution and the Neuroanatomy
of Religious Experience*   **268**
PART I: OBSERVATIONS AND SPECULATIONS   268
THE ANTIQUITY OF THE SOUL: MIDDLE PALEOLITHIC SPIRITUALITY   269
    The Amygdala, Temporal Lobe, and Religious Experience   273
    The Twilight of the Gods: Cro-Magnon and Upper Paleolithic Spiritual Evolution   277
THE LIMBIC SYSTEM AND THE SOUL   278
    The Hypothalamus, Sex, and Emotion   278
    The Amygdala and Emotion   279
    The Out-of-Body and "Near Death" Experience   280
    Fear and Out-of-Body Experiences   282
    Limbic Hyperactivation and Astral Projection   283
    Limbic System Hyperactivation, Hallucinations, and Near Death   283
    Out-of-Body, Heavenly, and "Otherworldly" Limbic Experiences   284
    Embraced By the Light: Temporal Lobe Epilepsy, "Death," and Astral Projection   284
THE FRONTAL LOBE, LIMBIC SYSTEM, MURDER, AND RELIGIOUS-SEXUAL
    EXPERIENCE   286
    Modern Religious Murderers   292
    Sex, God, and Religion   292
ETIOLOGICAL AND DIAGNOSTIC SPECULATIONS: SEXUALITY, RELIGIOUS
    EXPERIENCE, AND TEMPORAL LOBE HYPERACTIVATION   294
    Religion, Limbic System Hyperactivation, and Temporal Lobe Seizures   296
    Isolation, Limbic Hyperactivation, and Hallucinations   297
THE BIRTH OF GOD AND LIMBIC HYPERACTIVATION   298
    The Limbic System and the Transmitter to God   300
    Souls, Spirits, and Poltergeists   300
DREAMS AND THE ROYAL ROAD TO THE SPIRIT WORLD   301
    Animal Spirits and Lost Souls   301
    Soulful Dreams   301
    Dreams, Spirits, and Reality   302
    Right Hemisphere, Temporal Lobe Hyperactivation, and Dreaming   304
    Day Dreams and Foreseeing the Future   305

PART II: THOUGHT EXPERIMENTS AND SPECULATIONS   305
    In Search of the God Neuron: The God Within   305
    Evolution or the Guiding Hand of Planned Metamorphosis   308
    The Origins of Life   309
    Complexity and Evolutionary Metamorphosis   310
    Genetic Stability Versus "Random" Mutation and the Evolution of New Species   311
    God and Evolutionary Metamorphosis   312
    Exo-Biological Cerebral Organization: The Neocortex of the Gods?   314
EVOLUTIONARY METAMORPHOSIS   315
IN THE BEGINNING THERE WAS LIFE   316
    Death and the Body   317
PART III: ALPHA AND OMEGA   318

SECTION IV   *The Brainstem and Basal Ganglia*

9   Caudate, Putamen, Globus Pallidus, Amygdala, and Limbic Striatum: *Parkinson's Disease, Alzheimer's Disease, Psychosis, Catatonia, Obsessive-Compulsions, and Disorders of Movement*   **323**
    The Amygdala, Emotion, Memory, Psychosis, and Basal Ganglia   324
THE CORPUS STRIATUM   328
    The Caudate: Mania, Apathy, and Catatonia   330
CATATONIA AND THE FRONTAL-CAUDATE-AMYGDALA NEURAL NETWORK   333
    The Amygdala, Striatum, SMA, and Life-Threatening Fear and Arousal   333
    Implications Regarding Parkinson's Disease   335
PARKINSON'S DISEASE   336
    Striatal Imbalance and Parkinson's Disease   337
    The Supplementary Motor Areas and Parkinson's Disease   338
    The Hippocampus and Supplementary Motor Areas   338
    Dopamine, Acetylcholine, and Striatal Imbalance   339
PSYCHOSIS   339
    Obsessive-Compulsive Disorders   340
THE LIMBIC STRIATUM, MEMORY, AND ALZHEIMER'S DISEASE   342
    The Limbic Striatum, Dopamine, Serotonin, Norepinephrine, and Memory   342
    Memory and Alzheimer's Disease   344
    Widespread Neuronal Death and Alzheimer's Disease   345
NEOCORTICAL, THALAMIC, AND STRIATAL MOVEMENT FEEDBACK LOOPS   345
    The Supplementary Motor Areas, Putamen, and Globus Pallidus   345
    The Putamen and Medial and Lateral Globus Pallidus   346
    The Medial and Lateral Globus Pallidus   347
THE MOTOR THALAMUS   348
THE SUBTHALAMIC NUCLEUS   349
DOPAMINE AND GABA STRIATAL INTERACTIONS AND MOVEMENT DISORDERS   351
HUNTINGTON'S CHOREA   352
CONCLUDING STATEMENT: THE AMYGDALA   353

10   The Brainstem, Midbrain, DA, 5HT, NE, Cranial Nerves, and Cerebellum: *Motor Programming, Arousal, Psychosis, Coma, Sleep and Dreaming, Sleep Disorders, Vocalization, and Emotion*   **354**
    Reflexive Motor Actions and Hierarchical Functional Representation   356
    Sex and the Brainstem   357
    Multisensory-Motor Brainstem Neurons   358
THE MIDBRAIN   358
    The Midbrain, Brainstem and Speech   359
    Cries and Laughter   359

Contents   xv

THE MIDBRAIN SUPERIOR AND INFERIOR COLLICULI 360
The Superior Colliculus 361
The Inferior Colliculus 361
BRAINSTEM AUDITORY PERCEPTION 362
Auditory Transmission from the Cochlea to the Temporal Lobe 363
THE BRAINSTEM RETICULAR FORMATION 363
Coma, Lethargy, and the Brainstem 364
DOPAMINE, NOREPINEPHRINE AND SEROTONIN 365
Dopamine 366
Norepinephrine 366
Serotonin 367
NOREPINEPHRINE AND SEROTONIN INTERACTIONS: FEAR, PAIN, AND STRESS 369
Pain and Stress 369
Stress and Psychosis 369
Depression, Serotonin, and Norepinephrine 370
DREAMING, SLEEPING, AND RHYTHMIC AND OSCILLATING BRAINSTEM
ACTIVITY 371
REM-Off and REM-On Neurons 371
PGO Waves 372
Synchronized, Slow-Wave Sleep 373
Thalamic Contributions 374
Motor Inhibition During Sleep 374
BRAINSTEM SLEEP DISORDERS: NARCOLEPSY AND CATAPLEXY 374
Narcolepsy 374
CRANIAL NERVES OF THE MEDULLA: SHOULDER, HEAD, JAW, TONGUE
MOVEMENT, BREATHING, PHONATING, HEART RATE 376
Cranial Nerve XII: Hypoglossal 376
Cranial Nerve XI: Spinal Accessory 376
Cranial Nerve X: Vagus 376
Cranial Nerve IX: Glossopharyngeal 377
Lateral Medullary Syndrome 377
THE CRANIAL NERVES OF THE PONS 377
Cranial Nerve VIII and the Cochlear System: Tinnitus, Deafness, Dizziness,
Vertigo 377
The Eighth (Vestibular) Nerve 378
ORAL-FACIAL MOVEMENT, OLFACTION AND SENSATION 379
The First Cranial Nerve: Olfactory 379
The Seventh Cranial (Facial) Nerve 379
The Fifth Cranial Nerve: Trigeminal 380
EYE MOVEMENT 380
The Sixth Cranial Nerve: Abducens 380
CRANIAL NERVES OF THE MIDBRAIN 380
The Second Nerve: Optic 380
The Fourth Nerve: Trochlear 382
The Third Nerve: Oculomotor 382
Disturbances of Eye Movement 382
THE CEREBELLUM 383
Evolution 383
Structure and Cytoarchitecture 385
CEREBELLAR MOTOR CONTROL 385
Movement and Learning 385
CEREBELLAR DISTURBANCES OF GAIT AND VOLUNTARY MOVEMENT 386
THE CEREBELLUM AND VISION 387
EMOTION, COGNITION, PSYCHOSIS AND THE CEREBELLUM 387
The Paleo-Cerebellum and Emotion 387

SECTION V   *The Lobes of the Brain*

11   The Frontal Lobes: *Arousal, Attention, Perseveration, Personality,
Catatonia, Memory, Aphasia, Melodic Speech, Confabulation,
Schizophrenia, Movement Disorders, and the Alien Hand*   **393**

FUNCTIONAL OVERVIEW   393
MOTOR REGIONS OF THE FRONTAL LOBES   395
    Cellular Organization   395
CORTICOSPINAL PYRAMIDAL TRACT   396
    Primary Motor Area   396
    Premotor Neocortex   398
    The SMA and the Medial Frontal Lobes   399
CATATONIA AND THE MEDIAL FRONTAL LOBES   400
    Functional Anatomy of the Medial Walls   400
    Gegenhalten and Waxy Flexibility   401
    Catatonia   402
    Will and Apathy   402
    Forced Grasping, Compulsive Utilization, and the Alien Hand   402
    A Cytoarchitectural Continuum of Related Symptoms   404
    Anterior Cingulate   405
POSTERIOR FRONTAL CONVEXITY   406
    Frontal Eye Fields   406
    Visual Scanning Deficits and Neglect   407
    Exner's Writing Area   408
    Broca's Speech Area   409
CONFABULATION AND RIGHT FRONTAL EMOTIONAL AND PROSODIC SPEECH   410
    Emotional and Prosodic Speech   410
    Tangential and Circumlocutory Speech   411
    Confabulation   412
ORBITAL FRONTAL LOBES AND INFERIOR CONVEXITY   413
    Orbital Frontal Lobes   413
    Orbital Mediation of Limbic Arousal   413
    Autonomic Influences   414
    Emotional Unresponsiveness   415
    Emotional Disinhibition   415
    Attention   416
    Perseveration   416
LATERAL FRONTAL NEOCORTICAL REGULATION   419
    Frontal-Thalamic Control of Cortical Activity   420
    Lateral Frontal-Thalamic Arousal and Perceptual Steering   421
THE INFERIOR AND LATERAL FRONTAL LOBES AND ATTENTION   421
    Disinhibition and Response Suppression   421
    Attention   422
THE FRONTAL LOBES, THE DORSAL MEDIAL THALAMUS, AND MEMORY   423
    Memory Search and Retrieval   423
OVERVIEW   426
PERSONALITY AND BEHAVIORAL ALTERATIONS   427
    The Frontal Lobe Personality   427
    Urinary Incontinence   428
    Alterations in Appetite   428
    Disinhibition and Impulsiveness   429
    Uncontrolled Laughter and Mirth   430
    Disinhibited Sexuality   431
    Apathy and Depression   431
    Overview   432

RIGHT AND LEFT FRONTAL LOBES   432
    Neuropsychiatric Disorders   432
    Intellectual and Conceptual Alterations   436
    Right Frontal Dominance for Arousal   438
CONCLUSIONS   439

12   The Parietal Lobes: *The Body Image and Hand in Visual Space,*
Apraxia, Gerstmann's Syndrome, Neglect, Denial, and the Evolution
of Geometry and Math   **441**
PARIETAL TOPOGRAPHY   441
PRIMARY SOMESTHETIC RECEIVING AREAS   441
    Functional Laterality   443
    Somesthetic Agnosias   444
SOMESTHETIC ASSOCIATION AREA   445
    Hand Manipulation Cells   446
    The Body In Space   446
    Tactile Discrimination Deficits and Stereognosis   446
    Pain: Areas 5 and 7 and the Supramarginal Gyrus   447
    Pain and Hysteria   448
AREA 7 AND THE SUPERIOR-POSTERIOR PARIETAL LOBULE   448
    Three-Dimensional Analysis of Body-Spatial Interaction   448
    Visual-Spatial Properties   449
    Visual Attention   450
    Motivational and Grasping Functions   450
RIGHT AND LEFT PARIETAL LOBES: LESIONS AND LATERALITY   451
    Attention and Visual Space   451
INFERIOR PARIETAL LOBULE   452
    Multi-Modal Assimilation Area   452
    Language Capabilities   453
    Agraphia   453
    Lateralized Temporal-Sequential Functions   453
APRAXIA   454
    Sensory Guidance Of Movement   454
    Apraxia   455
    Ideomotor Apraxia   456
    Ideational Apraxia   456
    Left-Sided or Unilateral Apraxia (also called Callosal and Frontal Apraxia)   456
    Pantomime Recognition   457
    Constructional Apraxia   458
GERSTMANN'S SYNDROME: FINGER AGNOSIA, ACALCULIA, AGRAPHIA,
    LEFT-RIGHT CONFUSION   459
    The Knowing Hand   459
    Finger Agnosia   459
THE EVOLUTION OF GEOMETRY AND MATH   461
ACALCULIA   463
ATTENTION AND NEGLECT   464
    Left Hemisphere Neglect   466
    Delusional Denial   467
    Disconnection, Confabulation, and Gap Filling   468
    Delusional Playmates and Egocentric Speech   469
SUMMARY AND OVERVIEW   470

13   The Occipital Lobes: *Vision, Blind Sight, Hallucinations,*
Visual Agnosias   **471**
PRIMARY AND ASSOCIATION VISUAL CORTEX   471

PRECORTICAL VISUAL ANALYSIS   471
    Neocortical Columnar Organization   472
    Simple, Complex, Lower- and Higher-Order Hypercomplex Feature Detectors   472
STRIATE CORTEX: AREA 17   474
    Hallucinations   476
ASSOCIATION AREAS 18 AND 19   476
    Homonymous Hemianopsia and Quadrantanopsia   477
    Hallucinations   478
CORTICAL BLINDNESS   478
    "Blind Sight"   478
    Denial of Blindness   479
VISUAL AGNOSIA   479
PROSOPAGNOSIA   480
SIMULTANAGNOSIA   481
IMPAIRED COLOR RECOGNITION   481
OVERVIEW   482

14   The Temporal Lobes: *Language, Auditory, Visual, Emotional, and Memory Functioning, Form and Face Recognition, Aphasia, Epilepsy, and Psychosis*   **483**

TEMPORAL TOPOGRAPHY   483
    Functional and Evolutionary Considerations: Overview   483
AUDITORY NEOCORTEX   483
    Cortical Organization   484
AUDITORY TRANSMISSION FROM THE COCHLEA TO THE TEMPORAL LOBE   484
FILTERING, FEEDBACK, AND TEMPORAL-SEQUENTIAL REORGANIZATION   486
    Sustained Auditory Activity   486
PHONETICS, CONSONANTS, VOWELS, LANGUAGE   488
    Phonemes   488
    Consonants and Vowels   489
GRAMMAR AND AUDITORY CLOSURE   489
    Universal Grammars   490
LANGUAGE ACQUISITION: FINE TUNING THE AUDITORY SYSTEM   490
LANGUAGE AND REALITY   492
SPATIAL LOCALIZATION, ATTENTION, AND ENVIRONMENTAL SOUNDS   493
    Hallucinations   495
NEOCORTICAL DEAFNESS   495
    Pure Word Deafness   496
    Auditory Agnosia   496
THE AUDITORY ASSOCIATION AREAS   497
    Wernicke's Area   497
    Receptive Aphasia   499
THE MELODIC-INTONATIONAL AXIS   500
    The Amygdala   501
THE MIDDLE AND MEDIAL TEMPORAL LOBES   501
    The "Auditory" Middle Temporal Lobe   502
    Left AMT Language Capabilities   502
    Hallucinations   502
THE "VISUAL" MIDDLE TEMPORAL LOBE   503
    MTL Visual Functioning   503
INFERIOR TEMPORAL LOBE   504
    Visual Capabilities and Form Recognition   504
    Form and Facial Recognition   505
    Visual Attention   505
    Prosopagnosia and Visual Discrimination Deficits   506
MEMORY AND THE INFERIOR TEMPORAL LOBE   507

Memory, the Hippocampus, and the ITL   507
Temporal Lobe Injuries and Memory Loss   508
HALLUCINATIONS AND THE ITL   509
Hallucinations and the Interpretation of Neural "Noise"   509
Dreaming   510
TEMPORAL LOBE (PARTIAL COMPLEX) SEIZURES AND EPILEPSY   511
Auras and Automatisms   512
Running, Laughing, and Crying Seizures   512
Emotional Disturbances   513
TEMPORAL LOBE EPILEPSY AND RELIGIOUS EXPERIENCE   513
Out-Of-Body, Heavenly, and "Otherworldly Experiences"   513
SEIZURES, PERSONALITY, AND PSYCHIATRIC DISTURBANCES   514
Schizophrenia and Temporal Lobe Epilepsy   514
PSYCHOSIS, SPEECH, AND THE TEMPORAL LOBE   515
APHASIA AND PSYCHOSIS   516
Temporal Lobe Seizures, Psychosis, and Aphasia   516
Between-Seizure Psychoses   517

SECTION VI   *The Neuroanatomy of Psychotic and Emotional Disorders*

15   The Neuropsychology of Repression: *Hemispheric Laterality, Positive versus Negative Emotions, Corpus Callosum Immaturity, The Frontal Lobes, Trauma, Contextual Cues, and Recovered Memories*   **521**
VERBAL AMNESIA AND INFANT MEMORY   522
REPRESSION VERSUS AMNESIA FOR INFANT MEMORY   523
Corpus Callosum Immaturity and Interhemispheric Transfer   524
EMOTION, HEMISPHERIC LATERALITY, AND SPECIALIZATION   526
Right Cerebral Specialization and Emotion   527
Sex, Tactual Sensation, Pain, and the Body Image   528
POSITIVE EMOTIONS AND THE LEFT HALF OF THE BRAIN   530
VERBAL VERSUS EMOTIONAL MEMORY STORAGE   532
Verbal Memory and the Left Temporal Lobe   532
Right Temporal Lobe Emotional Memory   532
AMYGDALA HYPERACTIVATION AND EMOTIONAL MEMORY   533
Abuse and Amygdala Hyperactivation   533
Right Hemisphere Activation, Left Hemisphere Deactivation   534
CORPUS CALLOSUM AND UNILATERAL MEMORY STORAGE   534
EARLY ENVIRONMENTAL INFLUENCES AND REPRESSION IN CHILDREN   536
Childhood Amnesia versus Repression   536
MEMORY ISOLATION, NEOCORTICAL MATURATION, AND SYNAPTIC OVERLAY   536
CONTEXTUAL CUES AND EMOTIONAL MEMORY AND BEHAVIOR   538
AMNESIA AND RECOGNITION MEMORY   539
FUNCTIONAL AMNESIA   539
REPRESSION, RECALL, AND RECOGNITION MEMORY   541
DREAM RECOVERY   542
CONTEXT AND VISUAL RECOGNITION   543
Carol's Recall of Her Childhood Molestation   545
Speculations on Differential Memory Storage in Carol's Brain   545
AROUSAL, EMOTIONAL STRESS, AND MEMORY LOSS   546
MEMORY RECOVERY   549
THE FRONTAL LOBES: AROUSAL, ATTENTION, INHIBITION, AND REPRESSION   550
Orbital Frontal Mediation of Arousal   551
Arousal, Cortical Inhibition, and the Lateral Frontal-Thalamic System   551
Repression and the Right Frontal Lobe   553

REPRESSION AND PSYCHIC CONFLICTS   553
    Disconnection and the Failure to Integrate Emotional Experience   554
    Disconnection, Repression, and Differential Emotional Memory Storage 554
FUNCTIONAL OVERLAP, COPING STYLES, AND INDIVIDUAL DIFFERENCES   555
    Sex Differences   556
CAVEATS AND CAUTIONS   556
SUGGESTIBILITY, FALSE MEMORIES, AND CONFABULATION IN
   YOUNG CHILDREN   556
    Competing Memories   557
    Authoritative and Peer Pressure to Change Verbal Reports   558
    Verbal versus Visual Memory Distortion in Young versus Older Children 559
CONFABULATION   559
MEMORY CONFABULATIONS   560
MEMORY GAPS AND SUGGESTIBILITY   561
REPRESSION VERSUS CHILDHOOD AMNESIA   562
CONCLUSION   563

**16   The Neuroanatomy and Neurophysiology of Dissociation,
Repression, and Traumatic Stress:** *The Amygdala,*
Hippocampus, and Disconnection   **564**
THE CONSCIOUS AND UNCONSCIOUS MIND   564
DISCONNECTION SYNDROMES   565
    Blind Sight   566
    Disconnection and Dissociation   566
DISCONNECTION, DISSOCIATION, AND RIGHT CEREBRAL INJURIES   568
EMOTIONAL TRAUMA, DISSOCIATION, AND REPRESSION   570
TRAUMATIC STRESS, REPRESSION, DISSOCIATION   571
FUGUES, LOSS OF PERSONAL IDENTITY, AND AMNESIA   572
THE CEREBRAL HEMISPHERES, TEMPORAL LOBES, AND FUGUE   573
DISSOCIATION AND MULTIPLE PERSONALITY   574
    Multiple Personalities   574
    Fear and Splitting of the Ego: Out-of-Body Experiences   576
    The Hippocampus and Dissociation   577
    Sex Abuse, Dissociation, and Multiple Personalities   577
    Hypnosis and Multiple Personality Disorder   579
    Predisposing Factors: Sex, Age, Previous Trauma   580
SEX, FEAR, AND AMYGDALA HYPERACTIVATION   580
    The Amygdala and Emotional Memory   581
    Repression, Sex Abuse, and Amygdala Hyperactivation   581
    Sex, Satan, Alien Abductions, and Amygdaloid Hyperactivation   582
EMOTION, MEMORY, AROUSAL, AND REPRESSION   583
    Emotional Arousal and Memory Loss   583
    The Amygdala, Intense Fear, Stress, and Hippocampal Amnesia   585
    Opiates, Cortisone, and Hippocampal Amnesia   587
    Abnormal Neural Networks   588
    The Amygdala, Serotonin, Neural Plasticity, and Traumatic Stress Disorder 589
    Serotonin: Pain, Stress, Memory, Flashbacks, Startle, and Neural Circuits 590
NE, ABUSE, AND PSYCHOSIS   591
    Dopamine, Psychosis, and Kindling   592
SUMMARY   593

17   Neuroanatomy of Psychosis: *Depression, Mania, Hysteria,*
Obsessive-Compulsions, Hallucinations, Schizophrenia   **595**
FUNCTIONAL LOCALIZATION AND PSYCHOSIS   595
DEPRESSION   596
FRONTAL LOBE DEPRESSION   596

Sex Differences   597
THE TEMPORAL LOBES AND DEPRESSION   597
THE HYPOTHALAMUS AND DEPRESSION   598
THE SUPRACHIASMATIC NUCLEUS AND DEPRESSION   599
THE HYPOTHALAMUS-PITUITARY-ADRENAL AXIS   600
DEPRESSION, SEROTONIN, AND NOREPINEPHRINE   600
MANIA   601
    The Right Frontal and Right Temporal Lobes   601
HYSTERIA   603
    The Parietal Lobes and Hysteria   603
    Sex Differences, the Amygdala, and Hysteria   603
    Sex Differences and Collective, Socially Sanctioned Hysteria   604
    Mass "Female" Hysteria   605
    Sex Differences, Hysteria, and the Limbic System   608
OBSESSIVE-COMPULSIVE DISORDERS   609
    The Caudate and Frontal Lobes   609
HALLUCINATIONS   610
    Hallucinations and Hemispheric Laterality   611
SCHIZOPHRENIA   612
    The Frontal Lobes and Schizophrenia   612
    Lateral Frontal Psychosis and Neocortical Information Processing   613
THE CAUDATE AND PSYCHOSIS   614
    Catatonic Schizophrenia   614
    Dopamine and the Caudate   614
TEMPORAL LOBES AND PSYCHOSIS   615
THE AMYGDALA, ANTERIOR TEMPORAL LOBE, HALLUCINATIONS,
    AND PSYCHOSIS   616
    The Amygdala, DA, and Psychosis   616
    Aphasic and Psychotic Speech   617
    Between-Seizure Psychoses   617
SOCIAL-EMOTIONAL AGNOSIA   618
CONGENITAL AND EARLY ENVIRONMENTAL CONTRIBUTIONS TO PSYCHOSIS   619
    Abuse and Cerebral Development   621
    Schizophrenia, Early-Onset Frontal Lobe Injury, and Symptom Formation   621
CONCLUSIONS   622

SECTION VII   *Neuroanatomy and Pathophysiology of Head Injury, Stroke, Neoplasm, and Abnormal Development*

18   Neuroanatomy of Normal and Abnormal Cerebral Development: *Neuronal Migration Errors, Congenital Defects, Neural Plasticity, Environmental Influences, Trauma, Abuse, Psychosis*   **625**
ONTOGENY OF CEREBRAL AND CORTICAL DEVELOPMENT   626
OVERVIEW   626
    Ontogeny of Cerebral and Cortical Development   626
CORTICAL FORMATION   627
    Ontogeny and Evolution   627
    The Limbic System, Striatum, Neocortex   629
BRAINSTEM, CEREBELLUM, AND LIMBIC SYSTEM FETAL DEVELOPMENT   630
    Brainstem   630
    Midbrain   630
    Cerebellum   631
    Hypothalamus   631

Amygdala   631
Limbic Striatum   631
Corpus Striatum   631
Septal Nuclei   631
Hippocampus   631
THE TELENCEPHALON   632
ORIGIN AND ONTOGENESIS OF THE NEOCORTEX   632
Neural Migration and Cortical Layers I and VII: The Preplate   633
Layer I and Neocortical Genesis   633
Neural Migration and Cortical Layer VII   634
Neural Migration: Axons and Dendrites   635
Dendrites   635
Neural Networks: Local and Long-Distance Projecting Neurons   636
Fissures, Sulci, and Gyri   636
NEUROTRANSMITTERS AND BRAIN DEVELOPMENT   637
Norepinephrine   638
NE Influences on Neocortical Development   638
NE and 5-HT   639
Vascularization   639
Myelination   639
CONGENITAL MALFORMATIONS OF THE NERVOUS SYSTEM   640
Disturbances of Neuronal Migration: Migration Failure and Heterotopias   641
Neurodevelopmental Psychosis   641
Heterotopias   642
Heterotopias, Trophic Influences, and Neural Grafts   642
Heterotopias: Epilepsy and Psychosis   643
Late-Onset Symptoms   644
DEFECTS OF BRAINSTEM, SPINAL, AND FOREBRAIN DEVELOPMENT   644
Anencephaly   644
Encephalocele   645
Spina Bifida   645
Forebrain Growth Failure   646
Hemorrhage   649
Necrosis   650
Cysts   650
White Matter Tears   651
Hydrocephalus   651
ENVIRONMENTAL AGENTS AND NEURODEVELOPMENTAL ABNORMALITIES   652
Human Immunodeficiency Virus (HIV)   652
Maternal Malnutrition   652
Maternal Drug Abuse   653
Cocaine   653
Heroin   653
Maternal Alcohol Abuse   654
Maternal Trauma   654
Birth Trauma   654
EARLY COGNITIVE AND INFANT BRAIN DEVELOPMENT   654
REGIONAL DIFFERENCES IN NEOCORTICAL GROWTH AND MATURATION   655
The Corpus Callosum   655
The Lobes of the Brain   656
The Right and Left Hemispheres   656
Disconnection, Neocortical Immaturity, and Cross-Modal Cognitions   658
Cross-Modal Processing   659
Neocortical Maturation: Language   660
Neocortical Maturation: Memory   661

ABUSE AND EARLY ENVIRONMENTAL AND POSTNATAL DISTURBANCES OF
    NEUROANATOMICAL DEVELOPMENT, RECOVERY, AND NEURAL PLASTICITY   661
    Abnormal Environmental Experience and Neural Plasticity   664
FALLS, HEAD INJURIES, AND ABUSE   665
    The Battered Infant   665
    The Shaken Baby   667
    Repetitive Trauma and Aberrant Synaptic Recovery   669
GLIA, NE, FUNCTIONAL RECOVERY AND REORGANIZATION   669
    Glia, Amino Acids, Repetitive Injuries, and Limitations on Recovery   670
NOREPINEPHRINE, ABUSE, NEURAL RECOVERY, AND REORGANIZATION   671
    Stress, Emotional Abuse, and Limbic System Reorganization   672
    Neocortical Forebrain Plasticity and Susceptibility to Permanent Injury   673

19 **Cerebral and Cranial Trauma:** *Neuroanatomy and Pathophysiology of Mild, Moderate, and Severe Brain Injury*   **675**
    THE MENINGES   675
    THE SKULL   675
        Skull Injuries   676
    HEMATOMAS   682
        Intradural Hematomas   683
    HERNIATION   685
    CONTUSIONS   686
        Coup and Contre Coup Contusions   687
    ROTATION AND SHEARING FORCES   687
    DIFFUSE AXONAL INJURY   688
    HYPOXIA AND BLOOD FLOW   689
        Respiratory Distress   689
    COMA AND CONSCIOUSNESS   690
        Reticular Damage and Immediate Cessation of Consciousness   690
        Levels of Consciousness   691
    COMA   691
        The Glasgow Coma Scale   692
        Mortality   692
        Vegetative States   693
    MEMORY LOSS   694
        Retrograde Amnesia   695
        Post-Traumatic/Anterograde Amnesia   695
        Caveats: PTA and Long-Term Disorders   696
    PERSONALITY AND EMOTIONAL ALTERATIONS   696
    RECOVERY   698
        Age   699
    EPILEPSY   699
    CONCUSSION AND MILD HEAD INJURIES   700
        Concussion   700
        Mild and Classic Concussion   701
        Mild Head Injury   701
        Neuropsychological Deficits   702
    POSTCONCUSSION SYNDROME   703
        Emotional Sequela   703
    PREMORBID CHARACTERISTICS   706

20 **Stroke and Cerebral-Vascular Disease**   **707**
    FUNCTIONAL ANATOMY OF THE HEART AND ARTERIAL DISTRIBUTION   707
        Arterial Cerebral Pathways   708
    THE BLOOD SUPPLY OF THE BRAIN   709
        The Carotid System   709

Vertebral System   710
CEREBRAL-VASCULAR DISEASE   711
Atherosclerosis   713
CEREBRAL ISCHEMIA   716
THROMBOSIS   717
EMBOLISM   718
Cardiac Embolic Origins   718
The Carotid and Left Middle Cerebral Arteries   718
Focal Influences   719
Vertebral Basilar Artery   719
Onset   719
Embolic Hemorrhage   719
TRANSIENT ISCHEMIC ATTACKS   719
TIA and Stroke   720
LACUNAR STROKES   721
Multi-Infarct Dementia   721
HEART DISEASE, MYOCARDIAL INFARCTION, AND CARDIAC SURGERY   722
Ischemia and Heart Disease   722
Myocardial Infarction   722
Global Ischemia   722
HEMORRHAGE   724
Subarachnoid Hemorrhage   725
Cerebral/Intracerebral Hemorrhage   725
HYPERTENSION   726
ANEURYSMS, AVMS, TUMORS, AMYLOID ANGIOPATHY   727
Ischemia and Hemorrhagic Infarcts   727
Aneurysm   727
Cerebral Amyloid Angiopathy   728
Arteriovenous Malformations   728
Drug-Induced Hemorrhages   729
LOCALIZED HEMORRHAGIC SYMPTOMS   729
Hemorrhages and Strokes: Arterial syndromes   730
RECOVERY AND MORTALITY   734
SUMMARY   735

21 **Cerebral Neoplasms**   **736**
TUMOR DEVELOPMENT: ONCOGENES AND DEFECTIVE DNA EXCISION REPAIR   736
TELOMERASE, TELOMERES, "IMMORTAL CELLS," AND TUMOR GROWTH   737
THE ESTABLISHMENT AND GROWTH OF A TUMOR   737
Vascularization and Necrosis   737
Invasion and Dissemination   738
Metastasis   738
Malignancy   738
Tumor Grading   739
Extrinsic and Intrinsic Tumors   739
NEOPLASMS AND SYMPTOMS ASSOCIATED WITH TUMOR FORMATION   740
Fast- Versus Slow-Growing Tumors   740
GENERALIZED NEOPLASTIC SYMPTOMS   740
FOCAL SYNDROMES AND NEOPLASMS   741
Tumors of the Ventricles   743
Tumors of the Pineal Gland: Pinealomas   744
Tumors of the Meninges: Meningioma   745
Acoustic Neuromas and Schwannomas   745
ASTROCYTOMA, GLIOMA, GLIOBLASTOMA MULTIFORME,
AND OLIGODENDROGLIOMA   746
Abnormal Glia Development   746

Astrocytomas   746
Gliomas   747
Glioblastoma Multiforme   747
Oligodendroglioma   748
LYMPHOMAS, SARCOMAS, NEUROBLASTOMAS, AND CYSTS   748
Lymphomas   748
Sarcomas   749
Neuroblastomas   749
Cysts   749
UNILATERAL TUMORS AND BILATERAL DYSFUNCTION   749
HERNIATION   749
PROGNOSIS   751

References   *753*
Index   *835*

BRAIN PLATES

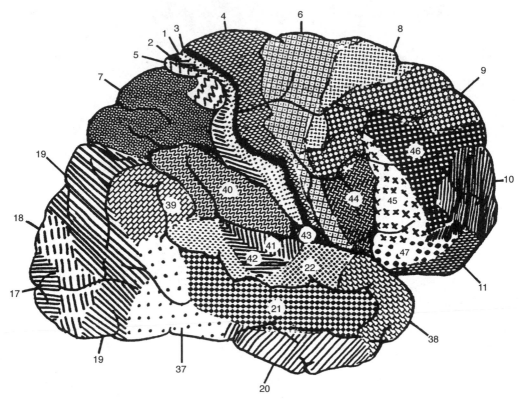

Plate 1   Lateral view of right hemisphere. Numbers refer to Brodmann's areas.

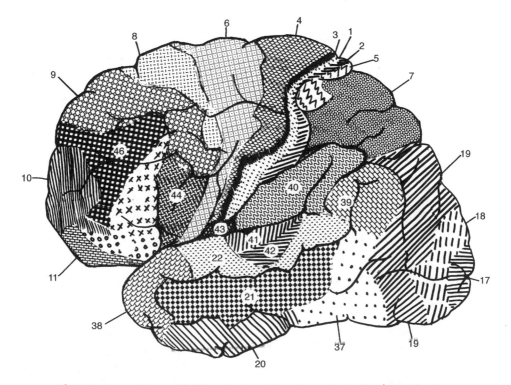

Plate 2   Lateral view of left hemisphere. Numbers refer to Brodmann's areas.

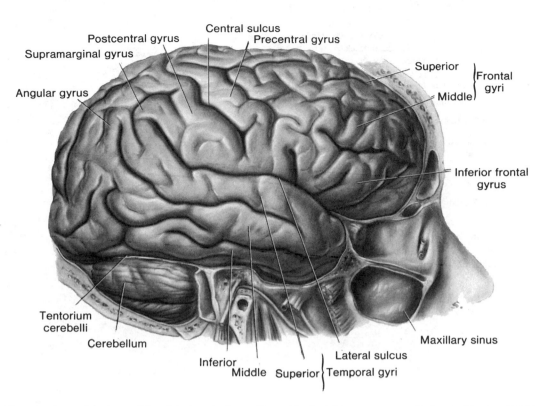

Plate 3   Lateral surface of the right hemisphere. (From Mettler's Neuroanatomy. St. Louis: Mosby, 1948.)

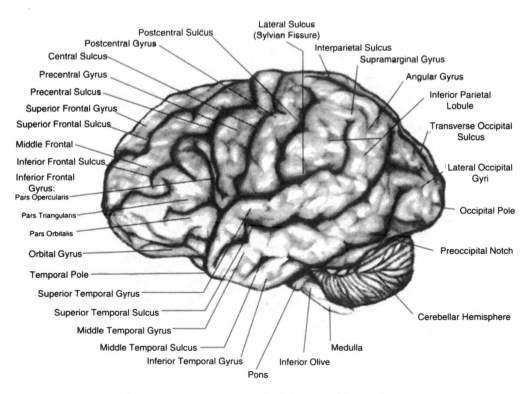

Plate 4   Lateral surface of the left cerebral hemisphere.

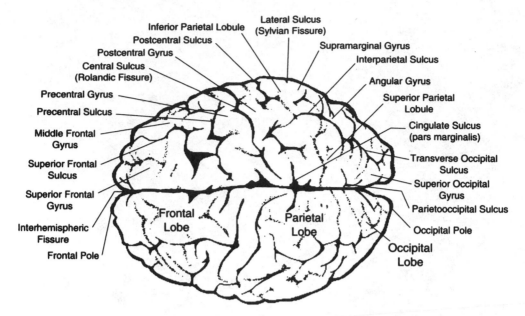

Plate 5   Superior surface of the brain.

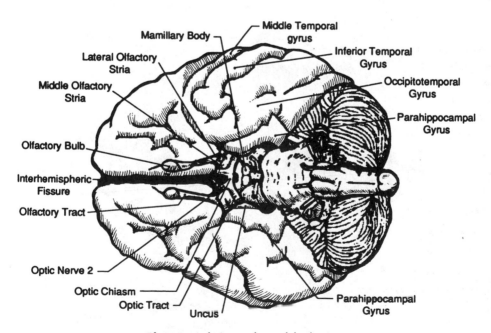

Plate 6   Inferior surface of the brain.

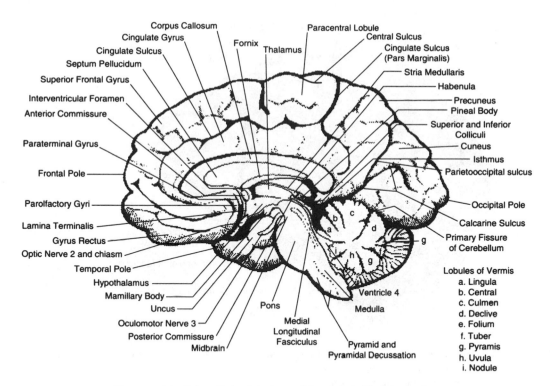

Corpus Callosum
Cingulate Gyrus
Cingulate Sulcus
Septum Pellucidum
Superior Frontal Gyrus
Interventricular Foramen
Anterior Commissure
Paraterminal Gyrus
Frontal Pole
Parolfactory Gyri
Lamina Terminalis
Gyrus Rectus
Optic Nerve 2 and chiasm
Temporal Pole
Hypothalamus
Mamillary Body
Uncus
Oculomotor Nerve 3
Posterior Commissure
Midbrain

Fornix   Thalamus

Paracentral Lobule
Central Sulcus
Cingulate Sulcus
(Pars Marginalis)
Stria Medullaris
Habenula
Precuneus
Pineal Body
Superior and Inferior
Colliculi
Cuneus
Isthmus
Parietooccipital sulcus
Occipital Pole
Calcarine Sulcus
Primary Fissure
of Cerebellum

Pons   Medulla
Medial
Longitudinal
Fasciculus
Pyramid and
Pyramidal Decussation

Ventricle 4

Lobules of Vermis
a. Lingula
b. Central
c. Culmen
d. Declive
e. Folium
f. Tuber
g. Pyramis
h. Uvula
i. Nodule

Plate 7   Medial (split-brain) view of the right half of the brain.

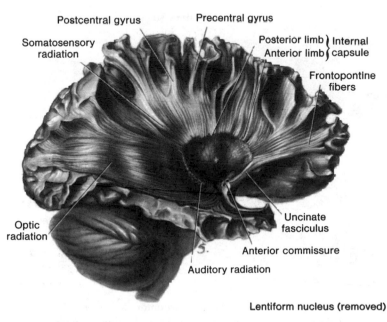

Postcentral gyrus   Precentral gyrus

Somatosensory
radiation

Posterior limb } Internal
Anterior limb { capsule

Frontopontine
fibers

Optic
radiation

Uncinate
fasciculus

Anterior commissure

Auditory radiation

Lentiform nucleus (removed)

Plate 8   Dissection of the lateral aspect of the right cerebral hemisphere depicting the corona ra- diata and optic radiation. (From Mettler's Neuroanatomy. St. Louis: Mosby, 1948.)

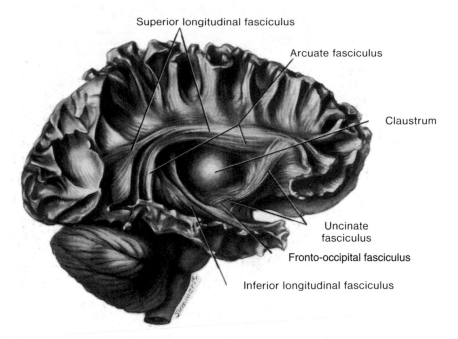

Superior longitudinal fasciculus

Arcuate fasciculus

Claustrum

Uncinate fasciculus

Fronto-occipital fasciculus

Inferior longitudinal fasciculus

Plate 9   Dissection of the lateral surface of the right hemisphere revealing the long association fibers that interconnect different cortical regions. (From Mettler's Neuroanatomy. St. Louis: Mosby, 1948.)

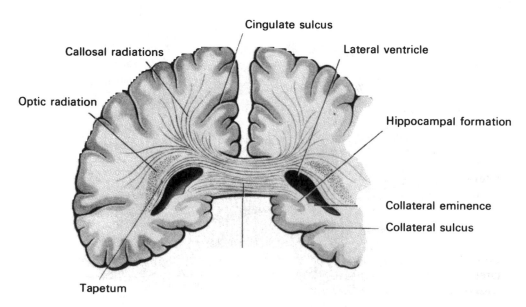

Cingulate sulcus

Callosal radiations

Lateral ventricle

Optic radiation

Hippocampal formation

Collateral eminence

Collateral sulcus

Tapetum

Plate 10   A frontal section at the level of the splenium of the corpus callosum. Tapetum refers to callosal fibers lateral to the ventricle. (From Carpenter M. Core Text of Human Neuroanatomy. Baltimore: Williams & Wilkins, 1991.)

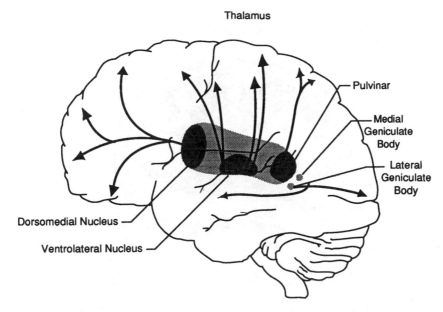

Plate 11   Thalamus and thalamic-neocortical projections.

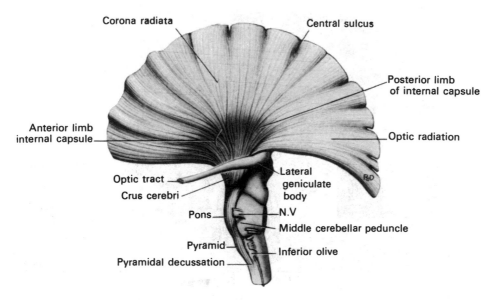

Plate 12   A dissection demonstrating the continuity and relationship of the internal capsule, corona radiata, crus cerebri, and medullary pyramids. (From Carpenter M. Core Text of Human Neuroanatomy. Baltimore: Williams & Wilkins, 1991.)

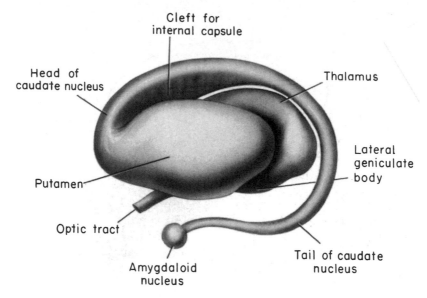

Plate 13   Striatum thalamus and amygdala of the left hemisphere. (From Carpenter M. Core Text of Human Neuroanatomy. Baltimore: Williams & Wilkins, 1991.)

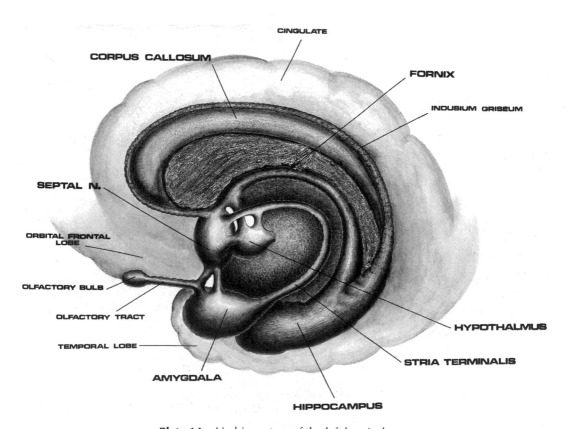

Plate 14   Limbic system of the left hemisphere.

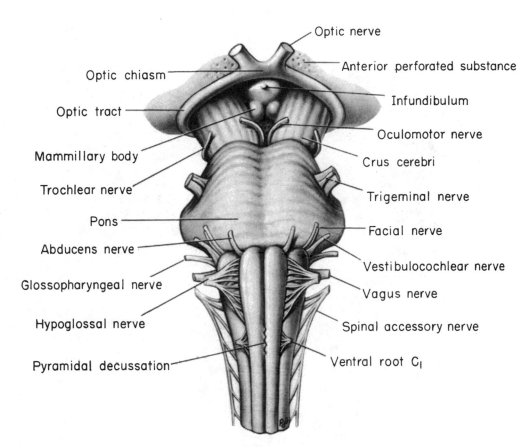

Optic nerve

Anterior perforated substance

Optic chiasm

Infundibulum

Optic tract

Oculomotor nerve

Mammillary body

Crus cerebri

Trochlear nerve

Trigeminal nerve

Pons

Facial nerve

Abducens nerve

Vestibulocochlear nerve

Glossopharyngeal nerve

Vagus nerve

Hypoglossal nerve

Spinal accessory nerve

Pyramidal decussation

Ventral root C_I

Plate 15   Anterior ventral view of the brainstem and midbrain. (From Carpenter M. Core Text of Human Neuroanatomy. Baltimore: Williams & Wilkins, 1991.)

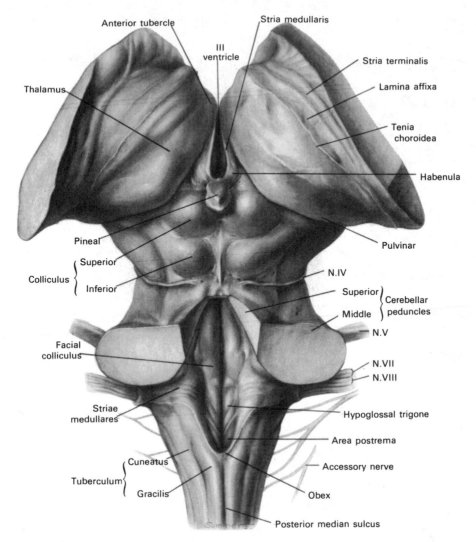

Plate 16   Dorsal view of the brainstem with the cerebellum removed. (From Carpenter M. Core Text of Human Neuroanatomy. Baltimore: Williams & Wilkins, 1991.)

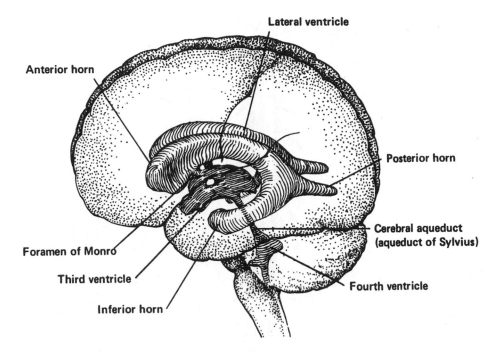

Plate 17   Ventricular system of the brain. (From Waxman and DeGroot. Correlative Neuroanatomy. East Norwalk: Appleton & Lange, 1994.)

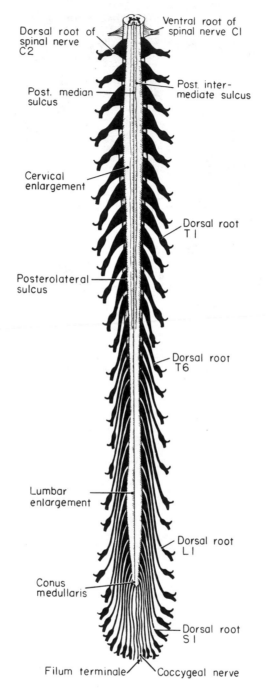

Dorsal root of
spinal nerve
C2

Ventral root of
spinal nerve CI

Post. median
sulcus

Post. inter-
mediate sulcus

Cervical
enlargement

Dorsal root
T I

Posterolateral
sulcus

Dorsal root
T6

Lumbar
enlargement

Dorsal root
L I

Conus
medullaris

Dorsal root
S I

Filum terminale

Coccygeal nerve

Plate 18   Posterior view of the spinal cord. Letters and numbers indicate corresponding spinal nerves. (From Carpenter M. Core Text of Human Neuroanatomy. Baltimore: Williams & Wilkins, 1991.)

SECTION I

Evolution

The Evolution of the Brain

The Neuron, Nerve Net, Limbic System, Brainstem, Midbrain, Basal Ganglia, and Telencephalon

It has been estimated that half a billion years after the Earth was formed (some 4.6 billion years ago), single-celled organisms began to proliferate within the crust of the Earth and throughout the nutrient-rich primeval ocean and salty seas. Presumably, these single-celled creatures possessed a single strand of DNA and practiced a crude form of photosynthesis such that ultraviolet and other wave lengths of light were absorbed, thus enabling them to survive (Bitter, 1991; Lovelock, 1979; Margulis, 1970; Schopf, 1978, 1992; Swimme and Berry, 1992).

Due to the harsh and poisonous nature of Earth's early environment, it is believed that these creatures were protected by a semipermeable membrane that allowed them to maintain their own internal atmosphere of fluids and essential elements. This porous membrane also enabled these organisms to discharge waste products or absorb chemical nutrients arising in the external environment as well as to secrete or detect material released by other single-celled creatures (Bitter, 1991; Hyman, 1942; Margulis, 1970; Schopf, 1978, 1993; Shore, 1994; Strauss and Wilson, 1990). Indeed, at a very rudimentary level these cells were able, presumably, to sense, acquire, store, activate, and exchange information in a meaningful fashion.

Possibly, these exceedingly ancient sensory and communicative capacities enabled single-celled creatures to engage in cooperative efforts. For example, an examination of the fossil records reveals huge colonies of these ancient single-celled organisms stacked together in towering sheets, referred to as stromatolites (Schopf, 1978, 1992). These creatures came to live together in an interdependent fashion over 3.5 billion years ago.

This initial means of single-celled cooperative communication presumably took place via chemical secretions, electrical discharges, and alterations in membrane polarity and electric currents (Bishop, 1956; Clayton, 1932). For example, it is well known that the surface membrane of a living cell is maintained by electrical forces and the interactions between macromolecules. Via alterations in these electrical-magnetic forces, molecular cellular structures are able to approach, position themselves, interact and even exchange material, and then separate. For example, positively charged areas on one molecule are attracted to the negatively charged surface of a different molecule. Moreover, alterations in membrane polarization result in alterations in excitability, which are translated into different membrane currents and discharge patterns. Complex and variable forms of information can thus be conveyed.

In that the most primitive means of true nerve cell communication also occurs via electrical and electrochemical interactions, whereas true neuronal actions and neuronal circuit activity are also determined by membrane currents (Bishop, 1956; Dowling 1992), it is possible that these first single-celled communicative and cooperative efforts were the harbingers of advances in communication yet to come: i.e., the evolution of the neuron. However, the first neurons did not arise until after the planet became oxygenated. When the single-celled prokaryotes ruled the planet there was little or no atmospheric oxygen and no sex (Bitter, 1991; Lovelock, 1979; Margulis, 1970; Schopf, 1978, 1992; Swimme and Berry, 1992). These cells merely divided and produced two identical copies of themselves and their DNA.

Oxygen and Single-Cell Diversification

Initially, and for almost a billion years, the Earth's atmosphere was composed of methane, ammonia, nitrogen, and carbon dioxide, with little or no oxygen present. Because the planet was also continually bombarded by life-neutralizing ultraviolet radiation, most forms of life, as we know it, remained at best only a possibility. Life forms dependent on oxygen had to wait almost 3 billion years for the atmosphere to alter enough to sustain them.

It is believed that the first single-celled life forms produced oxygen as a waste product. With the proliferation of life beneath the sea and over the ensuing billion years, the atmosphere was flooded by increasing amounts of oxygen (Lovelock, 1978; Shore, 1994; Swimme and Berry, 1992). This was a fortuitous event, at least in terms of evolution, for when oxygen is exposed to the sun's radiation, the resulting photochemical reactions create a tri-atomic form of oxygen called ozone. Ozone acts as a remarkable filter through which the rays of the sun must pass.

Over the ensuing hundreds of millions of years, the layer of ozone increased such that the life-neutralizing effects of ultraviolet solar radiation were filtered out. This allowed life to emerge from the sea and to diversify, and enabled oxygen-dependent life forms to take root. Those that could breath oxygen and produce carbon dioxide as a waste product and those that were dependent on carbon dioxide but which produced oxygen as a waste product prospered, evolved, diversified, and became increasingly interdependent. Carbon dioxide-breathing plants and oxygen-dependent animals soon came to rule and overrun the planet (Lovelock, 1979; Schopf, 1978, 1992; Swimme and Berry, 1992).

SINGLE AND MULTICELLULAR LIFE: NEURONS AND MICROTUBULES

Single-celled creatures, such as amoeba and paramecium, are able to sense and learn from their environment and are capable of navigating around obstacles, searching out food, and avoiding danger (Bitter, 1991; Clayton, 1979; McConnell et al., 1959, 1961). Hence, these single-celled organisms can behave like a very rudimentary motor-sensory neuron capable of movement; during embryonic development, neurons are also capable of movement via their pre-dendritic leading fiber process.

There are also other single cell and neuronal similarities. Like neurons, many single-celled creatures are surrounded by a protective cellular membrane, the cytoskeleton (Bitter, 1991). They also contain (in addition to other material) microtubules, which are exceedingly important in cellular regulation and information transfer and reception.

Microtubules are essentially hollow cylindrical tubes, organized in tiny bundles, that may run the length of the cell. However, each microtubule consists of two parts that are able to position themselves and to shift positions, a consequence of changes in electric polarization. Microtubules are, therefore, able to approach, position themselves, interact, exchange material, and then separate (Strauss and Wilson, 1990).

Within neurons, it is the microtubules that are responsible for both anterograde and retrograde transport of various molecules, including neurotransmitter substances that come to be released and taken up at the synaptic cleft (Dowling, 1992). Once released, the neurotransmitter substance is absorbed by a dendrite, and, via dendritic microtubules, is transported toward the cell body, much like the root of a plant absorbs nutrients. More specifically, it is the activity of two different proteins, dynein and kinesin, which essentially act to transport materials along the microtubules. Thus, it is via the aid of the microtubules, protein activity, and transport, coupled with alterations in resting potential, that nerve cells are able to communicate (Dowling, 1992).

Likewise, it is via microtubules that single-celled and multicellular creatures (who lack neurons) are also able to communicate and analyze environmental stim-uli (Bitter, 1991; Strauss and Wilson, 1990). However, microtubules also play a role in cell division, and may have contributed not only to the development of the neuron but to the evolution of sex and multicellu-lar creatures.

Sex, DNA, and Microtubules

It was probably not until about 1 billion years ago that the planet became suffi-ciently oxygenated (the cumulative result of 3 billion years of single-celled photosyn-thesis and oxygen waste production) that oxygen-breathing multicellular creatures were finally provided an opportunity to flourish and prosper and to indulge in sex-ual relations (Lovelock, 1979). Indeed, it was about a billion years ago that sex first became a widespread form of repro-duction, and this was first practiced by oxygen-breathing multicellular creatures who, in turn, became capable of indulging in more complex modes of communication.

Be it worm or human, the tissues of the body, perhaps like our ancestral multicellu-lar cousins, are derived from a single cell: i.e., the sexually fertilized ovum of the female. Essentially, the female ovum comes to be invaded by a spermatozoon. The sper-matozoon, in turn, immediately loses its tail, whereas its head rapidly increases in size and then disappears as it absorbs ovum protoplasm. As a consequence of this inva-sion, the single-celled ovum quickly under-goes meiosis and divides; this process is repeated by all its daughter cells, thus creating a complex multicellular organism. However, this process is also dependent on DNA and microtubular activity.

For example, the centrosome of the cell is connected to separate strands of DNA, via the microtubules. During the first stage of sexual cell division (i.e., meiosis) these DNA strands separate, and chunks of chromosome become detached and are swapped with their counterparts via the assistance of these microtubules. They are then paired together so as to make new chromosomes that are arranged in combi-nations different from those of the parent organisms. Hence, a tremendous variety of chromosomes and chromosomal combi-nations can be produced. In consequence, evolutionary metamorphosis is made pos-sible, due in part to microtubular activity.

The single-celled creatures that first pop-ulated this planet (bacteria and blue-green algae) were presumably without micro-tubules. However, a billion or so years ago, single-celled creatures may have been in-vaded by microtubules, forming a relation-ship that was maintained symbiotically. Together, microtubules and single-celled creatures formed a single organism that, in turn, was susceptible to further invasions (such as from sperm). Sperm enabled the single cell to divide sexually, thereby, over time, creating a variety of multicellular creatures. However, that is but one theory among many as to the origin of multicellu-lar (as well as single cellular) life (see Bitter, 1991; Lovelock, 1979; Irvine et al., 1980; Joseph, 1993; Reid et al., 1976).

Given the role of microtubules in cellular communication and active transport within neurons, over the course of evolution, cellu-lar development and communication was given tremendous developmental impetus. Indeed, in addition to the development of oxygen and sexual reproduction, it may well have been advances in communication secondary to microtubular symbiosis that gave rise, not only to multicellular crea-tures, but to the eventual evolution of the neuron.

THE NEURONAL KINGDOM OF LIFE

Being chemical creatures whose cellular surface also functioned at the behest of chemical and perhaps light-bearing mes-sengers, even the earliest primitive life forms maintained a rudimentary sensitiv-ity and capacity to respond to external chemical agents (via dendritic-like fibers), as well as secrete them via specialized cells that were externally located—the harbingers to the evolution of axons, plant leaves, and branches. Eventually, the ability to detect and search out food or mates based on the presence of chemicals excreted by other life forms came into being as motor cells and specialized receptors and appendages became especially adapted for this purpose.

Pheromones and Chemical Messengers

Chemically and light sensitive, albeit basically blind and deaf, later-appearing primitive and ancient creatures (such as sponges and primitive worms) were able to richly sample their world and detect and analyze other forms of life, even when separated by enormous distances (Applewhite, 1966; Behrens, 1961; Burnett and Diehl, 1964, Emson, 1966; McConnell, et al., 1959, 1961; Jacobson, 1963; Orton, 1924), via the detection of pheromonal cues. This is a capability that humans retain (see Chapter 5).

Specifically, the ability to detect externally originating chemicals (or pheromones) was first made possible by the evolution of specialized receptors that were located on various parts of the organism's body, such as along the legs or abdomen (Burnett and Diehl, 1964). Not only single-cellular and simple multicellular creatures but modern mammals, primates, and even humans, secrete as well as receive chemical and pheromonal messages via olfactory/pheromonal neurons located on the body surface; in humans, these centers have become concentrated within the nose and under the arms and genitals.

It is these externally located, nasal neurons which in turn gave rise to the olfactory lobe that would become part of the forebrain and limbic system—a series of nuclei that initially were concerned almost exclusively with feeding, fornicating, fighting, or fleeing (Joseph, 1990a, 1992a, 1994; MacLean, 1990). Indeed, this pattern of olfactory "neural" organization has not only been retained in all subsequent animal species, but it provided the foundation for what would become the telencephalon and the neocortex (e.g., Allman, 1990; Haberly, 1990).

Sponges and Neuronal Evolution

Although intelligent and capable of communication and interaction, the vast majority of multicellular creatures lack neurons. Only members of the Animal Kingdom of Life possess neurons. However, those animals that are believed to have evolved first, some 700 or so million years ago, such as sponges (which are a step below coelenterates: e.g., jelly fish, sea anemones, and Hydrozoa) are without true neurons (Ariens Kappers, 1929; Bishop, 1956; Burnett and Diehl, 1964; Emson, 1966; Lentz, 1968; Papez, 1967; Jacobson, 1963).

For example, the outer sensory surface of the sponge's body is very poorly differentiated and is without specialized receptors, and no nerves or neurons are found. Rather, they have only a very primitive organization of nervous-like tissue. This includes generalized sensory cells, the bulk of which are concentrated within and around their external orifices and pore sphincters, through which sea water freely circulates. Sponges also have very generalized motor cells, which is why young sponges are capable of ameboid movements. These sensory and motor cells presumably enable adult sponges to (very slowly) react to stimulation.

Sponges are capable of a very slow, protoplasmic form of information transmission, which is made possible via microtubular neuroid activity. In addition to microtubules, sponges contain a number of recognized neurotransmitters, such as serotonin, norepinephrine, epinephrine, and acetylcholinesterase (Lentz, 1968). These chemical transmitters are found in high concentrations within the human brain and are involved in memory, emotion, and movement. Hence, the nervous system of the sponge apparently employs these neurotransmitters and utilizes a very primitive type of neuroid transmission, from which true nervous conduction may have evolved.

THE NERVE NET AND THE EVOLUTION OF NEURAL GANGLIA

It was probably about 650 to 700 million years ago that the first true neurons appeared (Joseph, 1993). These first chemical, photosensitive, and motor neural cells most likely were not well developed, their axons and dendrites being rudimentary and plantlike. Presumably, these neural cells (or at least their sensors) were located outside the body and were especially responsive to light and to chemical and pheromonal messages. Nevertheless, it is likely that these first primitive neurons, via microtubular activity, absorbed, secreted, or discharged electrical and chemical substances that acted on other cells.

Later, neurons evolved single long, thin, axonal transmission fibers through which these same electrochemicals could be selectively secreted to a second neuron, via microtubular and protein transport. The dendrite of this second neuron would, in turn, absorb these chemicals molecule-by-molecule via a selective receptor surface located along its terminal junctions.

However, over the course of evolutionary metamorphosis and the ensuing eons of time, these original externally located sensory and motor neurons began to migrate inward and to eventually collect together within the body so as to form collections of nuclei: i.e., neural ganglia (Ariens Kappers, 1929; Papez, 1967; Lentz, 1968).

Hydra

It is with the fresh water flatworms (the hydra) where the first evidence of a central nervous system appears (Ariens Kappers, 1929; Colbert, 1980; Jarvik, 1980; Jerison, 1973; Papez, 1967; Romer, 1966). The flatworm is symmetrical and has an enlarged anterior region that corresponds to its head. Hydra are also capable of twisting, bending, swaying, and waving their tentacles, which respond to a variety of stimuli.

The hydra's sensory cells are located along the body surface or upon its tentacles, which send fiber processes to ganglion cells. When food comes into contact with the surface of the mouth, sensory cells are stimulated; as these act on motor cells, the mouth closes and the upper body contracts, forcing the food into the body cavity. In many respects, the hydra could almost be characterized as a floating stomach, for once the food has been taken, the creature twists and turns and contracts in various directions. Later, it will undergo a violent contraction, causing undigested material to be expelled through the mouth.

Hence, it is presumably within the outer epidermis that sensory cells arose (which were responsive to chemical, photo, and perhaps tactile stimuli), and within the contractile motor tissue that motor-effector neurons arose. A sensory (chemical/photosensitive) and motor nervous system were semi-independently evolving. However, these motor-effector neurons became increasingly specialized for the purpose of responding to discrete sensory messages, and for arousing and activating the body so as to induce movement; whereas, the sensory cells also became increasingly specialized, albeit dependent on external sensory sources, to induce excitation and arousal.

Photosensitivity and Activation

One primary energy source for inducing excitation is oxygen, which the original motor and sensory cells presumably consumed. However, of equal importance in energy production were photosensitive cells, which not only responded differentially to various shades of light but which, very early in their evolution, may have engaged in a rudimentary form of photosynthesis, like a plant. It is these photosensitive cells that would give rise to the retina, and to the visual midbrain, dorsal thalamus, epithalamus, and dorsal hypothalamus—nuclei that, over the course of evolution, would serve as a bridge linking the motor-sensory brainstem/spinal cord with the chemically sensitive and olfactory-sensitive forebrain-telencephalon.

The Nerve Net

Over the course of evolution, as the number of secreting and transmitting motor and sensory nerve cells that a creature possessed increased, a network of interlinked neurons, called the "nerve net," was soon fashioned (Ariens Kappers, 1929; Dowling, 1992; Papez, 1967; Lentz, 1968). With the establishment of the nerve net, a multicellular organism could now behave as a complete unit—in a controlled, highly coordinated, and directed

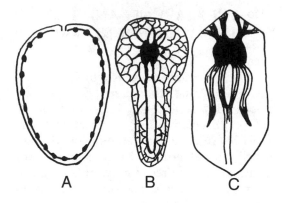

Figure 1.2 Stages in the evolution of the nerve net in three flatworms (Acoela, Polycladida, Rhabdoceoela). **A.** Epidermal nervous system and nerve cells. **B.** Nerve net with bilobed ganglia. **C.** Cephalization of brain with loss of nerve net. (Redrawn from Hyman LH. The invertebrates. Vol. II. New York: McGraw-Hill, 1951.)

Figure 1.1 The nerve net. The nervous system of the triclad tuberllarian (worm). The nerve net forms two pairs of ganglia in the anterior head region. A corpus of axons forms a commissure that connects the two ganglionic lobes. Dual nerve cords extend the length of the body. Eye cups and photosensitive nerves are located in the anterior ganglia. (From Lentz TL. Primitive nervous systems. New Haven: Yale University Press, 1968.)

manner—since different regions of the body could now communicate together almost simultaneously.

The evolution of the "nerve net" coincided with the appearance of the more advanced planaria and flatworms: the coelenterates. These are the most primitive members of the animal kingdom to possess not only neurons but a "nervous system," composed of distinct olfactory-chemical, photosensory, and sensory-motor neurons, including those which are unipolar, bipolar, and multipolar (Ariens Kappers, 1929; Papez, 1967; Lentz, 1968). They also contain a third type of cell called a ganglion cell (or protoneuron). These nerve cells also possess true synapses, which resemble those of the mammalian nervous system.

However, these creatures were without true brains. Rather, in addition to the nerve net they possessed a dual pair of nerve cords that ran lengthways within the body, and tiny neural ganglia located in the anterior head region. These nerve cords and ganglia, in turn, were connected to chemosensory neurons located externally in ciliated pits within the head area and along the body surface. They were also connected to photosensitive retinal cells found within their cephalic eyes, and to motor and tactual-sensory neurons located on the body surface.

Arthropod Versus Vertebrate Patterns of Neural Organization

The initial olfactory, visual, motor neuronal network first developed by the coelenterates was repeated and elaborated upon over the ensuing 50 million years, only to diverge into two different patterns of central nervous system organization. Among vertebrates, the nervous system would first form from a single hollow tube that runs the length of the back and expands into a brain (see Chapter 18). By contrast, the central nervous system of the arthropods would consist of a pair of nerve cords that run the length of the belly. It is from arthropods that spiders, insects, crabs, lobsters, and shrimps evolved.

By contrast, those creatures whose descendants were destined to become vertebrates increasingly adapted to a life of swimming and developed an internal skeleton of rigid bone and cartilage. Instead of shedding and then replacing the external skeleton (which is typical of arthropods), the skeletal system instead developed internally by accretion.

Like those who had come before them, these primitive and ancient prevertebrate animals maintained specialized neuroganglia sensitive to tactile, visual, and chemical/pheromonal (olfactory) stimulation, as well as ganglia involved in tactile-motor functions. As such, the first rudimentary features of what would become the olfactory forebrain, the visual midbrain, and the motor (and tactile-sensory) brainstem were probably first established almost 600 million years ago. By the time the first vertebrates and armored fish began to swim the oceans, around 500 million years ago, the first primitive lobes of the brain had become fashioned through the collectivization of these neural ganglia (Ariens Kappers, 1929; Colbert, 1980; Jarvik, 1980; Jerison, 1973; Joseph, 1993a; Papez, 1967; Romer, 1966; Sarnat and Netsky, 1981).

The Olfactory Limbic System (Rhinencephalon) and Memory

Originally, via the analysis of chemical secretions, a potentially edible substance might have been deemed good to eat and was consumed. Since it is not very adaptive merely to absorb, or to eat whatever comes into one's mouth, the ability to make distinctions, to recognize, and to make comparisons was also necessitated, which resulted in the creation of memory, and this, too, was accomplished through chemical analysis, smell (the olfactory-hippocampus, olfactory-amygdala-ventral hypothalamus), and, later, taste.

Memory may well have been originally derived from and based on olfactory information. Because of these olfactory roots, the amygdala and hippocampal (memory) regions of the brain are sometimes referred to as the rhinencephalon, or rather, the "nose brain." It is referred to as the nose brain because the nerve fiber pathway that leads from the olfactory mucosa of the nose to the olfactory bulbs terminates in these brain regions, which, in fact, evolved from the olfactory bulbs.

The olfactory forebrain gave rise to a very primitive ventral amygdala and rudimentary dorsal hippocampus, probably some 500 million years ago when the ancestors of the eel-like cyclostomes and the wormlike, burrowing, limbless Gymnophiona first slithered upon the scene. The modern-day descendants of these creatures are apparently little different from their ancestors and possess a well-developed hypothalamus, a rudimentary hippocampus, and a primitive ventral-striatum and amygdala (Ariens Kappers, 1929; Papez, 1967): nuclei collectively referred to as the limbic system, a term coined by Dr. Paul MacLean.

THE BRAINSTEM AND MIDBRAIN

By 500 million years ago, motor nerve neural networks and related neural ganglia became increasingly organized and united, thereby giving rise to the brainstem in species such as jawless tunicates and amphioxus (protochordates, the ancestors to true vertebrates).

The brainstem (which soon differentiated into a medulla and pons) was (and continues to be) concerned with monitoring and controlling heart rate, breathing patterns, cortical arousal, and sensory filtering, and with reflexively triggering specific motor reactions to visual, vestibular, painful, sexual, and edible stimuli (Donkelaar, 1990 (see Chapter 10). It is also from the brainstem that the posterior-ventral portion of the midbrain evolved (see Romer, 1966; Sarnat and Netsky, 1981). Indeed, the caudal midbrain is tightly linked to, and in many respects anatomically resembles, the brainstem. Spanning the length of both structures are reticular neurons.

PHOTOSENSITIVITY

In contrast, the anterior-dorsal midbrain and dorsal thalamus may have had a different origin (e.g., Butler, 1994). As noted, many primitive creatures possessed photosensitive cells, even though they lacked eyes. It is these cells that eventually evolved the capacity to distinguish form and shape and, thus, the ability to see. However, initially they may have been more concerned with extracting photochemicals from light and detecting various shades of illumination (including, perhaps, temperature).

These photosensitive cells were also dorsally located, as is the visual midbrain (and visual colliculi) and (dorsal) visual thalamus, and as the dorsal (visual) hippocampus initially was, before it migrated ventrally, forming portions of the superior parietal, occipital, and posterior temporal lobes in the process. Because this initial dorsal portion of the brain could detect light and shadow, it evolved the capacity to direct the brainstem and motor centers so as to approach (or avoid) light versus darkness.

VISUAL-SPATIAL MAPS

Eventually (and prior to the ventral migration of the hippocampus), some of these photosensitive neurons also gained the capacity to form a rudimentary "visual-spatial" map of their surroundings—a capacity greatly expanded in the modern hippocampus. For example, hippocampal cells greatly alter their activity in response to certain spatial correlates, particularly as an animal moves about in its environment (Nadel, 1991; O'Keefe, 1976; Olton, et al., 1978).

NEUROTRANSMITTER PRODUCTION

Unlike the hippocampus, the midbrain (and anterior pons) also evolved the capacity to manufacture its own chemicals: i.e., neurotransmitters (via the locus coeruleus, raphe nuclei, substantia nigra, midbrain tegmentum), which, in turn, is probably related to its photosensitive capabilities. This capacity enabled the upper brainstem to neurochemically control arousal, motor activity, and information processing within the forebrain. However, because the midbrain (and dorsal thalamus) also acted to funnel information to the forebrain and brainstem, and from the brainstem to the forebrain, it essentially became a bridge linking these structures, as well as a potential site for future cortical construction (see Chapter 18).

RHYTHM GENERATION

Large portions of the anterior-dorsal midbrain and dorsal thalamus (and dorsal hypothalamus) were initially derived from photosensitive cells. Because these cells are affected by the rhythmical nature of the planet's rotation (which induces rhythmical periods of light and dark), photosensitive neurons located in these tissues also became adapted to these alterations and are, thus, rhythmically sensitive (see Aronson et al., 1993; Morin, 1994, for related discussion). Eventually, these neurons became capable of inducing and enforcing rhythmic activities within the central nervous system, including the forebrain, via their control over arousal and the production of norepinephrine, serotonin, and dopamine—neurotransmitters that are manufactured by neurons located near the midbrain-brainstem junction (see

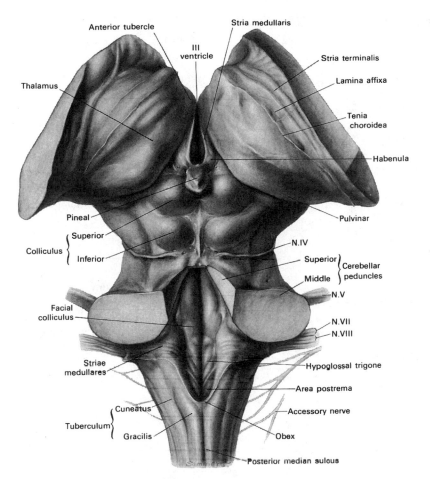

Figure 1.3   Dorsal view of human midbrain, visual (superior) and auditory (inferior) colliculi, and thalamus. The cerebellum and caudate/puta-men have been dissected away. (From Mettler FA. Neuroanatomy. St. Louis: CV Mosby, 1948.)

Chapter 10). Hence, the manufacture and release of these neurotransmitters is also rhythmical, which in turn effects the overall functioning and arousal of the brain. Moreover, because the brainstem (i.e., medulla, pons, midbrain) also controls breathing, heart rate, the sleep cycle, and dreaming, as well as the external musculature and thus swimming or walking, these functions, too, came to be governed by specific rhythms (see Skinner and Garcia-Rill, 1990; Chapter 10), long before jawed or even jawless fish had evolved some 500 million years ago.

Summary

The central nervous system of the proto-chordates consisted of a primitive olfactory-chemical-emotional-visual memory system (the limbic forebrain), photosensitive cells and transmitter-producing nuclei (midbrain), and a brainstem (Colbert, 1980; Jarvik, 1980; Jerison, 1973; Romer, 1966). However, there is no evidence of even a rudimentary cerebellum or neocortex.

The Dorsal Visual Thalamus

It was possibly around 500 million years ago that the thalamus began to rapidly evolve. Due to its dorsal superior position atop the ancestral brain, it received and was quite responsive to overhead ambient light. It also soon began receiving visual signals through retinal neurons located in a depression in the

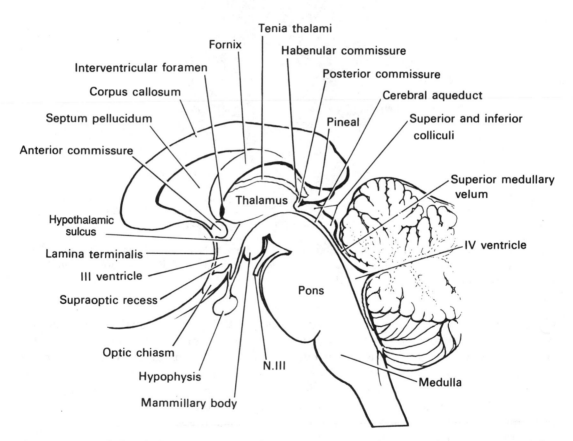

Figure 1.4   The brain of an armored fish (Macropetalichthys) that became extinct over 325 million years ago. Note parietal/pineal eye. (From Stensio E. Notes on the endocranium of a Devonian. Bulletin of the Geological Institute Upsala 1937; 27:128–144.)

Figure 1.5   Medial (split-brain) view of brainstem, midbrain, and thalamus. Note pineal. (From Carpenter M. Core text of neuroanatomy. Baltimore: Williams & Wilkins, 1991.)

superior surface of the skull, as did the midbrain (Butler, 1994). Hence, the thalamus was probably originally concerned predominantly with visual stimuli and perhaps relaying signals between the forebrain and brainstem.

THE THIRD EYE

Over the course of the ensuing 100 million years, the thalamus continued to undergo evolutionary metamorphosis and developed (via the epithalamus), a third photosensitive "eye," i.e., the pituitary/pineal eye, which first appeared about 400 million years ago with the evolution of the lamprey. Many ancient vertebrates possessed a parietal eye, including creatures ancestral to modern reptiles, amphibians, and fish (Colbert, 1980; Jarvik, 1980; Jerison, 1973; Romer, 1966), as well as some modern fish, frogs, and reptiles.

Among more advanced vertebrates, the third eye has completely disappeared. However, the thalamus evolved a specific nuclei, the geniculate nucleus, which continues to receive, process, and transfer visual information received from the visual tectum, and which became increasingly concerned with visual stimuli (Butler, 1994), receiving the bulk of retinal input (Casagrande and Joseph, 1978, 1980; Ebbesson, et al., 1972). However, in modern mammals, collaterals from the retinal visual system are still sent to the midbrain superior (visual) colliculus. Retinal signals are also transmitted via the optic radiations to the inferior temporal, superior parietal, and occipital lobes—neocortical regions that also receive widespread lateral geniculate and dorsal thalamic input as well.

JAWLESS AND CARTILAGINOUS FISH

With the evolutionary metamorphosis of jawless and cartilaginous fish (cyclostomes), the basic organization plan for the spinal cord, brainstem, midbrain, and olfactory forebrain/telencephalon had come to be established although there is no evidence of the cingulate gyrus or the six-layered neocortex. The thalamus was a major processing and relay center, and the olfactory forebrain consisted of a dorsal hippocampus and septum, and a primitive and undifferentiated olfactory-amygdala-striatum (Smeets, 1990); a portion of which would become the basal ganglia. As noted, it is these olfactory-derived nuclei (hippocampus, septal nuclei, amygdala) that gave rise to and provided the physiologic, morphologic, and pharmacologic circuitry of the neocortex (Haberly, 1990) and what would soon become the cingulate.

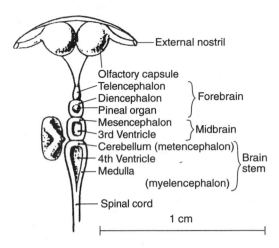

Figure 1.6   The brain of a jawless fish. (Redrawn from Tarlo H. A review of Upper Jurassic pliosaurs. Bulletin of the British Museum of Geology 1960;4:145–189.)

The brainstem, too, however, was destined to become even more complex, because the evolution of primitive jawless fish, 500–450 million years ago, coincided with the development of the vestibular system and the brachial nerves; these, in turn, were initially concerned with controlling the gills and movement of the tongue (Romer, 1966).

The tongue may well have been the first muscular appendage subject to fine motor control, and it is a structure that first appears in cyclostomes. Like the brainstem, however, the tongue became more fully developed with the evolution of amphibians (i.e., frogs) and reptiles and snakes (Colbert, 1980; Jarvik, 1980; Jerison, 1973; Romer, 1966).

However, it was the development of the jaw which added impetus to the evolution of the motor brainstem and what would later become cranial nerves as well as the middle ear (200 million years later).

Sharks

The first jawed armored fish evolved around 450 million years ago (Colbert, 1980; Jarvik, 1980; Jerison, 1973; Romer, 1966). Initially, there were at least two types of jawed species, one line of which diverged and gave rise to cartilaginous sharks and rays (elasmobranchs)—creatures that first appeared about 450 million years ago. Sharks possess a well developed brainstem, olfactory limbic forebrain (telencephalon), and large optic lobe (or tectum/colliculus) situated atop the midbrain (Haberly, 1990; Smeets, 1990).

With the evolution of sharks, in addition to olfactory and pheromonal input, the telencephalon became increasingly concerned with analyzing and integrating visual, auditory, mechanoreceptive, and electrical impulses relayed via the thalamus from the brainstem (Butler, 1994; Smeets, 1990). The hippocampus also increased in size and began to dominatethe roof (pallium) whereas the olfactory-striatum, primordial amygdala, and hypothalamus continued to dominate the floor (the ventral or subpallium) of the telencephalon, with the septal nuclei contributing to the development of the medial walls of the brain (Haberly, 1990; Smeets, 1990). However, these creatures are devoid of transitional (cingulaic) cortex or neocortex.

THE STRIATUM (BASAL GANGLIA)

With the evolution of sharks, the striatum appears to have become a repository for emotion-motoric-memories (or motor programs) that could be selectively activated by the olfactory-limbic system and coordinated by the extended (central-medial) amygdala, a nucleus now increasingly adapted for exerting some controlling influence on the brainstem and spinal nerves and, thus, over motor functioning.

However, in contrast to the more "hard wired" brainstem, the limbic-striatum became exceedingly concerned with motorically expressing the creature's emotional state in a manner that was much more flexible and less reflexive. Thus, in response to amygdala (and hippocampal) activation, the striatum could be stimulated in order to produce complex emotional displays and reactions (cf Heimer and Alheid, 1991; Mogenson and Yang, 1991). However, it is not until the evolution of amphibians and reptiles that the sirinium mushrooms in size and complexity.

AMYGDALA

Reflecting its increasing importance in analyzing and coordinating emotional-motoric activities, the amygdala and related olfactory tissue also began to insure that the two halves of the telencephalon

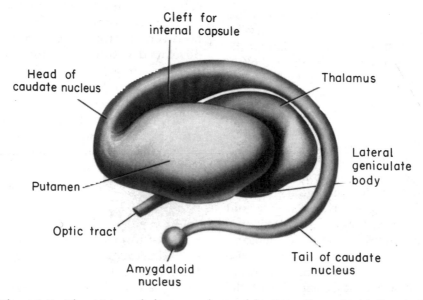

Figure 1.7 The striatum, thalamus, and amygdala. (From Carpenter M. Core text of neuroanatomy. Baltimore: Williams & Wilkins, 1991.)

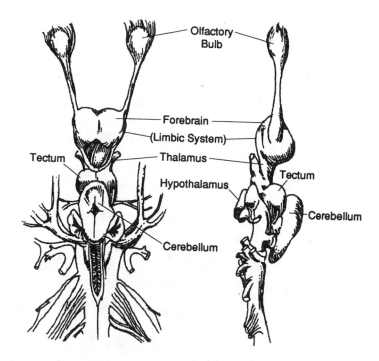

Figure 1.8 Dorsal and lateral view of a shark's brain.

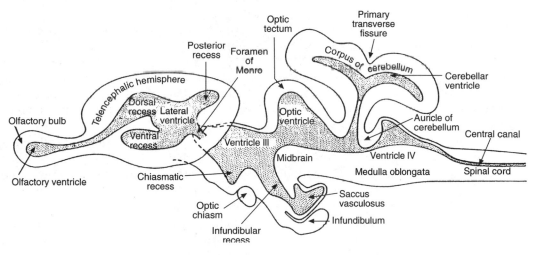

Figure 1.9 Medial (split-brain) view of a shark's brain. (Redrawn from Sarnat HB and Netsky MG. Evolution of the nervous system. New York: Oxford University Press, 1981).

acted as a coordinated unit. This was accomplished through the evolution of a thick rope of axonal fibers, the anterior commissure, around 400 million years ago. Large numbers of these axons also terminated in the thalamus and brainstem reticular formation (Butler, 1994; Smeets, 1990). Thus, in addition to its role in analyzing olfactory, visual, auditory, and tactile stimuli funneled to it (and the hippocampus) by the thalamus and olfactory nuclei, the amygdala increasingly became capable of coordinating and stimulating arousal and movement via bilateral influences exerted on both halves of the brain via the anterior commissure.

OPTIC AND AUDITORY TECTUM

With the evolution of sharks and other jawed creatures, the midbrain optic tectum had become increasingly concerned with processing and orienting toward visual as well as auditory stimuli. That is, the optic tectum began to differentiate into an auditory tectum. In sharks, auditory-vibratory information is received via receptors located in maculae sacculi (Smeets, 1990), whereas mechanical and electrical impulses are received via the lateral line system—all of which are then relayed to the brainstem and cerebellum, and then the midbrain and telencephalon via the thalamus.

THE FOREBRAIN/ TELENCEPHALON
Neocortex

As noted, over the course of evolutionary metamorphosis, the olfactory system, hippocampus, amygdala, and septal nucleus became more complex and began developing additional layers of nerve cells, which in turn contributed to the expansion of the forebrain/telencephalon: i.e., the neocortex. However, the first layers of neocortex did not appear until around 150 million years ago—100 million years after the evolution of the cingulate (transitional) cortex. Specifically, the septal nuclei and hippocampus eventually contributed to the development of the cortical medial walls, including the cingulate, with the hippocampus also providing tissue to portions of what would eventually become the superior parietal, occipital, and medial and posterior temporal lobes. The amygdala gave rise to the inferior and superior temporal neocortex as well as the striatum and portions of the orbital frontal lobe.

The midbrain, or phylogenetically old midbrain-like cortical tissue (Marin-Padilla, 1988a), may have also contributed to the evolution of the neocortex, perhaps by providing layers I and VII, between which are sandwiched neocortical layers II–VIa (see Chapter 18). Layers I and VII are phylogenetically old in structure and composition, having a unique horizontal organization, abundance of fibers, but scarcity of neurons—characteristics shared by all vertebrates except fish (Marin-Padilla, 1988a). These layers also appear to maintain an intimate relationship with the reticular activating system (Marin-Padilla, 1988a), as well as the norepinephrine transmitter system (see Chapter 18), perhaps acting as their most distal component. For example, layer I and VII neurons may provide a low threshold background of excitation (that can rapidly increase or decrease in magnitude), which acts to stimulate and maintain the functional and anatomical stability of newly arriving neurons during the course of neortical development.

THINKING, REASONING, REMEMBERING

Forebrain versus Brainstem

When coupled with its extensive thalamic and limbic input, the expanding neocortical forebrain/telencephalon became increasingly specialized for receiving, integrating, discriminating, and remembering multimodal sensory stimuli, as well as for expressing or transferring this data to other regions of the brain. However, in contrast to the brainstem, the limbic forebrain evolved the capacity to think and remember before acting.

Because chemical sensitivity in early creatures exceeded visual acuity, and as they were incapable of analyzing sound, the olfactory-limbic forebrain was given an intellectual and mnemonic evolutionary advantage over the midbrain and brainstem in regard to the development of thinking, planning, and memory. That is, since olfactory stimuli tend to arise in the distant environment, this gives the organism more time to analyze these signals before responding. Hence, being the recipient of this distant information, the forebrain was given time to "think," whereas the brainstem, in response to tactile, painful stimuli, had to react reflexively. Moreover, because the source of this chemical/pheromonal information might be far away and hidden, the forebrain had to retain this information in "memory" for long time periods, or for however long it took to approach or get away.

In that the brainstem and the thalamus also relayed sensory data to the forebrain (Butler, 1994; Donkelaar, 1990), the forebrain became increasingly specialized for interpreting and associating multiple sensory signals simultaneously including those already analyzed. It was able to "think" about what actions to take, and/or store this information in memory so as to "think" about or act on it later, rather than reflexively act in a stereotyped and unthinking manner, which is characteristic of the brainstem. In contrast to the forebrain/telencephalon, which is associated with emotion, memory, cognition, and the evolution of the neocortex, the brainstem has remained specialized for receiving information arising in the immediate environment (touch, taste, temperature, pain, balance); this allows for immediate reflexive responses by motor nuclei.

TELEOSTS

With the evolution of teleosts, and then the lobe-finned fish (Sarcopterygii), and lungfish (Dipnoi), around 400 million years ago, the forebrain/telencephalon increasingly became the recipient of visual, acoustic, and lateral line information which was relayed via the brainstem, midbrain, and thalamus (Butler, 1994; Nieuwenhuys and Meek, 1990a, b). The hemispheres of the telencephalon also became increasingly differentiated (Nieuwenhuys and Meek, 1990a, b) due to increased need for processing space. The amygdala limbic-(ventral)-striatum, and the corpus (dorsal) striatum, ballooned outward to become the caudate, putamen and nucleus accumbens, such that, in some respects, the striatum began to resemble the posterior portion of the mammalian striatum (i.e., the putamen and globus pallidus). Also, the first hints of what would eventually become a two-layered cortex began to form along the pallial surface of the telencephalon (Nieuwenhuys & Meek, 1990a, b). However, as detailed in Chapter 18, this two-layered

cortex (unlike the later-appearing cortical layers II–VIa) in some respects appears to be a derivative of the midbrain.

Brainstem Nuclei

With the evolution of jawed fish (the teleosts) about 400 million years ago, the arches between the brachial clefts (or gills) became increasingly modified and much smaller and concerned with supporting the jaw (Romer, 1966; Sarnat and Netsky, 1981). Indeed, over the course of phylogeny, the brachial arches eventually became the mandible and maxilla of the jaw.

The brachial nerves became increasingly specialized to form the vagal, glossopharyngeal, trigeminal, and facial nerves (Romer, 1966; Sarnat and Netsky, 1981), all of which terminate within the brainstem to form various subnuclei. These subnuclei are present-day mammalian brainstem features. Over the ensuing 100 million years, the acoustic nerve appeared as a derivative of portions of the vagal, glossopharyngeal, and facial nerves and the vestibular nuclei (Sarnat and Netsky, 1974). It is from the brainstem vestibular system that the mammalian auditory system evolved (Ariens Kappers, 1929; Papez, 1967).

LOBED-FINNED (SARCOPTERYGIAN) FISH

The sarcopterygian (lobed-finned) fish appeared during the Devonian age, some 400 million years ago, and possessed a cranium that was divided into two parts (a frontal-sphenoid portion and an occipital region) as well as a brain and cerebellum that, in many respects, is quite similar to amphibians and even reptiles (Nieuwenhuys and Meek, 1990b). Like sharks, lobed-finned fish never became extinct, as over 80 specimens have been caught off the coast of southeast Africa over the course of the last 50 years (Nieuwenhuys and Meek, 1990b).

These particular fish are unique in that, in addition to their dorsal fins, they possess two well-formed, "fleshy-lobed" paired fins: the internal skeleton of which includes a humerus, femur, radius, ulna, tibia, and fibula (Jerison, 1973; Nieuwenhuys and Meek, 1990b; Romer, 1966). It is from these lobed fins that legs would eventually "evolve" (Colbert, 1980; Jarvik, 1980; Jerison, 1973; Romer, 1966).

These lobed fins, although acting to improve swimming and maneuverability, also enabled these fish to probe and root among the fertile ocean floor and eat of its plant and other organic life (Jerison, 1973; Romer, 1966). The presence of lobed fins also increased and required greater telencephalic, striatal motor control, including the capacity to direct movement of these rudimentary extremities.

These animals, as well as other teleosts, were also equipped with the lateral line system, which enabled them to perceive vibration and rudimentary sound (Butler, 1994; Colbert, 1980; Jarvik, 1980; Jerison, 1973; Romer, 1966). The brainstem, midbrain, thalamus, and hippocampus, amyg-dala-striatal system therefore, underwent further modification in order to receive, process, and act on this information.

As in sharks and other teleosts, the olfactory bulb remained prominent and largely anatomically similar to that of previous (as well as forthcoming) animal life. However, in contrast to all previous life forms, the cerebral hemispheres had become increasingly differentiated, and the first two-layered cerebral (extended midbrain) cortex appeared (Nieuwenhuys and Meek, 1990b). However, there is still no evidence of the cingulaic gyrus or neocortex.

LUNG FISH

The lung fish (a subgroup of the lobed-finned fish), also appeared during the Devonian period, about 400 million years ago. These creatures had a long eel-shaped body with scales embedded in

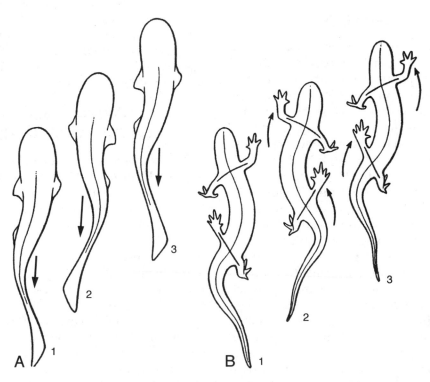

Figure 1.10 Rhythmic swimming motions of (**A**) three fish, and (**B**) three salamanders that utilize the same movement patterns, which are in sequence with the truncal muscle activity. The limbs are extensions of the axial muscles. (From Romer S. Vertebrate paleontology. Chicago: University of Chicago Press, 1966).

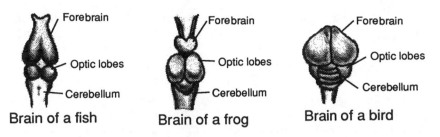

Figure 1.11 The brains of a fish, frog, and bird.

the skin. They are considered "fleshy-finned" fish (Nieuwenhuys and Meek, 1990b). The lung fish were also equipped with ventral lobe-like appendages; some investigators believe that from this group (i.e., sarcoptergia—lobe-finned, and dip-noans—lung fish) all terrestrial verte-brates evolved (Romer, 1966).

Because they possessed lungs and spiny lobed fins, these creatures were presumably capable of periodically leaving the water so as to venture along river banks and ocean fronts, and onto dry soil (Colbert, 1980; Jarvik, 1980; Jerison, 1973; Romer, 1966). Presumably, the lobed lung fish served as the intermediary species that led from fish and gave rise to amphibian-like, land-living vertebrates 400 to 350 million years ago (Colbert, 1980; Jarvik, 1980; Jerison, 1973; Romer, 1966). When this occurred, the basal ganglia mushroomed in size.

EUSTHENOPTERONS, ICHTHYOSTEGA, AND AMPHIBIANS

Those ground-hugging creatures that evolved from the lobed-finned lung fish, were capable of crawling as well as hopping from shallow pool to shallow pool when searching for food, including that which might be snacked upon between water holes, such as plants, worms, snails, and insects. To accommodate the organism's increased need for motor control, further alterations occurred within the brainstem, cerebellum, thalamus, and, in particular, the striatal forebrain; these increasingly worked in tandem to initiate and coordinate species-specific behavior patterns.

The visual and auditory system became modified, such that the optic lobe and tec-tum of the midbrain was now assisted by the evolution of auditory tectum (colliculi). Moreover, the midbrain auditory and visual colliculi also evolved feature-detecting cells that could recognize prey by sight or sound, as well as the ability to visually mediate prey-catching behaviors (Donkelaar, 1990).

Living on land also necessitated addi-tional evolutionary developments within the brainstem reticular formation, which at the point of reptilian evolution began to resemble that of mammals (Newman and Cruce, 1982). It is the evolution and exten-sion of the reticular formation as well as the optic tectum that gave rise to the inferior tectum, which in turn began to mediate ori-enting and arousal reactions in response to sound. However, not just the midbrain tec-tum, but also the thalamus became increas-ingly differentiated and specialized for processing auditory as well as visual and motoric information (Butler, 1994). In addi-tion, the hypothalamus was forced to develop new methods of thermoregulation in order to accommodate to life on dry land.

The lobed fins also underwent further modification due to the gravitational de-mands of a dry land environment on flexi-bility and the ability to move about. That is, the forces of gravity require strength in the supporting and moving of appendages; consequently, legs evolved. Soon, such creatures were able to expend a greater amount of time out of the water and began to diversify. The sirinium mushroomed in size and importance.

However, it was not the lobed-finned lung fish but their descendants that gave rise to amphibians. These amphibian-like creatures looked something like a cross between a fish and a big salamander, with flat heads, long tails, and short, stocky feet like a turtle. These include the eusthenopterons, as well as the ichthyostegas, which used four feet in order to move about (Colbert, 1980; Jarvik, 1980; Jerison, 1973; Romer, 1966). Hence, by 350 million years ago the lobe-finned fish presumably evolved into a fish with legs, the eusthenopterons and ichthyostegas, which in turn evolved into amphibians, some of which grew to 15 feet long.

Amygdala

As these creatures were now living on land, the vestibular nucleus, medial geniculate thalamus, amygdala, and inferior tectum became modified not only to perceive sounds but to trigger orienting and alarm reactions in response to transient auditory stimuli—characteristics these nuclei retain, even in modern primates (Edeline and Weinberger, 1991; Hitchock and Davis, 1991; Hocherman and Yirmiya, 1990). In addition, the lateral (cortical) amygdala (which is concerned with complex sexual and social-emotional functions) began to evolve (Neary, 1990).

Like the much more ancient central medial (and extended) amygdala, the lateral amygdala maintains rich interconnections with the hypothalamus and the preoptic area—nuclei concerned with sexual activities and male versus female sex-specific behavior (see Chapter 5). It is with the appearance of amphibians that a greater degree of complexity begins to typify gender-specific sexual behavior and social-emotional behavior. The amygdala, hypothalamus, and limbic (ventral) and dorsal striatum evolved accordingly.

The Striatum and Amygdala

In amphibians, reptiles, mammals, and humans, the striatum (basal ganglia) is intimately interconnected with the thalamus, midbrain, and brainstem as well as the medial and lateral amygdala (Butler, 1994;

Heimer and Alheid, 1991; Mogenson and Yang, 1991). Indeed, the striatum is the main motor output center of the forebrain (Donkelaar, 1990) in non-mammals, and be it amphibian or human, it receives dopamine via the nigrostriatal bundle and from the mesolimbic pathway—neurotransmitters that are associated with cognitive-motor and emotional motor expressiveness. As noted, the extended (central-medial) amygdala gives rise to, merges with and becomes coextensive with the ventral (limbic) striatum (Heimer and Alheid, 1991; Mogenson and Yang, 1991). However, the posterior medial-lateral amygdala gives rise to (i.e., grows a thick tail) and merges with the dorsal caudate and putamen, and the dorsal pallium (Neary, 1990).

In mammals and primates, these nuclei remain intimately interconnected as the tail of the caudate and putamen merges with the medial-lateral amygdala, whereas the extended (central-medial) amygdala and limbic striatum are in many respects indistinguishable. Hence, in this regard, by the time amphibians began to crawl upon the earth, the amygdala had helped to fashion the basal ganglia, and had essentially sandwiched the ventral-limbic striatum (nucleus accumbens substantia innominata) and the dorsal, (caudate and putamen) striatum together. In this manner, the amygdala is able to control and direct the quite different (albeit related) motoric and emotional-cognition functions subserved by these nuclei (see Chapter 9) so as to meet the increased social and emotional demands of living in a terestrial environment.

Amygdala control over the basal ganglia is still maintained in humans, and this nuclei is capable of exerting tremendous social-emotional influence over this motor system (see Carpenter and Jayaraman, 1986; MacLean, 1990; Mogenson and Yang, 1991). Indeed, there is some evidence to suggest that the amygdala is involved in the development of a variety of basal ganglia disorders, including Parkinson's disease, ballismus, and Alzheimer's disease (reviewed in Chapter 9).

Social-Emotional and Motivational Motor Programs and the Striatum

Smiling, kicking, punching, running, and complex sexual acts are mediated by the basal ganglia. In humans and in amphibians, the motor programs for engaging in species-specific social-emotional acts appear to be stored in this region (MacLean, 1990).

In humans, damage involving the dorsal caudate and putamen can cause the face to become frozen and emotionless and the body stiff (as in Parkinson's disease); with massive lesions the individual may become catatonic and cease to move (see Chapter 9). Conversely, lesions involving the limbic striatum or subthalamic nucleus can result in uncontrolled ballistic movements, such as kicking, hitting, and flailing—actions normally initiated when threatened or engaged in defensive or offensive attack.

Similarly, the destruction of the amphibian forebrain/striatum will disrupt species-specific and stereotyped social-emotional motor acts, such as dominance displays or even the desire to eat, and they may become catatonic (MacLean, 1990). With forebrain/striatal destruction, amphibians, such as frogs, will cease to respond to emotional or motivational stimuli, including the presence of food, and instead will simply sit. However, due to the preservation of the brainstem and cerebellum, if turned over the frog will right itself, or if thrown into the water it will swim until reaching a dry surface, where again it will simply sit. Indeed, if unmolested it will sit forever in the same spot, failing to demonstrate any desire to explore, eat, or have sex, until finally dying and becoming mummified. As noted, in some respects this is reminiscent of the catatonia and loss of will that occur with massive striatal lesions involving the medial frontal lobes (see Chapters 9 and 11).

REPTILES
Forebrain and Midbrain

With the evolution of amphibians, the forebrain underwent considerable expansion and began to surround the midbrain (Donkelaar, 1990; Colbert, 1980; Jarvik, 1980; Jerison, 1973; Romer, 1966; Sarnat and Netsky, 1981). However, with the evolution of reptiles the mammillary bodies began to evolve out of the hypothalamus, and the forebrain/telencephalon underwent yet further cortical expansion with the development of three-layered (albeit unmyelinated) cortices. This multilayered cortex began to spread dorsally and laterally over the surface of the hemisphere (Haberly,

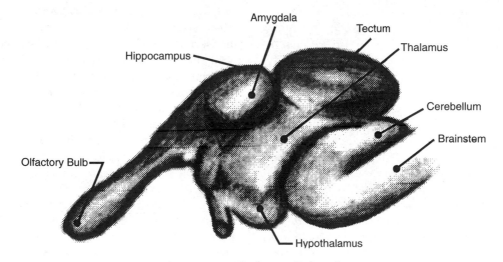

Figure 1.12   The brain of a lizard.

1990; Ulinski, 1990). The brainstem also underwent additional differentiation and elaboration, and, unlike all previous creatures, reptiles began demonstrating mammal-like sleep periods (Ulinski, 1990). However, there is no evidence that reptiles experience "slow wave" or dream sleep.

Cortex and Thalamus

The lateral region of the expanding cortex, although maintaining its rich limbic and olfactory interconnections, began to receive extensive visual, auditory, and somatosensory input from the increasingly differentiated thalamus (Butler, 1994). That is, the cortex began to receive sense data that had previously been directed primarily to the limbic system (Neary, 1990)—data which had originated within the midbrain and brainstem. However, with the exception of cortical layers I and VII, which is similar to midbrain tissue, since much of the cortex serves as an extension of the limbic system and in fact acts to funnel data back to these nuclei, the amygdala and hippocampus receive data that has now undergone an additional step of cortical analysis.

Motor Cortex and Striatum

Over the course of evolution, although reciprocal thalamic interconnections with the striatum were retained, the new motor areas in the dorsal cortex began receiving thalamic input as well (Neary, 1990). In reptiles, these cortical motor areas were tightly linked to the striatum, acting as a striatal (and hippocampal-cingulate-medial-rontal) extension. That is, initially (i.e., 300 million years ago), the motor cortex was subservient to the caudate, putamen, globus pallidus, and subthalamic nucleus, as there is no evidence in these creatures that the cortical motor areas project to the spinal cord or even to the brainstem (Ulinski, 1990). This does not occur until the evolution of mammals, at which point the motor cortex appears to become dominant over the striatum, and the cortico-spinal pathways are formed.

Nevertheless, as in amphibians, the reptilian (and primate) striatum controls the motoric aspects of species-specific, social-emotional, sexual, and signature displays (see MacLean, 1990). However, because reptiles live a somewhat more complex social life than amphibians, these nuclei became increasingly complex and exceedingly important in all aspects of social-emotional and motivational motor functioning.

Hence, as with amphibians, destruction of the reptilian forebrain (including the basal ganglia) results in the complete elimination of social-emotional behaviors, including signature displays. The ability to learn or remember or the capacity to engage in social actions is also eliminated (MacLean, 1990). However, reptiles devoid of forebrain can run or walk if vigorously stimulated, due to preservation of the brainstem, as can other creatures with forebrain/striatal destruction.

THE EVOLUTION OF AUDITORY COMMUNICATION

The first amphibians and reptiles lacked an inner or true middle ear but were capable of hearing low-level vibrations and sounds, such as croaking, tails thumped on the ground, certain distress calls, and signals of contentedness. Limbic language capabilities and the ability to engage in complex auditory communication was (and is) not well developed in these creatures, since sound production was, for the most part, reflexive and mediated by upper brainstem nuclei (the periaqueductal gray). Purposeful vocal communication would first require tremendous alterations in the ability to hear sound.

The development of hearing was a consequence of the evolutionary metamorphosis of the vibratory sense and the lateral line organs: structures that make their first rudimentary appearance in cyclostomes (van Bergeik, 1966). However, among teleosts, a well-developed closed canal system of lateral line organs developed beneath the scales and within the skin. It is the membrane that covers these

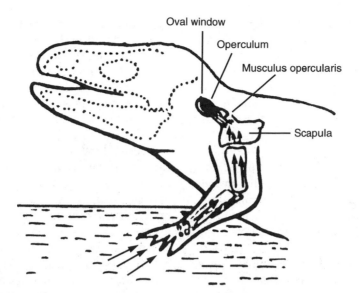

Oval window

Operculum

Musculus opercularis

Scapula

Figure 1.13 "Hearing" via bone conduction in a primitive amphibian, and the evolution of the "oval window." (From Tumarkin A. Evolution of the auditory conducting apparatus in terrestrial vertebrates. In: de Reuck AVS and Knight J, eds. Hearing mechanisms in vertebrates. CIBA Foundation Symposium. Boston: Little, Brown & Company, 1968).

canals which responds to and transmits vibrations picked up in sea water.

However, sharks and teleosts do not perceive sound per se (van Bergeik, 1966), at least in a manner analogous to mammals. Rather, in the progression from cartilaginous to lobed-finned fish, the lateral line system became increasingly modified so as to form the labyrinth that, in turn, was capable of detecting the animal's own movements as well as that of other creatures (van Bergeik, 1966). In more advanced vertebrates, the labyrinth forms and leads to the creation of the three semicircular canals that are responsible for perceiving motion, movement, and spatial orientation. Over the course of evolution, nerves from the semicircular canals, in turn, became enmeshed within the acoustic nerve. Together, they terminate within the medulla of the brainstem and send collaterals to the evolving cerebellum and the inferior auditory midbrain tectum (Brodal, 1981).

The mammalian auditory system, however, did not evolve from the lateral line but from a cluster of nerve cells located within the brainstem, the vestibular nucleus (Ariens Kappers, 1929; Papez, 1967), as well as from the jaw and scapula. However, in this regard, hearing initially remained tightly linked to vibratory perception, which, in turn, was also made possible via bone conduction.

Middle Ear/Midbrain Evolution

Once amphibians evolved and began to crawl upon dry land, some 300 million years ago, bone conduction replaced the lateral line system as a means of detecting vibrations and, thus, the rudiments of sound (Tumarkin, 1968). Because these creatures at first crawled upon the earth, they were able to detect vibrations through their legs as well as through the lower jaw. In consequence, the hyomandibular bone of the teleost evolved into the amphibian (and mammalian) stapes, whereas the vestibuloscapular bone located atop the scapula of the amphibian shoulder became modified to form the vestibuloscapular ossicle. This enabled vibrations to be perceived through the legs, even when the head and jaw were no longer on the ground (Tumarkin, 1968). Coinciding with these developments were further elaborations within the midbrain and brainstem, as well as within the forebrain.

With further refinements in the structure and support of the jaw, and with the evolution of the reptiles, repto-mammals, and then the dinosaurs some 250–225 million years ago, the articular and quadrate bones (which formerly supported the mandible) became modified, along with the stapes, to form the three ossicles of the middle ear. What had been the amphibian vestibuloscapular ossicle evolved into the oval window (Tumarkin, 1968). When this occurred, the dorsal midbrain (the inferior auditory tectum/colliculus) had become specialized for perceiving, analyzing, and orienting toward complex sounds, whereas the midbrain periaqueductal gray became specialized for coordinating the oral-facial, laryngeal, and respiratory muscles. The midbrain periaqueductal gray enabled these creatures to vocalize an increasing range of sounds.

The Evolution of Emotional-Vocal Perception and Communication

The most prominent nuclei of the midbrain include the superior (visual) and inferior (auditory) tectum, referred to as colliculi in the mammalian brain. As noted, the superior colliculus is concerned with analyzing visual signals and orienting the head and body toward the source of stimulation (via its extensive interconnections with the brainstem reticular formation and motor nuclei).

Once creatures began to adapt to a life on land, and with the evolution of hearing, the auditory tectum appeared. Being an extension of the brainstem, the auditory tectum was also concerned with arousal and orienting reactions in response to sound, and triggering specific motor programs when stimulated (see Donkelaar, 1990).

However, the types of auditory stimuli the organism was initially most concerned with were those indicating the presence of mates, food sources, and hidden or approaching predators—which, in turn, are associated with the production of transient sounds. It is these sudden, transient sounds that could alert the animal of the need to flee, freeze, or engage in further exploration. As such, these sounds were also relayed to and analyzed by the amygdala (see Chapter 5).

These adaptive auditory capabilities were further enhanced with the evolution of the repto-mammals (therapsids) and via the evolution of the anterior cingulate gyrus, and by further modifications within the dorsal and ventral striatum. As noted, it is the striatum in which associated species-specific motor programs (e.g., kicking, running in circles) are stored (in conjunction with the motor primitive motor subroutines maintained within the brainstem). In addition, with the evolution of the repto-mammals, the amygdala also become increasingly concerned with sound perception and the analysis of auditory nuances in order to trigger emotional, motivational, and alarm reactions, such as flight or fight (Joseph, 1992a, 1993; MacLean, 1990; Parham and Willot, 1990; Rosen et al., 1991). This is made possible via its rich interconnections with the basal ganglia, midbrain, and brainstem reticular formation.

THE REPTO-MAMMALS (THERAPSIDS)

Reptiles began to differentiate and evolve into the repto-mammals (the therapsids) some 250 million years ago (Brink, 1956; Broom, 1932, Crompton and Jenkins, 1973; Paul, 1988; Romer, 1966). Twenty-five million years later the first tiny dinosaurs (who diverged from a different line of reptiles, the theocondants) began to roam the Earth (Paul, 1988). Initially, the repto-mammals ruled the planet. However, their dominion was overturned by a series of catastrophic events that occurred when life-destroying meteors struck the planet about 225 million years ago (Raup, 1991; Stanley and Yang, 1994). Since larger creatures (i.e., the therapsids) were more severely affected, this event led to the ascendancy of the dinosaurs (Joseph, 1993), who soon grew to huge sizes.

This forced the remaining and smaller species of repto-mammals to adapt to a whole new way of living, including night-time foraging. This new life style, however, resulted in improved or greatly modified visual, auditory, and olfactory-pheromonal functional capacities, and the telencephalon began to expand with the addition of new layers of cortex. As such, these creatures also became more intelligent.

The initial evolution of the repto-mammals also coincided with and resulted in major biological alterations involving cranial and postcranial skeletal structure, mammillary development, thermoregulation, sexual reproduction, and hypothalamic function and structure (Bakker, 1971;

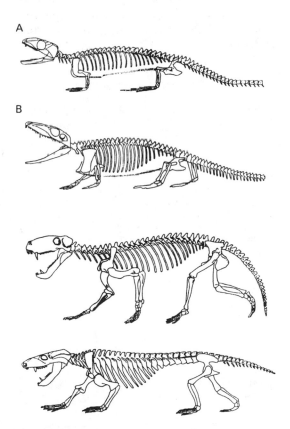

Figure 1.14 Skeletal structure of (**A**) a primitive reptile and (**B**) three repto-mammals. Note placement of legs has shifted from alongside the body to beneath the body. (From MacLean P. The evolution of the triune brain. New York: Plenum, 1990.)

Brink, 1956; Broom, 1932, Crompton and Jenkins, 1973; Duvall, 1986; Joseph, 1993; MacLean, 1990; Paul, 1988; Quiroga, 1980; Romer, 1966). Important physical alterations occurred, not just in their brains but in their body structure, stance, gait, and limb development (Colbert, 1980; Jarvik, 1980; Jerison, 1973; Romer, 1966), which coincided with further expansions in the striatum and motor cortex. For example, in contrast to lizards and amphibians, the legs of the therapsids were now located beneath rather than alongside the body. This enabled them to run for long periods while simultaneously breathing. Reptiles must stop in order to breathe since their legs, situated alongside their body, constrict the expansion of the lungs as they run.

The forebrain, brainstem, midbrain, thalamus, and amygdala (and evolving cingulate) became increasingly concerned with auditory sound perception and production. Coinciding with these developments, the ability to communicate expanded beyond posturing and olfactory signaling and now included the capacity to produce a variety of complex meaningful sounds (Joseph, 1993; MacLean, 1990), i.e., limbic language. Indeed, the evolution of these creatures coincided with tremendous advances in the ability to engage in audio-vocal communication, as well as the capacity to nurse the young, changes reflected in the expansion of the amygdala and the evolution of transitional cingulate cortex.

Over the next 150 million years, due to their new lifestyle and the competitive pressure and presence of the dinosaurs, the repto-mammals became smarter. Their brains became larger and more complex, until finally they evolved into the first true, neocortically equipped mammals, around 150 million years ago (Olson, 1970).

THE ANTERIOR CINGULATE, MATERNAL BEHAVIOR, AND THE SEPARATION CRY

With the appearance of the repto-mammals 250 million years ago, the middle ear began to undergo tremendous modification and the first rudiments of an inner ear developed (Broom, 1932; Brink, 1956; Crompton and Jenkins, 1973; Romer, 1966). The scent glands were modified and became mammillary glands—i.e., the maternal nipple—such that the capacity to nurse came into being (Duvall, 1986).

It was also at this time that the cingulate gyrus began to evolve and increasingly enshroud the anterior medial and dorsal surface of the forebrain (Joseph, 1993; MacLean, 1990), whereas the amygdala evolved additional cortex and nuclei. The evolution of the cingulate gyrus coincided with a stage in evolution when sound came to serve as a means of purposeful and complex communication, such as occurs not only between potential mates or predator and prey, but between mother and infant (Joseph, 1993, 1994; MacLean, 1990). In fact, the most recently evolved, five-layered transitional limbic cortex, the (anterior) cingulate gyrus, is exceedingly vocal, and can produce sounds that do not correlate with mood, indicating considerable flexibility and plasticity within this structure (Jurgens and Muller-Preuss, 1977).

The cingulate is not only capable of producing a complex medley of emotional sounds but also the separation cry, similar if not identical to that produced by an infant (MacLean, 1990; Robinson, 1967). Coupled with the newly evolved auditory capabilities of the amygdala, it is these particular sounds that promote intimate maternal behavior and watchfulness as well as maintain emotional attachments. Infantile separation cries and maternal behavior are both closely associated with the functional integrity of the cingulate (as well as the amygdala).

For example, among humans and lower mammals, anterior cingulate destruction can result in a loss of maternal responsiveness and mutism, and can significantly impair emotional-prosodic vocalizations (Barris and Schuman, 1953; Kennard, 1955; Laplane et al., 1981; Smith, 1944; Tow and Whitty, 1953). In primates, maternal behavior is also abolished, and the majority of infants whose mothers have suffered anterior cingulate destruction soon die from lack of care.

Limbic Language

The amygdala, as well as the septal nuclei, hypothalamus, and periaqueductal gray, became involved in sound production

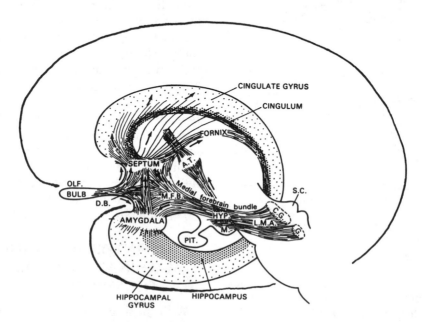

Figure 1.15 Internal view of human brain, depicting portions of the limbic system. (Adapted from MacLean P. The evolution of the triune brain. New York: Plenum, 1990).

long before the evolution of the cingulate gyrus. Moreover, each nuclei was (and is) concerned with different types of sound production (Jurgens, 1990; Jurgens and Muller-Preuss, 1977; Robinson, 1967, 1972). This is because different limbic nuclei, and, in fact, different divisions within these nuclei, subserve unique functions and maintain different anatomical interconnections with various regions of the brain (see Chapter 7 for a detailed discussion of limbic language).

For example, portions of the septal nuclei, hippocampus, medial amygdala, midbrain periaqueductal gray, and medial hypothalamus have been repeatedly shown to be generally involved in the generation of negative and unpleasant mood states (Joseph, 1990a, 1992a; Olds and Forbes, 1981, for review). Other limbic tissues, including the lateral hypothalamus, lateral amygdala, and portions of the septal nuclei, are associated with pleasurable feelings. Hence, areas associated with pleasurable sensations will produce pleasurable calls, whereas those linked to negative mood states will trigger shrieks and cries of alarm.

It is noteworthy, however, that vocalizations triggered by excitation of the amygdala, hypothalamus, or septal nuclei are usually accompanied by appropriate and mood-congruent behaviors (Gloor, 1960; Joseph, 1990a, 1992a; Jurgens, 1990; Robinson, 1967; Ursin and Kaada, 1960). In contrast, the cingulate gyrus is capable of considerable emotional flexibility and can trigger a variety of vocalizations that do not correspond to mood, including, for example, the production of sounds that may fool a predator into thinking its prey is injured; this directs the predator away from those who are truly helpless.

The Evolution of Social and Group Attachments

Sharks, teleosts, amphibians, and reptiles, although lacking a cingulate gyrus, possess a limbic system consisting of an amygdala, hippocampus, hypothalamus, and septal nuclei. It is these limbic nuclei that enable a group of fish to congregate and "school," and which make it possible for reptiles to form territories that include an alpha female, several subfemales, and a few juveniles (Joseph, 1993; MacLean, 1990). That is, these nuclei promote the formation of group attachments.

Nevertheless, these creatures do not develop long-term social or emotional attachments. In fact, amphibians and most, but not all, reptiles show little or no maternal care and will greedily cannibalize their infants, who, in turn, must hide from their parents and other reptiles in order to avoid being eaten (MacLean, 1990).

In contrast to reptiles and amphibians, the repto-mammalian therapsids were equipped with a recently evolved cingulate cortex, as well as a newly evolved cortex within the amygdala; they lived in packs or social groups and presumably cared for and guarded their young for extended periods lasting until the juvenile stage (Bakker, 1971; Brink, 1956; Crompton and Jenkins, 1973; Duvall, 1986; Paul, 1988; Romer, 1966). Being equipped with a more complex amygdala, septal nucleus, and cingulate gyrus, these animals developed long-term social and emotional relationships.

It is these limbic nuclei that enable human infants to indiscriminately seek social and physical contact, then, as these nuclei mature, to increasingly display fear of strangers and to form specific attachments. In humans and other mammals, these changes in attachment behavior correspond with the maturation of the amygdala, followed by the septal nuclei, and finally the cingulate gyrus (Joseph, 1982, 1990a, 1992a, 1993, 1994). However, it is the mammalian cingulate gyrus that not only made possible long-term social-emotional attachments, but which provided the impetus for what would become mammalian social-emotional and maternal behavior, including the pronounced tendency to engage in play (MacLean, 1990). It is also the cingulate, coupled with the amygdala, that would eventually give rise to the evolution of modern human language.

For example, the anterior cingulate gave rise to the medial frontal lobe and supplementary motor area (Sanides, 1968, 1970. The medial frontal lobes, in turn, evolved up and over the lateral convexity to form the motor areas, including Broca's speech area. Hence, it could be argued that Broca's area evolved from the anterior cingulate.

Over the course of the last half billion years, the amygdala became increasingly cortical in structure as well as multimodally responsive. As it evolved, the amygdala also increased its involvement in social-emotional perception and expression including feelings of affection, and the formation of emotional attachments. Moreover, as the amygdala (and hippocampus) became increasingly cortical, the medial and lateral temporal convexity began to balloon outward and upward. This gave rise to the insula and superior temporal lobe (Sanides, 1969, 1970), which contains the primary and secondary auditory receiving areas, and, eventually, portions of the inferior parietal lobule (see Chapter 4).

Hence, over the course of evolution but prior to the acquisition of modern human speech, the lateral amygdala evolved neocortical interconnections with the primary and association auditory areas, including Wernicke's area, the lateral and orbital frontal lobes, and the parietal lobule (Joseph, 1993). These neocortical interconnections enabled the lateral amygdala to extract as well as impart social-emotional and motivational significance to sound sources and expressive speech. Similarly, the anterior cingulate gave rise to, established, and has maintained rich interconnections with the left and right frontal convexity and medial frontal lobes, thereby influencing and modulating sound production (Joseph, 1993). However, with the evolution of the angular gyrus of the inferior parietal lobule, and with further functional elaborations within Wernicke's and Broca's areas, not only could limbic language be hierarchically expressed and produced by the neocortex, but it became subject to temporal sequencing such that modern grammatical utterances became the norm (see Chapters 2 and 4).

THE NEOCORTEX

As noted, the first rudimentary form of two-layered cortex appears with the evolution of the lobed-finned fish, and appears to have been derived from phylogenetically old, midbrain-like tissue. Over the course of evolution, these two layers became pushed apart as a third cortical layer appeared (associated with the evolution of reptiles [Haberly, 1990]), followed by the remaining layers II–VIa over the course of the last 150 million years. In general, two-layered cortex is referred to as archicortex, whereas the six-layered to seven-layered cortex is referred to as neocortex.

Specifically, the six layers of neocortex are:

I. Molecular (Golgi II cells, with small scattered horizontal cells);
II. External granular (densely packed small pyramidal, stellate, and granule cells);
III. Pyramidal (two sublayers of medium pyramidal cells);
IV. Internal granular (small pyramidal and stellate cells);
V. Ganglionic (large and medium pyramidal cells);
VI. Multiform (spindle-shaped) and pyramidal cells.

The thickness of these layers, the size of their neurons, and, thus, the size of the neocortex varies, however, depending on brain region, with a range of 1.3–4.5 mm.

The Evolution of the Lobes

Over the course of evolution, the lateral olfactory area came to correspond with the piriform (paleocortex) and entorhinal cortex and developed into the anterior-dorsal-lateral hippocampus (or archicortex) and amygdala. Large portions of the medial olfactory area gave rise to and became homologous with the septal nuclei and the medial-dorsal-posterior portion of the hippocampus (as well as the medial amygdala).

Presumably, during the transition from repto-mammals to true mammals, the cortex differentiated and ballooned outward. The hippocampus, as a whole, became displaced

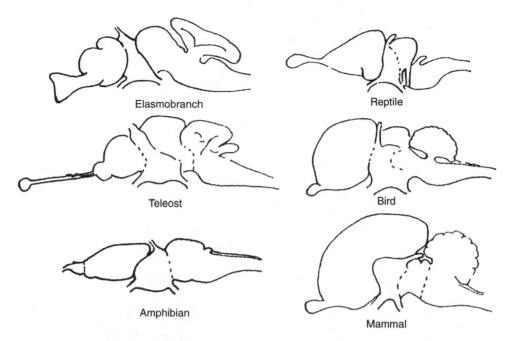

Elasmobranch

Reptile

Teleost

Bird

Amphibian

Mammal

Figure 1.17   The evolution of the brain, depicting expansion of forebrain/telencephalon. (Redrawn from Hebb C and Ratkovic D. Choline acetylase in the evolution of the brain in vertebrates. In: Richter D, ed. Comparative neurochemistry. New York: Macmillan, 1964.)

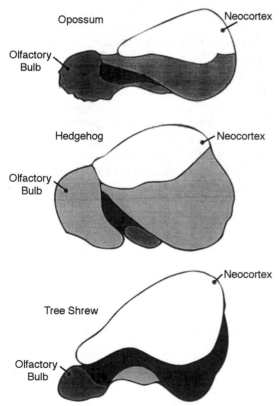

Opossum

Neocortex

Olfactory Bulb

Hedgehog

Neocortex

Olfactory Bulb

Neocortex

Tree Shrew

Olfactory Bulb

Figure 1.18   Expansion of the neocortex in two "primitive" mammals, and a prosimian primate (tree shrew).

and was increasingly pushed back, down, and into the medial temporal lobe (Sarnat and Netsky, 1981; Sanides, 1969, 1970), which, in turn, is a hippocampal-amygdala derivative. In that the hippocampus initially received visual input, its neocortical derivatives (e.g., portions of the superior parietal, occipital, and medial-posterior temporal lobes) also became visually responsive, with the superior parietal lobe also becoming concerned with body-image memories and the position of the body in space. In addition, as the dorsal hippocampus became increasingly ventrally located, it simultaneously began to produce the posterior cingulate gyrus, and (in conjunction with the septum) the posterior medial and posterior lateral walls of the cerebral hemispheres (Sanides, 1969). Presumably, in part, the anterior cingulate is a septal nucleus and amygdala-hippocampal derivative.

Neocortex

Initially (i.e., 150–165 million years ago), this neocortical shroud was quite smooth (lissencephalic), and the representation of visual, somatosensory, and auditory input was quite diffuse and crude (Sanides, 1970; Ulinski, 1990). Presumably, since the ancient neocortex was diffusely organized, the primary receiving areas may have been the last to evolve (Sanides, 1969), which makes sense in regard to fine motor function. However, some neuroscientists (see Allman, 1990) argue that the primary areas evolved first, followed by reorganization and functional elaboration in the secondary and association areas. In either case, defined topographic neocortical-functional maps came to be superimposed over the more diffuse pattern characteristic of reptiles and similar creatures (Allman, 1990), making fine perceptual analysis possible.

As repto-mammals, then mammals, evolved (around 150 million years ago), the neocortex continued to undergo evolutionary metamorphosis and became the recipient of topographically organized visual, auditory, and tactile input (Allman, 1990). The neocortex then funneled this highly processed information back to the limbic system as well as to the thalamus and brainstem. In this regard, large portions of the neocortex could be viewed, not only as a limbic derivative, but as an extension of the limbic system that evolved in order to engage in a much higher level of information processing.

Cortical Receiving Areas

Presumably, by 100 million years ago, convolutions and gyri began to form (at least in some mammals) as the neocortex began to fold up and under itself due to the limited amount of available space within the skull. Many regions within the neocortex became increasingly topographically and locally organized into visual, auditory, tactile, and motor areas, a pattern that has since become standard among all subsequent mammalian species. Hence, the posterior regions of the hemispheres are primarily concerned with sensory perception, analysis, and association, whereas the frontal areas are concerned with motor functioning and sensory integration, regardless of species.

THE ASCENDANCY OF MAMMALS

Once the primary, association, and multimodal tertiary neocortex had evolved, it began to provide the repto-mammals, and then mammals and primates, with a tremendously advanced capability to engage in multimodal sensory analysis and information storage. Neocortex provided mammals with the cognitive abilities to outthink their rivals. Therefore, when the mammals were given an additional, albeit temporary, competitive advantage when a huge meteor struck near the Gulf of Mexico some 65 million years ago (thus killing off large numbers of animals, particularly those of huge size, e.g., most dinosaurs), the remaining (smaller) dinosaurs were apparently unable to compete and were completely eradicated. Mammals, primates, and then humans soon came to rule the Earth.

2

Paleo-Neurology and the Evolution of the Human Mind and Brain

The Frontal and Inferior Parietal Lobes, Sex Differences, Language, Tool Technology, and the Cro-Magnon/Neanderthal Wars

By approximately 150 million years ago, mammals began to evolve from and replace the repto-mammal therapsids. Corresponding with the ascendancy of mammals was the continued expansion and development of the telencephalon and neocortex (Jerison, 1990). For example, the septal nucleus (and amygdala) contributed to the evolution of the anterior cingulate; from the anterior cingulate gyrus the medial and then the lateral (motor areas) of the frontal lobe were fashioned, whereas portions of the amygdala and hippocampus became increasingly cortical and gave rise to the inferior, medial, lateral, and superior temporal lobe (e.g., Sanides, 1969).

Hence, new layers of cortex (i.e., neocortex) began to enshroud the cerebral hemispheres; this, in turn, conferred upon these creatures extraordinary powers of intelligence, foresight, planning, and communication. Being much more intelligent, mammalian predators were becoming a serious competitive threat to the more dimwitted dinosaurs.

Presumably because of the competitive pressures exerted by the neocortically equipped mammals (Joseph, 1993), by 75–100 million years ago the number of dinosaurs began to decline throughout most parts of the world (cf. Schopf cited by Raup, 1991; Paul, 1988), with the possible exception of North America. However, about 65 million years ago a huge asteroid or meteor struck the planet near the gulf of Mexico (Alvarez, 1986) and a vast number of (large sized) species died out in response to this

cataclysm (Raup, 1991), particularly those who lived in North America: i.e., the dinosaurs. Nevertheless, mammals (including prosimian primates) and other smaller creatures were able to recover and apparently take competitive advantage of the situation. Hence, with the evolution of neocortically endowed mammalian Creodonta and related carnivores (Carnivora), the remaining dinosaurs were simply no match and were completely eradicated (Joseph, 1993). The Age of Mammals had begun.

PRIMATE EVOLUTION

With mammals and primates (who apparently diverged from the mammalian line some 70–100 million years ago [Colbert, 1980; Jarvik, 1980; Jerison, 1973; Romer, 1966]) in ascendancy—exploring, exploiting and gaining dominion over the Earth—the brain underwent further adaptive alterations and evolutionary advances in structure and organization. For example, as these creatures now ruled the day (as well as the night), there were tremendous expansions in visual cortex and in the size of the occipital and temporal lobes, especially among the primates some 50 million years ago (Allman, 1990; Jerison, 1990). In fact, 50 million years ago the primate brain was much larger than that of mammals of similar size.

It is believed that some of the first proto-primates may have been little rat-like creatures with long snouts and whiskers who devoured insects. In this regard, they were no match for the numerous mammalian predators, who lurked everywhere. On the other hand, it is just as likely that the first primates were, in fact, much larger, perhaps equivalent to a medium-sized dog. Nevertheless, it was presumably from one of these first primate stocks that "old world" monkeys, apes, and humans branched off and allegedly descended (Leakey and Lewin, 1977; Pfeiffer, 1985).

Some primate lines adapted to life on the ground, and these creatures flourished for almost 10 million years before dying out. By contrast, those primates who took to the trees rapidly adapted to living among the branches and the forest, and they flourished. They began to grow fingers, and their hands and feet became adapted for grasping. In consequence, tremendous alterations occurred in the parietal lobe and motor system, particularly in the frontal neocortex which was now significantly contributing to the control of the extremities as these creatures became increasingly adapted to a life in the trees. However, what has been (unfortunately) referred to as "prefrontal" cortex remained poorly developed (LeGros Clark, 1962).

The visual system was also modified in that the eyes had shifted from the sides of the head and had become frontally placed, which in turn resulted in stereoscopic vision, greater central visual acuity, and improved hand-eye coordination (Allman, 1990). Like other mammals, these primates also continued to rely on olfactory and pheromonal communication; and as in mammals, the olfactory bulb remained large (Jerison, 1973, 1990).

It was from this tree-loving stock that monkeys evolved in Africa about 40 million years ago. The apes evolved about 10 million years later (Leakey and Lewin, 1977; Pfeiffer, 1985). Although humans did not evolve from monkeys or apes, presumably it was from this same branch that the ancestral line leading to human beings diverged about 5–10 million years ago.

Evolutionary metamorphosis is most likely to occur when an organism is exposed to a multiplicity of environments, or where two divergent worlds meet. For the primates, the netherworld of change was found where the forest ended and the savanna and grasslands began. It was probably in such an environment that a new type of primate evolved and the line that would eventually give rise to Homo sapiens sapiens (the wise man who knows he is wise) appeared.

This pre-human may have been similar to Ramapithecus, who stood about three feet high, had a low forehead, flat wide nose, and face shaped like a muzzle. Although he was a seed eater and tended to grind his food, like his ape cousins, Ramapithecus (or like-minded males) probably occasionally hunted, captured, and killed by hand small game and possibly other primates, whereas females and their young obtained the brunt of their food by gathering (Leakey and Lewin, 1977; Pfeiffer, 1985). It is while on the ground that Ramapithecus or some other closely related primate-pre-hominid underwent further evolutionary metamorphosis, around 3–5 million years ago, when several subtypes of Australopithecus appeared in various parts of the world—all of whom coexisted for some time (Grine, 1988; Leakey and Lewin, 1977; Pfeiffer, 1985). Soon thereafter (about 2–3 million years ago), Homo habilis (the handy man) appeared.

AUSTRALOPITHECUS AND HOMO HABILIS

There is no general or widespread agreement as to the various possible phylogenetic relationships shared by the wide variety of Plio-pleistocene hominids so far discovered (Skelton and McHenry 1992), including Australopithecus and H. habilis. However, it has been proposed that H. habilis and the wide range of Australopithecines (A. aethopicus, A. africanus, A. robustus, A. boisei) that populated the Earth at that time, some 2–3 million years ago, may have shared a common ancestor; perhaps Australopithecus afarensis (see Grine, 1988; Leakey and Walker, 1988; Skel-

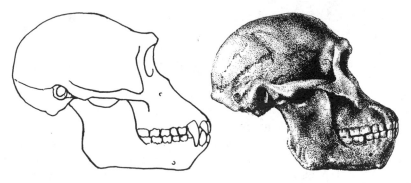

Figure 2.1 The skull of an Australopithecus transvallensis (right) and that of a chimpanzee (left). (From Clark G. The stone age hunters. London: Thames & Hudson, 1967.)

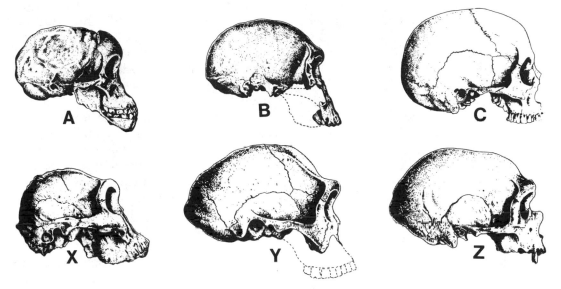

Figure 2.2 Two possible separate lines of human evolution as proposed by P. MacLean and as based on the enlargement of the frontal portion of the cranium: from Australopithecus africanus (**A**), to Homo habilis (**B**), to Cro-Magnon (**C**); versus Australopithecus robustus (**X**), to Homo erectus (**Y**), to Neanderthal (**Z**). (From MacLean P. The evolution of the triune brain. New York: Plenum, 1990.)

ton and McHenry, 1992 for a detailed discussion). Presumably, around 2.5 million years ago, H. habilis and the varying lines of Australopithecus branched off from this common Australopithecine ancestor, such that one or the other eventually gave rise to what would eventually become modern humans. Of course, it is also possible that H. habilis and many of the supposed different subtypes of Australopithecus may have given rise to separate branches of the human race.

For example, as based on enlargements in the frontal portion of the cranium, MacLean (1990) has proposed that modern humans (including the Cro-Magnon) may have descended from H. habilis, who he believes descended from Australopithecus africanus, whereas Neanderthal evolved from a distinct branch of H. erectus, who evolved from Australopithecus robustus. In truth, however, the evidence supporting any of the many competing hypotheses is so meager that arguments pro or con are little more than informed speculation.

In any case, unlike those who had come before them, Australopithecus and H. habilis were extremely advanced, for they had learned to stand on their hind legs and to walk in an upright manner (Johanson and Shreeve, 1989; Leakey, 1979; Leakey and Lewin, 1977). Similarly, their feet evolved so as to better accommodate standing, and they tended to walk and run on two legs rather than on all four, as do monkeys. In consequence, the arms and hands ceased to be weight bearers and were freed of the necessity of holding or hanging on to a tree branch in order to move about. In addition, the fingers and thumb underwent further modification and functional elaboration, and they gained the ability to not only hold and grasp, but to explore and manipulate objects.

For example, by the time H. habilis had come to wander the Earth, the thumb had become longer and stronger, which enabled H. habilis to engage in complex acts involving a refined precision grasp: e.g., the construction and utilization of complex tools (see Hamrick and Inouye, 1995; McGrew, 1995; Susman, 1995 for discussion of sup-

portive and contrary evidence). Correspondingly, the areas of the neocortex devoted to the representation of the hand increased (Richards, 1986; see also Darwin, 1871), which also improved versatility in fine motor control and communication.

Because the hands had become more versatile they could also be used for signaling among other primates so as to convey feelings of anger (such as by making a fist, pounding a tree, angrily swinging or throwing a tree branch) or a desire to share (such as begging for a small piece of meat), to indicate friendliness or the desire to form a social bond, or even to soothe an angry neighbor by grooming his coat. In this regard, the neocortex subserving the manipulation of the hand (the parietal lobe) became adapted for social-emotional communication as well as exploration and manipulation (Joseph, 1993).

Probably due in part to the increased efficiency in hand use, about 1.5 million years ago (Susman, 1995), these pre-humans began to shift from opportunistic scavenging to systematic hunting and gathering (Shipman, 1983), though, like male chimpanzees, they may have always been hunter-scavengers. It was also around this time that the brain doubled in size. However, as will be illustrated below, the evolution of modern human beings, and the transition from primitive Homo to modern sapiens sapiens, is not merely a consequence of increased brain volume and versatility in hand use but is due to the functional and anatomical evolution of the inferior parietal and anterior (pre-) frontal lobes (Joseph 1992b, 1993).

HOMO ERECTUS

Following on the heels of Homo habilis and Australopithecus were a wide range of quite different individuals collectively referred to as Homo erectus who lived throughout Africa and Eurasia from approximately 1.7 million ago until about 300,000 years ago. The Homo erectus were the first individuals to harness fire and the first to develop crude shelters and home

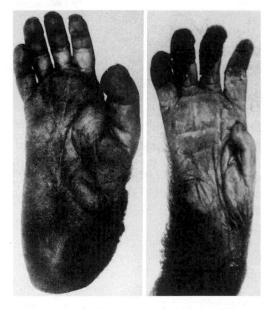

Figure 2.3 The foot (left) and hand (right) of a chimpanzee.

bases (Clark and Harris, 1985; Potts, 1984), around 500,000 years ago. They also utilized various earth pigments (ocher), possibly for cosmetic or artistic purposes. In this regard, perhaps such individuals were beginning to experiment with individual creative and artistic expression.

It was during the later stages of Homo erectus development, about half a million years ago, that the human brain appears to have become significantly enlarged (Rightmire, 1990; Tobias, 1971). It was also during this time that big game hunting had its onset.

As such, an additional divergence in the mind and brain of man and woman may have resulted, as males and females became increasingly specialized to perform certain tasks (see below). Males increasingly engaged in big game hunting, which coincided with improved visual-spatial perceptual and motor functions involving predominantly the right hand, whereas females continued to spend a considerable amount of time gathering, which coincided with improvement in bilateral fine motor and somesthetic perceptual skills, vocalizing, child rearing, scavenging or hunting small game, tool making, preparing food, and in maintaining the home base (see below).

The human females also probably ceased to have an estrus cycle around this time, so as to become sexually receptive at all times (Joseph, 1993). The loss of the estrus cycle and full-time sexual receptivity no doubt contributed significantly to the development of long-term male-female relationships, and so forth.

MULTIREGIONAL EVOLUTION AND REPLACEMENT

About 500,000 years ago, the first primitive and archaic Homo sapiens began to appear in increasing numbers in North Africa, the near Middle East, and even southern England. Wherever they wandered, they proved to be far more intelligent and resourceful as compared to the last remnants of Homo erectus, who over the next 200,000 years were eradicated or simply died out as a species.

It has been logically surmised that H. sapiens evolved from H. erectus. However, whether humans first "evolved" in Africa, Asia, or some other location is as yet an unresolved question. For example, considerable evidence suggests that not all H. sapiens evolved at the same time, in the same place, or from the same stock of H. erectus (Frayer et al., 1993; Moritz et al., 1987; Wolpoff, 1989). This has been referred to as the "multiregional" theory.

As argued by Frayer, Wolpoff, Thorne, Smith, and Pope (1993:17), "the best evidence of the fossil record indicates that modern humans did not have a single origin, or for that matter a number of independent origins, but, rather, it was modern human features that originated—at different times and in different places. This view also implies that certain features that distinguish some modern groups were developed very early in our history, after the exodus from Africa. *It is also important to recognize that not all features common to modern humans emerged at precisely the same time.*" (Emphasis added)

For example, there is some evidence that "archaic" Homo sapiens may have "evolved" from at least three different branches of H. erectus, each of which, in turn, gave rise to at least three branches of human beings who lived throughout Africa, Europe, the Middle East, and Asia. Of course, not all authorities would agree with the above (see Nitecki and Nitecki, 1994).

"Replacement"

Whereas the "multiregional" theorists (e.g., Frayer et al., 1993; Moritz et al., 1987; Templeton, 1992; Wolpoff, 1989) claim that the colonization of Asia and Europe may have begun over 800,000 years ago, the "replacement" theorists (e.g., Brauer, 1989; Mellars, 1989; Stringer, 1992) argue that modern humans originated in Africa, perhaps about 200,000 years ago. According to the replacement theory, in venturing out of Africa and invading the Middle East and

Europe, "modern" Homo sapiens sapiens "replaced," perhaps by killing off or out-competing, the more archaic H. sapiens (e.g., Neanderthals) who already dwelled in these coveted lands.

Presumably, due to population growth and other factors perhaps related to changes in culture, tool making, climate, or availability of game animals, "early modern" and anatomically "modern" H. sapiens (e.g., Cro-Magnons) eventually pushed their way up through Northeastern Africa, and then into the Middle East, and then into Europe, and the Northeastern Americas, and so on, eradicating "inferior" species as they went, such that by 50,000 to 40,000 B.P. the Cro-Magnon and "modern" H. sapiens had come to occupy a considerable part of the Earth (see Nitecki and Nitecki, 1994).

Hence (and broadly speaking), the "multiregional" view presents humans as evolving in multiple places from H. erectus, to archaic, to "modern." The "replacement" theorists argue that "modern" humans evolved in Africa, and in proliferating and invading other lands they eradicated those who had come before them.

However, as to the "replacement" and "Eve" hypothesis that all modern humans have descended from a single female ancestor in Africa about 250,000 years ago, the DNA evidence upon which this theory has been based has been shown to be invalid (Templeton, 1992). The recent notion that all humans have descended from a single male ancestor about 200,000 years ago is equally dubious.

"Multiregional Replacement"

Although the "replacement" versus the "multiregional" theories appear to be at odds, as will be detailed below, these theories are (at least somewhat) mutually supportive, particularly if one considers that the "replacement" theory may accurately depict events that occurred "multiregionally." For example, there is evidence suggesting that those "early modern" and anatomically "modern" H. sapiens sapiens who "evolved" in the Far East and Asia (versus those who "evolved" in Africa) radiated and ventured outward in all directions, invading the Middle East, Siberia, the Arctic, Japan, Australia, and the Americas (Barnes, 1993; Frayer et al., 1993; Moritz et al., 1987; Wolpoff, 1989), and "replaced," perhaps by killing off or out-competing, the more archaic H. sapiens who already dwelled in these lands. In fact, (as based on thermoluminescence and optical dating techniques), it has been recently reported (Rhys Jones, cited by Morell, 1995) that "modern" H. sapiens had probably arrived in Australia at least 60,000 years ago. Given that they would have had to sail almost 100 kilometers across open ocean in order to reach Australia (southern Asia being the closest land mass), and that there is evidence they utilized ocher for painting, considerable intellectual prowess and, thus, modernness is assumed.

By contrast, those who evolved in Africa radiated outward into Europe and the Middle East, and killed off the more archaic, more slowly evolving Neanderthals: a short (males 5'4" on average), heavily muscled, brutish-looking people who occupied Europe and parts of the Middle East for at least 500,000 years (e.g., 800,000 to about 35,000 years ago) and then quite suddenly disappeared from the face of the Earth. Presumably, modern H. sapiens sapiens (be they of African or Asian origin) were provided a comparative competitive advantage due to the anatomical and functional expansion of the frontal and superior-inferior parietal lobe and killed off the Neanderthals. In this regard, the "replacement" and "multiregional" views of evolutionary development are complementary in that both views entail competition and the replacement of "inferior" species

SUMMARY

Overall, the evidence suggests that the H. sapiens sapiens who appeared in Africa and the Middle East, and those who appeared in the Far East and Asia, probably semi-independently evolved from different lines of archaic H. sapiens, who, in turn, may have branched off from advanced forms of H. erectus over 500,000 years ago in Africa, Europe, and the Middle and Far East (Frayer et al., 1993; Moritz et al., 1987; Templeton, 1992; Wolpoff, 1989). It also appears that "early modern" and "modern" H. sapiens were provided a competitive neurological advantage over archaic humans, including Neanderthals: an advantage reflected in the functional evolution of the frontal and inferior parietal lobule (IPL)—which in part is derived from superior parietal and auditory neocortex in the temporal lobe. That is,

although Neanderthals or other archaic humans might have eventually evolved into modern humans, they were placed at a competitive disadvantage as these traits and functions differentially appeared in a different race of people at an earlier time.

Given that the nature of living things is to compete for survival, it would therefore be expected that neurologically advanced racial groups would replace those of inferior physical, cerebral, and intellectual stock: behaviors and evolutionary trends that are also consistent with the development and dispersion of other species (Darwin, 1859; Wallace, 1895) and which is

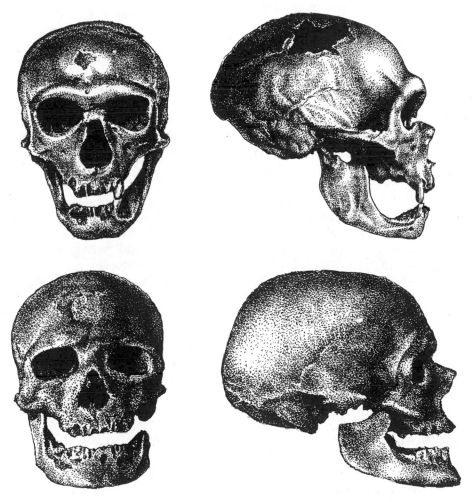

Figure 2.4 Neanderthal (top) and Cro-Magnon (bottom) craniums. Note expanded frontal region of skull in Cro-Magnon. (From Clark G. The stone age hunters. London: Thames & Hudson, 1967.)

known as "survival of the fittest," though in some cases, it is survival of the "lucky" (Raup, 1991).

"REPLACEMENT" AND THE NEANDERTHAL FRONTAL LOBE

The role of the frontal lobe and IPL in the evolution of modern H. sapiens sapiens has all but been ignored by anthropologists, who instead have concentrated on postcranial differences and evolutionary adaptations in tool-making. However, as will be detailed below, it is the evolution of the brain, particularly the anterior ("pre-") frontal lobe and IPL (i.e., angular gyrus) that not only accounts for the cultural differences between Middle Paleolithic and Upper Paleolithic peoples, but which accounts for the demise of the Neanderthals (Joseph, 1993).

From a "replacement" standpoint, it could be (and is) argued that Neanderthals died out and/or were killed off and thus replaced (see Mellars, 1989; Nitecki and Nitecki, 1994). Others have argued that despite the considerable differences between these peoples and the Cro-Magnon in height, stature, cranial development, and so on, that Neanderthals suddenly "evolved" into anatomically modern humans over the course of approximately 5,000 years. However, when considered not just from a cultural but from a neurological perspective, the evidence supports a "replacement" scenario. The Neanderthals were probably subject to widespread "ethnic cleansing" and were exterminated—a consequence of an inability to successfully compete due to inferior cultural, cognitive, and frontal and parietal lobe evolutionary development. As compared to the modern "Cro-Magnon" and even "early modern" H. sapiens, the Neanderthal was neurologically and cognitively inferior (Joseph, 1993) and were apparently eradicated as a species about 35,000 years ago (Mellars, 1989).

Unfortunately, as there are no Neanderthal brains in existence, this competitive disadvantage must be inferred or deduced neuropsychologically from the archaeological record, and as based on an examination

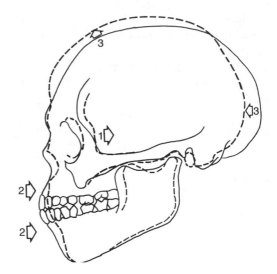

Figure 2.5 A "modern" (solid line) mesolithic cranium compared with a more ancient cranium (dotted line). Arrows indicate the main average changes in skull structure, including a reduction in the length of the occiput and an increase and upward expansion in the frontal cranial vault. (From Wolpoff MH. Paleo-Anthropology. New York: Alfred A Knopf, 1980.)

of cranial configuration. For example, it is apparent that the height of the frontal portion of the skull is greater in moderns and "early moderns" than Neanderthals (see also Wolpoff, 1980:Table 12.1), which suggests impoverished frontal lobe capacity in Neanderthals. Indeed, their characteristic "sloping forehead" was a limiting factor in frontal lobe development. (See Fig. 2.5)

Similarly, a gross photographic examination of various Neanderthal versus "modern" endocasts (taken from Smith, 1982:673; and from Tilney, 1928:919 and Figures 389–400 and 417–422), and the data provided by Tilney (1928:923) strongly suggests that the "modern" as contrasted with the Neanderthal (superior and inferior-orbital) frontal lobe has, in fact, expanded (Joseph, 1993; MacLean, 1990).

Although the Neanderthal brain is believed to be about 6% larger than the modern cerebrum, at least in overall volume (Holloway, 1985; Wolpoff, 1980), the maximum length and breadth is in the occipital and superior parietal areas

(Wolpoff, 1980)—regions concerned with visual analysis and positioning the body in space (see Chapters 12 and 13). As male and female Neanderthals spent a considerable amount of their time engaged in hunting activities (see below), scanning the environment for prey and running and throwing in visual space were more or less ongoing concerns. A large occipital and superior parietal lobe would reflect these activities.

It is recognized, of course, that one must not give undue credence to impressions based on skull size alone. Endocasts are also highly suspect, as these are molds of the inside of a skull and not the brain and can, therefore, provide only gross overall impressions—not anything resembling precise anatomical detail.

Nevertheless, it is interesting to note that the occipital region of the skull is much shorter in "moderns" (and the frontal por-

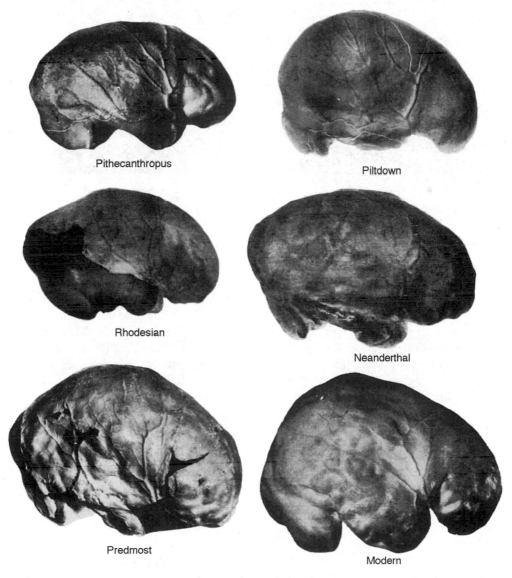

Pithecanthropus

Piltdown

Rhodesian

Neanderthal

Predmost

Modern

Figure 2.6   A comparison of endocranial casts of Neanderthal, "Predmost" (a contemporary of Cro-Magnon), and three H. erectus from various geo- graphical areas, as compared with a "modern" H. sapiens sapiens. (From Tilney F. The brain from ape to man. New York: Hoeber, 1928.)

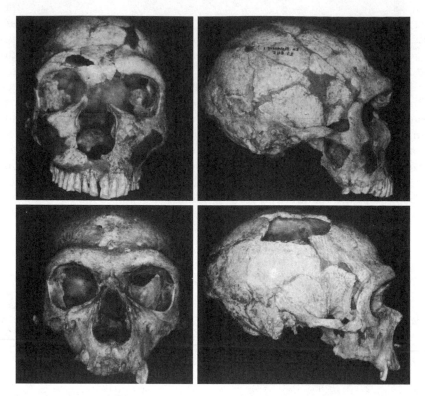

Figure 2.7 Two well-preserved crania from northern European male Neanderthals. (From Wolpoff MH. Paleo-Anthropology. New York: Alfred A Knopf, 1980.)

tion is larger). This suggests that the huge expanse of visual cortex has either shrunk—which is unlikely, given that the Cro-Magnon also spent a considerable amount of time hunting—or the visual cortex was displaced into the medial regions of the brain—which is typical of the modern human cerebrum (see Chapter 13). What would cause such a reorganization? Expansion of the parietal lobe, the inferior parietal lobe in particular (Joseph, 1993). As based on a functional and neuropsychological analysis of technological expertise and behavior, "modern" and even "early-modern" humans appear to be superiorly endowed with an inferior parietal lobule (i.e., angular gyrus), as well as enlarged frontal lobe—as compared to Neanderthals.

Given the functional significance of the anterior (nonmotor) frontal lobe (see Chapter 11), and its role in planning, foresight, and creative and intellectual activities, and the contribution of the IPL to tool making and language (see Chapter 4 and below), the Neanderthals, with their inferior cognitive and cultural capabilities and poorly developed frontal and inferior parietal lobule, would have been no match for the superiorly endowed Cro-Magnon. Hence, they were likely "replaced" about 35,000 years ago, at which point they simply disappeared from the scene.

However, if considered from a "multiregional" perspective, it could also be surmised that Neanderthals might have "evolved" into modern humans. That is, if they had not been exterminated by the Cro-Magnon, a people who stood as tall, lived as long, and, in fact, had a greater cranial capacity as compared to present-day humans.

Neanderthal Brain Development

Brain size alone tells us little about function, whereas action and behavioral accomplishment speak volumes as to cerebral, cognitive, and intellectual functional integrity and capability. Moreover, it could certainly be argued that what appears to be impoverished Neanderthal frontal lobe development is, in fact, an illusion created by their massive brow ridge. But this does not appear to be the case.

Even if we forgo the temptation to make dubious claims based on the size and shape of various endocasts, and ignore the somewhat flattened, sloping Neanderthal forehead and the evolutionary record, which indicates a comparative impoverishment in Middle Paleolithic cultural, tool making, and symbolic accomplishment, there is physical evidence which suggests Nean-

derthal brain development was characterized by significant limitations in plasticity and learning capability (e.g., Chapter 18 for related discussion). For example, based on Neanderthal pubic morphology and speculations on gestation length, it appears that their infants were born at an advanced stage of maturity and development (Rosenberg, 1988; Wolpoff, 1980). This would not be unusual, as our closest living primate relative, the chimpanzee, is born with a brain that is almost 50% of that of an adult, whereas newborn human brain volume and size is less than 25% of an adult (Passingham, 1982; Sacher and Staffeldt, 1974).

If Neanderthals were born at an advanced state of development, then like other

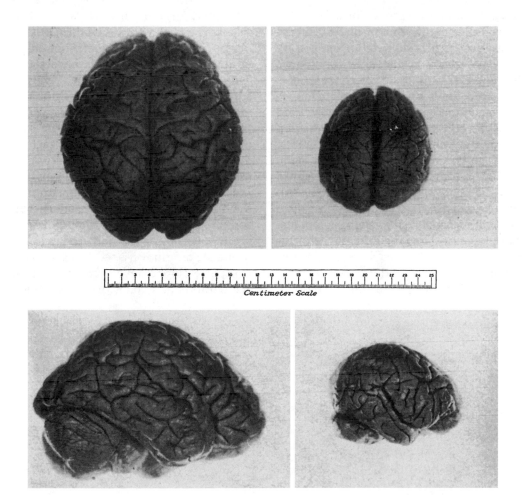

Figure 2.8 Comparison in size and configuration of the modern human brain with the chim-

panzee brain. (From Tilney F. The brain from ape to man. New York: Hoeber, 1928.)

rapidly maturing creatures their brains would have been more "hard wired" (however, see Trinkaus, 1987). Although this might have enabled them to respond to environmental emergencies without being hindered by prolonged helplessness, their ability to modify behavior patterns over the course of development would have been significantly hampered, at least as compared to modern humans.

Moreover, given that the frontal and inferior parietal lobes take well over 7–20 years to functionally mature in "modern" humans (see Goldman, 1971; A. Diamond, 1991; Huttenlocher, 1979; Joseph, 1982; Joseph and Gallagher, 1985; Joseph, et al., 1984), rapid brain development and maturity would preclude functional growth and elaboration in these regions. Consider, for example, that the more rapidly developing chimpanzee brain is completely lacking an angular gyrus (see Geschwind, 1965), whereas the majority of the frontal lobe is devoted to motor functions.

In fact, from an educational (enriched versus deprived) perspective alone, given that Neanderthals lived on average to age 40, compared with age 60 for Cro-Magnon, and the amount of time males and females spent engaged in hunting, the Neanderthal young would not have had aged relatives to care for and train them, and their mothers would have not have had the time to educate them. As detailed in Chapter 18, conditions such as these would have also exerted pronounced influences on the developing

brain, although individuals with rapidly developing and maturing brains would not be as profoundly affected.

However, even if the Neanderthal brain did not mature more rapidly than in modern humans, and we ignore their limited life spans and time spent hunting, it is still clear that they were not as well endowed neurologically, and thus cognitively and intellectually, and were therefore destined to lose the human race.

Paleoneurology and the Paleolithic Cultural Transition

Regardless of which theory one might favor, e.g., "replacement," "multiregional," or the possibility that Neanderthals evolved into modern humans, overall and ultimately it is the evolution of the frontal and inferior parietal lobule (the progression of which coincides with the transition from "archaic" to anatomically "modern" populations) that accounts for the major changes in tool making, cognitive development, language, and artistic achievement which characterize the Middle to Upper Paleolithic transition about 35,000 years ago.

Those who stress "cultural" vs neurological factors as responsible for this transition are simply ignoring (and/or are ignorant of) the obvious relationship between the brain and cognition and the restrictive influence of impoverished frontal and inferior parietal lobe development on cultural acquisition and expansion and even learn-

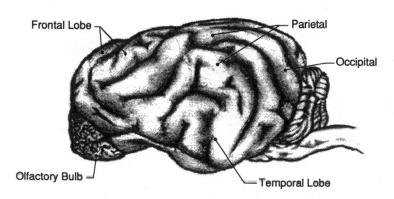

Figure 2.9   Lateral view of the dog brain.

ing and memory. For example, although a superior cultural environment and greater experiential opportunities can influence neuronal density, size, and function (as compared to those reared in a deprived environment, Chapter 18), a chimpanzee or a dog brain will remain exactly that, regardless of environment or "culture" (unless repeatedly experienced across multiple generations for many thousands of years); the same can be said of the Neanderthal versus "modern" brain.

Thus, as will be explained in detail below, the differential evolution of human cognitive and cultural development, particularly over the course of the last 100,000 to 6,000 years, is best understood in regard to differential frontal and inferior parietal lobule evolution; that is, as contrasted between Neanderthals, "early modern" and anatomically "modern" H. sapiens. Differences in cognition, culture, tool technology, etc., are a function of and reflect these neurological developments.

EVOLUTION OF "MODERN" AND NEANDERTHAL HOMO SAPIENS: THE FRONTAL LOBES

According to the interpretation of the data that will be presented here, as recently as 500,000 and perhaps as long as 800,000 years ago, variant populations of archaic humans occupied parts of Asia, the Middle East, Europe, and Africa (Frayer et al., 1993; Wolpoff, 1989). However, due to the extreme arctic cold and unique environmental conditions that characterized Middle Paleolithic glacial Europe and the middle East, a particular branch of the human race—i.e., Neanderthal—may have become geologically, genetically, and socially isolated and appear to have evolved differently from those archaic and "early modern" Homo sapiens who were living in Africa and Asia under different environmental and climatic conditions (Mellars, 1989).

As has now been well established, environmental influences, particularly during infancy and childhood, can exert drastic and significant influences on neurochem-

istry, neural development, and functional and structural interconnections throughout the brain, including the neocortex and limbic system (see Chapters 7 and 18). Behavior, emotional expression, perception, intelligence, neuronal growth and size, dendritic arborization, and neurochemistry are correspondingly effected (Bowlby, 1940, 1951; Casagrande and Joseph, 1978; 1980; Diamond, 1985; Denenberg, 1981; Ecknerode et al., 1993; Harlow and Harlow, 1965; Joseph, 1979; Joseph and Casagrande, 1980; Joseph and Gallagher, 1980; Joseph et al., 1978; Langmeier and Matejcek, 1975; Rosenzweig, 1971; Salzinger et al., 1993; see Chapter 18).

If generation after generation are exposed to similar adverse and intellectually limiting environmental influences, such as characterized the Neanderthal occupation of glacial Europe during the Middle Paleolithic, the impact would have been even more profound, effecting not just cranial and postcranial development but cerebral functioning across generations. By contrast, given that the racial-cultural groups who "evolved" in Africa and Asia some 800,000 years ago were exposed to different and changing environmental experiences and challenges for hundreds of thousands of years, it might be expected that their brains and thus their cultural and cognitive capabilities would have also "evolved," or at least developed, somewhat differently.

Frontal Cranial Evolution

These "racial" cerebral/cognitive differences are evident by 100,000 years B.P. and include postcranial anatomy, tool construction, hunting techniques, dental wear, the size and functional capacity of the frontal lobe, and the evolution of the IPL and angular gyrus.

For example, in Africa "archaic" H. sapiens, with their characteristic sloping frontal cranium, have been discovered buried in caves and rock shelters on the Klasies River of South Africa, and from Laetoli in Tanzania, from sites dated around 120,000 B.P. (Butzer, 1982; Grun et al., 1990; Rightmire, 1984). However, over the next 30,000 years the frontal cranium increases in size and "early modern" H. sapiens begins to appear.

By 90,000 B.P. these "early modern" peoples appear to have radiated northward, as Homo sapiens with much more advanced features, including individuals with rounded frontal cranium, have been unearthed from Qafzeh in the Middle East (Schwarcz et al., 1988). As these peoples apparently evolved from archaic H. sapiens, it is reasonable to assume the frontal lobes increased in size—as based on an examination of cultural record, as well as the frontal cranium.

However, these "early modern" H. sapiens were not only more advanced than "archaics" but they were superior to the Neanderthals who were living in Europe, as well as those living and occupying nearby sites (Mellar, 1989; Stringer, 1988). This is true culturally, cognitively, and as based on the expansion of the frontal cranium. Indeed, classic Neanderthal skulls with their large brows, flattened foreheads, and thus reduced frontal lobes have been found in the Middle East (i.e., Kebara cave)—sites dated as recently as 60,000 B.P. (Bar-Yosef, 1989). Therefore, the 90,000-year-old "early modern" Qafzeh skulls are more advanced than the 60,000-year-old Neanderthal skulls located nearby. Moreover, a Neanderthal skull from Saint-Cesaire in Western France, and dated from between 33,000 to 35,000 B.P., also displays the characteristically reduced frontal lobe cranial features (see Mellars, 1989: 351).

However, by 35,000 B.P., "early modern" H. sapiens apparently had evolved into "modern" H. sapiens: individuals who sported fully modern cranial and postcranial features and who now lived in Asia, Africa, the Middle East, and Europe. For example, also unearthed in western France are the 30,000- to 34,000-year-old skeletal remains of a Cro-Magnon displaying the bulging frontal cranium that is characteristic of "modern" humans (see Mellars, 1989). This strongly suggests that Cro-Magnon and Neanderthals occupied adjacent (or even the same) territories at about the same time, or that a tremendous expansion suddenly occurred in frontal lobe capacity such that Neanderthals "evolved" into modern humans within 5,000 years.

THE FRONTAL LOBES

The frontal lobes serve as the "Senior Executive" of the brain and personality, and together with the motor areas make up almost half of the cerebrum (see Chapter 11). In contrast, the "archaic" H. sapiens and Neanderthal frontal lobe makes up about a third, as a larger proportion of the brain appears to be occipital lobe and visual cortex.

Because the modern frontal lobe is so extensive and highly developed, and as different frontal regions have evolved at different time periods, they are concerned with different functions (see Chapter 11; see also Fuster, 1996; Van Hoesen, 1996). For example, about one-third of the frontal lobe, i.e., the motor areas, are concerned with initiating, planning, and controlling the movement of the body and fine motor functioning. It is this part of the "archaic" and Neanderthal frontal lobe that appears to be most extensively developed. The "orbital frontal lobes" act to inhibit and control motivational and emotional impulses arising from within the limbic system (Joseph, 1990d). Via orbital frontal interconnections with the limbic system it is possible for emotions to be represented as ideas, and for ideas to trigger emotions (Joseph, 1990a, d). An examination of the "archaic" H. sapiens and Neanderthal orbital area (i.e., endocasts) suggests a relative paucity of development.

The more recently evolved anterior (pre-) frontal lobe and the lateral frontal convexity are highly important in regulating the transfer of information to the neocortex, and they are involved in perceptual filtering and exerting steering influences on the neocortex so as to direct attention and intellectual processes (Como, Joseph, Fiducia, Siegel, 1979; Heilman and Van Den Abell, 1980; Joseph, et al., 1981; Joseph, 1986a, 1988a,

1990d). That is, the anterior half of the frontal lobes acts to mediate and coordinate information processing throughout the brain by continually sampling, monitoring, inhibiting and thus controlling and regulating perceptual, cognitive, and neocortical activity.

Moreover, social skills, planning skills, the formation of long-range goals, the ability to marshal one's resources so as to achieve those goals, the capacity to consider and anticipate the future rather than living in the past, and the ability to develop alternative problem-solving strategies and consider a multiple range of ideas simultaneously are capacities clearly associated with frontal lobe functional integrity. Hence, an individual is able to anticipate not only the future and the consequences of certain acts, but secondary goals that depend on the completion of one's planned actions. Indeed, the capacity to decide to do something later, to remember and do it later, and to dream, fantasize, and visualize the future as pure possibility are made possible via the frontal lobes (see Chapter 11).

Conversely, when the frontal lobes have been damaged, or when the "prefrontal" lobes have been disconnected from the rest of the brain (such as following prefrontal lobotomy), status seeking, social concern, initiative, and foresight are negatively impacted, as are emotional, motivational, intellectual, conceptual, problem-solving, and organization skills (Fuster, 1996; Joseph, 1986a, 1990d; Luria, 1980; Passingham, 1993).

Frontal lobe damage or surgical disconnection of the prefrontal lobe reduces one's ability to profit from experience, to anticipate consequences, or to learn from errors by modifying future behavior. There is a reduction in creativity, fantasy, dreaming, abstract reasoning, the capacity to synthesize ideas into concepts, or to grasp situations in their entirety. Interests of an intellectual nature seem to be diminished, or, with severe damage, abolished.

The frontal lobes also take well over 10 years to mature. Not surprisingly, following disconnection or frontal lobe injuries ante-

rior to the motor areas, patients may act without concern regarding long-term consequences, and/or behave in a very childish and puerile manner (Fuster, 1996; Joseph, 1986a, 1990d; Lishman, 1973; Luria, 1980).

Self-consciousness, concern regarding the future, clothing, personal adornments, status, competitiveness for material goods, the desire to impress others, and personal identity matter little. Like little children and chimpanzees, although sometimes capable of thinking about "tomorrow" or events hours into the future, those with poorly developed or injured frontal lobes tend to live in the "here and now."

THE FRONTAL LOBES AND THE NEUROPSYCHOLOGY OF THE MIDDLE-TO-UPPER PALEOLITHIC TRANSITION

It is noteworthy that the ability to consider problems from multiple perspectives, as well as the capacity to express personal identity or status such as via the creation and wearing of personal adornments, coincides with the evolution of modern Homo sapiens sapiens during the Upper Paleolithic (Chase, 1989; Chase and Dibble, 1987; Hayden, 1993; Joseph, 1993; Leroi-Gourhan, 1964; Mellars, 1989), i.e., 35,000 to 10,000 B.P. In contrast, Neanderthals, archaics, and other peoples of the Middle Paleolithic, i.e., 150,000 to 35,000 B.P. constructed and made essentially the same tools over and over again for perhaps 50,000 years, until around 35,000 B.P., with little variation or consideration of alternatives (Binford, 1982; Gowlett, 1984; Mellars, 1989)—a clear frontal lobe deficiency referred to as "perseveration."

Conversely, with the advent of the Upper Paleolithic and "modern" H. sapiens sapiens, the capacity to visualize multiple possibilities and to use natural contours and shapes in order to create not just tools but a variety of implements, decorations, and objects, came into being; items that could also be employed for multiple purposes (Leroi-Gourhan, 1964, 1982).

Neanderthals tended to create simple tools that served a single purpose (Plisson,

1988, cited by Hayden, 1993). There is also good evidence to indicate that the Neanderthals (but not the "early moderns") of the Middle Paleolithic tended to live in the "here and now," with little ability to think about or consider the distant future (Binford, 1973, 1982; Dennell, 1985; Mellars, 1989; however, see Hayden, 1993).

For example, in summing up some of the major behavioral differences between those of the Upper versus Middle Paleolithic, Binford (1982:178) argues that "early adaptations appear to be based on tactics which do not require much planning ahead, that is, beyond one or two days...and...the ability to anticipate events and conditions not yet experienced was not one of the strengths of" those of the Middle Paleolithic.

This is also apparent from a comparison of differences in hunting strategies employed by Neanderthals vs modern and even "early modern" humans. For example, Neanderthals, "early moderns" and modern H. sapiens sapiens, hunted large animals (Chase, 1989; Lieberman and Shea, 1994; Mellars, 1989; Rolland, 1989), which should not be surprising given that even male chimpanzees and baboons stalk, kill, dismember, and consume a variety of animals. However, just as there is evidence over the course of hominid evolution that large game animals increasingly served as human prey, there is also considerable data which indicate that the Cro-Magnon and "modern" H. sapiens (Upper Paleolithic), as well as "early moderns" utilized a far superior and effective strategy as compared to the Neanderthals (Chase, 1989; Lieberman and Shea, 1994).

For example, Lieberman and Shea (1994:317) argue that "early moderns" from Qafzeh utilized a pattern of "circular mobility" that involved moving residential camps in a recurrent annual cycle. They note that "such a strategy puts a premium on a group's ability to monitor the availability of distant resources," and that it increases their capacity to position themselves "near the highest-quality resources at the most advantageous time to collect them." In contrast, the Kebara Neanderthals who lived in the same region, albeit 30,000 years later, appear to have employed a much less efficient "radiating" strategy, which results in a "threefold increase in the frequency of hunting trips."

Similarly, Mellars (1989:357) argues the Neanderthals appear to have employed a "pattern of exploitation that was in certain respects significantly less logistically organized than that practiced by many of the later, Upper Paleolithic communities in the same regions" (see also Chase, 1989). The same could be said when comparing them to the "early moderns" (Lieberman and Shea, 1994). This reflects directly on frontal lobe functioning; i.e., "early moderns" and modern humans are superiorly endowed.

These differences associated with frontal lobe functional integrity are also apparent from analyses of the faunal assemblages from Upper versus Middle Paleolithic sites throughout Europe (Chase, 1989; Lieberman and Shea, 1994; Mellars, 1989; Rolland, 1989). According to Chase (1989) and Mellars (1989:357), as contrasted with the Neanderthals (Middle Paleolithic), the hunting strategies of the Cro-Magnon and other "moderns" "appear to reflect a high degree of planning, foresight, and logistic organization in the structuring of hunting activities, which involved a clear ability to predict and anticipate the movement of the reindeer herds."

As is well known, Upper Paleolithic (Cro-Magnon and other "modern" H. sapiens) social and organizational skills were exceedingly well developed and sophisticated (Chase and Dibble, 1987; Lindly and Clark, 1990; Mellars, 1989). In contrast, Neanderthal's social organization skills seem to have been exceedingly primitive, and they appear to have been somewhat socially isolated.

Neanderthals also tended to be quite mobile, occupying various sites for only short time periods, and their "home bases" tended to lack any form of sophisticated organization or structure (Mellars, 1989). Of course, population densities during the Middle Paleolithic were also quite low (Whallon, 1989), which in turn would have limited their opportunities for intersocial exchange, thus reducing selective pressures to evolve these particular skills—which are associated with the frontal lobes as well as the limbic system (see Chapters 5 and 11).

As based on these data and arguments, functionally the frontal lobes appear to be comparatively more well developed in Cro-Magnon and even "early modern" H. sapiens. If "moderns" and Neanderthals occupied similar or adjacent territories these differences in frontal lobe capacity would have conferred upon "moderns" a tremendous intellectual advantage over their less well-endowed competitors.

Progressive frontal lobe development would have also led to refinements and vast improvements in social skills and social organization, symbolic and artistic expression, and hunting and related technology. These are all characteristics associated with the Upper Paleolithic and much less so with the Middle Paleolithic period of evolutionary development. As neatly summed up by an ardent defender of Neanderthal cognitive capabilities (Hayden, 1993:139), "as a rule, there is no evidence of private ownership or food storage, no evidence for the use of economic resources for status or political competition, no elaborate burials, no ornaments or other status display items, no skin garments requiring intensive labor to pro-

duce, no tools requiring high energy investments, no intensive regional exchange for rare items like sea shells or amber, no competition for labor to produce economic surpluses and no corporate art or labor intensive rituals in deep cave recesses to impress onlookers and help attract labor."

Neanderthals tended to live in the here and now with little concern for the future, social status, the thoughts of others, or long-term consequences—typical features of reduced frontal lobe functioning (see Chapter 11).

ARCHAIC, "EARLY MODERN," AND NEANDERTHAL MORTUARY PRACTICES AND SYMBOLISM

Despite the poverty of frontal lobe development, archaic H. sapiens maintained a somewhat primitive cultural and social life and apparently developed spiritual beliefs regarding the dead and their souls. Indeed, it is these early humans who apparently first practiced complex mortuary rites as they buried infants, children, and adults with grave offerings and animal bones: a practice that would become increasingly complex with the evolution of "early moderns" over the ensuing 30,000 years (see Chapter 8 for a review of this and related literature).

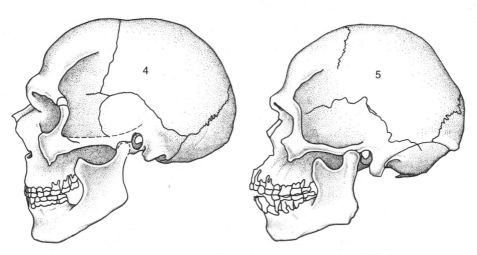

Figure 2.10 The crania of two "early modern" H. sapiens from Skhul. 90,000 to 98,000 B.P. (From Wolpoff MH. Paleo-Anthropology. New York: Alfred A Knopf, 1980.)

Thus, archaic H. sapiens and "early moderns" were carefully buried in Qafzeh, near Nazareth and in the Mt. Carmel Mugharetes-Skhul caves on the Israeli coast over 90,000 to 98,000 years ago (McCown, 1937; Smirnov, 1989; Trinkaus, 1986). This includes a Qafzeh mother and child who were buried together, and another infant who was buried holding the antlers of a fallow deer across his chest. In a nearby site equally as old (i.e., Skhul), yet another was buried with the mandible of a boar held in his hands whereas an adult had stone tools placed by his side (Belfer-Cohen and Hovers, 1992; McCown, 1937). Thus, it is quite clear that humans have been burying and presumably weeping over their dead, and perhaps preparing them for a journey to the Hereafter, for over 100,000 years.

However, in this regard, archaic and modern humans were and are little different from the Neanderthals who also engaged in complex religious rituals. For example, Neanderthals were buried in sleeping positions with the body flexed or lying on its side, surrounded by goat horns placed in a circle, with reindeer vertebrae, animal skins, stone tools, red ocher, and flowers, with large bovine bones above the head, with limestone blocks placed on top of the head and shoulders or beneath the head like a pillow, and with heads severed, coupled with evidence of ritual decapitation, facial bone removal, and cannibalism (Belfer-Cohen and Hovers, 1992; Binford, 1968; Harold, 1980; Smirnov, 1989; Solecki, 1971; also see commentary in Gargett, 1989). Moreover, Neanderthals presumably buried a bear at Regourdou, and at Drachenloch they buried stone "cysts" containing bear skulls (Kurten, 1976); hence, "the clan of the cave bear."

Of course, the fact that these Neanderthals were buried does not necessarily imply that they held a belief in God. Rather, what is indicated is that they had strong feelings for the deceased and were perhaps preparing them for a journey to the Hereafter or the land of dreams—hence, the presence of stone tools, the sleeping position, and stone pillows. Throughout the ages, dreams have been commonly thought to be the primary medium in which gods and humans interact (Campbell, 1988; Jung, 1945, 1964). Dreams, therefore, served as a doorway, a portal of entry to the spirit world.

The possibility that these ancient humans believed the dead (or their souls) might return and cause harm is also suggested by the (admittedly controversial) evidence of ritual decapitation, and the placement of heavy stones upon the body. This suggests they believed in ghosts, souls, or spirits, and a continuation of "life" after death, and therefore took necessary precautions to prevent certain souls from being released from the body or returning to cause mischief.

Similarly, the buried animal skulls and bones implies a degree of ritual symbolism, which when coupled with grave offerings and positioning of the body certainly seems to imply the Neanderthals were capable of very intense emotions and feelings, ranging from love to perhaps spiritual and superstitious awe (a function of the limbic system and inferior temporal lobe). When coupled with the evidence reviewed above (and below), there thus seems to be good reason to assume that Neanderthals maintained spiritual and mystical belief systems involving, perhaps, the transmigration of the soul and the horrors, fears, and hopes that accompany such feelings and beliefs.

Given that Middle Paleolithic peoples (archaic, "early moderns," Neanderthal) and those of the Upper Paleolithic ("moderns" and Cro-Magnon) all buried their dead with grave offerings indicates that all groups shared a certain commonality in regard to that region of the brain that has been implicated in the generation of fear, love, intense emotions, and religious and spiritual beliefs: the limbic system (e.g., amygdala) and inferior temporal lobe (Bear, 1979; d'Aquili and Newberg, 1993; Gloor, 1986, 1992; Halgren, 1992; Horowitz et al., 1968; Joseph, 1982, 1990a, 1992a; MacLean, 1990; Penfield and Perot, 1963; Rolls, 1992; Taylor, 1972, 1975; Weingarten et al., 1977; Williams, 1956).

It could thus be concluded that the inferior temporal lobe (as well as the amygdala) may have been as developed in "archaics," Neanderthals, and "early moderns" as in Upper Paleolithic (Cro-Magnon and other "modern") H. sapiens. The evolution of these cerebral nuclei, in turn, made it possible not only to experience but to attribute spiritual or religious significance to certain actions and objects (see Chapter 8).

For example, in addition to burial and mortuary practices, one of the first signs of exceedingly ancient religious symbolism is the discovery of an engraved "cross" (Vertes, 1964, cited by Mellars, 1989) that is perhaps between 60,000 to 100,000 years old. Regardless of time and culture, from the Aztecs, American Indians, Romans, Greeks, Africans, Cro-Magnons, and modern Christians, the cross consistently appears in a mystical context, and/or is attributed tremendous cosmic or spiritual significance (Campbell, 1988; Jung, 1964; Sitchin, 1990).

THE INFERIOR TEMPORAL LOBE:
Visual-Shape, Face, and "Cross" Neurons

Given that there are neurons that fire selectively to specific geometric visual shapes (e.g., faces, hands, triangles), and that they exist largely within the inferior temporal lobe (Desimone and Gross, 1979; Gross et al., 1972; Richmond et al., 1983, 1987), it could therefore be assumed that "cross" neurons as well as "mystical/religious" feeling neurons (or neural networks) had probably evolved by 100,000 years ago.

Indeed, along the neocortical surface of the inferior temporal lobe (and within the amygdala) are dense neuronal fields that contain neurons that fire selectively in response to visual images of faces, hands, eyes, and complex geometric shapes, including crosses (Gloor, 1960; Gross, et al., 1972; Richmond, et al., 1983, 1987; Rolls, 1984, 1992; Ursin and Kaada, 1950). The ability to recognize faces, geometric shapes, and social emotional nuances is dependent

on these specialized temporal lobe and amygdala neurons that respond selectively to these stimuli (Gross et al., 1972; Richmond et al., 1983, 1987; Rolls, 1984). However, since neurons in the amygdala and inferior temporal are also multimodally responsive and subserve almost all aspects of emotion, including religious feeling, it is possible for faces and geometric symbols to become infused with emotional, mystical, and religious significance. For example, heightened emotional activity within these nuclei could result in feelings of fear, foreboding, or religious awe, as well as activation of neural networks that respond selectively to crosses, such that emotional and spiritual significance is attributed to objects such as "crosses" (e.g., a dead tree taking the shape of a cross or a bird soaring with wings outstretched). Similar explanations could be offered in regard to the spiritual significance attributed to triangles (i.e., pyramids) and circles. In fact, along with crosses, triangles and circles were etched on Cro-Magnon cave walls over 30,000 years ago (Leroi-Gurhan, 1964).

In that the inferior temporal lobe (including the amygdala and hippocampus) are heavily involved in the formation of long-term memories, this would also suggest that these differing groups were equally endowed in this regard. That is, all probably possessed excellent powers of retention and recall. Those with a poorly developed frontal lobe, however, would be much more likely to live in the past rather than preparing for the distant future.

THE INFERIOR PARIETAL LOBULE AND TEMPORAL-SEQUENTIAL MOTOR CONTROL

THE INFERIOR PARIETAL LOBE

Although archaics, "early moderns," and Neanderthals all experienced an expansion in inferior temporal lobe functional development quite early in their history, the same does not appear to be the case in regard to IPL (or angular gyrus) development.

The evolution of the angular gyrus was yet another advantage enjoyed by the Cro-Magnon and other "modern" Homo sapiens, and one that the Neanderthals were probably denied. That is, Neanderthal inferior parietal and angular gyrus functional and anatomical development lagged behind that of "early modern" and fully modern humans. This would have exerted a limiting influence in regard to technological and linguistic development and competition for resources, and would be yet another factor leading to Neanderthal eradication.

Handedness

Among the majority of the population, motor control is dominated by the left cerebral hemisphere and the right hand (reviewed in Corbalis, 1991; Joseph, 1982, 1993; see also Kimura, 1973, 1979, 1993). Among lower primates, there is some evidence suggesting that the right hand is more frequently used for postural support; i.e., grasping, hanging, and holding tree branches and other objects (MacNeilage, 1993), whereas the left hand tends to be employed for gross manual manipulation tasks. However, among apes, the forelimbs are less involved in postural support, which in turn frees the right hand for other purposes.

MacNeilage (1993) argues that in "very general terms" handedness may have shifted from left to right over the course of primate evolution, such that prosimians tend to be left-handed when engaging in acts requiring manual prehension, whereas this preference is weaker in monkeys except in regard to tasks with a strong visual-spatial demands. According to MacNeilage (1993), right hand preference in monkeys and apes begins to be seen in tasks involving manipulation and auditory discrimination. He also notes that in lower primates the right hand is frequently used for postural support, but that in higher primates, since the forelimbs are less involved in postural support, the right hand was freed and could be utilized for other tasks,

thereby giving rise to right hand dominance, as is evident in humans.

However, contrary to the predictions of MacNeilage (1993), McGrew and Merchant (1992:118) have reported that chimpanzees show "no sign of a right-hand bias in object manipulation during tool use ... if anything, the pooled data suggest a leftward tendency. Conversely, for the visually guided movement of reaching, the tendency seems to be to the right and not the left." Indeed, it appears that apes, monkeys, and prosimians show little or no population bias toward using the right or left hand (see also Lehman, 1993). Nevertheless, across the span of hominid evolution, right hand control and dominance progressively increased (Corbalis, 1991).

It has been theorized that right-sided motor dominance, at least in humans, is a direct consequence of the comparatively earlier maturation of the left frontal motor areas and the fact that the left corticospinal tract descends the brainstem and crosses over at the medullary pyramids; this therefore establishes brainstem and spinal-motor interconnections in advance of the right hemisphere (Joseph, 1982). Given that the corticospinal (pyramidal) tract originates in the somatomotor areas of the frontal and parietal lobe, and as these areas program hand use and fine motor control, therefore, among the majority of "modern" humans, earlier left cerebral corticospinal tract development would confer right hand dominance for grasping, tool making, sewing, and communicating. Hence the majority of humans are more likely to spontaneously activate the right rather than the left half of the body when gesturing or speaking (e.g., Kimura, 1973, 1979; see also Corballis, 1991; Hewes, 1973). The right hand, in fact, appears to serve as a kind of motor extension of language and thought insofar as it often accompanies or acts at the behest of linguistic impulses and even emphasizes certain aspects of speech.

The Parietal Lobe

Motor functioning and hand use are dependent on sensory (e.g., proprioceptive, kinesthetic) and visual feedback, which in turn is provided by the parietal lobe (see Chapter 12). For example, if not for the sensory feedback of the muscles and joints (information which is transmitted to the parietal lobes and then relayed to the motor areas), movement would become clumsy and uncoordinated; i.e., a person would not know where their limbs were in space and in relation to one another. Since the parietal lobes and the frontal motor areas are richly interconnected, they serve in many ways as a single neurocortical unit, i.e., sensorimotor cortex (Joseph, 1990b, d; Luria, 1980).

The parietal lobes, however, are considered a "lobe of the hand" (see Chapter 12) and contain neurons that guide hand movements. Moreover, in contrast to the visual areas in the inferior temporal and occipital lobes, the parietal lobe responds to visual input from the periphery and lower visual fields—the regions in which the arms,

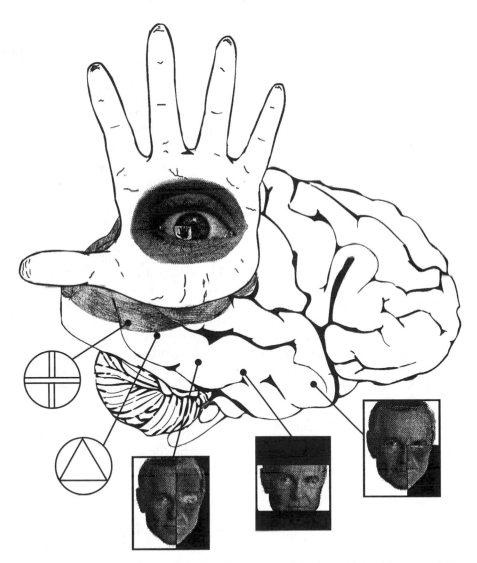

Figure 2.11 The parietal lobe has been referred to as a "lobe of the hand." Parietal neurons respond to visual information from the periphery and lower visual field and guide hand and reaching movements via sensory feedback to the motor areas of the frontal lobe. Neurons within the inferior temporal lobe respond to foveal and the upper visual field, and contain "facial" and "feature" detection neurons.

hands, and feet are most likely to come into view (Joseph, 1993). Therefore, when engaged in activities involving the hands, the parietal lobes become activated, not just due to hand movement but due to visual feedback. The parietal lobes are watching and visually guiding the hands, such as when gesturing or manipulating some object or constructing a tool. Because they also receive selective feedback, the parietal lobes are also uniquely situated so as to learn and memorize hand-movement-related behaviors. And, because a right-handed individual is more likely to use the right hand for tool making and the left hand for holding the tool, it is the left parietal lobe (which monitors the right lower visual field and controls the right hand) that would be more significantly affected by and involved in temporal-sequential activities. In consequence, the left (and right) superior parietal lobule greatly expanded in size, thereby creating a large portion of the multimodal neocortical tissue that comprises the angular gyrus and inferior parietal area.

In contrast, the temporal and occipital lobes are more concerned with identifying and observing whatever target the hands may be aiming toward or manipulating. In this manner, the parietal and temporal lobes interact with regard to aiming, throwing, and identifying relevant targets. For example, when throwing, the hand and upper arm often leaves the lower visual field and enters the upper visual field, which is also the domain of the temporal lobe and which, in turn, is observing the target.

Apraxia and Temporal Sequencing

Apraxia is a disorder of skilled temporal-sequential movement that in most cases is due to strokes or tumors that have invaded the inferior and superior parietal lobule of the left half of the brain (see Chapters 4 and 12). Affected individuals are generally unable to perform tasks that involve a series of interrelated steps and sequences.

Consider the simple steps necessary to make a pot of instant coffee. Obtaining the

coffee container, filling it, and heating the water, filling a cup with coffee grounds and then pouring the hot water into the cup are just a few of the many steps that must be performed in a highly interrelated sequence. Take just one of these steps and perform it out of order, and one destroys the overall integrity of what one was attempting to accomplish. This is exactly what occurs with apraxia.

With IPL injuries, the individual may not only be unable to make a pot of coffee or retrieve a cigarette and then strike a match in the correct sequence, but may even have trouble putting on his clothes, much less sewing them together. Similarly, individuals lacking an inferior parietal lobe, or those with injuries to this region, would be unable to fashion complex tools—much less utilize them in a complex temporal sequence.

Tool Making and the Inferior Parietal Lobe

Once the preference for right hand control became established, the left IPL (and the frontal motor areas) began to evolve and adapt as proficiency in right hand motor control increased; this, in turn, via natural selection and multigenerational influences, affected brain development (and vice versa). Tool making also evolved, as did the ability to make more efficient weapons and hunting implements.

For example, although simple stone tools were being struck by Homo habilis, Australopithecus, or both around 2.4 to 2.6 million years ago (see Hamrick and Inouye, 1995; McGrew, 1995; Susman, 1995), it was not until the Middle Paleolithic that H. sapiens were making a variety of very simple stone tools in abundance.

Even so, it was not until the Upper Paleolithic and the appearance of anatomically "modern" humans, including the Cro-Magnon, that tool making became literally an art, evolving beyond the use of rock and stone, such that complex, multifaceted features were incorporated in their construction (Leroi-Gourhan, 1964; Mellars, 1989). It is at this stage of evolutionary development that we have clear functional and neuropsychological evidence for the evolution of the angular gyrus of the IPL, which then continued to evolve and develop until as recently as 6,000 B.P. (and beyond), at which point reading and writing had evolved (Joseph, 1993)—functions also associated with the left inferior parietal lobule (see Chapters 4 and 12).

However, this does not mean to imply that this evolutionary acquisition occurred independently of cultural/environmental influences, and/or that all humans became similarly endowed at about the same time. That is, it is unlikely that cerebral "evolution" has ceased. Nor does it appear that all humans are equally endowed with inferior parietal and, in particular, frontal neocortical tissue or capabilities. Indeed, from a functional perspective, and as based on differences in impulse control, planning skills, and so on, it could be argued that some racial groups are probably differentially functionally and perhaps neuroanatomically endowed—a consequence of racial differences in environmental, climatic, geological, and cultural influences across time and generations.

Nevertheless, the evolution of the frontal lobe and, in particular, the angular gyrus and expansion within the IPL (which is also, in part, an evolutionary derivative of expanding limbic and superior parietal neocortex) provided the neurological foundations for tool design and construction, the ability to sew and even wear clothes, and the capacity to create art and pictorial language in the form of drawing, painting, sculpting, and engraving. Finally, it enabled human beings to not only create visual symbols but to talk about them and create verbal symbols in the form of written language (Joseph, 1993; Strub and Geschwind, 1983), which the IPL not only directs but directly observes. The parietal lobe, in fact, not only guides and observes hand movements but comprehends gestures, including those that produce written symbols (Joseph, 1993).

Again, however, it is important to note that although we possess two parietal lobes,

they do not perform the exact same functions. The left parietal is more concerned with temporal sequences, grammar, gestural communication, and language, including writing, spelling, and the production of signs, such as American Sign Language (Joseph, 1993; Kimura, 1993; Poizner et al., 1987; Strub and Geschwind, 1983).

The right parietal lobe is more concerned with guiding the body as it moves through space, and with the analysis, manipulation, and depiction of spatial relations such as through drawing, carpentry, masonry, throwing, aiming, and determining spatial relationships, as well as painting and art or even sewing together or putting on one's clothes (see Chapters 5 and 12). If the right parietal lobule were injured, an individual might suffer from constructional, visual-spatial, and dressing apraxia (see Chapter 3).

As such, the two parietal lobes are concerned with different aspects of language, motor control, visual-spatial analysis, and the art of gestural communication. Indeed, it was the evolution of these two differently functioning parietal tissues (as well as Broca's area in the left frontal lobe) and their subsequent harmonious interaction that made possible complex, grammatically correct, written and spoken language, as well as visual artistry and the ability to draw, as well as run, aim and throw with accuracy (e.g., move through visual space and dispatch a distant animal via bow or spear).

As noted, this was made possible via the expansion of the superior parietal and temporal lobes, which together (in conjunction with the occipital cortex) came to fashion the angular and marginal gyrus. Hence, the IPL is multimodal and concerned with hand movement and language.

The Inferior Parietal Lobule and the Middle-Upper Paleolithic Transition

As noted, apes and monkeys do not possess an angular gyrus (Geschwind, 1965). Rather (as based on an analysis of tool technology), the angular gyrus of the IPL may not have begun to truly evolve until about 100,000 B.P, and then only among a select group of perhaps early modern humans living in central Africa (e.g., modern-day Zaire) and in selected localities elsewhere.

For example, double-row barbed spears and bone points have been discovered in eastern Zaire from sites dated to approximately 75,000 to 90,000 B.P. (Brooks et al., 1995; Yellen et al., 1995). These bone tools not only required a complex series of steps for their construction but their manufacture indicates these peoples were able to recognize that bone could serve as a workable plastic medium that could be employed for a variety of purposes.

LIMITATIONS ON NEANDERTHAL ANGULAR GYRUS DEVELOPMENT

As noted, experience influences brain structure and functional organization. Neanderthals were not only more limited in regard to environmental opportunity, but they possessed relatively short distal limbs which, in turn, restricted arm movement. Coupled with limitations in the overall configuration of their glenohumeral joint surfaces, this reduced their ability to throw rocks or other hunting implements (Churchill and Trinkaus, 1990). These experiential limitations would have negatively impacted neurological development.

In contrast, Cro-Magnon and other "modern" humans of the Upper Paleolithic possessed a much rounder humerus, which is indicative of more frequent and efficient throwing. Hence, although Neanderthals may well have utilized throwing spears (Anderson-Gerfaud, 1989), these same weapons would have been wielded with considerably more efficiency and accuracy and at a farther distance, in the hands of "modern" H. sapiens, including Cro-Magnon (Churchill and Trinkaus, 1990; Joseph, 1993).

In addition, due to the relatively large pubic length in the Neanderthal population (Rosenberg, 1988; Trinkaus, 1984), they would not have been able to run as fast or move as rapidly or efficiently through space as compared to "moderns." These restrictions in functional flexibility and thus sensory feedback would have placed the Neanderthals at an additional developmental disadvantage.

In that the IPL of the right hemisphere is concerned with the hand and the movement of the body in space, and thus the analysis of body-spatial relationships, it could be inferred that these postcranial limitations on Neanderthal aiming, throwing, and running directly reflect on the functional capacity and evolutionary development of this region of the cerebrum. In that experience can in turn influence brain growth and development as well as gene selection, and because the brain functions in accordance with the "use it or lose it" principle, the Neanderthal parietal lobe would have been negatively impacted due to diminished experiential opportunity (see Chapter 18).

As compared to "modern" and Cro-Magnon peoples, these physical and experiential restrictions would have resulted in an accelerated rate of neuronal "pruning" and cellular atrophy, and related perceptual, behavioral, and cognitive deficiencies among the Neanderthal peoples, thus placing them at a significant disadvantage.

Tool Technology

Although there is some evidence that some Middle Paleolithic groups made simple bone tools (Hayden, 1993), particularly those who lived in Zaire (Brooks et al., 1995; Yellen et al., 1995), the Middle/Upper Paleolithic transition is characterized by the creation of complex bone tools, the appearance of the sewing needle, and the creation of personal adornments, such as carefully shaped beads of bone, ivory, and animal teeth, animal engravings, perforated shells that were presumably traded or transported over long distances, and statuettes, drawings, and paintings of animal and female figures (Leroi-Gourhan, 1964, 1982; Mellars, 1989; White, 1989); i.e., creative and utilitarian endeavors that are mediated by the inferior parietal lobe of the left and right cerebral hemispheres. The fact that art and tool making became exceedingly more complex during the Upper Paleolithic, and with the appearance of "modern" humans, can be directly attributed to the expansion of the left and right inferior parietal lobe, and the left frontal motor areas controlling fine hand movement.

However, this is not to suggest that Neanderthals were completely devoid of tool making capabilities, for their tool kit, albeit consisting largely of rocks, included "blade" technologies, and hafted points (see Hayden, 1994; Lieberman and Shea, 1994; Shea, 1989). Even so, other than the fact that a lot of effort was put into their construction there was nothing complex about their manufacture (Gowlett, 1984), and the same tools were made the same way over and over again for perhaps 50,000 years, until around 35,000 B.P., at which point the Neanderthals departed the scene.

Moreover, unlike the 90,000-year-old blades discovered in eastern Zaire, and particularly the tools associated with the Cro-Magnon peoples, Middle Paleolithic Neanderthal tools were predominately "use-specific" and thus served, for the most part, a unidimensional purpose (Plisson, 1988; cited by Hayden, 1994). In fact, similar to children, the Neanderthals tended to use their mouths for manipulative tasks (Molnar, 1972; Trinkaus, 1992).

As noted, it is not until the demise of the Neanderthals and the rise of the Upper Paleolithic that highly complex blade and completely new, diverse, and multifaceted tool (Aurignacian) technologies became the norm (Jelinek, 1989; Leroi-Gourhan, 1964; Mellars, 1989). Moreover, with the Upper Paleolithic peoples, the capacity to impose form, to visualize multiple possibilities, and to use natural contours and shapes in order to create not just tools but a variety of implements, decorations, and objects, came into being, including complex representational and mobile art, complex scaffolding to support cave artists, and the sewing needle (Leroi-Gourhan, 1964, 1982). All these things require an inferior parietal lobule and a

Figure 2.12 "Upper Paleolithic" art. The life-sized head and chest of a bull and the head of a horse. Aurignacian-Perigordian. Picture gallery. Lascaux. (Courtesy of Bildarchiv Preubischer Kulturbesitz.)

Figure 2.13   "Upper Paleolithic" art. Female engaged in ritual dance and surrounded by other dancers. (Courtesy of Bildarchiv Preubischer Kulturbesitz.)

Figure 2.14   "Upper Paleolithic" art. Hewn from and carved around the natural contours of a reindeer antler, a bison turns and licks its flank. This 4-inch ornament is thought to be part of a spear thrower. (From Prehistoire de l'Art Occidental. Editions Citadelles & Mazenod. Photo by Jean Vertut.)

motor cortex capable of controlling fine hand and finger movements; so that they may not only be fashioned but employed correctly.

In contrast, there is no evidence of a sewing needle or complex tool construction among Neanderthal populations during the Middle Paleolithic, and the capacity to visualize possibilities in regard to shape and form was comparatively absent as well (Binford, 1982; Mellars, 1989). The failure to invent complex tools, especially the sewing needle, certainly suggests an inability to utilize this implement, which in turn speaks volumes regarding their level of IPL development. Neanderthals utilized only the most simplistic of methods for hide or buck-

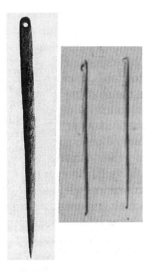

Figure 2.15   "Upper Paleolithic" sewing needles fashioned from bone.

skin preparation, and then only during the very end of the Middle Upper Paleolithic (Anderson-Gerfaud, 1989). As noted, the inability to make or even correctly put on clothing is directly associated with IPL abnormalities.

TRANSITIONAL HUMAN FORMS

Nevertheless, in addition to the manufacture of simple bone tools, an intermediate stage of tool construction also appears to characterize the Middle to Upper Paleolithic transition; i.e., the Chatelperronian. Skeletal remains intermediate between Neanderthal and "modern" H. sapiens and associated with the earliest phases of Aurignacian development have also been found (Allsworth-Jones, 1989).

There is also evidence that the Chatelperronian and Aurignacian tool making industries may have overlapped and existed simultaneously for a period of several centuries in various regions of northern Spain and western France (Harold, 1989; Mellars, 1989). Mellars (1989), however, argues that the Chatelperronian and Aurignacian industries are associated with two different populations. Moreover, since the Chatelperronian disappeared from France and Spain about 30,000 years ago, and was replaced by Aurig-

nacian technologies, Mellars (1989) argues that this may represent the extinction of Neanderthals and their replacement by Cro-Magnon.

Of course, the Chatelperronian could also represent a process of cultural exchange (Allsworth-Jones, 1989) and the interaction of Cro-Magnon and Neanderthal populations (which presumably began utilizing Chatelperronian technology as a consequence of this exchange), and/or the evolutionary metamorphosis of Neanderthals, or, more likely, "early moderns" into "moderns."

An Intermediate Stage of Frontal and Inferior Parietal Development

Regardless of which position we accept, the Chatelperronian technology would seem to represent an intermediate stage of frontal and parietal lobe development and thus continuity in archaic to "early modern" to "modern" human cerebral development. This progression is also evident from an examination of the paleoarchaeological evidence from Africa and the Middle East. For example, in addition to the 90,000-year-old barbed points from eastern Zaire (Brooks et al., 1995; Yellen et al., 1995), carefully shaped, incised and notched bones have been recovered from sites near the Klasies River Mouth (Singer and Wymer, 1982), and it is from South African sites that "early modern" H. sapiens have been discovered (Rightmire, 1984). In addition, a perforated Conus shell buried with an archaic H. sapiens infant in Border Cave on the Swaziland border, notched rib fragments, and seven split-tusk "daggers" have also been recovered (Volman, 1984). These sites have been dated from between 90,000 to 98,000 B.P. However, nothing similar has yet been recovered from any Middle Paleolithic (Neanderthal) sites in Europe.

In the Middle East, further advances in tool and blade technology, including the development of decorated personal ornaments have been discovered in sites associated with "modern" and "early modern" H. sapiens, which have been dated from between 40,000 to 47,000 B.P. (Clark and Lindly, 1989; Marks, 1989). However, it is not until around 38,000 to 35,000 B.P. that tools and related artistic and symbolic achievements appear to undergo a creative explosion (White, 1989), with even further advances occurring from 20,000 to 15,000 years B.P. (Leroi-Gourhan, 1964, 1982; Lindly and Clark, 1990) and, of course, up to the present.

Hence, there is evidence of a progression in mental and cognitive capability that is reflected in tool and artistic technology (Leroi-Gourhan, 1964, 1982), and which is also associated with progressive evolutionary advances in the inferior parietal as well as frontal lobes.

In that Upper Paleolithic tool technology is also characterized by complex and highly standardized blade and tool technologies (Jelinek, 1989; Leroi-Gourhan, 1964), and given the precision as well as the specific temporal-sequential steps that characterized tool construction (Leroi-Gourhan, 1964), it thus appears that the basic motoric neural templates that subserve not just temporal-sequential tool construction but language expression are also in evidence by this time period.

SEX ROLES, TOOL MAKING, GATHERING, AND LANGUAGE

The basic skills necessary in the gathering of vegetables, fruits, seeds, and berries and the digging of roots include the ability to engage in fine and rapid, often bilateral, temporal-sequential physical maneuvers with the arms, hands, and particularly the fingers. For almost all of human history, human females have been the traditional gatherers (Dahlberg, 1981; Martin and Voorhies, 1975; Murdock and Provost, 1981; Zilman, 1981).

However, in grubbing for roots and bulbs, the gatherers would need a digging stick, which they probably had to sharpen periodically by using stone flakes. In fact, the first tools created by hominids were probably digging sticks employed by female gatherers.

The gatherer would also carry a hammerstone for cracking nuts and for grinding the various produce collected during the day. In addition to food preparation, clothes had to be fashioned out of hides, and these, too, are tasks associated with women (Cremony, 1868; Kelly, 1962; Neithammer, 1977).

Thus, her duties probably included cleaning the hides via the use of a scraper, drying and curing the skin over the smoke of a fire, and then using a knife or cutter to make the general desired shape and then a punch to make holes through which leather straps or vines could be passed so as to create a garment that could keep out the cold. By Cro-Magnon times, they were weaving and using a needle to sew garments together.

Thus, in addition to gathering, women made tremendous use of tools and may have been the first tool makers. Indeed, in recent and current hunting and gathering groups (e.g., the American Indians), these and related "domestic" tasks are almost exclusively associated with "women's" work (Cremony, 1868; Kelly, 1962; Neithammer, 1977). With the exception of hunting implements (probably fashioned exclusively by males), it is the females and not the males who make and use tools (Niethammer, 1977). Similarly, among apes, female chimps generally use tools much more frequently than males (see McGrew and Marchant, 1992).

In tool making, technique comes to have preeminence. With the invention of technique, a linear, temporal, and sequential approach was developed over eons of time, and probably came to characterize the process. Certain tools are made a certain way with certain instruments with certain movements, and with a certain degree of muscular power and considerable precision. To make and utilize tools involving a precision grip requires that the manufacturer not only have a hand capable of such feats, but a brain that could control this hand and that could use foresight and planning in order to carry out the manufacture.

Due to selective pressures and the survival and breeding of those who were successful at these activities, the left half of the brain, which controls the right hand, became increasingly adapted for fine motor control, including temporal-sequencing, be it for the purposes of tool making or for gathering. In this regard, it is noteworthy that fine motor skills, such as those involving rapid temporal sequencing, are abilities at which females tend to excel as compared to males (Broverman, et al., 1968).

As the female has engaged in gathering much longer than the male, coupled with her possible role in tool manufacturing and tool use (e.g., skinning, clothes making, etc.), and the fact that speech production would not have been restricted but encouraged (as compared to the silent hunters), it might be assumed that the neural substrate for the temporal-sequential and grammatical aspects of what would become spoken language developed earlier and to a greater extent in the brains of women, particularly in the areas observing and controlling hand movements and somesthetic functioning (the parietal lobes).

Although women, through tool making and gathering, may have acquired and developed language and fine motor temporal sequencing abilities at an earlier point and to a greater extent than men (which is certainly true of modern women; see below), males would have also directly benefited from this acquisition via genetic inheritance and because the male would also have a mother who would talk to him and teach him language. Woman may well have provided him with the fruit of linguistic knowledge and what would become linguistic consciousness (see Chapter 16).

Limbic Language and Temporal Sequencing

Once the left IPL and the left frontal lobe and motor areas had become organized for imposing temporal sequences on motoric activities, this capacity soon expanded to include the temporal sequencing of auditory stimuli. That is, as the brain area representing the hand increased, internal as well as external stimuli could be manipulated and linked together. As detailed in Chapter 4, the angular gyrus acts with Broca's area for programming the speech neural-musculature, by organizing outgoing linguistic impulses into grammatically organized sequential units of speech (Joseph 1982, 1988a). That is, the left frontal and inferior parietal lobes act as a unit and, in fact, become simultaneously activated when vocalizing.

In this regard it appears that what had been limbic social-emotional, melodic speech (i.e., "limbic language") became subject to segmentation imposed by the left inferior parietal and Broca's area in the human brain. That is, as the amygdala is interlinked with the neocortical auditory and speech areas via the inferior arcuate fasciculus, and as the anterior cingulate is linked to the left (and right) frontal convexity (see Chapters 3 and 7), limbic language became hierarchically represented and produced by the left (and right) hemisphere, which imposed temporal sequences on vocal input and output, at least in modern humans.

Thus, once handedness and modern humans had evolved, left hemisphere speech was now superimposed over limbic language so as to create modern and grammatically correct language.

Presumably, in contrast, the language of Neanderthals and other archaic H. sapiens and pre-humans, remained largely social-emotional and limbic, similar to (albeit somewhat more complex than) that produced by chimpanzees and other apes. In this regard, the Neanderthal "speech" and "language" areas were probably also quite similar to chimps and other primates in neuroanatomical structural and functional organization as well.

PRIMATES, HOMO SAPIENS, AND APHASIA

In primates, destruction of the left frontal operculum (i.e., "Broca's area") does not effect vocalization rate, the acoustical structure of their calls, or social-emotional communication (Jurgens et al., 1982; Myers, 1976)— these features of "language" being, for the most part, a product of the limbic system: i.e., "limbic language" (Joseph, 1982, 1988a, 1990a, 1992a, 1993, 1994) and produced by the cingulate gyrus and other limbic tissue. By contrast, humans become severely aphasic. Unlike humans, however, primates do not require a Broca's area or fine motor control over the oral musculature (lips, tongue, larynx, etc.) in order to vocally express their needs, at least to the extent demonstrated by humans (Hauser and Ybarra 1994), whereas human articulation is exceedingly dependent on this capacity and brain area.

Although 90% of primate auditory cortex neurons can be activated by species-specific calls (Newman and Wollberg, 1973), as can amygdala nuclei (Gloor, 1960; Ursin and Kaada, 1960; see Chapter 5) destruction of the primate left superior temporal lobe and primary and association auditory areas (which in humans would include Wernicke's receptive speech area) does not render these creatures "aphasic," and their capacity to detect and recognize species-specific calls is not significantly affected (Dewson et al., 1975; Heffner and Heffner, 1984; Hupfer et al., 1977). Destruction of this area appears to only affect their ability to make fine auditory discrimination between similar sounds. This is because the primate primary auditory receiving area is responsible for perceiving different complex sounds and as this area maintains a reciprocal relationship with the amygdala. Hence, destruction of this tissue would in effect disconnect the amygdala, which would thus be restricted in the reception of vocal input. In fact, an analysis of the spectral response patterns of auditory cortex neurons in monkeys (Macaca mulatta) indicates a complete absence of pitch and pitch tone sensitivity (Schwarz and Tomlinson, 1990), which is not the case in humans. In the primate brain, therefore, these sounds are processed by subcortical nuclei, i.e., the limbic system (see Chapter 5).

Although primates are well endowed with neocortex, "language" in these creatures (like most of our hominid ancestors) is predominantly social and emotional (Erwin, 1975; Fedigan, 1992; Goodall, 1986, 1990; Joseph, 1993), and lacking in grammatical or temporal sequential organization (Premack and Premack, 1983), i.e., limbic language. "Language" functioning and comprehension are thus preserved with extensive destruction of the primate "Wernicke's" area, i.e., the left superior temporal lobe, but not with limbic lesions.

In contrast, destruction of the human left superior temporal lobe results in profound disturbances of linguistic comprehension, i.e., Wernicke's aphasia, and an impaired capacity to discern the individual units of speech and their temporal order (see Chapter 4). The same is not true in primates whose capacity to comprehend auditory stimuli and other primate vocalizations is not significantly affected. Primate "language" is not yet yoked to the neocortex or inferior parietal lobe, their speech sounds being produced irrespective of complex temporal sequences.

In contrast, the sounds of complex human language must be separated into discrete interrelated linear units or they will be perceived as a blur, or even as a foreign language. Hence, a patient with Wernicke's aphasia may perceive a spoken sentence such as the "pretty little kitty" as the "lip-kitterlitty" (Joseph, 1990e, 1993). He also may have difficulty establishing the boundaries of phonetic (confusing "love" for "glove") and semantic (cigarette for ashtray) auditory information. Grammatical speech is always severely impaired.

Laterality, Language, and Functional Specialization in Primates

Apes and monkeys do not utilize grammar and are not dependent on complex temporal sequencing in order to make their needs known. In fact, although apes and monkeys employ gestures in order to communicate, they are incapable of producing grammatically correct and sophisticated sign language (American Sign Language), despite extensive training (reviewed in Joseph, 1993; Premack and Premack, 1988).

American Sign Language is intimately linked with language expression and the human inferior parietal lobe, which, as noted above, is responsible for the imposition of auditory and motoric temporal sequences. Thus, in deaf humans, damage to the left parietal or frontal lobes can induce expressive or receptive gestural aphasia (Kimura, 1993; Poizner et al., 1987), whereas in speaking adults, similar destruction results in expressive and receptive aphasia.

Again, primates do not become aphasic with left cerebral neocortical injuries because limbic language has not yet been yoked to the neocortex and is not dependent on temporal-sequencing. Nonhuman primates are devoid of an angular gyrus.

Similarly, there is no evidence that Neanderthals, archaic H. sapiens, or any of the previous species of Homo or Australopithecus "evolved" an angular gyrus. As their tool making capabilities were also rather "primitive" it could be surmised that until "early modern" and anatomically "modern" and Cro-Magnon man and woman appeared on the scene, Neanderthal and other Homo brains were similar to those of other primates in this regard.

Nevertheless, in that left but not right cerebral destruction interferes with the capacity to make fine auditory discriminations in monkeys, there is thus some evidence of at least the beginning stages of left-sided dominance for language in nonhuman primates—which certainly suggests that Homo was also somewhat lateralized in this regard. Indeed, left-sided dominance for vocalizing has also been demonstrated in birds and even amphibians (e.g., Nottebohm, 1989). In large part this may well be a function of the greater concentrations of dopamine in the left hemisphere and amygdala (see Chapter 9), dopamine being directly associated with motor control, which in turn is linked to language (see Chapter 4).

However, in primates, and presumably in Australopithecus, and H. habilis, H.

erectus, and archaic H. sapiens, the right hemisphere also displays the rudiments of lateralized functional specialization, similar to that of modern humans. For example, baboons and rhesus macaques display a right hemisphere advantage in identifying and discriminating musical cords and pure tones, as well as for visual-spatial, form, facial, and facial-emotional discrimination, recognition, and expression (Hamilton and Vermeire, 1988; Hauser, 1993; Hopkins, et al., 1990; Pohl, 1983). Humans show a similar (albeit much more pronounced) right cerebral specialization (see Chapter 3). Moreover, primates with "language training" demonstrate a greater right hemisphere advantage in visual-spatial analysis than those without (Hamilton and Vermeire, 1988), suggesting a crowding effect, i.e., nonlanguage functions are crowded out of the left hemisphere and concentrated in the right.

Hence, although there is no convincing evidence for handedness or grammar in nonhuman primates, the rudiments of functional specialization and lateralization, including what would become left cerebral dominance for temporal-sequential language, is evident. With the evolution of humans, handedness, and the fingers and thumb, followed by systematic gathering and tool making, these natural lateralized neurological endowments would become more refined and exaggerated, as would other innate traits, including sex differences in cognition and behavior.

LANGUAGE, GATHERING, AND HUNTING SILENCE

Female gatherers (unlike silent male hunters) can chatter to their hearts content. Gathering, as well as tool making, is in fact directly related to the acquisition of the neurological foundations for language (Joseph, 1992b, 1993). By contrast, hunting does not promote linguistic development, as the successful hunter must be silent. Wolves, wild dogs, and lionesses spend a considerable part of each day tracking and hunting, and there is no evidence of speech among these creatures. Rather, although a hunter may throw his spear with the right hand (which is stronger and which can also be directed by the right hemisphere, via bilateral control over gross movements) he requires excellent visual-spatial skills and must maintain long periods of silence in order to be successful in these endeavors. Human speech and sound production is not promoted, however, except in regard to mimicking animal sounds.

Aspects of hunting would also put a premium on parietal lobe and right cerebral cognitive development as tracking, aiming, throwing, geometric analysis of spatial relationships, and environmental sound analysis are also directly related to the functional integrity of the right half of the brain (Guiard et al., 1983; Haaland and Harrington, 1990; Joseph, 1988a; see Chapter 3). In modern humans, it is the right half of the brain that mediates visual-spatial perceptual functioning, including the ability to aim and throw a spear, but not grammatical speech. And males consistently demonstrate superior visual-spatial, maze-learning, tracking, aiming, and related nonverbal skills, as compared to females (see Chapter 3).

HUNTING, GATHERING, LANGUAGE, AND THE MIDDLE/UPPER PALEOLITHIC TRANSITION

Among the Cro-Magnon and "modern" H. sapiens, where hunting was the center of religious and artistic life, 60–80% of the diet consisted of fruits, nuts, grains, honey, roots, and vegetables (Prideaux, 1973), which were probably gathered by the females. Even among the great majority of the very few modern hunting and gathering societies in existence today, women are the gatherers and main providers of food whereas spoils from the hunt account for only about 35% of the diet (Dahlberg, 1981; Martin and Voorhies, 1975; Murdock and Provost, 1981; Zilman, 1981).

In contrast, male and female Neanderthals, including their young children,

were predominately hunters and meat eaters as is apparent from analyses of the striation patterns along the surface of their teeth (Fox and Perez-Perez, 1993; Molnar, 1972), and as based on the faunal remains from their hunting sites (Lieberman and Shea, 1994). That is, a diet high in vegetation exerts a highly abrasive influence on the teeth, and this is not the case with Neanderthals. Moreover, glacial (Middle Paleolithic) Europe was very meager in plant food resources (Gamble, 1986). Therefore, silence and not speech production would have been emphasized among these Neanderthal hunting populations.

It is noteworthy, however, that although Neanderthals may have lived during a rather cool period, the early "modern" H. sapiens from Qafzeh lived during a rather warmer phase and did not spend as much time engaged in hunting (see Lieberman and Shea, 1994). Thus, these two populations would have adapted differently with the "early moderns" given a competitive advantage in regard to acquiring temporal-sequential and language skills, since there was less pressure to remain silent and more time available to devote to nonhunting activities.

Unlike "early modern" and Upper Paleolithic hunter gatherers, Neanderthal spent three times as much time hunting and scavenging for meat (Lieberman and Shea, 1994). Thus, they would have also had to practice protracted periods of silence, which would not contribute to the development of speech. Moreover, hunting was an activity that apparently consumed a considerable amount of Neanderthal female energy, as is evident from their relative lack of sexual dimorphism (as compared to males and females of the Upper Paleolithic) and as based on postcranial analysis of the upper extremities (Ben-Itzhaket al., 1988; Jacobs, 1985). That is, Neanderthal females engaged in activities similar to men—hunting for meat.

Hunting and scavenging for meat do not require language or temporal sequential skills but rely on the functional integrity of the right half of the brain. Given the lack of evidence for a well-developed inferior parietal lobe, the paucity of evidence to indicate complex cognitive or tool making capabilities, and the somewhat socially isolated manner in which they lived, it could be assumed that the Neanderthals, who lived predominantly as meat eaters and not gatherers, would not have been likely to develop temporal-sequential or language skills.

Coupled with Lieberman's (1992; Lieberman et al., 1992) analysis of the supralaryngeal airways, and his discussion of the contrary data provided by Arensburg et al. (1990) regarding the Kebara Neanderthal hyoid bone (which, if accurate, refutes his theory), there thus appears to be both anatomical as well as functional evidence that this population was not capable of complex language production.

As argued by Lieberman (1992:409), "Neanderthals lacked the anatomical prerequisites for producing unnasalized speech and the vowels [i], [u], and [a]. To speak any human language unimpaired requires the ability to produce these vowels and unnasalized speech, and Neanderthals did not have this ability. They would also have been unable to produce velar consonants such as [k] and [g]; these consonants are almost universally present in human languages."

It is noteworthy, however, that Lieberman and colleagues (1992) argue that the supralaryngeal airways of "early modern" H. sapiens, including those from Skhul V, are essentially similar to modern humans, which is consistent (in regard to temporal sequencing and thus language) with the notion that this population evolved an IPL in advance of Neanderthals. On the other hand, if an individual (or species) does not possess the neurological foundations for language production, then the shape and organization of their supralaryngeal pathways is irrelevant and tells us little about the evolution of human speech.

LANGUAGE AND ENDOCAST CAVEATS

A mold of the inside of a skull is referred to as an "endocast." It has been claimed, based on an analysis of an "endocast" of an H. habilis, that an area homologous to Broca's expressive speech area can be visualized (Falk, 1983a). That is, Falk (1983a) and others (e.g., Holloway, 1981; LeMay, 1976) claim that by examining the cast of the inside of an ancient and fragile skull, creating "stereoplotting" techniques and comparing these with chimpanzee or human brains, the "impression" of sulci and gyri left by the soft gelatinous brain (which is covered by a thick, tough outercovering, the dura) can be discerned, including Broca's area.

It has also been claimed, based on an analysis of endocasts, that Neanderthals possessed a Wernicke's area (LeMay, 1976). That is, it is claimed that the impression of the left sylvian fissure is larger than the impression of a fissure made on the right inside portion of the skull. In modern humans, left hemisphere dominance for language is associated with a significantly larger left-sided asymmetry in the planum temporale and parietal operculum (e.g., Geschwind and Galaburda, 1985; Habib et al., 1995).

Are these claims based on endocasts credible? Some think not and have disputed the findings (Jerison, 1990). Indeed, the "evidence" upon which these assumptions have been made—i.e., the discovery of supposed markings inside a hard inner skull left by a very soft gelatinous cerebral mass that is covered by a tough outer membrane (the dura)—is not exactly impressive.

Because these are molds of the smooth inside of the skull, human endocasts are exceedingly smooth. Although rendering information regarding gross size and position of the various lobes of the brain, it is exceedingly difficult to discern and precisely identify primary sulci and gyri, much less a "Broca's area" and certainly not a "Wernicke's area" (see Jerison, 1990 for related discussion). In fact, some would say it is impossible.

Even those involved in making these claims have exchanged accusations of improper methodology, shoddy scholarship,

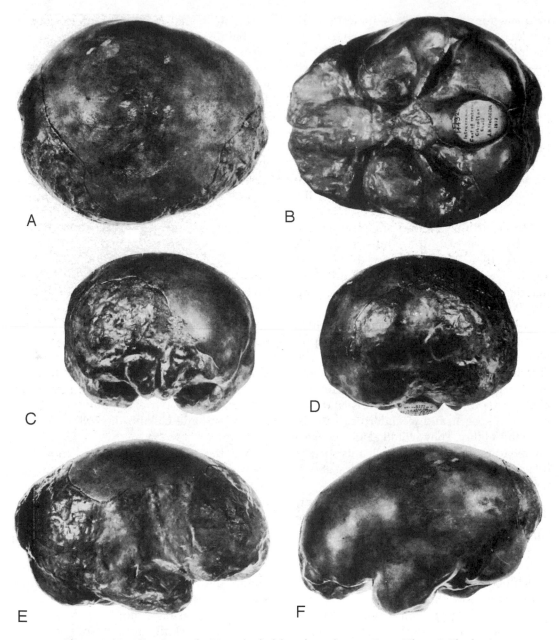

Figure 2.16   Six views of a Neanderthal female endocast. (From Tilney F. The brain from ape to man. New York: Hoeber, 1928.)

and arbitrary identification or even guesswork as to the position of various landmarks on these endocasts (see Falk, 1983b; Holloway, 1981). As stated by Holloway (1981), "Falk's use of L/H indices is unusual.... Twisting the arguments by injudicious use of indices" which "camouflages" important facts. Oddly, in attacking

Falk, Holloway (1981) claims that his data are based on an endocast collection which "is not carefully selected."

Even disregarding the questionable methodology and guesswork employed, it must be emphasized that these authors are claiming to identify secondary "sulci" and secondary "gyri" based on an examination of a mold conforming to the shape of the inside of a skull, and not the brain! Falk, Holloway, LeMay, and colleagues, have, in fact, never seen a cast of a Neanderthal, H. habilis, or Australopithecine brain and are merely speculating based on questionable methodology regarding the bumps on the inside of a skull. Dubious claims of mental and brain function based on an examination of the skull has a long history and in the past has been referred to as phrenology.

Given the above, there appears to be no credible evidence that H. habilis or Neanderthal possessed a linguistically functional Broca's or Wernicke's area and/or were capable of modern human temporal-sequential, complex grammatical speech (however, see Arensburg et al., 1990). Of course, one can also speculate otherwise.

Moreover, even if H. habilis, H. erectus, and Neanderthal possessed a Broca's and/or a Wernicke's area this does not indicate that these brain regions were linguistically functional, for similar landmarks can be discerned on chimpanzee and even dog brains. As detailed above, in nonhuman primates and other creatures, these brain areas serve a nonlanguage function.

CONCLUSION: THE CRO-MAGNON– NEANDERTHAL WARS

Clearly, from a review of the available functional, neuropsychological, cultural, technological, linguistic, and behavioral neurological data, it appears that the evolution of modern H. sapiens sapiens is characterized by advances in frontal lobe and inferior parietal/angular gyrus development. It also appears that Neanderthals flourished and

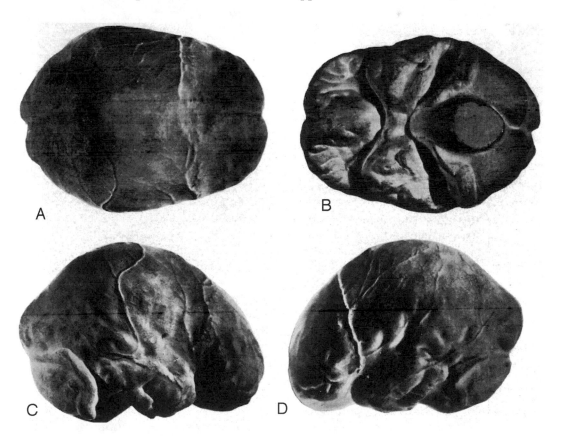

Figure 2.17 Four views of a modern human endocast which provide a gross indication of the position of some convolutions, fissures, and cerebral arteries. (From Tilney F. The brain from ape to man. New York: Hoeber, 1928.)

continued to proliferate in Europe, Palestine, and Iraq until about 35,000 years ago but were completely wiped off the face of the Earth, most likely by the Cro-Magnon, who in successive waves of humanity increasingly encroached, invaded, and probably raped, enslaved, and killed them off, ushering in the Upper Paleolithic and "replacing" Neanderthals in the process.

It is likely that a similar process of invasion and neurological competition took place among the "archaic" and "modern" H. sapiens sapiens in the Far East and Asia (Barnes, 1993). It is also possible that the Cro-Magnon and Asian "moderns" may have interbred in the Middle East and elsewhere (see Frayer et al., 1992, for related discussion), and that due to rape and interbreeding, recessive Neanderthal traits are still present in the modern gene pool.

THE EVOLUTION OF SEX DIFFERENCES IN LANGUAGE AND COGNITION

It was during the later stages of Homo erectus development, about half a million years ago, that the human brain appears to have become significantly enlarged, approximating that of a small-brained, modern-day human. It was also during this time that big game hunting had its onset.

As such, an additional divergence in the mind and brain of man and woman may have resulted over half a million years ago, as each became increasingly specialized to perform certain tasks (Joseph, 1992b, 1993). Males increasingly engaged in big-game hunting, which coincided with improved visual-spatial perceptual and motor functions, and thus frontal-parietal lobe and right hemisphere functioning.

However, because a bigger brain comes in a bigger head, this required an increase in the size of the pelvic opening of our ancient female ancestors, which in turn increased the width of the hips and pelvis, thus widening the birth canal in order to accommodate a big-brained baby. In consequence, the ability of females to run and maneuver about in space, at least as com-

pared to males, was negatively impacted. As a consequence of these changes in the hips and pelvis, the gait and balance of human females became altered such that they wiggle when they walk and are not able to run as fast as males because their mobility became slightly restricted as well (Joseph 1992b, 1993).

However, because of these sex-specific changes in the female pelvis and the positioning of her legs, the female's pelvis also became subsequently more fragile than that of a male and thus more subject to fracture if stressed by continuous hard running and jumping—a condition that plagues even modern women, including female soldiers, who suffer an unusually high incidence of pelvic and leg fractures, even though their training and duties are not nearly as strenuous or arduous as required of males (1995 U.S. Marine Corps Report, Parris Island).

All in all, these changes exerted limitations on the female as a swift and agile hunter, a condition also compounded by her reduced upper body strength. Thus, females continued to spend a considerable amount of time gathering (which coincided with improvement in bilateral, temporal-sequential, and fine-motor skills), vocalizing (cingulate gyrus, Broca's area), tool making (parietal lobe), food preparation, and maintaining the home base and engaged in child care (Joseph, 1992b, 1993). Indeed, a bigger brain requires more time to grow and mature, which in turn results in prolonged immaturity and child care.

Be it chimpanzee or human, males are far more likely to engage in hunting and related activities than females, who spend more time foraging and engaged in gathering as well as child care (Goodall, 1990; reviewed in Joseph, 1993). Because they engage in different tasks, not surprisingly, neurological and cognitive functioning is somewhat different for males versus females (see Chapters 3, 4, and 5); and this is true of many species, including dogs, chimpanzees, baboons, and rats.

For example, female monkeys and apes are more vocal—they engage in more social vocalizations, and, in fact, vocalize more often than males, who in turn are more likely to vocalize only when threatened or threatening or when engaged in dominance displays (Cross and Harlow, 1965; Erwin, 1975; Fedigan, 1992; Goodall, 1986, 1990; Mori, 1975; Mitchell, 1979). Similarly, as is well known, human females produce more social-emotional vocalization, and when in all-female groups, they tend to vocalize much more than males in all-male groups (Glass, 1993; Joseph, 1992b; Tanner, 1990). This suggests (like the visual-spatial superiorities in male rats and humans [Harris, 1978; Joseph, 1979, Joseph et al., 1978; Joseph and Gallagher, 1980]) that it has been a female tendency to vocalize more than males for well over 5 million years.

Gathering, and Why Women Like to Talk and Share Their Feelings

Over the course of human evolution, and in contrast to the silent, noncommunicative (or threatening) males, mothers and female gatherers were able to freely chatter with their babies or amongst themselves. Indeed, gathering not only fostered the development of language, but, like hunting for men, it also served as a social activity that enabled women to interact in a socially intimate manner (Joseph, 1992b, 1993).

Our ancient female ancestors probably gathered in large groups of seven or more individuals, as the typical size of a band is about 25 individuals on average. Some women were pregnant or probably carried infants who might be set on the ground here and there, or accompanied by young adolescents who would frolic about and play. Such gathering groups must have commonly been loud, noisy, and very gay affairs filled with the talk of the women and the sounds of the children's games and yells. Hence, unlike the men who must remain quiet for long time periods in order to not scare off game, the women are free to chatter and talk to their hearts' delight. Talking also served as a means of maintaining the location of the group, so that if a

gatherer or a child chanced to walk away, she (or her child) could always relocate the others by their hodgepodge of speech.

Talking, thus, became part of the gathering glue and served the purpose of keeping the group together and, thus, of bonding the group as a collective (a function similar to that performed by the anterior "maternal" cingulate gyrus). The women talked about their children, their mates, each other, and were thus exposed and allowed to expose their own feelings and thoughts to those who valued talking as much as they. For women, to socialize and be together means to talk, and to talk is a very social bonding element for women even today (Glass, 1993; Joseph, 1992b; Tanner, 1990)—a characteristic first forged by the evolution of the anterior "maternal" cingulate gyrus and the subsequent female tendency to vocalize with her young.

Female Language Superiorities

Due in large part to the evolution of the anterior cingulate, maternal vocalizing, her "female" ancestral heritage as gatherer and toolmaker, and sex differences in the organization of the limbic system (see Chapter 5) and Language Axis of the left hemisphere, human females display clear language, articulation, word knowledge, syntactic, and related linguistic superiorities (e.g., reading) over males (Bayley, 1968; Broverman, et al., 1968; Harris, 1978; Koenigsknecht and Friedman, 1976; Kimura, 1993; Levy and Heller, 1992; McGlone, 1980; McGuiness, 1976; Moore, 1967), who in turn suffer from more language-related disturbances, e.g., dyslexia and stuttering (Corballis and Beale, 1983; Lewis and Hoover, 1983).

In contrast to males, modern human females vocalize more, engage in more social speech, display superior linguistic skills, and excel over males on word fluency tests, for example, naming as many words containing a certain letter, or words belonging to a certain category (Broverman, et al., 1968; Harris, 1978; Levy and Heller, 1992). Females also vocalize more as infants, and speak their first words and develop larger vocabularies at an earlier age. Their speech as children is easier to understand, they improve their articulation and grammatical skills at a faster rate, and the length and complexity of their sentences is greater than that of males (Harris, 1978; Koenigsknecht and Friedman, 1976; Bayley, 1968; McGuiness, 1976; Moore, 1967). Boys tend to stutter more than girls (Corballis and Beale, 1983), and stuttering is largely a male problem for adults as well as children. Female language and related skills are, therefore, far superior to those of males, on average, and they appear to have more brain space devoted to linguistic functions (Joseph, 1992b, 1993; Levy and Heller, 1992).

Presumably, females demonstrate superior language skills because their brains are better adapted for producing temporal-sequential and social-emotional vocalizations (via the maternal cingulate gyrus), and because, as gatherers and tool makers, they were not under the same constraints as the silent male hunters. Males were required to maintain long periods of relative silence, at least while hunting. Their brains adapted accordingly.

However, it is not just expressive language skills, for girls learn how to read more quickly and more proficiently than males, who in turn suffer more difficulties learning how to read (Corballis and Beale, 1983). And by age 9 girls score higher than boys in reading comprehension and demonstrate superior writing and spelling skills (Lewis and Hoover, 1983). In fact, according to the 1993 report of the U.S. Department of Education, the writing skill of a 9-year-old girl in fourth grade is equal to that of a 13-year-old boy in eighth grade. As noted, reading and writing are directly associated with the functional integrity of the left inferior parietal lobule.

Overview: Sex Differences, the Parietal Lobe, Corpus Callosum, and Language

Superior female linguistic capacities appear to be related to the females' evolu-
tionary history as child bearers (see Chapter 7), gatherers, tool makers, hide preparers, and so on—tasks that require superior social-emotional capabilities (e.g., child care, socializing) and considerable temporal-sequential motor capability and somesthetic sensitivity, which in turn is associated with Broca's area and the parietal lobule in particular.

Gathering (and to a lesser extent tool making versus hunting), is far more likely to require bimanual activation of the hands, and thus simultaneous activation of both parietal lobes and the motor areas of the frontal lobe. These activities might also be expected to coincide with increased functional and metabolic activity in the parietal lobes, which in turn might lead to an expansion of this tissue (inferior parietal) and an increased need to intercommunicate. In this regard, it is noteworthy that the posterior portions of the corpus callosum, i.e., the rope of fibers that interconnect the right and left parietal lobes, may be thicker and larger in females than males (Holloway et al., 1994). However, contrary evidence abounds (reviewed in Holloway et al., 1994).

Nevertheless, as gathering and related activities are more likely to involve both hands (versus hunting and throwing with a single hand), it might be expected that the posterior (female) corpus callosum would be larger because the female right and left parietal lobe were probably simultaneously activated and utilized by females more than males for the last 100,000 or so years. On the other hand, one might equally expect that the corpus callosal interconnections linking the motor areas, particularly those involved in fine motor functioning, would be more developed in females.

Given that women have clear language superiorities and, unlike male hunters, are able to frequently and continually talk and chatter (if not with other unrelated women, then with their children, sisters, or mothers) and as the IPL is involved in word finding, sentence construction, and organizing Broca's area for grammatical speech production, it might also be expected that the female IPL is more involved in language production, as compared to males. Again, however, the same might be said of Broca's area.

It might also be expected that the female left parietal lobe would be more resistant to aphasic disturbances as compared to males, whose language representation is more sparse and fragile. In fact, males are far more likely to become severely aphasic with left parietal injuries than are females, who in turn more quickly recover from similar damage (Kimura, 1993). Conversely, females are far more likely to experience expressive aphasia with damage to Broca's area than males (Hier et al., 1994; Kimura, 1993). Again, however, they are also more likely to regain expressive speech functions.

On the other hand, it may be that sex differences in severity of initial language loss in anterior versus posterior lesions may be a function of sex differences in the vasculature and the origin and/or type of debris responsible for cerebral infarcts. In this regard, although females demonstrate clear language superiorities and have more neurological space in the right and left hemisphere devoted to linguistic and expressive functions (Joseph 1993), the exact nature and neurological foundations for these differences are certainly still open to debate.

OVERVIEW: UPPER PALEOLITHIC HOMO SAPIENS SAPIENS AND SPOKEN LANGUAGE

The ability to engage in complex conversational speech probably remained severely limited until the appearance of the Cro-Magnon and anatomically modern Homo sapiens sapiens in Europe, Africa, Asia and Australia, well over 60,000 years ago. Indeed, the Cro-Magnons stood as tall as and had a larger brain than present-day Homo sapiens sapiens, and probably possessed fully developed vocal capabilities (reviewed in Joseph, 1992b, 1993).

The Cro-Magnons and Asian moderns also believed in an afterlife, had complex supernatural beliefs and rituals which they regularly performed, and buried their dead with flowers, clothing, beads, head bands, necklaces, weapons, and offerings of food.

However, unlike those who came before them, the frontal lobes had increased significantly in size, and they evolved a new neocortical structure: i.e., the angular gyrus. Hence, due to these neurological developments, technology and innovation also mushroomed.

Some Cro-Magnons and other Upper Paleolithic peoples built large houses with stone foundations and lived in villages containing as many as 500 individuals, well over 25,000 years ago. Moreover, unlike those that came before them, they made and wore jewelry and personal decorations, and sewed their own clothes—an ability dependent not only on fine-motor skills but on evolutionary advances and developments in the inferior parietal lobe.

Hence, unlike those who came before them, the Cro-Magnon and other modern humans of the Upper Paleolithic were painting and sculpting in profusion as well as creating symbols that may well have been the forerunners to what would become written language. Indeed, the Cro-Magnon were accomplished artists and left thousands of paintings, drawings, etchings, and fine sculptures of bison, deer, wild horses, and bears in caves and on rocky cliffs everywhere they dwelled—including depictions involving temporal sequences.

These are all capabilities associated with the functional integrity of the inferior parietal lobule as well as developmental advances in the right and left cerebral hemisphere and the frontal lobes. However, it is the evolution of the inferior parietal lobule that not only gave rise to the eventual development of visual-pictorial language but written language as well, and which provided a major impetus to rearranging the functional organization of the right and left half of the brain (Joseph, 1992b, 1993). In part, this is because with the evolution of the angular gyrus of the inferior parietal lobule, the anterior (Broca's, anterior cingulate) and posterior (Wernicke's area, amygdala) language areas became interlinked

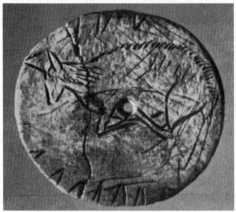

Figure 2.18 "Upper Paleolithic temporal-sequential picture writing. Carved on either side of a 1-inch disc of bone, on one side an animal is standing, and on the other an animal is lying on the ground, asleep or dead. (From Prehistoire de l'Art Occidental. Editions Citadelles & Mazenod. Photo by Jean Vertut.)

and locked together, thereby forming the Language Axis.

As argued in detail elsewhere, with the appearance of language and then reading and writing and math, older, nonlanguage functions formerly associated with the left half of the brain were displaced (Joseph, 1992b, 1993; see Chapter 4). This is because there is only so much neocortical space available and these new functions drove out the old (e.g., Joseph 1986b). This is not to imply that prior to this development the two halves of the brain were identical in functional representation, for this was certainly not the case (e.g., Hamilton and Ver-

meire, 1988; Hauser, 1993; Hopkins et al., 1990; Pohl, 1983).

For example, baboons and rhesus macaques display a right hemisphere advantage in identifying and discriminating musical cords and pure tones, as well as for visual-spatial, form, facial, and facial-emotional discrimination, recognition, and expression (Hamilton and Vermeire, 1988; Hauser, 1993; Hopkins et al., 1990; Pohl, 1983).

LINGUISTIC COMPETITION FOR NEOCORTICAL SPACE

Over the course of human evolution, and with the increasing adaption of the left hemisphere for fine-motor skills and language mediation, older, nonlinguistic functions were presumably displaced and diminished, and/or they lost their neocortical representation. That is, they were presumably crowded out by language (Joseph, 1986b, 1992b, 1993).

As per functional crowding, consider, for example, that those who are highly educated versus those of low socioeconomic and educational status show differential patterns of cerebral specialization, such that those with university training are more clearly lateralized (Alvarez and Fuentes, 1994). Moreover, primates with "language training" demonstrate a greater right hemisphere advantage in visual-spatial analysis than those without (Hamilton and Vermeire, 1988), suggesting a crowding effect. Presumably, this is due to the limited amount of neocortical space available. As language-related functions increase in importance, their neocortical representation increases, thereby giving rise to differential hemispheric specialties.

For example, it has been repeatedly demonstrated (e.g., Dennis and Whitaker, 1976; Feldman et al., 1992; Joseph, 1986b, Kurthen et al., 1992; Novelly and Joseph, 1983) that if the left hemisphere is damaged early in infancy (thus reducing neocortical space), language functions will migrate and take over vacant space within the right hemisphere. However, when this occurs, capacities normally associated with the right half of the cerebrum are diminished due to the effects of functional crowding. That is, language takes over cortical space devoted to later-appearing right cerebral functions, which in turn suffer due to lack of representation.

However, not all aspects of language will migrate, relocate, or functionally redevelop. Hence, the syntactic (temporal-sequential) components of speech may be deficient (Eisele and Aram, 1994; Dennis and Whitaker, 1976) and language development may lag (Feldman et al., 1992; Marchman et al., 1991). Presumably, this is due to innate hemispheric differences in motor control as related to language expression.

Conversely, if the right cerebrum is damaged early in life, many of its capacities can be, in part, acquired by the left hemisphere (Novelly and Joseph, 1983), in which case language begins to suffer (Feldman et al., 1992; Marchman et al., 1991). There is only so much cortical space available, and there is much competition for representation.

With the evolution and development of language, therefore, not just the human left hemisphere but the right half of the brain underwent further functional metamorphosis and developed and evolved. Thus, the right and left halves of the brain became more and more differentially specialized, which enabled humans to accomplish twice as much and which, in turn, gave birth to language, the art of creativity, and what has been referred to as "neuroses" (Joseph, 1992b). That is, with the appearance of the angular gyrus, and the evolution of the frontal lobe, language, profound artistic expression, self-consciousness, and right and left hemisphere functional specialization (in addition to that of the limbic system, midbrain and brainstem), a schism formed in the psyche of woman and man and between the right and left halves of the brain. The human mind was now subject to fragmentation and prone to the development of psychoses, neuroses, and intrapsychic conflicts.

SECTION II

The Cerebral Hemispheres

3

The Right Cerebral Hemisphere

Emotion, Language, Music, Visual-Spatial Skills, Confabulation,
Body Image, Dreams, and Social-Emotional Intelligence

Over the course of evolution, the limbic system and each half of the brain have developed their own unique strategies for perceiving, processing, and expressing information, as well as specialized neuro-anatomical interconnections that assist in mediating these functions (Bogen, 1969; Gazzaniga and LeDoux, 1978; Joseph, 1986b, 1988a, 1988b, 1992a; Levy, 1974, 1983; MacLean, 1990; Ornstein, 1972; Seymour et al., 1994; Sperry, 1966, 1982). Indeed, whereas the limbic system mediates the more primitive aspects of social-emotional intelligence, the neocortex and the cerebral hemispheres are organized such that two potentially independent mental systems coexist, literally side by side.

For example, expressive speech, linguistic knowledge and thought, mathematical and analytical reasoning, as well as the temporal-sequential and rhythmical aspects of consciousness, are associated with the functional integrity of the left half of the brain in the majority of the population. In contrast, the right cerebral hemisphere is associated with social and emotional intelligence; nonlinguistic environmental awareness; visual-spatial perceptual functioning, including analysis of depth, figure-ground, and stereopsis; facial recognition; the maintenance of the body image; and the perception, expression, and mediation of almost all aspects of emotionality.

LEFT HEMISPHERE OVERVIEW

The left cerebral hemisphere is associated with the organization and categorization of information into discrete temporal units, the sequential control of finger, hand, arm, gestural, and articulatory movements (Beaumont, 1974; Corina, et al., 1992; Haaland and Harrington, 1994; Heilman et al., 1983; Kimura, 1977, 1993; Mateer, 1983; McDonald et al., 1994; Wang and Goodglass, 1992), and the perception and labeling of material that can be coded linguistically or within a linear and sequential time frame (Efron, 1963; Lenneberg, 1967; Mills and Rollman, 1980). It is also dominant in regard to most aspects of expressive and receptive linguistic functioning, including grammar, syntax, reading, writing, speaking, spelling, naming, verbal comprehension, and verbal memory (Albert et al., 1972; Carmazza and Zurif, 1976; DeRenzi et al., 1987; Efron, 1963; Goodglass and Kaplan, 1982; Haglund et al., 1993; Hecaen and Albert, 1978; Heilman and Scholes, 1976; Kertesz, 1983a, 1983b; Milner, 1970; Njemanze, 1991; Vignolo, 1983; Zurif and Carson, 1970). In addition, the left hemisphere has been shown via dichotoic listening tasks to be dominant for the perception of real words, backwards speech, and consonants, as well as real and nonsense syllables (Blumstein and Cooper, 1974; Kimura, 1961; Shankweiler and Studdert-Kennedy, 1966, 1967; Studdert-Kennedy and Shankweiler, 1970).

As is generally well known, within the neocortical surface of the left hemisphere there is one area that largely controls the capacity to speak and another region that mediates the ability to understand speech. Specifically, Broca's expressive speech area is located along the left frontal convexity, whereas Wernicke's receptive speech area is found within the superior temporal lobe and becomes coextensive with the inferior parietal lobule.

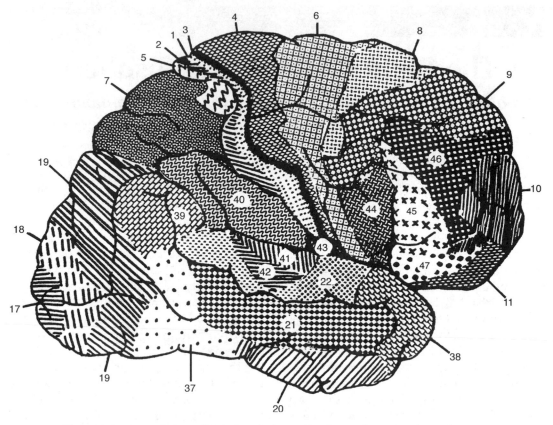

Figure 3.1   Lateral view of right hemisphere. Numbers refer to Brodmann's areas.

Broca's Aphasia

If an individual were to sustain massive damage to the left frontal convexity, his or her ability to speak would be curtailed dramatically. Even if the damage is only partial, disturbances that involve grammar and syntax, and reductions in vocabulary and word fluency in both speech and writing, result (Benson, 1977, 1979; Goodglass and Berko, 1960; Hofstede and Kolk, 1994; Milner, 1964). However, the ability to comprehend language is often (but not completely) intact (Bastiaanse, 1995; Tyler et al., 1995). This disorder is called Broca's (or expressive) aphasia (Benson, 1977, 1979; Goodglass and Kaplan, 1982; Levine and Sweet, 1983) and has also been referred to as "motor aphasia."

Individuals with expressive aphasia, although greatly limited in their ability to speak, nevertheless are capable of making emotional statements or even singing (Gardner, 1975; Goldstein, 1942; Joseph, 1988a; Smith, 1966; Smith and Burklund, 1966; Yamadori et al., 1977). In fact, they may be able to sing words that they cannot say. Moreover, since individuals with Broca's aphasia are able to comprehend, they are aware of their deficit and become appropriately depressed (Robinson and Szetela, 1981). Indeed, those with the smallest lesions become the most depressed (Robinson and Benson, 1981)—the depression, as well as the ability to sing, being mediated, presumably, by the intact right cerebral hemisphere (Joseph 1988a).

Wernicke's Aphasia

If, instead, the lesion were more posteriorly located along the superior temporal lobe, the patient would have great difficulty understanding spoken or written lan-

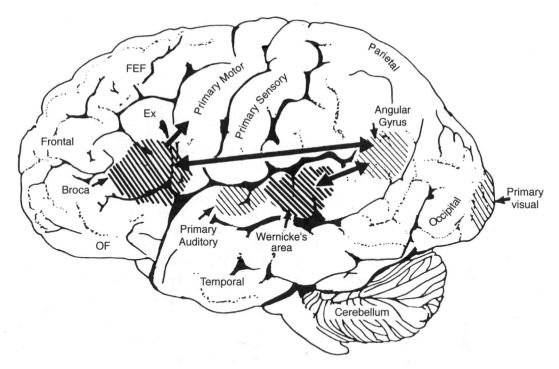

Figure 3.2 Lateral view of left hemisphere depicting the Language Axis, including frontal eye fields **(FEF)**, Exner's area **(Ex)**, orbital frontal lobes **(OF)**, and primary visual, auditory, motor, and somatosensory **(Sensory)** areas. Heavy lines and arrows indicate interactive neural pathways.

guage (Benson, 1993; Goodglass and Kaplan, 1982; Hecaen and Albert, 1978; Kertesz, 1983a). Presumably, this disorder is due in part to an impaired capacity to discern the individual units of speech and their temporal order. That is, sounds must be separated into discrete interrelated linear units or they will be perceived as a meaningless blur (Carmon and Nachshon, 1971; Christman, 1994; Efron, 1963; Joseph, 1993; Lackner and Teuber, 1973).

Wernicke's area (in conjunction with the inferior parietal lobule) acts to organize and separate incoming sounds into a temporal and interrelated series so as to extract linguistic meaning via the perception of the resulting sequences (Efron, 1963; Lackner and Teuber, 1973; Joseph, 1993; Lenneberg, 1967). When damaged, a spoken phrase such as the "big black dog" might be perceived as "the klabgigdod." This is referred to as Wernicke's aphasia. However, comprehension is improved when the spoken words are separated by long intervals.

Patients with damage to Wernicke's area are, nevertheless, still capable of talking (due to preservation of Broca's area and the fiber pathway linking these regions). However, because Wernicke's area also acts to code linguistic stimuli for expression, expressive speech becomes severely abnormal and characterized by non sequiturs, neologisms, paraphasic errors, sound and word order substitutions, and the omission of pauses and sentence endings (Christman, 1994; Goodglass and Kaplan, 1982; Hecaen and Albert, 1978; Hofstede and Kolk, 1994; Kertesz, 1983a). That is, temporal-sequential expressive linguistic encoding becomes disrupted.

For example, one patient with severe receptive aphasia responded in the following manner: "I am a little suspicious about what the hell is the part there is one part scares, uh estate spares, Ok that has a bunch of drives in it and a bunch of good googin...what the hell...kind of a platz goasted klack..." Presumably, since the coding mechanisms involved in organizing what humans are planning to say are the same mechanisms that decode what they hear, expressive as well as receptive speech becomes equally disrupted with left superior temporal lobe damage.

Nevertheless, a peculiarity of this disorder is that these patients do not always realize that what they say is meaningless (Maher et al., 1994). Moreover, they may fail to comprehend that what they hear is meaningless as well (cf. Lebrun, 1987). This is because when this area is damaged, there is no other region left to analyze the linguistic components of speech and language. The rest of the brain cannot be alerted to the patient's disability. Such patients are at risk for being misdiagnosed as psychotic.

Presumably, as a consequence of loss of comprehension, these patients may display euphoria, or in other cases, paranoia because there remains a nonlinguistic or emotional awareness that something is not right. That is, emotional functioning and comprehension remain intact (though sometimes disrupted due to erroneously processed verbal input). Hence, aphasic individuals are often able to assess to some degree the emotional characteristics of their environment, including the prosodic (Monrad-Krohn, 1963), stress contrasts (Blumstein and Goodglass, 1972), and semantic and connotative features of what is said to them, i.e., whether they are being asked a question, given a command, or presented with a declarative sentence (Boller and Green, 1972).

For example, many individuals with severe receptive (Wernicke's) aphasia can understand and respond appropriately to emotional commands and questions (e.g., "Say 'shit'," or "Do you wet your bed?" (Boller et al., 1979; Boller and Green, 1972). Similarly, the ability to read and write emotional words (as compared to nonemotional or abstract words) is also somewhat preserved among aphasics (Landis et al., 1982) due to preservation of the right hemisphere. Indeed, the capacity to identify emotional words and sentences is a capac-

ity at which the right hemisphere excels (Borod, 1992; Graves et al., 1981; Van Strien and Morpurgo, 1992).

Because these paralinguistic and emotional features of language are analyzed by the intact right cerebral hemisphere, the aphasic individual is able to grasp, in general, the meaning or intent of a speaker, although verbal comprehension is reduced. This, in turn, enables him to react in a somewhat appropriate fashion when spoken to.

For example, after I had diagnosed a patient as suffering from Wernicke's aphasia, her nurse disagreed and indicated the patient responded correctly to questions such as, "How are you this morning?" That is, the patient replied: "Fine." Later, when I reexamined the patient I used a *tone* of voice appropriate for "How *are you* today?," but instead said: "It's *raining* outside?" The patient replied, "Fine!" and appropriately smiled and nodded her head (Joseph, 1988a). Often, our pets are able to determine what we mean and how we feel by analyzing similar melodic-emotional nuances.

THE RIGHT CEREBRAL HEMISPHERE

It has now been well established that the right cerebral hemisphere is dominant over the left in regard to the perception, expression, and mediation of almost all aspects of emotionality (e.g., Borod, 1992; Cancelliere and Kertesz, 1990; Freeman and Traugott, 1993; Heilman and Bowers, 1995; Heilman et al., 1984; Joseph, 1988a; Tucker and Frederick, 1989; see below), including social-emotional intelligence). This emotional dominance extends to bilateral control over the autonomic nervous system, including heart rate, blood pressure regulation, galvanic skin conductance, and the secretion of cortisol in emotionally upsetting or exciting situations (Rosen et al., 1982; Wittling, 1990; Wittling and Pfluger, 1990; Yamour et al., 1980; Zamarini et al., 1990).

In part, it is believed that the right hemisphere dominance over emotional functioning is due to more extensive interconnections with the limbic system (Joseph,

1982, 1988a). Hence, over the course of evolution, limbic social-emotional functions have come to be hierarchically subserved by the right cerebrum. This includes the expression and representation of limbic language, and, as noted above, bilateral control over the autonomic nervous system.

COMPREHENSION AND EXPRESSION OF EMOTIONAL SPEECH

Although language is often discussed in terms of grammar and vocabulary, there is a third major aspect to linguistic expression and comprehension by which a speaker may convey and a listener discern intent, attitude, feeling, mood, context, and meaning. Language is both emotional and grammatically descriptive.

A listener comprehends not only the content and grammar of what is said but the emotion and melody of how it is said—what a speaker feels. Feeling—be it anger, happiness, sadness, sarcasm, empathy—often is communicated by varying the rate, amplitude, pitch, inflection, timbre, melody, and stress contours of the voice. When devoid of intonational contours, language becomes monotone and bland and a listener experiences difficulty discerning attitude, context, intent, and feeling. Conditions such as these arise after damage to select areas of the right hemisphere or when the entire right half of the brain is anesthetized (e.g., during sodium amytal procedures).

It is now well established (based on studies of normal and brain-damaged subjects) that the right hemisphere is superior to the left in distinguishing, interpreting, and processing vocal inflectional nuances, including intensity, stress and melodic pitch contours, timbre, cadence, emotional tone, frequency, amplitude, melody, duration, and intonation (Blumstein and Cooper, 1974; Bowers et al., 1987; Carmon and Nachshon, 1973; Heilman et al., 1975; Ley and Bryden, 1979; Mahoney and Sainsbury, 1987; Ross, 1981; Safer and Leventhal, 1977; Samson and Zatorre, 1988, 1992;

Shapiro and Danly, 1985; Tucker et al., 1977). The right hemisphere, therefore, is fully capable of determining and deducing not only what a persons feels about what he or she is saying, but why and in what context he is saying it—even in the absence of vocabulary and other denotative linguistic features (Blumstein and Cooper, 1974; DeUrso et al., 1986; Dwyer and Rinn, 1981). This occurs through the analysis of tone and melody.

Hence, if I were to say, "Do you want to go outside?," although both hemispheres are able to determine whether a question versus a statement has been made (Heilman et al., 1984; Weintraub et al., 1981), it is the right cerebrum that analyzes the paralinguistic emotional features of the voice so as to determine whether "going outside" will be fun or whether I am going to punch you in the nose. In fact, even without the aid of the actual words, based merely on melody and tone the right cerebrum can determine context and the feelings of the speaker (Blumstein and Cooper, 1974; DeUrso et al., 1986; Dwyer and Rinn, 1981). This may well explain why even preverbal infants are able to make these same determinations, even when spoken to in a foreign language (Fernald, 1993; Haviland and Lelwica, 1987). The left hemisphere has great difficulty with such tasks.

For example, in experiments in which verbal information was filtered and the individual was to determine the context in which a person was speaking (e.g., talking about the death of a friend, speaking to a lost child), the right hemisphere was found to be dominant (Dwyer and Rinn, 1981). It is for these and other reasons that the right half of the brain sometimes is thought to be the more intuitive half of the cerebrum.

Correspondingly, when the right hemisphere is damaged, the ability to process, recall, or even recognize these nonverbal and social-emotional nuances is greatly attenuated. For example, although able to comprehend individual sentences and paragraphs, such patients have difficulty understanding context and emotional connotation, drawing inferences, relating what is heard to its proper context, determining the overall gist or theme, and recognizing discrepancies such that they are likely to miss the point, respond to inappropriate details, and fail to appreciate fully when they are being presented with information that is sarcastic, incongruent, or even implausible (Beeman, 1993; Brownell et al., 1986; Foldi et al., 1983; Gardner et al., 1983; Kaplan et al., 1990; Rehak et al., 1992; Wapner et al., 1981).

Such patients frequently tend to be very concrete and literal. For example, when presented with the statement, "He had a heavy heart," and asked to choose several interpretations, right-brain-damaged (versus aphasic) patients are more likely to choose a picture of an individual staggering under a large heart rather than a crying person. They also have difficulty describing morals, motives, emotions, or overall main points (e.g., they lose the gestalt), although the ability to recall isolated facts and details is preserved (Delis et al., 1986; Hough, 1990; Wapner et al., 1981)—details being the province of the left hemisphere.

Although they are not aphasic, individuals with right hemisphere damage sometimes have difficulty comprehending complex verbal and written statements, particularly when there are features that involve spatial transformations or incongruencies. For example, when presented with the question "Bob is taller than George. *Who is shorter?*," those with right brain damage have difficulties due, presumably, to a deficit in nonlinguistic imaginal processing or an inability to search a spatial representation of what they hear (Carmazza et al., 1976).

In contrast, when presented with "Bob is taller than George. *Who is taller?*," patients with right hemisphere damage perform similar to normals, which indicates that the left cerebrum is responsible for providing the solution (Carmazza et al., 1976), given that the right hemisphere is injured and the question does not require any type of spatial transformation. That is, because the question "Who is shorter?" does not necessarily *follow* the first part of the statements

(i.e., is incongruent), whereas "Who is taller?" does, these differential findings further suggest that the right hemisphere is more involved than the left in the analysis of incongruities.

Right Hemisphere Emotional-Melodic Language Axis

Just as there are areas in the left frontal and temporal-parietal lobes that mediate the expression and comprehension of the denotative, temporal-sequential, and grammatical-syntactical aspects of language, there are similar regions within the right hemisphere that mediate emotional speech and comprehension (Gorelick and Ross, 1987; Heilman et al., 1975; Joseph, 1982, 1988a, 1993; Lalande et al., 1992; Ross, 1981; Shapiro and Danly, 1985; Tucker et al., 1977)—regions that become highly active when presented with complex nonverbal auditory stimuli (Roland et al., 1981).

For example, right frontal damage has been associated with a loss of emotional speech and emotional gesturing and a significantly reduced ability to mimic various nonlinguistic vocal patterns (Joseph, 1988a; Ross, 1981; Shapiro and Danly, 1985). In these instances, speech can become flat and monotone or characterized by inflectional distortions.

With lesions that involve the right temporal-parietal area, the ability to comprehend or produce appropriate verbal prosody or emotional speech, or to repeat emotional statements, is reduced significantly (Gorelick and Ross, 1987; Heilman et al., 1975; Lalande et al., 1992; Ross, 1981; Starkstein et al., 1994; Tucker et al., 1977). Indeed, when presented with neutral sentences spoken in an emotional manner, right hemisphere damage disrupts perception and discrimination (Heilman et al., 1975; Lalande et al., 1992) and the comprehension of emotional prosody (Heilman et al., 1984; Starkstein et al., 1994), regardless of whether it is positive or negative in content. Moreover, the ability to differentiate between different and even oppositional emotional qualities (e.g., "sarcasm" versus "irony" or

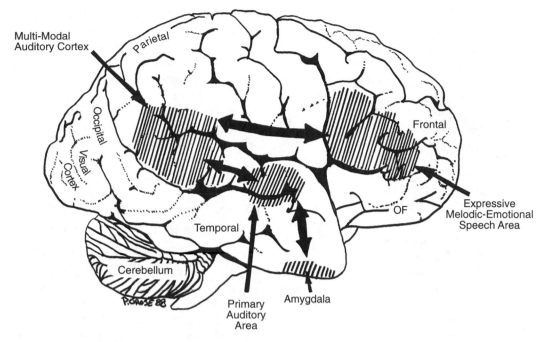

Figure 3.3   Lateral view of right hemisphere. Heavy lines and arrows indicate interactive neural pathways.

"love" versus "hate") can become distorted (Cicone et al., 1980; Kaplan et al., 1990), and the capacity to appreciate and comprehend humor or mirth may be attenuated (Gardner et al., 1975) and social and emotional intelligence is negatively impacted.

The semantic-contextual ability of the right hemisphere is not limited to prosodic and paralinguistic features, however, but includes the ability to process and recognize familiar, concrete, highly imaginable words (J. Day, 1977; Deloche et al., 1987; Ellis and Shephard, 1975; Hines, 1976; Joseph, 1988b; Landis et al., 1982; Mannhaupt, 1983), as well as emotional language in general. The disconnected right hemisphere also can read printed words (Gazzaniga, 1970; Joseph, 1986b, 1988b; Levy, 1983; Sperry, 1982; Zaidel, 1983), retrieve objects with the left hand in response to direct and indirect verbal commands, e.g., "a container for liquids" (Joseph, 1988b; Sperry, 1982), and spell simple three- and four-letter words with cut-out letters (Sperry, 1982). However, it cannot comprehend complex, nonemotional, written or spoken language.

In part, the right hemisphere dominance for vocal (and nonverbal) emotional expression and comprehension is believed to be secondary to hierarchical neocortical representation of limbic system functions (Joseph, 1982). It may well be dominance by default, however. That is, at one time both hemispheres may well have contributed more or less equally to emotional expression, but with the evolution of language and right handedness, the left hemisphere gradually lost this capacity, where it was retained in the right cerebrum (Joseph, 1993). Even so, without the participation of the limbic system, the amygdala, and cingulate gyrus in particular, emotional language capabilities would, for the most part, be nonexistent.

THE AMYGDALA AND CINGULATE

The amygdala appears to contribute to the perception and comprehension of emotional vocalizations, which it extracts and imparts to the neocortical language centers via the axonal pathway, the arcuate fasciculus, which links the frontal convexity, the inferior parietal lobule, Wernicke's area, the primary auditory area, and the lateral amygdala (Joseph, 1993). That is, sounds perceived are shunted to and fro between the primary and secondary auditory receiving areas and the amygdala, which then acts to sample and analyze them for motivational significance (see Chapters 5 and 7).

In addition, the anterior cingulate contributes to emotional sound production via axonal interconnections with the right and left frontal convexity (Broca's area). As noted in Chapters 1, 2, 7, and 11, over the course of evolution the anterior cingulate appears to have given rise to large portions of the medial frontal lobe and the supplementary motor areas, which in turn continued to evolve, thus forming the lateral convexity, including Broca's area. Via these interconnections, emotional nuances may be imparted directly into the stream of vocal utterances.

Conversely, subcortical right cerebral lesions involving the anterior cingulate, the amygdala, or these fiber interconnections can also result in emotional expressive and receptive disorders (Joseph, 1988a, 1993; see also Starkstein et al., 1994). Similarly, lesions to the left temporal or frontal lobe may result in disconnection, which in turn may lead to distortions in the vocal expression or perception of emotional nuances, that is, within the left hemisphere.

Sex Differences in Emotional Sound Production and Perception

As detailed in Chapters 1 and 7, the evolution of the anterior cingulate gyrus corresponded with the onset of long-term maternal care, and, presumably, the advent of mother-infant vocalization and intercommunication. The presence of an adult female, in fact, appears to promote language production in infants, as well as in

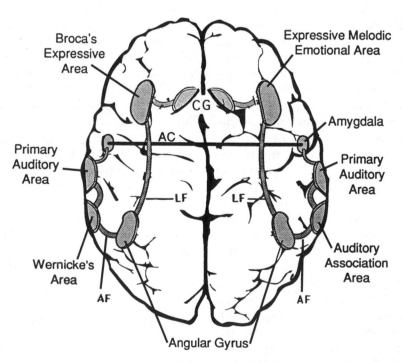

Figure 3.4 Schematic diagram of interactive pathways linking the amygdala and the cingulate gyrus **(CG)** with the "language" areas of the right and left hemispheres, including the longitudinal fasciculus **(LF)** and arcuate fasciculus **(AF)**, and the anterior commissure **(AC)**. (From Joseph R. The naked neuron: evolution and the languages of the body and brain. New York: Plenum, 1993.)

adult males and other females. Thus, non-threatening, complex, social-emotional vocalizations are, to some degree, strongly associated with female-maternal behavior and the desire to form social-emotional attachments (Joseph, 1993).

Almost regardless of mammalian species, females tend to produce a greater range of social-emotional (limbic) vocalization (Joseph, 1993). For example, human females tend to employ five to six different prosodic variations and to utilize the higher registers when conversing. They are also more likely to employ glissando or sliding effects between stressed syllables, and they tend to talk faster as well (Brend, 1975; Coleman, 1971; Edelsky, 1979). Men tend to be more monotone, employing two to three variations on average, most of which hover around the lower registers (Brend, 1975; Coleman, 1971; Edelsky, 1979). Even when trying to emphasize a point, males are less likely to employ melodic extremes but instead tend to speak louder. This is not, however, a function of sex differences in the oral-laryngeal structures, but is due to the greater capacity of the female right hemisphere (and limbic system) to express and perceive these nuances.

For example, it has been repeatedly demonstrated that females are more emotionally expressive and are more perceptive in regard to comprehending emotional verbal nuances (Burton and Levy, 1989; Hall, 1978; Soloman and Ali, 1972). This superior sensitivity includes the ability to feel and verbally express empathy (Burton and Levy, 1989; Safer, 1981) and the comprehension of emotional faces (Buck et al., 1974; Buck et al., 1972; see also Evans et al., 1995). In fact, from childhood to adulthood women appear to be much more emotionally expressive than males, in general (Gilbert, 1969; see Brody, 1985, and Burton and Levy, 1989, for review). Females, therefore, are superior to males in regard to social-emotional intelligence.

As argued in detail elsewhere (Joseph, 1992b, 1993), in addition to the evolution of the "maternal" anterior cingulate gyrus, this appears to be in part a consequence of

the differential activities engaged in by men (hunting) versus females (gathering, food preparation) for much of human history, and the possibility that the female right hemisphere has more neocortical space committed to emotional perception and expression (whereas males have more space devoted to visual-spatial functions; see below).

As noted, this emotional intellectual superiority is probably also a consequence of limbic system sexual differentiation, and the role of the female limbic system in promoting maternal care and communication (see Chapters 5 and 7). Thus, regardless of culture, human mothers tend to emphasize and even exaggerate social-emotional and melodic-prosodic vocal features when interacting with their infants (Fernald, 1992; Fernald et al., 1989), which in turn appears to greatly influence infant emotional behavior and attention (Fernald, 1991).

CONFABULATION

In contrast to left frontal convexity lesions that can result in *speech arrest* (Broca's expressive aphasia) and/or significant reductions in verbal fluency, right frontal damage sometimes has been observed to result in *speech release*—excessive verbosity, tangentiality, and in the extreme, confabulation (Fischer et al., 1995; Joseph, 1986a, 1988a). When secondary to frontal damage, confabulation seems to be due to disinhibition, and difficulty monitoring responses, withholding answers, utilizing external or internal cues to make corrections, or suppressing the flow of tangential and circumstantial ideas (Shapiro et al., 1981; Stuss et al., 1978). When this occurs, the language axis of the left hemisphere becomes overwhelmed and flooded by irrelevant associations (Joseph, 1986a, 1988a, 1990d). In some cases, the content of the confabulation may border on the bizarre and fantastical, as loosely associated ideas become organized and anchored around fragments of current experience.

For example, one 24-year-old individual who received a gunshot wound that resulted in destruction of the right inferior convexity and orbital areas attributed his hospitalization to a plot by the government to steal his inventions and ideas—the patient had been a grocery store clerk. When it was pointed out that he had undergone surgery for removal of bone fragments and the bullet, he pointed to his head and replied, "That's how they're stealing my ideas." Another patient with a degenerative disturbance that involved predominantly the right frontal lobe at times claimed to be a police officer, a doctor, or married to various members of the staff. When it was pointed out repeatedly that he was a patient, he at one point replied, "I'm a doctor. I'm here to protect people."

Gap Filling

Confabulation also can result from lesions that involve the posterior portions of the right hemisphere, immaturity or surgical section of the corpus callosum, or destruction of fiber tracts that lead to the left hemisphere (Geschwind, 1965; Joseph, 1982, 1986a, 1988a; Joseph et al., 1984). This results in incomplete information transfer and reception. As a consequence, because the language axis of the left hemisphere is unable to gain access to needed information, it attempts to fill the gap with information that is related in some manner to the fragments received. However, because the language areas are disconnected from the source of needed information, it cannot be informed that what it is saying (or, rather, making up) is erroneous, at least insofar as the damaged modality is concerned.

For example, in cases presented by Redlich and Dorsey (1945), individuals who were suffering from blindness or gross visual disturbances due to injuries in the visual cortex continued to claim that they could see even when they bumped into objects and tripped over furniture. Apparently, they maintained these claims because the areas of the brain that normally would alert them to their blindness (i.e., visual cortex) were no longer functioning.

Confabulation and delusional denial also often accompany neglect and body-image disturbances secondary to right cerebral (parietal) damage (Joseph, 1982, 1986a, 1988a). For example, the left hemisphere may claim that a paralyzed left leg or arm is normal or that it belongs to someone other than the patient. This occurs, in many cases, because somesthetic body information no longer is being processed or transferred by the damaged right hemisphere; the body image and the memory of the left half of the body have been deleted. In all these instances, however, although the damage may be in the right hemisphere, it is the speaking half of the brain that confabulates.

On the other hand, there is some evidence to suggest that when information flow *from* the left *to* the right hemisphere is reduced, a visual-imaginal-hypnogocic form of confabulation may result, i.e., dreaming (Joseph, 1988a). Dreaming is possibly one form of right hemisphere confabulation. Of course, many other factors are also involved (see below and Chapters 5 and 10). We will return to this issue.

MUSIC AND NONVERBAL ENVIRONMENTAL SOUNDS

Individuals with extensive left hemisphere damage and/or severe forms of expressive aphasia, although unable to discourse fluently, may be capable of swearing, singing, praying, or making statements of self-pity (Gardner, 1975; Goldstein, 1942; Smith, 1966; Smith and Burklund, 1966; Yamadori et al., 1977). Even when the entire left hemisphere has been removed completely, the ability to sing familiar songs or even learn new ones may be preserved (Smith, 1966; Smith and Burklund, 1966)—although in the absence of music the patient would be unable to say the very words that he or she had just sung (Goldstein, 1942). The preservation of the ability to sing has, in fact, been utilized to promote linguistic recovery in aphasic patients, i.e., melodic-intonation therapy (Albert et al., 1973; Helm-Estabrooks, 1983).

Similarly, there have been reports that some musicians and composers who were suffering from aphasia and/or significant left hemisphere impairment were able to continue their work (Alajoanine, 1948; Critchly, 1953; Luria, 1973). In some cases, despite severe receptive aphasia and/or although the ability to read written language (alexia) was disrupted, the ability to read music or to continue composing was preserved (Gates and Bradshaw, 1977; Luria, 1973).

One famous example is that of Maurice Ravel, who suffered an injury to the left half of his brain in an auto accident. This resulted in ideomotor apraxia, dysgraphia, and moderate disturbances in comprehending speech (i.e., Wernicke's aphasia). Nevertheless, he had no difficulty recognizing various musical compositions, was able to detect even minor errors when compositions were played, and was able to correct those errors by playing them correctly on the piano (Alajoanine, 1948).

Conversely, it has been reported that musicians who are suffering from right hemisphere damage (e.g., right temporal-parietal stroke) have major difficulties recognizing familiar melodies and suffer from expressive instrumental *amusia* (Luria, 1973; McFarland and Fortin, 1982). Even among nonmusicians, right hemisphere damage (e.g., right temporal lobectomy) disrupts time sense, rhythm, and the ability to perceive, recognize, or recall tones, loudness, timbre, and melody (Chase, 1967; Gates and Bradshaw, 1977; Milner, 1962; Yamadori et al., 1977). Right hemisphere damage also can disrupt the ability to sing or carry a tune and can cause toneless, monotonous speech as well as abolish the capacity to obtain pleasure while listening to music (Reese, 1948; Ross, 1981; Shapiro and Danly, 1985), i.e., a condition also referred to as *amusia*. For example, Freeman and Williams (1953) report that removal of the right amygdala in one patient resulted in a great change in the pitch and timbre of speech and that the ability to sing also was severely affected. Similarly, when the right hemisphere is anesthetized, the melodic aspects of speech and singing become significantly impaired (Gordon and Bogen, 1974).

It also has been demonstrated consistently in normals (such as in dichotic listening studies) and in brain-injured individuals that the right hemisphere predominates in the perception (and/or expression) of timbre, chords, tone, pitch, loudness, melody, meter, tempo, and intensity (Breitling et al., 1987; Curry, 1967; R. Day et al., 1971, Gates and Bradshaw, 1977; Gordon, 1970; Gordon and Bogen, 1974; Kester et al., 1991; Kimura, 1964; Knox and Kimura, 1970; McFarland and Fortin, 1982; Milner, 1962; Molfese et al., 1975; Piazza, 1980; Reese, 1948; Segalowitz and Plantery, 1985; Spellacy, 1970; Swisher et al., 1969; Tsunoda, 1975; Zurif, 1974)—the major components (in conjunction with harmony) of a musical stimulus.

In addition, Penfield and Perot (1963) report that musical hallucinations most frequently result from electrical stimulation of the right superior and lateral surfaces of the temporal lobe. Berrios (1990) also concluded from a review of lesion studies that musical hallucinations were far more likely following right cerebral dysfunction. Findings such as these have added greatly to the conviction that the right cerebral hemisphere is dominant in regard to the non–temporal-sequential aspects of musical perception and expression.

However, this does not appear to be the case with professional musicians, who in some respects tend to treat music as a mathematical language that is subject to rhythmic analysis. As such, some professional musicians may well demonstrate a left hemisphere dominance for musical stimuli.

Environmental Sounds

In addition to music, the right hemisphere has been shown to be superior to the left in discerning and recognizing nonverbal and environmental sounds (Curry, 1967; Joseph, 1988b; Kimura, 1963; King and Kimura, 1972; Knox and Kimura, 1970; Neilsen, 1946; Piazza, 1980; Roland et al.,

1981; Schnider et al., 1994; Spreen et al., 1965; Tsunoda, 1975). Similarly, damage that involves select areas within the right hemisphere not only disturb the capacity to discern musical and social-emotional vocal nuances, but may disrupt the ability to perceive, recognize, or discriminate between a diverse number of sounds that occur naturally within the environment (Joseph, 1993; Neilsen, 1946; Schnider et al., 1994; Spreen et al., 1965), such as water splashing, a door banging, applause, or even a typewriter; this is a condition that also plagues the disconnected left hemisphere (Joseph, 1988b).

The possibility has been raised that music, verbal emotion, and nonverbal environmental sounds are, in some manner, phylogenetically linked (Joseph, 1982, 1988a, 1993). For example, it is possible that right hemisphere dominance for music may be a limbic outgrowth and/or strongly related to its capacity to discern and recognize environmental acoustics as well as its ability to mimic these and other nonverbal and emotional nuances. That is, music may have been invented as a form of mimicry, and/or as a natural modification of what has been described as "limbic language" (a term and concept coined by Joseph, 1982).

For example, it is somewhat probable that primitive man and woman's first exposure to the sounds of music was environmentally embedded, for musical sounds are, obviously, heard frequently throughout nature (e.g., the singing of birds, the whistling of the wind, the humming of bees or insects). For example, bird songs can encompass sounds that are "flute-like, truly chime- or bell-like, violin- or guitar-like," and "some are almost as tender as a boy soprano" (Hartshorne, 1973:36).

Hence, perhaps our musical nature is related to our original relationship with nature and resulted from the tendency of humans to mimic sounds that arise from the environment—such as those which conveyed certain feeling states and emotions. Perhaps this is also why certain acoustical nuances, such as those employed in classical music, can affect us emotionally and make us visualize scenes from nature

(e.g., an early spring morning, a raging storm, bees in flight).

Music and Emotion

Music is related strongly to emotion and, in fact, may not only be "pleasing to the ear" but invested with emotional significance. For example, when played in a major key, music sounds happy or joyful. When played in a minor key, music often is perceived as sad or melancholic. We are all familiar with the "blues" and perhaps at one time or another have felt like "singing for joy," or have told someone, "You make my heart sing!"

Interestingly, it has been reported that music can act to accelerate pulse rate (Reese, 1948), raise or lower blood pressure, and, thus, alter the rhythm of the heart's beat. Rhythm, of course, is a major component of music.

Music and vocal emotional nuances also share certain features, such as melody, intonation, and so forth, all of which are predominantly processed and mediated by the right cerebrum. Thus, the right hemisphere has been found to be superior to the left in identifying the emotional tone of musical passages and, in fact, judges music to be more emotional as compared to the left cerebrum (Bryden et al., 1982).

LEFT HEMISPHERE MUSICAL CONTRIBUTIONS

There is some evidence to indicate that certain aspects of pitch, time sense, and rhythm are mediated to a similar degree by both cerebral hemispheres (Milner, 1962), perhaps more so by the left. Of course, time sense and rhythm are also highly important in speech perception. These findings support the possibility of a left hemisphere contribution to music. In fact, some authors have argued that receptive amusia is due to left hemisphere damage and that expressive amusia is due to right hemisphere dysfunction (Wertheim, 1969).

It also has been reported that some musicians tend to show a left hemisphere dominance in the perception of certain aspects of music (Gates and Bradshaw, 1977). However, as noted, it also appears that when the sequential and rhythmical aspects of music are emphasized, the left hemisphere becomes increasingly involved (Breitling et al., 1987; Halperin et al., 1973). In this regard, it seems that when music is treated as a type of language to be acquired, or when its mathematical and temporal-sequential features are emphasized (Breitling et al., 1987), the left cerebral hemisphere becomes heavily involved in its production and perception, especially in professional musicians.

However, even in nonmusicians the left hemisphere typically displays, if not dominance, then an equal capability in regard to the production of rhythm. In this regard, just as the right hemisphere makes important contributions to the perception and expression of language, it also takes both halves of the brain to make music.

MUSIC, MATH, AND GEOMETRIC SPACE

Pythagoras, the great Greek mathematician, argued over 2,000 years ago that music was numerical, the expression of number in sound (Durant, 1939; McClain, 1978). In fact, long before the advent of digital recordings, the Babylonians and Hindus, and then Pythagoras and his followers, translated music into number and geometric proportions (Durant, 1939). For example, by dividing a vibrating string into various ratios they discovered that several very pleasing musical intervals could be produced. Hence, the ratio 1:2 was found to yield an octave, 2:3 a fifth, 3:4 a fourth, 4:5 a major third, and 5:6 a a minor third (McClain, 1978). The harmonic system utilized in the nineteenth century by various composers was based on these same ratios. Indeed, Bartok utilized these ratios in his musical compositions.

These same musical ratios, the Pythagorians discovered, also were found to have the capability of reproducing themselves. That is, the ratio can reproduce itself within itself and form a unique geometrical configuration that Pythagoras and the ancient Greeks referred to as the "golden ratio" or "golden rectangle." The golden rectangle was postulated to have divine inspirational origins. Indeed, music itself was thought by early man to be magical, whereas musicians were believed by the ancient Greeks to be "prophets favored by the Gods" (Wormer, 1973).

This same golden rectangle is found in nature, i.e., the chambered nautilus seashell, the shell of a snail, and in the ear, the cochlea. The geometric proportions of the golden rectangle were also employed in designing the Parthenon in Athens, and by Ptolemy in developing the "tonal calendar" and the "tonal Zodiac" (McClain, 1978), the scale of ratios "bent round in a circle."

In fact, the first cosmologies, such as those developed by the ancient Egyptians, Hindus, Babylonians, and Greeks, were based on musical ratios (Durant, 1939; McClain, 1978). Pythagoras and Plato applied these same "musical proportions" to their theory of numbers, planetary motion, and to the science of stereometry, the gauging of solids (McClain, 1978). Indeed, Pythagorus attempted to deduce the size, speed, distance, and orbit of the planets based on musical ratios as well as estimates of the sounds generated (e.g., pitch and harmony) by their movement through space, i.e., "the music of the spheres" (Durant, 1939). Interestingly, the famous mathematician and physicist Johannes Kepler in describing his laws of planetary motion also referred to them as based on the "music of the spheres."

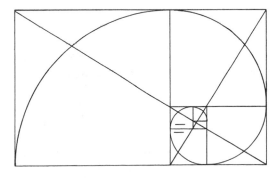

Figure 3.5 The "Golden Rectangle."

Thus, music seems to have certain geometric properties, such as are expressed via the ratio. Indeed, Pythagoras, the "father" of arithmetic, geometry, and trigonometry, believed music to be geometric. As we know, geometry is employed in the measurement of land, the demarcation of boundaries, and, thus, in the analysis of space, shape, points, lines, angles, surfaces, and configurations.

In nature, one form of musical expression—that is, the songs of most birds—also is produced for geometric purposes. That is, a bird does not "sing for joy" but to signal others of impending threat, to attract mates, to indicate direction and location, to stake out territory, and to warn away others who may attempt to intrude on his space (Catchpole, 1979; Hartshorne, 1973).

If we may assume that long before man sang his first song, the first songs and musical compositions were created by our fine-feathered friends (sounds that inspired mimicry by woman and man), then it appears that musical production was first and foremost emotional and motivational, and directly related to the geometry of space: the demarcation of one's territory. Emotion and geometry are characteristics that music still retains today; they are linked neurologically. This is also why training in music can improve visual-spatial (as well as mathematical) skills and vice versa.

CONSTRUCTIONAL AND SPATIAL PERCEPTUAL SKILLS

Based on studies of brain-injured, neurosurgical (e.g., temporal lobectomy, split-brain), and normal populations, the right cerebral hemisphere has been found to be dominant over the left in the analysis of geometric and visual space; the perception of depth, distance, direction, shape, orientation, position, perspective, and figure-ground; the detection of complex and hidden figures; the performance of visual closure; and the ability to infer the total stimulus configuration from incomplete information, route finding and maze learning, localizing targets in space, the performance of reversible operations, stereopsis, and the determination of the directional orientation of the body as well as body-part positional relationships (Benton, 1993; Butters and Barton, 1970; Carmon and Bechtoldt, 1969; DeRenzi and Scotti, 1969; DeRenzi et al., 1969; Ettlinger, 1960; Fontenot, 1973; Franco and Sperry, 1977; Fried et al., 1982; Hannay et al., 1987; Kimura, 1966; 1969, 1993; Landis et al., 1986; Lansdell, 1968, 1970; Levy, 1974; Milner, 1968; Nebes, 1971; Sperry, 1982).

In addition, the isolated right hemisphere has been found to be superior in "fitting designs into larger matrices, judging whole circle size from a small arc, discriminating and recalling nondescript shapes, making mental spatial transformations, sorting block sizes and shapes into categories, perceiving wholes from a collection of parts, and the intuitive perception and apprehension of geometric principles" (Sperry, 1982:1225).

Thus, it is the right hemisphere that enables us to find our way in space without getting lost, to walk and run without tripping and falling, to throw and catch a football with accuracy, to drive a car without bumping into things, to draw conclusions based on partial information, and to see the forest when looking at the trees. The right is also superior to the left in analyzing manipulo-spatial problems, drawing, and copying simple and complex geometric-like figures, and performing constructional tasks, block designs, and puzzles (Benson and Barton, 1970; Black and Strub, 1976; Critchly, 1953; DeRenzi, 1982; Gardner, 1975; Hecaen and Albert, 1978; Hier et al., 1983; Kertesz, 1983b; Levy, 1974; Luria, 1973, 1980; Piercy et al., 1960). It is for these and other reasons that the right brain is often viewed as the artistic half of the cerebrum.

The right hemisphere is also dominant over the left in regard to localizing and thus referencing the position of an object in space (Cook et al., 1994), as well as in aiming and closed-loop throwing accuracy (Guiard et al., 1983; Haaland and Harrington, 1990). Of course, most individuals use the right hand to throw (as well as to draw). Presumably, the right hemisphere is able to guide right limb and related axial movements (Rapcsak et al., 1993) via bilateral supplementary motor area and parietal lobe innervation of the basal ganglia and lower motor neurons (see Chapters 9 and 10).

Sex Differences in Spatial Ability

Right cerebral visual-spatial and geometric superiorities constitute skills that would enable an ancient hunter to visually track, throw a spear at, and dispatch various prey while maintaining a keen awareness of all else occurring within the environment (Joseph, 1992b, 1993). For most of human (and probably chimpanzee and baboon) history, males have typically been the hunters, whereas females (including female chimps) spend much more time gathering.

Not surprisingly, males are far superior to females in regard to visual-spatial functioning and analysis (Broverman, et al., 1968; Dawson et al., 1975; Harris, 1978; Joseph, 1992a, 1993; Kimura, 1993; Levy and Heller, 1992). This includes a male superiority in recalling and detecting geometric shapes, detecting figures that are hidden and embedded in an array of other stimuli, constructing three-dimensional figures from two-dimensional patterns, visually rotating or recognizing the number of objects in a three-dimensional array, and playing and winning at chess (which requires superior spatial abilities). Males also possess a superior geometric awareness, directional sense, and geographic knowledge, are better at solving tactual and visual mazes, and are far superior to females in aiming, throwing, and tracking, such as in coordinating one's movements in relationship to a moving target (reviewed in Broverman, et al., 1968; Harris, 1978; Joseph, 1992a, 1993; Kimura, 1993; Levy and Heller, 1992). In contrast, only about 25% of females, in general, exceed the average performance of males on tests of such abilities (Harris, 1978).

Moreover, some of these differences are present during childhood and have been demonstrated in other species (Dawson et al., 1975; Harris, 1978; Levy and Heller, 1992; Joseph, 1979; Joseph and Gallagher, 1980; Joseph et al., 1978). For example, male rats consistently demonstrate superior visual-spatial skills as compared to females (Joseph, 1979, Joseph and Gallagher, 1980; Joseph et al., 1978). In fact, these sex difference are ameliorated only when females are reared in a complex, socially and environmentally enriched environment and when males are placed in a restricted and impoverished environment, in which case males and females perform similarly (Joseph and Gallagher, 1980). When environments are similar (enriched or deprived), males outperform females. Hence, environmental influences do not cause these sex differences, which are obviously innate.

In large part, these sex differences in spatial ability are clearly a function of the presence or absence of testosterone during human fetal development (Joseph et al., 1978), as well as the differential activities of males versus females for much of human history: i.e., hunting versus gathering (Joseph, 1992b, 1993). Hence, over the course of evolution, natural male-versus-female superiorities have been enhanced.

However, in this regard, just as females appear to have more brain space devoted to language and social-emotional functions, it also appears that males have more neocortical space devoted to spatial-perceptual and related expressive functions (Joseph, 1993).

Visual-Perceptual Abnormalities

When the male or female right hemisphere is damaged, most aspects of visual-spatial and perceptual functioning can become altered, including nonlinguistic memory. For example, right temporal lobe damage impairs memory for abstract designs, tonal melodies, and visual mazes (Kimura, 1963; Milner, 1968). The result can be deficits in left-sided attention; deficits in the ability to make judgments that involve visual-figural relationships, to detect hidden, embedded, and overlapping nonsense figures, and to recognize or recall recurring shapes; and disturbances in the capacity to perceive spatial wholes and achieve visual closure (Bartoloemeo et al., 1994; Benton, 1993; Binder et al., 1992; DeRenzi, 1982; DeRenzi et al., 1969; Ettlinger, 1960; Gardner, 1975; Kimura 1963, 1966, 1969; Landis et al., 1986; Lansdell, 1968; 1970; Levy, 1974).

Such individuals may misplace things; have difficulty with balance and stumble and bump into walls and furniture; become easily confused, disoriented or lost while they are walking or driving; and have difficulty following directions or even putting on their clothes. Indeed, such patients easily can get lost while they are walking down familiar streets and even in their own homes (Benton, 1993; DeRenzi, 1982; DeRenzi et al., 1969; Ettlinger, 1960; Gardner, 1975; Landis et al., 1986; Lansdell, 1968; 1970; Levy, 1974). Right brain damage also can result in disorientation, problems in assuming different perspectives, and even dressing (Hecaen, 1962; Hier et al., 1983).

In some instances, the deficit can be quite subtle and circumscribed. For example, one patient's only complaint (3 months after he suffered a circumscribed blunt head injury that resulted in a subdural hemotoma over the right posterior temporal-parietal area) was that his golf game had deteriorated significantly and he was no longer as accurate when throwing wads of paper into the trash can in his office. Formal testing also indicated mild constructional and manipulo-spatial disturbances, with most other capacities in the high average to superior range.

Drawing and Constructional Deficits

The left hemisphere also contributes to visual-spatial processing and expression such that, when damaged, drawing ability can be affected (Kimura, 1993), albeit in a manner different from that of right cerebral injury (Mehta et al., 1987). For example, because the left is concerned with the analysis of parts or details, lesions result in sequencing errors, oversimplification, and a lack of detail in drawings such that details may be ignored, although the general outline or shape may be retained (Gardner, 1975; Joseph, 1988a; Kimura, 1993; Levy, 1974).

In contrast, the right cerebrum is more involved in the overall perceptual analysis of visual and object interrelations, including visual closure and gestalt formation. Thus, patients with right cerebral injuries have trouble with general shape and organization, although certain details may be drawn correctly. Drawings may also be grossly distorted and/or characterized by left-sided neglect.

Right-sided damage also can affect writing. When patients are asked to write cursively, writing samples may display problems with visual closure, as well as excessive segmentation due to left hemisphere release (Joseph, 1988a). That is, cursively the word "recognition" may be written "re cog n i tion," or letters such as "o" may be only partly formed.

Constructional deficits are more severe after right hemisphere damage (Arrigoni and DeRenzi, 1964; Black and Strub, 1976; Benson and Barton, 1970; Critchly, 1953; Hier et al., 1983; Piercy et al., 1960). However, lesions to either hemisphere can create disturbances in constructional and manipulo-spatial functioning, including performance on the WAIS Block Design and Object Assembly subtests (Arrigoni and DeRenzi, 1964; Cubelli et al., 1991; Kimura, 1993; Mehta et al., 1987; Piercy et al., 1960).

If the left hemisphere is damaged, performance may be impaired due to motor programming errors and/or an inability to transform the percept into a motor action with preservation of good spatial-perceptual functioning (Warrington et al., 1966); in this case, errors may be recognized by the patient. With right cerebral injuries, visual-spatial perceptual functioning becomes distorted (although motor activities per se are preserved) and the patient may not realize he has made an error (Hecaen and Assal, 1970).

Thus, visual motor deficits can result from lesions in either hemisphere (Arrigoni and DeRenzi, 1964; Bartoloemeo et al., 1994; Kimura, 1993; Piercy et al., 1960), though visual-perceptual disturbances are more likely to result from right hemisphere damage. Lesions to the left half of the brain may leave the perceptual aspects undisturbed, whereas visual motor functioning and selective organization may be compromised (Kim et al., 1984; Mehta et al., 1987; Poeck et al., 1973; Teuber and Weinstein, 1956) and attentional functioning may be disturbed (Bartoloemeo et al., 1994; Cubelli et al., 1991).

In general, the size and sometimes the location of the lesion within the right hemisphere have little or no correlation with the extent of the visual-spatial or constructional deficits demonstrated (Kimura, 1993), although right posterior lesions tend to be the worst of all. For example, in a restrospective study of 656 patients with unilateral lesions that employed a short form of the WAIS, Warringtonet al. (1986) found that those with right posterior (versus anterior or left hemisphere) damage had the lowest performance IQs as well as difficulty performing the block design and picture arrangment subtests.

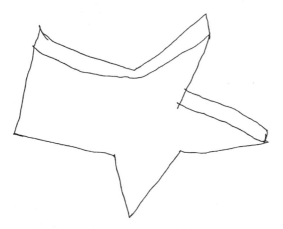

Figure 3.6 Patient's copy of a "star." Note left-sided distortion.

Conversely, visual-perceptual disturbances associated with left hemisphere damage are positively correlated with lesion size; and left anterior lesions are worse than left posterior (Benson and Barton, 1970; Black and Bernard, 1984; Black and Strub, 1976; Lansdell, 1970). The larger the lesion, the more extensive the deficit. Kimura (1993), however, argues that males are more likely to demonstrate these disorders with left posterior lesions, whereas in females the lesion tends to be more anterior.

As pertaining to WAIS performance IQ–verbal IQ differences, based on an extensive review of the literature, Bornstein and Matarazzo (1982) have confirmed what is now generally well known: that those with left hemisphere lesions have lower verbal IQs, whereas those with right-sided damage have lower mean performance IQs.

READING AND MATH

Because of its importance in visual orientation, the right hemisphere also participates in math and reading. Visual-spatial orientation is important when performing a variety of math problems, such as in correctly aligning numbers when adding multiple digits. Conversely, right-sided lesions may cause the patient to neglect the left half of digit pairs when adding or subtracting (Hecaen and Albert, 1978; Luria, 1980).

Moreover, via the analysis of position, orientation, and so forth, the right cerebrum enables a human to read the words on this page without losing his place and jumping haphazardly from line to line. Conversely, the patient with right cerebral injury may fail to attend to the left half of written words, or even the left half of the page (Critchly, 1953; Gainotti, et al., 1972, 1986).

Because the right half of the cerebrum can make conclusions based on partial information, e.g., closure and gestalt formation, humans need not read every or all of each word in order to know what they have read. For example, when presented with incomplete words or perceptually degraded written stimuli, there is an initial right hemisphere superiority in processing (Hellige and Webster, 1979); i.e., there is visual closure, which enables an individual to fill in these gaps and, thus, comprehend. Of course, sometimes people draw the wrong conclusions from incomplete perception (e.g., reading "word power" as "world power"), a problem that can become exaggerated if the right half of the brain has been damaged.

In addition, when the visual-figural characteristics of written language are emphasized, such as when large gothic scripts are employed (e.g., in tachistoscopic studies), the right hemisphere is dominant (Bryden and Allard, 1976; Wagner and Harris, 1994). Similarly, when presented with unfamiliar written words or a foreign alphabet, there is an initial right hemisphere perceptual dominance (Silverberg et al., 1979), apparently as the left hemisphere does not immediately recognize these stimuli as "words," whereas the right hemisphere attends to their shape and form.

Indeed, based on an extensive analysis of the evolution of written language, there is some evidence to suggest an initial right hemisphere dominance, particularly in that much of what was written was at first depicted in a pictorial, gestalt-like fashion (Joseph, 1993; see Chapter 4). The use of images preceded the use of signs (Campbell, 1988; Joseph, 1993; Jung, 1964).

INATTENTION AND VISUAL-SPATIAL NEGLECT

Unilateral inattention and neglect are associated most commonly with right hemisphere (parietal, frontal, thalamic, basal ganglia) damage, particularly following lesions located in the temporal-parietal and occipital junction (Bartoloemeo et al., 1994; Binder et al., 1992; Bisiach et al., 1983; Bisiach and Luzzatti, 1978; Brain, 1941; Calvanio et al., 1987; Critchly, 1953; DeRenzi, 1982; Ferroet al., 1987; Gainotti, et. al., 1986; Heilman, 1993; Heilman et al., 1983; Joseph, 1986a; Motomura et al., 1986; Neilsen, 1937; M. Roth, 1949; N. Roth, 1944; Watson et al., 1981). Such patients initially may fail to

respond to, recall, or perceive left-sided auditory, visual, or tactile stimulation, fail to comb, wash, or dress the left half of their head, face, and body, only eat food on the right half of their plate, write only on the right half of a paper, fail to read the left half of words or sentences (e.g., if presented with "toothbrush" they may see only the word "brush"), or on drawing tasks, distort, leave out details, or fail to draw the left half of various figures, e.g., a clock or daisy (Binder et al., 1992; Bisiach et al., 1983; Calvanio et al., 1987; Critchly, 1953; DeRenzi, 1982; Gainotti et al., 1972, 1986; Hecaen and Albert, 1978; Umilta, 1995; Young et al., 1992).

When shown their left arm or leg, such patients may deny that it is their own and claim that it belongs to the doctor or a patient in the next room. Indeed, "patients with severe unilateral neglect behave as if a whole system of beliefs had vanished, as if one half of the inner model of the environment were simply deleted from their mind (Bisiach et al., 1983:35). In the less extreme cases, patients may seem inattentive, and/or when their attention is directed to the left half of the environment, they are able to respond appropriately (see Jeannerod, 1987, and Umilta, 1995, for a detailed review).

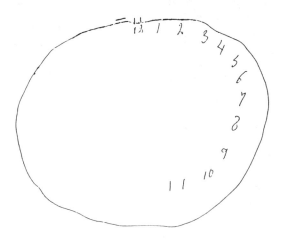

Figure 3.7 Left-sided neglect. The patient, a 54-year-old man with damage to the right frontal-parietal area, was instructed to "draw the face of a clock and put all the numbers in it, and make it say 10 after 11."

Figure 3.8 Examples of left-sided neglect and distortion. Note preservation of right-sided details. The patient, a 33-year-old woman with a right parietal injury, was instructed to copy the "cross" and the "star."

Imaginal and memory functioning also are disrupted such that patients may fail to attend to the left half of images recalled from memory. For example, Bisiach and Luzzatti (1978) found that when patients with right brain damage were asked to recall and describe a familiar scene from different perspectives, regardless of perspective (e.g., imagining a street from one direction and then from another), they consistently failed to report details that fell to their left—although the same details were recalled when imagined from the opposite direction, their right. Bisiach and Luzzatti (1978) suggest that visual images and scenes may be split into two images when conjured up, such that the right hemisphere images the left half of space and the left brain the right half of space. Similar results were presented by Meador et al. (1987).

Thus, right cerebral injuries can result in visual-spatial as well as (imaginal) representational neglect. In some severe cases, patients may demonstrate both forms, whereas in others visual-spatial (but not representational) neglect may be found in isolation (Bartoloemeo et al., 1994).

Neglect also can be influenced by the task demands (Bartoloemeo et al., 1994; Binder et al., 1992; Starkstein et al., 1993; Umilta, 1995) and may be differentially expressed in the horizontal and vertical spatial dimension and in near versus far peripersonal space (Mennemeir et al., 1992). For example, some patients may demonstrate neglect in the visual versus tactile modality (Umilta, 1995), whereas others may adequately draw simple figures, but fail to correctly place numerals in the right half of space when drawing a clock. Moreover, some authors have argued that letter cancellation tests appear to be more sensitive than line bisection tasks when evaluating (right cerebral) neglect (Binder et al., 1992).

Moreover, with the possible exception of letter cancellation, when the task is normally performed best by the damaged hemisphere (that is, before it was damaged), the neglect may be more pronounced (Leicester et al., 1969).

Thus, a patient with a left cerebral injury may not respond to a verbal question or command, or they may tend to ignore objects or written material that falls to their extreme right. Hence, left-brain-damaged patients also may show unilateral inattention or neglect (Albert, 1973; Bartoloemeo et al., 1994; Cubelli et al., 1991; Denny-Brown et al., 1952; Gainotti et al., 1986), albeit in a less severe form.

The Right Frontal Lobe: Arousal and Neglect

In general, there is some evidence to suggest that the right cerebral hemisphere may be involved more greatly in attention and arousal (Beck et al., 1969; Dimond and Beaumont, 1974; Heilman, 1993; Joseph, 1986a, 1990d), such that it may exert bilateral influences on cerebral and limbic activation (see Chapter 11). In consequence, right frontal damage can produce mixed extremes in arousal, such that with massive damage there results hypoarousal, bilateral reductions in reaction time, and, thus, diminished attentional functioning

(DeRenzi and Faglioni, 1965; Heilman et al., 1978; Heilman and Van Den Abell, 1979; Howes and Boller, 1975; S. Weinstein, 1978).

These patients, therefore, also tend to demonstrate variable degrees of left-sided neglect, and may also have considerable difficulty activating and correctly engaging in memory search (see Chapter 11). In this regard, Meador et al. (1987) found that by turning the head to the left (supposedly, more greatly activating the right hemisphere), although still deficient, patients were able to recall more left-sided objects and stimuli.

On the other hand, with smaller lesions involving the right frontal convexity, rather than a loss of arousal there can result a loss of control over arousal, and such patients can respond in a highly disinhibited fashion (Joseph, 1986a, 1990d). Such patients may also, however, demonstrate neglect. Presumably, due to the loss of counterbalancing right frontal input, the left frontal region may be unable to disengage from attending to the right half of auditory, visual, and physical space so as to explore the (neglected) left half of space. However, this may be partly overcome by requiring the patient to physically orient toward the left (e.g., Meador et al., 1987).

In addition, frontal injuries may also result in disconnection such that sensory input arriving in the posterior regions of the cerebrum are prevented from being transferred to the right hemisphere. Thus, input from the undamaged hemisphere continues to be processed (via the assistance of frontal lobe steering and activating influences) to the exclusion of data normally processed by the other half of the brain.

DISTURBANCES OF BODY IMAGE

In addition to having nonlinguistic, prosodic, melodic, emotional, and visual-spatial dominance, the right cerebrum has been shown to be superior to the left in processing various forms of somesthetic and tactile-spatial-positional information, including geometric, tactile-form, and Braille-like pattern recognition (Bradshaw et al., 1982; Carmon and Benton, 1969; Corkin et al., 1970; Desmedt, 1977; Dodds, 1978; Fontenot and Benton, 1971; Franco and Sperry, 1977; Hatta, 1978a; Hermelin and O'Connor, 1971; Hom and Reitan, 1982; Pardo et al., 1991; E. Weinstein and Sersen, 1961).

The right cerebrum is also dominant for two-point discrimination (S. Weinstein, 1978), pressure sensitivity (Semmes et al., 1960; S. Weinstein, 1978; E. Weinstein and Sersen, 1961), and processing tactual-directional information (Carmon and Benton, 1969; Fontenot and Benton, 1971). The right hemisphere may be more involved than the left in the perception of somesthetically mediated pain (Cubelli et al., 1984; Haslam, 1970; Murray and Hagan, 1973).

In addition, unlike the left, the right hemisphere is responsive to tactual stimuli that impinge on either side of the body (Desmedt, 1977; Pardo et al., 1991). Indeed, a somesthetic image of the entire body appears to be maintained by the parietal lobe of the right half of the brain (Joseph, 1988a), and not just a body image, but memories of the body, the left half in particular.

When the right hemisphere is damaged, somesthetic functioning can become grossly abnormal, and patients may experience peculiar disturbances that involve the body image (Critchly, 1953; Gerstmann, 1942; Gold et al., 1994; Hillbom, 1960; Joseph, 1986a; Miller, 1984; Nathanson et al., 1952; M. Roth, 1949; N. Roth, 1944; Sandifer, 1946; E. Weinstein and Kahn, 1950, 1952). These patients may fail to perceive stimuli applied to the left side; wash, dress, or groom only the right side of the body; confuse body-positional and spatial relationships; misperceive left-sided stimulation as occurring on the right; fail to realize that their extremities or other body organs are in some manner compromised; and/or literally deny that their left arm or leg is truly their own as the (right parietal lobe) memory of the body has been destroyed.

When confronted by their unused or paralyzed extremities, such patients may claim that they belong to the doctor or a

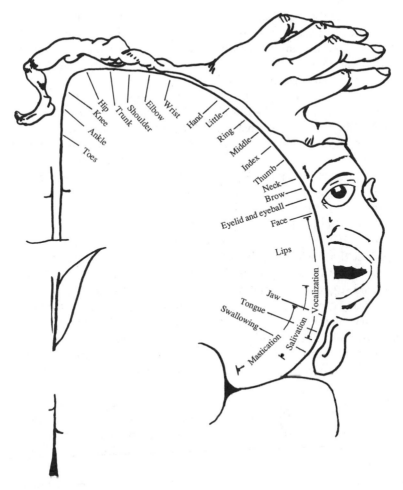

Figure 3.9 The motor "homunculus" of the frontal lobe. Body parts are represented in accordance with sensory/motor importance.

person in the next room or, conversely, seem indifferent to their condition and/or claim that their paralyzed limbs are normal, even when they are unable to comply with requests to move them.

"When asked why she could not move her hand she replied, 'Somebody has ahold of it.' Another patient, when asked if anything was wrong with her hand said, 'I think it's the weather. I could warm it up and it would be all right.' One woman, when asked whether she could walk, said, 'I could walk at home, but not here. It's slippery here.' Another patient, when asked why he couldn't raise his arm said, 'I have a shirt on'" (Nathanson et al., 1952:383).

EMOTIONAL REACTIONS

Many patients also appear indifferent and/or make inappropriate emotional remarks about their disability. Given right cerebral dominance for emotion and social-emotional intelligence, perhaps not surprisingly, many patients may appear and behave inappropriately and may even laugh and joke about being paralyzed.

In an extensive examination of these disturbances, E. Weinstein and Kahn (1950) found that of 22 patients (only 3 of whom were thought to have predominantly left hemisphere dysfunction), 15 were euphoric and manifested an air of serenity or bland unconcern about their condition, despite the fact that they were suffering from disorders such as hemiplegia, blindness, loss of memory, and incontinence. Ten of these individuals also behaved in a labile or transiently paranoid fashion.

Right cerebral lesions have been reported to slow the appearance of phantom limbs on either side of the body and can result in the loss of phantom limb pain (S. Weinstein, 1978). In contrast, left-sided lesions seem to have little effect.

In general, like neglect and inattention, lesions which result in body image disturbances tend to involve the right parietal or right frontal lobe (Critchly, 1953; Joseph, 1986a, 1990d; Stuss and Benson, 1986). In this regard, the fact that lesions of the left hemisphere rarely result in neglect or body image disturbances suggests that the right hemisphere maintains a bilateral representation and memory of the body, whereas the left cerebrum maintains a unilateral representation and a memory of only the right half of the body—memories that are stored in the parietal lobe, a portion of which evolved from the hippocampus, which is also concerned with memory, including memory of the body in space (see Chapters 1, 2, 5, and 12).

Hence, when the left hemisphere is damaged, the right parietal lobe continues to monitor (and remember) both halves of the body and there is little or no neglect—an impression supported by findings indicating that the right hemisphere electrophysiologically responds to stimuli that impinge on either side of the body, whereas the left hemisphere predominantly responds only to right-sided stimulation (Desmedt, 1977; Pardo et al., 1991).

Pain and Hysteria

In addition to body image distortions, parietal lobe injuries (particularly when secondary to tumor or seizure activity) also can give rise to *sensory misperceptions*, such as *pain* (Davidson and Schick, 1935; Hernandez-Peon et al., 1963; Ruff, 1980; Wilkinson, 1973; York et al., 1979). That is, in the less extreme cases, rather than failing to perceive (i.e., neglecting) the left half of

the body, patients may experience sensory distortions that concern various body parts due to abnormal activation of the right hemisphere and parietal lobe.

For example, one 48-year-old housewife complained of diffuse, poorly localized (albeit intense) pain in her left leg, which occurred in spasms that lasted minutes. She subsequently was found to have a large tumor in the right parietal area, which, when removed, alleviated all further attacks. Head and Holmes (1911) reported a patient who suffered brief attacks of "electric shock-like" pain that radiated from his foot to the trunk; a glioma in the right parietal area subsequently was discovered. McFie and Zangwill (1960) reported an individual who began to experience intense, extreme pain in a phantom arm after a right posterior stroke.

In another instance reported by York et al. (1979), a 9-year-old boy experienced spontaneous attacks of intense scrotal and testicular pain and was found to have seizure activity in the right parietal area. Ruff (1980) reports two patients who experienced paroxysmal episodes of spontaneous and painful orgasm, which was secondary to right parietal seizure activity. In one patient the episodes began with the sensation of clitoral warmth, engorgement of the breasts, tachycardia, and so forth, all of which rapidly escalated to a painful climax. Interestingly, in the normal intact individual, orgasm is associated with electrophysiological arousal predominantly within the right hemisphere (H. Cohen et al., 1976).

It is important to note, however, that although the predominant focus for paroxysmal pain is the right hemisphere, pain also has been reported to occur with tumors or seizure activity that involve the left parietal region (Bhaskar, 1987; McFie and Zangwill, 1960).

Unfortunately, when the patient's symptoms are not considered from a neurological perspective, their complaints with regard to pain may be viewed as psychogenic in origin. This is because the *sensation* of pain, stiffness, and engorgement, is, indeed, *entirely* "in their head" and based on distorted perceptual functioning at the level of the neocortex. Physical examination may reveal nothing wrong with the seemingly affected limb or organ—unless neocortical damage is extensive, in which case the patient may display paresis, sensory loss, and so forth. Even under these latter conditions, such patients may be viewed as hysterical or hypochondriacal, particularly in that right hemisphere damage also disrupts emotional functioning. For example, one female patient who was suspected of hysteria subsequently was found to have a right frontal-parietal and mid-temporal cyst (Joseph, 1988a). This same individual had difficulty recognizing or mimicking emotional and environmental sounds.

It is noteworthy that individuals suspected of suffering from hysteria are two to four times more likely to experience pain and other distortions on the left side of the body (Axelrod et al., 1980; Galin et al., 1977; Ley, 1980; D. Stern, 1977); these findings, in turn, suggest that the source of the hysteria may be a damaged right hemisphere.

This supposition is supported further by the raw MMPI data provided by Gasparrini, Satz, Heilman, and Coolidge (1978) in their study of differential effects of right versus left hemisphere damage on emotional functioning. That is, whereas depression (elevated MMPI scale 2) is more likely subsequent to left cerebral damage, a pattern suggestive of hysteria and conversion reactions (elevations on scales 1 and 3, reductions on scale 2; i.e., the Conversion "V") are more likely subsequent to right hemispheric lesions.

The findings of Gasparrini et al. (1978) were replicated based on a restrospective case analysis of individuals with long-standing right versus left hemisphere injury (verified by neuropsychological exam coupled with CT scan and/or EEG). In this study the MMPI Conversion V profile was found to be associated signficantly and almost exclusively with right hemisphere injuries (among males only), whereas elevated scale 2 was associated with left-sided damage (Joseph, unpublished data). However, not every patient demonstrated this pattern.

At least one investigator reporting on psychiatric patients, however, has attributed hysteria to left cerebral damage (Flor-Henry, 1983). In part, this may be a function of the statistical permutations utilized in analyzing his data. However, this also may reflect the nature of the population studied (i.e., psychiatric rather than neurological) and, thus, the differential effect of long-standing drug treatments and biochemical and congenital disturbances versus recently acquired neuroanatomical lesions (see Chapter 18 for related discussion).

FACIAL-EMOTIONAL RECOGNITION AND PROSOPAGNOSIA

Possibly due, in part, to the visual-spatial complexity as well as the social-emotional significance of the human face, the right hemisphere has been shown to be dominant in the perception and recognition of familiar and unfamiliar faces (Alvarez and Fuentes, 1994; Bradshaw et al., 1980; DeRenzi, 1982; DeRenzi et al., 1968; DeRenzi and Spinnler, 1966; Deruelle and de Schonen, 1991; Evans et al., 1995; Geffen et al., 1971; Hecaen and Angelergues, 1962; Levy et al., 1972; Ley and Bryden, 1979; Moreno, et al., 1990; Rizzolatti et al. 1971; Sergent et al., 1992) and to become more greatly activated when viewing faces as measured by positron emission tomography and regional cerebral blow flow (Sergent et al., 1992).

There is some indication, however, that the left hemisphere is involved in the recognition of famous faces (Marzi and Berlucchi, 1977; Rizzolatti, et al., 1971) and the differentiation of highly similar faces (presented in outline form) when analysis of fine detail is necessary (Patterson and Bradshaw, 1975).

Be it the face of a friend or that of a stranger, the right hemisphere superiority for facial recognition is augmented by the additional display of facial emotion (Ley and Bryden, 1979; Suberi and McKeever, 1977). Indeed, not only is it predominant in perceiving facial emotion, regardless of the emotion conveyed (Buchtel et al., 1978;

Dekosky et al., 1980; Landis et al., 1979; Strauss and Moscovitch, 1981; Suberi and McKeever, 1977), faces also are judged to be more intensely emotional when viewed exclusively by the right hemisphere (Heller and Levy, 1981).

Conversely, with right (but not left) cerebral injuries, patients tend to to demonstrate an overall impairment in recalling, imaging, identifying, and visualizing facial emotional expressions (Bowers et al., 1991; Young et al., 1995), and with right temporal atrophy, patients may suffer a progressive prosopagnosia (Evans et al., 1995).

In addition, the left side of the face has been found to be more emotionally expressive (Campbell, 1978; Chaurasis and Goswami, 1975; Moreno et al., 1990; Sackheim et al., 1978) and to be perceived as more intensely emotional as well (Borod and Caron, 1980; Sackheim and Gur, 1978). In response to emotional stimuli, the left half of the face becomes more activated, and a significant majority of individuals respond with conjugate lateral eye movements to the left (Schwartz et al., 1975; Tucker, 1981), the left half of the body being under the control of the right hemisphere.

Conversely, damage to the right (but not left) hemisphere significantly reduces facial emotional expressiveness (Blonder et al., 1993).

In fact, the right hemisphere superiority for facial recognition is augmented by the additional display of facial emotion (Ley and Bryden, 1979; Moreno et al., 1990; Suberi and McKeever, 1977), regardless of the emotion conveyed (Buchtel et al., 1978; Dekosky et al., 1980; Landis et al., 1979; Strauss and Moscovitch, 1981; Suberi and McKeever, 1977). Faces also are judged to be more intensely emotional when viewed exclusively by the right hemisphere (Heller and Levy 1981), and the right hemisphere is dominant with regard to memory for facial expression as well (Weddell, 1989). Hence, the right half of the cerebrum is dominant for visual, facial, and auditory modes of emotional expression and perception, including memory for faces and facial emotion.

Conversely, when the right hemisphere is damaged, particularly the occipital-temporal region, there can result a severe disturbance, not just in the capacity to perceive facial emotion but in the ability to recognize the faces of friends, loved ones, or pets (DeRenzi, 1986; DeRenzi et al., 1968; DeRenzi and Spinnler, 1966; Evans et al., 1995; Hecaen and Angelergues, 1962; Landis et al., 1986; Levine, 1978; Whiteley and Warrington, 1977; Young et al., 1995), i.e., *prosopagnosia*. Some patients may be unable to recognize their own face in the mirror.

For example, one patient could not identify his wife, and although tested for 7 hours by the same examiner was unable to recognize him at the end of the session (DeRenzi, 1986). Another patient was unable to recognize relatives or her pets; she was even unable to discriminate between people on the basis of sex but instead had to rely on the presence of details (such as lipstick, rouge, hair length, moustache) to make discriminations (Levine, 1978).

Delusions and Facial Recognition

As noted above, lesions to the right cerebrum may result in difficulty recognizing, distinguishing, or differentiating facial emotion (Bowers et al., 1991; DeKosky et al., 1980; Evans et al., 1995). That is, patients are unable to recognize or determine what others are feeling via facial expression.

Electrical stimulation of the posterior portion of the right middle temporal gyrus also results in an inability to correctly label the emotion shown in faces, whereas posterior right temporal stimulation disrupts visual-spatial memory for faces in general (Fried et al., 1982). Hence, the right hemisphere is clearly dominant for perceiving, recognizing, differentiating, expressing, and even recalling facial emotion.

In some instances, depending on the extent of damage, rather than a frank failure to recognize, patients may contend that friends, lovers, or their children look *different, strange, or unfamiliar*—perceptions that may give rise to a host of abnormal emotional reactions and upheavals, including frank paranoia, for example, fear that one's wife may have been replaced by an imposter.

Delusional misperception of familiar and unfamiliar individuals, as well as disturbances such as Capgras syndrome (delusional doubles; reduplication) or false identification, also can result from right hemisphere and/or (bilateral) frontal damage (Alexander et al., 1979; Benson et al., 1976; Hecaen, 1964; Jacocic and Staton, 1993). For example, one patient who was looking into a dark tachistoscope suddenly said in an emotional voice, "I see my daughter—oh, she's gone," and was unable to recognize ward personnel or relatives when present (Levine, 1978).

"POSITIVE" EMOTIONS AND THE LEFT HALF OF THE BRAIN

There is some (albeit, controversial) evidence that the left hemisphere processes positive emotions, and is likely to view negative and neutral emotions as positive (Davidson, 1984; Davidson et al., 1985, 1987; Dimond and Farrington, 1977; Dimond et al., 1976; Gainotti, 1972; Lee et al., 1990, 1993; Ostrove et al., 1990; Otto et al., 1987; Rossi and Rosadini, 1967; Sackeim et al., 1982; Schiff and Lamon, 1989; Silberman and Weingartner, 1986; Terzian, 1964). For example, when right cerebral influences are eliminated, such as due to right hemisphere damage or anesthetization, many patients are likely to view and report neutral and even negative events in a positive manner and to exhibit a positive mood, including laughter (Gainotti, 1972; Lee et al., 1990; Rossi and Rosadini, 1967; Sackeim et al., 1982; Terzian, 1964).

Rossi and Rosadini (1967), for example, injected sodium amytal into the left versus right hemisphere and found that 68% of those with left hemisphere inactivation reacted with depression (expressed presumably by the awake right hemisphere). In contrast, 84% of those with right-sided inactivation responded with euphoria (expressed presumably by the completely awake left hemisphere) and 16% responded

in a depressed fashion (see also Gainotti, 1972; Lee et al., 1990; Rossi and Rosadini, 1967; Terzian, 1964). These differences are not always observed, however. Indeed, I have observed approximately a dozen sodium amytal tests and never witnessed these changes in affect.

Nevertheless, when the left anterior region of the brain has been damaged or is dysfunctional, individuals are likely to respond with severe depression, or anger, irritability, and paranoia (Gainotti, 1972; Gruzelier and Manchanda, 1982; Hillbom, 1960; Joseph, 1990d; Lebrun, 1987; Robinson and Benson, 1981; Robinson and Szetela, 1981; Sherwin et al., 1982; Sinyour et al., 1986).

This suggests that when left cerebral ("positive") influences are negated, positive emotions are replaced by "negative" feeling states, which in turn are a consequence of right cerebral emotional dominance; i.e., the right half of the brain is accurately perceiving the consequences of the injury and is understandably upset and depressed. However, in some cases those with massive left posterior and temporal lobe damage may respond with euphoria, which in turn may be due to subsequent emotional disorganization in response to loss of comprehension.

Therefore, although the left plays a minuscule role in regard to emotions, there is evidence to suggest that it is inclined to view or express emotional material in either a neutral or positive light, regardless of its actual affective value. This may also explain why following massive right frontal injuries (at least in males), "positive" emotions are sometimes expressed indiscriminately and inappropriately. Indeed, affected patients may appear to be in a manic state (see below).

DISTURBANCES OF EMOTION AND PERSONALITY OVERVIEW

The right cerebral hemisphere appears to be dominant in regard to most aspects of somestheses, including the maintenance of the body image, visual-spatial-geometric analysis, facial expression and perception, and musical and paralinguistic, melodic-intonational processing. The right hemisphere also predominates in regard to almost all aspects of emotional functioning and social-emotional intellectual activity, and exerts bilateral influences on autonomic nervous system functioning. Moreover, norepinephrine (NE) concentrations are higher in the right thalamus (Oke et al., 1978), whereas, conversely, damage to the right hemisphere disrupts NE levels on both sides of the brain and similar damage to the left hemisphere only affects local NE levels (Robinson 1979). This is highly significant, given the role and importance of NE in emotions and arousal.

Hence, when the right cerebrum is damaged there can result a myriad of peculiar disturbances that involve a number of modalities. Patients with body image disturbances may seem emotionally abnormal and possibly hysterical rather than neurologically impaired. Those with facial agnosia may become paranoid and convinced that friends or lovers have been replaced by imposters. Individuals with intonational-melodic and emotional-linguistic deficiences may be unable to adequately vocally express their feelings, fail to recognize or misinterpret the feelings conveyed by others, as well as "miss the point" or fail to recognize discrepancies in speech, such as when presented with implausible information. Conversely, their own speech patterns and behavior may become abnormal, tangential, disinhibited, and contaminated by implausible, confabulatory, and delusional ideation.

Hence, in all instances, regardless of where within the right hemisphere damage occurs, social-emotional abnormalities may result. Indeed, emotional disturbances may be the dominant or only manifestation of a patient's illness. Unfortunately, if not accompanied by gross neurological signs, the possibility of right hemisphere damage may be overlooked.

Mania and Emotional Incontinence

In December of 1974, associate Supreme Court Justice William O. Douglas suffered a massive infarct in the right cerebral hemisphere that left him paralyzed and in pain for many years. As reviewed by Gardner et al. (1983):

> For all public purposes, Douglas acted as if he were fine, as if he could soon assume full work on the Court. He insisted on checking himself out of the hospital where he was receiving rehabilitation and then refused to return. He responded to seriously phrased queries about his condition with offhanded quips: "Walking has very little to do with the work of the Court; if George Blanda can play, why not me?" He insisted in a press release that his left arm had been injured in a fall, thereby badly denying the neurological cause of his paralysis. Occasionally, he acted in a paranoid fashion, claiming, for example, that he was the Chief Justice. During sessions of the Court, he asked irrelevant questions, and sometimes rambled on. Finally, Douglas did resign. But ... refused to accept that he was no longer a member of the Court. He buzzed for his clerks ... tried in inject himself into the flow of business, took aggressive steps to assign cases to himself, asked to participate in, author, and even publish separately his own opinions, and he requested that a tenth seat be placed at the Justices' bench.

In short, this formerly highly impressive and dignified man acted for a long time after his stroke in a highly unusual and bizarre manner.

Since the right hemisphere is dominant in the perception and expression of facial, somesthetic, and auditory emotionality, damage to this half of the brain can result in a variety of affective and social-emotional abnormalities, including indifference, lability, hysteria, florid manic excitement, pressured speech, ideas of reference, bizarre confabulatory responding, childishness, irritability, euphoria, impulsivity, promiscuity, and abnormal sexual behavior (Bear, 1977, 1983; Bear and Fedio, 1977; Clark and Davison, 1987; Cohen and Niska, 1980; Cummings and Mendez, 1984; Erickson, 1945; Forrest, 1982; Gardner et al., 1983; Gruzelier and Manchanda, 1982; House et al., 1990; Jamieson and Wells, 1979; Jampala and Abrams, 1983; Joseph, 1986a, 1990d; Lishman, 1968; Offen et al., 1976; Rosenbaum and Berry, 1975; Spencer et al., 1983; Spreen et al., 1965; Starkstein et al., 1987; Stern and Dancy, 1942).

For example, M. Cohen and Niska (1980) report an individual with a subarachnoid hemorrhage and right temporal hematoma who developed an irritable mood; shortened sleep time; loud, grandiose, tangential speech; flight of ideas; and lability, and who engaged in the buying of expensive commodities. Similarly, Oppler (1950) documented an individual with a good premorbid history who began to deteriorate over many years' time. Eventually, the patient developed flight of ideas, emotional elation, increased activity, hypomanic behavior, lability, extreme fearfulness, distractibility, jocularity, and argumentativeness. The patient was also overly talkative and produced a great deal of tangential-circumstantial ideation with fears of persecution and delusions. Eventually, a tumor was discovered (which weighed over 74 grams) and removed from the right frontal-parietal area.

Similarly, Spreen et al. (1965) describe a 65-year-old man who, following a right-sided stroke (with left hemiparesis), developed extremely unpredictable behavior and lability. Regardless of external circumstances he would begin crying at one moment and at the next demonstrate irritability, happiness, or extreme depression. Secondary mania also has been reported with right frontal encephalopathy accompanied by bioelectric epileptiform activity (Jack et al., 1983).

Over the course of the last dozen years, I have examined 16 male and 2 female patients who developed manic-like symptoms after suffering a right frontal stroke or trauma to the right hemisphere (some of whom are described in Joseph, 1986a, 1988a). All but 3 of the males had good premorbid histories and had worked steadily at the same job for 3–5 years. Upon recovering from their injuries, all developed delusions of grandeur, pressured speech, flight of ideas, decreased need for sleep, indiscriminate financial activity, extreme emotional lability, and increased libido.

One patient, formerly very reserved, quiet, conservative, and dignified with more than 20 patents to his name, and who had been married to the same woman for over 25 years, began patronizing up to four different prostitutes a day and continued this activity for months. He left his job, began thinking up and attempting to act upon extravagant, grandiose schemes, and camped out at Disneyland in an attempt to convince personnel there to finance his ideas for developing an amusement park on top of a mountain. At night he frequently had dreams in which either John F. or Robert Kennedy would appear and offer him advice—and he was a Republican!

CONSCIOUSNESS, AWARENESS, MEMORY, AND DREAMING
Right Hemisphere Mental Functioning

That the right brain is capable of conscious experience has now been well demonstrated in studies of patients who have undergone complete corpus callosotomies (i.e., split-brain operations) for the purposes of controlling intractable epilepsy. As described by Nobel Laureate Roger Sperry (1966:299), "Everything we have seen indicates that the surgery has left these people with two separate minds, that is, two separate spheres of consciousness. What is experienced in the right hemisphere seems to lie entirely outside the realm of awareness of the left hemisphere. This mental division has been demonstrated in regard to perception, cognition, volition, learning and memory."

For example, when split-brain patients are tactually stimulated on the left side of the body, their left hemispheres demonstrate marked neglect when verbal responses are required, they are unable to name objects placed in the left hand, and they fail to report the presence of a moving or stationary stimulus in the left half of their visual fields (Bogen, 1979; Gazzaniga, 1970; Gazzaniga and LeDoux, 1978; Joseph, 1988b; Levy, 1983; Seymour et al., 1994; Sperry, 1982). They (i.e., their left hemispheres) cannot verbally describe odors, pictures, or auditory stimuli tachistoscopically or dichotically presented to the right cerebrum, and have extreme difficulty explaining why the left half of their body responds or behaves in a particular purposeful manner (such as when the right brain is selectively given a command). In addition, they demonstrate marked difficulties in naming incomplete figures (and thus forming visual closure), as well as a reduced ability to name and identify nonlinguistic and environmental sounds (Joseph, 1986b, 1988b)—capacities associated with the functional integrity of the right hemisphere.

However, by raising their left hand (which is controlled by the right half of the cerebrum) the disconnected right hemisphere is able to indicate when the patient is tactually or visually stimulated on the left side. When tachistoscopically presented with words to the left of visual midline, although unable to name them, when offered mutiple visual choices in full field their right hemispheres are usually able to

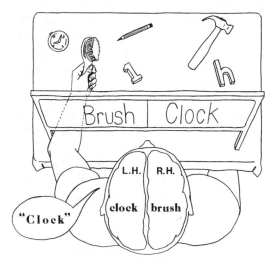

Figure 3.10 The split-brain patient is facing a screen upon which is flashed (via tachistoscope) the words "brush" and "clock." He is instructed to reach inside with his left hand to retrieve the object that corresponds to what he has seen. Because the word "brush" was viewed by the right hemisphere, he retrieves the brush with the left hand. However, because the word "clock" was viewed by the left (language-dominant) hemisphere, when asked what he saw he states "clock."

point correctly with the left hand to the word viewed.

In this regard, when presented with words like "toothbrush," such that the word "tooth" falls in the left visual field (and thus, is transmitted to the right cerebrum) and the word "brush" falls in the right field (and goes to the left hemisphere), when offered the opportunity to point to several words (i.e., hair, tooth, coat, brush, etc.), the left hand usually will point to the word viewed by the right cerebrum (i.e., tooth) and the right hand to the word viewed by the left hemisphere (i.e., brush). When offered a verbal choice, the speaking (usually the left) hemisphere will respond "brush" and will deny seeing the word "tooth."

Overall, this indicates that the disconnected right and left cerebral hemispheres, although unable to communicate and directly share information, are nevertheless fully capable of independently generating and supporting mental activity (Bogen, 1969, 1979; Gazzaniga, 1970; Gazzaniga and LeDoux, 1978; Joseph, 1986b, 1988b; Levy, 1983; Sperry, 1982). Hence, the right hemisphere a highly evolved visual spatial, environmental, and social-emotional intelligence (a term and concept developed by Joseph, 1992b) that accompanies in parallel what *appears* to be the "dominant" temporal-sequential, language-dependent stream of consciousness in the left cerebrum.

Moreover, as has been demonstrated by Sperry, Bogen, Levy, Gazzaniga, and colleagues, the isolated right cerebral hemisphere, like the left, is capable of self-awareness; can plan for the future; has goals and aspirations, likes and dislikes, social and political awareness; can purposefully initiate behavior, can guide responses, choices, and emotional reactions; and can recall and act upon certain desires, impulses, situations, or environmental events—without the aid, knowledge, or active (reflective) participation of the left half of the brain.

Right Brain Perversity

In that the brain of the normal as well as "split-brain" patient maintains the neu-

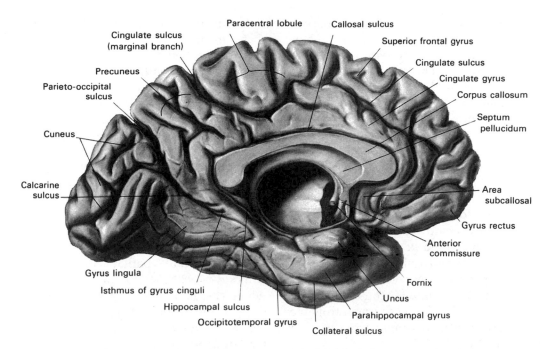

Paracentral lobule

Callosal sulcus

Cingulate sulcus
(marginal branch)

Superior frontal gyrus

Cingulate sulcus

Precuneus

Cingulate gyrus

Parieto-occipital
sulcus

Corpus callosum

Septum
pellucidum

Cuneus

Calcarine
sulcus

Area
subcallosal

Gyrus rectus

Anterior
commissure

Gyrus lingula

Fornix

Isthmus of gyrus cinguli

Uncus

Hippocampal sulcus

Parahippocampal gyrus

Occipitotemporal gyrus

Collateral sulcus

Figure 3.11   Medial (split-brain) view of the left hemisphere. (From Mettler
F.A. Neuroanatomy. St. Louis: CV Mosby, 1948.)

roanatomy to support the presence of two psychic realms, it is surprising that a considerable degree of conflict does not arise during the course of everyday activity. Frequently (such as in the case of the "split-brain" patient, LB, described below), although isolated, the right half of the brain is fully willing to assist the left in a myriad of activities. Presumably, such difficulties do not occur because both minds, having once been joined, share the same goals and interests. However, common experience seems to argue otherwise, for even in the intact individual, psychic functioning often is plagued by conflict.

In its most subtle manifestations, the disconnected right hemisphere may attempt to provide the left with clues when the left (speaking) hemisphere is called upon to describe or guess what type of stimulus has been secretly shown to the right (such as in a T-scope experiment). Because the corpus callosum has been severed, transfer and information exchange is not otherwise possible.

Hence, when a picture has been shown to the right, and the left has been asked to guess, the right hemisphere may listen and then nod the head or clear the throat so as to give clues or indicate to the left cerebrum that it has guessed incorrectly. In one case, the right hemisphere attempted to trace or write an answer on the back of right hand (e.g., Sperry et al., 1979). For example, when the right hemisphere was secretly shown a picture of Hitler, and the patient was asked to indicate his attitude toward it before verbally describing it, the patient (i.e., the right hemisphere) responded by signalling with his left hand, "thumbs down."

EX: "That's another 'thumbs-down'?"
LB: "Guess I'm antisocial."
EX: "Who is it?"
LB: "GI came to mind, I mean..." (Subject at this point was seen to be tracing letters with the first finger of the left hand on the back of his right hand.)
EX: "You're writing with your left hand; let's keep the cues out."
LB: "Sorry about that."

Nevertheless, the behavior of the right hemisphere is not always cooperative, and sometimes it engages in behavior that the left cerebrum finds objectionable, embarrassing, puzzling, mysterious, and/or quite frustrating. This is probably true for the normal as well as for the "split-brain" individual.

For example, Akelaitis (1945:597) describes two patients with complete corpus callosotomies who experienced extreme difficulties making the two halves of their bodies cooperate. "In tasks requiring bimanual activity the left hand would frequently perform oppositely to what she desired to do with the right hand. For example, she would be putting on clothes with her right and pulling them off with her left, opening a door or drawer with her right hand and simultaneously pushing it shut with the left. These uncontrollable acts made her increasingly irritated and depressed."

Another patient experienced difficulty while shopping; the right hand would place something in the cart and the left hand would put it right back again. Both patients frequently experienced other difficulties as well. "I want to walk forward but something makes me go backward." A recently divorced male patient noted that on several occasions while walking about town he found himself *forced* to go some distance in another direction. Later (although his left hemisphere was not *conscious* of it at the time) it was discovered (by Dr. Akelaitis) that this diverted course, if continued, would have led him to his former wife's new home.

Geschwind (1981) reports a callosal patient who complained that his left hand on several occasions suddenly struck his wife—much to the embarrassment of his left (speaking) hemisphere. In another case, a patient's left hand attempted to choke the patient himself and had to be wrestled away (Goldstein; cited by Geschwind, 1981).

Brion and Jedynak (cited by Geschwind, 1981) indicate that this type of independent left-sided (right hemisphere) activity was common in their split-brain patients and termed it the "alien hand."

In addition, Bogen (1979:333) indicates that almost all of his "complete commissurotomy patients manifested some degree of intermanual conflict in the early postoperative period." One patient, Rocky, experienced situations in which his hands were uncooperative; the right would button up a shirt and the left would follow right behind and undo the buttons. For years, he complained of difficulty getting his left leg to go in the direction he (or rather his left hemisphere) desired. Another patient often referred to the left half of her body as "my little sister" when she was complaining of its peculiar and independent actions.

A split-brain patient described by Dimond (1980:434) reported that once when she had overslept her "left hand slapped me awake." This same patient, in fact, complained of several instances where her left hand had acted violently. Similarly, Sweet (1945) describes a female patient whose left hand sometimes behaved oppositionally and in a fashion, which, on occasion, was quite embarrassing.

Similar difficulties plagued a split-brain patient on whom I reported (Joseph, 1988b). Indeed, after callosotomy, this patient (2-C) frequently was confronted with situations in which his left extremities not only acted independently, but engaged in purposeful and complex behaviors—some of which he (or rather, his left hemisphere) found objectionable and annoying.

For example, 2-C complained of instances in which his left hand would perform socially inappropriate actions (e.g., attempting to strike a relative) and would act in a manner completely opposite to what he expressively intended, such as turn off the TV or change channels, even though he (or rather his left hemisphere) was enjoying the program. Once, after he had retrieved something from the refrigerator with his right hand, his left took the food, put it back on the shelf, and retrieved a completely different item, "Even though," he said, "that's not what I wanted to eat!" On at least one occasion, his left leg refused to continue "going for a walk" and would only allow him to return home.

In the laboratory, he often became quite angry with his left hand; he struck it and expressed hate for it. Several times, his left and right hands were observed to engage in actual physical struggles. For example, on one task both hands were stimulated simultaneously (while out of view) with either the same or two different textured materials (e.g., sandpaper to the right, velvet to the left), and the patient was required to point (with the left and right hands simultaneously) to an array of fabrics that were hanging in view on the left and right of the testing apparatus. However, at no time was he informed that two different fabrics were being applied.

After stimulation, the patient would pull his hands out from inside the apparatus and point with the left to the fabric felt by the left and with the right to the fabric felt by the right. Surprisingly, although his left hand (right hemisphere) responded correctly, his left hemisphere vocalized: "That's wrong!" Repeatedly, he reached over with his right hand and tried to force his left extremity to point to the fabric experienced by the right (although the left hand responded correctly; his left hemisphere, however, didn't know this). His left hand refused to be moved and physically resisted being forced to point at anything different. In one instance a physical struggle ensued, the right hand grappling with the left.

Moreover, while 2-C was performing this (and other tasks), his left hemisphere made statements such as: "I hate this hand" or "This is so frustrating" and would strike his left hand with his right or punch his left arm. In these instances, there could be little doubt that his right hemisphere was behaving with purposeful intent and understanding, whereas his left brain had absolutely no comprehension of why his left hand (right hemisphere) was behaving in this manner.

Lateralized Goals and Attitudes

Why the right and left cerebral hemispheres behave, in some situations, cooper-atively and yet, in others, in an oppositional fashion is in part a function of functional lateralization and specialization and the differential representation of social-emotional, analytical abilities predominantly within the right hemisphere. Hence, because each hemisphere is concerned with different types of information, even when analyzing ostensibly the same stimulus each hemisphere may react, interpret, and process it differently and even reach different conclusions (Joseph, 1988b; Levy and Trevarthen, 1976). Moreover, even when the goals are the same, the two halves of the brain may produce and attempt to act on different strategies.

Functional lateralization may thus lead to the development of oppositional attitudes, goals, and interests. For example, Reuben Gur (reported in Joseph, 1988a) described one split-brain individual whose left hand would not allow the patient to smoke, and would pluck lit cigarettes from his mouth or right hand and put them out. Apparently, although his left cerebrum wanted to smoke, his right hemisphere didn't approve—apparently, he had been trying to quit for years.

As noted above, 2-C experienced conflicts when attempting to eat, watch TV, or go for walks, his right and left brain apparently enjoying different TV programs or types of food (Joseph, 1988b). Nevertheless, these difficulties are not limited to split-brain patients, for conflicts of a similar nature often plague the intact individual as well.

LATERALIZED MEMORY FUNCTIONING

Although a variety of neurochemical and neuroanatomical regions are involved in the formulation of memory (Gloor, 1992; Graff-Radford et al., 1990; Halgren, 1992; Murray, 1992; Rolls, 1992; Sarter and Markovitch, 1985; Squire, 1992; Victor et al., 1989), functional specialization greatly determines what type of material can be memorized or even recognized by each half of the cerebrum. This is because the code or form in which a stimulus is represented in the brain and memory is largely determined by the manner in which it is processed and the transformations that take place. Because the right and left cerebral hemispheres differentially process information, the manner in which

this information is represented also will be lateralized. Hence, some types of information only can be processed or stored by the right versus the left cerebrum.

For example, it is well known that the left hemisphere is responsible for the encoding and recall of verbal memories, whereas the right cerebrum is dominant in regard to visual-spatial, nonverbal, and emotional memory functioning (Barret al., 1990; Fried et al., 1982; Frisk and Milner, 1990; Hecaen and Albert, 1978; Kimura, 1963; Levy, 1983; Milner, 1962, 1968; Sperry et al., 1979; Squire, 1992; Suberi and McKever, 1977; Wechsler, 1973; Whitehouse, 1981). If the left temporal lobe were destroyed, verbal memory functioning would become impaired, since the right cerebrum does not readily store this type of information. Conversely, the left has great difficulty storing or remembering nonlinguistic, visual, spatial, and emotional information.

Specifically, left temporal lobectomy and seizures or lesions involving the inferior temporal areas can moderately disrupt immediate memory and severely impair delayed memory for verbal passages, and the recall of verbal paired-associates, consonant trigrams, word lists, number sequences, and conversations (Barr et al., 1990; Delaney et al., 1980; Kapur et al., 1992; Meyer and Yates, 1955; Milner, 1968; Milner and Teuber, 1968; Samson and Zatorre, 1992; Weingartner, 1968). Similarly, severe anterograde and retrograde memory loss for verbal material has been noted when the left anterior and posterior temporal regions (respectively) are electrically stimulated (Ojemann et al., 1968, 1971), lobectomized, or injured (Barr et al., 1990; Kapur et al., 1992).

In contrast, right temporal lesions or lobectomy significantly impair recognition memory for tactile and recurring visual stimuli, such as faces and meaningless designs, memory for object position and orientation, and visual-pictorial stimuli, and short-term memory for melodies (Corkin, 1965; Delaney et al., 1980; Kimura, 1963; Milner, 1968; Samson and Zatorre, 1988, 1992; Taylor, 1979). Similarly, memory for emotional material is also significantly impaired with right versus left cerebral lesions (Cimino et al., 1991; Wechsler, 1973) including the ability to recall or recognize emotional faces (DeKosky et al., 1980; Fried et al., 1982; Weddell, 1989). Individuals with right hemisphere damage also have more difficulty recalling personal emotional memories (Cimino et al., 1991).

Hence, it is the left hemisphere that is responsible for the encoding and recall of verbal, temporal-sequential, and language-related memories, whereas the right cerebrum is dominant in regard to visual-spatial, nonverbal, and social-emotional memory functioning. Each hemisphere stores the type of material that it is best at recognizing, processing, and expressing.

Unilateral Memory Storage

In the intact, normal brain, even nonemotional memory traces appear to be stored unilaterally rather than laid down in both hemispheres (Bures and Buresova, 1960; Doty and Overman, 1977; Kurcharski et al., 1990; Levy, 1974; Risse and Gazzaniga, 1979). Moreover, when one hemisphere learns, has certain experiences, and/or stores information in memory, this information is not always available to the opposing hemisphere; one hemisphere cannot always gain access to memories stored in the other half of the brain (Bures andd Buresova, 1960; Doty and Overman, 1977; Joseph, 1982, 1988a, 1988b, 1992b; Kurcharski et al., 1990; Levy, 1974; Risse and Gazzaniga 1979).

To gain access to these lateralized memories, one hemisphere has to activate the memory banks of the other brain half via the corpus callosum (Doty and Overman, 1977) or anterior commissure (Kurcharski et al., 1990). This has been demonstrated experimentally in primates. For example, after one hemisphere had been trained to perform a certain task, although either hemisphere could respond correctly once it was learned, when the commissures were subsequently cut, only the hemisphere that originally was trained was able to perform—i.e., could recall—it. The untrained hemisphere acted as though it never had been exposed to the task; its ability to retrieve the original memories was now abolished (Doty and Overman, 1977).

In a conceptually similar study, Risse and Gazzaniga (1979) injected sodium amytal into the left carotid arteries of intact patients so as to anesthetize the left cerebral hemisphere. After the left cerebrum was inactivated, the awake right hemisphere, although unable to speak, was still able to follow and behaviorally respond to commands, e.g., palpating an object with the left hand.

Once the left hemisphere had recovered from the drug, as determined by the return of speech and motor functioning, none of the eight patients studied was able to verbally recall what objects had been palpated with the left hand, "even after considerable probing." Although encouraged to guess, most patients refused to try and insisted that they did not remember anything. However, when offered multiple choices in full field, most patients immediately raised the *left* hand and pointed to the correct object!

According to Risse and Gazzaniga (1979), although the memory of touching and palpating the object was not accessible to the verbal (left hemisphere) memory system, it was encoded in a nonverbal form within the right hemisphere and was unavailable to the left hemisphere when normal function returned. The left (speaking) hemisphere was unable to gain access to information and memories stored within the right half of the brain. Nevertheless, the right brain not only remembered, but was able to act on its memories.

This indicates that when exchange and transfer is not possible, is in some manner inhibited, or if for any reason the two halves of the brain become functionally disconnected and are unable to share information, the possibility of information transfer at a later time is precluded (Bures and Buresova, 1960; Kurcharski et al., 1990; Risse and Gazzaniga, 1979), even when the ability to transfer is acquired or restored. The information is lost to the opposite half of the cerebrum.

Moreover, because some types of information are processed by the right and left hemispheres in a wholly different fashion, each hemisphere is unable to completely share or gain access to the data or even the conclusions reached by the other—as they are unable to process or recognize it—which in turn precludes complete interhemispheric transfer (Berlucchi and Rizzolatti, 1968; Hicks, 1974; Joseph, 1982, 1988a; Marzi, 1986; Merriam and Gardner, 1987; Miller, 1990, 1991; Myers, 1959, 1962; Rizzolatti et al., 1971; Taylor and Heilman, 1980); information is lost during the transfer process.

Nevertheless, although inaccessible or lost, these memories, details, and attached feelings can continue to influence whole brain functioning in subtle as well as in profound ways. That is, one hemisphere may experience and store certain information in memory and at a later time, in response to certain situations, act on those memories, much to the surprise, perplexity, or chagrin of the other half of the brain; one hemisphere cannot always gain access to memories stored in the other half of the brain.

DREAMING

Of course, complete functional deactivation is probably quite rare in the normal brain. However, there is some evidence to suggest that interhemispheric communication is reduced, for example, during sleep and possibly during dreaming (Banquet, 1983; reviewed in Joseph, 1988a).

Most dreaming occurs during REM, which possibly is associated with right hemisphere activation and low-level left hemisphere arousal (Goldstein et al., 1972; Hodoba, 1986; Meyer et al., 1987). It also becomes progressively more difficult to recall one's dreams as one spends time in or awakens during non-REM (Wolpert and Trosman, 1958), which is associated with high left hemisphere and low right brain activation (Goldstein et al., 1972). Thus, are dreams really forgotten, or are they locked away in a code that is not accessible to the speaking left hemisphere?

DREAMING AND HEMISPHERIC OSCILLATION

Although up to five stages of sleep have been identified in humans, for our purposes we will be concerned only with two distinct sleep states. These are the REM (rapid eye movement) and non-REM (NREM) periods. NREM occurs during a stage referred to as "slow-wave" or synchronized sleep. In contrast, REM occurs during a sleep stage referred to as "paradoxical sleep." It is called paradoxical, for electrophysiologically the brain seems quite active and alert, similar to its condition during waking. However, the body musculature is paralyzed, and the ability to perceive outside sensory events is greatly attenuated (reviewed in Hobson et al., 1986; Steriade and McCarley, 1990; Vertes, 1990).

Most individuals awakened during REM report dream activity approximately 80% of the time. When awakened during the NREM period, dreams are reported approximately 20% of the time (Foulkes, 1962; Goodenough et al., 1959; Monroe et al., 1965). However, the type of dreaming that occurs during REM versus NREM is quite different. For example, NREM dreams (when they occur) are often quite similar to thinking and speech (i.e., linguistic thought), such that a kind of rambling verbal monologue is experienced in the absence of imagery (Foulkes, 1962; Monroe et al., 1965). It is also during NREM that an individual is most likely to talk in his or her sleep (Kamiya, 1961). In contrast, REM dreams involve a consider-

able degree of visual imagery and emotion, and tend to be distorted and implausible to various degrees (Foulkes, 1962; Monroe et al., 1965).

REM is characterized by high levels of activity within the brainstem, occipital lobe, and other nuclei (Hobson, et al. 1986; Steriade and McCarley, 1990; Vertes, 1990). It also has been reported that electrophysiologically the right hemisphere becomes highly active during REM, whereas, conversely, the left brain becomes more active during NREM (Goldstein et al., 1972; Hodoba, 1986). Similarly, measurements of cerebral blood flow have shown an increase in the right temporal and parietal regions during REM sleep and in subjects who, upon waking, report visual, hypnogogic, hallucinatory, and auditory dreaming (Meyer et al., 1987).

Interestingly, abnormal and enhanced activity in the right temporal and temporal-occipital areas acts to increase dreaming and REM sleep for an atypically long period. Similarly, REM sleep increases activity in this same region much more than in the left hemisphere (Hodoba, 1986), which indicates that there is a specific complementary relationship between REM sleep and right temporal-occipital electrophysiological activity.

At least one group of investigators, however, has failed to find significant hemispheric EEG differences between REM and NREM (Ehrlichman et al., 1985).

Day Dreams, Night Dreams, and Hemispheric Oscillation

There is some evidence to suggest that during the course of the day and night the two cerebral hemispheres oscillate in activity every 90 to 100 minutes and are 180 degrees out of phase—a cycle that corresponds to changes in cognitive efficiency, the appearance of day dreams, REM (dream) sleep, and, conversely, NREM sleep (Bertini et al., 1983; Broughton, 1982; Gordon et al., 1982; cited by Hodoba, 1986; Klein and Armitage, 1979; Kripke and Sonnenschein, 1973; Levie et al., 1983, cited by

Hodoba, 1986). That is, like two pistons sliding up and down, it appears that when the right cerebrum is functionally at its peak of activity, the left hemisphere is correspondingly at its nadir.

Similarly, shifts in cognitive abilities associated with the right and left hemispheres have been found during these cyclic changes during the day and after awakening from REM and NREM sleep. That is, performance across a number of tasks associated with left hemisphere cognitive efficiency is maximal during NREM, whereas, conversely, right hemisphere performance (e.g., point localization, shape identification, orientation in space) is maximal after REM awakenings (Bertini et al., 1983; Gordon et al., 1982; Levie et al., 1983; cited by Hodoba, 1986). Moreover, Bertini et al. (1983) found that left hand motor dexterity (in right-handed subjects) was superior to the right when awakened during REM and that the opposite relationship was found during NREM, i.e., right hand superiority (see Hodoba, 1986, for review.)

Conversely, there have been reports of patients with right cerebral damage who have ceased dreaming altogether or dream only in words (Humphrey and Zangwill, 1951; Kerr and Foulkes, 1978, 1981). For example, defective dreaming, deficits that involve visual imagery, and loss of hypnogocic imagery have been found in patients with focal lesions or hypoplasia of the posterior right hemisphere and abnormalities in the corpus callosum (Botez et al., 1985; Kerr and Foulkes, 1981; Murri et al., 1984).

An absence or diminished amount of dreaming during sleep also has been reported after split-brain surgery, i.e., as reported by the disconnected left hemisphere (Bogen and Bogen, 1969; Hoppe and Bogen, 1977). Similarly, a paucity of REM episodes has been noted in other callosotomy patients, although these particular individuals continued to report some dream activity (Greenwood et al., 1977).

On the other hand, it has been reported that when the left hemisphere has been damaged, particularly the posterior portions (i.e., aphasic patients), the ability to verbally report and recall dreams also is greatly attenuated (Murri et al., 1984; Pena-Casanova and Roig-Rovira, 1985; Schanfald et al., 1985). Of course, aphasics have difficulty describing much of anything, let alone their dreams. Moreover, Language Axis disconnection from the right hemisphere would also account for this failure to verbally report dreams and related imagery.

In some respects, however, a parallel between these latter findings and those of Risse and Gazzaniga (1979) in their amytal studies may be explanatory regarding the failure to report dreams with left hemisphere lesions. That is, when the left hemisphere is damaged or at a low level of arousal, the ability to verbally recall or report events experienced or generated by the right hemisphere appears to be reduced; i.e., the left (speaking) half of the brain cannot remember because it cannot gain access to right cerebral memory centers.

Thus, it appears that the right hemisphere provides the physiological foundation from which dreams in part derive their source and origin (Goldstein, et al., 1972; Hodoba, 1986; Joseph, 1988a; Meyer et al., 1987). However, in some instances in which these dream centers are disconnected from the language-dominant left hemisphere, due to posterior right or left hemisphere lesions or after callosotomy, the ability to recall, report, and/or to produce vivid visual and hypnogogic dream imagery is attenuated (Joseph, 1988a).

However, also important in the capacity to engage in memory search and retrieval, or to dream and fantasize, are the frontal lobes (see Chapter 11). Likewise, frontal lobe damage and lobotomy also have been reported to abolish dreaming (Freeman and Watts, 1942, 1943).

Hallucinations

In addition to dream production, the right hemisphere also appears to be the dominant source for complex nonlinguistic hallucinations. Specifically, tumors or electrical stimulation of the right hemisphere or temporal lobe are much more likely to result in complex visual as well as musical and singing hallucinations, whereas left cerebral tumors or activation gives rise to hallucinations of words or sentences (Berrios, 1990; Halgren et al., 1978; Hecaen and Albert, 1978; Jackson, 1880; Mullan and Penfield, 1959; Penfield and Perot, 1963; Teuber et al., 1960). Conversely, LSD-induced hallucinations are significantly reduced following right but not left temporal lobe surgical removal (Serafetinides, 1965), and dreaming is sometimes abolished with right but not left temporal lobe removal (Kerr and Foulkes, 1981). In one study, however, it was reported that an alcoholic patient with a right subcortical injury and left-sided neglect experienced hallucinations only in the right (non-neglected) half of visual-space (Chamorro et al., 1990).

Imagery

It is important to note that some investigators believe the left hemisphere is responsible for dreaming and the production of images (see Greenberg and Farah, 1986, and Miller, 1990, for a detailed review). In fact, some investigators have claimed that the left (posterior) hemisphere is "dominant" for the production of imagery (Farah, 1989; Trojano and Grossi, 1994), and/or that it is faster at generating images from categorically stored (language-related) information (Findlay et al., 1994).

Unfortunately, those making these claims appear to have confused apraxia, agraphia, and disturbances of language as somehow indicative of imagery deficits, and/or the ability to quickly *verbally* describe a verbal "image" as somehow synonymous with imagery production.

For example, in some studies, patients were required to verbally describe and/or draw the object they have "visualized" (e.g., Farah et al., 1988). In other words, imagery production was not assessed, but drawing and verbal ability was. Those

imagery experiments with individuals with left posterior lesions are equally suspect in that a common abnormality would be difficulty in word finding. Thus, these left cerebral "imaginal" deficits appear to be more a function of naming and word-finding disturbances coupled with visual agnosia and/or apraxia—disturbances associated with left posterior (parietal) injuries (see also Trojano and Grossi, 1994).

Individuals with left hemisphere damage have difficulty accurately describing much of anything, let alone visual images, which is clearly more the productive province of the right hemisphere. Indeed, those who champion a left cerebral dominance for imagery have very little evidence to support their claims and instead appear to be confusing disconnection syndromes and apraxic and language disorders for imagery deficits (see also Sergent, 1990).

Nevertheless, the left cerebrum no doubt can produce a variety of images, especially in response to a *verbal* command (Goldenberg, 1989). However, as noted above, right (but not left) cerebral injuries can result in a failure or inability to generate half of visual images (Bartoloemeo, 1994), which indicates that in the absence of right cerebral input, the left hemisphere is producing only a partial image and is therefore lacking dominance in this regard.

Dream Stimulants

Since the right hemisphere utilizes a form of visual-spatial and emotional *language* that the left hemisphere does not speak, dream imagery often seems incomprehensible to the language-dependent half of the brain, even when it is providing the accompanying narrative or dialogue. Rather, the nontemporal, often gestalt nature of dreams appears to require that they be consciously scrutinized from multiple angles in order to discern their meaning, for the last may be first and what is missing may be just as significant as what is present (Freud, 1900; Jung, 1945, 1964). This is because the right hemisphere analyzes and expresses information in a manner much different from the left.

Moreover, in that the right hemisphere is at a higher level of activation during REM, it also tends to predominate when attempting to analyze internal and external sensory input during this stage of sleep. For example, although sensory perception is restricted, the right cerebrum may respond to a sensation experienced during sleep by creating a dream to explain it. However, when this occurs we sometimes dream backwards.

A case in point, "Trish" dreams she is walking in San Francisco, lugging large bags of gifts. Feeling tired, she sets them down on the sidewalk. She looks for a bus and sees a cable car coming. As it pulls up the conductor begins to ring its bell. The sound of the bell grows louder and then jolts her awake. Fully awake, she realizes that someone is ringing her doorbell. In this regard, the hearing of the bell seemed to be a natural part of the dream, and it is. What seems paradoxical, however, is that the dream seemed to lead up to the bell so that its ringing made sense in the context of the dream.

The dream did not lead up to the bell, however, for the bell initiated the dream. The dream was produced, via the unique language of the right hemisphere during sleep (as well as amygdala activation), so as to explain the sound of the bell. The bell was heard and the dream was instantly produced in explanation and association. The bell stimulated the dream (perhaps by startling the amygdala; see Chapter 5), which may have only lasted a second.

Since most dreams last only a few seconds (although they may seem to take place over the course of hours) and since the right hemisphere does not analyze in temporal and linear units, the sequence of events is not all that important, that is, to the right half of the brain.

Fortunately, backward dreams are the most easily comprehended because the left cerebrum recalls the dream from its ending forward, and then, like a reflection from a mirror, reverses all that is perceived so that it makes temporal-sequential sense. Backward dreams also tend to be devoid of hidden meanings.

For example, one individual (described by Freud, 1900) dreamt he was in 18th century France in the midst of the French Revolution. After a trial in which he was found guilty, he was led down a street lined with yelling and cursing French men and women. At the end of the street he could see the gallows, where the heads of various political criminals were being chopped off at the neck. Mad with fright, he saw himself being led up the stairs and felt his head being placed in the yoke of the chopping block. The executioner gave the signal, the crowd screamed its approval, he could hear and sense the blade falling, and with a loud crack it struck him across the neck. Indeed, it struck him with such a jolt that he awoke to find that his poster bed had broken and that a railing had fallen and struck him across the side and back of his neck.

DREAM PATTERNS

Although dreams probably serve a number of purposes, and at times are highly improbable and bizarre, they sometimes reflect something significant about the mental and emotional life of the dreamer, as well as other issues of concern to the right half of the brain. For example, when subjects are awakened repeatedly during REM over the course of several days, often an evolving thematic pattern, like an unfolding story, can be discerned (Cartwright et al., 1980). These patterns frequently reflect mental-emotional activity concerned with the solution of particular problems (Cartwright et al., 1980; Freud, 1900; Joseph 1992b; Jung, 1945, 1964; Kramer et al., 1964).

For example, one subject, a student, noted that "after being woken many times and seeing three or four dreams a night, I could realize there was a certain problem being worked out, like coping with responsibilities that were thrust upon me, but that weren't necessarily my own but I took on anyway. It was working out the feelings of resentment of taking somebody else's responsibility, but I met them well in my dreams. A good thing about spending time in the sleep lab was you could relate a common bond to some of the dreams" (Cartwright et al., 1980:277). Similar bonds and patterns were, of course, recognized by Freud (1900) and Jung (1945) many years ago.

Dreams and Emotional Trauma

Given right cerebral dominance for emotion and dream production, it is perhaps not surprising that emotional conflicts and traumas are often represented via dream imagery (Freud, 1900; Joseph, 1992b; Jung, 1945).

Consider, for example, Sara, who was desribed by her parents as a "very good and obedient" 6-year-old girl who suffered from "night terrors" (Joseph, 1992b). Specifically, she frequently woke up at night screaming about the "creek" that ran near her home. According to her mother, Sara was afraid because she had been told that hoboes lived under the bridge and to never go there without her mommy or daddy, because they would "get" her.

I had Sara tell me her dream:

> I'm walking on the sidewalk near the big creek. Then I go to the edge and stare down at the big rocks at the bottom. All at once the whole world starts to shake. Like it's turning upside down. It's trying to throw me into the creek. I get scared and start to be afraid and start grabbing at the trees and bushes to keep from falling into the creek and onto the big rocks. Sometimes I see this hole and I crawl in. Then everything is okay. Sometimes I fall and fall and fall and I can see the big rocks coming closer and I know I'm going to fall on them. When I fall on them I wake up 'cause it hurts.

Sara had this particular dream repeatedly. Was it really because she was afraid of the hoboes and the creek?

In Sara's case it is noteworthy that her parents fought, screamed, yelled, and argued almost nonstop and there were con-

stant threats of divorce. From talking with Sara it was also apparent that she was traumatized by her parents' constant fighting—although they were both ostensibly quite good to her. Nevertheless, Sara's world was being turned upside down and was in complete chaos, which was reflected in her dreams.

Sara also had another troubling and recurrent dream where she went riding on her bike and when she came back to her street, her house was gone. Every house on the street was the same, including the neighbors' house. But when she asked about her family, no one knew what she was talking about and no one recognized her.

Although the symbolic content of these dreams was not apparent to Sara or her mother, one need not be a psychiatrist to decipher their obvious meaning. Sara's emotional world was literally being turned upside down and was in chaos due to the horrible fighting engaged in by her parents. She was terrified of losing her home and the catastrophe she perceived as befalling her family. Her very identity and functional integrity as a person was at stake, for if she lost her family she lost her Self.

The dream imagery involving the hole that she climbs into is also very interesting in that it suggests the desire to return to the safety and security of the womb. On the other hand, the "hole" may have been exactly that: a hole. Indeed, dreams frequently mean exactly what they seem to mean (Jung, 1945).

Nevertheless, it might reasonably be asked, if dreams are of importance and not merely reflective of random and purely confabulatory ideation, why are they so difficult to recall? In part, as noted in the introduction to this section, this may be a function of lateralization, alterations in hemispheric arousal and activity, differential memory storage, and decreased interhemispheric communication during REM. Perhaps they only can be recalled by the right hemisphere. Of course, in some instances, dreams are probably nothing more than mundane and confabulatory *noise*.

LONG, LOST CHILDHOOD MEMORIES

For most individuals it is extremely difficult, if not impossible, to verbally recall events that occurred before the age of 3 years (Dudycha and Dudycha, 1933; Gordon, 1928; Wadvogel, 1948; White and Pillemer, 1979). There are several reasons for this (see Chapter 15). Information processed and experienced during infancy versus adulthood is stored via certain transformation and retrieval strategies that are quite different. As the brain matures and new information processing strategies are learned and developed, the manner in which information is processed and stored is altered. Although these early memories are stored within the brain, the organism no longer has the means of retrieving them, i.e., the key no longer fits the lock. That is, early experiences may be unrecallable because infants use a different system of codes to store memories whereas adults use symbols and associations (such as language) not yet fully available to the child (Dollard and Miller, 1950; Joseph, 1982, 1988a; Piaget, 1952, 1962, 1974). Much of what was experienced and committed to memory during early childhood took place prior to the development of linguistic labeling ability and was based on a prelinguistic or nonlinguistic code (Dollard and Miller, 1950; Freud, 1900). Hence, the adult, who is relying on more sophisticated and language-related coding systems, cannot find the right set of neural programs to open the door to childhood memories. The key does not fit the lock because the key and the lock have changed.

The inability to recall early memories is also a function of the immaturity of the corpus callosum in children. That is, nonverbal information perceived and processed by the right versus left hemisphere is generally stored in the right versus left hemisphere. Later, when the commissures mature, this information cannot be transferred except under special conditions (see Chapter 15). However, under conditions of traumatic memory loss (e.g., repression), not just the the right and left hemisphere, but the differential activation of the amygdala and hippocampus are contributory; and later the recall of such memories may be opposed by the frontal lobes, the right frontal in particular (see Chapter 16).

Emotion and Right Brain Functioning in Children

As is now well known, the developing organism is extremely vulnerable to early experience during infancy such that the nervous system, perceptual functioning, and behavior may be altered dramatically (Bowlby, 1982; Casagrande and Joseph, 1978; 1980; Diamond, 1985, 1991; Denenberg, 1981; Ecknerode et al., 1993; Harlow and Harlow, 1965; Joseph, 1979; Casagrande and Joseph, 1980; Joseph and Gallagher, 1980; Joseph et al., 1978; Langmeier and Matejcek, 1975; Rosenzweig, 1971; Salzinger et al., 1993; Sternberg et al., 1993). Interestingly, there is some evidence that the right cerebral hemisphere and the right amygdala may be more greatly affected (see Denenberg, 1981; Diamond, 1985).

Moreover, during these same early years our traumas, fears, and other emotional experiences, like those of an adult, are mediated not only by the limbic system, but also via the nonlinguistic, social-emotional right hemisphere. And, just as they are in adulthood, these experiences are stored in the memory banks of the right cerebrum. However, much of what was experienced and learned by the right half of the brain during these early years was not always shared or available for left hemisphere scrutiny (and vice versa). That is, a child's two hemispheres are not only functionally lateralized but limited in their ability to share and transfer information. In many ways, infants and young children have split-brains (Deruelle and de Schonen, 1991; Finlayson, 1975; Gallagher and Joseph, 1982; Galin et al., 1979a, 1979b; Joseph and Gallagher, 1985; Joseph et al.,

1984; Kraft et al., 1980; Molfese et al., 1975; O'Leary, 1980; Ramaeker and Njiokiktjien, 1991; Salamy, 1978; Yakovlev and Lecours, 1967).

Indeed, due to the immaturity of the corpus callosum and, in particular, the slow rate of axonal myelin within the callosum (Yakovlev and Lecours, 1967), communication is so poor that children as old as age 4 have mild difficulty transferring tactile, auditory, or visual information between the hemispheres, e.g., accurately describing complex pictures shown to the right brain (Joseph et al., 1984). That is, in addition to differential functional specialization, the slow development of the myelination process can in turn slow and disrupt axonal information transmission (e.g., Konner, 1991; Ritchie, 1984; Salamy, 1978).

Indeed, although *pain* can be transmitted and received via axons devoid of myelin, this type of data is completely lacking in complexity and is devoid of initial cognitive attributes. Pain-transmitting axons have very simple requirements. By contrast, as data complexity increases, so too does the complexity of those neurons that transmit these signals (Konner, 1991; Ritchie, 1984). For example, as axon diameter increases, so does the extent of myelination. Similarly, quantities of axoplasm and cytoplasm, nuclear diameter, and neuronal packing density are also correlated with myelination (Konner, 1991). Conversely, lack of myelin, or those neurons that have not yet myelinated, are associated with an increased susceptibility to conduction failure and interference due to extraneous influences, including signal modification by neighboring axons (Konner, 1991; Ritchie, 1984).

Functional Commissurotomies and Limited Interhemispheric Transfer

The corpus callosum is the gateway via which information may travel from one brain half to the other. However, it also acts to limit information exchange since almost 40% of the adult callosum lacks myelin (Selnes, 1974). Since myelin acts to insulate

and, thus, preserve information transmission by minimizing leakage and, increasingly, conduction velocity and integrity (Konner, 1991; Rogart and Ritchi, 1977; Ritchie, 1984), some information is lost and degraded, even when transfer is possible (Berlucchi and Rizzolatti, 1968; Hicks, 1974; Joseph et al., 1984; Marzi, 1986; Merriam and Gardner, 1987; Myers, 1959, 1962; Rizzolatti et al., 1971; Taylor and Heilman, 1980).

Moreover, particularly when one is dealing with complex or emotional information, situations probably sometimes arise in which one brain half has little or no knowledge as to what is occurring in the other (Dimond, 1980; Dimond and Beaumont, 1974; Dimond et al., 1972; Geschwind, 1965; Joseph, 1982, 1988a, 1992b; Joseph et al., 1984; Marzi, 1986; Myers, 1959). In part, this is a consequence of lateralized specialization. Certain forms of information can only be processed and, thus, recognized by the right or left half of the brain. Even information that is transferred may be subject to interpretation and misinterpretation (Gazzangia and LeDoux, 1978; Joseph, 1982, 1986a, 1986b, 1988a, 1988b; Joseph et al., 1984) and this includes even the learning of sequential and fine motor movements (Hicks, 1974; Taylor and Heilman, 1980; see also Parlow and Kinsbourne, 1989).

In addition, one brain half can be prevented from knowing what is occurring in the opposite half due to inhibitory or suppressive actions initiated by, for example, the frontal lobes, such that certain forms of information are suppressed or censored, and interhemispheric (as well as intrahemispheric) information transmission prevented (Hoppe, 1977; Joseph, 1982, 1988a, 1992b). Thus, there sometimes results a *functional commissurotomy* (Hoppe, 1977; see also Galin, 1974).

Therefore, these three conditions—lateralized specialization, frontal lobe inhibitory activity, and incomplete myelination of callosum axons—can reduce the ability of the two hemispheres to communicate among normal, intact individuals. Hence, in many ways the brain of even a normal adult is functionally split and disconnected, and for good reason. These conditions protect the brain and linguistic consciousness from becoming overwhelmed. As we have seen with frontal lobe damage, when communication is allowed to occur freely (due to disinhibition), the overall integrity of the brain to function normally is curtailed dramatically.

Nevertheless, a unique side effect of having two brains that are not always able to communicate completely and successfully is intrapsychic conflict. That is, we sometimes find ourselves feeling happy, sad, depressed, angry, and so forth without a clue as to the cause. In other instances, we actually may commit certain thoughtless, impulsive, overly emotional, or embarrassing actions and "have no idea" as to "what came over" us. To posit the notion that we have such experiences "simply because" is absurd. Nor, among "normals," are such experiences always due to biochemical fluctuations or the result of "unconscious" urges. Rather, unbeknownst to the left brain, sometimes the right perceives, remembers, or responds to some external or internal source of experience and/or to its own memories and, thus, reacts in an emotional manner. The left (speaking) hemisphere, in turn, only knows that it is feeling something but is unsure *what* or *why*, or, conversely, confabulates various denials, rationalizations, and explanations, which it accepts as fact.

Split-Brain Functioning in Children: The Ontology of Emotional Conflict

Thus, due in part to the slow pace of corpus callosum myelination, coupled with differential right and left cerebral specialization, the left hemisphere of a young child has, at best, incomplete knowledge of the contents and activities that are occurring within the right. This sets the stage for differential memory storage and a later inability to transfer this information between the cerebral hemispheres once the child reaches adulthood (Joseph 1982, 1988a, 1992b). Because of lateralization and limited exchange, the effects of early "socializing" experience can have potentially profound effects. "As a good deal of this early experience is likely to have unpleasant, if not traumatic moments, it is fascinating to consider the later ramifications of early emotional learning occurring in the right hemisphere unbeknownst to the left; learning and associated emotional responding that *later* may be completely inaccessible to the language centers of the left half of the brain. That is, although limited transfer in children confers advantages, it also provides for the eventual development of a number of very significant psychic conflicts—many of which do not become apparent until much later in life".

Moreover, due to the immaturity of the callosum, children frequently can encounter situations in which the right and left cerebrum not only differentially perceive what is going on, but are unable to link these experiences so as to understand fully what is occurring or to correct misperceptions (Galin, 1974; Joseph, 1982, 1992b).

Consider, for example, a young divorced mother with ambivalent feelings toward her young son (Galin, 1974; Joseph, 1988a, 1992b). Although she does not express these feelings verbally, she conveys them through her tone of voice, facial expression, and in the manner in which she touches her son. She knows that she should love him, and at some level she does. She wants to be a good mother and makes herself go through the motions.

However, she also resents her son because she has lost her freedom, he is a financial burden, and he may hinder her in finding a desirable mate. She is confronted by two opposing attitudes, one of which is unacceptable to the image she has of a good mother. Like many of us, she must prevent these feelings from reaching linguistic consciousness. However, this does not prevent them from being expressed nonlinguistically via the right brain.

Her son, of course, also has a right hemisphere, which perceives her tension and ambivalence. The right half of his brain notes the stiffness when his mother holds or touches him and is aware of the manner in which she sometimes looks at him. Worse, when she says, "I love you," his right hemisphere senses the tension and tone of her voice and correctly perceives that what she means is, "I don't want you," or "I hate you." His left hemisphere hears, however, "I love you" and notes only that she is attentive. He is in a "double bind" conflict, with no way for his two cerebral hemispheres to match impressions.

The right half of this little boy's brain feels something painful when the words "I love you" are spoken. When his mother touches him, he becomes stiff and withdrawn because his right hemisphere, via the analysis of facial expression, emotional tone, tactile sensation, and so forth is fully aware that she does not want him.

Later, as an adult, this same young man has one failed relationship after another. He feels that he can't trust women, often feels rejected, and when a girl or woman says "I love you," it makes him want to cringe, run away, or strike out. As an adult, his *left* hemisphere hears "love," and his right cerebrum feels pain and rejection.

Because the two halves of his cerebrum were not in communication during early childhood, his ability to gain insight into the source of his problems is greatly restricted. The left half of his brain cannot access these memories. It has "no idea" as to the cause of his conflicts.

In this regard, this asymmetrical arrangement of hemispheric function and maturation may well predispose the developing child in later life to come upon situations in which he finds itself responding emotionally, nervously, anxiously, or neurotically, without linguistic knowledge, or without even the possibility of linguistic comprehension as to the cause, purpose, eliciting stimulus, or origin of his behavior. As a child or an adult, he may find himself faced with behavior that is, for example, mysterious or embarrassing. His reaction:"I don't know what came over me."

OVERVIEW AND CONCLUDING COMMENTS

Overall, based on numerous studies conducted on normal subjects and brain-injured and neurosurgical patients, the right cerebral hemisphere has been shown to dominate in the perception and identification of environmental and nonverbal sounds (e.g., wind, rain, thunder, birds singing); somestheses; stereognosis; the maintenance of the body image; and the comprehension and expression of prosodic, melodic, and emotional features of speech, as well as in the perception of most aspects of musical stimuli, i.e., chords, timbre, tone, pitch, loudness, melody, intensity. Moroever, the right predominates in the analysis of geometric space and visual-space, including depth perception, orientation, position, distance, figure-ground, perspective, visual closure, and stereopsis.

It also appears to be more involved than the left in the production of certain forms of visual imagery (see Ehrlichman and Barrett, 1983; Trojano and Grossi, 1994, for contradictory evidence) and dreams during REM sleep, as well as day dreams during waking. Conversely, the left brain appears to be associated with non-REM sleep and the thinking-type of mentation that sometimes occurs during this stage. However, the left hemisphere probably provides much of the dialogue and commentary that accompany dream activity.

A considerable body of evidence indicates that the right hemisphere is dominant in the comprehension and expression of prosodic, melodic, and emotional features of speech, the expression and perception of visual, facial, and verbal affect, and the ability to determine a person's mood, attitude, and intentions via the analysis of gesture, facial expression, and vocal-melodic and intonational qualities. All aspects of social-emotional intellectual activity are mediated by the right hemisphere. However, the female right hemisphere appears to have more neocortical space devoted to these social-emotional and related language functions, whereas in males, more neocortical space is devoted to spatial perceptual processing.

Because the right half of the brain is dominant in regard to most, if not all, aspects of social-emotional intellectual functioning, when it is damaged a myriad of affective disturbances may result. These include mania, depression, hysteria, gross social-emotional disinhibition, euphoria, childishness, puerility, or, conversely, complete indifference and apathy.

Patients may become delusional, engage in the production of bizarre confabulations, and experience a host of somatic disturbances that range from pain and body-perceptual distortions to seizure-induced sexual activity and orgasm. They may fail to recognize the left half of their own bodies or, in other instances, fail to recognize the faces of friends, loved ones, or even their pets. In fact, individuals with right-sided lesions show less recovery and are more likely to die as compared to those with left-sided destruction (Denes et al., 1982; Hobhouse, 1936; Hurwitz and Adams, 1972; Knapp, 1959; Marquardsen, 1969).

In contrast, the range of emotional disturbances associated with left cerebral damage seems to be limited to apathy, depression, emotional blunting, and schizophrenia, although euphoria sometimes accompanies receptive aphasia and loss of comprehension (Gainotti, 1972; Gasparrini et al., 1978; Geschwind, 1981; Gruzelier and Manchanda, 1982; Hillbom, 1960; Robinson and Szetela, 1981; Robinson et al., 1984; Sherwin et al., 1982; Sinyour et al., 1986).

Although there is evidence of considerable functional overlap as well as inter-hemispheric cooperation on a number of tasks, it certainly appears that the mental system maintained by the right cerebral hemisphere is highly developed, social-emotional, bilateral, and in many ways dominant over the temporal-sequential, language-dependent half of the cerebrum. Indeed, the right cerebrum can independently recall and act on certain memories with purposeful intent; is the dominant source of our dreams, psychic conflicts, and desires, and is fully capable of motivating, initiating, as well as controlling behavioral expression—often without the aid or even active (reflective) participation of the left half of the brain.

4

The Left Cerebral Hemisphere

Language, Aphasia, Apraxia, Alexia Agraphia, Psychosis,
the Evolution of Reading and Writing, and
the Origin of Thought

Overview

Among nonhumans, vocal "language," for the most part, is subcortically and limbically mediated (medial frontal lobes, cingulate, amygdala, periaqueductal gray; see Chapters 2 and 5). Like language in nonhumans, motor functioning and movement of the extremities appears to be controlled by the medial regions, basal ganglia, cerebellum, and brainstem to a much greater degree than is apparent in humans.

Although human movement is also dependent on these brain structures, fine motor control involving the hands and fingers, as well as the neck, head, eyes, and lower extremities, is more the domain of the neocortex of the frontal and parietal lobes (see Chapters 11 and 12). Fine motor control is dependent on the neocortex. Fine motor control is also lateralized, the left hemisphere being dominant for fine motor functioning, including handedness in almost 90% of the population (reviewed in Corballis, 1991).

Similarly, as is now well known, the left hemisphere is dominant for most aspects of language perception and production (Albert, 1993; Barr et al., 1990; Benson, 1993; DeRenzi et al., 1987; Frisk and Milner, 1990; Goodglass and Kaplan, 1982; Joseph, 1990e; Kapura et al., 1992; Kertesz, 1983a; Kimura, 1993; Levin, 1993). As detailed in Chapter 2 (and below), language and motor functioning are both lateralized to the left hemisphere because modern human language is, in large part, an outgrowth and directly related to the acquisition of fine motor functioning, whereas right hand dominance is due to the earlier

maturation of the left corticospinal pyramidal tract (Joseph, 1982), which descends from the neocortical motor areas into the brainstem, and then crosses at the pyramidal decussation (Kertesz and Geschwind, 1971; Yakovlev and Rakic, 1966), and then descends the spinal cord in advance of those fibers from the right, thereby giving it a competitive advantage in establishing both motor control and thus dominance (Joseph, 1982).

By contrast, among nonhumans, there is little or no evidence that either hemisphere dominates in regard to "language," fine motor control, or handedness (see edited volume by Ward and Hopkins, 1993). Indeed, 50% of primates and mammals tend to favor the right hand (paw, leg) and 50% tend to utilize the left extremity (however, see MacNeilage, 1993). Presumably, the right and left corticospinal (pyramidal) tracts grow and mature at the same rate.

Left Hemisphere Dominance for Language

Among 80% of right-handers and over 50% of left-handers, the left cerebral hemisphere provides the neural foundation and mediates most aspects of expressive and receptive linguistic functioning, such as reading, writing, speaking, spelling, naming, and the comprehension of the grammatical, syntactical, and descriptive components of language, including time sense, rhythm, verbal concept formation, analytical reasoning, and verbal memory (Albert et al., 1972; Albert, 1993; Barr et al., 1990; Benson, 1993; Caramazza and Zurif, 1976; DeRenzi et al., 1987; Efron, 1963; Frisk and Milner,

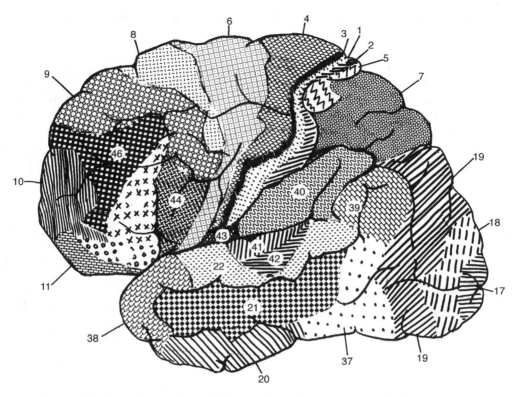

Figure 4.1 Lateral view of the left hemisphere. Numbers refer to Brodmann's areas.

1990; Goodglass and Kaplan, 1982; Hecaen and Albert, 1978; Heilman and Scholes, 1976; Joseph, 1990e; Kapura et al., 1992; Kertesz, 1983a; Kimura, 1993; Levin, 1993; Levine and Sweet, 1983; Luria, 1980; Milner, 1970; Vignolo, 1983; Zurif and Carson, 1970).

As is well known, these capacities are associated with Broca's expressive and Wernicke's receptive speech areas in the frontal and temporal lobes, and the inferior parietal lobule—areas that are linked together via the arcuate and longitudinal fasciculus. It is via these and associated nuclei that humans are able to verbally and consciously give form to and express their thoughts, ideas, and perceptions.

In this regard, it could be argued that because language is localized to the left hemisphere, those aspects of consciousness that are dependent on language in order to achieve comprehension and understanding are also associated with the left hemisphere, a position advocated by a number of independent neuroscientists (e.g., Albert et al., 1976; Bogen, 1969; Dixon, 1981; Galin, 1974; Hoppe, 1977; Ornstein, 1972; Miller, 1991; Popper and Eccles, 1977; see Chapter 16).

The human left hemisphere also dominates in the perception and processing of real words, word lists, rhymes, numbers, backwards speech, morse code, consonants, consonant/vowel syllables, nonsense syllables, the transitional elements of speech, and single phonemes (Blumstein and Cooper, 1974; Cutting, 1974; Kimura, 1961, 1993; Kimura and Folb, 1968; Levy, 1974; Mills and Rollman, 1980; Haglund et al., 1993; Papcunn et al., 1974; Shankweiler and Studdert-Kennedy, 1966, 1967; Studdert-Kennedy and Shankweiler, 1970). It is also dominant for recognizing phonetic, conceptual, and verbal (but not physical) similarities; for example, determining if two letters (g and p versus g and q) have the same vowel ending (Cohen, 1972; Levy, 1974; Moscovitch, 1973).

The perception, organization, and categorization of information into discrete temporal units or within a linear and sequential time frame are also left-hemisphere–mediated activities, including the sequential control of finger, hand, arm, and articulatory movements (Beaumont, 1974; Christman, 1994; Corina et al., 1992; Efron, 1963; Eisele and Aram, 1994; Heilman et al., 1983; Kimura, 1977, 1993; Mateer, 1983; Haaland and Harrington, 1994; Haglund, et al., 1993; Lenneberg, 1967; Wang and Goodglass, 1992). The left half of the brain is sensitive to rapidly changing acoustics, be they verbal or nonlinguistic, and is specialized for sorting, separating, and extracting in a segmented fashion the phonetic and temporal-sequential or articulatory features of incoming auditory information so as to identify speech units (Eisele and Aram, 1994; Joseph, 1993; Kimura, 1993; Luria, 1980; Shankweiler and Studdert-Kennedy, 1967; Studdert-Kennedy and Shankweiler, 1970).

Linguistic and motoric temporal sequencing is an exceedingly important capability that appears to be unique to humans and dependent on the inferior parietal lobule. For example, among humans, disturbances of linguistic comprehension (i.e., receptive aphasia) is frequently a consequence of an impaired capacity to discern the individual units of speech and their temporal order. Sounds must be separated into discrete interrelated linear units or they will be perceived as a blur, or even as a foreign language. Hence, a patient with Wernicke's aphasia may perceive a spoken sentence, such as the "big black dog" as the "blickbakgod".

In fact, the superiority of the left hemisphere regarding temporal sequencing includes even the visual domain, for the left is superior to the right in visual sequential memory and in detecting nonverbal sequences (Halperin, et al,. 1973; Zaidel, 1977). These capabilities, in turn, appear to be dependent on the inferior parietal lobule and the evolution of left hemispheric control over fine motor functioning.

Language and Motor Control

Modern human language is intimately associated with handedness and fine motor control (reviewed in Joseph, 1993; Kimura, 1993). Immediately adjacent to Broca's expressive speech is a huge expanse of neocortical tissue representing the hand (see Chapter 11). The hand and mouth areas are also richly interconnected. Hence, activation of the hand or Broca's area can result in spreading excitation, such that both neocortical areas become excited simultaneously (see below); this, in turn, can result in mutual interference (Kimura and Archibald, 1974; Kinsbourne and Cook, 1971).

There is also considerable evidence that the neural substrate and the evolution of language and linguistic thought are related to and are outgrowths of right hand and left hemisphere temporal-sequential motor activity (Bradshaw and Nettleton, 1982; Corballis, 1991; Corballis and Morgan, 1978; Faaborg-Anderson, 1957; Jacobson, 1932; Joseph, 1982, 1993; Kimura, 1980, 1993; MacNeilage, 1993; MacNeilage et al., 1987; McGuigan, 1978; Morgan and Corballis, 1978; Joseph, 1982, 1993; Hewes, 1973; Kimura, 1973, 1976. 1979, 1993). This linkage accounts for why individuals often gesture with the right hand when they speak and why Broca's aphasia is almost invariably accompanied by paralysis of the right upper extremity.

Indeed, among the majority of the population it is the right hand that is dominant for grasping, manipulating, exploring, writing, creating, destroying, and communicating. That is, although the left hand assists, it is usually the right which is more frequently utilized for orienting, pointing, gesturing, expressing, and gathering information concerning the environment. We predominantly use the right for waving good-bye, throwing a kiss, delivering a vulgar gesture, greeting, and so forth. The right hand appears to serve as a kind of motoric extension of language and thought in that it acts at the behest of linguistic impulses (Joseph, 1982) via parietal lobe programming. As detailed in Chapter 12, the parietal lobe is considered a "lobe of the hand."

As detailed below and in Chapter 2, with the evolution of the inferior parietal lobule, what had been limbic language became yoked to the neocortex. Briefly, the lateral frontal convexity, including Broca's area, may have evolved from the supplementary motor areas and medial frontal lobe, which in turn evolved from and is richly interconnected with the anterior cingulate (e.g., Sanides, 1969, 1970). The amygdala (and hippocampus) gave rise to the medial and inferior temporal lobes, the insula (see Sanides, 1969, 1970), followed by the superior temporal lobe, Wernicke's area, and by extension, portions of the inferior parietal lobule. The inferior parietal lobule, however, is also an evolutionary derivative of superior parietal tissue that expanded in accordance with and as represented by temporal sequential hand use and fine motor function involving the fingers and the thumb.

The evolution of the inferior parietal lobule, therefore, may have served as a nexus, interlocking, at a neocortical level, the cingulate-Broca pathways and the amygdala-Wernicke's pathway, thereby enabling limbic language impulses to become hierarchically represented as well as subject to temporal sequencing by neocortical neurons (Joseph, 1993). Prior to the evolution of the inferior parietal lobule, Broca's area presumably was unable to receive sufficient input from primary auditory receiving and Wernicke's areas (and the amygdala), and language thus remained, by and large, limbic and controlled by the anterior cingulate gyrus.

The impetus for inferior parietal and frontal lobe and Broca's area evolutionary development, however, appears to be twofold, being in part limbic derivatives (amygdala-hippocampus, amygdala-cingulate) and a function of the evolution of fine motor control involving the facial-oral musculature, vocalization, and especially the establishment of handedness. Given that the human left corticospinal tract matures earlier and crosses the medullary pyramids at an earlier age than fibers from the right (Kertesz and Geschwind, 1971; Yakovlev and Rakic, 1966), thereby presumably establishing synaptic control over the spinal and cranial motor nuclei in advance of the right as well, dominance for hand control and temporal sequential processing became the province of the left hemisphere (Joseph, 1982). With motor dominance, the left amygdala, cingulate gyrus, superior temporal lobe, inferior parietal and frontal lobes were reorganized accordingly.

Evolution of Handedness, Language, and Left Hemisphere Specialization

Fine motor control and handedness, as well as the grammatical and temporal-sequential, denotative aspects of language, appear to have been acquired rather gradually. For example, there is some evidence which suggests that by 2 or 3 million years ago up to 60–70% of Australopithecines were right-handed (Dart, 1949), whereas by 1.5 million years ago, about 80% of H. habilis were similarly inclined (Toth, 1985). However, it was not until about 150,000 years ago, that up to 90% of archaic humans had become right-handed (Cornford, 1986), which is similar to modern-day estimates of handedness.

There is no evidence, however, to suggest that by 150,000 years ago modern human language had appeared, for the neural substrate that would provide the motoric foundations for grammatical speech—i.e., the angular gyrus of the inferior parietal lobule—had probably not yet evolved (see Chapter 2). Moreover, until as recently as 100,000 years ago, with the exception of motor control, the left and right hemispheres were probably more alike than dissimilar in functional specialization—at least in regard to language. This is because the neocortex of the left hemisphere probably had not yet become as specialized or organized for expressing or comprehending modern human speech until around the late Middle Paleolithic Age.

This process of acquiring modern human speech and the accompanying alterations in the neural architecture of the left hemisphere may have begun in earnest around 100,000 years ago, and then may have undergone a rapid acceleration 50,000 years ago (with the evolution of the angular gyrus), followed by yet another possible rapid progression during the last 10,000 years with the invention of reading, writing, and math (Joseph, 1993). Human beings then had the ability to not only speak and think in words but to write them down.

Presumably, because of the increase in left hemisphere dominance over handedness and fine motor control, followed by language representation, former functions were displaced or minimized in importance and/or became the sole domain of the right hemisphere (Joseph, 1992b, 1993). That is, older functions were crowded out first by the development of fine motor control, and then again with the acquisition of language. The left and right hemispheres, therefore, became increasingly dissimilar in functional organization.

As per evolutionary changes in human neural motor organization, it appears that approximately 5 million years ago our pre-human ancestors had become increasingly adapted to walking and running on two legs rather than on all four, as do apes and monkeys. Thus, by 3.5 million years ago Australopithecus and H. habilis had become bipedal creatures and had learned to stand on their hind legs and to walk in an upright manner (Johanson and White, 1979; Leakey, 1979). In consequence, the arms and hands ceased to be used as feet and were freed of the need to hang or hold on to a tree branch for postural support and/or in order to move about. They ceased to be weight bearers. Subsequently, the hands were also increasingly utilized to explore and manipulate objects, and to make simple stone tools.

However, once the feet had become modified for walking, they ceased to be employed for more complex acts, such as grasping, and those areas of the neocortex devoted to foot control and sensory reception diminished in size (Richards, 1986; Falk, 1990). Correspondingly, neocortical areas subserving the fingers and hand increased in size and sensory-motor importance. In fact, more somatomotor neocortical space is devoted to representing the fingers and the human hand (in conjunction with the face and mouth) than the elbow, wrist, or forearm (see Chapters 11 and 12).

As detailed in Chapter 18, the brain is exceedingly plastic and is capable of undergoing tremendous functional reorganization, not just over the course of evolution but over the span of a few months (or years) of a single individual's lifetime (Feldman et al., 1992; Joseph, 1982; Juliono et al., 1994; Strauss et al., 1992; Weiller et al., 1993). However, if the forearm is repeatedly stimulated, corresponding increases in forearm representation are noted in the parietal lobe. Similarly, if a single digit is amputated, the remaining fingers will increase their neocortical representation and will take over the temporarily vacated space (Juliano et al., 1994).

Gradually, over the course of human evolution, the importance of the hands increased, as did their neocortical representation. Subsequently, tool technologies became more complex, as did those areas of the brain devoted to hand control, i.e., the inferior parietal lobule and frontal motor areas (Joseph, 1993). However, because the motor neocortex of the human left hemisphere matures before the right (Corballis and Morgan, 1978; Joseph, 1982; Morgan and Corballis, 1978), and as the left frontal-parietal pyramidal tract develops and establishes its spinal-motor interconnections in advance of the right hemisphere (Kertesz and Geschwind, 1971; Yakovlev and Rakic, 1966), the left hemisphere was (and is) given a competitive advantage in motor control, and, thus, right hand motor dominance which, in turn, led to the establishment of the grammatical and temporal-sequential foundations of modern human language.

Limbic Language and the Inferior Parietal Lobule

As noted, a large area of parietal lobe neocortex is devoted to hand and finger representation and the guidance of the hand and arm during reaching, throwing, and temporal-sequential and related movements (see Chapter 12). Indeed, motor functioning is dependent on sensory feedback, which in turn is provided by the parietal lobe (see Chapter 11). That is, if not for the sensory feedback provided by the muscles and joints (information that is transmitted to the parietal lobes), movement would become clumsy and uncoordinated, since a person would not know where his limbs were in space and in relation to one another. Since the parietal lobes and the motor areas in the frontal lobes are richly interconnected, they serve in many ways as a single neurocortical unit, i.e., sensorimotor cortex (Luria, 1980).

It is, in part, due to the interrelationship between sensory feedback, motor control, and gesture, that hand control and gesture are greatly dependent on the parietal area, the left inferior parietal lobe (IPL) in particular.

In addition, the motor engrams that make possible temporal and sequential motor acts (e.g., making a cup of coffee, fashioning a tool) appear to be localized within the inferior parietal lobe (Heilman et al., 1982; Kimura, 1993; Strub and Geschwind, 1983). It is the IPL which enables humans to engage in complex activities involving a series of related steps, create and utilize tools, produce and comprehend complex gestures, such as American Sign Language (ASL), and express and perceive grammatical relationships (Joseph, 1993, Kimura, 1993; see also Corina et al., 1992). Hence, when the left IPL has been injured, patients may be afflicted with apraxia (Heilman et al., 1982) and

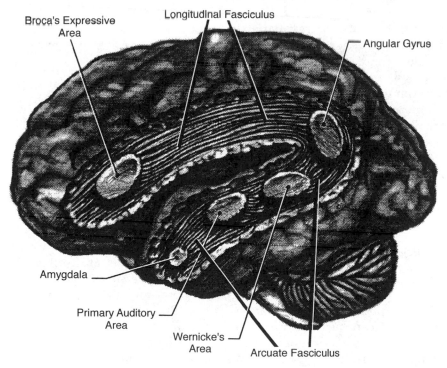

Figure 4.2 Schematic diagram of underlying axonal pathways linking Broca's and Wernicke's areas, the amygdala, and angular gyrus.

Figure 4.3 Schematic diagram indicating converges of sensory information within the inferior parietal lobule.

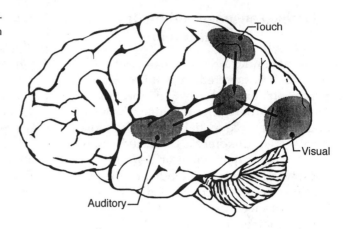

have difficulty with tasks requiring complex motor sequencing.

Broca and Wernicke's areas and, thus, left cerebral linguistic functioning are exceedingly dependent on the IPL and its capacity to impose rhythmic temporal sequences on auditory stimuli and motoric actions (Geschwind, 1965; Goodglass and Kaplan, 1982; Joseph, 1990b, 1993; Heilman et al., 1982; Kimura, 1993; Strub and Geschwind, 1983), including vocalizations that arise from the limbic system.

Presumably, when the inferior parietal lobule and the angular gyrus evolved and when human "hand" representation increased, humans acquired the capacity to segment incoming sounds and to hierarchically represent and punctuate social-emotional, limbic vocalizations so as to vocally express themselves in temporal and grammatical sequences. Thus, social-emotional vocalizations came to be governed by grammatical rules of organization, producing "modern" human language.

Multimodal Properties

In humans, the left IPL, being an indirect product of amygdala as well as hippocampal and superior parietal evolution (see Chapter 1), is capable of multimodal processing of auditory and visual, as well as tactile, impressions. Its unique position at the juncture of the auditory, visual, somesthetic, and motor neocortex also confers upon it the capability of analyzing, associating, and assimilating this divergent data in order to create multiple categories of visual, auditory, and tactile imagery and meaning (Geschwind, 1966; Joseph, 1982).

One can also integrate and assimilate these diverse sensory signals so as to abstract, classify, and produce multiple overlapping categories of experience and cross-modal associations (Geschwind, 1965; Joseph, 1990b, d, 1993).

THE ORGANIZATION OF LINGUISTIC THOUGHT ASSIMILATION AND ASSOCIATION WITHIN THE INFERIOR PARIETAL LOBE

The primary sensory receiving areas for vision, audition, and somethesis are located in the occipital, temporal, and parietal lobes, respectively. Adjacent to each primary zone is a secondary-association neocortical region where higher level information processing occurs and where complex associations are formed. Wernicke's region is one such zone, as is the middle temporal and the superior parietal lobe. The IPL (which includes the angular and supramarginal gyri) is located at the junction where the three secondary-association areas meet and overlap. In this regard, the inferior parietal region receives converging higher order information from each sensory modality and, in fact, makes possible the formation of multiple associations based on the assimilation of this divergent sensory input (Geschwind, 1965; Joseph, 1982, 1990b, 1993). One can thus feel an object while blindfolded and know what it would look like and be able to name it as well. Through its involvement in the construction of cross-modal associations, this region acts so as to increase the capacity for abstraction, categorization, differentiation, and the verbal as well as visual labeling of sensory-motor experience. One is, therefore, able to classify a single stimulus or event in multiple ways. In part, this is made possible because the inferior parietal lobule is the recipient of the simple and complex associations already performed in the primary and association cortices via the 10 billion axonal interconnections that occur in this region.

Stimulus Anchors and the Train of Thought

The angular gyrus and the left IPL of which it is part makes possible the assimilation of complex associations that have been constructed elsewhere so that multiple classifications, categorizations, and descriptions are possible. The IPL also acts to integrate and arrange them according to preestablished (gestural) temporal sequences and the requirements of what needs to be communicated. Moreover, via rich interconnections with Wernicke's area and the middle temporal lobe, the IPL is able to associate auditory/verbal labels with other sensory experiences such that we can describe things as "sticky, sweet, moist, red, lumpy," as well as use single word descriptions, e.g., "jelly."

This capability is particularly important in regard to reading and naming, as described in further detail in Chapters 12 and 14. For instance, when a word is read, the pattern of visual input is transmitted from the visual areas in the occipital and temporal lobes to the left IPL (which is coextensive with Wernicke's area), which then performs an auditory visual matching function. That is, it calls for and integrates the auditory equivalent of what is viewed or read so that we know what animals, objects, words, and letters look like as well as what their names sound like. If this area were damaged, reading ability would be lost.

THE TRAIN OF ASSOCIATIONS

In most instances in which the IPL is activated via internal or external sources of stimulation, multiple trains of inquiry are initiated via the numerous interconnections this areas maintains. Impressions, memories, ideas, and feelings that are in any manner associated with the initial stimulus probe are aroused in response (Joseph, 1982, 1986a, 1990e, 1993).

If a student is asked, "What did you do in school today?," a number of verbal and memory associations are aroused in parallel and integrated within the language axis, all of which are related in some manner to each element of the eliciting stimulus. Finally, in the process of associational linkage, those associations with the strongest stimulus value, which most closely match each element of the question in terms of internal and external appropriateness, and which therefore have the highest probability of being the most relevant, rapidly take a place in a hierarchical and sequential, grammatical arrangement that is being organized in a form suitable for expressing a reply.

To return to the question regarding "school," each speech segment and sound unit become triggers that first activate and then, like a magnet, draw associations accordingly. All aroused forms of mental imagery, verbal associations, and so on that are received in the IPL are then arranged, individually matched, and group matched such that the associations that correspond to all sources of relevant input with sufficient value of probability then act as templates of excitation that stimulate and attract other relevant ideas and associations. These, in turn, are assimilated and associated or are subsequently deactivated due to their low probability in contrast to the association already organized. Moreover, because the strength and value of closely linked associations change in correspondence to the developing sequential hierarchy (or the initial parallel hierarchies), previously aroused and assimilated material may subsequently come to have a lower value of probability or appropriateness within the the matrix of overall activity and may be deactivated (Joseph, 1982, 1986a, 1993).

Consider the question: "What is furry, small, loves milk and makes the sound *meow*?" At the level of the neocortex, each word—"furry," "small," "milk," and "meow"—acts to trigger associations (e.g., "furry = coat-animal-...," "milk = liquid-white-cow-..."). The grammatical linkage of these words also acts to trigger certain associations (e.g., "furry-milk-meow = animal-cat-...") while deactivating others (e.g., "cow"). Following the analysis and comprehension of these sounds and words in Wernicke's area, the angular gyrus, and the middle temporal lobe, the IPL continues to call forth associations so that a reply to the question can be generated.

So that the animal can be named, the IPL via its interactions with the temporal lobe activates the necessary phonemic elements (e.g., "k-a-t"), and then transfers this information to Broca's area and the question is answered: "Cat." If, instead, the individuals replies "tak," this would indicate a problem in organizing the correct phonemic elements once they were activated (see Chapter 14 for an extended and detailed discussion). The final product of this hierarchical, highly grammatical arrangement of mutually determining and parallel associational linkages is the train of thought or a temporal-sequential stream of auditory associations in the form of words and sentences. However, before this occurs, these verbal associations must receive a final temporal-sequential grammatical stamp, which is a consequence of the organization imposed on this material as it passes from

Broca's area to the oral-speech musculature (see Chapter 11 for further detail).

Confabulation and Gap Filling

Assimilation of input from diverse sources is a major feature of the language axis in general. However, if, due to an injury, the language axis is functionally intact but isolated from a particular source of information about which the patient is questioned, he may then suffer word finding difficulty, anomia, or agnosia (Davidoff and De Bleser, 1994; Geschwind, 1965; Shelton et al., 1994).

Following massive lesions of a brain area with which it normally communicates, the language axis sometimes begins to invent an answer or reply to questions based on the information *available*, despite the gaps in that data or the incongruent nature of what is being reported. Consider, for example, denial of blindness (following massive injuries to the visual neocortex) or denial or neglect of the left extremity, which may also be paralyzed (due to massive right cerebral injuries involving the motor and parietal neocortex). Patients will claim to have sight although they bump into objects or fall, or they may claim that their paralyzed left arm belongs to the doctor or a person in the next room (Joseph, 1986a).

To be informed about the left leg or left arm, the language axis must be able to communicate with the neocortical area that is responsible for perceiving and analyzing information about the extremities. For example, since the right parietal area maintains the somesthetic body image, as well as serves as the storage site for body image memories, when that area is destroyed the left half of the "body-image" and all associated memories are essentially "erased"—as if they never existed.

When no message is received by the Language Axis, due to destruction of the neocortical area responsible for that message or memory, and when the language axis is not informed that no messages are being received (because the brain area that would alert them is no longer functioning), the language zones instead rely on some other source, even when that source provides erroneous input (Joseph 1982, 1986a). Substitute material is assimilated and expressed and corrections cannot be made (due to loss of input from the relevant knowledge source), and the patient begins to confabulate (see Chapter 3). That is, the language axis fills the "gap" with erroneous material.

THE ORIGIN OF EGOCENTRIC AND LINGUISTIC THOUGHT
Thought

Nonverbal imagery and hallucinations, as well as visual-emotional dream activity, are associated with activity within the right hemisphere and inferior temporal lobe (Chapters 3 and 5), as well as within the brainstem (Chapter 9). Conversely, verbal thinking is associated with the left hemisphere (Joseph, 1982). Although an individual may utilize visual, emotional, olfactory, musical, or tactile "imagery" when he thinks, thinking often takes the form of "words" that might be "heard" or, rather, experienced, within one's own head (or mind).

Verbal thinking is clearly a form of communication and generally consists of an organized hierarchy of associations, symbols, and labels that appear before an observer, or which are heard by the thinker, within the mind's "ear" and "eye." Thought (i.e., verbal thinking) can be a means of deduction, clarification, plan and goal formation, and reality manipulation (Craik, 1943; Freud, 1900; Miller et al., 1960; Piaget, 1962). However, it is also a progression, an associative advance that leads from an inner or outer perception to linguistic-motor expression (Freud, 1900), and an elaboration, which some have argued appears with an initial or leading idea that is followed by a series of related verbal ideations, or, as originating developmentally from the inaccessible regions of the mind (Freud, 1900; James, 1961; Jung, 1954; Piaget, 1962). In the process of thinking "in words," one often acts to organize information that is "not thought out" and that is

not clearly understood, so that it may become thought out and, thus, comprehended (Ach, 1951; James, 1961; Joseph, 1982; Schilder, 1951).

On the other hand, sometimes the verbal "train of thought" emerges spontaneously and reflexively, as if related ideas simply become strung and attached together with no specific goal or purpose in mind. Moreover, in these instances, sometimes these thoughts rapidly alternate in content and fluctuate between *seemingly* unrelated ideas —as if triggered by a verbal domino effect where associated ideas become sequentially aroused, each subsequent idea triggering the next. Verbal thoughts are also triggered by agents external to the left half of the brain (such as the right hemisphere or limbic system). Sometimes, however, the production of *thought* reflects random neural activity. Indeed, sometimes it is exactly that: random and reflexive.

The Purpose of Verbal Thought

Directed, reflexive, or spontaneous, verbal thoughts always unfold before an observer and are heard within the mind's ear. It is a series of pseudo-auditory transactions, which are *experienced*, as well as produced sometimes, as a purposeful means of *explanation*. Thinking sometimes is experienced as a form of self-explanation through which ideas, impulses, desires, or thing-in-the-world may be understood, comprehended, and possibly acted upon. Paradoxically, it is often a process by which one *explains things to oneself*. Indeed, as a means of deduction or explanation, and as a form of internal language, it is almost as if one is talking to oneself inside one's head.

Nevertheless, the fact that one acts as both audience and orator raises a curious question: "*Who* is explaining what to *whom*?" A functional duality is thus implied in the production and reception of thought.

Assuming that the subject of thought originates in *me*, the thinker, and given that the organization of this often linear verbal arrangement is also a product of self-generated activity, then it should be expected in some instances that "I" should know what "I" am about to think prior to thinking it. "I" should also know the conclusion before it is communicated. In fact, often we do *know* (albeit nonverbally) before we think (and while we think). Sometimes we do not think, simply because the question-answer-implications are simultaneously understood without the aid of verbal thought. There is, thus, some redundancy built into the thinking process as well as an almost inescapable sense of duality in its production and reception.

In that thinking is often a form of communication, it seems that one aspect of the Self, or, rather, the brain, has access to the information that is to be verbally thought about, before it is thought about in a verbal form. That is, the source of what will become a thought is often within the Self. However, that source, or pre-thought, exists in a pre-verbal form, and must then be translated and be organized in a verbal linear sequence in order to be thought about and thus understood. In this regard, thinking is *sometimes* a form of communication through which one part of the brain gains access and an understanding regarding information or knowledge possessed in yet other brain regions; albeit, in nonlinguistic form.

Indeed, thinking often serves in part as a means of organizing, interpreting, and explaining impulses that arise in the nonlinguistic portions of the nervous system so that the language-dependent regions may achieve understanding (Joseph, 1982, 1990e). In fact, although thought may take various nonlinguistic forms, e.g., musical thought, visual-imagery, and so forth, one need only listen to one's own thoughts in order to realize that thinking often consists of an internal linguistic monologue, a series of words heard within one's own head. And, because these particular forms of thought are structured and perceived as *words* heard within one's head, then they must rely on the same neural pathways subserving the production and perception of language and speech sounds produced outside the head, i.e., the language axis.

THE DEVELOPMENT OF LANGUAGE AND THOUGHT
Three Linguistic Stages

There are three maturational stages of verbal development that correspond to the acquisition and development of language and thought (Joseph, 1982). Initially, linguistic expression in the infant is reflexive and/or indicative of generalized and diffuse feeling states. Vocalizations are largely emotional-prosodic in quality, and mediated by limbic and brainstem nuclei (see Chapter 7).

At approximately 3–4 months of age, the infant's utterances begin to assume meaningful as well as imitative qualities, are indicative of specific feeling states, and begin to become influenced by both the right and left cerebral hemispheres. It is during this time period that a second babbling stage develops and the child's prosodic utterances begin to assume temporal-sequential characteristics. That is, the left hemisphere begins to provide rhythm and specification to the melody and associated feeling states expressed by the right hemisphere and limbic system. From this point on, true language begins to develop.

However, it is not until a third stage of linguistic functioning makes its appearance that the child begins to not only speak in words, but to think them out loud. This final stage coincides with the expression and development of egocentric speech.

Limbic and Brainstem Cognitions

For some time after birth, due to the immaturity of the neocortex, its subcortical and thalamic interconnections, and degree of axonal myelination (see Chapter 18), most behavior is initially mediated by limbic, brainstem, and spinal nuclei (Churgani et al., 1987; Gibson, 1991; Joseph, 1982; Milner, 1967). For example, PET scan studies of glucose utilization in the newborn indicates high levels of brainstem activity but very low levels of neocortical activity (Churgani et al., 1987). It is not until about 1 year of age that infant neocortical glucose activity begins to significantly increase (Churgani et al., 1987) and not until ages 4–10 that the sensory and association cortical layers begin to become increasingly myelinated (Gibson, 1991; Lecours, 1975).

Therefore, because of neocortical immaturity, the psychic functioning of the newborn is probably no more than a vague, somewhat undifferentiated awareness, consisting of a multitude of excitatory and inhibitory neuronal interactions and a series of transient feeling states and emotional upheavals that correspond to the activation of specific and related subcortical and limbic nuclei.

The neonate is essentially internally oriented, its psychic attentional functions almost entirely directed to stimuli impinging on the body surface and sensations transmitted by the mouth (Milner, 1967). That is, although the newborn can cry and scream, turn his or her head to sounds, and within a few weeks imitate facial expressions, reach for objects, and show defensive reactions (Gibson, 1991; Meltzoff, 1990), these behaviors are under the control of the brainstem, limbic system, and basal ganglia.

Cognition, therefore, consists largely of generalized and diffuse feeling states that are aimed at the alleviation of displeasure or painful affect and, with the reactivation of experiences, associated with pleasurable sensations (reviewed in Chapter 18). In fact, from birth to 1 month, the infant displays only two attitudes–accepting and rejecting–and a very limited range of vocalization: crying and cooing (McGraw, 1969; Milner, 1967; Spitz and Wolf, 1946). These feeling states and vocalizations are largely mediated and expressed by the hypothalamus of the limbic system.

Indeed, as noted, the original impetus to speak springs forth from roots buried within the depths of the ancient limbic lobe and is bound with and tied to mood, impulse, feeling, desire, pleasure, pain, and fear. The infant cries, coos, and produces various prosodic inflectional variations that are without temporal-sequential organization and that serve only to communicate diffuse feelings. It is only over the course of the first few months that these prosodic-melodic utterances become associated with specific moods and emotions (Joseph 1982, Piaget, 1952). It is at this time that babbling makes its appearance (Brain and Walton, 1969).

Early and Late Babbling and Probable Speech

Although the diffuse prosodic-melodic utterances of the infant are limbic (and/or brainstem), much of the early babbling that occurs is a product of random and reflexive motor activity, motor overflow, and self-stimulation (McGraw, 1969). Early babbling is also representative of the neocortical maturational changes occurring within the language axis and the initial exercise and activation of these developing pathways.

By approximately 3–4 months of age the nonmotor neocortex of the right cerebral hemisphere leads the left in regard to dendritic growth and length (Scheibel, 1991). By this time, it has greatly increased its hierarchical acquisition of and synaptic control over various limbic nuclei (Joseph, 1982).

Prosodic vocal output begins to assume an imitative quality, which, in turn, is context-specific (Piaget, 1952), and the infant's largely intonational-melodic vocal repertoire becomes more elaborate and increasingly tied to certain feeling states, and begins to acquire probable meanings (Hoff-Ginsberg and Shatz, 1982; Joseph, 1982; Piaget, 1952). For example, by 3 months the infant may purposefully cry or call "ma ma," whereas initially "ma ma" signified nothing beyond the random universal babbling produced by all infants.

As cries, babbles, or, for example, the word "ma ma" first assumes probable meanings, they initially signify a variety of diffuse feelings and needs. Hence, "ma ma" might mean "mama, come here," "mama, I hurt," "mama, I thirst," "mama, I am hungry," and so forth (Joseph, 1982; Piaget, 1952; Vygotsky, 1962). Moreover, "ma ma," at times, may continue to signify nothing more than reflexive (early) babbling for the first several months.

Late Babbling and Temporal-Sequential Speech

By time the infant reaches 4 months of age, a second babbling stage develops, i.e., "late babbling." Whereas the first babbling stage reflects motor self-stimulation, late babbling heralds the first real shift from emotional-prosodic-melodic speech to temporal-sequential language (Joseph, 1982; Leopold, 1947), i.e., left hemisphere speech. Indeed, these developmental changes coincide with dendritic development in the language areas of the left hemisphere (Scheibel, 1991) and the maturation of the arcuate fasciculus (Lecours, 1975), the major pathways via which auditory-linguistic input is transmitted from Wernicke's area to the inferior parietal lobule to Broca's area for motoric sequencing and expression.

The initial appearance of left hemisphere speech and the late babbling stage also corresponds to the completion of the myelogenetic cycle within Heschl's gyri and the primary and association auditory receiving areas (Lecours, 1975). Moreover, during this later time period infants begin to behaviorally respond to specific speech sounds and their phonemic characteristics (Moffit, 1971; Morese, 1971; Treub, 1972). This is because Wernicke's area is now able to begin appropriately processing these signals.

Nevertheless, language comprehension (and production) remains largely emotional as infants between the ages of 10 weeks and 5 months continue to produce limbic (i.e., the melodic, prosodic, emotional) vocalizations and to listen selectively for similar speech sounds in order to discern, discriminate, and appropriately respond to social-emotional vocalizations conveying approval, disapproval, happiness, and anger (Fernald, 1993; Haviland and Lelwica, 1987). Moreover, 5-month-old (American) infants are able to make these discriminations based merely on melodic tone and in the absence of words and vocabulary, and in response to nonsense English and to German and Italian vocalizations (Fernald, 1993). Presumably, these vocal emotional nuances are comprehended (as well as produced) by the infant's right hemisphere and limbic system.

However, as the infant's left cerebral hemisphere begins to mature, and to respond to and influence auditory-linguistic stimuli, it begins to stamp and impose temporal sequences onto the stress, pitch, and melodic intonational contours that up until this time have characterized infant speech output (Joseph, 1982). This is part of what the late babbling stage heralds: the ability to sequence.

That is, syllabication is imposed on the intonational contours of the child's speech by the left hemisphere, such that the melodic features of generalized vocal expression come to be punctuated, sequenced, and segmented, and vowel and consonantal elements begin to be produced (Joseph, 1982, 1993; see also Berry, 1969; De Boysson-Bardies et al., 1980). Left hemisphere speech comes to be superimposed over limbic (and right hemisphere) melodic language output.

Thus, as the left hemisphere matures and wrests control of the peripheral and cranial musculature from subcortical and limbic influences, a second form of language emerges, one that arises through interactions and nominal associations with external stimulatory activities and via the production of sequences: i.e., left hemisphere speech. Left hemisphere speech is, therefore, grammatical, temporal-sequential, denotative, and social, and is closely bound with the eventual expression and development of verbal thought. Verbal thinking, however, does not appear until much later in development—an unfolding event which corresponds to the appearance of a third stage of language: egocentric speech.

Egocentric Speech

Egocentric speech is self-directed speech that consists of an explanatory monologue in which children comment on or explain their play and other actions, usually after the action has occurred. That is, the child essentially talks to himself but in an explanatory fashion. Egocentric speech is essentially speech for oneself (Joseph, 1982; Piaget, 1962; Vygotsky, 1962). It is a self-directed form of communication that heralds the first attempts at self-explanation via thinking-out-loud.

According to Vygotsky (1962), egocentric speech makes its first appearance at approximately 3 years of age. According to Piaget (1952, 1962, 1974), at its peak, egocentric speech comprises almost 40–50% of the preoperational child's language; the remainder consists of social speech (denotative, interactional, and emotional).

Social speech is produced so as to communicate with others. Egocentric speech is produced so as to communicate with no one other than the child who produces it. Prior to the development of egocentric speech, communication is directed strictly toward outside sources. There is no attempt to verbally communicate with the Self, for there is no internal dialogue. Verbal thought has not yet developed, and children do not talk to themselves about their ongoing behavior or feelings.

At around age 3, egocentric speech—the peculiar linguistic structure from which thought will arise—appears in the context of social-denotative vocalizations (Vygotsky, 1962). That is, part of the time the child engages in social speech, whereas the remainder of speech activities are egocentric and directed and produced for the sole benefit of the child who listens to his speech and external commentary as he or she plays.

While the child is engaging in egocentric speech, he does not appear concerned with the listening needs of his audience, simply because, to all appearances, his words are meant for his ears alone (Piaget, 1952, 1962, 1974; Vygotsky, 1962). The child is essentially thinking out loud in an explanatory fashion, commenting on and describing his or her actions (Joseph, 1982; Vygotsky, 1962).

When engaged in an egocentric monologue, there is no interest in influencing or explaining to others what, in fact, is being explained. The child will keep up a running verbal accompaniment to his actions, commenting on his behavior in an explanatory fashion, even while alone. Moreover, while engaged in this self-directed external monologue, the child appears oblivious to the responses of others to his statements (Piaget, 1952, 1962, 1974; Vygotsky, 1956). It is as if the child has no awareness that others hear him. In fact, many a child has been shocked when he later hears his mother (or a friend) repeat or comment upon something he assumed no one else could hear.

Egocentric speech is not simply talking out loud, but rather tends to be self-explanatory, serving as a form of commentary that is initially produced only after an action has been completed. The child observes what he or she has done and then comments on and/or explains what has taken place.

Egocentric speech presents us with a curious anomaly, for we must accept that the child knows what he or she has done without commenting or explaining her actions; moreover, she must know why she has performed certain behaviors without the need to explain them to herself. Nevertheless, the fact that she explains and comments upon her behaviors after they occur argues otherwise. Paradoxically, the child acts as both actor and witness, explainer and explained to. Clearly, *the child explains his actions to himself* (Vygotsky, 1962).

The Internalization of Egocentric Speech

Initially, egocentric speech is completely external and after the fact. Presumably, it is external and after the fact because the child is incapable of internally generating linguistic thoughts (Piaget, 1952, 1962, 1974; Vygotsky, 1956). This is probably due to the slow pace of myelination in the language axis and, in particular, the corpus callosum (Joseph, 1982). Because of these internal limitations, the child, therefore, thinks about his or her behavior, out loud.

It is important to emphasize, however, that egocentric speech is not random or pointless, but is largely explanatory. They are explaining their behavior to themselves. Since they are utilizing words, it is thus apparent that they are explaining their actions to their left hemisphere. The fact that the explanation occurs, initially, only after the behavior has been completed suggests that the left hemisphere did not have access to the behavioral plan, or the motivation behind it, until after the act was completed; this is then explained as a verbal commentary.

This suggests that the behavior being explained was, therefore, planned, initiated, or mediated, presumably, by the right hemisphere or limbic system (Joseph, 1982). Because of the slow pace of corpus callosum myelination, the left hemisphere of the child's brain cannot gain access to right cerebral impulses, memories, or plans, until after they are expressed and can be observed and, thus, commented on. However, as the callosum matures, the child's left hemisphere begins to receive earlier or advanced access and thus produces the egocentric commentary earlier and earlier as well.

For example, first a child will paint a picture and then explain it. As she ages, she will paint a picture and explain it while she is painting. Finally, she will announced what she is going to paint, and then paint it (Vygotsky, 1962). Hence, as the child grows older her comments and explanations occur earlier in the sequence of expression, until finally the child begins to explain her actions *before* they are performed instead of *after* they have occurred.

Essentially, as the child ages she appears to receive advanced warning of her intentions and actions, until finally this information is available before rather than after she acts (Joseph, 1982). At this later stage, however, egocentric speech has been greatly internalized as verbal thought.

According to Vygotsky (1962), after its initial appearance and elaboration, egocen-

tric speech also begins to occur internally and, in fact, becomes progressively more covert as the child grows older. At its overt maximum, when it appears to be fully developed (comprising by age 4 almost 50% of the child's speech), its traits and structures are simultaneously being internalized and strengthened and comprise a greater portion of the child's cognitive activities than may be witnessed (like an iceberg).

That is, egocentric speech never disappears but becomes completely internalized in the form of inner speech, i.e., thought. The child has now learned to think words as well as to speak them, and to think them in a temporal and organized sequence that retains its original and primary function— self communication.

Self-Explanation and Interhemispheric Communication

The essential feature of the external components of egocentric speech is that it is based on stimuli and actions that occur outside the child's immediate sphere of understanding and experience—at least, insofar as the language-dependent left hemisphere is concerned. In this regard, children, when they "misbehave," are probably sometimes telling the truth when they say they "don't know why" they did such and such. That is, the left hemisphere does not know why.

Although egocentric speech is a self-directed monologue, it is nevertheless a product of the left cerebral hemisphere. That egocentric speech appears initially only after an action has occurred indicates that the left hemisphere of the young child is responding to impulses and actions initiated outside its immediate realm of experiences and comprehension. It seems that the left hemisphere in the production of egocentric speech is not only "thinking out loud," but is attempting to interpret what it observes and experiences externally, thus creating a meaningful explanatory sequence, which it then linguistically communicates to itself.

As noted above, the appearance and eventual internalization of egocentric speech occur in response to several maturational changes in the central nervous system and parallels the myelination of the language axis and the corpus callosum and the increased ability for the cerebral hemispheres to communicate (Joseph, 1982). As has been demonstrated by a number of independent researchers, communication between the right and left halves of the brain is somewhat poor prior to age 3 and remains limited until approximately age 5 (Deruelle and Schonen, 1991; Finlayson, 1975; Gallagher and Joseph, 1982; Galin et al., 1977, 1979; Joseph, 1982; Joseph and Gallagher, 1985; Joseph et al., 1984; Kraft et al., 1980; Molfese et al., 1975; O'Leary, 1980; Ramaekers and Njiokiktjien, 1991; Salamy, 1978). Presumably, this is a function of the immaturity of the corpus callosal fiber connections between the hemispheres (Yakovlev and Lecours, 1967).

Essentially, egocentric speech appears to be a function of the left hemisphere's attempt to organize, interpret, and make sense of behavior initiated by the right hemisphere and limbic system (Joseph, 1982). Presumably, because interhemispheric communication is, at best, grossly incomplete, the left utilizes language to explain to itself the behavior in which it observes itself to be engaged. As the commissures mature and information flow within and between the hemispheres increases, the left hemisphere —gaining increased access to these impulses as they are formulated in the nonlinguistic portions of the brain—begins to linguistically organize what it experiences internally (rather than externally).

Essentially, increased commissural transmission allows the left hemisphere access to right hemisphere impulses-to-action before the action occurs, rather than forcing it to make sense of the behavior after its completion (which is typical of split-brain patients; see Chapter 3). Of course, as noted in Chapter 3, even in the "normal" intact adult, commissural transmission is not complete. As such, the adult left hemisphere sometimes finds itself witnessing and participating in behaviors that it did not initiate, and which it does not understand.

DISORDERS OF LANGUAGE AND THOUGHT
The Language Axis

When listening to someone speak, the information is transferred from the brainstem to the inferior colliculus and medial geniculate of the thalamus, and transferred to the amygdala and the primary auditory cortex, where the data are extensively analyzed (see Chapter 14). These auditory signals are then transferred to Wernicke's area (which merges with the angular gyrus), where the temporal-sequential, semantic, and related linguistic features are stabilized, extracted, analyzed, and labeled.

If a question has been asked, the message is transferred from Wernicke's to the inferior parietal lobule, where various ideational associations are aroused and organized via its vast interconnections with other cortical regions. Presumably, a series of interactions continue to occur between Wernicke's and the inferior parietal lobe (as these areas are coextensive) until a reply is formulated and properly organized for possible expression, at which point it is transferred to Broca's area (Geschwind, 1965; Joseph, 1982).

Specifically, the semantically correct and suitably chosen reply is transferred from Wernicke's region and the inferior parietal lobe, via the interlining axonal bundle, the *arcuate fasciculus*, to Broca's area, which then programs the speech musculature and neocortical motor areas so that the reply can be expressed.

Conduction Aphasia

Damage involving the arcuate fasciculus and/or the supramarginal gyrus of the left hemisphere can result in a condition referred to as *conduction aphasia* (Benson et al., 1973; Geschwind, 1965). Broca's area is essentially disconnected from the inferior parietal lobule and Wernicke's area, and although comprehension is intact, the patient cannot repeat words or read out loud. In these cases, the lesion may extend to the insula and auditory cortex and underlying white matter of the left temporal lobe, thereby destroying the axonal fibers of passage.

Individuals with this disorder have great difficulty communicating because a lesion in this vicinity disconnects Broca's area from the posterior language zones. Although a patient would know what he wanted to say, he would be unable to say it. Nor would he be able to repeat simple statements, read out loud, or write to dictation, although the ability to comprehend speech and written language would remain intact.

Nevertheless, these individuals are still able to talk. Unfortunately, most of what they say is contaminated by fluent paraphasic errors, phonetic word substitutions, and the telescoping of words due to impaired sequencing. Patients also tend to confuse words that are phonetically similar. Because they can comprehend, they are aware of their disturbances and will try to come up with the correct words via the generation of successive approximations (Marcie and Hecaen, 1979). Hence, speech may be circumlocutional, and seemingly tangential, as well as contaminated by paraphasic distortions. Sentences are usually short and are often unrelated to each other (Marcie and Hecaen, 1979).

When the patient's writing, grapheme formation is usually normal. However, because they produce the wrong words and/or misspell what they produce, patients may frequently cross out words and perform overwriting (i.e., writing over various letters with additional strokes). Patients are frequently very frustrated, irritable, and upset regarding their condition.

Broca's Aphasia

As noted, within the left frontal convexity there is a large region that is responsible for the expression of speech: i.e., Broca's area, which is named after Paul Broca, who, over 100 years ago, delineated the symptoms associated with damage to this area. Immediately adjacent to Broca's area is the portion of the primary motor area that subserves control over the oral-facial musculature and the right hand.

Various forms of linguistic information are transferred from the posterior portion of the language axis (via an axonal fiber bundle, the arcuate fasciculus) and converge upon Broca's area, where they receive their final sequential (syntactical, grammatical) imprint so as to become organized and expressed as temporally ordered motoric linguistic articulations, i.e., speech. These impulses are then transferred to the adjacent frontal motor areas, which in turn control the oral musculature. Hence, verbal communication and the expression of thought in linguistic form is made possible.

Anomia or Word Finding Difficulty

Because Broca's and Wernicke's areas and the inferior parietal lobe tightly interact in the production of language, and as these areas are also dependent on extensive interconnections with yet other neocortical areas and neurons, damage anywhere within the left hemisphere can therefore result in "word finding difficulty." Some people refer to this as "tip of the tongue," when they have trouble coming up with or remembering a particular word.

Every individual with aphasia has some degree of word finding difficulty, i.e., dysnomia (or if severe: anomia). Dysnomia is very common and can occur with fluent output, good comprehension, and the ability to repeat, and in the absence of paraphasias or other aphasic abnormalities. Although many *normal* individuals who have trouble finding a word sometimes experience it as "on the tip of the tongue," anomia is a much more pervasive abnormality and may involve naming objects, describing pictures, and so forth, even when hints are provided.

In general, anomic difficulties suggest left hemisphere functional impairment, and the condition is sometimes secondary to disconnection, or a deficit in activating the correct phonological sound-word patterns (Kay and Ellis, 1987). Frequently, if dysnomia is the only predominant problem, the patient may erroneously ascribe it to deficient memory. That is, the patient may complain of memory problems when in fact they have word finding difficulty. However, this is not due to mnemonic deficits. If provided with hints or even the initial letter there is little improvement (Goodglass and Kaplan, 1982). Moreover, once the word is supplied, the patient may again experience the same problem almost immediately. In general, the ability to generate nouns is more affected than verbs.

In severe cases, while conversing the patient may be so plagued with word finding difficulty that speech becomes "empty," and characterized by many pauses as the patient searches for words. This condition is sometimes referred to as anomic aphasia (Hecaen and Albert, 1978). These same patients may erroneously substitute phrases or words for the ones that cannot be found. For example, they may call a "spoon" the "stirrer," a "pencil" the "writer," or various objects "the whatchmacallit." This can lead to circumlocution as they tend to talk around the word they are after: "Get me the uh, the uh, thing over on the uh, on the top over there..."

Children are often plagued by such difficulties, which they outgrow; this is the result of immaturity within the inferior parietal lobe and the late establishment of the necessary axon fiber connections with other brain areas.

If accompanied by problems with reading and writing, the lesion is probably situated in the posterior portions of the left hemisphere, near the angular gyrus. However, lesions anywhere within the left hemisphere can result in anomic difficulties.

In some cases, however, depending on where the damage and thus the disconnection occurs, a person may not be able to name an item when it is shown to them (due to visual cortex-inferior parietal disconnection), but can name it if they touch it, or if it is described to them out loud; this is due to preservation of the auditory and tactile pathways, linking them with the inferior parietal lobe. For example, since the axonal fibers leading from Wernicke's area and the tactual association areas have not been injured, the person can name the ob-

ject if he touches it, if it is described, or if it makes a sound (e.g., by running the fingers across the teeth of a comb). If the lesion disconnected the inferior parietal lobe from the rest of the parietal lobe, the person would be able to name an object if he saw or heard it, but not by touch alone. If the lesion disconnected the inferior parietal lobe from Wernicke's area, or if the inferior parietal lobe were destroyed, he would be unable to name the object, regardless of modality of presentation (Geschwind, 1965; Joseph, 1993). If provided with hints or even the initial letter of the object's name, there is little improvement. Moreover, once the word is supplied, the patient may again experience the same problem almost immediately. In general, the ability to generate nouns is more affected than verbs.

Expressive Aphasia

Because Broca's area acts as the final common pathway by which language is vocally and temporally-sequentially expressed, damage to the left frontal convexity results in a dramatic curtailment of the ability to speak. Immediately following a massive stroke in this region, individuals often suffer a paralysis of the right upper extremity and initially are almost completely unable to speak: i.e., Broca's (or expressive) aphasia (Bastiannse, 1995; Benson, 1993; Goodglass and Kaplan, 1982; Haarman and Kolk, 1994; Hofstede and Kolk, 1994; Levine and Sweet, 1983). Comprehension, however, is generally intact.

Although symptoms differ depending on the severity of the lesion, in general, individuals with Broca's aphasia are very limited in their ability to articulate or repeat statements made by others. In severe cases, speech may be restricted to a few stereotyped phrases and expressions, such as "Jesus Christ," or to single words such as "fine," "yes," or "no," which are produced with much effort.

Even if they are capable of making longer statements, much of what these patients say is poorly articulated and/or mumbled, such that only a word or two may be intelligible (Bastiannse, 1995; Haarman and Kolk, 1994; Hofstede and Kolk, 1994). However, this allows them to make one-word answers in response to questions. Nevertheless, speech is almost always agrammatical (i.e., the production of some correct words but in the wrong order), contaminated by verbal paraphasias (e.g., "orrible" for "auto"), and/or marked by the substitution of semantically related words (e.g., "mother" for "father"), as well as characterized by the omission of relational words, such as those that tie language together (i.e., the propositions, modifiers, articles, and conjunctions).

Similarly, patients sometimes have difficulty comprehending these same grammatical features (Samuels and Benson, 1993; Tyler et al., 1995; Zurif et al., 1972), as well as related verbal material, such as demonstrated on the Token Test (DeRenzi and Vignolo, 1962). Their ability to repeat what is said to them, although grossly deficient, is usually not as severely reduced as is conversational speech. The ability to write is always affected. Similarly, their capacity to write to dictation is severely limited. However, the ability to copy is much better preserved. In addition, reading comprehension is usually intact although they cannot read aloud.

Patients with moderate and severe expressive abnormalities also have difficulty performing three-step commands although two-step requests may be performed without difficulty. However, they are better able to comprehend as well as verbalize semantically significant words. Even in moderate and mild cases, a consistent defect in the syntactical structure of speech is noted, including reductions in vocabulary and word fluency in both speech and writing (Bastiannse, 1995; Goodglass and Berko, 1960; Haarman and Kolk, 1994; Hofstede and Kolk, 1994; Levine and Sweet, 1983).

Individuals with expressive aphasia, although greatly limited in their ability to speak, are nevertheless capable of making emotional statements or even singing (Gardner, 1975; Goldstein, 1942 ; Joseph, 1988; Smith, 1966; Smith and Burklund, 1966; Yamadori et al., 1977). In fact, they may be able to sing words they cannot say.

Except for emotional speech, however, their language production is largely monotonic (Goodglass and Kaplan, 1982), or characterized by prosodic distortions such that, in some cases, they sound as if they are speaking with a foreign accent (Graff-Radford et al., 1986). This is often due to shifts in the enunciation of vowels; i.e., increased duration of the utterance and the pauses between words as they struggle to speak. However, in some instances this is secondary to deep lesions involving, perhaps, the anterior cingulate and other nuclei (see Chapter 11).

SEX DIFFERENCES

As discussed in Chapter 2, Broca's speech area appears to be more functionally developed in women than in men. That is, there is some evidence suggesting that expressive (and emotional) speech tends to be more clearly concentrated in the anterior regions of the female's brain (Kimura, 1993). In consequence, females are far more likely to become severely aphasic with left frontal injuries, whereas males become more severely aphasic with left parietal damage (Hier et al., 1994; Kimura, 1993).

However, there is some evidence that the posterior portions of the corpus callosum, i.e., the rope of fibers that interconnect the right and left parietal lobes, is thicker and larger in females than in males (Holloway et al., 1993). As gathering and related activities are more likely to involve both hands (versus hunting and throwing with a single hand), this findings might be expected, particularly in that the motor areas are dependent on the parietal somesthetic area for fine movement programming. Thus, it might be expected that the posterior female corpus callosum would be larger because the female right and left parietal lobes were probably simultaneously activated and utilized by females for the last 100,000 or so years—a consequence of woman's role as gatherer, tool maker, food and hide preparer; activities that promote temporal-sequential motor activities and the production of speech. Thus, females are more resistant to posterior injuries in regard to developing aphasic abnormalities, because they have an advantage over males in this regard (however, see Chapter 2).

It is, therefore, noteworthy that boys also tend to stutter more than girls (Corballis and Beale, 1983), and stuttering is largely a male problem for adults as well as children. This suggests either dysfunction or a lack of functional (temporal-sequential) development in the male inferior parietal lobule (although medial frontal, anterior cingulate, and/or abnormalities involving Broca's area and the anterior cingulate gyrus and medial frontal lobes may well be equally responsible). These sex differences in evolutionary history and, thus, language acquisition might also account for the many other "learning disabilities" that predominantly affect boys and men.

Depression and Broca's Aphasia

Individuals with left frontal damage and Broca's aphasia often become frustrated, sad, tearful, and depressed (Gainnoti, 1972; Robinson and Szetela, 1981). Presumably, this is because individuals with Broca's aphasia are fairly (but not completely) able to comprehend. Hence, they are aware of their deficit and become appropriately depressed (Gainnoti, 1972). Indeed, those with the smallest lesions become the most depressed (Robinson and Benson, 1981), the depression as well as the ability to sing being mediated, presumably by the intact right cerebral hemisphere.

Nevertheless, it has also been reported that psychiatric patients classified as depressed (who presumably have no signs of neurological impairment) often demonstrate, electrophysiologically, insufficient activation of the left frontal lobe (d'Elia and Perris, 1974; Perris, 1974). With recovery from depression, left hemisphere activation returns to normal levels.

SEX DIFFERENCES

Females are far more likely to experience depression and depressive episodes than males (DSM IV). Although this sex difference is no doubt related to differential stresses and hormonal factors affecting women versus men, this may also be secondary to the female's greater risk for anterior cerebral artery dysfunction, e.g., embolism (Hier et al., 1994) and to become depressed with even subtle injuries involving the left (and right) frontal cortices. That is, females (and males) with small left frontal strokes may be diagnosed as depressed when, in fact, they have suffered a cerebrovascular infarct.

Apathetic States

Depression and depressive-like features also occur with left frontal and medial lesions that spare Broca's area (Robinson and Szetela, 1981). However, rather than demonstrating depression per se (particularly when the frontal pole is damaged), such patients frequently appear severely apathetic, hypoactive with reduced motor functioning, and poorly motivated. Of course, they may also be depressed. Even so, when questioned, rather than worried or truly concerned about their condition, the overall picture may be one of bland confusion and disinterest. Patients also frequently seem motorically and emotionally blunted.

Presumably, these depressed and apathetic states are sometimes due to disconnection from the limbic system and/or right cerebral hemisphere. Unfortunately, with frontal pole injury, tumor, or degeneration, the underlying neurological precursors to their condition are not very apparent until late in the disease. Hence, misdiagnosis is likely. We will return to these issues in Chapter 11.

Psychosis and Blunted (Negative) Schizophrenia

Different forms of schizophrenia are also associated with destruction or dysfunction of specific cerebral nuclei; e.g., the temporal lobe, caudate-putamen, medial frontal lobe (see Chapter 17), and the dopamine system. However, the medial and left (and right) frontal and caudate-putamen and dopamine system interact in regard to a number of functions as a single unit. As such, when this unit becomes dysfunctional, obsessive-compulsive behaviors or a schizophrenic psychosis may ensue (see below; also, see Chapters 9 and 17). However, different regions of the brain contribute to different symptoms.

For example, with deep medial frontal lesions involving the supplementary motor areas, the individual may develop catatonic symptomatology (Joseph, 1990d; also, see Chapter 11). In contrast, apathetic, blunted, and "negative" forms of schizophrenia, coupled with "psychomotor retardation" reduced verbal and intellectual output and slowness of thought, tend to be associated with left frontal lobe dysfunction (Buchsbaum, 1990; Carpenter et al., 1993; Casanova et al., 1992; Leven, 1994; Weinberger, 1987; see also Crowe and Kuttner, 1991).

When the lesions or pattern of abnormality involve the lateral convexity, underlying white matter, and long distance axons of passage leading to the head of the caudate, there can also result a disconnection involving impulse and thought control. For example, a lesion involving the frontal-caudate-putamen "loop" may derail the entire system, thereby giving rise to retarded motor, cognitive, and verbal functioning such that the patient appears to be suffering from a blunted "schizophrenic" psychosis. These and related frontal lobe and basal ganglia disorders will be detailed in Chapters 9 and 11 (see also Chapter 17).

LEFT TEMPORAL LOBE: SOUND PERCEPTION, LANGUAGE, AND APHASIA
Pure Word Deafness

Located within the superior temporal lobe of both hemispheres are the primary auditory receiving areas. If these areas are destroyed bilaterally, the individual becomes cortically deaf. Although able to *hear*, these persons cannot *perceive* or comprehend nonverbal sounds or understand spoken language.

When the primary auditory area is destroyed, the auditory association areas (Wernicke's area in the left hemisphere) are disconnected from all sources of auditory input and thus cannot extract meaning from the auditory environment. When individuals are unable to perceive and identify linguistic and nonlinguistic sounds, the deficit is described as a *global* or *generalized auditory agnosia* (Schnider et al., 1994). Cerebrovascular disease is the most common cause of this abnormality.

These patients are not deaf, however, as this can be ruled out by testing pure-tone thresholds. In fact, even with complete bilateral destruction of the primary auditory cortex, there is no permanent loss of acoustic sensitivity (Rubens, 1993). This is because sounds continue to be received in thalamic and subcortical centers, including the amygdala. Hence, patients can still detect sounds; they just don't know what the sounds are (Schnider et al., 1994). Moreover, deficits in loudness discrimination and sound localization are common.

In some instances, damage may be unilateral and involve predominantly the primary auditory receiving area of the left hemisphere as well as the underlying white matter, such that Wernicke's region becomes disconnected from all sources of auditory input. Although the patient can hear and identify nonlinguistic and environmental sounds, he is unable to comprehend spoken language. These patients do not have Wernicke's aphasia, however, and are very aware of their deficits. Their own verbal output is normal (albeit, sometimes loud and dysprosodic). Frequently, they describe the sounds they hear as being "muffled," or they may complain that voices sound like "echoes" or "noise" (Tanaka et al., 1987).

Although they cannot repeat what is said to them, patients can read and write in a normal manner, and there are few or no aphasic symptoms (Schnider et al., 1994). Nor do they demonstrate difficulty understanding nonverbal or pantomimed actions.

Nevertheless, some patients, as they recover from Wernicke's aphasia, move to the stage of pure word deafness. However, in these instances there are remnants of aphasic abnormalities. Many such patients, like those who suffer some forms of deafness, may experience auditory hallucinations and/or exhibit paranoid ideation (Rubens, 1993). The hallucinations are likely a consequence of the attribution or erroneous extraction of meaning from randomly produced neural "noise." The paranoia is largely a normal reaction arising secondarily from the confusion associated with reduced verbal and social-linguistic contact. That is, patients become fearful and mistrustful,as they are not sure what is "going on."

In part, pure word deafness is sometimes due to disturbances in processing rhythm and temporal sequences (Tanaka et al., 1987), or making phonemic or semantic discriminations (Denes and Semenza, 1975; Schnider et al., 1994). Thus, slowing the rate of sound presentation or rate of speech can improve comprehension (Buchman et al., 1986).

Wernicke's Aphasia

Within the left superior temporal lobe, extending from the border zones of the primary auditory reception area toward the inferior parietal lobule, is located Wernicke's area. Wernicke's area—in conjunction with and aided by the inferior parietal lobule (Kimura, 1993)—acts to decode and encode auditory-linguistic information (be it externally or internally generated), so as to extract or impart temporal-sequential order and related linguistic features. In this manner, denotative meaning may be discerned or applied (Efron, 1963; Lackner and Teuber, 1973; Joseph, 1982, 1988a; Lenneberg, 1967). That is, this area acts to verbally label information transmitted from external sources as well as from other brain regions. For example, it may act to provide the audi-

tory equivalent of a visually perceived written word. In this manner, we know what the words we read sound like.

Receptive Aphasia

If the auditory association (Wernicke's) area is damaged, patients will have great difficulty comprehending spoken or written language (Goodglass and Kaplan, 1982; Kertesz, 1983a; Hecaen and Albert, 1978). Naming, reading, writing, and the ability to repeat or understand what is said are severely affected, the condition being labeled Wernicke's (receptive) aphasia. When patients prove capable of reading, then they are probably suffering from "pure word deafness" rather than receptive aphasia (to be explained).

Frequently, disturbances involving linguistic comprehension are due to an impaired capacity to discern the individual units of speech and their temporal order, due to destruction of the superior temporal lobe and adjacent tissue. Sounds must be separated into discrete interrelated linear units or they will be perceived as a blur, or even as a foreign language (Carmon and Nachshon, 1971; Efron, 1963; Joseph, 1982, 1988a; Lackner and Teuber, 1973). Hence, a patient with Wernicke's aphasia may *perceive* a spoken sentence such as the "big black dog" as "the klabgigdod." However, comprehension is improved if the spoken words are separated by long intervals.

Many receptive aphasics can comprehend frequently used words but have difficulty with those less frequently heard. Thus, loss of comprehension is not an all-or-none phenomenon. They will usually have the most difficulty understanding relational or syntactical structures, including the use of verb tense, possessives, and prepositions. However, by speaking slowly and by emphasizing the pauses between each individual word, comprehension can be modestly improved.

FLUENT APHASIA

Patients with damage to Wernicke's area are still capable of talking (due to preservation of Broca's area). However, because Wernicke's area also acts to code linguistic stimuli for expression (prior to its transmission to Broca's area), expressive speech becomes severely abnormal, lacking in content, containing neologistic distortions (e.g., "the razgabin"), and/or being characterized by non sequiturs, literal (sound substitution) and verbal paraphasic (word substitution) errors, a paucity of nouns and verbs, and the omission of pauses and sentence endings (Christman, 1994; Goodglass and Kaplan, 1982; Kertesz, 1983a; Hecaen and Albert, 1978). Patients may speak in a rush (e.g., press of speech) and what is said often conveys very little actual information: a condition referred to as *fluent aphasia*.

Patients also may have difficulty establishing the boundaries of phonetic (confusing "love" for "glove") and semantic ("cigarette" for "ashtray") auditory information. The speech of these patients may also be characterized by long, seemingly complex (albeit, unintelligible) grammatically correct sentences, such that speech is often hyperfluent and produced at an excessive rate. They, thus, have difficulty bringing sentences to a close and many words are unintelligibly strung together.

They also suffer severe word finding difficulty. This adds a circumlocutory aspect to their speech, which can deteriorate into jargon aphasia such that no meaningful communication can be made. (Christman 1994; Kertesz, 1983a; Marcie and Hecaen, 1979).

For example, one patient with severe receptive aphasia responded in the following manner: "....Oh hear but that was a long time ago that was when that when before I even knew that much about this place although I am a little suspicious about what the hell is the part there is one part scares, uh estate spares, Ok that has a bunch of drives in it and a bunch of good googin, nothing real big but that was in the same time I coached them I said hey stay out of the spear struggle stay out of trouble so don't get and my kids, uh, except for the body the boys are pretty good although lately they have become winded or some-

thing...what the hell...kind of a platz goasted klack..."

Presumably, because the coding mechanisms involved in organizing what they are planning to *say* are the same mechanisms that decode what they *hear*, expressive as well as receptive speech becomes equally disrupted. In fact, one gauge of comprehension can be based on the amount of normalcy in their language use. That is, if they can repeat only a few words normally, it is likely that they can only comprehend a few words as well. Nevertheless, in testing for comprehension it is important to insure that the patient's major difficulty is not apraxia or agnosia, rather than aphasia (to be discussed).

In addition, like speech, the ability to write may be preserved, although what is written is usually completely unintelligible, consisting of jargon and neologistic distortions. Copying written material is possible, although it is also often contaminated by errors.

Anosognosia

Although their speech is bizarre, in severe cases patients with receptive "fluent" aphasia do not realize that what they say is meaningless (Maher et al., 1994). Moreover, they may fail to comprehend that what they hear is meaningless as well (Lebrun, 1987). Nor can you tell them, since they are unable to comprehend. This is because when Wernicke's area is damaged, there is no other region left to analyze the linguistic components of speech and language. The brain cannot be alerted to the patient's disability. They don't know that they don't know; that they don't understand. However, they may be somewhat more capable of recognizing that their writing is abnormal (Marcie and Hecaen, 1979).

Aphasia and Emotion

Patients with Wernicke's aphasia, in some instances, may display euphoria due to the disorganizing effect of comprehension loss on emotional functioning, and/or due to involvement of the temporal lobe limbic nuclei (e.g., amygdala). That is, the patient's right hemisphere continues to respond to signals generated by the left, even though they are abnormal.

In some cases, patients become paranoid as there remains a nonlinguistic emotional awareness that something is not right; that what they hear and what they observe does not mesh or make sense. That is, although they are aphasic, emotional functioning and affective comprehension may remain intact, though it is sometimes disrupted due to erroneously processed verbal input (Boller and Green, 1972; Boller et al., 1979). As such, they are exceedingly sensitive to social-emotional nuances, which they may interpret and respond to correctly. Similarly, the ability to read and write emotional words (as compared to nonemotional or abstract words) is also somewhat preserved among aphasics (Landis et al., 1982). This is because the right hemisphere is intact.

Since these paralinguistic and emotional features of language are analyzed by the intact right cerebral hemisphere, the aphasic individual is able to grasp, in general, the meaning or intent of a speaker, although verbal comprehension is reduced. This, in turn, enables the individual to react in a somewhat appropriate fashion when spoken to. Unfortunately, this also makes aphasic persons appear to comprehend much more than they are capable of.

Similarly, Wernicke's aphasics also retain the melodic, prosodic, and intonation contours of speech. In fact, at times their speech can become hypermelodic as well as characterized by inappropriate abnormal fluctuations in melodic contour. In addition, these patients often show normal gestural and facial expressions (Marci and Hecaen, 1979).

"Schizophrenia"

Because individuals with receptive aphasia display unusual speech, have lost comprehension, and fail to comprehend that they no longer comprehend or "make sense" when speaking, they are at risk for being misdiagnosed as psychotic or suffering from a formal thought disorder, i.e., "schizophrenia." Indeed, individuals with abnormal left temporal lobe functioning sometime behave and speak in a "schizophrenic-like" manner (see Chapter 14). That they may also behave in a euphoric and/or paranoid manner only increases the likelihood of a misdiagnosis.

According to Benson (1993:42), "those with Wernicke's aphasia often have no apparent physical or elementary neurological disability. Not infrequently, the individual who suddenly fails to comprehend spoken language and whose output is contaminated with jargon is diagnosed as psychotic. Patients with Wernicke's aphasia certainly inhabited some of the old lunatic asylums and probably are still being misplaced."

Temporal lobe abnormalities and the development of psychotic disturbances are further detailed in Chapters 14, 16, and 17.

Global Aphasia

Global aphasia is essentially a total aphasia due to massive left hemisphere damage involving the entire language axis, i.e. the frontal, parietal, and temporal convexity. Comprehension is severely reduced as is the ability to speak, read, write, or repeat. Patients are usually but not always (Legatt et al., 1987) paralyzed on the right side due to damage extending into the motor areas of the frontal lobe. Frequently, this disturbance is secondary to cerebrovascular disease involving the middle cerebral artery. However, tumors and head injuries can also create this condition.

Isolation of the Speech Area (or Transcortical Aphasia)

Isolation of the speech area is a condition where the cortical border zones surrounding the language axis have been destroyed due to occlusion of the tiny tertiary blood vessels that supply these regions (Geschwind et al., 1968). That is, the language axis of the left hemisphere becomes completely disconnected from surrounding cortical tissue but remains (presumably) an intact functional unit. This is in contrast to global aphasia, in which the three major zones of language have been destroyed.

Nevertheless, like global aphasia, the rest of the cerebrum is unable to communicate with the language zones. As such, an individual is unable to verbally describe what they see, feel, touch, or desire. Moreover, because the language axis cannot communicate with the rest of the brain, linguistic comprehension is largely abolished. That is, although communication between Wernicke's, Broca's, and the inferior parietal lobule is maintained, associations from other brain regions cannot reach the speech center. Although able to talk, the patient has nothing to say. Moreover, although able to see and hear, the patient is unable to linguistically understand what they perceive. However, they are capable of generating automatic-like responses to well-known phrases, prayers, or songs.

In an interesting case described by Geschwind et al., (1968:343–346), a 22-year-old woman with massive destruction of cortical tissue due to gas asphyxiation was found to have a preserved language axis. It was noted, once the patient regained "consciousness," that "she sang songs and repeated statements made by the physicians." However, she would follow no commands, resisted passive movements of her extremities, and would become markedly agitated and sometimes injured hospital personnel. In all other regards, however, she was completely without comprehension or the ability to communicate.

The patient's spontaneous speech was limited to a few stereotyped phrases, such as "Hi, daddy," "So can daddy," "Mother," or "Dirty bastard." She never uttered a sentence of propositional speech over the 9 years of observation. She never asked for anything, she never replied to questions, and she showed no evidence of having comprehended anything said to her. Occasionally, however, when the examiner said, "Ask me no questions," she would reply, "I'll tell you no lies," or when told, "Close

your eyes" she might say, "Go to sleep." When asked, "Is this a rose?" she might say, "Roses are red, violets are blue, sugar is sweet and so are you." To the word "coffee" she sometimes said, "I love coffee, I love tea, I love the girls and the girls love me."

An even more striking phenomenon was observed early in the patient's illness. She would sing along with songs or musical commercials heard over the radio or would recite prayers along with the priest during religious broadcasts. When a record of a familiar song was played, the patient would sing along with it. If the record was stopped, she would continue singing correctly both words and music for another few lines, and then stop. If the examiner kept humming the tune, the patient would continue singing the words to the end.

She could learn new songs as well, as evidenced by her ability to sing a few lines of the new song correctly after the record was stopped. Furthermore, she could sing two different sets of words to the same melody. For example, she could sing "Let me call you sweetheart" with the conventional words, but also learned the parody beginning with "Let me call you rummy." Her articulation of the sounds and her production of melody were correct, although she might sometimes substitute the words "dirty bastard" for some of the syllables.

Nevertheless, although such patients are able to sing, curse, and even pray, it is not clear if these expressions are the product of right hemisphere activity or reflexive activation of the intact language zones (or both). I favor the former rather than the later explanations. In some cases, the isolation is only partial, involving either the anterior or posterior regions. In these instances, the disorder is referred to as *transcortical motor* or *transcortical sensory aphasia*, respectively (see Hecaen & Albert, 1978 for greater detail).

Individuals with a partial transcortical aphasia have reportedly been able to read out loud, to write to dictation, and to repeat simple verbal statements. However, although repetition is somewhat preserved, spontaneous speech is severely limited. In sensory transcortical disturbances, comprehension is largely lost, whereas with motor transcortical aphasia there is a greater preservation of comprehension.

LANGUAGE AND TEMPORAL-SEQUENTIAL MOTOR CONTROL

As discussed in detail above and in Chapters 2 and 12, there is considerable evidence that the evolution of language and linguistic thought are related to and are, in part, an outgrowth of right hand temporal-sequential motor activity. The right hand appears to serve as a kind of motoric extension of language and thought in that it acts at the behest of linguistic impulses (Joseph, 1982). In fact, the hand and oral-facial musculature are neuronally represented in adjacent cortical space and are intimately interconnected with Broca's area, which lies immediately posterior-lateral and anterior to these motor areas. Because the hand and oral-articulatory neural structures occupy adjoining neocortical space, sometimes dual and simultaneous activation of these two modalities results in competitive interference.

For example, while speaking, the ability to simultaneously track, manually sequence, position, or maintain stabilization of the arms and hands is concurrently disrupted (Bathurst and Kee, 1994; Hicks, 1975; Lomas and Kimura, 1976; Kimura and Archibald, 1974; Kinsbourne and Cook, 1971). And as the phonetic difficulty of the verbalizations increase (Hicks, 1975), motor control decreases—a function, presumably, of simultaneous activation of (and thus competition for) the same neurons.

Conversely, when the left hemisphere is damaged, the right extremity may become paralyzed, the individual may be plagued by apraxic disorders (see below), and performance on problems involving not only language but temporal order are selectively impaired (Carmon and Nachshon, 1971; Efron, 1963; Haaland and Harrington, 1994; Kimura, 1993; Lackner and Teuber, 1973), as is nonverbal manual or oral performance (Mateer, 1983; Mateer and Kimura, 1976; Heilman et al., 1975; Kimura, 1993),

the copying of meaningless movements (Kimura and Archibald, 1974) rapidly of limb and pursuit-rotor movements (Heilman et al., 1975; Wyke, 1968), and the analysis of temporal speech sequences (Joseph, 1990e, 1993; Lenneberg, 1967).

Temporal-sequencing is, of course, a fundamental property of language, as demonstrated by the use of syntax and grammar. That is, syntax is a system of rules that govern the positioning of various lexical items and their interrelations to one another. This allows us to do more than merely name but to describe and to analyze how various parts and segments of speech interrelate. We can determine what comes first or last (e.g., "point to the door after you point to the window"), and what is the subject and object.

When the left hemisphere is damaged, particularly the anterior portions, expressive and receptive aspects of syntactical information processing suffer. In contrast, the right hemisphere has great difficulty utilizing syntactical or temporal-sequential rules. For example, if the isolated right hemisphere is given a command, such as "pick up the yellow triangle and then the blue star," it may pick up the yellow star and the blue triangle (Zaidel, 1977). It responds to the first object in the sentence regardless of grammatical relationship. Hence, we find that the ability to extract denotative meaning from language is dependent on the ability to organize and coordinate speech into temporal and interrelated units—an ability at which the left hemisphere excels, and an ability that is, at least in part, an outgrowth or a function of motoric processing and the predominant use of the right hand for gathering, tool making, food preparation, and related temporal and sequential activities.

NAMING, KNOWING, COUNTING, FINGER RECOGNITION, AND HAND CONTROL

Ontogenetically, it is first via the hand that one comes to know the world so that it may be named and identified. For example, the infant first uses the hand to grasp various objects so they may be placed in the mouth and orally explored. As the child develops, rather than mouthing, more reliance is placed solely on the hand (as well as the visual system), so that information may be gathered through touch and manipulation.

As the child and its brain matures, instead of predominantly touching, grasping, and holding, the fingers of the hand are used for pointing and then naming the object indicated. It is these same fingers that are later used in counting and in the development of temporal-sequential reasoning; i.e., the child learns to count on his or her fingers, then to count (or name) by pointing at objects in space.

In this regard, counting, naming, object identification, finger utilization, and hand control are ontogenetically linked. In fact, these capacities seem to rely on the same neural substrates for their expression: i.e., the left inferior parietal lobule. Hence, when the more posterior portions of the left hemisphere are damaged, naming (*anomia*), finger recognition (*finger agnosia*), object identification (*agnosia*), arithmetical abilities (*acalculia*), and temporal-sequential control over the hands and extremities (*apraxia*) are frequently compromised.

It is relationships such as these which lend considerable credence to the argument that over the course of evolution the predominant use of the right hand enabled the left brain to develop neurons specialized for counting and naming, and for subserving the development of the temporal-sequential properties necessary for the mediation of grammatical-syntactical speech and language.

AGNOSIA, APRAXIA, ACALCULIA, AND ORIENTATION IN SPACE

As described in Chapter 3, disturbances involving spatial-perceptual motor functioning can occur with lesions to either hemisphere. However, the nature of the disturbance (as well as the severity) differs depending on which half of the brain has been compromised. For example, right cerebral injuries usually have a more pronounced effect on visual-spatial and related perceptual abilities, and more greatly disrupt the overall perception and expression of configurational relationships (e.g., disruption of the overall gestalt), as demonstrated on drawing or constructional tasks.

Individuals with left-sided damage tend to preserve spatial relations but show a reduction in the number of parts represented. Moreover, left hemisphere lesions tend to more severely effect motoric aspects of spatial-perceptual functioning. As such, left-hemisphere–injured patients tend to recognize their errors.

Nevertheless, it sometimes occurs that patients not only fail to recognize their errors, but fail to even recognize whatever object they are attempting to examine or reproduce. This condition is referred to as *agnosia* (a term presumably coined by Sigmund Freud during his studies in neurology).

Finger Agnosia

Finger agnosia is not a form of finger blindness, as the name suggests. Nor is it an inability to recognize a finger as a finger. Rather, the difficulty involves naming and differentiating among the fingers of either hand as well as the hands of others (Gerstmann, 1930, 1942). This includes pointing to fingers named by the examiner, or moving or indicating a particular finger on one hand when the same finger is stimulated on the opposite hand.

In addition, if you touch the patient's finger while her eyes are closed, and ask her to touch the same finger, she may have difficulty. Often, patients who have difficulty identifying fingers by name or simply differentiating between them nonverbally also suffer from receptive language abnormalities (Sanguet et al., 1971). However, disturbances in differentiating between the different fingers can occur independently of language abnormalities or with right parietal injuries (in which case the problem is seen only with the patient's left hand).

Usually, however, finger agnosia is associated with left parietal (as well as temporal-occipital) lobe lesions, in which case the agnosia is demonstrable in both hands. It is also part of the constellation of symptoms often referred to as *Gerstmann's syndrome*: i.e., agraphia, acalculia, left–right disorientation, and finger agnosia (Gerstmann, 1942). Gerstmann's symptom complex is most often associated with lesions in the area of the supramarginal gyrus and superior parietal lobule (Hrbek, 1977; Strub and Geschwind, 1983).

Visual Agnosia

Visual agnosia is a condition in which the patient loses the capacity to visually recognize objects, although visual sensory functioning is largely normal. That is, objects are detected but the ability to evoke or assign meaning is lost and the object cannot be identified (Critchly, 1964; Shelton et al., 1994; Teuber, 1968). The precept becomes stripped of its meaning.

For example, if shown a comb, the patient might have no idea as to what it is or what it might be used for. However, he may be capable of giving a rough estimate of its size, proportions, and other features. In general, this disturbance is associated with damage involving the medial and deep mesial portion of the left occipital lobe. However, the lesion may extend into the left posterior temporal lobe.

Nevertheless, this is not a naming disorder. Individuals with anomia have diffi-

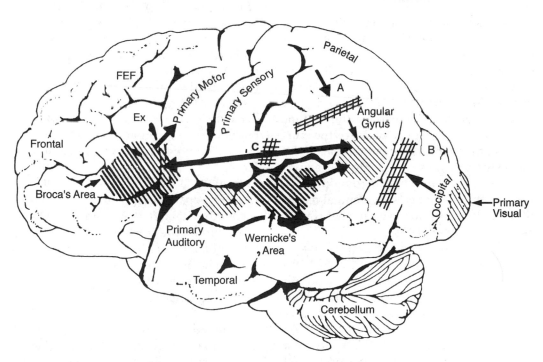

Figure 4.4 Schematic diagram of the Language Axis of the left hemisphere. Lesions disconnecting the angular gyrus of the inferior parietal lobe from **(A)** the superior parietal lobe results in an inability to name or verbally recognize objects held in the right hand. However, this same object can be named when it makes a characteristic sound or when the patient looks at it. Disconnection of the inferior parietal lobe from the visual cortex **(B)** results in an inability to name objects or written material that is viewed. However, the same object can be named if it is touched or if it makes a characteristic sound. Disconnection due to a small and deep lesion at **(C)** results in conduction aphasia. However, a massive lesion enveloping this same area and adjoining tissue results in global aphasia.

culty naming an object if it is presented by touch, by sound, via visual inspection, or through auditory description.

Individuals may have agnosia of multiple modalities or limited to a single modality (Davidoff and De Bleser, 1994; Shelton et al., 1994). If shown a comb, they may be unable to name or describe its use. However, if placed in their hand and encouraged to palpate and tactually explore it, they may be able to identify it without difficulty (Davidoff and Bleser, 1994). Conversely, patients who can recognize an item by sight but not by touch suffer from stereognostic abnormalities due to parietal lesions. Hence, anomia and agnosia can occur across modalities or may be limited to a single modality.

Simultanagnosia

Simultanagnosia is an inability to see more than one thing or one aspect of an object at a time. Although individual details are correctly perceived, the patient is unable to relate the different details so as to discern what is being viewed. For example, if shown a picture of a man holding an umbrella and a suitcase, he may see the suitcase and the man, and the umbrella, but be unable to relate these items into a meaningful whole. In fact, by surrounding the object with other objects, perceptual recognition may deteriorate even further (Luria, 1980; Shelton et al., 1994).

With severe damage, the patient may be unable to even recognize individual objects. Indeed, one patient I examined with bilateral posterior parietal-occipital damage (following stroke) could only identify parts of objects, but not the object itself. As described by Luria (1980), this is due to a breakdown in the ability to perform serial feature-by-feature visual analysis, and is sometimes accompanied by abnormal eye movements. Nevertheless, a variety of anatomical regions, when compromised, can give rise to this abnormality.

For example, simultanagnosia has been noted to occur with left, right, and bilateral superior occipital lobe lesions, or injuries involving the frontal eye fields (Kinsbourne and Warrington, 1962; Luria, 1980; Rizzo and Hurtig, 1987). Moreover, I have frequently observed patients with diffuse degenerative disturbances who demonstrate simultanagnostic deficiencies.

Right–Left Disorientation

Right–left disorientation (e.g., "show me your right hand") is usually associated with left hemisphere and left parietal-occipital damage. It occurs extremely rarely among individuals with right cerebral injuries (Gerstamann, 1930; McFie and Zangwill, 1960; Sanguet et al., 1971).

In general, these patients have difficulty differentiating between the right and left halves of their bodies or the bodies of others. This may be demonstrated in a number of ways, for example, by asking the patient: to touch or point to the side named by the examiner (e.g., "touch your left cheek," or "point to my right ear"); to point on his own body to the body part the examiner has pointed to on his or her body; or to perform crossed commands ("touch your left ear with your right hand"). In mild cases patients have difficulty only with the crossed commands (Strub and Geschwind, 1983).

In part, it seems somewhat odd that right–left spatial disorientation is more associated with left than right cerebral injuries, given the tremendous involvement the right half of the brain has in spatial synthesis and geometrical analysis. However, orientation to the *left* or *right* transcends geometric space as it relies on language. That is, "left" and "right" are designated by words and defined linguistically. In this regard, left and right become subordinated to language usage and organization (Luria, 1980). Hence, left–right confusion is strongly related to problems integrating spatial coordinates within a linguistic framework. It is perhaps for this reason that individuals with aphasic disorders generally perform the most poorly of all brain-damaged groups on left–right orientation tasks (Sauguet et al., 1971).

Acalculia

When an individual has difficulty working with numbers or performing arithmetical operations secondary to a brain injury, the manner and level at which the problem is expressed will differ depending on which part of the cerebrum has been compromised (e.g., Cohen and Dehaene, 1994; Joseph, 1990be). For example, an individual with a parietal-occipital lesion may suffer an *alexia/agnosia for numbers*, in which case she is unable to read or recognize different numerals. An individual with a posterior right cerebral injury may have difficulties with spatial-perceptual functioning and, thus, misalign numbers when adding or subtracting (referred to as *spatial acalculia*). Or she may suffer a very selective disconnection syndrome involving arithmetic long-term memory (Cohen and Dehaene, 1994). Hence, in many instances a patient may appear to have difficulty performing math problems when, in fact, the basic ability to calculate per se is intact. That is, the difficulty is due to spatial, linguistic, agnosic, or alexic abnormalities.

On the other hand, with left posterior lesions localized to the vicinity of the inferior parietal lobe, patients may have severe difficulty performing even simple calculations, e.g., carrying, stepwise computation, or borrowing (Boller and Grafman, 1983; Hecaen and Albert, 1978). When this occurs in the absence of alexia, aphasia, or visual-spatial abnormalities, and is accompanied by finger agnosia, agraphia, and right–left orientation, it is considered part of Gerstmann's syndrome (Gerstmann, 1930). This deficit is the purest form of acalculia and is due to an impairment in the ability to maintain order and to correctly plan in sequence; thus, there is a loss of the ability to appropriately manipulate numbers. Calculation disturbances will be discussed in greater detail in Chapter 12.

Apraxia

The left cerebral hemisphere has been shown to be superior to the right in the control of certain types of complex, sequenced motor acts (Haaland and Harrington, 1994; 1993; McDonald et al., 1994). If the left hemisphere is damaged, patients may be impaired in their ability to acquire or perform tasks involving sequential changes in the hand or upper musculature or those requiring skilled movement (Haaland and Harrington, 1994; Heilman et al., 1982; Kimura, 1979, 1980, 1993; McDonald et al., 1994). Indeed, the deficit may extend to well-learned and even stereotyped motor tasks, such as lighting a cigarette or using a key. This condition is referred to as *apraxia*.

Apraxia is a disorder of skilled movement in the absence of impaired motor functioning or paralysis (Kimura, 1993; Heilman et al., 1982; Heilman, 1993). Usually, apraxic patients show the correct intent but perform the movement in a clumsy fashion. Like patients with many other types of deficiencies following brain damage, apraxic patients and their families may not notice or complain of apraxic abnormalities. This is particularly true if they are aphasic or paralyzed on the right side. That is, clumsiness with either extremity may not seem significant. Hence, direct examination is necessary to rule out the presence of this problem, even in the absence of complaints.

Apraxia is usually mildest when objects are used, and performance deteriorates the most when patients are required to imitate or pantomime the correct action. For example, the patient may be asked to show the examiner "how you would use a key to open a door," or "hammer a nail into a piece of wood." In many cases, the patient may erroneously use the body, i.e., a finger, as an object (e.g., a key). That is, he may pretend his finger is the key (inserting it into a "key hole" and turning it) rather than have the finger and thumb hold an imaginary key. Although performance usually improves when they use actual objects (Kimura, 1993; Geschwind, 1965; Goodglass and Kaplan, 1982), a rare few may show the disturbance when using the real object as well (Heilman, 1993).

In addition, patients with apraxia may demonstrate difficulty properly sequencing their actions. For example, if you pretended to place a cigarette and matches in front of the patient and asked her to demonstrate, by pantomime, how she would light it and take a drag, the patient may pretend to hold up the match, blow it out, strike it, and then suck on the cigarette. That is, patients incorrectly sequence a series of acts. Individual acts, however, may be performed accurately.

Broadly speaking, there are several forms of apraxia, which, like many of the disturbances already discussed, may be due to a number of causes or anatomical lesions. These include *ideational apraxia, ideomotor apraxia, buccal facial apraxia,* and *dressing apraxia.* With the exception of dressing apraxia (which is due to a right cerebral injury), apraxic abnormalities are usually secondary to left hemisphere damage, in particular, injuries involving the left frontal and inferior parietal lobes (see Chapter 12).

The inferior parietal lobule IPL appears to be the central region of concern in regard to the performance of skilled temporal-sequential motor acts. This is because the motor engrams for performing these actions appear to be stored in the left angular and supramarginal gyri (Geschwind, 1965; Heilman, 1993; Kimura, 1993). These engrams assist in the programming of the motor frontal cortex where the actions are actually executed. If the inferior parietal region is destroyed, the patient loses the ability to appreciate when he or she has performed an action incorrectly. If the motor region is destroyed, although the act is still performed inaccurately (due to disconnection from the IPL), the patient is able to recognize the difference (Heilman et al., 1982; Heilman, 1993).

Thus, apraxic abnormalities secondary to left cerebral lesions tend to either involve destruction of the inferior parietal lobule or lesions resulting in disconnection of the frontal motor areas (or the right cerebral hemisphere) from this more posterior region of the brain. Apraxia will be discussed in more detail in Chapter 12.

Pantomime Recognition

Individuals with left cerebral and inferior parietal lobe lesions not only have difficulty performing motor acts but comprehending, recognizing, and discriminating between different types of motor acts performed and pantomimed by others (Bell, 1994; Corina et al., 1992; Heilman et al., 1982; Rothi et al., 1986; Wang and Goodglass, 1992). That is, in the extreme, if one were to pantomime the pouring of water into a glass versus lighting and smoking a cigarette, these individuals would have difficulty describing what was viewed or choosing which was which. Wang and Goodglass (1992) and Kimura (1993) argue that pantomime imitation and production are related to both apraxia and the ability to interpret purposeful movements, the engrams for which are located in the inferior parietal lobule (Heilman et al., 1982). However, with anterior lesions, although the ability to imitate may be impaired, the ability to comprehend pantomime is retained (Heilman et al., 1982).

Nevertheless, some scientists view pantomime and apraxic disorders as strongly related to language processing and aphasia (Bell, 1994; Corina et al., 1992; Kertesz et al., 1984). That is, if an individual cannot comprehend what is being asked, or what his own verbal intentions might be, he is going to fail to correctly perform associated movements and will fail to generate the correct verbal label so as to achieve linguistic comprehension as to what is being demonstrated via pantomime. Even so, most investigators are in agreement as to the common source of this deficit, the inferior parietal lobule (Heilman et al., 1982; Wang and Goodglass, 1992).

ALEXIA AND AGRAPHIA
The Evolution of Reading and Writing

There are a variety of theories that purport to explain the mechanisms involved in reading and the comprehension of written language. As detailed above, presumably, the evolution of these capacities is directly associated with the evolution of the inferior parietal lobule and is related to tool making, gathering, gesticulating, and gossiping for over 100,000 years.

The evolution of the inferior parietal lobule not only made language possible but provided the neurological foundations for tool design and construction, the ability to sew clothes, and the capacity to create art and pictorial language in the form of drawing, painting, sculpting, and engraving. Finally, it enabled human beings not only to create visual symbols but to create verbal ones in the form of written language (Joseph, 1993). In fact, with the exception of written language, all these abilities, including, perhaps, spoken language, may have appeared essentially within the same time frame, the first evidence of which is approximately 50,000 years old and was left in the deep recesses of caves and rock shelters throughout Europe, Africa, and perhaps Australia (see Chapter 2).

Although we have no recordings of what people may have sounded like, or any evidence regarding the presence of grammatical sequences in their speech, complex and detailed paintings left in the deep and forgotten recesses of ancient

Figure 4.5   Upper Paleolithic bow-and-arrow hunting scene. Drawn perhaps 15,000 years ago.

caves indicates that Homo sapiens sapiens of the Upper Paleolithic were capable of telling stories and making signs, which are still comprehensible and pregnant with meaning 40 millennia later.

However, it is likely that these peoples were speaking and sharing thoughts and desires since they first appeared on the scene, about 130,000 years ago. Although we can debate the merits and purposes for which these pictorial displays were created (magic, religious, instructive), they nevertheless represent the first form of written language. Body movements and gestures

Figure 4.6   A bison is wounded and impaled by arrows in this Upper Paleolithic cave drawing.

(From Préhistoire de l'Art Occidental Editions Citadelles & Mazenod. Photo by Jean Vertut.)

Figure 4.8   A lion is wounded and impaled by arrows. Detail from the "Great Lion Hunt."

Figure 4.7   Detail from the "Great Lion Hunt," depicting King Ashubanipal hunting lions with bow and arrow. Neo-Assyrian (Post-Babylonian). Carved approximately 3,000 years ago.

had become adapted for reproduction in the form of pictures, the result of well-crafted, delicate, and precise finger, hand, and arm movements.

The Evolution of Visual Symbols and Written Signs

When, where, how many times, and how many places the art of writing was invented, no one knows. The earliest preserved evidence detailing the evolution of written symbols comes from ancient Sumer, around 6000 or more years ago. In Sumer (a land where ancient Babylon would be established) as elsewhere, the first forms of "writing" were pictorial, a tradition that was already in use at least 20,000 years before this time, i.e., during the Upper Paleolithic.

Humans of the Upper Paleolithic, like modern westernized humans, initially utilized single pictures, such as a red hand to indicate a clear, readily understood message like "stop" or "do not enter." In most American cities, this same "hand" symbol is used at intersections to indicate if and when someone may cross a street.

Moreover, Upper Paleolithic humans also utilized abstract symbols, the meanings of which are not at all clear, though some scholars have associated them with sex or fertility, male versus female signs, and so on. However, these symbols may have actually been a primitive form of writing (see Figure 4.12).

However, the peoples of the Upper Paleolithic (e.g., Cro-Magnon) also liked to produce art for the sake of art. They were also the first people to paint and etch what today might be considered "cheesecake." That is, slim, shapely, naked, and nubile young maidens in various positions of repose (Joseph, 1993).

Like those of the Upper Paleolithic, Sumerians initially utilized single pictorial symbols to convey specific messages, such as a "lion" to indicate "watch out, lions around," or a "grazing gazelle" to inform others that good hunting could be found in the vicinity (Chiera, 1966; Kramer, 1981; Wooley, 1965). Although the Cro-Magnons sometimes used single two-step pictorial displays to indicate a sequence of action, as based on available evidence it was not until the time of the Sumerians and the Egyptians that people began to skillfully employ a series of pictures to represent, not just actions, but abstract as well as concrete ideas.

For example, prior to the Sumerians and the rise of the Egyptian civilization, a depiction of a lion or a man might indicate the creature itself, or the nature-God that was incarnate in its form. The Sumerians and Egyptians then took symbolism to its next evolutionary step: they strung these symbols together. For example, by drawing a "foot" or a "mouth" or an "eye," they could indicate the idea "to walk," "to eat,"

Figure 4.9   Victory tablet of an ancient Egyptian Pharaoh, possibly depicting the conquest of Sumerian territories some 6,500 or more years ago. Early stages of Egyptian "writing" utilized pictorial images organized in a combined temporal, linear, and gestalt fashion. At the top are depicted the heads of two bulls, which symbolize the power of the Pharaoh as well as possible gods. At top center the Pharaoh is marching in procession with his subjects and standard bearers and, in consequence, many men were decapitated. At bottom center two giant, long-necked leonine creatures have been captured (symbols of Sumerian royalty). Compare the long-necked, lion-headed beasts with Figure 4.10. At bottom, the Egyptian bull is besieging a city fortress.

and to "look" or "watch out." By combining these pictures they were able to indicate complex messages, such that "so and so" had to "to walk" to a certain place where "gazelles" might be found in order "to eat," but they would have to keep an "eye" out for "lions" (Chiera, 1966; Kramer, 1981; Wooley, 1965).

This was a tremendous leap, for earlier in Sumerian and Egyptian civilizations the indication of a complex message such as the above would have required elaborate pictorial detail of entire bodies engaged in particular actions. Although beautiful to behold, and easily understood by all, this was a very cumbersome and time-consuming process.

The next step in the evolution of writing was the depiction, not just of actions and ideas, but sounds and names. As argued by Edward Chiera (1966), if Sumerians wanted to write a common Sumerian name, such as "kuraka," they would draw objects that contained the sounds they wanted to depict, such as a "mountain," which was pronounced "kur," and "water," which was read as "a," and then a "mouth," which was pronounced "ka." By combining them together one was able to deduce the sound or name that the writer desired.

This was a tremendous leap in abstract thinking and in the creation of writing, for at this time visual symbols became associated with sound symbols; one could now not only look at pictures and know what was meant, but what the symbols sounded like as well. In this manner, writing, although still in pictorial form, became much more precise. Ideas, actions, and words,

Figure 4.10 Cylinder seals and impressions from the earliest Sumerian periods, 6,500 or more years ago. Note resemblance between the long-necked, long-tailed lions and those har- nessed by the victorious Egyptian soldiers in Figure 4.9. These ring cylinders were employed by Sumerian royalty as a form of signature.

Figure 4.11 Victory tablet of an ancient Egyptian Pharaoh, possibly depicting conquest of Sumerian territories (reverse of Figure 4.9) perhaps over 6,500 years ago. The victorious Pharaoh is ritually smiting a captured enemy. The hawk and rebus above the head of the captive basically reads, "Pharaoh, the incarnation of the hawk-god Horus, and his strong right arm...."

and, therefore, complex concepts could now be conveyed.

Nevertheless, although this was a highly efficient means of communication, it still remained quite cumbersome, which required that further steps in symbolic thought be invented. The Sumerians met this challenge by inventing wedge-form *cuneiform* characters, which gradually began to replace the older pictographs and ideographs (Chiera, 1966; Kramer, 1981; Wooley, 1965).

A B C D E F G H I
1
2
3
4
FEMALE
5
6
7
8
MALE
COUPLED

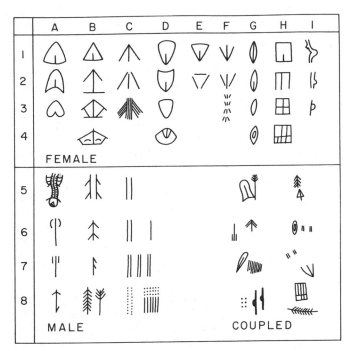

Figure 4.12 A tabulation of various "Cro-Magnon" (Upper Paleolithic) "signs" found in the caves of France. (From Clark G. The Stone Age Hunters. London: Thames & Hudson, 1967.)

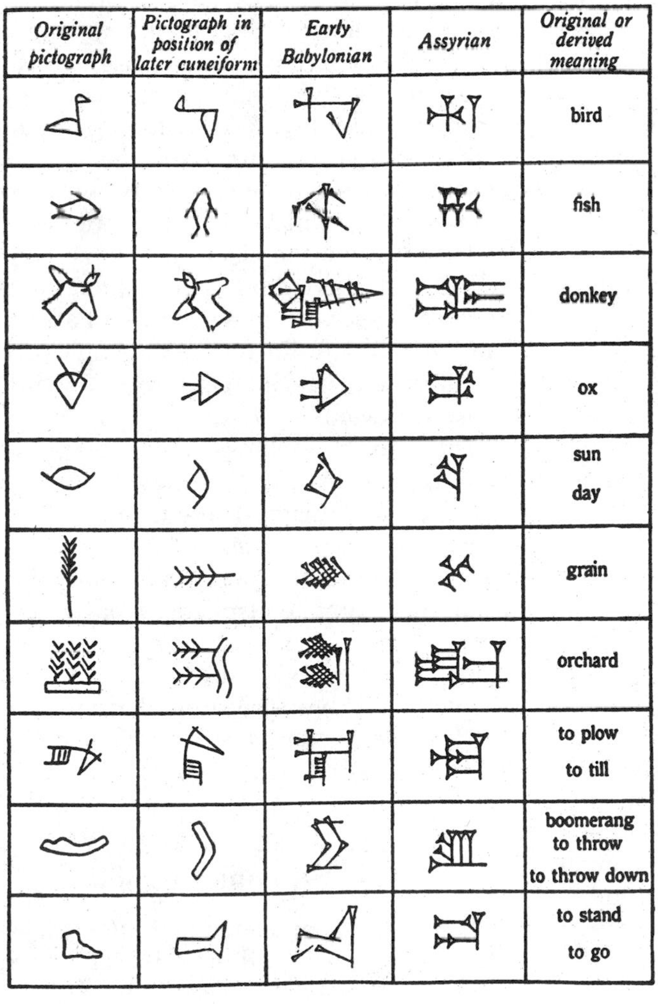

Original pictograph
Pictograph in position of later cuneiform
Early Babylonian
Assyrian
Original or derived meaning
bird
fish
donkey
ox
sun
day
grain
orchard
to plow
to till
boomerang
to throw
to throw down
to stand
to go

Figure 4.13 The origin and evolution of writing from pictures employed by the peoples of Sumer and then Babylon to develop cuneiform characters. The people of Sumer realized that pictures could represent, not only images and ideas, but sounds, which gave their writing greater precision in meaning. (From Chiera E. They wrote on clay. Chicago: University of Chicago Press, 1966.)

Initially these characters resembled the pictures they were destined to replace. However, as the use of cuneiform continued to evolve, pictorial details were gradually minimized, until finally these characters lost all pictorial relationship to that which they were meant to describe. Among the ancient Egyptians a similar process occurred, with the exception that with the invention of hieroglyphics, pictures remained an essential feature of their writing until the very end of their civilization about 2,000 years ago.

The Sumerians, Babylonians, and the later-appearing Assyrians (all of whom lived in what is today Iraq), although able to depict vowels, consonants, and complex ideas and sounds via cuneiform, did not develop an alphabet (Chiera, 1966). In fact, even when they were introduced to foreign-devised alphabets, they resisted this innovation. Even so, pictures came to be placed in temporal sequences, then the pictures were transformed and their transformed images became represented by sequences as well: a series of wedge-shaped lines that were read from left to right.

Be it the writing of the Egyptians, the Sumerians, or modern-day Americans, the process of reading and writing remains essentially similar. Both require the interaction of brain areas involved in visual and auditory analyses, as well as the evolution of the inferior parietal lobule, which enabled visual signals to be matched with sounds so that auditory equivalents could be conjured up. In this manner people are able to not only look at a visual symbol, but to recognize it as a "word" and know what it *sounds* like as well.

ALEXIA
Reading Abnormalities

From a neurodynamic perspective, the process of reading involves the reception of visual impulses in the primary visual receiving areas where various forms of perceptual analyses are initiated. This information is then transferred to the visual association cortex where higher-level information pro-cessing is carried out and visual associations formed. These visual associations are next transmitted to a variety of areas (see Chapters 12 and 14), including the inferior and middle temporal lobe, inferior parietal lobule, and Wernicke's area.

It is in these latter cortical regions where multimodal and linguistic assimilation take place so that the auditory equivalent of the visual stimulus may be retrieved. That is, via these interactions visual grapheme clusters become translated into phonological sound images (see Barron, 1980; Ellis, 1982). In this manner, we know what a written word looks and sounds like. It is also possible, however, to bypass this phonological transcoding phase so that word meanings can be directly accessed (i.e., lexical reading).

Although it is likely that most individuals utilize both lexical and phonological strategies when reading, in either case the temporal-parietal cortex (Rumsey et al., 1992) and the lingual, fusiform, and angular gyrus are involved (Gloning et al., 1968). With lesions involving the angular gyrus, or when damage occurs between the fiber pathways linking the left inferior parietal lobule with the visual cortex (i.e., disconnection), a condition referred to as *pure word blindness* sometimes occurs. Patients can see without difficulty but are unable to recognize written language. Written words evoke no meaning.

There are, however, several subtypes of reading disturbances that may occur with left cerebral damage. These include *literal, verbal,* and *global alexia,* and *alexia for sentences.* In addition, alexia can sometimes result from right hemisphere lesions, a condition referred to as *spatial alexia.* All these disorders, however, are acquired and should be distinguished from developmental dyslexia, which is present since childhood (see Njiokiktjien, 1988).

Pure Word Blindness:
Alexia Without Agraphia

Pure alexia, or *alexia without agraphia* (due to the preservation of the ability to write), is a condition in which patients are unable to read written words, or even recognize single letters. However, if words are spelled out loud, they have little difficulty with comprehension (due to intact pathways from Wernicke's to the angular gyrus). Moreover, they are able to speak and spell as well as write without difficulty. Nevertheless, although able to write, they are unable to read what they have written.

In general, the lesion appears to be between the left angular gyrus and the occipital lobe (in the arterial distribution of the left posterior cerebral artery), and extends to within the splenium of the corpus callosum (Benson, 1979; Vignolo, 1983); this prevents right hemisphere visual input from being transferred to the inferior parietal lobe of the left half of the brain.

Sometimes, this condition is due to a tumor or a head injury accompanied by a hematoma involving the white matter underlying the inferior parietal lobule (Greenblatt, 1983) and/or the temporal-parietal cortex (Rumsey et al., 1992). In these instances, the left angular gyrus is unable to receive visual input from the left and right visual cortex, and visual input cannot be linguistically translated. The patient cannot gain access to the auditory equivalent of a written word.

Although the patient is unable to read written language, this syndrome is not always accompanied by a visual field defect (hemianopsia). Moreover, objects may be named correctly (Hecaen and Kremin, 1976). However, patients may suffer from color agnosia (Benson and Geschwind, 1969), i.e., an inability to correctly name colors. There is often (but not always) an inability to copy written material (because of disconnection from the visual areas), and many patients have difficulty performing math problems. In some cases, the patient suffers not only pure word blindness but blindness for numbers as well.

This condition has been described as *global alexia* by some authors. However, global alexia is a more pervasive disorder in which the ability to write (agraphia) and name objects is also compromised.

Alexia for Sentences

Patients suffering from alexia for sentences are able to read letters and single words but are unable to comprehend sentences. However, patients may have difficulty with unfamiliar or particularly long words,however, easily understand more familiar material. Lesions are usually localized within the dominant occipital lobe but may extend into the inferior parietal area.

Verbal Alexia

When patients are able to read and recognize letters but are unable to comprehend or recognize whole words they are said to be suffering from verbal alexia. However, if presented with short words their reading ability improves. Hence, the longer the word, the greater the difficulty reading (Hecaen and Albert, 1978). Verbal alexia usually results from lesions involving the dominant medial occipital lobe (Hecaen and Albert, 1978; Hecaen and Kremin, 1977).

Literal/Frontal Alexia

When the patient is unable to recognize or read letters, this is referred to as *pure letter blindness* and *literal alexia*. Patients are usually unable to read by spelling a word out loud, and the ability to read numbers and even musical notes is often disturbed. Some authors have attributed literal alexia to left inferior occipital lesions (Hecaen and Kremin, 1977).

This disorder has also been referred to as *frontal alexia*. This is because individuals with Broca's aphasia or lesions involving the left frontal convexity have difficulty reading aloud, and have the most difficulty with single letters rather than whole words (Benson, 1977).

Spatial Alexia

Spatial alexia is associated predominantly with right hemisphere lesions. This disorder is due, in part, to visual-spatial abnormalities, including neglect and inattention. That is, with right cerebral lesions the patient may fail to read the left half of words or sentences, and may, in fact, fail to perceive or respond to the entire left half of a written page. Right posterior injuries may also give rise to spatial disorientation such that the patient is unable to properly visually track and keep place, his eyes darting haphazardly across the page, or, for example, skipping to the wrong line. Spatial alexia may also result from left cerebral injury, in which case the right half of letters, words, and sentences is ignored (Marshall and Newcombe, 1966).

AGRAPHIA

Just as one theory of reading proposes that visual graphemes are converted into phonological units (visual images into sound), it has been proposed that in writing one transcodes speech sounds into grapheme clusters: i.e., a phoneme to grapheme conversion (Barron, 1980; Ellis, 1982; Friederici et al., 1981). In the lexical route, there is no phonological step. Instead, the entire word is merely retrieved.

Regardless of which theory one adheres to, it appears that the angular gyrus plays an important role in writing. Presumably, the angular gyrus provides the word images (probably via interaction with Wernicke's area) that are to be converted to graphemes. These representations are then transmitted to the left frontal convexity (i.e., Broca's and Exner's areas) for grapheme conversion and motoric expression. However, it is possible that the initial graphemic representations are formed in the inferior parietal lobule (Roeltgen et al., 1983).

Nevertheless, it appears that there are at least two stages involved in the act of writing—a linguistic stage and a motor-expressive-praxic stage. The linguistic stage involves the encoding of information into syntactical-lexical units (Goldstein, 1948). This is mediated through the angular gyrus, which thus provides the linguistic rules that subserve writing.

The motor stage is the final step in which the expression of graphemes is subserved. This stage is mediated presumably by Exner's writing area (located in the left frontal convexity) in conjunction with the inferior parietal lobule. Hence, disturbances involving the ability to write can occur due to disruptions at various levels of processing and expression and may arise secondary to lesions involving the left frontal or inferior parietal cortices.

Thus, similar to alexia, there are several subtypes of agraphia that may become manifest, depending upon which level and anatomical region is compromised. These include *frontal agraphia, pure agraphia, alexic agraphia, apraxic agraphia,* and *spatial agraphia.*

Exner's Writing Area

Within a small area along the lateral convexity of the left frontal region lies Exner's writing area (see Hecaen and Albert, 1978). Exner's area appears to be the final common pathway where linguistic impulses receive a final motoric stamp for the purposes of writing. Exner's area, however, also seems to be very dependent on Broca's area, with which is maintains extensive interconnections. That is, Broca's area acts to organize impulses received from the posterior language zones and relays them to Exner's area for the purposes of written expression.

Lesions localized to this vicinity have been reported to result in disturbances in the elementary motoric aspects of writing, i.e., *frontal agraphia* (Penfield and Roberts, 1959), sometimes referred to as *pure agraphia*. However, pure agraphia has also been attributed to left parietal lesions (Basso et al., 1978; Strub and Geschwind, 1983). In general, with frontal agraphia, grapheme formation becomes labored, incoordinated, and takes on a very sloppy appearance. Cursive handwriting is usually more disturbed than printing.

The ability to spell, per se, may or may not be affected, whereas with parietal lesions spelling is often abnormal. Rather, there are disturbances in grapheme selection and patients may seem to have "forgotten" how to form certain letters, and/or they may abnormally sequence or even add unnecessary letters when writing (see Hecaen and Albert, 1978). With severe damage, the left hand is more affected than the right (due to disconnection from the left hemisphere language axis). However, with either hand patients may not be able to write or spell correctly, even if given block letters.

In addition, frontal agraphia can result from lesions involving Broca's area. These individuals are unable to write spontaneously or to dictation, and when they do write their samples may be contaminated by perseverations or the addition of extra strokes to letters (e.g., such as when writing an "m"). Often, such patients are able to write their names or other well-learned (automatic) items without difficulty. Hence, when tested they should be required to write spontaneously as well as to dictation.

Pure Agraphia

In pure agraphia, patients have difficulty in the motor control of grapheme selection and production, frequently misspell words, and may insert the wrong letters or place them in the wrong order or sequence when attempting to write (Marcie and Hecaen, 1979). In contrast, reading, oral speech, and the ability to name objects or letters are usually unimpaired.

Most commonly, pure agraphia is associated with lesions involving the superior and mid parietal regions of the left hemisphere—areas 5 and 7 (Basso, et al., 1978; Vignolo, 1983), and/or the inferior parietal region (Strub and Geschwind, 1983). Patients are unable to write because the area involved in organizing visual-letter organization (i.e., the inferior parietal lobe) is cut off from the region controlling hand movements in the frontal lobe (Strub and Geschwind, 1983). However, according to Vignolo (1983), the posterior superior left parietal lobule (area 7) is crucial for the sensorimotor linguistic integration needed for writing.

Alexic Agraphia

Alexic agraphia is a disturbance involving the ability to read and to write. It is a disruption of both the ability to decode and to encode written language. However, reading and writing may not be equally affected. Usually, recognition of letters is better than words, and the longer the word, the greater is the difficulty with recognition.

Patients have severe difficulty forming graphemes and spelling, even when block letters are employed (Marcie and Heilman, 1979). Although words are misspelled, syntactical sequencing is preserved such that what is written remains grammatically correct. Nevertheless, patients often display some degree of anomia and apraxia (due to inferior parietal involvement). Verbal-auditory comprehension is generally intact, as is expressive speech.

Often, alexia with agraphia is due to a lesion involving the left inferior parietal lobule and angular gyrus (Benson and Geschwind, 1969; Hecaen and Kremin, 1976), and for this reason has also been called *parietal alexia*. Hence, the destruction of the angular gyrus results in a massive disconnection such that the auditory areas cannot completely communicate with the visual areas. Nor can the frontal areas interact with the inferior parietal lobe.

There are, thus, two separate disconnections. Patients are unable to correctly perceive visual-linguistic symbols and are unable to reproduce them; i.e., they cannot read or write.

Apraxis Agraphia

It has been argued that the sensory motor engrams necessary for the production and perception of written language are stored within the inferior parietal lobule of the left hemisphere (Strub and Geschwind, 1983). When this region is damaged, patients sometimes have difficulty forming letters due to an inability to access these engrams (Strub and Geschwind, 1983). Writing samples are characterized by misspellings, letter omissions, distortions, temporal-sequential misplacements, and inversions (Kinsbourne and Warrington, 1964).

This type of agraphia is often seen in conjunction with *Gerstmann's syndrome*. Because the patient is also apraxic, the ability to make gestures or complex patterned movements, such as those involved in writing, is also deficient. That is, the ability to correctly temporally sequence hand movements, independent of writing, is affected (see section on apraxia). The patient no longer knows how to correctly hold or manipulate a pen, or how to move her hand when writing.

Spatial Agraphia

Right cerebral injuries can secondarily disrupt writing skills due to generalized spatial and constructional deficiencies. Hence, words and letters will not be properly formed and aligned, even when they are copied. There may be difficulty keeping lines straight, and letters may be slanted at abnormal angles. In some cases, the writing may be reduced to an illegible scrawl.

In addition, patients may write only on the right half of the paper such that as they write, the left-hand margin becomes progressively larger and the right side smaller (Hecaen and Marie, 1979). If allowed to continue, patients may end up writing only along the edge of the right-hand margin of the paper.

Patients with right hemisphere lesions may tend to abnormally segment the letters in words when writing cursively (i.e., "cur siv e ly"). This is due to a failure to perform closure as well as a release over the left hemisphere (i.e., left hemisphere release). That is, the left hemisphere, acting unopposed, begins to abnormally temporally-sequence and thus produce segments unnecessarily.

Also of note, Hecaen and Marcie (1979) report that the insertion of gaps or blanks into words can also occur following left hemisphere lesions.

Aphasia and Agraphia

Although every patient with agraphia does not necessarily suffer from aphasic abnormalities, every patient who has aphasia has some degree of agraphia. In some instances, the aphasia may be mild and the agraphic disturbance may be severe or, conversely, quite subtle and only revealed through special testing. In general, it is best to require the patient to write cursively rather than to print. Moreover, the more complex the writing task, the more likely will an agraphic disturbance be demonstrated. Hence, in testing one should not be satisfied with the writing of simple words or sentences if agraphia is truly to be ruled out.

Sex Differences

It is well known that males are far more susceptible to developing dyslexic and agraphic disorders. In addition, girls learn how to read more quickly and more proficiently than males, who in turn suffer more difficulties learning how to read (Corballis and Beale, 1983). By age 9, girls score higher than boys in reading comprehension and demonstrate superior writing and spelling skills (Lewis and Hoover, 1983). In fact, according to the 1993 report of the US Department of Education, the writing skills of a 9-year-old girl in 4th grade is equal to that of a 13-year-old boy in 8th grade.

Presumably (as described above and in Chapter 2), these sex differences are related to the differential activities engaged in by males and females for the last 100,000 years of human evolution—activities that conferred upon females language-related neuroanatomical and neuropsychological superiorities. However, as detailed in Chapters 2 and 7, it was also the evolution and elaboration of the anterior cingulate gyrus and the lateral amygdala that not only contributed to the female advantage in vocalizing capabilities, but which provided the motive source as well as the neocortical origins for the development of language, and for the desire to communicate in the first place.

SECTION III

The Limbic System

5

The Limbic System

Sex, Emotion, Emotional Intelligence, Pheromones, Sexual Differentiation, Attention, Memory, the Pleasure Principle, and Primary Process

Buried within the depths of the cerebrum are several large aggregates of limbic nuclei that are preeminent in the control and mediation of memory, emotion, learning, dreaming, attention, and arousal, and in the perception and expression of emotional, motivational, sexual, and social behavior, including the formation of loving attachments. Indeed, the limbic system not only controls the capacity to experience love and sorrow, but it governs and monitors internal homeostasis and basic needs, such as hunger and thirst (Bernardis and Bellinger, 1987; Joseph, 1992a, 1993, 1994; Gloor, 1992; LeDoux, 1992; MacLean, 1973, 1990; Rolls, 1984, 1992; Smith et al., 1990; Squire, 1992).

In general, the primary nuclei of the limbic system include the hypothalamus, amygdala, hippocampus, septal nuclei, and anterior cingulate gyrus. However, also implicated in the functioning of the limbic system are the limbic striatum (nucleus accumbens, ventral caudate and putamen), the orbital frontal and inferior temporal lobes, the midbrain monoamine system, and portions of the cerebellum.

However, as noted in Chapter 1, the orbital and inferior temporal lobes are, in fact, evolutionary derivatives of the limbic system, as is much of the neocortex. Specifically, it appears that the anterior cingulate and the amygdala have contributed to the development of the orbital frontal lobes. The cingulate also gave rise to much of the supplementary motor area and medial frontal lobes, whereas the hippocampus contributed to the evolution of the posterior cingulate, and portions of the parietal and temporal lobes. Similarly, the amygdala also gave rise to the inferior and superior temporal lobe as well as the auditory neocortex (in conjunction with the inferior colliculus and medial thalamus) and visual neocortex (in conjunction with the hippocampus, superior colliculus, and lateral thalamus).

Thus, much of the neocortex is an evolutionary derivative of the limbic system, with secondary contributions from the thalamus, and with portions of the midbrain possibly contributing layers I and VII (6b). The neocortex, therefore, is concerned with performing hierarchical analysis of stimuli, which was formerly the exclusive domain and concern of the limbic system (as well as the midbrain and thalamus). Once processed at the level of the neocortex, this information may then be shunted back to the thalamus and limbic system, as well as to other neocortical structures and/or the brainstem and spinal cord for motoric expression.

Therefore, whereas the limbic system remains largely concerned with emotion, motivation, attention, learning, and memory, and visual, sexual, tactile, and gross motor activities, so too is the neocortex. However, in contrast to the limbic system, what some have referred to as "the rational mind" is believed to be the exclusive domain and province of the new brain (neocortex).

The Rational and Irrational Mind

Although, over the course of evolution, a new brain (neocortex) has developed,

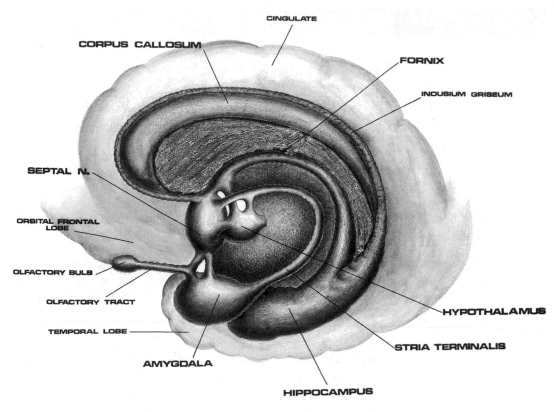

Figure 5.1   The limbic system of the left hemisphere.

Homo sapiens sapiens ("the wise man who knows he is wise") nevertheless remain creatures of emotion. We have not completely emerged from the phylogenetic swamps of our original psychic existence.

The old limbic brain has not been replaced, and is not only predominant in regard to all aspects of motivational and emotional functioning but is capable of completely overwhelming "the rational mind." Hence, humans not uncommonly behave "irrationally" or in the "heat of passion," and, thus, act at the behest of their immediate desires—sometimes falling "madly in love" and at other times acting in a blind rage such that even those who are "loved" may be murdered.

Indeed, emotion is a potentially powerful overwhelming force that warrants and yet resists control—as something irrational that can happen to someone ("you *make* me so angry") and that can temporarily snuff out or hijack the "rational mind." Indeed,

this schism between the rational and the emotional is real, and is due to the raw energy of emotion having its source in the nuclei of the ancient limbic lobe: a series of nuclei that first make their phylogenetic appearance long before humans walked upon this earth and that continue to control and direct human behavior.

AFFECTIVE ORIGINS: OLFACTION AND PHEROMONAL COMMUNICATION

Emotionality serves a protective function, either to promote survival of the individual (e.g., fight or flight) or of the species (e.g., sexual activity). The first and most primitive manifestations of emotion are elicited in response to olfactory (e.g., pheromones and other externally secreted chemical messengers) and tactile-vestibular sensory stimulation, which, in turn, are transmitted to the olfactory-limbic system

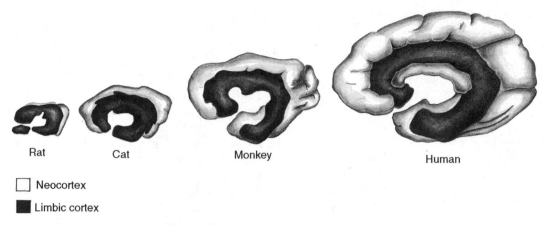

Rat   Cat   Monkey   Human

☐ Neocortex

■ Limbic cortex

Figure 5.2   The limbic system across four species. Not drawn to scale.

and the brainstem vestibular system, which also shunts this data to limbic nuclei.

Being embedded in the functioning of he limbic system, these initial, primitive emotional reactions are, therefore, expressed as withdrawal, "freezing," or aggressive reactions to pain or threat; they may also be expressed as approach or receptive responses to potential sexual partners or food (prey) and nourishment; i.e., fleeing, fighting, feeding, and fornicating.

Olfaction and Memory

Olfaction was originally extremely important in evolutionary and phylogenetic development, as it allowed the organism to be informed about the environment at a distance, and without the necessity of physical contact (Haberly, 1990). Via olfactory cues and the detection of pheromones, the organism was able to detect and track food, determine the intent and social-emotional status of conspecies, as well as signal its own intent, motivation, social position, and/or sexual availability (Ackerman, 1990; Blum, 1985; Joseph, 1993; Mayer and Mankin, 1985; Michael and Keverne, 1974; Tamaki, 1985; Wilson, 1980).

Moreover, through the olfactory system the creature was able to learn and remember these attributes, including their location in space; and this was made possible via the evolution of the amygdala and, in particular, the dorsal hippocampus, a nucleus derived from olfactory epithelium (Haberly, 1990). That is, initially the dorsal hippocampus was probably exceedingly responsive to light and may well have been responsible for computing the creature's position, as well as that of other creatures and objects, in visual-space—which is a function that it currently subserves in mammals; i.e., the analysis of place and position (Nadel, 1991; O'Keefe, 1976; O'Keefe and Nadel, 1974; Wilson and McNaughton, 1993).

However, long before the modern hippocampus gave rise to neocortex and shifted from a dorsal to a ventral position in humans, the primitive and ancient hippocampus (in conjunction with the amygdala) probably evolved the capacity to cross-correlate and match olfactory/pheromonal cues with visual (and other) cues. This created a spatial-olfactory map of the environment and provided the means to store and later recall that information when necessary (see Chapter 1). Learning, memory, and cognitive-spatial positional analysis are also attributes characteristic of the modern human amygdala and hippocampus, whereas the modern limbic system still remains responsive to pheromonal chemical signals.

For example, in humans, various odors affect learning and memory and can trigger vivid recollections of some far away and past event (Hirsh, 1993). Moreover, specific "positive" odors (e.g., peppermint) have been reported to improve learning, creativity, and work efficiency (Baron, 1990; Ehrlichman and Bastone, 1992) and even influence time spent gambling (Hirsh, 1993).

Conversely, individuals suffering from severe memory loss and Alzheimer's disease (which is associated with damage involving the substantia innominata, part of the limbic striatum) tend to suffer from severe odor detection deficits (Doty et al., 1987). Indeed, loss of smell may be one of the earliest indicators of Alzheimer's disease onset.

PHEROMONES

Pheromones are chemical substances that are secreted by the skin through specialized glands, and which can also be found in urine and feces. They are, likewise, perceived by specialized olfactory re-ceptors and are then transmitted to the olfactory bulb (or lobe is some species) and the olfactory-limbic system for analysis.

Chemical communication via pheromones is utilized by moths, social insects, dogs, cats, primates, and humans, as well as amphibians, sharks, and reptiles. Although among insects detection of pheromones is often accomplished via specialized chemoreceptors located on various parts of the body (Mayer and Mankin, 1985; Wilson, 1980), sharks, mammals, and primates rely upon olfactory receptors that are located in or along the snout; once received, this information is transmitted to the olfactory bulb and the telencephalon (Graeber, 1980; Haberly, 1990; Savage, 1980).

Among mammals as well as insects, olfactory cues are utilized for detecting food and potential mates, and for marking one's possessions and territory. For example, dogs will urinate on trees and bushes, whereas a stallion might urinate on the feces of his mare. Prosimian primates will

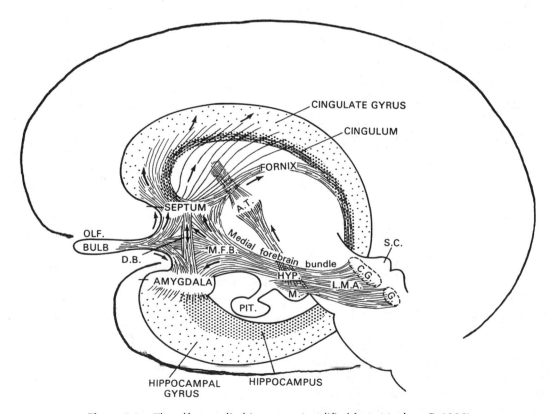

Figure 5.3   The olfactory-limbic system (modified from Maclean P, 1990).

urinate on their hands, rub the secretions over their body, thus marking everything they come into contact with, including their mates.

Conversely, male primates (as well as other mammals) rendered anosmic (via the cutting of the olfactory nerves) completely lose interest in sexually available females due to the inability to detect olfactory (pheromone) cues indicating sexual readiness (Bronson, 1968; Michael and Keverne, 1974). Interest in and the ability to enjoy food is also diminished.

Pheromones are produced by males and females and do much more than signal sexual availability or the potential tastiness of some edible treat. In a variety of species, pheromones indicate social status, aggressiveness, fear, threat, alarm, and the need to aggregate or disperse (Blum, 1985; Bronson, 1968; Wilson, 1971, 1980). In addition, there are pheromones that can stimulate mating, and others that inhibit it. There are those that attract males to a willing female, and others that are released by the first lucky consort, which has the effect of driving away other suitors (e.g., Cade, 1985; Mayer and Mankin, 1985; Tamaki, 1985).

Different pheromones are also employed to stimulate grooming, the disposal of the dead, regurgitative feeding, and emigration. Other chemical combinations promote individual recognition, determine caste assignment, inhibit ovarian function, promote the construction of enclosures where insect queens might be reared, and mark food sources.

Olfactory Blends: A Symphony of Smells

Various odors are comprised of a multitude of chemical agents. Indeed, some so-called connoisseurs have likened smell, particularly that of perfumes, to a symphony, for to make a perfume requires multiple blends that must harmonize together so as to form a particular, pleasing impression (Ackerman, 1990).

Similarly, when an insect or mammal releases a pheromone, it is not composed of a single olfactory chemical. Rather, like high-grade gasolines and premium whiskies, it is composed of blends of many different chemical entities (Cade, 1985; Mayer and Mankin, 1985; Tamaki, 1985). Thus, the release and perception of a pheromone can, therefore, exert a number of different influences because a variety of chemicals have been transmitted and received.

When these pheromone blends are perceived, a different pattern and number of dendrites may fire, depending on which chemicals are received. That is, if a blend of four different pheromone molecules is received on one occasion, and a blend of three different agents on another, and yet a blend of all seven agents on yet another occasion, a minimum of three different patterns of excitation and three different types of behavior will result, depending on the composition; it is as if these molecules were punching in a dendritic code. It is via these different patterns and codes that the identity of certain individual odors is indicated and the appropriate type of behavioral response is determined.

Since there are many different receptors that are sensitive to different pheromonal chemicals, once contact is made, different behaviors can then be triggered. Hence, a considerable degree of complex communication is possible as a number of different messages can be transmitted. However, only a fraction of whatever has been released is ever perceived, if at all.

Pheromonal-Olfactory Evolution/Involution

The ability to detect externally originating chemicals, i.e., pheromones, was initially made possible by specialized receptors that were located on external cells. Later, these specialized receptors became situated on various parts of the organism's body, such as within the abdomen or along the legs (Blum, 1985; Bruce, 1979; Mayer and Mankin, 1985; Wilson, 1980). These receptors enabled many ancient (and modern-day) species to communicate and sense pheromonal/chemical alterations in their environment—information that was transmitted to the olfactory-limbic system for analysis.

Although humans and other mammals no longer appear to have the ability to perceive pheromones or odors via the body surface, other than through the nose, the capacity to excrete these chemicals through the skin has been retained. Humans possess, for example, scent-producing glands, some of which are referred to as ancillary organs.

Ancillary organs are located under the arms and are maximally developed among females. Their purposes is presumed to be for sexual and social communication via the secretion of various scents, like a flower. Other mammalian scent glands are located in the anal and genital region and beside the teats, which enable a baby to recognize its mother and the source of its sustenance, so as to orient in order to feed.

Over the course of evolution, although these olfactory/pheromonal receptors have remained externally located, i.e., within the nose and snout, the actual cell bodies have migrated internally, some of which eventually formed ganglia, then the olfactory lobes, and then the forebrain (Haberly, 1990; Smeets, 1990; see Chapter 1).

Specifically, among humans and mammals, these tiny airborne chemical molecules are drawn into the nasal cavity where they make contact with microscopic hairs called cilia, which are located throughout the nasal mucosa. It is here where the initial dendritic receptor is located. The dendrite then leads to cell bodies located in the mucous-covered olfactory epithelium of the inner nose.

Axons arising from these nasal nerve cells collectively form the olfactory nerve, which terminates in the olfactory bulb. Within the olfactory bulb there is a considerable degree of convergence, with axons from many different cells forming synapses on the dendrite of a single neuron. This receiving neuron, in turn, acts to analyze and integrate the many messages received.

In some animals, particularly those where smell is essential to sex, social relations, and the detection of prey or predators, the olfactory bulb is quite large and is considered a lobe of the brain. For example, dogs, anteaters, sharks, and bottom-feeding fish possess large olfactory bulbs. By contrast, birds have a very poorly developed olfactory bulb, and some species of whales and porpoises have none, although they still possess an olfactory system (see Haberly, 1990 for related discussion).

In addition to species differences in olfactory-pheromonal sensitivity, there are sex differences. For example, among humans, men are supposedly not as sensitive to smell as are women (Doty et al., 1985). Moreover, female olfactory sensitivity is greatest at the time they ovulate (Mair et al., 1977), although exactly why that might be is not known. Nevertheless, given woman's historical and ancestral role in gathering, food preparation, and even the creation of herbal remedies, it might be expected that females would develop a more refined sense of olfactory perception; her survival and that of her family may well have depended on it.

Sex and Pheromones

Among humans it has been claimed that olfaction has lost its leading role in the signalling of motivationally significant information. Humans, however, are capable of detecting as many as 10,000 odors. About 1% of the human genetic blueprint remains devoted to olfactory functioning, and almost 1,000 different genes are responsible for the creation of odor-detecting nerve cells.

As for emotional and motivational functioning, one need only suffer a severe cold in order to appreciate the dominant function of smell in the ability to fully detect and appreciate the flavor of food and, thus, experience pleasure in eating. Conversely, it has been reported that specific blends of odorants can also induce weight loss and a reduced desire to eat (Hirsh and Gomez, 1995).

Moreover, olfactory cues are still employed for indicating sexuality (e.g., perfume), and odor exerts a powerful influence on what is considered socially acceptable—hence, the abundance of artificial chemicals designed to eliminate various body fragrances, or conversely, to smell "sexy."

Indeed, it is rather obvious that smell serves a sexual purpose. This is why so many animals spend time sniffing and licking each other's genitals or thoroughly investigating urine or feces. Moreover, just as some insects will manufacture a chemical that acts as a sexual attractant to other species (so that they can eat them when they arrive drunk with desire), the pheromones produced by many animals exert similar, albeit much less pronounced, effects on humans as well as on unrelated species of animals, acting to promote sexual arousal and related physiological responses (reviewed in Ackerman, 1990).

Certain animal odors can affect the menstrual cycle and, in some instances, can cause females to ovulate and to even conceive more easily (Bronson, 1968; Bronson and Whitten, 1968). However, among certain groups of mammals, such as mice, the odor of a strange male can act to terminate the pregnancy of the female (Bronson et al., 1969).

Similarly, a woman's scent can exert profound effects on the hormonal and sexual functioning of other women as well as men. It has sometimes been noted that female roommates who live together for long time periods begin to menstruate at the same time, and similar phenomena have been reported in all-female dormitories (reviewed in Ackerman, 1990). Women who are simply exposed to the sweat of other females on a daily basis for 5 minutes, within a few months begin to cycle at the same time (Ackerman, 1990). This effect is not limited just to humans but occurs among dogs, cats, and other social creatures who live in close proximity.

Although not always consciously realized, the sexual nature of these chemicals is why many women inundate themselves with a variety of perfumes, the purpose of which is not merely to smell pleasant, but to indicate sexual availability and to arouse members of the opposite sex. This is also why some of the most popular and expensive perfumes have such obvious names: "My Sin," "Tabu," "Decadence," "Opium," "Indiscretion."

One need not cover the body with artificial scents in order to elicit sexual arousal. Natural pheromones exuded by men and women often contain subtle sexual messages, including when a woman is most likely to become pregnant. For example, in one experiment it was reported that men find the vaginal smells of women most pleasant when they are at that point of their cycle when they are most likely to conceive (see Ackerman, 1990).

In most species the effects of many pheromones on the limbic system and sexual behavior are almost reflexive, if not overpowering. For example, in response to a sex pheromone being actively secreted by a nearby receptive female, the vast majority of male insects, cats, dogs, and so forth are compelled to respond and to travel long distances in order to mate, and they will persist in their attempts, even when actively rebuffed or faced with the threat of being killed or eaten.

In general, male humans, and to a lesser degree other apes and monkeys, are able to resist these "sexual" urges (such as in the presence of a dominant male), either via frontal lobe inhibition or by employing neocortically derived strategies that enable them to meet the woman in a socially acceptable manner and offer "pick-up lines," ask her for a date, and so forth, so that what is desired might be acquired.

Nevertheless, among humans, despite the tremendous expansion in the neocortex, particularly within the frontal lobe, emotional upheavals frequently occur, often with murderous or sexually inappropriate consequences. Among humans, these old limbic and rhinencephalic influences remain quite powerful and exert almost continual streams of influence that affect many different aspects of our lives, including not just sex and the consumption of food, but the manner in which we interact with strangers and loved ones alike. Although largely unconscious, we still employ and rely upon olfactory and pheromonal communication, the essence of which not only tells us about the world, but through which the world is told about us.

Somesthesis and Emotion

Ontogenetically, although influenced by olfactory stimuli, human infants most obviously experience and react emotionally in response to tactile sensations or rapid changes in body position (Emde and Koenig, 1969; Spitz and Wolf, 1946), which, in turn, activates the vestibular system and initiates a startle reaction within the amygdala, which is also responsive to tactile sensation and pain (see below). Pain, of course, is also first experienced in relation to the body and is thus somesthetically rooted. However, some have argued that pain is not an emotion.

Among human infants, the earliest smiles are induced through tactile stimula-

tion (e.g., light stroking or even blowing on the skin), whereas loss of support is the most powerful stimulus for triggering an emotional reaction in the newborn (Emde and Koenig, 1969; Spitz and Wolf, 1946). Of course, the earliest and most consistent manifestation of emotion in the infant consists of screaming and crying, whereas positive affect is limited to an attitude of acceptance and quiescence (Spitz and Wolf, 1946)—emotions that are first mediated by the hypothalamus, and which may, or may not be, *true* emotions at all.

In their journey from the external to the internal environment, olfactory and tactile input are transmitted to various limbic nuclei, such as the lateral hypothalamus and entorhinal area of the hippocampus (olfactory only) and the amygdala (olfaction and somesthesis). Indeed, it is via nuclei such as the amygdala (versus the hypothalamus) that the first true (or rather, *felt*) aspects of emotion appear to be generated. It is also because of the tremendous input of olfactory information to various limbic nuclei that this part of the brain at one time

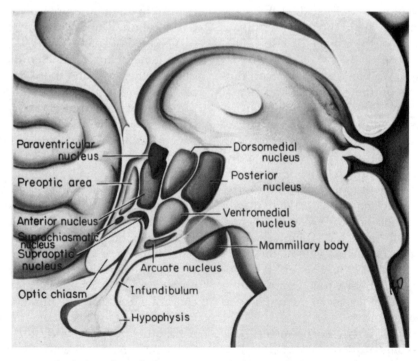

Figure 5.4   The human hypothalamus. (From Carpenter M. Core text of neuroanatomy. Baltimore: Williams & Wilkins, 1991.)

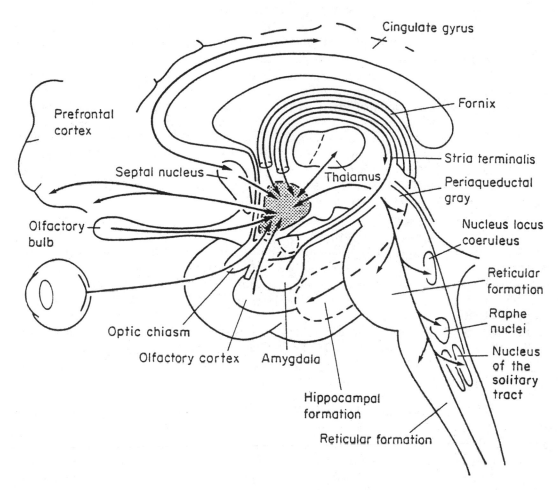

Cingulate gyrus

Fornix

Stria terminalis

Periaqueductal gray

Nucleus locus coeruleus

Reticular formation

Raphe nuclei

Nucleus of the solitary tract

Prefrontal cortex

Septal nucleus

Thalamus

Olfactory bulb

Optic chiasm

Olfactory cortex

Amygdala

Hippocampal formation

Reticular formation

Figure 5.5   Some of the main efferent and afferent connections of the hypothalamus. (After Bro-dal A. Neurological anatomy. New York: Oxford University Press, 1981).

was referred to as the rhinencephalon, or "nose brain."

HYPOTHALAMUS

The hypothalamus is an exceedingly ancient structure. Unlike most other brain regions, it has remained somewhat similar in structure throughout phylogeny and apparently over the course of evolution (Crosby et al., 1966). Located in the most medial aspect of the brain, along the walls and floor of the third ventricle, this nucleus is fully functional at birth and is the central core from which all emotions derive their motive force (Joseph, 1992a, 1994). Indeed, the hypothalamus is highly involved in all aspects of endocrine, hormonal, visceral, and autonomic functions and mediates or exerts controlling influences on eating, drinking, the experience of pleasure, rage, and aversion.

Although traditionally considered part of the diencephalon, the hypothalamus appears to have dual origins, i.e., visual and chemical-olfactory. It is likely the chemical/pheromonal component that gave rise to the ventral, medial, lateral, and preoptic hypothalamus, portions of which (like the amygdala) are also sexually dimorphic. That is, structurally and functionally, the hypothalamus of men and women is sexually dissimilar.

Sexual Dimorphism in the Hypothalamus

As is well known, sexual differentiation is strongly influenced by the presence or absence of gonadal steroid hormones during certain critical periods of prenatal development in many species, including humans. Not only are the external genitalia and other physical features sexually differentiated, but certain regions of the brain have also been found to be sexually dimorphic and differentially sensitive to steroids, particularly the preoptic area and ventromedial nucleus of the hypothalamus, as well as the amygdala (Bleir et al., 1982; Dorner, 1976; Gorski et al., 1978; Rainbow et al., 1982; Raisman and Field, 1971, 1973).

Indeed, it has now been well established that the amygdala and the hypothalamus (specifically, the anterior commissure, and the anterior-preoptic, ventromedial, and suprachiasmatic nuclei) are sexually differentiated and have sex-specific patterns of neuronal and dendritic development (Allen et al., 1989; Blier et al., 1982; Gorski et al., 1978; Rainbow et al., 1982; Raisman and Field, 1971, 1973; Swaab and Fliers, 1985).

This is a consequence of the presence or absence of testosterone during fetal development in humans, or soon after birth in some species, such as rodents. Specifically, the presence or absence of the male hormone, testosterone, during this critical neonatal period directly effects and determines the pattern of interconnections between the amygdala and hypothalamus, between axons and dendrites in these nuclei, and thus the organization of specific neural circuits. In the absence of testosterone, the female pattern of neuronal development occurs. Indeed, it is the presence or absence of testosterone during these early critical periods that appears to be responsible for neurological alterations that greatly effect sex differences in thinking, sexual orientation, aggression, and cognitive functioning (Barnett and Meck, 1990; Beatty, 1992; Dawson et al., 1975; Harris, 1978; Joseph, et al., 1978; Stewart et al., 1975).

For example, if the testes are removed prior to differentiation, or if a chemical blocker of testosterone is administered, thus preventing this hormone from reaching target cells in the limbic system, not only does the female pattern of neuronal development occur, but males so treated behave and process information in a manner similar to females (e.g., Joseph et al., 1978); i.e., they develop *female* brains and think and behave in a manner similar to females. Conversely, if females are administered testosterone during this critical period, the male pattern of differentiation and behavior results (see Gerall et al., 1992 for review).

That the preoptic and other hypothalamic regions are sexually dimorphic is not surprising in that it has long been known that this area is extremely important in controlling the basal output of gonadotropins in females prior to ovulation and is heavily involved in mediating cyclic changes in hormone levels (e.g., follicle-stimulating hormone, luteinizing hormone, estrogen, progesterone). Chemical and electrical stimulation of the preoptic and ventromedial thalamic nuclei also triggers sexual behavior and even sexual posturing in females and males (Lisk, 1967, 1971).

In primates, electrical stimulation of the preoptic area increases sexual behavior in males, and significantly increases the frequency of erections, copulations, and ejaculations, we well as pelvic thrusting followed by an explosive discharge of semen, even in the absence of a mate (MacLean, 1973, 1990). Conversely, lesions in the preoptic and posterior hypothalamus eliminate male sexual behavior and result in gonadal atrophy.

Although the etiology of homosexuality remains in question, it has been shown that the ventromedial and anterior nuclei of the hypothalamus of male homosexuals demonstrate the female pattern of development (LeVay, 1991; Swaab and Hoffman, 1990). When coupled with the evidence of male versus female and homosexual differences in the anterior commissure of the amygdala (see below), as well as the similarity between male homosexuals and women in regard to certain cognitive attributes (including spatial-perceptual capability), the possibility is raised that male homosexuals are in possession of a limbic system that is more "female" than "male" in functional as well as structural orientation.

Lateral and Ventromedial Hypothalamic Nuclei

Although consisting of several nuclear subgroups, the lateral and medial (ventromedial) hypothalamic nuclei play particularly important roles in the control of the autonomic nervous system, the experience of pleasure and aversion, eating and drinking, and raw (undirected) emotionality. They also appear to share a somewhat antagonistic relationship.

For example, the medial hypothalamus controls parasympathetic activities (e.g., reduction in heart rate, increased peripheral circulation) and exerts a dampening effect on certain forms of emotional/motivational arousal. The lateral hypothalamus mediates sympathetic activity (increasing heart rate, elevation of blood pressure) and is involved in controlling the metabolic and somatic correlates of heightened emotionality (Smith et al., 1990). In this regard, the lateral and medial region act to exert counterbalancing influences on each other.

Hunger and Thirst

The lateral and medial region are highly involved in monitoring internal homeostasis and motivating the organism to respond to internal needs, such as hunger and thirst (Bernardis and Bellinger, 1987). For example, both nuclei appear to contain receptors that are sensitive to the body's fat content (lipostatic receptors) and to circulating metabolites (e.g., glucose), which together indicate the need for food and nourishment. The lateral hypothalamus also appears to contain osmoreceptors (Joynt, 1966), which determine if water intake should be altered.

Electrophysiologically, it has been determined that the hypothalamus not only becomes highly active immediately prior to and while the organism is eating or drinking, but the lateral region alters its activity when the subject is hungry and simply looking at food (Hamburg, 1971; Rolls et al., 1976). In fact, if the lateral hypothalamus is electrically stimulated, a compulsion to eat and drink results (Delgado and Anand, 1953). Conversely, if the lateral area is destroyed bilaterally there results aphagia and adipsia so severe that animals will die unless force fed (Teitelbaum and Epstein, 1962).

If the medial hypothalamus is surgically destroyed, inhibitory influences on the lateral region appear to be abolished, such that hypothalamic hyperphagia and severe obesity result (Teitelbaum, 1961). Hence, the medial area seems to act as a satiety center—but a center that can be overridden.

Overall, it appears that the lateral hypothalamus is involved in the initiation of eating and acts to maintain a lower weight limit, such that when the limit is reached, the organism is stimulated to eat. Conversely, the medial region seems to be involved in setting a higher weight limit, such that when these levels are approached, the cessation of eating is triggered. In part, these nuclei exert these differential influences on eating and drinking via motivational/emotional influences they exert on other brain nuclei (e.g., via reward or punishment).

Pleasure and Reward

In 1952, Heath (cited by MacLean, 1969) reported what was then considered remarkable. Electrical stimulation near the septal nuclei elicited feelings of pleasure in human subjects: "I have a glowing feeling. I feel good!" Subsequently, Olds and Milner (1954) reported that rats would tirelessly perform operants to receive electrical stimulation in this same region and concluded that stimulation "has an effect which is apparently equivalent to that of a conventional primary reward." Even hungry animals would demonstrate a preference for self-stimulation over food.

Feelings of pleasure (as demonstrated via self-stimulation) have been obtained following excitation to a number of diverse limbic areas, including the olfactory bulbs, amygdala, hippocampus, cingulate, substantia nigra (a major source of dopamine), locus ceruleus (a major source of norepinephrine), raphe nucleus (serotonin), caudate, putamen, thalamus, reticular formation, medial forebrain bundle, and orbital frontal lobes (Brady, 1960; Lilly, 1960; Olds and Forbes, 1981; Stein and Ray, 1959; Waraczynski and Stellar, 1987).

In mapping the brain for positive loci for self-stimulation, Olds (1956) found that the medial forebrain bundle (MFB) was a major pathway that supported this activity. Although the MFB interconnects the hippocampus, hypothalamus, septum, amygdala, orbital frontal lobes (areas that give rise to self-stimulation), Olds discovered that in its course up to the lateral hypothalamus, reward sites become more densely packed. Moreover, the greatest area of concentration and the highest rates of self-stimulatory activity were found to occur not in the MFB but in the lateral hypothalamus (Olds, 1956; Olds and Forbes, 1981). Indeed, animals "would continue to stimulate as rapidly as possible until physical fatigue forced them to slow or to sleep" (Olds, 1956).

Electrophysiological studies of single lateral hypothalamic neurons indicate that these cells become highly active in response to rewarding food items (Nakamura and Ono, 1986). In fact, many of these cells will become aroused by neutral stimuli repeatedly associated with reward, such as a cue-tone—even in the absence of the actual reward (Nakamura and Ono, 1986; Ono et al., 1980). However, this ability to form associations appears to be secondary to amygdaloid activation (Fukuda et al., 1987), which in turn influences hypothalamic functioning.

Nevertheless, if the lateral region is destroyed, the experience of pleasure and emotional responsiveness is almost completely attenuated. For example, in primates, faces become blank and expressionless, whereas if the lesion is unilateral, a marked neglect and indifference regarding all sensory events occurring on the contralateral side occurs (Marshall and

Teitelbaum, 1974). Animals will, in fact, cease to eat and will die.

Aversion

In contrast to the lateral hypothalamus and its involvement in pleasurable self-stimulation, activation of the medial hypothalamus is apparently so aversive that subjects will work to reduce it (Olds and Forbes, 1981). Hence, electrical stimulation of the medial region leads to behavior that terminates the stimulation—apparently so as to obtain relief (e.g., active avoidance). In this regard, when considering behavior such as eating, it might be postulated that when upper weight limits (or nutritional requirements) are met, the medial region becomes activated; this in turn leads to behavior (e.g., cessation of eating) that terminates its activation.

It is possible, however, that medial hypothalamic activity may also lead to a state of quiescence such that the organism is motivated to simply cease to respond or to behave. In some instances, this quiescent state may be physiologically neutral, whereas in other situations medial hypothalamic activity may be highly aversive. Quiescence is also associated with parasympathetic activity, which is mediated by the medial area.

Hypothalamic Damage and Emotional Incontinence: Laughter and Rage

When electrically stimulated, the hypothalamus responds by triggering two seemingly oppositional feeling states, i.e., pleasure and displeasure/aversion. The generation of these emotional reactions in turn influences the organism to respond so as to increase or decrease what is being experienced. The hypothalamus, via its rich interconnections with other limbic regions, including the neocortex and frontal lobes, is able to mobilize and motivate the organism to either cease or continue to behave. Nevertheless, at the level of the hypothalamus, the emotional states elicited are very primitive, diffuse, undirected, and unrefined. The organism feels pleasure in general, or aversion/displeasure in general. Higher-order emotional reactions (e.g., desire, love, hate, etc.) require the involvement of other limbic regions as well as neocortical participation.

Emotional functioning at the level of the hypothalamus is not only quite limited and primitive, but is also largely reflexive. For example, when induced via stimulation, the moment the electrical stimulus is turned off the emotion elicited is immediately abolished. In contrast, true emotions (which require other limbic interactions) are not simply turned on or off but can last from minutes to hours to days and weeks before completely dissipating.

Nevertheless, in humans, disturbances of hypothalamic functioning (e.g., due to an irritating lesion, such as tumor) can give rise to seemingly complex, higher-order behavioral-emotional reactions, such as pathological laughter and crying that occurs uncontrollably. However, in some cases, when patients are questioned they may deny having any feelings that correspond to the emotion displayed (Davison and Kelman, 1939; Ironside, 1956; Martin, 1950). In part, these reactions are sometimes due to disinhibitory release of brainstem structures involved in respiration, whereas in other instances the resulting behavior is caused by hypothalamic triggering of other limbic nuclei.

Uncontrolled Laughter

Pathological laughter has frequently been reported to occur with hypophyseal and midline tumors involving the hypothalamus, aneurysm in this vicinity, hemorrhage, and astrocytoma or papilloma of the third ventricle (resulting in hypothalamic compression), as well as surgical manipulation of this nucleus (Davison and Kelman, 1939; Dott, 1938; Foerster and Gabel, 1933; Martin, 1950; Money and Hosta, 1967; Ironside, 1956; List, Dowman, and Bagheiv, 1958).

For example, Martin (1950:455) describes a man who, while "attending his mother's funeral was seized at the grave site with an attack of uncontrollable laughter which embarrassed and distressed him considerably." Although this particular attack dissipated, it was soon accompanied by several further fits of laughter, and he died soon thereafter. Postmortem, a large, ruptured aneurysm was found, compressing the mammillary bodies and hypothalamus.

In a similar case (Anderson, 1936; cited by Martin, 1950), a patient literally died laughing following the eruption of the posterior communicating artery, which resulted in compression (via hemorrhage) of the hypothalamus. "She was shaken by laughter and could not stop: short expirations followed each other in spasms, without the patient being able to make an adequate inspiration of air, she became cyanosed and nothing could stop the spasm of laughter which eventually became noiseless and little more than a grimace. After 24 hours of profound coma she died."

Because *laughter* in these instances has not been accompanied by corresponding feeling states, this pseudo-emotional condition has been referred to as "sham mirth" (Martin, 1950). However, in some cases, abnormal stimulation in this region (such as due to compression effects from neoplasm) has triggered corresponding emotions and behaviors—presumably due to activation of other limbic nuclei.

For example, laughter has been noted to occur, with hilarious or obscene speech—usually as a prelude to stupor or death—in cases where tumor has infiltrated the hypothalamus (Ironside, 1956). One group of neurosurgeons (Foerster and Gagel, 1932) has reported several instances in which, while surgeons swabbed the blood from the floor of the third ventricle, patients "became lively, talkative, joking, and whistling each time the infundibular region of the hypothalamus was manipulated." In one case, the patient became excited and began to sing.

Hypothalamic Rage

Stimulation of the lateral hypothalamus can induce extremes in emotionality, including intense attacks of rage accompanied by biting and attack upon any moving object (Flynn et al., 1971; Gunne and Lewander, 1966; Wasman and Flynn, 1962). If this nucleus is destroyed, aggressive and attack behavior is abolished (Karli and Vergness, 1969). Hence, the lateral hypothalamus is responsible for rage and aggressive behavior.

As noted, the lateral hypothalamus maintains an oppositional relationship with the medial hypothalamus. Hence, stimulation of the medial region counters the lateral area such that rage reactions are reduced or eliminated (Ingram, 1952; Wheately, 1944), whereas if the medial is destroyed there results lateral hypothalamic release and the triggering of extreme savagery.

In man, inflammation, neoplasm, and compression of the hypothalamus have also been noted to give rise to rage attacks (Pilleri and Poeck, 1965), and surgical manipulations or tumors within the hypothalamus have been observed to elicit manic and rage-like outbursts (Alpers, 1940). These appear to be release phenomenon, however. That is, rage, attack, aggression, and related behaviors associated with the hypothalamus appear to be under the inhibitory influence of higher-order limbic nuclei, such as the amygdala and septum (Siegel and Skog, 1970). When the controlling pathways between these areas are damaged (i.e., disconnected), sometimes these behaviors are elicited.

For example, Pilleri and Poeck (1965) described a man with severe damage throughout the cerebrum, including the amygdala, hippocampus, and cingulate but with complete sparing of the hypothalamus, who continually reacted with howling, growling, and baring of teeth in response to noise or a slight touch, or if approached. Hence, the hypothalamus being released responds reflexively in an aggressive-like, nonspecific manner to any stimulus. Lesions of the frontal-hypothalamic pathways have been noted to result in severe rage reactions as well (Fulton and Ingraham, 1929; Kennard, 1945).

Nevertheless, like "sham mirth," rage reactions elicited in response to direct electrical activation of the hypothalamus immediately and completely dissipate when the stimulation is removed. As such, these outbursts have been referred to as "sham rage."

Circadian Rhythm Generation and Seasonal Affective Disorder

As noted in Chapter 1, during the initial stages of cerebral evolution, the dorsal hypothalamus (like the dorsal thalamus, dorsal hippocampus, and dorsal midbrain) was likely fashioned, at least in part, from photosensitive cells located in the anterior head region. Given the daily and seasonal changes in light versus darkness, nuclei in the midbrain-pons, and in the hypothalamus, became sensitive to and capable of generating rhythmic hormonal, neurotransmitter, and motoric activities. It is the hypothalamus, however, the suprachiasmatic nucleus (SCN) in particular, which appears to be the "master clock" for the generation of circadian rhythms: rhythms that have a period length of 24 hours (Aronson et al., 1993; Morin, 1994).

In humans and other species, the SCN (and the midbrain superior colliculus) is a direct recipient of retinal axons. It also receives indirect visual projections from the lateral geniculate nucleus of the thalamus (see Morin, 1994). In this regard, the visual system appears to act to synchronize the SCN (and probably the midbrain-pons) to function in accordance with seasonal and day-to-day variations in the light/dark ratio. However, the SCN does not "see" per se, nor can it detect visual features, as its main concern is adjusting mood and activity in regard to light intensity as related to rhythm generation.

There is thus some evidence which suggests that when the SCN of the hypothalamus is deprived of (or unable to effectively respond to) sufficient light, although rhythm generation is not grossly effected (Morin, 1994), individuals may become depressed: a condition referred to as **seasonal**

affective disorder (SAD). That is, the hypothalamus (and midbrain-pons) appear to decrease those hormonal and neurochemical activities normally associated with activation and high (daytime) activity, thus resulting in depression.

For example, the hypothalamic-pituitary axis secretes melatonin in phase with the circadian rhythm. Phase-delayed rhythms in plasma melatonin secretion have been repeatedly noted in most (but not all) studies of individuals with SAD (see Wirz-Justice et al., 1993, for review). However, with light therapy, not only is the depression relieved but the melatonin secretions return to normal. This is significant, for melatonin is derived from tryptophan via serotonin, and low serotonin levels have been directly linked to depression (e.g., Van Pragg, 1982).

There is some evidence which suggests that the hypothalamus (and the midbrain) may act to regulate serotonin release (Chaouloff, 1993; however, see Morin, 1994), which in turn may explain why serotonin levels rhythmically fluctuate (e.g., such as during the sleep cycle), or become abnormal when denied sufficient light; i.e., the production of serotonin by the raphe nucleus (in the pons) is abnormally affected.

On the other hand, numerous studies have reported that SAD and major depression occurs most often during the spring and not the winter, and is not influenced by latitude (e.g., Magnusson and Stefansson, 1993; Wirz-Justice et al., 1993). There is also some suggestion that abnormal temperature perception, or aging within the SCN, may be responsible for the genesis of SAD and related depressive disorders. For example, age-related changes in the SCN have been noted to adversely effect circadian rhythm generation as well as metabolic and peptide activity (Aronson et al., 1993). In consequence, rest versus active cycles also become abnormal, with reductions in arousal and activity; i.e., the patient becomes depressed.

It is also possible, however, that although light therapy can assist in alleviating depressive symptoms associated with SAD, the deregulation of the SCN (and

melatonin/serotonin) might be unrelated to light, temperature, or aging, but may be a consequence of stress on the hypothalamus (Chauloff, 1993). For example, the hypothalamic-pituitary axis is tightly linked with and, in fact, mediates stress-induced alterations in serotonin (see Chauloff, 1993, for review) as well as norepinephrine (NE) (Swann et al., 1994), which has also been repeatedly implicated in the genesis of depression.

The Hypothalamus-Pituitary-Adrenal Axis

The hypothalamic, pituitary, adrenal (HPA) system is critically involved in the adaptation to stressful changes in the external or internal environment. For example, in response to fear, anger, anxiety, disappointment, and even hope, the hypothalamus begins to release corticotropin releasing factor (CRF); this activates the adenohypophysis, which begins secreting adrenocorticotropic hormone (ACTH), which stimulates the adrenal cortex, which secretes cortisol.

These events, in turn, appear to be under the modulating influences of NE. That is, as stress increases, NE levels decrease, which triggers the activation of the HPA axis. As is well known, low levels of NE are associated with depression.

Normally, cortisol secretion is subject to the tonic influences of NE, whereas cortisol can indirectly reduce NE synthesis. Thus, a feedback system is maintained via the interaction of these substances (in conjunction with ACTH). Moreover, cortisol and NE levels fluctuate in reverse, and thus maintain a reciprocal relationship with the circadian rhythm; i.e., in oppositional fashion they increase and then decrease throughout the day and evening.

Among certain subgroups suffering from depression, it appears that this entire feedback regulatory system, and thus the HPA axis, is disrupted (Carrol et al., 1976; Sachar et al., 1973). This results in the hypersecretion of ACTH and cortisol with a corresponding decrease in NE, which re-

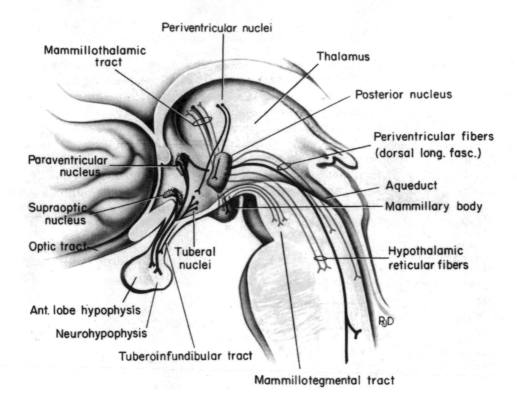

Figure 5.6 Connections of the hypothalamic-hypophysis system. (From Carpenter M. Core text of neuroanatomy. Baltimore, Williams & Wilkins, 1991.)

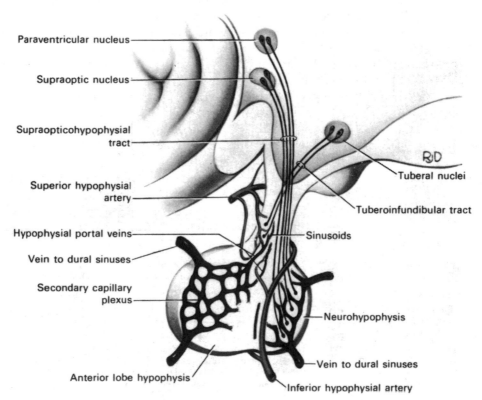

Figure 5.7 Diagram of the hypothalamic-hypophysis portal system. (From Carpenter M. Core text of neuroanatomy. Baltimore: Williams & Wilkins, 1991.)

sults in NE-induced depression. It was these findings which led to the development of the dexamethasone suppression test over 25 years ago.

Via the administration of dexamethasone (a synthetic corticosteroid), it was determined that many depressed individuals have excess cortisol, and an increased frequency of cortisol secretory episodes (Carrol et al., 1976; Sachar et al., 1973; Swann et al., 1994). Moreover, those who demonstrate excess cortisol were found to respond to NE potentiating agents, whereas those who were depressed but with normal cortisol responded best to serotonin-potentiating compounds (Van Pragg, 1982).

It is also noteworthy that dexamethasone nonsuppression rates are increased in mania; specifically, "mixed manic" states, which consist of lability, grandiosity, and lability superimposed over depression (see Swann et al., 1994). These "mixed manic" individuals also display elevated NE levels but respond poorly to lithium and show higher levels of cortisol during the depressed phase of their illness (Swann et al., 1994).

As noted, the hypothalamus may greatly influence circadian activities within the midbrain and pons, and thus the rhythmical secretion of various neurotransmitters. For example, corticotropin-releasing factor acts directly on the locus ceruleus (Valentino et al., 1983), which manufactures NE, and on the raphe, thereby influencing serotonin release. These findings suggest that a disturbance in circadian or rhythmical control of hypothalamic and midbrain-pontine activity can give rise to depression, or mixed mania in some individuals, particularly women.

LATERALIZATION

Although scant, there is some evidence which suggests that the right hypothalamus may be more heavily involved in the control of neuroendocrine functioning, particularly in females. Females are also far more likely to suffer from depression and from SAD. Moreover, right cerebral dysfunction can reduce NE levels in both the right and left hemisphere (Robinson, 1979). Greater right hypothalamic concentration of substances such as luteinizing hormone has also been reported (Gerendai, 1984), which in turn is a "female" hormone involved in lactation and pregnancy.

Psychic Manifestations of Hypothalamic Activity: The Id

Phylogenetically and from an evolutionary perspective, the appearance and development of the hypothalamus predates the differentiation of all other limbic nuclei, e.g., amygdala, septal nucleus, hippocampus (Andy and Stephan, 1961; Brown, 1983; Herrick, 1925; Humphrey, 1972). It constitutes the most primitive, archaic, reflexive, and purely biological aspect of the psyche.

Biologically, the hypothalamus serves the body tissues by attempting to maintain internal homeostasis and by providing for the immediate discharge of tensions in an almost reflexive manner. Hence, as based on studies of lateral and medial hypothalamic functioning, it appears to act reflexively, in an almost on/off manner so as to seek or maintain the experience of pleasure and escape or avoid unpleasant, noxious conditions.

Emotions elicited by the hypothalamus are largely undirected and short-lived, being triggered reflexively and without concern for or understanding of consequences—that is, unless the hypothalamus is chronically stressed or aroused. Nevertheless, direct contact with the real world is quite limited and almost entirely indirect, as the hypothalamus is largely concerned with the internal environment of the organism. Although it receives and responds to light, it cannot "see." It has no sense of morals, danger, values, logic, and so forth, and cannot feel or express love or hate. Although quite powerful, hypothalamic emotions are largely undifferentiated, consisting of feelings such as pleasure, displeasure, aversion, rage, hunger, thirst, and so forth.

As the hypothalamus is concerned with the internal environment, much of its activity occurs outside conscious awareness.

Moreover, being involved in maintaining internal homeostasis via, for example, its ability to reward or punish the organism with feelings of pleasure or aversion, it tends to serve what Freud (1911) has described as "the pleasure principle," and best corresponds to what has been described as the "ID."

The Pleasure Principle

The lateral and medial nuclei exert counterbalancing influences that serve to modulate activity occurring in the other. As described by Freud (1911), the pleasure principal not only serves to maximize pleasant experiences, but acts to keep the psyche as a whole free from high levels of excitation (be they pleasurable or unpleasant).

Like the hypothalamus, the pleasure principle is present from birth; and for some time thereafter, the search for pleasure is manifested in an unrestricted manner and with a great deal of intensity, as there are no oppositional forces (except those between the lateral and medial regions) to counter its striving. Indeed, higher-order limbic nuclei have yet to mature.

Functionally isolated, the hypothalamus at birth has no way of reducing tension or mobilizing the organism for any form of effective action. It is helpless. When tensions associated with immediate needs (e.g., hunger or thirst) become unpleasant, the only response available to the hypothalamus is to cry and make rage-like vocalizations. When satiated, the hypothalamus can only respond with a feeling state suggesting pleasure or at least quiescence.

Indeed, as is well known, for the first few months of life the infant's awareness largely consists of a very restricted matrix involving tactile, visceral (hunger), and kinesthetic sensations, where emotionally the infant is capable of screaming, crying, or demonstrating very rudimentary features of pleasure, i.e., an attitude of acceptance of quiescence (McGraw, 1969; Milner, 1967; Piaget, 1952; Spitz and Wolf, 1946).

It is only with the further differentiation and maturation of higher-order limbic nuclei (e.g., amygdala, septal nucleus, and hippocampus) that the infant begins to achieve some awareness of external reality and begins to form memories as well as differentiate and associate externally occurring events and individuals.

AMYGDALA

In contrast to the primitive hypothalamus, the more recently developed amygdala (the "almond") is preeminent in the control and mediation of all higher-order emotional and motivational activities and, in fact, serves as the seat of social and emotional intelligence. Via its rich interconnections with various neocortical and subcortical regions, amygdaloid neurons are able to monitor and abstract from the sensory array stimuli that are of motivational significance to the organism (Gaffan, 1992; Gloor, 1960, 1992; Joseph, 1992a; LeDoux, 1992; Rolls, 1992; Steklis and Kling, 1985; Ursin and Kaada, 1960). This includes the ability to discern and express even subtle social-emotional nuances, such as friendliness, fear, love, affection, distrust, anger, and so forth, and, at a more basic level, to determine if something might be good to eat. In fact, amygdaloid neurons respond selectively to the flavor of certain preferred foods, as well as to the sight or sound of something that might be especially desirable to eat (Fukuda et al., 1987; Gaffan, 1992; O'Keefe and Bouma, 1969; Ono et al., 1980).

Single amygdaloid neurons receive a considerable degree of topographic input, and are predominantly polymodal, responding to a variety of stimuli from different modalities simultaneously (Amaral et al., 1992; O'Keefe and Bouma, 1969; Perryman et al., 1987; Rolls, 1992; Sawa and Delgado, 1963; Schutze et al., 1987; Turner et al., 1980; Ursin and Kaada, 1960; Van Hoesen, 1981). The amygdala is also very sensitive to somesthetic input and physical contact, such that even a slight touch in a very circumscribed area of the body can produce amygdaloid excitation. Overall, because emotional, motivational, and multimodal assimilation of various sensory impressions occurs in this region, it is also involved in attention, learning, and memory.

Medial and Lateral Amygdaloid Nuclei

The amygdala is buried within the depths of the anterior-inferior temporal lobe and consists of several major nuclear groups, including what has been referred to as the "extended amygdala." This chapter will focus on the phylogenetically ancient anteromedial group (or medial amygdala), which is involved in olfaction and motor activity (via its interconnections with the striatum), and a relatively newer basolateral division (lateral amygdala), which is most fully developed in primates and humans (Amaral et al., 1992; Herrick, 1925; Humphrey, 1972; McDonald, 1992). Like the lateral and medial hypothalamus, these two amygdaloid nuclei subserve different functions and maintain different anatomical interconnections.

Embryologically, the medial amygdala is the first portion of the basal ganglia (limbic) striatal complex to appear during development, being formed via neuroblast migration from the epithelium of the lateral

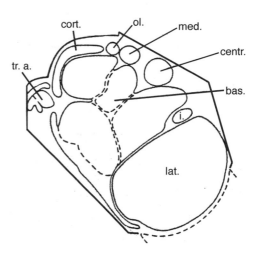

Figure 5.8 Diagram of the human amygdala. **bas**=nucleus basalis; **lat**=lateral nucleus; **centr**=central nucleus; **cort**=cortical nucleus; **med**=medial nucleus; **ol**=olfactory tract. (After Brodal A. Neurological anatomy. New York: Oxford University Press, 1981.)

ventricle (Humphrey, 1972). In addition, the tail of the caudate nucleus (as it circles in an arc from the frontal to temporal lobe) terminates and merges with the medial (and lateral) amygdala. Hence, the amygdala is, in fact, part of the basal ganglia and is heavily involved in motivating and coordinating gross, or whole body, motor activ-

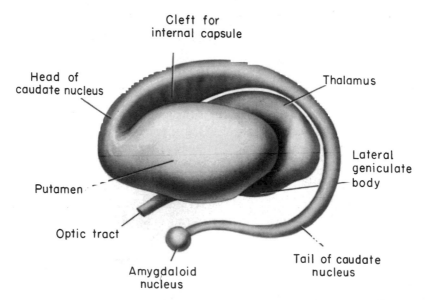

Figure 5.9 The amygdala, thalamus, and corpus striatum (caudate, putamen). (From Carpenter M. Core text of neuroanatomy. Baltimore: Williams & Wilkins, 1991.)

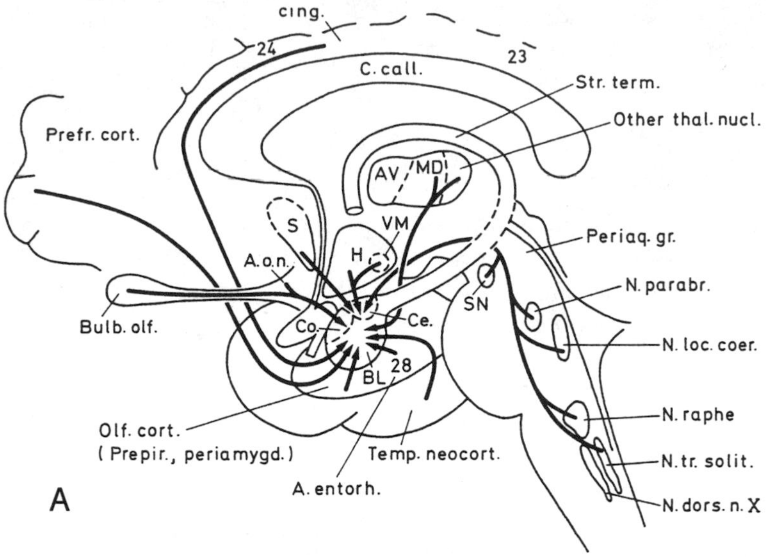

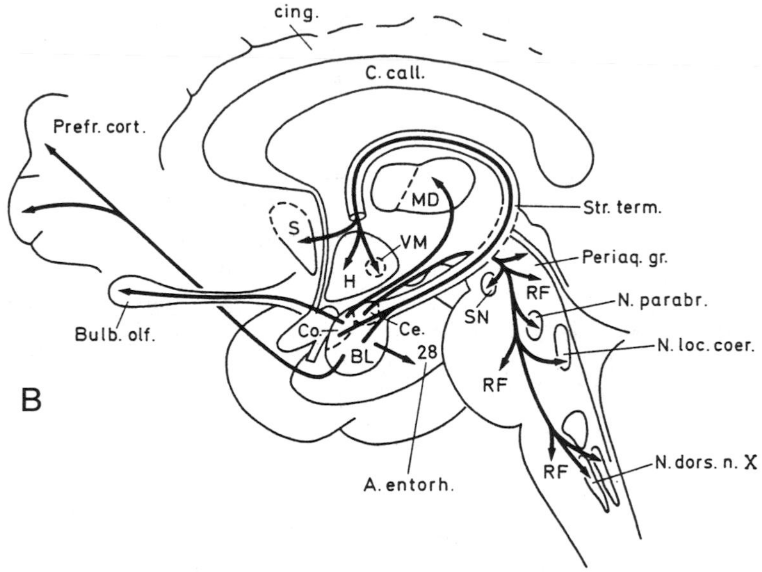

Figure 5.10   Diagram of some of the main afferent **(A)** and efferent **(B)** connections of the amygdala. Basolateral nucleus **(BL)**, central nucleus **(Ce)**, cortical nucleus **(Co)**, hypothalamus **(H)**, septal nucleus **(S)**, dorsal medial nucleus of the thalamus **(MD)**, substantia nigra **(SN)**. (After Brodal A. Neurological anatomy. New York: Oxford University Press, 1981.)

ity (Heimer and Alheid, 1991; Mogenson and Yang, 1991).

The medial amygdala receives fibers from the olfactory tract, and via a rope of fibers called the **stria terminalis** projects directly to and receives fibers from the medial hypothalamus (via which it exerts inhibitory influences) as well as the septal nucleus (Amaral et al., 1992; Carlsen et al., 1982; Gloor, 1955; McDonald, 1992; Russchen, 1982; Swanson and Cowan, 1979). The stria terminalis is significantly larger and thicker in males versus females (Allen and Gorksi, 1992), which suggests that information and impulse exchange (or inhibition) between the hypothalamus and amygdala is different in men than in women. In addition, the medial (and lateral) regions are rich in cells containing enkephalins, and opiate receptors can be found throughout the amygdala (Atweh and Kuhar, 1977a; Uhl et al., 1978). In this regard, the amygdala is capable of inducing extreme feelings of pleasure.

Lateral Amygdala

With the evolutionary ascent of primates, the lateral division of the amygdala progressively expands and differentiates. The lateral amygdala contributes fibers to the stria terminalis and gives rise to the amygdalofugal pathway, via which it projects to the lateral and medial hypothalamus (upon which it exerts inhibitory and excitatory influences, respectively), the dorsal medial thalamus (which is involved in memory, attention, and arousal), the limbic and corpus striatum, as well as other subcortical regions (Aggleton et al., 1980; Amaral et al., 1992; Carlsen et al., 1982; Dreifuss et al., 1968; Gloor, 1955, 1960; Klinger and Gloor, 1960; McDonald, 1992; Mehler, 1980; Russchen, 1982). It also receives fibers from the medial forebrain bundle, which in turn has its site of origin in the lateral hypothalamus (Mehler, 1980).

In general, whereas the medial amygdala is highly involved in motor, olfactory, and sexual functioning, the lateral division is intimately involved in all aspects of higher order emotional activity: hence, its rich interconnections with the lateral and medial hypothalamus and the neocortex.

The lateral amygdala maintains rich interconnections with the inferior, middle, and superior temporal lobes, as well as the insular temporal region, which in turn allows it to sample and influence the auditory, somesthetic, and visual information being received and processed in these areas, as well as scrutinize this information for motivational and emotional significance (Gloor, 1992; Herzog and Van Hoesen, 1976; Kling et al., 1987; Machne and Segundo, 1956; Mesulam and Mufson, 1982; O'Keefe and Bouma, 1969; Rolls, 1992; Steklis and Kling, 1985; Turner et al., 1980; Van Hoesen, 1981). Gustatory and respiratory sense are also re-represented in this vicinity (Amaral et al., 1992; Fukuda et

al., 1987; MacLean, 1949; Ono et al., 1980), as is the capacity to influence (via sensory analysis) food and water intake. The lateral division also maintains rich interconnections with cingulate gyrus, orbital and medial frontal lobes (Amaral et al., 1992; McDonald, 1992; Pandya et al., 1973), and the parietal cortex (O'Keefe and Bouma, 1969), through which it able to influence emotional expression and receive complex somesthetic information.

The lateral amygdala is highly important in analyzing information received and transferring information back to the neocortex, so that further elaboration may be carried out at the neocortical level. It is through the lateral division that emotional meaning and significance can be assigned to as well as extracted from that which is experienced so it may be analyzed intelligently.

The amygdala, overall, maintains a functionally interdependent relationship with the hypothalamus. It is able to modulate and even control rudimentary emotional forces governed by the hypothalamic nucleus. However, it also acts at the behest of hypothalamically induced drives. For example, if certain nutritional requirements need to be meet, the hypothalamus signals the amygdala, which then surveys the external environment for something good to eat or drink. On the other hand, if the amygdala, via environmental surveillance, discovers a potentially threatening stimulus, it acts to excite and drive the hypothalamus so that the organism is mobilized to take appropriate action. Hence, when the hypothalamus is activated by the amygdala, instead of responding in an on/off manner, cellular activity continues for an appreciably longer time period (Dreifuss et al., 1968; Rolls, 1992). The amygdala can tap into the reservoir of emotional energy mediated by the hypothalamus so that certain ends may be attained.

Attention

The amygdala acts to perform environmental surveillance and can trigger orienting responses, as well as mediate the maintenance of attention, should something of interest or importance appear (Gloor, 1955, 1960, 1992; Kaada, 1951; Rolls, 1992; Ursin and Kaada, 1960). Electrical stimulation of the lateral division can initiate quick and/or anxious glancing and searching movements of the eyes and head such that the organism appears aroused and highly alert, as if in expectation of something that is going to happen (Halgren, 1992; Ursin and Kaada, 1960). The EEG becomes desynchronized (indicating arousal), heart rate becomes depressed, respiration patterns change, and the galvanic skin response significantly alters (Bagshaw and Benzies, 1968; Kapp et al., 1994; Ursin and Kaada, 1960)—reactions that characteristically accompany the orienting response of most species.

Once a stimulus of potential interest is detected, the amygdala then acts to analyze its emotional-motivational importance and will act to alert other nuclei, such as the hypothalamus and basal ganglia, so that appropriate action may take place.

Fear, Rage, and Aggression

Initially, electrical stimulation of the amygdala produces sustained attention and orienting reactions. If the stimulation continues, fear and/or rage reactions are elicited (Gloor, 1992; Halgren, 1992; Ursin and Kaada, 1960; see also Cendes et al., 1994). When fear follows the attention response, the pupils will dilate and the subject will cringe, withdraw, and cower. This cowering reaction, in turn, may give way to extreme fear and/or panic such that the animal will attempt to take flight.

Among humans, the fear response is one of the most common manifestations of amygdaloid electrical stimulation and abnormal activation. Moreover, unlike hypothalamic on/off emotional reactions, attention and fear reactions can last up to several minutes after the stimulation is withdrawn.

In addition to behavioral manifestations of heightened emotionality, amygdaloid stimulation can result in intense changes in emotional facial expression. This includes crying and facial contortions, such as baring of the teeth, dilation of the pupils, widening or narrowing of the eyelids, flaring of the nostrils, as well as sniffing, licking, and chewing (Anand and Dua, 1955; Ursin and Kaada, 1960). Indeed, some of the behavioral manifestations of a seizure in this vicinity (i.e., temporal lobe epilepsy) typically include chewing, smacking of the lips and licking.

In many instances, rather than fear, there instead results anger, irritation, and rage, which seems to gradually build until, finally, the animal or human will attack (Egger and Flynn, 1963; Gunne and Lewander, 1966; Ursin and Kaada, 1960; Zbrozyna, 1963). Unlike hypothalamic "sham rage," amygdaloid activation results in attacks directed at something real, or, in the absence of an actual stimulus, at something imaginary. Moreover, rage and attack will persist well beyond the termination of the electrical stimulation of the amygdala. In fact, the amygdala remains electrophysiologically active for long periods, even after a stimulus has been removed (be it external-perceptual or internal-electrical), such that it appears to continue to process—in the abstract—information, even when that information is no longer observable (O'Keefe and Bouma, 1969). Hence, the amygdala is also responsible for generating and maintaining mood.

The amygdala, in addition to demonstrating sustained electrophysiological activity, has been shown to be heavily involved in the maintenance of behavioral responsiveness, even in the absence of an immediately tangible or visible object or stimulus (O'Keefe & Bouma, 1969). This includes motivating the organism to engage in the seeking of hidden objects or continuing a certain activity in anticipation of achieving some particular long-term goal. At a more immediate level, the amygdala is probably very important in object permanence (i.e., the keeping of an object in mind when it is no longer visible) and concrete or abstract anticipation. Anticipation is, of course, very important in the prolongation of emotional and mood states, such as fear or anger, as well as the generation of more complex emotions, such as anxiety. In this regard, the amygdala is important, not only in regard to emotion, but in the maintenance of mood states.

Fear and rage reactions have also been triggered in humans following depth electrode stimulation of the amygdala (Chapman, 1960; Chapman et al., 1954; Heath et al., 1955; Mark et al., 1972). Mark et al. (1972) describe one female patient who, after amygdaloid stimulation, became irritable and angry, and then enraged. Her lips retracted, and she displayed extreme facial grimacing, threatening behavior, and then rage and attack—all of which persisted well beyond stimulus termination.

Similarly, Schiff et al. (1982) describe a man who developed intractable aggression following a head injury and damage (determined via depth electrode) to the amygdala (i.e., abnormal electrical activity). Subsequently, he became easily enraged, sexually preoccupied (although sexually hypoactive), and developed hyper-religiosity and pseudo-mystical ideas. Tumors invading the amygdala have been reported to trigger rage attacks (Sweet et al., 1960; Vonderache, 1940).

Indeed, the amygdala appears capable of not only triggering and steering hypothalamic activity but acting on higher-level neocortical processes so that individuals form emotional *ideas*. Indeed, the amygdala is able to overwhelm the neocortex and the rest of the brain so that the person not only forms emotional ideas but responds to them. A famous example of this is Charles Whitman, who in 1966 climbed a tower at the University of Texas and began to indiscriminately kill people with a rifle.

Whitman had initially consulted a psychiatrist about his periodic and uncontrollable violent impulses but was unable to obtain relief. Prior to climbing the tower, he wrote himself a letter:

"I don't really understand myself these days. Lately I have been a victim of many unusual and irrational thoughts. These thoughts constantly recur, and it requires a tremendous mental effort to concentrate. I talked to a doctor once for about two hours and tried to convey to him my fears that I felt overcome by overwhelming violent impulses. After one session I never saw the Doctor again, and since then I have been fighting my mental turmoil alone. After my death I wish that an autopsy would be performed to see if there is any visible physical disorder. I have had tremendous headaches in the past."

Later he wrote:

"It was after much thought that I decided to kill my wife, Kathy, tonight after I pick her up from work ... I love her dearly, and she has been as fine a wife to me as any man could ever hope to have. I cannot rationally pinpoint any specific reason for doing this ... "

That evening he killed his wife and mother and wrote: "I imagine it appears that I brutally killed both of my loved ones. I was only trying to do a good thorough job ... " The following morning he climbed the University tower carrying a high-powered hunting rifle, and for the next 90 minutes he shot at everything that moved, killing 14, wounding 38. Autopsy of his brain revealed a glioblastoma multiforme tumor the size of a walnut compressing the amygdaloid nucleus (Sweet et al., 1969).

Docility and Amygdaloid Destruction

Bilateral destruction of the amygdala usually results in increased tameness, docility, and reduced aggressiveness in cats, monkeys, and other animals (Schreiner and Kling, 1956; Weiskrantz, 1956; Vochteloo and Koolhaas, 1987), including purportedly ferocious creatures, such as the agouti and lynx (Schreiner and Kling, 1956). In man, bilateral amygdala destruction (via neurosurgery) has been reported to reduce and/or eliminate paroxysmal aggressive and violent behavior (Terzian and Ore, 1955).

In some creatures, however, bilateral ablation of the amygdala has been reported, at least initially, to result in increased aggressive response (Bard and Mountcastle, 1948); if sufficiently aroused or irritated, even the most placid of amygdalectomized animals can be induced to fight fiercely (Fuller et al. 1957). However, these aggressive responses are very short-lived and appear to be reflexively mediated by the hypothalamus. Hence, these findings (and the data reviewed above) suggest that true aggressive feelings are dependent upon the functional integrity of the amygdala (versus the hypothalamus).

Social-Emotional Agnosia

Among primates and mammals, bilateral destruction of the amygdala significantly disturbs the ability to determine and identify the motivational and emotional significance of externally occurring events, to discern social-emotional nuances conveyed by others, or to select what behavior is appropriate, given a specific social context (Bunnel, 1966; Fuller et al., 1957; Gloor, 1960; Kling and Brothers, 1992; Klüver and Bucy, 1939; Weiskrantz, 1956). This is because the amygdala is primary in regard to all aspects of social-emotional intelligence (coined by Joseph 1992b). Bilateral lesions lower responsiveness to aversive and social stimuli, and reduce aggressiveness, fearfulness, competitiveness, dominance, and social interest (Rosvold et al., 1954). Indeed, this condition is so pervasive that subjects seem to have tremendous difficulty discerning the meaning or recognizing the significance of even common objects—a condition sometimes referred to as "*psychic blindness*," or the "Klüver-Bucy syndrome."

Thus, animals with bilateral amygdaloid destruction, although able to see and interact with their environment, may respond in an emotionally blunted manner, and seem unable to recognize what they see, feel, and experience. Things seem stripped of meaning. Like infants (who similarly are without a fully functional amygdala), individuals with this condition engage in extreme orality and will indiscriminately pick up various objects and place them in their mouths, regardless of its appropriateness. There is a repetitive quality to this behavior, for once they put the objects down they seem to have *forgotten* that they had just *explored* them, and will immediately pick them up and place them again in their mouths, as if the objects were completely unfamiliar.

Although ostensibly exploratory, there is thus a failure to learn, to remember, to discern motivational significance, to habituate with repeated contact, or to discriminate between appropriate versus inappropriate stimuli. Rather, when the amygdala has been removed bilaterally, the organism reverts to the most basic and primitive modes of object and social-emotional interaction (Brown and Schaffer, 1888; Gloor, 1960; Klüver and Bucy, 1939; Weiskrantz, 1956) such that even the ability to appropriately interact with loved ones is impaired.

For example, Terzian and Ore (1955) describe a young man who, following bilateral removal of the amygdala, subsequently demonstrated an inability to recognize anyone, including close friends, relatives and his mother. He ceased to respond in an emotional manner to his environment and seemed unable to recognize feelings expressed by others. He also demonstrated many features of the Klüver-Bucy syndrome (perseverative oral "exploratory" behavior and psychic blindness), as well as an insatiable appetite. In addition, he became extremely socially unresponsive such that he preferred to sit in isolation, well away from others.

When primates who have undergone bilateral amygdaloid removal are released from captivity and allowed to return to their social group, a social-emotional agnosia becomes readily apparent, as they no longer respond to or seem able to appreciate or understand emotional or social nuances. Indeed, they appear to have little or no interest in social activity and persistently attempt to avoid contact with others (Dicks et al., 1969; Jonason and Enloe, 1972; Kling and Brothers, 1992; Jonason et al., 1973). If approached they withdraw, and if followed they flee. Indeed, they behave as if they have no understanding of what is expected of them or what others intend or are attempting to convey, even when the behavior is quite friendly and concerned. Among adults with bilateral lesions, total isolation seems to be preferred.

In addition, they no longer display appropriate social or emotional behaviors, and if kept in captivity their dominance in a group or competitive situation will decline, even if they were formerly dominant (Bunnel, 1966; Dicks et al., 1969; Fuller et al., 1957; Jonason and Enloe, 1971; Jonason et al., 1973; Rosvold et al., 1954).

As might be expected, maternal behavior is severely affected. According to Kling (1972), a mother will behave as if her "infant were a strange object to be mouthed, bitten and tossed around as though it were a rubber ball."

Emotional Language and the Amygdala

Although cries and vocalizations indicative of rage or pleasure have been elicited via hypothalamic stimulation, of all limbic nuclei the amygdala is the most vocally active, particularly the lateral division (Robinson, 1967). In humans and animals, a wide range of emotional sounds have been evoked through amygdala activation, such as sounds indicative of pleasure, sadness, happiness, and anger (Robinson, 1967; Ursin and Kaada, 1960). Conversely, in humans, destruction limited to the amygdala

(Freeman and Williams, 1952, 1963), the right amygdala in particular, has abolished the ability to sing, convey melodic information, or properly enunciate via vocal inflection. Similar disturbances occur with right hemisphere damage (Joseph, 1988a). Indeed, when the right temporal region (including the amygdala) has been grossly damaged or surgically removed, the ability to perceive, process, or even vocally reproduce most aspects of musical and emotional auditory input is significantly curtailed (see Chapter 14).

Emotion and Temporal Lobe Seizures

The amygdala is buried within the depths of the anterior-inferior temporal lobe and maintains rich interconnections with areas throughout the temporal neocortex. Because of their intimate association, damage to the temporal lobe, particularly the anterior regions, often involves and disrupts amygdaloid functioning. In fact, because the amygdala and inferior-anterior temporal lobe have, of all brain regions, the lowest seizure threshold, and are minimally resistant and thus maximally vulnerable to developing abnormal seizure activity, even

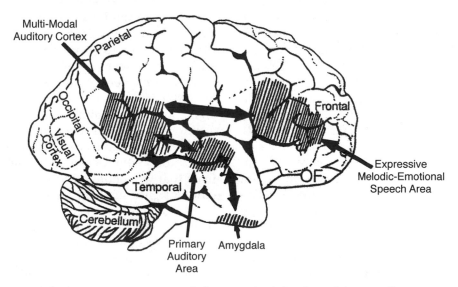

Figure 5.11 Schematic representation of the melodic-emotional language areas within the right hemisphere. Auditory information is received in the primary auditory area and the amygdala, where vocal emotional nuances are identified and analyzed.

mild injuries may result in kindling (i.e., abnormal activation), and thus disruption of their functional integrity. Indeed, damage to adjacent tissue has been known to spread, via kindling, to the amygdala and inferior regions (Cain, 1992). One consequence is temporal lobe epilepsy.

Personality, emotional, and sexual disturbances are a frequent complication of temporal lobe seizures in a significant minority of patients. Such individuals may develop paranoid, hysterical, or depressive tendencies, deepening of mood, hyposexuality, and other characteristics suggestive of affective disorders (Bear et al., 1982; Gibbs, 1951; Gloor, 1986, 1993; Herman and Chambria, 1980; Strauss et al., 1982; Williams, 1956). Immediately following or during the course of a seizure, 10% or more of such individuals experience a change in emotionality (Herman and Chambria, 1980; Strauss et al., 1982; Williams, 1956). In part, because the highest incidence of psychiatric disorders occurs in cases where the EEG spike focus is in the anterior temporal area (Gibbs, 1951), and because limbic nuclei such as the amygdala are frequently involved, it has been postulated that seizure activity sometimes hyperactivates these nuclei (Bear, 1979); this, in turn, distorts the affective meaning applied to afferent streams of visual, auditory, and somesthetic information (Gibbs, 1951). All aspects of emotional intellectual functioning are impaired.

Thus, during a seizure, individuals may be temporarily overwhelmed by feelings, such as fear. Or things they see or hear seem to become abnormally invested with emotional significance (Bear et al., 1979; Gibbs, 1951; Gloor, 1992; Gloor et al., 1982), presumably due to abnormal amygdala activation. Interestingly, one common symptom of temporal lobe epilepsy is an aura of tastes and, more often, odors that are usually quite unpleasant (e.g., like "burning wire," "burning shit," "burning rubber or tires," etc.)

Seizure-induced emotional changes tend to predominantly involve feelings of depression, pleasure, displeasure, or fear—with fear being one of the most common emotional experiences (Gloor, 1992; Gloor

et al., 1982; Halgren, 1992; Williams, 1956). More rarely, seizures involving sexual behavior, crying, laughing, or rage-like responses have been associated with temporal lobe epilepsy.

Some evidence suggests that certain emotional changes are more frequently associated with seizures originating in the right temporal lobe and/or amygdala, whereas disturbances involving thought (e.g., psychosis, schizophrenia) characterize left temporal lobe abnormalities (Flor-Henry, 1969, 1983; Offen et al., 1976; Joseph, 1988a; Sherwin, 1981; Schiff et al., 1982; Taylor, 1975; Weil, 1956; see also Chapters 3 and 4). For example, Gloor et al. (1982) in their depth electrode study of five individuals with seizure disorders, found that all feelings of fear and displeasure were associated with right temporal, right amygdala, or right hippocampal activation. These findings are consistent with the observations of Slater, Beard, and Glithero (1963), Bear and Fedio (1977), Flor-Henry (1969), and others, who have noted an association between right temporal seizures and affective disorders.

Other reports, however, have been less clear-cut—presumably due to the fact that seizures originating in the amygdala/temporal lobe can quickly spread from one hemisphere to another (via the anterior commissure), and the failure to employ depth electrodes to pinpoint the seizure foci. Hence, in some instances, emotions such as fear seem to arise regardless of which hemisphere is involved (Strauss et al., 1982). Nevertheless, even in these cases an emotional dichotomy is apparent.

For example, Herman and Chambria (1980) have described cases with right temporal foci who developed free-floating fears that were not tied to something specific but encompassed terrifying, "death-like," and "nightmarish" feelings, whereas one individual with a left temporal foci developed specific fears of certain individuals and situations. Similarly, Weil (1956) reports that the majority of his patients with right-sided foci developed intense fears but were unable to describe what they were afraid of.

Crying Seizures

As noted, stimulation of the amygdala can significantly alter facial emotional expression, including tearing. In a small minority of cases, right temporal seizures have been reported to cause paroxysmal attacks of weeping, with lacrimation and the making of mournful sounds, including sobbing and crying (Offen et al., 1976). However, crying as well as laughing seizures have also been noted to occur with left-sided involvement (Chen and Forster, 1973; Sethi and Raio, 1976). Nevertheless, without the benefit of depth electrodes, such as employed by Gloor and colleagues (1982), it is difficult to determine in which amygdala (right or left) and/or temporal lobe a seizure actually originates.

Sexual Seizures

The amygdala and regions of the hypothalamus are sexually dimorphic, and stimulation of either area can trigger sexual behavior. Similarly, sensations of sexual excitement, sometimes leading to orgasm, may also occur as a function of seizures originating in the temporal (Currier et al., 1971; Freemon and Nevis, 1969; Gloor, 1992; Remillard et al., 1983) and frontal lobes (Spencer et al., 1983). Seven of 10 patients with sexual seizures described by Remillard, et al. (1983) had right hemisphere foci. Similar findings were reported by Freemon and Nevis (1969), Penfield and Rasmussen (1950), and Spencer et al. (1983).

Sexual seizures due to temporal lobe abnormalities are often accompanied by actual sexual behavior. "The patient was sitting at the kitchen table with her daughter making out a shopping list. She stopped, appeared dazed, slumped to the floor on her back, lifted her skirt, spread her knees and elevated her pelvis rhythmically. She made appropriate vocalizations for sexual intercourse, such as, "it feels so good" and "further, further" (Currier et al., 1971:260).

Frontal Lobe Sexual Seizures

The frontal lobes, the orbital region in particular, maintain rich interconnections with the amygdala (Nauta, 1964) and receive olfactory projections as well (Tanabe et al., 1975). Epileptiform activity arising in the deep frontal (orbital) regions have also been associated with the development of "sexual seizures," including exhibitionism, genital manipulation, and masturbatory activity (Spencer et al., 1983).

Similar to that described regarding right temporal emotional and sexual disturbances, Spencer et al. (1983) found that three of their four patients with sexual automatism had seizures originating in the right frontal area, whereas the remaining patient had bifrontal disturbances. Other investigators have also noted that peculiar disturbances in emotion and personality are far more likely to arise following right versus left frontal damage (Joseph, 1986a, 1988a; Hillbom, 1960; Lishman, 1968).

Presumably, at least in part, emotional and sexual disturbances associated with right temporal and right frontal (orbital) dysfunction are due to activation of the limbic structures with which they are intimately interconnected. However, as discussed in Chapter 1, the inferior temporal and orbital frontal lobes are outgrowths of, and thus part of, the limbic system.

The Amygdala, the Anterior Commissure, Sexuality, and Emotion

When the amygdala or the bed nuclei for the anterior commissure of both cerebral hemispheres are damaged, hyperactivated, or completely inhibited, a striking disturbance in sexual and social behavior is evident (Brown and Schaeffer, 1888; Gloor, 1960; Klüver and Bucy, 1939; Terzian and Ore, 1955; Schriner and Kling, 1953). Specifically, humans, nonhuman primates, and felines who have undergone bilateral amygdalectomies will engage in prolonged, repeated, and inappropriate sexual behavior and masturbation, including repeated sexual acts with members of different species (e.g., a cat with a dog, a dog with a turtle, etc.). When activated from seizures, patients may involuntarily behave in a sexual manner and even engage in what appears to be intercourse with an imaginary

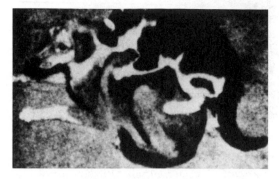

Figure 5.12   Social-emotional agnosia and inappropriate hypersexuality after bilateral destruction of the amygdala. Amygdalotomized cat attempting to have sex with a dog. (From Schriener LH, Kling A. Behavioral changes following rhinencephalic injury in the cat. J Neurophysiol 16:643,1953.)

partner. This abnormality is one aspect of the symptoms of Klüver-Bucy syndrome.

As noted, portions of the hypothalamus and amygdala are sexually dimorphic; i.e., there are male and female amygdaloid nuclei. For example, female primate amygdala neurons are more numerous and packed more closely together—a condition that would enhance electrical excitability, lower response thresholds, and increase susceptibility to kindling and, thus, hyperexcitation. In addition, electrical stimulation of the medial amygdala results in sex-related behavior and activity. In females, this includes ovulation, uterine contractions, and lactogenic responses, and in males, penile erections (Robinson and Mishkin, 1968; Shealy and Peele, 1957).

Moreover, the anterior commissure, the band of axonal fibers that interconnects the right and left amygdala/temporal lobe is sexually differentiated. Like the corpus callosum, the anterior commissure is responsible for information transfer as well as inhibition within the limbic system.

The female anterior commissure is 18% larger than in the male (Allen and Gorski, 1992). It has been argued that the increased capacity of the right and left female amygdala to communicate (via the anterior commissure), coupled with the more numerous and more densely packed neurons within the female amygdala (which in turn would

decrease firing thresholds and enhance communication) and the sex differences in the hypothalamus, predisposes females to be more emotionally intelligent and socially sensitive, perceptive, and expressive (Joseph, 1993). These limbic sex differences also induce her to be less aggressive and more compassionate, and affects her sexuality, feelings of dependency and nurturance, and desire to maintain and form attachments in a manner different from males.

In contrast, whereas the right and left female amygdala are provided a communication advantage not shared by males, the "male" amygdala, in turn, may be more greatly influenced by the (medial) hypothalamus via the stria terminalis, which is larger in men than in women (Allen and Gorski, 1992). The male limbic system would also be significantly impacted by high levels of circulating testosterone and other androgens.

Although environmental influences can shape and sculpt behavior and the functional organization of the brain (Chapter 18), most sex differences are innate and shared by other species—a direct consequence of the presence or absence of testosterone during adulthood and fetal development (see Gerall et al., 1992; Joseph, 1993; Joseph et al., 1978) and the sexual differentiation of the limbic system.

Because the limbic system is concerned with emotion, and since the female amygdala appears to be superior to that of males in this regard(whereas the male limbic system is more greatly influenced by the activating and aggression-inducing hormone, testosterone, as well as the medial hypothalamus), it is thus not surprising that women, and females of many other species, have been shown to be more nurturing, empathic, sympathetic, selfsacrificing, sensitive, and compassionate (Bakan, 1966; Belle, 1982; Blakemore, 1990; Berman, 1983; Eagly and Steffen, 1984; Fresbach, 1982; Graham, 1988; Hoffman, 1977). Human females also have a richer inner emotional life and are better able to understand, perceive, and express socialemotional nuances as compared to males

(Burton and& Levy, 1989; Brody, 1985; Buck, 1977, 1984; Buck et al., 1974, 1982; Card et al., 1986; Eisenberg et al., 1989; Fuchs and Thelan, 1988; Harackiewicz, 1982; Kemper, 1978; Lewis, 1983; Rubin, 1983; Shennum and Bugental, 1982; Soloman and Ali, 1972; Strayer, 1985).

Females are also more willing to express emotional issues and confide and discuss personal problems with others (Gilbert, 1969; Gilligan, 1982; Demos, 1975; Lutz, 1980; Pratt, 1985; Walker et al., 1987; Parelman, 1980; Lombardo and Levine, 1981), and are much more likely than males to cry (de Beauvoir, 1952; Thomas, 1993), or to freeze, panic, or run away in fear—functions and behaviors directly associated with amygdala activation.

Even female memories are more emotional and more inclined toward social and interpersonal concerns (Pratt, 1985; Walker et al., 1987; Friedman and Pines, 1991). Hence, women are superior to men in recalling emotional memories (Pratt, 1985; Walker et al., 1987; Friedman and Pines, 1991) and can recall events that some husbands or boyfriends swear did not even occur.

In contrast, males have difficulty discussing personal difficulties or expressing their emotions, other than anger and happiness (Balswick, 1982, 1988; Goldberg, 1976; O'Neil, 1982; Joseph, 1992b, 1993; Sattel, 1989) and are much more inclined to develop psychopathological conditions, such as sociopathy (Draper and Harpending, 1988). Many male criminals are sociopaths, and sociopaths, in general, have very little human regard, empathy, compassion, or concern for others.

As will be discussed in further detail below (and in Chapters 8 and 16), the activation of the amygdala-temporal lobes is also associated with the capacity to have religious experiences. Thus, females are not just more emotional, but women have more intense religious experiences, attend church more often, are more involved in religious activities, involve their children more in religious studies, hold more orthodox religious views, incorporate religious beliefs more often in their daily lives and activities, and pray more often, as well (Argyle and Beit-Hallahami, 1975; Batson and Ventis, 1982; De Vaus and McAllister, 1987; Lazerwitz, 1961; Lindsey, 1990; Sapiro, 1990). Presumably, this sex difference also is a consequence of sex differences in the structure and function of the limbic system, the amygdala and anterior commissure in particular (see Chapter 8).

It is tempting to speculate that these limbic system sex differences may also predispose females to become more easily emotionally stressed and upset—which in turn might affect the functioning of the hypothalamus and other nuclei, thereby giving rise to a greater incidence of depression (including SAD), as well as hysteria and related disorders, including "chronic fatigue syndrome"; conditions that are more common among women than men. Indeed, in almost all surveys of developed countries, females suffer from two to three times as many mood disorders as men (Mollica, 1989).

Male Aggression Is a Source of Male Pleasure

Due to sex differences in the size and functioning of the hypothalamus and amygdala, and given that these limbic nuclei control the ability to behave violently or experience pleasure or displeasure as well as sexual orgasm, some behaviors, like nurturance or aggression, are rewarding in themselves, depending if one possesses a male versus a female limbic system (Joseph, 1992b, 1993). That is, males and females experience pleasure in response to different experiences, e.g., holding and nurturing a baby versus watching a boxing match or playing a violent video game. This is because limbic neurons that respond to a baby's face versus a violent encounter are differentially activated or differentially generate pleasure or excitment depending on if the limbic system is "male" or "female."

For example, male humans, chimpanzees, birds, fowl, fish, mice, and rats will look for fights and opportunities to behave antagonistically or destructively because they experience pleasure when they behave aggressively. Males frequently seek aggressive competitive situations and the opportunity to behave combatively, even in otherwise neutral environments (Calhoun, 1950; Goodall, 1986; Hamburg, 1971; Johnson, 1972; Lorenz, 1966; Manning, 1972; McGrew and McGrew, 1970; Mitchell, 1979; Moyer, 1974).

In learning situations in which subjects are able to control their number of antagonistic encounters, males tend to chose situations where they can behave belligerently much of the time (Johnson, 1972). Even male mice will learn a maze when the only reward is the opportunity to attack another mouse (Tellegen et al., 1969). Females tend to have no such desire except under certain conditions, e.g., threat to her brood or the presence of another sexually receptive (attractive) female.

It is not uncommon for gangs of young human males to break and destroy things just for the pleasure of the destruction or to simply look for a fight or the opportunity to attack someone (McGrew and McGrew, 1970; Sears and Beach, 1965). Male chimpanzees often look for opportunities to attack and put much energy into creating encounters with strangers from neighboring troops. In fact, they often make raids into the core areas of neighboring groups just to fight and even kill one another. According to Goodall (1986), "These encounters are highly attractive to the participants. To some extent, all adult males show a keenness to participate in these exciting incidents." In fact, male chimps will become extremely aroused by violence and fights occurring among members of their own troop and may charge in and attack innocent bystanders, as well as join the general melee simply for the sake of fighting (Goodall, 1986). Human males responding likewise have been the fare of media depictions of barroom brawls since the inception of motion pictures.

Indeed, almost regardless of species, be it male deer, oxen, seals, walruses, wolves, sperm whales, hippopotamuses, or even Australian marsupials, males not only derive more enjoyment from aggression, but they are far more aggressive and violent than females in general (Elia, 1988; Fedi-

gan, 1992; Goodall, 1986; Hamburg, 1971; Johnson, 1972; Lorenz, 1966; Manning, 1972; Mitchell, 1979; Moyer, 1974). These differences are also characteristic of the majority of social primates, including humans, and are directly related to sex differences in the structure, organization, and interconnections between the amygdala and hypothalamus.

Over 80% of all violent crimes and murders are committed by males (Elliott et al., 1986). Similarly, male chimpanzees are much more violent and initiate attack five times as frequently as females (Van Lawick-Goodall, 1968), and adult male chimps engage in 90% of all aggressive behaviors (Bygott, 1979). Moreover, these differences are found among male baboons, gorillas, and rhesus monkeys, (Mitchell, 1979; Fedigan, 1992; Elia, 1988; Jolly, 1972) as well as chimps who are reared in a nursery with peers only (Nadler and Braggio, 1974) and those raised in social isolation (Sackett, 1974). Indeed, sex differences in violence and belligerence are apparent even in infancy, appearing as early as the first two months of life, even in chimps and other primates (Ransom and Rowell, 1972; Mitchell, 1979). Hence, male aggression is not learned, per se, but is an innate, limbic characteristic that is common among most species and which is directly related to sex differences in the limbic system. However, this is not to say that females are not aggressive, for this is not the case (Thomas, 1993).

The Limbic System and Testosterone

In large part, these and related sex differences in aggressiveness are also a consequence of the relatively higher concentrations of the activating hormone, testosterone, flowing through male bodies and brains. The overarching influence of neurological and hormonal predispositions are also indicated by studies showing that females who have been prenatally exposed to high levels of masculinizing hormones (i.e., androgens) behave similarly to males, even in regard to enhanced spatial abilities (Joseph et al., 1978; see Gerall et al., 1992).

They are also more aggressive and engage in more rough and tumble play than normal females (Money and Ehrhardt, 1972; Ehrhardt and Baker, 1974; Reinisch, 1974).

Similarly, female primates and other mammals who have been exposed to testosterone during neonatal development display an altered sexual orientation, as well as significantly higher levels of activity, competitiveness, combativeness, and belligerence (Mitchell, 1979). Nevertheless, it is important to reemphasize that it is generally the presence or absence of testosterone during the critical period of neuronal differentiation that determines if one is in possession of a "male" versus a "female" limbic system.

Sexual Orientation and Heterosexual Desire

As noted, the amygdala surveys the environment, searching out stimuli, events, or individuals that are emotionally, sexually, or motivationally significant and contains neurons which respond only to specific stimuli. Moreover, it contains facial recognition neurons, which are sensitive to different facial expressions and capable of determining the sex of the individual being viewed, and which become excited when looking at a male versus a female face (Leonard et al., 1985; Rolls, 1984). In this regard, the amygdala can act to discern and detect potential sexual partners and then motivate sex-appropriate behavior, culminating in sexual intercourse and orgasm.

That is, an individual who possesses a "male" limbic system is likely to view the female face, body, and genitalia as sexually arousing because the amygdala and limbic system respond with pleasure when stimulated by these particular features. Conversely, male physical features are likely to excite and sexually stimulate the limbic systems possessed by heterosexual females and homosexual males (Joseph, 1993). This happens because, at a very basic level, emotional, sexual, and motivational perceptual/behavioral functioning becomes influenced and guided by the anatomical sexual bias of the host.

THE HOMOSEXUAL LIMBIC SYSTEM

It is recognized that some men become homosexual or bisexual due to sexual abuse experienced as children, whereas in others there is a suggestion of a genetic contribution. Nevertheless, it has been reported that the hypothalamus (i.e., the ventromedial and anterior nuclei) in male homosexuals is organized in a manner similar to the female pattern of development (LeVay, 1991; Swaab and Hoffman, 1990). Moreover, the anterior commissure is not only larger in females but is 35% larger in homosexual males versus male heterosexuals (Allen and Gorski, 1992). Coupled with the evidence reviewed above, this raises the possibility that sex differences in male versus female and heterosexual male versus homosexual male sexual orientation, and thus the capacity to experience sexual pleasure when with a heterosexual or homosexual partner, may be determined by these nuclei.

For example, it could be assumed that homosexual males respond to males with feelings of sexual attraction and desire, as do heterosexual females, i.e., because they are in possession of a "female" limbic system, which responds to male physical and facial features with sexual arousal; i.e., the amygdala contains neurons that respond to faces and facial expressions, and which can determine the sex of the individual viewed (e.g., Leonard et al., 1985; Rolls, 1984). Conversely, heterosexual males, being in possession of a "male" hypothalamus and amygdala, not only respond to females with sexual arousal, but they behave and act more aggressively than females and homosexual males.

Moreover, one might expect homosexuals with a "female" limbic system to be inclined to behave in a manner similar to women. Indeed, homosexual males (in general) and females tend to be more alike than different in regard to social-emotional reactions and tendencies (Tripp, 1987). In some cases, these feminine tendencies are grossly exaggerated (Tripp, 1987): i.e., the "swishy" male with the exaggerated high-pitched voice.

A significant number of homosexuals, in fact, are psychologically similar to females in a number of ways, including showing a high interest in fashion and wearing apparel, and having a pronounced tendency to employ feminine body language and vocal tones, to shun sports and avoid fights, and to fear physical injury, particularly during childhood (Bell et al., 1981; Bieber et al., 1962; Van Den Aardweg, 1980; Tripp, 1987). Many also tend to maintain intense dependency relations with their mothers and to remain distant from strong male figures, including their fathers (Green, 1987); heterosexual males (including male primates; e.g., Fedigan, 1992; Goodall, 1986) tend to behave in a completely different fashion.

As children, homosexual males tended to prefer female companions and friends, to prefer girls' toys, activities, and often girls' clothes, and to behave in an effeminate manner (Bell et al., 1981; Saghir and Robins, 1973; Grellet et al., 1982; Green, 1987). Indeed, from 67% to 75% of homosexuals versus 2% to 3% of heterosexual males reported being "feminine" and more like girls than boys as children (Saghir and Robins, 1973; Green, 1987). Moreover, homosexual males, like heterosexual females, demonstrate comparatively inferior spatial perceptual capabilities as compared to heterosexual males (Gladue et al., 1990; Sanders and Ross-Field, 1986; Wilmot and Brierley, 1984). Homosexuals tend to perform similarly to females on these spatial tests.

It is tempting to speculate that because some homosexuals demonstrate an almost hyperdeveloped pattern in the structure of the anterior commissure (and thus, presumably, the amygdala, as well as the hypothalamus), this may account for what some might consider excessive, if not dangerous and sometimes "bizarre" (e.g., "fist fucking") sexual promiscuity and indiscriminate "orality," "anality," and group sex among 24% to 33% of this population (Arron, 1973; Gans, 1993; Pollak, 1993; Symons, 1979; Tripp, 1987)—manifestations of the Klüver-Bucy syndrome? It may also account for the high incidence of emotional disorders, including suicide, in homosexuals: i.e., the consequence of a dysfunctional and abnormal amygdala and limbic system.

This is not to negate the impact of social-environmental influences on the development of these "abnormalities," as the environment can have a significant impact on the amygdala (Diamond, 1985) as well as on one's self concept (Joseph, 1992b). Nor is it meant to imply that the majority of homosexuals engage in these excesses. Even so, the possibility of a limbic system "abnormality" in regard to homosexual self-destructiveness and indiscriminate and sometimes compulsive sexuality should not be ruled out.

Overview: The Amygdala

Over the course of early evolutionary development, the hypothalamus reigned supreme in the control and expression of raw and reflexive emotionality, i.e., pleasure, displeasure, aversion, and rage. Largely, however, it has acted as an *eye* turned inward, monitoring internal homeostasis and concerned with basic needs. With the development of the amygdala, the organism became equipped with an *eye* turned outward, so that the external emotional features of reality could be tested and ascertained. When signalled by the hypothalamus, the amygdala begins to search the sensory array for appropriate emotional-motivational stimuli until what is desired is discovered and attended to.

However, with the differentiation of the amygdala, emotional functioning also became differentiated and highly refined. The amygdala hierarchically wrested control of emotion from the hypothalamus.

The amygdala is primary in regard to the perception and expression of most aspects of emotionality, including fear, aggression, pleasure, happiness, sadness, and so forth, and, in fact, assigns emotional or motivational significance to that which is experienced. It is primary in regard to all aspects of social and emotional intelligence. It can thus induce the organism to act on something seen, felt, heard, or anticipated. The integrity of the amygdala is essential in regard to the analysis of social-emotional nuances, the organization and mobilization of the person's internal motivational status regarding these cues, as well as the mediation of higher-order emotional expression and impulse control. When it is damaged or functionally compromised, social-emotional intellectual functioning becomes grossly disturbed.

The amygdaloid nucleus, via its rich interconnections with other brain regions, is able to sample and influence activity occurring in other parts of the cerebrum and add emotional color to one's perceptions. As such, it is highly involved in the assimilation and association of divergent emotional, motivational, somesthetic, visceral, auditory, visual, motor, olfactory, and gustatory stimuli. Thus, it is very concerned with learning, memory, and attention, and can generate reinforcement for certain behaviors. Moreover, via reward or punishment, it can promote the encoding, storage and later retrieval of particular types of information. That is, learning often involves reward, and it is via the amygdala (in concert with other nuclei) that emotional consequences can be attributed to certain events, actions, or experiences, as well as extracted from the world of possibility so that it can be attended to and remembered.

Lastly, as is evident from studies of individuals with abnormal activity or seizures originating in or involving this nuclei, the amygdala is able to overwhelm the neocortex and thus gain control over behavior. As based on electrophysiological studies, the amygdala seems capable of literally *turning off* the neocortex (such as occurs during a seizure), at least for brief periods. That is, the amygdala can induce electrophysiological slow-wave theta activity in the neocortex, which indicates low levels of arousal (see below) as well as high-voltage fast activity. In the normal brain it probably exerts similar influences, such that, at times, individuals (i.e., their neocortex) "lose control" over themselves and respond in a highly emotionally charged manner.

HIPPOCAMPUS

The hippocampus ("Ammon's horn" or the "sea horse") is an elongated structure located within the inferior medial wall of the temporal lobe (posterior to the amygdala) and surrounds, in part, the lateral ventricle. In humans it consists of an anterior and posterior region and is shaped somewhat like a telephone receiver (or "sea horse").

There are three major neural pathways leading to and from the hippocampus. These include the fornix-fimbrial fiber system, a supracallosal pathway (i.e., the indusium griseum) that passes through the cingulate, and the entorhinal area, which is sometimes referred to as the gateway to the hippocampus.

It is via the entorhinal area that the hippocampus receives olfactory and amygdaloid projections (Amaral et al., 1992; Carlsen et al., 1982; Gloor, 1955; Krettek and Price, 1977; Murray, 1992; Steward, 1976) and fibers from the orbital frontal and temporal lobes (Van Hoesen et al., 1972). It is through the fornix and fimbrial pathways that the hippocampus makes major interconnections with the thalamus, septal nuclei, medial hypothalamus, and via which it exerts either inhibitory or excitatory influences on these nuclei (Feldman et al., 1987; Guillary, 1955; Poletti and Sujatanon, 1980).

SEPTAL INTERACTIONS

The hippocampus maintains a particularly intimate relationship with the septal nuclei. The septal nucleus partly serves as an interactional relay center, as it channels hippocampal influences to other structures, such as the hypothalamus and reticular formation (and vice versa), and as a major link through which the hippocampus and amygdala sometime interact (Hagino and Yamoaka, 1976).

AMYGDALA INTERACTIONS

The hippocampus is greatly influenced by the amygdala, which in turn monitors

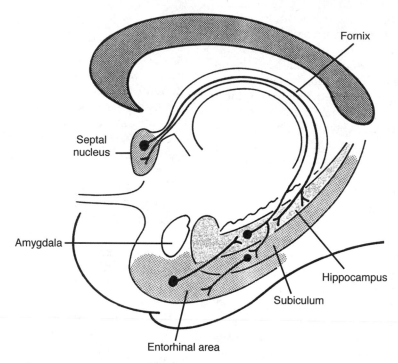

Figure 5.13   Depiction of some of the major connections of the hippocampus.

and responds to hippocampal activity (Gloor, 1955, 1992; Green and Adey, 1956; Halgren, 1992; Steriade, 1964). The amygdala also acts to relay certain forms of information from the hippocampus to the hypothalamus (Poletti and Sajatanon, 1980). Together, the hippocampus and amygdala complement and interact in regard to attention, the generation of emotional and other types of imagery, learning, and memory.

Hippocampal Arousal, Attention, and Inhibitory Influences

Various authors have assigned the hippocampus a major role in information processing, including memory, new learning, cognitive mapping of the environment, and voluntary movement toward a goal, as well as attention, behavioral arousal, and orienting reactions (Douglas, 1967; Eichenbaum et al., 1994; Frisk and Milner, 1990; Grastyan et al., 1959; Green and Arduini, 1954; Isaacson, 1982; Milner, 1966, 1970, 1971; Olton et al., 1978; Routtenberg, 1968; Squire, 1992; Victor and Agamanolis, 1990). For example, hippocampal cells greatly alter their activity in response to certain spatial correlates, par-

ticularly as an animal moves about in its environment (Nadel, 1991; O'Keefe, 1976; Olton et al., 1978; Wilson and McNaughton, 1993). It also develops slow-wave theta activity during arousal (Green and Arduini, 1954) or when presented with noxious or novel stimuli (Adey et al., 1960).

However, few studies have implicated this nucleus as important in emotional functioning per se, although responses such as "anxiety" or "bewilderment" have been observed when subjects are directly electrically stimulated (Kaada et al., 1953). Indeed, in response to persistent and repeated instances of stress and unpleasant emotional arousal, the hippocampus appears to cease to participate in cognitive, emotional, or memory processing (Joseph, 1990a, 1992a; see Chapters 6 and 16). Thus, the role of the hippocampus in emotion is somewhat minimal.

Arousal

HIPPOCAMPAL-NEOCORTICAL INTERACTIONS

Desynchronization of the cortical EEG is associated with high levels of arousal and

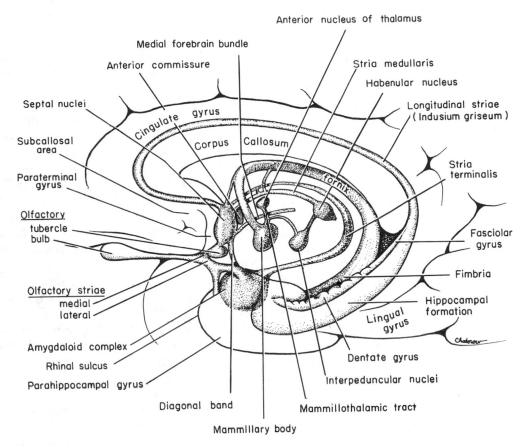

Figure 5.14 Schematic diagram of the limbic system. (From Carpenter M. Core text of neuroanatomy. Baltimore: Williams & Wilkins, 1991.)

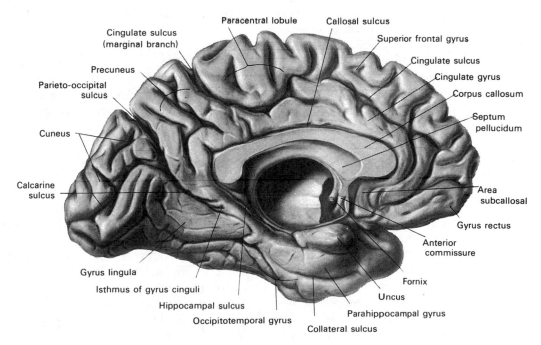

Figure 5.15 Medial surface of the left hemisphere. (From Mettler FA. Neuroanatomy. St. Louis: CV Mosby, 1948.)

information input. As the level of input increases, the greater is the level of cortical arousal (Como et al., 1979; Joseph et al., 1981). However, when arousal levels become too great, efficiency in information processing, memory, new learning, and attention become compromised as the brain becomes overwhelmed.

When the neocortex becomes desynchronized (indicating cortical arousal), the hippocampus often (but not always) develops slow-wave theta activity (Grastyan et al., 1959; Green and Arduni, 1954) such that it appears to be functioning at a much lower level of arousal. Conversely, when cortical arousal is reduced to a low level (indicated by EEG synchrony), the hippocampal EEG often becomes desynchronized.

These findings suggest that when the neocortex is highly stimulated the hippocampus, in order to monitor what is being received and processed, functions at a level much lower, in order not to become overwhelmed. When the neocortex is not highly aroused, the hippocampus presumably compensates by increasing its own level of arousal so as to tune in to information that is being processed at a low level of intensity.

Hence, in situations where both the cortex and the hippocampus become desynchronized, there results distractibility and hyperresponsiveness such that the subject becomes overwhelmed and confused, and may orient to and approach several stimuli (Grastyan et al., 1959). Attention, learning, and memory functioning are decreased. Situations such as this sometimes also occur when individuals are highly anxious or repetitively traumatized, emotionally or physically (see Chapter 16).

There is also evidence to suggest that the hippocampus may act so as to reduce extremes in cortical arousal. For example, whereas stimulation of the reticular activating system augments cortical arousal and EEG evoked potentials, hippocampal stimulation reduces or inhibits these potentials such that cortical responsiveness and arousal is dampened (Feldman, 1962; Redding, 1967). On the other hand, if cortical arousal is at a low level, hippocampal stimulation often results in an augmentation of the cortical evoked potential (Redding, 1967).

The hippocampus also exerts desynchronizing or synchronizing influences on various thalamic nuclei, which in turn augments or decreases activity in this region (Green and Adey, 1956; Guillary, 1955; Nauta, 1956, 1958). As the thalamus is the major relay nucleus to the neocortex, the hippocampus therefore appears able to block or enhance information transfer to various neocortical areas. Indeed, it may be acting to insure that certain precepts are stored in memory at the level of the neocortex (Squire, 1992).

It is, thus, likely that the hippocampus may act to influence information reception and storage at the neocortical level as well as possibly reduce extremes in cortical arousal (be they too low or high) via inhibition or excitation, and/or it may act so that the neocortex is not overwhelmed or underwhelmed when engaged in the reception and processing of information. That is, very high or very low states of excitation are incompatible with alertness and selective attention as well as the ability to learn and retain information (Joseph et al., 1981).

AVERSION AND PUNISHMENT

In many ways, the hippocampus appears to act in concert with the medial hypothalamus and septal nuclei (with which it maintains rich interconnections) so as to prevent extremes in arousal and thus maintain a state of quiet alertness (or quiescence). Moreover, similar to the medial hypothalamus, it has been reported that the subjective components of aversive emotion in humans have reportedly been correlated with electrophysiological alterations in the hippocampus and septal area (Heath, 1976); i.e., the hippocampus soon ceases to respond.

The hippocampus also appears to be heavily involved in the modulation of reactions to frustrations or mild punishment (Gray, 1970, 1990), particularly in regard to single-trial but not multiple-trial learning. For example, the hippocampus responds with trains of slow theta waves when presented with noxious stimuli but habituates or ceases to respond with repeated presentation. In consequence, learning may take place in the absence of hippocampal participation (see Chapters 6 and 16).

Attention and Inhibition

The hippocampus participates in the elicitation of orienting reactions and the maintenance of an aroused state of attention (Foreman and Stevens, 1987; Grastyan et al., 1959; Green and Arduini, 1954; Routtenberg, 1968). When exposed to novel stimuli or when engaged in active searching of the environment, hippocampal theta appears (Adey et al., 1960). However, with repeated presentations of a novel stimulus the hippocampus habituates and theta disappears (Adey et al., 1960). Thus, as information is attended to, recognized, and presumably learned and/or stored in memory, hippocampal participation diminishes. Theta also appears during the early stages of learning as well as when a person is engaged in selective attention and the making of discriminant responses (Grastyan et al., 1959).

When the hippocampus is damaged or destroyed, animals have great difficulty inhibiting behavioral responsiveness or shifting attention. For example, Clark and Issacson (1965) found that animals with hippocampal lesions could not learn to wait 20 seconds between bar presses if first trained to respond to a continuous schedule. There is an inability to switch from a continuous to a discontinuous pattern, such that a marked degree of perseveration and inability to change sets or inhibits a pattern of behavior, once it has been initiated (Douglas, 1967; Ellen et al., 1964). Habituation is largely abolished and the ability to think or respond divergently is disrupted. Disinhibition due to hippocampal damage can even prevent the learning of a passive avoidance task, such as simply ceasing to move (Kimura, 1963).

Hence, it appears that the hippocampus acts to possibly selectively enhance or diminish areas of neural excitation which, in turn, allows for differential selective attention and differential responding, as well as the storage and consolidation of information into long-term memory. When damaged, the ability to shift from one set of perceptions to another, or to change behavioral patterns, is disrupted and the organism becomes overwhelmed by a particular mode of input. Learning, memory, and attention are greatly compromised.

Learning and Memory: The Hippocampus

The hippocampus is most usually associated with learning and memory encoding, e.g., long-term storage and retrieval of newly learned information (Fedio and Van Buren, 1974; Frisk and Milner, 1990; Milner, 1966; 1970; Penfield and Milner, 1958; Rawlins, 1985; Scoville and Milner, 1957; Squire, 1992; Victor and Agamanolis, 1990), particularly the anterior regions. Hence, if the hippocampus has been damaged, the ability to convert short-term memories into long-term memories becomes significantly impaired in humans (i.e., anterograde amnesia) (MacKinnon and Squire, 1989; Squire, 1992; Victor and Agamanolis, 1990) as well as primates (Zola-Morgan and Squire, 1984, 1985, 1986). In humans, memory for words, passages, conversations, and written material is also significantly impacted, particularly with left hippocampal destruction (Frisk and Milner, 1990; Squire, 1992). Bilateral destruction of the anterior hippocampus results in striking and profound disturbances involving memory and new learning (i.e., anterograde amnesia). For example, one such individual who underwent bilateral destruction of this nuclei (H.M.) was subsequently found to have almost completely lost the ability to recall anything experienced after surgery. If you introduced yourself to him, left the room, and then returned a few minutes later he would have no recall of having met or spoken to you. Dr. Brenda Milner has worked with H.M.

for almost 20 years and yet she is an utter stranger to him.

Memory for events prior to his surgery is, in comparison, exceedingly well preserved. However, H.M. is so amnesic for everything that has occurred since surgery that every time he rediscovers that his favorite uncle died (actually a few years before his surgery) he suffers the same grief as the time he was first informed.

Although without memory for new (nonmotor) information, H.M. has adequate intelligence, is painfully aware of his deficit, and constantly apologizes for his problem.

"Right now, I'm wondering," he once said, "Have I done or said anything amiss? You see, at this moment everything looks clear to me, but what happened just before? That's what worries me. It's like waking from a dream. I just don't remember... Every day is alone in itself, whatever enjoyment I've had, and whatever sorrow I've had... I just don't remember." (Blakemore, 1977:96).

Presumably, the hippocampus acts to protect memory and the encoding of new information during the storage and consolidation phase via the gating of afferent streams of information and the filtering/ exclusion (or dampening) of irrelevant and interfering stimuli. When the hippocampus is damaged there results input overload; the neuroaxis is overwhelmed by neural noise, and the consolidation phase of memory is disrupted, such that relevant information is not properly stored or even attended to. Consequently, the ability to form associations (e.g., between stimulus and response) or to alter preexisting schemas (such as occurs during learning) is attenuated (Douglas, 1967).

Hippocampal and Amygdaloid Interactions: Memory

It has been argued that significant impairments involving memory (in humans) cannot be produced by lesions restricted to the hippocampus (Horel, 1978; see also commentary in Eichenbaum et al., 1994). Hence, in some instances of restricted lesions, good recall of new information is possible for at least several minutes (Horel, 1978; Penfield and Milner, 1958; Squire, 1992).

Rather, there is evidence which strongly suggests that the hippocampus plays an interdependent role with the amygdala in regard to memory (Gloor, 1992; Halgren, 1992; Kesner and Andrus, 1982; Mishkin, 1978; Murray, 1992; Sarter and Markowitsch, 1985). For example, it appears that the amygdala is responsible for storing the emotional aspects and personal reactions to events in memory, whereas the hippocampus acts to store the cognitive, visual, and contextual variables (see Chapter 6).

Specifically, the amygdala plays a particularly important role in memory and learning when activities are related to reward and emotional arousal (Gaffan, 1992; Gloor, 1992; Halgren, 1992; LeDoux, 1992; Kesner, 1992; Rolls, 1992; Sarter and Markowitsch, 1985). Thus, if some event is associated with a positive or negative emotional state, it is more likely to be learned and remembered, that is, by the amygdala.

The amygdala becomes particularly active when recalling personal and emotional memories (Halgren, 1992; Heath, 1964; Penfield and Perot, 1963), and in response to cognitive-determined and context-determined stimuli, regardless of their specific emotional qualities (Halgren, 1992). However, once these emotional memories are formed, the specific emotional or associated visual context are sometimes required to trigger their recall (Rolls, 1992; Halgren, 1992). If those cues are not provided or cease to be available, the original memory may not be triggered and may appear to be forgotten or repressed. However, even emotional context can trigger memory (see also Halgren, 1992) in the absence of specific cognitive cues.

Similarly, it is also possible for emotional and nonemotional memories to be activated in the absence of active search and retrieval, and thus without hippocampal or frontal lobe participation. Recognition memory may be triggered by contextual or emotional cues. Indeed, there are a small group of neurons in the amygdala, as well as a larger group in the inferior temporal lobe, which are involved in recognition memory (Murray, 1992; Rolls, 1992). Because of amygdaloid sensitivity to visual and emotional cues, even long forgotten memories may be evoked via recognition, even when search and retrieval repeatedly fail to activate the relevant memory store.

According to Gloor (1992), "a perceptual experience similar to a previous one can, through activation of the isocortical population involved in the original experience, recreate the entire matrix which corresponds to it and call forth the memory of the original event and an appropriate affective response through the activation of amygdaloid neurons." This can occur "at a relatively non-cognitive (affective) level, and thus lead to full or partial recall of the original perceptual message associated with the appropriate affect."

In this regard, it appears that the amygdala is responsible for emotional memory formation whereas the hippocampus is concerned with storing verbal-visual-spatial and contextual details in memory. Thus, in rats and primates, damage to the hippocampus can impair retention of context, and contextual fear conditioning, but it has no effect on the retention of the fear itself or the fear reaction to the original cue (Kim and Fanselow, 1992; Phillips and LeDoux, 1992; Rudy and Morledge, 1994). In these instances, fear-memory is retained due to preservation of the amygdala. However, when both the amygdala and hippocampus are damaged, striking and profound disturbances in memory functioning result (Kesner and Andrus, 1982; Mishkin, 1978).

Therefore, the role of the amygdala in memory and learning seems to involve activities related to reward, orientation, and attention, as well as emotional arousal and social-emotional recognition (Gloor, 1992; Rolls, 1992; Sarter and Markowitsch, 1985). If some event is associated with positive or negative emotional states, it is more likely to be learned and remembered. That is, reward increases the probability of attention being paid to a particular stimulus or consequence as a function of its association

with reinforcement (Gaffan, 1992; Douglas, 1967; Kesner and Andrus, 1982).

Moreover, the amygdala appears to reinforce and maintain hippocampal activity via the identification of motivationally significant information and the generation of pleasurable rewards (through action on the lateral hypothalamus). However, the amygdala and hippocampus act differentially in regard to the effects of positive versus negative reinforcement on learning and memory, particularly when highly stressed or repetitively aroused in a negative fashion. For example, whereas the hippocampus produces theta in response to noxious stimuli, the amygdala increases its activity following the reception of a reward (Norton, 1970).

LATERALITY

It is now well known that lesions involving the mesial-inferior temporal lobes (i.e., destruction or damage to the amygdala/hippocampus) of the left cerebral hemisphere typically produce significant disturbances involving verbal memory—particularly in contrast to those in individuals with right-sided destruction. Left-sided damage disrupts the ability to recall simple sentences and complex verbal narrative passages, or to learn verbal paired-associates or a series of digits (Frisk and Milner, 1990; Milner, 1966, 1970, 1971; Squire, 1992).

In contrast, right temporal destruction typically produces deficits involving visual memory, such as the learning and recall of geometric patterns, visual or tactile mazes, emotional sounds, or human faces (Corkin, 1965; Milner, 1965; Kimura, 1963). Right-sided damage also disrupts the ability to recognize (via recall) olfactory stimuli (Rausch et al., 1977), or recall emotional passages or personal memories (see Chapter 3).

It appears, therefore, that the left amygdala and hippocampus are highly involved in processing and/or attending to verbal information, whereas the right amygdala/hippocampus is more involved in the learning, memory and recollection of non-verbal, visual-spatial, environmental, emotional, motivational, tactile, olfactory, and facial information. These issues and the differing roles of these nuclei in memory formation, as well as amnesia and repression, will be discussed in greater detail in Chapters 6 and 16.

THE PRIMARY PROCESS
Dreams, Hallucinations, and The Amygdala and Pleasure

The amygdala maintains a functionally interdependent relationship with the hypothalamus in regard to emotional, sexual, autonomic, consumatory, and motivational concerns. It is able to modulate and even control rudimentary emotional forces governed by the hypothalamic nucleus. However, the amygdala also acts at the behest of hypothalamically induced drives. For example, if certain nutritional requirements need to be met, the hypothalamus signals the amygdala, which then surveys the external environment for something good to eat (Joseph, 1982, 1992a). On the other hand, if the amygdala, via environmental surveillance, discovers a potentially threatening stimulus, it acts to excite and drive the hypothalamus as well as the basal ganglia so that the organism is mobilized to take appropriate action.

When the hypothalamus is activated by the amygdala, instead of responding in an on/off manner, cellular activity continues for an appreciably longer period (Dreifuss et al., 1968). The amygdala can tap into the reservoir of emotional energy mediated by the hypothalamus so that certain ends may be attained (Joseph, 1982, 1992a).

Amygdala and Hippocampal Interactions During Infancy

HALLUCINATIONS

The amygdala-hippocampal complex, particularly that of the right hemisphere, is very important in the production and recollection of nonlinguistic and verbal-emotional images associated with past experience. In fact, direct electrical stimula-

tion of the temporal lobes, hippocampus, and particularly the amygdala (Gloor, 1990) not only results in the recollection of images, but in the creation of fully formed visual and auditory hallucinations (Gloor, 1992; Halgren, 1992; Halgren et al., 1978; Horowitz et al., 1968; Malh et al., 1964; Penfield and Perot, 1963), as well as feelings of familiarity (e.g., deja vu).

Indeed, it has long been known that tumors invading specific regions of the brain can trigger the formation of hallucinations that range from the simple (flashing lights) to the complex. The most complex forms of hallucination, however, are associated with tumors within the most anterior portion of the temporal lobe (Critchly, 1939; Gibbs, 1951; Gloor, 1992; Halgren, 1992; Horowitz et al., 1968; Tarachow, 1941); i.e., the region containing the amygdala and anterior hippocampus.

Similarly, electrical stimulation of the anterior lateral temporal cortical surface—particularly the right temporal lobe—results in visual hallucinations of people, objects, faces, and various sounds (Gloor, 1992; Halgren, 1992; Horowitz et al., 1968) (Halgren et al., 1978). Depth electrode stimulation and, thus, direct activation of the amygdala and/or hippocampus is especially effective.

For example, stimulation of the right amygdala produces complex visual hallucinations, body sensations, deja vu, and illusions, as well as gustatory and alimentary experiences (Weingarten et al., 1977), whereas Freeman and Williams (1963) have reported that the surgical removal of the right amygdala in one patient abolished hallucinations. Stimulation of the right hippocampus has also been associated with the production of memory-like and dream-like hallucinations (Halgren et al., 1978; Horowitz et al., 1968).

The amygdala also becomes activated in response to bizarre stimuli (Halgren, 1992). Conversely, if activated to an abnormal degree, it may in turn produce bizarre memories and abnormal perceptual experiences. In fact, the amygdala contributes, in large part, to the production of very sexual as well as bizarre, unusual and fearful memories and mental phenomena, including dissociative states, feelings of depersonalization, and hallucinogenic and dreamlike recollections (Bear, 1979; Gloor, 1986, 1992; Horowitz et al., 1968; Mesulam, 1981; Penfield and Perot, 1963; Weingarten et al., 1977; Williams, 1956). In addition, sexual feelings and related activity and behavior are often evoked by amygdala stimulation and temporal lobe seizures (Halgren, 1992; Jacombe et al., 1980; Gloor, 1986; Remillard et al., 1983; Robinson and Mishkin, 1968; Shealy and Peele, 1957), including memories of sexual intercourse (Gloor, 1990).

Moreover, intense activation of the temporal lobe and amygdala has been reported to give rise to a host of sexual, religious, and spiritual experiences; and chronic hyperstimulation (i.e., seizure activity) can induce some individuals to become hyper-religious or visualize and experience ghosts, demons, angels, and even God, as well as claim demonic and angelic possession or the sensation of having left their bodies (Bear, 1979; Gloor, 1986, 1992; Horowitz et al., 1968; MacLean, 1990; Mesulam, 1981; Penfield and Perot, 1963; Schenk and Bear, 1981; Weingarten et al., 1977; Williams, 1956).

LSD

As is well known, LSD (D-lysergic acid diethylamide) can elicit profound hallucinations involving all spheres of experience. After the administration of LSD, high-amplitude slow waves (theta) and bursts of paroxysmal spike discharges occur in the hippocampus and amygdala (Chapman and Walter, 1965; Chapman et al., 1963), but with little cortical abnormal activity. In both humans and chimps, when the temporal lobes, amygdala, and hippocampus are removed, LSD ceases to produce hallucinatory phenomena (Baldwin et al., 1959; Serafetinides, 1965). Moreover, LSD-induced hallucinations are significantly reduced when the right versus left temporal lobe has been surgically ablated (Serafetinides, 1965).

Overall, it appears that the amygdala, hippocampus, and the neocortex of the temporal lobe are highly interactionally involved in the production of hallucinatory experiences. Presumably, it is the neocortex of the temporal lobe that acts to interpret this material (Penfield and Perot, 1963) as perceptual phenomena. Indeed, it is the interrelated activity of the temporal lobes, hippocampus, and amygdala that not only produces memories and hallucinations, but dreams. In fact, the amygdala's involvement in all aspects of emotion and sexual functioning, including associated memories and the production of overwhelming fear as well as bizarre and dreamlike mental phenomenon, may well account for why this type of unusual stimuli, including personal and innocuous memories, also appears in dreams.

DREAMING

When hallucinations follow depth electrode or cortical stimulation, much of the material experienced is very dreamlike (Gloor, 1990, 1992; Halgren et al., 1978; Malh et al., 1964; Penfield and Perot, 1963) and consists of recent perceptions, ideas, feelings, and other emotions that are similarly illusionary and dreamlike. Indeed, the right amygdala, hippocampus, and the right hemisphere in general (Broughton 1982; Goldstein et al., 1972; Hodoba, 1986; Humphrey and Zangwill, 1961; Kerr and Foulkes, 1978; Meyer et al., 1987) also appear to be involved in the production of dream imagery as well as REM sleep (see Chapter 3).

For example, stimulation of the amygdala triggers and increases ponto-geniculo-occipital paradoxical activity during sleep (Calvo et al., 1987), which in turn is associated with REM and dreaming. In addition, during REM, the hippocampus begins to produce slow wave, theta activity (Jouvet, 1967; Olmstead et al., 1973). Presumably, during REM, the hippocampus and amygdala act as a reservoir from which various images, emotions, words, and ideas are drawn and incorporated into the matrix of dreamlike activity being woven by the right hemisphere. It is probably just as likely that the right hippocampus and amygdala serve as a source from which material is drawn during the course of a daydream.

THE RIGHT HEMISPHERE AND DREAMS

There have been reports of patients with right cerebral damage, hypoplasia, and abnormalities in the corpus callosum who have ceased dreaming altogether, suffer a loss of hypnagogic imagery or tend to dream only in words (Botez et al., 1985; Humphrey and Zangwill, 1951; Kerr and Foulkes, 1981; Murri et al., 1984). However, it has also been reported that when the left hemisphere has been damaged, particularly the posterior portions (i.e., aphasic patients), the ability to verbally report and recall dreams also is greatly attenuated (e.g., Murri et al., 1984). Of course, aphasics have difficulty describing much of anything, let alone their dreams.

Electrophysiologically, the right hemisphere also becomes highly active during REM, whereas, conversely, the left brain becomes more active during non-REM (NREM) sleep (Goldstein et al., 1972; Hodoba, 1986). Similarly, measurements of cerebral blood flow have shown an increase in the right temporal regions during REM sleep and in subjects who upon wakening report visual, hypnagogic, hallucinatory, and auditory dreaming (Meyer et al., 1987). Interestingly, abnormal and enhanced activity in the right temporal and temporal-occipital area act to increase dreaming and REM sleep for an atypically long period (Hodoba, 1986). Hence, it appears that there is a specific complementary relationship between REM sleep and right temporal electrophysiological activity.

Interestingly, daydreams appear to follow the same 90-minute to 120-minute cycle that characterizes the fluctuation between REM and NREM periods, as well as fluctuations in mental capabilities associated with the right and left hemispheres (Broughton, 1982; Kripke and Sonnen-schein, 1973). That is, the cerebral hemisphere tends to oscillate in activity every 90 to 120 minutes—a cycle that appears to correspond to the REM-NREM cycle and the appearance of day and night dreams.

FORGOTTEN DREAMS

Most individuals, however, have difficulty recalling their dreams. This may seem paradoxical, considering that hippocampal theta is being produced. However, this is theta punctuated by high levels of desynchronized activity, which is not conducive to learning. In this regard, theta activity may represent the reverberating activity of neural circuits formed during the day, such that the residue of daytime memories come to be inserted into the dream. Conversely, due to the high level of desynchronization occurring in the hippocampus (as it is so highly aroused), the hippocampus contributes images and the day's memories but does not participate in storing these dreamlike experiences into memory.

Consider the results from temporal lobe, amygdala, and hippocampal electrical stimulation on memory recall and the production of hallucinations. Although personal memories are often activated at low intensities of stimulation (memories that are verified not only by the patient but family), if stimulation is sufficiently intense, the memory instead will become dreamlike and populated by hallucinated and cartoon-like characters (Halgren et al., 1978). That is, at low levels of stimulation, memories are triggered, but these memories become increasingly dreamlike with high levels of activity.

Moreover, once these high levels of stimulation are terminated, patients soon become verbally amnesic and fail to verbally recall having had these experiences (Gloor, 1992; Horowitz et al., 1968). However, these memories can be later recalled if subjects are provided with specific contextual cues (Horowitz et al., 1968). The same can occur during the course of the day when a fragment of a conversation, or some other experience, suddenly triggers the recall of a dream from the previous night that had otherwise been completely forgotten. Presumably, it had seemingly been forgotten because the hippocampus did not participate in its storage and thus could not assist in its retrieval (Joseph, 1990a, 1992a; see Chapters 6 and 16).

There is also some evidence to suggest that different regions of the hippocampus show different levels of arousal during paradoxical sleep. For example, it appears that the posterior hippocampus becomes activated during paradoxical sleep and shows theta activity, whereas the more anterior portions become inhibited (Olmstead et al., 1973). Because the anterior portions are more involved in new learning (at least in humans), whereas the posterior hippocampus is more concerned with old and well established memories, this would suggest that the posterior hippocampus is contributing older or already established memories to the content of the dream. Conversely, the inhibition of the anterior region would prevent this dream material from becoming re-memorized.

DREAMS AND INFANCY

In the newborn, and up until approximately 6 to 9 months, there are two distinct stages of sleep that correspond to REM and NREM periods demonstrated by adults (Berg and Berg, 1978; Dreyfus-Brisac and Monod, 1975; Parmelee et al., 1967). Among infants, however, REM occurs during wakefulness as well as during sleep. In fact, REM can be observed when the eyes are open, when the infant is crying, fussing, eating, or sucking (Emde and Metcalf, 1970). Moreover, REM is also observed in infants within moments of engaging in nutritional sucking and appears identical to that which occurs during sleep (Emde and Metcalf, 1970).

The production of REM during waking in some respects seems paradoxical. Nevertheless, it might be safe to assume that, like an adult, when the infant is in REM, he or she is dreaming, or at least, in a dreamlike state. Possibly, this state corresponds to what Freud has described as the Primary Process. That is, when REM is produced while the infant is crying or fussing, the infant is *dreaming* of whatever relief it seeks. Correspondingly, REM that occurs while the infant is eating or sucking may have to do with the limbic structures that are involved, not only in the production of dreamlike activity but in the identification, learning, and retention of motivationally significant information (i.e., the amygdala and hippocampus).

Presumably this relationship is a consequence of REM, as well as eating and sucking, being mediated, in part, by the amygdala as well as other limbic nuclei, nuclei that are also concerned with forming motivationally significant memories. Hence, when one is hungry, the hypothalamus becomes aroused, which activates the amygdala, which is responsible for performing environmental surveillance so as to attend to, orient to, identify, and approach motivationally significant stimuli and eat (Joseph, 1990a, 1992a). However, because the infant's brain is so immature and as its resources for meeting its limbic needs are quite rudimentary, under certain conditions prolonged hypothalamus-induced amygdala activation results in the formation and recall of relevant memories, which may be experienced as hallucinations of the desired object. That is, previously formed neural networks become activated and the infant begins to dream about and hallucinate food; the infant will then suck and smack its lips as if eating or sucking when it is awake, in REM, and there is no food present.

The Primary Process

The hypothalamus, our exceedingly ancient and primitive Id, has an eye that only sees inward. It can tell if the body needs nourishment but cannot determine what might be good to eat. It can feel thirst, but has no way of slaking this desire. The hypothalamus can only say: "I want," "I need," and can only signal pleasure and displeasure. However, being the seat of pleasure, the hypothalamus can be exceedingly gracious in rewarding the organism when its needs are met. Conversely, when its needs go unmet it can respond, not only with displeasure or aversion, but with undirected fury and rage. It can cause the organism to cry out.

Nevertheless, the cry does not produce the immediately desired relief or reduction in tension. There is, thus, *pressure* on the limbic system and the organism to engage in environmental surveillance so as to meet the needs monitored by the hypothalamus.

Over the course of the first months of life, as the amygdala and then hippocampus develop, the organism begins to develop an eye that not only sees outward, but that can register and recall events, objects, people, and so forth associated with tension reduction, pleasure, and the satiety of the infant's internal needs (e.g., the taste, smell, and feel of mother's breast and milk, the experience of sucking and relief, etc.). This is called learning.

With the maturation of these two limbic nuclei, the infant is increasingly able to differentiate what occurs in the external environment based on hypothalamically monitored needs and the emotional/ motivational significance of that which is experienced. The infant can now orient, selectively attend, determine what brings satisfaction, and store this information in memory.

Primary Imagery

Although, admittedly, we have no direct knowledge as to the psychic interactions in the neonate, it does seem reasonable to assume that as the neocortex and underlying structures and fiber pathways mature, neural "programs," or networks, are formed that correspond to the repeated registration of experiences that are deemed significant (e.g., pleasurable). That is, neural pathways that are repetitively fired, deactivated, or activated in response to specific sensory and affective activities and experiences become associated with that activity, such that an associated neural circuit is formed (see Chapter 6); i.e., a memory is created.

Eventually, if this circuit is reactivated, the "learned" pattern is reexperienced: i.e., the organism remembers (Hebb, 1949; Joseph 1982).

Thus, infants as young as 2 days of age can learn to suck at the mere sight of a bottle (see Piaget, 1954), and in order to receive milk as a reinforcement, infants can even modify their sucking responses (Sameroff and Cameron, 1979). Hence, they are susceptible to classical conditioning (Sameroff and Cavanagh, 1979), although the possibility of operant conditioning has not been established. Nevertheless, the fact that they can recognize the bottle and suck (as well as cry and shed tears) indicates that various regions of the limbic system, especially that of the amygdala, is functional and that learning and the creation of specific, context-specific neural circuits have been formed very early in life.

Thus, when the amygdala/hippocampus are stimulated by a hungry hypothalamus, the events and images associated with past experiences of eating can not only be searched out externally, but recalled in imaginal form (Joseph 1982, 1990a, 1992a). For example, as an infant experiences hunger and stomach contractions as well as its own cries of displeasure, these states become associated with the sound, smell, taste, and so forth of mother and her associated movement and other stimuli that accompany being fed (cf Piaget, 1952:37, 407–408). Repetitively experienced, the sequence from hunger to satiety evokes and becomes associated with the activation of certain neural pathways and the creation of a specific neural network subserving that memory (see Chapter 6). Eventually, when the infant becomes hungry, if the hunger is prolonged there is the possibility that the entire neural sequence associated with hunger and feeding (i.e., hunger, mother, food, satiety), may become involuntarily triggered and activated (via association) such that an "image" of being fed is experienced. The activation of these rudimentary and infantile memory-images is probably what constitutes, at least in part, the primary process (Joseph 1982a, 1990a, 1992a).

Behaviorally, this is manifested by REM and via sucking and tongue movements as if eating, when in fact there is no food present (cf Piaget, 1952). That is, when hungry, the infant will begin to cry, rapid eye movement (REM) might be observed, and then the infant will stop crying, smack its lips, and make sucking movements (mediated by the amygdala) *as if* it were being fed. The infant experiences the experience of being fed in the form of a dream (Joseph, 1982) or hallucination, although it is awake.

In that the brain of the human infant is quite immature for, in fact, several years, which in turn restricts information reception and processing (see Chapters 3 and 15), and given the limited amount of reality contact infants are able to achieve, these rudimentary memories and images (even when occurring during waking, i.e., REM), are probably indistinguishable from actual experience simply because they are experience.

Like a dream that is replayed, the infant presumably reexperiences to some degree the sensations, emotions, and so forth originally linked to tension reduction. Thus, the young infant, as yet unable to distinguish between representation and reality, responds to the image as reality (Freud, 1900, 1911), even while awake—as manifested by REM. When hunger is prolonged, the associations linked to feeding are triggered and for a brief time the infant behaves as if its hunger has been sated. Reality is replaced by an image, or rather, a "dream." This is the primary process.

Since the hypothalamus (which monitors internal homeostasis) is not conscious that the dream images experienced are not real, it initially accepts the memory/dream images transmitted from the amygdala and hippocampus and ceases to cry, i.e., it re-sponds to the imagined sources of nourishment just as it responds to a cue-tone associated with a food reward (Nakamuar and Ono, 1986; Ono et al., 1980). However, the hypothalamus is not long fooled, for the primary process does not offer effective, long-lasting relief from tension. As the pain of hunger remains and increases, limbic activity is increased, and the image falls away to be replaced by a cry of hunger (Joseph, 1982). The amygdala and hippocampus are, therefore, forced to renew their surveillance of the environment in search of sources of tension reduction. Cognitive development is thus promoted.

"Whatever was thought of (desired) was simply imagined in an hallucinatory form, as still happens today with our dream-thoughts every night. This attempt at satisfaction by means of hallucination was abandoned only in consequence of the absence of the expected gratification, because of the disappointment experienced. Instead, the mental apparatus had to decide to form a conception of the real circumstances in the outer world and to exert itself to alter them...The increased significance of external reality heightened the significance also of the sense-organs directed towards the outer word, and of the consciousness attached to them; the latter now learned to comprehend the qualities of sense in addition to the qualities of pleasure and "pain" which hitherto had alone been of interest to it. A special function was instituted which had periodically to search the outer word in order that its data might be already familiar if an urgent need should arise; this function was attention. Its activity meets the sense-impressions halfway, instead of awaiting their appearance. At the same time there was probably introduced a system of notation, whose task was to deposit the results of this periodical activity of consciousness—a part of that which we call memory" (Freud, 1911:410–411).

The Hippocampus, Amygdala, Memory and Amnesia

Synaptic Potentiation and Cognitive and Emotional Neural Networks

AMNESIA

Memory loss may be secondary or related to a number of factors, including depression, aging, chronic emotional stress, degenerative disturbances, neurological disease or stroke, head injury, frontal or temporal lobe injury, intense fear, severe or repetitive physical trauma, rape, assault, and sexual molestation during childhood (Brier, 1992; Christianson and Nilsson, 1989; Donaldson and Gardner, 1985; Fisher, 1982; Frankel, 1976; Fredrikson, 1992; Grinker and Spiegel, 1945; Herman and Schatzow, 1987; Joseph, 1992b; Kuehn, 1974; Miller et al., 1987; Parson, 1988; Schacter et al., 1982; Shrouping et al., 1980; Williams, 1992). Even immersion in cold water and consensual sexual activity have been noted to give rise to transient amnesias (Miller et al., 1987; Shrouping et al., 1980), which is presumably a consequence of rapid depletion of serotonin and the release of opiate peptides, which in turn negatively influence the hippocampus (see Chapter 16).

Severe stress or emotional shock, as well as a blow to the head or a stroke or tumor, can induce sudden and drastic alterations in neurotransmitter and arousal levels as well as create structural alterations in the functional integrity of the brain, particularly the hippocampus—which is why memory may be affected. When the brain has been traumatized, particularly the hippocampus and the inferior temporal or frontal lobes, the ability to process information or retain, store, and recall ongoing events may be seriously altered and disturbed, thereby creating a functional amnesia (Donaldson and Gardner, 1985; Fisher, 1982; Grinker and Spiegel, 1945; Parson, 1988)—even when consciousness has not been lost (Clifford and Scott, 1978; Peters, 1991; Yarnell and Lynch, 1970).

However, depending on the nature of the trauma, the amnesia may be global and all encompassing (including a loss of personal identity if the amygdala and right inferior temporal lobe have been injured), or only for the events that occurred at the moment of impact or emotional upheaval, and for a few seconds before or thereafter. That is, the patient may suffer from an extensive and prolonged amnesia or a very brief period of amnesia that may include simple details that immediately preceded or followed the trauma. In general, global amnesia is associated with frontal-thalamic lesions (see Chapter 11).

In cases of very minor head injury or mild emotional shock, what is forgotten may be quite trivial. For example, Loftus and Burns (1982) reported that subjects who watched a 2.25-minute film of a violent (versus nonviolent) bank robbery (which ended with a 15-second sequence in which a boy is shot in the face) were unable to remember a number on the football jersey of a bystander. By contrast, those who had not seen the shooting could recall the number. Hence, emotion can negatively impact memory.

Clifford and Scott (1978) also found that the recall accuracy of witnesses viewing a video of an assault was significantly less than those viewing a nonviolent encounter, with women showing significantly poorer recall then men. Moreover, women rated the film as more violent than the men—a function of the differential structure and functioning of the female's limbic system and her greater sensitivity to emotional stimuli, which can interfere with memory storage and retrieval (see Chapter 5).

With mild head injury, such as often occurs among football players, the amnesia may relate to the previous or upcoming play (Lynch and Yarnell, 1973; Yarnell and Lynch, 1973)—even though consciousness was not lost. However, in these instances, it is not at all unusual for the forgotten memories to be recovered.

With more severe injuries, the memory loss may be much more enduring and extensive, and even personal identity may be temporarily forgotten. For example, when subjected to emotional and stressful extremes, and/or when the conditions are life-threatening, such as repeated frontline battlefield exposure or a brutal rape (Christianson and Nilsson, 1989; Donaldson and Gardner, 1985; Grinker and Spiegel, 1945; Janet, 1927; Myers, 1903; Nemiah, 1979; Parson, 1988; Prince, 1924), some individuals may become so amnesic they cannot recall their names, occupations, and so on; this, in turn, may be due to injury to the hippocampus and amygdala, or a temporary disconnection involving these nuclei (see Chapter 16). Even in less extreme instances involving non–life-threatening emotional trauma and stress, the victim may have no memory for the trauma that preceded or accompanied the development of the amnesia, and he or she may suffer a retrograde memory loss for events that occurred weeks, months, or even years beforehand (Janet, 1927; Myers, 1903; Nemiah, 1979; Prince, 1924). Amnesia victims such as these may also display a continuous anterograde amnesia such that they continue to forget events as they occur, including people whom they subsequently meet (Nemiah, 1979).

However, just as lost memories may be recalled over time following a mild or severe head injury, persons with emotionally induced traumatic amnesia may eventually recover much of this forgotten material, some of which may be retrieved while under hypnosis or via the assistance of a dream (Hilgard, 1977; Janet, 1927; Myers, 1903; Nemiah, 1979; Prince, 1939). Those who suffer amnesia following a head injury may also be exposed to forgotten details and other information, including the trauma itself, through their dreams and then suddenly remember. Indeed, regardless of the etiology of the amnesia, although it is often presumed that the memory (such as memories at the moment of impact) was not formed, and that other memories were erased, this is not always the case.

MEMORY GAPS

It has been argued that temporary amnesic states may be common, everyday events that most people just do not remember (Reed, 1979; Hodges and Warlow, 1990). For example, it is not at all uncommon for individuals to drive across twisting mountain roads, or through busy city streets, only to arrive home and have absolutely no recall of having driven their car or any other aspect of the journey. There are many such blank spots that fill the course of almost every day. One need only attempt to account for what he did over the course of the previous week to realize there may be a number of gaps—what Reed (1979) refers to as "time-gap" experiences.

It has also been suggested, however, that these transient "time gap" amnesias may be the result of the verbal half of the cerebrum simply ceasing to engage in or observe actions that it normally does not perform (Joseph, 1988a, 1992b) and as such are a form of *verbal* amnesia. Consider the "driving" example mentioned above. It is the right half of the brain and the right temporal and parietal lobes that are dominant for visual and depth perception and maneuvering through visual space. Therefore, the left hemisphere may not recall, for example, the driving experience since it was not engaged at the time and was possibly functioning at a lower level of arousal or engaged in other tasks. It is the right half of the brain that would process, store, and be responsible for recalling this information.

As noted, in previous chapters, it is this same differential organization of the brain, and the propensity for memories to be stored in one rather than in both hemispheres, that often gives rise to a *verbal* amnesia. This sometimes leads to the mistaken notion that an individual has either forgotten or failed to form certain memories, when in fact they are present but not available via verbal (or in some cases hippocampal) retrieval strategies.

Anesthesia and Unconscious Learning

It has been repeatedly demonstrated that patients in a deep state of unconsciousness due to anesthesia are capable of perceiving, storing in memory, and later recalling events that took place during surgery (Bennet, 1988; Furlong, 1990; Kilhstrom et al., 1990; Millar, 1987; Polster, 1993). Patients may even recover sooner when provided therapeutic suggestions while unconscious (Evans and Richardson, 1988; Furlong, 1990). Moreover, they may later respond to suggestions made to them while unconscious.

For example, in one study anesthetized patients received suggestions that they should touch their ears when interviewed after the surgery. And this is exactly what they did, although they claimed to have no knowledge or memory of having heard this suggestion (Bennet et al., 1981; cited by Schacter and Moscovitch, 1984).

Kilhstrom et al. (1990) reported that when word pairs were presented to unconscious patients during surgery they were more likely to free-associate these same words at a later time when presented with the other member of the pair. However, although learning and memory were demonstrated, the patients claimed to have no memory of having heard this material. Control subjects did not show this effect. Hence, in part, this represents a form of "state dependent learning."

Other investigators, however, have failed to find evidence of memory storage, even when recognition was employed (see Millar, 1987; Polster, 1993). In large part, this may be a function of the type of anesthesia as well as the type of surgery, for these studies are clearly in the minority.

Unconscious Knowledge: Verbal and Source Amnesia

Amnesic disturbances due to anesthesia, drugs, or dysfunction of the frontal lobe or dorsal medial thalamus or left hemisphere (see below) predominantly involve the *verbal* memory retrieval system, i.e., verbal material cannot be accessed or it is not verbally recognized as familiar. The visual and nonverbal recognition memory is much more robust and visual stimuli are more likely to be stored, although it, too, may be affected. For example, Adam (1973) found that subjects who inhaled nitrous oxide and who were then presented with words as well as nonverbal and melodic sounds, pictures, and geometric shapes were able to remember almost 100% of the visual and nonverbal acoustic stimuli but only about 50% of the words.

There have also been numerous examples of individuals who, although suffering from amnesia secondary to neurological dysfunction, are nevertheless able to demonstrate the possession of knowledge and information which they in fact deny possessing or learning (e.g., Claparede, 1911; see also Schacter et al.,1984). This includes recalling details from a short story read to them previously as well as names and addresses or little-known facts provided to them during their amnesic period.

For example, when amnesiacs were presented with words (which they quickly forgot), and were later presented with incomplete words that they were to fill in, the previously viewed (albeit forgotten) information facilitated performance, although subjects denied having heard this material before (Warrington and Weiskrantz, 1970, 1979; Tulving et al., 1982). Similarly, Graf, Squire and Mandler (1984) found that when amnesiacs were given a list of words and then later presented with incomplete stems (e.g., "inc___ or mot___"), those who were told to form the first word that comes to mind performed similarly to normals. That is, they were able to recall the words to which they had been exposed, although they denied having recently heard them.

In a similar study, an amnesic patient was told some very unusual stories about various pictures. When he was later shown the same pictures, he claimed to have no recollection of having seen them or having heard the story. However, when asked to pick a title for each picture, he picked those that mirrored the theme he had been told (Schacter and Moscovitch, 1984). Some individuals with amnesia can also learn motor skills and related tasks and will, for example, perform a jigsaw puzzle faster on the second trial, improving to the level of normals (Brooks and Baddeley, 1976). Amnesiacs have also proved capable of learning musical stimuli (Starr and Phillips, 1970). Similarly, some individuals suffering from profound amnesia are able to demonstrate recognition memory involving identification of visual objects and "picture puzzles," such that recognition time is reduced by over 100% on day two versus day one of training (Crovitz et al., 1981; Meudell and Mayes, 1981).

Many of the patients described above could, therefore, be considered to be suffering from "source amnesia"—an inability to recall how or in what context certain information was acquired. Hence, some amnesic syndromes are due to a source and contextual memory deficit (Stern, 1981; Winocur, 1982). Of course, in many instances, it is not the source but the information itself that amnesic individuals deny knowing or possessing. However, if also provided with forced choice testing situations, these same amnesiacs may be able to indicate they were in possession of that information all along. Again, it is important to note that although patients may *verbally* claim to have no verbal memory, and although they are clearly amnesic, the amnesias are often due to verbal access and retrieval failure, except, of course, in situations where the right hemisphere and right temporal lobe have been damaged. This is why these same amnesic individuals may retain and even learn information they *verbally* claim not to possess.

It is noteworthy that the retention and later recollection of forgotten (amnesic) memories is quite similar to instances of repression, in which a "forgotten" memory is suddenly reactivated after weeks or even years have elapsed. Although an individual does not verbally know, and may deny having certain memories and thus having had certain experiences, this may not be the case.

ANTEROGRADE AND RETROGRADE AMNESIA

Amnesic conditions are most frequently encountered by physicians and therapists after an injury to the brain, such as due to head trauma, stroke, neoplasms, or encephalitic conditions (Hodges and Warlow, 1990; Joseph, 1990f, Lynch and Yarnell, 1973; Miller, 1993; Russell, 1971; Yarnell and Lynch, 1970, 1973). Typically, following a significant head injury or stroke there may be a period of amnesia for events that occurred just before and just after the time of injury (Joseph, 1990f; Hodges and Warlow, 1990; Russell, 1971), even if consciousness is not lost (Yarnell and Lynch, 1970, 1973).

In more severe cases, such as those involving a brief or prolonged loss of consciousness, the brain may remain dysfunctional for some time, even after the trauma and the return of consciousness, and memory functioning and new learning may remain deficient for long periods (Joseph, 1990f; Miller, 1993; Russell, 1971). Indeed, following the recovery of consciousness patients may be unable to recall little or anything that occurred for days, weeks, or even months after their injury. This condition has been referred to as posttraumatic, anterograde amnesia (Russell, 1971).

Anterograde amnesia is a consequence of continued abnormal brain functioning. Because the brain is functioning abnormally, information is not processed or stored appropriately and cannot be accessed.

Post-Traumatic/Anterograde Amnesia

Individuals who suffer extended periods of emotional stress, particularly those who suffer unconsciousness and coma, such as following a head injury, may experience protracted periods of disorientation and confusion (Joseph, 1990f; Levin et al., 1982; Nemiah, 1979; Russell, 1971). Attention, new learning, and memory are necessarily compromised.

Frequently, with moderate or severe brain injuries there is a period of complete amnesia for continuously occurring events for some time after the return of consciousness (Russell, 1971) or following the cessation of the horrific emotional conditions that induced the disturbance (Donaldson and Gardner, 1985; Grinker and Spiegel, 1945; Parson, 1988).

That is, although consciousness has returned, and the patient can talk, respond to questions, and even perform simple arithmetical operations, because the brain is impaired or the effects of the emotional shock have not completely waned (such as in conditions of post-traumatic stress disorder), information processing and, thus, memory remain faulty for a variable length of time. The duration, however, may depend on the severity of the trauma, or the location and extent of the injury. This has been referred to as post-traumatic amnesia (PTA).

Nevertheless, PTA is not due to an inability to *register* information, for immediate recall may be intact whereas short- and long-term memory remain compromised (Joseph, 1990f; Yarnell and Lynch, 1970, 1973). Although patients are responsive and may interact somewhat appropriately with their environment, they may continue to have difficulty consolidating and transferring information from immediate to short-term to longer-term memory. However, the PTA may not be global, and may be selective for verbal versus visual material, or inclusive of both, whereas motor and emotional learning may remain intact.

It is also important to bear in mind that the first appearance of normal memory or personality or conscious functioning following an injury or an emotional trauma does not indicate the end of the amnesic period. That is, although behavioral functioning is ostensibly normal, brain functioning may not be. Hence, the "normality" may be temporary and followed by another period of PTA, such that the person may suffer repeated absences of memory (Hodges and Warlow, 1990; Joseph, 1990f; Myers, 1903). Memory functioning may remain abnormal even after 10 to 20 years (Schacter and Crovitz, 1977; Smith, 1974).

Retrograde Amnesia

Frequently, after a stroke or severe head injury, or in cases of extreme and prolonged emotional turmoil and trauma, the amnesia may include events that occurred well before the trauma or moment of impact. This retrograde amnesia (RA) may extend backwards for seconds, minutes, hours, days, months, or even years, depending on the severity of the injury, extent of degenerative damage (Blomert and Sisler, 1974; Hodges and Warlow, 1990) or degree of emotional trauma (Grinker and Spiegel, 1945; Janet, 1927; Myers, 1903; Nemiah, 1979; Prince, 1939; Southard, 1919).

Be it physically or emotionally induced, retrograde amnesia is often reflective of disturbances involving the posterior-ventral hippocampus, frontal or inferior temporal lobes, and/or the dorsal medial nucleus of the thalamus (see below) and possibly the amygdala. However, the RA may also be due to disconnection and dissociation, such that the Language Axis is no longer able to gain access to select neural networks and, thus, the missing information.

Retrograde amnesia is seldom inclusive for public facts or the entirety of an individual's life, for the remote past appears to be better preserved than the more recent past in these cases (Joseph, 1990f). Even when personal identity has been forgotten, facts, city names, the ability to read and write, and other forms of nonemotional memory may be retained. However, this also depends on the nature, location, and laterality of the trauma. For example, with traumatic injuries to the left hemisphere, public facts as well as reading, writing, math, and even the ability to speak may be severely compromised (see Chapter 4), whereas with right temporal lobe injuries, personal and emotional memories may be seemingly erased (see Chapter 3).

In general, the extent and severity of the RA is measured by determining the last series of consecutive events recalled prior to the trauma. That is, a patient aged 50 who has suffered a severe head injury may be able to remember events from late childhood and early adulthood, but have almost no memory for events that occurred between ages 30 and 40, and no memory for anything that occurred during the last 10 years. The more extensive the RA, the more severe the underlying brain damage.

However, be it due to physical or emotional trauma, patients may show islands of memory such that, in actuality, they may only be able to recall a few events from early adulthood, and only a few from middle age (see Squire et al., 1975; see Chapter 16). If they are not carefully examined, one might erroneously conclude that memory and brain functioning is not severely impaired. Moreover, individuals with amnesia, when repeatedly questioned at different times, can often recall information that, upon previous questioning, was not available to the speaking half of the brain (Squire et al. 1989). In this regard, these amnesiacs are somewhat like young children who report different details at different times, making them appear inconsistent. Unfortunately, this inconsistency may be viewed as malingering, when this is not the case.

Rather, what this indicates is that amnesia is variable and that when alternate neural networks and pathways become accessible, information or memories can be accessed when formerly they seemed to have been forgotten. In other words, although an individual at one time may appear to have no memory for certain events, and thus may be considered amnesic, at another time and under different circumstances these memories may suddenly become available, even in patients who are profoundly amnesic.

SHRINKING RETROGRADE AMNESIA

RA is not necessarily a permanent condition, and memory for various events may return. However, after head injury or stroke, usually the older memories return before more recent experiences. This has been referred to as a shrinking retrograde amnesia (Sisler and Penner, 1975; Squire et al., 1975). The shrinkage is not complete, as events occurring seconds or minutes before the trauma may be permanently forgotten (Sisler and Penner, 1975; Squire et al., 1975).

However, not all patients show a shrinking RA (Sisler and Penner, 1975), especially those with degenerative or functional disturbances. When there is secondary to degeneration, for example, the neurons making up various neural networks, and thus the networks themselves, may be destroyed, in which case there can be no recovery of memory. Moreover, due to degeneration, the RA will expand.

In emotionally induced amnesias, the forgotten material may be recovered in a haphazard, piecemeal fashion over a series of minutes, hours, or days (see chapters 15 and 16). Moreover, rather than a shrinking retrograde amnesia, in some cases (such as in fugue states) the forgotten memories (which may include almost all of the individual's prior life and childhood) may be quite suddenly recovered, almost in total (Janet, 1927; Myers, 1903; Nemiah, 1979; Prince, 1939; see also Chapters 15 and 16).

Thus, it is important to emphasize that even in well documented cases of amnesia secondary to verifiable brain injuries, or in those due to emotional trauma, some memory recovery is not only possible, but likely. However, even in instances where memories appear to have been permanently lost or erased, or perhaps never even formed, the neural network that supports the memory may simply be disconnected, dissociated, and unavailable to the Language Axis and the language-dependent regions of the conscious mind. With the neural network's recovery and reconnection, or if the patient is presented with associated contextual cues, the amnesia may clear and what had been forgotten may be suddenly remembered—sometimes to the shock and dismay of all concerned.

NEURAL NETWORKS

Almost a century ago it was proposed that learning was paralleled by physical changes in neuronal structure and function, as well as the establishment of new neuronal connections where before there had been none (Cajal, 1911, and Tanzi, 1893, cited by Cajal, 1954). Moreover, Cajal (1911) proposed that long-term memories were stored via the establishment of these new synaptic links.

A quarter of a century later, Hilgard and Marquis (1940) and Hebb (1949), basing their hypothesis in part on the neuroanatomical work of Lorente de No (1938) and Cajal (1911), proposed that perceptual activity continues in neurons even after the cessation of stimulus input, thereby inducing structural alterations, which in turn makes learning and memory possible. Hebb (1949) also proposed that different neurons are linked together via this activity so as to form reverberating neural circuits that can fire with minimal stimulation.

According to Hebb (1949:62), "when an axon of cell A is near enough to excite cell B or repeatedly or persistently takes part in firing it, some growth process or metabolic change takes place in one or both cells." He presumed that memories are represented via a specific pattern of activity across a network of neurons—a consequence of permanent reverberating structural changes (Hebb, 1949). Behaviorally, the establishment of these plastic, reverberating neural networks is represented by increases in learning efficiency and enlargement of the memory store.

In addition, like Cajal (1911), Hebb (1949) proposed that axonal-dendritic interconnections are likely to become more extensive in correspondence with the learning of events that are stored in long-term memory. He also postulated that memory storage may well take place in the same cortical regions where the information is processed (Hebb, 1949), such that those cells first involved continue to reverberate, creating these structural changes.

Because there are so many available synapses, cognitive complexity and memory capacity is potentially unlimited. Since the neocortex alone contains anywhere from 10^{14} to 10^{17} synapses, there are more than enough for all possible information storage requirements. Synapses make very low energy demands, so the system is energy efficient and requires little to maintain the memory stores. Therefore, as the number of neurons that come to be linked via perceptual and cognitive activity increases, so does complexity in mental functioning, which in turn is maintained via a vast system of interconnected memories and neural networks.

Moreover, via this correlated activity a new memory can be added to a neural network and/or the circuit can be modified so as to accommodate new learning. That is, a new neuron can be added to the network so long as its activity and interconnections become linked to the previously established pattern of excitation associated with a specific neural network. Behaviorally, this is expressed as learning.

A single dendrite may receive input from hundreds of axons, each of which may be concerned with different perceptual, emotional, behavioral, or cognitive functions. For example, dendrite A might receive input from axons 1 through 99. However, dendrite A and axon 1 may belong to one neural network, whereas dendrite A and axon 99 might belong to another. Moreover, depending on the variables involved, axon 50 could become part of circuit A-99, as well as circuit A-1, or belong to a number of different networks altogether.

Presumably, due to the creation of these neural circuits, complex actions can be initiated in an effortless and routine fashion due to the strength of the connections that maintain the pattern (Hebb, 1949). Presumably, these neural circuits can be associative, such that a variety of memories are linked as a whole and in parallel. However they may also be temporal-sequential, such that one neuron can predict or trigger the next step in the behavioral sequence by activating the next neuron, and so on. Based on Hebb's theory it could be predicted that memories can be activated by a variety of simple and even fragmentary cues, each of which can trigger activation of a portion or the entire circuit.

Over the course of the last 25 years these theories have found considerable empirical support (Barnes, 1979; Davies et al., 1989; Diamond, 1985; Gustafsson and Wigstrom, 1988; Lynch, 1986; Rosenzweig, 1971; W. Singer, 1990; Stevens, 1989). For example, when animals are reared in a complex versus simplified environment, not only are they found to have superior learning and memory capabilities (Joseph, 1979; Joseph and Gallagher 1980), but the cortex becomes more complex, with increases in synaptic density and postsynaptic thickness (Diamond, 1985, 1991; Greenough and Chang, 1988; Rosenzweig, 1971). As such, even neurological and perceptual functioning is altered (Casagrande and Joseph, 1978, 1980; Joseph and Casagrande, 1980).

Moreover, these synaptic changes have been found to occur in conjunction with the development of long-term synaptic potentiation in dendrites and axons (Gustafsson and Wigstrom, 1988; Lynch et al., 1990). Long-term synaptic potentiation has been correlated with learning and environmental input as well as with an increase in synaptic contacts and changes in the morphology of dendritic spines (Lynch et al., 1990).

This long-lasting synaptic activity, in turn, appears to bind the pre-and postsynaptic surfaces involved so that future neuronal activity is correlated as well. This would allow for the creation of widespread neuronal networks.

NEURAL CIRCUITS AND LONG-TERM POTENTIATION

These networking principles have been explored in a number of laboratories through the analysis of the establishment of long-term synaptic potentiation (Barnes, 1979; Lynch, 1986), a "reverberating" form of neural activity that has been noted to occur in both pre- and postsynaptic neurons. That is, long-term potentiation (LTP) is associated with what Hebb (1949) described as reverberation and the creation of long-lasting memories as represented by a neuronal circuit that is highly active.

An axon that is repeatedly utilized for information transmission will increase its supply of neurotransmitters. Conversely, a dendrite that is repeatedly stimulated not only becomes more complex, but each individual receptor surface (at the synaptic junction) may become more extensive so as to take advantage of the increased amount of neurotransmitter available. This may well set the state for the development of long-term potentiation. When certain synapses are repeatedly fired they develop a long-lasting increase in synaptic strength as measured, for example, by LTP.

LTP can generally be triggered by very brief periods of excitation, and may persist from hours to days to weeks to months (Barnes, 1979; Lynch, 1986). LTP also appears to interlock the pre- and postsynaptic junction (Davies et al., 1989; Kauer et al., 1988) of the various interacting neurons. However, it is during the course of the first half hour or so after learning that LTP develops in the presynaptic terminal. This may correspond to the development of long-term memory.

Nevertheless, temporal and spatial contiguity in axonal-dendritic synaptic activity has been postulated to be very important in establishing neural networks (Gustafsson and Wigstrom, 1988; Lynch, 1986; W. Singer, 1990). Presumably, different axons and dendrites must be highly active and active at the same time for these exclusionary networks to form.

That is, since a dendrite may receive input from hundreds of axons, all of which may fire at different times, in order for a neural circuit to be formed so as to maintain a particular memory, it has been hypothesized that the activity between a specific presynaptic and postsynaptic surface must be correlated so that others do not become linked in their place. A specific axon and dendrite must be active simultaneously, which in turn allows the pathways to consolidate (Singer, 1990b) and others to be excluded. For example, axon 1 and dendrite A (A-1) must be activated at the same time, so as to form one circuit, whereas the activity occurring between axon 99 and dendrite A must occur simultaneously but at a different point in time (as compared to A-1) so as to become incorporated into a second circuit (A-99).

For example, Lynch and colleagues (see Lynch et al., 1990), sequentially stimulated three different afferents that terminated at the same dendrite, such that each burst of activity overlapped somewhat with the next. They found that the degree of long-term potentiation (LTP) induced was greatest at the synapse that was first stimulated, intermediate for the second, and least of all for the third. Hence, LTP appears to bind together those neurons that initially share parallel activity.

As described by W. Singer (1990:225), "the integration interval during which presynaptic and postsynaptic activation must coincide in order to lead to stabilization of a pathway... (and)... excitatory and inhibitory inputs to the same dendrite must be activated and silenced respectively. The efficiency of stimuli to induce modifications of cortical circuitry will increase to the extent that the stimuli not only match the response properties, but also conform with the resonance properties of more distributed neuronal assemblies."

Hence, it appears that various neocortical and limbic regions may come to be linked via spatially and temporally correlated synaptic activity. Similarly, structural alterations and synaptic links may be induced by processing similar or associated perceptual experiences.

Due to LTP, not only might different neurons come to be activationally linked, but the neuron may come to have a lower firing threshold due to an increase in transmitter levels and activity as well as other factors. For example, LTP has been demonstrated in those synapses with glutamate receptors (see Stevens, 1989). These receptors open up, which enables ions to flow through into the postsynaptic neuron. According to Stevens (1989:461), "the glutamate that is released from the axon terminal... acts on the postsynaptic membrane to change the postsynaptic voltage and—under the right circumstances—permits an influx of calcium ions," which allows for the develop of postsynaptic LTP.

In this regard, LTP appears to be, at least in part, a function of increased transmitter levels (due to the release of more vesicles) as well as the changes in the postsynaptic receptors, which appear to increase in size, which allows them to absorb more of the excess transmitter. In this manner, the postsynaptic membrane becomes increasingly sensitive, which reduces the threshold for activation, thus allowing LTP to develop (see Davies et al., 1989). Hence, corresponding with the development of LTP, it is presumed that a neuron can fire more easily, particularly in response to input from the same exact source. That is, it is believed that LTP induces and/or is strongly related to the coupling of different neurons that repeatedly interact, thus forming a neural network that can be easily activated. As noted, neuronal sources that provide a different or later form of stimulation come to be excluded from this particular circuit and may be stored in a different neural network representing separate and distinct memories. LTP and the formation of neural networks appear to be formed based on strength as well as temporal sequence of stimulation.

Parallel, Sequential, and Isolated Neural Networks

Different memories are presumably represented by specific neuronal circuits, which, when later activated, may recreate the original cognitive, emotional, and perceptual experience (Joseph, 1982). Correspondingly, a complex memory may be triggered by a variety of stimuli as well as a single cue that initially activates only part of the neural network. When one region of the neural network is activated, associated neurons within the circuit are likely to become aroused in parallel or in sequence.

Not all neural networks, however, may be characterized by parallel processing. For example, neural networks may be temporal, such that activation of one neuron within the network will result in a stepwise pattern of sequential activation of various parts of the circuit. This would allow for one neuron to predict or determine what stimulus pattern or sequence will come next (Sejnowski and Tesauro, 1990). For instance, in getting dressed a whole sequence of associated actions take place which are so well learned that one need not even think about the different steps involved (e.g., putting on a shirt, fastening clothes, etc.) The entire circuit of experience is sequentially activated and occurs almost in reflex fashion. However, some neural circuits may be characterized by both parallel and sequential activation.

ISOLATED MEMORIES

Certain memories may be shared between circuits, or confined to a specific neural network. If there are few other memories associated with that circuit, then this memory can come to be isolated and not easily retrieved. That is, only a selective and quite narrow assortment of thoughts or associations or future experiences may trigger memory recall, in which case it may appear to be forgotten, or repressed (see Chapter 15). This is particularly likely during early infancy and childhood due to the immaturity of the neocortex (Joseph, 1982).

Moreover, over the course of development an earlier learned response may come to be superseded by a more complex learning experience. Or one neural circuit may act to suppress and inhibit the activation of a second network that is associated with certain behavioral acts that are no longer appropriate. With decreased activation, the original circuit may decay as neurons and dendrites drop out, and as LTP diminishes. The result is memory loss or amnesia for specific events and experiences.

THE PERMANENCE AND INSTABILITY OF MEMORY

Because of this neuronal complexity and the large numbers of synapses involved, it is presumed that individual neurons can die, or be eliminated from the circuit, without significantly disrupting the overall integrity of the pattern and associated memory (Gloor, 1990). As such, memories can be resistant to degradation. Moreover, this allows individual neurons to be involved in a variety of circuits, or to drop out of one and become an integral part of another without significantly affecting memory.

Presumably, it is because memories are stored in different neurons and in regions of the brain that different components may be retained or lost at different rates or as a consequence of different forms of interference, including brain damage and degeneration. It is also because there are different neural circuits that certain memories can come to be isolated, selectively forgotten, or repressed, or conversely, recalled forever with ease.

The Amygdala and Hippocampus: Monitoring, Inhibition, Activation

Reverberating neurons are presumably located in various regions of the neocortex, and are apparently bound together via the simultaneous activity and steering influences involving the frontal lobes, dorsal medial thalamus, and, in particular, the amygdala and hippocampus (Graff-Radford et al., 1990; Joseph, 1990a, d, 1992a; Lynch, 1986; Rolls, 1992; Squire, 1992), nuclei that are also interlinked and highly involved in attention, arousal, and memory functioning, and that presumably act to establish and maintain these neural circuits. That is, these different networks and neurons come to be linked via the steering influences exerted by the frontal lobes, and so forth, which can selectively activate or inhibit them in a coordinated fashion, and which can tie together certain perceptual experiences so as to form a complex multimodal memory.

For example, since the hippocampus is a prime location for the development of LTP and is significantly involved in many aspects of memory functioning, it is presumably able to exert steering influences on different neocortical sites with which it is also richly interconnected (see Chapter 5). Via LTP the hippocampus presumably acts to bind these divergent neocortical sites together so as to form a circuit of experience (Lynch, 1986; Squire, 1992).

However, the human hippocampus is important only in regard to certain aspects of memory, such as spatial, verbal, cognitive and recognition memory, whereas the amygdala is concerned with emotional memory. By contrast, the parietal lobes maintain body-image memories, whereas the frontal-thalamic system is more involved in selective attention and retrieval.

LTP is not an exclusive property of the hippocampus but also appears in the adjacent amygdaloid nucleus with which it is richly interconnected. Indeed, amygdaloid neurons show plasticity in response to learning (Lynch, 1986), and LTP has been induced in amygdala neurons (Chapman et al., 1990). Fear-induced neural plasticity in the form of LTP has been noted in amygdala neural pathways as well (Clugnet and LeDoux, 1990). This is presumably a consequence of the amygdala's involvement in most aspects of emotional experience, including the formation of cross-modal emotional associations and memories (Gloor, 1992; Halgren, 1992; Joseph, 1992a, 1994; Kesner, 1992; LeDoux, 1992; Rolls, 1992; however, see Murray and Gaffan, 1994).

The amygdala is also intimately interlinked with the anterior hippocampus and appears to exert reinforcing and modulating influences on this nuclei (see below and Chapter 5). Moreover, they both project to adjacent thalamic relay neurons, which raises the possibility that they act conjointly to form separate but closely aligned neural networks concerned with different aspects of memory.

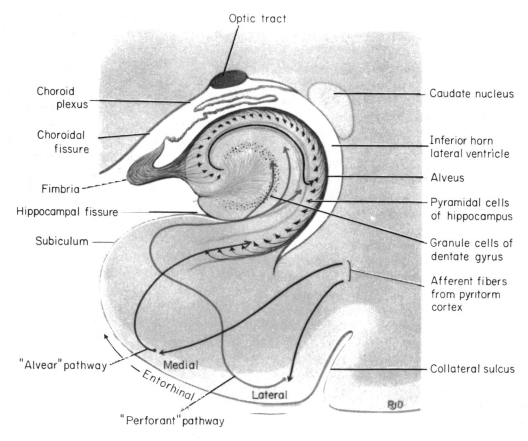

Figure 6.1 Semischematic diagram of the human hippocampus. (From Carpenter M. Core text of neuroanatomy. Baltimore: Williams & Wilkins, 1991.)

SHORT-TERM AND LONG-TERM MEMORY: THE ANTERIOR AND POSTERIOR HIPPOCAMPUS

In humans, the hippocampus, the anterior and ventral hippocampus in particular, is usually associated with learning and memory of cognitively relevant information, e.g., the long-term storage and retrieval of newly learned information (Fedio and Van Buren, 1974; Milner, 1966; 1970; Penfield and Milner, 1958; Rawlins, 1985; Scoville and Milner, 1957; Squire, 1992; see also commentary in Eichenbaum et al., 1994).

As noted, during learning activities LTP has been repeatedly found to occur within the hippocampus (Barnes and McNaughton, 1985; Lynch, 1986). Dendritic proliferation and the creation of specific neural circuits, as well as LTP, also occurs in the hippocampus during learning (Barnes, 1979; Lynch, 1986). It has also been demonstrated that hippocampal pyramidal cells undergo synaptic modification when flexible stimulus response associations are being formed (Rolls, 1987, 1988). Similar correlations between hippocampal LTP and learning have been found on tasks involving memory for visual-spatial relations (Barnes, 1979). Moreover, during acquisition, not only does LTP increase but so do EEG-evoked responses within the hippocampus (see Barnes and McNaughton, 1985).

Barnes and McNaughton (1985) found, however, that long-term hippocampus synaptic potentiation was more long-lasting and more quickly reached by young than old animals. Older animals also demonstrated slower learning and faster rates of forgetting spatial information.

Many of these synaptic and activational changes, in turn, are most apparent within the anterior regions of the hippocampus (Lynch, 1986), which maintain rich interconnections with the amygdala (Amaral et al., 1992). Moreover, this same region of the hippocampus will become electrophysiologically potentiated during learning tasks. In fact, long-term potentiation lasting up to several days has been noted in the hippocampus following successful learning trials (Lynch, 1986), which in turn may reflect the transition of information from short-term to long-term memory, at which point LTP ceases to be a factor in further memory maintenance.

These findings are consistent with the notion that the longer the potentiation, either at the cellular or hippocampal level, the stronger might be the memory, and the more likely it is to persist over time. Hence, the formation of short-term memories appears to be dependent on the anterior portion of the *human* hippocampus and the binding action of LTP, which creates links between different synapses, thus allowing for the transition from short-term to long-term memory. It is noteworthy, however, that the amygdala appears to act on the anterior hippocampus in order to emotionally reinforce as well as modulate its functional activity (Joseph, 1992a), and LTP occurs in both nuclei.

If the anterior hippocampus is injured or functionally suppressed, very little new cognitive learning occurs and potentiation does not appear. However, once these memories have been established in long-term memory, the role of the anterior regions of the human hippocampus appears to diminish. This would explain why long-term memory for long-ago events is spared with hippocampal destruction (see Chapter 5).

Squire argues (1992:222) that "the hippocampal formation is essential for memory storage for only a limited period of time. A temporary memory is established in the hippocampal formation at the time of learning in the form of a simple memory, a conjunction, or an index. The role of the hippocampus then gradually diminishes, and a more permanent memory is established elsewhere that is independent of the hippocampus…and…the neocortex alone gradually becomes capable of supporting usable, permanent memory. This reorganization could depend on the development of cortico-cortical connections between separate sites in neocortex, which together constitute the whole memory." On the other hand, there is some human evidence that the more posterior human hippocampus may be responsible for long-term memory access, which is why retrograde amnesia has been reported with damage to or electrical stimulation of this area (Fedio and Van Buren, 1974; Penfield and Mathieson, 1974). Indeed, Penfield and Mathieson (1974) suggested that memories might actually migrate over time along the length of the hippocampus in a posterior direction, and that the ability to retrieve these memories follows this posterior movement.

Short-Term Versus Long-Term Memory Loss, Retrieval, and Hippocampal Damage

When the hippocampus has been damaged, the ability to convert short-term memories into long-term memories (i.e., anterograde amnesia) becomes significantly impaired in humans (Eichenbaum et al., 1994; MacKinnon and Squire, 1989; Squire, 1992; Victor and Agamanolis, 1990) and primates (Zola-Morgan and Squire, 1984, 1985, 1986). Lesions of the hippocampus can also disrupt time sense and temporal sequencing, such as that involved in

timing tasks (Meck et al., 1984). Memory for words, passages, conversations, and written material is also significantly impacted, particularly with left hippocampal destruction (Frisk and Milner, 1990; Joseph, 1992a; Squire, 1992).

Moreover, spatial memory is significantly impaired among a variety of species with hippocampal lesions (Mishkin et al., 1984; Weiskrantz, 1987). Patients with right hippocampal destruction may demonstrate severe visual-spatial memory disturbances and may easily lose their way, or forget where they place items or where things are located (Joseph, 1990a; Squire, 1992). Nevertheless, initially individuals with even bilateral removal of the hippocampus demonstrate good initial retention and short-term memory (Horel, 1978; see commentary in Eichenbaum et al., 1994).

Bilateral Hippocampal Destruction and Amnesia

Given the powerful effect of unilateral lesions on memory, bilateral destruction of the anterior hippocampus results in striking and profound disturbances involving almost all aspects of cognitive and recognition memory and new learning (i.e., antero-

Figure 6.2 Schematic diagram of the right and left hippocampus, amygdala, and cingulate gyrus. (From Schwartz M. Physiological psychology. Englewood Cliffs, NJ: Prentice Hall, 1978.)

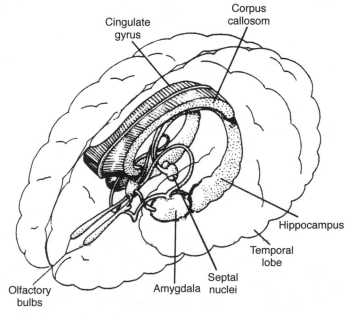

grade amnesia), as well as RA for events that may extend 2 to 3 years in time (Marslen-Wilson and Teuber, 1975; Milner, 1966; Murray, 1992; Squire, 1992), particularly if the posterior hippocampus has also been injured or removed.

For example, the father of an 18-year-old boy who sustained bilateral inferior-posterior temporal lobe and hippocampal destruction described his son's memory as like "a sieve." "You fill it up with information and 30 minutes later it's all leaked away. If you leave him somewhere alone for over 30 minutes, he will completely forget where he is, how he got there, and so on, and then he begins to panic, for he also forgets he was in a car accident and has lost his memory."

Nevertheless, his anterograde memory loss was not global, as he was capable of learning new motor routines and was able to get a job performing electronics assembly. However, he has also had to carry a note pad with a "reminder" to tell him where he is, etc. Similarly, monkeys with hippocampal destruction, including those with bilateral destruction, although demonstrating severe memory impairments are capable of learning motor skill tasks and of acquiring habits (Squire, 1992; Zola-Morgan and Squire, 1984, 1985a). Individuals with extensive hippocampal damage and who are amnesic also appear capable of describing and recalling remote memories and are little different from normals in this regard (Horel, 1978; MacKinon and Squire, 1989; Squire, 1992). In contrast, long-term memory for events following injury is severely disrupted: i.e., anterograde amnesia. However, as noted, among those with posterior hippocampal injuries, access to long-term memories stored prior to the injury may also be negatively impacted (Fedio and Van Buren, 1974); i.e., retrograde amnesia.

In addition, with hippocampal injuries, the creation of reverberating neuronal circuits and LTP is disrupted (Lynch, 1986). This may partly account for why long-term storage for certain cognitive perceptual experiences cannot be established.

Learning and Memory in the Absence of Hippocampal Participation

Considerable forms of learning and memory functioning are retained in the absence of the hippocampus (see Eichenbaum et al., 1994; Horel, 1978; Zola-Morgan et al., 1986). This includes the learning of skilled and coordinated motor programs, as these "memories" appear to be dependent on the basal ganglia (Heindel et al., 1988; Packard et al., 1989; Wang et al., 1990), cerebellum (Thompson, 1986), the inferior parietal lobule, the supplementary motor areas, and the lateral frontal motor areas (Joseph, 1990b, 1990d). The role of the hippocampus is minimal in these forms of learning (see Squire, 1992).

Classical conditioning is also independent of the hippocampus, which may be a function of the repeated nature of stimulus presentation—the hippocampus soon ceases to respond in repetitive experiences (see Chapter 16). In contrast, single learning associations are more dependent on the hippocampus (Squire, 1992).

The learning of emotional information also appears to occur independently of the hippocampus—being dependent on the amygdala (Joseph 1982, 1990a, 1992a). As noted, LTP has been induced in the amygdala (Chapman et al., 1990; Clugnet and LeDoux, 1990), which is presumably a function of its involvement in all aspects of emotional memory formation and the creation of emotional memory neuronal networks (Gloor, 1992; Halgren, 1992; Joseph, 1982; Kesner, 1992; LeDoux, 1992; Rolls, 1992).

In contrast, recognition memory appears to be more dependent on the hippocampus and adjacent nuclei (the inferior temporal lobes, amygdala). Recognition memory is severely disrupted with damage to this area but is spared when the hippocampus is intact (see Murray, 1992; Squire, 1992). Recognition memory is also spared when the medial thalamic nuclei and/or the frontal lobe destruction are the source of the amnesic disturbance (Graff-Radford et al., 1990; Jetter et al., 1986), in which case retrieval may be severely disrupted.

Hence, recognition memory may well depend on interactions involving the hippocampus, inferior temporal neocortex, and the amygdala (Mishkin, 1982; Murray, 1992; Squire, 1992), whereas retrieval may be more dependent on frontal-thalamic interactions (Aggleton and Mishkin, 1983; Graff-Radford et al., 1990; Squire, 1992; Victor et al., 1989) in conjunction with the posterior hippocampus. Emotional memory functioning, however, appears to be dependent on the amygdala whereas body-image memories are associated with the parietal lobe.

THE FRONTAL LOBES, GLOBAL AMNESIA, AND THE DORSAL MEDIAL AND ANTERIOR THALAMUS

The dorsal medial nucleus of the thalamus and the frontal lobes, are exceedingly important in memory functioning and information retrieval (Aggleton and Mishkin, 1983; Graff-Radford et al., 1990; Joseph, 1990d; Squire, 1992; Victor et al., 1989), and both nuclei are intimately linked. If severely injured, global amnesia may result (see Chapter 11).

The frontal lobes, in conjunction with the hippocampus (Goldman-Rakic, 1990), reticular activating system, the amygdala, and the dorsal medial thalamus (DMT), acts to gate and direct perceptual and cognitive activity occurring within the neocortex (Joseph, 1990d). That is, the frontal lobes, DMT, amygdala, and hippocampus interact when determining which neuronal circuits are created, linked together, inhibited, or activated.

The frontal lobes, in fact, are highly concerned with arousal, attention, activation, and neuronal inhibition as well as sustaining the creation of specific reverberating neural circuits. It is also particularly important in the capacity to retrieve previously stored information, as well as prevent access to those memories and perceptions deemed irrelevant or undesirable (see Chapters 11 and 15). Presumably, it can re-

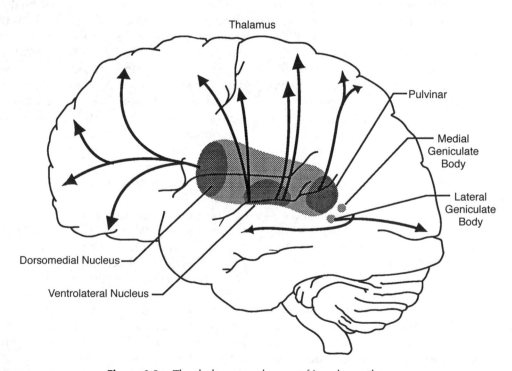

Figure 6.3   The thalamus and some of its relay pathways.

trieve or inhibit particular memories by acting directly on the neurons where they are maintained or indirectly via the thalamus and reticular formation, through which it can selectively effect arousal. In this manner, the frontal lobes can selectively sample, deactivate, or enhance neural activity, depending on the information it is searching for.

Conversely, damage to the frontal lobes can result in difficulty retrieving memories but may have little effect on recognition memory (Jetter et al., 1986) where search,

activation, and retrieval are not as important. Recognition memory can be activated by direct stimulation of the neurons and the neural network, which have stored similar information in the past. However, in some cases, recognition memory also fails and the patient may be globally amnesic.

The Dorsal Medial Thalamus

The frontal lobe, hippocampus, and amygdala appear to share major roles in regard to neocortical activation and thus

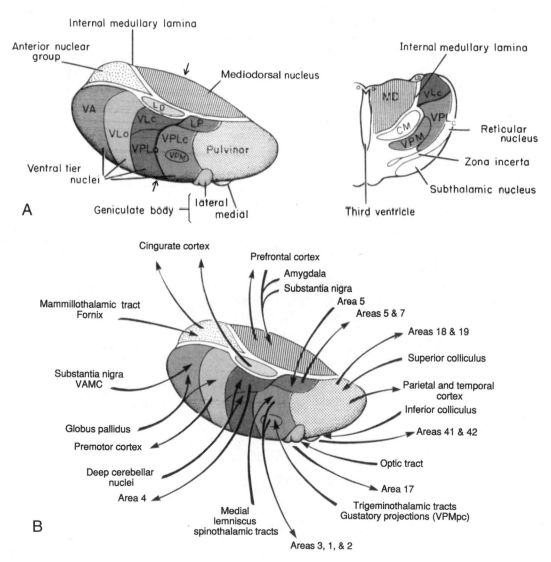

Figure 6.4 Schematic diagram of some of the major thalamic nuclei. (From Carpenter M. Core text of neuroanatomy. Baltimore: Williams & Wilkins, 1991.)

memory storage and retrieval, and all appear to partly rely on the DMT and anterior thalamus in this regard. The DMT (and the thalamus in general) are highly important in transferring information to the neocortex, and are involved in perceptual filtering and exerting steering influences on the neocortex so as to direct attention (see Chapter 11). When damaged, there can result a dense verbal and visual anterograde amnesia (Graff-Radford et al., 1990; von Cramon et al., 1985) as memories cannot be selectively stored or activated. Hence, patients may become globally amnesic.

The DMT and anterior thalamus are able to perform these functions via inhibitory and excitatory influences (mediated by the frontal lobe) on neurons located in the neocortex, limbic system, and reticular activating system. For example, the DMT and frontal lobes monitor and control, at the neocortical level, all stages of information analysis and insure that relevant data and associations are shifted from primary to association to multimodal association areas

so that further processing can occur. Conversely, the frontal-thalamic system can act to inhibit further processing and/or prevent this information from reaching the neocortex in the first place.

The Dorsal Medial Thalamus, Frontal Lobes, Korsakoff's Syndrome, Search and Retrieval

Through the widespread interconnections maintained both by the frontal lobes and thalamus, specific neuronal networks and perceptual fields can be selectively inhibited or activated so that information may be processed or ignored, and so that conjunctions between different neocortical areas can be formed. These nuclei, therefore, directly mediate information processing, memory storage, and the creation of or additions to specific neural networks via the control of neocortical activity (see Chapter 11, for further details). The frontal lobe and anterior and dorsal medial thalamus are also involved in search and re-

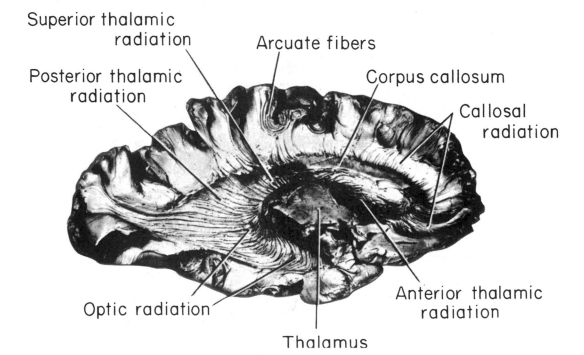

Figure 6.5 Dissection of the medial portion of the hemisphere, depicting the thalamic radiations (internal capsule). (From Carpenter M. Core text of neuroanatomy. Baltimore: Williams & Wilkins, 1991.)

trieval, for these aspects of memory are severely disrupted with damage to the frontal lobes and dorsal medial nuclei, whereas recognition memory may remain much more intact (Graff-Radford et al., 1990; Squire, 1992). However, recognition memory may be impaired as well. For example, one such patient was not only unable to recall that he had been a patient at the hospital for over a month, but repeatedly claimed to be "working" as a "janitor," and that he would be "going home at quitting time." However, when he was shown his hospital room and clothes, although he recognized the items as his, he persisted in claiming he was at work and that the reason his clothes were in the hospital room was so he could change before leaving (Joseph, 1986a).

Hence, when the frontal lobes and DMT and anterior thalamus are damaged, the ability to shift from one set of perceptions to another, or to selectively retrieve a specific memory, or to inhibit the activation of irrelevant ideas, associations, memories, and thoughts, is significantly impacted, and the patient may confabulate. The ability to alternate between, and thus selectively activate, specific neural networks is also disrupted, such that patients have difficulty shifting from one train of thought to another. That is, they may tend to become stuck in-set and to engage in recurrent perseveration (Joseph, 1986a, 1990d; Vikki, 1989) such that the same speech patterns or "memories" may be repeated, unless something occurs to arrest and shift their attention.

Hence, with frontal-thalamic damage, retrieval is impaired, as is the ability to selectively recall or shift between old versus recent memories and to keep track of the order in which they may have been formed (e.g., morning, evening, yesterday, last week, 5 years ago, and so on (see Graff-Radford et al., 1990; Squire, 1992; Talland, 1961). Therefore, a patient may deny possessing a certain memory of an event that occurred 5 weeks ago, because he is searching his memory store for events that were experienced 5 years ago or 5 days ago.

For example, one patient with severe frontal lobe damage claimed to have never been married and laughingly indicated that he never wanted to get married and that marriage was a bad idea. When pressed as to why, he exclaimed that "the wife may just up and die on you," and then began to laugh quite loudly and then just as suddenly began to sob and cry. When asked why he was crying, he claimed not to know why. However, when he was reminded that he had been married, he was suddenly able to recall that his wife (of 2 years) had been killed in the same accident that resulted in his injuries.

Another patient with anterior and dorsal medial thalamic damage erroneously stated that the president was Lyndon Johnson. However, when told the correct date ("1982") she correctly named Ronald Regan (Graff-Radford et al., 1990). Thus "time sense" and capacity to selectively search memory may be disrupted with frontal-dorsal medial dysfunction, a condition that is also characteristic of Korsakoff's syndrome. For example, Talland (1961:375) described one woman with Korsakoff's disease who "after ten year's hospitalization continually maintained that she has been brought in the previous day for observation. When questioned, she also argued that she still lived with her husband in a fancy hotel, where she had lived 'the life of a well-to-do lady of leisure.' She seemed unaware that the conditions of that very agreeable life had long since ceased to operate."

Individuals suffering from Korsakoff's syndrome demonstrate profound disturbances of memory, including a dense verbal and visual anterograde amnesia (Aggleton and Mishkin, 1983; Graff-Radford et al., 1990; Victor et al., 1989; von Cramon et al., 1985) and in some cases a more temporally graded memory loss for the remote past (Squire, 1992)—although the hippocampus is intact. Rather, in Korsakoff's syndrome, damage typically involves the dorsal medial nucleus of the thalamus and atrophy of the frontal lobes (Aggleton and Mishkin, 1983; Graff-Radford et al., 1990;

Mair et al., 1979; Victor et al., 1989; von Cramon et al., 1985; Wilkinson and Carlen, 1982). Nevertheless, recognition memory may be intact, and with selective reminders and reorientation, the retrieval deficits associated with frontal/dorsal medial damage can sometimes be partly overcome (Graff-Radford et al., 1990).

In some respects, this form of selective amnesia, where information that has been "stored" in memory cannot be retrieved, is also reminiscent of what occurs when the frontal lobes are overwhelmed by traumatic and emotional upheavals. That is, the frontal lobe and dorsal medial thalamic system enables an individual to avoid retrieving select memories, which is why these nuclei are the key to understanding purposeful repression (see Chapter 15).

Thus, by contrast, when the frontal-DMT system is damaged, the ability to shift from one storage depot (or neural network or train of thought) to another is disrupted such that, although the memory exists, the patient cannot gain access to it, even when provided assistance. In severe cases, they may demonstrate global amnesia. On the other hand, they may also recall events which they might prefer remained forgotten. For example, one patient with severe right frontal dysfunction suddenly recalled that she had been repeatedly molested as a child—traumas that were confirmed by family members (O'Connor, personal communication, 1994). However, even when aspects of the original memory are retrieved, closely aligned and associated neural networks may remain inaccessible and dissociated/disconnected.

For example, when an amnesic patient was stuck by a pin as she tried to shake hands with a physician, she later refused to shake hands although she had no memory of ever having met the doctor before. When asked why she refused the handshake, she answered that she didn't really know, but that sometimes people hid pins in their hands (Claparede, 1911; see also Kebeck and Lohaus, 1986; Meudell and Mayes, 1981).

This patient, thus, demonstrated what has been referred to as a source amnesia.

Moreover, the emotional link with the actual cognitive event associated with the original memory was ostensibly lost. That is, although she had an emotional memory of having been stuck with a pin when shaking hands, she could not recall when or where this occurred, or with whom. Hence, she experienced a handshaking aversion but was amnesic as to why. Repression, which involves the frontal lobes and dorsal medial nuclei (see Chapters 11 and 15), often gives rise to similar symptomatology, as does hypnosis (Evans, 1979; Hilgard, 1986; Myers, 1903).

THE HIPPOCAMPUS, DORSAL MEDIAL THALAMUS, AND NEOCORTEX

The frontal lobe and dorsal medial thalamic system probably act in concert with both the amygdala and hippocampus in memory storage and retrieval as well as in the temporal placement of events in regard to time and even place (see Graff-Radford et al., 1990; Talland, 1961). This is why retrieval and even memory storage may become abnormal with damage to any of these nuclei.

As discussed in Chapter 5, the hippocampus exerts desynchronizing or synchronizing influences on various thalamic nuclei, which in turn augments or decreases thalamic and neocortical activity (Green and Adey, 1956; Guillary, 1955; Nauta, 1956, 1958). As the thalamus is the major relay nucleus to the neocortex and is richly interconnected with the frontal lobes and amygdala, the hippocampus therefore appears able to act in concert with these nuclei so as to block or enhance information transfer to various neocortical areas where memories and perceptual experiences are presumably stored.

For example, when the neocortex becomes desynchronized (indicating cortical arousal), the hippocampus often (but not always) develops slow wave, synchronous theta activity (Grastyan et al., 1959; Green and Arduini, 1954) such that it appears to be functioning at a much lower level of arousal. Conversely, when cortical arousal is reduced to a low level (indicated by EEG synchrony), the hippocampal EEG often becomes desynchronized and thus highly aroused (Grastyan et al., 1959; Green and Arduini, 1954). However, when this occurs, theta activity disappears and learning and memory are also disrupted.

Neocortical interconnections with the hippocampus are bidirectional, which enables the hippocampus to not only activate select regions but to sample information after it has been partially processed. In this manner the hippocampus can also influence the processing that takes place. At a minimum, these reciprocal interconnections may enable the hippocampus to keep track of this activity so as to form conjunctions between different brain regions that process associated memories (Rolls, 1990a; Squire, 1992). By sampling the activity occurring in different regions and via its rich interconnections with these neocortical areas, the hippocampus may also be able to determine which perceptions are the most relevant and which should be stored versus ignored so that relevant memories need not compete with irrelevant sense data (Joseph, 1992a; Rolls, 1990a).

Presumably, the hippocampus acts to aid in the creation of these networks and to thus protect memory and the encoding of new information during the storage and consolidation phase, by the gating of afferent streams of information and the filtering/exclusion (or dampening) of irrelevant and interfering stimuli, i.e., by reducing arousal in select regions and via inhibitory and excitatory influences on the DMT. When the hippocampus is damaged or is overwhelmed, there is input overload, the neuroaxis is also overwhelmed, and the consolidation phase of hippocampal memory formation is disrupted such that cognitively relevant information is not properly stored or even attended to (see Chapters 5 and 16). Consequently, the ability to form nonemotional associations (e.g., between stimulus and response) and to create new neural networks, or to alter preexisting cognitive schemas and neural circuits, is attenuated (Douglas, 1967). These individuals appear to be amnesic. Again, however, this is a role the hippocampus also shares with the frontal lobes, which, in humans, are clearly dominant over the hippocampus in this regard. That is, the frontal lobes have taken over many functions that the hippocampus mediates in lower mammals.

Hippocampal Arousal and Memory Loss

As described in Chapter 5, when exposed to novel stimuli or when engaged in active searching of the environment, hippocampal theta appears (Adey et al., 1960), as does LTP (Lynch, 1986). There is, thus, a direct correlation between hippocampal theta and the development of hippocampal LTP (Lynch et al., 1990). However, with repeated presentations of a novel or familiar stimulus, the hippocampus habituates and theta disappears (Adey et al., 1960). Possibly, the reason the hippocampus and other brain structures appear to respond preferentially to novel stimuli and then to cease, at least during learning tasks, is because processing familiar stimuli in short-term memory would be a waste of energy and attentional space. Nevertheless, these findings suggest that when the neocortex is highly stimulated, the hippocampus (in order to monitor what is being received and processed), functions at a lower arousal level in order not to become overwhelmed (Joseph 1990a, 1992a). However, at extremely high levels of arousal, what is being experienced may not be learned, or it will be learned independently of the hippocampus due to diminished hippocampal activity. In situations where both the neocortex and the hippocampus become highly aroused and desynchronized, there results distractibility and hyperresponsiveness such that the subject becomes overwhelmed, confused, and may orient to and approach several stimuli (Grastyan et al., 1959)—a condition that also occurs following hippocampal lesions (Clark and Isaacson, 1965; Douglas, 1967; Ellen et al., 1964).

Under conditions of excessive arousal, the ability to think or respond coherently may be disrupted as is attention, learning, and memory functioning, i.e., memories are stored haphazardly, incompletely, or not at all.

Situations inducing high levels of arousal, perceptual disorganization, and memory loss sometimes also occur when individuals are highly anxious, frightened, or emotionally upset and traumatized (see Chapters 15 and 16), e.g., during a prolonged and brutal rape or physical assault, or during horrendous battlefield conditions—in which case, hippocampal participation in memory formation dramatically decreases as neocortical and limbic arousal increases.

However, under emotionally traumatic conditions it is not at all unusual for victims to profess a complete or partial amnesia for the event in question, which, in turn, may resolve or be accompanied by flashbacks, heightened startle reactions, and intrusive emotional images. In these instances of extreme emotional stress, other brain structures, such as the amygdala, probably play a more important role in memory and learning.

THE AMYGDALA AND EMOTIONAL MEMORY

The amygdala (although dependent on the hypothalamus) is preeminent in the control and mediation of all higher order emotional and motivational activities, including the formation of emotional memories (Gloor, 1992; Halgren, 1992; Joseph, 1982, 1992a; LeDoux, 1992; Rolls, 1992). Indeed, the amygdala becomes particularly active when recalling personal and emotional memories (Halgren, 1992; Heath, 1964; Penfield and Perot, 1963), and in response to cognitive- and context-determined stimuli, regardless of their specific emotional qualities (Halgren, 1992). The amygdala is also functionally related and richly (and reciprocally) interconnected with the hippocampus, inferior temporal lobe, and a variety of other limbic, brainstem, and neocortical regions that are also important in memory

functioning (Amaral et al., 1992; Gloor, 1992; Halgren, 1992; Rolls, 1992).

Amygdaloid neurons, therefore, are able to monitor and abstract from the sensory array stimuli that are of motivational significance to the organism (Gloor, 1992; Joseph, 1992a; Rolls, 1992; Steklis and Kling, 1985) as well as add emotional attributes to perceptual and cognitive activities. This includes the ability to discern and express even subtle social-emotional nuances, such as friendliness, fear, love, affection, distrust, anger, and so forth, and at a more basic level, to determine if something might be good to eat (Fukuda et al., 1987; O'Keefe and Bouma, 1969; Ono et al., 1980). The lateral amygdala also has high concentrations of acetylcholine (Woolf and Butcher, 1982), as does the hippocampus (Kuhar, 1975), and acetylcholine (ACh) activity within the amygdala as well as hippocampus is correlated with learning and memory (Todd and Kesner, 1978). Indeed, a portion of the "extended amygdala"—i.e., the substantia innominata (of the limbic striatum)—is a major source of neocortical ACh and is therefore exceedingly important in memory. Indeed, Alzheimer's disease is associated with degeneration in the substantia innominata (see Chapter 9).

Apparently, the amygdala acts on neocortical arousal in a very widespread manner via its influences on substantia innominata and brainstem ACh-containing neurons, which in turn project to the thalamus and onto the widespread areas of neocortex (see Chapter 9; Davis, 1992; Kapp et al., 1992; Steriade et al., 1990). This, in turn, increases the receptivity of the neocortex for processing sensory information (Steriade et al., 1990). This may also lead to a state of vigilance, attention, or emotional arousal, such as fear.

Single amygdaloid neurons receive a considerable degree of topographic input, and are predominantly polymodal, responding to a variety of stimuli from different modalities simultaneously (O'Keefe and Bouma, 1969; Perryman et al., 1987; Sawa and Delgado, 1963; Schutze et al., 1987; Turner et al., 1980; Ursin and Kaada,

1960; Van Hoesen, 1981). Hence, multimodal integration and assimilation occurs within single amygdala neurons. Amygdala neurons are also involved in recognition memory, and the detection of faces, facial expressions, and visual stimuli involving context (Gloor, 1992; Murray, 1992; Rolls, 1992). The amygdala, therefore, is also capable of forming cross-modal associations such that a neutral stimulus can come to be endowed with emotional attributes (Gaffan and Harrison, 1987; LeDoux, 1992; Rolls, 1992). This includes forming associations between primary as well as secondary reinforcers (Everitt et al., 1991; Gaffan and Harrison, 1987). In addition, there are single neurons in the amygdala that discriminate between stimuli that are rewarding or punishing (Rolls, 1992).

Because of its involvement in all aspects of social-emotional and motivational functioning, stimulation of the amygdala can also evoke highly personal and emotional memories, as it is highly involved in remembering emotionally charged experiences (Gloor, 1992; Halgren, 1981, 1992; Halgren et al., 1978; Rolls, 1992; Sarter and Markowitsch, 1985). In addition, the amygdala becomes particularly active when recalling personal and emotional memories (Halgren, 1992; Heath, 1964; Penfield and Perot, 1963), and in response to cognitive- and context-determined stimuli, regardless of their specific emotional qualities (Halgren, 1992).

Under some circumstances, however, including those involving extreme fear and repetitive emotional stress, the amygdala may act to form these memories in the absence of hippocampal participation. Moreover, because the amygdala becomes functional soon after birth (whereas hippocampal immaturity is more extended), early memories may be stored within or via the amygdala, tend to be created in the absence of verbal or cognitive attributes, and are likely to be emotional, gustatory, tactile, and visual (see Chapters 15 and 16). Therefore, these early memories can be quite difficult to access when one relies on hippocampal retrieval mechanisms, although they may be recalled in the form of bodily sensations or emotional disturbances (Courtois, 1988).

Fear, Anxiety, Startle, and Traumatic Stress

The amygdala is also the site where conditioned fear, startle, and autonomic reactions occur (Davis, 1992; Hitchcock and Davis, 1991; LeDoux, 1992)—reactions that may be stored in memory. Hence, in response to a certain stimulus that has been repeatedly associated with positive or negative outcomes, autonomic changes are triggered, which may be accompanied by fear as well as memory recall.

The amygdala directly influences breathing and heart rate in response to emotional stimuli, and in fact may create a circuit or neural network that triggers heart rate and autonomic changes, as well as neocortical desynchronization (low-voltage fast activity) in response to a neutral stimulus that has been repeatedly associated with highly arousing pleasant or aversive emotional outcomes (Kapp et al., 1992; Ursin and Kaada, 1960). That is, emotional learning of a well-defined conditioned response occurs within the amygdala, which in turn influences autonomic activity such that both classical and instrumental emotional conditioning occurs. However, this also makes possible the experience of anxiety and anticipatory fear in response to cues that might trigger the original emotional memory; in this regard, the amygdala contributes to the development of post-traumatic stress disorder (see van der Kolk, 1987).

In these particular instances, however, the more the cue is linked and associated with the original traumatic event, the more severe might be the reaction and the more upsetting the recalled memory. Kolb (1986:172), for example, in his discussion on the persistence of startle reaction in combat soldiers, noted that "it was remarkable 10, 15, or 20 years after the combat event to hear men tell about suddenly hearing a loud noise reminding them of battle sounds and then being in a panic state."

Kolb (1986:172) also reports that when they "played a tape consisting of combat sounds, including those of helicopters, mortars, and screaming wounded...the patients immediately returned to a critical event in the battlefield. Many then experienced and reenacted scenes of fight, flight, or rage." For most of these individuals, re-experiencing the traumatic memory is, in itself, extremely traumatic and they avoid all forms of stimulation that can reawaken these memories.

Hence, an individual may become severely stressed or repeatedly frightened, and later experience an increase in heart rate, changes in breathing patterns, and other autonomic alterations, in anticipation or in response to associated emotional and contextual cues, which in turn may activate the neural network in which the original emotional memory is stored. However, at least in regard to heart rate changes, this circuitry is, in part, hard-wired and involves a projection into the midbrain peri-aqueductal gray (as well as the basal ganglia) and then on to the lower brainstem into the vagal nucleus (Kapp et al., 1992).

According to Davis (1992:283), "much of the complex behavioral pattern seen during fear conditioning has already been hard wired during evolution. In order for a formerly neutral stimulus to produce the constellation of behavioral effects used to define a state of fear or anxiety, it is only necessary for that stimulus to now activate the amygdala, which in turn will produce the complex pattern of behavioral changes by virtue of its innate connections to different brain target sites." Hence, the amygdala need only learn what stimuli are the most arousing, threatening, fearful, pleasurable, and so on, and can then automatically trigger autonomic and behavioral responses that are already hard-wired into the nervous system.

Amygdala activation will also increase the frequency of respiration and will trigger behavioral freezing and an arrest of ongoing behavior (Kapp et al., 1992; Ursin and Kaada, 1960). As detailed in Chapter 9, the amygdala may also contribute to the frozen states common among those with Parkinson's disease or those who are extremely frightened.

Initially, however, freezing behavior is part of the attention response, which in turn may be followed by an orienting reaction, sometimes with anxious glancing about, and/or with an approach or withdrawal reaction (Gloor, 1960; Ursin and Kaada, 1960). If amygdala stimulation continues, fear and/or rage reactions are elicited (Ursin and Kaada, 1960).

In general, the right amygdala is more involved than the left in the production of fear and anxiety (Davis, 1987), particularly free-floating fears (see Chapter 5), and the right hemisphere is more greatly involved in the fear-potentiated startle response (Lang et al., 1990), as well as emotion in general (see Chapter 3).

Fear, Anticipation, and Repression

As noted, the amygdala can make fine discriminations between stimuli that are closely versus not so closely associated with fear, so that startle and fear reactions are elicited to stimuli that most closely match (Davis, 1992). The amygdala, therefore, can also anticipate fearful stimuli based on only fragmentary cues or sensory stimuli, some of which might trigger startle reactions (Davis, 1992; Kurtz and Siegel, 1966) or memory recall.

The amygdala is, thus, very important in the production of anticipatory anxiety (Davis, 1992) as well as emotional memory. Because it can anticipate based merely on fragmentary cues, individuals may come to feel upset, angry, frightened, and yet not know why or what might have triggered these emotions. This often occurs among individuals suffering from post-traumatic stress disorder, as well as among those with repressed memories, i.e., a consequence of a single cue triggering a complex emotional memory or associated feelings of anxiety (van der Kolk, 1987). Normally, however, these memories may seem to be completely forgotten.

This "forgetting" is a very adaptive arrangement, for if the individual were unable to inhibit or forget emotionally upsetting memories, their intrusion would cripple his or her ability to function. That is, perhaps it makes little evolutionary sense for an individual to recall extremely negative and emotionally upsetting memories, that is, except when in a highly aroused state. For these memories to seep or burst into consciousness at other times may exert a deleterious effect on functioning, which appears to be the case in post-traumatic stress disorder. Perhaps this is why the hippocampus sometimes ceases to participate in memory formation under extremely distressful conditions, and why only fragments of the emotional trauma may appear in consciousness under certain circumstances and conditions. Hippocampal deactivation reduces the likelihood of recalling or controlling the recall of negative memories. From the standpoint of survival it makes sense for these memories to be recalled only when in a similar state of high arousal or when presented with similar negative emotional cues. However, even under these conditions the forgotten memory may remain "repressed," and this too may be accomplished via the amygdala (as well as the frontal lobes).

For example, the amygdala in response to a partial cue may respond with feelings of anxiety and fear before the actual memory or associated neural network can be fully activated. In instances such as these, the actual memory might come to be inhibited, or repressed, in response to the anxiety that was triggered in advance. The individual is upset and does not know why or has only a partial "flashback."

Fear and Hippocampal Deactivation

In response to continued, ongoing fear, emotional stress, and anxiety, the amygdala begins to secrete large amounts of the amino peptide, corticotropin-releasing factor (CRF), which in turn potentiates the behavioral and autonomic reaction to fear and stress (see Davis, 1992). Cortisone levels also increase dramatically when stressed by restraint (see Davis, 1992). Moreover, high levels of cortisone exert an inhibitory influence on the hippocampus and eliminate hippocampal theta (see Spoont, 1992). Hence, in these instances, learning may occur in the absence of hippocampal participation.

When fear follows the attention response, the pupils dilate and the subject will cringe, withdraw, and cower, which is associated with high levels of amygdaloid activity. This cowering reaction, in turn, may give way to extreme fear and/or panic and the animal will attempt to take flight. This may be accomplished via amygdala activation of the locus ceruleus, which would release norepinephrine (NE), and activation of the raphe nucleus, which would release serotonin (Magnuson and Gray, 1990), both of which facilitate motor responsiveness (White and Neuman, 1980). However, if arousal levels continue to increase, the hippocampus may cease to participate in memory functioning such that whatever is experienced and learned may subsequently appear repressed or forgotten due to hippocampal deactivation (see Chapter 16 for further details). Among humans, the fear response is one of the most common manifestations of amygdaloid electrical stimulation (Gloor, 1990; Halgren, 1992; Williams, 1956). Moreover, unlike hypothalamic on/off emotional responses, attention and fear reactions can last up to several minutes after the stimulation is withdrawn. Apparently, this is associated with the development of LTP within amygdala neurons, which thus continue to reverberate, even when the subject is frightened or in pain.

In addition, when the (human) amygdala is abnormally or highly activated, it is able to possibly deactivate the anterior hippocampus as well as overwhelm the neocortex and the rest of the brain so that the person not only forms emotional ideas (that later cannot be fully remembered) but responds to them in an abnormal manner, which is also stored in emotional memory.

In fact, during periods of extreme fear, such as when a subject freezes, hippocampal theta disappears and is replaced by irregular electrophysiological activity (Vanderwolf and Leung, 1983). Prolonged stress also inhibits long-term potentiation in the hippocampus (Shors et al., 1989). Theta also disappears if a subject is placed in a situation where he is completely and hopelessly surrounded by a painful physical threat, such as an electrified grid floor (Vanderwolf and Leung, 1983). Similarly, under conditions of stress or when the body has been anesthetized (Green and Arduini, 1954), hippocampal theta disappears, as does LTP (Shors et al., 1989), and the threshold for hippocampal response is significantly raised. In part, it is overwhelmed and inhibited by the amygdala and the reticular activating system, which in turn reduces its participation in learning and memory. The hippocampus maintains a reciprocal relationship with the reticular formation in regard to the modulation of neocortical activity, except under conditions of high arousal, in which case it can be inhibited by the reticular activating system (Adey et al., 1957; Green and Arduini, 1954; Redding, 1967). This may occur in situations involving high levels of arousal and with repetitive stimulation (Redding, 1967), in which case theta and LTP may be prevented. Indeed, with persistent and repeated stimulation, instead of excitation there is synaptic depression with no evidence of LTP, even under neutral and nonemotional conditions (Lynch, 1986). Presumably this is related to rapid calcium depletion, for LTP and synaptic plasticity are affected by calcium levels.

This loss of hippocampal theta and LTP under repetitive, stressful, painful, or fearful conditions is not a function of lack of whole body movement, however. For example, theta appears when human subjects are sitting motionless but are mentally engaged in problem solving or learning tasks (Meador et al., 1991) and when the median raphe nucleus (which suppresses hippocampal theta and induces high states of arousal) is destroyed and subjects are immobilized (Vertes, 1981). Rather, this loss of theta activity is a function of overwhelming fear and high levels of arousal, for although the subject seems alert, the hippocampus is desynchronized; the subject

may seem frozen and will be unable to respond to environmental stimuli, be it threat or painful stimuli (Redding, 1967). In consequence, he is later unable to recall what occurred, although he may feel or act traumatized. as this data was stored by the amygdala in the absence of hippocampal participation.

Later, hippocampal retrieval mechanisms are unable to find the memory. The patient will appear amnesic when he is, in fact, relying on the wrong (albeit "normal") retrieval mechanisms, i.e., the hippocampus. Indeed, under conditions of high hippocampal arousal or inhibition, due to direct hippocampal stimulation, verbal amnesia results (Brazier, 1966; Chapman et al., 1967) and there is a loss or a significant reduction in theta (Redding, 1967). This occurs even when there has been no loss of consciousness, and while the subject seems fully alert, with no evidence of confusion. Normally, as described by Squire (1992:224, 208), "the possibility of later retrieval is provided by the hippocampal system because it has bound together the relevant cortical sites. A partial cue that is later processed through the hippocampus is able to reactivate all of the site and thereby accomplish retrieval of the whole memory.... In the absence of the hippocampus, representations that had been established in short-term memory are literally lost, become disorganized, or achieve some abnormal fate."

In this regard, memories formed by the amygdala in the absence of hippocampal participation, such as in instances of extreme emotional stress and neocortical arousal, may appear to be forgotten, as they cannot be accessed via normal hippocampal retrieval mechanisms. For example, in studies of individuals who have experienced high levels of emotional arousal, such as victims or witnesses of rape, murder, and assault, memory is significantly and negatively impacted (Deffenbacher, 1983; Kuehn, 1974; Williams, 1992). In one study, 38% of women who had been taken to an emergency room following a sexual assault had no memory of being at the hospital (Williams, 1992). Similarly, victims of rapes and physical assaults tend to provide poorer descriptions of their assailants than those who were victims of robberies (Kuehn, 1974). In one study of individuals who had committed murder and violent crimes, over 10% had no recollection and reported amnesia for their crimes (Taylor and Kopelman, 1984), whereas those who engaged in nonviolent crimes did not demonstrate a loss of memory. This follows the inverse U nature of arousal on memory (see Chapter 16).

It is possible, however, for these lost memories to be reactivated (see Chapters 15 and 16). Presumably, they are activated via retrieval strategies that are dependent on the amygdala and/or hypothalamus, if the memory is emotional (Halgren, 1992), or the basal ganglia, brainstem, or cerebellum, if it is motor routines that have been stored.

Moreover, although the hippocampal amnesia may extend backward in time from minutes to days and even weeks, the amnesia may shrink over the course of the next several hours, such that only those events that took place just prior and during the period of abnormal hippocampal excitation are seemingly forgotten (Brazier, 1966; Chapman et al., 1967), although, more likely, they too have been stored by the amygdala.

Even with the hippocampus destroyed, this amnesia does not extend to stimuli related to fear or pain (Seldon et al., 1991). In fact, aversive conditioning is completely unaffected (Seldon et al., 1991), particularly when the conditions are highly aversive and painful or highly emotional, as this form of learning is mediated by the amygdala (Cahill and McGaugh, 1990; Hitchcock and Davis, 1987; Kesner and DiMattia, 1987) as well as the hypothalamus. Rather, what may be forgotten are the place and the circumstances around which the pain was experienced (Seldon et al., 1991). However, when specific cues are provided, recall and memory recovery occurs (Seldon et al., 1991), perhaps due to direct activation of the memory store via recognition. Recognition memory and the neocortical storage site are activated and what was forgotten is suddenly recalled—sometimes to the disbelief of the family or treating physicians.

Memory Loss and Amygdala Dysfunction

As previously discussed in detail (see Chapter 5), the amygdala is involved in all aspects of social-emotional intelligence and memory functioning. Hence, destruction of the amygdala makes it exceedingly difficult to form associations between emotional, neutral, and reinforcing stimuli, regardless of their being positive or negative (Jones and Mishkin, 1972; LeDoux, 1992; Rolls, 1992; Weiskrantz, 1956). Amygdala destruction also results in a dense social-emotional agnosia and amnesia such that even loved ones are no longer recognized. Similarly, emotional learning becomes very limited, primitive, and dependent on the hypothalamus (Joseph, 1992a). For example, animals who have been amygdalectomized often demonstrate very poor performance on emotional learning and attention tasks (Kesner, 1992; Murray, 1992; Sarter and Markowitsch, 1985), as they are unable to assign emotional attributes to what they perceive. They also cannot learn associations between primary reinforcers and rewarded stimuli, including those where the subject has to approach or avoid stimuli that has either been rewarded or punished (Kesner, 1992; Murray and Mishkin, 1985; Rolls, 1985, 1992). Neurosurgical removal of the right or left amygdala also results in significant memory disturbances, particularly in the realm of information retrieval and social-emotional perception and contextual information storage (Andersen, 1978; Gloor, 1992; Rolls, 1992; Sarter and Markowitsch, 1985). Moreover, a complete amnesia can result following electrical stimulation of the amygdala (Chapman, 1958; Jasper and Rasmussen, 1958), and difficulty recognizing people and objects has been repeatedly noted following amygdalectomy in humans (Sawa et al., 1954).

Moreover, Alzheimer's disease as well as other dementias and encephalitic conditions not only result in severe memory loss, but often attack the amygdala and the anterior commissure (which links the right and left amygdala), creating both neurofibrillary changes and senile plaques (see Mann, 1992; Sarter and Markowitsch, 1985, for review). With normal aging there is also a progressive memory loss and a significant dropout of neurons in the amygdala (Herzog and Kemper, 1980), as well as in other structures, particularly the substantia innominata, which is part of the "extended amygdala" and which is directly implicated in the genesis of Alzheimer's disease.

The Amygdala, Hippocampus, and Dorsal Medial Nucleus

Both the hippocampus and amygdala maintain rich interconnections with the anterior thalamus (Amaral et al., 1992; Krettek and Price, 1977; Nauta, 1971; Graff-Radford et al., 1990). However, these two limbic pathways remain distinct and terminate in separate, albeit adjacent, thalamic areas (Graff-Radford et al., 1990). Indeed, a similar pattern of parallel organization characterizes amygdala and hippocampal input into the basal ganglia and other limbic and brainstem nuclei (see Chapters 9 and 10).

This suggests that the influences and contributions of the amygdala and hippocampus to memory, motor, and perceptual activity remain distinct within the brainstem, basal ganglia, and the neocortex such that perhaps two separate, albeit closely aligned, neural circuits are created in parallel—one concerning the emotional and the other the cognitive attributes of what has been attended to, learned, and stored in memory.

It is also noteworthy that although destruction of the anterior and DMT can induce severe verbal and visual amnesic disorders, it appears that both the amygdala and hippocampal pathways must be destroyed for this to occur (Graff-Radford et al., 1990). Presumably, this is because the

amygdala and hippocampus interact at the level of the limbic system, and again within the thalamus, and probably again within the neocortex to insure that certain memories are formed. Hence, damage to the DMT may result in a disconnection syndrome. That is, the interacting influences of the hippocampus and amygdala are no longer possible within the DMT, which negatively impacts memory formation within the neocortex.

In contrast, thalamic lesions that are more caudally located and that spare these limbic pathways do not induce amnesia (see Graff-Radford et al., 1990; von Cramon et al., 1985). Because these limbic-thalamic pathways have been spared, the hippocampus and amygdala are able to continue to exert tremendous influences on neocortical information processing, including the gating of perceptual activity and neocortical arousal and the filtering versus selective attention to events that are emotionally, motivationally, and cognitively significant.

These distinct, albeit codependent (emotional versus cognitive), memory traces are also probably mutually reinforcing and linked to some degree. However, because they are separate, it is possible to recall the cognitive attributes of an experience in the absence of its emotional significance, or conversely, to recall the emotional features of an event in the absence of memory for the event; e.g., an individual may recall having a horrible argument but be unable to remember what was said. If severely traumatized, these persons may become severely amnesic but they will remain upset.

For example, Christianson and Nilsson (1989) described a 23-year-old female who was raped and beaten while out jogging, after which she had no memory of her identity or that of her relatives, friends, boyfriend, or place of work. All recollections of her past life, including all aspects of the rape, were forgotten. However, when she returned with police to the jogging path she became extremely anxious and upset. Nevertheless, she remained "amnesic" until later, when specific cues associated with the rape and scene of the crime enabled her to suddenly recall what had happened.

Overview: The Amygdala, Hippocampus, and Memory

Based on the evidence reviewed above and in Chapters 5 and 16, there is, thus, considerable evidence to strongly suggest that the hippocampus plays an interdependent role with the amygdala (and the frontal-thalamic system) in regard to many aspects of memory (e.g., visual, verbal, spatial), and that the amygdala plays a primary role in regard to all aspects of emotional memory. That is, the amygdala (in conjunction with the hypothalamus) extracts and stores the emotional attributes as well as mediates the emotional reactions and feelings triggered by particular events. It also acts on the anterior hippocampus and DMT so that motivationally and emotionally significant information is attended to and stored via the creation of appropriate neural networks. The hippocampus performs likewise with verbal, visual-spatial, and cognitive material. Thus, these two nuclei work in concert.

The amygdala seems to reinforce and maintain hippocampal activity via the identification of social-emotional and motivationally significant information and the generation of pleasurable rewards—through interaction with the lateral hypothalamus and by "rewarding" the anterior hippocampus (see Chapter 5). That is, reward increases the probability of attention being paid to a particular stimulus or consequence as a function of its association with reinforcement (Douglas, 1967; Kesner and Andrus, 1982).

The same is true when an experience is personally significant and moderately emotionally arousing. It is perhaps for this reason that personal emotional events are often better recalled than those that are neutral and why learning and memory improves when rewarded; i.e., amygdala-hippocampal interactions are enhanced, which leads to the establishment of cognitive-emotional neural networks and, thus, long-term mem-

ory. Conversely, when both the amygdala and hippocampus are damaged, striking and profound disturbances in memory functioning result (Kesner and Andrus, 1982; Mishkin, 1978).

There is also evidence indicating that these emotional and cognitive memories may be stored in separate, albeit aligned, neural networks. As noted, the pathways leading from the amygdala and hippocampus to the anterior and DMT terminate in separate, albeit adjoining, thalamic neurons (Graff-Radford et al., 1990) as well as in adjacent basal ganglia and brainstem neurons. Thus, from the amygdala-hippocampus to the thalamus and then to the neocortex (as well as through any number of different routes) these two networks of emotional versus cognitive "memories" and associated neural pathways remain separate and may terminate on different dendrites and neurons. Hence, cognitive and associated emotional memories are maintained by semi-independent neural networks that might be activated or inhibited separately, partially, sequentially or in parallel, and conjointly.

Also, as noted, the amygdala and hippocampus act differentially in regard to the effects of positive versus negative reinforcement. For example, whereas the hippocampus produces theta in response to noxious stimuli, it soon ceases to participate in memory formation when negative conditions are continuous and ongoing; i.e., theta disappears (Vanderwolf and Leung, 1983) as does hippocampal LTP (Shors et al., 1989). Similarly, Heath (1977) reports that high emotional states, including fear and rage, are associated with high amplitude spindle EEG wave forms within the hippocampus. In contrast, the amygdala increases its activity following the reception of a reward and in response to stressful stimuli (see Gaffan, 1992; Halgren, 1992; Rolls, 1992). It also develops LTP under these conditions.

The amygdala also mediates behavioral responsiveness under conditions of high arousal, so as to maintain attention, to attack in anger (or hunger), or to run away in fear. Hence, whereas the hippocampus diminishes its contribution under these highly arousing conditions, the amygdala continues to subserve perceptual and behavioral activity, and continues to monitor and store the emotional attributes in memory, even in the absence of hippocampal participation. From the standpoint of survival, this makes evolutionary sense. Consequently, later this information may not be accessible via normal "hippocampal retrieval strategies" and may appear to be forgotten or repressed.

This "state-dependent" neurological relationship is quite adaptive, for in some life-threatening situations an immediate emotional reaction is called for and the recall and consideration of cognitively based alternatives might result in death. Thus, when in a similar state, only those memories that are relevant to survival are recalled. Hence, the hippocampus ceases to participate. Consequently, irrelevant information is not stored in memory, whereas when an individual is highly emotionally aroused, relevant memories might be suddenly recalled.

In fact, as individuals may not be capable of (verbally) thinking clearly under certain highly stressful and fearful conditions, the generation and storage of these thoughts might result in their latter recollection when a similar situation occurs (which, for almost all of our ancestry was probably likely); this, therefore, would disrupt the individual's ability to engage in effective action (such as fleeing or fighting) or to recall the last successful strategy.

However, these emotional memories may be suddenly recalled if the person is later confronted with similar emotional and contextual cues, or when in a similar emotional state, enabling him to act in the same manner that previously saved his life.

This may also account for the effect of certain contextual cues in the elicitation of traumatic stress reaction; i.e., hearing a car backfire and suddenly dropping to the floor in response to a "flashback" of being in battle. That is, the amygdala responds to associated cues to trigger the recollection of associated emotional memories and relevant (and life-saving) behaviors.

It is presumably due to these separate neural networks, and the differential contributions of the amygdala and hippocampus, that a cognitive memory might be retrieved or recalled, stripped of its original emotional attributes (which may be too troubling and upsetting to remember), or why emotional memories might be triggered in the absence of concrete auditory and visual cognitive cues. As such, an individual may feel exceedingly upset, anxious, angry, and so on, and not know why (Joseph, 1988a, 1992b), or, conversely, verbally recall certain traumatic or negative events that are expressed and verbally described in neutral or even positive terms (see Chapter 15), although the memory is actually quite negative, if not traumatic. However, in some cases the individual may "recall" nothing at all and may not even be aware that something unpleasant has been repressed.

Again, this is very "adaptive," for otherwise an individual may be plagued by memories that may best be forgotten. Indeed, although resulting in modern-day "neurosis" and paradoxical amnesias induced by brain damage, it is presumably this dual amygdala-hippocampal relationship and the capacity to separately store and selectively recall emotional versus cognitive information that has promoted, not just learning and memory, but individual and species survival (if not sanity) for much of human history.

7

Limbic Language, Social-Emotional Intelligence, Development, and Attachment

Amygdala, Septal Nuclei, Cingulate Gyrus, Maternal Language, Sex Differences, Emotional Deprivation, Contact Comfort, and Limbic Love

Infant Vocalizations and the Innate Languages of the Limbic System

It has been argued in detail elsewhere that phylogenetically and ontogenetically, the original impetus to vocalize springs forth from roots buried within the depths of the ancient limbic lobes, e.g. amygdala, hypothalamus, septal nucleus, cingulate gyrus (Joseph, 1982, 1992a, 1993, 1994; see also Jurgens, 1990; Jurgens et al., 1982; Jurgens and Muller-Preuss, 1977; Robinson, 1967, 1972, 1976; Ursin and Kaada, 1960). For example, although nonhumans do not have the capacity to speak, they still vocalize, and these vocalizations are often limbic and emotional in origin (Darwin, 1872; Jurgens, 1990; Jurgens and Muller-Preuss, 1977; Robinson, 1967, 1972, 1976; Ursin and Kaada, 1960). Characteristically, limbic vocalizations are evoked in situations involving sexual arousal, terror, anger, flight, helplessness, and separation from the primary caretaker when young.

The first vocalizations of human infants are similarly emotional in origin and limbically mediated, consisting predominantly of sounds indicative of pleasure and displeasure (Joseph, 1982, 1992a; Milner, 1967; Morath, 1979; Truby and Lind, 1965), vocalizations that are initially produced by the hypothalamus—a nucleus that could be likened to the Freudian Id. These sounds and cries are produced soon after birth, and even by infants born deaf and blind

(Eibl-Eibesfeldt, 1990), indicating that they are innate (Darwin, 1872).

Infants are exceedingly dependent on limbic language and the limbic system in order to communicate their needs and to discern the social-emotional intentions of others. Thus, although lacking denotative, grammatical language skills, because of limbic system expressive and perceptual activity, and its capacity to comprehend social-emotional nuances (see Chapter 5), infants are quite adept at distinguishing between different emotional vocalizations so as to determine the mood state and intentions of others (Fernald, 1993; Haviland and Lelwica, 1987).

For example, it has been demonstrated that preverbal infants between the ages of 10 weeks and 5 months are capable of appropriately discerning, discriminating, and responding to social-emotional vocalizations conveying approval, disapproval, happiness, and anger (Fernald, 1993; Haviland and Lelwica, 1987). Moreover, 5-month-old (American) infants are able to make these discriminations even when the words have been filtered, and in response to nonsense English, as well as to German and Italian vocalizations (Fernald, 1993). However, infants even younger are capable of producing these same limbic vocalizations (Joseph, 1982), which again indicates that these abilities are innate (Darwin, 1876).

Similarly, apes and monkeys reared in isolation or with surgically muted mothers

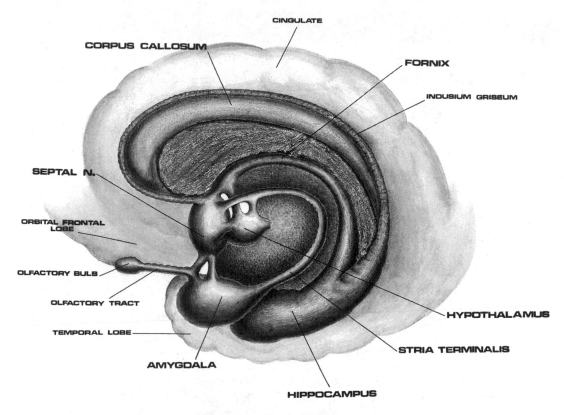

Figure 7.1   The limbic systems.

and, thus, with little or no "language" experience or training, are able to produce complex and appropriate calls and cries that accurately indicate fear or danger. And, they will react appropriately to these same calls the first time they are exposed to them (Winter et al., 1973). For example, squirrel monkeys reared in isolation respond appropriately with fear and anxiety in response to warning "yapping" calls (signifying the presence of a predator) the very first time they hear them (Herzog and Hopf, 1984). They will also produce an appropriate "yapping" cry when they are first exposed to a potential predator.

However, be it infant H. sapiens sapiens, or nonhuman primates and mammals (hereafter referred to as "primates" and "mammals"), different nuclei within the limbic system follow a somewhat different developmental course, which differentially affects the acquisition and expression of limbic language and social-

emotional intelligence. For example, the hypothalamus is almost fully functional at birth, whereas the amygdala, septal nuclei, and the cingulate gyrus (limbic nuclei exceedingly important in the production of maternal-infant social-emotional behavior, including feelings of love and attachment), continue to mature and develop over the course of the first year of life and beyond. This gives rise to increasing vocal-emotional complexity.

These differential limbic maturational events are also associated with developmental changes in the infants' generalized and intense desire for physical and social contact, their later fear of strangers, and then the development of specific emotional attachments (Joseph, 1982, 1992a). Over the course of development, limbic language and social-emotional intelligence become more complex and elaborate.

For example, initially infant emotional sound production appears to convey generalized meanings, e.g., pleasure and displeasure. However, gradually these vocalizations become modified, depending on context, and are then increasingly shaped and tied to specific mood states or events and social-emotional phenomena (Joseph, 1982, 1992a; Milner, 1966; Piaget, 1952). That is, the sounds produced and perceived become specifically tied to anger, sadness, joy, affection, sorrow, fear, and so on, rather than just pleasure and displeasure.

Similar developmental modifications in limbic vocalizations have been noted among primates. For example, vervet monkeys employ three distinct calls, which they differentially produce in the presence of eagles, snakes, and leopards (Cheney and Seyfarth, 1990). Experienced and normally reared monkeys respond to these calls by looking up ("eagle"), looking down ("snake"), or climbing up a tree ("leopard"), depending on which call is produced, even when played from a tape recorder (Seyfarth et al., 1980).

However, infants reared in isolation merely respond with generalized alarm when presented with these same calls, and are as likely to look up as down as climb a tree. That is, although they recognize the emotional significance of the call, they are not yet able to differentiate these sounds as signifying particular and quite specific social-emotional events. These abilities are increasingly acquired over the course of the first few months.

This limbic developmental course is not only innate, but similarly acquired across species (Darwin, 1876). However, because humans possess a similar limbic system, similar sounds exert similar effects, even across species. Thus, be it human, primate, or social mammal, certain sounds arouse fear, sadness, caution, or alarm (e.g., thunder, growling, rumbling tones of thunder) or, conversely, pleasure, gaiety, or peaceful conditions (e.g., soft, higher rolling, or smoother pitched tones and melodies) (see Chapter 3). It is because certain sounds can produce specific and universal mood states that movie and television programs are often accompanied by "mood" music.

Languages of the Limbic System

Emotional cries and warning calls have been produced via electrode stimulation of wide areas of the limbic system, including the septal nuclei, stria terminalis, the lateral, medial, and preoptic hypothalamus, and the periaqueductal gray located in the midbrain; these same areas often become activated in response to certain emotional sounds (see Jurgens, 1990; Jurgens et al., 1982; Jurgens and Muller-Preuss, 1977; Robinson, 1967, 1972; Ursin and Kaada, 1960). Indeed, the limbic system is more vocal than any other part of the brain (Jurgens, 1990; Robinson, 1967).

Nevertheless, the type of cry elicited, in general, depends upon which limbic nuclei has been activated (Jurgens, 1990; Jurgens and Muller-Preuss, 1977; Robinson, 1967, 1972). This is because different limbic nuclei, and in fact, different divisions within these nuclei, subserve unique functions and maintain different anatomical interconnections with various regions of the brain (see Chapter 5).

For example, portions of the septal nuclei, hippocampus, medial amygdala, and medial hypothalamus have been repeatedly shown to be generally involved in the generation of negative and unpleasant mood states (see Chapter 5; Olds and Forbes, 1981, for review). Other limbic tissues, including the lateral hypothalamus, lateral amygdala, and portions of the septal nuclei, are associated with pleasurable feelings. Hence, areas associated with pleasurable sensations often give rise, when sufficiently stimulated, to pleasurable calls, whereas those linked to negative mood states will trigger shrieks and cries of alarm.

Hierarchical Vocal Organization

When the hypothalamus is stimulated, not just emotional vocalization but complex and emotionally congruent behaviors can be elicited. On the other hand, once the stimulation is terminated, the behavior and vocalizations cease. Thus, the hypothalamus responds in an on/off fashion and is

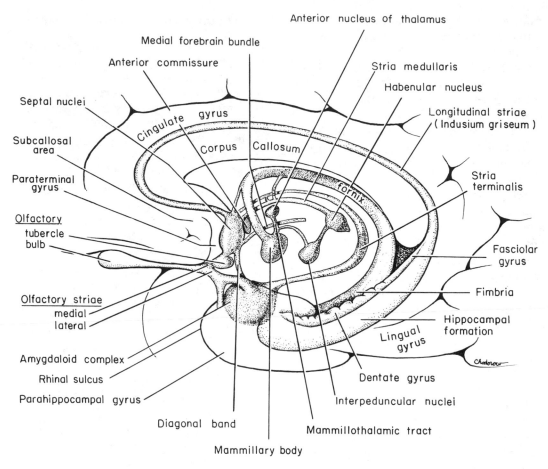

Figure 7.2   The limbic system. (From Carpenter M. Core text of neuroanatomy.
Baltimore: Williams & Wilkins, 1991.)

only indirectly involved in the production of complex and enduring emotional mood states (see Chapter 5). By contrast, activation of the more recently evolved amygdala and anterior cingulate produces sustained feeling and mood states. Moreover, these limbic nuclei appear to be hierarchically organized and therefore capable of a greater degree of complexity in function and emotional expression as compared with the more ancient hypothalamus. The same is true regarding the reflexive vocal motor centers located in the brainstem, which appear to be devoid of the capacity to experience specific emotions other than, perhaps, noxious and painful sensations (see Chapter 10). Instead, brainstem nuclei produce reflexive and stereotyped motor activities that normally accompany emotional expression when stimulated by the forebrain.

In this regard, a hierarchy of progressive complexity in emotional vocalization and feeling states could be said to begin in the brainstem (which would include periaqueductal gray), and come to include the hypothalamus, and then progressively expand so as to incorporate the amygdala followed by the cingulate gyrus. However, as the cingulate gyrus has given rise, over the course of evolution, to the medial frontal lobes and supplementary motor areas (SMA), these regions of the brain have also come to subserve the capacity to vocalize (see Chapters 2, 4, and 11). Moreover, as the evolutionary progression from cingulate to SMA continued up and over into the lateral convexity of the right and left frontal lobe, these neocortical areas (e.g., Broca's speech area) also came to subserve, not just emotional speech, but complex human language.

The advent of complex, grammatical human speech, however, is also reflected in the progressive expansion of the lateral amygdala and the medial and anterior temporal lobe, which expanded and evolved to include the insula, the superior temporal lobe (e.g., Sanides, 1968, 1969, 1970), and then portions of the inferior parietal lobule. With the evolution of the inferior parietal lobule, the anterior and inferior speech areas became interlocked at the neocortical level, thereby giving rise to the Language Axis, and modern human speech (see Chapters 2 and 4). Nevertheless, these nuclei and neocortical tissues remain dependent on the limbic system and the brainstem in order to vocalize and communicate.

PERIAQUEDUCTAL GRAY AND VOCALIZATION

Activity within the midbrain periaqueductal gray (which receives extensive input from the amygdala and other limbic nuclei) can trigger the production of a variety of sounds that are suggestive of exceedingly negative feelings (Larson et al., 1994; Zhang et al., 1994). The periaqueductal gray, in fact, appears to receive and respond to noxious and painful stimuli, as do other pontine-midbrain nuclei (see Chapter 10). However, if the periaqueductal gray is disconnected from the limbic system and neocortex (such as by a midbrain transection), stimulation of this nuclei will continue to evoke vocalization. Nevertheless, with the exception of facial contortions (produced by the fifth and seventh cranial nerves) and changes in breathing and, thus, vocalization, stimulation of the isolated periaqueductal gray is not accompanied by complex behavioral displays, and when the stimulation is removed, the vocalizations immediately cease. Moreover, as detailed below, patients with "bulbar palsy" (due to partial brainstem injury and disconnection), report that their vocalizations do not correspond to their actual feelings (see also Chapter 10).

This suggests that periaqueductal gray sound production, that is, at the level of the midbrain, is due to the activation of pre-programmed motor engrams that are stored within the brainstem. Moreover, this suggests that the periaqueductal gray responds either reflexively in response to painful stimuli (such as when an individual cries "ouch"), and in reaction to impulses transmitted via the amygdala and hypothalamus—nuclei with which it is intimately interconnected.

Specifically, the periaqueductal gray coordinates the activity of the laryngeal, oral-facial, and principal and accessory muscles of respiration and inspiration (Zhang et al., 1994) and appears to be the site where particular vocalization motor patterns are stored. The coordinated activity of these tissues and activation of these motor programs would enable an individual to laugh, cry, or howl, even if the rest of the brain (except the brainstem) were dead and there was no evidence of consciousness.

When the periaqueductal gray is activated by impulses received from the limbic system or neocortex, it activates the appropriate motor program and then organizes and coordinates the oral-laryngeal and respiratory muscles so that the appropriate sounds can be produced (Zhang et al., 1994). In this manner, the felt aspects of emotion are accompanied by appropriate sound production, as released and mediated by the brainstem and cranial nerves (see Chapter 10).

AMYGDALA SOCIAL-EMOTIONAL LIMBIC LANGUAGE AND BEHAVIOR

The amygdala is able to continually sample auditory as well as visual and tactual events so as to detect those that are of emotional and motivational significance (Fukuda et al., 1987; Gloor, 1960, 1986; O'Keefe and Bouma, 1969; Ono et al., 1980; Perryman et al., 1987; Rolls, 1992; Turner et al., 1980; Ursin and Kaada, 1960). Hence, long before the evolution of auditory neocortex, the amygdala (in conjunction with the inferior colliculus and auditory brainstem nuclei) received, analyzed, and responded to complex auditory stimuli, and this is a capacity it has retained even in humans. The amygdala is primary in regard to all aspects of emotion, including the capacity to feel love or hate and to form long-term emotional attachments (Joseph 1982, 1992a, 1994).

When motivationally relevant stimuli (e.g., food, sex partner) are detected, the amygdala can organize appropriate behavioral and vocal responses and can trigger startle reactions, as in response to transient sounds, or those typically made by predators, prey, or potential mates (Edeline and Weinberger, 1991; Hitchcock and Davis, 1991; Gloor, 1960, 1986; Hocherman and Yirmiya, 1990; Rolls, 1992; Ursin and Kaada, 1960). These behavioral reactions, in turn, are mediated by the limbic and corpus striatum, the periaqueductal gray, and lower brainstem, all at the behest of the amygdala.

Hence, when the amygdala is stimulated, a variety of complex vocalizations, emotions, behaviors, and mood states are triggered, including those indicative of pleasure, sadness, happiness, fear, anxiety, anger, and rage (Gloor, 1960, 1986; MacLean, 1990; Robinson, 1967; Ursin and Kaada, 1960). The amygdala (and the cingulate), in fact, is one of the most vocally active regions of the brain (Jurgens, 1990; Robinson, 1967). Moreover, once triggered,

these behaviors and moods will persist long after stimulus termination (Gloor, 1960, 1986; Ursin and Kaada, 1960)—which is not true of the hypothalamus or periaqueductal gray. In fact, the moods and emotions triggered may variably last hours or even days (see Chapter 14).

Conversely, in humans, destruction limited to the amygdala, the right amygdala in particular, can abolish the ability to sing, convey melodic information, or to enunciate properly via vocal inflection, and can result in great changes in pitch and the timbre of speech (Freeman and Williams, 1952, 1963). Similarly, if destruction is bilateral, the capacity to respond appropriately to emotionally significant visual or auditory stimuli is abolished (see Chapter 5).

The amygdala, therefore is primary in regard to the perception and expression of social and emotional nuances and in large part is responsible for the expression and comprehension of not just limbic language, but human speech, the sounds of which are shunted to and from the amygdala and neocortex via the inferior fasciculus (Joseph, 1993).

As noted (and described briefly below), portions of the auditory neocortex—which extends from the anterior and medial temporal lobe and beyond the insula to in-

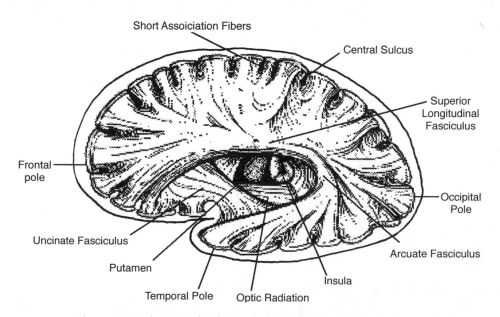

Figure 7.3   The axonal, white matter tracts of the left hemisphere.

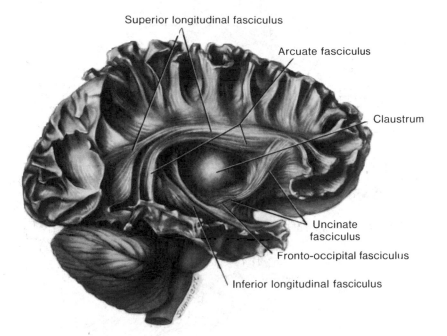

Figure 7.4   Dissection of the lateral surface of the right hemisphere, depicting the long association fibers that interconnect the frontal, inferior parietal, and superior and inferior temporal lobes. (From Mettler FA. Neuroanatomy. St. Louis: CV Mosby, 1948.)

clude the superior temporal lobe and the inferior parietal lobule—are, in part, an evolutionary derivative of the lateral amygdala (as well as the hippocampus; e.g., Sanides, 1969, 1970). In this regard, it could be argued that the primary, secondary, and auditory association areas, including Wernicke's area, have evolved (at least in part) from the amygdala, and, in fact, remain extensively interconnected with this nuclei via the inferior portions of the arcuate fasciculus.

In consequence, when the neocortical auditory areas are injured, the amygdala is sometimes disconnected and can no longer extract or impart emotional nuances to incoming or outgoing sounds, that is, at the level of the neocortex. For example, with right superior temporal lobe lesions, patients may suffer from a receptive auditory affective agnosia, as well as an agnosia for environmental sounds (see Chapters 3 and 14). They may have difficulty ascertaining the feelings of others or perceiving social emotional nuances. Hence, emotional perception and expression may become grossly disorganized and inappropriate.

By contrast, if the left amygdala is disconnected from the left superior temporal lobe, patients may verbally complain that they consciously feel cut off from their emotions (see Chapter 14). Moreover, with left temporal lobe dysfunction, speech and thought processes may come to be abnormally invested or devoid of emotion as well, and patients may be diagnosed as psychotic and/or paranoid.

Hence, although the left and right amygdala are functionally lateralized (see Chapters 5, 14, and 16), both contribute significantly to the perception and expression of language, and assist in maintaining the functional integrity of the neocortical auditory areas in the right and left temporal lobe. It is via these interconnections that limbic languages come to be hierarchically organized at the level of the temporal neocortex.

THE LANGUAGE AXIS

It is apparent that the thick band of (presumably) lateral amygdala-derived auditory neocortex continues in a belt-like

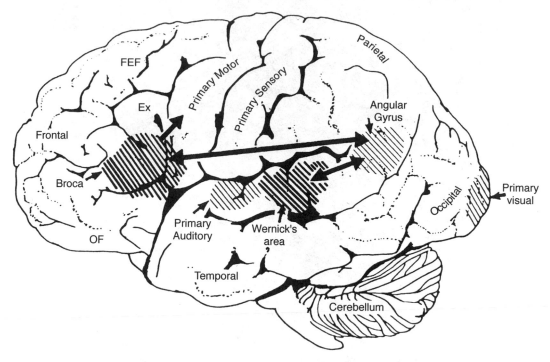

Figure 7.5 The Language Axis of the left hemisphere.

fashion beyond the inferior parietal lobule and extends and interconnects with Broca's area. However, Broca's area, and the hierarchical representation of limbic speech mechanisms in the frontal lobe, appear to be due, not to the amygdala per se, but to the anterior cingulate gyrus (Joseph, 1993). Moreover, it appears that Broca's area did not become fully functional as a neocortical speech center until after the more recent evolution of the inferior parietal lobule, which, in turn, appears to have acted as a nexus that interlocked the anterior and inferior speech centers and thus yoked language production to the neocortex, transforming language into modern human speech in the process (Joseph, 1993; see Chapters 2 and 5).

THE CINGULATE GYRUS

The five-layered cingulate gyrus sits atop the corpus callosum and can be broadly divided into two segments: the anterior cingulate (areas 24, 25, and 33), which is concerned with emotional functioning and regulating autonomic and endocrine activi-ties, and the posterior (area 23) cingulate, which is involved in visual-spatial and tactile analysis as well as motor output and memory.

The Posterior Cingulate

The posterior cingulate gyrus is richly interconnected with the superior parietal lobe (area 7), parahippocampal (inferior) temporal and superior temporal lobe (area 22), frontal lobe, caudate, putamen, substantia nigra, pulvinar of the thalamus, and dorsal hypothalamus (Baleydier and Maguiere, 1980; recently reviewed in Devinksy et al., 1995). In addition, the posterior cingulate projects to the red nucleus in the midbrain (which also receives frontal motor fibers) and to the spinal cord. Presumably, the posterior cingulate acts to integrate visual input with motoric output and is not concerned with emotional stimuli per se, with the possible exception of nociceptive functions (Devinksy et al., 1995). However, the posterior cingulate may also be involved in visual-spatial and memory-cognitive activities, particu-

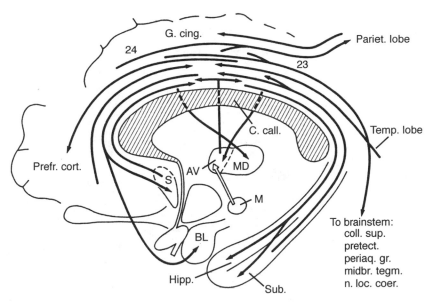

Figure 7.6   Diagram of some of the main connections of the cingulate gyrus. (After Brodal A. Neurological anatomy. New York: Oxford University Press, 1981.)

larly as relates to the body and movement—hence, the interconnections with the superior parietal lobe and the parahippocampal gyrus. The posterior cingulate may have, at least in part, evolved from the dorsal hippocampus (e.g., Sanides, 1964).

The Anterior Cingulate Gyrus

The anterior cingulate (areas 24, 25, and 33) is associated with processing and modulating the expression of emotional nuances, emotional learning and vocalization, the formation of long-term attachments, and maternal behavior, including the initiation of motivationally significant goal-directed behavior. It also has a role in influencing and, in part, regulating endocrine and autonomic activities (reviewed in Devinsky et al., 1995; MacLean, 1990). Hence, the anterior cingulate maintains rich interconnections with the septal nuclei (from which it may have, in part, evolved), amygdala, hypothalamus, mammillary bodies, hippocampus, dorsal medial nucleus of the thalamus, and periaqueductal gray (Baleydier and Maguiere, 1980; Powell, 1978; Powell et al., 1974; Muller-Preuss

and Jurgens, 1976), as well as the limbic striatum, caudate and putamen, and the frontal motor areas.

The anterior cingulate, thus, appears to be a supra-modal area that is involved in the integration of motor, tactile, autonomic, and emotional stimuli, as well as in the production of emotional sounds (see below) and the capacity to experience psychological "pain and misery." In fact, the cingulate has long been associated with the experience of psychic and even physical pain (e.g., identifying the affective attributes of noxious and psychic stimuli). Hence, during the 1930s and 1940s, bilateral cingulotomies were frequently performed to eliminate severe depressive and psychotic states as well as obsessive compulsive tendencies (Le Beau, 1954; Whitty and Lewin, 1957). However, following surgery, patients tended to become apathetic, emotionally blunted, and/or socially and emotionally inappropriate or unresponsive. Nevertheless, more recently it has been reported that 25% to 30% of patients with obsessive-compulsive disorder unresponsive to medication and behavioral treatment significantly improve following cingulotomy (Baer et al., 1995),

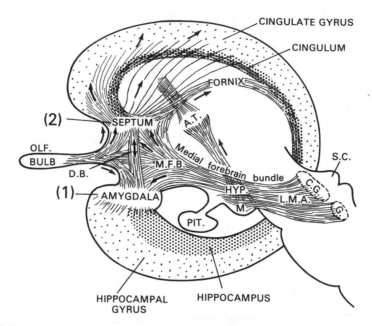

CINGULATE GYRUS

CINGULUM

FORNIX

(2)

SEPTUM

A.T.

OLF.
BULB

D.B.

Medial forebrain bundle

M.F.B.

S.C.

C.G.

(1)   AMYGDALA

HYP.

L.M.A

G.

M

PIT.

HIPPOCAMPAL   HIPPOCAMPUS
GYRUS

Figure 7.7   Schematic diagram of the limbic system and limbic system pathways.

The Cingulate Gyrus and Emotional Free Will

Electrical stimulation of the anterior cingulate can induce feelings of anxiety, pleasure, and fear (Meyer et al., 1973) as well as changes in heart rate, respiratory rate, and blood pressure, accompanied by pupil dilation, gonadal and adrenal cortical hormone secretion, penile erection, and aggression (Reviewed in Devinsky et al., 1995; MacLean, 1990). Stimulation also induces a wide range of divergent vocalizations, including growling, crying, high-pitched cackling, and sounds similar to an infant's separation cry.

It is noteworthy that vocalizations triggered by excitation of the amygdala, hypothalamus, or septal nuclei are usually accompanied by appropriate and congruent behaviors (Gloor, 1960; Jurgens, 1990; Robinson, 1967; Ursin and Kaada, 1960). Behavior, emotion, and vocalization are always highly correlated when produced by these nuclei, which apparently are exceedingly limited in the capacity to produce

behaviors or vocalizations that differ significantly from their true mood and emotional state.

In contrast, the most recently evolved five-layered transitional limbic cortex, the (anterior) cingulate gyrus, is not only exceedingly vocal, but is capable of producing emotional sounds that are not reflective of mood (Jurgens, 1990; Jurgens and Muller-Preuss, 1977). In addition, completely different emotional calls can be elicited from electrodes that are placed immediately adjacent (Jurgens, 1990). This includes the infant separation cry.

This suggests considerable flexibility, variability, and a high degree of voluntary control within the cingulate, and is reflective of its extreme importance in the evolution of complex maternal behavior, prolonged child care, extended social relations, establishment of long-term attachments, and what eventually became human language (Joseph, 1993, 1994; MacLean, 1990). Via the cingulate, a human can speak pleasantly while angry, or an animal can pretend to be hurt and produce appropriate (albeit false) vocalizations, so as to lure a predator away from their brood.

As per behavioral flexibility, it is noteworthy that some patients with anterior cingulate seizure activity have been reported capable of voluntarily modifying their automatisms so as to blend in with ongoing behavior (reviewed in Devinksy et al., 1995). This is completely in contrast to the complete loss of will and the automatisms associated with temporal, frontal, or parietal lobe epilepsy.

The anterior cingulate also assists in setting thresholds for vocalization (Jurgens and Muller-Preuss, 1977; Robinson, 1967), including modulating some of the prosodic and melodic features that characterize different speech patterns, e.g., happiness versus sadness. It accomplishes this via subcortical connections with the periaqueductal gray (Jurgens, 1990) and its axonal projections to the right and left frontal motor areas (Joseph, 1993). Hence, if the pathways linking the anterior cingulate

and periaqueductal gray were severed, the cingulate would cease to produce vocalization (Jurgens, 1979).

By contrast, abnormal cingulate-frontal activity may contribute to stuttering (see Devinksy et al., 1995). Lesions of the anterior cingulate (or their axonal connections to the frontal lobes) can also significantly impair emotional-prosodic vocalizations such that melodic-inflectional patterns become abnormal (Barris and Schuman, 1953; Kennard, 1955; Laplane et al., 1981; Smith, 1944; Tow and Whitty, 1953). However, due to the hierarchical organization of the speech centers, destruction of the pathway linking the cingulate with the speech areas located in the medial frontal lobes can also result in a loss of vocalization, i.e., *mutism* (see Chapter 11).

In that the anterior cingulate, like the medial frontal lobes, is implicated in what could be loosely defined as "free will" and

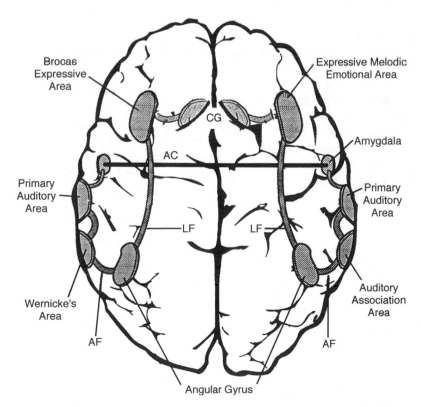

Figure 7.8 Schematic diagram of the pathways linking the amygdala and the anterior cingulate gyrus (**CG**) to the speech and vocalization areas of the right and left hemispheres. **AF**=arcuate fasciculus. **AC**=anterior commissure. **LF**=longitudinal fasciculus.

the voluntary and even deceptive control over behavior, abnormalities involving these tissues can also induce a loss of will, or a loss of the capacity to alter and modify behavior. One example of this is obsessive compulsions that are associated with lesions to the head of the caudate, medial frontal lobes, and anterior cingulate. As noted, anterior cingulate destruction has also been reported to relieve obsessive-compulsive behavior.

The Cingulate and the Evolution of Maternal Language

The cingulate is capable of producing a complex medley of emotional sounds, including the separation cry, similar if not identical to that produced by an infant (MacLean, 1990; Robinson, 1967). Moreover, abnormal activity in the anterior cingulate has, in some cases, induced infantile behavior, such as the assumption of the fetal position (Devinksy et al., 1995).

It has been argued that these particular anterior cingulate "infantile" sounds promote intimate maternal behavior and watchfulness and maintain emotional attachments between mother and infant. As noted, not just infantile separation cries but maternal behavior as well are closely associated with the functional integrity of the cingulate (Joseph, 1993; MacLean, 1990).

For example, among humans and mammals, anterior cingulate destruction can result in a loss of maternal responsiveness, as well as mutism and severe alterations in socially appropriate behavior (Barris and Schuman, 1953; Kennard, 1955; Laplane et al., 1981; Smith, 1944; Tow and Whitty, 1953).

Primates who have suffered cingulate destruction will also become mute, will cease to groom or show acts of affection, and will treat their fellow monkeys and infants as if they were inanimate objects. For example, they may walk upon and over them as if they were part of the floor or an obstacle rather than a fellow being (MacLean, 1990); i.e., their behavior becomes reptilian (reptiles are devoid of cingulate cortex). Mater-

nal behavior is also abolished, and the majority of infants whose mothers have suffered anterior cingulate destruction soon die from lack of care.

MATERNAL BEHAVIOR AND THE EVOLUTION OF INFANT SEPARATION CRIES

Sharks, teleosts, amphibians, and reptiles possess a limbic system consisting of an amygdala, hippocampus, hypothalamus, and septal nuclei (see Chapter 1). It is these limbic nuclei that enable a group of fish to congregate and "school," and that make it possible for reptiles to form territories that include an alpha female, several subfemales, and a few juveniles. These nuclei promote social attachment and interaction and provide the foundation for social-emotional intelligence.

It is these same nuclei that enable human infants to initially indiscriminately seek social and physical contact, and to then increasingly display fear of strangers and to form specific attachments—events that correspond with the maturation of the amygdala followed by the septal nuclei and then the cingulate gyrus (Joseph, 1990a, 1992a, 1994). However, it is the mammalian cingulate gyrus that not only made possible long-term and quite specific social-emotional attachments, but that probably provided the impetus for what would eventually become human language and the "family."

Evolution of Maternal-Infantile Vocalizations

The first amphibians and reptiles, like their modern counterparts, lacked the four- to five-layered cingulate gyrus. Nevertheless, although lacking an inner or true middle ear, amphibians and reptiles are attuned to low-level vibrations and sounds, such as croaking, tails thumped on the ground, a few distress calls, and those of contentedness. Limbic language capabilities are not well developed in these creatures.

As they also lack a cingulate gyrus, amphibians and most, but not all, reptiles show little or no maternal care, and rarely vocalize. They will also greedily cannibalize their infants, who in turn must hide from their parents and other reptiles, in order to avoid being eaten (MacLean, 1990).

When reptiles began to differentiate and evolve into the repto-mammals (the therapsids) some 250 million years ago, and then, 25 million years later, when the first tiny dinosaurs (who diverged from a different line of reptiles, the theocondants) began to roam the Earth, major biological alterations occurred involving cranial and postcranial skeletal structure, mammillary development, thermoregulation, sexual reproduction, and limbic system function and structure (Bakker, 1971; Brink, 1956; Broom, 1932; Crompton and Jenkins, 1973; Duvall, 1986; Paul, 1988; Quiroga, 1980; Romer, 1966)—all of which coincided with tremendous advances in the ability to engage in audio-vocal communication and the capacity to nurse the young (see Chapter 1).

It was not until the appearance of the therapsids that mammillary glands, and thus the capacity to nurse, came into being (Duvall, 1986). It was at this time that the middle ear began to undergo tremendous modification and the first rudiments of an inner ear developed (Broom, 1932; Brink, 1956; Crompton and Jenkins, 1973; Romer, 1966). Although the hypothalamus, septal nuclei, and amygdala continued to evolve, it was also around this time that the cingulate gyrus began to appear and increasingly enshroud the dorsal surface of the limbic forebrain (Joseph, 1993; MacLean, 1990)—an event that corresponded with the appearance of nursing nipples (Duvall, 1986) and the inner ear. When this began to occur (perhaps some 225 million years ago), sounds came to serve as a means of purposeful and complex communication, not only between potential mates or predator and prey, but between mother and infant (Joseph, 1993, 1994; MacLean, 1990). This ability, in turn, was probably made possible by the amygdala as well as through the evolution of the four- to five-layered transitional neocortex, the cingulate gyrus, the more anterior portions of this being derived, at least in part,

from the septal nuclei as well as from portions of the amygdala.

As noted, it is the limbic system and the interactions of limbic nuclei, such as the amygdala and the cingulate gyrus, that stimulate not only the desire to communicate but to form attachments, social groups, and eventually, the family. In fact, many of the repto-mammalian therapsids, and then later, many of the various dinosaurs, lived in packs or social groups and presumably cared and guarded their young for extended periods lasting until the juvenile stage (Bakker, 1971; Brink, 1956; Crompton and Jenkins, 1973; Duvall, 1986; Paul, 1988; Romer, 1966); presumably, long-term attachments were made possible via the evolution of the anterior cingulate.

As also noted, the first appearance of rudimentary nipples coincided with therapsid development. Hence, one of the hallmarks of this evolutionary transitional stage, some 250 million years ago, was the cingulate gyrus and the first evidence of nursing, maternal feeling, the creation of large social groups and hunting packs, and what would become the family.

Mother-Infant Vocalization

Among mammals and primates, the production of sound is very important in regard to infant care, for if an infant becomes lost, separated, or in danger, a mother could not quickly detect this by olfactory-pheromonal cues alone. These conditions would have to be conveyed via a cry of distress or a sound indicative of separation fear and anxiety, which would cause a mother to come running to the rescue. Conversely, vocalizations produced by the mother would enable an infant to continually orient and find its way back if it got lost or separated. Hence, the first forms of complex limbic social-emotional communication may well have been first produced in a maternal context.

As is well known, considerable vocalizing typically occurs between mothers and their infants (be they human, primate, or mammal); and the infants of many species will often sing along with or produce sounds in accompaniment to those produced by their mothers. These mutual interactions reinforce and promote mutual vocalization that is often initiated by the mother.

In fact, primate females are more likely to vocalize when they are near their infants versus non-kin, and infants are more likely to vocalize when their mother is in view or nearby (Bayart et al., 1990; Jurgens, 1990; Wiener et al., 1990). Similarly, infant primates will loudly protest when separated from their mothers so long as they are in view and will quickly cease to vocalize when isolated (Bayart et al., 1990; Wiener et al., 1990). However, adult males are also more likely to call or cry when in the presence of their mothers or an adult female versus an adult male (see Jurgens, 1990). It thus appears that the purpose of these vocalizations is to elicit a vocal response from mother, or an adult female, who in turn is likely to respond with soothing limbic language.

Hence, the initial production of emotional sounds appears to be limbically based, and increasingly, over the course of evolution, associated with maternal-infant care, and/or interactions with an adult female. In this regard, it is noteworthy that regardless of culture, human mothers tend to emphasize and even exaggerate social-emotional and melodic-prosodic vocal features when interacting with their infants (Fernald, 1992; Fernald et al., 1989), which, in turn, appears to greatly influence infant emotional behavior and attention (Fernald, 1991). Similarly, human infants prefer listening to and are more responsive to these exaggerated limbic vocalizations, as compared to "normal" adult speech patterns (Cooper and Aslin, 1990), particularly when produced by a female, i.e., by a female cingulate gyrus.

Female Superiorities in Limbic Language

Although produced by males and females alike, almost regardless of mammalian species, females tend to produce a greater range of limbic (social-emotional) vocalization. For example, female monkeys and apes are more vocal and engage in more social vocalizations and, in fact, vocalize more often than males; males, in turn, are more likely to vocalize when threatening or engaged in dominance displays (Cross and Harlow, 1965; Erwin, 1980; Fedigan, 1992; Goodall, 1986, 1990; Mori, 1975; Mitchell, 1979).

Similarly, human females not only engage in more social-emotional vocalizations than males (Glass, 1992; Joseph, 1993; Tannen, 1991) but they tend to employ five to six different prosodic variations and to utilize the higher registers when conversing. They are also more likely to employ glissando or sliding effects between stressed syllables, and they tend to talk faster as well (Brend, 1975; Coleman, 1971; Edelsky, 1979). Men tend to be more monotone, employing two to three variations on average, most of which hover around the lower registers (Brend, 1975; Coleman, 1971; Edelsky, 1979). Even when trying to emphasize a point, males are less likely to employ melodic extremes but instead tend to speak louder. Perhaps this is why men are perceived as more likely to bellow, roar, or growl, whereas females are perceived as more likely to shriek, squeal, and purr. Nevertheless, although influenced by sex differences in the oral-laryngeal structures, these differential capacities are also reflected in the greater capacity of the female brain to express and perceive these nuances.

For example, it has been repeatedly demonstrated that females are more emotionally expressive, and are more perceptive in regard to comprehending emotional verbal nuances (Burton and Levy, 1989; Hall, 1978; Soloman and Ali, 1972). This superior sensitivity includes the ability to feel and express empathy (Burton and Levy, 1989; Safer, 1981) and the comprehension of emotional faces (Buck et al., 1972, 1974). In fact, from childhood to adulthood women appear to be much more emotionally expressive than males in general (Brody, 1985; Burton and Levy, 1989; Gilbert, 1969).

Presumably females are superior in regard to most aspects of emotional intelligence and expressive and perceptive emotionality due to sex differences in the limbic system (see Chapter 5). For example, the hypothalamus and amygdala (as well as other limbic nuclei) are sexually differentiated and have sex-specific patterns of neuronal and dendritic development (Allen and Gorski, 1992; Allen et al., 1989; Blier et al., 1982; Gorski et al., 1978; Goy and McEwen, 1980; Raisman and Field, 1971; Swaab and Fliers, 1985; Swaab and Hoffman, 1988). Similarly, the rope of nerve fibers that interconnects the right and left amygdala and inferior temporal lobes (the anterior commissure) is 18% larger in females than males (Allen and Gorski, 1992).

Given the preeminent role of the amygdala (and inferior temporal lobe) in emotionality and sound production, as well as evidence indicating that this nuclei is sexually dimorphic, this latter finding of an enlargement in the female anterior commissure may be yet another reason why females are more emotionally expressive and receptive, and tend to employ a wider range of melodic pitch when they speak. That is, the two amygdala (and inferior temporal lobes) of the human female are better able to perceive, process, communicate, exchange, and express emotional information as compared to males.

In addition, given that the evolution of the cingulate gyrus appears to correspond with the advent of purposeful social-emotional communication and the development of long-term maternal-infant attachments, it is likely that it, too, is sexually differentiated. That is, in addition to the amygdala, hypothalamus, and anterior commissure, the sexually differentiated and, thus, the "female" anterior cingulate gyrus, probably confers upon females an advantage in the vocal expression of emotional nuances and in regard to all aspects of social-emotional intelligence.

Maternal Behavior, Attachment, and the Female Limbic System

It has been proposed that these limbic system sex differences are also responsible for and an evolutionary consequence of woman's role in bearing and rearing children and the female desire to form long-term attachments and to engage in maternal care and verbal communication (Joseph 1993). Female humans, primates, and other mammals apparently find these activities rewarding in themselves, due to these same limbic system sex differences.

That is, given that the anterior cingulate, at least in part, evolved in a maternal context and/or promoted the development of maternal feelings and long-term mother-infant attachment, the sexual differentiation of this nuclei may well account (when coupled with sex differences in the amygdala, anterior commissure, and hypothalamus) for why females differentially respond and desire to nurture, hold, cuddle, and stare at infants, much more so than males. Although abnormal early environmental experiences can attenuate or result in the abnormal expression of maternal behavior (or lack thereof), environment, per se, is not a factor in the differential expression of maternal and child care behaviors, or even the pronounced sex differences in dependency needs and the desire to form long-term attachments. They are innate and appear among all other mammalian species.

For example, female humans, chimps, baboons, and rhesus macaques are more dependent than males, cuddle more closely, and are cuddled more by their sisters, mothers, and other females (Jensen et al., 1968a; Hansen, 1966; Mitchell, 1968; Goodall, 1971, 1990).

Infant human or primate males are much more resistant to being held, and will kick, fuss, and actively attempt to escape their mothers much more so than females (Elia, 1988; Fedigan, 1992; Freedman, 1974, 1980; Mitchell, 1968, 1979; Goodall 1971, 1990; Kummer, 1971). Males are also less likely to cuddle, and this remains a male characteristic throughout adulthood. Moreover, these are not just primate characteristics, for even male dogs respond likewise (unpublished observations). In part, this reflects a struggle against potential physical domination, which most males find aversive (Joseph, 1993). Hence, be it a male dog, chimpanzee, baboon, or child, males are far more likely than females to resist attempts to hold them or pick them up, and may even respond as if they find it aversive.

Mothers are, therefore, more willing to hold female babies and for longer periods, as they are also easier to calm and are more fun to hold. Since females demonstrate greater social responsiveness and are more likely to employ facial, vocal, and social signals, mothers are more likely to physically, socially, and vocally interact with their infant daughters and vice versa (Moss, 1974). Being similarly socially inclined, mothers find it more socially rewarding and enjoyable to interact with their daughters.

Be it a female chimpanzee, baboon, rhesus macaque, or human, females also begin to demonstrate an extraordinary interest in babies and in play-mothering during even the earliest phases of their own childhood (Devore, 1964; Elia, 1988; Fedigan, 1992; Goodall, 1971, 1990; Jolly, 1972; Kummer, 1971, Mitchell, 1979; Strum, 1987; Suomi, 1972).

When girls play together, much of their fantasy and conversation concerns fashion, kissing, and making out with boys and revolves around adult relationships, including the raising of a family and the behavior and misbehavior of children (their dolls). Babies are of enormous interest to females, be they human, ape, or monkey; and social primates and female humans who have babies usually become tremendously popular and the center of attention (Elia, 1988; Fedigan, 1992; Jolly, 1972; Mitchell, 1979; Strum, 1987). Even among women enslaved in a harem, once a member becomes pregnant and has a child, her status is quickly and permanently elevated.

Mothers, grandmothers, young and adolescent females, and even women who describe themselves as "feminists" show much more interest in babies than do men, even when the babies are not their own (Zahn-Waxler et al., 1983; Berman, 1990; Berman and Goodman, 1984; Blakemore, 1981, 1985, 1990; Frodi and Lamb, 1978; Melson and Fogel, 1982; Nash and Fledman, 1981). Adolescent girls spend signifi-

cantly more time talking about new babies than do boys (Berman, 1990), and mothers spend more time talking about the babies with their daughters than their sons (Berman, 1990).

Girls not only talk more, but play and care for their infant sisters and brothers significantly more and show considerable amounts of nurturing interest in the baby's well-being (Blakemore, 1990), even when there has been no request or pressure to do so. Indeed, girls often demonstrate an intrusive interest in babies (Berman, 1983) and will give infants much more care than they require (Ainsworth and Wittig, 1969), as is often the case with new mothers (Stewart, 1990). Similarly, among nonhuman female primates, be it gorilla, chimpanzee, baboon, rhesus macaques, lemur, and so on, even those without infants will eagerly seek to groom, cuddle, and carry a child as much as the mother will allow (Jolly, 1972; Devore, 1964; Kummer, 1971; Strum, 1987; Suomi, 1972; Mitchell, 1979; Goodall, 1971). These primates may also spend all day passing them back and forth. Like human females, some will even steal these infants. Those female primates who show the greatest interest, however, are young females who have not yet had babies.

Moreover, among almost all social primates, the birth of a new baby has an extremely excitatory effect on all the other females of the troop who will gather around and touch, stare, hold, and cuddle it. This female interest, of course, is certainly quite adaptive, at least for those living in the dangerous conditions of the wild, for it insures that if a mother dies another will adopt her baby.

Such behavior is obviously not the result of sexist training, for it is typical of all primates. For example, boy chimpanzees show little interest in their younger infant siblings, whereas girl chimps become increasingly fascinated and will hold and cuddle them and will attempt to model their mother's interactions with the infant (Goodall, 1971). If a new mother dies but her baby has older male siblings, fewer than 25% of the males will adopt the little orphan, whereas females siblings are quite anxious and happy to take this role.

The Male Limbic System and Infant Care

With the exception of the baboon (Rowell et al., 1968; Kummer; 1968, 1971; Mitchell 1968, 1979; Fedigan, 1992), lack of interest in infants is characteristic of most social male primates and most male mammals, reptiles, amphibians, and fish. This includes human fathers, and men and boys in general. They have little or no interest in babies and generally provide little or no nurturing care for their own children or the children of others (Rossi, 1985; Gordon and Draper, 1982). Of course, there are always exceptions, particularly among males who may possess a "female" limbic system (see Chapter 5). Rather, like other social primates, boys seek boys for playmates and engage in considerable amounts of rough-housing, wrestling, and hitting—behavior that is completely inappropriate in regard to infant interactions. When boys or male primates begin to separate from their mothers, they show no interest in younger siblings but seek out adolescent and adult males to play with. Although they may on occasion seek nurturance, they seldom provide it in return.

Human males and fathers rarely behave in any manner that approximates normal female maternal behavior (Belsky et al., 1984; Clarke-Stewart, 1978; Frodi et al., 1982), as this is simply not an activity they find interesting, pleasurable, or rewarding. This is why, for example, child care professions and those jobs involving high levels of interactions with children, such as elementary school teachers, are overwhelmingly made up of women (Gordon and Draper, 1982)—a function not of pay but of lack of male interest (Blakemore et al., 1988), a function of possessing a "male" limbic system. Rather, fathers and adult heterosexual males tend to express interest in younger males and females only when they reach early adolescence, and this is also true of most male primates.

Given that these sex differences are obviously innate, it could therefore be argued that, in contrast to male humans, primates, and mammals who have little or no interest in child care, the female limbic system is designed to promote these interests. Just as the male limbic system rewards males for engaging in competitive and aggressive actions (see Chapter 5), the female limbic system probably generates rewarding feelings when females look at, hold, care for, and form attachments to their babies, infants, and young children, as well as to other adults, be they male or female. Although this has yet to be determined, the female limbic system probably contains nuclei, neural networks, and individual neurons that respond selectively to infant visual and auditory related stimuli, e.g., baby faces, infant cries.

Again, consider that the anterior cingulate, in part, evolved in a "maternal" context and acts to promote the development of maternal behavior and mother-infant communication. Indeed, sex-specific structural differences in the limbic system probably account, in large part, for most all sex differences in emotionality and related behavior, including child care, the desire to have and nurture babies, and the greater female propensity for developing affective and mood disorders. As noted, they probably also account for sex differences in emotional vocalization as well.

EMOTIONAL PROSODY AND HEMISPHERIC LANGUAGE SYSTEMS

Social-emotional and related contextual nuances—be they feelings of fear, love, anger, happiness, sadness, sarcasm, or empathy—are vocally and nonlinguistically communicated by varying the rate, amplitude, pitch, inflection, timbre, melody, and stress contours of the voice (Blumstein and Cooper, 1974; Dwyer and Rinn, 1981; Fernald, 1992; Joseph, 1988a; see Chapter 3), as well as through touch, facial expression, posture, and physical interaction (Eibl-Eibesfeldt, 1990; Ekman 1993; Joseph, 1993).

Although these social-emotional-vocal functions are associated with and mediated by specific limbic system nuclei, e.g. the amygdala, cingulate gyrus, hypothalamus, and septal nuclei (Freeman and Williams, 1952; Gloor, 1960, 1986; Joseph, 1982, 1992a, 1994; Jurgens, 1990; Jurgens et al., 1982; Jurgens and Muller-Preuss, 1977; Robinson, 1967, 1972, 1976; Ursin and Kaada, 1960), they are also hierarchically represented and expressed via the neocortex of the right frontal and temporal lobes (see Chapter 3). The right parietal lobe, being more concerned with visual-spatial motor activity (rather than with temporal-sequential actions), therefore, does not act to punctuate or segment emotional-melodic vocalizations—which, nevertheless, are subject to left hemisphere articulatory sequencing.

Thus, at the hierarchical level of the neocortex, limbic functions are predominantly processed, analyzed, interpreted, comprehended, and expressed by the right frontal and temporal-parietal areas of the right cerebral hemisphere. The right hemisphere not only hierarchically represents limbic system functions but is more richly interconnected with these nuclei, which are also lateralized in regard to functional and neurochemical activity (see Chapters 5, 6, 9, and 16).

Because these functions are lateralized at the level of the neocortex, the right hemisphere is capable of expressing and comprehending emotional vocalizations, even when a patient is suffering from severe left hemisphere damage and expressive (Broca's) and receptive (Wernicke's) aphasia (Boller et al., 1979; Boller and Green, 1982; Smith, 1966; Smith and Burklund, 1966). That is, although the patient with severe receptive aphasia is without the ability to comprehend human language, his intact right hemisphere and limbic system remains capable of perceiving and analyzing the emotional and melodic qualities of any words and sentences being said to him. The intact right hemisphere can then interact appropriately and sometimes guess at what he is being asked or what is being said.

For example, when a female patient with Wernicke's aphasia was asked, "How *are you* today?" she nodded her head and replied, "Fine, fine." However, when the next day I used an identical melody and

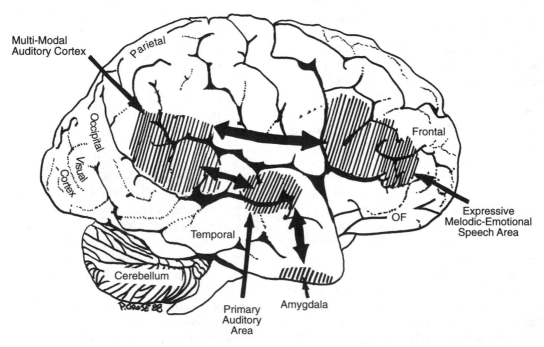

Figure 7.9   The emotional-melodic speech areas of the right hemisphere.

tone of voice but instead asked her, "It's *raining* outside?" she again nodded her head, smiled, and replied, "Fine..." Essentially, her right hemisphere and cingulate-amygdala correctly perceived the social-emotional qualities of my question and responded appropriately. In fact, it is via the limbic system and the right half of the brain that swearing, singing, and even praying is made possible, such that even severely aphasic individuals are capable of cursing and making music (Goldstein, 1944; Helm-Estrabrooks, 1983; Joseph, 1988a; Yamadori et al., 1977). Nevertheless, right neocortical lesions can also induce disconnection such that limbic system nuclei are prevented from receiving or transmitting data to the neocortex.

TEMPORAL SEQUENCING, GRAMMAR, AND THE LEFT HEMISPHERE

Evolution of Broca's Area, Superior Temporal and Inferior Parietal Lobes

The left cerebral hemisphere cannot sing, or carry a tune or melody, although, in all other respects, it certainly has "rhythm." In addition to rhythm, the denotative, syntactic elements of language, including reading, writing, spelling, naming, and so on, are associated with the left cerebral hemisphere, specifically, Broca's and Wernicke's areas, and the inferior parietal lobule (see Chapters 3 and 4). Hence, whereas the right hemisphere is dominant for emotional and melodic stimuli, the left dominates in regard to the nonemotional, temporal-sequential aspects of speech.

Broca's area and left cerebral linguistic functioning are exceedingly dependent on the inferior parietal lobule and its capacity to impose rhythmic temporal sequences on auditory stimuli (Geschwind, 1966; Goodglass and Kaplan, 1982; Joseph, 1990d, e, 1993; Kimura, 1993; Strub and Geschwind, 1983), including vocalizations that arise from the limbic system. However, Broca's area is also dependent on interconnections with the anterior cingulate, from which it may well have evolved.

For example, Sanides (1969, 1970) has argued that the frontal medial archicortex originated from the anterior cingulate. This "architectonic" evolutionary trend then continued its expansion, thus creating the lateral frontal convexity. This would include what would become "Broca's area" (Joseph, 1993). Hence, it could be argued that Broca's area evolved from the anterior cingulate, which gave rise to the medial frontal lobes, supplementary motor area (SMA), and then the lateral convexity.

However, not just Broca's area but a considerable proportion of motor neocortex, including that representing the hand, may be a cingulate derivative. As detailed in Chapters 2, 5, and 12, human language is intimately associated not just with the limbic system, but with handedness as well. Similarly, the motor areas of the frontal convexity and the medial frontal lobes are exceedingly important in fine motor functioning and control over the hand (see Chapter 11). It is via the internal representation of the human hand that internal stimuli can be manipulated and subject to sequencing. In that these tissues are presumably evolutionary derivatives of the cingulate, not surprisingly, abnormal activation of the anterior cingulate can also induce hand and finger movements, including the production of simple gestures as well as pressing, entwining, kneading, rubbing, and touching the hands and fingers—as well as sucking and lip puckering (reviewed in Devinksy et al., 1995).

Disconnection: Expressive Aphasia Versus Mutism

As detailed in Chapter 11, destruction of the medial frontal lobe—as well as the anterior cingulate—can result in mutism, which, however, is not the same as aphasia. Presumably, in part, medial frontal and SMA destruction results in a disconnection syndrome such that limbic impulses involved in the will to speak are no longer able to reach Broca's area or the right frontal emotional-melodic expressive areas; i.e., the patient becomes mute—but not aphasic.

For example, those with Broca's aphasia may complain (following some recovery) that they knew what they wanted to say, but couldn't find the words or say it. That is, although the Broca's and anterior cingulate connections may be intact, the axonal fibers linking these regions to the inferior parietal lobule and Wernicke's area are destroyed such that what is left of Broca's area no longer receives neocortical linguistic input. Nevertheless, patients can still cry, swear, and emotionally express themselves.

By contrast, those with medial frontal, SMA, and/or anterior cingulate destruction become mute and may (upon partial recovery) complain that nothing occurs to them to say, that "thoughts don't enter my head anymore." That is, the motivational foundations for the will to speak, or to think, or to call up and express one's feelings or thoughts, is abolished and Broca's and the right frontal melodic speech areas become deactivated—as if the primary circuit underlying speech production has been severed. The neocortex has been disconnected.

The Amygdala, Inferior Parietal Lobe, and Neocortical-Limbic Language Yoke

As noted, among primates destruction of the left frontal convexity, including what could be referred to as an area that might be homologous to "Broca's area" does not render these creatures expressively aphasic. Although the cingulate-frontal interconnections have been long established, the left frontal area is not organized for speech production. Rather, speech production (i.e., content, vocabulary, grammatical organization) is dependent on Wernicke's area and the angular gyrus of the inferior parietal lobule—a brain region possessed only by H. sapiens sapiens (Geschwind, 1965).

The inferior parietal lobule, however, has dual origins, being in part an expansive derivative of the superior parietal lobule, which in turn is associated with visual-spatial functions, including guiding the hand and body in visual space as well as tool

making, gesturing, and so forth. However, the inferior parietal lobule also processes auditory information and contains cells that are multimodally responsive.

These auditory and multimodal neurons are in large part an evolutionary derivative of the expanding Wernicke's area in the left hemisphere, and the auditory association areas, superior temporal lobule, and insula of the right cerebrum. However, the insula and superior temporal lobe, in turn, appear to have evolved from the amygdala (and to a lesser extent the hippocampus) via expansions in the medial and anterior-lateral temporal lobe (e.g., Sanides, 1969, 1970).

Specifically, over the course of evolution, as the lateral amygdala and portions of the hippocampus became increasingly cortical in structure, the medial and anterior-lateral temporal lobes ballooned outward and the neocortex began to form. According to Sanides (1969, 1970), this original paleocortex proceeded in a superior direction, thus forming the insula and then the superior temporal lobe. Eventually, this expansion continued and presumably contributed to the formation of the angular gyrus of the inferior parietal lobe. Hence, the angular gyrus and Wernicke's areas (and the auditory areas in the right hemisphere) remain extensively interlinked with the amygdala via the inferior and arcuate fasciculus.

Nevertheless, with the evolution of the angular gyrus, this expansion continued in an anterior direction, thus linking Broca's area with the posterior neocortex, including the amygdala. Presumably, it was at this point that the neuronal circuit that linked together the amygdala and anterior cingulate and the neocortical speech centers was established such that the Language Axis was formed and modern human speech came into being (Joseph, 1993).

With the evolution of the left angular gyrus, what essentially had been limbic language became yoked to the neocortex; i.e. limbic language impulses were now hierarchically represented as well as subject to temporal sequencing. Prior to the evolution of the angular gyrus, Broca's area pre-

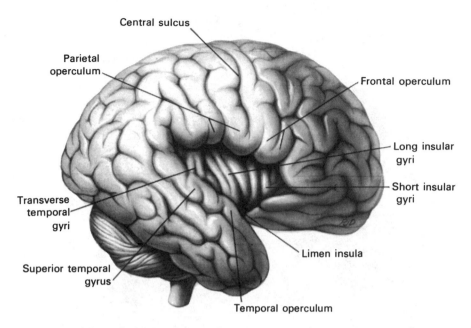

Figure 7.10   View of the right hemisphere with the insular cortex exposed. (From Carpenter M. Core text of neuroanatomy. Baltimore: Williams & Wilkins, 1991.)

sumably was unable to receive sufficient input from the primary auditory receiving and Wernicke's areas (and the amygdala), and language thus remained, by and large, limbic and under the control of the cingulate and amygdala in the absence of highly significant neocortical participation.

Temporal Sequencing

As detailed in Chapters 2 and 4, the left inferior-parietal lobe in conjunction with Wernicke's area acts to impose temporal order on incoming sounds. When this capability evolved, the capacity to vocally express these sounds in temporal and grammatical sequences unfolded next. In conjunction with Broca's area, the inferior parietal lobule imposed temporal sequences on auditory-linguistic stimuli destined for expression. Broca's area acts to match these linguistic-sound images to specific neural programs and muscle groups while organizing the secondary and primary oral-laryngeal motor areas in order to articulate speech. Broca's area also projects to the periaqueductal gray.

Therefore, at the level of the neocortex enveloping the left hemisphere, and due

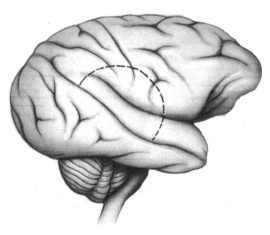

Figure 7.11   Auditory cortex of the rhesus monkey. Compare with figure 7.12. Note lack of angular gyrus and continuation of auditory area into the parietal lobe. (From Carpenter M. Core text of neuroanatomy. Baltimore: Williams & Wilkins, 1991.)

to the evolution of the left angular gyrus, social-emotional vocalizations come to be punctuated by temporal-sequences and governed by grammatical rules of organization, thus producing "modern" human language. That is, at the level of the neocortex of the left hemisphere, emotional, melodic, and prosodic limbic vocaliza-

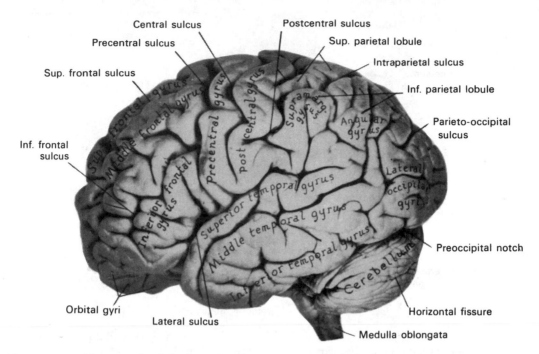

Central sulcus
Precentral sulcus
Sup. frontal sulcus
Inf. frontal sulcus
Postcentral sulcus
Sup. parietal lobule
Intraparietal sulcus
Inf. parietal lobule
Parieto-occipital sulcus
Preoccipital notch
Orbital gyri
Lateral sulcus
Horizontal fissure
Medulla oblongata

Figure 7.12 Photograph of the lateral surface of the left hemisphere. Note that the superior temporal gyrus extends from the anterior pole and then angles forward to include the inferior parietal lobule. (From Carpenter M. Core text of neuroanatomy. Baltimore: Williams & Wilkins, 1991.)

tions are fractionated into periodic and interrelated verbal units and are subject to classification into multiple categories of meaning (Joseph, 1982, 1986a, 1988a, 1993).

As described in Chapter 2 and elsewhere (Joseph, 1992b, 1993), presumably these motoric-linguistic alterations (and the angular gyrus of the left inferior parietal lobe) evolved in relation to gesturing, gathering, and tool making over the course of the last 100,000 years, such that by 50,000 B.P. what had been limbic speech had evolved into modern human speech. By contrast, social-emotional speech, i.e., "limbic language," has probably been employed for at least 225 million years; and it is from these collective and unconscious roots that modern human language may have sprung.

Overview

It is through the amygdala and anterior cingulate that emotional and motivational significance or nuances are extracted or im-

parted to much of auditory and perceptual experience. Via the rich interconnections maintained with the neocortex, auditory information processed by right temporal lobe and Wernicke's area, as well as the verbal stream transmitted to Broca's area for expression (via the arcuate fasciculus) becomes infused with emotion. However, it is through the cingulate, and its interconnections with Broca's area as well as the right frontal lobe, that melody and true and false emotions as well as genuine feelings of affection, joy, and other social-emotional (and environmental) sounds and vocalizations can be directly imparted into the stream of all that which an individual intends to say, including feelings of maternal concern. In fact, the motivation to speak may be localized within the cingulate, SMA, and medial frontal lobe.

Hence, via these interconnections, an individual is able to feign, hide, or truly express his or her feelings via the modulation of the melodic and emotional qualities of the voice, as well as discern sincerity, dishonesty, or true love in the voice of another. Of course, it is also via the sometimes inadvertent expression of limbic/ right cerebral vocalizations that some individuals betray feelings that might best remain hidden. Hence the admonishment, "It's not *what* you said, but the *way* you said it."

Fortunately, it is also due to these innate limbic roots that humans are able to perceive and convey warnings, threats, warmth, acceptance, love, and secret feelings, as well as a variety of social-emotional vocal nuances indicating mood, attitude, and intentions that are understood by infants, pets, friends and foes, and people from wholly different cultures who speak completely foreign dialects (Beier and Zautra, 1972; Fernald, 1992; Joseph, 1988, 1993; Kramer, 1964). Hence, the famous aside: "I don't know what they're saying, but I sure don't like the sound of it."

Indeed, as a universal language, limbic speech serves not just communication, but intraspecies and cross-cultural survival—even when the listener and speaker rely on completely different dialects or languages to communicate. This is because language is both emotional and grammatically descriptive. A listener comprehends not only the content and grammar of *what* is said but the emotion and melody of *how* it is said—what a speaker *feels*.

EMOTIONAL INTELLIGENCE: LIMBIC LOVE AND SOCIAL-EMOTIONAL ATTACHMENT

The limbic system provides the foundation for all aspects of emotional intelligence and makes it possible to experience and communicate social-emotional nuances via multiple modalities, such as is reflected in the evolution of emotional speech, including the ability to laugh, to cry, and to express sympathy and compassion, or the desire to form or maintain an emotional attachment. It is the evolution of these limbic nuclei (i.e., the amygdala, septal nuclei, cingulate gyrus) and their differential rates of maturation which, in fact, enable humans and other higher mammals to form changing and long-lasting emotional and loving

attachments, including the need and desire for contact comfort during infancy and early childhood (Joseph, 1982, 1990a).

Septal Nuclei

Phylogenetically, the septal nuclei develops at about the same time and rate as the hippocampus (Andy and Stephan, 1976; Brown, 1983; Humphrey, 1972), and eventually gives rise and contributes to the evolution of the medial portions of the hemispheres (Sanides, 1969, 1970), including portions of the anterior cingulate. It also increases in relative size and complexity as we ascend the ancestral tree, attaining its greatest degree of development in humans.

Specifically, the septal nuclei lies in the medial portions of the hemispheres, just anterior to the third ventricle near the hypothalamus and is comprised of the nucleus of the diagonal band of Broca and the nucleus of the medial septum. The septum projects heavily throughout the hypothalamus and maintains rich interconnections with all regions of the hippocampus (Mesulam et al., 1983; Siegel and Edinger, 1976) as well as the substantia innominata of the limbic striatum, a nucleus rich in cholinergic neurons that is implicated in memory (see Chapter 9).

In fact, cholinergic as well as GABAergic projections arise from the septal nuclei and project to the hippocampus in a topographic manner (Amaral and Kurtz, 1985; Panula et al., 1984), and septal neurons exhibit rhythmic, bursting activity that may serve as a pacemaker for the production of hippocampal theta, which is implicated in the mediation of neocortical arousal and perceptual activity (see Chapter 5). Hence, the septal nuclei appears to interact with the hippocampus in regard to memory and arousal. In fact, some septal cholinergic neurons are implicated in the generation of slow wave and paradoxical sleep (e.g., Sweeney et al., 1992).

The septal nuclei also contributes to and receives fibers from the amygdala via the stria terminalis (Swanson and Cowan,

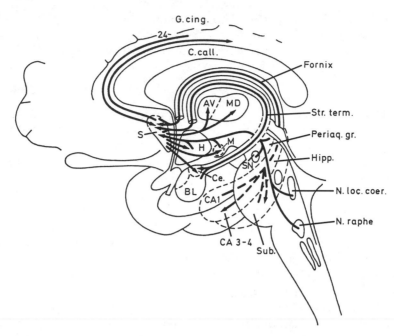

Figure 7.13   Schematic diagram of the main afferent and efferent pathways of the septal nuclei.

(After Brodal A. Neurological anatomy. New York: Oxford University Press, 1981.)

1979)—the amygdala also being implicated in memory, arousal, and dream production (see Chapters 5, 6, and 16). The septal nuclei receives and gives rise to portions of the anterior commissures and to one of the most massive roots of the medial forebrain bundle, via which it receives olfactory and ascending brainstem (reticular formation) fibers (Nauta, 1956). Hence, in addition to arousal, the septal nuclei is involved in emotional functioning. Moreover, like the amygdala and cingulate, the septal nuclei also produces emotional vocalizations.

Aversion and Internal Inhibition

In general, the septal nucleus appears to act in conjunction with the medial hypothalamus and hippocampus, particularly as related to internal inhibition and the exertion of quieting and dampening influences on arousal and limbic system functioning, which in turn aids in the maintenance of selective attention and memory.

For example, via impulses it receives and relays from the reticular activating system, the septal nuclei greatly influences hippocampal activity (Kolb and Whishaw, 1977; Petsche et al., 1965). In addition, the septal nuclei is able to exert facilitatory or inhibitory influences on medial versus lateral hypothalamic arousal (Mogenson, 1976).

However, in this regard, the septal nuclei appears to maintain a counterbalancing relationship with the amygdala (Joseph, 1992a). That is, this nuclei and the amygdala appear to enjoy largely an antagonistic relationship, particularly in regard to influencing the hypothalamus. In addition, the amygdala acts to either facilitate or inhibit septal functioning, whereas septal influences on the amygdala are largely inhibitory.

Moreover, whereas the lateral amygdala may activate the lateral hypothalamus, the septal nuclei may simultaneously facilitate the actions of the hippocampus and medial hypothalamus (Kolb and Whishaw, 1977; Mogenson, 1976; Petsche et al., 1965), which are associated with unpleasant mood states (see Chapter 5). In fact, electrophysiological alterations in septal activity that correspond to subjective feelings of aversion have been reported in humans (Heath, 1976).

Hence, in some respects the septal nuclei appears to serve as a motivational-emotional-arousal memory integrative interface that acts to modulate extremes in arousal as well as pleasurable and unpleasant sensation. However, whereas the amygdala and septal nuclei appear to exert the majority of their counterbalancing influences at the level of the hypothalamus, the dampening and mediating influences exerted by the septal nuclei on the hippocampus is, to a greater extent, independent of the amygdala.

QUIESCENCE

A primary activity of the septal nucleus appears to be that of reducing extremes in emotionality and arousal, and maintaining the organism in a state of quiescence and readiness to respond. Stimulation of the septal nuclei acts to reduce blood pressure and heart rate and induces adrenocortical secretion (Endroczi et al., 1963; Kaada, 1951; Ranson et al., 1935). It also counters lateral hypothalamic self-stimulatory activity (Mogenson, 1976).

RAGE

Electrical stimulation of the septal nuclei counters and inhibits aggressive behavior (Rubenstein and Delgado, 1963) and suppresses the expression of rage reactions following hypothalamic stimulation (Siegel and Edinger, 1976).

If the septal nucleus is destroyed, these counterbalancing influences are removed such that initially there results dramatic increases in aggressive behavior, including rage (Ahmad and Harvey, 1968; Blanchard and Blanchard, 1968; Brady and Nauta, 1953; King, 1958). Bilateral lesions, in fact, give rise to explosive emotional reactivity to tactile, visual, or auditory stimulation that can take the form of attack or flight. However, if the amygdala is subsequently lesioned, the septal rage and emotional reactivity are completely attenuated (King

and Meyer, 1958). Hence, septal lesions appear to result in a loss of the modulatory and inhibitory restraint that are normally exerted, in part, on the amygdala.

Septal Social Functioning

Eventually, within a few weeks or months, rage and aggressiveness due to septal lesions subside and/or completely disappear. In fact, it may well be that the septal rage reaction is due to the lesion encroaching on the hypothalamus. However, a generalized tendency to overrespond and a generalized failure to inhibit emotional responsiveness persists (McClary, 1961, 1966; Poplawsky, 1987), which also suggests orbital frontal and anterior cingulate dysfunction.

Due perhaps to functional recovery, reductions in swelling, and so forth, rage and irritability are soon replaced by indiscriminate socializing and an extreme need for social and physical contact (Jonason and Enloe, 1971; Jonason et al., 1973; McClary, 1961, 1966; Meyer et al., 1978). This may be related to the effects of septal lesions on the immediately adjacent anterior cingulate, and/or the loss of counterbalancing restraint on this nuclei and the amygdala (see Chapter 5).

Socialization and Contact Comfort

In contrast to amygdaloid lesions, which produce a severe social-emotional agnosia and social avoidance and withdrawal (see Chapter 5), septal lesions produce a dramatic and persistent increase in social cohesiveness and contact seeking (Jonason and Enloe, 1971; Jonason et al., 1973; Meyer et al., 1978). As detailed elsewhere (Joseph, 1992a), this suggests that the normal, intact amygdala appears to promote social behavior whereas the septal nucleus seems to counter generalized socializing tendencies and instead (in conjunction with the anterior cingulate) acts to promote selective attachments. Hence, when the septal nucleus, versus the amygdala nucleus, is destroyed, contact seeking rather than avoidance is triggered.

With complete bilateral destruction of the septal nuclei in animals, the drive for social contact appears to be irresistible such that persistent attempts to make physical contact occurs—even with species quite unlike their own. Presumably, this occurs because the amygdala (as well as other nuclei) are no longer opposed by the septal nucleus (and/or related to secondary abnormalities in the anterior cingulate). Hence, the amygdala (and cingulate) engages in indiscriminate contact seeking.

For example, Ritchie (1975, cited by Meyer et al., 1978) reported that septal-lesioned rats, unlike normals, will readily seek out mice (to which they are normally indifferent) or rabbits (which they usually avoid). If presented with a choice of an empty (safe) chamber or one containing a cat, septal lesioned rats persistently attempt to huddle and crawl upon this creature, even when the cat is acting perturbed. If a group of septally lesioned animals are placed together, extreme huddling results. So intense is this need for contact comfort following septal lesions that if other animals are not available, they will seek out blocks of wood, old rags, bare wire frames, or walls and then attempt to cuddle.

Among humans with right-sided or bilateral disturbances in septal functioning (such as due to seizure activity being generated in this region that may well involve the anterior cingulate), a behavior referred to as "stickiness" is sometimes observed. Such individuals seek to make repeated, prolonged, and often inappropriate contact with anyone who is available or who happens to be nearby so as to tell them stories or jokes, or to merely pass the time. They don't readily "take a hint" and are difficult to "get rid of." In hospital situations they can be found intruding on other patients and their families, hanging out by the nurses' station, or incessantly visiting other rooms to chat.

However, in some cases, similar behaviors can be triggered by amygdala as well as anterior cingulate hyperactivation. It is probably the abnormal interactions of these nuclei that account for stalking behaviors and the formation of delusional attachments to actresses, sports stars, coworkers, and so on.

Attachment and Amygdaloid-Septal Developmental Interactions

Broadly considered, the various nuclei comprising the limbic system functionally mature at different rates (Joseph, 1982, 1992a). Correspondingly, certain emotional behaviors and social tendencies appear at various times, overlay previous capacities, become differentiated and/or, in turn, become suppressed or eliminated.

At birth, the hypothalamus reigns supreme in regard to emotion. The infant freely expresses feelings of aversion and rage, or pleasure and quiescence, and demonstrates extreme orality and other behaviors similar to a state in which the amygdala is functionally nonexistent (e.g., the Klüver-Bucy syndrome). The "oral" stage, therefore, corresponds to the hypothalamic and immature amygdala state of neurological development.

As the amygdala and hippocampus begin to develop and mature, the organism becomes more reality oriented and more social as the ability to selectively attend to externally occurring events and to store this information in memory becomes more pronounced (see Chapters 5 and 6). The pleasure principle (although still dominant) begins to be served by the reality principle (Freud, 1911; Joseph, 1982, 1992a). That is, the amygdala and hippocampus not only begin to produce dreams in response to hypothalamic needs, but begin to learn behavioral strategies and to memorize certain acts and stimuli that meet those needs.

However, as the amygdala matures, the oral stage begins to be replaced by the phallic stage, and the infant increasingly seeks social, physical, and emotional contact. As the counterbalancing influences of the septal nuclei are comparatively reduced (due to functional immaturity), and as the cingulate is also slower to mature, infantile behavior is characterized by indiscriminate contact seeking (e.g., attachment behavior) with a variety of individuals, which is promoted by the amygdala. Specific and long-term attachments require the maturation and participation of the septal nuclei and cingulate.

Attachment

Attachment is not the same as dependency. Necessarily, the infant is dependent on its caretaker (e.g., mother). However, until the child is 6–7 months of age it will smile at the approach of anyone, even complete strangers. The child will also vigorously protest any form of separation from strangers (e.g., if they leave the room). This stage corresponds to the amygdaloid maturational period where septal and cingulate influences are still less well developed.

At about 7 months of age, the infant becomes more discriminating in his interactions, and it is during this period that a very real and specific attachment is formed, for example, to one's mother—an attachment that becomes progressively more intense and stable. This period represents septal and cingulate developmental influences such that global contact seeking becomes increasingly narrowed and restricted.

After these specific attachments have been formed, children begin to show anxiety, fear, and even flight reactions at the approach of a stranger (Spitz and Wolf, 1946). By 9 months, 70% of children respond aversively, whereas by 10 months they might cry out if a stranger appears (Schaffer, 1966; Waters, et al., 1975). By one year of age, 90% of children respond aversively to strangers (Schaffer, 1966). However, by this phase of development, long-term attachments with the primary caretaker have been well established and cingulate/septal influences (in conjunction with those of the amygdala) are much more pronounced.

Thus, during the amygdaloid maturational phase there is indiscriminate approach and contact seeking. During the septal-cingulate stage, indiscriminate social contact seeking is inhibited whereas specific attachments are narrowed, strengthened, reinforced, and maintained. The differential rates of amygdala and septal-cingulate development are, thus, crucial in promoting survival and social interaction with significant others.

Conversely, for these nuclei to mature and develop properly requires that physical and emotional contact be provided. If that contact is minimal or abnormal, then the development and neuronal integrity of these limbic nuclei are also likely to become abnormal (Joseph, 1982, 1992ab, 1993; see Chapter 18).

Indeed, if contact with others is abnormal or restricted during these early phases, then the ability to successfully interact socially at a later stage of development is retarded. This is even true among the so-called lower animals. For example, kittens that are not exposed to humans grow up to be "wild" and unapproachable. The phenomena of imprinting probably also requires similar interactions to take place and is dependent on the same limbic nuclei.

Contact Comfort

As is well known, for the first several years of life, physical and social interaction with others is critically important to psychological, neurological, and physical development. The nervous system of an infant, in fact, has been long adapted to receiving vigorous physical and vestibular stimulation. For much of human evolution, infants were often carried about by their mothers, who would bend, twist, walk, run, climb, and gather with a baby strapped or clutched tightly to their bodies: hence, the adaptive significance of the infant's grasp reflex—infants are provided the capacity to clutch back.

If diverse forms of stimulation are not provided, the growth of nerve cells, dendrites, and axons, and the establishment of certain neural circuits throughout the brain, become retarded due to cell atrophy or death and all subsequent behavior is severely affected (Casagrande and Joseph, 1980; Diamond, 1991; Greenough, 1976; Greenough et al., 1987; Joseph and Gallagher, 1980; Rosenzweig, 1971; see Chapter 18). Babies need to be held, touched, caressed, rocked, and carried about for a good part of every day, for these are the adaptive requirements of our 100 million-year-old mammalian nervous system. If denied this stimulation, infants have no way to provide it on their own and, in consequence, will either die or fail to thrive (Langmeier and Matejeck, 1975; Spitz, 1945).

As noted, so intense is this need for stimulation that during much of the first year of life children will indiscriminately seek contact and will smile at the approach of anyone, even complete strangers. Indiscriminate contact seeking and attachment during early development maximizes opportunities for physical-social interaction.

In fact, so intense is the need for physical and social contact that young animals raised in social isolation will form attachments to bare wire frames (Harlow, 1962), to television sets, to dogs that might maul them, to creatures that might kill them (Cairn, 1967)—and, among humans, to mothers that might abuse them. With removal of the septal nuclei, animals will even seek contact with creatures that might eat them.

Social-Emotional Deprivation in Humans

During the late 1930s, the Nazi Germans conducted an experiment designed to raise "supermen." Couples were selected according to strict physical, health, and racial criteria and were allowed to live in special secret camps until children were born. Soon after birth, the children were removed and reared in a special home designed to maximize their superiorities with the single exception that mothering was not provided, as it was thought undesirable to expose these impressionable youngsters to those who might instill in them womanly feelings of nurturance, compassion, and so forth. Within 2 years, 19 of the 20 children raised in this "superior" environment became severely developmentally and emotionally abnormal. They became socially withdrawn and unresponsive and would "lay like dead fish," and/or behaved like "idiots."

R.A. Spitz (1945) studied over 200 children reared in similar conditions. Half were raised in a prison nursery and the rest in a foundling home where mothering and social stimulation were quite minimal due to the high ratio of children to staff. Within one year, those reared in the foundling home became unresponsive to social stimulation and would lay passively on their beds. When approached, they would scream, and they made vigorous attempts to avoid strangers or novel objects or toys. Instead, these deprived youngsters spent hours on end engaged in repetitive, stereotyped, and bizarre movements designed to maximize self stimulation—i.e., rocking, head banging, or pinching precisely the same piece of skin until sores developed. Most of these children became permanently emotionally and socially scarred.

Of those who survived an infancy spent in institutions where mothering and contact comfort was minimized, signs of low intelligence, extreme passivity, apathy, as well as severe attentional deficits were characteristic (Dennis, 1960; Langmeier and Matejeck, 1975; Spitz, 1945). Such individuals have difficulty forming attachments or maintaining social interactions later in life.

Harlow (1962:138), in his famous series of experiments with monkeys, has also shown that those raised with surrogate terry cloth mothers develop extremely bizarre behaviors:

> The laboratory monkeys sit in their cages and stare fixedly into space, circle their cages in a repetitive stereotyped manner and clasp their heads in their hands or arms and rock for long periods of time. They often develop compulsive habits, such as pinching precisely the same patch of skin on their chest between the same fingers hundreds of times a day; occasionally such behavior may become punitive and the animal may chew and tear at its body until it bleeds.

Figure 7.14 During the 19th century and early 20th century, babies were warehoused in orphanages.

Figure 7.15 A baby monkey that was raised with a terry cloth surrogate mother. (Courtesy of the Harry Harlow Primate Laboratory, Wisconsin.)

Ronald Melzack (Melzack and Scott, 1956) performed what some might consider a rather sadistic experiment with puppies and young dogs, one group of which he raised in isolation. When they were approximately 7–10 months old he removed them from their isolation chambers and then stuck pieces of glass or burning matches into their snouts. Normally raised pups were also burnt and cut, but, as would be expected, they tried to bite him and flee. However, those who were raised in isolation were (presumably) so limbically starved for physical contact and social intimacy that, although they reflexively twisted away when burnt, they immediately stuck their noses right back into the flame and hovered excitedly next to him.

In this regard, not only are emotional and motivational functioning adversely effected by these conditions, but learning and intellectual capabilities, including the capacity to simply inhibit inappropriate behavior, is markedly disturbed (Joseph and Gallagher, 1980).

In fact, among humans, so pervasive is this need for physical interaction and social stimulation that when grossly reduced or denied, the result is often death. For example, in several well known studies of children raised in foundling homes during the early 1900s when the need for contact was not well recognized, the mortality rate for children less than 1 year of age was over 70%. Of 10,272 children admitted to the Dublin Foundling Home during a single 25-year period, only 45 survived (Langmeier and Matejcek, 1975).

Deprivation and Amygdala, Septal, and Cingulate Functioning

In addition to mediating all aspects of higher-order emotional intellectual functioning, the amygdala is particularly responsive to tactual stimulation such that single cells may respond to touch, regardless of where on the body the person is stimulated (see Chapter 5). When there is inadequate tactile and physical-social interaction during early development, the ability of neurons to adequately develop and function is significantly reduced (e.g., Diamond, 1985, 1991; Greenough et al., 1987). That is, the amygdala (as well as other nuclei) becomes environmentally damaged.

As has been repeatedly demonstrated in studies of the visual system as well as studies of cerebral development, one consequence of reduced environmental input during certain critical periods of development is cell death, atrophy, and functional

retardation, as well as inhibition by competing neuronal cell assemblies (Casagrande and Joseph, 1980; Diamond, 1991; Greenough, 1976; Greenough et al., 1987; Hubel and Wiesel, 1970; Rosenzweig, 1971; see Chapter 18). Consequently, gross behavioral and perceptual abnormalities result (Harlow, 1962; Hirsh and Spinelli, 1970; Hubel and Wiesel, 1970; Joseph and Casagrande, 1980). Moreover, visual functioning in the deprived eye and neuronal size and activity in the corresponding regions of the lateral geniculate of the thalamus also become abnormal, and animals so reared will fail to respond normally to visual input.

As noted, abnormalities in cellular development and function also occur within the amygdala, and probably the cingulate and septal nuclei, secondary to social-emotional deprivation. For example, Heath (1972) has reported that monkeys reared without mothers develop abnormal spiking in the septal region. Similarly, Heath (1954) found abnormal seizure-like discharges in the septum of withdrawn schizophrenics, noting that the severity of the abnormality was correlated with the severity of the psychosis. These individuals had no interest in social interactions, and if approached they would withdraw and become unresponsive, or, conversely, become agitated and irritable if contact was made.

In addition, it has been demonstrated that the right amygdala is larger than the left in animals reared in enriched and normal environments, whereas these differences are insignificant in animals reared in a restricted setting (Diamond, 1985), suggesting that deprivation differentially reduces right amygdaloid growth. This is significant because, among humans, the right cerebral hemisphere has been shown to be dominant in regard to most aspects of social-emotional intellectual functioning (see Chapter 3).

It is noteworthy that the maternal behavior of primates raised in isolation is also quite abnormal and, in fact, is similar to that of mothers whose amygdalas have been destroyed. As described by Harlow (1965:256–257, 259):

After the birth of her baby, the first of these unmothered mothers ignored the infant and sat relatively motionless at one side of the cage, staring fixedly into space hour after hour. As the infant matured desperate attempts to effect maternal contact were consistently repulsed...Other motherless monkeys were indifferent to their babies or brutalized them, biting off their fingers or toes, pounding them, and nearly killing them until caretakers intervened. One of the most interesting findings was that despite the consistent punishment, the babies persisted in their attempts to make maternal contact.

Mothers who have suffered bilateral amygdalectomies also abuse their infants, biting off toes and fingers, and bouncing the infants upon the floor like a rubber ball. It appears, therefore, that abnormal rearing conditions can drastically effect the amygdala, as well as related limbic nuclei.

Temporary Separation and Progressive Deterioration

Even temporary separation from the mother or primary caretaker can exert deleterious consequences, which become more severe and permanent as the separation continues. For example, children who are separated from their mothers between the ages of 1 and 2 years of age and placed, for example, in a hospital or childrens' home, eventually pass through three stages of emotional turmoil. The first is characterized by a protest period in which the child frequently cries and screams for his mother, and this may last several weeks or even months. This is followed by a stage of despair in which the child ceases to cry, loses interest in his environment, and withdraws, and this too can persist for months. In the final stage, he ceases to show interest in others, loses his appetite, fails to respond to the affection offered by others, and becomes quite passive and unresponsive. He may sit or lie for long periods with a frozen expression on his face, staring for hours at nothing. If the separation continues, he deteriorates further, becomes physically ill, and may die (Spitz, 1945). In general, males are more severely affected

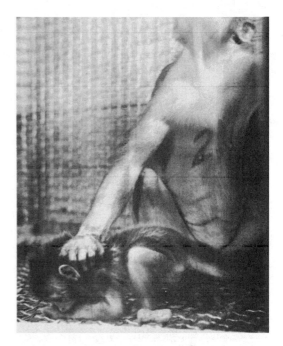

Figure 7.16 An "unmothered" mother rebuffs all attempts by her infant to establish physical contact. (Courtesy of the Harry Harlow Primate Laboratory, Wisconsin.)

than females (Bowlby, 1940, 1960; Sackett, 1974; Spitz, 1945), which is presumably due to sex differences in the limbic system.

As for those children who were only temporarily removed, once they were returned home they would desperately cling to their mothers, follow them everywhere, and become extremely fearful when left alone, even for short periods. Those who were deprived of maternal contact for 6 months or more behaved in a withdrawn, depressed manner and showed no interest in or were unable to reestablish their normal attachment to their mothers. According to J. Bowlby (1940, 1960), children who suffer long-term or repeated separations during the first 3–5 years of life are usually permanently affected.

These affects are presumably permanent due to the catastrophic effects that early deprivation induces within the limbic system and related brain structures, particularly the amygdala. That is, limbic neurons, being deprived of normal social-emotional input, become smaller, and throughout the nervous system there is cell death and atrophy. When the amygdala functions abnormally, subsequent social-emotional intellectual functions become similarly disturbed.

Deprivation and Abuse

The limbic system is organized so as to maintain survival and promote social-emotional interactions. As noted above, in cases of severe deprivation, the limbic system and subsequent behavior may become abnormal due to loss of normal input during critical developmental periods when axodendritic connections and related neural pathways are being formed (e.g., Chapters 16 and 18). However, deprivation is just one form of abnormal environmental input. Children who are reared in homes where they are verbally, physically, or sexually abused, or simply denied sufficient love and affection, can also be severely affected, not just emotionally, but in regard to the functional integrity of the limbic system and related nuclei (detailed in Chapters 6, 16, and 18).

As such, although these persons are seemingly "normal" and perhaps capable of obtaining employment, getting married, and having children, the abuse, neglect, and adverse patterns of interaction that characterized their upbringing may also characterize the structural and functional integrity of the limbic system such that they go through life forever emotionally crippled and/or feel incapable of love or happiness (Joseph, 1992b). As detailed in Chapters 15, 16, and 18, it is abusive, neglectful, and abnormal early environmental influences that not only create emotional-psychological and limbic system abnormalities but that contribute to and in many cases are responsible for the development of chronic psychotic disturbances.

IMPLICATIONS: LOVE, HATE, AND RELATIONSHIPS

The amygdala, septal nuclei, and cingulate gyrus are primary in regard to the development and maintenance of long-term social and emotional attachments, and this appears to be the case with humans, primates, mammals, therapsids (repto-mammals), and perhaps dinosaurs—creatures blessed with well developed limbic systems.

Similarly, among adult humans, it is presumably the interaction of these limbic nuclei (coupled with the neocortex) that give rise to attachment and bonding behavior, including feelings of love as well as lust and the experience of orgasm (see Chapter 5). Moreover, it is probably via the interactions mediated by these nuclei that emotions such as jealousy, rage, or fear of abandonment are also generated, as well as feelings of possessiveness, for a mate. It is these nuclei that probably account for instances of stalking and the formation of delusional attachments to actresses, sports stars, coworkers, and so on.

That some individuals respond with considerable grief, depression, anger and even uncontrollable rage when a "loved one" has ended their relationship probably can also be explained from a limbic (infantile hypothalamic/amygdala) perspective. Unfortunately, when the limbic system has been activated in this manner, feelings of rage may soon be manifested as acts of murder. Loss of love, such as occurs when a relationship ends, seconded only by jealousy and money issues, are prime elicitors of such murderous feelings and are due to the high involvement of the limbic system in all affairs of the heart.

Thus, if a person who has met another individual's primary needs for love, affection, and physical intimacy were to leave, or want to end the affair, the limbic system of the male or female being "abandoned" may respond in an uncharacteristic infantile fashion: i.e., with desperation, frustration, anger, and rage, or depression and despair—similar to the emotional stages demonstrated by children who are progressively deprived of mothering.

Indeed, although feelings of depression may predominate, many individuals feel angry or enraged if "abandoned" or if the relationship is broken off by someone whom they "need" and "love." These feelings, too, are generated by the limbic system. If the hurt and anger are sufficient, the frontal lobes and the rest of the brain may be overwhelmed and the person may act on his or her limbic attachment needs, either in an extremely dependent and despairing fashion ("Don't leave me or I'll kill myself") or in a violent and enraged manner ("Don't leave or I'll kill you").

It is noteworthy that these same limbic nuclei functionally mature during late infancy and early childhood, and in many respects they become (at least initially) functionally associated with their primary caretaker (e.g., their mother). Therefore, in some instances, adult relationships can activate, not just highly emotional and infantile behaviors, but the same feelings that were experienced when those childhood needs were no longer met by the mother.

For example, because some men may feel castrated as well as abandoned if their wives/girlfriends seek to leave, this may further activate those same limbic nuclei that may have raged when their mothers eventually ceased to meet their infantile desires and increasingly *abandoned* them for their fathers or new adult male lovers. They become hopelessly enraged. How-

ever, when this same boy (or girl) grows to adulthood, instead of simply raging helplessly when abandoned or neglected, he may threaten, stalk, beat, or even kill his former spouse or girlfriend and perhaps even his own children. Or, if it's his job he's lost, he may threaten, attack, or, as sometimes occurs in this society, kill his boss and coworkers.

Fortunately, only a small percentage of the population acts on such impulses. Nevertheless, even among those humans who maintain high levels of frontal inhibitory control in regard to all matters of the heart, the amygdala, cingulate, septal nuclei, hypothalamus, inferior temporal lobe...make possible not only intense feelings of emotion for a mate or lover, but correspondingly (at least in some people) an occasional "irrational" urge to throw that person in front of a train (metaphorically speaking, of course).

Indeed, because it provides the neurological foundation for emotion, attachment, jealously, and desire, the limbic system enables human beings not only to coo words of love and sorrow but to experience all the joys, lusts, warmth, thrills, romance, passion, sexual excitement, and "craziness" of "true love."

The Limbic System and the Soul

Evolution and the Neuroanatomy of Religious Experience

PART I: OBSERVATIONS AND SPECULATIONS

A belief in the transmigration of the soul, of an afterlife, of a world beyond the grave, may well have been a human characteristic for at least 100,000 years (Belfer-Cohen and Hovers, 1992; Butzer, 1982; McCown, 1937; Rightmire, 1984; Schwarcz et al., 1988; Smirnov, 1989; Trinkaus, 1986). Indeed, despite their primitive cognitive capabilities, even "archaic" human beings who wandered the planet over 120,000 years ago carefully buried their dead (Butzer, 1982; Rightmire, 1984); and like modern H. sapiens sapiens, they also, presumably, prepared the recently departed for the journey to the Great Beyond: across the sea of dreams, to the land of the dead, the realm of the ancestors and the gods. Thus, it was not uncommon for tools and hunting implements to be placed beside the body, even 100,000 years ago (Belfer-Cohen and Hovers, 1992; McCown, 1937; Trinkaus, 1986): a hunter in life, he was to be a hunter in death, for the ethereal world of the dead was populated by spirits and souls of bear, wolf, deer, bison, and mammoth (e.g., Campbell 1988; Kuhn, 1955). Food and water might be set near the head in case the spirit hungered or experienced thirst on its long sojourn to the Hereafter. And finally, flowers and red ocher might be sprinkled upon the body (e.g., Solecki, 1971), along with the tears of those who loved him.

Given the relative paucity of cognitive and intellectual development among Middle Paleolithic Neanderthal and "archaic" (as compared to modern) humans, and the likelihood that they had not yet acquired modern human speech (see Chapter 2), evidence of spiritual concerns among archaic and other Middle Paleolithic peoples (i.e., archaic, "early moderns," Neanderthals) may be somewhat surprising, if not unbelievable (Gargett, 1989). However, it appears, based on a gross photographic analysis of Neanderthal endocasts (see Chapter 2) as well as the evidence reviewed below, that "archaic," "early modern," and Neanderthal men and women possessed a well developed inferior temporal lobe and limbic system—brain areas directly implicated in the generation of religious experience.

In fact, a commonality in regard to the limbic system may account for why some other species (elephants, wolves, chimpanzees) have also been reported to sometimes bury their dead or hide their skeletal remains, to ritualistically howl at the moon (Lawrence, 1986), or engage in rhythmic ritualistic group activities during thunder storms—what Goodall (1971:67) referred to as a chimpanzee "rain dance." Like humans, these creatures also possess a limbic system and temporal lobe.

This is significant for, as will be detailed below, the amygdala, hippocampus, and inferior temporal lobe appear to subserve and provide the foundations for mystical, spiritual, and religious experiences (e.g., Bear, 1979; Daly, 1958; d'Aquili and Newberg, 1993; Gloor, 1986, 1992; Halgren, 1992; Horowitz et al., 1968; Jaynes, 1976; Joseph, 1982, 1990a, d, 1992a, b, 1993, 1994; MacLean, 1990; Mesulam, 1981; Penfield and Perot, 1963; Rolls, 1992; Schenk and Bear, 1981; Slater and Beard, 1963; Subirana and Oller-Daurelia, 1953; Trimble, 1991; Weingarten et al., 1977; Williams, 1956). However, in presumed contrast to other species, the spiritual realms associated with the human limbic system and temporal lobe activity (specifically, hyperactivity) include feelings of hyper-religiousness, cosmic wisdom, astral projection, and the perception, or rather, the "hallucination," of ghosts, demons, spirits, and sprites, and belief in demonic or angelic possession (Bear, 1979; Daly, 1958; d'Aquili and Newberg, 1993; Gloor, 1986, 1992; Horowitz et al., 1968; Jaynes, 1976; Joseph, 1990a, b; MacLean, 1990; Mesulam, 1981; Penfield and Perot, 1963; Schenk and Bear, 1981; Slater and Beard, 1963; Taylor, 1972, 1975; Trimble, 1991; Weingarten et al., 1977; Williams, 1956). When these nuclei are hyperactivated, "religious" experiences, although unusual, are not uncommon.

The limbic system is common to all peoples and appears to provide the foundations for religious feelings and related "hallucinations." This might explain why the belief in souls, spirits, haunted houses, and even angels or demons, or the capacity to have mystical experiences, including the sensation of being possessed by gods or devils and/or hearing their voices, is worldwide (Budge, 1994; Campbell, 1988; Frazier, 1950; Godwin, 1990; Harris, 1993; James, 1958; Jaynes, 1976; O'Keefe, 1982; Malinowkski, 1948; Smart, 1961; Wilson, 1951).

Presumably, because all humans possess a limbic system and a brain that is organized in a similar manner, they have similar religious and mystical experiences, what Jung (1964) referred to as "archetypes"—inborn tendencies to produce, create, dream of, and respond in a similar manner to specific images, symbols, and experiences. This commonality in "religious" or "archetypal" experience includes the capacity to experience "God" or the "Great Spirit" as well as the many vestiges or incarnations of what has been referred to as "the personal soul" or "ghost."

Indeed, it could be argued that the essence of God, and of our living soul, may be slumbering within the depths of the ancient limbic lobe, which is buried within the belly of the brain. And not just the soul or the Great Spirit, or the essence of the Lord God, for in the Upanishads and Tao it

is said, and as Buddha, Lao Tzu, Chuang Tzu, Jesus (St. Luke 17:21), the Sufis, and many Sumerian, Babylonian, Jewish, Arabic, Arian, Egyptian, Greek, Roman, Indian, Muslim, and Gnostic mystics have proclaimed, "The kingdom of God is within you."

THE ANTIQUITY OF THE SOUL: MIDDLE PALEOLITHIC SPIRITUALITY

The point at which humans first became aware of a "God" or "Great Spirit" cannot be determined. Nevertheless, the antiquity of religious beliefs extends well over the course of the last 100,000 years. Indeed, it has been well established that Neanderthals and other H. sapiens of the Middle Paleolithic (e.g., 150,000–35,000 B.P.) and Upper Paleolithic (35,000—10,000 B.P.) engaged in complex religious rituals. For example, Neanderthals (a people who lived in Europe and the Middle East from around 300,000—35,000 B.P.; see Chapter 2) have been buried in sleeping positions with the body flexed or lying on its side, surrounded by goat horns placed in a circle, accompanied by reindeer vertebrae, animal skins, stone tools, red ochre, and flowers, with large bovine bones above the head, with limestone blocks placed on top of the head and shoulders or beneath the head like a pillow, and with heads severed, coupled with evidence of ritual decapitation, facial bone removal, and cannibalism (Belfer-Cohen and Hovers, 1992; Binford, 1968; Harold, 1980; Smirnov, 1989; Solecki, 1971; see also commentary in Gargett, 1989). Moreover, Neanderthals presumably buried a bear at Regourdou, and at Drachenloch they buried stone "cysts" containing bear skulls (Kurten, 1976); hence, "the clan of the cave bear."

Of course, the fact that these Neanderthals were buried does not necessarily imply that they held a belief in God. Rather, what is indicated is that they had strong feelings for the deceased and were perhaps preparing them for a journey to the Hereafter or the land of dreams; hence, the presence of stone tools, the sleeping position, and stone pillows. Throughout the ages, dreams have been commonly thought to be the primary medium in which gods and humans interact (Campbell, 1988; Freud, 1900; Jung, 1945, 1964). Insofar as the ancients (and many "moderns") were concerned, dreams served as a doorway, a portal of entry to the spirit world.

The possibility that these ancient humans believed the dead (or their souls) might return and cause harm is also suggested by the (admittedly controversial) evidence of ritual decapitation, and placement of heavy stones upon the body. This suggests they believed in ghosts, souls, or spirits, and a continuation of "life" after death, and therefore took necessary precautions to prevent certain souls from being released from the body or returning to cause mischief.

Similarly, the buried animal skulls and bones implies a degree of ritual symbolism, which, when coupled with grave offerings and positioning of the body, certainly seems to imply the Neanderthals were capable of very intense emotions and feelings, ranging from love to perhaps spiritual and superstitious awe (a function of the limbic system). When coupled with the evidence reviewed above (and below), there thus seems to be good reason to assume that Neanderthals maintained spiritual and mystical belief systems involving, perhaps, the transmigration of the soul and the horrors, fears, and hopes that accompany such feelings and beliefs.

The Neanderthals, however, were not the first to practice mortuary rites. As noted, "early modern" and "archaic" Homo sapiens also buried infants, children, and adults with tools, grave offerings, and animal bones. For example, archaic H. sapiens and "early moderns" were carefully buried in Qafzeh, near Nazareth and in the Mt. Carmel Mugharetes-Skhul caves on the Israeli coast over 90,000–98,000 years ago (McCown, 1937; Smirnov, 1989; Trinkaus, 1986). This includes a Qafzeh mother and child who were buried together, and an-

other infant who was buried holding the antlers of a fallow deer across his chest. In a nearby site equally as old (i.e., Skhul), yet another was buried with the mandible of a boar held in his hands whereas an adult had stone tools placed by his side (Belfer-Cohen and Hovers, 1992; McCown, 1937). In this regard, it is quite clear that humans have been burying and presumably weeping over their dead, and perhaps preparing them for a journey to the Hereafter, for over 100,000 years.

However, it is with the rise of the Upper Paleolithic and dominance of the Cro-Magnon and Asian "moderns" that the quality and quantity of grave goods undergoes a creative and symbolic explosion (see Chapter 2). Dead were buried with jewelry, weapons, clothing, pendants, rings, necklaces, multifaceted tools, head bands, bracelets, all intricately fashioned with the care and talent equal to that of most modern artisans.

By 25,000 years ago the Cro-Magnon began painting, drawing, and etching bear and mammoth, dear and horse, and even pregnant females in the recesses of dark and dusky caverns (Bandi, 1961; Joseph, 1993; Leroi-Gurhan, 1964; Prideaux, 1973). The pregnant females include Venus stat-

Figure 8.1 A Neanderthal seemingly glowers from the grave at those who have disturbed his long rest. He had been purposefully buried and his body sprinkled with seven different types of flowers. From Solecki R. Shanidar: The first flower people. New York: Alfred A Knopf, 1971.)

Figure 8.2   Upper Paleolithic male buried with grave goods, necklaces, beads, and hunting implements. (Courtesy of the Novosti Press Agency, Moscow.)

uettes, some of which may be fertility and sex symbols or perhaps even representative of various goddesses.

However, in order to view many of these Cro-Magnon paintings and "religious" objects, one had to enter and crawl a considerable distance through a pitch black tunnel before reaching these Upper Paleolithic underground cathedrals. This is significant, for in the Egyptian & Tibetan Books of the Dead, and has been reported among many of those who have undergone a "near death" or "life after death" experience, being enveloped in a dark tunnel is commonly experienced soon after death and immediately prior to entering the "light" of "Heaven" or paradise, at which point "the recently dead" may be greeted by relatives, friends, and/or radiant human or animal-like entities (Eadie, 1992; Rawlings, 1978; Ring, 1980).

As is evident from their cave art and symbolic accomplishments, the nether world of the Cro-Magnon and other peoples of the Upper Paleolithic was also

Figure 8.3   Two Upper Paleolithic boys buried with grave goods, necklaces, beads, and hunting implements. (Courtesy of the Novosti Press Agency, Moscow.)

Figure 8.4 "Venus" figurines. Upper Paleolithic goddesses? (From Prehistoire de l'Art Occidental, Editions Citadelles & Mazenod, Paris. Photo by Jean Vertut.)

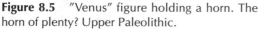

Figure 8.5 "Venus" figure holding a horn. The horn of plenty? Upper Paleolithic.

Figure 8.6 The entrance to Heaven. By Hieronymus Bosch, 1500.

haunted by the spirits and souls of the living, the dead, and those yet to be born, both animal and human (Brandon, 1967; Campbell, 1988; Kuhn, 1955; Prideaux, 1973). However, those of the Upper Pale-

olithic apparently believed these souls and spirits could be charmed and controlled by hunting magic, and the spells of sorcerers. Indeed, hundreds of feet beneath the earth, the likeness of one ancient shaman, attired in animal skins and stag antlers, graces the wall directly above the winding cavern en-

Figure 8.7 "The Shaman." Dressed in animal skins and stag antlers, the shaman was drawn above the entrance to the grand gallery at Les Trois-Freres Cave in Southern France. Upper Paleolithic. (From a copy made by Abbe Breuil.)

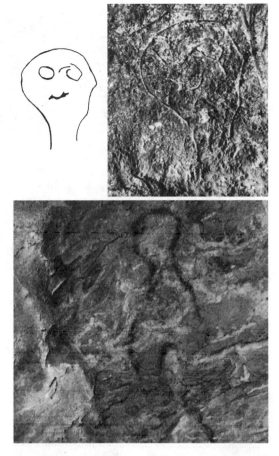

Figure 8.8 Ghostly vestiges drawn within an Upper Paleolithic cavern. Compare with Figure 8.9. (From Prehistoire de l'Art Occidental, Editions Citadelles & Mazenod, Paris. Photo by Jean Vertut.)

trance to the 20,000- to 25,000-year-old grand gallery at Les Trois-Freres in southern France (Prideaux, 1973). Galloping, running, and swirling about him are bison, stag, horse, deer, and, presumably, their souls.

The Amygdala, Temporal Lobe, and Religious Experience

The fact that Middle Paleolithic peoples (archaic, "early modern," Neanderthal) and those of the Upper Paleolithic (Asian and African "modern," and Cro-Magnon) all buried their dead with grave offerings and with the body placed in specific positions suggests that these peoples were capable of strong emotional feelings, ranging from love to fear to mystical and religious awe. This also indicates that all groups shared a certain commonality in regard to that region of the brain (and, in fact, the

only region of the brain) that has been implicated in the generation of fear, love, intense emotions, and religious and spiritual beliefs: the limbic system (e.g., amygdala and hippocampus) and inferior temporal lobe (Bear, 1979; Daly, 1958; d'Aquili and Newberg, 1993; Gloor, 1986, 1992; Halgren, 1992; Horowitz et al., 1968; Jaynes, 1976; Joseph, 1982, 1990a, d, 1992a, b, 1993, 1994; MacLean, 1990; Mesulam, 1981; Penfield and Perot, 1963; Rolls, 1992; Schenk and Bear, 1981; Slater and Beard, 1963; Subirana and Oller-Daurelia, 1953; Trimble, 1991; Weingarten et al., 1977; Williams, 1956).

The amygdala and inferior temporal lobe appear to have been as developed in

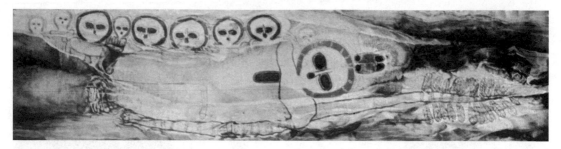

Figure 8.9 Wondjina figure with plum tree at its side. Wondjinas are spirits. Above the Wondjina are the "children of the wondjinas," which are the souls of men. (From Bildarchiv Preubischer Kulturbesitz, Baden-Baden, Germany.)

Figure 8.10 Souls ascending to Heaven? From Northwest Africa, Upper Paleolithic. These figures sometimes occur with "wondjina" figures, which are the "souls of men" (see Figure 8.9). (From Bildarchiv Preubischer Kulturbesitz, Baden-Baden, Germany.)

"archaics," Neanderthals, and early moderns as in Upper Paleolithic (Cro-Magnon and other "modern") H. sapiens. As will be detailed below, the evolution of these cerebral nuclei have made it possible, not only to experience but to attribute spiritual or religious significance to certain actions and stimuli.

For example, in addition to burial and mortuary practices, one of the first signs of exceedingly ancient religious symbolism is the discovery of an engraved "cross" (Vertes, 1964, cited by Mellars, 1989) that is perhaps 60,000–100,000 years old. Indeed, regardless of time and culture, from the Aztecs, American Indians, Romans, Greeks, Africans, Cro-Magnons, Egyptians (the key of life), and modern Christians, the cross consistently appears in a mystical context and/or is attributed tremendous cosmic or spiritual significance (Budge, 1994; Campbell, 1988; Jung, 1964; Sitchin, 1990).

Given that there are neurons that fire selectively to specific geometric visual shapes (e.g., faces, hands, triangles), and that they exist largely within the inferior temporal lobe (Desimone and Gross, 1979; Gross et al., 1972; Richmond et al., 1983; Richmond et al., 1987), it could, therefore, be assumed that "cross" neurons as well as "mystical/religious" feeling neurons (or neural networks) had probably evolved by 100,000 years ago—or longer, as such cross-like stimuli are common. For example, one need only walk in a forest to spy dead trees that take the form of a "cross," or look to the sky to view cross-like birds with extended wings.

Figure 8.11   A dead hunter and a bird's head. A disemboweled bison stands above him. Presumably, this scene depicts the death of a hunter and the flight of his soul, as symbolized by the bird. Bird heads were commonly employed by ancient peoples, including the Egyptians, to depict the otherworldly (see Figures 8.12 and 8.17). Eventually, bird heads were replaced by creatures with wings, e.g., angels. (From Bildarchiv Preubischer Kulturbesitz, Baden-Baden, Germany.)

Along the neocortical surface of the inferior temporal lobe (and within the amygdala) are dense neuronal fields that contain neurons that fire selectively in response to visual images of faces, hands, eyes, and complex geometric shapes, including crosses (Gloor, 1960; Gross et al., 1972; Richmond et al., 1983, 1987; Rolls, 1984, 1992; Ursin and Kaada, 1960); these are sometimes referred to as "feature detectors." The ability to recognize faces, geometric shapes, and social emotional nuances are dependent on these specialized temporal lobe and amygdala neurons and neural networks that respond selectively to these stimuli (Gross et al., 1972; Richmond et al., 1983, 1987; Rolls, 1984; see Chapter 14). However, since neurons in the amygdala and inferior temporal lobe are also multimodally responsive and subserve almost all aspects of emotion, including religious feeling, it is possible for

Figure 8.12   Goddess dressed with bird heads. From Libya, approximately 10,000 BP. The bird head symbolizes the capacity for flight and, thus, the ability to ascend to heaven. (From Bildarchiv Preubischer Kulturbesitz, Baden-Baden, Germany.)

Figure 8.14   Engraved cross. Sumerian, 6000 BP.

Figure 8.13   Example of a cross drawn by the Cro-Magnon people. Upper Paleolithic.

Figure 8.15   The Maya and Aztec God, Quetzalcoatl, the Plumed Serpent, was associated with the planet Venus, i.e., the bringer of rain, light, and knowledge. The sign of the cross was one of his many emblems, which is depicted in the center of his shield. (From Sitchin Z. The lost realms. New York: Avon Books, 1990.)

The Mayas and Aztecs utilized several different types of "calendars" for predicting celestial events, changes in the season, and impending catastrophes, including the expected return of Venus, or rather, Quetzalcoatl, the Plumed Serpent. This particular Venus "calendar" was based on multiples of 52 years, and it specifically predicted that in the year 1519 a comet (or star) would appear in the East and snake through the sky, and that the Plumed Serpent, carrying his sign of the cross, would return and overthrow the ruling god, to whom the Aztecs regularly sacrificed humans (see Figure 8.19). As predicted, in the year 1519 a comet, in fact, appeared in the East, and soon thereafter, the plumed and helmeted Cortez and his crew, flying the Spanish flag and the sign of the cross, appeared off the east coast of Mexico. And, as predicted, the old god, as well as the Aztec civilization, were destroyed.

A second "calendar" is based on a cycle of 5200 years, which corresponds to cycles of destruction and creation every 5200 years. According to the Mayan calendar, the last era of destruction began around 8313 BC, a time that corresponds with the ending of the last ice age and the onset of the great floods. According to Mayan cosmology and history, for the next 5000 years, civilization was all but abolished, and did not begin to reappear until the year 3113 BC—which corresponds with the development of civilization in Sumer and Egypt, followed by China, the Meso-Americas, India, and finally Greece, Rome, and elsewhere. According to this "calendar," the next cycle of destruction begins in the year 2012.

Figure 8.16   The God SEB supporting the Goddess Nut, who represents heaven. Note the repeated depictions of the key of life, i.e., a ring with a cross at the end.

faces and geometric symbols to become infused with emotional, mystical, and religious significance.

For example, heightened emotional activity within these nuclei could result in feelings of fear, foreboding, or religious awe, as well as activation of the neural networks that respond selectively to crosses, such that emotional and spiritual significance is attributed to objects such as "crosses." Similar explanations could be offered in regard to the spiritual significance attributed to triangles (i.e., pyramids) and circles. In fact, crosses, triangles, and circles were etched on Cro-Magnon cave walls over 30,000 years ago (see Leroi-Gurhan, 1964). Of course, when gazing at a cross, all four quadrants of the visual field (and thus the right and left temporal-parietal lobes) are activated.

The Twilight of the Gods: Cro-Magnon and Upper Paleolithic Spiritual Evolution

As the Cro-Magnon and other peoples of the Upper Paleolithic began to swarm across the earth, their spiritual belief systems soon outstripped those of their predecessors in complexity, originality, and artistic and symbolic accomplishments, due, in large part, to the tremendous advances that had occurred in frontal and inferior parietal lobe evolutionary development (see Chapter 2). As the brain and man and woman evolved, so too did their spiritual beliefs. Hence, the Cro-Magnon conception of the spirit world and their ability to symbolically express it was much more complex as well (Bandi, 1961; Leroi-Gurhan, 1964; Prideaux, 1973).

It appears that by the end of the Upper Paleolithic, many peoples believed that the Netherworld was populated, not just with the souls of trees, plants, animals and humans, but gods (Armstrong, 1994; Brandon, 1967; Budge, 1994; Campbell, 1988; Frazier, 1950; Harris, 1993; Smart, 1969; Wilson, 1951). There were forest gods, gods of the river, the sky, the clouds, and the seasons, gods of the day, and gods of the night. And then there were planetary gods, moon gods, sun gods, personal gods, and Lord Gods who created the

Figure 8.17   The pharaoh, dressed as the son of the god Horus (the hawk), holds the key of life in his right hand.

Heavens and the earth...gods who created man and woman in the image of the gods, and then a single God who gave of his own soul and spirit, thus awakening man and woman to their own spirituality through the breath of life...so claimed the ancients.

However, although the Heavens were thought to be the domain of the gods, some came to believe that the abode of God, like the spirit and the soul, was within each individual woman and man. Indeed, this Kingdom, this transmitter to God, may be none other than the limbic system.

THE LIMBIC SYSTEM AND THE SOUL

The Hypothalamus, Sex, and Emotion

The nuclei of the ancient limbic system include the hypothalamus, amygdala, cingulate gyrus, septal nuclei, and hippocampus (e.g., Gloor, 1960, 1986; Halgren, 1992; Joseph, 1990a, 1992a, 1993, 1994; MacLean, 1969, 1990; Rolls, 1992). It is these limbic system nuclei that are primary in regard to memory, the production of visual imagery, sexuality, and the expression and perception of most aspects of emotion, including love, sadness, grief, depression, fear, aggression, rage, pleasure, happiness, elation, and even sexual and religious ecstasy (see Chapters 5 and 7).

However, different limbic system nuclei, such as the hypothalamus, are more ancient and more primitive in emotional expression as compared to structures such as the more recently evolved amygdala (e.g., Halgren, 1992; Joseph, 1982, 1990a, 1992a, 1994; MacLean, 1969, 1990; Rolls, 1992). For example, the hypothalamus mediates the expression of very intense, rudimentary and transient emotional states that are common to most creatures, including reptiles, amphibians, fish, and even sharks (Joseph, 1993; MacLean, 1969, 1990). It monitors internal homeostasis, and mediates and controls the desire to eat, drink, attack, or have sex—i.e., the four Fs: feeding, fighting, fleeing, and fornicating. Indeed, sexuality and pleasure are of the utmost concern to the hypothalamus, and it is capable of generating orgasmic sensations of great intensity via the release of naturally occurring opiates (Joseph, 1990a, 1992a; Olds and Forbes, 1981; MacLean, 1969, 1990).

The Amygdala and Emotion

The amygdala (which is intimately interconnected with the hypothalamus) enables us to hear "sweet sounds," recall "bitter memories," and determine if something is spiritually significant, sexually enticing, or good to eat (Gloor, 1960, 1986; Halgren, 1992; Joseph, 1982, 1992a, 1994; Rolls, 1992; Ursin and Kaada, 1960). The amygdala makes it possible to experience the spiritually sublime, is concerned with the most basic animal emotions, and allows us to store affective experiences in memory or even to reexperience them when awake or during the course of a dream in the form of visual, auditory, or religious or spiritual imagery (Bear, 1979; d'Aquili and Newberg, 1993; Gloor, 1986; Joseph, 1982, 1990a,

1992a). The amygdala also enables an individual to experience emotions such as love and religious rapture, as well as the ecstasy associated with orgasm (via enkephalin secretion) and the dread and terror associated with the unknown.

In fact, the amygdala (in conjunction with the hippocampus) contributes in large part to the production of very sexual as well as bizarre, unusual, and fearful mental phenomenon, including dissociative states, feelings of depersonalization, and hallucinogenic and dreamlike recollections involving threatening men, naked women, sexual intercourse, religion, and the experience of god, as well as demons and ghosts and pigs walking upright dressed as people (Bear, 1979; Daly, 1958; d'Aquili and Newberg, 1993; Gloor, 1986, 1992; Halgren, 1992; Horowitz et al., 1968; Jaynes, 1976; Joseph, 1982, 1990a, d, 1992a, b, 1993, 1994; MacLean, 1990; Mesulam, 1981; Penfield and Perot, 1963; Rolls, 1992; Schenk and Bear, 1981; Slater and Beard, 1963; Subirana and Oller-Daurelia, 1953; Taylor, 1972, 1975; Trimble, 1991; Weingarten et al., 1977; Williams, 1956). Moreover, some individuals report communing with spirits, or receiving profound knowledge from the Hereafter, following amygdala (and hippocampal-temporal lobe) stimulation (MacLean, 1990; Penfield and Perot, 1963; Williams, 1956).

THE AMYGDALA, THE TEMPORAL LOBE, AND THE SOUL

According to d'Aquili and Newberg (1993), mystical states may be voluntarily or involuntarily induced and are dependent upon the differential stimulation and deafferentation of limbic system nuclei, including the hypothalamus, hippocampus, and amygdala, as well as the right frontal and right temporal lobe. However, it appears that these brain areas differentially contribute to religious and emotional experience.

For example, the hypothalamus is concerned with all rudimentary aspects of emotion and controls the hormonal and related aspects of sexual activity (including the capacity to experience an orgasm or the "nirvana" of a heroin "high" via enkephalin secretion). By contrast, it is the amygdala, in conjunction with the temporal lobe and hippocampus, that enables a human to have religious, spiritual, and mystical experiences (Bear, 1979; Daly, 1958; d'Aquili and Newberg, 1993; Horowitz et al., 1968; Mesulam, 1981; Penfield and Perot, 1963; Schenk and Bear, 1981; Slater and Beard, 1963; Subirana and Oller-Daurelia, 1953; Trimble, 1991; Weingarten et al., 1977; Williams, 1956).

The amygdala, hippocampus, and temporal lobe are richly interconnected and appear to act in concert in regard to mystical experience, including the generation and experience of dream states and complex auditory and visual hallucinations, such as may be induced by LSD (Broughton, 1982; Goldstein et al., 1972; Gloor, 1986, 1992; Hodoba, 1986; Horowitz et al., 1968; Joseph 1990a, e, 1992a; Meyer et al., 1987; Penfield and Perot, 1963; Weingarten et al., 1977; Williams, 1956). Intense activation of the temporal lobe, hippocampus, and amygdala has been reported to give rise to a host of sexual, religious, and spiritual experiences; and chronic hyperstimulation can induce an individual to become hyper-religious or visualize and experience ghosts, demons, angels, and even "God," as well as to claim demonic and angelic possession or the sensation of having left his body (Bear, 1979; Gloor, 1986, 1992; Horowitz et al., 1968; MacLean, 1990; Mesulam, 1981; Penfield and Perot, 1963; Schenk and Bear, 1981; Weingarten et al., 1977; Williams, 1956).

The amygdala and inferior temporal lobes are also highly involved in the generation of feelings of intense sexual arousal, fear, or conversely, rapture and euphoria— the latter being a consequence of the large quantities of enkephalins being released and the high concentrations of opiate receptors located throughout the amygdala (Atweh and Kuhar, 1977ab; Uhl et al., 1978). In response to pain, stress, shock, fear, or terror, the amygdala and other limbic nuclei begin to secrete high levels of opiates, which can induce a state of calmness as well as analgesia and euphoria.

As noted, if these neurons are hyperactivated, such as occurs during dream states, seizures, physical pain, terror, food deprivation, and social and sensory isolation, and under LSD (which disinhibits the amygdala by blocking serotonin), an individual might infuse his perceptions with tremendous religious and emotional feeling. Hence, under these conditions the individual may hallucinate, and ordinary perceptions, objects, or people may be regarded as spiritual in nature or endowed with special or religious significance. Hence, the individual may come to believe he or she is hearing, seeing, and interacting with gods, angels, and demons when in fact the individual is hallucinating and is excessively emotionally/religiously aroused and/or experiencing an "enkephalin" high, thereby giving rise to feelings of rapture or "nirvana."

This does not mean to say, however, that "gods," "angels," and "demons," do not exist, for conceivably this could be the case. Indeed, why would the brain have evolved neurons that enable members of the human race to hallucinate and/or believe in that which does not exist? Moreover, why would humans "evolve" neurons and neural networks that continue to fire, even after death, thereby causing them to think they have left their body and are euphorically basking in the light of Heaven?

The Out-of-Body and "Near Death" Experience

Some children and adults who have been declared "clinically" dead, but who subsequently return to life, have sometimes reluctantly reported that after "dying" they left their body and floated above the scene (Eadie, 1992; Rawling, 1978; Ring, 1980). Typically, they become increasingly euphoric as they float above their body (after which they may float away, become enveloped in a dark tunnel, and then enter a soothing radiant light); later, when they come back to life, they may even claim conscious knowledge of what occurred around their body while they were dead and floating nearby. Moreover, it is not uncommon for these individuals to see their entire past flash before their mind's eye—a "life review" which is no doubt due to activation of the amygdala-hippocampal memory centers. Indeed, similar experiences are even detailed in the Egyptian funerary texts and Book of the Dead almost 6000 years ago (Budge, 1994), as well as by otherwise completely "modern" and sophisticated humans.

"Lisa," for example, was a 22-year-old college coed with no religious background or formal spiritual beliefs. In 1983 Lisa was badly injured in an auto accident after the windshield collapsed and all but completely severed her arm. According to Lisa, when she got out of the car, blood was spraying everywhere, and she only walked a few feet before collapsing. Apparently, an ambulance arrived within minutes. However, the next thing she noticed was that part of the time she was looking up from the ground, and part of the time she was up in the air looking down; she could see the ambulance crew working, picking up her body, placing it on a gurney and into the ambulance.

According to Lisa, during the entire ride to the hospital it was like she was half in and half out of the ambulance, as if she were running along outside or just extending out of the vehicle, watching the cars and trees go by. When they got to the hospital, she was no longer attached to her body but was floating up and down the halls, watching the doctors, nurses, and attendants. One doctor, in particular, drew her attention because he had a big belt buckle with his name written on it. She could even read it: "Mike."

According to Lisa, she was "tripping out, bobbing up and down the halls, just checking everything out" when she noticed a girl lying on a gurney with several doctors and nurses working frantically. When she floated past and peered over the shoulder of one of the doctors to take a look, she suddenly realized the girl was

her and that her hair and face were very bloody and needed to be washed. At that point, Lisa realized she was floating well above her body and the doctors, and that she looked to be "dead." However, according to Lisa she did not feel afraid or upset, although the fact that her hair was dirty bothered her.

As detailed by Lisa, she soon floated up and outside the emergency room and was enveloped in total blackness, "like I was passing through a tunnel at the end of which was a vague light which became brighter and more brilliant, radiating outward." The light soon enveloped her body, which made her feel exceedingly happy and warm. A few moments later, she heard the voice of her grandmother who had died when Lisa was a young girl. Although Lisa had no memory of this grandmother, she nevertheless recognized her and felt exceedingly happy. However, as Lisa approached, her grandmother told her it was "too soon," that she would have to "go back." Lisa, however, didn't want to go back. But apparently, she had no choice. She then was drawn away from the light and felt herself falling, only to land with a painful thump in her own body. At this point, Lisa moved her hand, which alerted the emergency room staff that she was no longer dead.

It is noteworthy that Lisa had never heard of "near death experiences" (she was injured in 1983), and that after returning to life she only reluctantly explained what had happened when she was questioned by one of her doctors. Lisa also claimed that while she was dead and floating about the emergency room she saw, heard, and recalled everything that occurred up to the point when she was enveloped in darkness. She was able to accurately describe "Mike" as well as some of the staff who first attended her, the conversations that occurred around her, and some of the other patients.

Indeed, similar "after death" *claims* of not only leaving and floating above the body, but being able to see everything occurring below, are common (Eadie, 1992;

Moody, 1977; Rawling, 1978; Ring, 1980; Sabom, 1982; Wilson, 1987), and, as noted, are even reported in the 6000-year-old Egyptian Book of the Dead (Budge 1994), as well as the Tibetan Book of the Dead (the Bardo Thodol), which was composed over 1300 years ago (Evans-Wentz, 1960). Approximately 37% of patients who are resuscitated report similar "out of body" experiences (Ring, 1980).

Consider for example, the case of Army Specialist J. C. Bayne of the 196th Light Infantry Brigade. Bayne was "killed" in Chu Lai, Vietnam, in 1966. He was simultaneously machine-gunned and struck by a mortar. According to Bayne, when he opened his eyes he was floating in the air, looking down on his crumpled, burnt, and bloody body, and he could see a number of Vietcong searching and stripping him:

> I could see me...it was like looking at a manikin laying there...I was burnt up and there was blood all over the place.... I could see the Vietcong. I could see the guy pull my boots off. I could see the rest of them picking up various things...I was like a spectator.... It was about four or five in the afternoon when our own troops came. I could hear and see them approaching...I could see me...it was obvious I was burnt up. I looked dead...they put me in a bag...transferred me to a truck and then to the morgue. And from that point, it was the embalming process. I was on that table and a guy was telling a couple of jokes about those USO girls...all I had on was bloody undershorts...he placed my leg out and made a slight incision and stopped...he checked my pulse and heartbeat again and I could see that too...it was about that point I just lost track of what was taking place... [until much later] when the chaplain was in there saying everything was going to be all right...I was no longer outside. I was part of it at this point (reported in Wilson, 1987:113–114; and Sabom, 1982: 81–82).

It is noteworthy that some surgery patients, although ostensibly "unconscious" due to anesthesia, are also able to later describe conversations and related events that occurred during the operation (Furlong, 1990; Kihlstrom, et al., 1990; Polster, 1993). Hence, the notion that those who are

"clinically dead" or near death may also recall various events that occurred while they were ostensibly "dead" should not be dismissed out of hand. Moreover, some surgery patients also *claim* to have "left their bodies" while they were "unconscious" and *claim* to recall seeing, not just the events occurring below, but, as in one case, dirt on top of a light fixture (Ring, 1980): "It was filthy. And I remember thinking, 'Got to tell the nurses about that."

Did the above surgical patient or Lisa or Army Specialist Bayne really float above and observe their bodies and the events taking place below? Or did they merely transpose what they heard (e.g., conversations, noises, etc.) and then visualize, imagine, or hallucinate an accompanying and plausible scenario? This seems likely, even in regard to the "filthy" light fixture. On the other hand, not all those who have an "out-of-body" experience hear conversations, voices, or even sounds. Rather, they may be enveloped in silence.

> I was struck from behind. ...That's the last thing I remember until I was above the whole scene viewing the accident. I was very detached. This was the amazing thing about it to me...I could see my shoe, which was crushed under the car, and I thought: Oh, no. My new dress is ruined...I don't remember hearing anything. I don't remember anybody saying anything. I was just viewing things...like I floated up there...(Sabom, 1982:90).

Fear and Out-of-Body Experiences

The prospect of being terribly injured or killed in an auto accident or fire fight between opposing troops, or even dying during the course of surgery, is often accompanied by feelings of extreme fear. It is also not uncommon for individuals who experience feelings of terror to report perceptual and hallucinogenic experiences, including dissociation, depersonalization, and the splitting off of ego functions such that they feel as if they have separated from their bodies and floated away, or were on the ceiling looking down (Campbell, 1988; Courtois, 1988; Grinker and Spiegel,

1945; James, 1958; Neihardt and Black Elk, 1932; Noyes and Kletti, 1977; Parson, 1988; Southard, 1919; Terr, 1990). Consider the following:

> "The next thing I knew I wasn't in the truck anymore; I was looking down from 50 to 100 feet in the air." "I had a clear image of myself...as though watching it on a television screen." "I had a sensation of floating. It was almost like stepping out of reality. I seemed to step out of this world" (Noyes and Kletti, 1977).

Or as a close friend described his experience:

> I was shooting down the freeway doing about 100 or more in my Mustang when a Firebird suddenly cut me off. As I switched lanes to avoid him, he also switched lanes, at which point I hit the brakes and began to lose control. The Mustang began to slide and spin, and for a brief moment I felt real terror as I realized I was probably going to be killed. I didn't have a chance to think about being afraid, though, as I was trying to control the Mustang and avoid turning over, or hitting any of the surrounding cars or the guard rail. Instead, I became exceedingly calm, time seemed to slow down, and then I suddenly realized that part of my mind was a few feet outside the car looking all around; zooming above it and then beside it and behind it and in front of it, looking at and analyzing the respective positions of my spinning Mustang and the cars surrounding me. Simultaneously, I was inside trying to steer and control it in accordance with the multiple perspectives I was suddenly given by that part of my mind that was outside. It was like my mind split and one consciousness was inside the car, while the other was zooming all around outside and giving me visual feedback that enabled me to avoid hitting anyone or destroying my Mustang.

Limbic Hyperactivation and Astral Projection

As detailed in Chapters 5, 6, and 16, feelings of fear and terror are mediated by the amygdala, whereas the capacity to cognitively map, or visualize, one's position and the position of other objects and individuals in visual-space is dependent on the hippocampus (Nadel, 1991; O'Keefe, 1976; Wilson and McNaughton, 1993). The hippocampus contains "place" neurons that are able to encode one's position and movement in space.

The hippocampus, therefore, can create a cognitive map of an individual's environment and his movements within it. Presumably, it is via the hippocampus that individuals can visualize themselves as if looking at their bodies from afar, and can remember, and thus *see* themselves, engaged in certain actions, as if one were an outside witness. However, under conditions of hyperactivation (such as in response to extreme fear) it appears that the hippocampus may create a visual hallucination of that "cognitive map" such that the individual may "experience" himself as outside his body, observing all that is occurring.

In fact, it has been repeatedly demonstrated that hyperactivation or electrical stimulation of the amygdala-hippocampus-temporal lobe can cause some individuals to report they have left their bodies and are hovering upon the ceiling staring down (Daly, 1958; Jackson and Stewart, 1899; Penfield, 1952; Penfield and Perot, 1963; Williams, 1956). That is, their ego and sense of personal identity appears to split off from their body, such that they may feel as if they are two different people, one watching, the other being observed. As described by Penfield (1952), "It was as though the patient were attending a familiar play and was both the actor and audience."

Limbic System Hyperactivation, Hallucinations, and Near Death

Presumably, abnormal activation due to extreme fear or direct electrical stimulation induces an individual to think he is seeing himself from afar because the hippocampus is transposing and "hallucinating" one's image, similar to what occurs during normal remembering.

As noted, however, many patients who are diagnosed as "clinically dead" and then return to "life" report that after leaving their bodies they enter a dark tunnel and are then enveloped in a soothing radiant light. The same is reported in the Egyptian and Tibetan Books of the Dead.

Presumably, in that the hippocampus, amygdala, and inferior temporal lobe receive direct and indirect visual input and contain neurons sensitive to the fovea and upper visual fields, hyperactivation of this region also induces the sensation of seeing a radiant light. These brain regions would also account for the "life review" and hallucinations of seeing dead relatives, and so on, that are commonly reported by those who have "died." Similarly, the massive release of opiates (due to physical trauma leading to "death") would account for the immediate loss of fear and the experience of tranquility and joy.

Thus, the hyperactivation of these limbic nuclei would explain why those who have near death experiences report feelings of peace, rapture, and joy as they were "bathed by the light" and stood in the all-knowing-presence of God or other divine beings, including friends and relatives who had previously passed away. Indeed, these exact same feelings and related hallucinations can be induced by electrically stimulating the inferior temporal lobe and amygdala-hippocampal complex.

Out-of-Body, Heavenly, and "Otherworldly" Limbic Experiences

Penfield and Perot (1963) describe several patients who, during temporal lobe seizures, claimed they could see themselves in different situations. One woman stated that "it was though I were two persons, one watching, and the other having this happen to me," and that it was she who was doing the watching, as if she were completely separated from her body.

One patient had a sensation of being outside her body, watching and observing herself from the outside. Another neurosurgery patient alleged that, while outside her body, she was also overcome by feelings of euphoria and eternal harmony.

Other patients claim to have quite pleasant auras and describe feelings such as ela-

tion, security, eternal harmony, immense joy, paradisiacal happiness, euphoria, and completeness. Between 0.5 and 20% of such patients report these feelings (Daly, 1958; Williams, 1956). A patient of Williams (1956) claimed that his attacks began with a "sudden feeling of extreme well being involving all my senses. I see a curtain of beautiful colors before my eyes and experience a pleasant but indescribable taste in my mouth. Objects feel pleasurably warm, the room assumes vast proportions, and I feel as if in another world."

A patient described by Daly (1958) claimed his temporal lobe seizure felt like "a sunny day when your friends are all around you." He then felt dissociated from his body, as if he were looking down upon himself and watching his actions.

Williams (1956) describes a patient who during an aura reported that she experienced a feeling of being lifted up out of her body, coupled with a very pleasant sensation of elation and the feeling that she was "just about to find out knowledge no one else shares, something to do with the link between life and death."

Subirana and Oller-Daurelia (1953) described two patients who experienced ecstatic feelings of either "extraordinary beatitude" or of paradise, as if they had gone to heaven. Their fantastic feelings also lasted for hours.

Other patients suffering from temporal lobe seizures have noted that feelings and perceptions suddenly became "crystal clear," or that they had a feeling of clairvoyance, of having the truth revealed to them, or of having achieved a sense of greater awareness such that sounds, smells, and visual objects seemed to have a greater meaning and sensibility. Similar claims are made by those who have "died" and returned to tell the tale.

Embraced By the Light: Temporal Lobe Epilepsy, "Death," and Astral Projection

Some individuals (and their followers) claim to be able to voluntarily leave their bodies (e.g., Monroe, 1994), this includes any number of "mystics" and New Age spiritualists, as well as some priests, prophets, and shamans. Indeed, Monroe (1994) founded an Institute to study this phenomenon, and claims that others can learn this technique. Monroe, however, notes that when he had his first out-of-body experience he felt extremely frightened.

As noted, a few individuals suffering from (or at least demonstrating signs suggestive of) temporal lobe epilepsy also report "out-of-body experiences." One woman I evaluated claimed she not only could float on the ceiling, but in some instances, she could float outside and could see everything that was going on, including, on one occasion, a friend who was coming up the walkway. She also reported that by having a certain thought, she could propel herself to other locales, including the homes of her neighbors.

Some of those who have "after death" experiences (as well as those who claim to voluntarily leave their bodies, e.g., Monroe, 1994) have made similar claims. For example, Betty J. Eadie reports in her 1992 book, "Embraced by the Light," that after dying and then communing with three "ancient" men who appeared at her side and who "glowed," she suddenly thought of her husband and children whom she wanted to visit. "I began to look for an exit" and discovered that "my spiritual body could move through anything. ...My trip home was a blur. I began moving at tremendous speed...and I was aware of trees rushing below me. I just thought of home and knew I was going there. ...I saw my husband sitting in his favorite armchair reading the newspaper. I saw my children running up and down the stairs. ...I was drawn back to the hospital, but I don't remember the trip; it seemed to happen instantaneously" (pp. 33–35).

Compare Eadie's description with that of Black Elk (Neihardt and Black Elk, 1932), a Lakota Sioux Medicine Man and spiritual leader (born in 1863). During a visit to England (he was part of Buffalo Bill's Wild West Show), he suddenly fell out of his chair as if dead, and then experienced himself being lifted up. In fact, his companions thought he had died.

According to Black Elk: "Far down below I could see houses and towns and green land and streams. ...I was very happy now. I kept on going very fast...then I was right over Pine Ridge. I looked down (and) saw my father's and mother's teepee. They went outside, and she was cooking. ...My mother looked up, and I felt sure she saw me...then I started back, going very fast...then I was lying on my back in bed and the girl and her father and a doctor were looking at me in a queer way...I had been dead three days (they told him)...and they were getting ready to buy my coffin" (pp. 226–228).

This was not Black Elk's first out-of-body experience, however. In this regard, Black Elk (in my opinion) also demonstrated numerous behaviors and symptoms *suggestive* of temporal lobe epilepsy. Beginning even in childhood Black Elk repeatedly experienced "queer feelings" and heard voices, had visions, and suffered numerous instances of sudden and terrible fear and depression accompanied by weeping, as well as trance states in which he would fall to the ground as if dead.

Black Elk also had other visions similar to persons reporting "life-after-death" experiences, including the following incident that occurred during one of his trance and out-of-body states: "Twelve men were coming towards me, and they said, 'Our father, the two-legged chief, you shall see. ...' There was a man standing. He was not Wasichu (white) and he was not an Indian. While I was staring at him his body began to change and became very beautiful with all colors of light, and around him there was light. ..." (p. 245).

Similarly, Ms. Eadie (like many others who have experienced "life after death") came upon a man standing in the light, which "radiated all around him. As I got closer the light became brilliant...I saw that the light was golden, as if his whole body had a golden halo around it, and I could see that the golden halo burst out from around him and spread into a brilliant, magnificent whiteness that extended out for some distances" (pp. 40-41).

That so many people, regardless of culture or antiquity, have similar experiences (or hallucinations) after "dying" (e.g., the Tibetan and Egyptian Books of the Dead) or leaving their bodies is presumably due to all possessing a limbic system and temporal lobe that are organized similarly. The fact that, although ostensibly similar, many of these experiences are also colored by one's cultural background, can in turn be explained by differences in experience and cultural expectations. As explained in the Tibetan Book of the Dead: "It is quite sufficient for you (the deceased) to know that these apparitions are [the reflections of] your own thought forms."

DEATH OF THE TEMPORAL LOBE

Presumably, conditions involving extreme fear and/or traumatic injury, and some cases of temporal lobe epilepsy, result in hyperactivation of the amygdala and hippocampus, which in turn will begin to hallucinate and/or trigger a vision of brilliant light, as well as secrete opiate-like neurotransmitters that induce a state of euphoria and, thus, eternal peace and harmony. Given that similar experiences are reported by those who have been declared "clinically dead" also raises the possibility that the hippocampus and amygdala may be the first areas of the brain to be effected by approaching death, as well as one of the last regions of the brain to actually die. That is, as one approaches death and even after medical death, the amygdala and hippocampus may continue to function briefly and not only become hyperactivated, but produce a feeling of eternal peace and tranquility and a hallucination of floating outside the body and of meeting relatives and other religious figures, like a dream.

On the other hand (and as will be discussed in more detail in a later section of this chapter), it is curious that so many individuals have basically a very similar

"dream" and only under conditions suggestive of death. Moreover, it is exceedingly difficult to reconcile these experiences with the Darwinian notion of evolution. That is, what is the "evolutionary" adaptive significance of so many members of the human race having a dream of the "Hereafter" after they die?

THE FRONTAL LOBE, LIMBIC SYSTEM, MURDER, AND RELIGIOUS-SEXUAL EXPERIENCE

When the amygdala and hypothalamus act in a highly coordinated or lock step manner, it is usually in reaction to an exceedingly important emotional stimulus (e.g., fear), or in response to a specific limbic need, such as hunger, thirst, rage, or sexual desire (Joseph, 1982, 1990a, 1992a; MacLean, 1969, 1990). For example, in response to hypothalamically monitored needs (hunger, sexual desire), the amygdala may scan the environment until it determines that a particular food item or person has the necessary attributes (Gloor, 1960; Joseph 1982, 1990a, 1992a; Ursin and Kaada, 1960). In response to urgent hypothalamic desires, the amygdala might even assign sexual attributes to an individual that normally might not be viewed as sexually enticing.

It is also through hypothalamic and amygdala activity that a particular item or object (e.g., a banana) might be viewed as both a food item and a sexual object. Or conversely, certain individuals may be viewed as sexually exciting as well as aversive and hateful (e.g., one's husband or wife). Indeed, because the hypothalamus and amygdala are so concerned with sex, rage, fear, and hunger, not only may these attributes be assigned to one individual, animal, or object simultaneously (e.g., fear of the beast that one is going to enjoy killing and eating; hunger, and aversion regarding a high-caloric treat; hatred for a loved one) but they may be combined so as to give rise to exceedingly intense, albeit abstract, emotional states, e.g., religious awe. In this regard, feelings of religious awe may be based on fear (d'Aquili and Newberg, 1993), rage, extreme, hunger, or sexual arousal (see below). In fact, the God Yahweh depends on fear to reveal his presence and power.

"The beginning of wisdom is the fear of the Lord" (Proverbs 1:7, 9:10, 15:33). "And now, Israel, what does the Lord your god require of you, but to fear the Lord your god" (Deuteronomy 10:12). "God has come…in order that the fear of Him may be ever with you so that you do not go astray" (Exodus 20:17).

However, in addition to these nuclei (including the temporal lobe and hippocampus), d'Aquili and Newberg (1993) point out that the right frontal lobe also plays a significant role in the generation of mystical experience. It is, thus, noteworthy that the right frontal lobe can pray, swear, and curse God even when the (speaking) left cerebral hemisphere has been severely damaged and the patient is aphasic (Joseph 1982, 1986a, 1988a, 1993).

The right frontal and temporal lobes, hypothalamus, and amygdala also interact in regard to sexual arousal (Freemon and Nevis, 1969; Joseph, 1986a, 1988a, 1992a; MacLean, 1969, 1990; Remillard et al., 1983; Robinson and Mishkin, 1968; Spencer et al., 1983). This is a very important relationship, and in part explains why (although there are exceptions) religions tend to be quite sexual (see below) and/or exceedingly concerned with sexual mores and related activity. As is well known, fertility, female sexuality, pregnancy, and matters pertaining to birth control and abortion are of extreme concern to most modern as well as ancient religions (Campbell, 1988; Frazier, 1955; Parrinder, 1980; Smart, 1969)

However, the limbic system as well as the frontal and temporal lobes are also highly concerned with acting on or inhibiting aggression and murderous rage reactions, which also arise in the limbic system (Joseph 1986a, 1988a, 1990a, 1992a). This may also explain why many religious sects are so "righteously" belligerent and hateful and have employed torture and human or animal sacrifice, and sanction, if not encourage, the murder of nonbelievers: what could be referred to as limbic-religious blood lust. "Shed man's blood, by man be your blood shed" (Genesis 9:6).

For example, the God, Yahweh, repeatedly required that the Israelites undergo a blood ritual of submission (e.g., Exodus 24:1–14), and in fact proscribed a ritual of incredible bloodiness for the investiture of his priests (Exodus 29:1–46) as well as the slaughter and sacrifice of living creatures whose blood is splashed on his altar, and on his priests.

For ancient hunters, aggression and the murder of prey was a way of life, and a life-style that required hunting magic and related religious rituals. Religion and murder, like religion and sex, are linked by the limbic system and evolved accordingly. Consequently, when in the throes of religious excitement, torture and murder may even receive the blessing or be actively encouraged by one's God. Throughout history, many of the patriarchal Gods have been aggressive, jealous, conquering, angry, and warlike; i.e., "the Lord of hosts."

Indeed, these warrior gods, including Yahweh, were prone not just to extreme violent rages, but He would threaten and engage in the slaughter of enemies and believers alike, without mercy.

"Terror, and the pit, and the snare are upon you, O inhabitant of earth" (Isaiah 24:17). "And as the Lord took delight in doing you good and multiplying you, so the Lord will take delight in bringing ruin upon you and destroying you" (Deuteronomy 26:63). "The Lord will bring a nation against you from afar, from the end of the earth, which will swoop down like the eagle…a ruthless nation, that will show the old no regard and the young no mercy" (Deuteronomy 28:47–50). "It shall devour the offspring…you shall eat your own issue, the flesh of your sons and daughters…until He has wiped you out…leaving you nothing…until it has brought ruin unto you…"(Deuteronomy 28:50-55). These are emotions of the primitive hypothalamus and limbic system. Nonbelievers and those who sinned were not to be forgiven, but destroyed by the limbic God of war (the Lord of hosts).

> In the Name of God...by the Troops shall the unbelievers be driven towards Hell, until when they reach it, its gates shall be opened...for just is the sentence of punishment on the unbelievers. ...
>
> Koran, XXXIX

> Behold I send an angel before thee, to keep thee in the way. Beware of him and obey his voice, for I will be an enemy unto thine enemies, and an adversary unto thine adversaries, and I will cut them off...I will send my fear before thee, and will destroy all the people to whom thou shalt come...and I will drive them out from before thee, until thou be increased and inherit the land."
>
> Exodus 23:20–30

> When you approach a town, you shall lay seizure to it, and when the Lord your god delivers it into your hand, you shall put all its males to the sword. You may, however, take as your booty the women, the children, the livestock, and everything in the town—all its spoils—and enjoy the spoil of your enemy which the Lord your god gives you. ...In the towns which the Lord your god is giving you as a heritage, you shall not let a soul remain alive.
>
> Deuteronomy 20:12–16

> When Israel had killed all the inhabitants of Ai...and all of them, to the last man had fallen by the sword, all the Israelites turned back to Ai and put it to the sword...until all the inhabitants of Ai had been exterminated...and the king of Ai was impaled on a stake and it was left lying at the entrance to the city gate.
>
> Deuteronomy 8:24–29

Despite the commandment, "Thou shall not kill," the ancient Hebrews received special permission from God to murder non-Jews and Jewish nonbelievers, including women and children whom they slaughtered without mercy (e.g., Numbers 31:15–18; Numbers 34:50–53).

> And they warred...as the Lord commanded and slew all the males. And they slew the Kings...and they took all the women and their little ones...and they burnt all their cities wherein they dwelt, and all their goodly castles with fire. ... And Moses was wroth...and said unto them. Have ye saved all the women and the little ones alive? Now therefore kill every male among the little ones, and kill every woman that hath known man by lying with him. But all the women and female children that have not known a man, keep alive for yourselves.
>
> Numbers 31

Moreover, even the murder of innocent babies was encouraged. For example, throughout the Old Testament, it was a Jewish tradition to kill and slaughter not only non-Jewish males, in general, but first-born Jewish sons (a custom until the time of Moses, e.g., Bergmann, 1992).

"A blessing on him who seizes your babies and dashes them against rocks" (Psalm 137:9). "I polluted them with their own offerings, making them sacrifice all their first-born, which was to punish them, so that they would learn that I am Yahweh" (Ezekiel 20:25–36. See also Ezekiel 22:28–29). "This very day you defile yourselves in the presentation of your gifts by making your children pass through the fire of all your fetishes" (Ezekiel 20:31).

It was upon these images of the murderous warrior God, the Lord of hosts (armies), that Pope Urban II proclaimed that war, for the sake of God, was holy. In fact, the Catholic Popes instigated numerous Crusades and inquisitions, acting under the religious (or satanic) delusion that they were serving Jesus Christ, when in fact they were following the dictates of Peter, the "rock" of the church, a man who denied and betrayed Christ three times, and a man whom Jesus repeatedly castigated as "Satan," (e.g., Matthew 16:23; Matthew 17:24–25; Mark 8:33: "But he turned and said unto Peter, "Get thee behind me, Satan, thou art an offence unto me; for thou savourest not the things that be of God, but those of men.) Of course, in some respects, the Catholic Popes were also acting in accordance with the dictates of the "Old Testament" "Lord of hosts." In consequence, hundreds of thousands of Moslems and Jews, and women and children, were sexually tortured, slaughtered, spitted, and roasted alive, and their cities and villages pillaged and set ablaze. All in the name of God and Christ. However, so intense was their limbic blood lust that even Christians were murdered.

For example, in the 13th Century an army of some 30,000 Christian knights and Crusaders descended into southern France and attacked the town of Beziers in search of heretics. Over 13,000 Christians flocked to the churches for protection. However, when the Bishop, one of the Pope's representatives, was informed that the army was unable to distinguish between true believers and heretics, he replied, "Kill them all. God will recognize his own."

However, in order to recruit those willing to engage in these sadistic undertakings, the Pope had to appeal to murderers, rapists, molesters of children, and those who enjoyed the prolonged torture of their victims. "You oppressors of orphans, you robbers of widows, you homicides, you blasphemers, you plunderers of others' rights. ...If you want to take counsel for your souls you must go forward boldly as knights of Christ...": so proclaimed the Pope who offered "indulgences" and forgiveness to all those who would commit blasphemies in the name of God and Christ.

The limbic system is also concerned with sex. Unfortunately, an abnormal limbic system may abnormally link sex with murder, and among men, the sexual murder or torture of women.

Hence, when many of the men had marched off to Catholic Crusades or had been killed, the women were left unprotected; it was suspected that many had begun to practice their own religion and worship their own female Gods. In consequence, the Popes and the Catholic Church proclaimed them witches and declared war against women. Hence, in 1252, Pope Innocent IV issued the Ad Exitrpanda, which authorized the execution of heretics (e.g., wealthy landowners) and the seizure of their goods, and the prolonged sexual torture of women who were beautiful and wealthy, or old, ugly, and eccentric and/or who gathered in groups to converse and possibly worship pagan goddesses.

Indeed, when the Papal fathers and the Dominicans Heinrich Kramer and Johann Sprenger issued the infamous Papal Bull and the Malleus Maleficarum (witch's hammer), a blood lust regarding "woman the witch, healer and sorceress" was unleashed, and hundreds of thousands were burned and/or sexually tortured to death (Achterberg, 1991; Gies and Gies, 1978; Lederer, 1968). "For she is a liar by nature, so in her speech she stings while she delights us...for her voice is like the song of the Sirens, who with their sweet melody entice the passerby's and kill them..." (Malleus Maleficarum).

As noted, because many of the men (the Crusaders) had been killed or were serving in the army of the Catholic God, the women were often left unprotected. Sometimes whole villages were destroyed, or all the women in a given area were rounded up by the Catholic authorities. These females, particularly those who were exceedingly attractive or ugly, were then hideously tortured and then slaughtered by burning, boiling in oil, crushing, or subjected to whatever device the religious authorities felt appropriate or which suited their sick minds. Indeed, in Germany huge ovens were constructed for the purpose of mass female murder (Achterberg, 1991; Lederer, 1968).

However, it was not just beautiful females, for their numbers are limited, but those who were old, eccentric, or childless, and particularly women who owned pets, such as cats. Indeed, the cats would be tortured and murdered alongside the women. The "black plague" in fact was, in part, due to the denunciation and killing of cats, coupled with the sanctification of rats and mice (the proverbial church mouse), by the Catholic authorities.

As is well known, the Spanish and Catholic missionaries, acting at the behest of the Catholic Popes (and their Spanish/Catholic Sovereigns), continued these Satanic practices once they invaded the Americas during the 1500s and up through the 19th century. As the Catholic Dominican Bishop Bartolom de Las Casas reported to the Pope: the Aztec and Indian natives were hung and burnt alive "in groups of 13...thus honoring our Savior and the 12 apostles." (See Figure 8.18.)

Of course, the Aztecs did not practice a benign form of worship, for they tore the beating hearts from their victims in order to please their God (Carrasco, 1990) and they killed thousands if not hundreds of thousands in so doing. Similarly, many Indian tribes of the Mississippi valley practiced human sacrifice, as did the ancient Hebrews.

What is the origin of these sadistic religious practices? The human limbic system.

Nevertheless, despite the "infallible" proclamations of the Catholic Church, mass murder, torture, rape, pedophilia, and the castration of young boys were probably not

Figure 8.18   "A blessing on him who seizes your babies and dashes them against rocks" (Psalm 137:9). 15th century woodcut by De la Fray. According to Catholic Dominican Bishop Bartolom de Las Casas, the Indian natives were hung and burnt alive "in groups of 13... thus honoring our Savior and the 12 apostles." Because some of the various victims managed to live throughout the night, the Dominican priests ordered that sticks be shoved down their throats so the soldiers and priests would not be kept awake at night by their cries and moans.

what Jesus Christ had in mind when he preached his gospel, despite his own depressed, irritable, and labile character: "But love ye your enemies and do good and lend, hoping for nothing and your reward shall be great. Be ye merciful...judge not, and ye shall not be judged: condemn not and ye shall not be condemned: forgive and ye shall be forgiven" (Luke 7:35–37); "For the Son of man is not come to destroy men's lives, but to save them" (Luke 10:56; however, see Matthew 10:16 vs 34–35).

Figure 8.19 Illustration from Codex Magliabechiano, 16th century. The Aztecs regularly practiced human sacrifice by cutting out the hearts of captured warriors in honor of their god. A victim lies dead at the foot of the temple steps, whereas the warrior at top has his chest cut open by the temple priest, which allows his soul to ascend skyward, leaving behind a bloody trail.

Modern Religious Murderers

Murder of the innocents and the slaughter of infidels and nonbelievers are not antiquated religious customs. Cults and religious groups regularly arise in various lands and cultures and frequently indulge in similar practices (e.g., Jim Jones and "Jonestown" mass suicide, David Koresh and the fiery death of him and his followers in Waco, Texas).

Consider, the Japanese religious cult "Aum." Their leader, Shoko Asahara, and many top cult members were arrested and charged with murder in June of 1995 for releasing the nerve gas Sarin in five subway cars during rush hour, injuring over 5500 Japanese commuters (New York Times, 6/7/95).

Similarly, although the Islamic, Christian, and (modern) Jewish religions forbid it, many modern-day Middle Eastern and African Islamic fundamentalists, like some of their Jewish and Christian counterparts, regularly preach murder and hatred. Indeed, it has been reported that "militant rabbis in Israel had encouraged and condoned the assassination of Israel's Prime Minister, Yitzhak Rabin, and had issued a "pursuer's decree," which in effect morally required that he be killed (New York Times 11.11.95). And he was murdered by a student of religion, Yigal Amir, who claimed he acted upon God's instructions.

What is the source of these religious murderous feelings? The limbic system and the same cluster of nuclei that subserve sexuality and spirituality. It is the limbic system that enables human beings to respond with irrational and murderous blood lust in the name of God and religion.

Sex, God, and Religion

Sexuality is a major concern of most major religions (Lederer, 1968; Parrinder, 1980; Smart, 1969) as well as the limbic system. In fact, almost all major religions and their gods either act to promote sexuality ("be fruitful and multiply"), or to suppress it. This should not be entirely surprising, for religions are very sexual and many were originally concerned with the fertility of the fields and the abundance of prey (Campbell, 1988; Frazier, 1950; Harris, 1993; Kuhn, 1955; Malinowski, 1954; Parrinder, 1980; Prideux, 1973). Religious rituals evolved accordingly.

Many modern mystical and religious practices also involve the ritual control over sex and food. This includes many American Indian, Christian, Hebrew, and Moslem sects (Campbell, 1988; Parrinder, 1980; Smart, 1969). Thus, the commandment "thou shalt not. ..." These are limbic taboos, as eating and sexuality (like murder and violence) are under limbic control.

Many limbic taboos, however, promote survival, for example, by proscribing the eating of poisonous plants or unclean animals. Similarly, by forbidding anal or indiscriminate sex one was spared the wrath of God and whatever plagues he might send in the form of venereal disease or viruses. This is presumably what became of Sodom and Gomorrah where the anal-sex-crazed mobs attempted to sodomize even the Lord God himself (Genesis 19). Possibly, they all died due to a sexually transmitted plague.

Sex and food (along with fear and aggression) are probably the most powerful of all limbic emotions and motivators, and when harnessed or stimulated can completely overwhelm or control the brain and lead to limbic hyperactivation coupled with religious or spiritual sensations, or, at a minimum, complex dreams or hallucinations (see Chapter 5). Hence, hungry men, women, and infants will dream of food, and those who are sexually aroused (but are unfamiliar with the theories of Freud, 1900), will dream of sex. However, a parched and starving man will not just dream, he will hallucinate food and water and will attempt to slake his desires by consuming a hallucination.

Given that early (as well as modern) human populations were often concerned with obtaining food (as well as sex partners), many of their earliest religious beliefs and rituals were therefore concerned with increasing the abundance of game animals as well as preserving their own progeny (Armstrong, 1994; Campbell, 1988; Frazier, 1950; Harris, 1993; Kuhn, 1955; Parrinder, 1980; Prideux, 1973). As noted, many an ancient Upper Paleolithic cave was decorated with fertility and sex symbols, including pregnant women (Venus figures) and animals (Bandi, 1961; Joseph, 1993; Kuhn, 1955; Leroi-Gurhan, 1964), whereas Egyptian pyramids contain numerous paintings of food.

Thus, given our ancient hunter-gatherer (and then later, farming) heritage, many religions, both ancient and relatively modern, are highly concerned with fertility and food, or tend to be very sexual and limbic in orientation if not origin. This is also why there have always been gods who are associated with eating and drinking, especially alcohol (Campbell, 1988; Frazier, 1950; James, 1958; Parrinder, 1980; Smart, 1969). This also includes, for example, Osiris, and especially Dyonisus who was, among other things, a sex-crazed dancing god of the vine. (In fact, one of the first miracles performed by Jesus involved making wine from water.)

Even among the ancient religions of India and China, the sexual activity of the Gods and the promotion of similar sexual activities among the believers were widespread religious practices and beliefs (Campbell, 1988; Parrinder, 1980). For example, the ancient Vedas were greatly concerned, not only with the worship of various nature gods, but with the rituals of sexual union. Ancient Indian religious texts are filled with love charms and instructions as to how to win the love of a man or woman, or to protect against demons. In fact, temple prostitutes were quite common throughout India and the Middle East (as well as in Rome and Greece), and some temples employed so many girls that they were like giant brothel emporiums (Parrinder, 1980). As noted, sexuality and desire (like religious feeling) are directly mediated by the amygdala and hypothalamus.

In fact, sexual intercourse became a religious ritual among Hindus and Buddhists who practiced "tantra." Those who practiced tantra were inspired by visions of cosmic sex and were highly concerned with sexual energy. It was through tantra that one might be confronted with the cosmic mystery of creation as exemplified by another deity, Shakti, the divine mother. However, restrictions on where one could have sexual intercourse (not in public) and certain types of sexual acts, such as oral sex, were prohibited as well as sex with strange women or those of a lower caste (Parrinder, 1980). Nevertheless, the joys of sex were continually emphasized and embraced. Hence, the Kama Sutra, the "love text."

On the other hand, it was believed by some ancient Far Eastern sects that in order to gain power, one had to break taboos and, for example, have sex with women while they were menstruating and/or engage in sexual orgies. This was also a form of tantra, referred to as "left handed tantra." Those who followed the way of the left-handed tantra claimed that passion was nirvana and that adepts should cultivate all sexual pleasures (Parrinder, 1980). Moreover, both male and female deities, usually in the act of having sex, were worshiped.

Ancient Chinese and Taoist religions are also quite sexual (Parrinder, 1980). These beliefs are exemplified by the concepts of Yin and Yang, which appeared about 3000 years ago and which represented the male and female principles of the universe. In this regard, sexual intercourse was viewed as a symbolic union of the earth and heaven, which, during rainstorms, were believed to mate. It was in this manner that man and woman achieved harmony by following the example of the gods.

However, around 3000 years ago, at about the same time that the Judaic religion became more dominant in the Middle East, there occurred over the following thousand years a tremendous change in sexual thought, which continued to grow and prosper, enveloping the Roman Empire, and which eventually paralleled and coincided with the development of Christianity and Islam.

The Lord of the Gods (Genesis) gradually became the Lord God, the only God, and there is no hint of sexual duality in his personage. Moreover, the God of Abraham, and thus the God of the Hebrews, Christians, and Moslems, was not in any manner a sexual being, and in many ways, was opposed to sexual pleasure and required a sexual sacrifice. Indeed, as part of his covenant with Abraham and the Jews, it was ordered that every male child would suffer the amputation of the tip of his penis (which is densely innervated by fibers that yield intense sexual pleasure) and required a sexual sacrifice: "And ye shall circumcise the flesh of your foreskin; and it shall be a token of the covenant betwixt me and you" (Genesis 17:10–11). Nor does this God lust after or engage in sexual relations with women, which had been a common godlike (and "sons of God") behavior in the past.

Even Mary, the mother of the prophet Jesus Christ, became pregnant only indirectly, at least in the two stories that state Joseph was not the father of Jesus (e.g., Luke 1:26-35). Thus, after the angel Gabriel "came in unto her...and said unto her, And behold thou shall conceive in they womb, and bring forth a son...," Mary asked the angel, "How shall this be, seeing I know not a man? And the angel answered and said unto her, The Holy Ghost shall come upon thee, and the power of the Highest shall overshadow thee. ..."

Nevertheless, although this volatile and masculine seeming God was asexual, sexual behavior was of tremendous concern to "Him," for He commands sexual moral obedience and repeatedly tells his people, starting with Adam and Eve, "Be fruitful and multiply".

Why the concern regarding sex, pro or con, in religious thought? As noted, the relationship between fertility, the abundance of prey, and the evolution of hunting magic accounts in part for religious sexuality. In addition, sex, like religious experience or the ability to derive pleasure from eating and drinking, is mediated by the limbic system, i.e., the hypothalamus, amygdala, and temporal and frontal lobes (e.g., Freemon and Nevis, 1969; Joseph 1992a; MacLean, 1969, 1990; Remillard et al., 1983; Robinson and Mishkin, 1968).

Hence, in many ways violence (hunting) and religious-sexual beliefs are highly interrelated and appear to be derived from, or at least strongly associated with, the most ancient regions of the brain: the limbic system. As noted, activation of these structures (particularly the hypothalamus and amygdala) can give rise to rage reactions, sexual posturing, erection, ejaculation, orgasm, and hypersexuality, as well as hyper-religiousness, even in people who were not all that sexual or religious prior to their "conversion" or being "born again."

ETIOLOGICAL AND DIAGNOSTIC SPECULATIONS: SEXUALITY, RELIGIOUS EXPERIENCE, AND TEMPORAL LOBE HYPERACTIVATION

A not uncommon characteristic of high levels of limbic system and inferior temporal lobe activity are changes in sexuality as well as a deepening of religious fervor (Bear, 1979; Slater and Beard, 1963; Trimble, 1991; Taylor, 1972, 1975). It is noteworthy that not just modern-day evangelists, but many ancient religious leaders, including Abraham and Muhammad, tended to be highly sexual and partook of many partners (e.g., St. Augustine of Hippo: "Give me chastity, 'o lord, but just not yet"), or they shared their wives (Abraham), or they married women who were harlots (e.g., Hosea). Many also displayed evidence of the Klüver-Bucy syndrome, such as eating dung (Ezekiel), as well as temporal lobe limbic hyperactivation and epilepsy.

Muhammad, God's messenger, was apparently dyslexic and agraphic, and was known to lose consciousness and enter into trance states (Armstrong, 1994; Lings, 1983). In fact, he had his first truly spiritual-religious conversion when, as the story goes, he was torn from his sleep by the archangel Gabriel. Gabriel enveloped him in a terrifying embrace, so overpowering that Muhammad's breath was squeezed from his lungs. After squeezing and suffocating him repeatedly, Gabriel ordered Muhammad to speak the word of God, i.e., the qur'an. This was the first of many such episodes with the archangel Gabriel, who sometimes appeared to Muhammad in a titanic, kaleidoscopic, panoramic form.

In accordance with the voice of God or his angels, Muhammad not only spoke but he began reciting and chanting various themes of God in a random order over the course of the following 20 years, an experiencing he found quite painful and wrenching (Armstrong, 1994; Lings, 1983). However, in addition to his religious zest, Muhammad was reported to have the sexual prowess of 40 men, and to have bedded at least nine wives and numerous concubines (Lings, 1983). Women, in fact, tended to be his first and earliest converts (Armstrong, 1994).

Although Islam (which means "peace" or "surrender") is an exceedingly tolerant religion, and Muhammad was basically a kind and considerate man, he was also known to fly into extreme rages and to kill (or at least order killed) wealthy infidels and merchants and those who opposed him. These behaviors, when coupled with his increased sexuality, heightened religious fervor, trance states, mood swings, and possible auditory and visual hallucinations of a titanic angel, certainly point to the limbic system and inferior temporal lobe as the possible neurological foundation for these experiences. Indeed, Muhammad also suffered from horrible depressions, and on one occasion sought to throw himself from a cliff, only to be stopped by the archangel Gabriel.

Abraham (like his brother Lot), the patriarch of Jewish, Christian, and the Moslem religions, also experienced what could be considered visual as well as auditory hallucinations. Abraham also engaged in some unusual sex practices. For example, after Abraham left Babylon and before he arrived in Egypt, he told his wife Sara to pretend they were brother and sister because she was so beautiful and other men might wish to have sex with her.

"And it came to pass, that when Abram was come into Egypt, the Egyptians beheld the woman that she was very fair...and the woman was taken into Pharaoh's house." However, the Pharaoh thought Abraham and Sara were brother and sister, for he gave to Abraham, in order to pay for her sexual services, "sheep, and oxen, and asses and servants and camels" (Genesis 12:14–16). However, when Pharaoh found out she was married, he was so disgusted he threw them both out of Egypt (Genesis 12:18–20). But not before demanding an explanation. Abraham replied that he lied because he was "afraid."

However, when Abraham arrived at Gerar he repeated the lie and informed the King of the city that Sara "is my sister." Again, he offered her to those who wished to partake of her charms. "And Abimelech King of Gerar sent and took Sara. But God came to Abimelech in a dream by night and said to him, Behold thou art but a dead man, for the woman which thou has taken; she is a man's wife" (Genesis 20:2–3). The king was horrified and shocked that Abraham had behaved in so odd a manner.

Abraham, of course, also had sex with other women, banished his firstborn son and the handmaiden who bore him, and also attempted to murder his second son, Isaac. However, Abraham believed he was following the orders of God, for he heard a voice that instructed him to "Take now thy son, thine only son Isaac, whom thou lovest, and offer him for a burnt offering upon one of the mountains which I will tell thee of." However, after binding his son, laying him upon an alter, and picking up his knife, Abraham suddenly heard the voice of an angel who ordered him to let the boy go free (Genesis 22).

Abraham, therefore, heard voices, engaged in unusual sex practices, and was capable of extreme cruelty, including attempted murder. Is it possible that Abraham suffered from excessive limbic system activity, or even temporal lobe epilepsy? Consider that after hearing the voice of God, Abraham left ancient Babylon and changed his name (from Abram to Abraham), and, in a sense, his identity—as if he were experiencing a fugue state. Hence, he no longer recognized Sara as his wife and told the Pharoah, she is "my sister." Temporary fugue states are associated with abnormal temporal lobe activity (see Chapter 16), including temporal lobe epilepsy.

Epilepsy can be due to a number of different causes, such as head injury, heat stroke during infancy, and tumors. However, the predisposition to develop epilepsy can also be inherited.

Like his brother Abraham, Lot also saw angels and talked to God. It was God and his angels who warned Lot to leave Sodom, reportedly the most sexually corrupt city on Earth. However, once Lot escaped from Sodom, he celebrated by getting intoxicated and impregnating both his daughters, who willingly snuck into his bed on two separate nights (Genesis 20:33–38). In fact, even before they left Sodom, Lot had offered his daughters to some of the men of the city to do with as they pleased (Genesis 19:8).

We do not know if Lot followed Abraham's example and also let other men have sex with his wife. However, both Abraham and Lot clearly demonstrated signs of temporal lobe and limbic hyperactivation.

Religion, Limbic System Hyperactivation, and Temporal Lobe Seizures

Among a tiny minority of humans, the nuclei of the limbic system have a tendency to periodically become over-activated. When this occurs, emotions may be perceived or expressed abnormally, and the sensory and emotional filtering that normally takes place in these nuclei is reduced or abolished. Moreover, instead of being merely overly sensitive, those affected may suddenly experience extreme anger, rage, paranoia, depression, sexual desire, or even religious ecstasy or feelings of persecution. And they may hallucinate the presence of threatening people, animals, or even religious figures. Deepening of emotions, hallucinations, alterations in sex drive, and the development of extreme religious beliefs (i.e., hyper-religiousness) are not uncommon manifestations of limbic-temporal lobe

seizures and hyperactivation (Bear, 1979; Daly, 1958; Gloor, 1986, 1992; Horowitz et al., 1968; MacLean, 1990; Mesulam, 1981; Penfield and Perot, 1963; Schenk and Bear, 1981; Slater and Beard, 1963; Subirana and Oller-Daurelia, 1953; Trimble, 1991; Weingarten et al., 1977; Williams, 1956).

In fact, certain individuals who develop "temporal lobe epilepsy" and, thus, limbic hyperactivation may suddenly become hyper-religious and spend hours reading and talking about the Bible or other religious issues. Once this condition develops, they may spend hours every day preaching or writing out their mystical or religious thoughts, or engaging in certain actions they believe have religious significance. Many modern-day religious writers who also happen to suffer from epilepsy are, in fact, exceedingly prolific, and those who feel impelled to preach tend to do just that. (As noted in Chapter 1, the amygdala is directly interconnected with Wernicke's area.)

People who suffer from periodic episodes of limbic and temporal lobe hyperactivation, such as those with temporal lobe epilepsy, typically have seizures. It is not uncommon for these seizures to be preceded by an hallucination (Joseph, 1990a, d; Penfield and Perot, 1963; Williams, 1956). Patients can have any number of very odd hallucinations, such as smelling horrible odors, or hearing voices, music, or conversations. A rare few experience mystical awe, and might hallucinate religious entities, including angels, demons, ghosts, and God.

The great existential author, Feodor Dostoevsky, apparently suffered temporal lobe epilepsy. Dostoevsky alleged (via one of his characters) that when he had a seizure the gates of Heaven would open and he could see row upon row of angels blowing on great golden trumpets. Then two great golden doors would open and he could see a golden stairway that would lead right up to the throne of God.

As noted above, there is some evidence that many religious and spiritual leaders have had similar temporal lobe, limbic-system-induced religious experiences. For example, Moses may have suffered from temporal lobe seizures (Joseph, 1992b). Presumably, this was a consequence of being left, as an infant, for days to bake in the sun, after his mother abandoned him in a basket on a small stream. If that were the case, his brain could have become overheated and damaged by the scorching Egyptian sun.

If Moses subsequently developed temporal lobe epilepsy, this could explain his hyper-religious fervor, his rages, and the numerous murders he committed or ordered. Similarly, his speech impediment, hyper-graphia, and hallucinations, such as hearing the voice of God speaking to him from a burning bush, are symptoms not uncommonly associated with temporal lobe seizures and limbic hyperactivation.

Jesus Christ also was known to fly into violent and destructive rages, such as when he yelled, cursed, overturned the tables, and struck and chased the money lenders from the temple (John 2:14–15). He also frequently appeared irritable, sullen, gloomy, depressed, distrustful, and angry with his Disciples, whom he would sometimes curse, e.g., referring to Peter as "Satan" (e.g., Matthew 16:23). Jesus was also not beyond behaving in a petulant and sadistic manner, such as when he repeatedly refused the request of "a woman of Canaan" who "came and cried unto him, saying, Have mercy on me, O Lord; my daughter is grievously vexed with a devil." Instead, he refused by referring to her as a "dog." "It is not meet to take the children's bread and cast it to dogs" (Matthew 15:22–26). Indeed, he even cursed a fig tree for lacking fruit when he was hungry, "Let no fruit grow on thee henceforward forever" (Matthew 21:19). He was also apparently hyposexual, and although surrounded by unmarried female followers, many of whom had (presumably) been prostitutes, he apparently never succumbed to temptation, though he certainly enjoyed having Mary rub expensive oil on his body.

Depression, mood swings, episodic violence, and hyposexuality (as well as hyper-sexuality), are also associated with temporal lobe and amygdala hyperactivation. As per "hallucinations," Jesus did go alone into the wilderness for 40 days, and there he saw and spoke with "Satan." Jesus frequently sought solitude and isolation. Many religious figures have done likewise.

Isolation, Limbic Hyperactivation, and Hallucinations

It has been well established that even short-term social and sensory isolation lasting just a few days can induce emotionally and visually profound and complex hallucinations that can be so personally distressing that volunteers will refuse to discuss them (Bexton et al., 1954).

John C. Lilly (1972) combined LSD with prolonged water immersion and social and sensory isolation for about 7 hours on several occasions, and experienced and observed the presence of spiritual, god-like beings who beckoned to him (see also Eadie's similar description of "three men" noted above).

Isolation, as well as food and water deprivation, increased or decreased sexual activity, pain, drug use, self-mutilation, prayer, and meditation are common methods of attaining mystical states of religious and spiritual awareness, and have been employed world-wide, across time and culture (d'Aquili and Newberg, 1993; de Ropp, 1993; Frazier, 1950; James, 1958; Lehmann and Myers, 1993; Malinowski, 1948; Neihardt and Black Elk, 1989; Smart, 1969). These states also activate the limbic system.

For example, not only can pain or a desirable food item or sex partner result in limbic arousal, but when the limbic system is denied normal modes of input, be it sensory, emotional, social, or nutritional, it becomes hyperactive; stimuli normally deleted or subject to sensory filtering are instead perceived (Joseph, 1982, 1988a, 1992a). That is, limbic sensory acuity is increased, and in some respects what is perceived is not a hallucination but instead is the perception of overlapping sensory qualities that are normally filtered out. Sensory filtering is quite

Figure 8.20   An 18th century drawing by G Abana. Jesus, with Satan by his side, confronts some of the hordes of Hell. Jesus frequently sought solitude in order to commune with God.

common at the level of the amygdala, which contains neurons that are multimodally responsive a well as inhibitory (via serotonin). However, when this filtering is removed, hallucinations result.

LIMBIC HYPERACTIVITY AS AN INHERITED SPIRITUAL TRAIT

As pertaining to Jesus, although he may or may not have been hallucinating when he sought isolation, given the other features of his personality and religiosity, it could be argued that his amygdala and temporal lobe were highly active. Similarly, Abraham, Muhammad, Moses, and others capable of great spiritual faith demonstrate signs and behaviors associated with limbic and temporal lobe hyperactivation. Is this, perhaps, because they are, in fact, hyperactivating this region of the brain?

That is, a person who lives a highly spiritual or mystical life style might perpetually activate this region of the brain and achieve what others can only hope for via drugs, fasting, self-mutilation, and isolation/deprivation—i.e., access to God, or the spiritually sublime.

THE BIRTH OF GOD AND LIMBIC HYPERACTIVATION

In ancient Sumer (in southern Iraq around 6,000 years ago), it was believed that the Universe was ruled by a pantheon of gods (Armstrong, 1994; Kramer, 1956;

Figure 8.21   In seeking visions, the men of Arnhem Land expose themselves to the broiling sun and have stinging insects sprinkled repeatedly upon their bodies. Those with the short (versus the long) white sticks at their heads are those who have died during the ordeal. (New York Public Library.)

Wooley, 1965). However, many of the Sumerian people also worshipped household gods, including a personal god, which in some respects could be likened to a "guardian angel" or a spirit (totem) helper, as was common among the Plains Indians.

With the fall of Sumer and the rise of Babylon, many of these same gods, including these individual, personal gods, continued to be worshiped (Kramer, 1956; Wooley, 1965). This personal god served almost as a conscience and as a mediator between the head of the household and the great gods that ruled the cosmos (Joseph, 1992b).

Because this was a private, personal god, it was not uncommon for a believer to engage in prolonged and daily discussions with his deity (Kramer, 1956; Wooley, 1965). To this god, one could bear his heart and soul regarding sins, injustices, personal shortcomings, and hopes for the future. Hence, this god was indeed a personal god with whom one could "talk" and maintain a special personal relationship.

One day, however, something astounding and revolutionary occurred in the city of Ur of the Chaldees, in ancient Babylon, birthplace of Abram, a rich Babylonian prince. Abram began hearing voices. The voice was coming from his personal God and it gave him a command (Genesis 12): "Get thee out of thy country...and I will make of thee a great nation...and in thee shall all the families of the earth be blessed. ..."

And Abram and his personal god walked and talked, as God had not done since the time of Adam and Eve. Then one day this personal god came to a decision and said to Abram (Genesis 17), "Thy name shall be Abraham, for a father of many nations have I made thee. And I will make thee exceedingly fruitful...and I will make thee the father of many nations...and I will be their God."

Abraham both saw and heard his god on numerous occasions, both awake and dreaming, often falling on his face as God appeared. However, God and Abraham walked and spoke together during the heat of the day, and during the darkest hours of the night, his god making all types of grandiose promises and predictions.

Is it possible that Abraham was dreaming? Could this personal god from ancient Ur have been but a hallucination, and (given Abraham's odd sexuality and murderous actions) a product on temporal lobe epilepsy?

When we consider that this is the same god (at least in religious theory) who today is worshiped by Jews, Christians, and Moslems alike, is it possible he was merely hallucinating? Is it possible that Abraham (like Moses, Muhammad, and Jesus) may have been simply "blessed" with an over-active amygdala-temporal lobe, which not only would account for his spiritual-religious nature, but which may have enabled him to gain direct access to God (and vice versa)? Indeed, given that the prophecy of this personal god appears to have been largely fulfilled (as witnessed by the multitudes, i.e., Christians, Moslems, and Jews, who claim to worship this same god), the possibility of hallucinations, although quite plausible, seems unlikely. Of course, we could certainly argue otherwise.

The Limbic System and the Transmitter to God

There is, thus, some very speculative evidence to suggest that perhaps Abraham's hypersensitive amygdala and temporal lobe somehow enabled him to gain access to the spirit world of God, or conversely, that perhaps the presence of "God" triggered hyperactivity in Abraham's limbic brain. This would account for the fulfillment of the prophecy. That is, just as something frightening or sexual will activate limbic neurons, something *exceedingly* frightening, sexual, spiritual, or godlike might hyperactivate these same neurons.

Might this mean that maybe people who claim they are able to speak with God, or commune with spirits, are actually able to do so because these nuclei periodically become hyperactivated and "open" up a window leading to the Other Side? For the most part, this seems unlikely.

Consider Mary (described by Mesulam, 1981), a 26-year-old female, A-average college student. For several months she had been complaining of odd mystical experiences involving alterations in consciousness, accompanied by auditory and visual hallucinations as well as frequent experiences of déjà vu. These mystical experiences soon progressed to feelings of being possessed by the Devil. She was convinced the Devil was urging and trying to make her do horrible things to other people or to herself. She also claimed he would sometimes loudly cackle inside her head. Finally, a priest was brought in and a rite of exorcism was performed, as the Catholic hierarchy became convinced of the authenticity of her experiences, that she was possessed. However, her condition failed to improve. Finally an EEG was performed and abnormal activity was discovered to be emanating from both temporal lobes.

Another 44-year-old female college graduate suffering from temporal lobe abnormalities instead came to believe she was possessed by God and at times also thought she was the Messiah, who at the behest of God had a special mission to fulfill (Mesulam, 1981). At the urgings of "God" she ran for public office and almost won. However, she also engaged in some rather bizarre actions, including widespread and inappropriate sexual activity—another manifestation of limbic hyperactivation (see also Schenk and Bear, 1981; Trimble, 1991).

Souls, Spirits, and Poltergeists

As noted, another source of amygdala hyperactivation is extreme fear as well as extreme joy and ecstasy. In this regard, d'Aquili and Newberg (1993:194) note that "a combination of the experience of both fear and elation" is "usually termed religious awe." They also note that these feeling states are "almost always associated with religious symbols, sacred images, or archetypical symbols," which flow "from the inferior temporal lobe" and which "appear sometimes as monsters or gods." Indeed, angels, demons, and poltergeists may be experienced.

Most people find these experiences quite terrifying (Armstrong, 1994). They also frequently believe their perceptions are completely real and are not hallucinations.

"Cindy," a 22-year-old college student, was plagued by demons and ghosts for months until her right inferior/anterior temporal lobe was surgically removed. Cindy, however, had never been very religious, and had certainly never seen a ghost until after her auto accident. She had been thrown over 50 feet through the windshield of her car, suffering a fracture of the right temporal region of the skull and developing a subdural hematoma that was pressing on the temporal lobe, inducing herniation. This was surgically evacuated, and over the following weeks she seemed to quickly recover.

However, several days after her release from the hospital, she was startled, while watching television, when the arms, legs, hands, feet, and heads of the actors began protruding from the screen into the living room where she sat. Cindy said she first thought the television was broken and turned it off. But, as she stared back at the blank screen she saw what looked like her dead father staring back at her (which was probably her own reflection). As she backed away, the figure emerged from the television and began approaching and beckoning to her, as even more spooks and wraiths streamed from the picture tube. Crying for her mother, she raced for the bathroom and locked herself in. However, even as she hid within the inner sanctum of the washroom, spirits, sprites, and poltergeists streamed from the bathroom mirror and swirled about her. When she ran back into the living room she was even more horrified to observe a spirit enter and take possession of her mother.

Frightened and bewildered, Cindy ran into the street to flag down a police officer, who after investigating the scene brought her to the local hospital and psychiatry unit. Later, she decided what she had experienced were ghosts and lost souls of people who had either died in or had been entombed beneath her house.

Over the course of the next several weeks (until the temporal lobe was surgically removed), she also claimed to see "animal spirits" and complained that the "secret souls" of her mother's house plants were watching and observing her and that she could sometimes see filmy, soul-like entities traveling to and fro across the room and between different plants. And yet, in this regard, Cindy is not all that unusual.

DREAMS AND THE ROYAL ROAD TO THE SPIRIT WORLD
Animal Spirits and Lost Souls

Across time and culture, people have believed that not just humans and animals, but plants and trees, were alive, sensitive, and sentient, and were the abode of spirits, including the souls of dead ancestors (Campbell, 1988; Frazier, 1950; Harris, 1993; Jung, 1964; Malinowski, 1948). Because of this, among the ancients, before felling a tree the residing spirit sometimes had to be conjured forth to avoid harming it (Campbell, 1988; Frazier, 1950). However, be it animal or plant, souls were also believed capable of migrating to new abodes.

Among the ancients and many so-called primitive cultures, it was believed that souls are reflected in shadows, in streams, and pools of water (Campbell, 1988; Frazier, 1950; Harris, 1993; Jung, 1964; Malinowkski, 1948). However, because ghosts or demons sometimes attempt to abduct souls, one's shadow and reflection had to be protected. Indeed, even water spirits might try to capture a person's soul.

Moreover, the shadows and reflections of others had to be avoided so that one did not come into contact with the soul of a witch, sorcerer, or a demon. It was believed that the soul can be abducted by demons and witches as well as the recently departed. This is also why, in some cultures, people turn mirrors to the wall after a death and lay down pictures of the recently departed (Frazier, 1950). This insures that living souls are not stolen by the souls of the dead who are leaving this world for the next one.

Soulful Dreams

Souls were also believed by ancient humans to wander about while people sleep and dream (Brandon, 1967; Frazier, 1950; Harris, 1993; Jung, 1945, 1964; Malinowkski, 1948). That is, among many different cultures and religions the soul is believed to sometimes escape the body via the mouth or nostril during sleep. Moreover, during a dream, the soul may wander away from the body and may engage in certain acts or interact with other souls including those of the dear but long dead and departed.

Sometimes the soul is believed to take a form, such as that of a bird, deer, fox, rabbit, wolf, and so on. It could also hover about in humanlike, ghostly vestiges, at the fringes of reality, the hinterland where day turns into night (Campbell, 1988; Frazier, 1950; Jung, 1964; Malinowski, 1954; Wilson, 1951). However, sometimes the soul of an animal, such as a wolf or predatory bird, might take on various forms, including woman or man.

Hence, not just men but animals, too, had souls that had to be respected. However, it was believed that these souls could be influenced, their behavior controlled, and, in consequence, a good hunt insured. These beliefs gave rise to both animal worship and animal sacrifice, as well as the avoidance of certain animals, which were not to be killed or eaten at all, or killed or eaten only in a ritualized manner (Campbell, 1988; Frazier, 1950; Malinowkski, 1954; Smart, 1969).

Over the course of human cultural and cognitive evolution, these beliefs became increasingly complex. Specialists were required to interpret and minister the rituals and rites (Armstrong, 1994; Brandon, 1967; Campbell, 1988; Frazier, 1950; Smart, 1969; Wilson, 1951). Soon priests, prophets, and even the gods evolved. However, priests and prophets, as well as the common people, often experienced God as well as animal spirits and the souls of the dead during the course of a dream (Campbell, 1988; Frazier, 1950; Jung, 1945, 1964; Malinowkski, 1954).

Dreams (although mediated by brainstem nuclei) have their source in the amygdala (and hippocampus) and inferior temporal lobe (Joseph, 1982, 1988a, 1990a, b, 1992a, b). Indeed, activity within the amygdala may, in fact, trigger the first phase of dreaming (REM) sleep, which is heralded and then accompanied by what has been 7referred to as pontine-geniculate-occipital (PGO) waves (see Chapter 10). That is, the amygdala is active not only during REM, but amygdala activity triggers PGO waves (Calvo et al., 1987), which then lead to dream sleep.

In addition to amygdala activity during REM, the hippocampus (which is immediately adjacent to and also buried in the temporal lobe) begins to produce slow wave, theta activity (Jouvet, 1967; Olmstead et al., 1973). Presumably, during REM, the hippocampus and amygdala act as a reservoir from which various images, emotions, words, and ideas are drawn and incorporated into the matrix of dreamlike activity being woven by the right hemisphere (Joseph, 1982, 1988a, 1990a, b, 1992a, b). It is probably just as likely that the hippocampus and amygdala serve as a source from which material is drawn during the course of a daydream.

Dreams, Spirits, and Reality

When the limbic system becomes hyperactivated, it is not at all uncommon for an individual to experience a dream. Dreams, it has been proclaimed, are the royal road to the unconscious (Freud, 1900). It is also via dreams that gods frequently speak to men and women (Campbell, 1988; Jaynes, 1976; Jung, 1945, 1964), and it was via dreams that hunter-gatherers and ancient humans were able to gain access to the domicile of the soul (Frazier, 1950; James, 1958; Neihardt and Black Elk, 1989). Indeed, it has been argued that dreams (and thus the limbic system) enable an individual to come into contact with a different reality, the same reality shared and experienced by our ancestors and the Great Spirit (see Frazier, 1950; Jung, 1964; Neihardt and Black Elk, 1989).

Our ancient human ancestors lived in two realities, that of the physical and that of the spiritual, both of which were undeniable and experienced by enemies and friends alike (Frazier, 1950; Jung, 1945, 1964). One need only spend a night alone in the woods among the trees and the elements to become quickly convinced that one is not alone, but is being watched by various entities, both alive and supernatural, animal and spirit, benevolent and unkind.

Like modern-day humans, the ancients had dreams by which they were transported or exposed to a world of magic and untold wonders. It is as if one had been transported to a different world and a real-

ity that obeyed its own laws of time, space, and motion. It is through dreams that human beings came to believe the spiritual world sits at the boundaries of the physical, often where day turns to dusk—the hinterland of the mind, where imagination and dreams flourish and grow (Frazier, 1950; Jung, 1945, 1964; Malinowkski, 1954)—hence, the tendency to bury the dead in a sleeping position, even 100,000 years ago.

It is also via dreams that humans came to know that spirits and lost souls populated the night. The dream was real and so, too, were the gods and demons who thun-

Figure 8.22   Jacob's dream of the ladder to Heaven with angels going to and fro. Drawing by Hayley, 18th century.

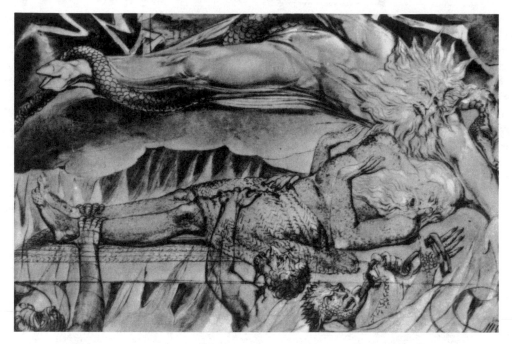

Figure 8.23   "With dreams upon my bed thou scarest me & affrightest me with visions."
Drawing by William Blake, 17th century.

dered and condemned, and the ghosts and phantoms that hovered at the edge of night. Although but a dream, like modern humans our ancient ancestors experienced this through the senses, much as the physical world is experienced. Both were real and were taken seriously.

Again, just as religious experiences can be secondary to amygdala and temporal lobe activity, dreams are also limbically produced. That is, neurons subserving spiritual experiences also give rise to dreams. Thus, the link between the world of dreams and the spirit land of gods and demons is the limbic system—i.e., the "transmitter to God," as unlikely as that may seem.

Right Hemisphere, Temporal Lobe Hyperactivation, and Dreaming

The amygdala and the neocortex of the temporal lobe, therefore, are highly interactionally involved in the production of religious and hallucinatory experiences, including dream states, the right temporal lobe and amygdala, in particular (Joseph, 1988a, 1992a).

Similarly, d'Aquili and Newberg (1993) argue that the right hemisphere (and right amygdala) is more involved than the left in the reception and production of religious imagery. This is likely, as the right hippocampus and amygdala, and the right hemisphere in general (Broughton, 1982; Goldstein et al., 1972; Hodoba, 1986; Humphrey and Zangwill, 1961; Joseph, 1988a, 1990a, b; Kerr and Foulkes, 1978; Meyer et al., 1987) also appear to be involved in the production of hallucinations and dream imagery, as well as REM during sleep.

For example, electrophysiologically the right temporal lobe becomes highly active during REM, whereas, conversely, the left temporal region becomes more active during NREM (Goldstein et al., 1972; Hodoba, 1986). Similarly, measurements of cerebral blood flow have shown an increase in the right temporal regions during REM sleep and in subjects who, upon wakening, report visual, hypnogogic, hallucinatory, and auditory dreaming (Meyer et al., 1987). Interestingly, abnormal and enhanced activity in the right temporal and temporal-occipital areas acts to increase dreaming and REM sleep for an atypically long time (Hodoba, 1986).

Conversely, LSD-induced hallucinations are significantly reduced when the right but not the left temporal lobe has been surgically ablated (Serafetinides, 1965). Similarly, it has been reported that dreaming is abolished with right but not left temporal lobe destruction (Bakan, 1977), though this has also been disputed (see Chapter 3). Nevertheless, it thus appears that there is a specific complementary relationship between REM sleep, hallucinations, mystical experiences, and right temporal (and, thus, right amygdala and hippocampus) electrophysiological activity.

Day Dreams and Foreseeing the Future

During dream states we see and experience events that are normally filtered from the conscious mind. We can also gain insight into problems that have plagued us, or gain access to knowledge of events that occurred in the past or that will occur in the future (Joseph, 1988a, 1990b, 1992b; Jung, 1945, 1964)—just as we can think about the future.

Consider the day dream. In addition to its images and memories, the fantasy produced also consists of anticipations regarding the future, and in this respect, day dreams could be considered an imaginal means of preparation for various possible realities. Interestingly, daydreams appear to follow the same 90- to 120-minute cycle that characterizes the fluctuation between REM and NREM periods, as well as fluctuations in mental capabilities associated with the right and left hemispheres (Broughton, 1982; Kripke and Sonnenschein, 1973). That is, the cerebral hemisphere tends to oscillate in activity every 90–120 minutes—a cycle that appears to correspond to the REM-NREM cycle and the appearance of day and night dreams, both of which may contain important information, not just regarding the past or the world of souls and spirits, but the future as well. As possible harbingers of the future, the intentions of the gods, and the future of self, friends and family, it has long been believed that dreams should be observed most carefully (Campbell, 1988; Frazier, 1950; Freud, 1900; Jung, 1945, 1964; Malinowkski, 1954).

In fact, among the ancients, the American Indians, and even the highly cultured ancient Romans, every once in awhile someone would have what is called "a big dream," which is of great importance to the whole clan, tribe, city, or nation. Often, the man or woman having the dream would gather the others together and announce it.

Given that dreams reflect mental activity, it is, thus, not terribly surprising that meaningful information might be derived, particularly in that the manner in which data are analyzed is so different and involves variables, as well as sensory stimuli, that are normally ignored or filtered out. During dream states, serotonin levels diminish (similar to what occurs under LSD) and multimodal neurons begin to fire, such that the brain becomes overwhelmed by sensory and ideational events that are normally filtered out (Joseph, 1990a, b). Hence, because the limbic system and temporal lobe are hyperactivated during dream states, not only does the brain become freed of inhibitory restraint, but one is presumably able to gain access to dreamlike alternate realities, including, perhaps, the spiritual reality of the Hereafter, that is if we wish to believe in such phenomena. Presumably, the same occurs when one is fasting, isolated, in pain, under LSD, in trance, or in the throes of religious ecstasy.

If the doors of perception were cleansed everything would appear... as it is, infinite...
William Blake ✳

PART II: THOUGHT EXPERIMENTS AND SPECULATIONS
In Search of the God Neuron: The God Within

Mystical, spiritual, and religious feelings, experiences, and beliefs are worldwide and have been in evidence for over 100,000 years. It is also clear that these beliefs and perceptions, including the capacity to dream and to experience the spiritual world through fasting, isolation, pain, drugs, dreams, and hallucinations, are dependent on specific and specialized neurons located in the limbic system (e.g., amygdala, hippocampus) and temporal lobe.

That there are neurons and neural networks that make it possible to perceive geometric patterns, forms, and faces, or (at least among some people) spirits or angels, would explain why hyperactivation of the limbic system and temporal lobe might result in dreams and hallucinations of faces, geometric shapes, colors, and so on, as well as gods, angels, and demons. During a dream, and/or due to temporal lobe and limbic system seizures or abnormalities, neurons subserving the perception of various visual and auditory stimuli become activated. The brain believes it is seeing a face or a demon, because "face" and other neurons have been activated and infused with intense emotion—i.e., the brain begins to hallucinate and dream.

Of course, it may be that the neural basis for the *perception* or hallucination of a ghost or demon is composed of input from a variety of different neurons, each of which contributes some feature to the resulting visual/auditory religious emotional hallucinogenic mosaic. That is, there are no "demon," or, for that matter, "God" neurons, but rather neural assemblies that interact under certain conditions to produce hallucinations and feelings of God and the spiritual Hereafter. Thus, the source is within the brain.

According to the Bardo Thodol (Tibetan Book of the Dead), "These realms are not come from somewhere outside thyself. They come from within...they exist from eternity within the faculties of thine own intellect...issuing from within thine own brain...reflections of thine own thought-forms. ..."

On the other hand, why are these spiritual states most likely to be experienced under conditions involving extreme fear (which may induce even a committed atheist to pray to God for help) or following death, or where the yoke of sensory inhibition and filtering has been removed, thus hyperactivating the limbic system? As noted, some of what is experienced, even under LSD, are not hallucinations per se, but are the result of disinhibition and multisensory neurons processing signals from divergent sources simultaneously. In consequence, one can see sound, feel colors, and so on: "real" stimuli that the brain can perceive but are normally filtered out. Is it possible that gods, demons, or angels are also filtered out?

Similarly, if religious and mystical experiences are hallucinations, and there is no Hereafter or spirit world, then why has our brain become adapted for perceiving and dreaming about what supposedly does not exist? Why would the limbic system evolve specialized neurons or neural networks that subserve the capacity to dream about, experience, or hallucinate spirits, angels, and the souls of the living and the dear departed, if these *entities* have no basis in reality?

That is, we are able to *hear* because there are sounds and voices that can be perceived and because we possess specialized brain tissue (e.g., auditory cortex) to analyze this information. First there were sounds, then specialized nerve cells that could initially analyze vibrations and, later, sounds.

We *see* because there are people and objects to view and because we possess neurons that code for various visual features and shapes. If there were nothing to visually contemplate, we would not have evolved eyes or a visual cortex that analyzes this information or we would have lost the capacity over time. (Of course, we are not able to "see" all wave lengths of light.) However, visual stimuli existed prior to the neurons that evolved in order to process these signals.

Shouldn't the same evolutionary principles apply to the limbic system and religious experience? Indeed, it could be argued that the evolution of this neuronal spiritual, mystical, religious capacity is the consequence of repeated and exceedingly intense perceptual and emotional experiences with "God" or the "Great Spirit" and the spiritually sublime over countless generations. Via, perhaps, the guiding influence of "God," or perhaps, after repeated experiences with gods, spirits, demons, angels, and lost souls, Homo sapiens evolved these "neurons" that enabled them to better cope with the unknown, as well as to perceive and respond to spiritual messages, which increased the likelihood of survival. A true scientist would not rule out such a possibility.

There are, however, other likely explanations that may even account for the belief in a life after death, and, thus, a world of spirits and souls. For example, the limbic system is exceedingly concerned with and desirous of maintaining life. Hence, perhaps it generates a desire to physically survive that is so intense that it has evolved specific neural networks that create dreams and hallucinations of souls and spirits of friends and relatives, so as to promote the promise of spiritual salvation and the illusion of eternal self-preservation, even after death.

The ability to experience God and the spiritually sublime is obviously an inherited "limbic" trait that is variably expressed by different individuals. However, from an evolutionary (Darwinian) standpoint, these hallucinations and illusions must contribute to the survival (or spiritual salvation) of those capable of experiencing these states. Presumably, the preservation of these traits is the result of natural selection and environmental influences on the survival and neural-biology of past generations. Those who did not possess these neurons were weeded out as "unfit." Those who possessed religious and spiritual capabilities were "selected for" and passed on these traits to their children. Those who developed and practiced limbic taboos and religious rituals, and who evolved a religious-moral conscience capable of redirecting and controlling the more dangerous limbic impulses, were more likely to survive, and, presumably, more likely to successfully breed.

Presumably, an evolutionary process such as this would have led to an exponential increase in the number of survivors and "religious" neural networks. Soon, religion and religious beliefs, as well as mystical experience and related "hallucinations," would have become worldwide and increasingly intense and profound. However, if that is the case, then the limbic system has evolved the capacity to not only regulate itself but to deceive itself via heightened emotional and opiate-induced religious euphorias, and the creation of false hopes and dreams, whose only purpose is to promote the survival of the species.

In this regard it could be argued that these mystical images, archetypes, and spiritual feeling states have always been *internally* generated, like a dream, and that they again have no basis in external reality but serve only the need and desire to survive. That is, there may not be an "out there" or "Heaven" or external God in some mystical space and alternate spiritual dimension, but rather, an internal heaven (or Hell) dominated, controlled, and produced by the limbic system. Thus, the limbic system and temporal lobes insure the survival of the *self* by dreaming and hallucinating ghosts, spirits, and avenging angels, and by promoting the illusion of perpetual and eternal survival (or damnation) if taboos, rituals, and the laws of God, are (not) obeyed.

Unfortunately, this argument does not explain why, even after death, individuals continue to dream and hallucinate, and why so many who return from the dead report similar religious and spiritual experiences that include being welcomed by the dear departed. That is, what is the adaptive significance of these neurons firing and creating hallucinations, *even after death* or *only after death*? A capacity such as this would represent a degree of limbic-evolutionary foresight that is almost too incredible to accept without positing some guiding intelligent force behind its design. For how could one pass on a trait that becomes functional *only* after death? Does that intelligence or foresight belong to the gods, and/or to the limbic system of woman and man?

Consider also the massive secretion of opiates, which guarantees most prey and other hapless creatures a "merciful death" as they fall and lie still while they are eaten alive by predators. What is the adaptive significance of a "merciful death?" Can a merciful loss of life promote the survival of the creature that is dying an otherwise horrible death? Obviously not. Rather, "mercy" such as this again raises the possibility of an intelligent force that purposefully and thoughtfully insured its "evolution."

Evolution or the Guiding Hand of Planned Metamorphosis

According to Darwin (1871): "A belief in all-pervading spiritual agencies seems to be universal, and apparently follows from a considerable advance in man's reason, and from a still greater advance in his faculties of imagination, curiosity, and wonder. I am aware that the assumed instinctive belief in God has been used by many persons as an argument for His existence. But this is a rash argument, as we should be, thus, compelled to believe in the existence of many cruel and malignant spirits, only a little more powerful than man; for the belief in them is far more general than in a beneficent Deity."

Charles Darwin, although championing and expanding upon the theory of evolution, was not the first to profess a belief in evolution, for similar theories were espoused 2600 years ago by Anaximander, a Greek philosopher. Anaximander argued that humans descended from fish. Over the centuries, others have not only come to similar conclusions but have written page after page that basically parallels the later work of Darwin (e.g., G.L.L. Buffon's 18th century treatise, "Natural History"). Indeed, when others pointed out the obvious similarities between the theories of Buffon and Darwin, Darwin agreed that "whole pages are laughably like mine."

A.R. Wallace, however, was probably not amused when informed that Darwin had also repeated many of his ideas on evolution (as laid out years before in his widely distributed, albeit unpublished, monograph). Wallace immediately contacted Darwin to set the record straight. It is for this reason that Wallace receives credit as the cofounder of the theory of evolution.

However, in contrast to Darwin, Wallace was repeatedly struck by the fact that various faculties and anatomical and neuroanatomical structures had somehow "evolved" and existed prior to the conditions that would make them necessary or useful. To Wallace (and others) this seemed to imply a degree of purposeful anticipation or planning. Indeed, according to Wallace (1895), the process of "natural selection" argues against a *mindless* or "random evolution," for the characteristic that is selected for must exist prior to its selection and, thus, prior to the conditions that require it. Adaptations made in advance of their utilization and, thus, before they were adapted or adaptive suggests the presence of some type of very long-ranged planning. As argued by Wallace (1895), this raises the possibility of a "guiding hand" or even that of a "divine intelligence" at work that has designed these features in advance and in anticipation of their later utilization.

Darwin was apparently furious with Wallace for pointing out the obvious, for he fired off a letter announcing his displeasure: "I hope you have not murdered too completely your own child and mine."

It is because of Wallace's insistence on the possibility of this guiding hand that subsequent evolutionary and genetic theorists have virtually ignored his work as well as the obvious implications of his observations, i.e., adaptive changes and modifications in structure prior to their usefulness suggests that their future employment was anticipated.

For example, those who are "selected" survive because they contain a DNA/RNA predisposition that enables them to develop or express the necessary traits in advance (or at least as possibility) and, thus, to cope or thrive in changing environments or under "new" conditions that were anticipated genetically. Because these specific creatures "survive," the possibility of further evolutionary developments in later appearing creatures is insured.

Conversely, it could be argued that those individuals or species who ceased to "anticipate," and were unable to continue genetically responding to changing environmental conditions, diseases, predators, and so on by failing to develop or activate certain DNA capacities or characteristics prior to actually needing them, were weeded out. Is it possible that these latter creatures somehow served their purpose (e.g., as a biological bridge leading to subsequent creatures) and their continued existence was no longer necessary?

The Origins of Life

The theory of a "random" evolution (versus a preplanned and anticipated evolutionary metamorphosis), is exceedingly dependent on the notion that life originated on this planet. Otherwise, one must take into account genetic traits or predispositions that were developed before life began on this planet, which precludes the notion of randomness. However, is it reasonable to assume that life accidentally originated on Earth?

One version of the prevailing theory of life's origins is that a single inert molecule suddenly sprung to life and began self-replicating after lightning repeatedly struck the hot, moist planet and the nutrient-rich cosmic soup that made up the Earth's primeval oceans, thereby giving rise to all subsequent forms of life (see Orgel, 1994, and Rebek, 1994, for related discussion). These and related notions could be referred to as the "Frankenstein theory of life." These theories are also based on the premise that the Earth is somehow isolated and sealed off from all other planets and the rest of the cosmos and, thus, could not have been contaminated by extraterrestrial life forms.

A competing theory is that the newborn Earth was repeatedly seeded by life-bearing meteors and galactic clouds of living protoplasm (Irving, 1980; Joseph, 1993; Reid et al., 1976) such that single and multicellular creatures, including bacteria and viruses (and their constituent RNA/DNA), were repeatedly hurled upon the face and seas of the young planet. Indeed, much of this extraterrestrial material would have survived entry into the Earth's early and very thin atmosphere without burning up.

In that *this* particular Universe may well have been in existence for *at least* 12–20 billion years (reviewed in Cowen, 1995; and Hellemans, 1995), whereas the Earth was created a scant 4.5 billion years ago, life has had ample time to "evolve" elsewhere. In fact, given that the age, size, complexity, and macro and molecular composition of the cosmos is unknown and may extend infinitely in time and space and interminably into the long ago, life has not just had ample time, but ample opportunity to appear other than on this speck of water, dust, and mud that we call Earth.

Like the various species who have lived and died on Earth, planets, solar systems, and galaxies exist for a limited time before dying, burning up, or collapsing into themselves. If any of the other stars and galaxies that had at one time swirled through the heavens had also swarmed with life, even if these planets or solar systems were somehow destroyed, isn't it possible that bacteria, viruses, and simple multicellular creatures might have survived (in dormant form) among the swirling debris? Tons of debris enter our atmosphere on a regular basis, including blocks of ice as big as a house (Frank, 1990). Isn't it conceivable that this debri might be contaminated, perhaps with dormant forms of life?

Consider, for example, that Raul Cano, of California Polytechnic State University, and researchers at Ambergene Corp, a biotechnology company, have recently reported (e.g., Cano and Borucki, 1995) successfully liberating microbes from fossilized amber, and bringing back to life over 1500 dormant bacteria, fungi, and yeast, ranging in age from 25–40 million years.

Given this and the fact that diverse forms of Earth-based-life can live and flourish in boiling, frigid, and seemingly "poisonous" environments, it therefore does not seem outlandish that some organisms may have "hitch-hiked" across space, attached to debris and encased in meteors and so forth, thereby subjecting the Earth to extraterrestrial contamination. If life were suddenly to appear on an otherwise desert island, would we not assume it washed to shore? The Earth, too, is an island, suspended in an ocean of space, and is constantly bombarded by debris.

Regardless of which theory one might subscribe to, it is generally believed that the earliest Earth-based life forms possessed RNA, or at least a single strand of DNA (see Chapter 1). Given that DNA/RNA provides the blueprint, or genetic memories, for the construction of all life forms, this would suggest that the earliest Earth-based life forms also contained these "blueprints," or "genetic memories," including those memories and instructions that enabled a small percentage of subsequent life forms to anticipate and "evolve" various structures in advance of their employment on this planet.

However, where did this "original" DNA/RNA and its storehouse of past and future genetic memories and predispositions come from? Did lightning just happen to randomly strike the correct inert molecule at the right time, causing DNA/RNA to suddenly spring forth, complete with instructions and data base? Or was the Earth infected by life-bearing debris, meteors, ice, and so forth, and their storehouse of genetic memories? Of course, if life (and its DNA/RNA) first appeared or evolved elsewhere, then the DNA/RNA genetic instructions of the first Earth-based life forms would be based on genetic knowledge acquired *elsewhere* as well.

Complexity and Evolutionary Metamorphosis

As pointed out by one of the true fathers of evolutionary theory, A. R. Wallace, natural selection is often characterized by anticipation of future environmental change such that some creatures evolved new (potential) capacities before they were functionally useful. In fact, some "adaptations" are not all that adaptive and do not promote the survival of that particular species. Rather, these anticipated adaptions are only maximally adaptive for those future species whose "evolution" depends on that particular adaptation in a previous species.

Consider, for example, the lung fish, which appeared over 400 million years ago, and the stunted appendages, i.e., lobed fins (see Chapter 1) that they "evolved." These creatures did not flourish or come to dominate any particular niche, as the environmental conditions under which they lived (and continue to live) do not require that a fish maintain the capacity to breath via lungs or move about on stunted appendages. However, the evolution of legs and feet and, thus, the capacity to "adapt" to living on dry land required that lungs and lobed fins "evolve" among a small and select group of ocean-dwelling fish, which in turn enabled amphibians, then reptiles, then repto-mammals, then mammals, primates, and Homo sapiens sapiens to "evolve," and become increasingly complex.

Could this obvious sequence of increasing complexity be due merely to chance? Under random conditions, complex systems eventually become increasingly chaotic, not more complex (Gleick, 1987).

What are we to make of the fact that although the Earth was initially populated by exceedingly simple single-celled organisms, and although almost 99% of all present-day life forms are also exceedingly simple (i.e., bacteria, viruses, etc.), a mere 1% have evolved and become increasingly complex? Moreover, many "unrelated" species (e.g., squid versus mammal) have somehow managed to evolve almost identical complex structures, such as the eye, or quite similar modes of communicating (reviewed in Joseph, 1993). Doesn't this suggest that even supposedly diverse organisms are striving toward similar modes of expression and being, as if following the same plan?

Numerous scientists have attempted to address these obvious weaknesses in evolutionary theory by proposing a variety of hypotheses to account for what is so difficult to explain: the seeming purposefulness and anticipation inherent in what is supposed to be a completely random process. These sometimes circumlocutional and circular theories and arguments include mutation, parallel evolution, exaptation, viral infection and transduction, jumping genes, and so on (Maynard Smith and Haigh, 1974; McClintock, 1984; Williams, 1966). Of course, these notions cannot account for why 99% of all life forms remain exceedingly simple and have not been similarly affected by these various agents and random forces.

Genetic Stability Versus "Random" Mutation and the Evolution of New Species

It has been repeatedly demonstrated that gene structure has been highly conserved during evolution, which in turn argues against the notion that any of the supposed agents (mentioned above) are responsible for this increased complexity and the evolution of various species. Moreover, as to the supposedly *random* genetic mutation involved in the evolution of new species, it has also been demonstrated that chromosomes and DNA have built-in genetic regulators that oppose "unplanned" and random alterations in genetic structure. Indeed, unplanned and random alterations result in malignant cellular formation, tumors, and death.

For example, tumors are sometimes believed to be due to a loss of restraint in regard to cellular growth and differentiation such that a malignant progression ensues (Kim et al., 1994; Modrich, 1994; Sancar, 1994). This is largely believed to be due to environmentally induced genetic lesions, the loss or inactivation of tumor suppressor genes, the abnormal presence of an enzyme referred to as telomerase (see Chapter 20), and/or the loss of regulatory proteins (pro-

duced by oncogenes) that normally control cellular growth (Gilbert, 1983; Shapiro, 1986; see Marx, 1994 for a recent review).

These same factors also act to control and stamp out genetic mutation, or unplanned and unmatched genetic alterations, for if they fail to do so, the result is not the development of a new adaptive trait, but disability and death.

For example, the aberrant expression (or inhibition) of oncogenes are associated with cancer-causing abnormalities in chromosomal structure. If these oncogenes become abnormal or are inappropriately expressed, they in turn induce malignant cellular formation, and not the creation of new species or adaptive modifications. Nevertheless, in conjunction with environmental influences, similar mechanisms supposedly account for Darwinian evolution, i.e., the creation of *adaptive* mutations.

Adaptive mutation (like its maladaptive counterpart) requires and is dependent upon altered oncogene expression, and chromosomal abnormalities involving unplanned deletions, duplications, and aberrant reciprocal translocations; otherwise, at least according to evolutionary theory, new species could not evolve. For example, depletions are associated with the removal of the constraints normally exerted on oncogenes, thus promoting the rapid and abnormal proliferation of cells as well as the possible loss of chromosomes during cell division (Rowley, 1983; Unis, 1983; Shapiro, 1986a, b). Although these mutations are maladaptive, again these are factors associated with "evolution," such that the weeding out process not only failed, but the "error" just randomly turned out to be beneficial.

Duplications are associated with the transposition, amplification, and hyperactivation of oncogenes (Alitalo et al., 1983), such that numerical abnormalities in chromosomal expression result as well as segregational errors involving the gain of chromosomes during cell division (Shapiro, 1986). Again, this is presumed to be secondary to a loss of regulatory restraint. Hence, if there is a recessive gene present, its duplication may result in a malignant overrepresentation and, thus, the expression of recessive malignant traits. The ability to suppress its expression has been overridden and a malignant hybrid cell is developed (Modrich, 1994; Shapiro, 1986).

However, according to evolutionary theory, these same forces result in adaptive (and unplanned) random alterations and, thus, the creation of new species. And yet, as stressed above, accept under abnormal conditions, these replication and recombination errors are corrected (Modrich, 1994; Sancar, 1994) such that the development of unplanned traits and cellular modifications and, therefore, the random evolution of new species is actively thwarted.

Evolutionary theory also requires that these "adaptive" translocation errors produce altered base pairings within the DNA helix. However, this violates the "Watson-Crick" pairing rules by creating unpaired bases within the helix. Normally, the "cellular mismatch repair system" recognizes these mismatches and eliminates them from the newly synthesized strands of DNA (Modrich, 1994). That is, they normally act to insure genetic stability by removing the damaged, unplanned, or altered nucleotide by replacing it with a normal base by inserting a complementary strand as a template; i.e., the duplex normally contains redundant information that can be reinserted when errors or mutations occur (Sancar, 1994). It is only when these errors go uncorrected that cells mutate and begin to accrue additional mutations at rates hundreds of times that of normal cells (Modrich, 1994). Again, evolutionary theory requires that these errors not only be allowed free expression, and that corrective mechanisms just happen to fail, but that they also be adaptive. Usually, however, these mutations give rise to cancer.

In summary, according to Darwinian and neo-evolutionary theory, genetic or environmentally induced *random* events have somehow not only overridden these protective and stabilizing genetic mechanisms, but the mutations produced were miraculously to the organism's advantage, thus giving rise to the creation of new species. Again, given the protective mechanisms described briefly above, this seems unlikely.

Rather, the fact that these presumed genetic *mutations* that gave rise to the evolution of new species were not corrected, did not induce cancer or other structural abnormalities, and just happened to occur at the right time so as to be adaptive, argues forcefully against the notion of randomness. Indeed, these alterations may well have been preplanned (i.e., existing as a potential genetic possibility or as genetic memory) and/or part of *normal* genetic expression, otherwise their expression would have been corrected and eliminated—if not genetically, then through death. Of course, one may also argue otherwise. ...

God and Evolutionary Metamorphosis

Most scientists are well aware of the infinite regression that characterizes the analysis of molecular and atomic structure; the building blocks of matter get smaller and smaller. However, humans assume that they are at one end of the *microscope*, the eyepiece (so to speak) looking down, whereas the quarks, gluons, and so forth that comprise the supposed building blocks of matter are at the other. But what if humans are actually in the middle of the *microscope*, and are not the crown of creation?

That is, from the human perspective, looking down through the eyepiece of the *microscope*, material forms can be broken down into subatomic particles (e.g., protons, neutrons), which are composed of yet smaller particles (e.g., quarks), which are held together by yet smaller entities (e.g., gluons). Hence, in one direction there is the possibility of an infinite regress.

However, there is also the likelihood of an infinite aggress. Consider, for example, that according to quantum chromodynamics, these same protons, and so forth, comprised the original state of the Universe within milliseconds of the supposed "big bang" creation of the cosmos. And yet, if considered from a macro-universal perspective, the constituent elements of the Universe, i.e., solar systems, galaxies, and so forth, could also be likened to macroparticles that are held together by yet smaller entities and cosmic forces.

From the human perspective, i.e., sitting at the eyepiece of the telescope, there is but one Universe. That Universe is composed of billions of separate galaxies. Galaxies are comprised of billions of stars, such as our sun, and perhaps billions of planets and individual solar systems. However, the supposed edge of our curved universe is defined by the most distant galaxies that are currently detectable or whose existence is inferred via analysis of varying wavelengths of light, radio emissions, the "Hubble constant" and the cosmic microwave background. Currently, the most distant galaxy yet discovered lies approximately 12–15 billion light years from Earth (reviewed in Hellemans, 1994), i.e., its radio and light emissions have taken 12–15 billion years to reach our planet.

However, this definition of the Universe does not take into account those stars and galaxies that lived and died so long ago that traces of their existence have not yet been found or recognized for what they are or simply no longer exist. Nor does it take into account the existence of galaxies that may be so distant that it is completely impossible to discern their existence. Indeed, this may well explain why astronomers are unable to account for the gravitational pull that guides the movements of all the varying galaxies comprising this (the "known") Universe. That is, there must be at least 10 times more matter than this Universe contains (reviewed in Hellemans, 1995). There is not enough matter in this universe to account for the gravitational influences that hold this universe together!

This suggests that there exists the equivalent of 10 more universes (each containing billions of galaxies), each of which is so distant that their existence is not detectable, at least from our end of the telescope. Possibly, these 10 additional (proposed) universes make up the matter necessary to account for cosmic gravitational forces that

hold *this* universe together. However, just as enormous distances separate the varying galaxies of *this* universe, the distance to the next universe (with its own galaxies, etc.) must border on the incomprehensible.

Most astronomers are not yet willing to seriously entertain such a possibility, however, as one of the most popular theories as to the origin of the Universe requires that there be only one universe, and that it was created by the "big bang"—which I believe might best be described as the "big deception."

Consider, for example, that it has been determined, based on how fast this universe is believed to be expanding, that the Universe was formed perhaps as recently as 8 billion years ago (Freedman, 1994). Nevertheless, even if there were a big bang, and even if it did occur 8 billion years ago, this cannot account for the existence of those stars and galaxies that are between 12–16 billion years old. That is, there are stars that are older than the "big bang." Moreover, the notion of the "big bang" also rests on measurements of the speed at which the Universe is supposedly expanding, i.e., the red shift infrared background and its intensity at different wavelengths, which indicates the recession velocity of a galaxy in relation to the Earth. However, not all galaxies provide a red shift background. Moreover, some galaxies appear to be flowing in the wrong direction. In fact, there appear to be rivers of galaxies flowing in various directions (e.g., Lauer and Postman, cited by and reviewed in Flamsteed, 1995).

Hence, there are stars that are older than the big bang, not all stars provide the appropriate redshift background (which suggests they are not moving away at the speeds predicted by a big bang), some galaxies are flowing in the "wrong" direction, and all the matter within the Universe is one-tenth of what is necessary to account for the gravitational forces that hold this universe together. Given the above evidence (as well as other data not discussed here), it certainly appears that the "big bang" cannot account for creation, and that the current conception of the size and extent of the Universe may be in error.

However, let us take this a step further. Again, just as we can engage in an infinite regress in the attempt to understand the foundations of matter, the possibility of an infinite aggress (or expansion) is equally likely. That is, the 10 extra universes that are required to explain the gravitational forces maintaining this universe would, in turn, and in all likelihood, require the existence of yet another 100 universes, and so on, to account for their gravitational forces. However, as this process of an infinite aggress and expansion continues, the Earth and its inhabitants become correspondingly minute and insignificant in size and composition; i.e., we become the equivalent (from a human perspective) of a subatomic particle and perhaps a potential object of study for whatever life forms whose existence is comprised of these infinite constellations and collections of universes. Indeed, it could be assumed that whatever is sitting at the eyepiece of *that* microscope (so to speak) would, thus, have godlike proportions.

Consider the possibility that if there is a God (or gods) or Great Spirit, it may well be impossible for humans to comprehend the nature of His/Her existence, form, plans, designs, and so forth. That is, just as a worm cannot comprehend a man (due to the substantial differences in neurological complexity) it might be similarly impossible for humans to comprehend life forms whose complexity dwarfs our own. Given that there are stars that are up to 16 billion years older than our own, then it is possible that any creatures that "evolved" in these more distant regions of the Universe may exceed humans in cerebral complexity. As compared to our 6 layers of neocortex, such craniums could contain 12 layers or more.

Exo-Biological Cerebral Organization: The Neocortex of the Gods?

If indeed there is life on other planets, then it could be surmised that evolution, or some variant of this process (e.g., evolutionary metamorphosis) is characteristic of that life, and that somewhere in the vastness of the cosmos, intelligent creatures may have also evolved. How they evolved, their physical appearance, the extent of their intellectual and creative capabilities, and so on, would in turn be determined genetically as well as environmentally and experientially over the eons of time.

On Earth, five Kingdoms of Life are recognized. If we grant the possibility of extra-terrestrial life, then it could be assumed that the same five Kingdoms are variably represented in various regions of the cosmos, though the actual number of Kingdoms may be in the thousands, millions, etc. It could also be surmised, that those extra-terrestrial creatures who belong to and/or who evolved in a fashion similar to those of the earth-based animal Kingdom of Life, probably have evolved neurons, and a central nervous system.

Given that Earth-based vertebrate and invertebrate morphological organization is basically identical at the neuronal and synaptic level (e.g., possessing cell bodies, dendrites, axons, chemical neurotransmitters), whereas a definite progressive sequence in cerebral structural and hierarchical organization is apparent in the progression from fish to primate brain, then it might be expected that on some planets cerebral evolutionary development might follow similar patterns. This would include alien brains organized in a sponge-like, reptilian, mammalian, or in a primate and thus, humanlike fashion, and with similar praxic, creative, technological and linguistic capabilities—again depending on the nature of their environment, for it is likely that such creatures may be able to process sensory information which the human brain may fail to perceive.

It is likely that the technological and cultural achievements of those exo-biological organisms with brains commensurate with that of the human may be highly advanced or quite primitive. However, given that *this* Universe has been in existence anywhere from 9 to 20 or more billion years (reviewed in Cowen, 1995; Hellemans, 1995), whereas the Earth has been in existence for a scant 4.6 billion years, it does not seem unreasonable to assume that those creatures living and evolving in the older regions of the cosmos may well have evolved beyond the present state of Homo sapiens sapiens before this planet was even formed, billions of years before life appeared on Earth.

If that is the case, then it could also be predicted that over the ensuing eons of time, these creatures may have acquired more complex neuronal capabilities, as well as increased association and assimilation (tertiary) neocortical processing space, and may have also evolved additional layers of neocortex, thus exceeding the 6- to 7-layered neocortex that is characteristic of primates and mammals. Since increasing complexity and progressive cerebral encephalization is characteristic of life on this planet, the same could be expected elsewhere. Hence, the neocortex of some exobiological brains may consist of 8, 10, 12, or more layers of neocortex, as well as enlarged neocortical associational and memory capacity, thus dwarfing the human brain in neuronal and cognitive capability.

In this regard, from the perspective of those exo-biological organisms who evolved in the older regions of the cosmos, and whose neocortical association and multimodal tertiary and assimilation areas have greatly expanded and evolved, and/or who possess 9, 12, or more neocortical layers, the human brain and mind may seem to be just one small step above a reptile; which, in many ways, it is. Conversely, the human ability to comprehend the intellectual and technological accomplishments and capabilities of an alien brain organized in this evolutionary advanced fashion might be analogous to a lizard's ability to comprehend a man. That is, the mental, intellectual, and technological capabilities of creatures who began to evolve 10 or 20 billion years ago may lie just beyond human understanding. They might appear as gods. Indeed, even the gods may have gods.

Consider also that Homo sapiens sapiens have recently begun creating "designer genes," and can manipulate, via genetic engineering, the structure and function of certain life forms, thus creating, for example, bacteria that can "eat" unrefined oil. Indeed, in May of 1995, theologians from almost all the major religions joined together and issued a proclamation warning against scientists "playing God."

Given the above, is it thus so difficult to consider at least the possibility (at least in the form of a thought experiment) that a "superior" life form that is more intelligent than human beings, and that lives or exists in a manner or physical dimension that is beyond our comprehension, or who has evolved 12 layers of neocortex, may have also been "playing God" with Earth-based life forms?

There are, of course, many other, even more plausible explanations...

EVOLUTIONARY METAMORPHOSIS

If life appeared elsewhere before the Earth was formed, it is not beyond reason to assume that if this planet were somehow scrubbed clean of all life, that simple life forms would soon repopulate the Earth. It might also be expected that the same sequence of "evolutionary" anticipation would take place, such that the progression leading from simple multicellular creatures to human (or humanlike) beings would again unfold.

That is, since the earliest Earth-based life forms contained RNA/DNA (and its genetic blueprints for life and, thus, the genetic potential for *all possible forms* of life), then if these genetic plans (or memories) "evolved" somewhere other than Earth, all *forms* of Earth-based life may not be a product of evolution and natural selection (per se) but of metamorphosis, i.e., evolutionary metamorphosis and a replication of previous life forms. That is, rather than a random evolution, and unlike the single seasonal metamorphosis that characterizes the transition from caterpillar to butterfly to baby caterpillar and so on, it is possible that humans (and/or the *neurological/*

genetic predispositions that were subsequently environmentally molded and shaped to create a human form) are an end product (or perhaps a midway product) of a process that takes approximately 1 billion years to unfold, a process that involves, not a two-step progression (caterpillar-butterfly) but a multi-step progression involving numerous successive species, each preprogrammed to give rise to the next. For example, from a particular single cell and DNA/RNA sequence, which is passed along and unfolds in order to become teleost—amphibian—reptile—...human.

The possibility of evolutionary metamorphosis would, in turn, account for the increasing complexity (rather than progressive simplicity) of specific organisms over the course of "evolution." The progression from simple to complex is preprogrammed.

Evolutionary metamorphosis (as well as most other current genetic theories, including "mutation") would also account for the "innate" resistance that members of many species demonstrate in response to chemicals (e.g., insects and DDT) or diseases to which they have supposedly never before been exposed. That is, they (i.e., the genes) have been exposed previously (prior to life on this planet), thereby creating a genetic inoculation that spares them extinction, that is, once these genetic memories are environmentally activated. Of course, there are other equally plausible explanations.

Metamorphosis and widespread genetic predisposition, coupled with environmental molding and shaping, would also explain the morphological and anatomical similarities of supposedly unrelated species that over the last 225–150 million years "evolved" in completely different environments, e.g., Australia versus the Americas, i.e., Tasmanian wolf/gray wolf, Koala/tree sloth, Western quoll/ocelot (e.g., Wilson, 1994). Because the same "genetic seeds" were planted on different continents, the "same" creature evolves, albeit in accordance with the molding influences of geography and environment over millions of years.

A preprogrammed metamorphosis would also account for the evidence that supports the "multiregional" theory of human evolution (see Chapter 2). That is, according to the "multiregional" view, modern Homo sapiens sapiens independently evolved from H. erectus in multiple places over multiple time periods. In other words, H. erectus was preprogrammed to eventually give rise to H. sapiens, such that archaic, "early modern," and modern humans independently evolved in Africa and Asia from different populations of H. erectus. If the "multiregional" view is correct, then it could also be argued that the Earth was seeded so as to "grow" humans (and other creatures).

On the other hand, even if life evolved "elsewhere" perhaps many billions of years before this planet (or even this universe) was formed, and although the evolution of complex Earth-based life forms may be a function of genetic predisposition, metamorphosis and the replication of previous extraterrestrial organisms, ultimately these organisms are shaped and molded by environmental influences, and their survival is determined by seemingly random or unexpected events and chance occurrences, including life-destroying catastrophic events.

Consider, however, if things had turned out differently and 100 million years ago some disease or catastrophic event selectively interfered with the evolutionary sculpting of what would become human beings. Since humans and other creatures are presumably descended from the same genetic seed (even if planted by the hand of God), then is it not possible, given the right environmental conditions, that some other animal would have filled this empty niche?

IN THE BEGINNING THERE WAS LIFE

Because the original genetic seed that is common to all life contains (at least as unrealized potential) the life plans for all life, different species also possess the genetic potential to assume multiple forms, depending on cross-generational environmental influences, geography, food sources, predators, and catastrophic events. Shape and form are functions of tens of thousands of years of cross-generational environmental influences acting on genetic potential, genetic predispositions, and the universal LIFE force that exists within us all.

Indeed, although it may seem mundane, it is noteworthy that all (known) life forms not only possess DNA/RNA, but life. Life is common to all life forms.

I would argue that LIFE cannot be created or destroyed, and that at The Beginning of Time and all being, there was life. In the Beginning there was life.

Although this proposition may appear to bear the stamp of metaphysics, it is, in fact, based on quantum mechanics and the same theorem proposed by Einstein to account for the indestructibility of energy: $E=mc^2$. Energy can neither be created nor destroyed. Is not life a form of energy?

Can life be destroyed? Consider a flowering plant. If we cut the plant in two, and nourish both halves so they live, and if the original half later dies, is the plant still alive? If we take that remaining half and cut it in half, and later the older half dies and the newly cut portion lives, is the plant still alive? If we take that half and divide it into fourths, eighths, sixteenths, and so on, and every portion dies but one, is the plant still alive? Of course. And does not this same principle extend to all other life forms, including humans, whose origins extend backward in time, interminably into the long ago? Does not a woman give birth not only to her own daughter but to the eggs in her daughter's ovaries and thus to her daughter's daughters? Are we not an interlinked continuation of those life forms which beget us, diverging again and yet again and again? Our ancestry extends backward in time, without interruption, to the very beginning—and in the beginning there was life. As is evident from the proliferation and undeniable existence of life and living things, life cannot be destroyed. Only the mask of individuality, and the husk that provides form to life, is subject to death and decay.

I also propose that the question regarding the origin of life is phrased incorrectly. Perhaps LIFE itself, and not inert matter, formed the heart of what we call "the beginning." Rather than considering life as some waste product of moist, nutrient-rich, electrified matter, perhaps matter is a nutrient-rich waste product of LIFE.

LIFE cannot be created or destroyed. LIFE can take many forms. Untold multiple physical manifestations of LIFE are shaped and molded by forces both internal and external. LIFE that takes the form of turtle, frog, woman, or dog remains LIFE; that is, alive. However, although life can be neither created nor destroyed, the structural organization that gives *form* to life is subject to disintegration and decay—what we call death.

Again, consider Einstein's theorem. Although energy cannot be destroyed or created, the destruction and transmutation of matter is something wholly different and is dependent on the organization and stability of the force field, which is energy. Although material forms may become unstable, disintegrate, or assume new organization, the constituent fabric that gives rise to matter and its manifest structure is physical energy, which cannot be destroyed. However, if that energy is liberated, its material form appears to disintegrate; e.g., it appears to die.

Death and the Body

In most instances, even "sudden" death is a gradual process, with some cells and tissues disintegrating in advance of others, and yet other tissues living for hours or even days before the body completely decays. Presumably, so long as the body (or at least the limbic system) lives, one's sense of a personal soul and identity remains intact (in the form of an out-of-body experience)—an ethereal existence and sense of personal identity that remains tethered to the body (or limbic system) until the body completely dies and decays.

Indeed, the linkage of the personal soul and individual immortality to the body were widespread beliefs and practices among the ancient Egyptians, which is why they expended so much effort to preserve the body via mummification. If the body could be preserved, so could one's personal soul and sense of individuality, leading to "immortality." Others, including the Tibetan Buddhists, sought just the opposite, to free the soul from the body so as to escape the "illusion" of individuality and personal existence.

Perhaps, too, what some experience as their personal soul upon death is but a gradual liberation of LIFE that at first retains its bodily (limbic) links, thus preserving one's sense of individuality—the shadow of one's previous form as the body dies. Thus, as the body is consumed, perhaps so, too, is the sense of individuality, freeing the soul, one's LIFE, to be embraced by the radiance of all LIFE, thereby becoming One with the Great Spirit and the Gods.

Life can neither be created nor destroyed.
In the beginning there was LIFE...

> Thine own consciousness, shining, void, and inseparable from the Great Body of Radiance, hath no birth, nor death, and is the Immutable Light—Buddha Amitabha [source of life and boundless light]. If all existing phenomena shining forth as divine shapes and radiances be recognized to be the emanations of one's own intellect, Buddhahood will be obtained at that very instant of recognition.
>
> Bardo Thodol (Tibetan Book of the Dead)

PART III: ALPHA AND OMEGA

The existence of God or Great Spirit, the "big bang," the plausibility of metamorphosis versus "Darwinian evolution," and the reality and validity of spiritual beliefs, religious beliefs, and mystical experiences, including "life after death," cannot be resolved here, nor was that my purpose. Rather, *my purpose* was *to speculate*...to pry open those windows and doors...and perhaps generate further dialogue, debate, and scientific inquiry.

It is an unfortunate tendency of some who proclaim to be scientists to ignore what they don't understand, to call it nonsense when it is pointed out, and to later state it was obvious when it is verified and documented as "real." When it comes to "religious" experience, however, some scientists tend to become hysterical and attempt to cloak themselves in the mantle of dogma (much like Temple Priests) so that the fictions they worship in the name of "science" may be preserved unchallenged. There is nothing scientific about this attitude, however, as it, in fact, prevents or hinders our ability to engage in true scientific inquiry. Those who dismiss or ridicule attempts to scientifically analyze religious belief—and the evidence that supports it—are in fact proclaiming: "Thou shall not know." Nevertheless, the theory of evolution is obviously flawed and we as yet have no clear understanding as to the nature of life, existence, or the structure and origins of the Universe, despite the almost religious fanaticism of those "scientists" who claim otherwise. Unfortunately, those who pretend otherwise rely on these fictions in order to ridicule those who seek to question the dogma that masquerades as "authority." As scientists we must question and we must doubt. Perhaps it is time that we, as scientists, cease to treat religious experience and belief in God as metaphysical mysteries but as subjects full worthy of hard-nosed scientific inquiry, as there is a factual basis for these experiences.

In this regard, despite the scientific or religious "truths" one might believe in or cling to, and in contrast to those who pretend that the theory of evolution is fact (when it is only a rather flawed theory), it is clear that there is a scientific and neurological foundation for religious and spiritual experience: i.e., the amygdala, hippocampus, and temporal lobe. Why that is, is yet to be determined.

Given the obvious role of the temporal lobe and limbic system in the generation and perception of these spiritual-feeling states, it could also be argued (at least at the level of metaphor) that the limbic system may well be the seat of the soul, and/or serve as the transmitter to "God." If that is indeed the case, then Buddha, Lao Tzu, Chuang Tzu, the Taoists, Sufis, and Jesus (like so many other Jewish, Arabic, Muslim, Indian, Babylonian, Sumerian, Egyptian, Greek, Roman, and Gnostic mystics) were correct when they proclaimed: "The kingdom of God is within you."

SECTION IV

The Brainstem and Basal Ganglia

9

Caudate, Putamen, Globus Pallidus, Amygdala, and Limbic Striatum

Parkinson's Disease, Alzheimer's Disease, Psychosis, Catatonia, Obsessive-Compulsions, and Disorders of Movement

Movement and motor functioning are dependent on the functional integrity of the basal ganglia, brainstem, cerebellum, spinal cord, and cranial nerve nuclei, as well as the thalamus, and the primary, secondary, and supplementary motor areas of the frontal lobes. These areas are all interlinked and function as an integrated system in the production of movement. For example, the basal ganglia provides input to the brainstem (via the "extra-pyramidal motor system") as well as to the motor thalamus and motor neocortex, which also projects to the brainstem and spinal cord (see Parent and Hazrati 1995).

The basal ganglia, however, is composed of several major nuclei that subserve different functions (DeLong, 1995; Parent, 1995; Parent and Hazrati, 1995). This includes the corpus (or dorsal) striatum ("striped bodies"), i.e., the caudate and putamen, which are extensively interconnected and which project to a variety of brain areas, including the immediately adjacent globus pallidus ("pale globe"). The dorsal globus pallidus (GP), although related to the midbrain, is, in many respects, coextensive and appears to merge with the putamen, giving the entire structure the appearance of a camera lens. Hence the globus pallidus and the putamen are referred to as the lenticular nucleus ("lens").

In contrast, the inferior *ventral* GP (also referred to as the ventral pallidum or substantia innominata) is part of the limbic (ventral) striatum and, in fact, eventually merges with the centromedial amygdala and receives extensive projections from the lateral amygdala, the olfactory tuber-

cle, and nucleus accumbens. The substantia innominata, nucleus accumbens, and olfactory tubercle constitute the limbic (or ventral) striatum, and are major nuclei of the basal ganglia.

In addition, the "motor thalamus," the orbital and medial frontal lobes and medial supplementary motor area, and hippocampus (as well as the central and medial amygdala), are richly interconnected with and constitute major components of the basal ganglia. However, as noted in Chapter 1, the basal ganglia (i.e., the corpus and limbic striatum and GP) evolved out of the olfactory-amygdala and in many respects could be considered part of the limbic system (Heimer and Alheid, 1991; MacLean, 1990).

Intimately linked and thus part of the basal ganglia is the substantia nigra and the midbrain tegmentum, which feed dopamine to the corpus and limbic striatum. Specifically, the nigrostriatal DA (A-9) cell group projects from the substantia nigra to the dorsal caudate and putamen and the medial frontal lobes (see Ellison, 1994; Fibiger and Phillips, 1986; Parent and Hazrati, 1995, for review), although some fibers also innervate the limbic striatum. It is the nigro-(dorsal)-striatal system that is thought to be related to motor functions, including the production of stereotyped and routine actions.

The mesolimbic DA system originates in the ventral midbrain tegmentum (A-10 DA cell group) and sends fibers to the amygdala, septal area, hippocampus, and frontal cortical areas, including the ventral caudate-putamen, nucleus accumbens, and substan-

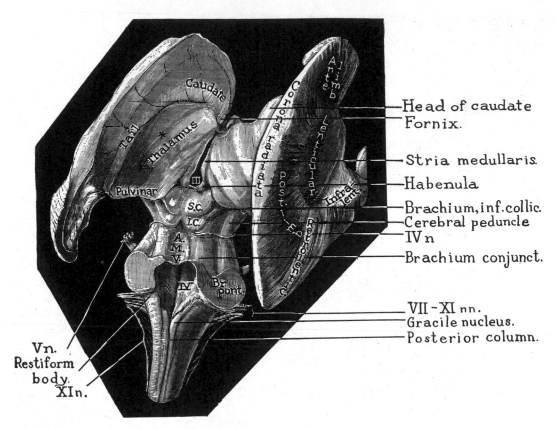

Figure 9.1 Dorsal view of the anatomical relationship between the caudate, thalamus, midbrain, and brainstem. (From Krieg W. Functional neuroanatomy. New York: Bakiston, 1952.)

tia innominata (Le Moal and Simon, 1991; Olton et al., 1991; Zaborszky et al., 1991), although some fibers also innervate the dorsal striatum. The mesolimbic DA system is believed to be related to emotion, mood, memory, and reward, including locomotor survival-related activities, such as running and even "galloping" (Ellison, 1994; Fibiger and Phillips, 1986; Le Moal and Simon, 1991).

The Amygdala, Emotion, Memory, Psychosis, and Basal Ganglia

As detailed in Chapter 1, considerable evidence indicates the limbic (ventral) striatum evolved out of the "extended" (centromedial) amygdala. As noted, the substantia innominata (ventral GP/nucleus basalis) merges with the centromedial amygdala and receives extension projections from the lateral amygdala.

In addition, the more recently evolved corpus striatum is a derivative of the ventral/posterior corticomedial-lateral amygdala with which it remains intimately linked via the "tail" of the caudate. However, in that the amygdala is the first portion of the basal ganglia to appear over the course of evolution, as well as during embryological development, being formed via neuroblast migration from the epithelium of the lateral ventricle (Humphrey, 1972), the caudate could be considered the bulbous "tail" end of the amygdala (and not vice versa).

Indeed, the human corpus and limbic striatum are not only derivatives of the amygdala (and to some respect the hippocampus), but are sandwiched between the posterior/ventral medial/lateral "tail" and the extended centromedial "nose" (or internal shoulder) of this nucleus. The amygdala (as well as the anterior cingulate,

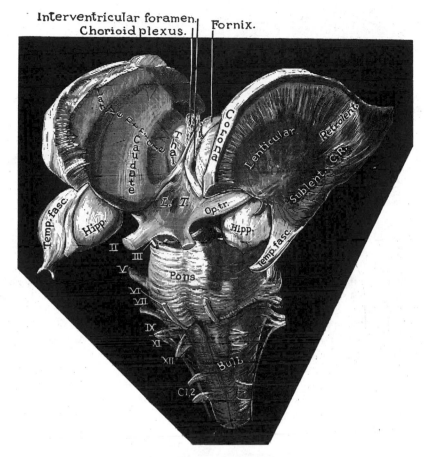

Figure 9.2   Anterior ventral view of the anatomical relationship between the caudate, thalamus, midbrain, and brainstem. (From Krieg W. Functional neuroanatomy. New York: Bakiston, 1953.)

lateral hypothalamus, and hippocampus), therefore, is able to exert considerable influence on the basal ganglia, which appears to have evolved in order to serve as an emotional-motor interface so that limbic needs and impulses may be acted on in a flexible manner (MacLean, 1990; Mogenson and Yang, 1991). These are functions the basal ganglia (the limbic striatum, in particular) continues to perform in humans as well as other creatures.

For example, the basal ganglia is exceedingly important in the stereotyped and species-specific motoric expression of social and emotional states, such as running away in fear or biting defensively, and as displayed in ballistic movements (hitting, kicking) and manifested through facial expression, posture, muscle tone, or gesture (Mogenson and Yang, 1991; MacLean, 1990;

Rapoport, 1991). Because all humans basically possess the same basal ganglia and limbic system, when happy, sad, angry, and so on, the facial and body musculature assumes the same readily identifiable emotional postures and expressions, regardless of culture or racial orgin (Ekman, 1993; Eibl-Ebesfeldt, 1990; however, see Russell, 1994).

Because of this similarity in basal ganglia functional architecture, regardless of culture or race (and in many respects, mammalian species), if frightened, angry, or in the process of being assaulted, animals or humans may similarly engage in (ballistic) hitting and kicking, biting, and/ or all of the above, depending on context, mood, and situational variables (including play). However, in contrast to the brainstem and cerebellum, which provides re-

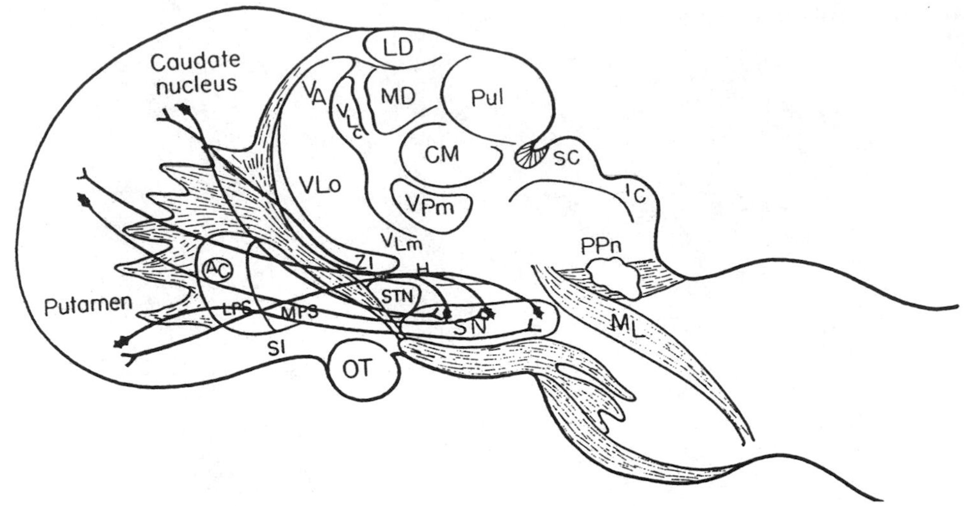

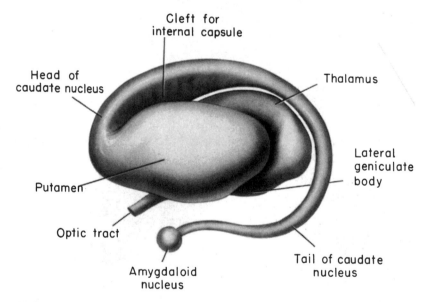

Figure 9.3   Schematic diagram of the nigrostriatal system viewed from a sagittal plane. **(AC)** anterior commissure; **(CM)** centromedian nucleus; **(H)** Forel's field; **(IC)** inferior colliculus; **(SC)** superior colliculus; **(LD)** lateral dorsal nucleus; **(LPS)** lateral pallidal segment; **(MD)** mediodorsal nucleus; **(ML)** medial lemniscus; **(MPS)** medial pallidal segment; **(OT)** optic tract; **(Pul)** pulvinar; **(STN)** subthalamic nucleus; **(ZI)** zona incerta. (From Carpenter M. Core text of neuroanatomy. Baltimore: Williams & Wilkins, 1991.)

Figure 9.4   Caudate, putamen, thalamus, and amygdala. (From Carpenter M. Core text of neuroanatomy. Baltimore: Williams & Wilkins, 1991.)

flexive and stereotyped motor programs that can be performed without thinking, the basal ganglia is capable of considerable flexibility in regard to motor-emotional expression, and is exceedingly responsive to the organism's motivational and emotional state—via its extensive interconnections with the limbic system.

Thus, the striatum contains cells that selectively respond to motivationally significant stimuli, including novel and familiar variables that are rewarding or punishing

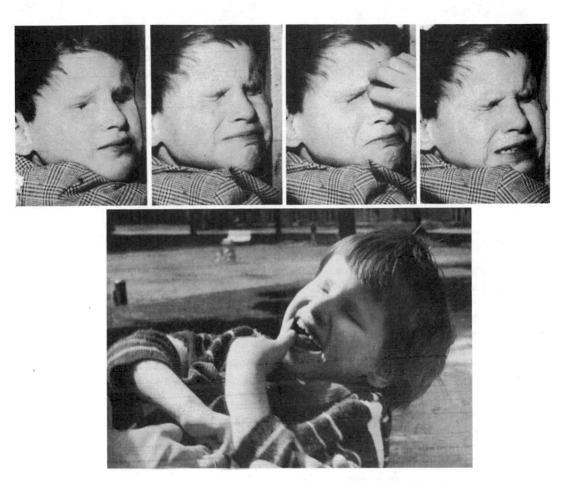

Figure 9.5   Emotional expressions are innate. A young girl, blind and deaf since birth, weeping and laughing. (From Eibl-Eibesfeldt I. Ethology. New York: Holt, 1975.).

(Rolls and Williams, 1987; Schneider and Lidsky, 1981). Striatal neurons can also react differently to familiar stimuli depending on their reinforcement properties, and many neurons will, in fact, increase their responsiveness as stimuli approach the mouth, and/or touch the face and mouth (Rolls and Williams, 1987; Schneider and Lidsky, 1981). Some striatal neurons also respond during tongue and lip movements (e.g., licking) and during arm movements toward a food item (Rolls and Williams, 1987).

These findings suggest the striatum is involved in orienting and guiding movements toward the mouth, presumably so that a desired object can be licked, sucked on, chewed, and consumed. In this regard, the striatum could be considered a primary motor center that enables various limbic desires, needs, and impulses, such as hunger, to be satisfied.

The limbic striatum (e.g., the nucleus accumbens) and the mesolimbic DA system that projects to these nuclei also appear to be highly involved in mediating feelings of pleasure, including the rewarding effects of amphetamines, cocaine, and opiates (see Ellison, 1994; Hakan et al., 1994; Koob et al., 1991). These effects (including the aversiveness of opiate withdrawal) are probably due, not only to the presence of opiate, DA, and related receptors, but to the rich interconnections maintained with the amygdala as well as the lateral hypothalamus (Kelsey and Arnold, 1994; Olton et al., 1991; Zaborszky et al., 1991), which constitute part of the "pleasure circuit" maintained by the medial forebrain bundle (Olds and Forbes, 1981; see also Chapter 5).

However, animals will also work in order to self-administer opiates directly into the accumbens (Koob et al., 1991; Olds and Forbes, 1981). Conversely, lesions to the nucleus accumbens may disrupt the capacity to experience pleasure or the rewarding effects of opiates and cocaine (Koob et al., 1991), or to engage in complex coordinated defensive acts (see below).

CATATONIA, PARKINSON'S DISEASE, AND PSYCHOSIS

Lesions to the corpus striatum and lenticular nucleus (putamen and GP) can attenuate one's capacity to motorically express his or her emotions via the musculature; e.g., the face may become frozen and mask-like. These latter motor disturbances are well-known symptoms associated with Parkinson's disease, a disturbance directly linked to dopamine deficiency and related neuronal degeneration, not only in the putamen (Goto et al., 1990; Kish et al., 1988), but within the limbic striatum, i.e., the nucleus accumbens (see Rolls and Williams, 1987), as well as in the supplementary motor areas and medial frontal lobe, which maintains rich interconnections with the striatum. However, when chemical or structural lesions extend beyond the basal ganglia and come to include the medial frontal lobe, not only might an individual suffer motor rigidity, he or she may become catatonic and experience extreme difficulty responding to external or internally mediated impulses (Joseph, 1990d). Various aspects of this symptom complex also characterize those with Parkinson's disease (see below).

Other disturbances associated with striatal abnormalities include Huntington's chorea, ballismus, sensory neglect and apathy, obsessive compulsive disorders, mania, depression, "schizophrenia," and related psychotic states (Aylward et al., 1994; Baxter et al., 1992; Caplan et al., 1990; Castellanos et al., 1994; Chakos et al., 1994; Davis, 1958; Deicken et al., 1995; Ellison, 1994;

Rauch et al., 1994; Richfield et al., 1987). Severe memory loss and social-emotional agnosia and an inability to recongize friends or loved ones is also characteristic of striatal abnormalities, particularly disturbances involving the limbic striatum.

Thus, although the basal ganglia is often viewed and described as a major motor center (detailed below), the functional capacities and symptoms associated with this group of nuclei are quite diverse, and vary depending on the nuclei and chemical neurotransmitters involved as well as the laterality, location, and extent of any lesion.

THE CORPUS STRIATUM

PATCHES AND MATRIX

The caudate and putamen are tightly interlinked, and in some respects are indistinguishable and possess a similar internal compartmental structure of patches and matrix (Graybiel, 1986; Gerfen, 1984). This is why a gross analysis of the the mammalian caudate and putamen reveals a striated (patchlike) appearance.

The patches and matrix are biochemically distinct and receive projections from different regions of the neuroaxis (Gerfen, 1984, 1987; Graybiel, 1986). For example, the patches contain dense concentrations of opiate receptors (Graybiel, 1986), receive projections from the amygdala, hippocampus, and other limbic tissue, and maintain interconnections with the DA neurons in the substantia nigra. The patches, in fact, form a continuous labyrinth that snakes throughout the striatum.

The surrounding matrix also receives projections from the cingulate gyrus and the motor thalamus, and from throughout the neocortex, and maintains interconnections with GABA and DA neurons in the substantia nigra (Gerfen, 1987). The matrix also contains large amounts of acetylcholinesterase.

MOTOR FUNCTIONS

Although tightly linked and similar in structural organization, the caudate appears to exert more influence and provide more input to the putamen, than vice versa. However, like the caudate, the putamen also receives considerable input from the medial frontal, supplementary, secondary, and primary motor cortex, as well as from areas 5 and 7 of the parietal lobe (Jones and Powell, 1970; Pandya and Vignolo, 1971).

As per motor functioning, presumably the putamen, in conjunction with the caudate, transmits this information to the GP, which in turn projects to the motor thalamus and brainstem reticular formation, as well as to the motor neocortex, thus creating a very elaborate feedback loop (see Mink and Thach, 1991; Parent and Hazrati, 1995) whose origin may begin in the medial frontal lobes (Alexander and Crutcher, 1990; Crutcher and Alexander, 1990) or perhaps the limbic system (Joseph, 1990a, 1992a), e.g., anterior cingulate and amygdala.

For example, when an individual is anticipating or preparing to make a movement, but prior to the actual movement, neuronal activity will first begin and then dramatically increase in the medial supplementary motor areas (SMA); activity will follow in the secondary and then the primary motor areas (see Chapter 11), and then the caudate, and last of all the putamen and GP (Alexander and Crutcher, 1990; Mink and Thach, 1991).

Hence, these areas, including the "motor" thalamus, in many respects act in a step-wise, coordinated fashion so as to mediate purposeful movement. As noted in Chapter 1, this was a principle role of the basal ganglia long before the evolution of the neocortex and frontal motor areas.

Some investigators have argued there are at least five different motor circuits involving the basal ganglia that are segregated to varying degrees (Alexander et al., 1990). The dorsal striatal motor circuit also consists of at least two separate systems involving the putamen and medial (internal) GP, and the putamen and lateral (external) GP (reviewed in Marsden and Obesco, 1994; Parent and Hazrati, 1995).

FUNCTIONAL DIFFERENTIATION OF THE CAUDATE AND PUTAMEN

Over the course of evolutionary metamorphosis, the corpus striatum and the motor thalamus began to develop in tandem and became increasingly interlinked in order to subserve motoric and related activities, including the processing and analysis of sensory information and the expression of feeling states via specific motor activities. That is, the corpus striatum initially served not only the motor functions and related information requirements of the limbic system, but in many respects performed (at a rudimentary level) some of the same "analytical" and perceptual functions that would later be subsumed by the neocortex.

With the continued expansion of the the neocortex and the exponential increase in the capacity to analyze and respond to divergent sensory information, the corpus striatum essentially was split in two by the tremendous proliferation of thalamic axons (i.e., the internal capsule, or rather, the thalamic radiations) which not only terminated on striatal dendrites, but which swept forward and radiated outward to innervate the frontal lobes (Kemp

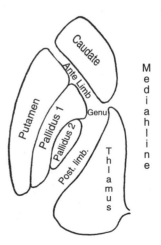

Figure 9.6 Diagram depicting the splitting apart of the striatum (caudate and putamen) in relationship to the thalamus. (Redrawn from Krieg W. Functional neuroanatomy. New York: Bakiston, 1953.)

and Powell, 1970). Thus, the putamen and caudate nucleus were formed and began to receive differential input from the thalamus as well as from the neocortex, and therefore began to subserve somewhat different functions. For example, although both nuclei contain significant amounts of dopamine (DA), 5-hydroxytryptamine (serotonin), adrenal cortical hormone, and γ-aminobutyric acid (GABA) (reviewed in Ellison, 1994; Parent and Hazrati, 1995; Stoof et al., 1992) and receive input from the amygdala and hippocampus, as well as motor and somatosensory perceptual data (Haber et al., 1985; Van Hoesen et al., 1981; Whitlock and Nauta, 1956), the putamen is the recipient of considerable bilateral and topographical input, such that a motor and sensory map of the body (particularly the face, mouth, leg, and arm) is maintained in this region (Delong et al., 1983; Parent and Hazrati, 1995). Axonal projections from the motor and sensory neocortex that are concerned with the arm converge in one area of the putamen, whereas those concerned with the leg converge in another.

From these observations, in conjunction with the symptoms and experiments described below, it is suspected that the putamen is concerned with integrating sensory with intended motor actions and coordinating the movement of the limbs and body in visual space via projections maintained with the medial and lateral GP, as well as the parietal lobe.

In contrast, the caudate nucleus is dominated by axons from association cortices, including the inferior temporal lobe and anterior cingulate (Percheron et al., 1987), the amygdala (Amaral et al., 1992; Heimer and Aheid, 1991) and the frontal motor areas. The caudate appears to be more involved in multimodal motor, emotional, and sensory integration, analysis and inhibitory functions. Consequently, lesions to the caudate can produce sensory neglect and unresponsiveness, or, conversely, loss of inhibitory control over the musculature, depending on the extent and laterality of the lesion.

The Caudate: Mania, Apathy, and Catatonia

The head of the caudate nucleus is concerned with multimodal information processing and inhibition. Via inhibition, the caudate is able to exert modulatory effects on motor activity and facial-gestural posture and expression, and aids in the maintenance of selective motoric attention, e.g., standing still and observing. In consequence, in primates and humans, lesions, destruction, or shrinkage of the head of the caudate can result in sensory neglect, agitation, hyperactivity, distractibility, and, in some cases, what appears to be a manic or "schizo-affective" psychosis (Aylward et al., 1994; Caplan et al., 1990; Castellanos et al., 1994; Chakos et al., 1994; Davis, 1958; Richfield et al., 1987), depending on the extent and laterality of the destruction.

For example, Richfield et al. (1987:768) report a 25-year-old female honor student (soon to be married) who, after complaining of headaches and nausea, disappeared for 3 days. "When found, she had undergone a dramatic personality change manifested by alterations in affect, motivation, cognition, and self-care." These changes were largely permanent. "Her abnormal behaviors included vulgarity, impulsiveness, violent outbursts, enuresis, indifference, hypersexuality, shoplifting, and exposing herself. She was inattentive and uninterested in her surroundings but could be encouraged to concentrate for short periods of time. She would frequently lie down to sleep. Her affect was flat." CT scan indicated bilateral damage to the head of the caudate nuclei.

Some patients will alternate between hyperactivity and apathy—a function, perhaps, of the laterality and extent of the lesion as well as associated biochemical alterations. For example, injuries or abnormalities restricted to the right caudate are more likely to result in a manic-like psychosis (Castellanos et al., 1994)—quite similar to what occurs after right frontal lobe injury (Joseph 1986a, 1988a, 1990d). Similarly, right caudate hypermetabolism and

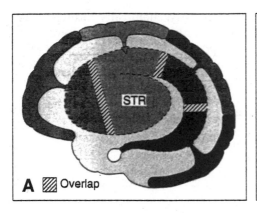
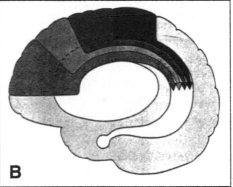

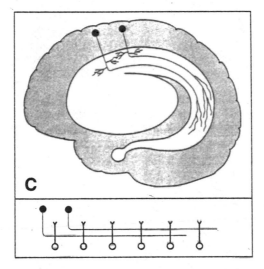

Figure 9.7   Schematic diagram depicting three different views of the corticostriatal projection system. **(A)** As advocated by Kemp and Powell (1970), who suggest a point-to-point relationship between the striatum and cortex. **(B)** As advocated by Goldman-Rakic et al. (1982), who propose that terminal fields of the cortical areas occupy distinct domains. **(C)** Hypothetical view suggested by A. Parent and L-N. Hazrati that axons from cortical neurons *(black circles)* follow a linear and poorly branched trajectory. (From Parent A and Hazrati L-N. Functional anatomy of the basal ganglia. I. II. Brain Research Reviews 20:91–154, 1995.)

blood flow has also been associated with obsessive compulsive disorders (Baxter et al., 1992; Rauch et al., 1994), whereas frontal abnormalities may induce perseverative disturbances (see Chapter 11) as well as obsessive-compulsions (Rauch et al., 1994).

Conversely, left caudate injuries (particularly those that extend to the mesial and left frontal lobe) may induce severe apathy as well as speech disturbances. As noted, left frontal injuries are associated with similar abnormalties (see Chapter 5 and 11).

With extensive left or bilateral caudate injuries it is not uncommon for patients to appear agitated or apathetic, with decreased spontaneous activity and slowed, delayed, dysarthric (or stuttering) and emotionally flat speech, with some patients responding to questions only after a delay of 20–30 seconds (Caplan et al., 1990). This condition is particularly likely if the medial frontal lobes have been compromised as well.

In fact, due perhaps to its extensive interconnections with the medial frontal lobes and SMA—a region that, when destroyed, can give rise to catatonia (Joseph 1990d), massive bilateral lesions to the caudate and anterior putamen and surrounding tissue has been shown to produce catatonic or "frozen" states, where animals show a tendency to maintain a single posture or simply stand, unmoving, for weeks at a time (Denny-Brown, 1962).

Similarly, chemically-induced lesions of the caudate can produce complete catatonia, posturing, and a cessation of all movement (Spiegel and Szekely, 1961). However, if the amygdala-striatal pathway and/or the amygdala is destroyed prior to lesioning the caudate, these frozen catatonic states can no longer be induced (Spiegel and Szekely, 1961).

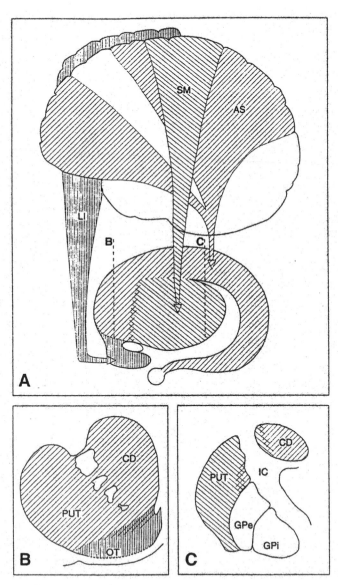

Figure 9.8 Schematic diagram depicting the projections of the limbic **(LI)**, sensorimotor **(SM)**, and association **(AS)** cortical areas. Hatched areas depict zones of overlap between limbic, sensorimotor, and associative striatal territories. Dotted lines in A indicate the levels of the transverse sections depicted in B and C. (From Parent A and Hazrati L-N. Functional anatomy of the basal ganglia. I. II. Brain Research Reviews 20:91–154, 1995.)

CATATONIA AND THE FRONTAL-CAUDATE-AMYGDALA NEURAL NETWORK

As noted, the head of the caudate is extensively interconnected with the frontal lobes. Thus, some of the symptoms associated with caudate destruction (e.g., mania, apathy, catatonia) parallels those associated with frontal lobe injury (Joseph, 1990d). For example, whereas lesions to the frontal lobe may produce perseverative abnormalities, caudate lesions and/or increased metabolism or regional blood flow may induce obsessive-compulsive behaviors (Baxter et al., 1992; Rauch et al., 1994). In part, this may represent a loss of frontal lobe control over the caudate, and/or the loss of striatal inhibitory influences over the frontal lobe.

However, the caudate and frontal lobes are also intimately linked to the amygdala (Ammaral et al., 1992; Joseph, 1990a, 1992a), a nucleus that, under conditions of extreme fear and arousal, can induce a complete cessation of movement, i.e., catatonic-like frozen panic states (see Chapters 5 and 16). The amygdala is able to accomplish this via interconnections with the basal ganglia, and brainstem, as well as the medial frontal lobes.

If the medial frontal lobes are damaged or electrically stimulated, there sometimes results an inability to initiate a voluntary movement, and the "will" to move or speak may be completely attenuated and abolished (Hassler, 1980; Laplane et al., 1977; Luria, 1980; Penfield and Jasper, 1954; Penfield and Welch, 1951).

Similarly, lesions of the medial frontal lobes can cause subjects to simply sit or stand motionless, as if frozen, and display waxy flexibility, or "gegenhalten" or counterpull—i.e., involuntary resistance to movement of the extemities (Brutkowski, 1965; Hasslet, 1980; Laplane et al., 1977; Luria, 1980; Mishkin, 1964). Posturing is also noted, and patients might remain in odd and uncomfortable positions for exceedingly long periods, making no effort to correct the situation (Freeman and Watts,

1942; Rose, 1950), as if they were dead and in the first stages of rigor mortis.

Similar reactions (see below) occur after disasters and in instances of extreme fear and terror. Affected individuals may display all the classic symptoms of catatonia, which, in many respects, could be likened to death feigning or "playing possum."

Presumably, in response to extreme fear the amygdala acts on the striatum and SMA (as well as the brainstem), thereby inducing not only rigidity and a frozen panic state, but a classic catatonic state accompanied by waxy flexibility, and a complete paralysis of the will. The SMA, in fact, is very susceptible to the disruptive influences of stress (Finlay and Abercrombie, 1991).

The Amygdala, Striatum, SMA, and Life-Threatening Fear and Arousal

In situations involving exceedingly high levels of arousal coupled with extreme fear, the individual may simply freeze; attentional functioning may become so exceedingly narrow that little or nothing is perceived, and cognitive activity may be almost completely (albeit temporarily) abolished (see Chapter 16). These behaviors are apparently under the control of the amygdala, which can trigger a "freezing" reaction and a complete arrest of ongoing behavior (Gloor, 1960; Kapp et al., 1992; Ursin and Kaada, 1960) via brainstem/striatal interconnections. This is part of the amygdala attention response, which, at lower levels of excitation, may be followed by anxious glancing about, an increase in respiration and heart rate, pupil dilation, and perhaps cringing and cowering or flight (Gloor, 1960; Ursin and Kaada, 1960). Among humans, the fear response is one of the most common manifestations of amygdaloid stimulation (Gloor, 1990; Halgren, 1992; Williams, 1956). However, if arousal levels continue to increase, subjects do not merely freeze in response to increased fear, they may become catatonic—a condition that may be secondary to dopamine and serotonin depletion and amygdaloid influ-

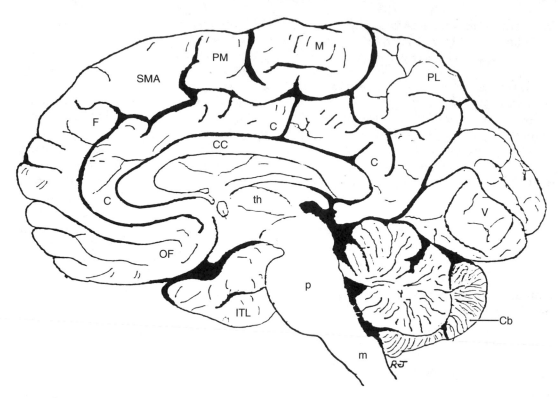

Figure 9.9   Medial view of the right hemisphere depicting the supplementary motor areas **(SMA)** and medial frontal lobe.

ences on the SMA as well as the striatum, nuclei that are intimately interconnected.

For example, in response to extreme fear, "one tendency is to remain motionless, which reaches its extreme form in death-feigning in certain animals and sometimes produces the waxy flexibility of catatonics" (Miller, 1951). The affected individual becomes psychologically and emotionally numb and unresponsive, which is coupled with a complete blocking off of cognition. Moreover, the individual may resist and fail to respond to attempts at assistance (Krystal, 1988; Miller, 1951; Stern, 1951). The airline industry has referred to this as "frozen panic states" (Krystal, 1988), a condition sometimes seen in air and sea disasters. For example, in mass disasters, 10–25% of the victims will become frozen, stunned, and immobile, and will fail to take any action to save their lives, such as attempting to evacuate a burning or sinking craft, even though they are not injured (see Krystal, 1988).

According to Krystal (1988), with increasing fear "there is also a progressive loss of the ability to adjust, to take the initiative or defensive action, or act on one's own behalf...that starts with a virtual complete blocking of the ability to feel emotions and pain, and progresses to inhibition of other mental functions" (Krystal, 1988:151).

EVOLUTIONARY SIGNIFICANCE OF RIGIDITY

Catatonic panic states are prevalent in the animal kingdom, and constitute a life-preserving reaction that is apparently mediated by the amygdala and striatum. That is, when the animal is freezing and not moving, predators may fail to take note of their presence. During war, soldiers who are lost behind enemy lines are also most likely to escape detection by lying still and not moving.

Catatonia, coupled with emotional and "psychological" numbing, also represents a total surrender reaction, usually as a pre-lude (and hopeful guarantee) of a painless death when attacked by predators or in-vaders. That is, the prey may cease to run or fight and may simply stand still or lie down and allow predators to literally eat them alive. Or, in the case of humans, as sometimes occurs during war and genoci-dal mass murders, individuals may pas-sively allow themselves to be marched into a ditch and shot. According to Krystal (1988:144), "thousands of European Jews obeyed orders in an automatonlike fashion, took off their clothes, and together with their children descended into a pit, lay down on top of the last layers of corpses, and waited to be machine-gunned," all the while seemingly almost petrified with fear and/or completely numb as to what was going on around them. Presumably, this numbing is made possible via the massive secretion of opiates within the amygdala and basal ganglia, whereas the rigidity and loss of the will to resist is a consequence of overwhelming fear and hyper-amygdala influences on the medial frontal lobe and corpus and limbic striatum.

Some animals, however, instead of run-ning in fear, will simply "freeze," fall to the ground, and lie stiff, rigid, and motionless, as if dead.   Unless exceedingly hungry, many predators will avoid eating creatures that appear to be already dead (i.e., un-responsive). Again, this is a very adaptive response that can promote survival. As sometimes occurs to potential victims dur-ing mass killings (e.g., the 1994 Rwanda civil war between the Hutus and the Tutsis) humans too, will sometimes fall down as if dead and may remain frozen, stiff, and and unmoving for long periods, even though they may not have been harmed (Krystal, 1988). Indeed, sometimes these individuals are believed to be dead even by rescuers or those who are clearing away and burying bodies.

Presumably, it is via connections with the basal ganglia and medial frontal lobes that the amygdala is able to induce these catatonic states, which in part is also de-pendent on dopamine. For example, it has been demonstrated that under extremely stressful conditions the striatal and frontal lobe DA system is adversely affected (see Le Moal and Simon, 1991).

Implications Regarding Parkinson's Disease

As noted, many of the functions origi-nally associated with the basal ganglia have been subsumed by the neocortex, and damage to the frontal lobes can produce essentially similar symptoms as seen fol-lowing caudate destruction, including stiff-ness, rigidity, and difficulty initiating movement (see Chapter 11).

Given that the medial frontal lobes, cor-pus and limbic striatum, and amygdala are extensively interconnected, and given the powerful influences of the limbic system on all aspects of behavior, it thus appears that when exceedingly aroused or emo-tionally stressed, the amygdala is able to inhibit (or overactivate) the frontal-striatal motor centers, which (in addition to the amygdala) are simultaneously undergoing DA depletion (which in turn results in hy-peractivation of these nuclei, including the amygdala; see Le Moal and Simon, 1991). When this occurs, the organism may fall and cease to move, blink, or even breathe (or breathe only shallowly and slowly). The creature, therefore, appears to be in a state of rigor mortis and, thus, dead (i.e., catatonic).

Overall, these amygdala-basal ganglia-medial frontal lobe-fear induced frozen and catatonic states are exceedingly adap-tive—that is, unless the hapless victim is in a burning airplane or sinking ship.

Although not as dramatic or severe, sim-ilar states occur with biochemical abnor-malities involving the dopamine pathways from the substantia nigra, which not only feed the caudate but the medial frontal lobes and the amygdala, as well. Specifi-cally, loss of DA (or excessive amygdaloid arousal) results in motor neuron hyperac-tivity and tonic EMG activity and, thus, limb and facial rigidity—conditions that also afflict those with Parkinson's disease.

In that some of these same "semi-frozen" and akinetic states are present in many of those with Parkinson's disease. Given that those with Parkinson's are sometimes described as excessively aroused and/or unable to relax, and to suffer from heightened autonomic nervous system activity (Stacy and Jankovic, 1992), it is possible that the amygdala and related limbic nuclei may significantly contribute to the development of this disorder. Indeed, as noted, destruc-

tion of the amygdala prior to chemically lesioning the corpus striatum prevents the development of Parkinsonian symptoms.

PARKINSON'S DISEASE

The basal ganglia plays a major role in controlling and facilitating specific movements, as well as inhibiting unwanted movements (Marsden and Obesco, 1994). The basal ganglia is also directly implicated

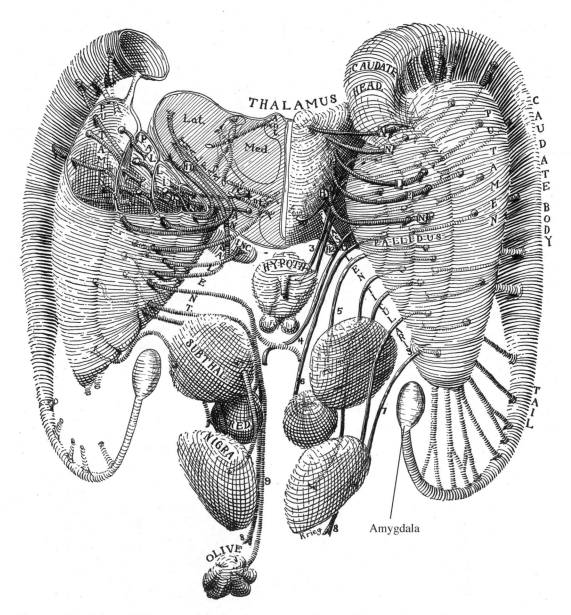

Figure 9.10   Schematic diagram of the basal ganglia, including the thalamus, subthalamic nucleus, and amygdala. (Redrawn from Krieg W. Functional neuroanatomy. New York: Bakiston, 1953.)

in the generation of a variety of movement disorders, including chorea, ballismus, rigidity, stiffness, and Parkinson's disease. Parkinson's disease is a progressive degenerative disorder characterized by rigidity and shuffling gait, stooped posture, generalized slowness and stiffness of movement, and a loss of facial emotional expression, as well as a loss of spontaneity and flexibility in making postural adjustments when eating, going to the toilet, or having sex. Moreover, many Parkinson's patients not only experience rigidity and akinesia but suffer from episodic freezing of movement (Dietz et al., 1990; Pascual-Leone et al., 1994), a tendency to easily fall (Stacy and Jankovic, 1992), an impairment of "righting reflexes" (Calne, 1994), and a reduced capacity to blink (Freedman, 1992, and even to breathe (Stacy and Jankovic, 1992), similar to frozen panic states and induced catatonia.

Hypophonia (reduced voice volume) dysarthria and a tendency to speak in a monotone, micrographia (small handwriting), and a 4–8 c/sec ("pill rolling") resting tremor (exacerbated by stress) involving antagonistic muscles are common (Freedman, 1992; Stacy and Jankovic, 1992). Depression, changes in personality, and slowed thought processes are also not unusual among Parkinson's patients.

Parkinsonism can result following neurological trauma or vascular abnormalities (Koller, 1987; Murrow et al., 1990) or from isolated lesions to the substantia nigra (Stern, 1966). Its precipitants, therefore, are numerous and varied, including infection and toxic exposure. Parkinson's disease may also overlap with other striatal disturbances, such as Alzheimer's disease (Calne, 1994). Indeed, sigificant cognitive deficits, i.e., subcortial dementia, are not unusual among those with Parkinson's disease (Freedman, 1992; Rajput, 1992).

Parkinson's disease usually begins after age 60 (though the first signs may appear during the early 40s), affects about 2% of the population (Koller, 1987), and is usually characterized by a massive loss of up to 80–85% of the dopamine neurons in the substantia nigra and 80% of striatal DA,

with DA depletion greatest in the putamen (Goto et al., 1990; Kish et al., 1988). By contrast, the mesolimbic pathways and limbic striatum are only mildly affected. However, DA depletion also occurs within the SMA, which is also implicated in the genesis of Parkinson's disease (see below).

Striatal Imbalance and Parkinson's Disease

The differential involvement of the nigrostriatal versus mesolimbic DA system in Parkinson's disease raises the possibility that, whereas the corpus striatum and medial frontal lobes are negatively impacted, the amygdala and limbic striatum may continue to function normally—due to preservation of the mesolimbic DA system. However, because the normal balance between these various nuclei is disrupted, amygdala influences received within the SMA and corpus striatum may, in fact, overwhelm and massively inhibit (or overactivate) these nuclei. This, thereby, gives rise to Parkinsonian symptoms (see Le Moal and Simon, 1991, for related discussion): i.e., rigidity and a tendency to fall, coupled with reduced blinking, disturbed righting reflexes, and so forth.

In addition to DA in the genesis of Parkinsonian symptoms, significant reductions in opiate receptors within the putamen as well as the GP have also been reported (Goto et al., 1990), such that neurons involved in the experience of "reward" are affected.

This selective loss of "reward" neurons suggests a reduction in the capacity of the basal ganglia to receive pleasurable or positive emotional input as might be provided not only by opiates, but via the lateral amygdala and lateral hypothalamus.

Possibly, because striatal neurons associated with negative feeling states may be selectively preserved in Parkinson's disease, the basal ganglia (but not the amygdala, hypothalamus, or neocortex) may respond as if in a highly aroused negative state (e.g., fearful, stressed), even when that is not the case. Adding to this imbalance would be the relative preservation of the mesolimbic DA system, and the continued input from the medial amygdala into the corpus and limbic striatum—the medial amygdala being more involved in the generation of unpleasant, including fearful, mood states (see Chapter 5). Indeed, loss of corpus striatal DA and/or excessive amygdaloid arousal is directly associated with the development of tonic EMG activity, motor neuron hyperactivity, and, thus, excessive tonic excitation of the musculature and limb rigidity—i.e., Parkinson's symptoms. Conversely, excess striatal DA can result in chorea and excessive movement, as well as psychosis.

The Supplementary Motor Areas and Parkinson's Disease

It has been reported that many patients with Parkinson's disease display reductions in norepinephrine (NE) as well as DA throughout the upper layers of the motor, premotor, and medial frontal lobes (Gaspar et al., 1991; Hornykiesciz, 1982). NE (and 5HT) depletion also occurs following heightened emotional states, including extreme fear, prolonged stress, and heightened arousal, and with amygdala hyperactivation. As noted, individuals with Parkinson's disease are also sometimes described as excessively aroused and/or unable to relax, and to suffer from heightened autonomic nervous system activity and gastrointestinal disturbances (Stacy and Jankovic, 1992)—disturbances that also implicate the amygdala and excessive limbic system arousal. Conversely, the autonomic nervous sytem of those who suffer from Alzheimer's disease has been characterized as hypoactive (Borson et al., 1992), the limbic striatum being implicated in Alzheimer's disease.

The SMA typically becomes functionally activated prior to movement (see Chapter 11) and well before the caudate or putamen with which it is interconnected. When Parkinson's patients attempt to make repetitive movements, SMA functional activity is abnormally reduced, as is putamen activity (Playford et al., 1992).

This suggests that, in addition to the corpus striatum, one of the major regions of the neuroaxis involved in the "will" to move and the capacity to prepare to move (i.e., the SMA) is dysfunctional in those suffering from Parkinson's disease. Thus, the SMA and medial frontal lobes, as well as the corpus striatum, DA system, and amygdala, are implicated in the genesis of this disorder (e.g., Playford et al., 1992).

Consider, for example, that Parkinson's patients have difficulty using both hands to perform two different movements (e.g., squeeze a ball with one hand, draw a triangle with the other). Rather, they tend to perform one action, then switch to the other (Caligiuri et al., 1992). SMA dysfunction can result in similar disturbances (Joseph, 1990d; see Chapter 11). Similarly, the ability of Parkinson's patients to engage in sequential and repetitive movements (Playford et al., 1992), or to simultaneously engage in separate movements such as walking, turning, and talking (Stacy and Jankovic, 1992), is severely affected—a disorder that is also associated with frontal lobe abnormalities.

The Hippocampus and Supplementary Motor Areas

As noted, the striatum is also richly innervated by the hippocampus, which is involved, not only in memory, but in spatial and cognitive mapping of the environment (see Chapters 5 and 6). Thus, the corpus striatum presumably relies on hippocampal (as well as parietal lobe) input in order to coordinate movement in visual-space.

It is noteworthy that Parkinson's patients find it easier to perform movements that are cued or guided by external visual stimuli, whereas movements that are internally generated are more difficult to perform (Dietz et al., 1990). Hence, perhaps those striatal neurons that normally receive and integrate hippocampal input may also be dysfunctional.

However, the role of the hippocampus appears to be minimal in the development of Parkinson's disease, at least in those cases without memory loss or dementia. Instead, the beneficial effect of external cues again raises the specter of significant medial frontal lobe and SMA involvement.

For example, lesions to the medial frontal lobes can result in an inability to internally generate movements, whereas external environmental and visual cues can exert an almost "magnetic" effect on limb activation. Patients may involuntarily respond and react to external stimuli by reaching, grasping, or utilizing whatever objects may be near (see Chapter 11). In contrast, the lateral frontal lobes are involved in externally driven movements (see Chapter 11).

Hence, Parkinson's patients are possibly better able to "move" in response to visual stimuli, presumably because of preserved lateral frontal lobe functioning and the loss of inhibitory SMA control, which (in part) accounts for their difficulty internally "willing" purposeful and fluid movements.

Dopamine, Acetylcholine, and Striatal Imbalance

The DA and cholinergic, ACh system appear to exert counterbalancing influences. For example, loss of nigrostriatal DA results in ACh, neuron hyperactivity (see Aghanjanian and Bunney, 1977; Bloom et al., 1965), and tonic EMG activity and, thus, limb rigidity, which can be reversed by anticholinergic drugs (Klockgether et al., 1987). That is, drugs that decrease ACh and those that increase DA can ameliorate Parkinson's symptoms, which suggests these two transmitters play oppositional and balancing roles (see Stoof et al., 1992). Moreover, ACh can be increased by drugs that decrease DA, and can be decreased by drugs that increase DA (reviewed by Stoof et al., 1992).

DA normally inhibits ACh release and increases GABA activity. The DA system appears to exert controlling influences, not only on GABA but on ACh neurons in the striatum (Le Moal and Simon, 1991). Hence, reductions in DA result in decreased GABA activity and increased lim-

bic striatal arousal and ACh activity, and reduced corpus striatal arousal and reduced GABA activity (see Stoof et al., 1992). Under these latter conditions the subject becomes rigid.

However, under conditions where striatal DA is depleted, the excessive release of ACh creates excessive neural activity (Aghanjanian and Bunney, 1977; Bloom et al., 1965), which, in turn, may be manifested in the form of excessive motor actions—i.e., ballismus and chorea, if the medial and ventral GP-subthalamic circuit is affected, or hypomovement and Parkinson's symptoms if the corpus striatum and dorsal GP are affected.

For example, fluctuations in DA levels can either act to excite or inhibit motor and related cognitive-memory activity within the limbic striatum (Le Moal and Simon, 1991; Mogenson and Yang, 1991). However, with high levels of limbic striatal DA (and/or amygdala) activity, and reduced GABA, animals may appear hyperactive and engage in running, "galloping," and biting or related movements.

Presumably, these latter actions are due to inhibitory release (a consequence of striatal dysfunction or hyperactivation) and, thus, the triggering of brainstem motor programs that subserve these specific behavioral acts. These same stimulus (predator) released actions, across evolution and across most animal species, are directly related to the functional integrity of the limbic system and striatum, and the capacity to survive when living in close proximity to predators, the elements, and cospecies. Reflexive actions without thought are, however, the province, not of the basal ganglia, but the brainstem, which is why these and related motor programs are stored within brainstem nuclei (see Chapter 10).

PSYCHOSIS

A major function of the head of the caudate (particularly the caudate nucleus of the right hemisphere) appears to be inhibition. Similarly, the principle effect of striatal DA is inhibition (Ellison, 1994; Le Moal and Simon, 1991; Mercuri et al., 1985). Hence, excessive and abnormal amounts of corpus striatal DA can result in increased levels of caudate-putamen activity such that limbic striatal, and the reception of amygdaloid/emotional input, may be severely diminished and/or abnormally processed. Thus, excessive corpus striatal activity might also result in emotional blunting or distortion as well as cognitive abnormalities. Conversely, because benzodiazepine receptors are located throughout the striatum, antipsychotic agents, due to their dopamine properties, can reduce this abnormal activation.

That is, abnormal levels of corpus striatal DA activity can interfere with selective attention and the reception of neocortical and limbic impulses within the striatum, as can increases in mesolimbic DA activity (Le Moal and Simon, 1991). However, with increases in striatal DA, cognitive, social, and emotional integrative functions may become abnormal and distorted, and the individual may become psychotic (see Castellanos et al., 1994; Chakos et al., 1994; Ring et al., 1994; Snyder, 1972).

As is well known, individuals diagnosed as paranoid and schizophrenic with psychomotor retardation have been found to have elevated DA levels and increased DA receptor binding (Crow, 1979; Ellison, 1994; Matthyse, 1981; Ring et al., 1994). Presumably, this is why dopamine blockers, such as the phenothiazines, are effective in reducing psychotic behavior; i.e., they reduce excessive DA binding and excessive striatal activity, which in turn restores the capability of the striatum (and related nuclei) to engage in cognitive-emotional integrative activities.

THE AMYGDALA, TEMPORAL LOBE, AND DOPAMINE

Psychotic disturbances, including "schizophrenia," need not involve the basal ganglia or DA dysfunction, as disorders of thought and mood may also appear following frontal or temporal lobe damage (see Chapter 4). However, with frontal lobe

damage the basal ganglia and SMA are often affected, whereas the amygdala is implicated in psychotic states related to the temporal lobe (see Chapters 16 and 17). Indeed, abnormal DA as well as norepinephrine (NE) and serotonin (5HT) have been reported in the amygdala of those diagnosed as psychotic (Spoont, 1993; Stevens, 1992). In addition, a gross asymmetry in the postmortem levels of mesolimbic DA has been noted in the brains of those diagnosed as schizophrenic (reviewed in Le Moal and Simon, 1991). Specifically, an increase in DA has been noted in the central amygdala and in the striatum of the left hemisphere, whereas DA levels within the right cerebral hemisphere are similar to those in normal persons.

In fact, the left amygdala and left temporal lobe (Flor-Henry, 1969; Perez et al., 1985; Stevens, 1992) have long been thought to be major components in the pathophysiology of psychosis and schizophrenia (Heath, 1954; Stevens, 1973; Torey and Peterson, 1974). For example, abnormal activity as well as size decrements have been noted in the left amygdala (Flor-Henry, 1969; Perez et al., 1985) as well as the left inferior temporal lobe in schizophrenic patients (see Stevens, 1992). Spiking has also been observed in the amygdala of psychotic individuals who are experiencing emotional and psychological stress (see Halgren, 1992). In this regard, it has been suggested that stress, particularly when experienced prenatally, may be responsible, in some cases, for the development of DA abnormalities (see Le Moal and Simon, 1991), whereas adverse early environmental experiences and trauma may contribute to the development of kindling, spiking, and abnormal amygdala functioning (see Chapter 16). As noted, Parkinson's symptoms, in some respects, also resemble the effects of chronic stress and an inability to relax.

Obsessive-Compulsive Disorders

Stress and emotional turmoil can be induced via external agents, as well as via one's personal and private thoughts and inclinations—particularly if, at a "conscious" level, the individual finds these impulses objectionable. However, whether internally or externally generated, the amygdala (and hypothalamus), as well as the frontal-striatal complex, is often involved and/or affected. This, in turn, can influence motor functioning in a number of ways, including the development of obsessive compulsive disturbances.

Obsessive-compulsive disorders are characterized by the repetitive experience of unwanted and recurrent, perseverative thoughts (obsessions) and/or a compulsion to repetitively perform certain acts. The worldwide prevalence rate is about 2.3% (see Okasha et al., 1994). Some patients tend to suffer from compulsions whereas others experience obsessions, with the majority experiencing both (DSM-IV).

Not uncommonly, obsessions tend to be religious, or concerned with contamination coupled with cleaning and washing rituals. The experience of intrusive and unpleasant (e.g., violent and/or sexual) images is not uncommon (Lipinski and Pope, 1994).

In some instances, obsessive compulsions serve as a means of coping with unwanted or troubling impulses; i.e., by perseverating in thinking certain thoughts or engaging in certain actions, the individual is able to avoid focusing on and recognizing thoughts, actions, or impulses that he or she finds disturbing.

For example, consider "Bob," a 30-year-old socially inept, single, male with a BA degree in English who, for several years, voiced homophobic and right-wing, extremist views and who experienced and engaged in a compulsion to work on his car and/or dust and wash the engine late in the evening, even after midnight. After neighbors repeatedly complained about the noise, he was forced to cease these acts, underwent a noticeable personality change, and was soon frequenting bars late at night, taking midnight drives, and having sex with men he met while intoxicated.

As noted, the basal ganglia is intimately associated with the hypothalamus, amyg-

dala, and orbital frontal lobes—nuclei that are also associated with sexual activity and the mediation of emotional functioning, with the orbital areas acting to gate and inhibit limbic system activity (Joseph, 1990d). Hence, the orbital frontal lobes are also implicated in the development of obsessive compulsive disorders (Baxter et al., 1988; Rapoport, 1991; Rauch, et al., 1994), as is the amygdala, which, in some cases, may be the source of the unwanted impulse.

In that obsessive-compulsive disorders are also associated with hypermetabolism within the caudate nucleus (Baxter et al., 1992; Rapoport, 1991; Rauch et al., 1994)—as well as reduced 5HT activity (Goodman et al., 1990; Zohar and Insel, 1987), which in turn results in disinhibition—it is possible that the inhibitory circuit maintained by the orbital frontal lobes and basal ganglia, in some instances, may be deactivated or overwhelmed.

Given that it is via the frontal lobes that thoughts come to be integrated and infused with emotion, and/or inhibited and suppressed, this also suggests that caudate/orbital dysfunction may result in the inability to extinguish certain thoughts and actions, even in the absence of an underlying emotional problem (see Baxter et al., 1992, and Rauch et al., 1994, for a different, albeit related, explanation). The individual may perseverate on a single thought or action, which in turn may interfere with his capability to engage in alternative modes of response, due to neurological dysfunction involving this circuit.

Conversely, as is suggested by the actions of "Bob" referred to above, it is possible that this same frontal-inhibitory circuit may be employed so as to prevent an individual from engaging in certain behavioral acts that the "conscious" mind finds objectionable or "dirty" (see Chapters 11 and 15). The individual will perseverate on a certain theme, thought, or action to the exclusion of others. Therefore, when "Bob" was feeling homosexually aroused, rather than "cruising" for sex partners, a different (albeit related) motor program was initiated and he would *wash and dust* his car and even its engine (which simultaneously and symbolically *cleansed* him of his *dirty* thoughts and/or at least prevented him from thinking and acting on them). However, in some cases, these disorders—when mild—are due to memory disturbances. The individual, therefore, checks again and again to see, for example, if the door is locked. As noted, the frontal lobes are instrumental in memory search and activation as well as impulse inhibition and memory avoidance (e.g., repression), whereas the limbic striatum is involved in memory.

THE LIMBIC STRIATUM, MEMORY, AND ALZHEIMER'S DISEASE
The Limbic Striatum, Dopamine, Serotonin, Norepinephrine, and Memory

The limbic striatum consists of the nucleus accumbens, olfactory tubercle, the extended (centromedial) amygdala, as well as the ventral aspects of the caudate, putamen, and GP (substantia innominata).

The nucleus accumbens is located immediately beneath the anterior portion of the caudate and from a microscopic level appears to be part of the ventral caudate (Olton et al., 1991; Zaborszky et al., 1991). However, it maintains extensive interconnections with the amygdala as well as the hippocampus via the fimbria-fornix fiber bundle (DeFrance et al., 1985), which implicates this nuclei in memory functioning and probably the learning of visual-spatial and perhaps social-affective relationships. Like the accumbens, the substantia innominata (SI) is also important in memory functioning and has been implicated as one of the principle sites for the initial development of Alzheimer's disease (see Olten et al, 1991; Zaborszky et al., 1991, for a review of related details).

The dorsal portion of the substantia innominata ("great unknown") merges (dorsally) with the ventral GP (and ventrally with the centromedial amygdala), and

maintains interconnections with the accumbens, hippocampus, lateral amygdala, and dorsal medial (DM) nucleus of the thalamus (Young et al., 1984; Zaborszky et al., 1991)—the DM being involved in regulating neocortical arousal and information reception, as well as memory (see Chapters 6 and 11).

The nuclei of the limbic striatum are also interconnected with the lateral and medial hypothalamus, brainstem reticular formation, and the frontal and inferior temporal lobes (Everitt and Robbins, 1992; Groenewegen et al., 1991; Heimer and Alheid, 1991; Kelly et al., 1982; Mogenson and Yang, 1991; Olton et al., 1991; Parent and Hazrati, 1995; Van Hoesen et al., 1981; Zaborszky et al., 1991). The limbic striatum is able, therefore, to exert widespread effects on neocortical and subcortical structures.

As noted, the limbic striatum receives DA from the mesolimbic system, as does the amygdala. The amygdala also sends axons that terminate in striatal neurons immediately adjacent to those innervated by mesolimbic DA neurons (Kelly et al., 1982; Yim and Mogenson, 1983, 1989). Thus, a complex interactional loop is formed between these nuclei, the integrity of which, in part, is dependent on the mesolimbic DA system, which can act on the amygdala, hippocampus, and limbic striatum simultaneously so as to modulate striatal reception of amygdala (Maslowski-Cobuzzi and Napier, 1994) and hippocampal excitatory signals, and regulate the transmission of accumbens input into the SI. Alterations in mesolimbic DA activity, therefore, can significantly influence limbic striatal, amygdala, and hippocampal activity as well as the motoric expression of limbic impulses.

For example, depletions in mesolimbic DA can disrupt motor and social-emotional memory functioning, a condition compounded by DA influences on ACh neurons, and the effects of striatal DA on the reception of hippocampal, amygala, and neocortical input. Indeed, the limbic striatum (and limbic system) contain high densities of ACh neurons, which are involved in memory as well as motor functioning (Olton et al., 1991; McGaugh et al., 1992; Zaborszky et al., 1991). In fact, of all striatal nuclei, densities of DA and ACh neurons are highest within the (SI) and nucleus accumbens (Meredith et al., 1989), which in turn greatly influences the SI (which is a major source of neocortical ACh), as does the amygdala (Yim and Mogenson, 1983) and hippocampus; i.e., these signals converge on the SI.

Like the striatum, the hippocampus (and the amygdala) also receives mesolimbic DA input and contains D1 and D2 receptors (Camps et al., 1990) as well as ACh. ACh influences on neuronal activity can be excitory or inhibitory, depending on the resting membrane potential of the receiving neuron (Ajima et al., 1990) as well as concurrent DA activity. Specifically, signals mediated by mesolimbic DA are relayed from the accumbens to ACh neurons in the SI, which in turn project to the DM and the neocortex, the frontal lobes in particular (Mogenson and Yang, 1991; Olton et al., 1991; Zaborszky et al., 1991). However, the DA and cholinergic, ACh system appear to exert counterbalancing influences.

For example, ACh is usually inhibited by DA. By contrast, DA depletion results in increased ACh and neuron hyperactivity (see Aghanjanian and Bunney, 1977; Bloom et al., 1965; Klockgether et al., 1987), which can greatly disrupt cognitive and memory functioning as well as motor activities, unless reversed by anticholinergic drugs. In fact, damage to this DA/cholinergic system can result in Alzheimer's disease (Olton et al., 1991; Zaborszky et al., 1991).

As per memory functioning, it appears that mesolimbic DA modulates the reception of hippocampal (motor-spatial) input into the accumbens, which projects to the SI (which in turn distributes these influences, perhaps via ACh to the neocortex). DA accomplishes this via inhibitory influences on ACh and facilitation of the GABA system, which appears to exert inhibitory influences within the striatum and the transmission of impulses from the nucleus

accumbens (and amygdala/hippocampus) to the SI. Therefore, if the inhibitory influences of GABA and ACh on the SI are dampened, the SI becomes activated and memory functioning may be enhanced or disrupted at the neocortical level, as this nucleus exerts widespread ACh influences on the cerebrum. Thus, fluctuations in DA levels can either act to excite or inhibit cognitive-memory activity within the limbic striatum (see Mogenson and Yang, 1991) as well as the corpus striatum (Packard and White, 1991) and the neocortex.

However, also of importance in this memory-striatal neural network is 5HT (McLoughlin et al., 1994; Mogenson and Yang, 1991) and NE (Roozendaal and Cools, 1994). For example, NE levels have been shown to fluctuate within the amygdala and nucleus accumbens when novel stimuli are presented (Cools et al., 1991) and during information acquisition. Presumably, NE levels within the accumbens can regulate or influence the reception of amygdala and hippocampal impulses within the accumbens and SI, which in turn projects to the neocortex and, in this regard, may act to shunt amygdala-hippocampal impulses to discrete or wide areas of the cerebrum.

Specifically, high alpha NE activity is associated with reduced amygdala input, whereas low beta NE may reduce hippocampal input into the striatum. Conditions such as these, however, are most likely to result when stressed or traumatized, in which case it is possible for amygdala (and thus emotional) input to be received, learned, and expressed by the striatum in the absence of hippocampal participation (see Rozzendaal and Cools, 1994). It is conditions such as these that can give rise to amnesia with preserved learning (see Chapters 6 and 16).

However, in some cases, it is also possible for both the alpha and beta NE systems to be affected simultaneously such that amygdala and hippocampal input to the limbic striatum are inhibited, in which case profound memory loss may result (see Rozzendaal and Cools, 1994). Similar disturbances are associated with the 5HT system (McLoughlin et al., 1994).

Depletion of 5HT can significantly affect the capacity to inhibit irrelevant sensory input (at the level of the brainstem, amygdala, basal ganglia, and neocortex). Hence, 5HT abnormalities or depletion typically results in confusion, sensory overload, and, in some instances, the production of hallucinations (see Chapter 5).

Thus, disruption of the mesolimbic DA system and/or severe disruptions in the 5HT system, in turn, interfere with the hippocampal-limbic striatal memory system (Packard and White, 1991) and can create neuronal hyperactivity—a condition that interferes with perceptual and hippocampal-memory-SI-amygdala functioning (see Chapters 6 and 16). In this regard, it is noteworthy that the central 5HT system is severely disrupted among those with severe memory loss and Alzheimer's disease (McLoughlin et al., 1994), and that calcification of the GP/SI can induce visual and auditory hallucinations as well as cognitive deterioration (Lauterbach et al., 1994).

Memory and Alzheimer's Disease

As detailed in Chapter 6, the amygdala and hippocampus provide massive input to the limbic striatum as well as the dorsal medial nucleus of the thalamus (DM), the frontal lobes, and the reticular activating system, as does the accumbens and SI (see Heimer and Alheid, 1991; Koob et al., 1991; Mogenson and Yang, 1991; Zaborszky et al., 1991). As noted, these nuclei play a significant role in memory and the gating of information destined for the neocortex and appear to be part of a massive neural network designed to control information processing and to establish memory-related neural networks (see also Chapter 11). Presumably, the nucleus accumbens acts to integrate hippocampal and amygdala input, which is then transmitted to the SI, which is also the recipient of limbic impulses concerned with cognitive, memory, and motoric activities (see Mogenson and Yang, 1991).

In part, it appears that the accumbens and SI play an inhibitory (filtering) role on information processing, and may exert inhibitory (and counterbalancing) influences on the the medial frontal lobes, the DM, and corpus striatum (see Mogenson and Yang, 1991). The amygdala exerts similar influences on these nuclei (see Chapter 5) and is also able to inhibit the SI and accumbens.

However, the role of the limbic striatum in cognitive and memory-related activity also includes the learning of reward-related and aversion processes, the facilitation of approach and withdrawal responses, and the memorization of where a reward or aversive stimulus was previously received (Everitt et al., 1991; Kelsey and Arnold, 1994). In this regard, the nucleus accumbens and SI are dependent on the medial and lateral amgydala, especially in the learning of negative experiences (Kelsey and Arnold, 1994).

As noted, the SI is the primary source of neocortical cholinergic innervation (Carpenter, 1991) and appears to be concerned with integrating social-emotional and cognitive input with motor memories, and then storing them perhaps within the SI as well as within the neocortex. Hence, if the SI is lesioned, the cholinergic projection system is disrupted, and cognitive as well as social-emotional memory functioning is negatively impacted.

Since the SI merges with and becomes coextensive with the centromedial amygdala (Heimer and Alheid, 1991), and is an amygdala derivative, not suprisingly, a significant loss of neurons, neurofibrillary changes, and senile plaques have been found in the amygdala (and hippocampus; Rapoport, 1990), as well as in the SI, in patients with Alzheimer's disease and those suffering from degenerative disorders and memory loss (Herzog and Kemper, 1980; Mann, 1992; Sarter and Markowitsch, 1985). Degeneration in the amygdala would also account for the social-emotional agnosia and prosopagnosia that is common in the advanced states of Alzheimer's disease; i.e., failure to recognize or remember loved ones (see Chapter 5).

Widespread Neuronal Death and Alzheimer's Disease

It has been argued that Alzheimer's disease and associated cognitive and memory disturbances are due to a "global cortico-cortical disconnection" syndrome (see Morrison et al., 1990; Rapoport, 1990); i.e., a loss of neurons in and interconnections with association neocortex, which in part would account for the cognitive deterioration. Indeed, among those with Alzheimer's disease, widespread atrophy, plaques, tangles, and metabolic disturbances have been noted in a variety of cortical areas, with relative preservation of the motor and primary receiving areas (Rapoport, 1990).

Neuronal loss has also been reported in the entorhinal and parahippocampal areas of the inferior temporal lobe (see Morrison et al., 1990; Rapoport, 1990), within which is buried the hippocampus and amygdala. Destruction of this tissue would also result in memory, social-emotional, facial recognition, and related visual abnormalities.

Therefore, given the likelihood that the limbic striatum and limbic nuclei may be selectively involved in the early stages of Alzheimer's, and given the fact that the SI provides cholinergic input to the neocortex, then the progressive loss of SI (and amygdala) neurons might result in a progressive deterioration and cell death within the neocortex such that otherwise healthy neurons are killed. That is, since the SI contains high concentrations of cholinergic neurons, which in turn project to widespread areas throughout the neocortex (reviewed in Carpenter, 1991), perhaps the initial cell loss within the SI (also referred to as the nucleus basalis) may trigger further cell death in healthy neurons (which project to or receive fibers from the affected cells), which essentially become "pruned" and drop out from disuse.

DEFECTIVE AXONAL TRANSPORT

There is evidence to indicate that, perhaps due to head injury, drug or toxic exposure, or the loss of synaptic junctions (from the death of unhealthy cells and dendritic retraction), axonal transport becomes defective due to the death of its target neuron. However, if a healthy cell cannot discharge and exchange information it too may die, thus leading to a domino effect and, thus, widespread cell death (see Burke et al., 1992). That is, owing to the loss of terminal synaptic junctions (due to cell death) or to other chemical abnormalties, including defects in microtubule assembly (which participates in neuronal transmission), axonal transport becomes dysfunctional after the receiving cell and its dendrites have died. Hence, there is a buildup of toxic oxidative metabolites and naturally occurring neurotoxins in the healthy cell body that project to the dead cell, which causes the normal cell to die as well. This would result in a progressive loss of neurons such that widespread areas of the cerebrum soon become affected. Presumably, this is what may occur if the limbic striatum becomes abnormal and cells within the SI and accumbens begin to die. As the disturbance and deterioration spreads, cognitive, emotional, memory, and related abnormalities, including Alzheimer's disease and Parkinsonian symptoms, begin to appear and become progressively worse as neocortical, striatal, and limbic neurons die and drop out.

NEOCORTICAL, THALAMIC, AND STRIATAL MOVEMENT FEEDBACK LOOPS
The Supplementary Motor Areas, Putamen, and Globus Pallidus

As noted, some studies have indicated that those with Parkinson's disease demonstrate the greatest amounts of DA depletion and related DA neuronal degeneration within the putamen. Hence, many authors have argued that Parkinson's symptoms are a consequence of damage to these nuclei (e.g., Goto et al., 1990), which in turn can result in the production of unwanted movements, such as tremor. Presumably, in these instances, the medial (internal) GP (which receives putamen input) ceases to inhibit irrelevant motor activity.

The putamen receives much of its input from the caudate, the SMA, the secondary and primary motor cortex, and areas 5 and 7 of the parietal lobe (Jones and Powell, 1970; Pandya and Vignolo, 1971).

Presumably, the putamen, in conjunction with the caudate, transmits this information to the medial and lateral GP, which in turn projects to the motor thalamus, and brainstem. However, like the caudate, the putamen and GP also project back to the motor neocortex, thus creating a very elaborate feedback loop (see Mink and Thach, 1991; Parent and Hazrati, 1995), whose origin may begin in the SMA (Alexander and Crutcher, 1990; Crutcher and Alexander, 1990) or perhaps the limbic system (see Chapter 15).

For example, when anticipating or preparing to make a movement, but prior to the actual movement, neuronal activity will first begin and then dramatically increase in the SMA; this is followed by activity in the secondary and then the primary motor area (see Chapter 11), and then the caudate and, last of all, the putamen-GP (Alexander and Crutcher, 1990; see also Mink and Thach, 1991).

Ignoring for the moment the role of the limbic system, this indicates that impulses to move first appear in the SMA and that other motor regions are temporally-sequentially recruited in a step-wise fashion: i.e., SMA—premotor—primary motor —caudate—putamen—GP—motor thalamus/frontal motor areas—brainstem. ...

However, the caudate nucleus also contains neurons that become active prior to, and in anticipation of, making body movements, and in response to associated auditory and visual environmental cues (Rolls et al., 1983; Rolls and Williams, 1987). This caudate neuronal activity precedes, or occurs simultaneously with, excitation in the lenticular nucleus, which also results (via feedback) in SMA activation. This is because the GP and the putamen not only re-

ceive caudate and neocortical afferents, but they project back to the frontal motor areas (as well as to the motor thalamus). Hence, parallel processing also occurs.

That is, activity in these regions quickly begins to overlap such that neurons in the motor neocortex and basal ganglia often remain activated simultaneously (Alexander and Crutcher, 1990). Moreover, the motor areas all independently send axons to the brainstem, with axons from different areas converging on the same motor neuron.

Therefore, activity in the motor areas is characterized by both temporal-sequential and parallel processing that, in turn, is made possible via feedback neural circuitry. Because these different networks are intimately linked and mutually interactive, they make coordinated and goal-directed movements possible.

Indeed, multiple feedback loops are probably also necessitated by the numerous variables and body—spatial—visual—kinesthetic—motor references, and so forth, that need to be computed in order to make a planned movement. Thus, both temporal-sequential and parallel processing is necessitated, as each area is functionally specialized to analyze specific types of information and to perform certain actions in semi-isolation as well as performing other functions in parallel with yet other motor areas.

The Putamen and Medial and Lateral Globus Pallidus

Just as different regions of the caudate appear to subserve different functions, a variety of neuronal types also characterize the lenticular nucleus. For example, preparatory and movement-related neurons are segregated within the putamen (Alexander and Crutcher, 1990). This suggests that the putamen employs two different neural networks, one which executes movements, the other which prepares to make the movement—information that is normally transmitted to the GP as well as back to the motor neocortex.

The putamen appears to act at the behest of impulses arising in the neocortical motor areas (as well as the limbic system) and then only secondarily acts to prepare for, and then to participate in, the guidance of movement (Mink and Thach, 1991). In conjunction with the caudate, the putamen accomplishes this via signals transmitted to the medial and lateral GP, the motor thalamus, and the brainstem reticular formation (Marsden and Obesco, 1994), and then back to the motor neocortex (see also Parent and Hazrati, 1995, for a related review).

As noted, the putamen and GP are also coextensive (the lenticular nucleus) and in many respects function as a prepatory motor unit involved in the guidance, and perhaps even the learning of various motor activities (Crutcher and DeLong, 1984; Kimura, 1987). For example, alterations in neuronal activity have been demonstrated in the GP and the putamen during tasks involving learned body movements (Crutcher and DeLong, 1984). Lenticular neurons also become highly active when learned facial or limb movements are triggered in response to a particular auditory or visual stimulus associated with that movement (Kimura, 1987).

Some putamen and GP neurons also fire in response to reward but not to movement, and vice versa (Kimura et al., 1984). However, these same neurons do not respond when the same learned and rewarded movements are made spontaneously and in the absence of associated cues, or when they are no longer rewarded. These findings raise the possibility that specific neurons within these nuclei are responsive to motivational cues associated with movement, and that these neurons utilize these cues in preparation for movement (Kimura, 1987).

Neurons in the GP also selectively respond and change their activity when making ballistic movements, particularly those that are visually guided (Mink and Thach, 1991). However, activation occurs too late for these neurons to be involved in the initiation or planning of the movement (Mink and Thach, 1991).

Conversely, destruction or massive inhibition of the GP can result in difficulty turning off a movement (see Mink and Thach, 1991), thus giving each movement a ballistic quality. Similarly, destruction or massive inhibition of the GP can make it exceedingly difficul to rapidly alternate between movements and thus switch from an ongoing to a different motor program (Hore and Vilis, 1980). This loss of control over motor programming with GP impairment extends even to attempts to make purposeful ballistic reaching or stepping movements (Horak and Anderson, 1984; Hore and Vilis, 1980), and the velocity and amplitude of these movements may in fact be significantly slowed and reduced (Mink and Thach, 1991).

The Medial and Lateral Globus Pallidus

The medial and lateral GP appear to play different and counterbalancing roles in the execution, inhibition, and excitation of different motor programs (Crossma, et al., 1987; Marsden and Obesco, 1994). For example, the medial GP is believed to provide positive/excitatory feedback to the neocortical motor areas, whereas the lateral GP provides indirect "negative" and inhibitory feedback so that unwanted movements are prevented. Moreover, DA inhibits the medial GP but excites the lateral GP (reviewed in Marsden and Obesco, 1994).

In consequence (at least in theory), reduced nigrostriatal DA levels can result in an overexcitation of the medial GP, which inhibits cortically mediated or initiated movements, thereby producing akinesia (loss of movement), bradykinesia (reduction in movement), and hypokinesis (slowness of movement) coupled with tremor.

On the other hand, excessive lateral GP (and limbic striatal) activity can also result in gross involuntary and excessive hyperkinetic and ballistic movements, usually involving the limbs on the contralateral

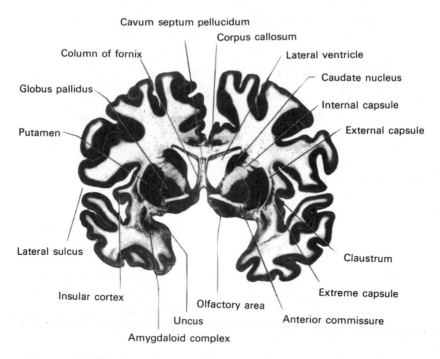

Cavum septum pellucidum
Corpus callosum
Column of fornix
Lateral ventricle
Caudate nucleus
Globus pallidus
Internal capsule
Putamen
External capsule
Lateral sulcus
Claustrum
Insular cortex
Extreme capsule
Olfactory area
Uncus
Anterior commissure
Amygdaloid complex

Figure 9.11   Photograph of the frontal section of the brain, depicting the caudate, putamen, globus pallidus, and amygdala. (From Carpenter M. Core text of neuroanatomy. Baltimore: Williams & Wilkins, 1991.)

side of the body, along with choreiform movements and tremors. However, surgical destruction of the fiber tracts leading to and from the lateral and medial GP can significantly diminish such disturbances, including hemiballismus, tremor, dystonia, chorea, and athetosis.

As noted above, lesions to the amygdala prior to the development of Parkinsonian and related symptoms can also prevent the development of these motor abnormalities, which in part is due to the extensive interconnections maintained with these regions and the fact that the amygdala becomes coextensive with the ventral GP.

In addition, because the putamen projects to the GP, lesions restricted or localized to the putamen can also result in abnormal competition between antagonist and agonist muscles, due to the release of the GP (and associated neural circuitry) from putamen control. Patients or primates so effected suffer from severe dystonia as well as rigidity and increased muscle tone some-

times accompanied by chorea (Segawa et al., 1987). Similarly, among those with Parkinson's disease, activity within the putamen is significantly reduced (Playford et al., 1992), and reduced functional activity is also seen in the caudate, which merges with the putamen.

THE MOTOR THALAMUS

The motor thalamus evolved in tandem with the basal ganglia and is richly interconnected with the caudate nucleus and GP in particular (Crossman et al., 1987; Parent and Hazrati, 1995; Powell and Cohen, 1956; Royce, 1987). It is an integral aspect of the motor circuit and becomes activated when making a variety of movements and in response to kinesthetic and proprioceptive stimuli (Vitek et al., 1994).

The motor thalamus consists of the ventromedial, ventrolateral, ventralis intermedius, centromedian, and parafascicular (posterior) intralaminar thalamic nuclei, receives input from the brainstem, cerebellum, and neocortex, and maintains reciprocal projections with the GP and SMA (Carpenter, 1991; Brodal, 1981; Kemp and Powell, 1970; Narabayashi, 1987; Royce, 1987; Vitek et al., 1994). These thalamic "motor" nuclei also receive input from the facial, leg, and arm regions of the motor and somatosensory cortex (Kunzle, 1976; Vitek et al., 1994) and maintain reciprocal interconnections with the amygdala, cingulate gyrus, substantia nigra, and superior colliculus (Jones et al., 1979; Mesulam et al., 1977; Royce, 1987; Vogt et al., 1979).

Thus, the motor thalamus is intimately associated with limbic and motor nuclei throughout the brain and is able to influence as well as receive multiple inputs from a variety of nuclei, which in turn are interlinked. For example, some intralaminar thalamic nuclei send collateralizing axons that project to both the striatum and the neocortex, whereas some motor neocortical axons project to both the striatum and the intralaminar nuclei (Royce, 1987). Hence, a richly interconnected neural circuit is maintained by these nuclei so as to control and guide motor functions.

In some respects, the motor thalamus appears to act as a nexus where multiple forms of input are integrated so as to regulate motor activity. Therefore, when the motor thalamus is abnormally inhibited or activated, significant motor abnormalities result; e.g., rigidity, tremor, ballismus, catatonia, and catalepsy. For example, unilateral infarcts and hemorrhages involving the thalamus can induce unilateral thalamic ataxia and apraxia (Nadeau et al., 1994; Soloman et al., 1994).

Conversely, injuries or surgical destruction of the specific thalamic nuclei can, in some instances, eliminate or reduce the influences of abnormal activity received from other regions within the motor circuit (Marsden and Obesco, 1994; Narabayashi, 1987). For example, patients suffering from Parkinsonian symptoms appear to derive the greatest benefit from thalamic lesions, which abolish contralateral tremor and rigidity (see Marsden and Obesco, 1994) via disruption of the GP-motor thalamus-frontal motor area motor circuit. Specifically, some evidence indicates that destruction of the ventrolateral (VL) motor thalamic nuclei can reduce rigidity (if the lesion is more anterior), or tremor (if the lesion is posterior). Neurosurgical destruction of the GP-to-motor thalamus projection fibers can also significantly decrease rigidity, whereas section of the thalamic-to-brainstem/cerebellum pathway also reduces tremor (see Narabayashi, 1987). Moreover, VL lesions can significantly reduce choreic and ballistic movements, such as those secondary to trauma or encephalitis.

In addition, neurosurgical destruction or electrical stimulation of these nuclei (e.g., ventralis intermedius) can eliminate tremors (see Narabayashi, 1987). Indeed, neurons in the motor thalamus not only fire in tandem with tremor, but electrical stimulation of this nuclei at a frequency similar to the tremor increases the frequency and amplitude of the tremor. However, high levels of thalamic stimulation reduces or abolishes tremor (Narabayashi, 1987).

THE SUBTHALAMIC NUCLEUS

The subthalamic nucleus is a small but densely innervated component of the basal ganglia-thalamocortical-limbic motor circuit. It maintains a very important reciprocal relationship and is richly interconnected with the medial and lateral GP (Crossman et al., 1987; Parent and Hazrati, 1995; Wichman et al., 1994) and merges medially with the lateral hypothalamus, with which it is intimately linked. The subthalamic nucleus also receives extensive and topographic projections from the neocortical motor areas, including inhibitory axonal fibers from the frontal lobes (Parent and Hazrati, 1995). It also projects to the caudate, putamen, SI and brainstem reticular formation, and provides excitatory influences to the substantia nigra and other target nuclei (Klockgether et al.,

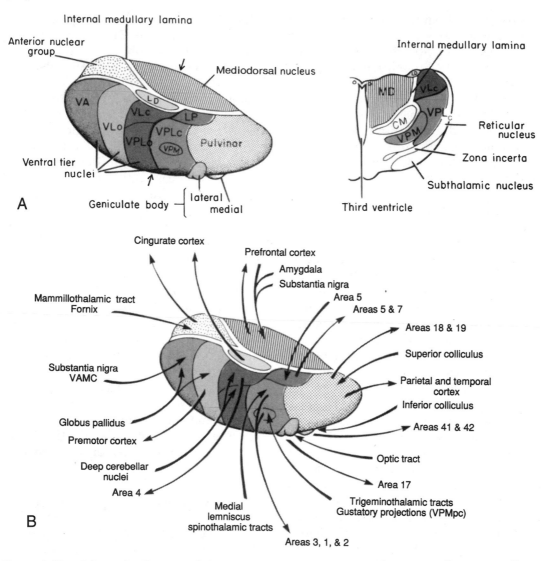

Figure 9.12 Schematic diagram of the major thalamic nuclei. (From Carpenter M. Core text of neuroanatomy. Baltimore: Williams & Wilkins, 1991).

1987; Parent and Hazrati, 1995), and thus influences the nigrostriatal dopamine system and a wide variety of brain areas via separate pathways (Wichman et al., 1994).

Due to its interconnections with so many motor-related areas of the brain (as well as the limbic system), the subthalamic nucleus is able to exert significant, albeit indirect, as well as direct influences on motor expression, which is accomplished in conjunction with the nigrostriatal DA system and the GP; this, in turn, modulates subthalamic activity and its reception

of cortical input (Parent and Hazrati, 1995; Wichman et al., 1994). Hence, the subthalamic nucleus appears to also exert modulating influences on movement (although it also serves non-movement functions), especially those involving the proximal limbs.

Abnormalities localized to or involving the subthalamic nucleus can, therefore, result in significant motor disturbances, including hemiballismus and chorea (Crossman et al., 1987) due presumably to interuption of the normal GABAinergic reciprocal

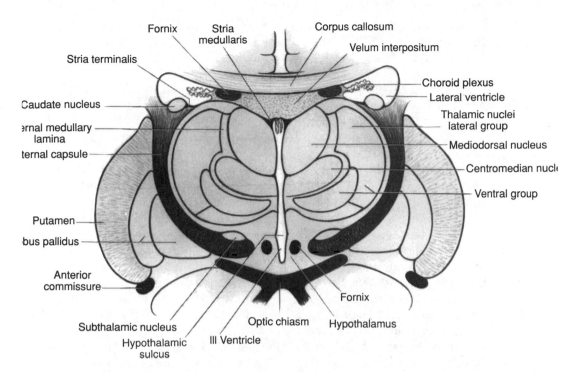

Figure 9.13   Diagram of a frontal section of the brain, depicting the major thalamic nuclei. (From Carpenter M. Core text of neuroanatomy. Baltimore: Williams & Wilkins, 1991.)

relationship it maintains with the caudate, putamen, and in particular the GP (McGeer and McGeer, 1987; Parent and Hazrati, 1995). However, the subthalamic nucleus may also respond abnormally if it receives abnormal signals.

For example, Parent and Hazrati (1995) argue that inhibitory influences exerted on the subthalamic nucleus by the GP can result in akinesia if GP activity is reduced, or hyperkinesia if enhanced. Similarly, Crossman et al. (1987) argue that in hyperkinetic states the medial GP and subthalamic nucleus are underactive and the lateral GP is overactive. In Parkinson's and akinetic disorders, the medial GP and subthalamic nucleus are overactive, whereas the lateral GP is underactive (Crossman et al., 1987).

Conversely, lesions of the subthalamic nucleus can result in increased activity within the GP and motor thalamus, such that the motor thalamus, SMA, and brainstem become hyperactivated, which results in excessive movement, including ballismus. However, one case has been re-ported in which a hemorrhage involving the subthalamic nucleus resulted in the amelioration of a patient's Parkinson's symptoms (see Wichmann et al., 1994 for detailed discussion).

DOPAMINE AND GABA STRIATAL INTERACTIONS AND MOVEMENT DISORDERS

The GP receives about two-thirds of its input from the putamen, and one-third from the caudate (as well as some fibers from the SMA). Presumably, activation of the SMA-primary motor-caudate-putamen loop results in differential activation of the lateral and medial GP, the subthalamus, the motor thalamus, and the brainstem—nuclei that project back to the SMA and primary motor areas, from which arises a massive rope of nerve fibers (the corticospinal/pyramidal tract) that directly innervates the brainstem and spinal cord. Presumably, these complex feedback loops insure that coordinated and smooth continuous movements are planned, coordinated, and finally carried out. As noted, these interactions are also dependent on DA activity.

For example, Parkinson's disease, being characterized by reduced DA, is associated with a reduction in striatal inhibitory activity and thus increased GP output to the motor thalamus and brainstem, which would produce tonic EMG activity, akinesia, and rigidity (Kockgether et al., 1986; Starr and Summerhayes, 1983), i.e., hypomovement due to excessive tonic excitation.

Conversely, excessive corpus stritatal DA would result in increased striatal GABA activity and, thus, inhibition of the GP; this would lead to a disinhibition of the motor thalamic and subthalamic nucleus and, thus, to increased activation of brainstem and spinal motor neurons. Movements would become hyper, ballistic, and chorea may develop—as frequently occurs with prolonged neuroleptic treatment.

Specifically, GABA activity within the GP and subthalamus is significantly affected by DA activity within the corpus striatum. For example, decreases in striatal DA result in decreased striatal GABA neural activity (which is normally inhibitory). As these GABA neurons project to the GP, the loss of this GABA inhibitory input would increase tonic activity within the GP, which in turn would result in increased inhibitory GP influences on the motor thalamus and brainstem, thereby decreasing their activity. Hence, the modulatory influences of these latter nuclei on motor functioning would be eliminated, resulting in heightened activity levels and, thus, overactivation of this aspect of the feedback circuit. Brainstem and spinal motor neurons would become tonically activated, thereby creating rigidity and stiffness of movement.

Conversely, the loss of GABA influences could induce choreic movements. In this regard, it is noteworthy that significant GABA loss and abnormalities have been noted in the substantia nigra and basal ganglia of those with Huntington's chorea

(Bird and Iversen, 1974), along with degeneration of the nigrostriatal DA projection pathway.

HUNTINGTON'S CHOREA

Chorea ("dance") is characterized by jerky, writhing, twisting, and unpredictable movements of the extremities. The two main types are Sydenham's chorea (St. Vitus dance) and Huntington's chorea. Sydenham's chorea involves choreiform movements of the face, tongue, and extremities, and is accompanied by loss of nerve cells in the caudate and putamen, as well as within the cerebral cortex, substantia nigra, and subthalamic nucleus.

Huntington's chorea is a progressive, deteriorative, inherited genetic disorder passed on by an autosomal dominant gene located on chromosome 4. It is characterized by an insidious onset that may begin during childhood or old age, with the illness beginning earlier in those who have an affected father (reviewed in Folstein et al., 1990; Young, 1995). Cognitive decline, however, is gradual. Affected individuals tend to suffer from memory and visual-spatial deficits, depression, and reduced verbal fluency, although aphasia is not typical. Difficulty with motor coordination, planning skills, decision making, and a reduced capacity to consider alternate problem solving strategies or to shift from one mental set to another is not uncommon (reviewed in Folstein et al., 1990). Hence, in some respects this disorder is suggestive of frontal lobe abnormalities (see Chapter 11).

This syndrome is also associated with widespread neuronal loss in the caudate, putamen, brainstem, spinal cord, cerebellum, and atrophy in the GP (see Vonsattel et al., 1987; Young, 1995). Degeneration is predominantly of small striatal neurons whereas larger neurons remain intact. Opiate neurons located in the striatum are also significantly affected (Reiner et al., 1988). It is believed that the degeneration of these corpus striatal neurons, as well as the loss of GABA influences, results in reduced striatal control over the GP (see Narabayashi, 1987), thereby producing excessive movement.

Of note, some reports indicate that the posterior caudate and putamen are more severely affected than the anterior regions (Vonsattel et al., 1987). Indeed, the posterior caudate and its tail is usually the earliest and most severely affected part of the brain—which implicates the amygdala as a factor in the development of chorea. Indeed, in the early stages of Huntington's chorea, atrophy and degeneration begin in the tail and spread dorsally and anteriorally, thus affecting the striatum and lenticular nucleus (Young, 1995).

In this regard, it is noteworthy that affective disorders and personality and mood changes are prominent early signs suggesting amygdala involvement. Indeed, disturbances of emotion may precede any motor or cognitive decline by as much as 20 years, with some patients displaying mania, depression, and antisocial tendencies (reviewed in Folstein et al., 1990). As noted, neural degeneration tends to begin in the amygdaloid tail of the corpus striatum (Young, 1995).

As per disturbances of movement, those with Huntington's disease tend to suffer from either or both voluntary and involuntary abnormalities. The involuntary aspects include the jerking and unpredictable movements of the limbs, trunk, and face that may occur when at rest, walking, or while actively engaged in some task, such that these individuals may appear to be intoxicated and/or attempting to dance about.

Voluntary disorders include rigidity, slowed or clumsy movement, or difficulty initiating movement. Those who suffer from voluntary movement disorders are the most likely to demonstrate cognitive decline (Folstein et al., 1990).

CONCLUDING STATEMENT: THE AMYGDALA

The basal ganglia motor circuit, as conceptualized and described in this chapter, consists of multiple neural networks involving the caudate, putamen, globus pallidus, limbic striatum, motor thalamus, subthalamus, brainstem, supplementary motor areas, pre-motor and primary motor areas, and the amygdala. Unfortunately, the role of the amygdala in movement is often ignored by investigators (and almost all textbooks) although all aspects of the motor circuit are directly or indirectly innervated by this nuclei and respond to amygdala-mediated impulses. Indeed, although the amygdala can be injured bilaterally without significantly affecting normal motor functioning, in many instances the amygdala is implicated in the genesis of abnormal motor activities and may play a significant role in striatal dysfunction and the development of Parkinson's and related diseases. Obviously, much more research on this issue is warranted.

10

The Brainstem, Midbrain, DA, 5HT, NE, Cranial Nerves, and Cerebellum

Motor Programming, Arousal, Psychosis, Coma, Sleep and Dreaming, Sleep Disorders, Vocalization, and Emotion

The brainstem consists of the medulla, the pons and midbrain, the reticular formation, and monoaminergic neurotransmitter systems as well as the cranial nerves and associated nuclei. Straddling the brainstem is the cerebellum, which is an outgrowth of the vestibular system.

The nuclei and fiber pathways of the brainstem subserve a variety of divergent as well as interrelated sensory and motor functions. These include the mediation and control of attention, arousal, the sleep cycle, heart rate, breathing, balance, gross axial movement, the initiation and coordination of eye, jaw, tongue, and head movements; as well as visual, somesthetic, gustatory, and auditory perception (Cowie et al., 1994; Davidson and Bender, 1991; Erickson et al., 1994; Johnson et al., 1994; Masino, 1992; Rhode and Greenberg, 1994). Many brainstem functions, however, are performed in a reflexive, rhythmic, or stereotyped manner, and in a fashion that does not necessitate thinking or planning: e.g., stepping, walking, chewing, and breathing.

Given its exceedingly long and ancient evolutionary history, not surprisingly, many brainstem functions occur without the aid of consciousness or awareness, or even forebrain/neocortical participation. That is, the motor programs that subserve many of these basic brainstem functions are essentially hardwired and need only to be activated in order for the associated motor acts to be performed. Moreover, many brainstem activities are rhythmical and wax and wane in accordance with their own intrinsic rhythms, e.g., sleep and dreaming.

GROSS ANATOMY

The brainstem consists of the medulla (myelencephalon), the pons (metencephalon), and the midbrain (mesencephalon). Straddling the dorsal surface of the pons is the cerebellum, and spanning the length of the brainstem are the nuclei for cranial nerves III–XII (see Carpenter, 1991; Parent, 1995), which will be briefly reviewed below.

The most rostral portion of the brainstem (i.e., the midbrain) merges with the diencephalon, which in turn is surrounded by the cerebral hemispheres. The more posterior portion of the midbrain consists of the superior (visual) and inferior (auditory) colliculi, which are directly related to the thalamus, and the red nucleus, which receives motor fibers from the cerebellum and frontal and parietal lobes. The dorsal roof of the midbrain is referred to as the tegmentum (which produces dopamine), whereas more centrally located is the substantia nigra, a major source of corpus striatal dopamine.

The pons represents the most rostral portion of the "hindbrain" and is separated from the midbrain via the superior pontine sulcus, and from the medulla by the inferior pontine sulcus. The two components of the vestibulocochlear nerve (VIII), and the facial (VII), abducens (VI) (which also lies in the floor of the fourth ventricle), and the trigeminal nerve (V) are the cranial nerve nuclei associated with the pons.

The medulla is the most caudal portion of the brainstem. The transition from spinal cord to medulla, however, is not obvious, though located at the medulla-spinal junction are the paired medullary pyramids

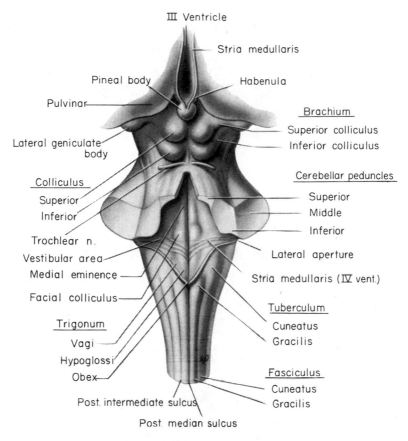

Figure 10.1   Dorsal view of the brainstem. (From Carpenter M. Core text of neuroanatomy. Baltimore: Williams & Wilkins, 1991.)

through which course the corticospinal/pyramidal tract as they cross over before descending into the spinal cord. Basically, the medulla extends from the level of the foramen magnum to the inferior pontine sulcus. The hypoglossal (XII), accessory (XI), vagus (X), and glossopharyngeal (IX) cranial nerves are associated with the medulla, whereas the vestibulocochlear (VIII) nerve emerges at the junction of the medulla, pons, and cerebellum.

Spanning the medial and lateral lengths of the brainstem is a complex reticulum of richly interconnected cells with long ascending and descending axons, collectively referred to as the reticular activating system. The pontine tegmentum and the more dorsal portion of the midbrain tegmentum are also considered part of the reticular formation through which course ascending sensory fibers and descending motor fibers.

EVOLUTIONARY CONSIDERATIONS

As detailed in Chapter 1, the brainstem in large part evolved from motor-sensory cells that had collected together to form like-minded ganglia within the caudal regions of the head. Some of these ganglia came to constitute the portions of the spinal cord where purely reflexive acts may be triggered (e.g., flexion or extension), whereas yet others formed the lower brainstem (and brainstem-spinal junction) and came to mediate a variety of vital vegetative functions, which for the most part occur completely outside of conscious control: e.g., thermoregulation, salivation, cardiovascular functioning, swallowing, and respiration. This includes lower brainstem (medulla) motor nuclei that coordinate and program the movement of the tongue, palate, pharynx, larynx, and jaw, as well as

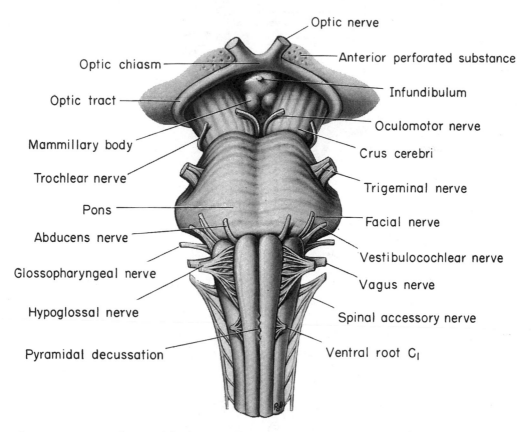

Optic nerve

Optic chiasm

Anterior perforated substance

Optic tract

Infundibulum

Mammillary body

Oculomotor nerve

Trochlear nerve

Crus cerebri

Pons

Trigeminal nerve

Abducens nerve

Facial nerve

Glossopharyngeal nerve

Vestibulocochlear nerve

Hypoglossal nerve

Vagus nerve

Pyramidal decussation

Spinal accessory nerve

Ventral root C$_I$

Figure 10.2   Ventral view of the brainstem. (From Carpenter M. Core text of neuroanatomy. Baltimore: Williams & Wilkins, 1991.)

swallowing, breathing, phonating, and "vocalization." These latter oral-facial-glutorial functions are mediated by brainstem nuclei and cranial nerves XII, XI, X, IX, VII, and V (see Carpenter, 1991; Parent, 1995; Waxman and De Groot, 1994).

Over the course of evolution, as motoric capabilities became more extensive and increasingly complex, the brainstem also evolved specific nuclei that could reflexively trigger and organize these movements. However, those nuclei concerned with the simpler aspects of movements are also found within the lower brainstem.

Reflexive Motor Actions and Hierarchical Functional Representation

When specific brainstem nuclei (and associated neural networks) are stimulated, sucking, chewing, swallowing, swimming, stepping, walking, and running movements can be induced (Barman, 1990; Cohen et al., 1988; Klemm, 1990; Mogenson, 1990; Steriade and McCarley, 1990; Skinner and Garcia-Rill, 1990; Vertes, 1990). However, these movements decrease in complexity as the stimulation site moves in a caudal direction from the upper pons to the lower brainstem; i.e., they are hierarchically organized, with those requiring the least amount of organizational or conscious control being localized near the brainstem-spinal border. Hence, at the level of the spinal cord, movement programs are exceedingly simplified and usually consist of only fragments of the entire motor display.

Conversely, as one ascends from the spinal cord through the brainstem, the degree and extent of these various motor programs and behaviors becomes more complex. For example, at the most caudal regions of the medulla, neurons, when stimulated, can induce stepping motions, whereas stimulation of the more anterior midbrain pontine junction (at the level of the inferior colliculus), can initiate eye movements, head turning, controlled walking, and running (Cowie and Robinson, 1994; Cowie et al., 1994; Skinner and Garcia-Rill, 1990). Conscious and/or forebrain/ telencephalic participation is not required for these motor responses.

In addition, brainstem activation can also induce involuntary, stereotyped, and routine motor acts, such as yawning, opening and closing the mouth, wiping the face, or lifting the legs, taking steps, walking, or running (Cohen et al., 1988; Klemm, 1990), even when forebrain influences have been eliminated via transection (Skinner and Garcia-Rill, 1990). For example, if the brainstem is completely severed from the midbrain and forebrain (by a lesion below the superior colliculi at the midbrain-pons transition) mammals may respond to facial stimulation by raising one or both paws and wiping the face (see Klemm, 1990). Or, following transection, direct stimulation of the brainstem can induce walking and running movements. Because the early maturing brainstem stores these motor programs, infants and even anencephalic "monsters" are capable of making stepping, walking, and swimming movements if properly stimulated.

Brainstem motor control is maintained via the cranial nerves and descending fibers that terminate on spinal motor neurons in the dorsal horns. As such, components of the various motor programs that mediate these activities are located within the spinal cord as well as the brainstem. Hence, in response to stimulation, the brainstem can act on the spinal and cranial nerves in order to trigger these complex motor acts in the absence of forebrain influences. Even the stereotyped motor components involved in sexual posturing and behavior appear to be under direct brainstem control.

Sex and the Brainstem

Portions of the caudal medulla (as well as the spinal cord) have been implicated in the motor control over sexual posturing (Benson, 1988; Rose, 1990). Specifically, there are medulla neurons that respond to hormonal influences, neurons that react to tactile-genital contact transmitted via the spinal cord, and neurons that integrate these messages as well as those transmitted by the amygdala and ventromedial hypothalamus and basal ganglia—nuclei that are also involved in sexual activities. Many of these same sexual neurons are located near cranial nerve nuclei that control phonation and respiration. Hence, when these latter neurons are activated, the subject may scream or cry out in orgasmic pleasure (see Rose, 1990).

Moreover, there are neurons in the medulla that, when directly stimulated, can induce females to assume the lordosis (or a receptive) posture. Similarly, there are brainstem nuclei, including serotonin (5HT) neurons that act to enhance or suppress male sexual behavior and arousal and the capacity to sustain an erection—due to 5HT suppressive influences on related spinal motor neurons and the amygdala.

However, female sexual posturing appears to be more reflexive and "hard wired" as compared to male sexual behavior, which is more complicated. That is, the female need only to stand, crouch, or lie still while the male must pursue, stimulate, and mount the female, insert his penis, and discharge semen—activities that are highly dependent on sensory feedback and guidance, and more greatly under the influence of the limbic forebrain. It is because male sexual behavior is more complicated and more sensitive that it can be more easily disrupted, not only by psychological factors but also by lesions involving a wide variety of brainstem and forebrain nuclei.

Overall, it appears that these and related sexual motor acts are stored in the brainstem, and do not require forebrain participation or voluntary or conscious control for their elicitation. However, most of the motor actions and "programs" stored in the brainstem can be activated by forebrain nuclei, in particular the amygdala and hippocampus, the basal ganglia, dorsal medial thalamus, and the motor areas of the frontal neocortex.

Multisensory-Motor Brainstem Neurons

Some brainstem neurons respond only to a single sensory system; others respond almost regardless of the nature of the stimulus. As one ascends from the spinal cord to the lower brainstem, and then from the medulla to the midbrain, neurons and nuclei become increasingly responsive to a number of different sensory modalities (Grillner and Shik, 1973; Scheibel, 1980).

For example, fibers from different sensory modalities sometimes terminate on the same neuron, indicating that many brainstem and reticular neurons have multimodal properties. Indeed, via these converging sensory fibers, brainstem neurons are able to map multisensory space so as to create almost a three-dimensional sensory representation of the environment (Scheibel, 1980), whereas others create a motor map that may be juxtaposed over the sensory map (Grillner and Shik, 1973).

Via the creation of these brainstem sensory-motor maps, the organism can respond in a complex and reflexive manner to a variety of sensory stimuli, particularly those that are threatening or aversive, in which case reticular neurons induce generalized, or in some instances, selective arousal in response to specific or a combination of sensory signals. However, these brainstem sensory-motor neurons may also preprocess stimuli, prior to activating and transferring these signals to appropriate regions of the neocortex (see Klemm, 1990; Steriade and McCarley, 1990).

Prior to the evolution of neocortex, these signals were transferred to the midbrain as well as the basal ganglia (and limbic system); the basal ganglia reciprocated by transmitting its motoric signals back to the midbrain. Initially the midbrain acted to assimilate and organize these data and, thus, the organism's behavioral reaction.

THE MIDBRAIN

The midbrain is the least differentiated and the smallest segment of the brainstem and can be divided into three portions: a medial segment (the tegmentum), which includes dopamine (DA) manufacturing neurons and which represents the anterior continuation of the reticular formation (described below); the most recently evolved ventral segment, through which pass descending cortical fibers; and the tectum, which is dorsally located and includes the superior (visual) and inferior (auditory) colliculi. Located between and below the ventral segment and the DA-producing (mesolimbic) tegmentum is the substantia nigra, which also manufactures dopamine. However, the most obvious structure of the midbrain is the pinkish colored "red nucleus," which receives descending motor fibers from the frontal lobe and gives rise to the rubrospinal tract, which facilitates flexor muscle tone.

THE PERIAQUEDUCTAL GRAY

Also located within the midbrain is the periaqueductal gray, a nucleus that receives extensive input from the motor areas, the amygdala and other limbic nuclei and is implicated in the motoric-vocal aspects of emotional expression. For example, stimulation of the periaqueductal gray can trigger the production of a variety of sounds that are suggestive of exceedingly negative moods (Larson et al., 1994; Zhang et al., 1994). If the periaqueductal gray is disconnected from the limbic system and neocortex (such as by a midbrain transection), stimulation of this nuclei will continue to evoke vocalization. Hence, these "sounds" are preprogrammed as motor engrams located in the periaqueductal gray.

Specifically, the periaqueductal gray coordinates the activity of the laryngeal, oral-facial, and principal and accessory muscles of respiration and inspiration (Zhang et al., 1994). The coordinated activity of these tissues enables an individual to laugh, cry, or howl, even if the rest of the brain (except the brainstem) is dead and there is no evidence of consciousness.

It thus appears that rather than being a center for emotional vocalizations, the midbrain periaqueductal gray instead appears to be the site where particular vocalization motor patterns are stored, as direct stimulation of this area produces vocalization in the absence of complex emotional-behavioral displays. Hence, when the periaqueductal gray is activated by impulses received from the limbic system or neocortex (or via spinal nuclei involved in transmitting noxious stimuli), it activates the appropriate motor program and then organizes and coordinates the oral-laryngeal and respiratory muscles so that the appropriate sounds can be produced (Zhang et al., 1994). Presumably, these neural vocal motor programs are stored within the periaqueductal gray.

The Midbrain, Brainstem and Speech

Brainstem mechanisms involved in vocalization include control over the respiratory muscles; control over the vocal cords so that one can phonate; control over the pharyngeal, oral, and nasal passages, which effect resonance; and control over the palate, tongue, lips, and mandible, which control articulation. However, articulation also requires a mixture of nonvoiced and voiced (plosive) sounds (e.g., "puh, guh, kuh")—which require a strong puff of air. In part, these functions are under the control of the midbrain periaqueductal gray, as well as cranial nerves XII, X, IX, V, and VII.

In order to expel air through the mouth, the soft palate must elevate to close off the nasopharynx from the oropharynx. If the subject articulates plosives poorly, with air escaping through the nose, the palate is not properly elevating, which suggests a lower brainstem lesion. Speech may sound hypernasal, or, conversely, hyponasal if there is insufficient air pushed into the nose (as if the nose has been pinched closed during speech), which may also be due to nasal obstruction.

Absence of any phonation is called *aphonia*. Faulty articulation is a result of a forebrain, brainstem, or muscular lesion and is called *dysarthria*. If the individual cannot speak due to a nerve or muscular lesion it is called *anarthria*. Because nerve X is associated with the palate (articulation), pharynx (resonance), and larynx (phonation), disorders of speech and swallowing may be due to a lesion to this nerve or its bed nucleus.

However, not just phonation and articulation but swallowing and respiration would be affected by lesions in this vicinity. As such, individuals may articulate, phonate, and inhale and exhale inappropriately while speaking; they may even sound as if they are laughing or crying when, in fact, inspiration and expiration are affected. This latter condition has been referred to as a bulbar palsy.

However, with severe bilateral lesions of the corticobulbar fibers, patients may lose the ability to speak or swallow and become mute and comatose. In the recovery phase, or with gradual lesions of the corticobulbar tracts (due to tumor, small strokes), the patient shows a characteristic, virtually pathonomic syndrome termed *pseudobulbar palsy*: he initiates speaking or swallowing very slowly, slurring words, with the voice having a peculiar strained pitch and quality. In addition, there is much emotional lability, crying one moment and laughing the next. The patient's face, most of the time, appears immobile, as though it were a wooden mask. However, when he laughs or cries, the facial movements return and become greatly exaggerated and prolonged. However, when questioned, the patient will claim that he does not feel the emotion being conveyed.

Cries and Laughter

As noted, the midbrain periaqueductal gray coordinates the activity of the laryngeal, oral-facial, and principal and accessory muscles of respiration and inspiration (Zhang et al., 1994). The coordinated activity of these tissues enables an individual to laugh, cry, or howl, even if the rest of the brain (except the brainstem) were dead and there were no evidence of consciousness.

Laughter and crying often involve activation of the entire musculature of the face, including the muscles of the throat, jaw, diaphragm, and to a lesser extent, the neck, chest, and abdomen. Laughter is also accompanied by respiratory expiration, whereas crying is associated with inspiration, such that rhythmic waves of air rush over the larynx, giving rise to the phonation of laughter, or the sound of sobbing.

Laughter and crying are, therefore, dependent on the functional integrity of the periaqueductal gray, as well as the pons, medulla, and cranial nerves XII, X, IX, V, and VII. However, the correspondence between the motoric-expressive and the felt aspects of emotion are maintained and dependent on two supranuclear pathways: the corticobulbar, which arises in the frontal motor areas, and the fronto-limbic-pontine-medullary pathway, which arises in the orbital frontal lobes and comes to include the medial frontal lobes and anterior cingulate, hypothalamus, amygdala, periaqueductal gray, and brainstem. It is this latter pathway that is responsible for conveying the felt aspects of emotional expression, such as is involved in laughter and crying (see Chapters 5 and 11).

When the functional integrity of these interconnections is disrupted, such as due to a progressive bulbar (pontine-medullary) palsy, brainstem hemorrhage, tumor, amyotrophic lateral sclerosis (ALS), or related injuries to the hypothalamus and deep orbital area, pathological laughing and/or crying may be triggered or reflexively released via brainstem activation.

Specifically, crying is conveyed via prolonged bursts of respiratory inhalation, and laughter via sudden bursts of expiration. That is, what appears to be "laughter" is actually an uncontrolled and sustained clonus accompanied by an explosion of air passing over the larynx, which repeats itself in a series of bursts. If the expiration is sustained, the sound produced becomes distorted and the patient may sound as if he is wailing or even screaming. Essen-

tially, the periaqueductal gray and the lower brainstem have been released and disconnected from forebrain control such that normal brainstem activities, e.g., vocalization, become uncontrolled. Moreover, as the brainstem nuclei mediating facial expression are also affected, facial expression is also out of control. The face becomes contorted and seemingly expressive of extreme happiness or grief, although patients will deny experiencing these feelings—and they may feel distressed and embarrassed by what is happening.

As described by a patient with ALS: "I begin to smile, but the smile becomes an exaggerated grin, which attaches itself to my face...[then]...I become angry, humiliated, and subject to uncontrolled tears. You have no idea how terrible it is when the crying is fully triggered and takes hold like a seizure. I can't control any of it. I simply disintegrate" (Lieberman and Benson 1977:717–718).

THE MIDBRAIN SUPERIOR AND INFERIOR COLLICULI

MIDBRAIN EVOLUTION

As pointed out in Chapter 1, the brainstem may have evolved in a posterior-caudal and an anterior direction so as to merge with the visual (midbrain) areas that probably initially supplied photopically derived energy to run the motor system and to arouse and energize the brain. As also detailed in Chapter 1, the anterior most segment of the midbrain and the anterior pons appear to have evolved from sensory-motor and photosensitive cells.

Due presumably, in part, to its photosensitive origins, and thus its concomitant ability to extract energy, the midbrain-pontine region of the brainstem developed the capacity to respond to light versus darkness, and to manufacture various neurotransmitters, e.g., serotonin, dopamine, and norepinephrine. In this manner the brainstem is able to exert widespread influences on cerebral arousal and neural transmission.

Over the course of evolutionary development, the capacity to respond to visual input increased beyond the detection of light versus shadow, and soon came to include motion, as well as the ability to rotate the eyeballs and focus the retina via cranial nerves II, III, IV, and VI. These complex functions, in turn, became subject to the control of the superior colliculus (or tectum).

However, with the evolution of "hearing," the midbrain became increasingly modified to subserve complex auditory functioning as well, which in turn became associated with the evolution of the inferior (auditory) colliculus. In this regard, just as the control and governance of motor functioning becomes more complex and hierarchically (caudal to anterior) organized, so too does sensory analysis and related sensory-motor integrative functions that take place within the midbrain. Multimodal processing and complex behavioral reactions, and the integration and analysis of auditory, visual-tactile, and even motor stimuli, are characteristics of the visual (superior) and auditory (inferior) colliculi of the midbrain.

The Superior Colliculus

The superior colliculi are located in the anterior half of the midbrain. Although the midbrain per se is not well differentiated, these nuclei consist of gray and white layers and resemble cerebral cortex (Carpenter, 1991). As noted in Chapter 1, neocortical layers I and VII (between which are sandwiched layers II–VIa), in turn, appear to be related to and/or an outgrowth of the midbrain.

The superior colliculi can be functionally divided into superficial and deep layers that subserve somewhat different functions. (Cowie and Robinson, 1994; Davidson and Bender, 1991; Weller, 1989). For example, the superficial layers receive considerable input from the retina as well as temporal and occipital visual cortex, and respond to moving stimuli. The superficial layers also project to vision-related cranial nerve nuclei.

By contrast, the intermediate and deeper layers receive converging motor, somesthetic, auditory, visual, and reticular input, and in fact serve as an extension of the reticular formation, and maintain interconnections with the caudal medulla and those cranial nerves associated with movement of the head. Hence, these layers serve as multimodal assimilation areas that are concerned with orienting toward external stimuli and movements of the head and eyes during gaze shifts (Cowie and Robinson, 1994; Davidson and Bender, 1991; Keller and Edelman, 1994).

The deep and superficial portions are also intimately interconnected (Carpenter, 1991; Parent, 1995) and both receive extensive optic input from the contralateral eye, and contralateral visual input from the ipsilateral eye. Moreover, they receive ipsilateral projections from the neocortical visual areas regarding the contralateral half of visual space. Hence, each colliculus is concerned with either the right or left half, i.e., the contralateral half of visual space. In addition, the colliculi receive input from the frontal eye fields. It is the superior colliculi which make "blind sight" (following neocortical blindness) possible.

As noted, cells within the colliculi are responsive to visual motion and many tend to respond only to movement in a certain direction (Davidson and Bender, 1991). Most do not respond to stationary stimuli. Given that reptiles, amphibians, and fish are devoid of higher cortical centers, it thus appears that the colliculi evolved so as to detect the presence of prey (or predators) and to guide orienting reactions and, thus, movement related to escape or food procurement.

The Inferior Colliculus

The inferior colliculus (IC) is predominantly concerned with detecting and analyzing auditory stimuli and, in fact, is tonotopically organized; i.e., neurons are arranged in a laminar pattern that represents different auditory frequency bands (reviewed in Carpenter, 1991; Brodal, 1981). The IC can also respond to sounds arriving from either ear. This enables the IC to analyze and localize the source of various sounds and to correlate them with their various spatiotemporal characteristics. In

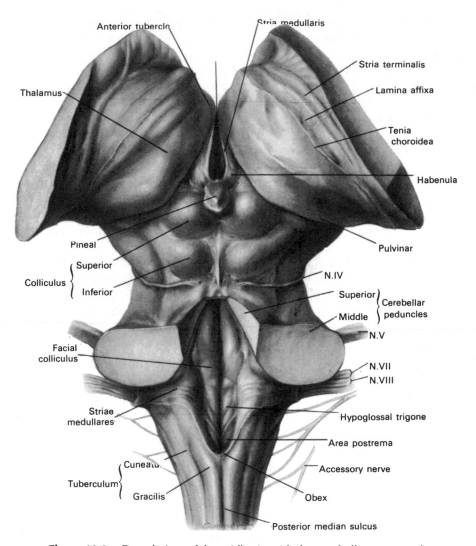

Figure 10.3   Dorsal view of the midbrain with the cerebellum removed. (From Mettler FA. Neuroanatomy. St. Louis: CV Mosby, 1948.)

this manner, sounds can be identified as coming from a certain direction and from a specific source.

The IC sends fibers to the pons and medulla as well as the superior colliculus, and relays auditory impulses to the spinal cord and to the nuclei subserving the neck and facial musculature (Henkel and Edwards, 1978; Wright and Barnes, 1972). Hence, via the IC, auditory impulses can trigger head and body turning and thus orientation toward sound sources. In mammals, the IC also acts to relay auditory signals received from the lateral lemniscus to the medial (auditory) geniculate of the thalamus; in this regard, it serves not only to analyze and orient toward auditory stimuli, but as a major auditory relay nucleus. However, in this regard it is completely dependent on cranial nerve nuclei subserving the vestibular-cochlear system: cranial nerve VIII, which is located at the pontine, medulla junction.

BRAINSTEM AUDITORY PERCEPTION

Long before amphibians left the sea to take up life on land and in the air, fish had already learned to detect and analyze sounds made in the water. This was made possible by a structure located along both sides of the fish's body, called the lateral line. The lateral line is very sensitive to vibrations, including those made by sound. The mammalian auditory system, however, did not evolve from the lateral line but from the vestibular nucleus (see Chapter 1).

The human brainstem vestibular nucleus is very sensitive to vibrations and gravitational influences and helps to mediate balance and movement, such as in swimming and running. Over the course of evolution, it is through the vestibular nucleus (and the amygdala) that the first neuronal rudiments of "hearing" and the analysis of sound also took place. Among all mammals, this is a function the vestibular nuclei continues to serve via the signals it receives from the inner ear: the brain's outpost for detecting certain vibratory waves of molecules and their frequency of occurrence.

Initially, however, the auditory system also evolved so that organisms could turn, orient toward, and localize unexpected sound sources. This was made possible by messages transmitted to the midbrain inferior colliculus. This includes "things that go bump in the night," such as the settling of the house and the creaking of the floor, all of which can give rise to alarm reactions in the listener (in the amygdala). This is a sense humans and their ancestors have maintained for several million years, its main purpose being to maximize survival via the detection of approaching predators. In fact, the inferior colliculus (as well as the amygdala of the limbic system) was the main source of auditory analysis 200 million years before the evolution of the neocortex some 100 million years ago (see Chapter 1). This is a function it continues to perform among modern-day reptiles, amphibians, and sharks, as well as mammals.

Auditory Transmission from the Cochlea to the Temporal Lobe

Within the cochlea of the inner ear are tiny hair cells that serve as sensory receptors. These cells give rise to axons that form the cochlear division of the eighth cranial nerve, i.e., the auditory nerve. This rope of fibers exits the inner ear and travels to and terminates in the cochlear nucleus, which overlaps and is located immediately adjacent to the vestibular-nucleus from which it evolved within the brainstem.

Among mammals, the cochlear nucleus, in turn, projects auditory information to three different collections of nuclei. These are the superior olivary complex, the nucleus of the lateral lemniscus, and, as noted, the inferior colliculi.

A considerable degree of information analysis occurs in each of these areas before being relayed to yet other brain regions, such as the amygdala (which extracts those features that are emotionally or motivationally significant), and to the medial geniculate nucleus (MGN) of the thalamus, where yet further analysis occurs.

Among mammals, from the MGN and amygdala, auditory signals are transmitted to the primary auditory receiving area located in the neocortex of the superior temporal lobe: Heschl's gyrus. Here, auditory signals undergo extensive analysis and re-analysis, and simple associations begin to be formed. However, by the time it has reached the neocortex, auditory signals have undergone extensive analysis by the MGN of the thalamus, amygdala, and the other ancient structures mentioned above.

THE BRAINSTEM RETICULAR FORMATION

The pontine and medullary brainstem significantly contributes to cortical arousal, attention, sensory and motor facilitation and inhibition, and the orienting reaction. This is made possible through ascending limbic, thalamic, and neocortical influences, and descending inhibitory influences on brainstem, spinal sensory, and motor nuclei. In this manner the brainstem is able to arrest movement, while simultaneously filtering out distracting or irrelevant sensory stimuli (presumably via 5HT influences) so that attention may be selectively directed and maintained (see Klemm, 1990; Steriade and McCarley, 1990).

Brainstem motor and sensory filtering and inhibition is also made possible via reticular neuronal modulatory control over arousal: i.e., the reticular activating system. The brainstem reticular formation consists of a number of morphologically distinct nuclei and neurons (see Steriade and McCarley, 1990; Vertes, 1990, for review), which differ in regard to size, shape, number, and orientation of dendrites, and axonal patterns of projection. It is via these nuclei that the reticular formation can exert widespread influences on motor and sensory functioning, as well as arousal.

For example, cranial nerve and sensory fibers originating in the spinal cord give off collaterals as they ascend or terminate in the brainstem reticular formation (Brodal, 1981; Pompeiano, 1973). This includes auditory and visual fibers originating in the midbrain tectum (Edwards, 1980). The more lateral portions of the reticular formation consist of short axons that richly innervate adjacent brainstem nuclei and are connected to the collaterals of the various sensory tracts.

Specifically, the ascending reticular activating system consists of a multineuronal, polysynaptic core that conveys, via long central axons with numerous laterally projecting collaterals, nonspecific impulses to the forebrain. These long axons ascend via the central tegmental tract, which projects to the intralaminar nuclei of the thalamus.

Many of these core reticular neurons are concerned with motor activation and inhibition as well as with alerting and arousal functions, so that motor functioning is coordinated with forebrain sensory processing. By contrast, fibers from the medial reticular formation project to the limbic system.

Coma, Lethargy, and the Brainstem

The ascending sensory tracts, including those involved in transmitting noxious stimuli, are exceedingly important in maintaining neocortical arousal. If these sensory tracts are severed at the brainstem level, patients may become exceedingly lethargic, and even apathetic. Similarly, destruction of the olfactory, auditory, and visual pathways, even at a peripheral level, can result in prolonged lethargic states (see Steriade and McCarley, 1990). This indicates that sensory input, from the periphery to the brainstem and up through the thalamic level, is exceedingly important in maintaining arousal and cortical alertness, as well as the waking state.

However, destruction or interruption of these long, ascending sensory pathways (i.e., the lemniscal systems) that span the spinal cord and terminate in the thalamus does not eliminate the cortical arousal response. This is because forebrain arousal is dependent on the medial reticular formation, which receives collateral sensory fibers from these ascending systems.

Thus, whereas lateral brainstem injuries may create lethargic and reduced states of arousal, destruction of the core brainstem reticular formation can result in a loss of forebrain arousal and create a permanent comatose state such that the patients or animals may not even respond to noxious stimuli (French and Magoun, 1952; Lindsley et al., 1950).

However, this is not only a consequence of a loss of reticular input to the forebrain, but a loss of forebrain (and frontal lobe) and midbrain input to the brainstem reticular formation (Bremer, 1975; see also Steriade and McCarley, 1990). Just as the complexity of motor programming increases as one ascends the brainstem, so too does control over arousal.

For example, if the transection is in the upper brainstem (i.e., at the pons, trigeminal level) well below the tectum, the subject continues to demonstrates EEG signs of alertness, as well as visual tracking eye movements (reviewed by Steriade and McCarley, 1990), although he is otherwise in a coma. However, if the lesion is more anterior at the midbrain level, the subject becomes completely comatose and unresponsive—which indicates an important role for the midbrain reticular formation in maintaining cortical and forebrain arousal.

However, if the lesion is anterior to the midbrain, rather than coma and a loss of arousal, there results a loss of control over arousal such that environmental stimuli may be ignored or neglected. For example, if the immediately adjacent thalamus is lesioned (which is the terminal junction of many reticular fibers), the result may be hemi-neglect of auditory, visual, and tactual space, depending on the laterality of the damage. Presumably, this is a consequence of an inability to transmit or relay sensory stimuli or to activate the neocortex, for with massive lesions to the thalamus, cortical EEG desynchronization is usually absent, regardless of brainstem activity (Steriade and McCarley, 1990).

Conversely, if the lower brainstem is lesioned, the patient not only become comatose but cardiovascular and respiratory disturbances will result. Hence, the patient may die.

FRONTAL-THALAMIC AND AMYGDALOID CONTROL OVER AROUSAL

When thalamic output increases, the amplitude of cortical responses also increases (Steriade and McCarley, 1990). However, the thalamus, in turn, is also under the control of the lateral and orbital frontal lobes, which also exert significant modulatory influences on the brainstem reticular formation while it simultaneously monitors neocortical and thalamic activity (Joseph, 1990d). Via the frontal lobe, the thalamus and the reticular formation can be activated or inhibited selectively or globally; in this way, specific sensory modalities are attended to while others are filtered or suppressed so that the organism can engage in selective attention and information processing.

Similarly, the amygdala is capable of inducing arousal as well as EEG desynchronization (Kriendler and Steriade, 1964; see Chapter 5). This is because the amygdala is also intimately interlinked with the upper brainstem reticular formation (Takeuchi et al., 1982), as well as the frontal lobe and

thalamus. Hence, when stimulated, the amygdala can induce alerting reactions and arousal (see Chapter 5) as well as PGO waves (pontogeniculo-occipital waves, or rapid eye movements) (Calvo et al., 1987), which in turn are associated with the onset of paradoxical sleep and brainstem activity.

In addition, the medial and lateral portions of the posterior hypothalamus also project to the midbrain reticular formation (Steriade and McCarley, 1990), as do the preoptic regions of the hypothalamus (Swanson et al., 1987)—a region intimately involved in sexual behavior and sexual posturing (see Chapter 5). The hypothalamus is also intimately linked with the orbital frontal lobes and the amygdala. Therefore, the hypothalamus, being exceedingly involved in all aspects of rudimentary motivational and emotional functioning, is also important in arousal and the waking state (Ranson, 1939), and if the posterior regions are damaged there can result severe apathetic states.

Hence, these findings implicate the forebrain (i.e., the amygdala, hippocampus, frontal lobes, thalamus, hypothalamus) in maintaining forebrain arousal. Nevertheless, even in this regard, the forebrain is dependent on the midbrain and brainstem. It is the brainstem that contains nuclei and cell clusters that manufacture specific neurotransmitters directly responsible for maintaining and promoting behavior, and emotional, cognitive, and cerebral arousal, including sleep and dreaming.

DOPAMINE, NOREPINEPHRINE AND SEROTONIN

The midbrain and pons contain neurons and nuclei that are responsible for manufacturing three principle monamine neurotransmitters: i.e., dopamine (DA) and norepinephrine (NE)—also referred to as catecholamines—and serotonin (5HT), which is an indolamine. DA is produced predominantly within the midbrain substantia nigra and tegmentum, whereas NE is manufactured and distributed by neurons located predominantly in the locus ceruleus, which is located in the pons. 5HT-producing neurons and cell groups are found in both the pons and medulla (e.g., the dorsal and median raphe nucleus).

Dopamine

Dopamine is believed to play a dominant role in cognition and motor functioning, and disturbances in the DA transmitter systems are closely related to the development of Parkinson's disease and psychotic abnormalities (see Chapter 9). For example, reductions in corpus striatal DA are associated with catatonia and Parkinson's disease, whereas excess DA has been linked to paranoid and psychotic disturbances (see Chapter 9). However, the nature of the disturbance is also dependent on which DA system is affected, i.e., substantia nigra-corpus striatal or tegmentum-limbic striatum.

For example, the midbrain ventral tegmentum (A-10 DA cell group) provides dopamine to the amygdala, the limbic striatum, and the ventral caudate and putamen, and is believed to be related to emotion, mood, and reward (referred to as the mesocorticolimbic/mesolimbic DA system). The mesolimbic DA system can, therefore, enhance or depress emotional-motoric expression.

The substantia nigra DA system is related to motor functions, including the production of stereotyped and routine actions (for related details see Chapter 9, and/or Fibiger and Phillips, 1986; Robins and Everitt, 1992). Prominent projection sites include the dorsal caudate, putamen, and medial frontal lobes.

The nigrostriatal and mesolimbic DA systems sometimes act in an oppositional as well as complementary and mutually supportive manner. For example, the release of DA occurs in response to specific biologically significant stimuli, which in turn may activate the limbic and corpus striatum and thus trigger motor activity, such as approach or withdrawal (Fibiger and Phillips, 1986; Robins and Everitt, 1992). However, if one (or both) of these two DA systems becomes dysfunctional, counterbalancing influences are removed, which may result in excessive activity within the supposedly remaining "normal" DA system and related nuclei. Hence abnormalities in these DA systems may induce hyperactivity and mania or, conversely, a poverty of movement (e.g., Parkinson's symptoms and catatonia; see Chapter 9).

As noted in Chapter 1, midbrain neurons, being derived in part from photosensitive cells, also display circadian and natural rhythms that may well have been induced via adaptions to changes in the light–dark cycle. Similarly, DA neurotransmission also displays circadian variations, which in turn are reflected by cyclic motor and DA activity within the caudate nucleus, and less so within the limbic striatum (Paulson and Robinson, 1994).

Norepinephrine

The NE neuron system is composed of three main neuronal clusters located in the pons and medulla: i.e., the dorsal medullary (A-2 cells), the locus ceruleus (LC, A-4, A-6 cells), and the subceruleus (A-1, A-3, A-5, A-7 cells). These nuclei give rise to ascending and descending monosynaptic projections that make diffuse and specialized selective contact with a variety of brain areas via their highly branching axons (Dahlstrom and Fuxe, 1965; Fuxe, 1965; Levitt and Moore, 1979). It has been estimated that, via their diffuse projections, a single NE neuron can make contact with up to 75,000 other neurons and can contribute fibers to both the cerebral cortex, limbic system, and cerebellum (Cooper et al., 1974). However, this has also been disputed.

Broadly considered, the NE system projects to the midbrain colliculi and lateral and medial geniculate of the thalamus, the amygdala (which contains very high concentrations of NE), and throughout the brainstem, where it influences auditory and visual functioning and is a major transmitter involved in facial expression. Indeed, the NE system subserves a wide range of functions, including regulation of cardiovascular activity, micturition, respiration, sleep and dreaming, motivation, emotional expression, and mobilization in response to pleasure, fear, or stress (reviewed in Chapters 6 and 16).

The locus ceruleus (LC)–NE system is clearly implicated in emotional functioning and contributes fibers (via the medial forebrain bundle) to the hypothalamus, amygdala, septal nuclei, hippocampus, thalamus, and cingulate gyrus (Dalhstrom and Fuxe, 1965; Fuxe, 1965) as well as the midbrain, including the lateral geniculate and pulvinar of the thalamus, and the frontal and occipital lobes (see Foote et al., 1983; Morrison and Foote, 1986; Sakaguchi and Nakamura, 1987). However, it maintains predominantly ipsilateral connections, including those that diffusely project to the neocortex. There is also some evidence to indicate that, whereas DA is more greatly concentrated in certain left cerebral nuclei and brain regions, NE concentrations are greater in the right hemisphere, particularly within the right somesthetic nucleus of the thalamus (Oke et al., 1978). In this regard, it is noteworthy that, whereas left cerebral lesions will reduce left cerebral NE levels within the areas of injury, right hemisphere lesions can induce diffuse and bilateral NE reductions (Robinson, 1979).

Locus ceruleus–NE neurons become highly active in response to novel or motivationally significant somesthetic, visual, and auditory stimuli, and NE levels are affected and rapidly metabolized by stressful and emotional experience (Foote et al., 1980; see also Chapters 6 and 16), especially fear. Indeed, the neural circuitry involving the NE system makes it "uniquely situated to subserve all the known physiological correlates of fear" (Redmond and Huang, 1979:2154).

Hence, electrical stimulation of the LC can trigger severe anxiety and fear reactions. Conversely, ablation of the locus ceruleus results in a significant decrease in fear and anxiety reactions, even in response to threatening stimuli (Redmond and Huang, 1979). Similarly, NE reuptake blockers (tricyclics) can reduce feelings of anxiety, whereas NE activators, such as piperoxan, can induce anxiety reactions in humans (Klein et al., 1978).

The NE system is also involved in the mediation of pleasure and reward, and is rapidly synthesized when subjects are engaged in pleasurable activities (Bliss and Zwanziger, 1966). Conversely, low levels of NE are associated with a lack of pleasurable feeling, i.e., depression. Hence, NE is essential to the maintenance of the "pleasure circuit."

The NE system is also highly involved in arousal, including the sleep–wake cycle, and the generation of REM, PGO activity, and dreaming (Chu and Bloom, 1974). NE levels, therefore, fluctuate in cycles and demonstrate specific circadian rhythms that are also related to sleep and dreaming. For example, in humans and other primates, NE levels are lowest at 3 AM and gradually increase throughout the day until around 3 PM (Ziegler et al., 1976), which in some cultures corresponds to "nap" or "siesta" time.

Serotonin

Serotonergic (5HT) neurons exert widespread and often tonic and inhibitory influences on a variety of brain areas (Applegate, 1980; Jacobs and Azmita, 1992; Soubrie, 1986; Spoont, 1992). This includes the modulation of appetite, fear, pain, and emotional states, such as depression; the inhibition of incoming sensory input; and the modulation of motor expression. That is, 5HT restricts perceptual and information processing and, in fact, increases the threshold for neural responses to occur at both the neocortical and limbic levels.

For example, in response to arousing stimuli, 5HT is released (Auerbach et al., 1985; Roberts, 1984; Spoont, 1992), which aids attentional and perceptual functioning so that the most salient stimuli are attended to. That is, 5HT appears to be involved in learning not to respond to stimuli that are irrelevant and not rewarding (see Benninger, 1989). These signals are filtered out and suppressed.

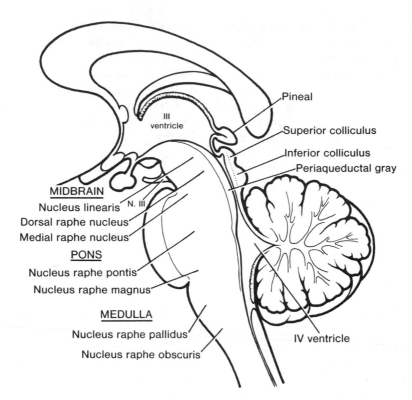

Figure 10.4   Schematic diagram of the brainstem, depicting raphe nuclei. (From Carpenter M. Core text of neuroanatomy. Baltimore: Williams & Wilkins, 1991.).

It has also been demonstrated that 5HT acts to suppress activity in the lateral (visual) geniculate nucleus of the thalamus and synaptic functioning in the visual cortex as well as the amygdala and throughout the neocortex (Curtis and Davis, 1962; Marrazzi and Hart, 1955). By contrast, substances that block 5HT reception—such as LSD—result in increased activity in the sensory pathways to the neocortex (Purpura, 1956), which induces complex hallucinatory experiences.

As will be detailed below, 5HT is also implicated in the generation of sleep and dreaming, as well as the production of motor paralysis during sleep. For example, 5HT, via descending influences, facilitates motorneuronal activity while simultaneously decreasing or inhibiting sensory input (Morrison and Pompeiana, 1965). In this manner, selective attention (via sensory filtering) can be maintained while the organism is engaged in goal-directed motor behavior. However, while sleeping and dreaming, internal sensory filtering is reduced (due to decreased 5HT) whereas motor functioning is inhibited—which prevents the subject from walking about and acting out his or her dreams (see below).

THE DORSAL AND MEDIAN RAPHE-5HT NUCLEI

Serotonin (5HT) is produced by the raphe nuclei located in the pons and medulla, with the highest percentage located in the dorsal raphe (79%) and the bulk of the remainder produced by the median raphe. Thus, the main source for limbic and neocortical 5HT is the dorsal (DR) and median raphe (MR) nuclei (Azamita, 1978; Azamita and Gannon, 1986; Wilson and Molliver, 1991a, b). Specifically, the amygdala, hypothalamus, basal ganglia, primary and association receiving areas, and frontal lobe are innervated by the DR, whereas the hippocampus, cingulate gyrus, and septum receive their 5HT from the MR nuclei (Azamita, 1978; Azamita and Gannon, 1986; Wilson and Molliver, 1991a, b).

The MR, however, is diffusely organized and appears to exert a nonspecific and global influence on arousal and excitability (Wilson and Molliver, 1991a, b). The DR is much more discretely organized, can exert highly selective inhibitory or excitatory influences, and plays a role in the coordination of excitation in multiple functionally related areas, including the frontal lobes and amygdala (Wilson and Molliver, 1991a, b). Because of the manner in which they are organized, the DR and MR can exert select inhibitory influences so as to engage in perceptual filtering in one or a variety of areas. It does this while simultaneously exciting yet other multiple regions, which in turn aid selective attention, and by creating specific neuronal networks.

NOREPINEPHRINE AND SEROTONIN INTERACTIONS: FEAR, PAIN, AND STRESS

The 5HT and NE systems appear to interact in regard to a number of different functions, including arousal, sensory, and emotional processing, and sleep and dreaming, and (in conjunction with brainstem enkephalins) the experience and modulation of painful experience, including depression. For example, it has been repeatedly and widely reported that altered brain 5HT turnover and reduced 5HT levels are often discovered postmortem among individuals who have committed suicide (e.g., Arranz et al., 1994). Moreover, some individuals commit suicide because of feelings of chronic stress and emotional turmoil, and an inability to experience pleasure in living—which, in turn, implicates the NE as well as 5HT systems.

Pain and Stress

Somatic and visceral sensations are transmitted to the cerebrum by way of the spinal cord and brainstem. At the spinal level, these signals are transmitted via the dorsal horns, where collateral fibers from the peripherally located somatic receptors enter and then terminate near the substantia gelatinosa. At this level they form and give rise to the lateral spinal tract, which transmits pain and thermal impulses to the somesthetic (VPL) thalamus, which contains high concentrations of NE, particularly the right VPL (Oke et al., 1978).

However, descending LC-NE and raphe-5HT fibers also terminate in the substantia gelatinosa, which is also rich in opiate receptors (Dalhstrom and Fuxe, 1965). In response to nociceptive impulses, both NE and 5HT (as well as the brainstem enkephalins) act to reduce and inhibit the transmission of these impulses (Headley et al., 1978). Similarly, stimulation of the raphe nucleus induces analgesia, and direct application of 5HT depresses dorsal horn reactivity to noxious stimuli.

In this regard, not only does NE and 5HT serve to counteract the reception of painful stimuli, they also act to arouse and mobilize the organism when it experiences pain or fear. As noted, stimulation of the LC can induce fear reactions as well as defensive behavior, including flight (Redmond and Huang, 1979). In this manner, fear can also become associated with stimuli that induce pain, and both can be associated with escape. However, in response to chronic fear, pain, or stress, 5HT and NE levels begin to rapidly decrease.

Stress and Psychosis

In response to highly stressful, fearful, or painful stimuli, 5HT levels, at least initially, are increased, which results in numbing, analgesia, and a loss of pain perception (Auerbach et al., 1985; Roberts, 1984; Spoont, 1992). Under conditions of prolonged stress, pain, or fear, 5HT is rapidly depleted. In consequence, the individual may begin to feel overwhelmed and unable to appropriately process incoming sensory stimuli. Soon thereafter (i.e., following chronic stress) he or she may also demonstrate a heightened fear and startle response (Davis, 1984) as well as social withdrawal (Raleigh et al., 1983).

With continued high stress, sensory filtering is reduced, and neurons in the amygdala, hippocampus, and inferior temporal lobe cease to be inhibited, which in turn can result in dreamlike states, hallucinations (Joseph, 1988a, 1992a; Spoont, 1992), depersonalization (Sallanon et al., 1983), and associated psychotic and emotionally disturbed states (see Chapter 16).

Under conditions of chronic emotional stress, 5HT begins to be depleted. However, not just trauma, but repeated and chronic sexual activity is also associated with significant reductions and depletions in 5HT (Spoont, 1992; Zemlan, 1978); this might explain why individuals who are repeatedly sexually molested or assaulted may develop post-traumatic stress disorder, dissociative and amnesic states, and psychosis (see Chapter 16).

In addition, stress, be it due to combat, exposure to cold, prolonged swimming, or pain, is associated with the depletion of NE (Bliss et al., 1968). With the exception of strenuous physical activity, fear and stress can reduce NE concentrations by up to 50%.

In fact, even a single trauma may result in alterations in NE and 5HT neurotransmitter turnover and release, which in turn can influence and increase the size of both the presynaptic and postsynaptic substrates (Goelet and Kandel, 1986; Krystal, 1990) and even gene expression. This may induce, not only alterations in learning and memory, but increase emotional reactivity and alarm responses that may persist indefinitely (Charney et al., 1993; Goelet and Kandel, 1986).

Under chronic conditions, affected neurons and associated neural networks may also become adapted to reduced 5HT and NE levels, thereby increasing the likelihood of enhanced startle reactions, social withdrawal, feelings of depersonalization, hallucinatory states, sleep disturbances, increased REM latency, and other associated symptoms associated with a chronic post-traumatic stress disorder (Carlson and Rosser-Hogan, 1991; Cassiday et al., 1992; Charney et al., 1993; Foa et al., 1991; McNally et al., 1990; Putnam, 1985; Wilcox

et al., 1991; van der Kolk, 1987). It is, thus, noteworthy that blockage of 5HT reuptake (which increases the amount of 5HT in the synapse) acts to reduce the symptoms of post-traumatic stress disorder (Hollander et al., 1990).

Depression, Serotonin, and Norepinephrine

As is now well known, major factors in the pathogenesis of depression include disturbances in the LC–NE and 5HT systems (Arranz et al., 1994; Delgado et al., 1994). For example, a disturbance in NE synthesis and metabolism, regardless of its cause, will disrupt limbic and cortical functioning and arousal, and reduce neuroendocrine activity and, thus, the organism's ability to mobilize its defenses in response to stress, adverse experience, or even self-defeating thoughts. Every component of social–affective expression will be altered, especially that related to the experience of pleasure and reward—a condition conducive to depression.

Similarly, when 5HT levels are reduced, individuals may become depressed. There is also an increased tendency to respond to unrewarding situations (Soubrie, 1986) and to continue to respond regardless of punishment (Spoont, 1992). Affected individuals may fail to avoid or may even be drawn to abusive or potentially frightening or traumatic situations, and may seem helpless to alter their behavior, e.g., learned helplessness (see Charney et al., 1993; Krystal, 1990). Indeed, reduced 5HT has been repeatedly noted in the brains of those who have committed suicide ,particularly in a violent fashion (e.g. Brown et al., 1982; Cronwell and Henderson, 1995).

Although it is difficult to classify a depressive episode or psychosis strictly in terms of a biochemical abnormality, it has been known for over two decades that at least two subgroups differ in regard to depressive symptoms and NE and 5HT activity levels (Goodwin et al., 1979; Van Dongen, 1982; Van Praag, 1982). For example, those with an NE depression may behave in a more emotionally labile and expressive manner, whereas those with a 5HT depression may demonstrate a much more profound motor retardation coupled with social withdrawal and confusion due to sensory overload.

Moreover, both groups respond differently to pharmacological treatment. For example, some respond best to imipramine, which inhibits NE reuptake (thus increasing NE levels in the synapse), whereas others react best to amitriptyline, which potentiates 5HT (Goodwin et al., 1978; Van Praag, 1981) and inhibits NE reuptake. This also suggests that among some individuals, NE and 5HT may simultaneously contribute to a distinct form of depression.

It has also been argued that reduced 5HT does not cause depression per se and is not linearly related to the level of depression (Delgado et al., 1994). Rather, 5HT may be a predisposing factor in depression, which in turn may be related to a postsynaptic deficit in 5HT utilization or binding affinity (Arranz et al., 1994; Delgado et al., 1994).

DREAMING, SLEEPING, AND RHYTHMIC AND OSCILLATING BRAINSTEM ACTIVITY

As noted, many of the functions performed by the brainstem are not only reflexive and stereotyped but characterized by specific oscillating rhythms. For example, the sex act involves the temporal-sequential insertion and movement of the male penis and rhythmic motion of the hips, whereas walking involves the rhythmic control of the extremities. As noted in Chapter 1 (and above), over the course of neuronal evolution dorsally situated neurons, including those that formed the pineal body, and portions of the thalamus and midbrain, were highly responsive to light and in consequence developed circadian and rhythmic cycles of activation.

Hence, a considerable number of neurons located in the midbrain and pons display a bursting, oscillating rhythm—a pattern that also accompanies controlled

locomotion and a host of other cyclic activities. This includes 5HT and NE neurons, which are also involved in controlling a wide variety of rhythmic functions, including respiration, chewing, and micturition (Skinner and Garcia-Rill, 1990) as well as sleep and dreaming. Presumably, these interconnected monoamine, cholinergic, and catecholaminergic neuronal clusters are out of phase, which induces cyclic oscillations (Skinner and Garcia-Rill, 1990), including those involved in sleep and dreaming.

REM-Off and REM-On Neurons

It is now well known that the visual–emotional hallucinatory aspects of dreaming occur during REM, whereas more thought-like and verbal ideational patterns are produced during non-REM (NREM) sleep. A variety of nuclei and brain regions appear to be involved in the production of REM, i.e., the amygdala, hippocampus, right temporal lobe, and, especially, brainstem nuclei located in the lateral and medial pons.

Specifically, cholinergic neurons located in the lateral pons, and neurons located in the medial pontine reticular formation, appear to be the locus for REM-on neurons that initiate and/or maintain the production of REM sleep (see Steriade and McCarley, 1990; Vertes, 1990). That is, during the production of REM and paradoxical sleep, there is increased cholinergic activity in these regions (REM-on neurons), whereas with the termination of REM, these same neurons greatly reduce their activity (REM-off neurons).

Similarly, when cholinergic agonists are injected into the medial pons, REM sleep is produced (Steriade and McCarley, 1990). When cholinergic agonists are injected into the lateral pons, muscle atonia is produced—a common characteristic of paradoxical sleep. Thus, REM-on cells tend to be cholinergic and become very active during paradoxical sleep and induce muscle atonia, but markedly decrease their activity during slow wave and NREM sleep, at which point atonia also tends to disappear.

In contrast, REM-off neurons, which tend to be located in the medial raphe nucleus (which contain 5HT neurons) and in the locus ceruleus (located at the midbrain-pons junction and which contain NE neurons) are highly active during waking. However, they begin to reduce their activity during slow-wave sleep and then significantly decrease their activity with the initiation and onset of REM and the production of pontine, lateral geniculate, occipital activation, i.e., PGO waves (see Steriade and McCarley, 1990; Vertes, 1990). Conversely, experimental cooling (and thus deactivation) of the LC consistently and repeatedly induces REM sleep (Cespuglio et al., 1982).

Hence, these REM-off neurons appear to suppress REM and PGO activity and interfere with the onset of dreaming and paradoxical sleep, whereas REM-on neurons initiate the opposite sleep phase, in which case the brain appears to be highly active (e.g., the amygdala, temporal lobe, pons-geniculate-occipital lobe) and the individual begins to dream.

REM-on and REM-off neurons, therefore, appear to oscillate in a rhythmic fashion, thus inducing sleep, dreaming, and waking. When the REM-off neurons cease to fire, the REM-on neurons (which are predominately cholinergic) become highly active until a REM episode is produced. However, as REM-on neuron activity decreases, REM-off activity increases, thus setting into motion a continuous 90-minute cycle of REM–NREM sleep alteration.

PGO Waves

Paradoxical sleep and the production of dream states are associated with the development of high levels of activity within the pons, the lateral geniculate nucleus of the thalamus, and the occipital lobes—activity collectively referred to as PGO waves (see Hobson et al., 1986). Presumably, as these neurons are activated, visual imagery is produced as these regions are concerned with analyzing visual input as well as visual scanning of the environment.

However, PGO waves, although generated in a variety of brainstem reticular neuronal clusters, are also induced by amygdala activation, which, in concert with the hippocampus, contributes emotional imagery and experiential impressions to the dream state (Joseph, 1982, 1992a, 1994). Indeed, via the extensive interconnections the amygdala maintains with the brainstem and the geniculate and occipital lobes, and its capacity to induce high levels of arousal, this nucleus is ideally situated to initiate, if not produce (again in conjunction with the hippocampus), the auditory-visual imagery characteristic of the dream state, including all aspects of related emotion (see Chapter 5). Moreover, the amygdala also receives 5HT and NE from the brainstem (with which it is intimately interconnected), thereby creating an elaborate feedback circuit.

The lateral pontine reticular formation is also involved in producing hippocampal theta during REM (Vertes, 1984). As detailed in Chapter 5, hippocampal theta is associated with arousal, and theta activity may represent the introduction and transformation of hippocampal-based memory–imagery into dream imagery (Joseph, 1992a). However, the production of hippocampal theta may also help insure that whatever is experienced during REM sleep is stored in such a fashion that dream stimuli do not become committed to memory.

Nevertheless, in regard to dream sleep, the amygdala and hippocampus appear to be largely dependent on the brainstem, for when disconnected (see below), although the brainstem continues to demonstrate a sleep–wake cycle, the forebrain will cease to demonstrate paradoxical sleep activity. Moreover, although the amygdala can induce PGO activity, in general, neurons in the upper medial pontine reticular formation begin to fire well in advance of the development of PGO waves (McCarley and Ito, 1983). It is also these pontine neurons (in conjunction with the cranial nerves and midbrain) that initiate eye movements during REM. Typically, PGO waves reach their maximum amplitude in the lateral geniculate nucleus of the thalamus, which (like the amygdala) are also cholinergically responsive (Steriade and McCarley, 1990).

It is noteworthy that PGO neurons also become active during waking, when eye movements are made (Nelson et al. 1983). Therefore, PGO waves are not strictly a phenomenon associated with sleep, but can occur when the subject is emotionally aroused and perhaps looking anxiously around for the source of stimulation—which again implicates the amygdala in their production.

Characteristically, however, PGO waves and their transfer to the thalamus have two distinct states (Steriade and McCarley, 1990). That is, they first appear during the last stages of synchronized sleep just prior to REM, and then again, accompanying the development of desynchronized sleep and REM.

THE MIDBRAIN RETICULAR FORMATION AND REM

Although NE and 5HT neurons appear to make a major contribution to the development of sleep and dreaming, REM is also dependent on activity within the midbrain reticular formation—activity that, presumably, includes excitation of the superior (visual) colliculus, which in turn can trigger eye movements and orienting reactions to sensory stimuli.

Midbrain reticular activity during waking and REM is twice that during slow-wave sleep (Steriade and McCarley, 1990). Midbrain reticular neurons also markedly increase their activity during the transition from slow-wave to REM sleep.

Conversely, if the midbrain is transected, REM does not appear in the forebrain, although all associated components are preserved in the brainstem, including REM (although reduced), muscle atonia, and the pontine contribution to PGO waves (Hobson et al., 1986; Steriade and McCarley, 1990; Vertes, 1990). Transection of the lower (medulla) brainstem does not significantly affect REM sleep (Siegel et al., 1986), which again implicates the midbrain and pons as

the dream–sleep generators. However, they also contribute to the waking state as well.

For example, midbrain reticular neurons tend to significantly reduce their activity immediately prior to spindle production as well as during the transition from wakefulness to sleep and during drowsiness (Steriade and McCarley, 1990). However, reticular activation can also abolish the thalamic production of spindles, such as occurs during waking and paradoxical sleep. Hence, brainstem stimulation or activity can either inhibit or promote spindle wave production and induce paradoxical as well as slow-wave, synchronized sleep.

Synchronized, Slow-Wave Sleep

The defining feature of synchronized (or quiet, NREM) sleep is large amplitude rhythmic EEG waves of varying frequencies (from slow to fast). Even during waking, however, and when highly aroused or alert, some cortical areas may display synchronized EEG activity, whereas even during REM sleep some cortical areas display synchronized activity (reviewed in Steriade and McCarley, 1990). Presumably, that some cortical areas demonstrate slow-wave activity whereas yet others are highly aroused is a reflection of selective attention and discrete neocortical activation. Some cortical regions become suppressed, thus narrowing the range of incoming stimuli, whereas other specific regions remain activated in order to process incoming information and/or for the purposes of contributing to whatever mental process may be taking place.

Steriade and McCarley (1990) propose that the development of synchronized sleep is dependent on the removal or lessening of brainstem input and is a passive rather than an active process. That is, fluctuations and decreases in cholinergic and noncholinergic activity within the reticular formation decreases thalamic activity, thus producing sleep spindles in the reticular thalamic nucleus followed by synchronous activity. However, with increases in cholinergic and noncholinergic activity, thalamic activity increases, and synchronous EEG activity is replaced by desynchronization and the production of REM (see also Hobson et al., 1986; Vertes, 1990).

Thalamic Contributions

Given that the thalamus acts to relay sensory input to the neocortex, not surprisingly thalamic neurons are also involved in synchronized and paradoxical sleep. Specifically, thalamic neurons inhibit sensory transfer to the neocortex during the synchronized sleep stage as well as during the transition from wakefulness to sleep and when drowsy; otherwise the individual would keep waking up due to neocortical activation.

Some thalamic neurons, however, act to enhance sensory transfer during desynchronized sleep (Steriade and McCarley, 1990) as well as during waking. Presumably, this allows for sensory stimuli to become incorporated into and to, in fact, induce dream states (see Chapter 5).

Specifically, the reticular thalamic nucleus appears to also act as a "pacemaker" center for the development of sleep spindles (i.e., EEG waves that wax and wane between 7 and 14 Hz), which herald the onset of sleep (Steriade and McCarley, 1990). Spindles are typically associated with loss of consciousness and thalamic inhibition and, thus, loss of information transfer to the cortex. Moreover, spindles originate in the reticular thalamus, which in turn projects to other thalamic neurons. These thalamic nuclei are indirectly under the control of the frontal lobe and dorsal medial thalamus (see Chapter 11).

Motor Inhibition During Sleep

The frontal lobe as well as brainstem motor areas becomes activated during REM (see Steriade and McCarley, 1990). However, although motor commands may be initiated and even transmitted from the neocortical motor centers, they cannot be acted out during REM due to the production of muscle atonia. Muscle atonia is produced, in part, via descending 5HT influences on the spinal cord (see above).

Specifically, during the transition from slow-wave to paradoxical sleep, motor neurons in the brainstem and the spinal cord come to be actively inhibited (Chandler et al., 1980; Chase and Morales, 1985). The result is muscle atonia, or what has also been referred to as *sleep paralysis*. This condition is almost exclusively associated with REM sleep (Chase and Morales, 1985). Presumably, these inhibitory influences arise within the pontine and bulbar brainstem reticular formation and via descending 5HT influences so as to prevent an individual from motorically acting out his dreams and possibly injuring himself (e.g., sleep walking). Indeed, the dorsal lateral portion of the pontine tegmentum has been repeatedly implicated as responsible for muscle atonia during sleep (see Steriade and McCarley, 1990).

By contrast, lesions of the pontine reticular formation (with sparing of the locus ceruleus) abolish muscle atonia (Hendricks et al., 1982; Jouvet, 1979). In consequence, during REM sleep, the subject will move about and engage in semipurposeful behaviors, including walking, orienting movements, and attack—as if they were acting out their dream (Hendricks et al., 1982).

BRAINSTEM SLEEP DISORDERS: NARCOLEPSY AND CATAPLEXY

Narcolepsy

Narcolepsy is a syndrome that is often characterized by excessive daytime sleepiness and may be accompanied by partial or complete and sudden attacks of REM sleep, even though the patient may have been quite awake just moments before (Broughton, 1990; Guilleminault, 1989). In some instances, patients may also experience sleep paralysis, hypnagogic hallucinations, and cataplexy. These daytime attacks sometimes follow the 90-minute cycle characteristic of REM sleep. Abnormalities involving rapid eye movements (REM) and motor disturbances, such as muscle twitching and persistence of muscle tone, are not uncommon (Schenck and

Mahowald, 1992), and the latency of REM sleep is often very short.

This condition is often treated with amphetamines or with tricyclic antidepressants. It is also believed to be secondary to abnormal concentrations of NE and DA. However, narcolepsy has several different components and features that may be variably expressed. These are described below.

Excessive Daytime Sleepiness

Excessive daytime sleepiness (EDS) involves the sudden sleep onset of REM , including dreaming. These sleep episodes tend to last from a few minutes to half an hour.

Sleep Paralysis

Sleep paralysis (SP) usually occurs upon awakening, such that the person experiences a paralysis of the voluntary muscles. The person cannot open his eyes, move his arms or legs, or speak. Some are terrified by these experiences. Presumably, many people have experienced this in a mild form at least once. The condition tends to last just a few minutes.

Presumably, SP is due to the abnormal prolongation of activity within the dorsal lateral portion of the pontine tegmentum, a region that is normally responsible for producing muscle atonia during sleep (see above).

Hypnogogic Hallucinations

Some patients experience hypnogogic hallucinations upon awakening and while seemingly completely conscious. These can be exceedingly vivid and realistic and involve auditory, kinesthetic, and somesthetic as well as visual hallucinations. The affected individual may see what he believes to be ghosts or even intruders attempting to attack him. Hypnogogic hallucinations (and sleep paralysis) appear to represent the intrusion of REM into wakefulness.

SOMNAMBULISM

Somnambulism, or sleep walking, is most common among children; presumably, up to 15% of all children sleep walk at some point (see Kavey et al., 1990), which is probably related to brainstem immaturity. In contrast, fewer than 1% of adults sleep walk.

Somnambulism occurs in the first third of the night during NREM sleep. Usually, patients have no recollection or only a vague memory of walking about (Kavey et al., 1990). While sleep walking, some patients may behave in an aggressive or agitated manner, scream and engage in fighting or fleeing behaviors, and fall, strike, or injure themselves. Fragmentary memories may include being chased.

Adults who sleep walk often have a history of sleep walking during childhood (Kavey et al., 1990). Some patients demonstrate epileptiform EEGs and/or suffer from atypical nocturnal complex partial seizures (Pedley and Guilleminault, 1977).

REM Sleep Behavior Disorder

REM sleep behavior disorder is associated with narcolepsy and REM sleep (Schenck and Mahowald, 1992) and is a form of motor disinhibition accompanied by often violent behaviors. Patients appear to be acting out their dreams. It usually affects men over 60, though it is also seen in children with obvious brainstem dysfunction (see Schenck and Mahowald, 1992). However, when it accompanies narcolepsy, this disorder tends to have its onset in men in their late 20s (Schenck and Mahowald, 1992).

Presumably, somnambulism and REM sleep behavior disorder are a consequence of disinhibition of the brainstem and spinal motor neurons that are normally inhibited during REM. That is, it is likely that the dorsal lateral portion of the pontine tegmentum, a region that is normally responsible for producing muscle atonia during sleep, fails to induce muscle atonia such that the person begins to walk about and act out his dreams. As noted above, destruction of this nuclei also results in motor disinhibition during REM such that the subject begins to move about as if actively engaged in purposeful behavior.

CATAPLEXY

Cataplexy involves a sudden loss or decrease of muscle tone: atonia. Hence, the affected individual may suddenly fall to the ground, sometimes injuring himself. In mild cases, the jaw may suddenly become slack and the knees may weaken and buckle. In severe cases, the patient immediately falls to sleep and enters REM, which is accompanied by dream mentation. Often, cataplexic attacks are induced by sudden surprise or attacks of emotion, including anger or laughter. Cataplexy is generally considered a component of narcolepsy. The production of REM and the loss of muscle tone suggests a brainstem loci for this disorder. However, the influence of emotion in inducing cataplexic attacks and the fact that some individuals become rigid and stiff, almost catatonic, suggests that the amygdala, the basal ganglia, and perhaps, indirectly, the medial supplementary motor areas (SMA) of the frontal lobe may also be contributory. That is, in some respects this disorder is reminiscent of abnormalities within the basal ganglia and SMA—nuclei that are significantly influenced by the amygdala (see Chapter 6).

CRANIAL NERVES OF THE MEDULLA: SHOULDER, HEAD, JAW, TONGUE MOVEMENT, BREATHING, PHONATING, HEART RATE
Cranial Nerve XII: Hypoglossal

The hypoglossal nucleus located in the caudal medulla gives rise to the twelfth nerve, which controls tongue movements via innervation of the somatic skeletal muscles. A lesion, tumor, or infarct invading this area can give rise to weakness or atrophy of the tongue. For example, if the tongue is weak and deviates to the right, the ipsilateral lower motor neuron (LMN) is involved, indicating that the right genioglossus muscle is weak. Weakness can be tested by asking the patient to place his tongue in one side of his cheek and to press against your finger (on the outside of

the cheek). The clinical and EMG signs of an LMN lesion of the twelfth nerve are hemiatrophy, unilateral fasciculation or fibrillation, and, usually, severe paralysis with obvious deviation to the paralytic side when the tongue is protruded.

Cranial Nerve XI: Spinal Accessory

Cranial nerve XI consists of two distinct segments: a cranial portion that, along with the vagus nerve, forms the inferior laryngeal nerve, which innervates the muscles of the larynx; and a spinal portion, which innervates the sternocleidomastoid and upper trapezius muscles and therefore aids in turning the head and elevating the shoulders. Injuries to this nerve may result in a sagging of the shoulder on the affected side, or weakness in turning the head, particularly when resistance is applied.

Cranial Nerve X: Vagus

The vagus nerve is actually a complex mix of nerves that innervate a variety of structures, including the trachea, larynx, pharynx, esophagus, abdominal and thoracic viscera, epiglottis, and the external auditory meatus. Hence, this nerve (in conjunction with the ninth nerve, both of which are derivatives of the skeletal muscles that originally formed the brachial (gill) arches, is important in regard to oral activities, including swallowing, breathing, speaking, pharyngeal constriction, and, thus, with movement of the palate, the pharynx and larynx. The vagus nerve is, therefore, responsible for swinging the soft palate upward and backward to contact the posterior wall of the pharynx, sealing off the oropharynx from the nasopharynx when one swallows, whistles, or speaks.

An injury to this region can result in severe and enduring palatal weakness as well as a condition referred to as pseudobulbar palsy. Unless the soft palate elevates properly, liquid will escape into the nose when one drinks and when one speaks, resulting in nasal speech. Indeed, damage to these nuclei can severely affect speech.

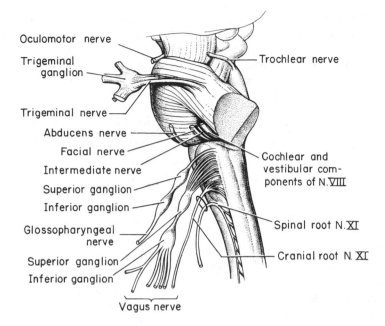

Figure 10.5 Schematic diagram depicting the cranial nerves and peripheral ganglia. (From Carpenter M. Core text of neuroanatomy. Baltimore: Williams & Wilkins, 1991.)

Cranial Nerve IX: Glossopharyngeal

The 9th nerve is closely related to the vagus; both have similar functional components, including general somatic and visceral afferents and efferents. The glossopharyngeal nerve, thus, receives tactile, thermal, and pain sensations from the tongue, and contributes fibers to the solitary nucleus, thereby forming the gustatory nucleus. In addition, the glossopharyngeal nerve receives fibers from the carotid sinus that transmit impulses regarding increases in carotid arterial pressures. The 9th nerve then transmits this data to the solitary nucleus, which in turn contributes fibers to the vagus nerve. In this manner, the tenth and ninth nerve can induce reductions in arterial blood pressure and heart rate.

Lesions to the ninth nerve generally results in a loss of taste and sensation in the posterior third of the tongue, and a loss of the gag reflex and the carotid sinus reflex. In some cases, intense pain may be triggered by swallowing or coughing.

Lateral Medullary Syndrome

The lateral medullary syndrome is typified by dysphonia (hoarse voice), persistent hiccups, disordered equilibrium, vertigo, nausea, loss of thermal sense, and pain involving the contralateral half of the body and the ipsilateral half of the face. Usually, this condition is due to vascular abnormalities involving the vertebral arteries, or occlusion of the posterior inferior cerebellar artery.

THE CRANIAL NERVES OF THE PONS

Cranial Nerve VIII and the Cochlear System: Tinnitus, Deafness, Dizziness, Vertigo

A portion of the eighth nerve arises from the cochlear nuclei so as to form the auditory branch of the eighth nerve, which in turn bifurcates and innervates the ampulla, utricle, saccule, and cochlear duct of the semicircular canals of the inner ear. It is via the acoustic nerve that auditory stimuli are relayed to the ventral and dorsal cochlear nuclei for further transduction prior to transfer to the inferior colliculus of the midbrain, and the medial geniculate nucleus of the thalamus. Hence, lesions involving this nerve or the cochlear nucleus can give rise to significant hearing problems, including deafness and tinnitus—ringing in the ears, which may be reported as buzzing, humming, whistling, roaring, hissing, or clicking.

TINNITUS

There are two general types of tinnitus—vibratory and nonvibratory. Nonvibratory tinnitus is actually quite common, being due to contractions of the muscles of the inner and middle ear, but it is usually masked and may only be heard at night or in a very quiet environment. In contrast, vibratory tinnitus is most often due to disorders of the inner ear, ossicles of the middle ear, tympanic membrane, or eighth nerve, and may be accompanied by deafness in one ear.

DEAFNESS

If a patient complains of deafness, it must be determined if this is due to otosclerosis, chronic otitis, or occlusion of the external auditory canal or eustachian tube, in which case the problem is conductive and is not related to nerve damage. By contrast, disease of the cochlear, of the cochlear division of the eighth nerve, or of its central connections results in sensorineural (or nerve) deafness.

These two distinct disorders can be easily differentiated with the help of a tuning fork. When struck, the fork is placed at the center of the forehead or on the mastoid bone behind the ear so that the vibration can bypass the middle ear and can be mechanically conveyed to the inner ear so as to excite auditory impulses. If the condition is due to nerve deafness, the sound is localized in the normal ear and fails to be perceived by the affected ear. If the disorder is conductive, the sound is heard as louder in the defective ear. If there is an obstruction or a middle ear abnormality (versus a nerve disorder), then the sound is heard louder in that ear due to dampening of noise via the air conduction block. If there is a lesion of the nerve, then the sound is heard in the opposite or normal ear only.

The Eighth (Vestibular) Nerve

The vestibular division of the eighth nerve arises from the vestibular nerve, which innervates the labyrinth and the maculae of the saccule and utricle and the ampullae of the semicircular canals. As noted in Chapter 1, the vestibular system was derived from the lateral line system, and in turn gave rise to the cerebellum.

The major concern of the vestibular division is not hearing or vibratory perception, however, but determination of the body's position in visual–space so as to maintain equilibrium during movement. All sensory receptors for the vestibular system are located in the membranes of the labyrinth of the internal ear and in the vestibular ganglion, which is in the internal auditory canal. Thus, it can determine changes in position via alterations in fluid balance within the semicircular canals.

Lesions of the vestibular receptors, their nerves, or their central connections cause abnormal sensations of movement, vertigo, nausea, tendencies to fall, dizziness, motion sickness, and so forth.

The vestibular system also provides information about the position of the head and correlates head and eye movements with somatic muscle activity. Together with the descending medial longitudinal fasciculus (MLF) and vestibulospinal tracts, the vestibular system is able to mediate the postural reflexes. In part, it is able to accomplish this via rich interconnections with those cranial nerve nuclei (via the MLF) that subserve eye movement (nerves VI, IV, and III). Disease involving these tissues can, therefore, cause nystagmus.

Following injuries to this system, patients may complain of to–and–fro or up–and–down movements of body or floors, and may note that the walls seem to tilt, sink, or rise. When walking, there may be feelings of unsteadiness such that they veer to one side. Or there may be a feeling of being pulled or drawn—a feeling of impulsion. There also may be a disinclination to walk (particularly during an attack), and a tendency to list to one side, and the condition may be aggravated by riding in a vehicle. Some disturbances may occur only for a few seconds, or after lying down or sitting up, turning, and so forth. When less severe the patient may merely veer to one side while walking.

The vestibular system is also concerned with eye movement, for it is also via ocular signals that the position of the head and body in space can be determined. Hence, vestibular dysfunction can include difficulty focusing or fixating on objects while walking, or when the object is moving. This is due to a loss of stabilization of ocular fixation by the vestibular system during body movement and is caused by an inability to integrate visual with vestibular input. These functions are normally made possible through rich interconnections (via the MLF) between the vestibular system and the sixth, fourth, and third nerves, which subserve eye movement, as well as the bilateral interconnections between these regions and the cerebellum.

ORAL–FACIAL MOVEMENT, OLFACTION AND SENSATION
The First Cranial Nerve: Olfactory

The olfactory nerve is not a brainstem nerve, but a complex axonal pathway associated with the forebrain and limbic system. Nevertheless, the olfactory nerve is associated with brainstem functions and, in particular, the seventh and fifth nerves, as originally it was via the olfactory system that information regarding not just smell but taste were derived.

The olfactory nerve originates via bipolar cells located in the olfactory epithelium, which in turn is composed of very primitive cells that have a life span that lasts only days. New axons and new synapses are continually forming. These nerve fibers perforate and pass through the cribriform plate (making them very susceptible to injury and shearing—see Chapter 20) and project directly to the olfactory bulb. From the olfactory bulb the olfactory tract is formed, which projects to pyriform, periamygdaloid, and entorhinal areas—which in turn constitute the primary olfactory cortex. Projections continue to the amygdala, hippocampus, thalamus, orbital frontal lobes, and insula.

Following a head injury the cribriform plate may fracture, the olfactory nerve may be severed, and the meninges may rupture. If this occurs, in addition to loss of smell, cerebrospinal fluid may continually drip or gush into the nose—which suggests a cerebrospinal fluid fistula. To determine if nasal drip is due to CSF leakage or normal mucous secretion, the fluid can be subject to a glucose test tape (as used for urinalysis). CSF contains glucose, whereas mucous does not. If glucose is indicated, call a neurosurgeon.

However, in some cases, a head injury may cause these nerves to be severed although the cribriform plate remains intact. In these instances, loss of smell may result: anosmia. However, this loss may be unilateral or bilateral (involving both nostrils). If unilateral, the patient will not notice the loss, in which case each nostril must be assessed separately.

Dysosmia refers to a perversion of the sense of smell, which in turn may be due to partial injuries of the olfactory bulbs, or due to tumor of the nasal sinuses or the temporal lobes, in which case food may also be said to have an extremely unpleasant odor and/or taste. Similarly, olfactory hallucinations are associated with tumors, seizure activity, as well as head injuries involving the inferior temporal lobes.

The Seventh Cranial (Facial) Nerve

The seventh nerve, and the brainstem nuclei that it innervates, is concerned with facial movement, including elevation of the eyebrows, retraction of the lips, and closure of the auditory canals, as well as with gustatory sensation. Injuries to the seventh nerve can, therefore, produce a lip retraction, eyebrow lifting, or eyelid closure paralysis, i.e., Bell's palsy. Patients have difficulty or are unable to wrinkle their forehead, purse their lips, and show their teeth, and the corner of the mouth may droop.

In addition, whereas cranial nerves IX and X innervate the taste buds of the posterior third of the tongue, cranial nerve VII innervates the anterior two-thirds. Hence, a lesion of the seventh nerve can result in a disturbance of taste sensation.

In addition to the muscles of the face, the seventh nerve also innervates the stapedius muscle, which acts to dampen excessive sound via inhibition of the movement of the ossicles. If the seventh nerve is injured, the stapedius may become paralyzed and the patient may report that sounds are uncomfortably loud. However, disturbances of hearing are most usually associated (at least at the brainstem level) with damage involving the eighth cranial nerve.

As to oral and facial movement, including swallowing and speaking, the functional integrity and participation of the fifth nerve is also important, for it controls the jaw. In this regard, the fifth, seventh, ninth, tenth, and twelfth nerves and associated nuclei frequently act in concert regarding oral-facial, jaw, head, and shoulder movement, and are richly interconnected.

The Fifth Cranial Nerve: Trigeminal

The fifth nerve is the largest of the cranial nerves and innervates the trigeminal nucleus within the medulla. It is concerned with jaw closure, as well as chewing, grinding, and lateral movement of the jaw. In this regard, the fifth nerve also acts in concert with the seventh nerve which innervates all the muscles involved in facial expression. These muscles control the size of every facial aperture, including the auditory canals. A lesion involving this nucleus, therefore, can result in difficulty chewing, and if severe, atrophy and complete paralysis of the left or right temporal and masseter muscles. Paralysis and atrophy are always ipsilateral to the lesion.

The general somatic afferent components of the fifth nerve also mediate the general sensory modalities for the face, teeth, and mouth, and the mucous membranes of the nose, cheek, tongue, and sinuses. These general sensory functions include proprioception, touch, pain, and temperature.

EYE MOVEMENT
The Sixth Cranial Nerve: Abducens

The abducens is a motor nerve that innervates the lateral rectus muscle of the eye, and is part of a collection of fibers found with the loop of fibers that also form the facial nerve. However, fibers from the abducens nuclei also ascend the brainstem and terminate on neurons belonging to the oculomotor complex, which innervates the medial rectus muscle. It is, therefore, responsible for horizontal eye movements. In large part, the pontine center for lateral gaze and the abducens nuclei form a single entity (Carpenter, 1991). In addition, the abducens is linked to the pontine/midbrain center for vertical gaze (the rostral interstitial nucleus of the MLF).

Thus abducens nerve and nuclei are the pontine centers that control lateral eye movements outward to the right or to the left. Injuries to the sixth nerve can, therefore, produce a lateral gaze paralysis as well as a paralysis of the lateral rectus muscle that results in double vision (horizontal diplopia).

CRANIAL NERVES OF
THE MIDBRAIN
The Second Nerve: Optic

The rods and cones of the retina project to horizontal and bipolar cells, which in turn project to X, Y, and W ganglion cells. Axons of the retinal ganglion cells run in parallel along the surface of the retina and converge at the optic disc (the blind spot) where they gather together and punch through the retina, thereby forming the optic nerve.

The right and left optic nerves project into the cranial cavity via the optic foram-

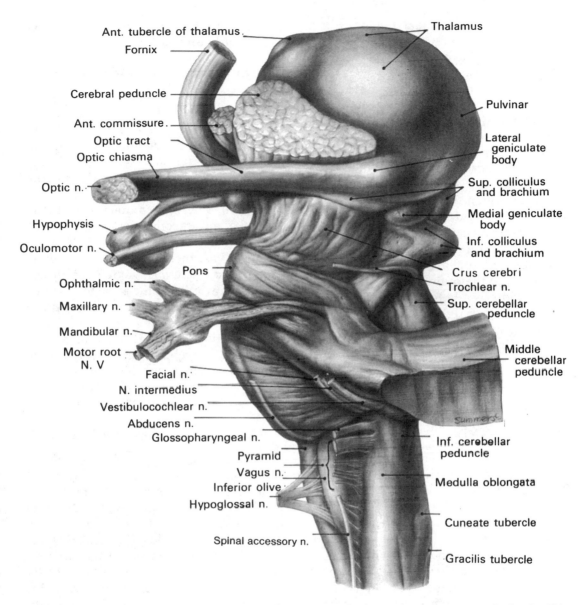

Figure 10.6   Lateral view of brainstem with cerebellum removed, depicting the site of entrance and emergence of most cranial nerves.

(From Mettler FA. Neuroanatomy. St. Louis: CV Mosby, 1948.)

ina and then unite to form the optic chiasm where a decussation occurs, thereby forming the optic tracts. Subsequently, visual input from the left half of visual space projects to the right hemisphere via the optic tracts, which in turn project predominantly to the lateral geniculate nucleus of the thalamus, and to a much lesser extent, to the superior colliculus, and probably the hypothalamus. Hence, the optic nerve is not a true cranial nerve—at least in mammals—although some fibers (via the optic tract) are received in the midbrain visual colliculi.

From the lateral geniculate nucleus these visual fibers form the optic radiations, which project predominantly to the striate cortex within the occipital lobe, and give off collaterals to the inferior temporal and superior parietal lobe where the upper and lower visual fields are also represented.

Injuries to the optic pathways necessarily produce visual defects and are referred to as homonymous if they are restricted to either the right or left visual field, or heteronymous if both visual fields are disrupted to some degree. Heteronymous defects suggest a lesion involving both cerebral hemispheres, or a lesion to retina or the optic nerve before it crosses over at the decussation.

If the visual defects are homonymous, then the lesion can be localized to one side of the optic tract or radiations and thus to the right or left hemisphere. Complete destruction of the optic tract or its terminal zones induces a homonymous hemianopsia to the left or right, whereas a partial injury may induce a quadratic homonymous defect: upper quadrant defects being associated with temporal lobe defects, and lower quadrant defects associated with the superior parietal lobe.

EYE MOVEMENT

Movement of the eye is dependent on the functional integrity of the posterior parietal neocortex, the frontal eye fields of the lateral frontal lobes, the midbrain visual colliculi, the cerebellum, and the cranial nerves and nuclei of the upper pons (cranial nerve VI) and midbrain (nerves IV and III). All pathways mediating saccadic, pursuit, and vestibulo-ocular movements, including those originating in the forebrain and midbrain, then converge onto the pontine centers for horizontal gaze.

The Fourth Nerve: Trochlear

As noted, the sixth (abducens) nerve and nuclei are the pontine centers that control lateral eye movements outward to the right or to the left. By contrast, the fourth (trochlear) nerve and nucleus are located just caudal to the inferior colliculi, and innervate the superior oblique muscle of the eye, a muscle that has three actions: depression, abduction, intorsion. The fourth nerve, therefore, assists in moving the eye downward or inward.

The Third Nerve: Oculomotor

The third (oculomotor) nerve is responsible for rotating the eye upward and downward as well as inward, and innervates all ocular rotary muscles (except for the lateral rectus and superior oblique). These include the medial, superior, inferior recti, and inferior oblique.

The third nerve also innervates the intraocular and smooth muscles of the pupil, i.e., the ciliary and pupilloconstrictor muscles. Lesions, therefore, may result in an inability to rotate the eye upward, downward, or inward, and the pupil (on the side of the lesion) may fail to respond to direct light.

In addition, the third nerve innervates the levator palpebrae muscle that elevates the eyelid. Injury to the third nerve may result in levator palpebrae weakness and, thus, ptosis (drooping) of the eyelid.

Disturbances of Eye Movement

Eye movement nuclei are all located in the pontine brainstem and receive input from (and transmit information to) the superior (visual) colliculus within the midbrain (Grantyn and Grantyn, 1976), and the frontal eye fields (Leichnetz et al., 1984), which also project to the superior colliculus (Segraves and Goldberg, 1987). As is well known, the frontal eye fields and the superior colliculus become activated immediately prior to the onset of an eye movement, and stimulation of these regions elicits eye movements and visual orienting reactions as well.

Hence, damage to these cortical areas or to any of these nerves or nuclei can result in significant visual disturbances. For example, strabismus (or squint) is due to a brainstem lesion that results in muscle imbalance and, thus, improper alignment of the visual axes of the two eyes. A complete lesion of the oculomotor nerve causes ptosis and an inability to rotate the eye upward, downward, or inward. Lesions of the fourth nerve result in weakness of downward movement of the affected eye; patients may complain of difficulty reading or walking downstairs. Head tilting to the opposite shoulder is especially characteristic. Lesions of the sixth nerve result in paralysis of lateral or outward eye movements.

Most common causes of damage to the third, fourth, and sixth nerves are tumors of the base of the brain, trauma to the head, ischemic infarction of one of these nerves, and aneurysms of the circle of Willis. Sixth nerve palsies in children are due to neoplasm, i.e., pontine glioma. The fourth nerve is most commonly injured by head trauma. Third nerve damage is often due to compression by aneurysm, tumor, or temporal lobe herniation; enlargement of the pupil is an early sign.

As noted, cranial nerves III, IV, and VI are linked to the eighth (vestibular) nerve via the MLF. It is due to these rich interconnections that a lesion involving these regions and the MLF can cause vertigo, dizziness, nystagmus, and impairment of vertical fixation and pursuit. However, disturbances such as these also raise the possibility of cerebellar dysfunction (reviewed in Stein and Glickstein, 1992).

THE CEREBELLUM
Evolution

The cerebellum evolved out of the vestibular nuclei and is derived from the rhombic lip and ectodermal thickenings around the cephalic borders of the fourth ventricle (see Chapters 1 and 18). As an evolutionary outgrowth of the vestibular system, a primary concern of the cerebellum has been and continues to be stabilizing the body and providing information about the position and movement of the head in relation to gravity.

Initially, however, during the early stages of evolution, as primitive creatures were without limbs, they possessed only a small nubbin of cerebellum—referred to as the flocculonodular lobe—what would become the paleocerebellum as well as the archicerebellum. This tissue presumably acted to coordinate the axial muscles with the position of the head, trunk, and eyes, and probably also acted to integrate these movements in response to motivational commands transmitted by the limbic system. In modern creatures, including humans, disequilibrium of stance and gait with little or no extremity dystaxia is usually due to lesions in the flocculonodular lobe (caudal vermis).

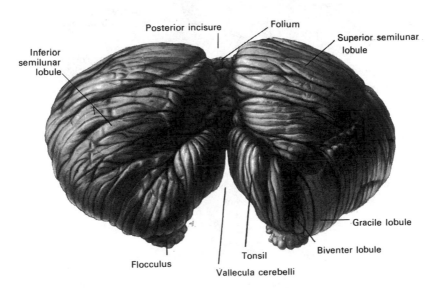

Figure 10.7   Superior *(top)* and posterior-inferior view *(bottom)* of the cerebellum. (From Carpenter M. Core text of neuroanatomy. Baltimore: Williams & Wilkins, 1991.)

With the evolution of legs, the cerebellum was forced to assume new roles, including the coordination of axial (trunk) and appendicular (limb) muscles, which it influences via feedback from the spinocerebellar tracts. The anterior lobes of the cerebellum began to evolve in order to meet these demands. Thus, disturbances of this region can result in dystaxia predominantly in the legs. Disturbances like these are often the consequence of alcoholism and nutritional deficiency (Adams and Victor, 1994).

With the emergence of bipedal posture, increasingly complex demands were placed on the anterior and the emerging posterior cerebellum in order to coordinate gait and upper as well as lower limb movements. This corresponded with the evolution of the neocerebellum (including the dentate gyrus and the cerebellar hemispheres), which in turn paralleled neocortical expansion in the forebrain and the development of temporal-sequential and fine motor functioning.

Thus, the dentate gyrus (of the neocerebellum) controls voluntary actions and direction of movements involving multiple joints (Kane et al., 1989) and the upper limbs. Dentate neurons also tend to respond or alter their response prior to changes in the EMG or neuronal activity in the motor cortex (Thach, 1978), indicating they are involved in preprogramming.

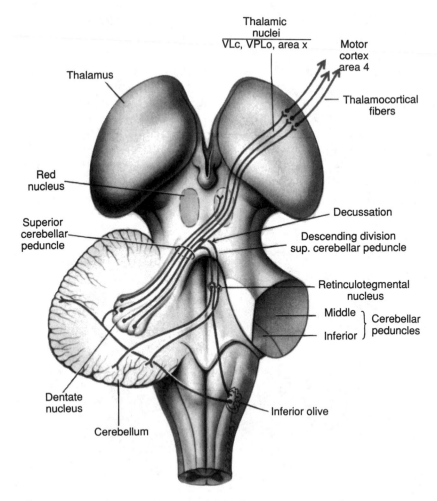

Figure 10.8   Schematic diagram of the efferent fibers of the dentate nucleus. (From Mettler FA. Neuroanatomy. St. Louis: CV Mosby, 1948.)

Disturbances of the neocerebellum (and the dentate gyrus and posterior lobes) can give rise to ataxia of the upper musculature as well as the lower half of the body.

Structure and Cytoarchitecture

Despite its long and varied evolutionary history, the cerebellum is strikingly homogeneous in appearance. Structurally, it can be divided into the cerebellar vermis (located medially), and into two lobes, the cerebellar hemispheres. Phylogenetically and functionally, however, the cerebellum can be divided into three parts, the archicerebellum (which includes the flocculonodular lobe), the paleocerebellum (or anterior lobe) which influences autonomic and emotional activity, and the most recently evolved neocerebellum which constitutes the posterior lobe and which influences motor, somatic, and cognitive activities. The neocerebellum appears to be involved in neocortical information processing (see Leiner et al., 1986).

Enshrouding the cerebellum is a three-layered cerebellar cortex (i.e., a molecular, Purkinje, and granular layer), beneath which course thick sheets of axonal fibers (medullary substance), thereby forming the white matter. Within the white matter are embedded four pairs of intrinsic nuclei.

Information is shunted primarily to the cerebellar cortex via fiber tracts that course through the middle cerebellar peduncles which anchor the cerebellum to the brainstem. This includes afferents that convey tactile, auditory, and visual impulses, including those that essentially reconstruct and maintain somesthetic images of the body within the different lobules (reviewed in Brodal, 1981; Carpenter, 1991).

CEREBELLAR MOTOR CONTROL
Movement and Learning

The cerebellum has been implicated in emotional functioning; the pathogenesis of schizophrenia and autism (Bauman and Kemper, 1985; Gaffney et al., 1987; Heath, 1977; Heath et al., 1982; Taylor, 1991; Wein-

berger et al., 1979, 1980); cognitive processing, including speech (Silveri et al., 1994; Van Dongen et al., 1994; Wallesch and Horn, 1990); learning and memory (Fiez et al., 1992; Lavond et al., 1990); and visual functioning, including the visual guidance of movement (Bloedel, 1992; Stein and Glickstein, 1992). When injured, patients may, therefore, suffer from tremor, nystagmus, gait disturbances, incoordination, and postural instability, as well as emotional and speech disturbances.

In general, the cerebellum is associated with motor functioning, classical conditioning, and the learning of new motor programs. For example, cerebellar climbing fibers become activated during early learning stages but then decline (Watanabe, 1984), and lesions abolish the acquisition and retention of conditioned responses (Lavond et al., 1990). Similarly, when subjects are asked to imagine motor movements, cerebellar blood flow increases (Dacety et al., 1990). This suggests a role in learning and memory. On the other hand, it has also been argued that what the cerebellum appears to learn are habits, and that it does not act to form memories per se (see Bloedel, 1992).

Motor learning and motor habit formation are associated with the functional integrity of the neocerebellum, which maintains rich interconnections with forebrain motor areas. Specifically, the neocerebellum projects to the motor thalamus, which relays these impulses to motor areas 4 and 6 in the frontal lobe and the somatosensory areas of the parietal lobe (Sasaki, 1979). In general, neocortical information is relayed to the neocerebellum via the pons, which in turn acts to relay information from the neocerebellum to the neocortex directly and via thalamic relays (Miyata and Sasaki, 1983).

In addition, the neocerebellum (i.e., the dentate gyrus) appears to be involved in planning motor movements, including the direction of movement. For example, neuronal modulation occurs in relationship to ongoing movement and in response to activation of the muscles that generate those movements (see Bloedel, 1992).

CEREBELLAR DISTURBANCES OF GAIT AND VOLUNTARY MOVEMENT

Typically, disturbances of the cerebellum involve voluntary and skilled motor functions. Patients may display gait ataxia, dysarthria, vertigo, hypotonia, and intention tremor, as well as abnormalities in the force, accuracy, range, and rate of goal-directed voluntary movements (Brodal, 1981; Carpenter, 1991; Holmes, 1917, 1939). Similarly, visually guided tracking movements and, thus, determination of movement trajectory are disrupted (Beppu et al., 1984; Miall et al., 1987)—e.g., past pointing, in which distances are incorrectly judged and the patient may fall short or go too far when attempting to touch or grasp an object.

CEREBELLAR GAIT

One of the most easily recognized cerebellar disturbances is what has been referred to as "cerebellar gait." Patients typically have a wide-based gait, and their steps are characteristically unsteady, irregular, uncertain, and of variable length. If dysfunction is mild, these disturbances may only be noticeable when the patient is tired. However, if severe the patient may not be able to stand without assistance. Loss of muscle tone, incoordination of volitional movements, intention tremor, minor degrees of muscle weakness, fatigability, and disorders of equilibrium, including nystagmus, are common features. Due to these motor disturbances, it is not uncommon for afflicted individuals to be erroneously perceived as being severely intoxicated. On the other hand, alcohol abuse has long been known to injure the cerebellum.

TIMING, TRACKING, AND SEQUENTIAL MOVEMENTS

The lateral cerebellum appears to be involved in regulating the timing of sequential movements, and cerebellar patients have difficulty with timing and rhythm (Ivry and Keele, 1989). Thus, abnormalities in the rate, range, and force of movement are apparent, as demonstrated via deficiencies in finger-to-nose testing. For example, when asked to touch the physician's finger and then his own nose, the patient's arm and hand may sway and miss on both counts.

Irregularities in acceleration and deceleration of movement are also common, which may be due to an inability to alter movement strategy, perhaps because of programming errors (Sanes et al., 1990). For example, when the patient is reaching, extension of the limb may be arrested prematurely, such that the target/objective is attained by a series of jerky movements. When asked to touch his nose, the patient may do it in two stages: by lifting the arm to nose level, then by bringing the fingers to the nose.

It has been proposed that the cerebellum translates motor-spatial concepts into time, that is, it is involved in timing muscle activity so as to produce smooth and accurate rapid movements (Kornhuber, 1971). Hence, damage to the cerebellum can induce a loss of tracking accuracy and create disjointed responses that are performed too fast (Beppu et al., 1984; Miall et al., 1987); such movements may appear ballistic. They make their movements too quickly, and fail to slow their movements as quickly as normals. Or, the limb overshoots the mark and the error is corrected only by a series of secondary movements.

Lateralized cerebellar signs limited to one half of the body are due to lesions (infarct, neoplasm, or abscess) affecting only one of the cerebellar hemispheres. Bilateral cerebellar signs are likely a result of toxic-metabolic, demyelinating, or other degenerative diseases.

ATAXIC SPEECH

Cerebellar-related disorders of movement and disturbances of motoric coordination all affect speech. That is, speech may become dysarthric, and in some cases with rapid and acute onset, patients may become transiently mute (van Dongen et al., 1994).

Cerebellar dysarthric speech may be of two types. Speech may become slowed and slurred, particularly when required to repeat sounds ("ga ga ga"). Or the patient may suffer from a scanning dysarthria with variable intonations as words are broken up into syllables, some of which are explosively uttered (i.e., ballistic speech).

THE CEREBELLUM AND VISION

Patients with cerebellum lesions experience more difficulty with motor tasks when the visual cues utilized to guide movement are eliminated. This suggests that it utilizes visual cues and integrates this with motor output (Bloedel, 1992; Stein and Glickstein, 1992). However, if a patient performs more poorly due to the loss of visual input (closing the eyes) rather than because of a loss of visual guidance (removing visual cues), then the disturbance is due to sensory abnormalities ("sensory dystaxia") and loss of visual and proprioceptive input rather than cerebellar damage. Hence, one must distinguish between the two in making a proper diagnosis.

If the disorder gets worse with the eyes closed, then the disorder is sensory in nature (referred to as "dorsal column, sensory dystaxia"). This is because the patient with sensory/dorsal column dystaxia substitutes visual guidance for proprioceptive information. The cerebellum can equilibrate only if it has proper proprioceptive information.

If the proprioceptive system is interrupted at the level of the dorsal or vestibular roots of the spinal cord, the patient will suffer disequilibrium that is exacerbated by the loss of visual input. If proprioceptive information is lacking, and the patient closes his eyes, the movement becomes more dystaxic, which indicates a dorsal column abnormality. However, if the dystaxia remains the same with eyes open or closed, the patient is suffering from cerebellar dystaxia.

Because it is linked with the vestibular system as well as the cranial nerve nuclei subserving eye movements, lesions to the cerebellum can affect eye movement. For example, conjugate lateralized eye movements or attempts at visual tracking may be characterized by a series of jerky eye movements.

EMOTION, COGNITION, PSYCHOSIS AND THE CEREBELLUM

It is apparent that the neocerebellum contributes to and is concerned with cognitive functioning. For example, the right cerebellum (which is interconnected with the left hemisphere) becomes activated when patients are asked to produce verbs in response to nouns (Petersen et al., 1988). Conversely, it has been reported that those with cerebellar lesions may display disturbances on verbal-paired associates tests as well as spatial tasks (Bracke-Tolkmitt et al., 1989). As noted, some patients may initially display mutism (van Dongen et al., 1994).

Patients with lesions in the left neocerebellum (which is interconnected with the right hemisphere) have also been shown to have mild difficulty performing cognitive-spatial operations in three-dimensional space (Wallesch and Horn, 1990).

Presumably, these disturbances are a function of the extensive interconnections that link the right cerebral hemisphere with the left cerebellar hemisphere, and the left cerebral hemisphere with the right half of the cerebellum. That is, they may be secondary to the disrupting influences of cerebellar lesions on neocortical processing, and/or on neocortical afferents as they pass through the brainstem in route to the cranial nerve nuclei.

The Paleo-Cerebellum and Emotion

Having evolved out of the vestibular system, the cerebellum continues to straddle the medulla-pontine junction, and is anchored to the pons via the three cerebellar peduncles. Indeed, it is via the pons that the cerebellum receives and transmits information (Brodal, 1981; Carpenter, 1991). However, it is not just motor and vestibular activities with which it is concerned, but those aspects of emotional experience that may require movement.

Via fibers passing through the pons, the paleocerebellum (including the fastigial nucleus and vermis) is directly linked with the amygdala, hippocampus, temporal lobe, hypothalamus, septal nuclei, nucleus accumbens, and substantia nigra (Harper and Heath, 1973, 1974; Heath and Harper, 1974; Snider and Maiti, 1976). Hence, stimulation of the paleocerebellum can induce violent rage and emotional reactions, including threat and attack (Bharos et al., 1981), as well as autonomic changes, including alterations in arterial pressure and heart rate, piloerection, dilation of the pupils, urination, gastrointestinal changes, and the production of sleep-like EEG spindles (Ban et al., 1980; Chambers, 1947; Rasheed et al., 1970). Moreover, activation of the anterior cerebellum will increase blood pressure, heart rate, and respiration, and will inhibit gastromotility, whereas the posterior cerebellum has the exact opposite effect. However, if its connnections with the amygdala and hypothalamus are severed, cerebellum stimulation ceases to have an influence on these functions (Sawyer et al., 1960), which indicates that it is not exerting these influences independently via the brainstem.

Given its intimate linkage with the limbic system and emotional functioning, it is not surprising that cerebellar dysfunction is associated with emotional abnormalities. For example, it has been reported that up to 50% of patients diagnosed as psychotic or schizophrenic display cerebellar abnormalities, including atrophy of the vermis or tumors (Heath, 1977; Heath et al., 1982; Taylor, 1991; Weinberger et al., 1979, 1980). Heath et al. (1982) also argue that approximately 50% of those who are psychotically depressed show a similar pattern of cerebellar abnormality. Moreover, cerebellar hypoplasia has been impli-

cated in autism (Bauman and Kemper, 1985; Gaffeny et al., 1987).

Indeed, it has been repeatedly noted that stimulation of the paleocerebellum (i.e., the vermis) can induce electrophysiological desynchronization in the thalamus, anterior cingulate, amygdala, hippocampus, orbital frontal lobes, midbrain reticular formation, hypothalamus, and the ventral striatum (Anand et al., 1958; Rasheed et al., 1970)—nuclei that have been implicated in the genesis of psychotic and emotional disturbances (see Chapters 5, 7, 9, 14, and 16).

Conversely, Heath (1977) has also reported that electrical stimulation of the vermis of the cerebellum can relieve chronic and intractable psychotic disturbances in over 90% of those treated. Hence, these latter findings suggest that the cerebellum may maintain not just a reciprocal relationship with the limbic system and ventral striatum, but that it may exert an inhibitory influence on these nuclei—even when they are reacting abnormally. Thus, it has been reported that cerebellar stimulation can decrease limbic system seizure activity, whereas ablation will result in enhanced limbic system seizures (Dow, 1974; Snider and Maiti, 1975). Conversely, cerebellar atrophy has been noted in patients with severe epilepsy (Botez et al., 1988).

In large part, however, although linked to the limbic and autonomic nervous system, rather than mediating or controlling emotional expression, the cerebellum may merely be providing the guidance for motor subroutines that make emotionally motivated actions possible. That is, the cerebellum may not be related to emotion and motivation except indirectly via its influences on the motor system.

For example, stimulation of the fastigial nucleus, vermis, and superior cerebral peduncle can trigger biting, chewing, swallowing, and lip and tongue movements (Bernston et al., 1973). However, stimulation of the cerebellum does not trigger searching for food. Hence, rather than motivating hunger, the motor program associated with eating is being triggered, for animals will approach and pick up and swallow pieces of food as well as nonnutritive substances.

Similarly, although stimulation of the paleocerebellum can trigger rage-like reactions, they are undirected and seem at best semi-purposeful (i.e. "sham rage"). Therefore, although the motor programs for these behaviors may be stored in the cerebellum, this nuclei appears to interact with and respond to feedback from forebrain structures, for removal of the cerebellum does not eliminate these and related emotional behaviors (Bernston et al., 1973). Obviously, much more research needs to be conducted in this area.

SECTION V

The Lobes of the Brain

11

The Frontal Lobes

Arousal, Attention, Perseveration, Personality, Catatonia, Memory, Aphasia, Melodic Speech, Confabulation, Schizophrenia, Movement Disorders, and the Alien Hand

The frontal lobes serve as the "Senior Executive" of the brain and via the assimilation and fusion of perceptual, volitional, cognitive, and emotional processes, modulate and shape character and personality. When damaged there can result excessive or diminished cortical and behavioral arousal, disintegration of personality and emotional functioning, difficulty planning or initiating activity, abnormal attention and ability to concentrate, severe apathy or euphoria, disinhibition, and a reduced ability to monitor and control one's thoughts, speech, and actions. Paralysis of the extremities, severe unilateral neglect of visual-auditory space, or conversely, compulsive utilization of tools or other objects can occur.

The reason for such a wide range of potential disturbance is that rather than a single pair of frontal lobes there are several "frontal" regions that differ in regard to embryology, phylogeny, cellular composition, functional specificity, and interconnection and interactions with other brain areas. Each frontal region is concerned with somewhat different functions (Cummings, 1993; Fuster, 1996; Joseph, 1986a; Luria, 1980; Passingham, 1993; Selemon et al., 1995; Strub and Black, 1993; Van Hoesen, 1996). Moreover, damage is seldom restricted to one specific frontal area but commonly disrupts adjoining "frontal" tissues as well. Hence, a host of divergent abnormalities can result, sometimes simultaneously, depending on the extent, site, depth, and laterality of the lesion.

Broadly speaking, the frontal lobes can be subdivided into four major functional-anatomical regions. These include the central-lateral motor areas (Brodmann's areas 4, 6, 8, 44, 45), the lateral convexity (areas 9, 10, 11, 45, 46, 47), orbital regions (10, 11, 12, 13, 14), and the medial inner walls of the hemispheres that overlap with the motor, lateral convexity, and orbital areas. However, the right frontal lobes also appear to be functionally lateralized, such that the right frontal region performs activities that in many respects are profoundly different from those involving the left frontal lobe.

FUNCTIONAL OVERVIEW

Broadly considered, from a symptomatic perspective, the right lateral frontal lobe appears to be more involved in regulating arousal and attentional functions, and in the expression and modulation of the emotional-melodic aspects of human speech. Injury to the right frontal lobe may result in periods of manic-like excitement, coupled with pressured and/or confabulatory speech that may be melodically distorted.

In some respects, orbital frontal dysfunction is similar to right frontal abnormalities in that patients may behave in a labile, impulsive, and emotionally inappropriate fashion. The orbital area is intimately associated with the anterior cingulate and the amygdala.

By contrast, left frontal symptoms are associated with apathy, depression, schizophrenic-like cognitive distortions, reduced speech output, and expressive aphasia. Similarly, medial frontal lobe abnormalities are associated with reduced speech output and mutism, severe apathy and/or catatonia.

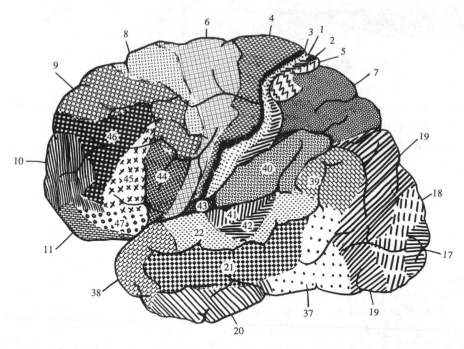

Figure 11.1   The lateral surface of the left cerebral hemisphere with Brodmann's areas indicated.

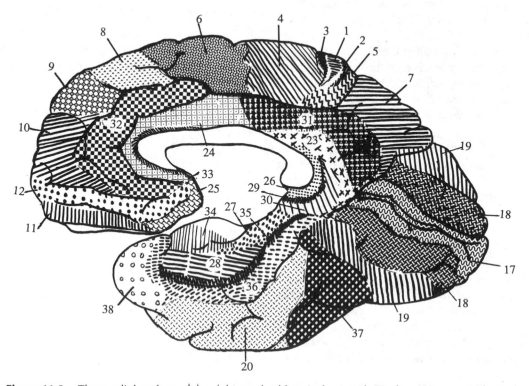

Figure 11.2   The medial surface of the right cerebral hemisphere with Brodmann's areas indicated.

REGULATION OF AROUSAL

An additional and major distinction between the orbital and (the right and left) lateral convexity concerns their differential influences on the control and mediation of limbic (and subcortical) arousal versus neocortical arousal and perceptual activity. For example, the frontal convexity receives overlapping projections from the primary, association, and multi-modal visual, auditory, and somesthetic areas; via the thalamus, it can exert a steering influence on neocortical activity and arousal. In this manner the frontal convexity aids in the suppression or enhancement of perceptual and ideational activity, and assists in the storage and retrieval of this data.

By contrast, the orbital region exerts inhibitory and modulatory influences on limbic system functioning and emotional expression. Like the lateral frontal lobe, the orbital area accomplishes this, in part, via rich interconnections with the dorsal medial thalamus.

MOTOR REGIONS OF THE FRONTAL LOBES

As detailed in Chapter 9, motor functioning and movement are dependent on the functional integrity of the basal ganglia, brainstem, cerebellum, spinal cord and cranial nerve nuclei, motor thalamus, and the primary, secondary, and supplementary motor areas (SMA) of the frontal lobes. Indeed, these areas are all interlinked and function as an integrated system in the production of movement (Mink and Thach, 1991; Parent and Hazrati, 1995).

However, with the evolution of the neocortex and the frontal motor areas, the flexibility and capability to engage in complex movements, including fine motor activities, was made possible. This is largely accomplished via the interactions of the various neocortical motor areas in conjunction with the more ancient motor centers.

Specifically, the frontal neocortical motor regions consist of three major subdivisions: the primary motor cortex, located along the precental gyrus (area 4), the premotor cortex (area 6), and the supplementary motor cortex, which is located along the medial wall of the hemispheres and includes portions of area 6. The frontal eye fields (area 8 with overlap from areas 9 and 6), Exner's writing area, and Broca's expressive speech area (areas 45, 46) are also obvious motor areas involved in eye and hand movement, and the expression of speech.

Also of significance in the discussion of frontal lobe functional neuroanatomy is the head of the caudate nucleus, which is buried within the depths of the frontal lobes and which is related to the lateral convexity and motor areas (as is the anterior cingulate). The caudate is discussed in detail in Chapter 9.

Similarly, the anterior cingulate gyrus (area 24) maintains rich interconnections with the medial frontal lobes (as well as with the limbic striatum and amygdala), and in some respects eventually merges with and essentially becomes indistinguishable from the medial frontal cortices. Although the cingulate is discussed in Chapter 7, those aspects associated with frontal lobe motor activity are briefly reviewed below.

Cellular Organization

The neocortex consists of six to seven layers: I, the molecular (Golgi II cells, with small scattered horizontal cells); II, external granular (densely packed small pyramidal, stellate, and granule cells); III, pyramidal (two sublayers of medium pyramidal cells); IV, internal granular (small pyramidal and stellate cells); V, ganglionic/pyramidal (large and medium-sized pyramidal cells); VIa, multiform (spindle-shaped cells) with VIb (VII) resembling midbrain cortex. The thickness of these layers, the size of their neurons, and thus the size of the neocortex, varies depending on brain region, with a range in size of 4.5 mm to 1.3 mm.

The neocortex of the motor area (areas 4, 6, 8, and 44) is relatively thick and "agranular" in structure. This is because internal "granular" layer IV is obscured by the increased size of layers III and V, which is due to increases in the size and number of pyramidal cells, including the giant cells of Betz. By contrast, in somesthetic areas 3, 1, and 2, layers III and V are much smaller, whereas layer IV is much thicker.

Corticospinal neurons are found predominantly in layer V, and tend to vary greatly in size. Layer V also gives rise to the corticobulbar, corticopontine, and corticorubral fibers. Corticothalamic fibers arise from layer VIa.

CORTICOSPINAL PYRAMIDAL TRACT

Axonal projections from the motor and somesthetic regions all give rise to the massive corticospinal (pyramidal) tract, a rope of fibers that descend to the brainstem and spinal cord to make contact with cranial nerve and sensory and spinal motor neurons (Brodal, 1981; Kuypers and Catsman-Berrevoets, 1984). Approximately 31% of the corticospinal tract arises from the pyramidal cells of agranular area 4, with the remainder arising from areas 6, 8, and somesthetic areas 3, 1, and 2.

LATERAL AND VENTRAL CORTICOSPINAL TRACTS

By definition, the corticospinal/pyramidal tract consists of all axonal fibers that turn and cross over longitudinally in the pyramids of the lower medulla, and then descend into the spinal cord.

Approximately 90% of corticospinal axons range from 1 to 4 µm in diameter, with the giant Betz cells giving rise to the larger axons. About 60% are myelinated. Specifically, the axons of the corticospinal (and related) tracts arise from large pyramidal neurons located in layers III and V of the neocortex and descend via the internal capsule to the midbrain, some of which make contact and terminate in the red nucleus, the periaqueductal gray, and a variety of other brainstem motor nuclei. Collectively these latter fibers are actually referred to as the corticobulbar tract, and

many of its axons arise in the orbital and medial frontal lobe as well as the inferior lateral convexity.

By contrast, corticospinal axons descend through the pons where they separate into tiny nerve bundles before regrouping within the medulla to form the medullary pyramid. However, at the spinal-medulla border, about three-fourths of these axons cross over at the midline of the medulla to form the pyramidal decussation, with the crossed fibers forming the lateral corticospinal tracts and the remaining uncrossed axons forming the ventral corticospinal tract. The lateral tract projects to the lateral motor nuclei of the ventral horn and to intermediate zone interneurons, whereas the ventral tract projects bilaterally to the medial cell column, which is concerned with the axial muscles. Most of the ventral (uncrossed) axons originate in Brodmann's area's 4 and 6 and the SMA, whereas the lateral tract originates in areas 4, 6, and 3, 1, 2.

Primary Motor Area

Functionally the primary motor cortex appears to extend well beyond the confines of the precentral gyrus (area 4) and includes portions of area 6 and the somatosensory (1, 2, 3, 5) regions (Brodal, 1981). These areas are richly interconnected (Jones et al., 1978; Jones and Powell, 1970; Kuypers and Catsamn-Berrevoets, 1984), the somatosensory projections providing information important in the sensory guidance of movement (Godschalk et al., 1981; Lebedev et al., 1994; see Chapter 12). As noted, axonal projections from the motor and somesthetic regions all give rise to the massive cortico-spinal (pyramidal) tract, which innervates cranial nerve and sensory and spinal motor neurons (Brodal, 1981; Kuypers and Catsman-Berrevoets, 1984). For these reasons some authors have referred to the somesthetic and motor regions as the sensorimotor cortex.

The primary motor area is concerned with the coordination and expression of gross and fine motor functioning, including finger movements (Luria, 1980; Rao et al., 1995; Shibasaki et al., 1993; Woolsey, 1958), and serves as a neocortical nodal point where impulses organized in other brain areas are transferred for expression. Indeed, it is likely that the primary motor area may have been the last neocortical motor region to evolve (Sanides, 1969, 1970)—an event that corresponded with the evolution of facial expression and of the fingers, thumb, and fine motor control.

Hence, unlike the sensory areas where information is first received in the primary zones before transmission to the association areas, motor impulses begin their organizational journey in the SMA (Alexander and Crutcher, 1990; Crutcher and Alexander, 1990) and/or the limbic system (see Chapter 6) prior to transmission to the primary regions where they are acted upon. Because the impulse to move originates elsewhere, direct electrical stimulation of the primary motor cortex does not give rise to complex, coordinated, or purposeful movements (Penfield and Jasper, 1954; Penfield and Rasmussen, 1950; Rothwell et al., 1987), although some gross movement of single muscle groups may be elicited, including twitching of the lips, flexion or extension of a single finger joint, protrusion of the tongue, and elevation of the palate.

However, if electrical stimulation is applied when the patient is attempting to move, the result is paralysis (Penfield and Jasper, 1954; Penfield and Rasmussen, 1950). Presumably the reception of impulses to move (which are initiated and organized elsewhere) are blocked by primary motor electrical stimulation.

MOTOR HOMUNCULUS

As noted, electrical stimulation of discrete points within the motor cortex while the subject is at rest can trigger contractions and movements of tiny muscle groups on the opposite side of the body. In fact, an almost one-to-one correspondence between single motor neurons and particular muscles is evident, such that the entire musculature of the body is neuronally represented in the motor cortex. However, since certain muscle groups play a proportionately greater role in the performance of complex (versus simple) movements, a relatively greater number of neurons are involved in their representation. Thus, the fingers have extensive cortical representation, whereas a smaller neuronal field is concerned with the elbow. The motor homunculus, therefore, is quite distorted.

Specifically, represented deep in the inferior medial portion of area 4 and moving anterior toward the superior medial frontal lobe are the toes, ankle, knee, hip, anal sphincter, and genitals. Circling up and over the superior surface to the lateral convexity are the shoulder, elbow, wrist, and hand, with extensive areas of neocortex devoted to the fingers and thumb, followed by the brow, eyelids, larynx, lips, jaw, and tongue, which is located at the most inferior opercular portion of the precentral gyrus. Like the somesthetic neocortex (see Chapter 12), which maintains a double representation of the body surface (e.g., a double body image), there is some possibility that the primary motor area may contain multiple representations of the body's musculature.

PARALYSIS

Damage to the motor areas or to the descending cortico-spinal tract initially results in a flaccid hemiplegia such that the muscles are completely without tone contralateral to the lesion (Adams and Victor, 1994; Brodal, 1981). If the examiner were to raise and release an affected arm, it would drop like that of a limp rag doll.

Over the course of the next several days the muscles develop increased tone and there is resistance to passive movements. Reflexes become very brisk and spasticity and hyperreflexia are manifest. With massive lesions extending into the medial regions (where the leg is represented), the leg will become permanently extended and the arm will assume a flexed position (Adams and Victor, 1993; Brodal, 1981). After several weeks or months, very limited gross

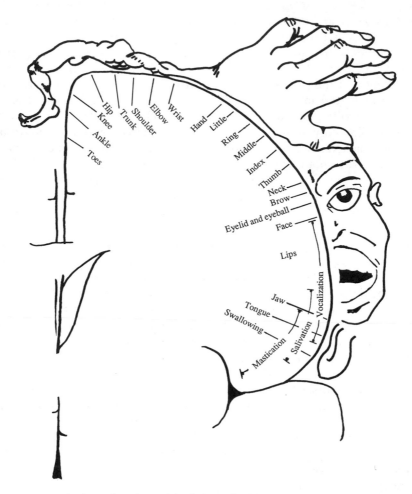

Figure 11.3   The lateral surface of the left cerebral hemisphere with relative location of body parts indicated along the primary motor cortex.

movements become possible. Fine movements are usually permanently lost (Brodal, 1981).

Premotor Neocortex

The premotor cortex (area 6) is intimately interconnected with the primary motor area, both of which continue on to the medial wall of the hemisphere. Although the premotor area does not contain giant Betz cells (which are found in area 4), it contributes almost one-third of the fibers of the corticospinal tract.

The premotor cortex sends axons to area 4 and receives projections from this and the SMA (Jones et al., 1978; Jones and Powell, 1970). It also receives information directly from the primary and secondary somesthetic and visual cortices (areas 17, 18, 19) (Jones and Powell, 1970; Pandya and Kuypers, 1969), and is heavily involved in the guidance and refinement of movement via the assimilation of sensory information provided by the sensory areas (Godschalk et al., 1981; Porter, 1990) and the basal ganglia, motor thalamus, and SMA (Alexander and Crutcher, 1990; Crutcher and Alexander, 1990; Mink and Thach, 1991; Parent and Hazrati, 1995).

Whereas neurons in the primary motor region become active during movement, excitation in the premotor cortex precedes cellular activation of the primary region (Weinrich et al., 1984). Moreover, cells in the premotor cortex become activated before movements are even initiated. These and other findings suggest that the premotor area may be modulating and exerting controlling influences on impulses that are to be transmitted to the primary region for expression. Indeed, the premotor area appears to be highly involved in the programming of various gross and fine motor activities, and becomes highly active during the learning of new motor programs (Porter, 1990; Roland et al., 1981). Moreover, electrical stimulation elicits complex patterned movement sequences as well as stereotyped and gross motor responses such as head turning or torsion of the body (Fulton, 1934; Passingham, 1981, 1993).

Unlike the primary area, damage limited to the pre-motor cortex does not result in paralysis but disrupts fine motor functioning and dexterity, including simple activities such as finger tapping (Luria, 1980). With extensive damage fine motor skills are completely lost and phenomena such as the grasp reflex are elicited (Brodal, 1981), i.e., if the patient's hand is stimulated it will involuntarily clasp shut.

The SMA and the Medial Frontal Lobes

The SMA is located along the medial walls of the hemispheres but has no clearcut anatomical boundaries. Its anterior portion abuts the medial overlap of the primary area and it extends downward along the medial wall where it meets the anterior cingulate gyrus (area 24). The SMA contains a crude neuronal representation of the body (Goldberg, 1985).

The SMA receives axonal projections from the primary and association somatosensory areas (Jones and Powell, 1970; Pandya and Vignolo, 1969) and shares rich interconnections with primary motor cortex, the anterior cingulate (Devinksy et al., 1995; Jones et al., 1978), and the basal ganglia (Alexander and Crutcher, 1990; Crutcher and Alexander, 1990).

The SMA appears to be concerned with the general problem of guiding and moving the extremities through space. Electrical stimulation has produced complex semipurposeful movements, vocalization (Penfield and Jasper, 1954), and postural synergies involving the trunk and extremities bilaterally (Van Buren and Fedio, 1976). Moreover, single cell recordings (Brinkman and Porter, 1979; Tanji and Kurata, 1982) and studies of blood flow (Orgogozo and Larsen, 1979; Shibasaki et al., 1993) and movement-related evoked potentials (Ikeda et al., 1992) indicate increased activity in this area while a subject performs and even imagines complex movements of the fingers and hands. This region also becomes highly active during the modification, learning, and establishment of new movement programs (Brinkman and Porter, 1979; Roland et al., 1980; Tanji et al., 1980). Moreover, activity begins in the SMA well before movements are initiated and prior to activation within the premotor and primary motor areas (Alexander and Crutcher, 1990; Crutcher and Alexander, 1990; Mink and Thach, 1991).

For example, when anticipating or preparing to make a movement, but prior to the actual movement, neuronal activity will first begin and then dramatically increase in the SMA, followed by activity in the secondary and then the primary motor area, and then the caudate and last of all the putamen and globus pallidus (Alexander and Crutcher, 1990; Mink and Thach, 1991). Presumably the putamen, in conjunction with the caudate, transmits this information to the globus pallidus, which in turn projects to the motor thalamus, brainstem reticular formation, and the motor neocortex, thus creating an elaborate feedback loop (Mink and Thach, 1991; Parent and Hazrati, 1995), which is also influenced by the anterior cingulate and amygdala.

SMA DAMAGE

Paralysis or paresis (in the classic sense) does not result with damage to the SMA, although the body may become stiff and movements tend to be slow and incoordi-

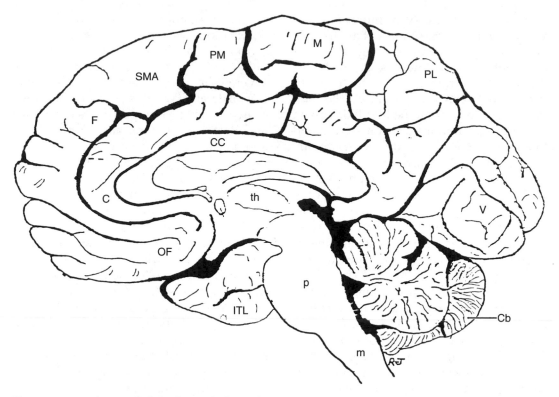

Figure 11.4 The medial surface of the right cerebral hemisphere. The pons (p) and medulla (m) of the brainstem are indicated as well as the thalamus (th), corpus callosum (CC), cingulate (C), orbital frontal lobes (OF), supplementary motor area (SMA), premotor (PM) and the primary motor (M) areas, parietal lobule (PL), visual cortex (V), and cerebellum (Cb).

nated—a condition also seen with mild lesions (Penfield and Jasper, 1954). However, with extensive and massive injuries patients may become mute (McNabb et al., 1988; Watson et al., 1986) and so stiff and unmoving that they appear to demonstrate all the classic signs of catatonia, including gegenhalten and waxy flexibility.

More typically, however, patients demonstrate clumsiness, severe agraphia, impairments of bimanual coordination, and difficulty performing rapid or alternating movements (Brinkman, 1981; Gasquione, 1993; Goldberg and Bloom, 1990; Goldberg et al., 1981; McNabb et al., 1988; Penfield and Jasper, 1954; Travis, 1955; Truelle et al., 1995; Watson et al., 1986). Patients may walk with short steps and suffer disturbances involving posture, balance, and gait. Initially mutism may be observed (McNabb et al., 1988; Watson et al., 1986).

CATATONIA AND THE MEDIAL FRONTAL LOBES

Functional Anatomy of the Medial Walls

The medial walls of the anterior portion of the hemispheres contain the SMA, the anterior cingulate (area 24), portions of the frontal eye fields (area 8), and areas of overlap from the primary and supplementary motor cortices, lateral convexity (areas 9, 10), and orbital frontal lobe (areas 11, 12). It maintains massive interconnections with all these regions as well as the hippocampus, lateral amygdala, lateral hypothalamus, basal ganglia, and reticular formation, and it receives axonal projections from the sensory cortices (Jones and Powell, 1970, Leichnetz and Astruc, 1976; Van Hoesen, 1995).

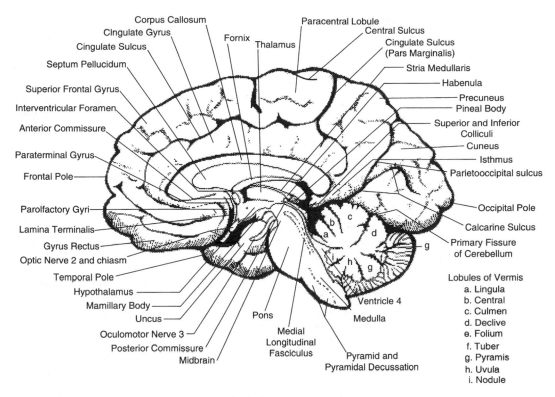

Figure 11.5 A medial view of the right hemisphere.

As is suggested by these rich interconnections with so many diverse brain regions, the medial frontal lobes are involved in the synthesis of motor, sensory, and emotional information. Moreover, as limbic and neocortical input intersects within the medial frontal lobes, ideas and thoughts can become emotionally charged, and motivational significance may be assigned to external perceptions or particular motor activities.

Hence, if damaged, disturbances of "will" (e.g., apathy) as well as peculiar abnormalities involving motor activity (e.g., gegenhalten, waxy flexibility) are likely. Neglect, indifference, catatonic-like features (Hassler, 1980; Joseph, 1990d; Laplane et al., 1977; Luria, 1980; Penfield and Jasper, 1954; Penfield and Welch, 1951), or conversely, compulsive activation of the limbs, including in the extreme a syndrome referred to as the "alien hand," may result (Freeman and Watts, 1942; Gasquoine, 1993; Goldberg and Bloom, 1990; Travis, 1955).

Gegenhalten and Waxy Flexibility

As noted, lesions involving the medial frontal lobes and SMA are sometimes associated with gegenhalten (counterpull), i.e., involuntary resistance to movement of the extremities (Travis, 1955). If a physician attempts to move an affected arm, it will suddenly stiffen and become increasingly rigid as pressure to move it increases. Although aware, the patient cannot decrease the resistance.

Waxy flexibility has been attributed to deep SMA and medial lesions as well as following surgical destruction of the underlying SMA white matter. If the examiner moves the arm, resistance appears after completion of the movement such that the extremity will remain in whatever position or posture it is placed and then only very slowly return to a normal resting position.

Posturing is also noted. Such patients, for example, might remain in odd and uncomfortable positions for exceedingly long time periods and make no effort to correct the situation (Freeman and Watts, 1942; Rose, 1950). Usually this condition will resolve within a few weeks. Nevertheless, the condition sometimes is mistaken for catatonia.

Catatonia

Although a variety of neural pathways and structures have been implicated in the pathogenesis of schizophrenia (see Chapters 4, 9, 14, 17, 18), there is some evidence that deep medial lesions of the frontal lobe sometimes give rise to emotional blunting, posturing, and what appears to be catatonic-like symptoms. For example, Freeman and Watts (1942) described one person who developed waxy flexibility, catatonia, and related symptoms after a gunshot wound (suffered during the war) in which the bullet passed completely through the frontal lobes.

"The patient [laid] in a catatonic-like stupor for two months, always upon one side with slightly flexed arms and legs, never changing his uncomfortable position; if he were rolled into some other position, he would quickly get back into his former one. He did not obey commands, but if food and drink were given to him, he swallowed them naturally. He was incontinent, made no complaints, gazed steadily forward and showed no interest in anything. He could not be persuaded to talk, and then suddenly he would answer quite correctly about his personal affairs and go back to mutism. From time to time he showed a peculiar explosive laugh, especially when his untidiness was mentioned." Incredibly, the patient "was eventually returned to active duty" (Freeman and Watts, 1942:46–47).

Similarly, Hillbom (1951) reported that among a large sample of persons who had suffered traumatic missile wounds, those who developed catatonic features (or hebephrenia) commonly had frontal lobe injuries.

One person with no previous psychiatric history, after having been beat around the head and suffering frontal subdural hematomas (which required the drilling of burr holes for evacuation), within days developed catatonic behaviors, resisted the efforts of others to move him, and would sit motionless and unresponsive for hours in odd and uncomfortable positions. Interestingly, the patients symptoms seemed to wax and wane such that he demonstrated some periods of seeming normality (Joseph 1990d).

Will and Apathy

Stimulation of the SMA sometimes results in a functional disconnection between this and other areas such that the ability to initiate or complete a voluntary movement is disrupted, and the "will to move or speak" is completely attenuated and abolished (Hasslet, 1980; Laplane et al., 1977; Luria, 1980; Penfield and Jasper, 1954; Penfield and Welch, 1951). Upon partial recovery or with extensive prodding, such patients have complained that thoughts do not enter their head (Luria, 1980). That is, they are unable to think or generate ideas, and experience a motivational-ideational void (Brutkowski, 1965; Hasslet, 1980; Laplane et al., 1977; Luria, 1980; Mishkin, 1964).

Patients may bump into objects, not blink in response to threats, stand about motionless, or demonstrate waxy flexibility. Some patients, like the soldier described by Freeman and Watts above, may initially seem to be in a catatonic-like stupor. Similar abnormalities (excluding catatonia) are sometimes seen with convexity lesions bordering on the frontal pole or motor areas, and with anterior cingulate damage (Devinksy et al., 1995).

Forced Grasping, Compulsive Utilization, and the Alien Hand

In other instances of medial frontal lobe dysfunction, semi-purposeful uncontrolled motoric responses may become superimposed on the patients otherwise seemingly apathetic and confused condition. That is, with massive or bilateral destruction of the SMA and portions of the medial walls, there can result semi-purposeful albeit reflexive motoric abnormalities such as forced grasping and "magnetic" groping (Denny-Brown, 1958; McNabb et al., 1988; Travis, 1955).

In some cases patients seem "stimulus bound" and involuntarily respond to or even compulsively utilize objects or stimuli with which they come in contact (Denny-Brown, 1958; Gasquoine, 1993; Goldberg and Bloom, 1990; Lhermitte, 1983; Travis, 1955). Presumably this is due to a loss of internal motivational controls (i.e., disconnection) and is partially a release phenomenon such that patients appear reflexively or magnetically directed solely by external stimuli that trigger involuntary motor reactions.

FORCED GRASPING AND GROPING

With extensive destruction and/or as the lesion extends mesially, the mere visual presence of an object near the hand triggers groping movements as well as grasping. Similarly, if an object touches the palm of the hand, the patient grips it involuntarily and cannot let go. Denny-Brown (1958) has referred to this as "magnetic apraxia" and "compulsive exploration." He notes that touching the hand will elicit orienting movements to bring the object into the palm. Once it is grasped the patient cannot release his grip.

This aberration is often accompanied by gegenhalten. Immediately prior to actually grasping the object the entire arm will stiffen and one is met with resistance if an attempt is made to move the extremity. If a patient were to attempt to write, the hand stiffens and becomes seemingly "stuck" to the paper. If they try to walk, their feet seem to stick to the floor as if glued, and steps are made with great difficulty (Denny-Brown, 1958).

Denny-Brown (1958) believed that magnetic groping was due in part to a perseveration of contractual reactions (as well as parietal lobe disinhibitory release), for so long as stimulation is applied to the skin, the deficit, including gegenhalten, not only persists but becomes more intense.

This abnormality is also triggered visually. As if stimulus bound, the patient's eyes and head may compulsively follow moving objects. If an object is brought near the face the patient may compulsively reach out and take it; if the object is brought near the lips the patient may attempt to mouth or suck it. Such behavior is completely involuntary and will occur even if the patient "willfully" attempts to oppose it.

COMPULSIVE UTILIZATION

With extensive mesial damage involving the medial and medial-orbital areas, patients not only magnetically grope but may demonstrate a compulsive involuntary tendency to use whatever objects may be nearby (Lhermitte, 1983). For example, if an experimenter were to place a hammer and nail close by, the patient might impulsively begin to hammer the nail into the testing table or the wall—even if instructed not to. If the examiner places a pair of spectacles on the table, the patient compulsively puts them on. One such patient put three pair of spectacles on simultaneously (Lhermitte, 1983). If food and water are near, he might eat and drink even when repeatedly told to refrain and in the absence of hunger or thirst. Hence, damage is this area can cause patients to respond and use whatever objects or tools may be close even when they understand that they are not to do so (Lhermitte, 1983).

THE ALIEN HAND

In rare instances, this compulsive groping and utilization behavior can become confined to one limb and involve complex and seemingly purposeful actions (Goldberg, 1987; Goldberg and Bloom, 1990). This may occur when the lesion has predominantly destroyed either the right or left SMA and medial portion of the hemisphere as well as the anterior corpus callosum (McNabb et al., 1988; Gasquoine, 1993; Goldberg and Bloom, 1990; Goldberg et al., 1981), such that the right and left frontal lobe have become partially or fully disconnected.

For example, McNabb et al. (1988: 219, 221) describe one woman with extensive damage involving the medial left frontal lobe and anterior corpus callosum, whose right "hand showed an uncontrollable tendency to reach out and take hold of objects and then be unable to release them. At times the right hand interfered with tasks being performed by the left hand, and she attempted to restrain it by wedging it between her legs or by holding or slapping it with her left hand. The patient would repeatedly express astonishment at these actions." A second patient frequently experienced similar difficulties. When "attempting to write with her left hand the right hand would reach over and attempt to take the pencil. The left hand would respond by grasping the right hand to restrain it" (p. 221).

Similar problems, however, have also plagued some patients following complete (surgical) destruction of the corpus callosum (Joseph, 1988a, 1988a, b). In many of these instances, however, the independent "alien" behaviors demonstrated were often purposeful, intentional, complex, and obviously directed by an intelligence maintained by the disconnected right hemisphere. For example, one patient's left hand would not allow him to smoke and would pluck lit cigarettes from his mouth, whereas another patient's left hand (right brain) preferred different foods and even television shows and would interfere with the choices made by the right hand (left hemisphere). In one instance the patient was pulled to the television and the left hand changed the channel.

However, in some instances patients display alien movements in both hands (Gasquoine, 1993; Goldberg and Bloom, 1990) as well as alien vocalization of thoughts. For example, a patient described by Gasquoine (1993) had a propensity to reach out and touch female breasts, as well as novel objects and persons within reach. He reported this caused him great embarrassment and that he would typically attempt to take hold of his right with his left hand or voluntarily grasp objects, such as his lap tray, so that he would not sponta-

neously reach out and grab someone. Moreover, although he had difficulty initiating speech, early in his illness he would spontaneously vocalize his thoughts.

Internal Utilization

As the locus of the lesion becomes even more inferior and posterior, encroaching on the orbital-temporal (amygdaloid) area, external utilization is replaced by attempts at internal (oral) utilization. Instead of utilizing an object by motorically interacting with it, the person orally interacts and attempts to consume the object. Everything that is seen and touched is placed indiscriminately in the mouth and orally "explored."

This latter disturbance, extreme orality, has been referred to as "psychic blindness" and the Klüver-Bucy syndrome, and is a consequence of destruction of the amygdaloid—a nucleus with massive interconnections with the orbital/inferior medial area (see Chapter 5).

A Cytoarchitectural Continuum of Related Symptoms

Although seemingly disparate, a continuum of related aberrations can result following damage to the medial and motor areas of the frontal lobe. With deep inferior orbital-temporal (amygdaloid) destruction the patient may engage in internal utilization (e.g., heightened orality and internal utilization behavior). With more medial destruction the patient may instead display external compulsive utilization of objects or, in the extreme, complex uncontrollable acts limited to one extremity. With more superior medial damage involving the SMA the behavioral displays are less complex and may involve involuntary activation of the fingers and extremities, including forced grasping and groping. However, damage involving the premotor area results in a loss of fine motor skills, whereas primary motor destruction results in paralysis.

Hence, with primary motor damage a patient cannot move although he may want to, whereas with medial and SMA damage the patient may compulsively move although he does not want to. In some instances (such as following partial disconnection of the anterior regions due to corpus callosum involvement) a complex syndrome of motoric responses, referred to as the alien hand syndrome, may result.

These findings raise the possibility that the motor areas, although guided (or even triggered) by external sensory impressions, often act at the behest of impulses originating in or beyond the medial and orbital-amygdaloid area. When these internal influences are blocked (because of disconnection) but external perceptual activity is preserved, compulsive responding to external stimuli results (Denny-Brown, 1958; Goldberg and Bloom, 1990). The motor areas become released from internal guidance or from inhibitory influences mediated by the opposing hemisphere. However, with massive medial damage (where internal influences becomes integrated with external sources of input) the patient loses the ability to respond to either internal or external forces and becomes mute and sometimes seemingly catatonic.

Anterior Cingulate

The SMA and portions of the lateral convexity appear to have evolved from the anterior cingulate gyrus (Sanides, 1969, 1970, 1972). That is, over the course of evolutionary development the cingulate has grown up and over the medial wall to become, in modified form, part of the SMA and lateral convexity. Indeed, the SMA and portions of the lateral convexity are not only structurally related to the anterior cingulate but maintain massive interconnections with this area (Sanides, 1969, 1970, reviewed in Devinksy et al., 1995).

The anterior cingulate (Brodmann's area 24) also maintains extensive interconnections with the lateral amygdala, septum, hippocampus, anterior hypothalamus, caudate and putamen, dorsal medial nucleus of the thalamus, and inferior parietal lobule, as well as the lateral convexity, medial wall, and orbital frontal lobes (Baleydier

and Maguiere, 1980; Pandya and Kuypers, 1969; Powell, 1978; Powell et al., 1974). In addition, the cingulate projects to the spinal cord as well as to the motor neocortex and the striatum (Devinksy et al., 1995). The cingulate appears to maintain an intermediate functional-anatomical position between the limbic system and neocortex and probably serves as an interface linking volitional, motoric, cognitive, and emotional impulses (Devinksy et al., 1995; Goldberg, 1985).

Electrical stimulation of the anterior cingulate gyrus (hereafter referred to simply as the cingulate) in humans has produced primitive motor subroutines, or rather, motor fragments that if linked together would constitute an entire movement (Devinksy et al., 1995; Goldberg, 1985). Conversely, injuries to the anterior cingulate can induce forced grasping and utilization behavior (Degos et al., 1993), which is probably secondary to medial frontal lobe dysfunction. Hence, the cingulate is involved in and appears to aid in performing complete movements or movement sequences, including finger movements (Shibasaki et al., 1993).

The cingulate may also be involved in setting thresholds for vocalization (Jurgens and Muller-Preuss, 1977), including modulation of some of the prosodic (melodic) features that characterize different speech patterns (e.g., happiness versus sadness). For example, among primates, when this area is electrically stimulated, "limbic" vocalizations indicative of alarm, fear, or "separation anxiety" have been triggered (Robinson, 1967). Conversely, destruction of the anterior cingulate and adjacent medial frontal tissue can abolish spontaneously vocalizations such as the "separation cry" (MacLean, 1990). Moreover, all cortical areas that give rise to vocalization maintain direct interconnections with area 24 (Jurgens and Muller-Preuss, 1977).

Among humans and primates, electrical stimulation has also resulted in feelings of anxiety, fear, and pleasure (Meyer et al., 1973). However, these effects may be secondary to indirect involvement of the hypothalamus, amygdala, septal nuclei, or even the orbital area.

Similar to damage involving the SMA, massive lesions of the cingulate trigger a variety of motor abnormalities, including paralysis of will, magnetic apraxia, compulsive utilization, and mutism (Barris and Schuman, 1953; Devinksy et al., 1995; Kennard, 1955; Laplane et al., 1981; Smith, 1944). Damage or surgical removal has also resulted in generalized emotional dampening and unresponsiveness, inability to make active avoidance responses, hypokinetic states, severe social indifference and apathy, and decreased awareness of the environment (Barris and Schuman, 1953; Glees et al., 1950; Kennard, 1955; Laplane et al., 1981; Pechtel et al., 1958; Smith, 1944; Tow and Whitty, 1953).

For example, Barris and Schuman (1953) describe one patient who rapidly developed akinetic mutism and indifference following bilateral damage to the anterior cingulate and portions of the SMA. After returning home from work he sat down on the sofa, held up a newspaper as if reading (i.e., utilization behavior), became incontinent of urine (about which he was totally unconcerned), was unresponsive, and was unable to reply to questions. The next morning family members found him sitting on the floor polishing a cigarette lighter although he was unable in all other respects to respond intelligibly.

Laplane et al. (1981) describe a similar patient who, following bilateral infarcts to the cingulate, became indifferent, docile, and incontinent, and demonstrated both magnetic groping and utilization behavior. However, rather than become mute she developed confabulatory tendencies and distractibility.

In addition, initially following surgical anterior cingulotomies patients have complained of difficulties distinguishing between events occurring internally and externally such that thoughts and waking experience seem dream-like (Whitty and Lewin, 1957). Confabulation and delusions have been noted, including tangentiality and disinhibited speech (Whitty and Lewin,

1957; 1960). These latter symptoms, however, are probably secondary to damage involving adjacent frontal tissues (Joseph, 1986a).

EVOLUTION

As noted, the medial frontal lobes appear to be an evolutionary extension of the anterior cingulate. For example, Sanides (1969, 1970) has argued that the frontal medial archicortex originated from the anterior cingulate. As detailed above, destruction of the medial frontal lobe can result in mutism (which, however, is not the same as aphasia). Nevertheless, via expansions and evolutionary alterations in the anterior cingulate, the SMA was fashioned. This "architectonic" evolutionary trend then continued its expansion, thus creating the lateral frontal convexity and the motor areas, including what would eventually become the inferior frontal lobe and "Broca's area." Hence, it could be argued that Broca's area evolved from the anterior cingulate.

POSTERIOR FRONTAL CONVEXITY
Frontal Eye Fields

The frontal eye fields (FEF) are located along the superior lateral convexity immediately adjacent and anterior to the premotor area, and encompass all of area 8 as well as portions of areas 9 and 6. The FEF receives projections from the primary and association visual cortices in the occipital lobe (17, 18, 19), the auditory association (22) and multimodal visual association areas (20) in the temporal lobe (Barbas and Mesulam, 1981; Jones et al., 1978; Jones and Powell, 1970), and the somatosensory association area (Crowne, 1983). It also shares interconnections with the caudate, superior colliculus, and oculomotor nucleus (Astruc, 1971; Knuezle and Akert, 1977; Segraves and Goldberg, 1987). Hence, the FEF receives information concerning the auditory, tactual, and visual environment and is multimodally responsive.

The FEF is heavily involved in coordinating and maintaining eye and head movements, and thus orienting and attentional reactions in response to predominantly visual, but also tactile and auditory, stimuli (Barbas and Dubrovsky, 1981; Denny-Brown, 1966; Gottlieb et al., 1994; Latto and Cowey, 1971a,b; Pragay et al., 1987; Segraves and Goldberg, 1987; Wagman et al., 1961). It is also involved in focusing attention on certain regions within the visual field, particularly the fovea (Wurtz et al., 1980; Segraves and Goldberg, 1987), as well as in making smooth pursuit movements (Gottlieb et al., 1994) and perhaps reading and the guidance of the hand during writing (Ritaccio et al., 1992).

In addition to visual tracking and supporting focused visual attention, neurons in the FEF demonstrate anticipatory activity (Gottlieb et al., 1994; Pragay et al., 1987); that is, firing before a response is made. In fact, these neurons will continue to fire at a high rate until the moment when behavior is initiated. Yet other cells begin to fire only when the waiting period becomes prolonged (Pragay et al., 1987). In this regard, they probably exert a countering influence so that attention does not drift.

Electrical stimulation of the FEF results in complete saccades of the eyes (Barbas and Dubrovsky, 1981; Wagman et al., 1981), as well as pupillary dilation. Moreover, cells in the FEF will fire selectively in response to stationary and moving stimuli, to objects within arm's reach, and to tactual stimuli applied to the hands and/or mouth (Rizzolatti et al., 1981a, b). In fact, as an object approaches the face and mouth, some of these cells correspondingly increase their rate of activity (Rizzolatti et al., 1981a, b). Hence, cells within the FEF are highly involved in mediating sustained attention and orienting reactions of the head and eyes, maintaining visual fixation and modulating visual scanning, and coordinating eye-hand and hand-to-mouth as well as smooth pursuit eye movements.

Visual Scanning Deficits and Neglect

Damage to the FEF can cause abnormalities in fixation, decreased sensitivity to stimuli throughout the visual field, slowed visual scanning and searching (Latto and Cowey, 1971a, b; Teuber, 1964), inattention and neglect, and mislocation of sounds (Denny-Brown, 1966; Welch and Stuteville, 1958). With massive lesions, searching and responsiveness become so profoundly reduced that a complete unilateral neglect and failure to attend to any and all stimuli falling to one side of the body results (Heilman and Valenstein, 1972). Like confabulation, neglect is more frequently seen after right cerebral damage (see Chapter 3).

Even with less severe destruction, performance on tasks involving visual search is disrupted (Teuber, 1964), as is attention to visual detail (Luria, 1980). If shown and asked to describe a complex picture, patients may focus on only one detail, neglecting the remainder, or look about haphazardly (Luria, 1980). It is probable that individuals with certain types of dyslexia suffer from similar abnormalities involving visual search and synthesis.

With large lesions (particularly when involving the right frontal area), they may tend to make leaps of judgment and impulsively guess and describe the meaning of the whole based on the perception of a fragment. For example, focusing only on a "drummer boy" in a battle scene they may describe the picture as being about "musicians" or "a rock band." Hence, such patients may erroneously extrapolate from an isolated detail around which they construct and confabulate a conclusion (see Chapter 3).

In the extreme, rather than a true analysis they may produce irrelevant associations because they have difficulty not only in analyzing and synthesizing the different components of a visual presentation (Luria, 1980) but in correcting their impressions via search and feedback (Joseph, 1986a, 1988b). In these instances the lesion typically extends well beyond the confines of the FEF.

With extensive damage involving the FEF and convexity, not only confabulation but Capgras syndrome (false identification) and reduplicative paramnesia have been reported, particularly if the damage is bilateral (Alexander et al., 1979; Benson et al., 1979; Hecaen, 1964).

Exner's Writing Area

Exner's writing area lies within a small region along the lateral convexity, near the foot of the second frontal convolution of the left hemisphere, occupying the border regions of Brodmann's areas 46, 8, 6. Although some authors have denied the existence of Exner's area, this region appears to be the final common pathway where linguistic impulses receive their final motoric stamp for the purposes of writing, i.e., the formation of graphemes and their temporal sequential expression.

Exner's area, however, depends on Broca's expressive speech area, with which it maintains extensive interconnections. In fact, Exner's writing center extends to and appears to become coextensive with Broca's area (Lesser et al., 1984). Broca's area possibly acts to organize and relay impulses re-

ceived from the posterior language zones to Exner's area when written expression is desired. Exner's area, in turn, transfers this information to the secondary and primary motor areas for final expression.

Electrical stimulation of this vicinity in awake moving patients has resulted in the arrest of ongoing motor acts, including the capacity to write or perform rapid alternating movements of the fingers (Lesser et al., 1984). In some instances, writing and speech arrest were noted.

AGRAPHIA

Lesions or seizure activity localized to this vicinity lead to deficiencies involving the elementary motoric aspects of writing, i.e., agraphia (Penfield and Roberts, 1959; Ritaccio et al., 1992; Tohgi et al., 1995). Grapheme formation becomes labored and incoordinated, and takes on a very sloppy appearance. Cursive handwriting is usually more disturbed than printing. In cases of well circumscribed lesions, usually there are no gross deficiencies of motor function-

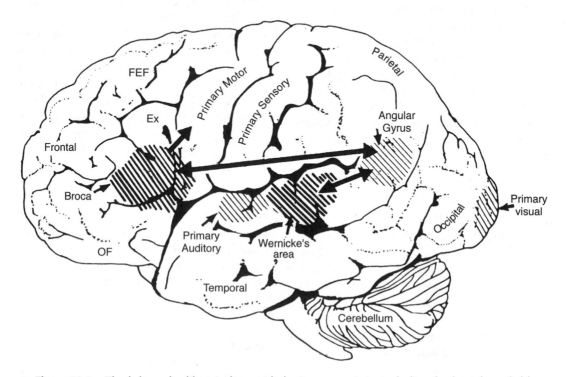

Figure 11.6   The left cerebral hemisphere with the Language Axis, including the frontal eye fields (FEF) and Exner's writing areas (ex) depicted.

ing or speech, although mild articulatory disturbances may be observed (e.g., lisping) as well as abnormalities involving fine motor control (cf. Lesser et al., 1984; Levine and Sweet, 1983).

In cases of pure (frontal) agraphia, spelling may or may not be affected, whereas with parietal lesions spelling as well as writing is often abnormal. Rather, with left frontal lesions more frequently there are disturbances of grapheme selection such that the patient may seem to have "forgotten" how to form certain letters and/or may misplace or even add unnecessary letters when writing (Hecaen and Albert, 1978; Tohgi et al., 1995). When spelling orally or typing, the ability to spell is often better preserved.

Damage localized to this vicinity can be secondary to perinatal trauma, tumors, or vascular abnormalities. Disturbances involving constructional or manipulospatial functioning are not apparent. In fact, one such patient whose damage was secondary to birth injury, although able to write or print only with great difficulty, was able to draw and paint with some professional acumen. However, his ability to copy letters was severely affected. Hence, disturbances secondary to lesions localized to Exner's area are limited to abnormalities involving linguistic-symbolic grapheme motor control.

Broca's Speech Area

Broca's speech area is located in the general vicinity of the posterior-inferior region of the left frontal area (i.e., third frontal convolution), and includes portions of areas 45, 6, and 4, and all of area 44. This region is multimodally responsive and receives projections from the auditory, visual, and somesthetic areas (Geschwind, 1965; Jones and Powell, 1970), as well as massive input from the inferior parietal lobule and Wernicke's area via a rope of nerve fibers referred to as the arcuate fasciculus (which also links these areas to the amygdala). In addition, Broca's area receives fibers from and projects to the anterior cingulate.

Broca's speech area is a final converging destination point through which thought and other impulses come to receive their final sequential (syntactical, grammatical) imprint so as to become organized and expressed as temporally ordered motoric articulations (i.e., speech). Verbal communication, the writing of words (via transmission to Exner's area), and the expression of thought in linguistic form is made possible.

Hence, via this neuronal field impulses transmitted from the posterior language zones come to be temporally-sequentially prepared for transmission to the adjacent primary motor neurons that mediate oral-facial muscular activity (i.e., lips, tongue, jaw) and thus the actual expression of speech (Geschwind, 1965, 1979; Joseph, 1982; Kimura, 1993).

BROCA'S (EXPRESSIVE) APHASIA

With injuries to Broca's area the individual loses the capacity to produce fluent speech. Output becomes extremely labored, sparse, and difficult, and the patient may be unable to say even single words, such as "yes" or "no." Often, immediately following a large stroke patients are almost completely mute and suffer a paralysis of the upper right extremity as well as right facial weakness (since these areas are neuronally represented in the immediately adjacent area 4). Patients are also unable to write, read out loud, or repeat simple words. Interestingly, it has been repeatedly noted that almost immediately following stroke some patients will announce "I can't talk," and then lapse into frustrated partial mutism.

With less severe forms of Broca's (also referred to as expressive, motor, nonfluent, verbal) aphasia, speech remains labored, agrammatical, fragmented, extremely limited to stereotyped phrases ("yes," "no," "shit," "fine") and contaminated with syntactic and paraphasic errors, e.g., "orroble" for auto, "rutton" for button (Bastiaanse, 1995; Goodglass and Kaplan, 1982; Haarmann and Kolk, 1994; Hofstede and Kolk, 1994; Levine and Sweet, 1983; Tramo et al.,

1988). Writing remains severely affected, as are oral reading and repetition. Such patients also have mild difficulties with verbal perception and comprehension (Hebben, 1986; Maher et al., 1994; Tramo et al., 1988; Tyler et al., 1995), including the ability to follow three-step commands (Luria, 1980). Commands to purse or smack the lips, lick, suck, or blow are often, but not always, poorly executed (DeRenzi et al., 1966; Kimura, 1993), a condition referred to as buccal-facial apraxia. Nonspeech oral movements are seldom significantly affected (Goodglass and Kaplan, 1982; Hecaen and Albert, 1978; Kimura, 1993; Levine and Sweet, 1983).

With mild damage, patients may still demonstrate severe confrontive naming and word-finding difficulties (anomia), as well as possible right facial, hand, and arm weakness. Speech is often characterized by long pauses, misnaming, paraphasic disturbances, and articulatory abnormalities. Stammering and the omission of words may also be apparent.

Similarly, electrical stimulation of this region results in speech arrest (Lesser et al., 1984; Ojemann and Whitaker, 1978) and can alter the ability to write and/or perform various oral-facial movements.

It is noteworthy that even with anterior lesions or surgical frontal lobectomy sparing Broca's area a considerable impoverishment of spontaneous speech can result (Luria, 1980; Milner, 1971; Novoa and Ardila, 1987). Disturbances involving grammar and syntax and reductions in vocabulary and word fluency in both speech and writing have been observed with frontal lesions sparing Broca's area (Benson, 1967; Crockett et al., 1986; Goodglass and Berko, 1960; Milner, 1964; Novoa and Ardila, 1987; Petrie, 1952; Samuels and Benson, 1979; Tow, 1955). In word fluency tests, however, simple verbal generation (e.g., all words starting with "L") is usually more severely impaired than semantic naming (e.g., all animals that live in the jungle)—which is presumably a function of semantic processing being more dependent on posterior language areas.

CONFABULATION AND RIGHT FRONTAL EMOTIONAL AND PROSODIC SPEECH

Although unable to discourse fluently, individuals with severe forms of expressive aphasia may be capable of swearing, making statements of self-pity, praying, singing, and even learning new songs—although in the absence of music they would be unable to say the very words they had just sung (see Chapter 3). Presumably the ability to produce nonlinguistic and musical/emotional sounds is due to these functions being mediated by the undamaged right hemisphere and limbic structures (see Chapters 3, 5, 7).

Emotional and Prosodic Speech

Although language is usually discussed in regard to grammar and vocabulary, it is also emotional, melodic, and prosodic—features that enable a speaker to convey and a listener to determine intent, attitude, feeling, and meaning (Joseph, 1988a, 1993). A listener comprehends not only what is said, but how it is said—what a speaker feels.

Feeling and attitude are conveyed through the melody (musical qualities), inflection, intonation, and prosody of one's voice, and by varying the pitch, timbre, stress contours, melody, and the rate and amplitude of speech—capacities predominantly mediated by the right half of the cerebrum (see Chapter 3).

Indeed, just as there is a large region within the left frontal convexity (i.e., Broca's area) that subserves the syntactical, temporal-sequential, motoric, and grammatical aspects of linguistic expression, there appears to be a homologous region within the right frontal area that mediates the expression of emotional/melodic speech (Gorelick and Ross, 1987; Joseph, 1982, 1988a, 1993; Ross, 1981; Shapiro and Danly, 1985).

With massive damage involving the right frontal language area, speech may become flat and monotonous; conversely, the

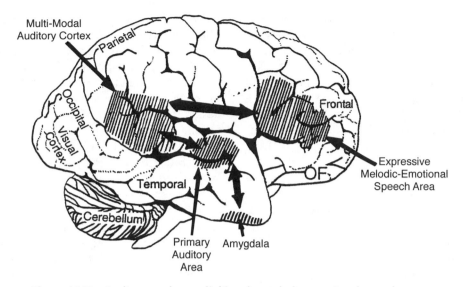

Figure 11.7   Auditory pathways linking the melodic-emotional speech areas.

ability to alter or convey melodic/prosodic elements can become exceedingly abnormal and distorted.

One such patient, following a bullet wound damaging part of the right orbital and convexity region, frequently lost control over his voice and at times sounded as if he were crying, wailing, or screeching. He also suffered some paralysis involving the left extremities. Such patients may lose the ability to engage in vocal mimicry or to accurately repeat various statements in an appropriate emotional manner (Gorelick and Ross, 1987; Joseph, 1988a; Ross, 1981).

With mild damage, rather than severe distortions or a loss of melody, the intonational qualities of the voice can become mildly abnormal and patients may seem to be speaking with an odd Midwestern-like accent, particularly with deep lesions of the right frontal area, perhaps involving the cingulate or basal ganglia. Prosodic distortion in the form of an unusual accent is sometimes seen in seizure disorders involving deep right frontal or frontal-temporal areas.

On the other hand, with left frontal lesions some patients develop what sounds like an unlearned foreign accent, as if they were from Germany, France, etc. (Blumstein et al., 1987; Graff-Radford et al., 1986).

This is due, in part, to distortions involving the pronunciation of vowels.

When damage is limited to this right frontal emotional-motor speech area, the ability to comprehend and understand prosodic-emotional nuances appears to be somewhat intact (Gorelick and Ross, 1987; Joseph, 1988a; Ross, 1981). Nevertheless, since there have been so few studies concerned with these issues, much more research is needed before truly definitive statements can be made.

Tangential and Circumlocutory Speech

With bilateral or right frontal lobe damage, tangentiality is frequently observed in speech (Joseph, 1986a, 1988a). For example, when a patient with severe right orbital damage was asked if his injury affected his thinking, he replied, "yeah—it's affected the way I think—it's affected my senses— the only thing I can taste are sugar and salt—I can't detect a pungent odor—ha ha—to tell you the truth it's a blessing this way" (Blumer and Benson, 1975:197).

When asked what he received for Christmas, one frontal patient replied, "I got a record player and a sweater." (Looking down at his boots.) "I also like boots, westerns, popcorn, peanuts, and pretzels." When asked in what manner an orange and a banana were alike, another right frontal patient replied, "fruit. Fruitcakes— ha ha—tooty fruity." When asked how a lion and a dog were alike, he responded, "They both like fruit—ha ha. No. That's not right. They like trees—fruit trees. Lions climb trees and dogs chase cats up trees, and they both have a bark."

Tangentiality is in some manner related to impulsiveness as well as circumlocution. However, patients with circumlocutious speech often have disturbances involving the left cerebral hemisphere and frequently suffer from word-finding difficulty and sometimes receptive or expressive dysphasia. They experience difficulty expressing a particular idea or describing some need as they have trouble finding the correct words. Thus they seem to talk around the central point and only through successive approximations are able to convey what they mean.

Patients with tangential speech lose the point altogether. Instead, words or statements trigger other words or statements that are related only in regard to sound (e.g., like a clang association) or some obscure and ever-shifting semantic category. Speech may be rushed or pressured and the patient may seem to be free-associating as they jump from topic to topic.

Confabulation

In contrast to left frontal convexity damage, which can result in speech arrest and/or significant reductions in verbal fluency, right frontal damage has frequently been observed to result in speech release, verbosity (i.e., "motor mouth"), tangentiality, and in the extreme, confabulation (Joseph, 1982, 1986a, 1988a, 1993). Reductions in fluency, perhaps secondary to motor or frontal pole involvement, can also occur to a mild degree in some cases. That is, spontaneous speech per se may be reduced. However, once speech is initiated, confabulatory or tangential tendencies may be observed.

When secondary to right (or deep and/or bilateral) frontal damage, confabulation seems to be due to disinhibition. Difficulties in monitoring responses, withholding answers, utilizing external or internal cues to make corrections, or suppressing the flow of tangential and circumstantial ideas (Fischer et al., 1995; Joseph, 1986a, 1988a, 1993; Kapur and Coughlan, 1980; Shapiro et al., 1981; Stuss et al., 1978) such that the language axis of the left hemisphere becomes overwhelmed and flooded by irrelevant associations. Frontal lobe confabulators also sometimes demonstrate marked perseveratory tendencies or difficulty in shifting response sets or in maintaining a coherent line of reasoning. In some cases the content of their speech may be bizarre and fantastic, as loosely associated ideas become organized and anchored around fragments of current experience.

For example, one 24-year-old individual who received a gunshot wound that resulted in destruction of the right inferior convexity and orbital areas, attributed his hospitalization to a plot by the government to steal his ideas (Joseph, 1986a). When it was pointed out that he had undergone surgery for removal of bone fragments and the bullet, he pointed to his head and replied, "that's how they are stealing my ideas."

Another patient, formerly a janitor, who suffered a large right frontal subdural hematoma (which required evacuation) soon began claiming to be the owner of the business at which he formerly worked (Joseph, 1988a). He also alternatively claimed to be a congressman and fabulously wealthy. When asked about his work as a janitor he reported that as a congressman he had been working under cover for the CIA. Interestingly, this patient stated that he realized what he was saying was probably not true. "And yet I feel it and believe it though I know it's not right."

Although diverse forms of cerebral injury may trigger confabulatory responding, this type of disturbance is frequently seen with memory loss (Fischer et al., 1995; Talland, 1961), Korsakoff's psychosis, and the accompanying amnesic syndrome. It is thus noteworthy that persons with Korsakoff's disease and memory loss who were brought to autopsy, were found to have neuronal damage involving predominantly the dorsal medial nucleus of the thalamus (Victor et al., 1989)—a nucleus with major inter-relaying fiber pathways that project to and from the frontal lobes (Pribram et al., 1953; Walker, 1940; Whitlock and Nauta, 1956). Moreover, the dorsal medial nucleus, when damaged, often gives rise to memory disturbances (see Chapter 6) as well as bizarre, delusional, and confabulatory speech (Bogousslavasky et al., 1988).

Similarly, as will be detailed below, the frontal lobes are involved in memory storage and retrieval, including the capacity to temporarily set an idea to the side (so to speak) with the intention to recall and act on it later (Baron et al., 1994; Cockburn, 1995; McAndrews and Milner, 1991; Moscovitch, 1995; Selemon et al., 1995). This includes maintaining visual and verbal representations as well as temporal order.

As a function of memory loss, it appears that when the individual is not consciously aware that he or she no longer remembers, he or she fills the gap in memory with associations and ideas that are in some manner (often tangentially) linked to what is available or to the question being asked (Joseph, 1982, 1986a, 1988a; Talland, 1961). Confabulation due to "gap filling" is also associated with corpus callosum immaturity (Joseph et al., 1984) and right parietal injuries (Joseph, 1986a) such that the left hemisphere is denied relevant input.

As noted, in some respects injuries involving the orbital frontal lobes can result in symptoms similar to those with right frontal injuries, including the production of confabulatory ideation. However, in contrast to right (or bilateral) frontal injuries, which may result in the production of fantastic spontaneous confabulations in which contradictory facts are ignored or simply incorporated, confabulatory responses associated with orbital injuries tend to be more restricted and transitory, and in some cases must be provoked (Fischer et al., 1995).

ORBITAL FRONTAL LOBES AND INFERIOR CONVEXITY
Orbital Frontal Lobes

The orbital frontal lobes receive higher-order sensory information from the sensory association areas throughout the neocortex and maintain rich interconnections with the lateral convexity, anterior cingulate, inferior temporal and inferior parietal lobes, medial walls, parahippocampal gyrus (Fuster, 1980, 1996; Johnson et al., 1968; Jones and Powell, 1970; Pandya and Kuypers, 1969; Van Hoesen, 1996; Van Hoesen et al., 1975), and hypothalamus and amygdala (Van Hoesen, 1996).

The orbital cortices also send fibers directly to the reticularis gigantocellularis of the reticular formation, including the reticular inhibitory regions within the medulla and the excitatory areas throughout the pons (Kuypers, 1958; Rossi and Brodal, 1956; Sauerland et al., 1967). There are also rich interconnections with the medial magnocellular dorsal medial nucleus of the thalamus (Pribram et al., 1953; Siegel et al., 1977), a major relay nucleus involved in the gating and filtering of information destined for the neocortex (Skinner and Yingling, 1977; Yingling and Skinner, 1977) and limbic system (see Chapter 6). That is, this portion of the thalamus appears to exert modulating influences on information reception and processing as well as arousal.

Orbital Mediation of Limbic Arousal

The medial portion of the magnocellular dorsal medial thalamic (MDMT) nucleus (like the orbital region) also receives fibers from the reticular formation and amygdala (Chi, 1970; Krettek and Price, 1974; Siegel et al., 1977). However, this portion of the dorsal medial thalamus is

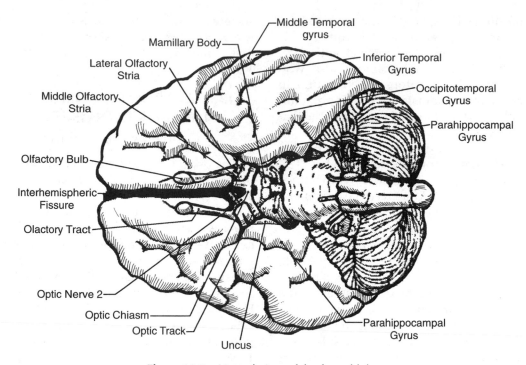

Figure 11.8 Ventral view of the frontal lobes.

subject to orbital control (Joseph, 1990d). Indeed, the orbital region, via its interconnections with the reticular formation, limbic system, MDMT, is able to exert considerable influence on the interactions that take place in these nuclei, including control over various forms of limbic, behavioral, and emotional arousal.

For example, electrical stimulation of the orbital area can cause a hungry animal to stop eating, walk away from its dish, lie down, and even fall into slow-wave synchronized sleep (Lineberry and Siegel, 1971). It can inhibit monosynaptic spinal reflexes (Clemente et al., 1966; Sauerland et al., 1967), as well as reduce and inhibit arousal throughout the neocortex (Lineberry and Siegel, 1971) and limbic system (Steriade, 1964), including the reticular formation (Lineberry and Siegel, 1971; Siegel and Wang, 1974). Indeed, the orbital region appears to exert hierarchical control over the MDMT, reticular formation, limbic system, and autonomic nervous system, thus mediating generalized arousal throughout the neuroaxis. As pertaining to emotional arousal, it has also been postulated that the

orbital area exerts a major influence on the experience of anxiety (Gray, 1981).

Autonomic Influences

Electrical stimulation of the orbital cortex can slow or arrest respiration, alter arterial blood pressure, inhibit pyloric peristalsis, increase salivation, dilate the pupils, decrease gastric motility, and increase skin temperature (Bailey and Sweet, 1940; Chapman et al., 1950; Kaada, 1951, 1972; Livingston et al., 1948; Wall and Davis, 1951). Hence, when the orbital cortex is severely damaged, the autonomic nervous system is liberated from higher-order inhibitory control (Rinkel et al., 1950). Blood pressure and skin temperature are lowered, sweating is increased, and widespread disturbances involving gastrointestinal activities, salivation, micturition, diuresis, and sexual arousal can occur (Chapman et al., 1950; Delgado and Livingston, 1948; Langworthy and Richter, 1939; Mettler et al., 1936; Rinkel et al., 1950). Urinary bladder control is also diminished, resulting in a disturbance of micturition.

It is thus easy to see how activity in the orbital area, i.e., the "Senior Executive" of the subcortical and limbic brain, can affect gastric functioning, including the possible development of ulcers (Freeman and Watts, 1942), as well as contributing to many of the symptoms that underlie feelings of anxiety.

Emotional Unresponsiveness

Usually immediately follow orbital or surgical destruction of the frontal lobes (e.g., frontal lobotomy), there result severe reductions in activity and emotional/motivational functioning, including apathy, indifference to loud noises or threats, and extreme reductions in arousal and motor functioning, (Butter et al., 1970; Freeman and Watts, 1942, 1943). With extensive orbital destruction involving portions of the medial wall and anterior tip of the cingulate, reduced responsiveness persists, and patients will sit quietly and silently, nearly motionless, making little or no attempt to communicate as if mute. The ability to respond socially and emotionally seems abolished. Rather, if sufficiently stimulated, humans and animals seem capable of reacting only in an irritable and aversive manner (Butter et al., 1968), and many may appear depressed (Grafman et al., 1986).

In free-ranging monkeys as well as those reared in enclosed settings, global social disintegration is observed following orbital destruction. Animals cease to groom or produce appropriate vocalizations, adults attempt to completely avoid members of their social group, and mothers neglect and rebuff their infants (Myers et al., 1973; Raleigh, 1976, cited by Kling and Steklis, 1976). Similarly, human females with surgical destruction of this vicinity will neglect and/or strike or beat their children without provocation (Broffman, 1950).

In the free-ranging situation, animals with orbital destruction in fact ran away from their social groups and remained solitary until their deaths (Myers et al., 1973). In contrast, lateral convexity lesions do not result in changes involving social proximity, grooming, or bonding (Kling and Mass,

1974). Hence, complete orbital destruction results in complete abolition of most forms of emotional and social behavior.

Emotional Disinhibition

With less extensive damage, or as swelling and ischemia become reduced so that neighboring structures are not affected, this initial unresponsive phase may pass. Rather than a loss of emotion there is a loss of emotional control and the subject becomes disinhibited, hyperactive, euphoric, extroverted, labile, and overtalkative, and develops perseveratory tendencies (Butter, 1969; Butter et al., 1970; Dax et al., 1948; Greenblatt, 1950; Kennard et al., 1941; Kolb et al., 1974; Reitman, 1946, 1947; Ruch and Shenkin, 1943). Patients are frequently described as markedly irresponsible, antisocial, lacking in tact or concern, and having difficulty planning ahead or foreseeing consequences, and they suffer from generalized disinhibition. There can result tendencies toward impulsive actions, to laugh inappropriately and make trivial jokes, or to behave in a demanding or transiently aggressive manner. Proneness to criminal behavior, promiscuity, grandiosity, and paranoia have also been observed (Blumer and Benson, 1975; Benson and Geschwind, 1971; Lishman, 1973; Luria, 1980; Stuss and Benson, 1984). In this regard it is noteworthy that murderers show reductions in frontal lobe glucose metabolism (Raine et al., 1994).

Much of this inappropriate behavior is a consequence of a loss of control over reticular and limbic structures. That is, the orbital region appears as a kind of "censor" or even "conscience" such that the removal of these "superego"-like influences results in heightened levels of generalized and emotional arousal and the patient responds in a childish, inappropriate, and unrestrained manner (Joseph, 1986a).

With less extensive damage, or when the lesions are confined to one hemisphere, the long-term effects are less drastic. However, right orbital damage seems to result in the most severe alterations in mood and emotional functioning (Grafman et al., 1986).

Attention

Among primates and mammals, disturbances involving visual, spatial, auditory, tactile, and olfactory discrimination have consistently been observed with orbital lesions (Fuster, 1980, 1996; Oscar-Berman, 1975) as well as with frontal lesions in general. That is, because of heightened generalized arousal levels (resulting from orbital and inferior convexity release of the reticular formation and other nuclei), competing perceptions do not seem to stand out enough to attract and shift attention, i.e., there is a reduced capacity to deactivate whatever perceptual and behavioral activity is predominant such that the ability to differentiate and attend to salient versus irrelevant stimuli is reduced (Como et al., 1979).

Moreover, the ability to shift attention may be reduced. That is, in a manner analogous to magnetic motor behaviors following medial lesions, orbital destruction may lead to attentional stimulus binding such that the subject appears "stuck in set" and may demonstrate perseverative attentional activity and difficulty shifting responses. However, if the patient or orbitally lesioned animal is purposefully distracted, e.g., via a novel stimulus, this pattern of perseverative attention is momentarily halted and the ability to shift response and attention is briefly regained (Mishkin, 1964; Pribram et al., 1964). Nevertheless, lesions confined to the orbital area do not result in distractibility per se.

Perseveration

Lesions involving the orbital frontal cortex and inferior convexity have consistently resulted in increased activity levels and an abnormal tendency to repeat previous responses in a repetitive, perseverative fashion even when the context is no longer appropriate or rewarded and/or the response is punished and the individual realizes that his or her responses are incorrect (Butter, 1969; Butter et al., 1963; Iversen and Mishkin, 1970; Jones and Mishkin, 1972; Kolb et al., 1974; Luria, 1980; Mishkin,

1964)—i.e., the capacity to shift responses is attenuated. Thus, once a behavior is completed, particularly if it is repeatedly performed, the pattern continues to be involuntarily executed such that the ability to change to a different pattern of activity is disrupted. Similar disturbances occur with inferior medial damage.

Perseverative, repetitive abnormalities can affect motor behavior. For example, a common feature of large lesions involving the neocortical regions is macrographia and expansiveness, which can be seen in drawings that are made abnormally large. Conversely, deep frontal lesions may produce micrographia, such that drawings become abnormally small. However, in some instances, micro- or macro-graphia may be also be contaminated with perseverative abnormalities.

Perseverative tendencies also influences speech. For example, patients may tend to repeat phrases: "Doctor, can I look at this, can I look at this, Doctor, can I look at this." Or once a topic has changed they may reintroduce and again repeat certain statements or words. During the administration of the Vocabulary subtest from the WAIS-R, for example, an orbitally damaged patient managed to use the word "summer" in five different definitions.

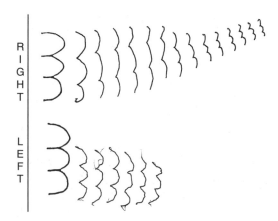

Figure 11.9 Example of micrographia associated with deep frontal lesions. Patient is instructed to draw a figure identical to and the same size as the "sample" and to then draw it over and over again.

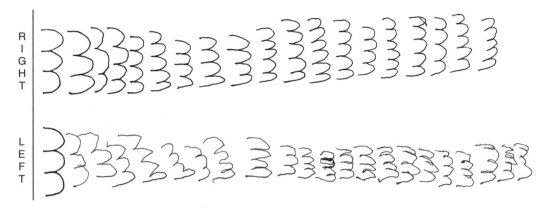

Figure 11.10   Micrographia coupled with perseveration.

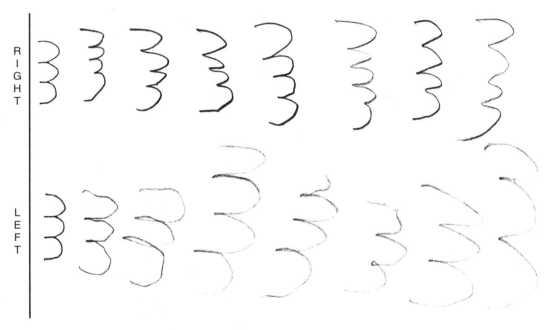

Figure 11.11   Macrographia coupled with perseveration. Associated with neocortical lesions of the frontal lobe.

Another patient with damage and suspected subclinical seizures involving the mesial orbital cortex and possibly the basal ganglia demonstrated a striking perseveration of movement. If asked to draw a star he would dash off four or five. When he attempted to sit, the downward motion continued in a repetitive, machine-like fashion and he would slide to the floor.

In severe cases it is frequently found that once a patient has performed a required action, such as drawing a particular figure

several times, he or she will continue to draw the first when asked to draw a different figure. If asked to draw a star, a cross, and a square, the patient may draw the first or even the second figure correctly but then impulsively draw the star again. In less extreme cases, patients may simply inappropriately introduce components of a previous response in their next set of actions.

In one instance, a patient with bilateral damage was asked to draw "a pair of spectacles" and did so correctly. When asked to

Figure 11.12   "The M and N test." Patient is instructed to write "m and n and m and n" until they reach the end of the paper. Note perseverations.

Figure 11.13   Examples of macrographia coupled with perseveration.

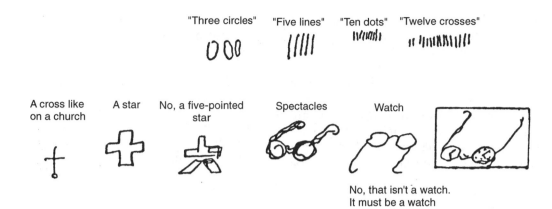

Figure 11.14   Examples of perseveration in a patient following damage to the frontal lobes. (From Luria A. Higher Cortical Functions in Man. New York: Basic Books, 1980.)

draw a watch, he again drew the spectacles. When the mistake was pointed out, he again drew a pair of spectacles but this time drew a watch in the center of one of the lenses (Luria, 1980).

Conceptually, patients may have difficulty considering or even recognizing alternatives such that they seem to become locked into a particular mode of thinking or activity. For example, Luria (1980) describes "a patient with a wound of the frontal lobes, who, when working in the carpenter's shop of the hospital, inertly went on planing a piece of wood until nothing of it remained. Even then he did not stop but continued to plane the bench" (p. 294).

Although orbital patients may have difficulty inhibiting emotionality, the perseverative abnormality is not always due to disturbances involving impulsivity or the withholding of a behavioral response. For example, across tasks requiring delayed responding, orbital lesions have little or no effect on performance (Brutkowski et al., 1963; Rosenkilde, 1979). Rather, the problem is in shifting sets and in inhibiting the recurrence of a previous response when the next action is initiated. Once a behavior occurs it tends to persist and contaminates the performance of unrelated actions.

Perseverative reactions, however, may occur with lesions situated well outside the orbital areas. For example, perseveration of speech (of which there are several sub-

types) can also occur among persons with aphasia and with lesions localized to the right or left frontal lobe or damage situated throughout the left hemisphere (Joseph, 1986a; Pietro and Rigdrodsky, 1986; Sandson and Albert, 1987). In these instances, as patients grope for words, the same word may be repeated in a number of erroneous contexts in a perseverative fashion.

Again it must be stressed, however, that lesions are seldom confined to a single quadrant of the frontal lobe, even following "frontal lobotomy"—which during the 1940s and 1950s was often performed as an in-office procedure and was simply quite sloppily performed. Moreover, damage in one frontal area may result in abnormal activity in yet a different region of the frontal lobe because of disconnection and loss of input or inhibitory restraint. Hence, symptoms associated with orbital destruction may also appear with medial or lateral convexity injuries. In addition, it appears that the right orbital region performs somewhat different functions than the left orbital areas such that different symptoms may appear with right versus left lesions (Grafman et al., 1986; see below).

LATERAL FRONTAL NEOCORTICAL REGULATION

With the exception of olfactory information, which via the olfactory tracts projects to the limbic system and is relayed to the orbital region (Cavada, 1984), all sensory impulses are first transferred to the thalamus before being transmitted to the primary auditory, visual, and somesthetic receiving areas. From the primary zones this information is sent to three separate major locations: to the immediately adjacent sensory association area, back to the thalamus, and to the motor cortex of the frontal lobes.

The motor area then relays this information to the lateral convexity, which simultaneously receives fiber projections from the sensory association areas (Cavada, 1984; Jones et al., 1978; Jones and Powell, 1970; Pandya and Kuypers, 1969) and the inferior parietal lobule. Hence, the frontal cortex is "interlocked" with the posterior sensory areas via converging and reciprocal connections with the first, second, and third levels of modality specific analysis, including the multimodal associational integration performed by the inferior parietal

lobule. It can therefore sample activity within all cortical sensory/association regions at all levels of information analysis.

Frontal-Thalamic Control of Cortical Activity

The role of the lateral convexity is not limited to sampling, but also involves regulation of information flow to and within the neocortex. This is accomplished, in part, via projections linking the frontal lobes with the dorsal medial thalamic nucleus.

RETICULAR THALAMUS

Fibers passing to and from the thalamus and the cortical sensory receiving areas give off collaterals to the reticular thalamic nucleus—which in addition sends fibers that envelop and innervate most of the other thalamic nuclei (Scheibel and Scheibel, 1966; Updyke, 1975). The reticular thalamus acts to selectively gate transmission from the thalamus to the neocortex

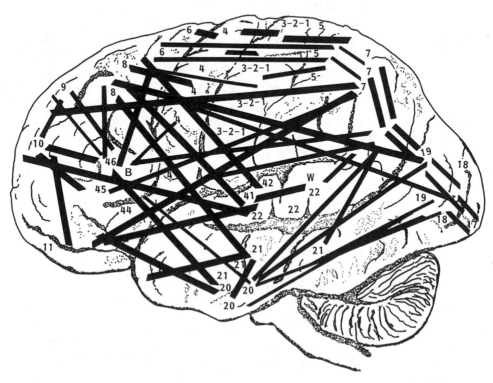

Figure 11.15   Schematic depiction of pathways interlinking the various regions of the brain to the frontal lobe. Numbers refer to Brodmann's areas.

and continually samples thalamic-cortical activity (Skinner and Yingling, 1977; Yingling and Skinner, 1977).

Lateral Frontal-Thalamic Arousal and Perceptual Steering

The reticular thalamus is controlled by the lateral convexity of the frontal lobes, the reticular formation, and the lateral portion of the dorsal medial thalamus with which it maintains dense interconnections (Skinner and Yingling, 1977; Yingling and Skinner, 1977).

The convexity and lateral dorsal medial nucleus (LDM) are also richly interconnected and together exert significant steering influences on the reticular thalamus. That is, the lateral frontal convexity appears to exert specific influences on the LDM so as to promote or diminish the flow of information to the cortex and thus modulate specific perceptual and cognitive activities occurring within the neocortex— activity that it is simultaneously sampling. This is in contrast to the orbital region with its connections to the reticular formation and the medial magnocellular segment of the dorsal medial thalamus, and its influences on generalized arousal and limbic activation/inhibition.

To recapitulate, the lateral frontal system can influence cognitive/perceptual cortical functioning via the sampling of activity occurring throughout the neocortex at all levels of informational analysis, and via its modulating influences on the lateral portion of the dorsal medial and reticular thalamic nuclei. The lateral frontal region thus can act at any stage of processing, from initial reception to motor expression, so as to facilitate or inhibit further analysis, selectively acting to determine exactly what type of processing occurs throughout the neocortex.

Via integration and inhibitory action and through its neocortical and thalamic links, the lateral convexity is able to coordinate interactions between various regions of the neuroaxis so as to organize, mobilize, and direct overall cortical and behavioral activity and to minimize conflicting demands, impulses, distractions, and/or the processing of irrelevant information.

When damaged, depending on the site (e.g., inferior versus posterior convexity) or laterality of the lesion, there can result behavioral disinhibition, flooding of the sensory association areas with irrelevant information, hyperreactivity, distractibility, memory loss, impulsiveness, and/or apathy, reduced motor-expressive activities (e.g., speech arrest), and sensory neglect (Como et al., 1979; Joseph, 1986a, 1988a, 1990a; Joseph et al., 1981). Similar disturbances can result when the dorsal medial nucleus or the bidirectional pathways linking the thalamus and frontal lobe are severed (Graff-Radford et al., 1990; Skinner and Yingling, 1977; Victor et al., 1989).

Hence, in summary, the orbital region exerts modulating influences on subcortical and generalized limbic arousal, whereas the lateral regions are more concerned with the control of information processing, attention, and arousal, as well as information storage and retrieval at the neocortical level. In consequence, when the lateral regions are injured, selective attention and memory functioning may become impaired such that patients may become distractible and disinhibited and have difficulty keeping their mind on certain tasks and/or recalling and acting upon events planned for the future.

THE INFERIOR AND LATERAL FRONTAL LOBES AND ATTENTION
Disinhibition and Response Suppression

The inferior and lateral convexity appears to be highly involved in the inhibition of behavior and the ability to withhold or delay responses, which in turn is in part a function of its involvement in controlling neocortical perceptual activity and arousal (Como et al., 1979; Joseph 1990d; Joseph et al., 1981). Hence, electrophysiological analysis of cellular activity within the lateral convexity indicates that many neurons alter their discharge rates when a subject is required to wait before responding to a signal. Yet others increase or decrease their activity as the time interval between the onset of the delay and the release of the response increases (Fuster et al., 1982). Most of these delay neurons are found within the inferior convexity. However, a number of neurons in the superior convexity show similar properties (Pragay et al., 1987).

Similarly, high-frequency electrical stimulation of the lateral and inferior convexity has been shown to disrupt the ability to inhibit, delay, and withhold responses (Goldman et al., 1970; Gross and Weiskrantz, 1964; Stamm and Rosen, 1973), whereas low-frequency stimulation actually improves performance on delayed response tasks and enhances behavioral inhibition (Wilcott, 1974, 1977). Interestingly, electrical stimulation of the right frontal region, as compared to the left, more greatly disrupts delayed response performance.

Conversely, when the lateral convexity and inferior convexity are damaged there results a consistent disturbance across tasks requiring the withholding and delay of a response (Brutkowski et al., 1963; Gross and Weiskrantz, 1964; Mishkin and Pribram, 1956; Stepien and Stamm, 1970). That is, subjects become disinhibited and impulsive—disturbances that in turn affect all aspects of behavior. Patients may spontaneously speak or make comments "without thinking" and act on sudden impulses without regard for consequences. Depending on the extent of the lesion, both emotion and cognitive activity can be affected (to be discussed).

In addition, humans and animals tend to become hyperreactive and may demonstrate increased activity levels (Bradford, 1950; French, 1959; Fuster, 1996; Joseph, 1986a, 1988a; Latto and Cowey, 1971a; Rose, 1950). For example, noise, threat, or novel stimuli result in significantly heightened activity in frontal animals coupled with distractibility (French, 1959). Among humans a perpetual shifting of attention may result as they are inordinately distracted by noises in the hall or even specks on the testing table. However, if potential distractors are re-

moved or the subject is placed in a darkened room, activity and hyperresponsiveness declines (French, 1964; French and Harlow, 1955).

With more restricted or lateralized lesions, long-term effects are more distinct. For example, right-sided damage is more frequently associated with motor decontrol, such as edginess (Grafman et al., 1986) and even mania (to be discussed). In addition, right-sided injuries may give rise to profound attentional and memory deficits, including transient global amnesia (Baron et al., 1994).

Attention

As noted, the frontal lobes are highly involved in attentional functioning (Como et al., 1979; Crowne, 1983; Fuster, 1980, 1996; Joseph, 1986a, 1988a; Joseph et al., 1981; Knight et al., 1981; Luria, 1980; Pragay et al., 1987), and attentional disturbances are frequently associated with frontal lobe lesions. Some patients, although fully alert and oriented, are easily distracted and show wandering or a perpetual shifting of attention (Bianchi, 1922; Stuss and Benson, 1984). They may seem distracted by noises in the hall, specks on the testing table, or extraneous objects around the room.

Others may seem easily overwhelmed by complexity or behave as if their sensory-perceptual capacities were significantly narrowed (Yarcorzynski and Davis, 1942). In severe cases attention may be focused for only short time periods. For example if asked to count, they may stop after reaching 10 or 15 (Rose, 1950) and then must be prodded to continue.

On the other hand, some patients seem remarkably able to maintain directed attention, at least when performing simple tasks. Hence, some frontal patients can perform tasks such as digit span without difficulty (Benson et al., 1976; Partridge, 1950; Petrie, 1952; Stuss et al., 1978). Indeed, patients may seem to be locked into this as if all potentially interfering stimuli were completely blocked out, e.g., perseverative attention. Nevertheless, although repetition of digits may be normal, or even well above average, when required to recite digits backwards performance often is abnormal (Partridge, 1950; Petrie, 1952). Hence, although able to attend (or at least echo what has been said), these patients often have a disturbed ability to maintain concentration.

As might be expected, a secondary consequence of attentional abnormalities are disturbances of memory. A patient who is not paying attention is not going to remember.

Memory is not always significantly affected, however (Delaney et al., 1980), which in turn may be due to the location and extent of the lesion. For example, lesions that destroy not only frontal tissue but encroach upon the dorsal medial thalamus may produce profound memory disturbances, including transient global amnesia (Baron et al., 1994; Graff-Radford et al., 1990; Victor et al., 1989).

THE FRONTAL LOBES, THE DORSAL MEDIAL THALAMUS, AND MEMORY

The dorsal medial nucleus of the thalamus, as well as the frontal lobes, are exceedingly important in memory functioning and information retrieval (Aggleton and Mishkin, 1983; Graff-Radford et al., 1990; Joseph, 1990d; Squire, 1992; Victor et al., 1989), and both nuclei are intimately linked (see Chapter 6).

As noted above and detailed in Chapter 6, the frontal lobes, in conjunction with the hippocampus (Goldman-Rakic, 1990), the reticular activating system, the amygdala, and the dorsal medial thalamus (DMT), act to gate and direct perceptual and cognitive activity occurring within the neocortex. That is, the frontal lobes, DMT, amygdala, and hippocampus interact when determining which neuronal circuits are created, linked together, inhibited, or activated.

Moreover, the frontal lobes are highly concerned with activation and neuronal inhibition as well as sustaining the creation of specific reverberating neural circuits. They are also particularly important in the

capacity of retrieving previously stored information, as well as preventing access to those memories and perceptions deemed irrelevant or undesirable (see Chapter 15).

Presumably the frontal lobes can retrieve or inhibit particular memories by acting directly on the neurons where they are maintained, or the frontal lobes may act indirectly via the thalamus and reticular formation through which they can selectively affect arousal. In this manner, the frontal lobes can selectively sample, deactivate, or enhance neural activity depending on the information they are searching for (Joseph, 1990d). Moreover, in this manner future intentions and plans may be temporarily stored (such as when a person is distracted or engaged in a different task) and then later recalled and acted upon (Cockburn, 1995).

Conversely, damage to the frontal lobes can result in difficulty in retrieving memories but may have little effect on recognition memory (Graff-Radford et al., 1990; Jetter et al., 1986; Squire, 1992) wherein search, activation, and retrieval are not as important. Nevertheless, this is not always the case and even recognition memory may remain impaired.

For example, one such patient was not only unable to recall that he had been a patient at the hospital for over a month, but repeatedly claimed to be working as a janitor, and that he would be "going home at quitting time." However, when he was shown his hospital room and clothes, although he recognized the items as his, he persisted in claiming he was at work and that his clothes were in the hospital room so he could change before leaving for home.

Memory Search and Retrieval

The frontal lobe, hippocampus, and the amygdala appear to share major roles in regard to neocortical activation and thus memory storage and retrieval, and all appear to partly rely on the DMT and anterior thalamus in this regard. The DMT (and the thalamus in general) is highly important in transferring information to the

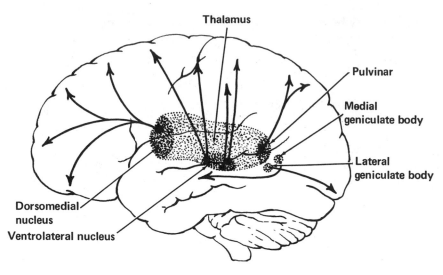

Figure 11.16   Diagram of principle thalamocortical projections. (From Waxman SG, De Groot J. Correlative Neuroanatomy, 22nd Ed. New Jersey: Appleton & Lange, 1994.)

neocortex, and is involved in perceptual filtering and exerting steering influences on the neocortex so as to direct attention. When the DMT is damaged, there can result a dense verbal and visual anterograde amnesia (Graff-Radford et al., 1990; von Cramon et al., 1985) as memories cannot be selectively stored or activated.

The DMT and anterior thalamus are able to perform these functions via inhibitory and excitatory influences (mediated by the frontal lobe) on neurons located in the neocortex, limbic system, and reticular activating system. As noted, the DMT and frontal lobes monitor and control all stages of information analysis, and ensure that relevant data and associations are shifted from primary to association to multimodal association areas to permit further processing. Conversely, the frontal-thalamic system can act to inhibit further processing and/or prevent this information from reaching the neocortex in the first place.

Through the widespread interconnections maintained both by the frontal lobes and the thalamus, specific neuronal networks and perceptual fields can be selectively inhibited or activated, so that information may be processed or ignored, and so that conjunctions between different neocortical areas can be formed. These nuclei therefore directly mediate information processing, memory storage, and the creation of or additions to specific neural networks via the control of neocortical activity.

Hence, when the frontal lobes, DMT, and anterior thalamus are damaged, the ability to shift from one set of perceptions to another, or to selectively retrieve a specific memory, or to inhibit the activation of irrelevant ideas, associations, memories, and thoughts is significantly affected and the patient may confabulate. The ability to alternate between and thus selectively activate specific neural networks is also disrupted, such that patients have difficulty in shifting from one train of thought to another. That is, they may tend to become stuck in set and to engage in recurrent perseveration (Joseph, 1986a, 1990d; Vikki, 1989) such that the same speech patterns or "memories" may be repeated unless something occurs to arrest and shift their attention.

Hence, with frontal-thalamic damage, retrieval is impaired, as is the ability to selectively recall or shift between old and recent memories, and to keep track of the order in which they may have been formed, e.g., morning, evening, yesterday, last week, five years ago, and so on (see Graff-Radford et al., 1990; Squire, 1992; Tal-

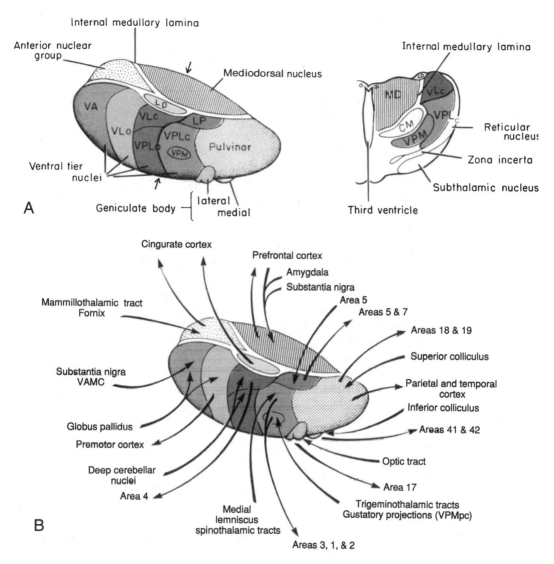

Figure 11.17   Schematic diagram of the major thalamic nuclei and their principal efferent and afferent projections. (From Carpenter M. Human Neuroanatomy. Baltimore: Williams & Wilkins, 1991.)

land, 1961). Hence, a patient may deny possessing a certain memory of an event that occurred five weeks ago, because he is searching his memory store for events that were experienced five years ago.

For example, one patient with severe frontal lobe damage claimed to have never been married and laughingly indicated that he never wanted to get married and that marriage was a bad idea. When pressed as to why, he exclaimed that "the wife may just up and die on you," and

then began to laugh quite loudly and then just as suddenly began to sob and cry. When asked why he was crying, he claimed not to know why. However, when he was reminded that he had been married, he was suddenly able to recall that his wife (of two years) had been killed in the same accident that resulted in his injuries. Another patient with anterior and dorsal medial thalamic damage erroneously stated that the president was Lyndon Johnson. However, when told the correct date ("1982"), she correctly named Ronald Regan (Graff-Radford et al., 1990).

Thus "time sense" and the capacity to selectively search memory may be disrupted with frontal-dorsal medial dysfunction—a condition that is also characteristic of Korsakoff's syndrome. For example, Talland (1961:375) described one woman with Korsakoff's disease who "after ten year's hospitalization continually maintained that she has been brought in the previous day for observation." When questioned she also argued that she still lived with her husband in a fancy hotel, where she had lived "the life of a well-to-do lady of leisure. She seemed unaware that the conditions of that very agreeable life had long since ceased to operate."

Indeed, persons suffering from Korsakoff's syndrome demonstrate profound disturbances of memory, including a dense verbal and visual anterograde amnesia (Aggleton and Mishkin, 1983; Graff-Radford et al., 1990; Victor et al., 1989; von Cramon et al., 1985) and in some cases a more temporally graded memory loss for the remote past (Squire, 1992)—although the hippocampus is intact. Rather, in Korsakoff's syndrome damage typically involves the dorsal medial nucleus of the thalamus and atrophy of the frontal lobes (Aggleton and Mishkin, 1983; Graff-Radford et al., 1990; Mair et al., 1979; Victor et al., 1989; von Cramon et al., 1985; Wilkinson and Carlen, 1982).

In summary, when the frontal-DMT system is damaged, the ability to shift from one memory storage depot (or neural network, or train of thought) to another is disrupted such that, although the memory exists, the patient may not be able to gain access to it—even when provided assistance.

OVERVIEW

Lesions of the supplementary, secondary, and primary motor areas give rise respectively to compulsive utilization and magnetic groping and grasping, diminished fine motor functioning, and paralysis. Damage

to the posterior (particularly the left) lateral convexity involving Broca's speech area, Exner's writing area, or the frontal eye fields reduces motor responsiveness and behavioral expression, i.e., aphasia, agraphia, apraxia, and deficiencies involving visual search and orienting. With inferior convexity and right frontal lesions there is increased responsiveness and impulsivity. Perseverative aberrations are found with orbital, inferior medial, and inferior convexity lesions, whereas deep medial-orbital damage can also result in compulsive utilization behavior. Medial lesions involving the corpus callosum may result in complex involuntary behavior acts, i.e., the "alien hand." Hence, regardless of where a frontal lesion occurs, motor functioning is in some manner affected.

Damage to the medial wall produces symptoms that in some respects parallels those associated with convexity damage. For example, lesions to the primary motor strip cause paralysis of movement, whereas damage to the medial wall and cingulate can result in paralysis of will. Damage involving the inferior convexity is associated with disinhibition and impulsiveness, whereas destruction of the inferior medial wall is associated with compulsive utilization. When the orbital area is also compromised, disinhibition, manic excitement, and internal utilization behaviors (increased sexuality, orality) may occur.

Disturbances involving emotion can result with medial cingulate or convexity damage. With complete destruction of the orbital area, emotional and social functioning is abolished, but with less extensive damage, rather than a loss of emotion there is a loss of emotional control. These findings suggest that the orbital region acts as the senior executive of the social-emotional brain and exert tremendous inhibitory as well as expressive influences on emotion and generalized arousal through its massive interconnections with various limbic nuclei, the dorsal medial nucleus of the thalamus, and the reticular formation.

The orbital region also seems to coordinate and integrate emotional and perceptual activity. For example, fibers from the sensory association areas, including the inferior parietal lobule, project into the orbital frontal cortex. Hence, the orbital region is continually informed as to the perceptual processing being performed in the neocortex and can thus integrate and assign emotional-motivational significance to cognitive impressions; the association of emotion with ideas and thoughts. It is in this manner that thoughts can come to be upsetting or emotionally arousing. It is also thus that disturbances such as anxiety can result (Gray, 1987).

The medial frontal cortex and cingulate are also involved in these processes. Hence, with complete destruction of these areas emotion can no longer be assigned to external events, thoughts, or ideas, and the patient seems apathetic and blunted.

The lateral convexity is much less concerned with emotional and motivational functioning. Rather, the domain of the convexity appears to involve the integration and coordination of perceptual and cognitive processes with volitional-expressive activities. In addition, the lateral convexity is highly involved in monitoring and exerting steering influences on neocortical perceptual activity and in this regard influences attention, as well as memory storage and retrieval.

PERSONALITY AND BEHAVIORAL ALTERATIONS
The Frontal Lobe Personality

Lesions involving the nonmotor frontal regions are usually not accompanied by gross disturbances of visual, auditory, and tactile sensation. However, attentional functioning may become grossly compromised, behavior may become fragmented, and initiative and the general attitude toward the future may be lost. The range of interests may shrink and patients may be unable to adapt to new situations, carry out complex, purposive, and goal-directed activities, or evaluate their attempts (Fuster, 1996; Freeman and Watts, 1942; Girgis, 1971; Hecaen, 1964; Joseph, 1986a; Luria,

1980; Passingham, 1993; Petrie, 1952; Rylander, 1939; Tow, 1955).

With massive trauma, stroke, neoplasm, or surgical destruction (i.e., frontal lobotomy), patients may show a reduction in activity and take very long to achieve very little. They may be unconcerned about their appearance and their disabilities, and demonstrate little or no interest in self-care or the manner in which they dress, or even if their clothes are soiled or inappropriate (Bradford, 1950; Broffman, 1950; Freeman and Watts, 1942; Petrie, 1952; Strom-Olsen, 1946; Tow, 1955). Although often demonstrating restlessness, they may tire easily and show careless work habits and a desire to get things over and done with quickly. They may in fact immediately develop a tendency just to lie in bed unless forcibly removed (Broffman, 1950; Freeman and Watts, 1942, 1943; Rylander, 1948; Tow, 1955), a condition that may pass with recovery.

Even patients with mild and subtle damage may seem to take hours to get dressed, to finish their business in the bathroom, or to shop and purchase simple items. A curious mixture of obsessive compulsiveness and passive aggressiveness may be suggested by their behavior.

In severe cases compulsive utilization may occur, as well as distractibility and perseveration. As described by Freeman and Watts (1943) regarding their lobotomy cases, patients may pick up and play with various objects for long periods or spend hours in the bathtub playing with the bubbles. "Sometimes a pencil and a piece of paper will be enough to start an endless letter that may end up with the mechanical repetition of a certain phrase, line after line and even page after page" (Freeman and Watts, 1943:801). Not all patients behave this way, however.

Even with mild injuries to the frontal lobes patients may initially demonstrate periods of tangentiality, grandiosity, irresponsibility, laziness, hyperexcitability, lability, personal untidiness and dirtiness, poor judgment, irritability, fatuous jocularity, and tendencies to spend funds extravagantly. Failure to be concerned about consequences, tactlessness, and changes in sex drive and even hunger and appetite (usually accompanied by weight gain) may occur. Disturbances of attention, perseverative tendencies, and a reduction in the ability to produce original or imaginative thinking, as well as fantasy, seem to be associated with mild or severe cases, at least initially following injury.

Urinary Incontinence

Urinary and (in rare instances) fecal incontinence are associated with frontal lobe damage, as are excessive eating, drinking, and smoking (Bianchi, 1922; Fulton et al., 1932; Freeman and Watts, 1942; Langworthy and Richter, 1939; Greenblatt, 1950; Rose, 1950; Watts and Fulton, 1934). Patients may freely urinate upon themselves while lying in bed, sitting in front of the television, or while at the dinner table, and seem totally unconcerned about their condition. One patient, standing at her front door, emptied her bladder while waving good-bye to a friend. She stopped on request but completed the act as she walked to the toilet (Rose, 1950).

Apparently urinary bladder capacity is significantly reduced after frontal damage, and with severe destruction (such as following lobotomy) the bladder may become spastic and hypertonic (Rinkel et al., 1950). This is due to a loss of frontal inhibitory control. However, even in mild cases patients may report occasional instances of involuntary urination.

Alterations in Appetite

With bilateral or anterior orbital damage appetite may become exceedingly excessive, and patients (or animals) may eat two or three times their normal amount (Anand et al., 1958; Bianchi, 1922; Fulton et al., 1932; Langworthy and Richter, 1939). One patient I examined who suffered a series of small bilateral strokes gained over 150 pounds within 8 months. A second patient with a degenerative disturbance involving the inferior medial and orbital area, in addition to developing perseveration and compulsive utilization, gained over 200 pounds within a year and a half. Indeed, it has been reported that in some patients the craving for food became so intense that they repeatedly crammed their mouths with food, became cyanotic, and died with their mouth still full (Greenblatt, 1950).

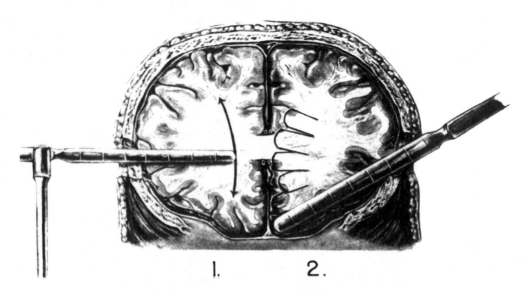

Figure 11.18 Frontal lobotomy. Destruction of the lateral fibers via a "lateral approach." (From Freeman W, Watts JW. Psychosurgery in the Treatment of Mental Disorders and Intractable Pain. Springfield, IL: Charles C Thomas, 1942.)

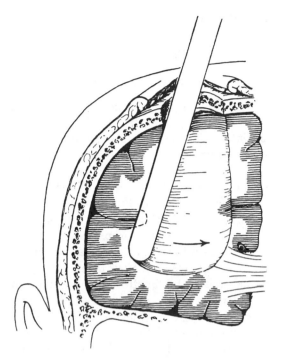

Figure 11.19   Frontal lobotomy. Destruction of the medial fibers via a "superior" approach. (From Poppen JL. Technic of prefrontal lobotomy. J Neurosurg 5:514, 1948.)

In other instances patients might eat non-nutritive objects such as cigarette butts or drink enormous amounts of fluid. Hence, hyperphagia, extreme orality, even copro-phagia may occur (Butter et al., 1970; Butter and Snyder, 1972). Similar disturbances result from hypothalamic/amygdala damage (Joseph, 1992a).

Disinhibition and Impulsiveness

Following massive frontal injuries patients may become emotionally labile, irritable, euphoric, aggressive, and quick to anger, and yet be unable to maintain a grudge or sustain other emotions and mood states (Bradford, 1950; Greenblatt, 1950; Joseph 1986a; Rylander, 1939; Strom-Olsen, 1946). They may become unrestrained, overtalkative, and tactless, saying whatever "pops into their head" with little or no concern as to the effect their behavior may have on others or what personal consequences may result (Bogous-slavasky et al., 1988; Broffman, 1950;

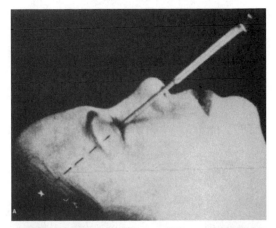

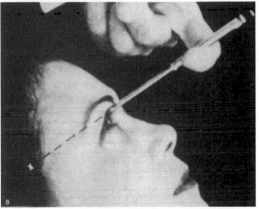

Figure 11.20   Frontal lobotomy. Destruction of orbital fibers via an inferior transorbital approach. Freeman often coupled this type of lobotomy with electroconvulsive shock. According to Freeman the "shock treatment disorganizes the cortical patterns that underlie the psychotic behavior, and by severing the connections between the thalamus and frontal pole the pattern is prevented from reforming." (From Freeman W. Transorbital leucotomy. Proc R Soc Med 42:8, 1949.)

Freeman and Watts, 1943; Joseph, 1986a; Luria, 1980; Miller et al., 1986; Partridge, 1950; Rylander, 1939, 1948; Strom-Olsen, 1946).

Patients may seem inordinately disinhibited and influenced by the immediacy of a situation, buying things they cannot afford, lending money when they themselves are in need, and acting and speaking "without thinking." Seeing someone who is obese they may call out in a friendly manner, "Hey, fatty" and comment on their presumed eating habits. If they enter a

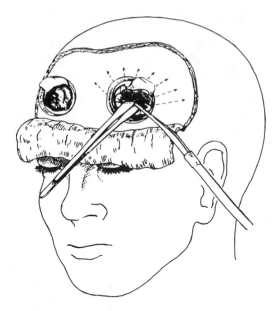

Figure 11.21 Orbital undercutting. The frontal lobes were lifted upward via a spatula and an incision was made so as to cut the medial orbital fibers. (From Asenjo A. Neurosurgical Techniques. Springfield, IL: Charles C Thomas, 1963.)

room and detect a faint odor, it's "Hey, what stinks" or "Who farted?"

In many instances following severe injuries there may be periods of gross disinhibition that may consist of loud, boisterous, and grandiose speech, singing, yelling, beating on trays, or even destruction of furniture and the tearing of clothes. Some patients may impulsively strike doctors, nurses, or relatives and thus behave in a thoroughly callous and irresponsible manner (Freeman and Watts, 1942, 1943; Strom-Olsen, 1946). One patient, following resection of a tumor involving the right frontal area, attempted to throw a fellow patient's radio through the window because he did not like the music. He also loudly sang opera in the halls. Indeed, during the course of his examination he would frequently sing his answers to various questions (Joseph, 1986a).

Disturbances involving impulsiveness can also be quite subtle. Luria (1980:294), describes one patient with a slowly growing frontal tumor "whose first manifestation of illness occurred when, on going to

the train station, got into the train which happened to arrive first, although it was going in the opposite direction."

Uncontrolled Laughter and Mirth

Frontal lobe patients can act in a childish and puerile manner, laughing at the most trivial of things, making inappropriate jokes, teasing, and attempting to engage total strangers in hilarious conversation (Ackerley, 1935; Freeman and Watts, 1942, 1943; Kramer, 1954; Luria, 1980; Petrie, 1952; Rylander, 1939, 1948). Some may initially appear almost completely emotionally unrestrained. Pathological laughter, joking, and punning may occur superimposed upon a labile effect and many seem vastly amused by their own jokes (Ironside, 1956; Kramer, 1954; Martin, 1950). Indeed, such patients can be quite funny, but often they are not! In part, this humor is a function of tangentiality and disinhibition such that loosely connected ideas or events come to be strung together in an unusual fashion. The tendency to exaggerate and to impulsively comment upon whatever happens to draw their attention is also contributory.

Nevertheless, rather than funny, many seem crude and inappropriate. In other instances, they may laugh without reason and with no accompanying feelings of mirth. Kramer (1954) for example, describes four cases of uncontrollable laughter after lobotomy and noted that these persons could not stop their laughter on command or by their own volition. The laughter would come on like spells, occurring up to a dozen times a day and/or continue into the night, requiring sedation in some cases. In these instances, however, the laughter had no contagious aspects but seemed shrill and "frozen." When questioned about the laughter the patients either confabulated a reason or seemed completely perplexed as to the cause. Similar disturbances have been noted with damage to the hypothalamus, a nucleus with major interconnections with the orbital regions (see Chapter 5).

Focal tumors of the orbital regions have also been reported to give rise to gelastic seizures (Chen and Forster, 1973; Daly and Moulder, 1957; Loiseau et al., 1971)—that is, seizures that induce uncontrolled laughter. Often the damage appears more extensive on the right.

In addition to laughter, punning, and "Witzelsucht" (puerility), language might become excessively and inappropriately profane, and the patient may seem inordinately inconsiderate, outspoken, and obstinate (Broffman, 1950; Partridge, 1950; Strom-Olsen, 1946).

Disinhibited Sexuality

Frontal lobe damage sometimes results in a loss of control over sexual behavior and patients may engage in inappropriate sexual activity (Brutkowski, 1965; Freeman and Watts, 1942, 1943; Girgis, 1971; Lishman, 1973; Miller et al., 1986; Strom-Olsen, 1946). One patient, after a right frontal injury, began patronizing up to four prostitutes a day, whereas his premorbid sexual activity had been limited to Tuesday evenings with his wife of 20 years (Joseph, 1988a).

Another patient with a right frontal stroke began propositioning nurses and once, when a particularly large-breasted woman chanced to be near, he reached out and fondled her. It is thus not unusual for a hypersexual, disinhibited person with frontal lobe injury to become forceful and indiscriminate One person who was described as quite gentle, concerned, caring, and loving prior to his injury became exceedingly forceful afterwards and raped several women.

Similar behavior has been described regarding certain persons following lobotomy (Freeman and Watts, 1943). "Sometimes the wife has to put up with some exaggerated attention on the part of her husband, even at inconvenient times and under circumstances which she may find embarrassing. Refusal, however, has led to one savage beating that we know of, and to an additional separation or two" (p. 805). Curiously, in these situations Freeman and Watts (1943:805) have suggested that "spirited physical self-defense is probably the best strategy of the woman. Her husband may have regressed to the cave-man level, and she owes it to him to be responsive at the cave-woman level. It may not be agreeable at first, but she will soon find it exhilarating if unconventional."

Seizure activity arising from the deep frontal regions have also been associated with increased sexual behavior, including sexual automatisms, exhibitionism, genital manipulation, and masturbation (Spencer et al., 1983). In most instances, "sexual" seizures are associated with right frontal or right temporal seizure foci (Joseph, 1988a). Nevertheless, in many cases patients may become hyposexual (Greenblatt, 1950; Miller et al., 1986).

Apathy and Depression

In many instances, however, patients may appear severely apathetic, indifferent, and lethargic, and they may develop bradykinesia, inertia, and mutism (Blumer and Benson, 1975; Freeman and Watts, 1942, 1943; Girgis, 1971; Hecaen, 1964; Luria, 1980; Strom-Olsen, 1946). Periods of impulsiveness, tangentiality, perseveration, and irritability may also be evident although pseudodepressive features predominate. There is a loss of drive and initiative, inactivity, a narrowing of interests, untidiness, excessive sleep, confusion, and defects of memory and attention. Indeed, even those who initially behave in a floridly disinhibited fashion over time tend to behave in a generalized apathetic manner.

For example, one patient who "prior to his accident requiring amputation of the left frontal pole, had been garrulous, enjoyed people, had many friends, was active in community affairs [and was described as having] true charisma" became quiet and remote, spent most of his time sitting alone smoking, and was frequently incontinent of urine and occasionally of stool following his head injury. He remained unconcerned and was frequently found soaking wet, calmly sitting and smoking. When asked, he would deny illness (Blumer and Benson, 1975:196).

Bilateral frontal damage in monkeys has also been reported to result in a severe generalized apathetic and disinterested/indifferent state. As described by Kennard (1939) they would sit with their head sunk between their shoulders, neither blinking nor turning their heads in response to noise, threats, or the presence of intruders, but would stare absently straight ahead with no facial expression. In a similar study, following massive frontal destruction a monkey who was formerly quite active and the dominant leader of his group became inactive, indifferently watched others, failed to respond emotionally, and seemed to have lost all interest and ability to engage in complex social behavior (Batuyev, 1969).

According to Freeman and Watts (1943), inertia and apathy usually immediately follow surgical destruction of the frontal lobes. "This is characterized by more or less complete blankness. Whoever has charge of the patient will have to pull him out of bed, otherwise he may stay there all day. It is especially necessary since he won't get up voluntarily even to go to the toilet."

In some patients this initial state of inertia disappears, whereas in others it becomes a lasting or even progressively severe disturbance. "The previously busy housewife who has always been a dirt-chaser, and who has kept her fingers perpetually busy with darning, crocheting, knitting, and so on, sits with her hands in her lap watching the 'snails whiz by.' Like a child she must be told to wipe the dishes, to dust the sideboard, to sweep the porch" and even then the patient completes only half the task as there is no longer any interest or initiative (Freeman and Watts, 1943:803).

In his summary of two large-scale frontal tumor studies Hecaen (1964) noted that most patients seemed confused, disorganized, apathetic, hypoactive, and to be suffering from inertia and feelings of indifference. However, puerility was also common among these patients, and many demonstrated decreased judgment with

either total or partial unawareness of the environment.

Overview

Overall, two extremes in the qualitative features of abnormal personality and emotion apparently may result from frontal injury. One group of patients may tend to display disinhibition, hyperactivity, and impulsivity, including tangentiality and confabulatory tendencies, whereas a second group may demonstrate apathy, hypoactivity, confusion, lethargy, and depression. Many patients, however, demonstrate both extremes coupled with periods of intensification, amelioration, and islands of seeming normality. In part, the type and pattern of abnormality depends on whether the lesion involves the orbital, medial, inferior, or lateral convexity, or the frontal pole, as well as the laterality of the destruction.

RIGHT AND LEFT FRONTAL LOBES
Neuropsychiatric Disorders

In general, symptoms associated with frontal lobe lesions, particularly those involving the convexity, are often differentially expressed depending on the laterality of the lesion. For example, in one study of 670 patients with missile wounds, "a specific association" between emotional-psychiatric disability and right frontal wounds was noted (Lishman, 1968). In particular, the so-called "frontal lobe personality" (e.g., euphoria, childishness, egoism, irritability) was almost exclusively associated with right but not left frontal wounds.

Similarly, confabulation, hypersexuality, tangentiality, and manic-like behaviors seem to be more frequently associated with right frontal dysfunction (Joseph, 1986a, 1988a). In contrast, reduced responsiveness is more frequently linked to lesions of the left convexity.

DEPRESSION

Patients with severe forms of Broca's expressive aphasia often become frustrated, tearful, sad, and depressed (Robinson and

Benson, 1981). This is because comprehension is usually relatively intact and they are painfully aware of their deficit (Gainotti, 1972). In other words, depression in these cases appears to be a normal reaction and as such is mediated by normal tissue, i.e., the undamaged right hemisphere (Joseph, 1982, 1988a).

Depressive-like features, however, also seem to result with left anterior damage sparing Broca's area such as when the frontal pole of either hemisphere is compromised (Robinson et al., 1984; Robinson and Szetela, 1981; Sinyour et al., 1986). In part, this is probably due to its close proximity and interconnections with the medial region, an area that when damaged induces hypokinetic and apathetic states. Hence, although these patients look depressed, often they are severely apathetic, indifferent, hypoactive, and poorly motivated. When questioned, rather than worried or truly concerned about their condition the overall picture is that of confusion, disinterest, and blunted emotionality (Kleist, 1934 cited by Freeman and Watts, 1942: Hacaen, 1964), i.e. there is a lack of worrisome thoughts or depressive ideation.

On the other hand, it has been reported that psychiatric patients classified as depressed (who presumably show no obvious signs of neurological impairment) demonstrate a significant lower integrated amplitude of the EEG evoked response over the left versus the right frontal lobe (d'Elia and Perris, 1973; Perris, 1974), i.e., the left frontal region is insufficiently activated. Based on EEG and clinical observation, d'Elia and Perris have argued that the involvement of the left hemisphere is proportional to the degree of depression. They have also noted that with recovery from depression the amplitude of the evoked response increases to normal right and left hemisphere levels. Reduced bioelectric arousal over the left frontal region has also been reported following depressive mood induction (Tucker et al., 1981).

In part, apathetic and depressive features may result from left frontal convexity and frontal pole damage due to a severance of fibers that link emotional impulses (such as those being transmitted via the orbital and medial region) with external sources of input or cognitive activity that are transmitted to the convexity (i.e., disconnection).

For example, areas 9 and 10 of the frontal lobe receive converging sensory fibers from the auditory, visual, and somesthetic cortices (Jones and Powell, 1970; Pandya and Kuypers, 1969), the inferior parietal lobule, and the anterior cingulate and medial wall. Areas 9 and 10 then project into as well as receive fibers from the orbital regions that share projections with the hypothalamus, amygdala, and septal nuclei.

It was the recognition that the frontal lobes acted as a bridge between emotion and idea that led to the wide-scale use of frontal lobotomy, i.e., surgical destruction of interlinking fibers—a technique, which when used in the 1940s and 1950s, often involved little more than blindly swishing a "surgical ice pick" inside somebody's brain!

Moreover, convexity lesions, like medial damage, may result not only in a disconnection between cognitive-perceptual and emotional activity, but would prevent limbic system output from reaching the motor areas such that emotional-motivational impulses cannot become integrated with neocortical motor activities. The patient is thus motorically hypoemotionally aroused (i.e., depressed) and appears to be demonstrating psychomotor retardation.

Just as left frontal convexity motor damage can result in left-sided apraxia (due to right hemisphere disconnection from left parietal temporal-sequential output), the reverse can also occur. That is, with left frontal damage, linguistic impulses not only fail to become expressed, but emotional output from the right hemisphere and limbic system fail to become integrated with linguistic ideation (i.e., thought). Ideas no longer come to be assigned emotional significance. In the extreme, the motivational impetus to even engage in thought production is cut off.

As pertaining to laterality, left frontal (versus right frontal) lesions are associated with reductions in intellectual and conceptual capability (to be explained), which often leads to confusion and a reduced ability to appreciate and appropriately respond to the external or internal environment. In these instances, one possible consequence is apathy, indifference, hyporesponsiveness, and depressive-like symptoms.

On the other hand, it is possible that among psychiatric patients and otherwise normal, albeit depressed individuals, the left frontal region appears relatively inactive because the right frontal area is preoccupied with being depressed; the right frontal region is excessively aroused. That is, excessive right frontal arousal leads to massive left frontal inhibition, i.e., the bilateral arousal system of the right hemisphere inhibiting the left (to be explained).

Depressive and apathetic disturbances, however, also occur with right hemisphere lesions, particularly those involving the parietal region and the frontal pole (Joseph, 1988a).

OBSESSIVE-COMPULSIVE DISORDERS

There is now considerable evidence that frontal lobe dysfunction may give rise to obsessive-compulsive (Ob-C) disorders— particularly (among psychiatric patients) when the left frontal region (Flor-Henry et al., 1979) or orbital frontal areas are compromised (Baxter et al., 1988; Rappoport, 1991; Rauch et al., 1994). It is noteworthy that Ob-C disturbances are also frequently accompanied by strong feelings of depression and anxiety (Goodwin and Guze, 1979)—affective states linked to alterations in the functional integrity of the anterior regions of the brain.

Similarly, abnormal activation of the frontal lobes has been reported to elicit recurrent and intrusive ideational activity ("forced thinking"), as well as compulsive urges to perform aberrant actions, e.g., shouting (Penfield and Jasper, 1954; Ward, 1988). A more frequent complication, however, is perseveration of speech and motor functioning.

The mixture of perseveration and compulsions, when subtle, may take on the appearance of obsessions. Of course, perseveration is not due to an obsessive disorder. Nevertheless, obsessions are perseverative in nature, often involving intrusive recurring thoughts, feelings, or impulses to perform certain actions (Goodwin and Guze, 1979). This disinhibition of verbal representation, i.e., forced, repetitive, and intrusive thinking, as pointed out by Miller (1990), strongly reinforces the impression of an Ob-C frontal lobe association.

Motorically obsessive compulsions may involve repetitive, stereotyped acts including the perseverative manipulation and touching of objects (Goodwin and Guze, 1979)—abnormalities also associated with certain forms of frontal lobe damage.

Hence, in many ways there appears to be a convergence of symptoms that suggests that some individuals suffering from certain subtypes of severe Ob-C irregularities may be victims of abnormalities involving the frontal lobes and/or the frontal-basal ganglia-amygdala circuit.

For example, right caudate hypermetabolism and blood flow has also been associated with Ob-C disorders (Baxter et al., 1992; Rauch et al., 1994) as well as reduced 5-HT activity (Goodman et al., 1990; Zohar and Insel, 1987), which in turn results in disinhibition. As detailed in Chapter 9, the inhibitory circuit maintained by the orbital frontal lobes and basal ganglia, in some instances, is possibly deactivated or overwhelmed, thus giving rise to obsessive acts or thoughts.

Given that it is via the frontal lobes that thoughts and emotions come to be integrated and infused with emotion, and/or inhibited and suppressed, this also suggests that caudate/orbital dysfunction may result in the inability to extinguish certain thoughts and actions, even in the absence of an underlying emotional problem. Hence, the person may perseverate on a single thought or action, which in turn may interfere with his or her ability to engage in alternative modes of response.

SCHIZOPHRENIA

It has also been reported that some persons classified as schizophrenic demonstrate EEG and other abnormalities suggestive of bilateral or left frontal dysfunction or hypoarousal (Akbarian et al., 1993; Ariel et al., 1983; Ingvar and Franzen, 1974; Kolb and Whishaw, 1983; Levin, 1984). It is likely, however, that only certain subpopulations of schizophrenics actually suffer from frontal lobe dysfunction, such as those with catatonia, posturing, mannerisms, and emotional blunting coupled with reduced speech output and apathy.

For example, patients displaying unusual mannerisms, catatonia, "emotional blunting", apathy, and/or "hebephrenia" (blunting associated with puerile, silly childishness as well as obsessive-compulsiveness; cf. Hamilton, 1976) are certainly "frontal lobe" candidates. Indeed, Hillbom (1951) notes exactly this association among persons with head trauma and missile wounds who later developed schizophrenic-like symptoms, including catatonia and hebephrenia, i.e., left frontal patients are more likely to develop these symptoms. However, it is also likely that lesions involving the left temporal lobe or dysfunctional nuclei buried deep within the frontal lobe, i.e., the caudate and other portions of the basal ganglia, may in fact be the causative agents in some instances.

For example, Richfield and coworkers (1987:768) report a 25-year-old female honor student who "had undergone a dramatic personality change manifested by alterations in affect, motivation, cognition, and self-care which included vulgarity, impulsiveness, violent outbursts, enuresis, indifference, hypersexuality, shoplifting, and exposing herself." Computed tomography (CT) indicated bilateral damage to the head of the caudate nuclei.

In some respects this patient's behavior suggested the acute onset of a "schizoaffective" disorder. However, repeated psychi-

atric hospitalizations and treatment with major tranquilizers over the course of the next 2 years were of no assistance. And no wonder, her problem was neurologic, not psychiatric (as illustrated by Richfield et al., 1987). And yet, the wards of many psychiatric hospitals are probably filled with such unfortunates, their neurological abnormalities unrecognized.

MANIA

When the right frontal, orbital, and portions of the inferior lateral convexity are damaged, behavior often becomes inappropriate, labile, and disinhibited. Patients may become hyperactive, distractible, hypersexual, tangential, and confabulatory (Joseph 1986a, 1988a). However, although laughing and joking one moment, these same patients can quickly become irritated, angered, enraged, destructive, or conversely tearful with slight provocation. This pattern is one of mania and/or hypomanic excitement.

Mania and manic-like features have been reported in many patients with injuries, tumors, and even seizures involving predominantly the frontal lobe and/or the right hemisphere (Bogousslavsky et al., 1988; Clark and Davison, 1987; Cohen and Niska, 1980; Cummings and Mendez, 1984; Forrest, 1982; Girgis, 1971; Jack et al., 1983; Jamieson and Wells, 1979; Joseph, 1986a, 1988a; Lishman, 1973; Miller et al., 1986; Oppler, 1950; Rosenbaum and Berry, 1975; Starkstein et al., 1987; Stern and Dancy, 1942). For example, one patient described as a very stable, happily married family man became very talkative, restless, grossly disinhibited, and sexually preoccupied; he extravagantly spent money and recklessly purchased a business that soon went bankrupt (Lishman, 1973). Frontal lobe destruction was indicated.

In another case, a woman of 46 was admitted to the hospital and observed to be careless about her person and room, to be incontinent of urine and feces, to sleep very little, and to act in a hypersexual manner. Her symptoms had developed several months earlier when she began accusing a neighbor of taking things she had misplaced and on other occasions stripping in front of him. She began going about in just her nightdress, informing people she was descended from queens, was very rich, and that many men wanted to divorce their wives and marry her. During her hospitalization she was frequently quite loud, disoriented as to time and place, and extremely tangential, jumping from subject to subject. After several years she died, and a meningioma involving the orbital surface of her frontal lobes was discovered (Girgis, 1971).

Oppler (1950), reported one person who developed flight of ideas, increased activity, and emotional lability, including elation, extreme fearfulness, distractibility, puerility, and argumentative behavior. Subsequently a tumor was discovered and removed from the right frontal-parietal area.

Over the course of the last dozen years, I have examined 16 male and 2 female patients who developed manic-like symptoms after suffering a right frontal stroke or trauma to the right hemisphere (some of whom are described in Joseph, 1986a, 1988a). All but three of the males had good premorbid histories and had worked steadily at the same job for over 3–5 years.

Following their injuries all developed delusions of grandeur, pressured speech, flight of ideas, decreased need for sleep, indiscriminate financial activity, extreme emotional lability, and increase libido. However, in all cases the mania subsequently subsided after a few weeks or months of the injury and/or after placement on lithium.

Intellectual and Conceptual Alterations

It has frequently been claimed that intelligence is not affected even with massive injuries to the frontal lobes. However, this view is completely erroneous, for even in mild cases, although intelligence per se may not seem to be reduced, the ability to effectively employ one's intelligence is al-

most always compromised to some extent.

Frontal lobe damage reduces one's ability to profit from experience, to anticipate consequences, or to learn from errors (Bianchi, 1922; Drewe, 1974; Goldstein, 1936–37, 1944; Halstead, 1947; Joseph, 1986a; Milner, 1964, 1971; Nichols and Hunt, 1940; Petrie, 1952; Porteus and Peters, 1947; Rylander, 1939; Tow, 1955). There is a reduction in creativity, fantasy, dreaming, abstract reasoning, and the capacity to synthesize ideas into concepts or to grasp situations in their entirety. Intellectual interests seem to be diminished, or, with severe damage, abolished (Bianchi, 1922; Freeman and Watts, 1942, 1943; Goldstein, 1936–37, 1944; Partridge, 1950; Tow, 1955).

For example, Freeman and Watts (1943:803) note that "patients who formerly were great readers of good literature will be interested only in comic books or movie magazines. Men of considerable intellectual achievement will read avidly the sports page, and when discussion turns on the great events of the day will pass off some clichés as their own opinions."

In mild or severe cases thinking may be contaminated by perseverative intrusions of irrelevant and tangential ideas, randomly formed associations, and illogical intellectual activity. These patient are also often affected by the immediacy of their environment and have difficulty making plans or adequately meeting long-term goals in an appropriate manner.

For example, one patient, formerly an executive at a local electronics firm, arrived at my office with a toothbrush, toothpaste, hair brush, and wash rag sticking out of his shirt pocket. When I asked, pointing at his pocket, "What's all that for?", he replied with a laugh, "Just in case I want to wash my face and brush my teeth," and in so saying he quickly drew the toothbrush from his pocket and began to demonstrate. During the course of the exam he laughingly wanted to show me how the hole in his head (from the craniotomy and bullet wound) could bulge in or out when he held his breath or held his head upside-down and climbed up on my desk so that I could better see his head when held up-side-down.

Although this person had suffered a bullet wound damaging the orbital area and mesial convexity of the right frontal lobe (he had shot himself in the mouth), his overall WAIS-R IQ was above 130 (98% rank: "Very Superior"). Nevertheless, throughout the exam he behaved in a silly, puerile manner, often joking and laughing inappropriately.

Frontal lobe patients also may have difficulty thinking up or considering alternative problem-solving strategies and thus developing alternative lines of reasoning. For example, Nichols and Hunt (1940) dealt a patient five cards down including the ace of spades, which always fell to the right on two successive deals and then to the left for two trials. The patient's task was to learn this pattern and turn up the ace. The patient failed to master this after 200 trials.

Some frontal lobe patients may also have extreme difficulty sorting even common everyday objects according to category (Rylander, 1939; Tow, 1955), such as sorting and grouping drinking containers (glasses) with other drinking containers (mugs) or tools with tools. Similarly, they may have difficulty performing the Wisconsin Card Sorting Task (Crockett et al., 1986; Drewe, 1974; Milner, 1964, 1971), which involves sorting geometric figures according to similarity in color, shape, or number. However, how a patient fails on this task depends on the locus and laterality of the damage.

For example, patients with orbital damage seem to have relatively little difficulty performing this category sorting task (Drewe, 1974; Milner, 1971). Similarly, patients with right frontal damage, although they tend to make perseverative-type errors (i.e., persisting in a choice pattern which is clearly indicated as incorrect), perform significantly better than those with left frontal damage (Drewe, 1974; Milner, 1964, 1971). Thus overall, patients with left medial and convexity lesions perform most poorly and have the most difficulty with thinking in a flexible manner or developing alternative response strategies.

IQ

In a number of studies of conceptual functioning in which either the Raven's Progressive Matrices or Porteus Mazes were administered both before and after surgical destruction of the frontal lobes, significant declines have resulted consistently (Petrie, 1952; Porteus and Peters, 1947; Tow, 1955). As with most tests, the usual pattern is improvement with practice. Hence, these results (and those mentioned above) indicate that frontal lobe damage disrupts abstract reasoning skills, verbal-nonverbal pattern analysis, learning and intellectual ability, and the capacity to anticipate the consequences of one's actions or to profit from experience.

Significant disturbances and reductions in intellectual ability have also been reported with administration of the Wechsler Intelligence Scales (Petrie, 1952; Smith, 1966a). In one study in which patients undergoing frontal leucotomy for intractable pain were administered the Wechsler test both before and after surgery, a 20-point drop in the IQ was reported (Koskoff, 1948, cited by Tow, 1955). Again, however, the effects of frontal damage on IQ depends on the locus of the damage.

For example, left frontal patients show lower Wechsler IQs than those with right frontal lesions (Petrie, 1952; Smith, 1966a). In fact, 17 of 18 patients with left frontal damage reported by Smith (1966a) scored lower across all subtests compared to those with right frontal lesions. Indeed, patients with left-sided destruction perform as poorly as those with bilateral damage (Petrie, 1952).

In analyzing subtest performance, Smith (1966a) notes that left frontal patients scored particularly poorly on Picture Completion (which requires identification of missing details). This is presumably a consequence of the left cerebral hemisphere being more concerned with the perception of details (or parts, segments) versus wholes (Joseph, 1982, 1988a). Petrie (1952), however, reports that performance on the Comprehension subtests (i.e., judgment, common sense) was most significantly impaired among left frontals.

In contrast, persons with severe right frontal damage have difficulty performing Picture Arrangement—often leaving the cards in the same order in which they are laid (McFie and Thompson, 1972). This may be a consequence of deficiencies in the capacity to discern social-emotional nuances, a function at which the right hemisphere excels (see Chapter 3).

Nevertheless, since so few studies have been conducted it is probably not reasonable to assume that lesions lateralized to the right or left frontal lobe will always affect performance on certain subtests, particularly if there is a mild injury. It is also important to consider how lateralized effects on IQ may be contributing to or secondary to reduced motivation and apathy since bilateral and left frontal damage often give rise to this constellation of symptoms. If patients are apathetic they are not going to be motivated to perform to the best of their ability.

ATTENTION

As noted, a tremendous amount of evidence indicates that the frontal lobes are highly involved in attentional functioning (Crowne, 1983; Fuster, 1980, 1996; Luria, 1980; Knight et al., 1981; Pragay et al., 1987). Nevertheless, the type of attentional disturbance depends in part on the laterality of the lesion as well as the extent to which it involves the convexity or orbital zones (already discussed). For example, some right frontal patients are impaired on sustained attention tasks when stimuli are presented at the rate of one stimulus per second. However, if these patients are presented with seven stimuli per second, performance is improved (Wilkins et al., 1987). This suggests that right frontal patients may be understimulated and show wandering attention if not fully engaged.

The left frontal lobe appears to be more concerned with verbal attentional functioning and the monitoring of temporal-sequential and detailed events (Milner, 1971; Petrides and Milner, 1982), whereas the right is more attentive to nonverbal auditory, visual, tactual, and social-emotional stimuli (Joseph, 1988a). For example, damage involving either the right or left frontal lobe may impair performance on the Picture Completion subtest of the WAIS-R (i.e., detecting a missing detail such as a dog leaving no footprints), but for different reasons. Left frontals may do poorly because of inattention to detail (e.g., "There's nothing wrong with this picture"). Right frontals may perform deficiently because of impulsive tendencies to say the first thing that comes to mind ("The dog doesn't have a leash"). Higher false-positive rates also occur when the frontal lobes are damaged.

Right Frontal Dominance for Arousal

The right cerebral hemisphere is clearly dominant in terms of mediation and control over most aspects of social-emotional intellectual functioning (see Chapter 3). A variety of findings also strongly suggest that the right frontal lobe exerts bilateral influences on arousal (DeRenzi and Faglioni, 1965; Heilman and Van Den Abell, 1979, 1980; Joseph, 1982, 1986a, 1988a; Tucker, 1981). For example, the intact, normal right hemisphere is quicker to react to external stimuli and has a greater attentional capacity compared to the left (Dimond, 1976, 1979; Heilman and Van Den Abell, 1979; Jeeves and Dixon, 1970; Joseph, 1988a, b). In split-brain studies the isolated left hemisphere tends to become occasionally unresponsive, suffers lapses of attention, and is more limited in attentional capacity (Dimond, 1976, 1979; Joseph, 1988a, b). The right frontal lobe is also larger than the left, suggesting a greater degree of interconnection with other brain tissue.

Moreover, it has been demonstrated that visual and somesthetic stimuli, or active touch exploration with either the right or the left hand, elicits evoked EEG responses preferentially and of greater magnitude over the right hemisphere (Desmedt, 1977). The right hemisphere also becomes desynchronized (aroused) following left- or right-sided stimulation (indicating that it is bilaterally responsive), whereas the left brain is activated only with unilateral (right-sided) stimulation (Heilman and Van Dell Abell, 1980).

It is also possible that the right and left frontal lobes exert different influences on arousal (Tucker, 1981). The right frontal lobe may exert predominantly bilateral inhibitory influences (or bilateral excitatory and/or inhibitory influences). The left frontal region may be more involved in unilateral excitatory (expressive) activation. Thus when the left frontal region is damaged, the right acts unopposed and there may be excessive inhibition as manifested, for example, by speech arrest, depression, and/or apathy. However, with lesions involving the right frontal lobe, not only is there a loss of inhibitory control, but the left may act unopposed such that there is excessive excitement, e.g., as manifested by speech release, confabulation, and disinhibited behavior.

Nevertheless, because left cerebral excitatory influences are predominantly unilateral, with massive right cerebral damage, although the left hemisphere is aroused, the left cannot activate the right half of the brain. This may result in unilateral inattention and neglect of the left half of the body and space (Heilman and Valenstein, 1972; Joseph, 1986a, 1988a, b). That is, the patients' (undamaged) left hemispheres may ignore their left arms or legs if their neglected extremities are shown to them, they may claim that they belong to the doctor or a person in the next room.

That such disturbances occur only rarely with left frontal or left hemisphere damage further suggests that the right hemisphere is able to continue to monitor events occurring on either side of the body. Thus, although the damaged left hemisphere is hypoaroused (or inhibited by the right), there is little or no neglect.

In contrast, with partial right frontal damage, rather than a loss of arousal, there can result a loss of inhibitory restraint, and the patient may demonstrates confabulatory tendencies (speech release), heightened sexuality, manic-like excitement, and a host of other disinhibitory disturbances that may wax and wane in severity.

CONCLUSIONS

The frontal lobes serve as the senior executive of the brain, and together with the motor area make up almost half of the cerebrum. They act to mediate information processing throughout the neuroaxis via cortical sampling, thalamic gating, and inhibitory control over various nuclei, including the limbic system and reticular formation. For example, the orbital regions appear to exert controlling influences over limbic arousal, whereas the convexity seems more concerned with influencing information processing throughout the neocortex. Hence, via these and other interactions, the frontal lobes act to maintain and shift attention, exert organizational control over all aspects of expression (e.g., thought, speech, motor), anticipate consequences, consider alternatives, plan and formulate goals, shape, direct, and modulate personality and emotional functioning, and act to integrate ideas, emotions, and perceptions. Capacities such as motor, emotion, and personality functioning may become compromised to some extent, regardless of where within the frontal lobe a lesion may occur.

Massive damage to the medial walls of the frontal lobes (including the SMA and cingulate gyrus) often initially result in reduced speech output, mutism, and variable motor abnormalities and mannerisms, including agraphia, forced groping, compulsive utilization, gegenhalten, waxy flexibility, and, in the extreme, catatonia. Patients may seem apathetic, indifferent, and/or severely depressed. Left convexity lesions are also associated with reduced intellectual functioning, and possibly some forms of schizophrenic and Ob-C abnormalities.

Regardless of laterality, lesions involving the orbital frontal lobes, if partial, can give rise to a loss of emotional control and social-emotional restraint coupled with disinhibition and irresponsible behavior. However, if such lesions are massive, rather than a loss of emotional control, there is a loss of emotional expression. Moreover, orbital lesions (as well as inferior medial and inferior convexity) are often associated with perseverative abnormalities.

Inferior convexity lesions also often result in a loss of restraint, including hyperexcitability and perseveration. Similarly, right frontal convexity lesions may initially give rise to disinhibitory states; in the extreme, patients may seem delusional and manic-like.

In many respects, whereas the left convexity seems to resemble the medial frontal lobes, the right frontal region seems to be more closely associated with the orbital areas. In part this may be due to the greater right frontal involvement with emotional functioning as well as its more intimate connections with and control over the limbic system, of which the orbital region is a part.

It must be emphasized, however, that not only is there considerable overlap in functional representation, lesions are seldom confined to one particular region of the frontal lobe. Hence, based on symptoms alone, one cannot with complete assurance localize damage to a particular quadrant of the anterior half of the cerebrum. Moreover, uterine or birth trauma, brain damage experienced during childhood, or even early rearing experiences can lead to considerable functional reorganization (Goldman, 1971; Joseph, 1982, 1986b, 1988b; Novelly and Joseph, 1983) (see Chapter 18). Not all brains are alike.

Moreover, long-term effects of frontal lesion are often quite different from what is seen initially or during the first year following injury. In part this is a function of the initial loss and then regrowth of white matter axonal interconnections versus permanent destruction of neocortical nerve cells. Some frontal patients become progressively more sluggish and apathetic in their behavior—even those who were floridly manic when first examined. Others may become less disinhibited and more appropriate as the white matter fiber tracts regenerate. Conversely, if the damage is mild, the disturbances may be quite subtle and or may be attributed to abnormalities involving personality and social and emotional maturity.

12

The Parietal Lobes

The Body Image and Hand in Visual Space, Apraxia,
Gerstmann's Syndrome, Neglect, Denial, and
the Evolution of Geometry and Math

The parietal lobes are commonly thought to be predominantly concerned with processing somesthetic, kinesthetic, and proprioceptive information. However, like the large expanse of tissue in the frontal portion of the brain, the parietal lobes are not a homologous tissue but consist of cells that respond to a variety of divergent stimuli, including movement, hand position, objects within grasping distance, audition, eye movement, and complex and motivationally significant visual stimuli (Burton et al., 1982; Cohen et al., 1994; Dong et al., 1994; Lebedev et al., 1994; Lin and Sessle, 1994; Lin et al., 1994; Mountcastle et al., 1975; Previc, 1990; Pred'Homme and Kalaska, 1994; Stein, 1992).

Damage to the parietal lobe can therefore result in a variety of disturbances, including abnormalities involving somesthetic sensation, the body image, visual-spatial relations, temporal-sequential motor activity, language, grammar, numerical calculation, emotion, and attention, depending on which area has been lesioned as well as the laterality of the damage.

PARIETAL TOPOGRAPHY

The parietal lobe may be subdivided into a primary receiving area (Broadmann's areas 3,1,2) within the postcentral gyrus, an immediately adjacent somesthetic association area (Broadmann's area 5), a polymodal (visual, motor, somesthetic) receiving area located in the superior-posterior parietal lobule (area 7), and a multi-modal-assimilation area within the inferior parietal lobule (areas 7, 39, 40) that encompasses the angular and supramarginal gyrus.

The primary somesthetic (as well as portions of the association area) contribute almost one-third of the fibers that make up the cortical-spinal (pyramidal) tract. Hence, this region is highly involved in motor functioning. Moreover, the primary motor and somesthetic regions are richly interconnected (Jones and Powell, 1970), for in order to make motoric responses with some precision, there must be tremendous sensory feedback concerning proprioception, the positions of the various joints and tendons, etc.—information that is provided by the somesthetic cortices (Cohen et al., 1994; Dong et al., 1994; Lebedev et al., 1994; Pred'Homme and Kalaska, 1994). Together, the motor and somesthetic areas comprise a single functional unit that some have referred to as the sensorimotor cortex (Luria, 1980).

PRIMARY SOMESTHETIC RECEIVING AREAS

The primary receiving area subserving general, superficial, and deep somatic sensation is located in the postcentral gyrus. These cortical tissues receive thalamic input from the ventral posterior thalamic nuclei (VPL and VPM). Specifically the VPL and VPM act to relay impulses received from the ascending trigeminothalamic, spinothalamic, and medial lemniscus pathways to cortical layer IV. This input is somatotopically organized, and the entire body surface comes to be spatially represented in the parietal neocortex.

However, the gyrus actually consists of three narrow strips of tissue (areas 3ab, 1, 2) that differ histologically, in architectural composition, and in sensory input. Specifically, area 3a receives input from the muscles spindles (group IA muscle afferents) and area 3b receives cutaneous stimuli such that, together, areas 3a and 3b appear to maintain a cutaneous map and can also signal muscle length (e.g., flexion or extension). However, almost all of the cells in area 3ab receive input only from the contralateral half of the body. Hence, only half the body is represented.

Information received and processed in area 3 is relayed to the immediately adjacent areas, 1 and 2, each of which also con-

tains a specialized spatial map of the body (Kaas et al., 1981; Lin et al., 1994; Sur et al., 1982). For example, area 1 appears to maintain an overlapping cutaneous-joint body map (Evarts, 1969; Mountcastle and Powell, 1959; Schwartz et al., 1973). Area 2 maintains a map of the joint receptors and can signal the position and posture of the limbs based on input from the muscle spindles. Hence, the somesthetic cortex maintains multiple maps of the body.

Moreover, within this tiny expanse of tissue there is a sequential hierarchical convergence of input from area 3 to area 1 onto area 2. That is, information is analyzed and then passed from area 3 to area 1, and from areas 3 and 1 onto area 2.

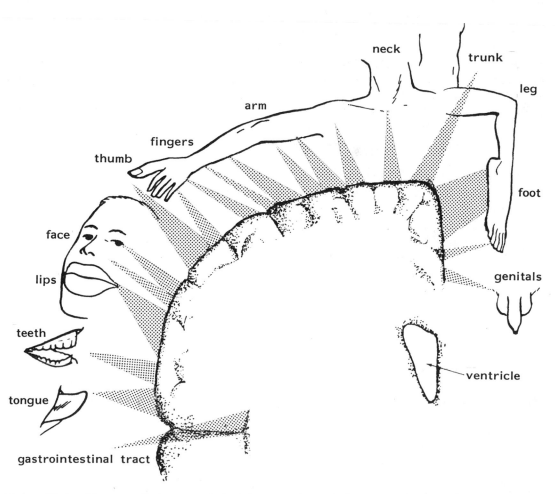

Figure 12.1. The sensory homunculus of the somesthetic cortex. (From Milner PM. Physiological psychology. New York: Holt, Rinehart and Winston, 1970.)

Therefore, a single neuron in area 2 receives multiple input from several cells in area 3, as well as vestibular input. Together these three strips of tissue comprise an interactional functional unit and are responsive to touch, texture, shape, motion, and the direction of stimulus movement, including temporal-sequential patterning, and can directly monitor the position and movement of the extremities (Cohen et al., 1994; Lebedev et al., 1994; Levitt and Levitt, 1968; Lin et al., 1994; Prud'Homme and Kalaska, 1994; Mountcastle, 1957; Warren et al., 1986a; Whitsel et al., 1972). Many cells are also responsive to changes in temperature as well as the presence of noxious stimuli applied to the skin.

Because most of these neurons receive input concerning pressure, light touch, vibration, the movement of joints, and muscular activity (Cohen et al., 1994; Lebedev et al., 1994; Levitt and Levitt, 1968; Mountcastle, 1957; Prud'Homme and Kalaska, 1994) they can signal and determine whatever posture or position the body is in as well as the amount of force or pressure being exerted by the limbs (Jennings et al., 1983), i.e., if the body is carrying or lifting some object. Conversely, via the reception and analysis of this input, a person can detect an insect crawling up or down the leg and the direction in which it is moving, as well as determine the position of the arms and legs without looking at them.

Nevertheless, predominantly elementary and simple contralateral somesthetic information is processed in this region (Lin et al., 1994; Prud'Homme and Kalaska, 1994). Electrical stimulation of the primary somesthetic area gives rise to simple, albeit well localized sensations on the opposite half of the body (Penfield and Jasper, 1954; Penfield and Rasmussen, 1950) such as numbness, pressure, tingling, itching, tickling, and warmth.

BODY IMAGE REPRESENTATION

The primary receiving areas for somesthesis continue up and over the top of the hemisphere and along the medial wall, where the lower half of the body is represented. Specifically the rectum, genitals, foot, and calf are located along the medial wall, the leg along the superior surface of the hemisphere, and the shoulder, arm, hand, and then face along the lateral convexity (Penfield and Jasper, 1954; Penfield and Rasmussen, 1950). Body parts are also represented in terms of their sensory importance, i.e., how richly the skin is innervated. For example, more cortical space is devoted to the representation of the fingers and the hand than to the elbow (Warren et al., 1986). Because of this the cortical body map is very distorted.

In summary, the primary receiving area receives precise information regarding events occurring anywhere along the internal or external body and responds to converging inputs from muscle spindles, cutaneous and joint receptors, and proprioceptive and vestibular stimuli. In this manner, not only the body but the global properties of objects held in the hand can be determined (Iwamura and Tanaka, 1978), i.e., stereognosis.

Functional Laterality

As detailed in Chapter 3, there is clear evidence that the right parietal area is dominant in regard to many aspects of somesthetic information processing. Hence, neurons in this half of the brain appear to be more sensitive to, more responsive to, and more closely involved in monitoring events occurring on either half of the body, but particularly the left. In fact, this relationship was noted over 150 years ago by Weber. According to Weber (1834/1978), the left half of the body exceeds the right in regard to most forms of tactual sensitivity. The left hand and the soles of the left foot, as well as the left shoulder, are more accurate in judging weight and have a more delicate sense of touch and temperature, such that "a greater sense of cold or of heat is aroused in the left hand" (p. 322). That is, the left hand judges warm substances to be hotter, and cold material to be colder as compared to the right

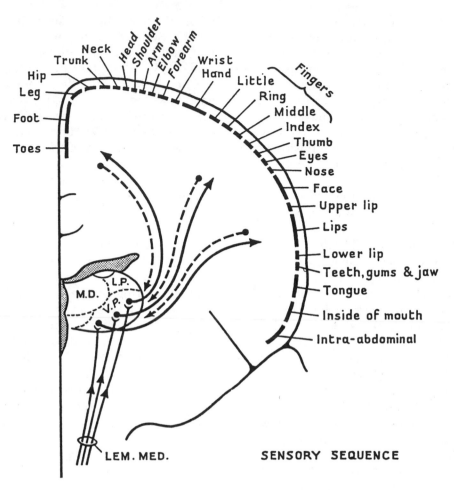

Figure 12.2. Sensory representation of the different regions of the body along the somesthetic cortex of the left hemisphere. The afferent pathways for discriminative somatic sensation are indicated by the unbroken lines. (From Bannister R. Brain's Clinical Neurology. Oxford: Oxford University Press, 1975.)

hand, even when both hands are simultaneously stimulated.

Somesthetic Agnosias

Surgical or other forms of destruction involving the primary somesthetic receiving areas results in a complete, albeit temporary, loss of sensation from the entire half of the body (Russell, 1945). Longer-term effects include elevation of sensory detection thresholds; loss of position and pressure sense, including two-point discrimination; and a greatly reduced ability to detect movement of the fingers. In addition, the capacity to determine texture, shape, and temporal-sequential patterning, or to recognize objects

by touch or to discriminate among different forms or their properties (e.g., size, texture, length, shape, and stereognosis) is significantly attenuated (Corkin et al., 1970; Curtis et al., 1972; LaMotte and Mountcastle, 1979; Randolf and Semmes, 1974). Passive (non-movement) sensation is less impaired. In some instances, over time a remarkable recovery of somesthetic discrimination sense may be observed (Semmes, 1973).

Nevertheless, even with complete removal of the postcentral gyrus, stimuli applied to the face are much better perceived than the same stimuli applied to the hand. Conversely, lesions that spare the hand area of the postcentral gyrus but that destroy the remaining tissue result in mild or no per-

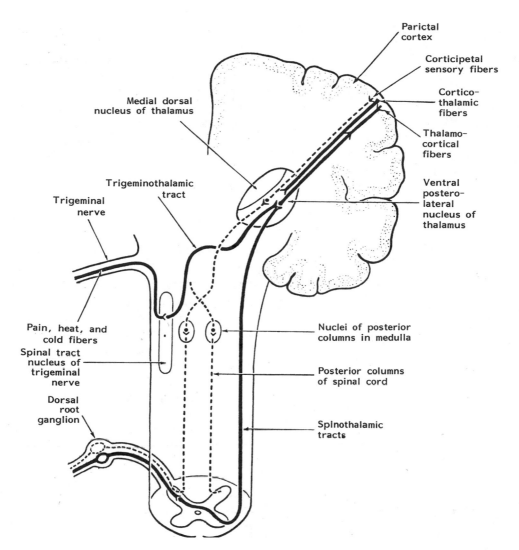

Figure 12.3. The principle sensory pathways leading from the spinal cord and cranial nerves to the thalamus and somesthetic cortex. (From Ban- nister R. Brain's Clinical Neurology. Oxford: Oxford University Press, 1975.)

manent sensory deficits when the hands are tested (with the exception of stereognosis) (Semmes, 1965). However, sensation across other body parts will be impaired. Hence, when testing for parietal lobe dysfunction, one must examine not only the face and hands, but other body parts as well. As noted, this region is also highly concerned with motor functioning as it is extensively interconnected with the primary motor area and gives rise to almost one-third of the corticospinal tract. Hence, damage to this region can give rise to motor disturbances such as paresis accompanied by hypotonia and/or produce inaccuracy and reduced

speed of movement (Cole and Glees, 1954; Luria, 1980). The ability (or will) to initiate movement may also be reduced.

SOMESTHETIC ASSOCIATION AREA

In addition to receiving information already analyzed in the immediately adjacent primary somesthetic cortices, some cells in the association area (Broadmann's area 5) also receive input from the contralateral primary zone (via the corpus callosum) as well as from the motor association areas (area 6) in the frontal lobes

(Dong et al., 1994; Jones and Powell, 1970). Hence, both halves of the body, the truncal area in particular (Robinson, 1973), are represented in this region. Indeed, the two halves of the body appear to be superimposed such that the body is bilaterally represented (Whitsel et al., 1969). However, as based on behavioral and electrophysiological data from intact and brain-damaged persons, bilateral representation is predominantly maintained within the right half of the human brain.

Hand Manipulation Cells

A small percentage of cells in area 5 also appear to be concerned with more complex activities such as the movement of the hand and arm and the manipulation of objects (Cohen et al., 1994; Mountcastle et al., 1975). Indeed, a detailed representation of the cutaneous surface of the body, particularly the hand (and face), is maintained here (Burton et al., 1982; Lin et al., 1994). These cells are referred to as "hand-manipulation cells." In fact, electrical stimulation of area 5 can result in limb movements (Hyvarienen, 1982).

Others neurons in area 5 are especially responsive to particular temporal-sequential patterns of sensation (LaMotte and Mountcastle, 1979) and can determine direction and rhythm of movement. It is presumably through the activity of these cells that one can "hear" via the detection of vibrations (such as reported by the deaf).

The Body In Space

Signals from joint and cutaneous receptors are transmitted to association neurons (Sakata and Iwamura, 1978). Many association cells also receive converging input from primary neurons concerned with different body parts (Dong et al., 1994; Sakata et al., 1973) and thus can determine positional interrelationships. For example, a single association neuron may receive information regarding the elbow and the shoulder, and become activated only when these two body parts are simultaneously stimulated or in motion. A considerable number of cells are especially sensitive to the pos-

ture and position of the trunk and extremities during movement (Hyvarienen, 1982). By associating this convergent input these cells can thus monitor, coordinate, and guide limb movement (Cohen et al., 1994) as well as determine the position of the body and objects in space.

The integrative and associative activities of the cell assemblies within area 5 maintain an interactional image of the body. In this manner, a person can ascertain the position of the body and the limbs at rest and in motion (Gross et al., 1974). In part, this may be accomplished through comparisons with a more stable image of the body, which is possibly maintained via the combined interactions of neurons in areas 3, 1, 2. That is, a stable body image (or body image memories) are stored in these tissues. Hence, when the body has moved, this new information (received and processed in area 5) can be compared to the more stable trace (or memory maintained in the primary regions) so that the new position of the limbs and body can be ascertained. In this regard it could be argued that body-related memories are stored in the parietal lobe. Nevertheless, to determine position, sensation per se is not sufficient. Rather, sensation must be combined with input regarding movement or positional change (Gandevia and Burke, 1992). It is for this reason that in the absence of movement (and in the absence of visual cues, such as when one wakes up in the middle of the night), one usually cannot tell where or in what position his or her arms or legs may be in. However, with a slight movement we can immediately determine position.

Tactile Discrimination Deficits and Stereognosis

Destruction of the somesthetic association area results in many of the same disturbances that follow lesions of the primary region. These include abnormalities involving two-point discrimination, position sense, and pressure sensitivity. However, detection threshold is not altered (LaMotte and Mountcastle, 1979). For example, a patient may recognize touch, but be unable to localize what part of the body has been stimulated. For example, following a right parietal stroke, one patient made gross errors of localization, naming his elbow when I touched his leg, and his shoulder when his wrist was touched. He in fact expressed considerable astonishment when I allowed him to open his eyes after each test so as to see where the stimulation was actually being applied.

In addition, with destruction of this tissue, although a patient may be able to recognize that he is holding something in his hand, he may be unable to determine what it might be (astereognosis). In such cases, however, the cortical area representing the hand must be compromised. With small lesions involving only a particular part of the somesthetic cortex (e.g., the area representing the arm and shoulder), the deficit will be only manifested when that part of the body represented is stimulated. For example, the patient may be able to localize touch and determine the direction of a moving stimulus when it is applied to the hand, face, or leg, but be unable to do so along the shoulder or arm if this area of the cortex has been compromised.

The laterality of the lesion is also important, right-sided damage having more drastic effects on somesthesis and stereognostic functioning than left parietal lesions. In addition, although lesions to either parietal lobe can give rise to astereognosis (an inability to recognize objects tactually explored), lesions to the right parietal lobe are likely to give rise to bilateral abnormalities, whereas left parietal injuries generally effect only the right hand (Hom and Reitan, 1982). Again, however, for there to be astereognostic deficits requires that the somesthetic area representing the hand (see below) be compromised (Roland, 1976).

Larger lesions extending into the posterior parietal lobe, area 7, also decrease the ability to discriminate size, roughness, weight, and shape (Blum et al., 1950; Denny-Brown and Chambers, 1958; Garchas et al., 1982; Ridley and Ettlinger, 1975; Semmes and Turner, 1977)

Pain: Areas 5 and 7 and the Supramarginal Gyrus

Some neurons located in areas 5 and 7 of the parietal lobe demonstrate pain sensitivity; some area 7 neurons responding exclusively to thermal and nociceptive stimuli (Dong et al., 1994). Hence, in some instances, such as when the more inferior portion of areas 5 and 7 or the supramarginal gyrus (Broadmann's area 40) has been destroyed, patients may demonstrate a lack of emotional responsiveness to painful stimuli, become indifferent, develop an increased pain threshold, tolerate pain for an unusually long time, and fail to respond even to painful threats (Berkley and Parmer, 1974; Biemond, 1956; Geschwind, 1965; Greenspan and Winfield, 1992; Hyvarinen, 1982; Schilder, 1935)—particularly with right parietal destruction (Cubelli et al., 1984). However, disturbance or lack of pain sensation has occurred with lesions to either hemisphere (Hecaen and Albert, 1978). Frequently, although pain responsiveness may be diminished or absent, elementary sensation is intact and the ability to differentiate, for example, between dull and sharp is retained. The deficit is usually bilateral.

Some researchers have claimed that pain sensitivity sometimes is lost when the lesion involves the frontal-parietal cortex (Hecaen and Albert, 1978). However, the supramarginal gyrus of the inferior parietal lobule (Geschwind, 1965; Hyvarinen, 1982; Schilder, 1935) and area 7 of the superior parietal lobule (Dong et al., 1994; Greenspan and Winfield, 1992) are the most likely candidates for this condition—particularly because a second somesthetic area is located here, as well as yet another image of the human body (Penfield and Rasmussen, 1950). In this regard, Schilder (1935) has argued that the loss of reaction to pain is due to disturbances in the image of the body. That is, the experience or threat of pain is no longer related to the body image. Geschwind (1965), however, raises the possibility that this condition is due to disconnection from the limbic system (see Cavada and Goldman-Rakic, 1989). If this were the case, somesthetic (painful) sensation would no longer be assigned emotional significance. Loss of sensation or an inability to react to pain may also occur from subcortical lesions, especially within the thalamus.

Pain and Hysteria

Whereas destruction of the inferior portions of areas 5, 7, and 40 may result in loss of pain sensation, when the injury is secondary to tumor or seizure activity patients may instead report experiencing pain (Davidson and Schick, 1935; Hernandez-Peon et al., 1963; Ruff, 1980; Wilkinson, 1973; York et al., 1979). In addition, patients may experience sensory distortions that concern various body parts due to abnormal activation of the parietal neocortex.

For example, one 48-year-old housewife complained of diffuse, poorly localized (albeit intense) pain in her left leg, which occurred in spasms that lasted minutes. She subsequently was found to have a large tumor in the right parietal area, which, when removed, alleviated all further attacks. Head and Holmes (1911) reported a patient who suffered brief attacks of "electric shock"-like pain that radiated from the foot to the trunk; a glioma in the right parietal area subsequently was discovered. Sometimes the pain may be related to abnormal sexual or genital sensations.

For example, one 9-year-old boy with seizure activity in the right parietal area experienced spontaneous attacks of intense scrotal and testicular pain (York et al., 1979).

Ruff (1980) reports two patients who experienced paroxysmal episodes of spontaneous and painful orgasm, which was secondary to right parietal seizure activity. In one patient the episodes began with the sensation of clitoral warmth, engorgement of the breasts, tachycardia, etc., all of which rapidly escalated to a painful climax.

It is important to note, however, that although the predominant focus for paroxysmal pain is the right hemisphere, pain also has been reported to occur with tumors or seizures, activity that involves the left parietal region (Bhaskar, 1987; McFie and Zang-

will, 1960). Unfortunately, when the patient's symptoms are not considered from a neurological perspective, their complaints with regard to pain may be viewed as psychogenic. This is because the sensation of pain, stiffness, and engorgement, is, indeed, entirely "in their head" and based on distorted neurological perceptual functioning. Physical examination may reveal nothing wrong with the seemingly affected limb or organ. Thus such patients may be viewed as hysterical or hypochondriacal, particularly in that right hemisphere damage also disrupts emotional functioning.

AREA 7 AND THE SUPERIOR-POSTERIOR PARIETAL LOBULE

POLYMODAL INFORMATION PROCESSING

Information processed and analyzed in the primary and association somesthetic areas is then transmitted to area 7, where polymodal analyses take place (Dong et al., 1994; Previc, 1990). Jones and Powell (1970) considered area 7 to be concerned with the highest levels of somesthetic integration, as it receives massive input from the visual receiving areas in the occipital and middle temporal lobe, from the motor and nonmotor areas in the lateral frontal convexity, and from the inferior parietal lobule (Cavada and Goldman-Rakic, 1989; Dong et al., 1994; Jones and Powell, 1970; Previc, 1990; Wall et al., 1982). Cells in area 7 also have auditory receptive capacities, including the ability to determine sound location (Hyvarinen, 1982). Hence, area 7 is heavily involved in the analysis and integration of higher-order visual, auditory, and somesthetic information, and single neurons often have quite divergent capabilities.

Three-Dimensional Analysis of Body-Spatial Interaction

A single area 7 neuron via the reception of converging input from the primary and association somesthetic regions can monitor activities occurring in many different body parts simultaneously—for example, the position and movement of the arms, trunk, and legs (Leinonen et al., 1979). Via the reception of auditory and visual input, area 7 thus can also create a three-dimensional image of the body in space (Lynch, 1980).

Moreover, cells in this area not only receive information about body part interrelationships (such as are maintained by area 5), but the interaction of the body with external objects and events (Stein, 1992). Indeed, many cells in this vicinity become highly active when the hand is moved toward an object or while reaching for and/or manipulating objects (Mountcastle et al., 1975; Robinson et al., 1978; Yin and Mountcastle, 1977). These cells also act to coordinate and guide whole-body positional movement through visual and auditory space. As stated by Mountcastle and coworkers (1980), "the parietal lobe, together with the distributed system of which it is a central node, generates an internal neural construction of the immediately surrounding space, of the location and movements of objects within it in relation to body position, and of the position and movements of the body in relation to that immediately surrounding space. The region appears in general to be concerned with continually updating information regarding the relation between internal and external coordinant systems" (p. 522).

Visual-Spatial Properties

As noted in Chapter 1, the parietal lobe, in part, evolved from the hippocampus, and maintains rich interconnections with this nucleus. Given that the hippocampus is concerned with the body in space and with the position of various objects and stimuli in visual space, it is not surprising that the parietal lobe performs similar functions.

Although cells in area 7 receive and transmit data to the hippocampus, these neurons also accomplish these functions via the convergence of somesthetic information from area 5, visual input from the visual association areas, and the reception of midbrain and vestibular signals (Cavada and Goldman-Rakic, 1989; Gross and Graziano, 1995; Hyvarinene, 1982; Kawano

and Sasaki, 1984; Previc, 1990). Like some hippocampal neurons, single parietal neurons are also involved in the representation of space (Gross and Graziano, 1995; Stein, 1992).

Area 7 neurons also have quite large visual receptive fields, sometimes occupying a whole quadrant, a hemifield, or the entire visual field (Robinson et al., 1978). However, the receptive visual fields of these neurons do not usually include the fovea and are more sensitive to objects in the periphery and lower visual fields (Motter and Mountcastle, 1981; Previc, 1990), i.e., where the hands, feet, and ground are more likely to be viewed. In this regard, many of these cells are not concerned with the identification of form but rather place, position, and reaching distance as well as depth perception and the coordination of the body as it moves through space.

Because many of the neurons in area 7 receive highly processed visual (as well as limbic) input, including information regarding ocular movement and direction of gaze, they are also responsive to and can determine a variety of visual-object qualities and interrelationships, such as motivational significance, direction of movement, distance, spatial location, figure-ground relationships, and depth, including the discrimination and determination of an object's three-dimensional position in space (Andersen et al., 1985; Gross and Graziano, 1995; Kawano et al., 1984; Lynch, 1980; Previc, 1990; Sakata et al., 1980; Stein, 1992). They are largely insensitive to velocity or speed of movement. In addition, many cells respond most to stimuli that are within grasping distance, whereas others respond most to stimuli that are just beyond arm's reach.

Neurons in this supramodal region are able to accomplish this by responding to somesthetic positional information provided by area 5, visual input from areas 18 and 19 and the inferior and middle temporal lobe and hippocampus, and from extraretinal signals regarding convergence and accommodation of the eyes, and the position and movement of the eyes while tracking (Gross and Graziano, 1995; Previc,

1990). Indeed, electrical stimulation of this area elicits eye movements as well as convergence, accommodation, and pupil dilation (Jampel, 1960). By integrating these signals, these cells are able to monitor and mediate eye movement and visual fixation, map out the three-dimensional positions of various objects in visual space, and determine the relation of these objects to the body and to other objects. Thus the visual analysis performed by many of these cells is largely concerned with visual spatial functions (Robinson et al., 1978; Sakata et al., 1980). Moreover, this area maintains extensive interconnections with two other regions highly concerned with visual functions and eye movements, i.e., the frontal eye fields and superior colliculus (Jones and Powell, 1970; Kawamura et al., 1974; Knuezle and Akert, 1977; Pandya and Kuypers, 1969; Previc, 1990) as well as visual areas 18 and 19 (Jones and Powell, 1970).

Conversely, when this area is damaged, depth perception, figure-ground analysis, and the ability to track objects or to correctly manipulate objects in space (e.g., constructional and manipulospatial skills) are compromised.

Visual Attention

Many neurons in area 7 can act so as to increase or decrease visual fixation, direct attention to objects of motivational significance, and promote the maintenance of visual grasp such that a moving object continues to be visually scanned and followed (Lynch et al., 1977; Previc, 1990). Electrical stimulation of this area induces lateral eye movements because of interconnections maintained with the frontal eye fields, visual cortex, and subcortical visual centers.

In contrast, lesions to this area often disrupt attentional functioning, such that in the extreme (e.g., following a right parietal lesion) patients fail to attend to the left half of not only visual space, but the left half of the body; i.e., neglect. Neglect is most common with right cerebral lesions but sometimes initially follows massive left cerebral destruction (see Chapter 3; see also Umilta, 1995).

As detailed below, with circumscribed parietal lesions involving area 7, neglect is more pronounced for the lower visual fields. The lower visual fields, in turn, are a predominant sensory domain of the parietal lobes, for this is the area of visual space in which the hands and feet are most likely to be viewed—therefore making movement of the body in space more efficient and coordinated. The lower visual field is also the area in which objects are most likely to be physically explored by the hands, which in turn are guided by the parietal lobe.

Motivational and Grasping Functions

A number of cells in area 7 have been described as exerting "command" functions (Mountcastle, 1976; Mountcastle et al., 1980), especially those located along the inferior lateral convexity. These cells are motivationally responsive, can direct visual attention, become excited when certain objects are within grasping distance, and can motivate and guide hand movements, including the grasping and manipulation of specific objects (Hyvarinene and Poranen, 1974; Lynch et al., 1977; Mountcastle, 1976). Most of these cells cease to fire when the object fixated upon is actually grasped, suggesting that they may be exerting some type of driving force or at least an alerting function so that objects of desire or of general (versus specific) interest will be attended to (Rolls et al., 1979).

It has been argued that many neurons in area 7 actually execute a matching function between the internal drive state of the subject and the object that is being attended to. That is, by responding to signals transmitted from the limbic system (Cavada and Goldman-Rakic, 1989; Mesulam et al., 1977), e.g., the cingulate gyrus as well as the middle and inferior temporal lobe, these cells in turn direct visual attention to objects of potential interest, and when detected, act so as to maintain visual grasp (Lynch et al., 1977). In other words, when an object is recognized as being of motivational significance (determined by the limbic system and visual form recognition neurons in the temporal lobe), this information is relayed to neurons in area 7. Although not concerned with form recognition, these cells will guide as well as monitor eye movement so that the object of interest is fixated upon. These cells then exert motor command functions so that the hand is guided toward the object until it is grasped.

RIGHT AND LEFT PARIETAL LOBES: LESIONS AND LATERALITY
Attention and Visual Space

Lesions involving the superior as well as the inferior parietal lobule (of which area 7 is part) and the parietal-occipital junction can greatly disturb the ability to make eye movements, maintain or shift visual attention, and visually follow moving objects; in the extreme such lesions result in oculomotor paralysis (Hecaen and De Ajuriaguerra, 1954; Previc, 1990).

Right parietal lesions are associated with deficiencies involving depth perception and stereopsis, including the ability to determine location, distance, spatial orientation, and object size (Benton and Hecaen, 1970; Ratcliff and Davies-Jones, 1972). Visual constructional abilities may also be compromised (see Cowey, 1981; Critchley, 1953), and many patients suffer from visual-spatial disorientation and appear clumsy. Persons with right parietal lesions show defective performance on line orientation tasks (Benton et al., 1975; Warrington and Rabin, 1971), maze learning (Newcombe and Russell, 1969), the ability to discriminate between unfamiliar faces (Milner, 1968), or to select from the visual environment stimuli that are of importance (Critchley, 1953). With right parietal-occipital injuries, patients may be deficient on tasks requiring detection of imbedded figures (Russo and Vignolo, 1967).

Others may also have severe problems with dressing (e.g., dressing apraxia) and may become easily lost or disoriented even in their own homes. One patient I examined with a gun shot wound involving predominantly the right superior posterior parietal area could not find his way to and from his hospital room (although he had been an in-patient for over 3 months) and on several occasions had difficulty finding his way out of the bathroom. Indeed, in one instance he was discovered feeling his way along the walls in his attempt to find the door.

LOCALIZATION OF OBJECTS IN SPACE

Patients with right parietal lesions perform defectively on visual localization tasks (Hannay et al., 1976). However, Ratcliff and Davies-Jones (1976) found that the localization of stimuli within grasping distance is disrupted equally by lesions to the posterior region of either hemisphere. Hence, the right parietal lobe appears to play an important role in generalized localization, whereas the left exerts influences in regard to objects that may be directly grasped and manipulated.

APRAXIA

Because of the involvement of the parietal region in mediating hand, arm, and body movements in space (Joseph, 1993; Kimura, 1993; Lynch et al., 1977), as well as in temporal sequencing, damage to this vicinity can also result in apraxia such that individuals have difficulty controlling or temporally sequencing the extremities. This is especially evident with left rather than right parietal injuries. There can also result gross inaccuracies as well as clumsiness when making reaching movements or when attempting to pick up small objects in visual space (Kimura, 1993; Lynch, 1980). Tendon reflexes may be slowed, and hypotonia coupled with a paucity of and/or a slowness in movement initiation may result (Denny-Brown and Chambers, 1958; LaMotte and Acunam, 1978; Lynch, 1980). Left parietal damage may result in difficulty in visually recognizing objects (agnosia), and left-right orientation may be grossly deficient.

EMOTION

Some of the effects of lesions in this region also include altered emotional-motivational functioning, body and visual-spatial neglect, and clumsiness and visual-spatial disorganization.

With massive right parietal lesions involving area 7, many patients are often initially hypokinetic and seem very passive, inattentive, and unresponsive, and they take very little interest in their environment (Critchley, 1953; Heilman and Watson, 1977). Moreover, when their disabilities are pointed out (e.g., paresis, paralysis), they may seem indifferent or conversely euphoric (Critchley, 1953).

Areas 7 and 5 (and the inferior parietal lobule) receive auditory information (Cavada and Goldman-Rakic, 1989; Hyvarinen, 1982; Roland et al., 1977) and can discern the emotional-motivational significance of this input as well as differentiate between different vocal-emotional characteristics—especially right parietal neurons. Hence, when the right parietal region is damaged, patients not only seem unconcerned about their disability, but they may have difficulty in perceiving and differentiating between different forms of emotional speech (Tucker et al., 1977).

INFERIOR PARIETAL LOBULE

Developmentally, of all cortical regions, the inferior parietal lobule is one of the last to functionally and anatomically mature (Blinkov and Glezer, 1968; Joseph and Gallagher, 1985; Joseph et al., 1984). Hence, many capacities mediated by this area (e.g., reading, calculation, the performance of reversible operations in space) are late to develop, appearing between the ages of 5 and 8.

Sitting at the junction of the temporal, parietal, and occipital lobes, the inferior region (which includes the angular and supramarginal gyri) has no strict anatomical boundaries, is partly coextensive with the posterior-superior temporal gyrus, and includes part of area 7. It maintains rich interconnections with the visual, auditory, and somesthetic association areas, the superior colliculus via the pulvinar, and the lateral geniculate nucleus of the thalamus, as well as massive interconnections with the frontal lobes, inferior temporal region, and other higher-order assimilation areas

throughout the neocortex (Bruce et al., 1986; Burton and Jones, 1976; Geschwind, 1965; Jones and Powell, 1970; Seltzer and Pandya, 1978; Zeki, 1974).

Multi-Modal Assimilation Area

As noted in Chapters 2 and 4, over the course of evolution the amygdala, hippocampus, and medial temporal lobe began to balloon outward and upward, giving rise to the superior temporal lobe, and then continued to expand in a posterior direction, forming part of the angular and marginal gyrus. Hence, this portion of the inferior parietal lobule has auditory and thus (in the left hemisphere) language capabilities. However, with the evolution of the thumb and the capability of precise grasping coupled with tool-making and related temporal-sequential tasks, the superior parietal lobule also expanded, thereby giving rise to inferior parietal neocortical tissue.

Given its location at the border regions of the somesthetic, auditory, and visual neocortices, and containing neurons and receiving input from these modalities, the inferior parietal lobule became increasingly multimodally responsive as it evolved—a single neuron simultaneously receiving highly processed somesthetic, visual, auditory, and movement-related input from the various association areas. Hence, many of the neurons in this area are multispecialized for simultaneously analyzing auditory, somesthetic, and spatial-visual associations, and have visual receptive properties that encompass almost the entire visual field, with some cells responding to visual stimuli of almost any size, shape, or form (Bruce et al., 1982, 1986; Hyvarinen and Shelepin, 1979).

Inferior parietal neurons are involved in the assimilation and creation of cross-modal associations and act to increase the capacity for the organization, labeling, and multiple categorization of sensory-motor and conceptual events (Geschwind, 1965; Joseph, 1982). One can thus create visual, somesthetic, or auditory equivalents of objects, actions, feelings, and ideas, simultaneously—for example, conceptualizing a "chair" as a word, a visual object, or in regard to sensation, usage, and even price.

Language Capabilities

Because the left angular gurus is involved in language functions damage can result in anomia, i.e., severe word-finding and confrontive naming difficulty. Such patients have difficulty naming objects, describing pictures, etc. Moreover, with lesions involving the angular gyrus, or when damage occurs between the fiber pathways linking the left inferior parietal lobule with the visual cortex, pure word blindness also can result. This is due to an inability to receive visual input from the left and right visual cortices and to transmit this information to Wernicke's area so that auditory equivalents may be called up. Such patients thus cannot read and suffer from alexia.

Because the inferior parietal lobule also acts as a relay center where information from Wernicke's region can be transmitted, via the arcuate fasciculus, to Broca's area (for expression), destructive lesions, particularly to the supramarginal gyrus of the left cerebral hemisphere, can result in conduction aphasia (see Chapter 4). Although comprehension would be intact and a patient would know what she wanted to say, she would be unable to say it. Nor would she be able to repeat simple statements, read out loud, or write to dictation. This is because Broca's area is disconnected from the posterior language zones.

Agraphia

It has been argued that the sensory motor engrams necessary for the production and perception of written language are stored within the parietal lobule of the left hemisphere (Strub and Geschwind, 1983). In fact, given that the parietal lobes are concerned with the hands and lower visual fields, they not only guide and observe hand movements, but learn and memorize these actions, including those involved in writing. Hence, when these areas are lesioned, patients sometimes have difficulty writing and forming letters because of an inability to access these engrams (Strub and Geschwind, 1983; Vignolo, 1983), i.e., they suffer from agraphia, an inability to write (detailed in Chapter 4). Writing samples may be characterized by misspellings, letter omissions, distortions, temporal-sequential misplacements, and inversions (Kinsbourn and Warrington, 1964). Sometimes agraphia is accompanied by alexia, the inability to read (Benson and Geschwind, 1969; Hecaen and Kremin, 1977).

Lateralized Temporal-Sequential Functions

Because the inferior parietal lobe receives so much information and aids the rest of the brain in various forms of analysis, one of its functions is to maintain track of input and output so that information may be organized appropriately in either a sequential (i.e., first, middle, last) or a spatial frame-

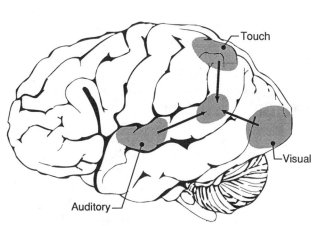

Figure 12.4. Schematic diagram of converging sensory flow to the inferior parietal lobule.

work. Hence, another side effect of lesions localized to the inferior parietal lobule is a disruption of visual-spatial functioning and temporal sequencing ability (e.g., apraxia), as well as logic, grammar, and the capacity to perform calculations, depending on which hemisphere is compromised.

Individuals with lesions involving the inferior-parietal-occipital border of either hemisphere may have difficulty in carrying out spatial-sequential tasks—for example, drawing "a square beneath a circle and a triangle beneath a square" (Luria, 1980). Often they draw the objects in the order described (i.e., square, circle, triangle, square). That is, they have difficulty in conceptualizing how to place the objects in relation to each other.

Patients with left inferior parietal lesions have trouble with more obvious sequential-grammatical relationships (Luria, 1980). For example, they may be unable to understand the question: "John is taller than Jim but shorter that Pete. Who is tallest?" In part, this is not only a function of left parietal dysfunction but of the right hemisphere's difficulty in dealing with temporal-sequential and grammatical relations.

Because the right brain does not understand grammatical relationships, a sentence that starts with the name "John" is interpreted by the right parietal area as all about "John," i.e., the first word of the sentence is understood by the right brain as the "agent" regardless of semantics or grammar (Chernigovaskaya and Deglin, 1986). In this manner, if presented with the sentence, "give me the book after you give me the pencil," the right brain would respond to the order of presentation rather than the grammatical relationship and would thus present the book, then the pencil. When left parietal input is abolished, proper temporal-sequential and grammatical programming and comprehension suffers.

APRAXIA
Sensory Guidance Of Movement

The parietal lobe is highly concerned with the mediation of movement. As noted in Chapter 11, the primary motor cortex ex-

tends well beyond area 4 and includes portions of the somatosensory regions, which in turn contribute almost one-third of the fibers that make up the pyramidal tract. These areas are in fact richly interconnected (Jones and Powell, 1970). Together, the motor and somesthetic regions comprise a single functional unit, i.e., the sensorimotor cortex. Nevertheless, it is important to emphasize, as pointed out by Luria (1980), that every voluntary movement is in fact composed of a series of movements that are spatially organized in accordance with successively changing input from other modalities (Kimura, 1993). That is, in order for a movement to be correctly planned and carried out, signals must be directed to the right muscle groups as based on efferent streams of visual, somesthetic, and auditory input. This includes information regarding the position of the body and limbs in space. Indeed, movement becomes extremely difficulty without sensory feedback and guidance. Because of this, parietal lesions can result in unilateral paresis and even wasting (i.e., parietal wasting). Hence, the somesthetic cortex is important in the guidance of movement; in fact, some neurons fire prior to making a movement (Lebedev et al., 1994).

Although the entire parietal lobule makes important contributions, the superior and inferior parietal lobules of the left hemisphere appear to be the central region of concern regarding the performance of skilled temporal-sequential motor acts. This is because the motor engrams for performing these acts appear to be stored in the left angular and supramarginal gyri (Geschwind, 1965; Heilman, 1993)—a consequence, in part, of this brain area's unique ability to guide, visually observe, and thus selectively learn hand movements, gestures, and complex temporal sequential actions such as those involving tool construction (Joseph, 1993). Related memories are therefore stored in this cortex.

Conversely, these hand movement related memories assist in the programming of the motor frontal cortex, where the actions are actually executed. However, the inferior parietal lobule in turn depends on input from the primary and association somesthetic areas.

There is some evidence of laterality in regard to all of the above; the right half of the brain may be more concerned with movement of the trunk and the lower extremities. This would include navigational movement through space, running, certain types of dancing, and actions requiring analysis of depth and balance. The left cerebral hemisphere exerts specialized influences on the upper extremities, including the control of certain types of complex, sequenced motor acts such as those requiring alterations in the orientation and position of the upper limbs (Haaland and Harrington, 1994; Kimura, 1982, 1993).

Apraxia

If the left inferior parietal region is destroyed, the patient loses the ability to perform actions in an appropriate temporal sequence or to even appreciate when they have been performed incorrectly. Such patients may also be impaired in their ability to acquire or perform tasks involving sequential changes in the hand or upper musculature (Kimura, 1979, 1982, 1993), including well learned, skilled, and even stereotyped motor tasks such as lighting a cigarette or using a key.

Apraxia is a disorder of skilled movement in the absence of impaired motor functioning or paralysis. Apraxic patients usually show the correct intent but perform the movements clumsily. As with many other types of disturbances, patients and their families may not notice or complain of apraxic abnormalities. This is particularly true if they're aphasic or paralyzed on the right side. That is, clumsiness with either extremity may not seem significant. Hence, this condition is something that requires direct evaluation.

Performance deteriorates the most when a patient is required to imitate or pantomime certain actions, including the correct usage of some object (McDonald et al., 1994). For example, the patient may be asked to show the examiner "how you would use a key to open a door" or "hammer a nail into a piece of wood." In many cases, the patient uses the body, i.e., a finger, as an object (e.g., a key) rather than the finger and thumb holding the key. Although performance usually improves when such patients use the real objects (Geschwind, 1965; Goodglass and Kaplan, 1972), a rare few show the disturbance when using the real object as well (Heilman, 1993; Kimura, 1993).

In addition, patients with apraxia may demonstrate difficulty in properly sequencing their actions. For example, they may pretend to stir a cup of coffee, then pretend to pour the coffee into the cup, and then take a sip. However, the individual acts may be performed accurately.

Broadly speaking, there are several forms of apraxia, which like many of the disturbances already discussed may be due to a number of causes or anatomical lesions. These include ideational apraxia, ideomotor apraxia, buccal facial apraxia, constructional apraxia, and dressing apraxia. With the exception of dressing and constructional apraxia, apraxic abnormalities are usually secondary to left hemisphere damage, particularly injuries involving the left frontal and inferior parietal lobes.

For example, Kimura (1982, 1993) found that the ability to perform meaningless oral or hand movements was related to the frontal or posterior nature of the lesion, such that persons with frontal lesions were impaired on oral tasks, whereas those with parietal lesions had the most difficulty in making hand postures or complex movements of the extremities (Kolb and Milner, 1981). Thus, apraxic abnormalities secondary to left cerebral lesions tend to involve either destruction of the inferior parietal lobule or lesions resulting in disconnection of the frontal motor areas (or the right cerebral hemisphere) from this more posterior region of the brain.

If the inferior parietal region is destroyed, the patient loses the ability to appreciate when he or she has performed an action incorrectly. If the motor region is destroyed, although the act is still performed inaccurately (because of disconnection from the inferior parietal lobule), the patient is able to recognize the difference (Heilman, 1993; Kimura, 1993).

Ideomotor Apraxia

Ideomotor apraxia is usually associated with lesions within the inferior parietal lobe of the left hemisphere. Rather than problems with temporal-sequencing of motor acts per se, these patients tend to be very clumsy when performing an act and/or they may perseverate and erroneously perform a previous movement. These individuals also tend to be very deficient when attempting to perform an action via pantomime or when engaged in meaningful imitation, meaningless imitation, and the meaningful or meaningless use of actual objects (Goodlass and Kaplan, 1972; Heilman, 1973; Kimura and Archibald, 1974; McDonald et al., 1994). This is presumably due to destruction of the engrams important in motor performance. Patients will demonstrate apraxic abnormalities in both the right and the left hand.

In addition, patients with ideomotor apraxia tend to have difficulty with simple versus complex movements, although various elements within a complex action may be performed somewhat abnormally (Hecaen and Albert, 1978; McDonald et al., 1994). Hence, actions such as waving good-bye, throwing a kiss, and making the "sign of the cross" may be performed deficiently. Moreover, many patients tend to uncontrollably comment on their actions, i.e., "verbal overflow." In other words, when asked to "wave good-bye they may say "good-bye" while waving even when instructed to say nothing.

It has been suggested that ideomotor apraxia can occur in the absence of ideational apraxia, but that the converse is not true (Hecaen and Albert, 1978). In this regard, ideomotor apraxia may be a less severe form of ideational apraxia (Kimura, 1993).

Ideational Apraxia

This form of apraxia is usually due to severe disturbances in the temporal sequencing of motor acts. That is, the separate

chain of links that constitute an entire movement become dissociated, such that the overriding idea of the movement in its entirety is lost. Hence, these patients commit a number of temporal and spatial errors when making skilled movements, although the individual elements, in isolation, may be preserved and performed accurately (Hecaen and Albert, 1978; Luria, 1980). For example, patients (via pantomime) may rotate their hand before inserting the key, drink from a cup before filling it from a pitcher of water, or puff from a cigarette and then lighting it. Thus they incorrectly sequence a series of acts. Both hands are effected.

Because of conceptual, ideational abnormalities, they may also have difficulty in using actual objects correctly. During pantomime they may use a body part as object such as an index finger for a key. Even so, their actions are out of temporal sequence. Hence, these patients seem to be unable to access the motor engrams (or "memories") that would allow them to perform appropriately (Luria, 1980). In this regard, patients are sometimes hesitant to perform a task as they have difficulty in understanding what has been asked of them. However, they can often describe verbally what they are unable to perform (Heilman, 1993).

Left-Sided or Unilateral Apraxia (also called Callosal and Frontal Apraxia)

Patients with unilateral (callosal/frontal) apraxia are unable to imitate or perform certain movements with their left (but not right) hand and are clumsy in their use of objects. Left-sided apraxia is sometimes due to a lesion of the anterior corpus callosum or left frontal motor cortex. This is because lesions of the corpus callosum or premotor and motor region of the left hemisphere can result in a disconnection syndrome, i.e., the motor areas of the right hemisphere cannot gain access to the motor engrams stored within the left inferior parietal lobe. Thus, a left frontal lesion results in apraxia of the left hand and paralysis of the right. Often this is secondary to strokes within the distribution of the anterior cerebral artery such that the anterior portion of the corpus callosum is destroyed (Geschwind, 1965).

It is noteworthy that patients also may show deficient finger tapping performance in the left hand due to apraxic abnormalities secondary to left hemisphere injury (Heilman, 1975). In these instances, reduced finger tapping is bilateral.

DRESSING APRAXIA

Dressing apraxia is usually secondary to right hemisphere lesions involving the inferior parietal region, and as the name implies, patients have difficulty putting on their clothes. For example, a patient may attempt to put a shirt on upside-down, then inside-out, and then backwards. Severe spatial-perceptual abnormalities as well as body image disturbances are usually contributing factors.

APHASIA AND APRAXIA

Many patients who are aphasic also appear to be apraxic because they have severe difficulty in comprehending language and understanding motor commands. That is, a patient may fail to perform a particular action because he or she doesn't comprehend what is being asked.

To distinguish between receptive aphasic abnormalities and apraxia, one must ask "yes" and "no" questions ("are you in a hospital?"), require patients to perform certain actions via pantomime ("show me how you would throw a ball" or "show me how a soldier salutes"), and require pointing response ("point to the lamp"). If they can answer appropriately "yes" or "no" or point to objects named but cannot execute commands, they have apraxia. It is important to note that in severe cases apraxic patients may have difficulty even with pointing.

Pantomime Recognition

Persons with damage involving the left parietal lobule not only make errors when performing motor acts but when comprehending, recognizing, and discriminating between different types of motor acts such

as those demonstrated via pantomime (Heilman et al., 1982; McDonald et al., 1994). Moreover, persons with lesions in the left inferior occipital lobe have also been shown to have difficulty in verbally understanding, describing, or differentiating between pantomimes (Rothi et al., 1986). That is, in the extreme, if such patients were to pantomime the pouring of water into a glass versus lighting and smoking a cigarette, they would have problems in describing what they have viewed or in choosing which was which.

Deficits in pantomime recognition occur frequently among persons with aphasia (Varney, 1978). Moreover, this disturbance is also significantly correlated with reading comprehension (Gainotti and Lemmo, 1976; Varney, 1978). In this regard, patients with alexia frequently suffer from pantomime recognition deficits as well. Because of this relationship it has been suggested that the ability to read may be based on or derived from the ability to understand gestural communication, i.e., the reading of signs (Joseph, 1993; Varney, 1982).

Wang and Goodglass (1992) and Kimura (1993) argue that pantomime imitation and production are related to both apraxia and the ability to interpret purposeful movements, the engrams for which are located in the inferior parietal lobule (Heilman et al., 1982; Joseph, 1993; Wang and Goodglass, 1992). In contrast with left anterior lesions, which may impair motor functioning and the capacity to imitate, the ability to comprehend pantomime is retained (Heilman et al., 1982).

As noted, the inferior and superior parietal lobules receive considerable visual input, particularly from the periphery and lower visual field—the area in which the hands are most likely to be viewed. Hence, this area of the brain views, manipulates, guides, and mediates hand-object coordination and reaching movements, including the comprehension of hand movements, i.e., gestures (Joseph, 1993). Hence, when the left superior and inferior parietal lobule is destroyed, gestural comprehension, including the understanding of (as well as the capacity to execute) complex gestures, including American Sign Language (ASL) and the capacity to engage in complex temporal sequential acts, is significantly affected (reviewed in Joseph, 1993; see Chapter 4). If the right parietal area is destroyed, these deficits may also include constructional apraxia (see Chapter 3).

Constructional Apraxia

Constructional apraxia is by no means a unitary disorder (Benton, 1969; Benson and Barton, 1970) and may be expressed in a number of ways. On a drawing or copying task, this expression may include the addition of unnecessary or nonexistent details or parts, misalignment of or inattention to details, disruptions of the horizontal and vertical axes with reversals or slight rotations in reproduction, and scattering of parts. For example, in performing the Block Design subtest from the WAIS-R, the patient may correctly reproduce the model but angle it incorrectly. In drawing or copying figures, the patient may neglect the left half, draw over the model, and misalign details.

Moreover, although constructional deficits are more severe after right hemisphere damage (Arrigoni and DeRenzi, 1964; Black and Strub, 1976; Benson and Barton, 1970; Critchley, 1953; Hier et al., 1983; Joseph, 1988a; Kimura, 1993; Piercy et al., 1960), disturbances involving constructional and manipulo-spatial functioning can occur with lesions to either half of the brain (Arrigoni and DeRenzi, 1964; Mehta et al., 1987; Piercy et al., 1960). Hence, depending on the laterality as well as the extent and site of the lesion, the deficit may also take different forms. For example, following posterior right cerebral lesions, rather than apraxic, the patient is spatially agnosic, i.e., suffering from constructional agnosia and a failure to perceive and recognize visual-spatial and object interrelationships. In other cases, such as following left cerebral injury, the disturbance may be secondary to a loss of control over motor programming (Kimura, 1993; Warrington, 1969; Warrington et al., 1966).

Although visual motor deficits can result from lesions in either hemisphere (Arrigoni and KDeRenzi, 1964; Piercy et al., 1960; Kimura, 1993), visual-perceptual disturbances are more likely to result from right hemisphere damage. In contrast, lesions to the left half of the brain may leave the perceptual aspects undisturbed, whereas visual motor functioning and selective organization may be compromised (Kim et al., 1984; Mehta et al., 1987; Poeck et al., 1973). As such the patient is likely to recognize that errors have been made.

In general, the size and sometimes the location of the lesion within the right hemisphere has little or no correlation with the extent of the visual-spatial or constructional deficits demonstrated, although right parietal lesions tend to be worst of all. With right parietal involvement patients tend to have trouble with the general shape and overall organization, the correct alignment, and the closure of details, and there may be a variable tendency to ignore the left half of the figure or to not fully attend to all details. Moreover the ability to perceive (or care) that errors have been made is usually compromised.

Conversely, constructional disturbances associated with left hemisphere damage are positively correlated with lesion size, and left anterior lesions are worse than left posterior lesions (Benson and Barton, 1970; Black and Bernard, 1984; Black and Strub, 1976; Kimura, 1993; Lansdell, 1970). This is because the capacity to control and program the motor system has been compromised. The larger the lesion, the more extensive the deficit.

Moreover, because the left hemisphere is concerned with the analysis of parts or details and engages in temporal-sequential motor manipulations, lesions result in oversimplification and a lack of detail, although the general outline or shape may be retained (Gardner, 1975; Levy, 1974). However, in some cases, when drawing, a patient tends to more greatly distort the right half of the figure with some preservation of left-sided details (see Fig. 12.5).

GERSTMANN'S SYNDROME: FINGER AGNOSIA, ACALCULIA, AGRAPHIA, LEFT-RIGHT CONFUSION

The Knowing Hand

It has been said that the parietal lobule is an organ of the hand. As noted, the hand appears to be more extensively represented than any other body part; parietal neurons respond to hand movements and manipulations, mediate and visually observe temporal-sequential hand movements, and become highly activated in response to objects that are within grasping distance. Indeed, given that the parietal lobe receives lower visual field input, the area in which the hands are most likely to be viewed, it is therefore highly sensitive to and concerned with the somesthetic as well as the visual guidance of hand (and lower extremity) movements.

It is also via the hand that the parietal lobe gathers information regarding various objects, i.e., stereognosis, and about the Self and the World, so that things and body parts come to be known, named, and identified. Ontogenetically, the hand is in fact primary in this regard. That is, the infant first uses the hand to grasp various objects so that they may be placed in the mouth and orally explored. As the child develops, rather than mouthing, more reliance is placed solely on the hand (as well as the visual system) so that information may be gathered through touch, manipulation, and visual inspection.

As the child and its brain matures, instead of predominantly touching, grasping, and holding, the fingers of the hand are used for pointing and then naming—hand activities that are most likely to be viewed (and later remembered) by the parietal lobe. It is these same fingers that are later used for counting and the development of temporal-sequential reasoning—i.e., the child learns to count on his or her fingers, then to count (or name) by pointing at objects in space.

In this regard, counting, naming, object identification, finger utilization, and hand control are ontogenetically linked and seem to rely on the same neural substrates for their expression, i.e., the left inferior and superior parietal lobules. In this regard, the parietal lobe, because of its control over the hand, and its chronic surveillance of the lower visual field, is therefore the unique recipient and observer of specialized hand-related activities and has thus become organized accordingly in regard to the guidance, as well as the learning and memory, of hand- and finger-related activities.

Hence, when the more posterior portions of the left hemisphere are damaged, naming (anomia), object and finger identification (agnosia), arithmetical abilities (acalculia), and temporal-sequential control over the hands (apraxia) are frequently compromised.

A variety of symptoms are therefore associated with left inferior parietal lobe damage, some of which occur as a related constellation of disturbances, i.e., finger agnosia, acalculia, agraphia, and left-right disorientation: Gerstmann's syndrome (Gerstmann, 1930, 1942; Strub and Geschwind, 1983). Gerstmann's symptom complex is most often associated with lesions in the area of the supramarginal gyrus and superior parietal lobule (Hrbek, 1977; Strub and Geschwind, 1983). However, because the symptoms of this complex do not always occur together, some authors have argued that Gerstmann's syndrome, per se, does not exist. We will not take sides on this controversy but instead will focus on those aspects of Gerstmann's syndrome that have not yet been discussed (finger agnosia, acalculia, left-right disorientation).

Finger Agnosia

Finger agnosia is not a form of finger blindness, as the name suggests. Rather, the difficulty involves naming and differentiating among the fingers of either hand as well as the hands of others (Gerstmann, 1940). This includes pointing to fingers named by the examiner, or moving or indicating a particular finger on one hand when the same

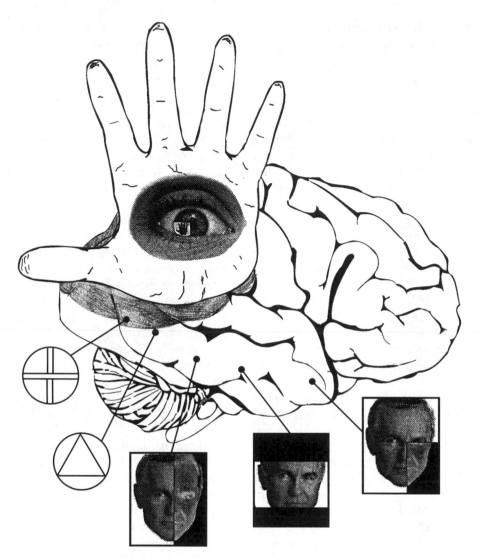

Figure 12.5. The parietal lobe has been considered a "lobe of the hand." The parietal lobule guides and comprehends hand movements and gestures. Facial and complex geometric recognition neurons are located within the inferior temporal lobe.

finger is stimulated on the opposite hand.

For example, if you touch the finger of such patients while their eyes are closed, and ask them to touch the same finger, they may have difficulty. If the disturbance is subtle the examiner may wish to stimulate two fingers and then ask the patient to indicate, in order, the same fingers. Many patients have difficulty on these tests regardless of whether they are administered in a verbal (naming) or a nonverbal (touching) format (Kinsbourne and Warrington, 1962). In general, the middle three fingers

are hardest to recognize, and the agnosia is demonstrable in both hands.

Although finger agnosia is rare among patients who have right hemisphere lesions (Hecaen, 1962), many patients also demonstrate some visual-constructive disability (Kinsbourne and Warrington, 1962). Hence, in testing for this disorder, both verbal and nonverbal forms help to determine the side of the lesion.

Often patients who have difficulty identifying fingers by name or simply differentiating between them nonverbally also suffer from receptive language abnormalities (Sanguet et al., 1971). Nevertheless, this disorder is not merely a manifestation of aphasia because finger agnosia may appear in the absence of language abnormalities (Strub and Geschwind, 1983).

In part, a good way of determining if the disorder is secondary to a right versus a left hemisphere injury is by noting if patients have more problems recognizing fingers on the right versus the left hand, or in transferring from the right to the left hand (or vice versa)—that is, by stimulating a finger (or fingers) on the right hand (while it or they are out of sight) and then having the patient indicate the same finger or fingers on the left hand. One must rule out deficient attentional functioning in making this diagnosis. Of course, one should not base a diagnosis of brain damage merely on poor performance of this one index but should look at the overall pattern of deficiency.

THE EVOLUTION OF GEOMETRY AND MATH

Three hundred thousand years ago, someone took a piece of red ocher pigment and sharpened it, presumably to mark something (Pfeiffer, 1985). On what surface it drew on and what the nature of the composition may have been, we do not know. We can only guess that it served some symbolic purpose, or it may have merely served only to make a mark.

Three hundred thousand years ago someone, took the rib of an ox and carved a series of geometric double arches on it (Pfeiffer, 1985). Was he or she just doodling, or was this a common form of artistic expression even in those lost days and forgotten nights? Again, we do not know.

Sixty thousand years ago, Neanderthals were painting their caves red; by the time they were overrun by the Cro-Magnon people twenty thousand years later, geometric patterns, designs, and doodles graced many a wall and cavern (Leroi-Gourhan, 1964; Prideaux, 1973).

However, it was not until about twenty thousand years ago that people began leaving marks on rocks and walls that suggested that may have been keeping track of, or counting, something. Perhaps the phases of the moon, or the number of animals killed? No one knows. Just as we have no idea when the first complex sentence was spoken, or when the first words were written, the point at which human beings first began to count or to measure the geometric properties of the land or the surrounding universe remains a mystery.

Geometry and the first forms of spoken and written pictorial language appear to be naturally related to the functional integrity of the right half of the brain (see Chapters 3 and 4) and the parietal lobe, which is exceedingly concerned with visual-spatial relations. It is likely, however, that the first (temporal-sequential) mathematical concepts were promulgated by the left cerebral hemisphere and parietal lobe, and like writing, were related to hand use (Joseph, 1993). That is, one first counts on his or her fingers, and then learns to count by pointing with one's fingers at objects that one wishes to add together, and then later to grasp a pen or pencil and make marks and signs that indicate the numbers used and their summations—actions that are guided and observed by the parietal lobes.

The decimal system is clearly an outgrowth on this reliance on the fingers for counting, for this system is based on the concept of tens. Even the decimal systems employed by the ancients of Meso-America was digitalized, with the exception that they used a base of 20 as they apparently counted their toes.

It has been postulated that human beings first became concerned with geometry and numbers with the advent of agriculture (around ten thousand years ago, after the last great flood) in order to count their crops as well as survey their fields. In addition, geometry may well have first been employed to survey the heavens: visual space. It was because of geometric-heavenly concerns that many of the ancients considered geometry to be the math of the gods and of divine origin. Perhaps this is why almost all ancient temples and buildings were not only constructed according to complex geometric principles, but oriented in regard to certain celestial configurations, including the temples of ancient Sumer.

The Sumerians were exceedingly knowledgeable about complex geometric principles (Kramer, 1981; Wooley, 1965), as were the Egyptians (Breasted, 1909; Gardner, 1961; Wilson, 1951). Although it is apparent that the Sumerians were also familiar with and utilized a decimal system, and that by 4000 years ago the Babylonians had developed the fundamental laws of mathematics, both cultures nevertheless relied on a sexagesimal system for their complicated calculations because it was far superior to the decimal (Chiera, 1966; Kramer, 1981; Wooley, 1965). For example, whereas the decimal system can be factored by 2, 5, and 10, the 60-unit sexagesimal system could be factored by 2, 3, 4, 5, 6, 9, 10, 12, 15, and so on.

The sexagesimal system is also clearly related to the geometry of space and the partition of what the ancients considered the cosmic, or divine, circle—an activity that many ancient and recent cultures have indulged in and also considered divine. Thus, when the cosmic circle or the heavens are equally divided into four quadrants, e.g., North, East, West, South, this forms the sign of a "cross." This is the same cross that most cultures have also deemed to be divine and celestial in origin (see Chapter 8).

However, the creation of the "divine" four, or three, represents only the most rudimentary features of the sexagesimal system. For example, a circle can be divided into 360 degrees and so on, and these principles can be applied not just to surveys of the heavens but to architecture. Via the complicated permutations made possible via the sexagesimal system, the Sumerians, the Babylonians, the Egyptians, the Greeks, and those living in ancient Meso-America were able to make precise calculations of angular, object, and mathematical relations, and to create temples and buildings, the likes of which today could only be designed, built, and fitted together using

extremely precise tools and advanced, computerized measuring devices.

It is this same sexagesimal system that is employed in the measurement of time, i.e., 60 seconds and 60 minutes. Similarly, the first calendars were created in the same manner, the Sumerians dividing the circle into 12 parts in accordance with their beliefs regarding the sacred celestial nature of the number "12." That is, the Sumerians, Egyptians, and Babylonians were well aware that the sun, and not the Earth, was at the center of the solar system. They also realized that the Earth was one of several planets and that all traveled around the sun. They postulated the presence of 10 planets (one of which may have been a moon), plus the moon that circles the Earth, which, when coupled with the sun, equaled the sum of 12 (Sitchin, 1990).

It is this same Sumerian "12" that makes the 12 hours of the day and the night (the 24-hour day), and is retained in the form of the 12 months and the 12 houses of the Zodiac (Kramer, 1981; Wooley, 1965). The ancient Egyptians essentially adapted this system for designing their own calendar, and in Meso-America an almost identical calendar system was devised by the Mayas.

However, the Babylonians (and probably the Sumerians before them) took the decimal and 60-unit sexagesimal system one step further and invented a way to write these numbers in a temporal sequence, the grammatical order of which revealed the value of the sum (Kramer, 1981; Wooley, 1965). In this manner, thanks to the Sumerian-Babylonians, when one writes 4254, it is clear that the first "4" is a thousand times greater than the last "4." Finally, when the ancient Hindus appeared on the scene, the concept of "nothing" was formulated, and thus "zero" came into being. Just as written language soon came to be organized in a nonpictorial series of temporal sequences, so to did the understanding of the cosmos, geometry, time, and numbers.

These tremendous intellectual and creative achievements, however, like language, depend on the functional integrity of the inferior parietal lobe (as well as other neural structures such as those located within the frontal and temporal lobes and the thalamus), for with the destruction of this tissue, one's sense of space, geometry, written language, and math may be abolished.

ACALCULIA

Problems working with numbers or performing arithmetical operations can be secondary to a number of causes and may result from injuries involving different regions of the brain. For example, a person may suffer from alexia/agnosia for numbers, amnesia for arithmetical facts (Cohen and Dehaene, 1994), or difficulties with spatial-perceptual functioning that causes misalignment of numbers when adding or subtracting (referred to as spatial acalculia). Hence, a patient often appears to have difficulty in performing math problems when in fact the basic ability to calculate per se is intact. That is, the apparent difficulty may in fact be due to spatial, linguistic, agnosic, or alexic abnormalities.

However, in many instances, patients who are no longer able to perform calculations demonstrate a number of deficiencies. They may erroneously substitute one operation for another, e.g., misreading the sign "+" as "×", such that they multiply rather than add. Or they may reverse numbers (e.g., "16" as "61"), substitute counting for calculation (e.g., 21 + 6 = 22), or inappropriately group (e.g., 32 + 5 = 325).

On the other hand, with left posterior lesions localized to the vicinity of the inferior parietal lobe, patients may have severe difficulty in performing even simple calculations, e.g., carrying, stepwise computation, borrowing (Boller and Grafman, 1983; Hecaen and Albert, 1978). When this occurs in the absence of alexia, aphasia, or visual-spatial abnormalities, and is accompanied by finger agnosia, agraphia, and right-left disorientation, it is considered part of Gerstmann's syndrome (Gerstmann, 1930). It has also been referred to as "anarithmetria" (Hecaen and Albert, 1978), or pure acalculia when not accompanied by other abnormalities.

PURE ACALCULIA/ANARITHMETRIA

Acalculia is an isolated impairment of calculation in the absence of alexia or agraphic or spatial organization problems and involves a disturbance of basic math processes, e.g., carrying, stepwise computation, and borrowing (Boller and Grafman, 1983; Levin, 1979). It is manifested as an impairment in the ability to maintain order, to correctly plan in sequence, and to appropriately manipulate numbers, and/or it may be related to an amnesia for arithmetical facts (Cohen and DeHaene, 1994).

Indeed, one patient, a former accountant who had suffered a small circumscribed left parietal subdural hematoma in an auto accident, was unable to add past 10 (e.g., 7 + 4) although she could read, write, speak appropriately, and recognize objects. However, she did demonstrate severe finger agnosia, and in fact the finger agnosia appeared to be directly related to her inability to perform calculations. In this regard, I was able to work with this patient for a number of months and by emphasizing finger recognition tasks, including finger calculations (e.g., taking two fingers on her right hand and four on her left, and saying "two times four is?), was able to raise her math ability to the high school level. Of course, the normal tendency to partially recover lost functions certainly played a part.

ALEXIA/AGNOSIA FOR NUMBERS

Alexia for numbers and digits is found in over 80% of persons with left temporal-occipital lesions, and in less than 10% of those with right hemisphere lesions. As the name implies, the patient is unable to recognize numbers. Usually these patients also suffer from generalized or literal alexia (Hecaen, 1962), i.e., an inability to recognize letters.

ALEXIA/AGRAPHIA FOR NUMBERS

In some cases acalculia is associated with alexia and/or agraphia for numbers, as well as aphasic abnormalities (referred to as aphasic acalculia) (Benson and Weir, 1972). Patients with this disorder are unable to recognize or properly produce numbers in written form. For example, they may be unable to write out or point to the number "4" versus the number "7" or the letter "B." The lesion is usually in the left inferior parietal lobule and localized within the angular gyrus. Not all patients are aphasic, however (Levin, 1993).

SPATIAL ACALCULIA

As described in Chapter 3, the right cerebral hemisphere is quite proficient in performing geometrical analysis. However, its (i.e., the isolated right hemisphere's) basic arithmetical abilities are limited to the performance of addition, subtraction, and multiplication of simple sums and numbers below 10 (Levy-Agresti and Sperry, 1968). However, it can also visually recognize correct answers, for small sums, when given a visual choice (Dimond and Beaumont, 1974). Nevertheless, problems with arithmetical reasoning per se are not usually due to right hemisphere lesions.

However, right hemisphere damage may result in difficulties in calculation due to visual-spatial disturbances (Hecaen and Albert, 1978). For example, figures and digits may not be properly aligned, arranged, or organized on the page when writing out a problem. Or the patient may ignore the left half of numbers when adding, subtracting, etc. If given the opportunity to perform the same calculation verbally, frequently little difficulty is demonstrated, indicating that the basic ability to calculate is intact.

In general, most persons with spatial acalculia have right hemisphere lesions. However, bilateral disturbances may be present (Boller and Grafman, 1983; Hecaen and Albert, 1978).

RIGHT-LEFT DISORIENTATION

Right-left disorientation (e.g., errors in response to "show me your right hand") is usually associated with left hemisphere and left parieto-occipital damage. It occurs only extremely rarely among persons with right cerebral injuries (Gerstmann, 1930; McFie and Zangwill, 1960; Sanguet et al., 1971).

In general, these patients have difficulty in differentiating between the right and left halves of their body or the bodies of others. This may be demonstrated by asking the patient to touch or point to the side named by the examiner, e.g., "touch your left cheek" or, "point to my right ear"; to point on the patient's own body to the body part the examiner has pointed to on his or her body; or in performing crossed commands, such as "touch your left ear with your right hand." In mild cases only the crossed commands may be performed deficiently (Strub and Geschwind, 1983). Nevertheless, patients with aphasic disorders generally perform most poorly of all brain-damaged groups (Sauguet et al., 1971). Interestingly, among presumably neurologically intact adults, approximately 18% of females and 9% of men perform deficiently on right-left orientation tasks (Wolf, 1973).

In part, it seems somewhat odd that right-left spatial disorientation is more associated with left than with right cerebral injuries, given the tremendous involvement of the right half of the brain in spatial synthesis and geometrical analysis. However, orientation to the left or right transcends geometric space as it relies on language. That is, "left" and "right" are designated by words and defined linguistically. In this regard, left and right become subordinated to language usage and organization (Luria, 1980). Hence, left-right confusion is strongly related to problems integrating spatial coordinates within a linguistic framework.

ATTENTION AND NEGLECT

Data from a variety of studies have indicated that the parietal lobes are heavily involved in directing and maintaining various aspects of motoric, visual, and somesthetic attentional functioning (reviewed above). This includes the maintenance of visual fixation, the guidance of hand and manipulatory activities, or the detection and monitoring of a stimulus moving across the body surface. In part, this is accomplished via interconnections with the various association areas, frontal lobes, and subcortical structures such as the superior colliculus and the reticular formation—all of which are heavily involved in attention and/or arousal.

In general, however, the parietal lobes of the right and left cerebral hemisphere do not exert identical or equal influences. For example, although many neurons in the secondary somesthetic areas receive contralateral and ipsilateral input, whereas many visual neurons in area 7 respond to both halves of the visual field, bilateral responsiveness appears to be more characteristic of the right parietal region. The right seems to contain a greater number of bilateral cells. Thus, visual and somesthetic stimuli exert greater EEG evoked responses over the right half of the brain (Beck et al., 1969; Schenkenberg et al., 1971), and the right cerebral hemisphere becomes activated by stimuli applied to the right or left half of the body (Desmedt, 1977; Heilman and Van Den Abell, 1980). Conversely, the left hemisphere becomes aroused predominantly in response to unilateral (right-sided) input.

Reaction times to visual stimuli are also more greatly reduced following right versus left cerebral injuries (Howes and Boller, 1975). Similarly, among split-brain patients, the right cerebral hemisphere is able to maintain attention for appreciably longer time periods (Dimond, 1976, 1979), whereas the left hemisphere tends to demonstrate attentional lapses as well as unilateral spatial-conceptual neglect (Joseph, 1988b). Perhaps because of the greater right cerebral monitoring ability, across tasks requiring sustained motor-visual attention performance in the left half of space (with either hand) is superior to performance in the right half of space (Heilman et al., 1987).

As is well known, lesions, or even surgical removal of the right parietal lobe (particularly the inferior regions extending into the second occipital convolution or the frontal lobe), can result in unilateral neglect of the left half of visual, somesthetic, and auditory space (Critchley, 1953; Hecaen et al., 1956;

Heilman, 1993; Heilman and Valenstein, 1972; Heilman et al., 1987; Joseph, 1986a, 1988; Nielsen, 1937; Roth, 1944, 1949; Umilta, 1993). In the extreme, these patients may fail to become consciously aware that half the body is in some way dysfunctional, or even that it exists (Bisiach et al., 1979; Critchley, 1953; Denny-Brown et al., 1952; Gold et al., 1994; Heilman, 1991; Joseph, 1986a, 1988a; Levine, 1990; Roth, 1944, 1949; Sandifer, 1946; Schilder, 1935). They may dress or groom only the right half of their body, eat only off the right half of their plates, etc. Indeed, the left half of the environment is ignored even when the body is aligned in the vertical axis (Calvanio et al., 1987) or when conjuring up mental imagery, i.e., the left half of the image disappears.

As attention may be directed to at least three dimensions of visual space, radial, vertical, horizontal, including near and far peripersonal space (see Mennimeir et al., 1992), affected persons may show neglect of only the lower quadrants—the domain of the parietal lobes—or in the upper quadrants—the domain of the temporal lobe (Shelton et al., 1990). Patients may also selectively neglect stimuli oriented toward the body (Shelton et al., 1990).

However, depending on the extent of the lesion, the neglect may involve multiple frames of reference. For example, patients with bilateral parietal-occipital lesions have been shown to neglect vertical (inferior) and near (radial) peripersonal space, and in both the tactile and the visual modalities (Rapcsak et al., 1988). Mennemeir and coworkers (1992) have reported similar findings (see also Umilta, 1995).

Neglect may take the form of hypoarousal or inattentiveness and may involve different modalities, e.g., touch versus vision (see Umilta, 1995). For example, the patient may not completely ignore the left half of space, but instead may fail to respond to left-sided stimuli only under certain conditions, such as when the patient is fatigued or if simultaneously stimulated bilaterally (e.g., to the left and right half of the face, or the right ear and left hand) i.e., extinction.

In addition, neglect may encompass the left half of an object rather than the left half of visual-somesthetic space, even when the entire object falls to the right (Bisiach et al., 1979). For example, if the word "toothbrush" is presented well to the patient's right, she may report only seeing the word "brush." If the word is presented to her left, she may state that she sees "nothing." This is probably because these patients begin scanning from the right and proceed only a short distance leftward before they stop or are pulled back toward the right.

Not surprisingly, patients with neglect may also display visual recognition deficits, such as an inability to recognize faces or complex objects (Young et al., 1992). However, this should not be taken as evidence that the parietal lobe contains facial recognition neurons. Rather, with parietal lesions the ability to search and explore visual, tactile, and even motoric aspects of the environment is disrupted such that only half of a face or tactile stimulus may be explored.

Similarly, patients may neglect the left half of images that they consciously conjure up (Bisiach and Luzzati, 1978; Umilta, 1995) or the left half of words or sentences when reading (Barbut and Gazzaniga, 1987; Kinsbourne and Warrington, 1962). This suggests that neglect is for both internal and external events, and greatly involves the ability to internally generate bilateral perceptions.

Neglect, however, encompasses more than internal and external inattention, but a failure to attend to gravitational influences as well (Gazzaniga and Ladavas, 1987). As noted, the parietal lobule receives and processes information concerning body-positional relationships and integrates this information with visual input regarding objects in space. Via the analysis of proprioceptive and other forms of input, the parietal lobule also takes into account gravitational influences—the position of the body in space.

As pointed out by Gazzaniga and Ladavas (1987), when the eyes are looking straight forward and the head is in an upright position, the visual frame of reference coincides with the gravitational frame. However, if the head is turned to the side, the left and right side of the gravitational field (which is invariant) no longer coincide with the visual reference; what lies to the right may be groundward, and that to the left may be skyward. If the head is tilted to the right, the left visual field encompasses that which is up, and the right visual field that which is below. When the head is tilted, however, persons with parietal damage and neglect ignore not only the left side of visual space, but the left side of gravitational space as well (Gazzniga and Ladavas, 1987). For example, if the head is tilted to the left, the patient would ignore everything that is downward as well as everything falling to the vertical left. Only the right upper quadrant would continue to be perceived and responded to.

Neglect from parietal lesions is probably also secondary to a disconnection of this region from the frontal lobes. That is, the frontal lobes, failing to receive input, cease to exert activational influences (via their connections with the thalamus, reticular formation) such that information normally processed by the damaged parietal lobes are no longer activated and attended to.

Left Hemisphere Neglect

Although not as common nor as severe, inattention and neglect has also been shown to occur following left-sided lesions (Albert, 1973; Ogden, 1987). However, whereas neglect induced by the right hemisphere is more profound, attentional disturbances following left cerebral damage tend to be more subtle, e.g., failing to attend to small figures on the right. Or neglect may only be manifested under procedures employing extinction—when the patient is stimulated simultaneously on the right and left halves of the body. Similarly, the right half of drawings, although not neglected per se, may tend to be more distorted and incomplete.

Neglect following left cerebral injuries, however, is more likely with left anterior rather than left parietal injuries (Kimura,

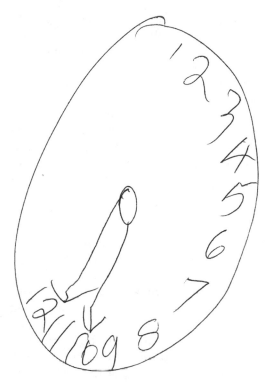

Figure 12.6. Left-sided neglect. Patient with a right parietal stroke was requested to "draw a clock, put all the numbers in it, and make it say ten after eleven."

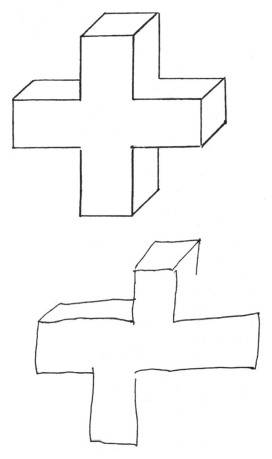

Figure 12.7. Right-sided neglect and distortion due to left parietal injury in a right-handed man.

1993; Ogden, 1987). Sometimes, however, this is due to motoric abnormalities, such as gaze paralysis, or difficulty moving the head or right arm toward the right, i.e., hypokinesis. That is, patients may seem to not fully attend to the right half of space because of damage to motor neurons and thus an inability to motorically respond. Moreover, because of a loss of counterbalancing influences, the right hemisphere acting unopposed also causes such patients to favor the left half of space. In addition, if verbal report is required, the left hemisphere, being damaged, is necessarily at a disadvantage such that what appears to be neglect may in fact be impaired language functioning.

Delusional Denial

Frequently, patients with right parietal lesions, when confronted with their unused or immobile limbs, may (at least initially) deny that it belongs to them or swear there is nothing wrong (Gold et al., 1994; Heilman, 1991; Joseph, 1986a, 1988a). More often, however, they tend to ignore their left half. In some cases, however, patients may perceive the left half of their body but refer to it using ego-alien language, such as "my little sister," "my better half," "my friend Tommy," "my brother-in-law," "spirits," etc. For example, Gerstmann (1942) describes a patient with left-sided hemiplegia who "did not realize and on being questioned denied that she was paralyzed on the left side of the body, did not recognize her left limbs as her own, ignored them as if they had not existed, and entertained confabulatory and delusional ideas in regard to her left extremities. She said another person was in bed with her, a little Negro girl, whose arm had slipped into the patient's sleeve" (p. 894). Another declared (speaking of her left limbs), "That's an old man. He stays in bed all the time."

One such patient engaged in peculiar erotic behavior with his "absent" left limbs, which he believed belonged to a woman. A patient described by Bisiach and Berti (1987:185) "would become perplexed and silent whenever the conversation touched upon the left half of his body; even attempts to evoke memories of it were unsuccessful." Moreover, although "acknowledging that all people have a right and a left side, he could not apply the notion to himself. He would affirm that a woman was lying on his left side; he would utter witty remarks about this and sometimes caress his left arm."

Some patients may develop a dislike for their left limbs, try to throw them away, become agitated when they are referred to, entertain persecutory delusions regarding them, and even complain of strange people sleeping in their beds because of their experience of bumping into their left limbs during the night (Bisiach and Berti, 1987; Critchley, 1953; Gerstmann, 1942). One patient complained that "the person" tried to push her out of the bed and then insisted that if it happened again she would sue the hospital. Another complained about "a hospital that makes people sleep together." A female patient expressed not only anger but concern lest her husband should find out; she was convinced it was a man in her bed.

Disconnection, Confabulation, and Gap Filling

In some respects it seems quiet puzzling that patients with right parietal injuries deny what is visually apparent and what they should easily be able to remember, i.e., the presence of the left half of their body. However, it is possible that when the language-dominant left hemisphere denies ownership of the left extremity it is in fact telling the truth. That is, the left arm belongs to the right, not the left, hemisphere. Indeed, in that the parietal lobe is in part an evolutionary derivative of the hippocampus (which also contains neurons that code

for the position of the body and objects in space), not just a body image and awareness of the body in space, but the memory of the body is maintained in this tissue.

Moreover, as discussed in Chapters 3, 6, and 15, memories are sometimes unilaterally stored, i.e., the left hemisphere maintains a somesthetic memory image of the right half of the body, the right cerebrum maintaining perhaps bilateral representations. In this regard, the left brain may in fact have no memory regarding the left half of the body. Of course, all this seems preposterous, and yet patients (i.e., the speaking half of their brain) will deny what is obvious (the existence or ownership of the left half of their body).

Inevitably, in order for a person to confabulate, erroneous information must become integrated somehow so that the confabulated response can be expressed. When the frontal lobes are compromised there is much flooding of the association and assimilation areas with tangential and irrelevant information, much of which is amplified completely out of proportion to more salient details (Fischer et al., 1995; Joseph, 1986a, 1988a, 1993; Stuss et al., 1978). Consequently, salient, irrelevant, highly arousing, and fanciful information is expressed indiscriminately. The normal filtering process is disrupted. However, when the parietal lobes are compromised, rather than flooding, there results a disconnection, and information received in the language axis is incomplete and riddled with gaps (Joseph, 1986a).

As noted, assimilation of input from diverse sources is a major feature of left and right parietal (i.e., inferior parietal) activity. Hence, when this area is damaged, errors abound in the assimilation of perceptions and ideas as the language axis can no longer access all necessary information. That is, when the language axis is functionally isolated from a particular source of information about which the patient is questioned (such as in cases of denial), it begins to make up a response based on the information available. To be informed about the left leg or left arm, it must be able to communicate with the cortical area (i.e., the parietal lobe) that is responsible for perceiving and analyzing information regarding the extremities. When no message is received and when the language axis is not informed that no messages are being transmitted, the language zones instead rely on some other source, even when that source provides erroneous input (Joseph, 1986a); substitute material is assimilated and expressed and corrections cannot be made (because of loss of input from the relevant knowledge source). The patient begins to confabulate. According to Geschwind (1965), when the speech area is disconnected from a site of perception, the speech area will be unable to describe what is going on at that site. This is because "the patient who speaks to you is not the 'patient' who is perceiving—they are in fact, separate."

In these instances delusions and confabulatory responses result from an attempt by the language axis to fill the gaps in the information received with associations and ideas that are somehow related to the fragments available (Joseph, 1982, 1986a; Joseph et al., 1984; Talland, 1961). In this regard, confabulatory-delusional statements, although erroneous, can contain some accurate elements around which erroneous, albeit related, ideations are anchored. Hence, a patient may see his left leg or arm and then state that it belongs to the doctor. In general, these disturbances are most common when the right frontal or right parietal lobe is damaged. However, neglect, denial, and delusional confabulation may also infrequently result from left parietal injuries (Joseph, 1986a).

Delusional Playmates and Egocentric Speech

Young children not only produce egocentric speech, in which they comment upon, explain, and describe their own behavior, but sometimes ascribe their behavior to others (such as their parents) using ego-alien language. They may claim not to know why they performed certain behaviors, may claim that someone else actually performed the deed for which they are accused, and develop imaginary friends with

whom they share secrets, play games, or upon whom they may place the blame for some untoward incident. These same "imaginary" friends sometimes urge them to commit certain acts or explain or inform them of the actions and motives of others.

Not all children, of course, develop elaborate imaginary friends. However, all (or almost all) children develop egocentric speech and at times employ ego-alien descriptions when confronted with their own disagreeable behavior.

Much of this behavior is secondary to corpus callosal immaturity and a reaction of the left hemisphere is response to impulses and behaviors initiated by the right cerebrum and/or limbic system, specifically hyperactivation of the hippocampus and adjoining tissue (which in turn has contributed to the evolution of the parietal lobe). However, among this same age group (up to age 7 and even 10), the parietal lobes are also quite immature, the inferior parietal area in particular.

Following destruction of the inferior parietal lobe and adjoining tissue, patients sometimes uncontrollably comment upon their actions, e.g., "Now I'm waving goodbye" (such as when given a command by their physician: "Wave your hand"), whereas in other instances they may claim that the person performing the action in someone other than themselves. As noted earlier, they may in fact completely deny that the left half of their body is their own and claim it belongs to another, which also occurs in some cases of early trauma-induced injury to the hippocampus (see Chapter 16), a nucleus which has contributed to the evolution of the parietal lobe.

Although a lesioned parietal lobe is not the same as parietal lobe immaturity (particularly in that immaturity is bilateral and damage is usually unilateral), there remains a curious similarity in the behavior of these adults and children, including those who suffer dissociative states secondary to hippocampus-amygdala hyperactivation (see Chapter 16).

Perhaps this is because the ego and Self are first identified with the body, whereas the image of the body is maintained by the parietal lobe. When one parietal lobe (because of damage or immaturity) is unable to communicate with the other half of the brain, the "normal" brain half is unable to recognize a continuity of Self. Consequently, behaviors initiated by the opposite (usually right) half of the brain, or the half of the body controlled by right brain, are recognized by the speaking half of the cerebrum only from a disconnected (i.e., alien, dissociated) perspective. When the speaking (i.e., left) hemisphere is questioned about this behavior, or about its left limbs, they are described as initiated by or belonging to someone else, such as "an old man," "my brother-in-law," or an imaginary friend.

SUMMARY AND OVERVIEW

The parietal lobes, although commonly associated with the mediation of somesthetic stimuli, are also concerned with motor and attentional functioning and the perception of spatial relations, including depth, orientation, location, and the identification of motivationally significant auditory, somesthetic, and visual stimuli. It is via the integrative interaction of the parietal lobes that a person can feel an object and not only determine its physical qualities (e.g., shape, size, weight, texture), but can visualize it, verbally label it, and even write out its name. It is also via the interaction of these various cell assemblies that a person can attend to specific objects in space as well as reach out and manipulate them. Indeed, the parietal lobe is important for mediating movement in visual space. That is, in order for a movement to be correctly planned and carried out, signals must be directed to the right muscle groups as based on efferent visual, somesthetic, and auditory input. Moreover, one requires sensory information as to the position of the body and limbs in space; otherwise movements will be clumsy. This is accomplished through the dense interconnections linking the primary somesthetic with the motor

areas in the frontal lobe and via impulses transmitted down the cortico-spinal tract. Moreover, parietal neurons appear to guide and monitor movements as they occur in visual space. Although the entire parietal lobule makes important contributions to movement, the inferior parietal lobule of the left hemisphere appears to be the central region in the performance of skilled temporal-sequential motor acts. These engrams assist in the programming of the motor frontal cortex, where the actions are actually executed. If the inferior parietal region is destroyed, the patient loses the ability to perform actions in an appropriate temporal sequence or to even appreciate when they have performed an action incorrectly. This condition is referred to as apraxia. In contrast, right parietal injuries are associated with severe disturbances of emotion, constructional deficiencies, and a host of visual-spatial perceptual abnormalities, including left-sided inattention and neglect. For example, with severe lesions in this area, patients may demonstrate a profound inattention to all forms of stimuli falling to their left. When drawing pictures they may fail to draw the left half of an object; when writing or reading they may ignore the left half of words or the left half of the page; and they may fail to perceive and respond to persons standing to their left, or even the left half of their own body. This condition is largely secondary to a destruction of neurons that are sensitive to various forms of visual and somesthetic input. Nevertheless, it is important to emphasize that lesions to the parietal lobe (or anywhere within the brain) are seldom localized to one particular quadrant (e.g., inferior, superior), or even restricted to the parietal lobe. That is, damage may be parietal-occipital, parietal-temporal, frontal-parietal, or even bilateral (such as damage due to cerebrovascular disease or compression from a unilateral tumor). Therefore, an afflicted person may display agraphia but normal reading, stereognosis in the absence of apraxia, or conversely a wide mixture of seemingly unrelated symptoms.

13

The Occipital Lobes
Vision, Blind Sight, Hallucinations, Visual Agnosias

Like the frontal, temporal, and parietal lobes, which respond to and process information from a number of modalities, the occipital lobe contains neurons that respond to vestibular, acoustic, visual, visceral, and somesthetic input (Beckers & Zeki, 1994; Horen et al., 1972; Jung, 1961; Morrell, 1967; Pigarev, 1994; Sereno et al., 1995). However, although over half the primate neocortex is also concerned with visual functions, the primary neocortical destination for visual input is the occipital lobes. Simple and complex visual and central/foveal analysis is one of the main functions of the occipital lobe (Kaas & Krubitzer, 1991; Sereno et al., 1995).

As noted, neocortical layers I and VII (6b) appear phylogenetically ancient in organization and structure and may well be an extension of midbrain tissue (Marin-Padilla, 1988a). As detailed in Chapter 1, it is likely that some of the visual cortex is derived from the hippocampus, and to a lesser extent, the amygdala and the midbrain superior colliculus. As detailed in Chapter 1, early in the course of evolutionary history the dorsal midbrain, as well as the dorsal hippocampus (and later the dorsal thalamus), were exceedingly responsive to light. In this regard, the modern human visual neocortex (layers II–VIa) is sandwiched between two cellular layers that initially were (and presently are) concerned with photoanalysis and that may be midbrain in origin.

As noted in Chapters 12 and 14, the superior parietal lobe also contains visual neocortex and receives peripheral and lower visual input via the optic radiations, whereas the middle and inferior temporal lobes receive foveal and upper visual field input via these fibers. And, as in the frontal neocortex, the neocortical layers of these tissues also contain phylogenetically old layers I and VII. Thus, the visual cortex extends well beyond the occipital lobe and includes the posterior third of both cerebral hemispheres as well as portions of the frontal lobes (e.g., frontal eye fields).

PRIMARY AND ASSOCIATION VISUAL CORTEX

The primary visual receiving area (i.e., striate cortex, area 17) is located predominantly within the medial walls and floor of the calcarine sulcus and extends around the lateral convexity. Area 17 is also characterized by rather thin cortical layers, particularly layer II, and by its striped appearance, which is due to the structure and composition of layer IV. That is, layer IV is divided into three sublayers, with the middle layer containing a rather thick band of cortex (the band of Baillarger/Gennari), which is visible to the naked eye.

The association cortices (areas 18 and 19) are also located medially and along the lateral convexity, extending into the superior parietal (area 7) and inferior and middle temporal lobes (Kaas and Krubitzer, 1991; Nakamura et al., 1994; Sereno et al., 1995; Tovee et al., 1994).

PRECORTICAL VISUAL ANALYSIS

There is much processing of visual input prior to its reception in the occipital lobe. Initially, visual information analysis takes place in the receptor cells (rods and cones) within the retina and then undergoes successive hierarchical stages of analysis as the information is passed through the sequential cell layers of which the retina is composed, i.e., horizontal cells, bipolar cells, amacrine cells, and ganglion cells. Information is then relayed via the optic nerve to the lateral geniculate nucleus of the thalamus (Casagrande and Joseph, 1978; Kaas and Krubitzer, 1991), where yet further forms of analysis are performed. From the lateral geniculate, visual stimuli are then transmitted via the optic radiations to the primary visual receiving area, the striate cortex. In addition, visual information is transmitted to the superior colliculus and then the pulvinar of the thalamus, and is then relayed to the visual cortex (Doty, 1983; Kaas and Krubitzer, 1991; Snyder and Diamond, 1968; Tigges et al., 1983).

In general, a strict topographical relationship is maintained throughout the visual projection system and the visual cortex. Within the visual cortex, immediately adjacent groups of neurons respond to visual information from neighboring regions within the retina (Kaas and Krubitzer, 1991).

In general, at the neocortical level information is first received in the primary visual receiving area, 17, with medial geniculate fibers predominantly innervating layer IV. From the visual cortex this information is then transferred back to the lateral geniculate nucleus of the thalamus (from which it was first relayed) to the superior colliculus (Kawamuara et al., 1974) and to the association areas 18 and 19 as well as the middle temporal lobe (Doty, 1983; Kaas and Krubitzer, 1991; Lin et al., 1982; Sereno et al., 1995), where higher-order processing occurs (Tovee et al., 1994).

This same arrangement seems to characterize information reception and transfer in the auditory and somesthetic cortices as well. That is, information transferred from the thalamus to the neocortex is then transferred back to the thalamus as well as to the adjacent association areas. In this manner a feedback loop is constructed so that information transmission from the thalamus to the cortex can be enhanced, diminished, or altered depending on neocortical requirements.

Once visual stimuli are relayed to the association areas 18 and 19 (where further analyses take place) this information is transmitted back to area 17 and to the temporal and parietal lobes, where polymodal associations are performed (Beckers and Zeki, 1995; Martinez-Millan and Hollander, 1975; Tigges et al., 1983; Tovee et al., 1994; Zeki, 1978b).

Neocortical Columnar Organization

Throughout the striate cortex, neurons with similar receptive properties are stacked in columns. Indeed, one column of cells may respond to a certain visual orientation and the cells in the next column to an orientation of a slightly different angle. Moreover, columns exist for color (Zeki, 1974), location, movement, etc. In addition, since certain cells respond predominantly to input from one eye, there are ocular (eye) dominance columns as well (Hubel and Wiesel, 1968, 1974). A similar columnar arrangement in regard to somesthetic input is maintained in the parietal lobe.

Nevertheless, although these neurons seem to communicate predominantly with those in the same or immediately adjacent columns, a considerable amount of parallel communication is likely (Dow, 1974; Kaas and Krubitzer, 1991). That is, information is analyzed vertically and horizontally so as to create a series of superimposed mosaics of the visual word.

Simple, Complex, Lower- and Higher-Order Hypercomplex Feature Detectors

The visual cortex is made up of a variety of cell types, each of which is concerned with the analysis of different visual features (Hubel and Wiesel, 1959, 1962, 1968; Kaas and Krubitzer, 1991; Sereno et al., 1995). These include simple, complex, and (higher- and lower-order) hypercomplex cells that are distributed disproportionately throughout areas 17, 18, and 19.

To briefly summarize, simple cells appear to be involved in the initial analysis of incoming visual cortical input, and are most sensitive to moving stimuli. They are found predominantly within area 17. Some

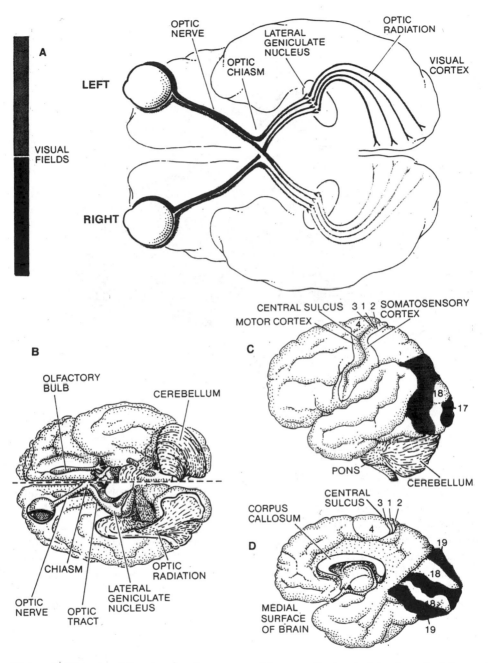

Figure 13.1. A, Diagram of the visual pathways. The left side of each retina projects to the left lateral geniculate nucleus of the thalamus and then to left visual cortex. B, Visual pathways in partially dissected brain, depicting the optic nerve, chiasm, optic tract, lateral geniculate nucleus, and optic radiations. C, Lateral view of visual cortex. D, Medial view of visual cortex. (From Kuffler SW, Nicholls JG. From neuron to brain. Sinauer Associates, 1976.)

are sensitive to stimuli moving in one direction, whereas others may respond to stimuli moving in any direction. In addition, simple cells are responsive to the particular position and orientation a stimulus may take. However, for a simple cell to fire, a stimulus must assume a specific orientation and position.

Simple cells relay this processed information to the far more numerous complex

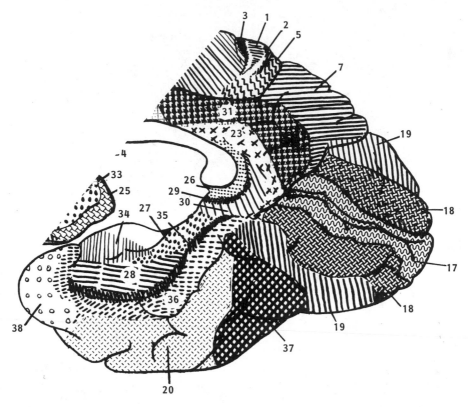

Figure 13.2.   Medial view of the posterior portion of right cerebral hemisphere. Numbers refer to Brodmann's areas. Note extensive medial representation of the visual areas.

cells. Each complex cell receives input from several simple cells. Complex cells are also concerned with orientation of the stimulus. However, these cells are more flexible and will respond to and analyze a stimulus regardless of its particular orientation. These cells, via the combined input from simple cells, are probably involved in the earliest stages of actual form perception, i.e., the determination of the outline of an object. A considerable number of complex cells receive converging input from both eyes, the remainder being monocular. Complex cells are found predominantly within area 18.

Hypercomplex cells are concerned with the analysis of discontinuity, angles, and corners, as well as movement, position, and orientation. That is, these cells respond selectively to certain visual configurations and thus act to determine precise geometric form. It is also via the action of these cells (in conjunction with visual neurons in the temporal lobe) that the first

stages of visual closure are initiated. This in part requires that the functional activity of these cells be suppressed such that when presented with an incomplete figure, these cells are overridden and the brain is able to "fill in the gaps" in stimuli perceived. This is also why one does not notice his or her "blind spot"; it is filled in. Hypercomplex cells are found predominantly within area 19.

STRIATE CORTEX: AREA 17

The primary visual cortex, area 17, is located predominantly within the medial walls of the cerebral hemispheres, extending only minimally along the lateral convexity. This area is often referred to as "striate" because the incoming fibers from the optic radiations form a stripe along the cortical surface that can be seen by the naked eye. Areas 18 and 19 do not have this striped appearance.

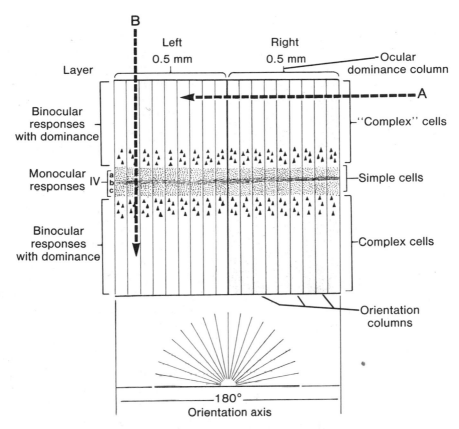

Figure 13.3. Schematic representation of ocular dominance and orientation column organization in the striate cortex. Ocular dominance columns (9.25 to 0.5 mm) are an order of magnitude larger than orientation columns. A pair of left-right ocular dominance columns (0.5 to 1 mm) are equal to orientation columns representing a complete cycle of orientations of 180 degrees. Lateral geniculate input to layer IV is monocular and consists of parallel and alternating stripes. Most of the cells in (blue) layer IV are "simple" cells. "Complex" cells in layers below and above layer IV are binocular. Arrow A, A recording electrode inserted tangential to the pail surface will detect responses to stimuli with different orientations. Arrow B, Electrode inserted vertically responds to stimuli presented in only one axis of orientation. (From Carpenter M. Human neuroanatomy. Baltimore: Williams & Wilkins, 1991.)

As with the primary motor and somesthetic cortices, cellular representation is greater for those areas that are the most densely innervated and of the most sensory importance, i.e., the fovea (Daniel and Whitteridge, 1961; Hubel and Wiesel, 1979; Kaas and Krubitzer 1991). Indeed, the central part of the retina has a cortical representation that is 35 times more detailed than that of the periphery. This is particularly important in that the fovea contains cells that are most sensitive to the detection and representation of form.

Although all neuron types are found within the striate cortex, simple and complex cells predominate. Hence, the primary receiving area is predominantly involved in the analysis of color, movement, position, and orientation, i.e., the most elementary aspects of form perception.

The primary visual cortex, however, receives fibers from nonvisual brain areas as well. These include brainstem nuclei, the pontine and mesencephalic reticular formation, the lateral amygdala, and the lateral hypothalamus (Doty, 1983; Tigges et al., 1983). Processing in the primary region can thus be enhanced or diminished via reticular influences and emotional-motivational concerns. In this manner, if a stimulus is emotionally significant, greater visual attention will be directed at the object.

Conversely, bilateral destruction of area 17 can result in loss of visual recognition capabilities—even with sparing of the association areas (Humphrey and Weiskrantz, 1967; Weiskrantz and Cowey, 1963). However, awareness of moving objects and visual-spatial orientation is preserved. Visual preservation with primary occipital destruction has been referred to as "blind sight" (see below).

Hallucinations

Electrical stimulation, tumors, seizures, or trauma involving the striate cortex may produce simple visual hallucinations, such as sparks, tongues of flames, colors, and flashes of light (Penfield, 1954; Tarachow, 1941). Objects may seem to become exceedingly large (macropsia) or small (micropsia), blurred in terms of outline, or stretched out in a single dimension, or colors may become modified or even erased (Hecaen and Albert, 1978). Sometimes simple geometric forms are reported. Usually the hallucination is restricted to one-half of the visual field. That is, if the seizure is in the right occipital lobe, the hallucination will appear in the left visual field.

Although elementary hallucinations are usually associated with abnormalities involving the occipital lobe they may occur with temporal lobe lesions or electrical stimulation (Penfield and Rasmussen, 1950; Tarachow, 1941).

ASSOCIATION AREAS 18 AND 19

Areas 18 and 19 are involved in the translation and interpretations of visual impressions transmitted from area 17. Although simple and complex cells are found in the association cortex, this region is predominantly populated by hypercomplex (both higher- and lower-order) neurons— most of which are concerned with the determination of precise geometric form as well as the assimilation of signals transmitted from the primary cortex (Kaas and Krubitzer, 1991).

In contrast to the neurons within area 17, many of the cells within area 18 receive binocular input and can be activated by either eye (Hubel and Wiesel, 1970). This same pattern of binocularity is evident in the parietal association area. It is via the action of these cells that one is able to gather information regarding distance and discrepancies in stimulus location and thus determine depth and achieve stereoscopic vision (Blakemore, 1970; Hubel and Wiesel, 1970). Indeed, some association neurons will only fire when a target is a definite distance from the eye.

Many of the neurons in this region, particularly area 19, receive higher-order converging input from the parietal and temporal lobes. For example, in addition to visual input, neurons in the superior portions of area 19 respond to tactile and proprioceptive stimuli, whereas those in the inferior portions respond to auditory signals (Morrell, 1967). It is probably in this manner (in conjunction with subcortical connections with, for example, the superior colliculus) that one is able to orient toward and gaze at an auditory stimulus as well as maintain stabilization of the head (via proprioceptive vestibular input) while engaged in visual search.

Hence, overall, the visual association area appears to be involved in the initial analysis of form, distance, and depth perception, as well as the performance of visual closure. It is thus heavily involved in the association of various visual attributes so that a variety of qualities may be ascertained. This would include an objects shape, length, thickness, and color (Sereno et al., 1995).

It is important to emphasize that the visual association areas also maintain intimate relationships with the parietal visual regions (area 7) as well as the visual areas in the middle and inferior temporal lobes. The temporal visual areas are in turn reciprocally interconnected with area 7. Hence, a complex interactional visual loop is maintained, the inferior-medial temporal lobe being concerned with form perception and the analysis of emotional-motivational sig-

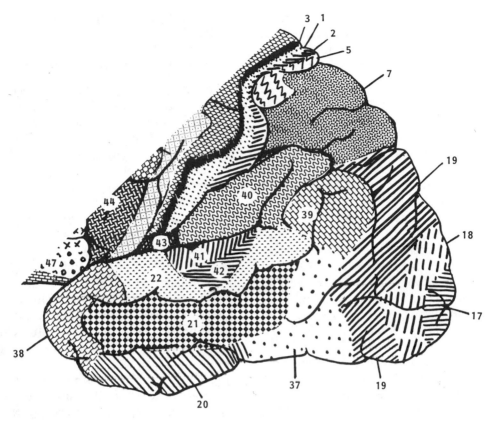

Figure 13.4. Depiction of the visual cortex of the lateral surface of the temporal, occipital, and parietal lobes.

nificance; area 7 being involved in visual attention, visual fixation, and the analysis of distance, depth, and objects within grasping distance; and areas 18 and 19 providing (and receiving from) both required visual input.

In this regard, when an object is recognized (via the interaction of the visual areas in the temporal-occipital lobe), the parietal visual areas are alerted so that distance may be determined and so that, once the object is within grasping distance, it may be grasped and tactually explored. If it is a potential food item (signaled by limbic structures within the temporal lobe), it can then be orally explored and eaten.

Hence, whereas the temporal lobes perform complex form recognition and central and upper visual field analysis, the parietal lobes observe the periphery and lower visual field (where the hands and feet are most likely to be viewed), and to-

gether these cortical regions compute and make possible eye-hand or hand-object coordination.

Homonymous Hemianopsia and Quadrantanopsia

Massive unilateral destruction of the visual cortices results in blindness in the contralateral temporal visual field and ipsilateral nasal visual field. Hence, a left visual cortex lesion produces a right homonymous hemianopsia. However, these visual disturbances may also result from destruction of the optic radiations or optic tract. It is noteworthy that patients are often unaware of having lost half their sight, particularly with right occipital lesions. Nevertheless, patients may complain of bumping into people and objects they cannot see.

As noted above, the parietal lobes are concerned with the lower visual fields, whereas the temporal lobes receive massive upper visual field projections. Thus, lesions to the inferior or superior occipital lobe can result in an upper or lower visual field homonymous quadrantanopsia.

Hallucinations

Electrical stimulation of or lesions involving areas 18 and 19 can produce complex visual hallucinations (Foerster, 1929, cited by Brodal, 1981 and Hecaen and Albert, 1978; Tarachow, 1941), such as images of men, animals, various objects and geometric figures, and lilliputian-type persons, including micropsias and macropsias (see Luria, 1980; Hecaen and Albert, 1978, for review). Sometimes objects may seem to become telescoped and far away, whereas in other situations, when approached, objects may seem to loom and become exceedingly large.

Complex hallucinations are usually quite vivid and fully formed and the patient may think that what he sees is real (Hecaen and Albert, 1978). Forester (1928, cited by Heacen and Albert, 1978) reported a patient who hallucinated a butterfly, then attempted to catch it when area 19 was electrically stimulated. Another patient hallucinated a dog and then called to it, denying the possibility that it was not real.

Complex hallucinations, although usually associated with tumors or abnormal activation of the visual association area, have also been reported with parietal-occipital, (Russell and Whitty, 1955), occipital-temporal, or inferior-temporal damage (Mullan and Penfield, 1959; Tarachow, 1941; Teuber et al., 1960), or with lesions of the occipital pole and convexity (Hecaen and Albert, 1978).

LATERALITY

According to Hecaen and Albert (1978), their review of the international literature shows that, although simple hallucinations are likely following damage to either hemisphere, complex hallucinations are usually associated with right rather than left cerebral lesions (Hecaen and Albert,

1978; Mullan and Penfield, 1959; Teuber et al., 1960).

CORTICAL BLINDNESS

Lesions of the occipital lobe, especially if the entire visual cortex is ablated, result in cortical blindness, such that pattern and form vision is lost. Rather, what is left is the ability to discriminate only between different fluxes in luminous energy, i.e., lightness and darkness (Brindley et al., 1969; Brodal, 1981; Hecaen and Albert, 1978; Weiskrantz, 1963). If damage is restricted to the occipital lobe of only one of the hemispheres, patients will lose patterned vision for the opposite half of the visual field (i.e., a hemianopsia). This is not the same as unilateral neglect or inattention. However, if the lesion is sufficiently large and involves the right parietal area as well, the patient may suffer from both hemianopsia and neglect.

Nevertheless, if only a portion of the visual cortex is destroyed, vision is lost only for the corresponding quadrant of the visual field (referred to as a scotoma). However, in cases of partial cortical blindness, patients are able to make compensatory eye movements and are not terribly troubled by their disability (Luria, 1980). Indeed, patients frequently are unaware that they have lost a quadrant or even half of their visual field. Hence, this condition must be tested for.

"Blind Sight"

Although somewhat controversial, it has been reported that some persons, although blind (because of destruction of the primary visual cortex), can indicate the presence or absence of a stimulus within the "blind" portion of their visual field, and even differentiate between various objects, although they do not recognize them (Humphrey and Weiskrantz, 1967; Weiskrantz and Cowey, 1963). Using forced choice procedures, "blind" patients can localize visual targets and perceive moving objects.

In one case, a patient who denied visual perception and was diagnosed with cortical blindness was able to correctly name objects, colors, famous faces, and facial emotions, as well as read various single words with greater than 50% accuracy (Hartmann et al., 1991).

Visual preservation following area 17 lesions has been referred to as blind sight (Weiskrantz, 1987; Weiskrantz et al., 1974). Possibly, although cortically blind, patients are able to accomplish this via intact subcortical nuclei involved in visual orientation, i.e., via the so called "second (retino-collicular-pulvinar-extrastriate) visual system" (Schneider, 1969; Weller, 1988).

However, it has also been argued that among those with blind sight, "useful visual function is preserved only when a critical amount of area 17 is spared (Celesia et al., 1991). For example, using PET and SPECT studies, Celesia and colleagues (1991) found that among those with visual cortical destruction and blind sight, islands of preserved area 17 functioning were observed. Although this finding is no doubt true, it does not mean that these preserved areas are responsible for the patient's "seeing" what he or she verbally claims cannot be seen—unless, of course, these lesions have resulted in a disconnection syndrome (Hartmann et al., 1991) (see Chapter 16). On the other hand, Celesia and coworkers (1991) report that patients with complete area 17 destruction do not demonstrate signs of blind sight.

Nevertheless, it is noteworthy that in the case presented by Hartmann and associates (1991), the patient, although denying visual perception, was able to name objects, colors, and so on when the stimuli were placed in the upper visual field. The upper visual fields are associated with the inferior temporal lobe, which are the recipients of fibers from the "second visual system" as well as from the visual neocortex. Hence, in cases of disconnection and blind sight due to occipital lesions, complex visual input may still reach the temporal lobe, and may therefore be directed to the auditory areas via a secondary route so that objects can be named, although the patient (relying on the lesioned and disconnected occipital route) continues to deny visual perception.

Denial of Blindness

More frequently, however, patients with cortical blindness (such as follows massive lesions of the visual cortex) seem initially quite confused and indifferent regarding their condition, and they report a variety of hallucinatory experiences that may be complex or elementary. Moreover, frequently these patients will initially deny that they are blind (Redlich and Dorsey, 1945)—a condition referred to as Anton's syndrome. For example, a number of patients described by Redlich and Dorsey (1945), although bumping into furniture and unable to recognize objects held before them, invented elaborate excuses for their errors and failure to see, e.g., claiming that it is a little dark and they need their glasses, or conversely, that they see better at home. That is, these patients confabulate.

As discussed in detail elsewhere (see Chapter 3) (Joseph, 1986a, 1988a), these confabulatory abnormalities are sometimes due to a disconnection such that the language axis, failing to receive information from the visual cortex (i.e., the fact that it cannot see), responds instead to associations from intact areas that concern "seeing." That is, the language axis does not know that it is blind because information concerning blindness is not being received from the proper neural channels.

It is also possible that these patients deny being blind because subcortically they are still able to see. Hence, although at a neocortical level there is no sight, subcortically there remains an unconscious awareness of the visual world.

VISUAL AGNOSIA

Visual agnosia is a condition in which the patient loses the capacity to visually recognize objects, although visual sensory functioning is largely normal. This condition often arises with lesions involving the inferior medial occipital lobe. In general, objects are detected, but they no longer evoke meaning and cannot be correctly identified or named (Critchley, 1964; Davidoff and De Bleser, 1994; Shelton et al., 1994; Teuber, 1968). The percept becomes stripped of its meaning. For example, if shown a comb, the patient might have no idea as to what it is or what it might be used for. If asked to guess they may call it a harmonica or a tiny box. If shown a pair of spectacles, they may call it a "bicycle" or "two spoons." A picture of a dial telephone may be described a "clock," etc. (Luria, 1980).

Many patients are unable to sort pictures or objects into categories or to match pictures with the actual object such that there appears to be a deficit not only in the ability to recognize but to classify visual percepts (Hecaen et al., 1974). Moreover, in severe cases, patients are unable to point to objects that are named.

Nevertheless, this is not a naming disorder (e.g., Davidoff and De Bleser, 1994), because regardless of modality, anomics continue to have word-finding and naming difficulties. In contrast, patients with agnosia show enhanced recognition if an object is presented via a second intact modality (e.g., if they palpate it by hand). Thus agnosia can often be limited to a single input channel, i.e., visual versus tactual. Moreover, if an object is used in context, recognition can be greatly enhanced (Rubens, 1993).

Some patients may complain that objects change while they are looking at them and/or that they disappear a condition that suggests optic ataxia. Usually, however, the deficit is conceptual rather than perceptual.

Although recognition is abolished, patients may be able to accurately describe, draw, or copy various aspects of the object they are shown (Albert et al., 1975; Mack and Boller, 1977; Rubens and Benson, 1971). However, they fail to correctly draw the entire object. Moreover, if asked to trace rather than copy, these patients may trace over and over the outlines of objects or drawing but cannot recognize where they started. Thus they seem unable to synthesize visual details into an integral whole

(Luria, 1980). That is, patients may recognize an isolated detail (e.g., the dial of a phone) but are unable to relate it to, for example, the headset (see Shelton et al., 1994). Hence, if asked what they have been shown, they may erroneously extrapolate from the detail perceived and thus confabulate a concept.

With increasing complexity, if an object is surrounded by other objects, if it is presented in pictorial (versus actual) form, or if unnecessary lines are drawn across the picture, the ability to recognize the object deteriorates even further (Luria, 1980; Rubens and Benson, 1971).

Agnosic patients also often (but not always—see Davidoff and De Bleser, 1994) have difficulty with reading (Albert et al., 1975; Mack and Boller, 1977; Rubens and Benson, 1971) and may suffer from prosopagnosia and/or impaired color naming (Mack and Boller, 1977; Shelton et al., 1994). Interestingly, in some cases visual memory may be intact (Rubens, 1993).

Like alexia, agnosia can occur following lesions to the medial and deep mesial portion of the left occipital lobe. The left inferior temporal lobe and posterior hippocampus may also be damaged in some cases (Shelton et al., 1994).

In some respects, it is likely that agnosia is due not so much to tissue destruction as to tissue disconnection. That is, if the visual form recognition neurons in the temporal lobe are no longer able to receive input from the visual association areas, this particular region becomes "cortically blind" and form recognition is prevented. However, as with some other disconnection syndromes, if a different input channel is employed, i.e., if the object is verbally described or tactually explored, recognition is enhanced.

PROSOPAGNOSIA

Prosopagnosia is a severe disturbance in the ability to recognize the faces of friends, loved ones, or pets (DeRenzi, 1986; DeRenzi et al., 1968; DeRenzi and Spinnler, 1966; Evans et al., 1995; Hecaen and Angelergues, 1962; Landis et al., 1986; Levine, 1978; Whiteley and Warrington, 1977). Some patients may in fact be unable to recognize their own face in the mirror. Nevertheless, they usually realize that a face is a face; they just don't know who the face belongs to. In this regard, prosopagnosia is not a visual agnosia as described above, wherein the patient cannot recognize that a chair is a chair or a clock a clock.

A number of authors have erroneously argued that prosopagnosia is due to bilateral injuries involving the inferior and medial occipital lobe visual association areas. Nevertheless, although patients may suffer from bilateral injuries, the lesions frequently are restricted to the right hemisphere and involve the occipital and inferior temporal regions (DeRenzi, 1986; DeRenzi et al., 1968; DeRenzi and Spinnler, 1966; Evans et al., 1995; Hecaen and Angelergues, 1962; Landis et al., 1986; Levine, 1978; Whiteley and Warrington, 1977; Young et al., 1995), although frontal lesions (Braun et al., 1994) and right parietal lesions (Young et al., 1992) may also disrupt facial recognition. By contrast, prosopagnosia does not usually result from lesions restricted to the left half of the brain, whereas loss of facial recognition with frontal and parietal lesions is more likely due to visual neglect and visual scanning abnormalities.

Although prosopagnosia could be explained from a disconnection perspective, it is likely that the actual face identification neurons have been destroyed, whereas neurons involved in the recognition of facial parts have been preserved (see Chapter 14).

SIMULTANAGNOSIA

Simultanagnosia occurs with left hemisphere damage and is an inability to see more than one thing, or all aspects of an item, at a time (Kinsbourne and Warrington, 1962, 1964; Rizzo and Robin, 1990). For example, some patients complain of seeing things only in a piecemeal fashion such that objects look fragmented. In fact, when an object is surrounded by other objects,

perceptual recognition deteriorates even further.

This condition is sometimes accompanied by abnormal eye movements (Luria, 1980). These patients often have difficulty in shifting gaze and/or performing visual search tasks such that their ability to scan and visually explore the environment is drastically reduced (Rizzo and Robin, 1990). As described by Luria (1980), this is due to a breakdown in the ability to perform serial feature-by-feature visual analysis. Visual attention is often largely limited to the central visual field, whereas the periphery is ignored (Hecaen and Ajuriaguerra, 1954; Luria, 1973, 1980). However, patients complain that even objects in the central visual field tend to disappear as they stare at them (Rizzo and Robin, 1990).

Simultagnosia has been described following lesions to the frontal eye fields and following bilateral superior occipital lobe lesions (Rizzo and Hurtig, 1987). In many cases the lesion is localized to the superior occipital-parietal region (area 7). Hence, the patient is no longer able to maintain visual fixation and cannot adequately focus on an object or explore its parts. This disorder has also been referred to as Balint's syndrome as well as optic ataxia, paralysis of gaze, and concentric narrowing of the visual field.

IMPAIRED COLOR RECOGNITION

In this condition, although patients can sometimes correctly name objects, they cannot correctly name, match, and identify colors or point to colors named by the examiner. No, this is not due to color blindness. Frequently persons with color imperception also display prosopagnosia (Green and Lessel, 1977; Meadows, 1974b).

DeRenzi and Spinnler (1967) found that 23% of those with right cerebral damage and 12% of those with left-sided destruction had difficulty with color matching (Lhermitte et al., 1969). On the other hand, some investigators note that impairments of color perception are frequently secondary to bilateral inferior occipital lobe damage (Green and Lessell, 1977; Meadows, 1974a; Shelton et al., 1994). In addition, almost 50% of patients with aphasia demonstrate deficient color naming and color identification (DeRenzi and Spinnler, 1967; DeRenzi et al., 1972). However, color perception per se is largely intact among aphasic patients.

OVERVIEW

The primary visual cortex is located predominantly within the medial walls of the cerebral hemispheres and is concerned with the elementary aspects of form perception. Damage limited to this area will usually affect foveal vision and/or give rise to simple hallucinations.

From the primary area, information is then relayed to the association areas, 18 and 19, where complex analysis, including form recognition, positioning, and analysis of depth take place. Damage involving these areas and the primary region can cause cortical blindness or hemianopsia if only one hemisphere is lesioned. Destruction of or abnormal activity in areas 18 and 19 is associated with the formation of complex hallucinations.

Visual information is next relayed to area 7 in the parietal lobe and to the inferior temporal lobule, where higher-order analysis and multimodal processing occurs. Damage to the parietal-occipital borders may result in abnormalities involving depth and form perception as well as visual neglect. Destruction of the temporal-occipital regions can give rise to visual agnosias and an inability to recognize complex objects and faces.

The occipital lobes also appear to be lateralized in regard to certain capabilities such as facial recognition. For example, destruction of the right occipital region is associated with prosopagnosia, and abnormal activity in this area is more likely to give rise to complex visual hallucinations.

14

The Temporal Lobes

Language, Auditory, Visual, Emotional, and Memory Functioning, Form and Face Recognition, Aphasia, Epilepsy, and Psychosis

TEMPORAL TOPOGRAPHY

The temporal lobes subserve a wide variety of functions including memory and complex visual, auditory, linguistic, and emotional functioning. Of course, different regions are involved with different functions, the superior temporal lobe being concerned with auditory-linguistic activity and the inferior region with emotional, visual, as well as auditory memory and related perceptual and visceral activity.

Functional and Evolutionary Considerations: Overview

As detailed in Chapter 1, over the course of evolution the dorsally situated hippocampus became displaced and progressively assumed a ventral position, during the course of which it also contributed to the neocortical development of portions of the parietal, occipital, and temporal lobes. Similarly, the lateral amygdala became increasingly cortical in structure, and together with the hippocampus, contributed to the evolution of the anterior, medial, superior, and lateral temporal lobes.

Given the role of the amygdala and hippocampus in memory, emotion, attention, and the processing of complex auditory and visual stimuli, the temporal lobe became similarly organized. As noted, the anterior and superior temporal lobes are concerned with complex auditory and linguistic functioning (Edeline et al., 1990; Geschwind, 1965; Joseph, 1993; Keser et al., 1991), whereas the inferior temporal lobe (ITL) harbors the amygdala and hippocampus, performs complex visual integrative activities including visual closure, and contains neurons that respond selectively to faces and complex geometric and visual stimuli (Gross and Graziano, 1995; Nakamura et al., 1994; Rolls, 1992; Tovee et al., 1994). The temporal lobes, however, also receive extensive projections from the somesthetic and visual association areas, and they process gustatory, visceral, and olfactory sensations (Jones and Powell, 1970; Previc, 1990; Seltzer and Pandya, 1978), including the feeling of hunger (Fisher, 1994).

AUDITORY NEOCORTEX

Although the various cytoarchitectural functional regions are not well demarcated via the use of Brodmann's maps, it is possible to very loosely define the superior-temporal and anterior-inferior and anterior middle temporal lobes as auditory cortex, although visual, visceral, and other sensations are received in this area. These regions are linked together by local circuit neurons (interneurons) and via a rich belt of projection fibers, which include the arcuate and inferior fasciculus. It is via the arcuate and inferior fasciculus that the ITL, as well as the amygdala, contributes and transfers (as well as receives) complex auditory information to the primary and secondary auditory cortex, which simultaneously receives auditory input from the medial geniculate of the thalamus and (sparingly) the midbrain. As will be detailed below, it is within the primary and neocortical auditory association areas that linguistically complex auditory signals are recognized, analyzed, and reorganized so as to give rise to complex, grammatically correct, human language.

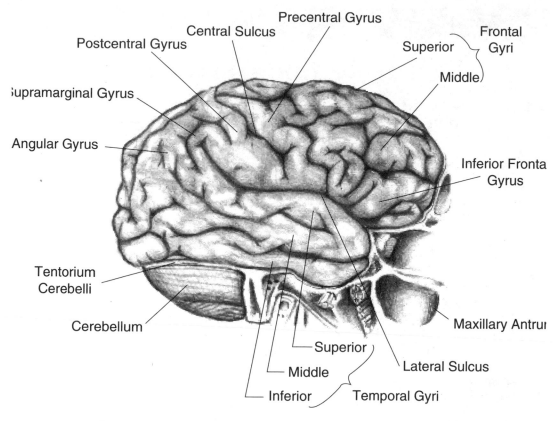

Figure 14.1.   Lateral view of the brain still partly encased by the skull.

Cortical Organization

The functional and columnar neural architecture of the auditory cortex is similar to that of the somesthetic and visual cortex. Like the other primary receiving areas, the auditory cortex, therefore, is basically subdivided into discrete columns, which extend from layers VI and VII to the pial surface (layer I). Likewise, neurons located in the same vertical columns have similar functional properties and are activated by the same type or frequency of auditory stimulus.

AUDITORY NEOCORTEX

Among humans, the primary auditory neocortex is located on the two transverse gyri of Heschl, along the dorsal-medial surface of the superior temporal lobe. The center-most portion of the anterior transverse gyri contains the primary auditory receiving area (Brodmann's area 41) and neocor-

tically resembles the primary visual (area 17) and somesthetic cortices (area 3). The posterior transverse gyri consists of both primary and association cortices (areas 41 and 42, respectively). The major source of auditory input is derived from the medial geniculate nucleus of the thalamus.

In humans, the auditory cortex appears to be larger on the left temporal lobe (Geschwind and Galaburda, 1985; Geschwind and Levitsky, 1968; Habib et al., 1995; Wada et al., 1975). Geschwind, Galaburda, and their colleagues have in fact argued that the larger left planum temporale is a significant factor in the establishment of left hemisphere dominance for language.

AUDITORY TRANSMISSION FROM THE COCHLEA TO THE TEMPORAL LOBE

Within the cochlea of the inner ear are tiny hair cells that serve as sensory recep-

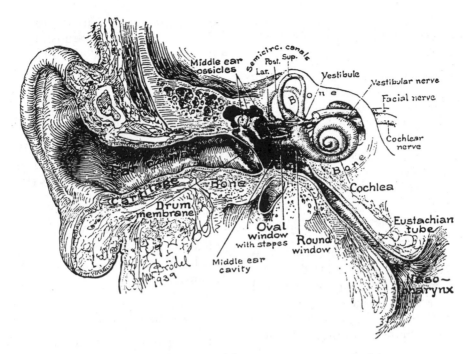

Figure 14.2. Diagram of the inner ear. From M. Brodal.

tors. These cells give rise to axons that form the cochlear division of the 8th cranial nerve, i.e., the auditory nerve. This rope of fibers exits the inner ear and travels to and terminates in the cochlear nucleus, which overlaps and is located immediately adjacent to the vestibular-nucleus from which it evolved within the brainstem (see Chapter 1).

Among mammals, once auditory stimuli are received in the cochlear nucleus, there follows a series of transformations as this information is relayed to various nuclei, i.e., the superior olivary complex, the nucleus of the lateral lemniscus of the brainstem, the midbrain inferior colliculus, and the medial geniculate nucleus of the thalamus, as well as the amygdala (which extracts those features which are emotionally or motivationally significant) and the cingulate gyrus (Carpenter, 1991; Devinksy et al., 1995; Edeline et al., 1990; Parent, 1995). Auditory information is then relayed from the medial geniculate nucleus of the thalamus as well as via the amygdala (through the inferior fasciculus) to Heschl's gyrus, i.e., the primary auditory receiving area

(Brodmann's areas 41 and 42), which is located within the superior temporal gyrus and buried within the depths of the Sylvian fissure.

Here auditory signals undergo extensive analysis and reanalysis and simple associations are established. However, by the time they have reached the neocortex, auditory signals have undergone extensive analysis by the medial thalamus, the amygdala, and the other ancient structures mentioned above.

Unlike the primary visual and somesthetic areas located within the neocortex of the occipital and parietal lobes, which receive signals from only one-half of the body or visual space, the primary auditory region receives some input from both ears, and from both halves of auditory space. This is a consequence of the considerable interconnections and cross-talk that occurs between different subcortical nuclei as information is relayed to and from various regions prior to transfer to the neocortex. Predominantly, however, the right ear transmits to the left cerebral neocortex and vice versa.

AUDITORY RECEIVING AREA: NEUROANATOMICAL-FUNCTIONAL ORGANIZATION

There is some indication that the primary auditory area is tonotopically organized (Edeline et al., 1990; Merzenich and Brugge, 1973; Pantev et al., 1989; Woolsey and Fairman, 1946), such that different auditory frequencies are progressively anatomically represented. However, it may be that the auditory cortex is "cochleotopically" rather than tonotopically organized (Schwarz and Tomlinson, 1990). Nevertheless, high frequencies are received and analyzed in the anterior-medial portions, and low frequencies are received and analyzed in the posterior-lateral regions of the superior temporal lobe (Merzenich and Brugge, 1973).

Neurons located in the primary auditory cortex can determine and recognize differences and similarities between harmonic complex tones and demonstrated auditory response patterns that vary in response to lower and higher frequency and to specific tones (Schwarz and Tomlinson, 1990). Some display "tuning bandwidths" for pure tones, whereas others are able to identify up to seven components of harmonic complex tones. In this manner, pitch can also be discerned (e.g., Pantev et al., 1989).

FILTERING, FEEDBACK, AND TEMPORAL-SEQUENTIAL REORGANIZATION

The old cortical centers located in the midbrain and brainstem evolved long before the appearance of neocortex and have long been adapted and specialized for performing a considerable degree of information analysis (Buchwald et al., 1966). This is evident from observing the behavior of reptiles and amphibians, who are devoid of auditory neocortex.

Moreover, as with the neocortex and thalamus, many of these old cortical nuclei also project back to each other such that each subcortical structure might hear and analyze the same sound repeatedly (Brodal, 1981). In this manner different regions

of the brain may heighten or diminish the amplitude of various sounds via feedback adjustment (Joseph, 1993; Luria, 1980). In fact, the actual order of the sound elements perceived can be rearranged when they are transferred back and forth between nuclei.

This same process continues at the level of the neocortex, which has the advantage of being the recipient of signals that have already been highly processed and analyzed by older brain structures (Buchwald et al., 1966; Edeline et al., 1990; Schwarz and Tomlinson, 1990). However, it is within the neocortex and in this manner that language-related sounds begin to be organized and recognized (e.g., Schwarz and Tomlinson, 1990).

Sustained Auditory Activity

One of the main functions of the primary auditory neocortical receptive area appears to be the retention of sounds for brief time periods (up to a second) so that temporal and sequential features may be extracted (Luria, 1980) and discrepancies in spatial location identified, i.e., so that we can determine where a sound may have originated (see Mills and Rollman, 1980). This also allows comparisons to be made with sounds that were just received and those which are just arriving.

Moreover, via their sustained activity, these neurons are able to prolong (perhaps via a perseverating feedback loop with the thalamus) the duration of certain sounds so that they are more amenable to analysis (Luria, 1980). In this manner, even complex sounds can be broken down into components that are then separately analyzed. Hence, sounds can be perceived as sustained temporal sequences.

Although it is apparent that the auditory regions of both cerebral hemispheres are capable of discerning and extracting temporal-sequential rhythmic acoustics (Milner, 1962), the left temporal lobe apparently contains a greater concentration of neurons specialized for this purpose, as the left half of the brain is clearly superior in this capacity.

For example the left hemisphere has been repeatedly shown to be specialized for sorting, separating, and extracting, in a segmented fashion, the phonetic and temporal-sequential or linguistic-articulatory features of incoming auditory information so as to identify speech units. It is also more sensitive to rapidly changing acoustic cues, be they verbal or nonverbal, as compared to the right hemisphere (Shankweiler and Studdert-Kennedy, 1967; Studdert-Kennedy and Shankweiler, 1970). Moreover, via dichotic listening tasks, the right ear (left temporal lobe) has been shown to be dominant for the perception of real words, word lists, numbers, backwards speech, Morse code, consonants, consonant vowel syllables, nonsense syllables, the transitional elements of speech, single phonemes, and rhymes (Blumstein and Cooper, 1974; Bryden, 1967; Cutting, 1974; Kimura, 1961; Kimura and Folb, 1968; Levy, 1974; Mills and Rollman, 1979; Papcun et al., 1974; Shankweiler and Studert-Kennedy, 1966, 1967; Studdert-Kennedy and Shankweiler, 1970).

In part the association of the left hemisphere and left temporal lobe with performing complex temporal-sequential and linguistic analysis is due to its interconnections with the inferior parietal lobule (see Chapters 2, 4, 7, 13). As noted in Chapters 2 and 4, the inferior parietal lobule is in part an outgrowth of the superior temporal lobe but also consists of somesthetic and visual neocortical tissue and is highly concerned with programming movement of the hand. Because it also represents the hand, internal as well as external stimuli may be manipulated. Hence, the inferior parietal lobule acts to manipulate and to impose temporal sequences on incoming auditory stimuli, as well as visual and somesthetic stimuli. It also serves to provide (via its extensive interconnections with surrounding brain tissue) and to integrate related associations, so as to create multiple representations and categories, thus making complex and grammatically correct human language possible.

However, the language capacities of the left temporal lobe are also made possible via feedback from "subcortical" auditory neurons, and via sustained (as well as diminished) activity and analysis. That is, because of these "feedback" loops, the importance and even order of the sounds perceived can be changed, filtered, or heightened (Joseph, 1993; Luria, 1980). In this manner sound elements composed of consonants, vowels, and phonemes and morphemes can be more readily identified, particularly within the auditory neocortex of the left half of the brain (Cutting, 1974; Shakweiler and Studdert-Kennedy, 1966, 1967; Studdert-Kennedy and Shankweiler, 1970).

For example, normally a single phoneme may be scattered over several neighboring units of sounds. A single sound segment may in turn carry several successive phonemes. Therefore a considerable amount of reanalysis, reordering, or filtering of these signals is required for comprehension (Joseph, 1993). These processes, however, presumably occur both at the neocortical and the old cortical levels. In this manner a phoneme can be identified, extracted, analyzed, and placed in its proper category and temporal position (see edited volume by Mattingly and Studdert-Kennedy, 1991 for related discussion).

Take, for example, three sound units, "t-k-a," which are transmitted to the superior temporal auditory receiving area. Via a feedback loop the primary auditory area can send any one of these units back to the thalamus, which again sends it back to the temporal lobe, thus amplifying the signal and/or allowing for the rearrangement of their order, "t-a-k" or "k-a-t." A multitude of such interactions are in fact possible so that whole strings of sounds can be arranged or rearranged in a certain order (Joseph, 1993). Mutual feedback characterizes most other neocortical-thalamic interactions as well, be it touch, audition, or vision (Brodal, 1981; Carpenter, 1991; Parent, 1995). Via these interactions and feedback loops sounds can be repeated, their order can be rearranged, and the amplitude on specific auditory signals can be enhanced, whereas others can be filtered out.

It is in this manner, coupled with neural plasticity, learning, and experience (Edeline et al., 1990; Diamond and Weinberger, 1989), that fine tuning of the nervous system occurs so that specific signals are attended to, perceived, processed, committed to memory, and so on so that a specific language can be learned over the course of development. Indeed, a significant degree of neural reorganization and plasticity in response to experience as well as auditory filtering occurs throughout the brain in regard to not only sound, but visual and tactual experience as well (Greenough et al., 1987; Hubel and Wiesel, 1970; Juliano et al., 1994). Moreover, the same process occurs when organizing words for expression.

This ability to perform so many operations on individual sound units has in turn greatly contributed to the development of human speech and language. For example, the ability to hear human speech requires that temporal resolutions occur at great speed so as to sort through the overlapping and intermixed signals being perceived. This requires that these sounds are processed in parallel, or stored briefly and then replayed in a different order so that discrepancies due to overlap in sounds can be adjusted for (see Mattingly and Studdert-Kennedy, 1991 for related discussion), and this is what occurs many times over via the interactions of the old cortex and the neocortex. Similarly, when we are speaking or thinking, sound units must also be arranged in a particular order so that what we say is comprehensible to others and so that we may understand our own verbal thoughts.

PHONETICS, CONSONANTS, VOWELS, LANGUAGE

Humans are capable of uttering thousands of different sounds, all of which are easily detected by the human ear. And yet, although own vocabulary is immense, human speech actually consists of about 12–60 units of sound, depending, for example, on whether one is speaking Hawaiian or English. The English vocabulary consists of several hundred thousand words, which are based on the combinations of just 45 different sounds.

Animals too, however, are capable of a vast number of utterances. In fact monkeys and apes employ between 20 and 25 units of sound, whereas a fox employs 36. However, these animals cannot string these sounds together so as to create a spoken grammatical or complex language. Most animals tend to use only a few units of sound at one time, which varies depending on their situation, e.g., lost, afraid, playing.

Humans combine these sounds to make a huge number of words. In fact, employing only 13 sound units, humans are able to combine them to form five million word sounds (see Fodor and Katz, 1964; Slobin, 1971).

Phonemes

The smallest unit of sound is referred to as a phoneme. For example, p and b as in pet versus bet are phonemes. When phonemes are strung together as a unit, they in turn make up morphemes. Morphemes such as "cat" are composed of three phonemes, "k,a,t". Hence, phonemes must be arranged in a particular temporal order in order to form a morpheme.

Morphemes in turn make up the smallest unit of meaningful sounds such as those used to signal relationships such as "-er" as in "he is older than she." All languages have rules that govern the number of phonemes, their relationships to morphemes, and how morphemes may be combined so as to produce meaningful sounds (Fodor and Katz, 1964; Slobin, 1971). Each phoneme is composed of multiple frequencies, which are in turn processed in great detail once they are transmitted to the superior temporal lobe.

As noted, the primary auditory area is tonotopically organized, such that related albeit differing auditory frequencies are analyzed by adjoining cell columns. As also noted, however, it is the left temporal lobe that has been shown to be dominant for the perception of real words, word lists, numbers, syllables, and the transitional elements of speech, as well as single phonemes, consonants, and consonant-vowel syllables.

Consonants and Vowels

In general, there are two large classes of speech sounds: consonants and vowels. Consonants by nature are brief and transitional and have identification boundaries that are sharply defined. These boundaries enable different consonants to be discerned accurately (Mattingly and Studdert-Kennedy, 1991). Vowels are more continuous in nature and, in this regard, the right half of the brain plays an important role in their perception. Consonants are more important in regard to left cerebral speech perception. This is because they are composed of segments of rapidly changing frequencies, which includes the duration, direction, and magnitude of sound segments interspersed with periods of silence. These transitions occur in 50 msec or less, which in turn requires that the left half of the brain take responsibility for perceiving them.

In contrast, vowels consist of slowly changing or steady frequencies with transitions taking 350 or more msec. In this regard, vowels are more like natural environmental sounds, which are more continuous in nature, even those which are brief such as a snap of a twig. They are also more likely to be processed and perceived by the right half of the brain, though the left cerebrum also plays a role in their perception (see Chapter 3).

The differential involvement of the right and left hemispheres in processing consonants and vowels is a function of their neuroanatomical organization and the fact that the left cerebrum is specialized for dealing with sequential information. Moreover, the left temporal lobe is able to make fine temporal discriminations with intervals between sounds as small as 50 msec. However, the right hemisphere needs 8–10 times longer and has difficulty discriminating the order of sounds if they are separated by less than 350 msec.

Hence, consonants are perceived in a wholly different manner from vowels. Vowels yield to nuclei involved in "continuous perception" and are processed by the right (as well as left) half of the brain. Consonants are more a function of "categorical perception" and are processed in the left half of the cerebrum. In fact, the left temporal lobe acts on both vowels and consonants during the process of perception so as to sort these signals into distinct patterns of segments via which these sounds become classified and categorized (Joseph, 1993).

Nevertheless, be they processed by the right or the left brain, both vowels and consonants are significant in speech perception. Vowels are particularly important when we consider their significant role in communicating emotional status and intent.

GRAMMAR AND AUDITORY CLOSURE

Despite claims regarding "universal grammars" and "deep structures," it is apparent that human beings speak in a decidedly nongrammatical manner with many pauses, repetitions, incomplete sentences, irrelevant words, and so on. However, this does not prevent comprehension since the structure of the nervous system enables Homo sapiens sapiens to perceptually alter word orders and sound so they make sense, and to even fill in words that are left out (auditory closure).

For example, when subjects heard a taped sentence in which a single syllable (gis), from the word "legislatures," had been deleted and filled in with static (i.e., le...latures), the missing syllable was not noticed. Rather, all subjects heard "legislatures" (reviewed in Farb, 1975).

Similarly, when the word "tress" was played on a loop of tape 120 times per minute (tresstresstress....") subjects reported hearing words such as dress, florists, purse, Joyce, and stress (Farb, 1975). In other words they organized these into meaningful speech sounds, which were then coded and perceived as words.

The ability to engage in gap-filling, auditory closure, sequencing, and the imposing of temporal order on incoming (supposedly grammatical) speech is important because human speech is not always fluent, as many words are spoken in fragmentary form and sentences are usually four words or less. Much of human speech also consists of pauses and hesitations, "uh" or "err" sounds, stutters, repetitions, and stereotyped utterances ("you know," "like").

Hence, a considerable amount of reorganization as well as filling in must occur prior to comprehension. Therefore, these signals must be rearranged in accordance with the temporal sequential rules imposed by the structure and interaction of Wernicke's area, the inferior parietal lobe, and the nervous system—what Noam Chomsky (1957) referred to as "deep structure"—so that they may be understood.

Consciously, however, most people fail to realize that this filling in and reorganization has even occurred, unless directly confronted by someone who claims that she said something she believes she didn't. Nevertheless, this filling in and process of reorganization greatly enhances comprehension and communication.

Universal Grammars

Regardless of culture, race, environment, geographical location, parental verbal skills, or attention, children the world over go through the same steps at the same age in learning language (Chomsky, 1972). Unlike reading and writing, the ability to talk and understand speech is innate and requires no formal training. One is born with the ability to talk, as well as the ability to see, hear, feel, and so on. However, one must receive training in reading, spelling, and mathematics, as these abilities are acquired only with some difficulty and much effort. On the other hand, just as one must be exposed to light or he will lose the ability to see complex forms, one must be exposed to language or he will lose the ability to talk or understand human speech.

In his book Syntactic Structures, Chomsky (1957) argues that all human beings are endowed with an innate ability to acquire language as they are born able to speak in the same fashion, albeit according to the tongue of their culture, environment, and parents. They possess all the rules that govern how language is spoken and they process and express language in accordance with these innate temporal-sequential motoric rules that we know as grammar.

Because they possess this "deep structure," which in turn is imposed by the structure of our nervous system, children are able to learn language even when what they hear falls outside this structure and is filled with errors. That is, they tend to produce grammatically correct speech sequences, even when those around them fail to do so. They disregard errors because they are not processed by their nervous system, which acts to either impose order even where there is none, or to alter or delete the message altogether. It is because humans possess these same Wernicke-inferior parietal lobe "deep structures" that speakers are also able to realize generally when a sentence is spoken in a grammatically incorrect fashion.

It is also because apes, monkeys, and dogs and cats do not possess these deep structures or an angular gyrus that they are unable to produce grammatically complex sounds, sequences, or gestures. Hence, their vocal production does not become punctuated by temporal-sequential gestures imposed on auditory input or output by the Language Axis.

LANGUAGE ACQUISITION: FINE TUNING THE AUDITORY SYSTEM

By time humans have reached adulthood, they have learned to attend to certain sounds and to ignore the rest. Some sounds are filtered and ignored as they are irrelevant or meaningless. Humans also lose the ability to hear various sounds because of actual physical changes, such as deterioration and deafness, that occur within the auditory system.

At birth and continuing throughout the first decade of life, however, there is a much broader range of generalized auditory sensitivity. It is this generalized sensitivity that enables children to rapidly and more efficiently learn a foreign tongue, a capacity that decreases as they age (Janet et al., 1984). This is because auditory neurons drop out if they are not employed and other neurons become sensitive to only certain sounds as a function of early experience. It is due, in part, to these same early neuronal experience changes that generational conflicts regarding what constitutes "music" frequently arise. As pertaining to language, since much of what is heard is irrelevant and is not employed in the language the child is exposed to, the neurons involved in mediating their perception either drop out or die from disuse, which further restricts the range of sensitivity. This also further aids the fine tuning process so that, for example, one's native tongue can be learned (Janet et al., 1984; Joseph, 1993).

For example, no two languages have the same set of phonemes. It is because of this that to members of some cultures certain English words, such as pet and bet, sound exactly alike. Non-English speakers are unable to distinguish between or recognize these different sound units. Via this fine tuning process, only those phonemes essential to one's native tongue are attended to.

Phonemes and Neuronal "Fine Tuning"

Language differs not only in regard to the number of phonemes, but the number of phonemes that are devoted to vowels versus consonants and so on. Some Arabic dialects have 28 consonants and 6 vowels. By contrast, the English language consists of 45 phonemes that include 21 consonants, 9 vowels, 3 semivowels (y, w, r), 4 stresses, 4 pitches, 1 juncture (pauses between words), and 3 terminal contours that are used to end sentences (Fodor and Katz, 1964; Slobin, 1971).

It is from these 45 phonemes that all sounds are derived that make up the vast multitude of utterances that make up the English language. However, learning to attend selectively to these 45 phonemes, as well as to specific consonants and vowels, requires that the nervous system become fine tuned to perceiving them while ignoring others. In consequence, those cells which are unused die.

Children are able to learn their own as well as foreign languages with much greater ease than adults because initially infants maintain a neural sensitivity to a universal set of phonetic categories (e.g., Janet et al., 1984). Because of this they are predisposed to hearing and perceiving speech and anything speech-like regardless of the language employed.

These neural sensitivities are either enhanced or diminished during the course of the first few years of life so that those speech sounds which the child most commonly hears becomes accentuated and more greatly attended to, such that a sharpening of distinctions occurs (Janet et al., 1984). However, this generalized sensitivity in turn declines as a function of acquiring a particular language and, presumably, the loss of nerve cells not employed (see Chapter 17 for related discussion). The nervous system becomes fine tuned so that familiar language-like sounds become processed and ordered in the manner dictated by the nervous system, i.e., the universal grammatical rules common to all languages.

Fine tuning is also a function of experience, which in turn exerts tremendous influence on nervous system and neuronal network development and cell death. Hence, by the time most people reach adulthood they have long learned to categorize most of their new experiences into the categories and channels that have been relied on for decades.

Nevertheless, in consequence of this filtering and neuronal reorganization and dropout, sounds that arise naturally within one's environment can be altered, rearranged, suppressed, and thus erased. Fine tuning the auditory system so as to learn culturally significant sounds and so that language can be acquired occurs at a sacrifice. It occurs at the expense of one's natural awareness of the environment and its orchestra of symphonic sounds. In other words (at least from the perspective of the left hemisphere), the fundamental characteristics of reality are subject to language-based alterations (Sapir, 1966; Whorf, 1956) and the perceptual sensitivity and organization of the neural networks that subserve language.

LANGUAGE AND REALITY

As detailed in Chapter 16 and elsewhere (Joseph, 1982, 1988a, 1992b, 1993), language is a tool of consciousness, and the linguistic aspects of consciousness are dependent on the functional integrity of the left half of the brain. Because left cerebral conscious experience is, in large part, dependent on and organized in regard to linguistic and temporal-sequential modes of comprehension and expression, language and the neural networks that subserve language can shape as well as determine the manner in which perceptual reality is perceived and conceived—at least by the left hemisphere.

According the Edward Sapir (1966), "Human beings are very much at the mercy of the particular language which has become the medium of their society...the real world is to a large extent built up on the language habits of the group. No two languages are ever sufficiently similar to be considered as representing the same social reality."

According to Benjamin Whorf (1956), "language...is not merely a reproducing instrument for voicing ideas but rather is itself the shaper of ideas.... We dissect nature along lines laid down by language." However, Whorf believed that it was not just the words we used but grammar that has acted to shape human perceptions and thought. A grammatically imposed structure forces perceptions to conform to the mold that gives them not only shape, but direction and order.

Hence, the distinctions imposed by the neural networks that subserve language and temporal-sequential processing and the employment of linguistically based cat-egories and verbal labels and verbal associations can shape and give form to one's thoughts and perceptual reality.

For example, Eskimos possess an extensive and detailed vocabulary that enables them to make fine verbal distinctions between different types of snow, ice, and prey, such as seals. To a man born and raised in Kentucky, snow is snow and all seals may look the same. But if that same man from Kentucky has been raised around horses all his life he may in turn employ a rich and detailed vocabulary to describe them, e.g., appaloosa, paint, pony, stallion, and so on. However, to the Eskimo a horse may be just a horse and these other names may mean nothing to him. All horses look the same.

Moreover, both the Eskimo and the Kentuckian may be completely bewildered by the hundreds of Arabic words associated with camels, the 20 or more terms for rice used by different Asiatic communities, or the 17 words the Masai of Africa use to describe cattle.

Nevertheless, these are not just words and names, for each cultural group is also able to see, feel, taste, or smell these distinctions as well, a consequence of experiential alterations in the neural assemblies subserving language and linguistic expressive and receptive perceptual experience. An Eskimo sees a horse but a breeder may see a living work of art whose hair, coloring, markings, stature, tone, height, and so on speak volumes as to its character, future, and genetic endowment.

Through language and via fine tuning of the neural assemblies subserving perceptual experience, one can teach others to attend to and to make the same distinctions and to create the same categories. In this way, those who share the same language and cultural group learn to see and talk about the world in the same way, whereas those speaking a different dialect may in fact perceive a different reality as their nervous system has been organized differently. Thus, the linguistic foundations of conscious experience, and thus linguistic-perceptual consciousness, can be shaped by language, early linguistic experience, and the neural networks that subserve language and perceptual activity.

For example, when A. F. Chamberlain (1903) visited the Kootenay and Mohawk Indians of British Columbia during the late 1800s, he noted that they heard animal and bird sounds differently from him. For example, when listening to some owls hooting, he noted that to him it sounded like "tu-whit-tu-whit-tu-whit," whereas the Indians heard "Katskakitl." However, once he became accustomed to their language and began to speak it (thereby relying on a different and newly developed neural network), he soon developed the ability to hear sounds differently once he began to listen with his "Indian ears." When listening to a whippoorwill, for example, he noted that instead of saying "whip-poor-will," it was saying "kwa-kor-yeuh."

Observations such as these thus strongly suggest that changing languages and thus developing different language-based neural networks might change one's perceptions and even the expression of one's conscious thoughts and attitudes. Consider, for example, the results from an experiment reported by Farb (1975). Bilingual Japanese-born women married to American serviceman were asked to answer the same question in English and in Japanese. The following responses were typical. "When my wishes conflict with my family's...it is a time of great unhappiness" (Japanese). "I do what I want" (English). "Real friends should...help each other" (Japanese). "Be very frank" (English).

Obviously, language does not shape all attitudes and perceptions. Moreover, language is often a consequence of these differential perceptions, which in turn requires the invention of new neural networks and linguistic labels so as to describe these new experiences. Speech of course is also filtered through the personality of the speaker, whereas the listener is also influenced by her own attitudes, feelings, beliefs, prejudices, neural architecture, and so on, all of which can affect what is said, how it is said, and how it is perceived and interpreted (Joseph, 1993).

In fact, regardless of the nature of a particular experience, be it visual, olfactory, or sexual, language not only influences perceptual and conscious experiences but even the ability to derive enjoyment from them, for example, by labeling them cool, hip, sexy, bad, or sinful, and by instilling guilt or pride. In this manner, language serves not only to label and filter reality, but affects one's ability to enjoy it.

SPATIAL LOCALIZATION, ATTENTION, AND ENVIRONMENTAL SOUNDS

In conjunction with the inferior colliculus, and because of bilateral auditory input, the primary auditory area plays a significant role in orienting to and localizing the source of various sounds (Sanchez-Longo and Forster, 1958)—for example, by comparing time and intensity differences in the neural input from each ear. A sound arising from one's right will reach and sound louder to the right ear as compared to the left ear.

Indeed, among mammals, a considerable number of auditory neurons respond or become highly excited only in response to sounds from a particular location (Evans and Whitfield, 1968). Moreover, some of these neurons become excited only when the subject looks at the source of the sound (Hubel et al., 1959). Hence, these neurons act so that location may be identified and fixated on. These complex interactions probably involve the parietal area (7), as well as the midbrain colliculi and limbic system. According to lesion studies in humans, the right temporal lobe is more involved than the left in discerning location (Penfield and Evans, 1934; Shankweiler, 1961).

There is also some indication that certain cells in the auditory area are highly specialized and will respond only to certain meaningful vocalizations (Wollberg and Newman, 1972). In this regard they seemed to be tuned to respond only to specific auditory parameters so as to identify and extract certain meaningful features,

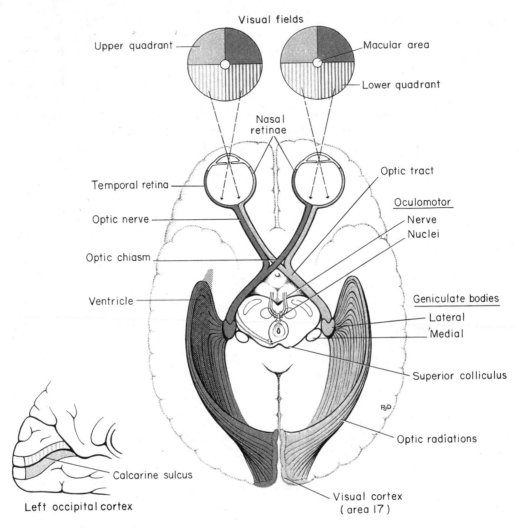

Figure 14.3. Schematic diagram of the visual pathways viewed from ventral portion of the brain. From Carpenter M. Human Neuroanatomy. Baltimore: Williams & Wilkins, 1991.

i.e., feature detector cells. For example, some cells will respond to cries of alarm and sounds suggestive of fear or indicating danger, whereas others will react only to specific sounds and vocalizations.

Nevertheless, although the left temporal lobe appears to be more involved in extracting certain linguistic features and differentiating between semantically related sounds (Schnider et al., 1994), the right temporal region is more adept at identifying and recognizing acoustically related sounds and nonverbal environmental acoustics (e.g., wind, rain, animal noises), prosodic-melodic nuances, sounds that convey emotional meaning, as well as most

aspects of music, including tempo and meter (Heilman et al., 1975, 1984; Joseph, 1988a; Kester et al., 1991; Schnider et al., 1994; Spinnler and Vignolo, 1966) (see Chapter 3).

Indeed, the right temporal lobe's spatial-localization sensitivity coupled with its ability to perceive and recognize environmental sounds no doubt provided great survival value to early primitive humans. That is, in response to a specific sound (e.g., a creeping predator), one is able to immediately identify, localize, locate, and fixate on the source and thus take appropriate action. Of course, even modern humans rely on the same mechanisms to avoid cars when walking across streets or riding bicycles, or to ascertain and identify approaching individuals, etc.

Hallucinations

Electrical stimulation of Heschyl's gyrus produces elementary hallucinations (Penfield and Jasper, 1954; Penfield and Perot, 1963). These include buzzing, clicking, ticking, humming, whispering, and ringing, most of which are localized to the opposite side of the room. Tumors involving this area also give rise to similar, albeit transient, hallucinations, including tinnitus (Brodal, 1981). Patients may complain that sounds seem louder and/or softer than normal, closer and/or more distant, strange, or even unpleasant (Hecaen and Albert, 1978). There is often a repetitive quality that makes the experience even more disagreeable.

In some instances the hallucination may become meaningful. These include the sound of footsteps, clapping hands, or music, most of which seem (to the patient) to have an actual external source. These hallucinations are due to the abnormal activation of neurons that normally respond to the sounds.

LATERALITY

Auditory verbal hallucinations seem to occur with right or left temporal destruction or stimulation (Hecaen and Albert, 1978; Penfield and Perot, 1963; Tarachow, 1941)—although left temporal involvement is predominant. With fewer temporal lobe abnormalities the hallucination may involve single words, sentences, commands, advice, or distant conversations that can't quite be made out. According to Hecaen and Albert (1978), verbal hallucinations may precede the onset of an aphasic disorder, such as one due to a developing tumor or other destructive process. Patients may complain of hearing "distorted sentences," "incomprehensible words," etc.

Penfield and Perot (1963) report that electrical stimulation of the superior temporal gyrus, the right temporal lobe in particular, results in musical hallucinations.

Similarly, patients with tumors and seizure disorders, particularly those involving the right temporal region, may also experience musical hallucinations. Frequently the same melody is heard over and over. In some instances patients have reported hearing singing voices, and individual instruments may be heard (Hecaen and Albert, 1978).

Conversely, it has been frequently reported that lesions or complete surgical destruction of the right temporal lobe significantly impairs the ability to name or recognize melodies and musical passages. It also disrupts time sense and the perception of timbre, loudness, and meter (Chase, 1967; Joseph, 1988; Kester et al., 1991; Milner, 1962; Shankweiler, 1966).

NEOCORTICAL DEAFNESS

In some instances the primary auditory receiving areas of the right or left cerebral hemisphere and/or the underlying "white matter" fiber tracts may be destroyed (e.g., following a middle cerebral artery stroke). This results in a disconnection syndrome such that sounds relayed from the thalamus cannot be received or analyzed by the temporal lobe. In some cases, however, the damage or strokes may be bilateral. When both primary auditory receiving areas and their subcortical axonal connections have been destroyed, the patient is said to be suffering from cortical deafness (Tanaka et al., 1991). The neocortex no longer receives auditory input.

However, sounds continue to be processed subcortically by the amygdala and inferior colliculis. Hence, the ability to hear sounds per se is retained. Nevertheless, since the sounds which are heard are not received neocortically and thus cannot be transmitted to the adjacent association areas; sounds become stripped of meaning. That is, meaning cannot be extracted or assigned by the neocortex and the patient becomes agnosic for auditory stimuli (Albert et al., 1972; Schnider et al., 1994; Spreen et al., 1965). Rather, only differences in intensity are discernible—much like cortical blindness.

When superior temporal lobe lesions are bilateral, patients cannot respond to questions, do not always show startle responses to loud sounds, lose the ability to discern the melody for music, cannot recognize speech or environmental sounds, and tend to experience the sounds they do hear as distorted and disagreeable, e.g., like the banging of tin cans, buzzing and roaring, etc. (Albert et al., 1972; Auerbach et al., 1982; Earnest et al., 1977; Kazui et al., 1990; Mendez and Geehan, 1988; Reinhold, 1950; Tanaka et al., 1991).

Commonly such patients also experience difficulty discriminating sequences of sound, detecting differences in temporal patterning, and determining sound duration, whereas intensity discriminations are better preserved. For example, when a 66-year-old right-handed male suffering from bitemporal subcortical lesions "woke up in the morning and turned on the television, he found himself unable to hear anything but buzzing noises. He then tried to talk to himself, saying 'This TV is broken.' However, his voice sounded like noise to him. Similarly, the patient could hear his wife's voice, he could not interpret the meaning of her speech. He was also unable to identify many environmental sounds" (Kazui et al., 1990:477).

Nevertheless, individuals who are cortically deaf are not aphasic (e.g., Kazui et al., 1990), though with massive bilateral lesions, both conditions can occur together. Although "cortically deaf," they can read, write, speak, and comprehend pantomime, and are fully aware of their deficit. However, afflicted individuals may also display auditory inattention (Hecaen and Albert, 1978) and a failure to respond to loud sounds. Nevertheless, although they are not aphasic per se, speech is sometimes noted to be hypophonic and contaminated by occasional literal paraphasias.

In some instances, rather than bilateral, a destructive lesion may be limited to the primary receiving area of just the right or left cerebral hemisphere. These patients are not considered cortically deaf. However, when the auditory receiving area of the left temporal lobe is destroyed the patient suffers from a condition referred to as pure word deafness. If the lesion is in the right temporal receiving area, the patient is more likely to suffer a nonverbal auditory agnosia (Schnider et al., 1994).

Pure Word Deafness

With a destructive lesion involving the left auditory receiving area and underlying white matter, Wernicke's area becomes disconnected from almost all sources of acoustic input and patients are unable to recognize or perceive the sounds of language, be it sentences, single words, or even single letters (Hecaen and Albert, 1978). All other aspects of comprehension are preserved, including reading, writing, and expressive speech.

Moreover, because of sparing of the right temporal region, the ability to recognize musical and environmental sounds is preserved. However, the ability to name or verbally describe them is impaired because of disconnection and the inability of the right hemisphere to talk.

Pure word deafness, of course, also occurs with bilateral lesions in which case environmental sound recognition is also effected (e.g., Kazui et al., 1990). In these instances the patient is considered cortically deaf.

Pure word deafness, when due to a unilateral lesion of the left temporal lobe, is partly a consequence of an inability to extract temporal-sequential features from incoming sounds. Hence, linguistic messages cannot be recognized. However, pure word deafness can sometimes be partly overcome if the patient is spoken to in an extremely slow manner (Albert and Bear, 1974). The same is true of those with Wernicke's aphasia.

Auditory Agnosia

An individual with cortical deafness due to bilateral lesions suffers from a generalized auditory agnosia involving words and nonlinguistic sounds. However, in many instances an auditory agnosia, with preserved perception of language, may occur with lesions restricted to the right superior temporal lobe. In these instances, an individual loses the capability to correctly discern environmental sounds (e.g., birds singing, doors closing, keys jangling) and acoustically related sounds, emotional-prosodic speech, and music (see Chapter 3).

These problems are less likely to come to the attention of a physician unless accompanied by secondary emotional difficulties. That is, most individuals with this disorder, being agnosic, would not know that they have a problem and thus would not complain. If they are their families notice (for example, if a patient does not respond to a knock on the door), the likelihood is that the problem will be attributed to faulty hearing, attentional deficits, or even forgetfulness.

However, because such individuals may also have difficulty discerning emotional-melodic nuances, it is likely that they will misperceive and fail to comprehend a variety of paralinguistic social-emotional messages. This includes difficulty discerning what others may be implying, or in appreciating emotional and contextual cues, including variables such as sincerity or mirthful intonation. Hence, a host of emotional and behavioral difficulties may arise (see Chapter 3).

For example, a patient may complain that his wife no longer loves him, and that he knows this from the sound of her voice. In fact, a patient may notice that the voices of friends and family sound in some manner different, which, when coupled with difficulty discerning nuances such as humor and friendliness, may lead to the development of paranoia and what appears to be delusional thinking. Indeed, unless appropriately diagnosed, it is likely that the patient's problem will feed on and reinforce itself and grow more severe.

It is important to note that rather than being completely agnosic or word-deaf, patients may suffer from only partial deficits. In these instances they may seem to be hard of hearing, frequently misinterpret

what is said to them, and/or slowly develop related emotional difficulties.

THE AUDITORY ASSOCIATION AREAS

The auditory area, although originating in the depths of the superior temporal lobe, extends in a continuous belt-like fashion posteriorly from the primary and association (e.g., Wernicke's) area toward the inferior parietal lobule, and via the arcuate fasciculus onward toward Broca's area. Indeed, in the left hemisphere, this massive rope of interconnections forms an axis such that Wernicke's area, the inferior parietal lobule, and Broca's area, together, are able to mediate the perception and expression of most forms of language and speech.

The rope-like arcuate and inferior and superior fasciculi are bidirectional fiber pathways, however, that run not only from Wernicke's area through to Broca's area but extend inferiorly deep into the temporal lobe where contact is established with the amygdala and deep into the frontal lobe and the anterior cingular gyrus. In this manner, auditory input comes to be assigned emotional-motivational significance, whereas verbal output becomes emotionally-melodically colored. Within the right hemisphere, these interconnections, which include the amygdala and cingulate, appear to be more extensively developed.

Wernicke's Area

Following the analysis performed in the primary auditory receiving area, auditory information is transmitted to the immediately adjacent association cortex (Brodmann's area 22), where more complex forms of processing take place. In both hemispheres this region partly surrounds the primary area and then extends posteriorly, merging with the inferior parietal lobule with which it maintains extensive interconnections. Within the left hemisphere, the more posterior portion of area 22 and part of area 42 correspond to Wernicke's area.

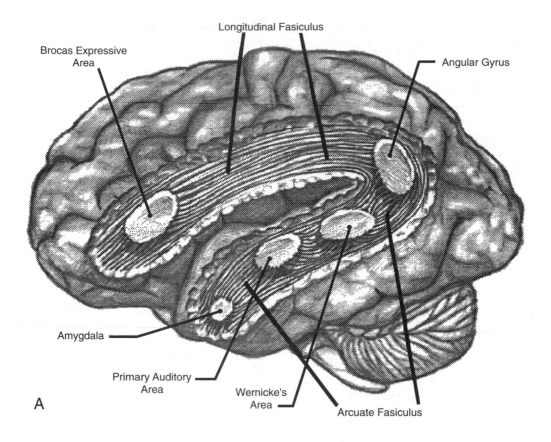

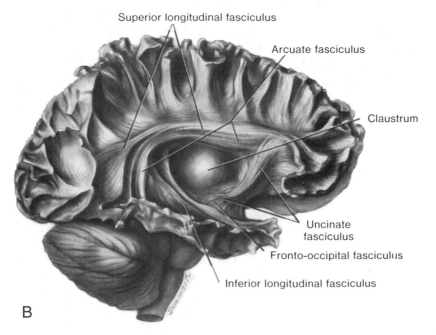

Figure 14.4. Top, Schematic diagram of the fiber pathways (the arcuate and inferior and superior fasciculus) linking the auditory neocortex with the amygdala. Bottom, dissection of the lateral surface of the right hemisphere revealing the inferior, superior, and arcuate fasciculus. From Carpenter M. Human Neuroanatomy. Baltimore: Williams & Wilkins, 1991.

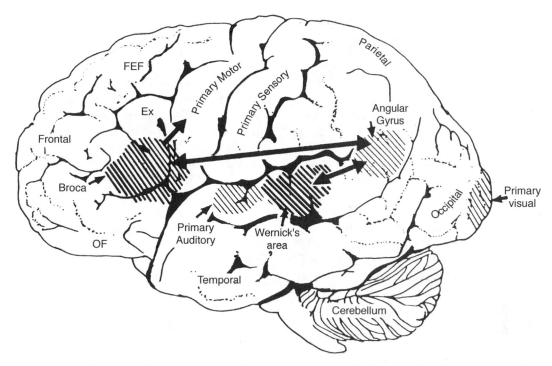

Figure 14.5. The left cerebral hemisphere with the Language Axis, including depictions of the frontal eye fields (FEF) and Exner's (Ex) writing areas.

Wernicke's area and the corresponding region in the right hemisphere are not merely auditory association areas as they also receive a convergence of fibers from the somesthetic and visual association cortices (Jones and Powell, 1970; Zeki, 1978b). Hence, this region is involved in multimodal as well as auditory associational analyses. The auditory association area (area 22) also receives fibers from the contralateral auditory association area via the corpus callosum and anterior commissure, and projects to the frontal convexity and orbital region, the frontal eye fields, and cingulate gyrus (Jones and Powell, 1970).

Hence, a considerable number of multimodal linkages are maintained by the auditory association area. In fact, this region appears to be responsible for the integration and assimilation of a diverse number of informational variables (Luria, 1980) and probably acts in close conjunction with the frontal and inferior parietal lobe in this regard. Indeed, the richness of language and the ability to utilize a multitude of words and images to describe a single object or its use is very much dependent on these interconnections.

Receptive Aphasia

When the posterior portion of the left auditory association area is damaged patients typically suffer from a severe receptive aphasia, i.e., Wernicke's aphasia. Individuals with Wernicke's aphasia, in addition to severe comprehension deficits (see Chapter 4 for details), usually suffer abnormalities involving expressive speech, reading, writing, repeating, word finding, etc. Spontaneous speech, however, is often seemingly fluent, and sometimes hyperfluent such that these persons speak at an increased rate and seem unable to bring sentences to an end—as if the temporal-sequential separations between words have been extremely shortened or in some instances abolished. Hence, this disorder has also been referred to as fluent aphasia. In addition, speech may be contaminated by neologistic and paraphasic distortions. As such, what they say is often incomprehensible (Christman, 1994; Goodglass and Kaplan, 1982; Hecaen and Albert, 1978).

Individuals with Wernicke's aphasia, however, may be able to perceive the temporal-sequential pattern of incoming auditory stimuli to some degree (because of perseveration of the primary region). Nevertheless, they are unable to perceive spoken words in their correct order, cannot identify the pattern of sound presentation, and become easily overwhelmed (Hecaen and Albert, 1978; Luria, 1980). Patients with Wernicke's aphasia, in addition to other impairments, are necessarily "word deaf." As these individuals recover from their aphasia, word deafness is often the last symptom to disappear (Hecaen and Albert, 1978).

THE MELODIC-INTONATIONAL AXIS

It has been consistently demonstrated among normals (such as in dichotic listening studies) that the right temporal lobe (left ear) predominates in the perception of timbre, chords, tone, pitch, loudness, melody, and intensity—the major components (in conjunction with harmony) of a musical stimulus (Joseph, 1988a). However, both the right and left hemispheres seem to process the rhythmic features of music with equal facility.

When the right temporal lobe is damaged (e.g., right temporal lobectomy, right amygdalectomy), time sense; rhythm; and the ability to sing, carry a tune, and perceive, recognize, or recall tones, loudness, timbre, and melody may be disrupted. Similarly, the ability to recognize even familiar melodies and the capacity to obtain pleasure while listening to music is abolished or significantly reduced—a condition referred to as amusia.

In addition, lesions involving the right temporal-parietal area have been reported to significantly impair the ability to perceive and identify environmental sounds; to comprehend or produce appropriate verbal prosody and emotional speech; or to repeat emotional statements (see Chapter 3). Indeed, when presented with neutral sen-

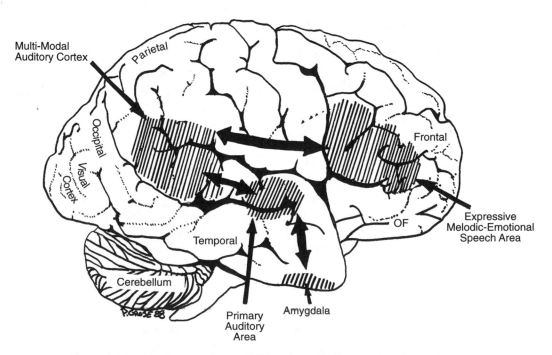

Figure 14.6.   Auditory pathways linking the melodic-emotional speech areas.

tences spoken in an emotional manner, persons with right temporal-parietal damage have been reported to experience disruptions in the perception and comprehension of emotional prosody, regardless of its being positive or negative in content (see Chapter 3).

Hence, the right temporal-parietal area is involved in the perception, identification, and comprehension of environmental and musical sounds and various forms of melodic and emotional auditory stimuli, and probably acts to prepare this information for expression via transfer to the right frontal convexity, which is dominant regarding the expression of emotional-melodic and even environmental sounds. Indeed, it appears that an emotional-melodic-intonational axis, somewhat similar to the Language Axis in anatomical design, is maintained within the right hemisphere (Gorelick and Ross, 1987; Joseph, 1982, 1988a; Ross, 1981).

When the posterior portion of the Melodic-Emotional Axis is damaged, the ability to comprehend or repeat melodic-emotional vocalizations in disrupted. Such patients are thus agnosic for nonlinguistic sounds. With right frontal convexity damage, speech may become bland, atonal, monotone, or distorted melodically.

The Amygdala

As detailed in Chapters 3 and 7, the right cerebral hemisphere appears to maintain more extensive as well as bilateral interconnections with the limbic system (Joseph, 1982). Indeed, the limbic system also appears to be lateralized in regard to certain aspects of emotional functioning such that the right amygdala, hippocampus, and hypothalamus seem to exert dominant influences (see Chapter 5).

As noted, the arcuate fasciculus extends from the amygdala (which is buried within the anterior-inferior temporal lobe) through the auditory association area and inferior parietal region and into the frontal convexity. It is through these interconnections that emotional colorization is added

to neocortical acoustic perceptions and to speech or musical sounds being prepared for expression. Hence, when the right amygdala has been destroyed or surgically removed, the ability to sing as well as to properly intonate is altered (see Chapter 5). In this regard, the amygdala should be considered part of the melodic-intonational axis of the right hemisphere (and the language axis of the left) as it not only subcortically responds to and analyzes environmental sounds and emotional vocalizations but (in addition to the anterior cingulate) imparts emotional significance to auditory input and output processed and expressed at the level of the neocortex.

To a lesser extent, emotional-prosodic intonation may be imparted directly to speech variables as they are being organized for expression within the left hemisphere. That is, although the right hemisphere is dominant in regard to melodic-emotional vocalizations, the two amygdalas are in direct communication via the anterior commissure, whereas callosal fibers connect the right and left cingulate. Hence, although originating predominantly within the right amygdala/ hemisphere, these influences can be directly transmitted to the left half of the brain via the cingulate, left amygdala, and anterior commissure and then via the arcuate fasciculus into the linguistic stream of the Language Axis.

THE MIDDLE AND MEDIAL TEMPORAL LOBES

The middle temporal region is a complex multifunctional zone that is functionally lateralized and has both extensive visual and auditory capabilities. In general, the anterior middle and medial regions are more involved in auditory functioning, whereas the posterior medial and middle temporal lobes (MTL) are more intimately concerned with the processing and recall of higher level visual information.

As detailed in Chapter 2, over the course of evolution the hippocampus became increasingly cortical and was dis-

placed from its superior position to an inferior position in the temporal lobe. However, in the course of its displacement it
gave rise to neocortical tissues (both medially and laterally) that, like the original
hippocampus, became responsive to visual
stimuli. The middle, medial, and portions
of the posterior ITL, therefore, are also visually responsive, being derivatives of the
hippocampus.

However, as the amygdala also became
increasingly cortical it also gave rise to neocortical tissues, including the anterior inferior, middle, medial, and superior temporal
lobes. Like the amygdala, these neocortical
tissues also became responsive to auditory,
emotional, and visual stimuli.

The "Auditory" Middle Temporal Lobe

The "auditory" medial and middle temporal region (AMT) maintains extensive
interconnections with the superior and inferior temporal lobes, auditory association
area, frontal convexity, and orbital region
(Jones and Powell, 1970; Pandya and
Kuypers, 1969). The inferior arcuate fasciculus, in its journey from the amygdala to
Wernicke's area, also passes through this
neocortical auditory territory. However, as
noted, the hippocampus also contributed
to the development of this tissue, thus involving it in visual as well as memory
functioning.

The anterior medial and AMT, therefore,
play a role in auditory memory and the
discrimination and organization of speech
and other auditory sounds (Dewson et al.,
1969; Weiskrantz and Mishkin, 1958), as
well as in visual functioning (Felleman and
Kaas, 1974; Heit et al., 1990; Maunsell and
Van Essen, 1983).

Left AMT Language Capabilities

As based on the electrical stimulation
studies of Penfield and Rasmussen (1950),
it appears that the auditory-linguistic capabilities of the medial and AMT are more extensively represented within the left
hemisphere and that these areas play a sig-

nificant role in speech output. Thus when
the medial or MTL is stimulated there can
result aphasic abnormalities (Penfield and
Rasmussen, 1950).

Lesions involving these areas are also
associated with subtle disturbances involving language, including word-finding
difficulty, confrontive naming deficits,
abnormalities in the maintenance of temporal order and sequence, and verbal
memory impairments (Luria, 1980). Patients may have difficulty recalling word
lists or order of word presentation, such
that the longer the list, the greater the difficulty. Hence, if a list of four words were
read, the patient cannot repeat the correct
order even after repeated presentations,
although individual words may be recalled correctly (Luria, 1980). However, if
the same material is presented visually,
i.e., in a written format, the patient is able
to perform without difficulty.

Hence, this disorder is due to a failure to
store auditory-linguistic rather than visual-
linguistic material and is presumably secondary to disconnection. That is, a lesion
involving the medial or middle temporal
region results in a severance of the fibers
linking the memory centers in the inferior
temporal region (e.g., the amygdala and
hippocampus) with the auditory receiving
areas. However, by presenting material visually an intact pathway may be utilized
and the region of disconnection may be
circumvented.

In that the inferior temporal lobule is
also specialized for the visual recognition
of forms, disconnection may result in an
inability to associate verbal labels with visual images—hence, confrontive naming
difficulties are associated with lesions to
the left MTL.

Hallucinations

Tumors involving the medial and MTL
have been associated with the development of auditory and visual hallucinations,
dreamy states, and alterations in emotional
functioning—particularly as the lesion en-
croaches on the inferior regions (Luria,

1980). Electrical stimulation also initiates the development of complex hallucinations and alterations in consciousness (Penfield and Roberts, 1959).

THE "VISUAL" MIDDLE TEMPORAL LOBE

The MTL, like other structures throughout the brain, appears to be functionally lateralized such that the right temporal region is more involved in visual and nonverbal auditory functioning, whereas the left MTL is more concerned with auditory-linguistic capabilities. Hence, whereas the auditory-linguistic zones appear to be more extensive within the left temporal lobe (and include portions of the posterior temporal region), the right MTL seems to be more associated with visual responsiveness. In general, however, the middle temporal gyrus of both hemispheres appear to contain visually responsive neurons (Beckers and Zeki, 1995; Felleman and Kaas, 1974; Heit et al., 1990; Maunsell and Van Essen, 1984; Selemon et al., 1994). Nevertheless, since there have been relatively few published studies of the effects of lesions or the functional capabilities of the MTLs in humans, notions regarding the extent of auditory versus visual lateralized representation are admittedly speculative.

MTL Visual Functioning

Visual cells in the medial and MTL receive direct projections from the striate cortex, area 17, and the association area 18 (Wall et al., 1982). This area in turns projects to the MTL of the opposite hemisphere via the corpus callosum; ipsilaterally to areas 17, 18, and 19; and the inferior temporal and the parietal-occipital cortices (Tigges et al., 1981; Wall et al., 1982), and maintains interconnections with the pulvinar, superior colliculus, and pontine nuclei of the brainstem (Walls et al., 1982). Hence, this region has connections with areas concerned with visual and visual attentional capabilities.

MTL neurons are sensitive to speed and direction of stimulus movement, motion, orientation, width, and disparity (Felleman and Kaas, 1984; Maunsell and Van Essen, 1983). However, most MTL neurons are not particularly sensitive to the form of a stimulus—most preferring narrow stimuli (Albright et al., 1984; Maunsell and Van Essen, 1983). A large number of cells have binocular capabilities and respond to stimuli from either or both eyes. Hence, these cells are significantly involved in stereoscopic vision as well as in determining distance.

Although not specialized for perception of form per se, via the combined analysis of width, orientation, speed, and direction of

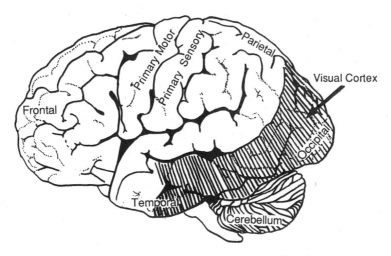

Figure 14.7.   Schematic representation of visual cortex within the parietal, occipital, and temporal regions of the left cerebral hemisphere.

movement, these cells can process visual stimuli in a three-dimensional fashion, particularly in regard to trajectory, depth, and position, i.e., if an object is near or far, approaching or withdrawing (Maunsell and Van Essen, 1983).

Via its interconnections with the parietal-occipital lobe (an area involved in depth perception, visual fixation, and the visual-somesthetic guidance of movement), the MTL appears to provide visual guidance in regard to interactional object and body movement. That is, via the analysis of the flow of movement, local direction, and speed, as well as the relative distances of objects, these neurons (in conjunction with area 7 cells) are important in guiding the movements of the body as well as the limbs toward the object (Manunsell and Van Essen, 1983) as in throwing a spear at running prey.

The medial and MTL maintain extensive interconnections with the inferior region and thus with limbic nuclei. It is probably via these connections that objects or species of motivational significance can be discerned and attended to. That is, not only are the speed, movement, distance, etc. of a stimulus determined, but its emotional attributes (e.g., should it be feared, chased, eaten).

MEMORY

Perhaps because of its rich interconnections with the ITL, amygdala, and hippocampus, the medial and MTL are also involved in and become activated during recognition memory tasks, including those involving words and faces (see Heit et al., 1990). As noted, the MTL may serve as a storage depot for "words" (in the left hemisphere) and perhaps "faces" and shapes in the right. Hence, left MTL lesions result in recognition deficits for words (Milner and Teuber, 1968), whereas those involving the right MTL disrupt memory for visual forms and faces (e.g., Hanley et al., 1990). As argued by Heit and colleagues (1990), "MTL firing patterns may contribute to the reactivation of neocortical circuits encoding a particular stimulus-context gestalt" which in turn makes recognition and retrieval possible.

INFERIOR TEMPORAL LOBE

As noted, it is possible to very loosely define the anterior-inferior temporal lobes as auditory cortex, although neurons in this vicinity receive visual as well as somesthetic input. Hence, the ITLs can respond to auditory as well as emotionally and visually significant stimuli.

That the ITL processes complex auditory, visual, and emotional stimuli is probably a function of it being a derivative of the amygdala and hippocampus (see Chapter 1) as well as the visual cortex (Diamond, 1973). Visual and auditory functions also reflect its reception of extensive (upper visual field) projections from the optic radiations, highly processed input from the visual association areas in the occipital lobes of both cerebral hemispheres (via the corpus callosum), and fibers from the superior colliculus by way of the pulvinar of the thalamus (Chow, 1950; Kaas and Krubitzer, 1991; Kuypers et al., 1965; Previc, 1990; Rocha-Miranda et al., 1975).

Visual Capabilities and Form Recognition

The neocortex of the ITL is specialized for receiving, analyzing, discriminating, recognizing, and recalling complex visual information and is involved in attention and visually guided behavior (Gross and Graziano, 1995; Gross et al., 1972; Tovee et al., 1994), including the recollection and learning of visual discriminations (Gross, 1972). Cells in the ITL have very large, bilateral visual receptive fields that include the fovea, and many are sensitive to the direction of stimulus movement, color, contrast, size, shape, and orientation and are involved in the perception of three-dimensional objects and the supramodal analysis of information already processed in the association areas (Eskandar et al., 1992; Gross and Graziano, 1995; Gross et al., 1972; Nakamura et al., 1994; Rolls, 1992;

Sergent et al., 1990). Thus ITL neurons appear to take part in the last stages of visual analysis for form recognition and receive terminal fibers from the primary and association visual areas (Gross et al., 1972; Ungerleider and Mishkin, 1982). Indeed, a single neuron can respond to a combination of these features, and many will fire selectively in response to particular shapes and faces (Eskandar et al., 1992; Gross et al., 1972; Nakamura et al., 1994; Richmond et al., 1983; Rolls, 1992; Sergent et al., 1990).

Form and Facial Recognition

Overall, the ITL appears to be involved in the highest level of visual integration, containing highly developed neurons that seem to be the end station of a hierarchical system that mediates the perception and recognition of particular shapes and forms (Desimone and Schein, 1987; Gross and Graziano, 1995; Nakamura et al., 1994; Richmond et al., 1983; Rolls, 1992). Indeed, in addition to the optic radiations, there is a pathway beginning in the primary visual cortex that passes through areas 18 and 19 and terminates in the ITL (Kuypers et al., 1965; see also Previc, 1990). As information is passed from the primary to these association areas, various features important in the identification of specific objects become progressively and hierarchically analyzed, and increasingly complex associations are formed.

In fact, based on single cell recordings, some of these ITL neurons have been found to become particularly excited when presented with two-dimensional patterns or three-dimensional objects such as hands, brushes, and, in particular, faces (Desimone and Gross, 1979; Gross et al., 1972; Nakamura et al., 1994; Richmond et al., 1983, 1987). A variety of different feature detectors are found. Some cells are responsive to particular facial orientations, such as a profile, some respond to only parts of the face, such as eyes or a mouth, whereas others respond only to the entire face, i.e., a correctly organized facial gestalt. Interestingly, the ITL also contains neurons that will fire even to a scrambled face, so long as all the features are present (Gross et al., 1984).

Some of these cells are selective to faces, hands, etc., whereas others respond to other objects, including those with high emotional value. Hence, a minority of ITL neurons are facial recognition neurons, hand neurons, etc. and are activated only when these items are viewed.

The majority of cells in this area, however, are also selective for aspects of shapes, colors, or texture (Gross et al., 1984; Nakamura et al., 1994). These particular neurons probably act collectively so as to code and assemble a particular shape, including the formation of gestalts and thus the performance of visual closure, particularly the right temporal region (see Chapter 3). Via closure an individual can detect a partial stimulus and recognize that it is a face versus a rabbit.

Visual Attention

Attention and visual fixation involves the activation of neurons in the superior parietal and frontal eye fields, the midbrain colliculi, and thalamic nuclei—regions with which the ITL maintains interconnections (Gross and Graziano, 1995; Previc, 1990). Attention, however, generally requires that visual fixation be focused. Since many ITL neurons maintain large bilateral visual

Figure 14.8. An incomplete figure requiring visual closure for recognition. (From Gollin ES. Developmental studies of visual recognition of incomplete objects. Perceptual and Motor Skills 1960;11:289–298, ©Southern Universities Press, 1960.)

fields, when engaged in visual fixation ITL cells appear to become partly suppressed. That is, the area of the visual field they respond to becomes restricted to the fovea (Previc, 1990; Richmond et al., 1983). In this manner, ITL neurons become less responsive to other objects such as those in the periphery as the receptive field contracts around the object attended to.

Hence, ITL neurons seem to scan the entire visual field so as to alert the organism to objects of interest or motivational importance (via interconnections with limbic nuclei). When such stimuli are detected, the frontal eye fields, the superior parietal lobe, and the subcortical neurons are activated, triggering visual fixation. Simultaneously, ITL visual form recognition neurons are activated, whereas those with wide nonspecific visual fields are inhibited. In this manner, objects of interest are detected and fixated on (e.g., Previc, 1990).

Of adjunctive importance is the MTL, which in turn can analyze the velocity and direction of the object's movement so that the individual may approach, and, via interaction with cells in area 7, grasp and manipulate the object.

Prosopagnosia and Visual Discrimination Deficits

Damage or removal of the ITL results in severe disturbances involving visual discrimination learning and retention (Braun et al., 1994; Gross and Mishkin, 1977; Mishkin, 1972), and severe difficulty performing visual closure and recognizing incomplete figural stimuli (Lansdell, 1968, 1970). For example, primates with lesions in this vicinity have severe difficulty learning to discriminate between different shapes and patterns and objects that differ in regard to size or color, although visual acuity is normal.

With damage to the right temporal-occipital region, a severe disturbance in the ability to recognize the faces of friends, loved ones, or pets may result (Braun et al., 1994; DeRenzi, 1986; DeRenzi et al., 1968; DeRenzi and Spinnler, 1966; Hanley et al.,

1990; Hecaen and Angelergues, 1962; Landis et al., 1986; Levine, 1978; Whiteley and Warrington, 1971) or to discriminate and identify even facial affect (Braun et al., 1994)—a condition referred to as prosopagnosia. In fact, with gradual deterioration and degeneration of the right ITL, patients may suffer a progressive prosopagnosia (Evans et al., 1995). Some patients may in fact be unable to recognize their own face in the mirror. For example, one patient was unable to even discriminate between people on the basis of sex but instead had to rely on the presence of details, such as lipstick, rouge, hair length, or a mustache to make discriminations (Levine, 1978).

Presumably, the inability to recognize faces is due to the destruction of facial recognition neurons, including those within the amygdala (Young et al., 1995) and hippocampus, which becomes activated when memorizing faces (Kapur et al., 1995). However, because, at a neocortical level, global facial recognition cells appear to be in a minority whereas those specialized for analyzing facial parts are more numerous, the ability to recognize facial details is more likely to be preserved following inferior temporal destruction—particularly if the left temporal lobe is spared.

LATERALIZATION AND FACIAL RECOGNITION

Although patients with prosopagnosia often suffer from bilateral injuries, in many cases the lesions are restricted to the right hemisphere (DeRenzi, 1986; DeRenzi et al., 1968; DeRenzi and Spinnler, 1966; Evans et al., 1995; Hanley et al., 1990; Hecaen and Angelergues, 1962; Landis et al., 1986; Levine, 1978; Whitely and Warrington, 1971). Disturbances involving facial recognition do not usually occur with isolated left cerebral lesions. Indeed, the right hemisphere appears to be dominant in regard to the recognition of both familiar and unfamiliar faces, as has been well demonstrated in numerous studies of brain-injured as well as normal, intact individuals (Bradshaw et al., 1980; DeRenzi, 1982; DeRenzi et al., 1968; DeRenzi and Spinnler, 1966; Evans et al., 1995; Geffen et al., 1971; Hecaen and Angelergues, 1962; Levy et al., 1972; Ley and Bryden, 1977). The left hemisphere, however, is involved in the recognition of famous faces (Marzi and Berlucchi, 1977; Rizzolatti et al., 1971; see also Hanley et al., 1990) and facial details.

Among neurosurgical patients, it has also been reported that electrical stimulation of the posterior right temporal gyrus disrupts visual-spatial memory for faces in general (Fried et al., 1982). Electrical stimulation of the posterior portion of the right middle temporal gyrus also results in an inability to correctly label emotion faces.

AGNOSIAS

When the ITL is damaged, in addition to prosopagnosia there may occur difficulty identifying various familiar stimuli and objects, e.g., utensils and cars, as well as differentiating among similar visual stimuli. Many patients also have difficulty with color recognition (Green and Lessell, 1977; Meadows, 1974b). Frequently these types of agnosic disturbances are related to left cerebral or bilateral dysfunction.

MEMORY AND THE INFERIOR TEMPORAL LOBE

The ITL is able to code for and learn new stimuli, can recall the code of the previous stimuli, and can make comparisons between what was perceived on one occasion versus another (Eskandar et al., 1992). This enables ITL neurons to make judgments regarding temporal order: whether the stimulus was first, second, or last (Baylis and Rolls, 1987; Eskandar et al., 1992). Thus, ITL neurons can perform sequential analysis and can compare initial with previous visual representations (Eskandar et al., 1992). Hence, neurons in the ITL are involved in form and facial recognition, discrimination, temporal-sequencing, and thus learning and memory for a variety of stimuli.

In addition, ITL neurons are directly connected with the entorhinal cortex (Jones

and Powell, 1970; Van Hoesen and Pandya, 1975)—the "gateway to the hippocampus"—as well as to the amygdala (Amaral et al., 1987). Thus, neurons located in the ITL are also involved in the maintenance of short-term emotional, visual, and cognitive memory, and many display heightened activity during delay periods while one is learning (Fuster and Jevey, 1981; Miyashita and Chang, 1988). Moreover, ITL neurons are also highly concerned with the behavioral context in which learning occurs (Baylis and Rolls, 1987; Gross et al., 1979; Riches et al., 1991).

For example, Eskandar and associates (1992) found in their studies of electrophysiological activity that neurons in the inferior temporal gyrus convey information concerning current versus previous stimulus patterns and their behavioral context. ITL neurons are therefore involved in remembering this information and matching it with previously learned information, and are capable of simultaneously transmitting these impressions (including contextual details) to other brain areas. However, ITL neurons are also capable of discriminating between stimuli independent of context (Eskandar et al., 1992).

Hence, ITL neurons are involved in both encoding, storage, and recall and interact with the amygdala and hippocampus in regard to learning, memory, and recognition. It is in this manner and through these interconnections that ITL neurons are involved in emotional as well as nonemotional cognitive processing and memory storage. Conversely, with injuries to the ITL, emotional, visual, and verbal memory functions suffer (see Chapters 3 and 4).

Memory, the Hippocampus, and the ITL

The entorhinal, perirhinal, and parahippocampal neocortices are located in the inferior and medial temporal lobe and are adjacent to and richly interconnected with the hippocampus, which is buried within its depths. Like the hippocampus and amygdala, these tissues are involved in memory functioning (Gloor, 1990; Murray, 1992; Squire, 1992) and may in fact serve as neuronal memory depots.

When the ITL is electrically stimulated, exceedingly vivid personal memories may be triggered and recalled (Gloor, 1990, 1992; Halgren, 1992; Halgren et al., 1978; Penfield, 1954; Penfield and Perot, 1963). Patients may report seeing a complex scene from their past or early childhood, including hearing conversations, seeing faces, and experiencing somesthetic sensations and related events. Curiously, however, these scenes do not move forward in time and are otherwise quite static (See Gloor, 1990; Halgren, 1992), like a wide angle, multisensory snapshot.

In that the inferior temporal region also contains neurons that are selectively sensitive to particular sensory features and will fire in response to faces, hands, and geometric patterns and objects (Gross et al., 1972, 1979; Sergent et al., 1990), it is possible that not only are auditory and visual memories stored within the ITL, but that it may act as an associational warehouse (so to speak) from which particular auditory-visual images can be activated, retrieved, and compared. In this manner, the hippocampus and amygdala can gain access to particular perceptual images as well as store associated information via the aid and guidance of the neural circuits maintained in the ITL, through which information is transmitted to both nuclei.

For example, the entorhinal cortex acts as a gateway via which information from the immediately adjacent perirhinal cortex (which sits above and is richly interconnected with the amygdala) and the parahippocampal gyrus (which receives visual, auditory, and tactual neocortical information) is analyzed and is then transmitted into the hippocampus (see Horel et al., 1987; Issausti et al., 1987; Squire, 1992). It is also via the entorhinal area that the hippocampus receives amygdaloid projections (Carlsen et al., 1982; Gloor, 1955; Krettek and Price, 1976; Steward, 1977) and fibers from the orbital frontal and temporal lobes (Van Hoesen et al., 1972). These neocortical regions are thus highly important in memory.

Squire (1992:202) in fact argues that "the cortical structures adjacent to the hippocampus (entorhinal, perirhinal, and parahippocampal cortex) appear to participate with the hippocampus in a common memory function." He bases this argument on his review of numerous studies (many of which were conducted by Squire and Zola Morgan and colleagues) that indicate that severe memory impairments can result from lesions restricted to these cortical areas.

Temporal Lobe Injuries and Memory Loss

Although a variety of neurochemical and neuroanatomical regions are involved in the formulation of memory, it has long been known that damage or the neurosurgical removal of the temporal lobes can produce profound disturbances in the learning and recollection of verbal and visual stimuli (Kimura, 1963; Milner, 1958; Squire, 1987, 1992). For example, left temporal lobectomy or seizures or lesions involving the inferior temporal areas can moderately disrupt immediate and severely impair delayed memory for verbal passages and the recall of verbal paired-associates, consonant trigrams, word lists, and number sequences (Delaney et al., 1980; Meyer, 1959; Meyer and Yates, 1955; Milner, 1958, 1968; Milner and Teuber, 1968; Weingartner, 1968). Similarly, severe anterograde and retrograde memory loss for verbal material has been noted when the anterior and posterior temporal regions (respectively) are electrically stimulated (Ojeman et al., 1968, 1971).

In contrast, right temporal lesions or lobectomy significantly impairs recognition memory for tactile and recurring visual stimuli such as faces and meaningless designs, as well as memory for object position and orientation and visual-pictorial stimuli (Corkin, 1965; Delaney et al., 1980; Evans et al., 1995; Kimura, 1963; Milner, 1968; Taylor, 1969). Electrical stimulation of the right anterior and posterior temporal region also causes, respectively, severe anterograde and retrograde memory loss for designs and geometric stimuli, and it impairs memory for faces (Fried et al., 1982; Ojeman et al., 1968).

Bilateral removal of the inferior temporal region results in a condition that has been variably referred to as "psychic blindness" and the "Klüver-Bucy syndrome." However, as explained in Chapter 3, this is due to destruction of the amygdala. If the mesial regions are removed, severe memory disturbances involving visual and auditory stimuli result such that the patient suffers a permanent anterograde amnesia, including facial processing impairments (Young et al., 1995).

Based on lesion, temporal lobectomy, and electrical stimulation studies, it appears that the anterior temporal region is more involved in initial consolidation storage phase of memory, whereas the posterior region is more involved in memory retrieval and recall. Overall, however, it appears that the amygdala and hippocampus are in some manner compromised and/or at least disconnected from the source of input for these deficiencies involving memory to arise (Milner, 1974; Mishkin, 1978).

HALLUCINATIONS AND THE ITL

The functional integrity of the temporal lobes, the inferior regions in particular, is highly important to the memorization and recollection of various auditory, visual, olfactory, and emotional experiences. When the lobes are destroyed, disconnected from sources of input, or compromised in some fashion, the ability to store information and to draw visual-verbal mnemonic imagery from memory is severely attenuated.

Conversely, when the temporal lobes and/or the limbic nuclei buried within their depths (i.e., the amygdala and hippocampus) are artificially or abnormally activated, it sometimes occurs that visual-auditory imagery as well as a variety of emotional reactions are evoked involuntarily. These may take the form of complex hallucinations, dream-like states, or confusional episodes, or they may involve the abnormal attribution of emotional significance to otherwise neutral thoughts and external experiences.

Hallucinations and the Interpretation of Neural "Noise"

Hallucinations may occur secondary to tumors or seizures involving the occipital, parietal, frontal, and temporal lobes, or arise secondary to toxic exposure, high fevers, general infections, exhaustion, starvation, extreme thirst, partial or complete hearing loss including otosclerosis, and partial or complete blindness such as that due to glaucoma (Bartlet, 1951; Flournoy, 1923; Lindal et al., 1994; Pesme, 1939; Rhein, 1913; Ross et al., 1975; Rozanski and Rosen, 1952; Semrad, 1938; Tarachow, 1941). Interestingly, when hallucinations are secondary to peripheral hearing loss, frequently individuals report hearing certain songs and melodies from their childhood—melodies that they had usually long forgotten. In addition, individuals suffering from cortical blindness, i.e., Anton's syndrome (Redlich and Dorsey, 1945), and deafness (Brown, 1972), as well as those recovering from Wernicke's aphasia, frequently experience hallucinations.

In general, hallucinations secondary to loss of visual or auditory input appear to be secondary to the interpretation of neural noise. That is, with loss of input, various brain regions begin to extract or assign meaningful significance to random neural events or to whatever input may be received. Thus we find that subjects will hallucinate when placed in sensory-reduced environments or even when movement is restricted (Lilly, 1956, 1972; Lindsley, 1961; Shurley, 1960; Zuckerman and Cohen, 1964).

Conversely, hallucinations can be due to increased levels of neural noise as well. For example, if an area of the neocortex is abnormally activated, that area in turn may act to interpret its own neural activity. However, the complexity or degree of interpretative activity depends on the type of processing performed in the region involved. In this regard we find that hallucinations become increasingly complex as the disturbance expands from primary to association areas and as involvement moves from the occipital to anterior temporal regions (Critchley, 1939; Penfield and Perot, 1963; Tarachow, 1941)—which is one of the major interpretive regions of the neocortex (Gibbs, 1951; Gloor, 1990, 1992; Halgren, 1992; Penfield and Perot, 1963). That is, in the primary regions, neural noise is given a simple interpretation (simple hallucinations), whereas in the association and multiassociational areas, neural noise is given a complex interpretation.

For example, tumors or electrical stimulation of the occipital lobe produce simple hallucinations such as colors, stars, spots, balls of fire, and flashes of light, whereas with superior temporal involvement the patient may experience crude noises, such as buzzing, roaring sounds, bells, and an occasional voice or sounds of music. However, with anterior and inferior temporal abnormalities, the hallucinations become increasing complex, consisting of both auditory and visual features, including faces, people, objects, animals, etc. (Critchley, 1939; Penfield and Perot, 1963; Tarachow, 1941). However, the anterior-inferior temporal region may give rise to the most complex forms of imagery because cells in this area are specialized for the perception and recognition of specific forms.

Indeed, it has frequently been reported that as compared to other cortical areas, the most complex and most forms of hallucination occur secondary to anterior temporal lobe involvement (Critchley, 1939; Horowitz et al., 1968; Malh et al., 1964; Penfield and Perot, 1963; Tarachow, 1941) and that the hippocampus and amygdala (in conjunction with the temporal lobe) appear to be the responsible agents (Gloor, 1990, 1992; Gloor et al., 1982; Halgren et al., 1978; Horowitz et al., 1968). For example, Bancaud and coworkers (1994), Halgren and associates (1978), and Horowitz and colleagues (1968) note that hippocam-

pal stimulation was predominantly associated with fully formed and/or memory-like hallucinations including feelings of familiarity, and secondarily with dream-like hallucinations. However, stimulation limited to the neocortex had relatively little effect in this regard (Gloor et al., 1982). Thus it appears that limbic activation is necessary in order to bring to a conscious level percepts that are being processed in the temporal lobes.

LSD

As per hallucinations, this does not mean, however, that neocortical involvement is not necessary, for frequently it is the interpretive interaction of the temporal lobe that gives rise to certain types of hallucinations, i.e., bringing them to a conscious level. For example, it is well known that the ingestion of LSD will trigger the formation of vivid and complex auditory and visual hallucinations. Following LSD administration, electrophysiological abnormalities are noted in the amygdala and hippocampus (Chapman et al., 1963). However, if the temporal lobes are surgically removed there is a significant decrease (with unilateral removal—Serafetinides, 1965) or complete abolition (with bilateral lateral removal) of LSD-induced hallucinatory activity (Baldwin et al., 1955)—even when the amygdala and hippocampus are spared. Again, this is presumably a function of form-recognition neurons being located and activated within the inferior temporal region. Interestingly, the hallucinatory effect of LSD appears to be greatest in the right temporal lobe (Serafetinides, 1965).

Dreaming

Vivid, visual-auditory, and sometimes intensely emotional hypnotic dream imagery is clearly associated with REM sleep (Foulkes, 1962; Goodenough et al., 1959; Monroe et al., 1965). As discussed in detail in Chapter 3, electrophysiological studies or measures of cerebral blood flow have indicated that the right hemisphere becomes highly active during REM, whereas conversely the left brain becomes more active during non-REM sleep (Goldstein et al., 1972; Hodoba, 1986; Meyer et al., 1987).

It has also been reported that abnormal and seizure-like activity in the right temporal and temporal-occipital areas acts to increase dreaming and REM sleep for an atypically long time period. Similarly, REM sleep increases activity in this same region much more than in the left hemisphere (Hodoba, 1986), indicating that there is a specific complementary relationship between REM sleep, dreaming, and right temporal-occipital electrophysiological activity.

In this regard, although not conclusive, there seems to be a convergence of evidence that suggests that the right temporal lobe may be more involved than the left in the production of visual-auditory (nonverbal) hallucinations and dream-like mental states whether they occur naturally during sleep or be they produced secondary to LSD, electrical stimulation, or abnormal seizure activity.

As pertaining to right versus left cerebral involvement in the production and recollection of hallucinations and dreams, it is interesting to note that Horowitz and associates (1968) reported that electrically induced hallucinated events were usually forgotten by patients within 10 to 15 minutes after the experience, and when patients were questioned the next day memory was not improved. Similar forgetting patterns are characteristic of memory for normal sleep-induced dreams as well (Joseph, 1988a). That is, it becomes progressively more difficult to recall one's dreams as one spends time in or awakens during non-REM sleep (Wolpert and Trosman, 1958)—which is associated with high left temporal lobe and low right temporal lobe activity.

TEMPORAL LOBE (PARTIAL COMPLEX) SEIZURES AND EPILEPSY

The hippocampus and amygdala have the lowest seizure thresholds of all brain structures, and temporal lobe epilepsy (TLE) is associated with atrophy and sclerosis of the hippocampus in about 50% of such patients. Incisural herniation at the time of birth is a major cause of hippocampal sclerosis.

Head injuries also cause injuries to the temporal region due to contusion from the bony structures in which the temporal region is encased, which may later give rise to seizure activity. Also, middle ear infections are associated with disturbances involving the temporal lobe, which in turn may result in TLE.

In addition, as detailed in Chapter 16, severe and repeated early emotional trauma can in fact injure the immature hippocampus and temporal lobe, giving rise to a propensity to develop kindling as well as abnormal neural networks. This would make these individuals more likely to develop psychotic and severe emotional and dissociative disorders and put them at risk for developing TLE.

SEIZURES

Usually, during the course of a temporal lobe seizure, there are no abrupt and drastic alterations in motor activity such as tonic-clonic spasms. However, some patients may simply cease to respond and stare blankly straight ahead, make licking or smacking movements of the lips, and/or fiddle with their clothing, as if picking up pieces of lint. Although conscious awareness is lost, these patients are not unconscious. Rather, their mental state is one of absence. In fact, they may appear awake and conscious, although unable to speak or respond to questions, and behaviorally their actions seem semipurposeful, though later they have no memory for what occurred.

Hence, it is not always obvious to an observer when someone is experiencing a temporal lobe seizure. In some cases an inexperienced observer may only have the impression that the person is acting somewhat oddly. Nevertheless, unless the patient has several different seizure foci, the same characteristic behavioral manifestations are elicited every time a patient has a seizure. The patient does not simply stare on one occasion and on the next begin rolling his eyes and cry out.

Auras and Automatisms

Some patients are aware of and note an aura immediately prior to seizure onset. This may involve feelings of fear or anxiety, alterations in gastric motility, or unpleasant tastes or, in particular, strange odors (e.g., burning rubber or feces)—i.e., an olfactory aura. Presumably, the experience of olfactory hallucinations is due to abnormal activation of the rhinencephalon (the "nose brain") and thus limbic nuclei such as the amygdala.

Following the aura, the patient's seizure may consist of semipurposeful movements or licking and smacking of the lips. Possibly the licking and smacking movements are also due to activation of the amygdala and other limbic structures associated with food consumption. The semipurposeful movements may be due to limbic seizure activation of the basal ganglia. Changes in gastric motility may be secondary to insular activation and/or limbic participation.

Automatisms associated with TLE occur in up to 75–95% of patients, which is twice what occurs with lesions outside the temporal lobe. These include staring, searching, groping, lip smacking, spitting, groping, and salivation. Also included are laughing, crying, hissing, gritting or gnashing of teeth, clenching the fist, confused talking, screaming, shouting, standing, running, walking, kissing, and so on.

Common visceral reactions associated with TLE auras include feelings of a racing or fluttering heart, including pounding or throbbing. Also there may be feelings of smothering or choking. Body sensation include numbness, tenseness, pressure, and heaviness. There may also be visual abnormalities, such that things look abnormally large or small.

Feelings of strangeness or familiarity are also common. Deja vu occurs in up to 20% of such patients, usually with right cerebral

seizures within the temporal lobe. Some claim to have feelings like a desire to be alone or of wanting something but not knowing what.

As noted, olfactory hallucinations are not uncommon and are usually quite disagreeable and include smells like burning meat, fish, lime, acid fumes, and feces. There may also be gustatory sensations that are also usually disagreeable coupled with very bad and bitter or metallic and sour tastes.

Fear is one of the most common feelings associated with temporal lobe auras and seizures, which include feelings of panic, terror, or of something horrible about to happen. Some patients have experienced fear coupled with sexual sensations, including fear associated with a feeling of sexual climax, or feelings of being sexually penetrated following by orgasm then a loss of consciousness. One patient stated it was like "a red hot poker being inserted into her vagina" followed by vomiting or orgasm.

Running, Laughing, and Crying Seizures

In some instances patients with temporal lobe seizures (triggered perhaps by tumor) may display extremely odd and bizarre behaviors during the course of the seizure (Chen and Forster, 1973; Trimble, 1991). For example, a number of authors have presented case reports of individuals who developed laughing (gelastic epilepsy), crying (dacrystic epilepsy), and/or running seizures (cursive epilepsy)—running apparently being triggered via amygdala-striatal activation (see Chapter 9).

One patient's seizure consisted of suddenly bursting into laughter, rubbing his upper abdomen, and running wildly about the room with an expression of fear on his face (Sethi and Rao, 1976). Another individual displayed paroxysmal attacks of weeping, sobbing, and mournful moaning (Offen et al., 1976). These behaviors were involuntary, however, and part of the seizure. Once consciousness returned, the patients would have no recollection of their actions.

Emotional Disturbances

The most common emotional reactions and sensations that occur secondary to or during the course of a temporal lobe seizure include feelings of fear, anxiety, depression, depersonalization, pleasure, displeasure, and familiarity (Bear, 1979; Cendes et al., 1994; Herman and Chambria, 1980; Gloor et al., 1982; Perez and Trimble, 1980; Slater and Beard, 1963; Strauss et al., 1982; Trimble, 1991; Weil, 1956; Williams, 1956), fear being the most frequently experienced. Many of these same feelings are also triggered by electrical stimulation of the temporal lobes, amygdala, and hippocampus (Gloor, 1972; Gloor et al., 1982; Heath, 1964; Mullan and Penfield, 1959).

Depression (lasting from hours to weeks) may occur as an immediate sequela to the seizure, many patients also experiencing confusion. A depressive aura may also precede and thus herald the coming of a seizure by hours or even days (Trimble, 1991; Williams, 1956).

Rather than increased emotionality, some patients complain of emotional blocking and feelings of emptiness: "Feelings don't reach me anymore" (Weil, 1956). Presumably this is a consequence of limbic disconnection. That is, the seizure foci (or lesion) acts to disconnect the limbic areas from the temporal (or orbital frontal) lobes. In consequence, percepts and thoughts no longer come to be assigned emotional or motivational significance.

TEMPORAL LOBE EPILEPSY AND RELIGIOUS EXPERIENCE

It has been reported by a number of neuroscientists that some patients who experience temporal lobe epilepsy (TLE), seizures, or related abnormalities sometimes experience very sexual as well as bizarre, unusual, and fearful mental phenomenon including dissociative states, feelings of depersonalization, and hallucinogenic and dream-like

recollections involving threatening men, naked women, sexual intercourse, religion, the experience of God, as well as demons and ghosts and pigs walking upright and dressed as people (Bear, 1979; Daly, 1958; Gloor, 1986, 1992; Halgren, 1992; Horowitz et al., 1968; Penfield and Perot, 1963; Slater and Beard, 1963; Taylor, 1972, 1975; Trimble, 1991; Weingarten et al., 1977; Williams, 1956). Moreover, some individuals report communing with spirits or receiving profound knowledge from the Hereafter following abnormal temporal lobe activation (Daly, 1958; MacLean, 1990; Penfield and Perot, 1963; Williams, 1956).

Intense activation of the temporal lobe, hippocampus, and amygdala has been reported to give rise to a host of sexual, religious, and spiritual experiences, and chronic hyperstimulation can induce an individual to become hyper-religious or visualize and experience ghosts, demons, angels, and even God, as well as claim demonic and angelic possession or the sensation of having left the body (Bear, 1979; Daly, 1958; Gloor, 1986, 1992; Horowitz et al., 1968; MacLean, 1990; Mesulam, 1981; Penfield and Perot, 1963; Schenk and Bear, 1981; Slater and Beard, 1963; Subirana and Oller-Daurelia, 1953; Trimble, 1991; Weingarten et al., 1977; Williams, 1956). Indeed, as detailed in Chapter 8, the temporal lobe, amygdala, and hippocampus enable humans to have religious, spiritual, and mystical experiences (Bear, 1979; Daly, 1958; d'Aquili and Newberg, 1993; Mesulam, 1981; Trimble, 1991).

Out-Of-Body, Heavenly, and "Otherworldly Experiences"

One patient had a symptom of being outside her body and watching and observing her body from the outside. Penfield and Perot (1963) describe several patients who during a seizure claimed they could see themselves in different situations. One woman stated that "it was though I were two persons, one watching, and the other having this happen to me," and that it was she who was doing the watching as if she was completely separated from her body. According to Penfield, "It was as though the patient were attending a play and was both actor and audience."

Other patients claim to have quite pleasant auras and describe feelings such as elation, security, eternal harmony, immense joy, paradisiacal happiness, euphoria, and completeness. Between 0.5% and 20% of such patients claim such feelings (Daly, 1958; Williams, 1956). One patient of Williams (1956) claimed that his attacks began with a "sudden feeling of extreme well being involving all my senses. I see a curtain of beautiful colors before my eyes and experience a pleasant but indescribable taste in my mouth. Objects feel pleasurably warm, the room assumes vast proportions, and I feel as if in another world."

A patient described by Daly claimed his seizure felt like "a sunny day when your friends are all around you." He then felt dissociated from his body, as if he were looking down on himself and watching his actions.

Williams (1956) describes a patient who claimed that during an aura she experienced a feeling that she was being lifted up out of her body, coupled with a very pleasant sensation of elation and the sensation that she was "just about to find out knowledge no one else shares, something to do with the link between life and death."

Subirana and Oller-Daurelia (1953) described two patients who experienced ecstatic feelings of either "extraordinary beatitude" or of paradise as if they had gone to heaven and noted that their fantastic feelings lasted for hours.

Other patients have noted feelings of things suddenly becoming "crystal clear," a feeling of clairvoyance, or a feeling of having the truth revealed to them or of having achieved a sense of greater awareness such that sounds, smells, and visual objects seemed to have a greater meaning and sensibility.

Hyper-religiousness, or hyperspirituality, coupled with a need to preach or write (hypergraphia) about these experiences

and feelings is thus not uncommon with TLE. Similarly, because of limbic system hyperactivation, hypo- or hyper-sexuality, as well as significant mood and emotional changes, is likely.

SEIZURES, PERSONALITY, AND PSYCHIATRIC DISTURBANCES

Personality and psychiatric disorders have been frequently reported as a common associated complications among a small percentage of individuals (between 4.5% and 7%) with seizures disorders localized to the temporal lobe (reviewed in Trimble, 1991). Such patients may appear depressed, paranoid, hysterical, or suffering from schizophrenic-like symptomology (However, see Gold et al., 1994).

Some authors have argued that, as with hallucinations, the nearer the seizure foci to the anterior temporal lobe, the greater the probability of significant psychiatric abnormality (Gibbs, 1951; Gloor, 1992; Trimble, 1991). Presumably this is a function of hyperactivation of underlying limbic structures and thus abnormal attribution of emotional significance to different afferent streams of perceptual experience (Bear, 1979).

Schizophrenia and Temporal Lobe Epilepsy

It has been repeatedly reported that a significant minority of individuals with TLE suffer from prolonged and/or repeated instances of major psychopathology (Bruton et al., 1994; Buchsbaum, 1990; Falconer, 1973; Flor-Henry, 1969, 1983; Gibbs, 1951; Jensen and Larsen, 1979; Perez and Trimble, 1980; Sherwin, 1977, 1981; Slater and Beard, 1963; Taylor, 1972; Trimble, 1991). Conversely, those with affective disturbances, e.g., mania, have been found to suffer from seizures involving the right temporal lobe (Bear et al., 1982; Falconer and Taylor, 1970; Flor-Henry, 1969; Trimble, 1991) or right frontal lobe (Joseph, 1986a, 1988a), although depression may be more likely with left versus right hemisphere injuries (see Chapter 4).

Estimates of psychosis and affective disorders among individuals suffering from TLE, be it a single episode or chronic psychiatric condition, have varied from 2% to 81%. However, as noted, probably a more realistic figure is 4–7% (see Trimble, 1991, 1992).

There are several distinct neurological explanations for the association between temporal lobe damage and psychotic and affective disorders. As discussed in detail in previous chapters, the right hemisphere is dominant for most aspects of emotional functioning. Hence, when it is damaged or abnormally activated, such as from a seizure or chronic subclinical seizure activity, emotional functioning becomes altered and disturbed. Moreover, since the amygdala and hippocampus are often secondarily aroused (or hyperaroused), the possibility of abnormal emotionality is enhanced even further. In this regard, Gloor and colleagues (1982) in a depth electrode stimulation study of patients suffering from epilepsy presented evidence indicating that all experiential and emotional alterations were encountered following right amygdala and hippocampal activation.

As discussed in Chapter 3, the limbic system seem to be lateralized and the right amygdala and hippocampus appear to be more involved in emotional functioning. Hence, abnormal activation of these nuclei (such as during the course of a temporal lobe seizure) are more likely to give rise to abnormal emotional reactions, with the exception of depression, which is common with left frontal and left temporal lobe injuries.

In addition, some individuals with temporal lobe seizures may complain of emotional blunting. Hence, rather than a limbic hyperconnection, these individuals suffer from a disconnection, similar to what occurs with a left frontal injury.

PSYCHOSIS, SPEECH, AND THE TEMPORAL LOBE

It has been repeatedly demonstrated in histological studies that the temporal lobes of schizophrenics are characterized by excessive focal temporal lobe neuronal damage including gliosis (Akbarian et al., 1993; Bruton, 1988; Bruton et al., 1990; Stevens, 1982; Taylor, 1972). Hence, these patients may experience auditory hallucinations and are more likely to demonstrate formal thought disorders and aphasic thinking, including the production of neologisms and syntactical speech errors, as well as emotional (or thought) blocking or blunting due to limbic system-temporal lobe disconnection (Perez and Trimble, 1980; Slater and Beard, 1963). As often occurs with receptive aphasia and left temporal lobe dysfunction, individuals suffering from schizophrenia may be unaware of their illness (Amador et al., 1994; Wilson et al., 1986).

Hence, certain subgroups of schizophrenics display significant abnormalities involving speech processing, such that the semantic, temporal-sequential, and lexical aspects of speech organization and comprehension are disturbed and deviantly constructed (Chaika, 1982; Clark et al., 1994; Flor-Henry, 1983; Hoffman, 1986; Hoffman et al., 1986; Rutter, 1979). Indeed, significant similarities between schizophrenic discourse and aphasic abnormalities have been reported (Faber et al., 1983). It is often difficult to determine what a schizophrenic individual may be talking about, particularly in that some of these individuals complain that what they say often differs from what they intended to say (Chapman, 1966). Temporal-sequential (i.e., syntactical) abnormalities have also been noted in their ability to reason. The classic example is "I am a virgin; the Virgin Mary was a virgin; therefore I am the Virgin Mary."

Hence, there is considerable evidence of a significant relationship between abnormal left hemisphere and left temporal lobe functioning and schizophrenic language, thought, and behavior. As detailed in Chapter 11, the left frontal region and medial frontal cortices are also implicated in the genesis of schizophrenic-like abnormalities, including catatonia.

However, in contrast to those with frontal lobe schizophrenia who are more likely to present with "negative" symptoms and emotional blunting, temporal lobe schizophrenics are likely to present with "positive" symptoms, including retention of the capacity to experience and express "warmth" (Crow, 1980; Crowe and Kuttner, 1991). Emotional expression, however, tends to be abnormal, inappropriate or incongruent, and/or exaggerated (Bear and Fedio, 1977; Crow, 1980; Crowe and Kuttner, 1991).

It has also been repeatedly reported that irritative lesions involving the lateral convexity of the left temporal lobe are the most likely to result in schizophrenic-like disturbances (Dauphinais et al., 1990; DeLisis et al., 1991; Flor-Henry, 1983; Perez et al., 1985; Rossi et al., 1990, 1991; Sherwin, 1981). Given the role of the temporal lobe in language (and the fact that the superior temporal lobe is derived from amygdala and hippocampal tissue), it is also not uncommon for these individuals to demonstrate aphasic abnormalities in their thought and speech (Chaika, 1982; Clark et al., 1994; Flor-Henry, 1983; Hoffman, 1986; Hoffman et al., 1986; Perez and Trimble, 1980; Rutter, 1979; Slater and Beard, 1963).

However, the type and form of temporal lobe schizophrenia may differ depending on which regions have been compromised. In some instances the amygdala and hippocampus are involved (Bogerts et al., 1985; Brown et al., 1986; Falkai and Bogerts, 1986), in which case memory and emotional functioning may be most severely affected, and patients may experience visual hallucinations. Temporal lobe schizophrenics with limbic involvement may behave violently and aggressively (Crowe and Kuttner, 1991; Lewis et al., 1982; Taylor, 1969). They may also experience extreme feelings of religiousness (Bear, 1979, Trimble, 1991, 1992), fear, paranoia, and depression (Crowe and Kuttner, 1991; Perez and Trimble, 1980; Trimble, 1991, 1992), as well as hallucinations.

APHASIA AND PSYCHOSIS

Since the left temporal lobe is intimately involved in all aspects of language comprehension as well as the organization of linguistic expression, altered neocortical activity involving this region is likely to result in significant disruptions in language functioning and the formulation of linguistic thought. As described in Chapter 4, individuals with left temporal lobe damage and Wernicke's aphasia are likely to be diagnosed as suffering from a formal thought disorder due to their fluent aphasic language disturbances and their failure to comprehend.

In this regard, several authors have noted significant similarities between schizophrenic discourse and receptive aphasia (Faber et al., 1983; Hillbom, 1951). In both conditions the semantic, temporal-sequential, and lexical aspects of speech organization and comprehension are often disturbed and deviantly constructed (Chaika, 1982; Flor-Henry, 1983; Hoffman, 1986; Hoffman et al., 1986; Rutter, 1979). Similar language and comprehension defects have been reported for individuals suffering post-traumatic schizophrenic-like symptoms following head injury and missile wounds (Hillbom, 1951). Moreover, just as some receptive aphasic individuals may be unaware of their disorder, similar findings have been noted with schizophrenics and those with schizoaffective disorders (Amador et al., 1994).

Temporal Lobe Seizures, Psychosis, and Aphasia

Patients with left temporal seizures commonly become globally aphasic during the course of their seizure. When spoken to or if their name is repeatedly called, they may make only fleeting eye contact, grunt, or utter partial words such as "huh?" Some individuals remain aphasic for seconds to minutes after seizure termination as well.

Language impairments, verbal memory disorders, and associated linguistic deficiencies are usually apparent even between seizures if adequately assessed. Moreover, individuals with left-sided foci tend to have lower (WAIS-R) Verbal IQs as compared to their Performance IQs and as compared to those with right-sided foci. This is not surprising as language is represented in the left hemisphere. However, in some cases, excessive activity within the left (or right) temporal (or frontal) lobe can actually result in what appears to be enhanced functional activity. That is, they may perform better than normal because of increased neuronal activity in these regions.

Overall, there is considerable evidence that abnormal activity in the left temporal lobe, as well as the presence of an active lesion, can result in significant alterations in not only language, but the organization of linguistic thought such that the patient appears psychotic. In this regard, abnormal thought formation may become a characteristic pattern among those with left temporal foci. In part, this is also due to the involvement of the amygdala, which is directly linked with the language axis. Conversely, if the left temporal region is periodically disconnected from centers mediating emotion, i.e., the amygdala, patients may demonstrate what appears to be emotional blunting as well as a formal thought disorder.

Between-Seizure Psychoses

The range of disturbances among those with TLE includes a high rate of sexual aberration, such as hyper- or hypo-sexuality, aggressiveness, paranoia, depression, deepening of emotion, intensification of religious concerns, disorders of thought, depersonalization, hypergraphia, complex visual and auditory hallucinations, and schizophrenia (Bear, 1982; Bear et al., 1982; Flor-Henry, 1969, 1983; Schiff et al., 1982; Sherwin, 1977, 1981; Stevens et al., 1979; Taylor, 1975; however, see Gold et al., 1994).

Frequently, seizure-related disturbances in emotional functioning wax and wane, such that some patients experience islands of normality followed by islands of psychosis. Possibly these changes are a function in variations in subclinical seizure activity and kindling. That is, abnormal activity may wax and wane.

In some instances, the psychiatric disturbance may develop over days as a prelude to the actual onset of a seizure, due presumably to increasing levels of abnormal activity, until the seizure is triggered. In other cases, particularly in regard to depression, the alterations in personality and emotionality may occur following the seizure and may persist for weeks or even months (Trimble, 1991).

In yet other instances, patients may act increasingly bizarre for weeks, experience a seizure, and then behave in a normal fashion for some time ("forced normalization,"—Flor-Henry, 1983), only to again begin acting increasingly bizarre. This has been referred to as an interictal (between-seizure) psychosis and is possibly secondary to a build-up of abnormal activity in the temporal lobes and limbic system (Bear, 1979; Flor-Henry, 1969, 1983) Hence, once the seizure occurs, the level of abnormal activity is decreased and the psychosis goes away only to gradually return as abnormal activity again builds up.

CAVEATS

Some authors have argued that there is no significant relationship between these psychotic disorders and TLE (Gold et al., 1994; Stevens and Hermann, 1981), whereas others have drawn attention to possible social-developmental contributions (Matthews et al., 1982). That is, growing up with a seizure disorder, feeling victimized, not knowing when a seizure may strike, loss of control over one's life, etc. can independently create significant emotional aberrations.

It is also important to emphasize that it is probably only a subpopulation of individuals with TLE who come to the attention of most researchers, e.g., those with the most serious or intractable problems. Thus one should not immediately view an individual with TLE (or any type of epilepsy) as immediately suspect for emotional-psychotic abnormalities. Nevertheless, it is noteworthy that not only is there an association between the temporal lobe and certain forms of schizophrenia, but that temporal lobe abnormalities have been noted across generations of individuals and families afflicted with psychotic disturbances (Honer et al., 1994).

SECTION VI

The Neuroanatomy of Psychotic and Emotional Disorders

15

The Neuropsychology of Repression

Hemispheric Laterality, Positive versus Negative Emotions, Corpus Callosum Immaturity, The Frontal Lobes, Trauma, Contextual Cues, and Recovered Memories

For most individuals it is extremely difficult if not impossible to verbally recall events that occurred before the age of 3 1/2 (Dudycha and Dudycha, 1933; Gordon, 1928; Wadfogel, 1948; White and Pillemer, 1979). For example, in a study of college students, Dudycha and Dudycha (1933) found that the mean age for first recollection among men was 3 years 8 months, and that for women it was 3.6. In a later study they determined that the average age at first recollection was 3.67 for adolescent males and 3.50 for females (see Dudycha and Dudycha, 1941). Similarly, Gordon (1928) and Wadfogel (1948) found that the average age was 3.64 for males and 3.4 and 3.23 for females. For the most part these memories were of highly significant emotional events such as fires, illnesses, births, and so on.

It is noteworthy, however, that some individuals claim memories from before the age of 2 and even 1 (Dudycha and Dudycha, 1941; White and Pillemer, 1979). Of course, for most individuals verbal access to memories even after the age of 3 and for many years thereafter is quite spotty. In part, this is due to language limitations.

That is, verbal memory for childhood events corresponds with the development of verbal ability (Wadfogel, 1948), which in turn is correlated with the myelination and maturity of the axonal interconnections linking the language areas within the left hemisphere (LeCours, 1975). Thus, Wadfogel (1948) reported that when verbal "memories were plotted according to the age of their origin, it was found that there was an increment from year to year which

took the form of an ogive and seemed to parallel the growth of language during childhood."

Investigators have also found that the earliest recollections tend to be emotional and visual rather than verbal, with fear, joy, and positive emotions predominating (Dudycha and Dudycha, 1933, 1941; Wadfogel, 1948). Wadfogel (1948) found that pleasant memories constituted about 50% of the total with unpleasant ones making up 30% and neutral memories constituting the rest (see also Colgrove, 1899; Hall, 1899; Potwin, 1901).

Similarly, in most studies of emotional recollections, positive memories predominate (Jersod, 1931; Kihlstrom and Harackiewicz, 1982; Meltzer, 1930, 1931; Stagner, 1931). In fact, when questioned in a neutral setting, "normal" adults tend to report negative or traumatic memories in a positive manner (Kihlstrom and Harackiewicz, 1982).

In a neutral setting, positive emotions are also easier to verbally recall for adults (Blaney, 1986; Rholes et al., 1987) and children (Casey, 1993; Cole, 1986; Saarni, 1984; Strayer, 1993). Children, however, find it quite difficult to verbally recall, describe, and identify negative versus positive emotions and memories regardless of their current mood (Feshbach and Roe, 1968; Mood et al., 1978; Strayer, 1980; see below).

Adults have little difficulty in identifying and describing negative or positive recollections, the only exception being that while adults are in a positive mood, pleasant memories are recalled more quickly, whereas when in a depressed mood retrieval latency tends to lag for positive but not negative recollections (see Blaney, 1986; Rholes et al., 1987; Singer and Salovey, 1988). However, adults tend to forget negative memories more quickly than positive ones (Meltzer, 1931; Stagner, 1931), though not nearly at the highly rapid rate typical for children.

As will be discussed, these differences in emotional forgetting and recollection and the tendency to assign "positive" emotional coloration to early memories, are in part determined by the differential functional capabilities of the left and right hemisphere and the fact that they are limited in their ability to intercommunicate during early childhood, and to a lesser extent, during adulthood.

As detailed in Chapter 3, the right half of the brain is involved in the perception and expression of almost all aspects of emotion (Borod, 1992; Cancelliere and Kertesz, 1990; Dekosky et al., 1980; Gorelick and Ross, 1987; Heller and Levy, 1981; Heilman et al., 1975, 1984, 1985; Heilman and Bowers, 1996; Joseph, 1988a, 1990e; Ley and Bryden, 1979; Ross, 1981; Shapiro and Danly, 1985; Strauss and Moscovitch, 1981; Suberi and McKeever, 1977; Tucker and Frederick, 1989; Tucker et al., 1977), whereas the left hemisphere (the anterior regions in particular) is biased toward perceiving and reporting emotional stimuli as neutral or positive (Davidson, 1984; Davidson et al., 1985, 1987; Dimond et al., 1976; Ostrove et al., 1990; Sackeim et al., 1982; Silberman and Weingartner, 1986), even when by its explicit dimensions, the memory reported is in fact negative or even traumatic (Kihlstrom and Harackiewicz, 1982).

VERBAL AMNESIA AND INFANT MEMORY

It thus appears that the earliest verbalizable memories appear around age 3 ½ and constitute verbalized interpretations of earlier visual and emotional feeling states. This is the age at which verbal capabilities begin to rapidly develop, i.e. a stage of life marked by an emotional-verbal transition in psychological and cognitive functioning. Hence, adults and older children can verbally recall some early memories, but not many, and those that occurred prior to the onset of verbal language development, and those which are stored nonverbally, appear to be wholly forgotten. This is because verbal memories are for the most part nonexistent prior to ages 2–3 (Howe and Courage, 1993).

This does not mean to imply that young children are without memory, as even infants can process, store, and recall emotional, visual, tactual, or auditory events (see Fivush et al., 1987; Howe and Courage, 1993; Rovee-Collier and Shyi, 1992). In fact, some memories formed as early as 6 months, and certainly those formed before the age of 2, can be triggered by contextual and visual and auditory cues, and demonstrated nonverbally, months and even years after they were formed (Howe and Courage, 1993; Meyer et al., 1987; Nelson, 1984; Perris et al., 1990; Pillemer and White, 1989; Rovee-Collier and Shyi, 1992; White and Pillemer, 1979). There is even some suggestion that "long term memory is functioning prenatally" (Howe and Courage, 1993).

However, insofar as these early memories are visual, emotional, and are stored nonverbally, later they cannot be recalled and accessed verbally. This verbal access failure plagues adults and even older children. For example, in one study it was found that 4-year-olds were able to verbally recall a considerable amount of information about the recent birth of a sibling and surrounding events. However, if they were questioned about any siblings born before they were age 3, they had absolutely no verbal memory of the event (Sheingold and Tenney, 1982). In contrast, there is some evidence that children below age 3 can demonstrate memory for events that occurred 1–2 years earlier, including events that occurred when they were only a few months old (Meyers et al., 1987).

In part, early experiences may be unrecallable for older versus younger individuals because infants use a different system of codes to store memories and are more likely to rely on emotions and feelings, whereas adults and older children use strategies, symbols, and associations, such as language, not yet fully available to younger children (Brown, 1975; Dollard and Miller, 1950; Freud, 1900, 1916; Hagen et al., 1975; Joseph, 1982, 1988a, 1992b; Piaget, 1952, 1962, 1974; Pillemer and White, 1989; Shantz, 1983; Vygotsky, 1962). Much of what is experienced and committed to memory during early childhood takes place prior to the complete development of linguistic labeling ability and is based on a prelinguistic or nonlinguistic code (Dollard and Miller, 1950; Freud, 1900, 1916; Joseph, 1982, 1988a, 1992b; Pillemer and White, 1989; Shantz, 1983). Therefore, since older persons are relying on more mature and sophisticated coding systems, such as language, they cannot find the right set of neural programs to open the door to early childhood memories (Dollard and Miller, 1950; Neisser, 1967; Schachtel, 1947; Underwood, 1969). The key does not fit the lock because the key has changed.

In that very young children demonstrate long-term memory capabilities, it seems that early memories appear to be forgotten or repressed when in fact they simply can no longer be verbally retrieved. In part this is a consequence of the maturation of a secondary memory system associated with language that appears much later in development, whereas the initial, primary memory system is for the most part emotional, visual, tactual, and auditory and stored in the nonverbal regions of the brain.

REPRESSION VERSUS AMNESIA FOR INFANT MEMORY

It is the development of the secondary language system that Freud (1916) believed acted in part to disguise early memories. Freud (1905) made a distinction, however, between childhood amnesia, which he believed hid the "earliest beginnings of childhood up to the sixth or eighth year," and repression.

Amnesia for early childhood experiences was a passive force, he argued, whereas repression is active and prevents conscious realization of unpleasant and emotionally traumatic experiences that are sometimes of a sexual or threatening nature. Indeed, Freud (1896) at one time argued that repression and female hysteria was a consequence of having been sexually molested during childhood. Although he later downplayed the idea as a product of female fantasy (Freud, 1909), he continued to note that incestuous relations were quite common and subject to repression (Freud, 1916, 1931, 1937).

Freud (1899) also argued that early events that are subject to emotional disorganization and not easily verbalized may also be subjected to repressive forces and that suggestibility and fantasy may play a role in their disguise, misinterpretation, and distortion. The process of translating these nonverbal memories into language results in considerable distortion (Joseph, 1982, 1988a, 1992a).

Thus, Freud (1916) argued that insofar as an adult is able to provide complex narratives of these early events, he or she is not accessing memories laid down by the child. The young child is too verbally immature and unsophisticated to create such complex verbal memories. Hence, these recollected early memories are reconstructions and in this regard may be completely false or serving as a screen (Freud, 1899) to block out of view the original, sometimes traumatic, memory.

Of course, one might ask, "Why should one aspect of the psyche attempt to hide information so as to deceive yet a different region of the mind?" Moreover, what is responsible for this psychic cleavage? (see Sartre, 1956).

Presumably, this information is hidden for the purposes of protection, so as to not disrupt the psychological and cognitive functioning of that aspect of the mind responsible for language—the same regions that control most aspects of fine motor functioning, the left cerebral hemisphere (Joseph, 1982, 1990b). In part, however, the process of repression is also made possible via the frontal lobes, which exert inhibitory control over neocortical information processing and interhemispheric transfer (Como et al., 1979; Davidson, 1984; Hoppe, 1977; Joseph, 1982, 1986a, 1988a, 1990d, 1992b; Joseph et al., 1981).

Secondly, this "psychic cleavage" is a consequence of the differential functional organization of the left versus right cerebral hemisphere (which is more concerned with visual, tactual, and social-emotional processing) and the fact that some information is processed in a wholly different fashion and cannot be recognized by one versus the other half of the brain, which in turn precludes complete interhemispheric transfer (Berlucchi and Rizzolatti, 1968; Hicks, 1974; Joseph, 1982, 1988a; Marzi, 1986; Merriam and Gardner, 1987; Miller, 1990, 1991; Myers, 1959, 1962; Rizzolatti et al., 1971; Taylor and Heilman, 1980). One hemisphere cannot always gain complete access to information stored in the other, and this includes even the learning of sequential and fine motor movements (Hicks, 1974; Taylor and Heilman, 1980; see also Parlow and Hicks, 1989). In addition, early memories stored in the right half of the brain are not completely shared with the left because of the immaturity and slow pace of myelination of the corpus callosum, conditions that give rise to verbal amnesia for early experience in general.

By contrast, traumatic memories, particularly those based on repetitively induced life-threatening terror, emotional trauma, and physical and sexual assaults, are mediated by the limbic system and right hemisphere. However, trauma and repetitive stress not only activates various limbic nuclei, it may damage them, particularly the hippocampus (see below). As detailed in Chapters 6 and 16, this can result in subsequent hippocampal retrieval failure and thus verbal amnesia (i.e., repression).

Many therapists and researchers indicate that repetitive, painful, life-threatening, and traumatic sexual and physical memories are the most resistant to immediate verbal recollection in both children and adults (Briere and Conte, in press, cited by Batterman-Faunce and Goodman, 1993; Courtois, 1988; Christianson and Nilsson, 1989; Donaldson and Gardner, 1985; Fisher, 1982; Fredrickson, 1992; Grinker and Spiegel, 1945; Herman and Schatzow, 1987; Kuehn, 1974; Parson, 1988; Southard, 1919; Summit, 1983; Williams, 1992). Therefore, differential limbic system and right hemisphere activation is implicated in repression versus amnesia.

For example, the right hemisphere becomes highly aroused when emotionally stimulated or the body is physically or painfully stimulated or engaged in sexual activity, and is responsible for the maintenance of the somesthetic body image (Cohen et al., 1976; Cubelli et al., 1984; Desmedt, 1977; Freemon and Nevis, 1969; Haslam, 1970; Murray and Hagan, 1973; Joseph, 1988a; Penfield and Rasmussen, 1950:27; Remillard, et al., 1983; Ruff, 1980; Spencer et al., 1983; York, et al., 1979). When this occurs, there sometimes results a disconnection syndrome such that the left hemisphere, being at a lower level of arousal, is unable to fully participate or monitor the events being processed and learned by the right hemisphere (Joseph 1988a) (see Chapter 3). In consequence, such persons may appear to be verbally amnesic as associated memories cannot be verbally accessed. Instead they appear to be repressed.

Corpus Callosum Immaturity and Interhemispheric Transfer

Since the brain matures over time, the manner in which information is committed to memory and shared between the cerebral hemispheres undergoes changes as well. In fact, for the first six years of life, the cycle of myelination for the corpus callosum, the interhemispheric fiber pathway that enables information to be transferred and shared, is so immature (Yakovlev and Lecours, 1967) that the exchange of tactual, auditory, and visual information is limited and incomplete (Deruelle and Schonen, 1991; Finlayson, 1975; Galin et al., 1977,

1979; Gallagher and Joseph, 1982; Joseph, 1982; Joseph and Gallagher, 1985; Joseph et al., 1984; Kraft et al., 1980; Molfese et al., 1975, see methods; O'Leary, 1980; Ramaekers and Njiokiktjien, 1991; Salamy, 1978). Even when some of this information is transferred to the left half of the brain, much of it is distorted and/or immediately forgotten.

For example, when complex pictures were selectively transmitted by tachistoscope to the brains of 4-year-olds, they were able to accurately describe what had been viewed by the left hemisphere. By contrast, verbal descriptions of pictures shown to the right cerebrum were sometimes fragmented, filled with information gaps, and contaminated by confabulatory embellishments (Joseph et al., 1984). Similar gaps and recall fragmentation occur when children are asked to describe affective-laden television shows (Collins, 1970, 1978; Collins et al., 1978; Hayes and Casey, 1992), and confabulatory tendencies are observed in a variety of situations involving memory reproduction in children (Hudson, 1990; Nelson and Gruendel, 1986).

Similarly, adults who have undergone complete or partial surgical sectioning of the corpus callosum (corpus callosotomy) respond with gaps and confabulatory embellishments when asked to verbally describe pictures shown only to the nonverbal half of their brain (Joseph, 1986a, 1986b, 1988b; Gazzaniga and LeDoux, 1978), as do children as old as 7 (Joseph et al., 1984).

In normal persons, however, in contrast to those with "split-brains," verbal, visual, and tactual memory for information transferred from the right to the left hemisphere progressively improves with age (Finlayson, 1975; Galin et al., 1977, 1979; Joseph et al., 1984; O'Leary, 1980; Ramaekers and Njiokiktjien, 1991; Salamy, 1978) and the likelihood of confabulating or presenting with information or memory gaps correspondingly decreases (Joseph et al., 1984). Hence, initially it appears that infants and young children function as if their right and left cerebral hemispheres were in part disconnected and are unable to fully share

certain types of information. In addition, the left, language-dependent region of the cerebrum respond as if it immediately forgets those details transmitted to it by the right hemisphere.

This lack of verbal memory or consciousness for right cerebral perceptions and memory in children has also been demonstrated in regard to complex cognitive processes involving spatial reasoning, nonverbal knowledge, and logic (Gallagher and Joseph, 1982; Joseph and Gallagher, 1985; Kraft et al., 1980; Ramaekers and Njiokiktjien, 1991). For example, as has been demonstrated by Piaget (1952, 1962, 1974) and others, conservation of space, length, and volume appears to require the ability to perform spatial analysis, internal spatial reversals, and the integration of an object's identity in separate and changing spatial contexts so that relational constancies may be ascertained (Gallagher and Joseph, 1982; Joseph and Gallagher, 1985). Young children supposedly lack these capabilities. Hence, if water is poured from one of two similarly sized and equally filled glasses into a tall and thin glass, the child will verbally state that the thin glass has more because the water volume is higher. They will do this even if a little water is first removed before it is transferred to the taller, thinner glass (Gallagher and Joseph, 1982; Joseph and Gallagher, 1985).

However, these abilities appear to depend on the intact right cerebral hemisphere (Gallagher and Joseph, 1982; Joseph and Gallagher, 1985; Kraft et al., 1980), which is nonverbal. Because of callosum immaturity, it is thus not surprising that children fail to demonstrate a verbal understanding of these spatial relationships. The standard test design requires verbal responses so that conservation can be demonstrated. However, if the conservation problem and question (e.g., "Which has more?") is altered so that emotional and motivational judgments are called for ("If you were thirsty and this was your favorite drink, which would you want?"), and the same paradigm is employed in which a little water is first removed, the majority of supposedly non-conserving children will subsequently demonstrate a tacit understanding of these principles and will show conservation, even when verbally they continue to make the same error. In other words, when asked to make a judgment based on their feelings and emotional desires, young children were suddenly able to demonstrate knowledge of information that is not available to the speaking half of the brain (Gallagher and Joseph, 1982; Joseph and Gallagher, 1985). They relied on their feelings to guide their responses, rather than employing language to guess at information that is otherwise not available to the left hemisphere. As such, they pointed to the correct glass, indicating that they knew which had more, despite the illusion of appearances (see also Wheldall and Poborca, 1980).

INFORMATION TRANSFER IN "SPLIT-BRAIN" PATIENTS

Similar modes of emotionally guided information transfer has been demonstrated in persons who have undergone complete corpus callosotomy (e.g., Gazzaniga and LeDoux, 1978; Sergent, 1990; Sperry et al., 1979). Corpus callosotomy involves severing the callosal fibers, which in turn disconnects the two cerebral hemispheres. When the right hemisphere is shown pictures or words, the left cannot verbally describe what has been viewed. However, if the stimulus is emotionally significant, limbic pathways (or any remaining callosal fibers) appear to allow for emotional transfer. For example, when pictures of family members or even of Adolf Hitler were selectively and secretly transmitted to the right half of the brain (via tachistoscope), the disconnected left half of the brain at first had no knowledge of what had been viewed, but was able to make progressively more accurate guesses (e.g., "G.I. came to mind."). However, these guesses were reinforced by the emotional tone generated by the right half of the brain and cues provided by the experimenters. In these split-brain patients, the left half of their brains soon gained ac-

cess to the images still stored within the right hemisphere and eventually came up with the correct response when provided cues and other forms of reinforcement—e.g., "Hitler," (Sperry et al., 1979).

Unfortunately, what is not originally shared usually remains unshared even when information transfer is later greatly improved or as the commissures and the brain develop (Bures and Buresova, 1960; Doty and Overman, 1977; Joseph, 1982, 1988a, 1992b; Risse and Gazzaniga, 1979). In consequence, the right half of the brain sometimes can store, recall, and even act on its own memories much to bewilderment of the left hemisphere, which is sometimes unable to even recognize, much less comprehend, the well-kept secrets of the opposite half of the brain.

EMOTION, HEMISPHERIC LATERALITY, AND SPECIALIZATION

LEFT HEMISPHERE SPECIALIZATION

It is now well known that expressive speech, linguistic knowledge and verbal thought, mathematical and analytical reasoning, and the temporal-sequential and rhythmical aspects of consciousness are associated with the functional integrity of the left half of the brain in most of the population. As detailed in Chapter 4, the left cerebral hemisphere is associated with the organization and categorization of information into discrete temporal units; the sequential control of finger, hand, arm, and articulatory movements; and the perception and labeling of material that can be coded linguistically or within a linear and sequential time frame. It is also dominant for most aspects of expressive and receptive linguistic functioning, including grammar, syntax, reading, writing, speaking, spelling, naming, verbal comprehension, and verbal memory (see Chapter 4). In addition, the left hemisphere has been shown via dichotic listening tasks to be dominant for the perception of real words, consonants, and real and nonsense syllables.

Right Cerebral Specialization and Emotion

The right cerebral hemisphere is associated with social-emotional and nonverbal environmental awareness; visual-spatial perceptual functioning, including analysis of depth, figure-ground, and stereopsis; facial recognition and the maintenance of the body image; and the perception, expression, and mediation of almost all aspects of emotionality (see Borod, 1992; Cancelliere and Kertesz, 1990; Heilman et al., 1985; Heilman and Bowers, 1996; Joseph, 1988a; Ross, 1981; Tucker and Frederick, 1989). This emotional dominance extends to bilateral control over the autonomic nervous system, including heart rate, blood pressure regulation, galvanic skin conductance, and the secretion of cortisol in emotionally upsetting or exciting situations (Rosen et al., 1982; Wittling, 1990; Wittling and Pfluger, 1990; Yamour et al., 1980; Zamarini et al., 1990).

Moreover, according to studies of brain-injured and normal persons, the right hemisphere is superior to the left in distinguishing, interpreting, expressing, and processing auditory and vocal-emotional inflectional nuances, including intensity, stress and melodic pitch contours, timbre, cadence, emotional tone, frequency, amplitude, duration, and intonation (Blumstein and Cooper, 1974; Bowers et al., 1987; Carmon and Nachshon, 1973; Heilman et al., 1975; Ley and Bryden, 1979; Mahoney and Sainsbury, 1987; Ross, 1981; Safer and Leventhal, 1977; Samson and Zatorre, 1988, 1992; Shapiro and Danly, 1985; Tucker et al., 1977). Because of this, the right hemisphere is fully capable of determining and deducing not only what a person feels about what he or she is saying, but why and in what context it is being said—even in the absence of vocabulary and other denotative linguistic features. In fact, even without the aid of the actual words, based merely on melodic tone the right hemisphere can determine context and the feelings of the speaker (Blumstein and Cooper, 1974; DeUrso et al., 1986; Dwyer and Rinn, 1981). Indeed, infants are able to make

these same determinations (Fernald, 1993; Haviland and Lelwica, 1987).

For example, although lacking denotative, grammatical language skills, infants between the ages of 10 weeks and 5 months are capable of appropriately discerning, discriminating, and responding to social-emotional vocalizations conveying approval, disapproval, happiness, and anger and can thus distinguish between different emotional vocalizations so as to determine the mood state and intentions of others (Fernald, 1993; Haviland and Lelwica, 1987). Moreover, 5-month-old (American) infants are able to make these discriminations even in the absence of words and vocabulary, and in response to nonsense English, as well as to German and Italian vocalizations (Fernald, 1993). Presumably, these vocal emotional nuances are comprehended by the infant's right hemisphere and limbic system (see Chapters 3 and 8).

The right hemisphere is thus very sensitive to social-emotional contextual information (Brownell et al., 1986; Cicone et al., 1980; Foldi et al., 1983; Gardner et al., 1975). The left hemisphere has great difficulty in correctly analyzing these variables, and in fact is quite socially dense and concrete, lacking even a sense of humor (Brownell et al., 1986; Cicone et al., 1980; Foldi et al., 1983; Gardner et al., 1975). Similarly, the capacity to identify emotional words and sentences is a capacity at which the right hemisphere excels (Borod et al., 1992; Graves et al., 1981; Van Strien and Morpurgo, 1992).

In addition, the left side of the face has been found to be more emotionally expressive (Campbell, 1978; Chaurasis and Goswami, 1975; Moreno et al., 1990; Sackheim et al., 1978) and to be perceived as more intensely emotional (Borod and Caron, 1980; Sackheim and Gur, 1978). In response to emotional stimuli, the left half of the face becomes more activated and a significant majority of persons respond with conjugate lateral eye movements to the left (Schwartz et al., 1975; Tucker, 1981), the left half of the body being under the control of the right hemisphere.

In fact, the right hemisphere superiority for facial recognition is augmented by the additional display of facial emotion (Ley and Bryden, 1979; Moreno et al., 1990; Suberi and McKeever, 1977), regardless of the emotion conveyed (Buchtel et al., 1978; Dekosky et al., 1980; Landis et al., 1979; Strauss and Moscovitch, 1981; Suberi and McKeever, 1977). Faces also are judged to be more intensely emotional when viewed exclusively by the right hemisphere (Heller and Levy, 1981), and the right hemisphere is dominant in regard to memory for facial expression (Weddell, 1989). In contrast, the left hemisphere mediates or greatly participates in the recognition of famous faces (Marzi and Berlucchi, 1977; Rizzolatti et al., 1971).

Similarly, the right half of the brain labels pictures and movies as more emotional, and is likely to view positive and negative images and emotions as more negative than the left. When the left hemisphere is selectively stimulated (or in the absence of right cerebral input such as occurs with brain damage or sodium amobarbital injection), it is more likely to verbally rate these stimuli as neutral or positive (Davidson, 1984; Davidson et al., 1985, 1987; Dimond and Farington, 1977; Dimond et al., 1976; Gainott, 1972; Ostrove et al., 1990; Schiff and Lamon, 1989; Silberman and Weingartner, 1986) and is more likely to produce positive affects such as laughter (Lee et al., 1990; Rossi and Rosadini, 1967; Sackeim et al., 1982; Terzian, 1964).

Nevertheless, although the left is biased toward making positive emotional judgments, studies of brain-injured and normal populations have repeatedly demonstrated that the right half of the brain is dominant in the expression, recognition, and perception of all aspects of emotion, be they negative or positive (e.g., Borod, 1992; Cancelliere and Kertesz, 1990; Dekosky et al., 1980; Gorelick and Ross, 1987; Heller and Levy, 1981; Heilman et al., 1975, 1984, 1985; Heilman and Bowers, 1996; Joseph, 1986a, 1988a; Ley and Bryden, 1979; Ross, 1981; Shapiro and Danly, 1985; Strauss and Moscovitch, 1981; Suberi and McKeever, 1977; Tucker and Frederick, 1989; Tucker et al., 1977). Moreover, right frontal dominance for the expression of emotional and even negative vocalizations such as crying is apparent by 10 months of age (Davidson and Fox, 1989). Hence, when the right half of the brain has been damaged, emotional expression and perception are significantly affected (see Chapter 3).

Sex, Tactual Sensation, Pain, and the Body Image

In addition to nonlinguistic, prosodic, melodic, emotional and visual-spatial dominance, the right cerebrum has been shown to be superior to the left in processing, perceiving, recognizing, and recalling somesthetic and tactile information, including tactile-form, position, direction, and pressure, as well as Braille-like patterns (Bradshaw et al., 1982; Corkin et al., 1970; Desmedt, 1977; Dodds, 1978; Fontenot and Benton, 1971; Franco and Sperry, 1977; Hatta, 1978; Hermelin and O'Connor, 1971; Hom and Reitan, 1982). The left half of the body (and thus the right hemisphere) is also more sensitive to tactile stimuli, including detecting hot or cold, and has a lower threshold for somesthetic pain perception (Carmon and Benton, 1969; Cubelli et al., 1984; Fontenot and Benton, 1971; Haslam, 1970; Murray and Hagan, 1973; Semmes et al., 1960; E. Weinstein and Sersen, 1961; S. Weinstein, 1978). In contrast, right cerebral destruction has been reported to result in the loss of pain sensation, including phantom limb pain (S. Weinstein, 1978).

In addition, unlike the left, the right hemisphere responds to tactual stimuli that impinge on either side of the body (Desmedt, 1977; Pardo et al., 1991). Indeed, a somesthetic image of the entire body appears to be maintained by the right half of the brain (see Chapter 3). Moreover, right hemisphere dominance for tactual processing is apparent by 2 years of age (Rose, 1984).

When the right hemisphere is damaged or abnormally activated, somesthetic functioning can become grossly abnormal, and patients may experience peculiar disturbances that involve the body image (Critchley, 1953; Hillbom, 1960; Joseph, 1986a, 1988a; Miller, 1984; Nathanson et al., 1952; M. Roth, 1949; N. Roth, 1944; Sandifer, 1946; E. Weinstein and Kahn, 1950, 1952). Like those with split-brains, these patients may fail to perceive tactile stimuli applied to the left side, or in severe cases, wash, dress, or groom only the right side of the body; confuse body-positional and spatial relationships; misperceive left-sided tactual stimulation as occurring on the right; fail to realize that their extremities or other body organs are in some manner compromised; and/or literally deny that their left arm or leg is truly their own (Critchley, 1953; Bisiach and Geminiani, 1991; Joseph, 1986a, 1990c, d). In contrast, left-sided lesions seem to have relatively little effect in this regard.

The fact that lesions of the left hemisphere rarely result in severe neglect or body image disturbances suggests that the right hemisphere maintains a bilateral representation of the body and is much more sensitive to tactile input, whereas the left cerebrum maintains a unilateral representation. Therefore, when the left hemisphere is damaged, the right brain continues to monitor both halves of the body and there is little or no neglect—an impression supported by findings that indicate that the right hemisphere electrophysiologically responds to stimuli that impinge on either side of the body, whereas the left hemisphere predominantly responds only to right-sided stimulation (Desmedt, 1977; Pardo et al., 1991). Hence, when the left hemisphere is disconnected from right-sided tactual and body image information, it verbally denies any memory or knowledge regarding half the body and may even fail to notice or correctly localize right-sided stimulation. Insofar as the left hemisphere is concerned, these memories, and thus the left half of the body, does not exist.

PAIN AND SEX

In addition to body image distortions and neglect, right parietal lobe abnormalities (particularly when secondary to tumor or seizure activity) can give rise to sensory misperceptions such as physical pain (Davidson and Schick, 1935; Hernandez-Peon et al., 1963; Ruff, 1980; Wilkinson, 1973; York et al., 1979). That is, because of heightened and neocortical seizure activity, rather than failing to verbally report or perceive (i.e., neglecting) the left half of the body, patients may experience sensory distortions that concern various body parts due to abnormal activation of the right hemisphere (see Chapter 3). This includes sensations of painful orgasm or testicular pain (Ruff, 1980; York et al., 1979).

For example, Ruff (1980) reports two patients who experienced paroxysmal episodes of spontaneous and painful orgasm, which were secondary to right parietal seizure activity. In one patient the episodes began with the sensation of clitoral warmth, engorgement of the breasts, tachycardia, etc., all of which rapidly escalated to a painful climax. In normal intact persons, orgasm is also associated with electrophysiological arousal predominantly within the right hemisphere (Cohen et al., 1976). Moreover, seizure activity in the right frontal and temporal lobes has also been reported to result in increased sexual activity, including exhibitionism, masturbation, and the acting out of coitus in the absence of a partner (Freemon and Nevis, 1969; Joseph, 1988a, 1990d, f; Penfield and Rasmussen, 1950:27; Remillard et al., 1983; Spencer et al., 1983).

There is thus considerable evidence that indicates that the right half of the brain is dominant in regard to the perception and expression of visual, tactual, and auditory emotions, the experience of tactual and painful body sensations, and the experience and expression of sexual feelings and behavior. However, when the left half of the brain is prevented from gaining access to this information, it may respond by developing dissociative reactions, creating alternative personalities that it attributes to the left half of the body, and/or by verbally denying or neglecting body sensations and the left half of the body.

It is thus noteworthy that in addition to developing temporary imaginary play-mates, most children will fail to verbally report when they have been touched or have undergone genital examination. In one study, 56% of 33 children tested failed to report that a stranger had touched and rubbed their head, and only 32% recalled this accurately (Peters, 1991). Moreover, it has been found that children up to age 7 who had undergone genital and anal touching during a medical checkup also failed to report it unless specifically asked and provided cues (Goodman and Clarke-Stewart, 1991; Saywitz et al., 1991). This suggests that the left half of their brain is unable to initially gain verbal access to this tactile-sexual information.

POSITIVE EMOTIONS AND THE LEFT HALF OF THE BRAIN

There is some evidence that the left hemisphere processes positive emotions (Davidson, 1984; Davidson et al., 1985, 1987; Dimond and Farrington, 1977; Dimond et al, 1976; Gainotti, 1972; Lee et al., 1990; Ostrove et al., 1990; Rossi and Rosadini, 1967; Sackeim et al., 1982; Schiff and Lamon, 1989; Silberman and Weingartner, 1986; Terzian, 1964), and that the left half of the brain is likely to view negative and neutral emotions as positive, whereas those with right hemisphere damage are more likely than normal persons to view neutral events as more positive, presumably because of the removal of right cerebral influences.

In addition, when right cerebral influences are eliminated, as occurs with right hemisphere damage or anesthetization, patients are likely to view and report neutral and even negative events in a positive manner and to exhibit a positive mood, including laughter (Gainotti, 1972; Lee et al., 1990; Rossi and Rosadini, 1967; Sackeim et al., 1982; Terzian, 1964).

For example, Rossi and Rosadini (1967) injected sodium Amytal into the left versus right hemisphere and found that 68% of those with left hemisphere inactivation reacted with depression (expressed presumably by the awake right hemisphere). In contrast, 84% of those with right-sided inactivation responded with euphoria (expressed presumably by the completely awake left hemisphere) and 16% responded in a depressed fashion (see also Gainotti, 1972; Lee et al., 1990; Rossi and Rosadini, 1967; Terzian, 1964). These differences are not always observed, however.

When the left anterior region of the brain has been damaged or is dysfunctional, persons are likely to respond with severe depression, or anger, irritability, and paranoia (Gainotti, 1972; Gruzelier and Manchanda, 1982; Hillbom, 1960; Joseph, 1990b, e; Lebrun, 1987; Robinson and Benson, 1981; Robinson and Szetela, 1981; Sherwin et al., 1982; Sinyour et al., 1986), although those with massive left posterior and temporal lobe damage may respond with euphoria (in which case bilateral dysfunction may be suspected). This suggests that it is the right and not the left half that mediates these negative feeling states, which are expressed as a consequence of reduced left cerebral input. However, unlike the abnormal emotions triggered by right cerebral injuries, it appears that with left hemisphere damage, the right half of the brain is accurately perceiving the consequences of the injury and is understandably upset and depressed.

Therefore, although the left hemisphere plays a minuscule role in regard to emotions, there is much evidence to suggest that it is inclined to view or express emotional material in either a neutral or positive light, regardless of its actual affective value. In part, its tendency to be overly positive may be for survival benefits so as to not create social gaffs, which in turn is a consequence of its difficulty in making inferences and ascertaining social emotional nuances, including humorous material (Brownell et al., 1986; Cicone, et al., 1980; Foldi et al., 1983; Gardner, 1975; Gardner et al., 1975, 1983; Ostrove et al., 1990). Depressive emotions also hinder language and behavioral expression, functions that the left half of the brain dominates.

The left half of the brain may also have difficulty gaining access to "negative" feeling states, not only because it is ill equipped to process them, but because they fall within the domain of the right hemisphere. That is, just as the right half of the brain cannot engage in grammatical discourse and has no reason to receive this information from the left, it also has little reason to transfer emotional information to the left hemisphere except under certain circumstances and conditions. This does not preclude transfer, however, as many persons dwell on depressive thoughts and feelings.

Given the immaturity of the callosum (which partially disconnects the left hemisphere from the right) and the left hemisphere's tendency to err in the direction of positive perception and verbal reports, young children are more likely to verbally label emotional experiences as positive, even when they are not (Casey, 1993; Cole, 1986; Saarni, 1984). Young children (that is, their left hemispheres) have difficulty in correctly verbally identifying negative emotion expressions in themselves or others, and tend to give verbal reports that misidentify these feelings as positive (Casey, 1993; Cole, 1986; Saarni, 1984). Adults do not have these difficulties. Children are also more likely to verbally recall positive versus negative experiences (Strayer, 1980). In fact, children find it much easier to verbally describe and identify positive versus negative emotions (Feshbach and Roe, 1968; Mood et al., 1978; Strayer, 1980), whereas similar difficulties do not plague adults. Events that occur during or depicting positive mood states are also easier for children to verbally recall as well (Forgas et al., 1988).

Moreover, young children are less likely to verbally recall events that occurred when they were in a negative mood (Hill, 1972; Masters et al., 1979; Peters, 1987) or when observing adults or children behaving in a negative manner (Blugental et al.,

1992; Hoffman, 1983). Even extraneous contextual details that occurred in association with negative, versus positive or neutral conditions, are also seemingly forgotten by young but not older children (Blugental et al., 1992). Young children are also likely to confabulate responses when asked to recall this material (Blugental et al., 1992). In fact, in some experiments young children verbally recalled less than 5% of emotional reactions observed, as compared with adults, who tended to recall approximately 40% (Hayes and Casey, 1992; Hayes and Kelly, 1985; Mandler, 1978). Moreover, although children are able to recognize these emotions (Gnepp, 1983; Hayes and Casey, 1992), emotional events quickly seem to be forgotten, for within minutes or following a short delay children have great difficulty in verbally recalling them (Hayes and Casey, 1992). The same occurs with matching and recalling negative faces, and even neutral events appear to become repressed if experienced in a negative setting. In contrast, positive memories and stimuli and events following a negative incident or mood are recalled much more correctly (Goodman et al., 1991), presumably because they are not associated with a negative context and are stored separately in the left half of brain.

The fact that children are relying on verbal labels suggests that the left half of the brain cannot gain access to right cerebral–mediated negative affective states. This is also largely a function of the immaturity of the corpus callosum, which prevents transfer in many modalities up until the age of 7 and beyond (Joseph et al., 1984). In fact, children's verbal recall of visual emotional material (e.g., facial matching ability) does not approach adult levels until children reach age 14 (Kolb et al., 1992, see methods and results).

However, this does not appear to be a function of lack of ability per se, for children as young as 5 also demonstrate a right cerebral advantage for facial recognition (Levine and Levy, 1986; Young and Bion, 1980), and happy faces are recognized and matched as well by children as by adults. Rather, this appears to be a consequence of limitations in interhemispheric information transfer. Intact adults do not display these deficits.

It is also noteworthy that young children have difficulty in recalling central points (Collins, 1970; Collins et al., 1978) and have more difficulty in recalling contextual or inferring implicit versus explicit motives and details (Collins et al., 1978; Paris, 1975; Thompson and Meyers, 1985) as compared to adults. Removal of right cerebral input in adults (such as following injury) often results in similar deficits (Brownell et al., 1986; Cicone et al., 1980; Foldi et al., 1983; Gardner et al., 1983), as the left hemisphere is no longer able to gain access to this information.

Although children have difficulty in verbally recalling recent emotional and negative experiences (Hayes and Casey, 1992; Hayes and Kelly, 1985), the ability to verbally describe one's emotions and emotional memories improves with age (Bretherton et al., 1986; Campos et al., 1983; Donaldson and Westerman, 1986; Harter and Whitesell, 1989; Lewis et al., 1979; Ridgeway et al., 1985; Russell and Bullock, 1986; Strayer, 1993). Presumably this corresponds to increases in corpus callosum maturity and the greater capacity to transfer and share information.

Nevertheless, even over the course of development it is negative feeling states that are the most resistant to verbal recall in children. In fact, the capacity to verbally recall negative emotional memories developmentally lags far behind positive memory description (Hayes and Casey, 1992; Taylor and Harris, 1983).

VERBAL VERSUS EMOTIONAL MEMORY STORAGE
Verbal Memory and the Left Temporal Lobe

Although a variety of neurochemical and neuroanatomical regions are involved in the formulation of memory (see Chapters 5 and 6), it has long been known that damage to

the amygdala and hippocampus coupled with the neurosurgical removal of the inferior temporal lobes can produce profound disturbances in the learning and recollection of verbal and visual stimuli (Barr et al., 1990; Frisk and Milner, 1990; Joseph, 1990f, 1992a, 1994; Halgren, 1992; Kapura et al., 1992; Kimura, 1963; Milner, 1968, 1970; Mishkin, 1978; Murray, 1992; Rolls, 1992; Sarter and Markowitsch, 1985; Squire, 1992). For example, left temporal lobectomy, seizures, or lesions involving the inferior temporal areas can moderately disrupt immediate and severely impair delayed memory for verbal passages, as well as the recall of verbal paired-associates, consonant trigrams, word lists, number sequences, and conversations (Barr et al., 1990; Delaney et al., 1980; Kapur et al., 1992; Meyer and Yates, 1955; Milner, 1968; Milner and Teuber, 1968; Samson and Zatorre, 1992; Weingartner, 1968). Similarly, severe anterograde and retrograde memory loss for verbal material has been noted when the left anterior and posterior temporal regions (respectively) are electrically stimulated (Ojemann et al., 1968, 1971), lobectomized, or injured (Barr et al., 1990; Kapura et al., 1992).

Right Temporal Lobe Emotional Memory

Right temporal lesions or lobectomy significantly impairs recognition memory for tactile and recurring visual stimuli such as faces and meaningless designs, memory for object position and orientation and for visual-pictorial stimuli, and short-term memory for melodies (Corkin, 1965; Delaney et al., 1980; Kimura, 1963; Milner, 1968; Samson and Zatorre, 1988, 1992; Taylor, 1979). Similarly, memory for emotional material is also significantly impaired with right versus left cerebral lesions (Cimino et al., 1991; Wechsler, 1973) including the ability to recall or recognize emotional faces (DeKosky et al., 1980; Fried et al., 1982; Weddell, 1989). Persons with right hemisphere damage also have more difficulty recalling personal emotional memories (Cimino et al., 1991).

Hence, the left hemisphere is responsible for the encoding and recall of verbal memories, whereas the right cerebrum is dominant in regard to visual-spatial, nonverbal, and emotional memory functioning. Each hemisphere stores the type of material that it is best at recognizing, processing, and expressing.

AMYGDALA HYPERACTIVATION AND EMOTIONAL MEMORY

It is important to emphasize that the amygdala (in conjunction with the orbital frontal lobes) is primary in regard to all aspects of emotional and social-personality functioning, including the capacity to form complex multimodal emotional memories (see Gloor, 1986, 1992; Halgren, 1992; Joseph, 1992a, 1994; Kesner, 1992; Kling and Brothers, 1992; LeDoux, 1992; Murray, 1992; Rolls, 1992). The amygdala is buried within the depths of the anterior-inferior temporal lobe, adjacent to the hippocampus, and is functionally related and richly (and often reciprocally) interconnected with these and a variety of other brainstem, limbic, and neocortical regions (see Chapter 5). It is via these reciprocal interconnections that it can sample, extract, and add emotional attributes to perceptual and cognitive activities.

In this regard, Gloor (1992), Halgren (1992), and Rolls (1992) argue that the amygdala adds to the emotional stream of cognitive and perceptual experience, including the formation of memories, by associating aversive or rewarding emotions to the flow of overall experience. According to Gloor (1992:523), "affective significance will also be attached if what is perceived occurs in the context of psychoaffective constellations related to affiliative behavior (for instance, kinship ties) or other forms of social behaviors, familiar surroundings, or earlier aversive experiences.... When the amygdala is activated under such circumstances, three streams of amygdaloid input will be set into motion, one directed to the hypothalamus and brainstem...[a] second to the hippocampus [that] serves to reinforce

consolidation of perceptual information...a third to association cortex neurons," within which the memories may be stored and that feeds back to the amygdala.

The amygdala also becomes particularly active when recalling emotional memories (Halgren, 1992; Heath, 1964; Penfield and Perot, 1963) and in response to cognitive and context-determined stimuli regardless of their specific emotional qualities (Halgren, 1992). However, once these emotional memories are formed, the amygdala sometimes requires the specific emotional or associated visual context to trigger their recall (Halgren, 1992; Rolls, 1992). In this regard, the amygdala is implicated in "repression." Other emotional feelings elicited by amygdala stimulation include terror and extreme fear, disgust, guilt, sadness, depression, and loneliness (see Chapter 5). The amygdala also becomes activated in response to stress as well as bizarre stimuli (Halgren, 1992), whereas in response to repetitive negative experiences, the hippocampus rapidly habituates (see below) and apparently ceases to participate in memory functioning (Joseph, 1990a, 1992a) (see Chapters 6 and 16).

In addition, sexual feelings and related activity and behavior are often evoked by amygdala stimulation and temporal lobe seizures (Gloor, 1986; Halgren, 1992; Jacome et al., 1980; Remillard et al., 1983; Robinson and Mishkin, 1968; Shealy and Peele, 1957). For example, males may experience penile erection, whereas females may ovulate, experience uterine contractions and lactogenic responses, and produce vaginal-vulva secretions in conjunction with orgasm and sexual feelings in the genital areas. In one case a patient experienced a memory flashback that involved a vivid recollection of her first experience with sexual intercourse (Gloor, 1986, 1992; see also Penfield and Perot, 1963).

Abuse and Amygdala Hyperactivation

Children often respond to sexual abuse by pretending that it is not taking place

(Courtois, 1988; Summit, 1983), such as by "playing possum." Many victims will lie in bed in a state of frozen fear and hypervigilance (Courtois, 1988), frozen fear states and hypervigilance being mediated by the amygdala (see Chapter 5). It is also not unusual for children to initially hyperventilate and then to hold their breath and pretend they are sleeping while the abuse occurs.

It is noteworthy that the temporal lobes and the amygdala-hippocampal neural complex are highly susceptible to hypoxia, which in turn can result in loss of memory or conversely create hallucinatory and dissociative states (detailed in Chapter 16). Hence, in cases of sexual abuse, whereas the amygdala may be highly activated because of feelings of fear and terror, in some cases it may be simultaneously denied sufficient oxygen.

In fact, some children may also be denied sufficient oxygen by having an adult lie across them. In some cases, sexually abused children suffer compression injuries to the chest and will die from lack of oxygen. Moreover, some sadistically abusive adults will hold their hands over the child's mouth as a form of torture or so as to prevent them from crying out while they are being assaulted.

Thus it is not surprising that some abused children feel as if they have detached from and left their bodies and experienced a "splitting of the ego" due to these influences (see Chapter 16). It is also likely that with repeated instances of abuse, coupled with oxygen depletion, that the temporal lobes and amygdala/ hippocampal complex might be injured by these experiences and forever function abnormally (see Braun, 1984, for related discussion). One additional consequence might be a form of amnesia for these experiences, particularly in that the hippocampus (in contrast to the amygdala) appears to cease to participate in the formation of memories when highly aroused (Joseph, 1992a) and in response to repeated negative and punished experiences (see Chapters 6 and 16), thereby inducing hippocampal memory retrieval and storage failure.

Right Hemisphere Activation, Left Hemisphere Deactivation

As noted, although both halves of the brain appear equally capable of experiencing fear, the right half of the brain is more likely to be involved in sexual and painful perception and emotional memory storage. It is also the right half of the brain and the amygdala, hippocampus, and temporal lobes that are more likely to produce visual-emotional and hallucinogenic experiences, including those induced by LSD or hyperactivation (see Chapters 3 and 5). Hence, when these feelings and images are produced or experienced, the right half of the brain is at a much higher level of arousal than the left which in turn appears to be partly disconnected from the right, as it is at a comparatively lower level of activation, the memory is stored only in the right hemisphere.

Not surprisingly, persons experiencing these dissociated states, or who are producing high-impact hallucinatory perceptions and memories, such as following a temporal lobe seizure or amygdaloid activation, soon become verbally amnesic and fail to verbally recall these experiences (Gloor, 1992; Horowitz et al., 1968). However, these memories and the hallucinations can be later recalled if the person is provided with specific contextual cues (Horowitz et al., 1968), or they may be reproduced in the form of fragmented and terrifying dreams (Donaldson and Gardner, 1985; Kramer et al., 1987) that in turn may be forgotten (Kaminer and Lavie, 1991; Kramer et al., 1987) by the left cerebral hemisphere.

CORPUS CALLOSUM AND UNILATERAL MEMORY STORAGE

In the intact, normal brain, even nonemotional memory traces appear to be stored unilaterally rather than laid down in both hemispheres (Bures and Buresova, 1960; Doty and Overman, 1977; Levy, 1974; Risse and Gazzaniga, 1979). Moreover, when one hemisphere learns, has certain experiences, and/or stores information in memory, this information is not always available to the opposing hemisphere; one hemisphere cannot always gain access to memories stored in the other half of the brain (Bures and Buresova, 1960; Doty and Overman, 1977; Joseph, 1982, 1988a, 1988b; Levy, 1974; Risse and Gazzaniga, 1979). These learning and memory transfer deficits include even the learning of sequential and fine motor movements (Hicks, 1974; Taylor and Heilman, 1980).

Presumably, to gain access to these lateralized memories, one hemisphere (perhaps via the frontal lobes) has to activate the memory banks of the other brain half via the corpus callosum or anterior commissure (which links the right and left amygdala and inferior temporal lobes). However, if the corpus callosum is prevented from transferring this information, the other half of the cerebrum will be almost completely unaware of its presence. This has been demonstrated experimentally in primates and humans.

For example, in a series of experiments performed by Doty and Overman (1977) one hemisphere had been trained to perform a perceptual-motor task. Once it was learned and both hemispheres were able to respond correctly, the commissures were subsequently cut. Following the corpus callosotomy, only the hemisphere that originally was trained was able to perform, i.e., could recall it. The untrained hemisphere acted as though it never had been exposed to the task; its ability to retrieve the original memories was now abolished (Doty and Overman, 1977) as it was now unable to gain access to them.

In a similar study, Risse and Gazzaniga (1979) injected sodium Amytal into the left carotid arteries of human patients so as to anesthetize the left cerebral hemisphere. After the left cerebrum was inactivated, the awake right hemisphere, although unable to speak, was still able to follow and behaviorally respond to commands, e.g., palpating an object with the left hand, looking at pictures.

Once the left hemisphere had recovered from the drug, as determined by the return of speech and motor functioning, none of the eight patients studied were able to verbally recall what objects had been palpated with the left hand, "even after considerable probing." Although encouraged to guess, most patients refused to try and insisted that they did not remember anything. However, when offered multiple choices in full field, most patients immediately raised the left hand and pointed to the correct object! Once this occurred, the patients then immediately verbally claimed to have suddenly remembered what the right half of the brain had experienced. Hence, information confined to the right half of the brain was suddenly made available to that of the left when provided sufficient cues, whereas before it had remained a "well kept secret" (Risse and Gazzaniga, 1979).

According to Risse and Gazzaniga (1979), although the memory of touching and palpating the object was not accessible to the verbal (left hemisphere) memory system, it was encoded in a nonverbal form within the right hemisphere. However, it remained unavailable to the left hemisphere when normal function returned. The left (speaking) hemisphere was unable to gain access to information and memories stored within the right half of the brain. Nevertheless, the right hemisphere not only remembered, but was able to act on its memories.

This indicates that when exchange and transfer is not possible or is in some manner inhibited, or if for any reason the two halves of the brain become functionally disconnected and are unable to share information, the possibility of information transfer at a later time is precluded (Bures and Buresova, 1960; Doty and Overman, 1977; Risse and Gazzaniga, 1979)—even when the ability to transfer is acquired or restored. The information is usually lost to the opposite half of the cerebrum, though recognition memory can be triggered if visual and other cues are provided (Joseph, 1992b).

Nevertheless, although lost, these memories and attached feelings can continue to influence whole brain functioning in subtle as well as in profound ways. The right hemisphere may experience and store certain information in memory and later, in response to certain situations, act on those memories, much to the surprise, perplexity, or chagrin of the left half of the brain; one hemisphere cannot always gain access to memories stored in the other half of the brain.

EARLY ENVIRONMENTAL INFLUENCES AND REPRESSION IN CHILDREN

As is now well known, the developing organism is extremely vulnerable to early experience during infancy such that the nervous system, perceptual functioning, and behavior may be altered dramatically (Bowlby, 1982; Casagrande and Joseph, 1978; 1980; Denenberg, 1981; Diamond, 1985; Ecknerode et al., 1993; Harlow and Harlow, 1965; Joseph, 1979; Joseph and Casagrande, 1980; Joseph and Gallagher, 1980; Joseph et al., 1978; Langmeier and Matejcek, 1975; Rosenzweig, 1971; Salzinger et al., 1993; Sternberg et al., 1993) (see Chapter 18). Interestingly, there is some evidence that the right cerebral hemisphere and the right amygdala may be more greatly affected (see Denenberg, 1981; Diamond, 1985).

Moreover, during these same early years our traumas, fears, and other emotional experiences, like those of an adult, are mediated not only by the limbic system, but also via the nonlinguistic, social-emotional right cerebral hemisphere. And, just as they are in adulthood, these experiences are stored in the memory banks of the right cerebrum.

However, as demonstrated by the experiments of Doty, Overman, Bures, Buresova, Risse, and Gazzaniga, when information transfer is for any reason prevented, even when transfer is later made possible, that information can remain confined to that half of the brain that originally perceived and processed it. Since information transfer between the two halves of the brain in young children is limited because of corpus callosum immaturity, the left hemisphere

of a young child has at best incomplete knowledge of the contents and activity that are occurring within the right (Deruelle and de Schonen, 1991; Finlayson, 1975; Gallagher and Joseph, 1982; Galin et al., 1979a, 1979b; Joseph, 1982; 1988a, 1988b, 1992b; Joseph and Gallagher, 1985; Joseph et al., 1984; Kraft et al., 1980; Molfese et al., 1975; O'Leary, 1980; Ramaeker and Njiokiktjien, 1991; Salamy, 1978). This sets the stage for differential memory storage and a later inability to transfer information between the cerebral hemispheres. Since traumatic and unpleasant as well as most other emotional experiences are more likely to be stored in the right half of the brain, emotional childhood memories, particularly those which are negative or which involve the body or sexuality, are thus even more likely to become well kept secrets, i.e., repressed.

Childhood Amnesia versus Repression

Children are better able to verbally describe and recall positive and nonemotional memories because the language-dominant left hemisphere is provided direct access and is better able to process them, whereas the converse is not true regarding those events stored in the right cerebrum.

However, insofar as these early memories are not traumatic or unpleasant, this failure to transfer is not due to repression. Rather, this passive failure to gain access and transfer information gives rise to verbal amnesia for early childhood experiences. Repression is a more active force that serves to prevent transfer and recollection in both the right and left halves of the brain. As will be discussed in more detail below, repression proper requires the arousal and inhibitory activity of the frontal lobes.

MEMORY ISOLATION, NEOCORTICAL MATURATION, AND SYNAPTIC OVERLAY

Several other factors, however, may be responsible for or at least contribute to repression and loss of memory of early childhood experiences. Although limbic nuclei have established much of their synaptic subcortical connections soon after birth, the infant is not fully capable at a neocortical level of processing much information, as many regions have not fully myelinated or developed. Neuronal associations and synaptic interconnections with other brain areas is quite limited in infants and children versus adults (see Blinkov and Glezer, 1968; Farber and Njiokiktjien, 1993; Joseph, 1982; Spear, 1979); these are processes that can continue for several years, and in some regions of the brain, beyond the first decade (Lecours, 1975; Milner, 1967; Yakovlev and Lecours, 1967) (see Chapter 18). Unmyelinated axons, although able to transmit primitive sensations such as pain, are slower to transmit and are not well adapted for complex information exchange.

Thus, over the course of development, islands of neocortical memories may be formed and then become progressively isolated as they are overlaid by new memories and the establishment of new myelinated neuronal circuits linked by common learning experiences that are multimodal and associated with different brain regions (Farber and Njiokiktjien, 1993; Joseph, 1982; Spear, 1979). Later memories have the advantage of numerous synaptic interconnections, which increases the likelihood that many different types of cues or probes might activate the memory by association. Infants and young children are not so well endowed, and thus only highly specific contextual and associated cues may trigger recollection.

This has been referred to as the "synaptic overlay hypothesis" (see Spear, 1979), which suggests that early memories become inaccessible because of new growth in the brain and synaptic morphological development. Once a person becomes an adult these early isolated memories are no longer accessible but come to be disconnected or increasingly isolated. Moreover, with new neural development, these early memories may also come to be grossly misinterpreted when passing through new

brain regions that then attempt to process and translate them.

Because the adult brain is more myelinated and richly interconnected via synaptic, dendritic, and axonal proliferation, it is likely that adult memories are retained in multiple regions. It is because of these multiple interconnected cortical associations and neural networks that, if we think about the word "chair," we are able to visualize it, name it, describe it, and imagine a multitude of chairs differing in every possible dimension. Diverse regions of the brain can create a circuit of experience by their activation (Gloor, 1992; Joseph, 1982; Halgren, 1992; Hebb, 1949; Kesner, 1992; Lynch, 1986; Penfield, 1954; Penfield and Perot, 1963; Rolls, 1992).

In this regard, the greater efficiency in adult retention may be directly related to myelinated and the multiple ways in which different attributes of a memory may be stored by diverse brain regions. A single memory in the adult (versus the child) thus has multiple representations and can be triggered by a variety of cues. Younger organisms, having less developed brains, will store these experiences in fewer regions, such that later these memories are less easily activated.

Nevertheless, it is clear that the adult's and child's limbic system is capable of perceiving, analyzing, and expressing complex social-emotional stimuli (Joseph, 1982, 1992a) and is certainly capable of expressing, responding to, and storing in memory negative as well as positive emotional states. As any parent or preschool teacher can easily testify, children and even infants are able to express a full range of complex positive and negative emotions. It is when children are asked to verbally describe negative emotions or memories that they begin to have significant difficulty.

Hence, although localized differential neocortical maturation and myelination accounts for changes in cognitive development (Farber and Njiokiktjien, 1993), as well as the difficulty in accessing early memories in general, differences in the expression and memorization of positive versus negative and traumatic emotions is probably not a function of maturational changes per se, except to a limited degree and in regard to the corpus callosum and perhaps the hippocampus, which in turn would influence the formation of cognitive memories (see Chapter 6).

For example, "myelination of the hippocampus and its key relay zones shows a curvilinear increase in the extent of myelination from the first to sixth decades of life" (Benes et al., 1994). Hence, a hippocampal contribution to childhood amnesia as well as repression is thus implicated—particularly if the child is repeatedly traumatized, i.e., an immature hippocampus is at an increased risk for injury (see Chapter 16).

On the other hand, amygdala and hippocampal synaptogenesis in primates has been shown to be complete soon after birth (Ecknhoff and Rakic, 1981), whereas in humans the hippocampus is at least 40% mature by birth and almost fully mature by 15 months of age (Kretschmann et al., 1986). In addition, the inferior temporal lobe has reached 85% of its adult size by age 2 (Blinkov and Glezer, 1968). However, even so, these nuclei have not completely matured (at least in regard to myelination) until much later in life (e.g., Benes et al., 1994; Lecours, 1975).

Thus in large part it appears that verbal amnesia for early experiences is largely a consequence of split-brain functioning in children, amygdala and right hemisphere hyperactivation, hippocampal immaturity and or injury, frontal lobe inhibitory activity, and the different ways in which adult versus early childhood experiences are coded and stored.

CONTEXTUAL CUES AND EMOTIONAL MEMORY AND BEHAVIOR

Although a person may be verbally amnesic for early childhood experiences or those repressed during adulthood, these memories can also be triggered by the appropriate contextual cues. This is also why,

be it a child or adult, it is always possible for these supposedly forgotten or repressed memories to be acted on and even recalled—that is, by the right half of the brain (Joseph, 1992b). However, sometimes these memories will be triggered but will remain confined to the right hemisphere, if it is opposed by the frontal lobes. However, depending on the extent of frontal lobe inhibition, the right hemisphere may then act on them.

Thus, because of contextual activation coupled with frontal lobe inhibition and right hemisphere transfer failure, these same repressed memories can also alter mood and behavior in subtle and profound ways, even when the individual is unable to verbalize or consciously determine what is affecting them. Consider for example a child who has witnessed a murder (Bergen, 1958) or who has been sexually molested. Although she may never talk about it, and may deny it if asked, she may incorporate the traumatic event into her play (Bergen, 1958), or go to school and pull up her dress, pull down her panties, play with the genitals of her friends, or allow them to play with hers or act out sexually with dolls, stuffed, animals, or their friends, or even adults in a highly aggressive fashion (Conte and Schuerman, 1987; Joseph, 1992b; Pynoos and Eth, 1985; Terr, 1988, 1990).

In other words, she acts out the memories and experiences selectively stored within and recalled by the right half of her brain. Because her right hemisphere cannot talk about what has happened, it acts it out instead. Sometimes, however, these traumas are played out emotionally within the privacy of a person's own head (Joseph, 1992b).

It is important to note, however, that although children who have been repeatedly sexually brutalized (Herman and Schatzow, 1987) or molested are reported to be the most likely to become amnesic, and to repress all associated memories (Brier, 1992; Brier and Conte, in press; Courtois, 1988; Fredrikson, 1992; Herman and Schatzow, 1987; Summit, 1983), not everyone so treated will do so. There are important individual differences that come to play a significant role in repression.

AMNESIA AND RECOGNITION MEMORY

There have been numerous examples of persons who, although suffering from verbal amnesia, nevertheless demonstrate the possession of knowledge and information that they in fact denied possessing (Claparede, 1911; Crovitz et al., 1981; Meudell and Mayes, 1981; Schacter et al., 1984), including details from a short story read to them previously as well as names and addresses or little known facts provided during their amnesic period. Similarly, persons suffering from profound amnesia are able to demonstrate recognition memory involving identification of visual objects and "picture puzzles," such that recognition time is reduced by over 100% on day 2 versus day 1 of training, although they deny having seen the material before (Crovitz et al., 1981; Meudell and Mayes, 1981). This has been referred to this as "source amnesia." Of course, in these and other instances, it is not just the source but the information itself that they deny. Nevertheless, because of findings such as these, some have argued that amnesia is a reversible, task-dependent retrieval deficit (Warrington and Weiskrantz, 1970; Weiskrantz, 1978).

These and related findings indicate that knowledge and information can be present and influence behavior although the person verbally denies knowledge of its presence, for this is exactly what occurs with amnesiacs (see Cermak et al., 1973; Corkin, 1965; Milner et al., 1968). For example, in one study an amnesic patient was told some very unusual stories about various pictures. When he was later shown the same pictures, he claimed to have no recollection of having seen them or hearing the story. However, when asked to pick a title for each picture, he picked those which mirrored the theme that he had been told (Schacter and Moscovitch, 1984).

In another instance, an amnesic patient was introduced to and shook hands with a doctor who had hidden a pin in his hand, sticking the patient in the process. Later, when the doctor had left and then returned, although the patient claimed to have no memory of having met the doctor before, the patient refused to shake hands. When asked why, the patient stated that sometimes people hid pins in their hands (Claparede, 1911).

Hence, although a patient is unable to verbally report these memories and actively deny having had certain experiences, this information can nevertheless be attended to, stored, and demonstrated even when the patient has no verbal recollection of having heard or seen it before. In this regard, even in amnesia, it appears that loss of memory may not due to a storage deficit per se, but to a failure or inability to employ the correct retrieval strategy.

This is also why recognition memory is not as severely affected as verbal memory in amnesic patients (Hirst et al., 1988), particularly those in whom the amygdala and hippocampus has been spared (see Murray, 1992; Squire, 1992) and in whom the medial thalamic nuclei and/or the frontal lobe destruction is the source of the amnesic disturbance (Jetter et al., 1986). Recognition memory may well depend on interactions involving the amygdala, hippocampus, and inferior temporal neocortex (Mishkin, 1982; Murray, 1992; Rolls, 1992; Squire, 1992), the right hemisphere mediating visual recognition memory, and the left being involved in recognizing verbal stimuli.

FUNCTIONAL AMNESIA

Amnesia and memory loss may be secondary or related to a number of different factors, including post-traumatic stress disorders, degenerative disturbances and neurological disease or stroke, and disconnection syndromes (such as corpus callosum immaturity) as well as head injury, intense fear, emotional shocks, and physical traumas, including rape, assault (Christianson and Nilsson, 1989; Donaldson and

Gardner, 1985; Fisher, 1982; Frankel, 1976; Grinker and Spiegel, 1945; Kuehn, 1974; Parson, 1988; Schacter et al., 1982; Southard, 1919; Williams, 1992), and sexual molestation during childhood (Brier, 1992; Brier and Conte, in press; Courtois, 1988; Fredrikson, 1992; Herman and Schatzow, 1987; Joseph, 1992b; Summit, 1983). Indeed, life-threatening and repetitive emotional, sexual, and physical traumas have been repeatedly noted to give rise to dissociation and fugue states and temporary amnesias. Fugue states are also associated with the development of "hysterical amnesias."

Functional amnesia can last from seconds to minutes to hours and even days, weeks, and years. In some cases the amnesia may be circumscribed and limited to certain events and time frames. In others, the condition is so severe that not only do patients suffer profound retrograde amnesia, but they may even lose personal identity.

For example, Schacter and coworkers (1982) describe a 21-year-old patient (hereafter called "P.N.") suffering from functional retrograde amnesia. He had no idea as to his name, address, or any other personal information and had been wondering the streets of Toronto until he approached a police officer and was taken to a hospital. It was only when his picture was placed in a newspaper that a cousin recognized him, although P.N. claimed not to recognize her. According to his cousin, his grandfather had died the week before. However, P.N. did not remember the death or the funeral or anything about his grandfather who his cousin said he loved dearly. It is noteworthy, however, that "P.N.'s amnesia cleared on the next evening while he watched an elaborate cremation and funeral sequence in the concluding episode of the television series Shogun. PN reported that as he watched the funeral scene, an image of his grandfather gradually appeared in his mind. He then remembered his grandfather's death as well as the recent funeral. During the next few hours, the large sections of his personal past that had been inaccessible for the previous 4 days returned" (p. 524). It is noteworthy in this case that a CT scan revealed a right temporal lobe injury suggestive of gliosis (Schacter et al., 1982). Gliosis, particularly when involving the temporal lobe, is highly likely to give rise to abnormal activity as well as seizure disorders, which in turn will hyperactivate the amygdala (Bear, 1979; Gloor, 1992; Joseph, 1990a, f).

It is also noteworthy that Schacter and colleagues report that an "island" of "positive" affective memories was discovered to be intact when P.N. was examined during his amnesic state. Later when he recovered from his amnesia, he continued to report this positive island of positive memories, whereas "he described most of his other experiences in either neutral or negative terms" (p. 531).

What these findings suggest is that the right hemisphere, and perhaps, specifically, the right temporal lobe in this case, was in fact responsible for the fugue state and the functional amnesia. That is, the right hemisphere, being already dysfunctional, may have become hyperactivated by the emotional trauma of the funeral, and the left half of the brain, which mediates positive affect and nonpersonal and related memories, continued to function, albeit in a disconnected state. This would also account for the preservation of positive affect and the verbal amnesia. The right hemisphere is dominant for all aspects of emotion including the self-image, whereas the left is more the domain for impersonal and verbal facts, including the recollection of positive emotions.

However, as noted, not just personal shocks, but physical assaults and rape can also lead to functional amnesia and memory loss (Christianson and Nilsson, 1989; Donaldson and Gardner, 1985; Kuehn, 1974). For example, Christianson and Nilsson (1989) described a 23-year-old female who was raped and beaten while out jogging. When found by a policeman she had no memory of her identity or that of her relatives, friends, boyfriend, or place of work, and all recollection of her past life, including all aspects of the rape, were forgotten. However, she recognized the police officer because he had a "famous face" as he was a well-known hockey player. As noted, memories for famous faces are under the dominant control of the left half of the brain, as is memory of verbal facts such as the names of cities and so on, which this patient had no difficulty recalling.

After approximately 2 weeks, when she returned with police to the jogging path to discover the scene of the rape (which she still could not recall), she became extremely anxious and upset. This occurred when she noticed some crumbled bricks and when she walked over toward a meadow, at which point she began to cry "hysterically." Later, when she came across yet another pile of crumbled bricks, images of the rape began to return. "Thereafter, isolated memory-pictures from the traumatic evening started to return in a nonchronological manner, and within a time space of 10–20 minutes she was able to reconstruct the whole episode" (p. 291), including the fact that she was attacked near the crumbled bricks and raped in the nearby meadow. It is noteworthy that this person also had a history of having been sexually abused as a child, which again raises the question of predisposing factors.

Similarly, Frankel (1976) described a black male who, after being accused of stealing money from his former employer, fell down a stairs, struck his head, lost consciousness, and awoke suffering from complete amnesia regarding all aspects of his past, including his identity and that of friends and relatives. He was also found to be suffering from left hemiparesis and impaired left-sided sensation, which suggests a right cerebral injury. However, over the course of the following week and with hypnosis, he was able to gradually recall his past.

These cases are not all that unusual. In one study, 38% of women who had been taken to an emergency room following a sexual assault had no memory of being at the hospital (Williams, 1992). Victims of rapes and physical assaults tend to provide poorer descriptions of their assailants than those who are victims of robberies (Kuehn, 1974). Similarly, Briere and Conte (cited in Batterman-Faunce and Goodman, 1993)

found that over 50% of 450 men and women who had been sexually abused as children had suffered periods of total or partial amnesia for the event (see also Brier, 1992; Courtois, 1988; Fredrikson, 1992; Herman and Schatzow, 1987; Summit, 1983; Terr, 1988, 1990). In a study of persons who had committed murder and violent crimes, over 10% had no recollection and reported amnesia for their crimes (Taylor and Kopelman, 1984; see also Christianson and Nilsson, 1989), whereas those who engage in nonviolent crimes or victims who are not physically assaulted do not demonstrate a loss of memory.

Traumatic memory loss also occurs among children (Bergen, 1958; Brier, 1992; Brier and Conte, in press; Courtois, 1988; Fredrikson, 1992; Herman and Schatzow, 1987; Summit, 1983; Terr, 1988, 1990). For example, Terr (1988, 1990), in her examination of children who have been sexually abused or traumatized by the death of a parent or a severe injury, found that although they were able to accurately reenact the event, their verbal memories tended to be fragmented, incomplete, and quite poor.

REPRESSION, RECALL, AND RECOGNITION MEMORY

It is noteworthy that the common denominator to many (but not all) of these traumatically induced amnesic syndromes is often severe and repeated emotional trauma and/or body contact in the form of sexual assault or physical injury. However, over time, adult victims are able to recall bits and pieces of what occurred, and memory may be triggered by contextual variables and other cues. Although considered controversial by some, and completely rejected by professional "witnesses for the defense," adults who have been repeatedly sexually abused as children also sometimes repress and recollect their traumas (Brier, 1992; Courtois, 1988; Fredrikson, 1992; Herman and Schatzow, 1987; Joseph, 1992b; Summit, 1983).

As noted above, emotional, sexual, and tactual body impressions, memories, and experiences are mediated and stored pre-dominantly by the right half of the brain. However, because the right cerebrum is also dominant in regard to nonverbal visual memories, the presence of contextual visual variables and other cues (such as the bricks or the funeral described above) can trigger recognition memory via the right temporal lobe, and eventually, if transfer occurs, verbal recall.

Similarly, among children even nontraumatic and nonverbal memories can be triggered by visually observed reenactment, or by visual observation of visual cues alone, or if they return to the original setting (e.g., a walk in the park) and are allowed to reenact the scene (Pipe et al., 1993; Wilkinson, 1988), even when they initially claim no verbal memory for these events.

As noted in the cases described by Risse and Gazzaniga (1979) and Sperry and colleagues (1979), once recognition occurred in the right hemisphere and certain actions or feelings were observed and experienced by the left, the speaking half of the brain gained access to these memories as well, stating, "Now I remember." The same principle seems to hold for repressed and even nontraumatic childhood memories that seem to have been wholly forgotten.

Indeed, this is what presumably occurs in cases such as that of the 1990 highly publicized conviction of a retired California fireman, George Franklin Sr., who received a life sentence for the 1969 murder of an 8-year-old girl, Susan Nason. Mr. Franklin was convicted in San Mateo Superior Court on the testimony of his daughter, Eileen Franklin-Lipsker, who reported to police in 1989 that she had suddenly recalled memories that had been repressed and forgotten for 20 years. Specifically she recalled watching her father rape and then kill the victim, her girlfriend, by smashing her skull with a large rock. She claimed that after almost 20 years the memory had come to her when certain actions and facial expressions of her own daughter suddenly triggered their recollection. However, she also reports that her father had molested her as well, which again raises the question

as to predisposing factors being involved in repression versus simple forgetting and subsequent verbal amnesia for early experiences. It is these latter experiences, however, that are most easily triggered via visual contextual cues.

Female friends and patients frequently have told me of instances of recalling sometimes unpleasant and sometimes pleasant but otherwise completely forgotten memories from their childhood that were triggered by observing, listening, or interacting with their own children. It may be a look on the child's face, something he or she said or did, or maybe the mother's own reaction to her child's antics that triggers this flood of visual and often emotional associations, all of which are usually immediately recognized as long-lost childhood memories (Joseph, 1992b).

Hall (1899), for example, reported how numerous forgotten memories from his childhood were triggered when he visited his old school. The triggering of memories such as these are commonplace. It is the nature of memory to be triggered by internal or external sources of input that may only indirectly or directly have a bearing on what we recall, and which may be related to these memories only in regard to contextual variables. In fact, it is not at all uncommon for persons to observe a scene or hear a fragment of conversation and be suddenly reminded of a dream from the night before that had seemingly been forgotten. As reviewed and discussed in detail in Chapter 3 the visual-emotional content of dream material appears to be produced and stored predominately by the right half of the brain, and it is within the domain of the dream that traumatic memories are often recalled (Freud, 1900), only to be sometimes forgotten upon waking.

DREAM RECOVERY

It is now well known that the visual-emotional hallucinatory aspects of dreaming occur during REM, whereas more thought-like and verbal ideational patterns are produced during non-REM (NREM). It is also during NREM that a person is most likely to talk in his or her sleep (Kamiya, 1961). In turn, REM has been shown to be associated with right hemisphere activation and low-level left hemisphere arousal (Goldstein et al., 1972; Hodoba, 1986; Meyer et al., 1987; see also Bakan, 1977, 1978; Fischman, 1983; for opposing views, see Ehrlichman et al., 1985). Similarly, measurements of cerebral blood flow have shown an increase in the right temporal lobe during REM sleep and in subjects who upon wakening report visual, hypnagogic, hallucinatory, and auditory dreaming (Meyer et al., 1987).

Abnormal and enhanced activity in the right temporal area, including the amygdala and hippocampus, also acts to increase dreaming and REM sleep for an atypically long time, whereas REM sleep increases activity in this same region much more than in the left hemisphere (Hodoba, 1986). Likewise, tumors or electrical stimulation of the right temporal lobe is much more likely to result in complex visual as well as musical and singing hallucinations, whereas left cerebral dysfunction or activation gives rise to hallucinations of words, sentences, or conversations (Halgren et al., 1978; Hecaen and Albert, 1978; Jackson, 1880; Mullan and Penfield, 1959; Penfield and Perot, 1963; Teuber et al., 1960). Conversely, LSD-induced hallucinations are significantly reduced following right but not left temporal lobe surgical removal (Serafetinides, 1965), and dreaming is sometimes abolished with right but not left temporal lobe removals (Kerr and Foulkes, 1981).

In addition, performance across a number of tasks associated with left hemisphere cognitive efficiency is maximal during NREM, whereas, right hemisphere performance (e.g., point localization, shape identification, orientation in space) is maximal after REM awakenings (Bertini et al., 1983; Gordon et al., 1982; Levie et al., 1983, cited by Hodoba, 1986). Left hand motor dexterity is also superior to that of the right hand when persons are awakened during REM, and the opposite relationship is found during NREM, i.e., right hand superiority (Bertini et al., 1983).

Of course, it is unlikely that the right hemisphere is solely responsible for the content of dreams. Rather, the left probably contributes some visual and ideational content as well as providing any and all verbal monologues and speech that is produced in accompaniment (Foulkes, 1978; Joseph, 1988a; Kerr and Foulkes, 1981). That is, although the right is clearly dominant, it takes both hemispheres to dream.

Although there have been conflicting reports as to the effect of right versus left cerebral lesions on the ability to recall dreams (see Miller, 1991), it is important to note that persons with massive left hemisphere lesions have difficulty describing much of anything, let alone their dreams (Joseph, 1988a). Moreover, it would be expected that when the left half of the brain has been damaged, its level of arousal would be lower than that of the right, in which case it would be unable to gain access to right cerebral memory centers. The two halves have become partially disconnected. In this regard it is interesting to note that it becomes progressively more difficult to recall one's dreams as one spends time in or awakens during NREM (Wolpert and Trosman, 1958), which is associated with high left hemisphere and low right brain activation, and that interhemispheric communication is reduced during dreaming (Banquet, 1983).

Thus, are dreams really forgotten, or are they locked away in a code that is not accessible to the speaking left hemisphere? However, as noted, it is not at all uncommon for persons to suddenly be reminded of an otherwise completely forgotten dream, the memory of which may be triggered by a closely associated visual or auditory experience the following day.

CONTEXT AND VISUAL RECOGNITION

Forgotten memories are often triggered by present-day experiences. In part this is because memory serves as a reference via which one can make judgments and take appropriate action and know what to do given one's current circumstance. When presented with cues that create a similar context, the contextual information appears to act as a cue that triggers recall or recognition (Dalton, 1993; Gartman and Johnson, 1972; Goodman and Aman, 1990; King and Yuille, 1987). These contextual cues can be composed of props or actual items or may be provided by returning to the actual environment in which the event to be recalled occurred.

As argued by Tulving (1983) and other researchers (Pipe et al., 1993) those cues that are suited for triggering these memories are those which are most similar to these memories. Tulving (1983) has referred to this as the "encoding specificity principle." These stimuli act as retrieval cues.

When something is experienced only or repeatedly within the same context, the context becomes part of the memory trace. In these instances recognition and recall may be bound to the context (see Dalton, 1993). For example, Gartman and Johnson (1972) report that subjects exposed to particular words and then tested for recall in the same environmental context were better able to remember the words than were persons exposed to words and tested in different contexts. Similarly, Smith and colleagues (1978) found a context-dependent effect for memory of common words.

In addition, Godden and Baddeley (1975) found that persons who learned word lists underwater versus above water were better able to recall those words when tested in the same environmental context in which they were first exposed. However, recognition memory could be triggered in either environment if cues were provided (Godden and Baddeley, 1980).

Indeed, it is not at all uncommon for a persons to be sitting, for example, in his or her living room, and in response to a thought, impulse, desire, or memory rise, walk out of the room to retrieve the item or act on the memory, and then completely forget what it is he or she intended to do. However, once the person returns to the context in which the thought or memory occurred, and then sits down, he or she suddenly is able to remember what was sought (Joseph, 1992b).

Hence, context plays an important role in memory, for it may not only be a retrieval cue and used for resolving ambiguity, but part of the memory trace. For example, as noted, the amygdala provides an emotional context that is stored along with the cognitive memory trace, whereas the recreation of a similar emotional context can trigger recall (Blaney, 1986; Bruhn, 1990).

Context may also provide a bridge linking the formation and storage of a series of separate memories that, although independently formed, now share similar contextual attributes. These multicontextual memories are easy to retrieve, as they may be triggered by so many different stimuli (Joseph, 1992b). Therefore, when events or stimuli are experienced in a variety of contexts, any number of contextual cues may trigger memory (Smith, 1988). However, if a single memory is linked to a single context, recognition and recall will be hindered when contexts are significantly different. However, if one is presented with a quite similar context and the appropriate cues, a single memory that had seemingly been long forgotten may be suddenly recalled.

Contextual factors are also a factor in infant memory, for recognition has been repeatedly found to be superior under conditions in which the context remains similar (see Fagan, 1984). According to Fagan (1984), "variations in context during encoding and retrieval influence the recognition memory performance of infants much as such variations change memory performance in adults."

Be it infant or adult, the more the cue matches or compliments information contained in memory the greater is the likelihood of retrieval and recollection (Tulving, 1983). Context also becomes important in recognition memory when the material to be recalled is actively opposed by other concerns such as personality and emotion and self-image. However, in this regard, one's current mood can also act to trigger selective memory recall (see Blaney, 1986; Bruhn, 1990), particularly when appropriate cues and contextual details are present.

Carol's Recall of Her Childhood Molestation

In 1988 Carol came to my office to determine if she had "gone crazy." She reported that while watching television she suddenly recalled bits and pieces of a terrible memory that seemed to explain her problems with men, sex, and her former boyfriend.

According to Carol, she had been in bed with her boyfriend and they had just finished making love for the first time, which, she noted, made her feel upset. However, he then began to stroke and run his fingers through her hair while simultaneously exerting downward pressure on her head, directing her to his genitalia. She began to panic, became quite hysterical, and started crying and trying to strike him. Grabbing up her clothes and quickly getting dressed, she ran from his apartment.

For the next several weeks she refused to talk to him, hung up when he called her, began to feel an overwhelming aversion toward men, and was troubled at night by nightmares of a man chasing her with a gun. She sought counseling, but to no avail.

It was only many months later while watching television that the entire memory of what had happened to her, so many years before, unraveled. In the show, a father, who was a police detective wearing a gun, walked into his crying daughter's bedroom, turned her around to face him, and began to brush and run his fingers through her hair as he pulled her toward him. Immediately Carol began to feel angry and upset, and then fragmented mental images began to flood her mind: a scary man, the gun, the fellatio, his fingers in her hair.

And then she remembered those nights when her mother briefly worked, and how her mother's boyfriend, for almost a year beginning when she was 4 or younger, had molested her until he moved away. Vividly she began to recall everything: his running his hands through her hair as he sat facing the television, how he frightened and intimidated her into performing fellatio, sometimes making her cry in the process. In addition she had troubling memories of this man sexually abusing her mother and terrifying her with a gun. Somehow she managed to forget all about this until many years later.

At my urgings, Carol discussed this memory with her mother who confirmed that this man (who lived in the house when she was between the ages of 4 and 6) had been abusive and that he had used a gun to terrorize her mother. However, her mother did not wish to discuss these episodes and was unable to confirm that he had molested Carol. It is also important to note that I explored the possibility that although her memory of her mother's abuse was seemingly accurate, that perhaps her memory of personal sexual abuse might be based on her fears, as a child, of what this man might do rather than on what he actually did, and that she might not have been molested. Carol adamantly rejected this and became very upset that I repeatedly raised this possibility.

Speculations on Differential Memory Storage in Carol's Brain

According to Terr (1990), "Traumatized children may later forget...[or] they may become so accustomed to chronic, long standing, or repeated traumas that they begin to shut-off from their consciousness.... In an attempt to see no evil, hear no evil, speak no evil, and feel nothing, the youngster starts ignoring what is at hand. His senses go numb and he guards against thinking.... Children struggle not to think trauma-related ideas and not to feel trauma related feelings. They fight any mental picture that might create new upsurges of feeling" (pp. 78–79).

This is particularly likely for events experienced while a person is isolated and where what happens is secret, frightening, and the person feels that he or she cannot discuss what occurred (Courtois, 1988; Summit, 1983). In contrast, those who are sufficiently verbal and who have a supportive audience who also know of and acknowledge the trauma may talk about it incessantly. This is not exactly what occurs when children are being secretly molested and repeatedly traumatized, as was the case with Carol.

Presumably, the memory of what this man had done had been stored in the memory banks of the limbic system and the right half of Carol's brain and had been forgotten by both cerebral hemispheres. Years later, when her boyfriend began to stroke her hair while physically encouraging her to perform fellatio, the emotional association between sex, fellatio, and having her hair fondled was reactivated. The right half of her brain then began recalling the visual and emotional images of what had happened when she was a child, and she became terribly upset, at which point the right frontal lobe acted, via inhibition, to prevent any further recollection. Her left hemisphere, having little linguistic memory of the tactual, sexual act and its associated emotions, had "no idea" as to why she was behaving in this manner. Nevertheless, her right hemisphere prevailed and she ran away in hysterics. Periodically thereafter, she was troubled by right hemisphere dreams.

It was only many months later that the left hemisphere was clued in to what had triggered her hysterical reaction. When both the right and left hemisphere observed on TV the father, who was wearing a gun, brush and fondle the little girl's hair, Carol's limbic system and right hemisphere became upset and the images of what had happened to her spilled forth. Looking at the television movie and feeling the revulsion, the left cerebrum suddenly was given a tremendous amount of information as to what was bothering her and was then offered access to the horror that had so long been locked away, seemingly forgotten. Transfer was made possible because both halves of her brain were observing the same event and feeling the same feeling:

upset. As noted earlier, similar means of emotional transfer leading to verbal recognition has been demonstrated in split-brain patients whose right cerebrums were selectively shown pictures of family members and, in one instance, Adolf Hitler. Moreover, contextual cues have repeatedly been observed to trigger the recall of hidden and even traumatic memories.

AROUSAL, EMOTIONAL STRESS, AND MEMORY LOSS

Some psychologists, however, argue that high levels of stress and emotional arousal are in fact good for memory, and that high-impact, emotionally upsetting events, such as the Challenger spacecraft explosion or the assassination of President Kennedy, literally become engraved in the mind's eye (Brown and Kulick, 1977; Rubin and Kozin, 1984; however, see also Ceci and Bruck, 1993; Christianson, 1992; Deffenbacher, 1983). Presumably these memories are so vivid that they have an almost "live…perceptual" quality, that is extremely resistant to forgetting. One might argue, therefore, that children should be able to recall traumatic memories because they would be so emotionally arousing.

Nevertheless, there have been a number of conflicting reports as to the effects of stress and arousal on memory functioning (see Bohannon, 1988; Ceci and Bruck, 1993; Deffenbacher, 1983; Goodman, 1991; Neisser and Harsch, 1992; Peters, 1991; Terr, 1988, 1990). In large part this is due to some experimenters failing to differentiate between types of stress as well as what constitutes high levels of emotional arousal. That is, hearing that Kennedy has been killed or having a mother take her reluctant daughter to a pediatrician may induce stress and emotional arousal, but this is much different from being brutalized, tormented, raped, or molested in a dark room or back alley where there are no witnesses and the acts are painful, frightening, humiliating, and conducted in secrecy. As noted, temporary functional amnesias are not uncommon.

Indeed, high arousal and emotional levels are not really assessed by these studies. For example, Deffenbacher (1983), in his review of 21 studies, argues that most subjects in laboratory experiments are not aroused to the levels typical of criminal or real-life situations. This is certainly true of actual victims, including witnesses.

For example, witnesses of real-life crimes, although resistant to misleading questions (Yuille and Cutshall, 1986), often produce a number of errors in their verbal reports and are initially verbally amnesic for some details. For example, Yuille and Cutshall (1986) interviewed 13 witnesses to a gun shooting incident, all of whom had initially been interviewed by police soon after. They noted that over a third of their sample produced a number of errors, and that many reports were riddled with errors. Moreover, over 60% of the information provided was completely new and had not been previously reported, which suggests that initially the witnesses had been amnesic for these details.

Similarly, a variety of studies on memories of the Challenger explosion or personal emotional experiences have indicated that these high-impact memories are subject to much forgetting and even distortion (e.g., Neisser and Harsch, 1992; McCloskey et al., 1988; Larsen, 1992). For example, Neisser and Harsch, (1992) had subjects fill out a questionnaire regarding where they were and how they heard about the Challenger accident and so on. When these subjects were questioned again 32–34 months later, only 25% recalled filling out the questionnaire. They also found that many of the subjects on later questioning had in fact changed their recollections. Moreover, many were tremendously surprised when shown their original statements and answers. According to Neisser and Harsch (1992), "As far as we can tell, the original memories are just gone." Moreover, there was no correlation between vividness and verbal accuracy, and confidence in accuracy often proved to be completely unrelated.

McCloskey and colleagues (1988) also found significant forgetting and inaccuracy

over a 9-month interval. Of course, subjects were asked to provide verbal reports (whereas the original experience was visual and emotional), context was not reinstated, and no one felt terrorized. Hence, although verbally amnesic, these memories may not have been forgotten. That is, they may have simply been repressed or no longer accessible to the left hemisphere because of differences in hemispheric specialization. This would be especially likely in children.

In fact, the younger the subject when first exposed to this high-impact, emotionally upsetting information, the greater the degree of verbal forgetting and verbal distortion. For example, in examining memory for national traumatic events such as the death of JFK and Robert Kennedy, Winograd and Killinger (1983) found a steep gradient of forgetting that became more profound for memories formed between the ages of 1 and 7. Hence, for those adults who had been 1–7 years of age when Kennedy had been killed, only approximately 50% of those who had been 4.5 years old or older could verbally recall the news and provide at least one verbal detail. Those who were younger than 3 had no verbal recollections regarding context or associated events or information sources, and only a few who were younger than 5 were able to demonstrate detailed verbal knowledge or memories when questioned as adults.

In yet another study of the recollections of children regarding the Challenger space craft explosion, Warren and Swartwood (1992) found that children of age 5 were also less accurate and more likely to delete features over time in describing their recollections. When tested 2 years later, those who had been the youngest as well as the most emotionally upset at the time were less likely to provide extensive verbal narratives and verbally deleted and left out more features.

Hence, even highly emotional and arousing public events have negative influences on verbal recall for those who are young. Indeed, these recollections, even during childhood, tend to become riddled with gaps that are filled with confabulatory ideas and fantasies.

In this regard, however, it must be emphasized that the subjects in these studies usually were required to provide verbal narratives, which in turn may be completely inappropriate for recalling emotional memories. That is, although verbal memory appears to be studded with gaps in these studies, the same may not be true regarding the nonverbal, emotional aspects of what has been learned and stored in memory. Moreover, in these laboratory studies, contextual cues associated with the original situation in which these events were experienced, were not reinstated, in which case, rather than memory loss, what is being demonstrated is verbal retrieval failure.

Nevertheless, given the amount of verbal forgetting and repression with these public events, it does not seem unreasonable, particularly when considered in respect to the facts detailed above, that extremely upsetting and traumatic memories that are life-threatening and/or entirely personal and shameful may be repressed, stored in the absence of hippocampal input, rather than forgotten.

Unfortunately, these studies not only fail to recreate context, but are unable to demonstrate the full range of verbal forgetting and repression that may occur. The problem with many of these cognitive studies on stress and arousal (at least insofar as repression and childhood amnesia are concerned) is that they focus on emotionally shocking events that occur a single time and that are shared and acknowledged by others. This is not what occurs when a person is placed in a life-threatening situation or front line battlefield conditions in which friends are horribly killed and death is a constant companion, or where a child is brutalized, tormented, or molested or raped in private and repeatedly traumatized. In the case of children, not only are they frightened, but they may be told that what they are doing is secret and that what is happening is not in fact happening, or that they may be killed if they

tell, and that no one will believe them anyway or will think they are "crazy" (Courtois, 1988; Summit, 1983). Moreover, rather than a single instance of terror and abuse, this may go on for years.

Indeed, similar forms of memory loss, including depersonalization, can also plague adults who are terrorized over time, and this includes hardened soldiers and aviators (Donaldson and Gardner, 1985; Grinker and Spiegel, 1945; Parson, 1988; Southard, 1919), as well as women who are held captive and raped. For example, Donaldson and Gardner (1985) describe a woman who was kidnapped and repeatedly raped over a period of weeks. She developed a dissociative reaction and was able to recall only bits a pieces of what occurred until nightmares and flashbacks and therapeutic assistance enabled her to remember.

However, there have also been instances of global amnesia following a single, albeit prolonged episode of terror. For example, Fisher (1982) reports that a druggist who had been terrorized, handcuffed, and robbed by two thieves subsequently became amnesic and lost all memories of what occurred. In another case, a woman became amnesic for 18 hours after seeing her husband die right before her eyes (Fisher, 1982).

Even young children who are not abused but who witness the death and murder of parents may be initially unable to verbally describe their experience (Eth and Pynoos, 1985) and as such are verbally amnesic and may not be able to verbally recall what transpired until weeks, months, and even years after the event (Bergen, 1958)—and then only if provided therapy, or encouragement and emotional support by a trusted friend or loved one. For example, Eth and Pynoos (1985) report of one instance in which a week passed before a little girl was able to recall and tell her great-grandmother about the killing of her mother, implicating her father in the process. Even hardened soldiers who witness the death of friends and who suffer repeated traumas involving their own near-death experiences may suffer memory loss, dissociative reactions, and "hysterical" amnesias (Grinker and Spiegel, 1945; Parson, 1988; Southard, 1919).

Young children are especially susceptible to developing these disorders (Wolfenstein, 1966), particularly if they have no one to talk to, such as in instances of physical or sexual abuse (Courtois, 1988; Summit, 1983). Rather, their experience may be surrounded by the denial of adults as well as their own shame and denial. In these instances, their memory loss may be much more prolonged.

Spear (1979) argues that immature organisms forget more and at a faster rate than adults, and that they are more susceptible to the disruptive influences of retroactive interference, particularly those of a negative or emotionally unpleasant nature. Moreover, repeated traumas exert a more deleterious influence on memory than single traumas (see also Terr, 1990). This is true in rats and other animals as well (see Spear, 1979). Moreover, Feigley and Spear (1970) found that what is retained are feelings of aversion and heightened emotionality. Hence, heightened emotionality coupled with feelings of fear, aversion, and pain can give rise to amnesia in adults as well as infants and leave them emotionally traumatized, even when they cannot recall what had happened (Feigley and Spear, 1970)—that is, unless recognition memory is triggered, a process that is mediated by the inferior temporal lobe in conjunction with the amygdala. Just because a person initially appears verbally amnesic does not mean that the memories have ceased to exist.

Nevertheless, such memories are also quite difficult to retrieve because of the fact that with repeated instances of punishment or under conditions of high arousal, the hippocampus may soon cease to participate in memory formation.

MEMORY RECOVERY

As noted, in a number of experimental studies it has been reported that although

verbal memory is nonexistent for certain details and objects associated with experimentally induced traumatic events, these memories are nevertheless stored nonverbally and can be reactivated by cues or as demonstrated on recognition tests. This is particularly true of unpleasant and negative emotional material (e.g., grotesque facial injuries) that are not as easily recalled as neutral material in free verbal recall situations unless retrieval cues are present (Christianson and Nilsson, 1984; Davis, 1990; Wagenaar, 1986). When the test is recognition versus verbal recall, these differences disappear (Christianson, 1992b; Christianson and Nilsson, 1984). In addition, over time memory begins to return.

For example, in one study it was found that emotional scenes or disgusting words (e.g., vomit), work against accurate immediate recall (Christianson, 1984; Kleinsmith and Kaplan, 1964; Kebeck and Lohaus, 1986), but after a 1-week time period or following a long interval, the emotional material was better retained than the neutral when subjects were appropriately cued.

Similarly, memory for violent and murderous acts that are witnessed, such as in experimenter-constructed videotapes, will progressively increase such that subjects will recall more details over time when tested, questioned, and thus cued every 12 hours (Scrivner and Safer, 1988). Similarly, in rape cases, if the victim is provided a supportive, understanding environment, any initial amnesia for the assault will gradually wane such that increasingly more details can be provided.

Overall, these studies indicate that verbal amnesia for emotionally upsetting or traumatic material is a consequence not of disturbances involving memory storage per se but the manner in which these experiences have been encoded and retrieval has been attempted. And the more personal, shocking, life-threatening, shameful, sexual, and physically painful the experience, the greater is the likelihood the person will initially appear verbally amnesic. This also indicates that although a person may claim to have no verbal memory for specific events, these memories can nevertheless be triggered by environmental and contextual cues, in which case recognition in the absence of verbal memory can be demonstrated. Lastly, highly arousing emotional and traumatic experiences, particularly those which are repetitive, life-threatening, or sexual or physically painful, are the most likely to be seemingly forgotten and repressed.

THE FRONTAL LOBES: AROUSAL, ATTENTION, INHIBITION, AND REPRESSION

There is yet another important factor in the maintenance of repression, and that is the right frontal lobe. Indeed, it has been well established that the frontal lobes are highly involved in all aspects of attentional functioning and arousal (Como et al., 1979; Crowne, 1983; Fuster, 1980, 1996; Joseph et al., 1981; Luria, 1980; Knight et al., 1981; Pragay et al., 1987), including orienting and attentional reactions to visual, tactile, and auditory stimuli (Barbas and Dubrovsky, 1981; Denny-Brown, 1966; Latto and Cowey, 1971a, b; Pragay, et al., 1987; Segraves and Goldberg, 1987; Wagman et al., 1961; Wurtz et al., 1980; Segraves and Goldberg, 1987). Hence, when the right frontal lobe is damaged, patients become easily distracted and disinhibited, and they show wandering or a perpetual shifting of attention and they may spontaneously free-associate (Bianchi, 1922; Como et al., 1979; Joseph, 1986a, 1990d; Rose, 1950; Stuss and Benson, 1984; Wilkins et al., 1987; Yarczynski and Davis, 1942). However, the type of attentional disturbances depends on the laterality of the lesion.

The left frontal lobe appears to be more concerned with left cerebral and verbal attentional functioning and the monitoring of temporal-sequential and detailed events (Milner, 1971; Petrides and Milner, 1982). In contrast, the right is more attentive to nonverbal auditory, visual, tactual, and social-emotional stimuli, and appears to exert bilateral influences on arousal (DeRenzi and Faglioni, 1965; Dimond, 1980; Dimond and Beaumont, 1974; Heilman and Van Den Abell, 1979, 1980; Joseph, 1986a, 1988a, 1990d; Pardo et al., 1991; Tucker, 1981). As such, the entire right half of the brain may become more aroused than the left in certain situations.

For example, the intact, normal right hemisphere is quicker to react to external stimuli, and has a greater attentional capacity compared to the left (DeRenzi and Faglioni, 1965; Dimond, 1976, 1979; Heilman and Van Den Abell, 1979, 1980; Howes and Boller, 1975; Jeeves and Dixon, 1970; Joseph, 1988b; Whitehead, 1991). In split-brain studies the isolated left hemisphere tends to become occasionally unresponsive, suffers lapses of attention, and is more limited in attentional capacity (Dimond, 1976, 1979; Dimond and Beaumont, 1974; Joseph, 1988b). The right hemisphere also becomes desynchronized (aroused) following left- or right-sided stimulation (indicating that it is bilaterally responsive), whereas the left brain is activated only with unilateral (right-sided) stimulation (Desmedt, 1977; Heilman and Van Dell Abell, 1979, 1980; Pardo et al., 1991).

Hence, the right frontal lobe is presumably able to exert bilateral inhibitory or excitatory influences, whereas the left frontal region may be more involved in unilateral excitatory (expressive) activation (Davidson, 1984; Joseph, 1986a, 1990d; Tucker, 1981). Thus when the left frontal region is damaged, the right acts unopposed and there is excessive inhibition—for example, as manifested by speech arrest, depression, and/or apathy (Joseph, 1988a, 1990b, e; Robinson and Benson, 1981; Robinson and Szetela, 1981; Tucker, 1981). However, with lesions involving the right frontal lobe, not only is there a loss of inhibitory control, but the left may act unopposed such that there is excessive excitement, e.g., as manifested by speech release and confabulation (Lishman, 1973; Joseph, 1986a, 1988a, 1990d; Stuss and Benson, 1984; Stuss et al., 1978). Repressive influences might also be removed, and patients may engage in the unrestricted production of verbal associations, i.e., they free-associate (Joseph, 1986a)—

what Freud associated with the "talking cure." Indeed, with right frontal injury, a person may come to recall previously repressed memories, including verified incidents involving sexual molestation when he or she was a child (O'Connor, personal communication, 1994).

Orbital Frontal Mediation of Arousal

It has been proposed that the frontal lobes are able to exert these influences on information processing via thalamic gating, reticular activation and inhibition, and exerting steering influences on neocortical activity (Joseph, 1982, 1990d). For example, the frontal lobes receive projections from the primary and association visual, auditory, and somesthetic neocortices in the occipital, temporal, and parietal lobes, respectively (Barbas and Mesulam, 1981; Crowne, 1983; Levin, 1984; Jones and Powell, 1970; Jones et al., 1978; Pandya and Kuypers, 1969), as well as the limbic system (see Chapter 11). Massive interconnections are also maintained with the thalamus and reticular formation (Kuypers, 1958; Rossi and Brodal, 1956; Sauerland et al., 1967).

Hence, the frontal cortex is "interlocked" with the posterior sensory areas via converging and reciprocal connections with the first, second, and third levels of modality-specific analysis, including the multimodal associational integration performed by the inferior parietal lobule and thalamic relay nuclei. The frontal cortex is therefore able to sample activity within all neocortical sensory/association regions at all levels of information analysis, as well as prior to their reception in the neocortex (Joseph, 1990d). However, different frontal areas maintain different interconnections with various cortical and thalamic regions and exert different influences on arousal.

For example, the orbital frontal lobes also receive higher-order sensory information from the sensory association areas throughout the neocortex and maintain rich interconnections with the lateral convexity, anterior cingulate, inferior temporal

and inferior parietal lobes, medial walls, parahippocampal gyrus, and "gateway" to and from the hippocampus, the entorhinal region of the temporal lobe (Fuster, 1980; Johnson et al., 1968; Jones and Powell, 1970; Pandya and Kuypers, 1969; Van Hoesen, 1982; Van Hoesen et al., 1975).

The orbital regions are also linked with the hypothalamus and amygdala (see Chapter 11). The orbital cortices also sends fibers directly to the reticularis gigantocellularis of the reticular formation, including the reticular inhibitory regions within the medulla and the excitatory areas throughout the pons (Kuypers, 1958; Rossi and Brodal, 1956; Sauerland et al., 1967). There are also rich orbital frontal interconnections with the (MDMT) medial magnocellular dorsal medial nucleus of the thalamus (Pribram et al., 1953; Siegel et al., 1977), a major relay nucleus involved in the gating and filtering of information destined for the neocortex (Skinner and Yingling, 1977; Yingling and Skinner, 1977) and a structure intimately involved in memory (Graff-Radford et al., 1990; Victor et al., 1989). This portion of the thalamus appears to exert modulating influences on neocortical information reception and processing as well as arousal.

The MDMT (like the orbital region) also receives fibers from the reticular formation and amygdala (Chi, 1970; Krettek and Price, 1974; Siegel et al., 1977). However, this portion of the dorsal medial thalamus is subject to orbital control and projects to the amygdala (Joseph, 1990d). Indeed, the orbital region, via its interconnections with all three regions (i.e., reticular formation, limbic system, MDMT), is able to considerably influence the interactions that take place in these nuclei, including control over various forms of limbic, behavioral, emotional, and autonomic arousal (see Chapter 11).

Arousal, Cortical Inhibition, and the Lateral Frontal-Thalamic System

Fibers passing to and from the thalamus and the cortical sensory receiving areas give off collaterals to the reticular thalamic nucleus, which also sends fibers that envelop and innervate most of the other thalamic nuclei (Scheibel and Scheibel, 1966; Updyke, 1975). The reticular thalamus acts to selectively gate transmission from the thalamus to the neocortex and continually samples thalamic-cortical activity (Skinner and Yingling, 1977; Yingling and Skinner, 1977). However, the reticular thalamus is controlled by the lateral convexity of the frontal lobes (Joseph, 1990d), the reticular formation, and the lateral portion of the dorsal medial thalamus, with which it maintains dense interconnections (Skinner and Yingling, 1977; Yingling and Skinner, 1977).

The frontal convexity and lateral dorsal medial nucleus of the thalamus (LDMT) are also richly interconnected and together exert significant steering influences on the reticular thalamus. That is, the lateral frontal convexity appears to exert specific influences on the LDMT so as to promote or diminish the flow of information to the cortex and thus modulates specific perceptual and cognitive activities occurring within the neocortex—activity that it is simultaneously sampling (Joseph, 1990d). This is in contrast to the orbital region, with its connections to the reticular formation and the MDMT and its influences on generalized arousal and limbic activation and inhibition. However, these frontal regions also act in concert.

It is thus via these multiple interconnections and executive inhibitory or excitatory functions that the frontal lobes are able to influence cognitive and perceptual cortical functioning via the sampling of activity occurring throughout the neocortex at all levels of informational analysis, as well as via its modulating influences on the lateral portion of the dorsal medial and reticular thalamic nuclei as well as the reticular formation (Joseph, 1990d). The frontal lobes are therefore able to act at any stage of perceptual processing, from initial reception to motor expression, so as to facilitate or inhibit further analysis, selectively acting to determine exactly what type and the extent of processing that occurs throughout the

neocortex (see Chapter 11). Via inhibitory action and through their neocortical and thalamic links, the frontal lobes are able to coordinate interactions between various regions of the neuroaxis so as to organize, mobilize, and direct attention and overall cortical and behavioral activity and to minimize conflicting demands, impulses, distractions, and/or the attention to or further processing of irrelevant or undesirable data. Information that is irrelevant, or unpleasant, may therefore be filtered and inhibited, perhaps even at the thalamic or hippocampal level, before it reaches the neocortex. The result is repression.

In this regard it is noteworthy that Weinberger and coworkers (1979) have found that persons identified as repressors (based on psychometric analysis) demonstrate higher levels of arousal when presented with aggressive and sexual cues, as compared to low anxious subjects (but similar to high anxious subjects) and their own physiological baseline. Moreover, although highly aroused, the "repressor" group downplayed their feelings of discomfort and anxiety when presented with these stimuli and increased their tendency to claim that they were not upset (see also Bonanno and Singer, 1990; Schwartz, 1990; Weinberger, 1990). This was measured by self-report versus heart rate, skin resistance, and EMG. The frontal lobes and the orbital regions, in conjunction with the amygdala and the right hemisphere in general (Heilman et al., 1978; Morrow et al., 1981; Rosen et al., 1982; Schrandt et al., 1989; Wittling, 1990; Wittling and Pfluger, 1990; Yamour et al., 1980; Zamarini et al., 1990), directly influence if not control these aspects of autonomic arousal.

Hence, when this system is damaged, depending on the site (e.g., orbital versus inferior versus posterior convexity) or laterality of the lesion, or if it is overwhelmed by limbic and emotional upheavals, there can result increased levels of emotional responsiveness; arousal; behavioral disinhibition; and flooding of the sensory association areas with irrelevant and even unpleasant information, ideas, and associations, as well as hyperreactivity, distractibility, impulsiveness, and/or apathy, reduced motor-expressive activities (e.g., speech arrest), and sensory/attentional neglect (see Fuster, 1980, 1995; Joseph, 1986a, 1990d; Stuss and Benson, 1984). Because of this disinhibition and the subsequent removal of repressive influences, some persons may also come to recall molestation experienced during childhood (O'Connor et al., in preparation).

Repression and the Right Frontal Lobe

Unlike the left, the right frontal lobe appears to maintain bilateral inhibitory and probably excitatory control over information processing in both halves of the brain and limbic system as well as suppressing and modulating emotional expression. Presumably it is the right frontal lobe that acts to prevent information that has been emotionally recognized or recalled by the right half of the brain from spreading across the corpus callosum, thereby preventing the disruption of information processing in the left half of the brain (see Chapter 11). Moreover, the right frontal lobe may act so as to prevent the activation of particularly painful or unpleasant memories within the confines of the right hemisphere as well through its inhibitory influences on the hippocampus and amygdala. It is in this manner, via the actions of the frontal lobe and the right frontal lobe in particular, that persons can protect themselves from gaining verbal access to unconscious traumatic memories. Indeed, many functions, ideas, thoughts, images, feelings, and memories may come to be actively inhibited by the right frontal region regardless of where they originate. In this manner a person can "know" (right hemisphere), yet "not know" (left hemisphere).

Although the right frontal lobe may exert bilateral repressive influences on memories, this is not to say that the left may not attempt to do the same regarding left (or right) cerebral memories as well (Davidson, 1984; Hoppe, 1977; Joseph,

1982). However, as the left half of the brain is more likely to label emotional material positive, or at least neutral, when in fact it is negative or mildly unpleasant, this is less likely, at least in regard to "unconscious" repression. Nevertheless, the left frontal lobe may play a significant role in suppression: the rejection, inhibition, and forgetting of information that is consciously recognized as undesirable. In this regard, once transfer begins, the left frontal lobe may then begin to oppose further information exchange, or it may act on undesirable verbal memories stored within the left half of the brain.

Hence, active and purposeful repression, in part, is a function of the frontal lobes in general and the right frontal lobe in particular, which can act to inhibit the recall of unpleasant emotional memories or, once such memories have been recalled, may attempt to inhibit their transfer to the left half of the brain. Essentially this is what may occur following a rape or physical assault. These painful memories are purposefully hidden and repressed. This also explains how persons can become aware of painful or unpleasant information and then hide it from themselves, i.e., from their left hemisphere.

REPRESSION AND PSYCHIC CONFLICTS

Because of frontal lobe inhibition, amygdala hyperactivation, lateralized specialization, and limited information exchange, the effects of early "socializing" experience can have potentially profound effects on the behavior of adults as well children. As a good deal of this early experience is likely to have unpleasant if not traumatic moments, it is fascinating to consider the later ramifications of early emotional learning occurring in the right hemisphere unbeknownst to the left, learning and associated emotional responding that later may remain completely inaccessible to the language centers of the left half of the brain (Joseph, 1982). That is, although limited interhemispheric transfer in children confers advantages (such as by reducing competitive interference), it also provides for the eventual development of a number of significant psychic conflicts—many of which do not become apparent until much later in life.

Disconnection and the Failure to Integrate Emotional Experience

Because of frontal lobe inhibition, partial interhemispheric disconnection, and the immaturity of the callosum, children frequently encounter situations in which the right and left hemispheres not only differentially perceive what is going on, but are unable to link these experiences so as to understand fully what is occurring or to correct misperceptions (Galin, 1974; Joseph, 1982, 1988a, 1992b). As such, children have difficulty in reconciling and assimilating divergent emotions and discrepant situations, such as when a person is acting happy in an unhappy situation (Gnepp, 1983; Harter and Buddin, 1987). Older children, in contrast, are able to internalize the feeling states of others and can describe their own feelings as well without relying on descriptions of overt actions.

Indeed, young children have difficulty in internalizing and verbally describing the emotions of others and instead tend to describe these feelings in regard to externally and observable situations and physical displays, even when they are incongruent (Harris et al., 1981; Hayes and Casey, 1992; Livesely and Bromley, 1973; Nannis and Cowan, 1987). It is also by observing their own actions that children are sometimes able to verbally guess at their own motivations, and it is because of partial intercerebral disconnection as well as their limited verbal skills that they tend to explain many of their actions after, rather than during or before, they occur, as demonstrated by the content and their production of egocentric speech (Joseph, 1982, 1990b). That is, it is the left hemisphere that observes and describes what is being acted out by the right half of the brain and that explains to itself, via the egocentric monologue, what is occurring.

Adults lacking right hemisphere input because of cerebral dysfunction demonstrate similar difficulties in making inferences about intent or motivation or in making nonliteral context-independent judgments (Brownell et al., 1986; Cicone et al., 1980; Foldi et al., 1983; Gardner et al., 1983), such as when a person's expressed mood does not correspond with externally occurring events (Kaplan et al., 1990; Molloy et al., 1990). Moreover, like children, adults with damage to the right hemisphere are more likely to make positive judgments about such observations and fail to attend appropriately to negative connotations (Kaplan et al., 1990).

Children, however, are not suffering from cerebral dysfunction but corpus callosum immaturity, which reduces the ability of the left hemisphere to receive information from the right. Lacking the ability to integrate right and left hemisphere perceptions, children are likely to be faced with difficulties in integrating divergent and contradictory observations, which in turn sometimes results in unfortunate consequences in regard to mental health.

Disconnection, Repression, and Differential Emotional Memory Storage

Consider, for example, a young divorced mother with ambivalent feelings toward her young son (cf. Galin, 1974; Joseph, 1982). Although she does not express these feelings verbally, she nevertheless may convey them through her tone of voice, facial expression, and the manner in which she touches her son. She knows that she should love him, and at some level she does. She wants to be a good mother and makes herself go through the motions. Hence, she is confronted by two opposing attitudes, one of which is unacceptable to the image she has of a good mother. Unfortunately, both attitudes are expressed.

Her son, of course, being in possession of an equally astute and sensitive right cerebral hemisphere, is able to perceive her tension and ambivalence. His right cerebrum notes the stiffness when his mother holds or touches him and is aware of the manner in which she sometimes looks at him. Worse, when she says, "I love you," the right half of his brain is able to sense the emotional tone of her voice and perceives that what she means is, "I don't want you" or "I hate you. " His left hemisphere hears only the positive: "I love you," and notes only that she is attentive. He is in a "double bind" conflict, with no way for his two cerebral hemispheres to match impressions.

This little boy's right cerebrum feels something painful when the words "I love you" are spoken. When his mother touches him, he becomes stiff and withdrawn because his right hemisphere, via the analysis of facial expression, emotional tone, tactile sensation, etc., is fully aware that she does not want him. However, the left half of his brain has labeled all of these experiences in a positive fashion even though they are in fact unpleasant and negative.

Later, as an adult, this same young man has one failed relationship after another. He feels that he can't trust women, often feels rejected, and when a girl or woman says "I love you," it makes him want to cringe, run away, or strike out. As an adult, his left hemisphere hears "love," and his right hemisphere feels pain and rejection. Because the two halves of his cerebrum were not in communication during early childhood, his ability to gain insight into the source of his problems is greatly restricted. The left half of his brain cannot verbally access these "repressed" memories. It has "no idea" as to the cause of his conflicts.

In this regard, this curious asymmetrical arrangement of function and maturation may well predispose the developing child in later life to come upon situations in which it finds itself responding emotionally, nervously, anxiously, or neurotically, without linguistic knowledge or without even the possibility of linguistic comprehension as to the cause, purpose, eliciting stimulus, or origin of its behavior. As a child or an adult, it may find itself faced with behavior that is mysterious, embarrassing, etc. "I don't know what came over me." Essentially such persons are unable to gain verbal access to these emotional and nonverbal memories that are stored in the limbic system and/or right half of the brain. They are thus verbally amnesic for these early traumas. However, sometimes, under certain conditions, they are able to suddenly gain access to these hidden memories, which are triggered by contextual cues similar to those associated with the original trauma.

FUNCTIONAL OVERLAP, COPING STYLES, AND INDIVIDUAL DIFFERENCES

It must be stressed that there is considerable overlap in functional representation such that hemispheric functioning is not completely dichotomized. In addition, there is individual variation in the functional organization of the brain. Some individuals (such as those suffering from early left cerebral injuries) have reversed lateralization, wherein the right hemisphere mediates language and the left is concerned with emotional perception and expression (Joseph, 1986b; Novelly and Joseph, 1983) (see Chapters 4 and 18).

Individual differences in brain functioning, personality (see Bonanno and Singer, 1990; Weinberger, 1990), previous traumatic experiences, or perhaps sex differences in the presence of a neocortical or limbic abnormality may in fact account for why some persons are more likely than others to repress emotional memories or to suffer from prolonged functional amnesia. Of course, as noted, early environmental influences can dramatically alter the structure of the brain as well. Nevertheless, these early experiences and/or neurological structural alterations in turn may well give rise to maladaptive behavior patterns, or conversely to particular personality styles that enable them to cope.

As to personality and coping, it is noteworthy that persons who are identified as "repressors" (via psychometric analysis) tend to display personality styles suggestive of interhemispheric disconnection (Davidson, 1984). For example, they are far less likely than other groups to engage in introspection or self-disclosure or to correctly assess their internal emotional state. They have fewer nightmares or aggressive or sexual dreams, take longer to perceive emotional words, tend to rate events and experiences more positively, have difficulty in recalling personal emotional experiences and childhood memories, report less intrapsychic conflict, and are lacking in the ability to make inferences or determine social nuances or implications (see Bonanno and Singer, 1990; Davis, 1990; Weinberger, 1990 for a detailed review).

Sex Differences

In addition to individual differences, there are significant sex differences in personality, emotion, and the organization of the limbic system and the right and left halves of the brain that I have not addressed here (see Chapters 3, 4, and 5). For example, the anterior commissure, the interhemispheric axonal pathway that links the right and left amygdala and inferior temporal lobe, is significantly larger in females than in males (Allen and Gorski, 1992). This presumably allows for a female advantage in emotional expression and perception and predisposes them to being more affected by these variables. These sex differences are significant in the creation of multiple personalities, repression, and the recollection of emotional memories.

CAVEATS AND CAUTIONS

It should be noted that this author does not advocate the use of hypnosis as a technique for discovering repressed memories (due to increased susceptibility to suggestion). Nor should this chapter be viewed as offering unequivocal support for every claim of recovered memory. As has been repeatedly emphasized, the left hemisphere and the linguistic aspects of consciousness are not always capable of gaining complete access to memories and related knowledge

sources stored within the right cerebrum (Joseph 1982, 1988a, 1992b). Material that is recovered is often subject to interpretation and confabulatory gap filling. Hence, what is recalled may have only a fragmentary relationship to what was originally repressed.

For example, consider "Carol" described above. It is noteworthy that a few aspects of her story varied somewhat over the three sessions I saw her. This indicates that her recollections were not pure reproductions of what had happened, but were, in part, reconstructions contaminated by time and the process of interhemispheric translation, which even nontraumatic memories and experiences are subject to.

In addition, from a therapeutic standpoint, one should never assume that memory loss or repression, or the presence of psychological disturbances, are indicative of sexual abuse, because any number of non-sex-related traumas may be involved. Moreover, given that some girls and boys fantasize about sex with their parents (Freud, 1900, 1905), and that some young females may act in a sexually suggestive manner with their fathers and step-dads (Friday, 1977; however, see Courtois, 1988; Krieger et al., 1980; Rosenfeld, 1979) whereas some older females may have rape and related fantasies sometimes even involving sex with animals (Friday, 1977, 1991), one must be particularly cautious in regard to what is being recalled and remembered. In some cases it may be nothing more than guilty secrets and repressed fantasies and may have little to do with the actual cause of their psychological torment. Indeed, it is well recognized that some persons remember things that never happened (as many a husband or wife might testify) as well as purposefully forget things that did.

SUGGESTIBILITY, FALSE MEMORIES, AND CONFABULATION IN YOUNG CHILDREN

There is absolutely no convincing evidence in the absence of brain injury, hypnosis, drugs, or aggressive brain washing techniques, that adult memory may be significantly altered by suggestion. Although a few "experts" in witness memory have made rather dubious claims in this regard, real-life witnesses to actual crimes about which experimenters have no "authoritative control" completely resist attempts to mislead them and refuse to incorporate "suggestions" (Yuille and Cutshall, 1986).

Nevertheless, young children who are sexually abused may not only find the experience frightening and surreal, but the adult who is abusing them may attempt to convince them that what is occurring is not really happening, that others will not believe them, that it didn't happen, that they are confused, lying, crazy, and so on. Moreover, the abusive parent may act as if nothing happened the following morning, and the child may therefore feel exceedingly confused as to the validity of the experience. As such, they may incorporate this "false" information, and/or rely on it to the exclusion of their actual memory of the actual event.

In this regard it has been demonstrated that when very young children are exposed to misleading suggestions or questions by persons in positions of authority, this may influence their ability to correctly verbally recall, or at least verbally describe, their memories (Ceci et al., 1987; Zaragoza, 1987). This is not the case with older children and adults.

Thus children around age 3 tend to be much more susceptible and more likely to incorporate suggestions than older children (ages 7 and up) (Ceci and Bruck, 1993; Ceci et al., 1987; Zaragoza, 1987)). Nevertheless, even by age 4, children tend to be resistant to incorporating false memories and suggestions, including those concerned with sexual abuse (Goodman and Clarke-Stewart, 1991), such as "He took your clothes off, didn't he?" when no clothes had been removed such as following a pediatric exam. In fact, most young children appear to be resistant to incorporating false suggestions, at least in regard to memory recall (Goodman and Clarke-Stewart, 1991; Zaragoza, 1987).

Of course, children can be bullied, coerced, or persuaded via persistent questioning so as to change their verbal reports, and in this regard, they may also change their verbal memories. However, in some instances, they may seemingly change their verbal reports (and thus seemingly their memories) because they fear the consequences of telling the truth, or because they really can't remember, or because they sense that they are playing some kind of game with the questioner and thus respond accordingly.

Competing Memories

In those children who supposedly change their verbal memories (or at least their verbal reports) in response to false suggestions, the misleading information presumably competes with impressions made during the original experience, such that the underlying memory seems to become altered and transformed, or conversely suppressed. However, another view of suggestibility is that the child has access to memories of the original event as well as memories of the inaccurate suggestions, but cannot distinguish the two.

That is, the tendency of some children to report misinformation instead of their original memory may not be due to forgetting or alteration but competition between the original and the misleading verbal memories, threats, and suggestions, which makes the original material difficult to verbally retrieve (Bekerian and Bower, 1983; Christiaansen and Ochalek, 1983; McCloskey and Zaragoza, 1985). The original memory remains indelible albeit nonexpressed. If for any reason the original memory is not accessible, the verbal memory for the inaccurate information is more likely to be reported (see Lindsay, 1990).

Ceci and Bruck (1993) propose that once the suggestion is received and memorized, it may come to have an equal or even dominant status as compared to the accurate memory. Presumably this is a consequence of the development of a separate neural circuit based on suggestion, which in turn is more easily accessed because of recency and potentiation (see Chapter 6). When intruding questions or misinformation is provided, it makes a stronger impression such that it is "more likely to be subsequently recalled because there is no strong coexisting trace for the original event to compete with" (Ceci and Bruck, 1993:413).

There is also the possibility that two different memories (what one was told versus what one observes) can interact to produce a single false memory involving minor details (Linberg, 1991). In this instance, more neurons are added to the neural network, thus altering the memory—essentially this is what occurs while learning; the original material becomes increasingly modified (see Chapter 6). That is, the tendency of some children to report misinformation instead of their original memory is not due to forgetting or alteration but due to competition between the original and the misleading memories or the right and left hemisphere, which makes the original material difficult to retrieve (Bekerian and Bower, 1983; Christiaansen and Ochalek, 1983; McCloskey and Zaragoza, 1985). Or, they are capable of only reporting the verbal distortions, whereas the original memory remains indelible albeit nonexpressed. If for any reason the original memory is not accessible, the memory for the inaccurate information is more likely to be reported (see Lindsay, 1990). Hence, the original memories remain unaffected, although the ability to verbally retrieve them versus the incorrect memories is affected. Presumably this is a consequence of hemispheric specialization and the development of a separate neural circuit based on suggestion in addition the suggestion may be more easily accessed because of recency and potentiation (see Chapter 6).

For example, Ceci and Bruck (1993) propose that the inaccurate memory may come to have an equal or even dominant status as compared to the accurate memory. When intruding questions or misinformation is provided, it makes a stronger impression such that it is "more likely to be subsequently recalled because there is no strong coexisting trace for the original event to compete with" (p. 413).

Authoritative and Peer Pressure to Change Verbal Reports

It is likely that what is being measured in the above studies is not suggestibility and memory at all, but the effects of social persuasion by persons in positions of authority. That is, children who alter their verbal reports may in fact be attempting to please the experimenter or acquiescing to authority, i.e., such children may be telling the experimenter what they believe the experimenter wishes to hear. Similarly, children who are abused may attempt to please their abuser by pretending the abuse is not occurring and/or by repeating the lies told to them.

For example, when children are asked by an adult stranger to lie about what they observed him do (i.e., keeping it a secret that he had "played" with a doll), almost 50% complied (Goodman and Clarke-Stewart, 1991). Moreover, children are likely to change their stories when they feel intimidated by the questioner or those making the request (Goodman et al., 1991; Goodman and Clarke-Stewart, 1991). In many instances of so-called memory distortion it is clear that the children are in fact being bullied into changing their stories (see Ceci and Bruck, 1993), though not necessarily their memories.

Children may also change their story (and not their memories) because they are able to discern through the examiner's repeated questions that their answers are not acceptable and that a different answer is desired regardless of its truthfulness. Hence, memories of observed actions can be affected by the manner in which children are questioned, particularly if they are repeatedly grilled and subjected to strong suggestions on the part of questioners.

Sometimes children incorporate false suggestions or "lie" so as to avoid the embarrassment of "not knowing" and thus appearing stupid. For example, "when asked nonsensical questions, such as 'is milk bigger than water?' most 5- and 7-year olds replied 'yes' or 'no'; they rarely responded 'I don't know'"(Ceci and Bruck, 1993, citing a study by Hughes and Grieve, 1980). Moreover, as occurs with adults, peer pressure, that is, hearing other children give false answers, can also lead to children changing their reports and thus seemingly altering their memories (Davies, 1991) when this is not the case at all. Like adults, children sometimes respond to suggestion and change their verbal reports simply because they have learned to lie.

Overall, with the exception of young children, it is extremely questionable whether memory can be significantly affected by suggestion (versus brain washing). In fact, when social and peer pressures are lacking, and when children are not repeatedly questioned, bullied, or persuaded to "please" the experimenters, these effects are not particularly robust (see Zaragoza, 1991). Ceci and colleagues (1987) counter, however, by pointing out that the memories of very young children "can be distorted through post event suggestions, not that they inevitably will be."

Nevertheless, in regard to "repressed memories" and early traumatic experiences, if there is any validity to these studies on suggestion, it could well be expected that children could suggest to themselves that what happened to them did not happen, so as to forget it. Conversely, an adult who molests or abuses and threatens while simultaneously lying to the child as to what is occurring may also interfere with memory formation and recall. That is, the child will incorporate the suggestion made by the tormentor and may change his or her memory accordingly or even verbally "forget" what happened.

Verbal versus Visual Memory Distortion in Young versus Older Children

It is noteworthy that the effect of suggestion or being misled is much more pronounced when the original memory is verbal (Ceci et al., 1987) versus visual (Zaragoza, 1987). That is, being misled verbally appears to have more effect on verbal rather than visual memories, which would

suggest that verbal memories maintained by the left hemisphere are more likely to be affected by suggestion, at least in young children, whereas right cerebral memories are much more robust and indelible.

Hence, as children age and with increases in verbal skills the amount of verbal information stored in verbal memory correspondingly increases (Fivush, 1993), whereas their susceptibility to suggestions and tendency to confabulate significantly decreases (Ceci and Bruck, 1993; Joseph et al., 1984). As noted, between the ages of 3 and 7 confabulatory tendencies markedly decrease, as do the gaps and deletions in verbal reports and descriptions of right cerebral perceptions (Joseph et al., 1984).

CONFABULATION

Because children (i.e., their left hemisphere) have incomplete verbal access to right cerebral motives and memories, they sometimes fill these gaps with what seems most relevant and familiar. For example, children may respond to fragmentary verbal memories for past events by recalling familiar events and everyday routines that they use to fill in these gaps (Fivush, 1993; Myles-Worsely et al., 1986), as do confabulators (see Chapters 3, 11).

However, like adult confabulators suffering from disconnection syndromes and memory gaps, young children may also respond to irrelevant details and confabulate and embellish responses by relying on extraneous or misleading props so as to fill any gaps in their verbal memory (O'-Callaghan and D'Arcy, 1990; Piaget, 1965). That is, they confabulate when requested information is not completely available to the language axis of their left hemisphere.

Among adults, confabulation is often a consequence of disconnection, memory failure, right frontal and parietal injuries, and dysfunction or immaturity of the corpus callosum (see Chapter 3). Right frontal injuries result in speech release and a flooding of the speech centers with disinhibited vocalizations and ideas (Joseph, 1986a, 1990d), whereas right parietal and corpus callosum disturbances result in disconnection syndromes such that the language axis is disconnected from sources of input. The same is thought to occur when confabulation accompanies certain memory disorders (Talland, 1961).

Among children, confabulation is often a consequence of verbal and corpus callosum immaturity (Joseph, 1982; Joseph et al., 1984) as well as the slow pace of frontal and inferior parietal lobe and neocortical development and myelination, which is not complete until the end of the first decade of life (Blinkov and Glezer, 1968; Lecours, 1975). In children, neocortical and cognitive functioning is characterized by a ariety of maturationally induced disconnection syndromes.

As a consequence of the multiple disconnections that characterize cerebral functioning, when children are questioned, depending on the cues available and the context, they sometimes inadvertently lie or confabulate regarding their behavior, yet do not realize they are in fact telling a lie. According to Piaget (1965:157), "without actually for the sake of lying, i.e., without attempting to deceive anyone, and without even being definitely conscious of what he is doing, he distorts reality in accordance with his desires and his romancing."

However, by age 5 most children realize the difference between a lie and the truth (Peterson et al., 1983). Even so, some continue to lie, most typically about misdeeds (Stouthamers-Loeber, 1987, cited in Bussey et al., 1993), which, not infrequently, they blame on imaginary playmates. However, as these misdeeds may be under the direction and control of the right hemisphere or limbic system, children may lie because their left cerebrum cannot gain verbal access to these knowledge sources and thus the motivational origins of certain aspects of their behavior (Joseph, 1982). Information reception in the left hemisphere of young children is often plagued by gaps.

MEMORY CONFABULATIONS

All memories and perceptions are influenced to varying degrees by personality, self-concept, and emotional status, including what we find objectionable, pleasurable, and so on. Moreover, our perceptions, and thus our memories, are affected by expectations, stereotypes, guilt, embarrassment, social circumstances, and so on, and persons sometimes leave out or distort various details and concentrate and even confuse events and conversations, including, for example, what was said yesterday versus today, and/or the order of events and the length of time between them.

For example, it sometimes happens in criminal trials that witnesses recall seeing a particular person and may identify him or her as a suspect or participant in a crime, when in fact they had seen this person on a different day but perhaps at the same time. In one case a gas station attendant who had been robbed and beaten by two men later positively identified one man as the culprit, when in fact he had an airtight alibi as to his whereabouts when the crime occurred (he was in court). It turned out, however, the misidentified suspect had stopped at the station the previous day, at the same time, in the company of a friend.

Similarly, in the bombing of the Federal building in Oklahoma in 1995, it was subsequently determined that "John Doe number 2" had indeed been at the same truck rental garage as the primary suspect, Timothy McVay. However, it had been the previous day.

Sometimes even two discrepant memories concerning minor details can be blended together so that a compromise in recall occurs. For example, when subjects were shown a film of an accident involving a green car and were later asked a question with the erroneous cue "blue" car embedded, although some subjects subsequently recalled the car as blue, many others recalled seeing a "blue-green" car. Hence, verbal cues can effect verbal as well as visual memory, at least when reported by the language-dominant half of the brain. However, if subjects are allowed to point to the color of the car, the original visual cues are more successfully and accurately recalled.

Related to this is interference by other verbal factors when trying to recall people or incidents. For example, consider a crime situation in which a witness is interviewed again and again, first by police, then other police, then attorneys and so on. One might expect, because of rehearsal, that memory for the incident in question would improve. But that is not the case, for if the witness is asked if the "suspect" did or wore or said such and such, and then a week later is asked similar questions by different people, there will be a ring of familiarity to it and the witness may say, yeah, he may have been wearing white socks, and I think he did look angry, when in fact this is not true.

Similar forms of contamination can occur just by thinking and remembering. As many husbands and wives have discovered following an argument, often their spouse will later recall details and other items that the other spouse is positive did not occur and were not said. This is because by thinking and rehearsing, other variables sometimes come to be considered and incorporated. "I should have said..." "I bet he was going to say..." and so on. Unfortunately, this information often comes to be associated with the original memory, which then becomes contaminated by erroneous material.

However, in this regard it must be stressed that these distortions and contamination involve minor details, or confusions regarding time or place, and do not involve the invention of complex acts that did not occur. Again, there is no evidence to suggest, in the absence of neurological dysfunction, that complex false memories can be created in this fashion—at least in adults. For erroneous details to become assimilated into memory requires that they be details and not main points. Main points or major events that did not happen are not erroneously accepted into memory, and there is no evidence to the contrary—again, except in cases of neurological dysfunction, drug intoxication, prolonged "brain washing," and so on.

Hence, minor aspects of memory can be altered if a person does not notice discrepancies during subsequent periods when talking, visualizing, or thinking about the event, or during the process of rehearsal and reconsideration. The central-most aspects of memory and experience are more resistant to change, particularly aspects that are visual, nonverbal, and emotional. Misinformation concerning major points will be identified as erroneous and will not be incorporated into memory.

MEMORY GAPS AND SUGGESTIBILITY

Although adults and children can be "brain washed," bullied, or threatened until they change their minds and presumably their memories, memory is most likely to be significantly and profoundly altered when it is incomplete and riddled with gaps (Joseph, 1982, 1986a, 1988a; Joseph et al., 1984; Talland, 1961). When memory is fragmented and a person does not realize that what is being recalled is incomplete, he or she may begin to add new details that are tangentially or in some manner related to the fragments received. This in turn may result in the creation of a "new memory." However, more often than not this represents confabulation (Joseph 1982, 1986a, 1988a). Nevertheless, if the patient, child or adult, is aware of the gap in memory, he or she will usually state that he or she does not remember and will resist suggestions to the contrary.

Confabulation, therefore, is often a secondary consequence of an attempt of the language axis of the left hemisphere to fill gaps in the information available with plausible details that may be in some way related to the fragments received. Inaccurate information is accepted because there is no disconfirming feedback.

However, be it confabulation due to memory failure and incomplete information, or a consequence of suggestibility, it is the verbal memory system that may come to be altered (Ceci et al., 1987), whereas the visual, nonverbal (Zaragoza, 1987) and emotional memory stores (see Chapter 6) are more resistant to modification and external suggestive influences.

However, once this visual and nonverbal information is transferred from the language-dominant half of the brain, it too can come to be riddled with gaps and subject to confabulatory embellishment (Joseph, 1986a, 1982b, 1988a, b, 1993; Joseph et al., 1984). And yet, in these instances, it is the verbal report that is inaccurate, and not the nonverbal (right cerebral/limbic) memory it is based on. This also appears to be the case in young children.

However, in young children, disconnections also characterize left cerebral functioning, such that even left hemisphere knowledge sources may be riddled with gaps and filled with confabulatory embellishments (e.g., Joseph et al., 1984). Apparently this is why suggestion can sometimes significantly influence and change verbal memory, but only in very young children under certain conditions.

In summary it must therefore be stressed that adults as well as children resist suggestions to change or conjure up new complex memories—unless subjected to drugs or horrendous emotional conditions (see Chapter 17). In this regard there is also absolutely no convincing evidence that adults who suddenly recall a forgotten and repressed memory are actually the victims of therapists who coerced and/or planted and suggested these complex and horrific "memories."

Again, although persons may have created false verbal memories as children and/or repressed any emotional traumas they were subjected to, nonverbal, visual, and emotional (right cerebral) memories are resistant to distortion via verbal suggestion. Although verbal memories may be distorted, nonverbal and visual memories are more stable, which thus increases the likelihood that if and when they are recalled during adulthood, they are by and large accurate descriptions of actual childhood trauma. Of course, memory, be it verbal, emotional, or visual, is subject to decay as well as erasure due to brain damage.

REPRESSION VERSUS CHILDHOOD AMNESIA

Memory loss and amnesia for early childhood are a normal consequence of the differential experiential capabilities of the right and left half of the brain, corpus callosum immaturity, lateralized memory storage, and the increasing reliance on words and vocabulary to retrieve and describe one's experience. These early memories have not necessarily been extinguished, but may be triggered and recalled, or, in some instances, actively suppressed by the frontal lobes, thus preventing left hemisphere, language axis acquisition.

Hippocampal Amnesia. As detailed in chapter 16, trauma induced repression and memory loss is associated with amygdala, right cerebral hyperactivation, and hippocampal injury or shutdown, thereby producing a hippocampal amnesia. The human brain is chemically complex and exceedingly fragile and may be injured by stress, fear, sexual abuse and emotional trauma, even when the brain has not been physically impacted. When traumatized or threatened the brain mobilizes its defenses and prepares for flight or fight. Under conditions of repetitive emotional trauma or sexual or physical abuse, neurotransmitters such as norepinephrine and serotonin may be depleted, whereas enkephalins, cortisol and glucocorticoids continue to be released which damages the hippocampus. Hippocampal neurons atrophy, shrink, lose connections, and may die, resulting in memory loss, amnesia, and repression, as well as related emotional abnormalities. However, the felt aspects of the emotional trauma are stored by the amygdala, sometimes dooming those afflicted with PTSD, and long term emotional problems. The victim may feel traumatized but not know why.

Sex, Satan, Snow White, Alien Abductions

Those who are severely traumatized or subject to excessive fear may experience sensory distortions and hallucinations that are exceedingly frightening or even demonic. Victims may even hallucinate Satanic and demonic entities, and, in some instances, these sensory distortions and hallucinations may be committed to memory. Later, the victim may recall the "hallucination" and believe they were abused in Satanic rituals, or abducted and sexually assaulted by demonic aliens.

Consider the Disney version of "Snow White." When the woodcutter, who'd been ordered to cut out her heart, urged Snow White to flee for her life, she flew into the darkening forest, panicked and near hysteria. And, as she ran and stumbled darting in tears here and there, the trees became demonic, growing eyes and mouths, and gnarled hands and arms which stretched threateningly toward her. Overcome with terror, she collapsed to the forest floor, sobbing uncontrollably.

Now, perchance, had Snow White later recalled this frightening misadventure, she may well explain to skeptical listeners that demons had emerged from the forest, and threatened to snatch her away. And, she may truly believe this happened, for it is what she experienced, and what she now remembers. Of course, the skeptics would have us believe that some evil therapist implanted this false memory. Unfortunately, a poorly trained therapist may mistake the remembered hallucination as a real Satanic event, or an alien abduction, and may then provide treatment for an otherworldly condition.

Those reporting alien abductions, often have a history of sexual molestation or severe emotional trauma, or temporal lobe epilepsy (Mack, 1994). Consider, fear and sex triggers the limbic system. Neurosurgical stimulation or seizures involving the amygdala and hippocampus can induce bizarre hallucinations, including feelings of being lifted up, placed in a room of vast proportions, receiving special knowledge, or being confronted by demonic, spiritual, or Satanic-alien-like beings.

False Memory Syndrome

The belief in thought and memory insertion is a common paranoid delusion that's been reported for centuries. According to this delusion, evil psychotherapists, the Devil, witches, the CIA, FBI, KGB, or a next door neighbor, are insidiously implanting complex memories into the minds of unsuspecting adults who accept the "false memory" as real. According to this paranoid view, everything becomes "false." Do you remember what you had for dinner last night? No you don't. Someone implanted a false memory.

Elizabeth Loftus has claimed to have convinced a female coed that she became lost while shopping with a relative as a child. A common childhood memory! Of undergraduate males, 58%, and of females, 38% report becoming lost while shopping with a relative (N=100). Moreover, fears of becoming lost, and admonitions to stay close or else, are common childhood experiences, whereas being raped or repeatedly sexually molested is not. Nor is there evidence that demonic or alien-abduction memories have been "implanted." Rather, these memories are likely based on terror or seizure induced hallucinations.

CONCLUSION

Those who have been raped, physically assaulted, molested, or repetitively placed in traumatic and life threatening situations may suffer severe and significant verbal memory loss and amnesia. Accurate recall of repetitive and severe trauma-related memories occurs infrequently. I've examined dozens of severely traumatized patients with no memory of their childhood or early teens.

Repressed memories are usually repressed for a reason, and should probably remain so, unless a skilled therapist is available to help navigate the victim through the remembered trauma. Many deteriorate when traumas are recalled.

Unfortunately, a repressed traumatic memory can still influence brain functioning. Those molested or sexually abused, or severely traumatized as adults, often demonstrate disturbed behavior and emotional patterns their entire lives. Some are plagued by hidden memories, distressing emotions, or intrusive images and disturbing dreams, the residue of forgotten horrors and "well kept secrets" of the limbic system and the right half of the brain.

16

The Neuroanatomy and Neurophysiology of Dissociation, Repression, and Traumatic Stress

The Amygdala, Hippocampus, and Disconnection

THE CONSCIOUS AND UNCONSCIOUS MIND

It is now well established that in most of the population the left cerebral hemisphere subserves the capacity to read, write, name, spell, as well as speak and comprehend the denotative and grammatical aspects of language (reviewed in Chapter 4). Moreover, it is also well known that the comprehension and expression of language and related temporal-sequential capacities are associated with Broca's expressive and Wernicke's receptive speech areas in the frontal and temporal lobes and the inferior parietal lobule. These neural regions are linked together by the thalamus and via the arcuate and longitudinal fasciculus (which also contributes fibers to the amygdala and cingulate respectively), thereby forming a "language axis" (Joseph, 1982, 1993).

Language is a tool of consciousness. The denotative and grammatical features of language enable an individual to verbally organize and articulate his or her thoughts, plans for the future, personal identity, and sense of self, as well as form abstractions and categorize personal experience so that it may be communicated to others in name and in words. Via language one may form an abstraction of one's sense of Self so that it may be scrutinized from multiple verbal perspectives, as well as modified or even disguised in accordance with an idealized self-concept (Joseph 1992b, 1993).

This does not mean that a personal identity or the ability to reflect upon the Self depends on (denotative, grammatical) language, but rather that these aspects of verbal expression are a dominant mode employed by humans in this regard. That is, humans often think in words, and it is through the grammatical and denotative aspects of language and verbal thought that Homo sapiens are able to manipulate the world, make predictions about the future, and symbolize aspects of the past in verbal memory and in written form. Via these multiple linguistic modalities humans are able to reflect upon and analyze and describe the world and the Self in a multimodal, multidimensional fashion. Thus language is not only a tool of consciousness but in some respects is indistinguishable from human consciousness, i.e., linguistic consciousness.

Consciousness, and grammatical/denotative language and linguistic thought are not only intimately interrelated, but are supported and maintained by the left half of the brain (Joseph, 1982, 1986a, 1988a, b). Because these aspects of nonemotional language depend on the functional integrity of the left cerebral hemisphere, the language-dependent aspects of consciousness are also associated with the left hemisphere, a position advocated by a number of independent neuroscientists (e.g., Albert et al., 1976; Bogen, 1969; Dixon, 1981; Eccles, 1981; Galin, 1974; Hoppe, 1977; Miller, 1991; Popper and Eccles, 1977; Rossi and Rosadini, 1967; Schwartz, 1967; Serafetinides et al.,

1967; Terzian, 1964). It is the left hemisphere that controls the ability to talk and think in words and that analyzes spoken and written (nonemotional) language. The ability to produce linguistic knowledge and thought, to engage in mathematical and analytical reasoning, or to process and express information in a temporal-sequential, grammatical, and rhythmical fashion, are associated with the functional integrity of the left half of the brain in most of the population. It is the left hemisphere that we associate with linguistic consciousness.

By contrast, the right hemisphere (and limbic system) is the domain of a highly evolved nonlinguistic (and thus unconscious) intelligence that is predominantly social-emotional, visual-spatial, and somesthetic, and which employs emotional and melodic sounds for expression (Joseph 1982, 1986a, 1988a, b, 1992b). Indeed, in the right hemisphere we deal with a second form of awareness that accompanies in parallel what appears to be the "dominant" temporal-sequential, language-dependent stream of consciousness in the left cerebrum (reviewed in Chapter 3).

As has been repeatedly demonstrated by Sperry, Bogen, Levy, and Gazzaniga and their colleagues, the isolated right and left cerebral hemispheres are each capable of self-awareness; can plan for the future; can have goals and aspirations, likes and dislikes, and social and political awareness; and can independently and purposefully initiate behavior, guide responses, choices, and emotional reactions, as well as recall and act on certain desires, impulses, situations, or environmental events—sometimes without the aid, knowledge, or active (reflective) participation of the other half of the brain (see Chapters 3 and 15).

Nevertheless, from the perspective of the language-dependent conscious mind, in some respects the mental system maintained by the right hemisphere could also be likened to the unconscious (Galin, 1974; Hoppe, 1977; Joseph, 1982, 1988a, b, 1992b; Miller, 1991), particularly in regard to social-emotional nuances, nonverbal and emotional memories, feelings, and related impulses. The mental system associated with the right hemisphere does not rely on written language or the grammatically descriptive, denotative, and temporal-sequential aspects of self-expression that is characteristic of the left cerebrum. Hence, mental activities that are associated with the right cerebrum sometimes cannot be understood or comprehended by the left, which in turn may be puzzled or even completely "unconscious" and unaware as to the origin, meaning, and even presence of these nonlinguistic knowledge sources. Each hemisphere is specialized to perform certain tasks, and just as a human may be unaware of certain sounds or odors that might be perceived, for example, by a dog, the same applies to some of the capacities associated with the other half of the brain, i.e., one hemisphere may be completely unaware of what is occurring in the other.

DISCONNECTION SYNDROMES

As detailed in Chapters 3, 4, and 15, although the right and left hemispheres and their respective "mental systems" are interconnected via the corpus callosum and limbic system, some types and forms of information are processed in a wholly different fashion and/or cannot be recognized by one versus the other half of the brain. That is, although there is considerable overlap in functional representation and although many functions are differentially processed in parallel, there are some types of stimuli (e.g., arithmetical) that may be incompletely transferred (or received) and that cannot be analyzed or correctly comprehended by the other (e.g., Berlucchi and Rizzolatti, 1968; Joseph, 1982, 1988a, 1990a, 1992a; Marzi, 1986; Merriam and Gardner, 1987; Miller, 1990, 1991; Myers, 1959, 1962; Rizzolatti et al., 1971). In consequence, the left (or right) hemisphere may be ignorant of, or may even misinterpret, what has been perceived or processed in the other half of the brain, as it does not have complete access to this data.

Under some conditions (e.g., stroke, tumor) the language axis (i.e., Broca's and Wernicke's areas, the inferior parietal lobule) may come to be disconnected anatomically (Geschwind, 1965; Joseph, 1986a, 1988a, b 1990a, 1993) and functionally (Galin, 1974; Hoppe, 1977; Joseph, 1982, 1988a, 1992a) from important sources of input not only within the right, but within the left hemisphere as well. From the perspective of the language axis, and thus the language-dependent conscious mind, those functions associated with the disconnected tissue may continue to be processed and even acted on "unconsciously."

Blind Sight

Consider, for example, the phenomenon of "blind sight." That is, in some cases certain aspects of visual perception may be retained following destruction of the pathways and neocortical tissue subserving visual analysis (see Chapter 13). Although these patients "verbally" claim to be blind, and are essentially cortically blind (because of destruction of visual neocortex), they may avoid obstacles when they walk and/or correctly retrieve objects that they desire. Patients with blind sight are capable of detecting movement and different gratings (but not different patterns or shapes— i.e., "X" versus a "triangle"), and they can correctly reach for various objects although they claim no conscious awareness of the visual stimulus (see Weiskrantz, 1986). The language axis has been disconnected and claims that it cannot see as it is no longer the recipient of these perceptions because of destruction of visual cortex.

Specifically, those who are "cortically" blind but demonstrate blind sight are able to accomplish these acts because thalamic, midbrain, and cortical tissues that are involved in visual functioning and that have not been damaged, continue to function normally, although they are completely dissociated from the "dominant" stream of verbal consciousness. That is, these visual signals are processed and responded to by "subcortical" tissues, but are not transferred to the neocortex or the language axis. Hence, the language axis, in failing to receive visual signals, therefore claims to have no knowledge of the visual world. Nevertheless, visual processing continues outside of linguistic conscious-awareness.

Disconnection and Dissociation

In some cases, although a lesion may disconnect and thus isolate the language axis from important sources of input, the "broken off" (lesioned) fragment of the psyche (or neocortex) may continue to act in a semi-independent manner. That is, similar (in a very limited respect) to what can occur following "split-brain" surgery, sometimes complex and purposeful behaviors can be initiated and acted out in the absence of left hemisphere and linguistic conscious participation. However, in some cases, the language axis may be disconnected from neocortical tissue also located in the left hemisphere, which in turn may act independently and outside conscious-awareness.

For example, if an individual suffers a discrete lesion that disconnects the language axis (i.e., Wernicke's area and the angular gyrus) from the visual cortex (such as by destroying the axonal connections that link these tissues), although the patient can see (because the visual cortex is spared) he or she may be unable to verbally recognize or name objects that are shown to them. Thus, such patients may call a glass of water a "clock" or a "folder" or they may call a comb a "harmonica" or "toothbrush," and so on, and they may fail to realize that an error has been made (Freud, 1891; Geschwind, 1965). However, although unable to name or identify a glass of water, or explain its function or utility, once they became thirsty they might pick up the glass and drink from it (Geschwind, 1965). Nevertheless they may remain unable to name the item even during the course of utilizing it (depending on the extent of disconnection).

If instead the lesion is disconnected the language axis from the left superior parietal lobule and the primary receiving areas

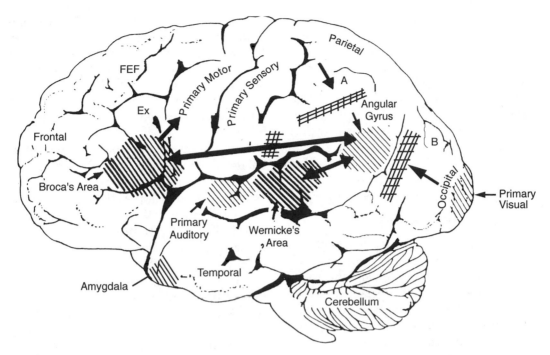

Figure 16.1 Schematic diagram of lesion sites that may induce disconnection syndromes and dissociative phenomenon. A, Disconnection involving somesthetic area and language axis. B, Disconnection involving primary and secondary visual areas and language axis. Complete destruction of these visual areas will result in neocortical blindness. However, patients may demonstrate blind sight due to preservation of thalamic and midbrain visual areas. C, A lesion of the arcuate fasciculus disconnects Broca's area from the posterior language areas, thereby producing conduction aphasia.

for somesthesis, although able to correctly name what they see, these patients would be unable to name whatever object might be secretly placed in their right hand, e.g., a comb. In fact, they may not consciously (verbally) realize that something was placed in their hand. However, they may be able to demonstrate its use by combing their hair.

These unusual disturbances are not due to a perceptual or "word finding" problem or disability (Freud, 1891; Geschwind, 1965). Rather the language axis and verbal-consciousness have been disconnected and thus dissociated from the area of perception, a result of a lesion that destroys the axonal interconnections that normally link these neocortical tissues. Hence, if provided the correct word or descriptive phrase—i.e., "Is it a comb or a toothbrush?"—such patients may still chose the wrong word or confabulate (see Chapters 3, 11). They are unable to match the word with the visual or somesthetic image because of disconnection.

However, as demonstrated above, the dissociated area of perception can sometimes continue to act in isolation and can maintain a fragmentary nonverbal awareness that "it" may act on independently (Geschwind, 1965). That is, the brain area that has been disconnected from the language axis may respond "normally." Hence, if asked to draw the object, or to show how it is used, they can do so, even though consciously they cannot verbally describe or recognize the object.

Thus, under these conditions, that aspect of the mind and personality that drinks from the glass or demonstrates the use of a comb represents a broken-off and disconnected fragment of the "mind" that is no longer attached to the dominant verbal stream of consciousness. Although a portion of the patient's mind and ego has essentially split off, it can act independently and in a purposeful and intelligent manner. As stated by Geschwind (1965), under conditions such as these "we are dealing with more than one 'patient.' The 'patient' that speaks to you is not the 'patient' who is perceiving—they are, in fact separate."

DISCONNECTION, DISSOCIATION, AND RIGHT CEREBRAL INJURIES

As discussed and detailed in Chapter 3, the right hemisphere is dominant for almost all aspects of emotion and the establishment of emotional and personal memories, including memories associated with the body image. Hence, in some instances following right cerebral injuries, the left hemisphere and language-dependent aspects of consciousness may be denied access to these memories, including memories regarding the left half of the body. In consequence, the left hemisphere may be unable to recognize, remember, or report or describe information that is stored exclusively within the right hemisphere, including knowledge regarding the left half of the body, the existence of which the left half of the brain may actively deny (Joseph, 1986a, 1988a). This is known as neglect and denial (see Chapter 3) and if presented with his or her left arm or leg, the patient (i.e., the left hemisphere) may claim it belongs to a relative or even the doctor.

However, depending on the severity of the lesion, the left half of the body (and thus the right hemisphere) may act independently. When this occurs, the patient (i.e., the left hemisphere) may produce delusional notions regarding the left half of the body, which they may believe belongs to someone else.

As reviewed in Chapters 3 and 11, "one such patient engaged in peculiar erotic behavior with his 'absent' left limbs which he believed belonged to a woman. He claimed that a woman was lying on his left side; he would utter witty remarks about this and

sometimes caress his left arm. One woman who was confronted with her paralyzed left arm said it belonged to another person whom she thought was in bed with her, and whose arm had slipped into her sleeve. Another declared (speaking of her left limbs), 'That's an old man. He stays in bed all the time.'"

In some instances, when the left hemisphere is denied input from the right (such as occurs with a massive right cerebral lesion), it may develop a dislike for the left limbs, try to throw them away, become agitated when they are referred to, entertain persecutory delusions regarding them, and even complain of strange people sleeping in their beds because of "bumping into" the left limbs during the night. One patient complained that this person on her left tried to push her out of the bed and threatened that she would sue the hospital if it happened again. Another patient, after bumping into her left arm and leg all night, complained about "a hospital that makes people sleep together." She expressed not only anger but concern lest her husband should find out; she was convinced it was a man in her bed.

Moreover, the left hemisphere of some split-brain patients may verbally claim to hate the left half of the body and attribute to it disagreeable personality traits or claim that it has engaged in behavior that the speaking half of the brain finds unpleasant, strange, objectionable, embarrassing, or contrary to its wishes (Joseph, 1988a, b), as if it is possessed by a second, independent personality. Sometimes this second personality engages in destructive acts. For example, Akelaitis (1945:597) describes a female split-brain patient who became quite depressed and irritated about the actions of the left half of her body. On several occasions it tried to slam a drawer on her right hand, and on a number of instances the left hand (right hemisphere) attempted to take her clothes off, even though that is not what she (i.e., the left hemisphere) desired to do. Geschwind (1981) reports a callosal patient who complained that his left hand on several occasions suddenly struck his wife—much to the embarrassment of his left (speaking) hemisphere. In another case, a patient's left hand attempted to choke the patient himself and had to be wrestled away.

A split-brain patient described by Dimond (1980:434) reported that once when she had overslept, her left hand began slapping her across the face until she (her left hemisphere) woke up. This same patient, in fact, complained of several instances where her left hand had acted violently. Another split-brain patient's left hand would not allow the patient to smoke, and would pluck lit cigarettes from his mouth or right hand and put them out. Another split-brain patient ("2-C"), complained of instances in which his left hand would perform socially inappropriate actions (e.g., attempting to strike a relative) and would act in a manner completely opposite to what he expressively intended, such as turn off the TV or change channels, even though he (or rather his left hemisphere) was enjoying the program (Joseph, 1988b). Once, after he had retrieved something from the refrigerator with his right hand, his left took the food, put it back on the shelf and retrieved a completely different item "even though that's not what I wanted to eat!" On at least one occasion, his left leg refused to continue "going for a walk" and would only allow him to return home. In the laboratory, he often became quite angry with his left hand, he struck it and expressed hate for it. Several times, his left and right hands were observed to engage in actual physical struggles. For example, on one task both hands were stimulated simultaneously (while out of view) with either the same or two different textured materials (e.g., sandpaper to the right, velvet to the left), and the patient was required to point (with the left and right hands simultaneously) to an array of fabrics that were hanging in view on the left and right of the testing apparatus. However, at no time was he informed that two different fabrics were being applied. After stimulation the patient would pull his hands out from inside the appara-

tus and point with the left to the fabric felt by the left and with the right to the fabric felt by the right. That is, the left hand (right hemisphere) did not mirror the behavior of the right hand, but acted with correct and purposeful intent. Surprisingly, although his left hand (right hemisphere) responded correctly, his left hemisphere vocalized: "That's wrong!" as it apparently believed that the fabric rubbed on the right hand was also rubbed on the left hand. Repeatedly he reached over with his right hand and tried to force his left extremity to point to the fabric experienced by the right. His left hand (right hemisphere) refused and physically resisted being forced to point at anything different. In one instance a physical struggle ensued, the right grappling with the left.

Goldberg (1987:290) described a 53-year-old right-handed woman, "B.D." who, while at work, was overcome with a "feeling of nausea and began to notice that her left leg felt 'as if it did not belong to me.' This feeling of being dissociated from her body spread to the rest of her left side. At home, her symptoms began to worsen." Subsequent CT scan and MRI indicated an infarct to the medial frontal lobe and damage to the body of the corpus callosum. "While sleeping one night a few days after admission to the hospital, she woke up suddenly and noticed her own left hand scratching her shoulder.... She would frequently look down to find the hand doing something that she had no idea it had been doing. She found this very disturbing and was convinced that she was going crazy."

Another patient described by Goldberg (1987:295) reported "an incident in which she was lying in bed with the window open when suddenly the impaired limb reached down and pulled up the covers, functioning entirely in the alien mode. She concluded that 'it' must have felt cold and needed to cover her up. She felt that frequently the 'alien' did things that were generally 'good for her.'"

In these instances it is clear that the language-dependent aspects of consciousness are not only completely disconnected from the mental activity occurring within the right, but that the left hemisphere views right cerebral mental and physical actions as completely alien and thus dissociated from its own realm of personal identity. From the perspective of the disconnected left hemisphere, what occurs within the right is completely outside of its personal consciousness.

Hence, when persons who have undergone complete corpus callosotomy are tactually stimulated on the left side of the body, their left hemispheres may fail to realize they have been touched. They are also unable to name objects placed in the left hand, and they fail to verbally report the presence of a moving or stationary stimulus in the left half of their visual fields (Bogen, 1979; Gazzaniga, 1970; Gazzaniga and LeDoux, 1978; Joseph, 1988a, b; Levy, 1983; Sperry, 1982). They (i.e., their left hemisphere) cannot verbally describe odors, pictures, or auditory stimuli tachistoscopically or dichotically presented to the right cerebrum, and they are often completely mystified when the left half of their body responds or behaves in a particularly purposeful manner (such as when the right hemisphere is selectively given a command). However, by raising their left hand (which is controlled by the right half of the cerebrum), the disconnected right hemisphere is able to indicate when the patient is tactually or visually stimulated on the left side. When tachistoscopically presented with words to the left of visual midline, although unable to name them, when offered multiple visual choices in full field the right hemisphere is usually able to point correctly with the left hand to the word viewed. Thus although disconnected from "verbal" consciousness, right cerebral mental activity continues unabated.

EMOTIONAL TRAUMA, DISSOCIATION, AND REPRESSION

One need not suffer a gross physical injury to the brain in order to suffer from disconnection and dissociative disorders, for the same abnormalities may occur follow-ing severe emotional trauma (Courtois, 1988; Devinksy et al., 1989; Yates and Nasby, 1993; Zlotnick et al., 1994; see also edited volumes by Kluft, 1985; J. Singer, 1990; and Spiegel, 1993), which, however, may injure the brain.

As summed up in DSM-IV (1994:477), dissociation is due to "a disruption in the usually integrated functions of consciousness, memory, identity, or perceptions of the environment"—disturbances that are sometimes associated with temporal lobe, amygdala, and hippocampal abnormalities, which in turn may be associated with emotional trauma (see below and Chapter 14). If the trauma is sufficiently severe, these nuclei may be injured, thereby inducing loss of memory, a dissociative state, and a fragmentation of the ego.

Indeed, Janet (1927) argued that under certain traumatic conditions, the ego may fracture and split apart, such that a fragmentary (and traumatized) aspect of the ego (including associated ideas, perceptions, feelings, and memories) comes to be dissociated from the dominant stream of mental activity. Because it has been fractured, the ego loses its ability to control or bind together all aspects of the psyche. In consequence, the traumatized and broken-off aspects of the ego may also act independently as if under the control of a separate personality.

Moreover, Janet (1927) argued that these dissociated elements of the ego, under some conditions, can essentially take on a life of their own and may in fact independently develop and grow (see also Jung, 1964 regarding complexes). Nevertheless, because of the dissociation and disconnection, the dominant personality is often completely agnosic and amnesic regarding the existence of the other.

Freud (1915) and Breuer and Freud (1893) also argued that various aspects of the psyche could come to be split off from consciousness. However, in contrast to Janet, Freud believed that it was the emotion or impulse associated with an idea (or memory) that came to be disconnected and dissociated, i.e., repressed. Thus the emotion or impulse and the associated memory disappear into the unconscious.

Via this expulsion from consciousness, painful ideas, memories, impulses, and feelings could come to be repressed and dissociated from the ego. Nevertheless, according to Freud, in some instances this disconnection and dissociation are incomplete, and repressed material and related ideas can seek entry or seep into consciousness in the form of fragmented images and "reminiscences" (Breuer and Freud, 1893) and thus achieve a partial conscious realization.

Hence, rather than emphasizing the fragmentation of the ego, Freud argued for the existence of a completely different mental domain that became the source and depository for these rejected and dissociated impulses. However, in developing his ideas and theories, Freud downplayed the significance of dissociation and instead embraced and developed the concept of the "unconscious" and "repression," which essentially became the "cornerstone" of psychoanalysis (Freud, 1900, 1915a). That is, the unacceptable impulse is either rejected by the conscious mind and thus purposefully suppressed, or it is denied entry into consciousness by the unconscious mind (via the unconscious censors) and remains repressed.

Although these two clinicians essentially championed major constructs that somewhat differed from that of the other (i.e., dissociation versus repression), and although Freud argued that repression may be triggered in response to unacceptable impulses, Janet (1927) and Freud (1896, 1900, 1905, 1916, 1931, 1937) basically agreed as to the role of emotional and psychic trauma as a primary cause of repression and dissociative states. That is, emotional trauma can create a condition wherein the experience and associated memories become disconnected and dissociated from the dominant stream of verbal consciousness and thus repressed. However, although engaged in repression and/or dissociation, the patient may continue to suffer the ill effects of what is otherwise an ongoing emotional disturbance.

As will be detailed below, emotional trauma is not just a psychological event but a physiological experience. Emotional trauma can induce cerebral dysfunction, neurochemical abnormalities, neuronal degeneration, and thus neuroanatomical disconnection.

TRAUMATIC STRESS, REPRESSION, DISSOCIATION

In some instances an individual may be severely traumatized and unable to forget what happened (Charney et al., 1993) and may wish to talk about it incessantly (Terr, 1988, 1990). Others suffer memory loss to varying degrees (Courtois, 1988; Grinker and Spiegel, 1945; Joseph, 1992b; Parson, 1988; Southard, 1919; Terr, 1990). However, like those who can remember, those who may be wholly or partially amnesic may experience flashbacks and hallucinations as well as severe emotional upset. Sometimes these flashbacks are triggered by environmental stimuli (e.g., engine backfire) and may seem so vivid and real that the person believes he or she is experiencing the actual event and then act accordingly (DSM IV, 1994; Krystal, 1990).

The fact that those who suffer memory disturbances can sometimes vividly recall or at least dream about the forgotten event indicates that the traumatic memory has been "stored" but is not completely available to the language-dominant aspects of consciousness.

In addition, those with preserved as well as partial or complete loss of memory may suffer autonomic nervous system dysfunction, heightened arousal and startle reactions, and other behavioral and emotional problems as well as the entire constellation of post-traumatic symptomology. That is, post-traumatic stress disorders (PTSDs) are common among those who are traumatized regardless of degree of memory loss or retention (Charney et al., 1993; DSM IV, 1994; Krystal, 1990; van der Kolk, 1987). In this regard, the neural circuitry that is associated with the experience and reaction to the trauma has also been preserved.

Also common to those with preserved as well as partial or complete loss of memory is the possibility of experiencing a dissociative reaction. Correspondingly, persons may fully recall what occurred during their dissociative states, or they may be amnesic ("Dissociative Amnesia" DSM IV, 1994). The extent and nature of the memory loss, however, may come to include even personal identity.

FUGUES, LOSS OF PERSONAL IDENTITY, AND AMNESIA

Janet (1927) has provided several volumes and numerous examples of severe memory loss associated with dissociative phenomena (also referred to as hysterical, functional amnesia). Typically such attacks are short-lived and are triggered by emotional shocks and turmoil that simultaneously may induce convulsive movements, recurrent hallucinations, and delusions—such as might occur with a temporal lobe seizure.

Insofar as the condition involves memory loss, the patient may have no memory for the trauma that preceded or accompanied the development of the amnesia, and sometimes cannot recall events that occurred several weeks, months, or even years before-hand—retrograde amnesia (Courtois, 1988; Grinker and Spiegel, 1945; Parson, 1988; Southard, 1919; Terr, 1990). Some patients also display a continuous anterograde amnesia such that they continue to forget events as they occur, or people whom they subsequently meet and so on (Nemiah, 1979). Such conditions sometimes arise with chronic amygdaloid kindling and hippocampal deactivation or injury, which may also be induced by emotional trauma (see below).

In prolonged and profound fugue states, patients suffer a complete loss of identity and a total loss of memory, may wander for days or weeks, and may travel great distances and may even begin a new life with little or absolutely no concern for the huge gap in their past (though some sense that they are escaping from something unpleasant).

In extreme and rare instances of fugue,

they may even take on a completely different identity and occupation (Miller, 1991; Nemiah, 1979). And then, weeks or months later, they may suddenly remember their previous identify and have no recollection for all that occurred during their fugue state. Indeed, they may be extremely upset to discover that they are sleeping in a strange bed and interacting with strange people.

Consider the classic case of Ansel Bourne, an itinerant preacher who one day disappeared and could not be found despite considerable publicity and police efforts (Hodgson, 1892, cited by Nemiah, 1979). Two months later Mr. Bourne awoke one morning lying in a strange bed:

> On pulling back the curtains and looking out the window he was alarmed to find his surroundings quite unfamiliar to him and was immediately overcome with fear, knowing that he was in a place where he had no business to be. Feeling a considerable amount of distress he ventured into the hall and knocked on the door to another room which was opened by one of the boarders of the house (Mr. Earle) who greeted him by saying, "Good-morning Mr. Brown."
>
> Mr. Bourne: "My name isn't Brown. Where am I?"
>
> Mr. Earle: "Norristown.... In Pennsylvania."
>
> Mr. Bourne: "What time of the month is it?"
>
> Mr. Earle: "The 14th."
>
> Mr. Bourne: "Does time run backward here? When I left home it was the 17th."
>
> Mr. Earle thought that Mr. "Brown" was out of his mind and sent for a doctor, to whom Mr. Bourne told the following:
>
> Mr. Bourne: "These persons tell me I am in Norristown and that I have been here for six weeks, and that I have lived with them all the time. I have no recollection of ever having seen of them before this morning."

Indeed, his last recollection was from January 17 when he was on his way to visit his sister. What may have happened to him (e.g., head injury, emotional shock) so as to induce the fugue state was not discovered. Nevertheless, once he remembered his former identity and recalled old memories he then became completely amnesic for his temporary (now discarded) identity and all that transpired during the last few weeks—suggesting a state-dependent form of learning.

However, once he was placed under hypnosis he was able to recall all his memories while he had been "Mr. Brown," and these were found to correspond with the memories and accounts supplied by others.

It is noteworthy that Mr. Bourne had not only changed identities but had been running a variety store (as Mr. Brown), indicating that although personal and all emotional memories had been lost during the course of the fugue, facts, data, names, reading, writing, and other abilities associated with the left hemisphere had been maintained—similar to what often occurs following traumatic amnesia, in which a person may forget their personal identity although he or she fails to take on a new one (see Chapters 6 and 15). However, in fugue states there may be a change in identity, or merely a loss of identity and personal memories coupled with automatic travel and preservation of nonpersonal memory right (or bilateral) amygdala dysfunction is often the causative agent.

THE CEREBRAL HEMISPHERES, TEMPORAL LOBES, AND FUGUE

The above is suggestive of (but not entirely consistent with) what may occur with severe right cerebral and amygdala injury and hemispheric disconnection. As noted in Chapter 3, right temporal lobe dysfunction, for example, can result in a loss of personal memories. As per the temporal lobe, it is also noteworthy that fugue states are sometimes associated with seizure activity or related abnormalities involving the temporal lobe, amygdala, and hippocampus. As will be detailed below, abnormalities involving these structures, particularly abnormal levels of activation, may induce not just fugues, but dissociative states in which personal identity is not only lost, but a new identity may be temporarily formed.

It is also noteworthy that like the left hemisphere of persons who have suffered a loss of right cerebral input and who may demonstrate a profound unconcern for their paralysis or hospitalized condition, the preserved aspect of the personality in fugue states also shows no concern for what is missing or absent in their life, such as their personal identity. Moreover, with severe right cerebral injuries, personal and emotional memories may no longer be accessible (see Chapter 3).

According to Nemiah (1979:307), "As the fugue state begins, the patient loses all memory whatsoever for the events of his entire past life. His origins, his family, his upbringing, wife, children, friends, occupation, all dissolve into the mists of forgetfulness, and the patient assumes a new name and life without any evident awareness or concern over the internal upheaval. Driven by often half-veiled inner urgings, he wanders far from his familiar surroundings to start a fresh existence, his conscious mind a virtual tabula rasa. Basic functions such as language, general knowledge, and the skills of coping with the tasks of everyday living remain under his command."

In other words, those aspects of emotional and personal identity, including emotional feelings and knowledge regarding loved ones, are lost—functions associated with the right half of the brain and that are mediated by the amygdala (see Chapters 3 and 5). What is preserved are those aspects of speech, impersonal facts, and so on that are associated with the left half of the brain.

However, these fugue states are not a simple function of right and left cerebral disconnection, for split-brain patients do not respond in this manner, though the two halves of their body may attempt to engage in completely opposite and conflicting actions. In this regard, the temporal lobes, amygdala, and hippocampus are again implicated.

Thus in fugue states it is almost as if the specific neural circuits involved in personal identity and associated aspects of emotion have become isolated and are no longer retrievable by the left half of the brain. As detailed below, and as suggested above, this may be due to abnormal activity (or injury) involving the limbic system, the amygdala, and hippocampus in particular. If these nuclei become abnormal, both the right and left hemispheres may be unable to gain access to personal, emotional, and related memories.

In consequence, the left hemisphere, having absolutely no access to this information, cannot be alerted to the disturbances, including their loss of personal identity—much as might occur with left-sided neglect, Wernicke's aphasia, or denial of blindness, in which case patients do not know that they are disabled. Being unaware of this gap in personal identity, those in a fugue state are unaware that anything is amiss, wander about, travel long distances, and may even begin a new life without knowing that they are doing anything different, as they have lost this personal reference. Apparently, those neural networks which are established during the course of the fugue, including those concerning newly experienced personal and emotional material, also remain dissociated from those previously established—as in state-dependent learning.

DISSOCIATION AND MULTIPLE PERSONALITY
Multiple Personalities

In some respects fugue states coupled with amnesia for one's personal identity bear some resemblance to multiple personality disorders. In cases of multiple personality, the main (or presenting) personality generally has no awareness of the existence of the alternative or other personalities. However, the converse is not always the case.

According to Putnam (1986:110), "some of the alternate personalities report being able to see and hear everything that occurs when another personality is in control of the body. These alternate personalities are said to be co-conscious. Other alternate personalities report no awareness of what transpires when they are not 'out' and manifest amnesia for these periods."

However, when the secondary personality retains access to the memories and experiences of the main personality, it usually views them quite differently and displays a completely different attitude regarding them and may even view the main personality with contempt. For example, Thigpen and Cleckley (1957), in describing their famous case involving "Eve," noted that when the alternate personality "Eve Black" took over, "Eve White remained functionally in abeyance, quite oblivious of what acts the coinhabitant of her body engaged in. However, Eve Black preserved an awareness and retained access to the memories of Eve White while absent. Invisibly alert at some unmapped post of observation, she was able to follow the actions and the thoughts of her spiritually antithetical twin."

According to Thigpen and Cleckley (1957), "The devil-may-care Eve Black, knows and can report what the other does and thinks, and describes her feelings. Those feelings, however, are not Eve Black's own. She does not participate in them." In fact, Eve Black reported that most of Eve White's concerns and thoughts were "pretty corny." For example, "Eve White's genuine and natural distress about her failing marriage is regarded by the other as silly. Eve White's love and deep concern for her only child, a little girl of four...[are] trite, bothersome, and insignificant" insofar as Eve Black was concerned.

Hence, in some instances of multiple personality, the alternate personality appears to maintain a bilateral awareness as to the psychological processes occurring in the "main" personality, whereas the "main" personality is fairly limited to a unilateral and quite restricted consciousness and is completely amnesic and disconnected from the mental activities occurring in the other. However, this disconnection is not always complete, for the main personality might be plagued by strange thoughts or impulses, or even hear voices that urges it to do things it finds objectionable.

Consider the case of Mrs. Martha G., a 36-year-old woman described by Frankel (1976). Martha was very depressed, with multiple aches and pains, and had developed an "hysterical" paralysis following a car accident. In addition to her aches and listlessness she complained of hearing voices that threatened to take control over her. When asked to "let it take over," she responded by clenching her fists, rolling her head, and writhing about. She then suddenly opened her eyes and smiled and seemed quite cheerful. She announced that her name was "Harriet" and was very pleased to have gotten "rid of that other one who stays sick all the time.... She and that husband of hers, all they do is go to church and read the Bible and go to church," she scornfully added, and then indicated that she was quite different. "I like to go drinking and dancing and night-clubbing, and she likes to go to church. But I make her miserable!"

When she was asked how she could go dancing given Martha's inability to walk, Harriet replied, "I'm not sick. She's the one that stays sick. I'm an altogether different person." And with that she was able to demonstrate her ability to move about without paralysis.

Later, when Harriet was asked to let Martha return, she reluctantly complied and there followed the same writhing movements accompanied by cries, after which she relaxed, opened her eyes, and sighed: "Oh, I've been asleep on you." She had absolutely no memory for what had transpired, or for those periods in which Harriet took over. In contrast, Harriet was aware of Martha's every thought and action. Moreover, she intensely disliked Martha and enjoyed causing her unhappiness by flooding her with evil thoughts.

It is also not uncommon for several personalities to vie with one another for control over consciousness, and each may have some awareness of the existence of the others (Nemiah, 1979; Prince, 1924; see also edited volumes by Kluft 1985; Spiegel, 1993). According to Nemiah (1979), these secondary personalities can have their own experiences and their own thoughts, which may become increasingly complex over time because of their ongoing and continual contact (even when inactive) with the environment, as if living a separate existence. Nevertheless, although vying for control, not all the alternates are knowledgeable about the presence of the others.

Consider a patient described by Prince (1924) who had three separate personalities (A, B, and C). The A and C personalities were amnesic for B and for each other, whereas B was aware of the thoughts, memories, and feelings of both A and C, and claimed to have her own complete consciousness even when A and C were in control. Per Dr. Prince's request, B wrote an account of her life as a co-conscious, albeit subordinate personality ("B," 1908).

According to "B" (1908), "When I am not here as an alternate personality, my thoughts still continue although A and C are not aware of them. I think my own thoughts, which are different from theirs, and at the same time I know their thoughts and what they do."

In many respects, however, the consciousness demonstrated by "B" was different from that of A or C, in that she was much more aware of environmental sounds and visual stimuli, as well as the nonverbal, social-emotional cues and nuances expressed by others. "The rustle of a paper, the cracking of a stick in the fire, the sound of a bird chirping, the smile or frown on the face of a person whom we meet, the gleam of their teeth...into my stream of consciousness most of these perceptions are absorbed" whereas A and C were lacking in this regard, and in many respects could be considered socially-emotionally as well as environmentally agnosic. The left cerebral hemisphere of split-brain patients, as well as the left hemisphere of most normal persons, is similarly lacking in social-emotional and environmental awareness, as this is the domain of the right half of the brain (see Chapter 3).

It is noteworthy that the presenting personality in those with multiple personality disorder (MPD) tends to be socially-emotionally inept, whereas the alternate personalities are frequently quite emotionally expansive and expressive as well as much more socially-emotionally astute. Indeed, it has frequently been noted, perhaps beginning with Janet (1927), that the secondary personality is often much more psychologically healthy than the primary personality, which is obviously quite emotionally "sick."

As noted, this social-emotional awareness of the secondary personality is often coupled with heightened awareness of environmental sounds and sights and emotional nuances. As described by "B" (1908), "Dr. Prince comes into the room and C is conscious of nothing but a sense of relief at seeing him; but I perceive things like this: Dr. Prince's hand is cold; he looks tired; he is nervous today, etc. These perceptions become my thoughts. C does not take them into her consciousness at all. She does not notice the sounds in the house or out of doors, but I do. I hear the blinds rattle, I hear the maid moving about the house, I hear the telephone ring, etc. She hears none of those things.... When she is talking with anyone, I often disagree with what she says. We rarely have the same opinion ...."

It is noteworthy that among some split-brain patients, the right hemisphere not only may have different opinions from the left, but is able to speak, write, and comprehend spoken language without significant difficulty (Joseph, 1986b; Novelly and Joseph, 1983). However, among the majority of split-brain and normal persons, the right hemisphere is devoid of significant language capabilities, whereas in MPD, each personality can understand and produce spoken and written language without difficulty. Thus one cannot simply ascribe MPD to a functional split in right and left hemisphere interaction, particularly in that those with MPD may have dozens of personalities.

However, among those split-brain patients whose right and left hemispheres have acquired language capabilities, there is evidence of developmental or neurological disturbances that occurred during early childhood. Presumably this early neurological trauma somehow resulted in abnormal right and left hemisphere functional and language development.

As will be detailed below, those with MPD typically have a history of early childhood trauma. As noted, it is trauma that creates the split and dissociation of the ego into separate personalities. In this regard it should be stressed that most "normal" persons also have multiple personalities that they differentially employ or activate depending on circumstances (e.g., talking with a girlfriend versus a police officer or one's employer). However, these alternate personalities are coextensive, whereas the alternate personalities among those with MPD are often fragmented, mutually amnesic, and possibly represented by isolated neural networks that were briefly created during a horribly traumatic series of incidents.

Fear and Splitting of the Ego: Out-of-Body Experiences

It is not uncommon for persons subject to sexual abuse or exposed to terrifying situations to report perceptual and hallucinogenic experiences, dissociation, depersonalization, and the splitting off of ego functions (Courtois, 1988; Grinker and Spiegel, 1945; Noyes and Kletti, 1977; Parson, 1988; Southard, 1919; Terr, 1990). For example, Courtois (1988) reports that "many survivors describe a process of dissociation. They separate their minds from their bodies in order not to feel their bodily sensations and not to be present." Some terrified and sexually abused persons (see Courtois, 1988; Fredrickson, 1992; Grinker and Spiegel, 1945; Noyes and Kletti, 1977; Summit, 1983) claim that they felt as if they left their bodies and floated away, or were on the ceiling looking down.

Noyes and Kletti (1977), for example, described several such persons who believed that they were about to die and then experienced dissociation: "I had a clear image of myself...as though watching it on a television screen." "The next thing I knew I wasn't in the truck anymore; I was looking down from 50 to 100 feet in the air." "I had a sensation of floating. It was almost like stepping out of reality. I seemed to step out of this world."

Similarly, persons with amygdaloid/hippocampal hyperactivation sometimes also experience out-of-body and out-of-this-world experiences, as do children in terrifying and abusive situations. In some respects these phenomena are also quite similar to the initial phases of what some describe as "after death" experiences (see Chapter 8).

The Hippocampus and Dissociation

As detailed in Chapters 5 and 6, the amygdala can act on the hippocampus, which concurrently becomes hyperactivated and thus deactivated under conditions of extreme emotional arousal. One consequence is "hippocampal" amnesia.

However, in addition to memory functioning, the hippocampus (probably in conjunction with neocortical areas such as the parietal lobe) is also responsible for determining one's place and position and the position of other objects and persons such that a visual-spatial cognitive map of the body and environment is maintained in this region (Nadel, 1991; O'Keefe, 1976; Wilson and McNaughton, 1993). The hippocampus presumably enables normal persons to "visualize" or remember themselves engaged in some act as if witnessing it from a different perspective. Hence, we can "see," remember, and dream about ourselves engaged in some act as if we were both actor and audience. However, when coupled with emotional trauma, the "actor" may "split off" and dissociate, whereas the "audience" later forgets that the trauma even happened—similar to what occurs when one forgets a dream upon waking.

Thus it is presumably hippocampal (in conjunction with the temporal lobe and amygdala) hyperactivation that induces the "sense" of dissociation and thus the "experience" of leaving the body and thus "visualizing" one's body and/or surrounding objects and people as if hovering from above, or from multiple perspectives. A dissociative reaction is induced, coupled with a sense of a "splitting of the ego."

Indeed, it has been reported by a number of independent investigators that electrical or hyperactivation of the amygdala, hippocampus, and temporal lobe will cause some patients to report that their ego and sense of personal identity has split off and separated from their body, such that they may feel as if they are two different people, one watching, the other being observed (Daly, 1958; Jackson and Stewart, 1899; Penfield, 1952; Penfield and Perot, 1963; Williams, 1956). As described by Penfield (1952), during electrical stimulation of this area, "it was as though the patient were attending a familiar play and was both the actor and audience."

Sex Abuse, Dissociation, and Multiple Personalities

Unlike what occurs in the otherwise safe environment of a hospital, under conditions of extreme terror or repetitive instances of physical or sexual abuse, this splitting of the ego may become more or less complete such that the dissociated aspect of the personality may continue to exist (and act) in isolation. That is, when coupled with extreme and repetitive physical or sexual abuse, a person may not just dissociate, but form an alternate, or numerous alternative, personalities that contain memories of the abuse and that consist of isolated neural networks associated with personal reactions to the abuse.

Indeed, it has been repeatedly reported that persons identified as demonstrating "multiple personalities" have histories of horrible emotional trauma and repeated instances of intense fear, sexual abuse, and even torture (Bliss, 1980; Coons and Milstein, 1986; Coons et al., 1988; Devinksy et al., 1989; Kluft, 1984; Mathew et al., 1985; Putnam, 1985; Putnam et al., 1986; Rosenbaum and Weaver, 1980; Schreiber, 1974; Wilbur, 1984). In most cases, the alternate personalities were created during early childhood (Bliss, 1980; Devinksy et al., 1989; Greaves, 1980; Schreiber, 1974), a time in which the amygdala, hippocampus, and temporal lobe may be most susceptible to promoting these disturbances because of immaturity.

However, in instances of repeated and severe sexual and physical abuse there may be repeated instances of dissociation. This in turn can give rise to the creation of numerous disconnected memories of (and neural networks associated with) each traumatic event and one's observations and reactions. In some instances, these neural constellations may constitute a fragmentary "personality," each fragment of which may be created (if even only for a few minutes) during periods of hyperactivation of the amygdala-hippocampus-temporal lobe, a consequence of the splitting off and detachment of the ego from the body.

Multiple instances of abuse coupled with multiple instances of dissociation would account for reports describing some individuals with dozens or more personalities—each probably created and then disconnected following a severe abusive episode (much like a child may create an "imaginary friend" when lonely). That is, each "personality" may be associated with the creation of a specific neural network. Although disconnected from the language axis and associated conscious mind, these isolated neural networks can be reactivated and may act independently if cues associated with the original trauma are reexperienced. When stressed or presented with a similar abusive situation, the dissociated "personality" (or imaginary friend) may be reactivated or created.

Consider for example "Kathy," a 29-year-old white married female with several personalities (Salama, 1980). She was first molested by her father at age 4 and later felt guilty, became tearful, and then engaged in denial, attributing what had happened to a person other than herself, whom she called "Pat." For the next 5 years it was "Pat" and not "Kathy" who engaged in oral sex with her father. However, when Kathy was 9 her mother began to molest her and she again engaged in denial and attributed these acts to a new person named "Vera." Vera engaged in sexual acts with her mother for the next 5 years. At age 14 she was raped by her father's best friend and developed a complete dissociative reaction coupled with trance-like symptoms.

Most abusers, however, do not wish to molest an unresponsive body, but a personality that is either excited or tormented and terrified by what is occurring. According to Courtois (1988), it is the abuser's desire "to achieve total control over the victim and her responses and for the victim to become a willing participant and to enjoy the abuse. To achieve this end, many incest offenders take great pains to sexually stimulate their victims to arousal." However, they may also pinch, cut, and hurt them to obtain the desired reaction.

In this regard, although these particular children may have "split off" from their body, they are nevertheless forced to respond, as a personality, to the abuse (T. Jobe, personal communication), in which case the "personality" may act in accordance with the manner in which it was created. Hence, some alternative personalities are highly sexual, or homosexual, or angry and self-destructive, mirroring either the role they were forced to play or the personality and behavior of the perpetrator.

ABNORMAL NETWORKS AND BRAIN INJURY

As has been repeatedly demonstrated, adverse early rearing and environmental influences can exert profound influences on the structure and function of the brain. Similarly, repeated instances of emotional stress and thus amygdala and hippocampal hyperactivation may also induce structure changes, including neuronal death (Charney et al., 1993; Uno et al., 1989). With severe and repeated instances of trauma, the very fabric and structural and functional integrity of the brain can be altered, which in turn provides a foundation that fosters and promotes repeated dissociative episodes. This suggests that individuals with multiple personalities may have sustained damage or developed abnormalities in the temporal lobes at a time period in which these nuclei were most vulnerable to plastic changes including injury, i.e., childhood.

In fact, not only are these disturbances rooted in childhood trauma, but in many cases of multiple personality dissociative disorder, EEG or blood flow abnormalities involving the temporal lobe have been demonstrated (Benson et al., 1986; Drake, 1986; Fichtner et al., 1990; Mathew et al., 1985; Mesulam, 1981; Schnek and Bear, 1981). Moreover, with increases in temporal lobe activity, and following a major seizure, affected individuals may shift from one personality to another (Benson et al., 1986; Shenk and Bear, 1981). Similarly, it has been reported that heightened emotional distress (which would activate the amygdala—see Chapters 5 and 6) may also precede the appearance of alternate personalities (Bliss, 1980; Greaves, 1980; Putnam, 1985).

Hypnosis and Multiple Personality Disorder

It has been reported that individuals with MPD tend to be easily hypnotized (Nemiah, 1979; Putnam, 1986). Similarly, some children also tend to slip into hypnoid or trance states when being assaulted (Courtois, 1988; Putnam, 1986), and are reported to be highly hypnotizable as adults (Hilgard, 1974; Nash and Lynn, 1985; Putnam, 1986). In fact, even "normal" non-abused children tend to be susceptible to hypnosis (London and Cooper, 1969). Possibly, because of temporal lobe immaturity and because some children are easily hypnotized and can in fact induce self-hypnosis, when repeatedly abused their risk for developing MPD is increased. That is, temporal lobe immaturity and the capacity to easily enter into trance states appear to be conducive to the development of dissociation and the creation of alternative personalities when children are repeatedly abused (Decker, 1986; Prince, 1890; Putnam, 1986).

It has also been reported that adults who are easily hypnotized also demonstrate a propensity for developing MPD. According to Prince (1890), if patients are repeatedly hypnotized under a variety of conditions and situations, a second personality can develop and become quite extensive. Some individuals who are highly hypnotizable can demonstrate MPD characteristics including amnesia for events that occurred during the trance state (see Putnam, 1986). Moreover, alternate personalities created under hypnosis may become more complex (e.g., by adding neurons to the neural network supporting that personality), and can be instructed to interfere with the mental functioning of the original personality, and may in fact be able to prevent it from hearing, seeing, or remembering something even after the session has been completed (Hilgard, 1977; Orne, 1959), i.e., via post-hypnotic suggestions.

Not surprisingly, individuals with MPD are highly susceptible to developing new personalities while under hypnosis (Decker, 1986). According to Decker (1986:43), "Under hypnosis a third personality could appear. The investigator himself, consciously or unconsciously, could prompt the multiplication and development of personalities. If an unconscious personality was given a name, it showed itself more clearly."

Given that some adults who are hypnotizable are reportedly susceptible to developing some form of alternative personality, it is thus likely that children would be even more so. Indeed, children are neurologically immature and more susceptible than adults not only to the effects of hypnosis but to being adversely effected by repeated traumas. Children are also far more likely to develop imaginary friends as a normal aspect of development ("friends" that can become alternate personalities).

In addition, individuals who were repeatedly physically and sexually abused as children tend to be more fantasy prone than those who were not (Hilgard, 1974; Rhue et al., 1991), which in turn may explain why they tend to be more prone to the effects of hypnotism as adults (Hilgard, 1974; Nash and Lynn, 1985; Putnam, 1986). That is, those who tend to live in a world of fantasy are not necessarily more creative, but tend to escape or to drift into dreamy, semi-hallucinatory states—which is associated with the development of dissociative reactions and depersonalization as well as with decreased serotonin (5-HT) and abnormalities within the temporal lobe and amygdala.

As detailed below, these neurochemical and neuroanatomical alterations are frequent complications of repeated traumatic and extremely emotionally arousing conditions in both adults and children. Hence, traumatic conditions, particularly when experienced early in life, may predispose an individual to develop emotional and personality disturbances, including psychotic reactions and MPD, or a propensity to be easily hypnotized—which in turn may be associated with some form of temporal lobe abnormality.

Predisposing Factors: Sex, Age, Previous Trauma

In addition to a previous trauma, head injury, or temporal lobe abnormalities, sex and age may also differentially contribute or predispose certain individuals to develop fugue states and dissociative disorders. For example, females are far more likely to suffer dissociative disorders than men (Coons et al., 1988), a function perhaps of sex differences in the limbic system (see Chapter 5). Younger children are also more likely than those who are older to suffer from these abnormalities (DSM IV, 1994), which may be due to temporal lobe and hippocampal immaturity. Indeed, the most severe forms of dissociative disorder are often directly linked to childhood trauma (DSM IV, 1994).

Freud also emphasized childhood trauma, as did Janet. Moreover, although Freud stressed the importance of repression in preventing the recollection or conscious realization of dangerous, painful, embarrassing, and unacceptable impulses,: often of a sexual nature, he and Janet were in agreement as to the importance of sexual trauma in the development of these disorders.

SEX, FEAR, AND AMYGDALA HYPERACTIVATION

According to Freud (1896, 1900, 1905, 1915, 1916, 1931, 1937), repression acts to purposefully prevent conscious realization of unpleasant, fearful, painful, and embarrassing impulses and emotionally traumatic sexual experiences and memories, including sexual molestation during childhood. Janet (1927) made similar arguments regarding fugue states and dissociation.

As is well known, the amygdala and hypothalamus, and to a lesser extent the orbital frontal lobes, are highly involved in the regulation, expression, control, and mediation of sexual feelings and behavior (Freemon and Nevis, 1969; Gloor, 1960, 1986; Halgren, 1992; Joseph, 1990a,c,e, 1992a, 1994; Lisk, 1971; MacLean, 1973, 1990; Remillard et al., 1983; Robinson and Mishkin, 1968; Schiff et al., 1982; Spencer et al., 1983; Ursin and Kaada, 1960). Sexual feelings and related activity and behavior are often evoked by amygdala stimulation and temporal lobe seizures (Gloor, 1986; Halgren, 1992; Jacombe et al., 1980; Remillard, et al., 1983; Robinson and Mishkin, 1968; Shealy and Peele, 1957).

For example, electrical activation of the amygdala can induce sex-related behavior, including penile erections, ovulation, uterine contraction, and lactogenic responses, and produce vaginal-vulva secretions in conjunction with orgasm and sexual feelings in the genital areas. In some cases patients may experience memory flashbacks involving sexual intercourse (Gloor, 1986, 1992; see also Penfield and Perot, 1963).

Conversely, destruction of the amygdala can disrupt sexual behavior or produce abnormal sexual activity, whereas sexual activity, as well as emotional stress, can induce amygdala activity (Gloor, 1960, 1986; Halgren, 1992; LeDoux, 1992; Ursin and Kaada, 1960). Conversely, sexual stimuli can trigger amygdala activation.

In addition, electrical stimulation of the amygdala can produce sustained feelings of fear and terror (Gloor, 1960, 1986, 1992; Ursin and Kaada, 1960) and subjects will cringe, withdraw, and cower. This cowering reaction in turn may give way to extreme fear and/or panic such that they will attempt to escape and take flight (reviewed in Chapter 5).

Among humans the fear response is one of the most common manifestations of amygdaloid electrical stimulation (Chapman, 1960; Chapman et al., 1954; Gloor, 1960, 1986, 1992; Williams, 1956). Moreover, unlike hypothalamic on/off emotional reactions, fear reactions can last for up to several minutes after the stimulation is withdrawn (Joseph, 1990a, 1992a). Thus the capacity to experience fear (and sex pleasure/arousal) is dependent on the functional integrity of the amygdala, which becomes exceedingly aroused in response to fearful (and sexual) stimuli.

The Amygdala and Emotional Memory

It has been repeatedly demonstrated that the amygdala is primary in regard to all aspects of emotional functioning, including the capacity to experience fear, terror, and sexual arousal, or form complex multimodal emotional memories (Gloor, 1960, 1986, 1992; Halgren, 1992; Joseph, 1990a, 1992a, 1994; Kesner, 1992; Kling and Brothers, 1992; LeDoux, 1992; Murray, 1992; Rolls, 1992; Ursin and Kaada, 1960).

The amygdala also becomes particularly active when recalling emotional memories (Halgren, 1992; Heath, 1964; Penfield and Perot, 1963) and in response to cognitive- and context-determined stimuli, regardless of their specific emotional qualities (Halgren, 1992). However, once these emotional memories are formed, their recall and amygdala activation sometimes requires the specific emotional or associated visual context (Halgren, 1992; Rolls, 1992); otherwise they will seem to have been forgotten.

Conversely, electrical stimulation of the amygdala (and/or the overlying temporal neocortex) can trigger the recollection of recent and forgotten memories, including incidents involving sexual intercourse, fear, or terror, or conversely, joy and happiness (Gloor, 1986, 1992; Halgren, 1992; Penfield and Perot, 1963). Indeed, it appears that the amygdala is responsible for the formation and establishment of sexual-social-emotional memories and in fact contributes to the emotional stream of cognitive and perceptual experience, by associating aversive or rewarding emotions to the flow of overall experience, including that which is stored in memory (Gloor, 1992; Halgren, 1992; Rolls, 1992) .

According to Gloor (1992:523), "affective significance will be attached if what is perceived occurs in the context of psychoaffective constellations related to affiliative behavior (for instance, kinship ties) or other forms of social behaviors, familiar surroundings, or earlier aversive experiences.... When the amygdala is activated under such circumstances, three streams of amygdaloid input will be set into motion, one directed to the hypothalamus and brainstem...[a] second to the hippocampus [that] serves to reinforce consolidation of perceptual information...[and] a third to association cortex neurons," within which the memories may be stored and fed back to the amygdala.

As per emotional memory, the amygdala appears to act in conjunction with the hippocampus (and temporal lobe) so that emotional memories are stored alongside associated cognitive attributes (Gloor, 1992; Joseph, 1990a, 1992a). That is, in addition to the specific neural networks created by the hippocampus, which ensures the consolidation and storage of cognitive experience, the amygdala appears to create a semi-independent neural network that stores emotional information (detailed in Chapter 6). Moreover, the amygdala appears to exert a steering influence on the hippocampus so as to reinforce and thus ensure that certain memories are formed (see Chapters 5, 6).

Repression, Sex Abuse, and Amygdala Hyperactivation

The sexual abuse of children by adults is not uncommon. Often this occurs under conditions of threat, fear, and coercion. For example, it has been estimated that of those who were sexually molested, approximately 29–31% were physically or aggressively forced and were pinned or held down, whereas from 2% to 5% were subjected to substantial violence such as hitting or slapping or beating (e.g., Courtois, 1988; Russell, 1986).

Children often respond to sexual abuse by pretending that it is not taking place (Courtois, 1988; Summit, 1983), such as by "playing possum." Many victims will lie in bed in a state of frozen fear and hypervigilance (Courtois, 1988)—frozen fear states and hypervigilance being mediated by the amygdala throughout its interconnections with the basal ganglia, brainstem, and medial frontal lobes (see Chapters 5 and 9). It is also not unusual for children to initially hyperventilate and then to hold their breath and pretend they are sleeping while the abuse occurs.

The amygdala, hippocampus, and temporal lobe neural complex are highly susceptible to hypoxia, which in turn can result in loss of memory, or conversely create hallucinatory and dissociative states. This is particularly true during childhood as the brain is still immature and plastic and consumes almost 50% of total body oxygen and glucose, compared with 25% in adults. Hence, in cases of sexual abuse, whereas the amygdala may be highly activated because of sexual activity coupled with feelings of fear and terror, in some cases it (and the hippocampus) may be simultaneously denied sufficient oxygen (because of breath holding), which it requires even more than it usually does (because it is highly active).

In fact, some children may also be denied sufficient oxygen by having an adult lie across them. In some cases, sexually abused children suffer compression injuries to the chest and will die from lack of oxygen. Moreover, some sadistically abusive adults will hold their hands over the child's mouth as a form of torture or to prevent him or her from crying out during the assault.

Given that the amygdala and the hippocampus in particular is highly involved in the experience of bodily dissociation, perhaps it is not surprising that some abused children feel as if they have detached from and left their bodies and experienced a "splitting of the ego" (Courtois, 1988; Fredrickson, 1992; Summit, 1983) due to these influences on these nuclei.

It is also likely that with repeated instances of abuse, coupled with oxygen depletion, the still immature temporal lobes and amygdala/hippocampal complex might be injured by these experiences and forever function abnormally (see Braun, 1984; Charney et al., 1993, for related discussion). Given the role of these nuclei in memory, related disturbances would also be likely.

Hence, one additional consequence of trauma might be a form of amnesia for these experiences that is dependent on context and emotional state, particularly in that the hippocampus (in contrast to the amygdala) appears to cease to participate in the formation of memories in response to repeated negative and punished experiences (Joseph, 1990f; see below). That is, since these memories were formed during an unusual state of emotional arousal, these memories may not be retrieved unless the traumatized person is aroused to the same degree or similar contextual cues are provided (Gloor, 1992; Joseph, 1992), i.e. state-dependent learning (Eich, 1987).

In the absence of these cues, the victim may remain amnesic such that all associated memories are repressed, and/or these troubling memories may be reproduced in the form of fragmented and terrifying dreams (Donaldson and Gardner, 1985; Kramer et al., 1987) that in turn may be forgotten (Kaminer and Lavie, 1991; Kramer et al., 1987) by the left cerebral hemisphere upon waking (Joseph, 1988a, 1990).

Sex, Satan, Alien Abductions, and Amygdaloid Hyperactivation

Repeated instances of intense physical sexual abuse (and associated fear states) are therefore linked to amygdala and temporal lobe activation. Given the amygdala's (and hippocampus') propensity for creating hallucinatory images, including those of a demonic, religious, and sexual nature, it is perhaps not surprising that some children who were sexually abused sometimes report that they were subject to bizarre sexual rituals that involved demonic (Satanic) activities.

In fact, the amygdala-hippocampal complex contributes in large part to the production of very sexual as well as bizarre, unusual, and fearful memories, "perceptions," and mental phenomena including dissociative states, feelings of depersonalization, and hallucinogenic and dream-like recollections and experiences involving threatening men, naked women, sexual intercourse, religion, the experience of god, as well as demons and ghosts and pigs walking upright dressed as people (Bear, 1979; Daly, 1958; Gloor, 1986, 1992; Horowitz et al., 1968; MacLean, 1990; Mesulam, 1981; Penfield and Perot, 1963; Schenk and Bear, 1981; Slater and Beard, 1963; Subirana and Oller-Daurelia, 1953; Trimble, 1991; Weingarten et al., 1977; Williams, 1956). Traumatic stress not uncommonly induces hallucinations.

For example, combat vets with PTSD may hallucinate hearing cries for help or battle sounds (Mueser and Butler, 1987; Wilcox et al., 1991). Similarly, individuals who are subject to repeated instances of terrifying and painful sexual abuse as children not uncommonly report hearing voices (Courtois, 1988).

Although it is apparent that some children are sometimes employed in "religious" sexual rituals, and others are abused by "Satanists," it is likely that the vast majority of these reports, including those involving alien abduction (e.g., Mack, 1994) are based on a combination of stress, intense and painful sexual arousal, breath holding, oxygen depletion, and terror-induced hallucinations and/or amygdala-temporal lobe seizure activity. Hence, demons or "aliens" might be hallucinated.

That is, the fear and terror coupled with abuse (and perhaps hyperventilation followed by breath holding) will hyperactivate the amygdala and hippocampus, thus producing these bizarre mental experiences, which are not only experienced as possibly real, but which are then stored in accompaniment with the actual abuse or sexual trauma. The involvement of the amygdala and hippocampus in dream states (reviewed in Joseph 1990a, 1992a) (see Chapters 3 and 5) might also explain why these intrusive images, including actual memories of abuse, may later appear in their dreams.

It is also possible that breath holding (or a preexisting structural lesion in this area—e.g., Horner et al., 1994) may account, in part, for why some individuals become amnesic and remember only the terror they felt as their abusive mother or father approached with all else seemingly a blank. In these instances, they may suspect but be verbally amnesic for sexual and physical abuse (Courtois, 1988; Fredrickson, 1992), their memories ending with mom or dad standing in the shadows watching or slowing approaching.

EMOTION, MEMORY, AROUSAL, AND REPRESSION

Emotional Arousal and Memory Loss

It is well known that events and material that are emotionally arousing are usually much more easily learned and recalled than neutral (or boring) stimuli. However, it has been repeatedly demonstrated that as arousal increases beyond a certain level, memory and learning correspondingly decrease and the functional integrity and capability of the brain to process information is correspondingly reduced (Berlyne, 1960; Deffenbacher, 1983; Gold, 1992; Joseph et al., 1981; McGaugh, 1989), particularly if the information is negative and unpleasant.

For example, Christianson (1992:193) reports that "memory for information associated with negative emotional events, that is, information preceding and succeeding such events...is found to be less accurately retained."

Moreover, it has been repeatedly demonstrated that as arousal increases, attention and retention begin to contract such that increasingly fewer details are recalled (Christianson, 1992b; Easterbrook, 1959; Kebeck and Lohaus, 1986; Kramer et al., 1990). However, at even higher levels of stress and arousal the field of attention may narrow to such an extent that all perceptions are seemingly eclipsed. Consider, for example, the severe memory loss and amnesia of some combat veterans, rape victims, and assault victims as well as the those children who have been sexually assaulted and molested (Grinker and Spiegel, 1945; Kuehn, 1974; Terr, 1990; Williams, 1992).

Memory loss due to high arousal has also been demonstrated in laboratory animals. For example, epinephrine will improve learning and memory at low and intermediate doses, but high doses result in an inverted U-shaped learning curve such that cognitive and memory functioning rapidly decreases and may completely disappear (see Berlyne, 1960; Gold, 1992; McGaugh, 1989).

Hence, Christianson and Nilsson (1984) found that high states of emotional arousal (as indicated by cardiac activity) result in lower recall and poorer recognition memory when subjects are asked to recall names, occupations, and so on listed below pictures of horribly disfigured versus neutral faces. Moreover, amnesia persisted even with cued recall but not with cued recognition. Similarly, Eysenck (1979), in his review of studies examining the relationship between performance and arousal level (as measured by anxiety), indicated that those who were highly aroused showed impaired memory and learning on complex and difficult tasks.

Nevertheless, those who have been personally traumatized may demonstrate even more severe memory disturbances. For example, victims or witnesses of rape, murder, and assault tend to suffer severe and significant memory loss (Deffenbacher, 1983; Kuehn, 1974; Williams, 1992). In one study, 38% of women who had been taken to an emergency room following a sexual assault had no memory of being at the hospital (Williams, 1992). Similarly, victims of rapes and physical assaults tend to provide poorer descriptions of their assailants than do those victims of robberies (Kuehn, 1974).

Be it human or animal, the higher the arousal or degree of emotional trauma, the poorer the memory, until at exceedingly high levels, the subject may simply freeze and no further learning occurs (see Chapter 9). In fact, these same U-shaped relationships between memory and arousal have been demonstrated even in regard to nonthreatening emotional events.

CHILDREN AND INDIVIDUAL DIFFERENCES

Some individuals have no difficulty recalling traumatic incidents, whereas others may suffer partial or complete memory loss. In part his may be due to predisposing factors (such as previous traumas, temporal lobe abnormalities) as well as sex differences in emotionality and individual differences in arousal. Some individuals and animals function at a much higher or lower level of arousal—as measured via EEG and evoked visual and auditory potentials—compared to other humans or animals of the same species (Como et al., 1979; Joseph et al., 1981). Those functioning at a higher level of arousal tend to become more easily overwhelmed.

In addition, attentional capacity is much narrower in children versus adults (Dempster, 1981). Thus children may become more easily overwhelmed by high-impact emotional events—a consequence related to the immaturity of the neocortex and the limbic system.

For example, Peters (1991) found that children who observed a theft (versus those who did not) and who became the most excited and upset (as measured by pulse rate) experienced a considerable degree of memory loss. That is, as pulse rate went up, eyewitness recognition memory declined such that those with the highest pulse rate had the most difficulty correctly identifying the faces of the thief. This was particularly evident when the children were later placed face to face with the "thief" (i.e., live line-up condition), thus inducing high levels of arousal, such that 58% made incorrect identifications.

Moreover, the younger the subject when first exposed to this high-impact emotional information, the greater is the degree of verbal forgetting and distortion. For example, in examining memory for national traumatic events such as the death of JFK and Robert Kennedy, Winograd and Killinger (1983) found a steep gradient of forgetting that became more profound for memories formed between the ages of 1 and 7. Hence, for those adults who had been 1–7 years of age when Kennedy was killed, only approximately 50% of those who had been 4.5 years old or older could verbally recall the news and provide at least one detail. Those adults who were younger than 3 had no verbal recollections regarding context or associated events or information sources, and only a few who were younger than 5 were able to demonstrate detailed verbal knowledge or memories when questioned as adults.

In yet another study regarding memory for the Challenger spacecraft explosion, Warren and Swartwood (1992) found that children age 5 were also less accurate and more likely to delete features over time in describing their recollections. When tested 2 years later, those who had been the youngest as well as the most emotionally upset at the time were less likely to provide extensive narratives and deleted more features.

THE AMYGDALA AND HIPPOCAMPUS

As noted above, even nonthreatening, highly emotional and arousing public events have a negative influences on verbal recall for both adults and children. Because the amygdala is preeminent in the perception and expression of fear and the formation of emotional memories, whereas the hippocampus is predominately responsible for the establishment of nonemotional and cognitive memories (Joseph, 1990a, 1992a; Squire, 1992), these studies suggest that

with high level of emotional arousal, the amygdala as well as the hippocampus may be negatively affected.

Indeed, under conditions of intense fear, arousal, or terror, in some individuals the amygdala may trigger a "freezing" reaction and a complete arrest of ongoing behavior (Gloor, 1960; Kapp et al., 1992; Ursin and Kaada, 1960), whereas the hippocampus may not only cease to participate in memory functioning (Joseph, 1990a, 1992a) but may suffer neuronal degeneration (Uno et al., 1989). This may result in a preserved "emotional" memory (that may be recalled only with the provision of associated cues) accompanied by a hippocampal "cognitive" amnesia (see Chapters 5 and 6). However, if arousal levels and/or feelings of terror continue to increase, these subjects may not merely freeze in response, they may become catatonic, and the amygdala as well as the hippocampus may become exceedingly overactivated and abnormally participate in memory functioning.

The Amygdala, Intense Fear, Stress, and Hippocampal Amnesia

The amygdala becomes aroused in response to emotional and fearful stimuli, and even under conditions of stress continues to participate in the establishment of associated emotional neural networks (see Chapter 6). In contrast to the hippocampus, long-term synaptic potentiation (LTP) and dendritic plasticity in response to fearful stimuli have been noted in amygdaloid neurons (Chapman et al., 1990; Clugnet and LeDoux, 1990).

This is significant, for it has been theorized that LTP is associated with the development of a long-lasting increase in synaptic strength such that various interacting neurons come to be interlocked, thus forming widespread neuronal networks (see Chapter 6 and Barnes, 1979; Davies et al., 1989; Kauer et al., 1988; Lynch, 1986).

In that LTP has been repeatedly demonstrated in hippocampal neurons (and in view of the well-established role of the hippocampus in the development of long-term memory), it has been postulated that this LTP corresponds to the establishment of long-term (rather than short-term) memories. Moreover, it is during the course of the first half-hour or so after learning that LTP develops, thus presumably interlocking the presynaptic and postsynaptic membranes. Hence, LTP within hippocampal and amygdala neurons and neural pathways may correspond to the development of long-term cognitive and emotional memory and the linkage of neurons that subserve and maintain these memories (see Chapter 6).

Hence, fear-induced LTP within amygdala neurons is probably a reflection of the amygdala's involvement in most aspects of emotional experience, including the formation of cross-modal emotional associations and memories (Gloor, 1992; Halgren, 1992; Joseph, 1990a, 1992b, 1994; LeDoux, 1992; Rolls, 1992) and associated neural networks.

In contrast, during periods of extreme fear or arousal, or in the course of high-impact emotional learning situations, hippocampal LTP and theta activity disappear (Redding, 1967; Shores et al., 1989; Varnderwolf and Leung, 1983) and are replaced by irregular electrophysiological activity (Vernderwolf and Leung, 1983). In addition, prolonged stress inhibits the development of LTP in the hippocampus (Shores et al., 1989).

Similarly, if a subject is placed in a situation in which he or she is surrounded by painful physical threat, such as an electrified grid floor, theta activity disappears (Vanderwolf and Leung, 1983). In addition, under conditions of stress or when the body has been anesthetized (Green and Arduini, 1954), hippocampal theta activity disappears, as does LTP (Shores et al., 1989). In other words, when repetitively stressed and highly emotionally aroused, the hippocampus appears to become so aroused that it is essentially deactivated (following the well-established inverse U-curve of arousal) and ceases to participate in memory formation. As such, it is disconnected, and whatever memories and neural networks are formed are established separately from this nuclei. In fact, under conditions of extreme stress the hippocampus may be damaged with subsequent neural degeneration (Uno et al., 1989).

Because the hippocampus may become deactivated or injured, later when an individual recovers, he or she may be unable to recall what occurred, even though this information has probably been stored via the amygdala. That is, because memory retrieval is normally dependent on hippocampal retrieval mechanisms (Squire, 1992), and since high-impact emotional memories were presumably formed and stored by the amygdala in the absence of hippocampal participation, the memory essentially becomes "lost," isolated, and disconnected, i.e., it is "repressed." This then could be classified as hippocampal amnesia, a condition that results from hippocampal hyperactivation (see Brazier, 1966; Chapman et al., 1967).

In fact, with hippocampal hyperactivation/deactivation, the amnesia may extend backward in time from minutes to days and even weeks (Brazier, 1966; Chapman et al., 1967). However, this retrograde amnesia will generally shrink over the course of the next several hours such that only those events which took place just prior to and during the period of excitation are seemingly forgotten (Brazier, 1966; Chapman et al., 1967). That is, they cannot be retrieved via the hippocampus.

Nevertheless, although the hippocampus presumably ceases to participate during periods of intense and repetitive stress and emotional arousal, and may be injured in consequence (Uno et al., 1989), the amygdala continues to function, and, as noted, even under fearful conditions will produce LTP. In consequence, although a person is seemingly amnesic, painful and distressful memories are formed by the amygdala (in the absence of hippocampal participation) and may be reactivated and recalled if the amygdala is aroused by similar contextual cues (Gloor, 1992; Halgren, 1992; Seldon et al., 1991). When such a patient is presented with associated contextual or emotional cues, what was disconnected, dissociated, and repressed may be suddenly remembered (see Chapter 15).

Indeed, even if the hippocampus is destroyed, although new learning and associated cognitive attributes may be forgotten, this amnesia does not extend to fear- or pain-related stimuli (Seldon et al., 1991). In fact, aversive conditioning is completely unaffected (Seldon et al., 1991), particularly when the conditions are highly negative, painful, or emotional as this form of learning is mediated by the amygdala (Cahill and McGaugh, 1990; Hitchcock and Davis, 1987; Kesner and DiMattia, 1987) as well as the hypothalamus (see Chapter 5).

Although disconnected and lost, the emotional distress associated with the trauma may continue to be experienced, although in some cases the individual may not consciously understand why, as these associated cognitive attributes are not accessible to linguistic consciousness (because of hippocampal deactivation and the dissociation of the memory). Rather, although the emotional memory may be preserved (or completely dissociated and repressed), what is forgotten is the place and the circumstances around which the pain was experienced (Seldon et al., 1991), whereas the emotional consequences continue to be felt. This in turn would give rise to symptom formation (Joseph, 1982), the possible experience of flashbacks and associated nightmares, and what is now recognized as PTSD (Charney et al., 1993; DSM IV). Presumably those with PTSD continue to experience intrusive images and flashbacks because the neural network associated with the original trauma remains primed or in a partially activated state (Charney et al., 1993; McNally et al., 1987).

Nevertheless, it is important to emphasize that a single high-impact emotional experience is not sufficient to induce hippocampal deactivation and the loss of theta activity and LTP—that is, unless the victim is plagued by a predisposition to overreact and/or has been exposed to a previous trauma. Rather, traumatic amnesia requires

that high levels of arousal be repeatedly experienced over a short period of time, in which case hippocampal theta activity and LTP may be prevented. Indeed, persistent and repeated stimulation will result in hippocampal synaptic depression with no evidence of hippocampal LTP (Lynch, 1986). Presumably this is related to rapid calcium depletion, for LTP and synaptic plasticity are affected by calcium levels.

Opiates, Cortisone, and Hippocampal Amnesia

Although the experience of sex per se may have little or no influence on hippocampal functioning, sexual activity and pain can induce the release of enkephalins (within the amygdala, hypothalamus, brainstem, and striatum), opiate-like substances that in turn can induce hippocampal deactivation and thus hippocampal amnesia.

For example, the amygdala is rich in cells containing enkephalins, and opiate receptors can be found throughout this nucleus (Atweh and Kuhar, 1977a, b; Uhl et al., 1978). In response to pain, stress, shock, and fear, the amygdala and other limbic nuclei begin to secrete high levels of opiates, which can induce a state of calmness as well as analgesia. Under nonpainful conditions, the release of opiates can increase feelings of pleasure as well as act as a reinforcer and a reward in response to certain experiences, including those which are purely sexual.

However, if opiate activity increases beyond a certain level, they can greatly interfere with memory and learning (Gold, 1992; McGaugh, 1989), particularly in situations involving pain or terror. This is because opiates and opioid peptides result in hyperactivation of the hippocampus and hippocampal pyramidal cells (Gahwiler, 1983; Henriksen et al., 1978). In consequence, hippocampal epileptiform and seizure activity can develop (albeit in the absence of convulsive seizures), which may be accompanied by abnormal high-voltage EEG paroxysmal waves that can last from

15 to 30 minutes (Gahwiler, 1983; Henriksen et al., 1978). In addition, LTP and theta activity is completely abolished (Gahwiler, 1983; Shores et al., 1989), as is the ability of the hippocampus to learn and remember (Gold, 1992; McGaugh, 1989). Hippocampal memory functioning is severely disrupted. Presumably, in consequence, learning and memory is again mediated by the amygdala and other nuclei, in the absence of hippocampal input.

In response to continued fear, emotional stress, and anxiety, the amygdala (and other limbic nuclei) also begin to secrete large amounts of the amino peptide, corticotropin-releasing factor (CRF), which potentiates the behavioral and autonomic reaction to fear and stress (Davis, 1992). Cortisone levels also increase dramatically when the subject is stressed by restraint. However, CRF and cortisone have an inhibitory effect on hippocampal functioning and can eliminate hippocampal theta activity. In consequence, hippocampal participation in memory functioning is again eliminated, altered, or reduced.

However, in addition to interfering with hippocampal memory formation, the massive secretion of cortisone and enkephalins possibly create a state-dependent memory. In consequence, these memories may be recalled and retrieved only under certain "painful" or neurochemical conditions that reproduce the original "state" and/or when an individual is presented with contextual cues or experiencing high levels of emotional stress—which in turn is associated with amygdala activation and opiate and cortisone release, i.e., state dependency (Eich, 1987). However, when stressed, not just the traumatic memory, but any personal reactions (i.e., fragmentary personality features) may also be recalled because of activation of the associated neural network (Charney et al., 1993; Krystal, 1990).

Abnormal Neural Networks

In addition to possible state-dependent effects of trauma-induced amygdala activation and neurotransmitter and peptide release, repetitive traumatic states might induce the development of abnormal neuronal networks, which in turn might produce and maintain permanent abnormal emotional states (see Chapter 4), as manifested by heightened startle reactions, neurotic abnormalities, and PTSD.

For example, in hearing a car backfire in the middle of the night, a traumatized former vet might suddenly spring from his bed, search for his gun, and/or crawl about the room in search of or in fear of enemies (Charney et al., 1993; Krystal, 1990). However, in addition to the flashback and the original feelings of fear, enkephalins and related substances might also be released.

That is, once the trauma is experienced and an associated neural network is created, associated feelings and/or the experience of the trauma itself may be later reactivated by associated cues. The entire circuit of experience may be involuntarily recalled. Unfortunately, even benign cues may trigger hyperarousal, traumatic feelings, and even defensive postures in situations unrelated to the original stressful event—even in the absence of obvious stressors or provoking situations (Charney et al., 1993; Krystal, 1990).

On the other hand, the existence of the abnormal neural network may in itself act as a source of potential stress. That is, although the victim might not be able to consciously recall the traumatic memory, or even if they wish to forget what had happened, the associated neural network may remain potentiated and continue to exert significant disruptive influences as if the individual were constantly subject to trauma.

THE UNCONSCIOUS PERMANENCE OF TRAUMATIC MEMORIES

One consequence of traumatic experience and the establishment of traumatic memories and associated neural networks is the development of PTSD. PTSD is often characterized by an extremely heightened startle response (Charney et al., 1993) that is mediated by the amygdala (see Davis, 1992; Hitchock and Davis, 1991).

Krystal (1990:10) argues that "PTSD could be associated with fundamental and long-lasting neuronal modifications, including alterations in neuronal structure and gene expression. This suggests that many PTSD symptoms may have the indelible qualities of long-term memory," which in turn may result in "long-lasting sensitization of alarm systems with transient habituation, with a subsequent reemergence of symptoms when the habituation fades. From the perspective of evolutionary biology, the resistance of traumatic learning to modification probably reflects the fundamental importance of avoiding catastrophic situations at all costs."

On the other hand, the inability to sometimes recall these traumatic memories also confers certain adaptive advantages. That is, the individual (or animal) need not be tormented by repeatedly recalling something that is terrifying. However, when these memories are triggered, as noted, many individuals feel they are vividly reexperiencing the original trauma without modification.

As noted, although verbal memories are subject to modification (see Chapter 6), emotional and visual memories appear to be much more indelible, particularly those formed by the amygdala (LeDoux, 1992). Given that LTP and synaptic plasticity in the amygdala can be induced by fear, and that opiates can alter the presynaptic and postsynaptic substrates (Goelet and Kandel, 1986; Krystal, 1990), trauma-related memories may well be permanent, although dependent on related stressors, contextual or emotional cues, or drugs for their retrieval (Charney et al., 1993).

In this regard, however, not only memory, but brain functioning and emotional and personality may be irreversibly altered by repeated traumas. If these events occur during childhood when neural pathways are still being formed, the consequences of traumatic learning and memory formation can be even more drastic and may result in permanent alterations in brain structure, synaptic organization, and neurochemistry.

The Amygdala, Serotonin, Neural Plasticity, and Traumatic Stress Disorder

The main source for limbic and neocortical 5-HT is the dorsal raphe (DR) and median raphe (MR) nuclei (Azamita, 1978; Azamita and Gannon, 1986; Wilson and Molliver, 1991a, b). The amygdala, hypothalamus, basal ganglia, primary and association receiving areas, and frontal lobe are innervated by the DR, whereas the hippocampus, cingulate gyrus, and septum receive their 5-HT from the MR nuclei (Azamita, 1978; Azamitia and Gannon, 1986; Wilson and Molliver, 1991a, b). The MR, however is diffusely organized and appears to exert a nonspecific and global influence on arousal and excitability (Wilson and Molliver, 1991a, b). The DR is much more discretely organized and can exert highly selective inhibitory or excitatory influences and plays a role in the coordination of excitation in multiple functionally related areas including the frontal lobes and amygdala (Wilson and Molliver, 1991a, b). Because of the manner in which they are organized, the DR and MR can exert select inhibitory influences so as to promote perceptual filtering in one or a variety of neocortical and subcortical areas while simultaneously exciting yet other multiple regions, which in turn aid selective attention and in the creation of neuronal networks.

However, under conditions of chronic emotional stress, 5-HT begins to be depleted, which can result in amygdala and hippocampal disinhibition. However, not just trauma, but repeated and chronic sexual activity is also associated with significant reductions and depletions in 5-HT (Spoont, 1992; Zemlan, 1978), which might explain why individuals who are repeatedly sexually molested or assaulted, may develop PTSD, dissociative and amnesic states, and psychosis.

Specifically, reductions in 5-HT can decrease hippocampal functioning, but increase amygdala activation and thus the increased secretion of opiates, cortisone, and dopamine. The possibility of amygdala kindling would be enhanced, as well as the development of stress-induced LTP and synaptic plasticity. It is likely that these neurons and associated neural networks would also become adapted to reduced 5-HT levels, which in turn would increase the likelihood of an enhanced startle reaction, social withdrawal, feelings of depersonalization, hallucinatory states, sleep disturbances, increased REM latency, and other associated symptoms associated with a chronic PTSD.

Increased startle, disturbances in sleep onset, increased REM latency and decreased REM, hypersensitivity to noises, hypnagogic and auditory hallucinations, intrusive images and emotions, depersonalization and dissociation, nightmares, and difficulty sleeping have been repeatedly noted in studies of combat veterans, concentration camp and holocaust survivors, victims of sea disasters and natural catastrophes, individuals who are physically assaulted, raped, or sexually abused as children, and adults who witness the murder of children, relatives, friends, parents, husbands, or wives (Charney et al., 1993; Grinker and Spiegel, 1945; Kaminer and Lavie, 1991; Mueser and Butler, 1987; Parson, 1988; van Kammen et al., 1990; Van Putten and Emory, 1973; Wilcox et al., 1991). Indeed, these disturbances (collectively referred to as PTSD, dissociation, and depersonalization) appear to be a universal reaction to traumatic experiences (Carlson and Rosser-Hogan, 1991; Cassiday et al., 1992; Charney et al., 1993; Foa et al., 1991; McNally et al., 1990; Putnam, 1985; Wilcox et al., 1991; Van der Kolk, 1987). They also appear to be directly related to permanent neurochemical (i.e., 5-HT, norepinephrine [NE], DA, and opiates) and neurological alterations involving the amygdala and hippocampus.

In fact, even a single trauma may result in alterations in NE and 5-HT neurotransmitter turnover and release, which in turn can influence and increase the size of both the presynaptic and postsynaptic substrates (Goelet and Kandel, 1986; Krystal,

1990) and even gene expression. This may induce not only alterations in learning and memory, but increased emotional reactivity and alarm responses that may persist indefinitely (Charney et al., 1993; Goelet and Kandel, 1986).

Serotonin: Pain, Stress, Memory, Flashbacks, Startle, and Neural Circuits

One of the major roles of 5-HT is inhibition of neural activity (Applegate, 1980; Soubrie, 1986; Spoont, 1992). It restricts perceptual and information processing and in fact increases the threshold for neural responses to occur at both the neocortical and limbic levels.

For example, in response to arousing stimuli, 5-HT is released (Auerbach et al., 1985; Roberts, 1984; Spoont, 1992), which acts to aid attentional and perceptual functioning so that stimuli that are the most salient are attended to. That is, 5-HT appears to be involved in learning not to respond to stimuli that are irrelevant and are not rewarded (see Beninger, 1989), which are in turn filtered out and inhibited.

In response to highly arousing and stressful, fearful, or painful stimuli, 5-HT (as well as opiate) levels, at least initially, are increased, which results in numbing, analgesia, and a loss of pain perception (Auerbach et al., 1985; Roberts, 1984; Spoont, 1992). However, under conditions of prolonged stress and fear, 5-HT is rapidly depleted. In consequence, the individual may begin to feel overwhelmed and unable to appropriately process incoming sensory stimuli. Soon thereafter he or she may also demonstrate a heightened fear and startle response (Davis, 1984) as well as social withdrawal (Raleigh et al., 1983).

With continued high stress, sensory filtering is reduced and neurons in the amygdala, hippocampus, and inferior temporal lobe cease to be inhibited, which in turn can result in dream-like states, hallucinations (Joseph, 1988a, 1990, 1992a; Spoont, 1992), depersonalization (Sallanon et al., 1983), and increased social fear.

Serotonin reductions are associated with increased sexual activity, fear, stress, and the onset of paradoxical sleep and dreaming (Sallanon et al., 1983)—which might account for why some individuals begin to hallucinate when exceedingly frightened. Moreover, LSD acts to inhibit 5-HT production and blocks synaptic uptake within the amygdala, hippocampus, and temporal lobe (see Chapter 5), which induces hallucinations.

DEPRESSION, APATHY, AND STRESS

When 5-HT levels are reduced, individuals may become depressed. There is also an increased tendency to respond to nonrewarding situations (Soubrie, 1986) and to continue to respond regardless of punishment (Spoont, 1992). Affected individuals may fail to avoid or may even be drawn to abusive or potentially frightening or traumatic situations, and may seem helpless to alter their behavior, e.g., learned helplessness and addiction to trauma (see Charney et al., 1993; Krystal, 1990).

These conditions are most likely to occur behaviorally in chronically emotionally stressful situations wherein feelings of helpless rage and emotional disorganization have been induced and wherein the individual feels continually traumatized and depressed. Hence, even when no longer stressed, their brain continues to respond as if exceedingly emotionally aroused, although they are otherwise severely depressed. Indeed, reduced 5-HT has been repeatedly noted in the brains of those who have committed suicide and taken their lives in a violent fashion (e.g., Brown et al., 1982; Cronwell and Henderson, 1995).

On the other hand, some individuals may seek traumatic or highly arousing or dangerous experiences so as to induce the release of naturally occurring opiates. They may become addicted to trauma, addicted to drugs, or addicted to unpleasant interpersonal relationships (see Charney et al., 1993; Krystal, 1990). It is thus noteworthy that individuals who have been repeatedly traumatized (such as those with PTSD) not uncommonly abuse opiates and other substances that decrease 5-HT and NE activity as well as reduce anxiety (Krystal, 1990). It is thus noteworthy that blockage of 5-HT reuptake (which increases the amount of 5-HT in the synapse) acts to reduce the symptoms of PSTD (Hollander et al., 1990).

NE, ABUSE, AND PSYCHOSIS

As suggested above, in some cases of childhood sexual and physical abuse, the amygdala and hippocampal temporal lobe complex may be injured following repeated instances of traumatic stress and/or via breath holding or temporary suffocation and/or as a consequence of high levels of opiate activity.

Opiate receptors appear to be located presynaptically on NE cells (Llorens et al., 1978), and opiate peptides appear to exert an inhibitory effect on the release of NE (Izquierdo and Graundenz, 1980). Stress increases NE turnover in the amygdala (Tanaka et al., 1982) and increases the production of opiates (Krystal, 1990). The repeated release of these substances, coupled with fear-induced LTP and synaptic plasticity, can in turn significantly and permanently alter the functioning of the amygdala (Cain, 1992; Racine, 1978).

When the amygdala is repeatedly and continually stimulated by excitatory neurotransmitters, or via pharmacological agents including opiates, or through direct electrical activation, an abnormal form of neuronal plasticity and a lowered threshold of responding results. These changes are associated with increases in the size of the evoked potential amplitudes and can give rise to epileptiform after discharges, seizures, and convulsions, as well as induce kindling (subseizure activity) within the amygdala (Cain, 1992; Racine, 1978).

Normally, the release of NE retards kindling in the amygdala. However, with repetitive stress and high levels of arousal, opiates will inhibit NE release. This can lead to permanent structural and functional alterations within amygdala neurons, affecting their postsynaptic densities, and in the size of the presynaptic terminals as well as their capacity to process and transmit information (Cain, 1992; Racine, 1978). Indeed, the threshold for amygdala activation and information transmission may be reduced as well as become abnormal. However, as noted, the hippocampus may also be injured by repetitive and traumatic stress (Uno et al., 1989).

Given the involvement of the amygdala and hippocampus in the development of dissociative states, it therefore seems likely that these changes and reduced threshold for activation might well predispose an affected individual to repeatedly experience similar dissociative states when emotionally stressed. Indeed, because of kindling, abnormal plasticity, and other changes, the amygdala, hippocampus, and temporal lobe may become abnormal and hypersensitive and potentially capable of inducing psychotic episodes (see Halgren, 1992; Horner et al., 1994; Stevens, 1992), leading to the reemergence of the "abused" personality and/or the continuation of a post-traumatic stress reaction (Charney et al., 1993).

Moreover, even when other brain areas are repeatedly and highly activated, kindling begins to occur within the amygdala, and in fact builds up more rapidly than in any other brain region (Cain, 1992; Racine, 1978). In contrast, the hippocampus is one of the slowest regions to develop kindling, with the anterior hippocampus being more likely to do so than the posterior regions (Cain, 1992; Racine, 1978).

Kindling in the amygdala can potentiate neuronal transmission and can induce neural plasticity, which in turn significantly affects learning and memory and the manner in which kindled circuits transmit nonseizure-related activity and information (Cain, 1992; Racine, 1978). That is, high activity-induced kindling can affect the actual creation and functioning of neural circuits as well as their manner of transmitting and processing information, particularly in that it also enhances inhibitory neurotransmission (Cain, 1992). For example, it has been found that kindling induces

increases in postsynaptic densities and in the size of the presynaptic terminals (Cain, 1992). Overall, the consequences can include the development of not just abnormal temporal lobe activity, but a susceptibility to develop seizures.

Dopamine, Psychosis, and Kindling

Amygdala kindling can also have a direct effect on personality, including increasing social withdrawal and defensiveness (Cain, 1992) as well as increasing sensitivity to dopamine. It is noteworthy that the amygdala receives extensive dopamine projections and has a higher concentration of dopamine than the caudate nucleus (Stevens, 1992). These amygdala dopamine neurons project to wide areas of the neocortex, including the frontal lobes, and are believed to play an inhibitory role in neocortical information processing (Stevens, 1992), which in turn is important in selective attention.

In this regard, under conditions of continual high-level emotional stress and arousal, opiates may be released, amygdala neural circuitry can be altered, the hippocampus may be damaged, NE production can be suppressed, and dopamine sensitivity and dopamine levels can be increased, particularly within the amygdala (see Krystal, 1990). Hence, not only might emotional responsivity be altered, but wide areas of the neocortex can be abnormally affected because of alterations in amygdaloid neurochemical and dopamine functioning.

Altered dopamine sensitivity and transmission has long been associated with the development of schizophrenia and psychosis. Specifically, individuals with schizophrenia contain a higher density of dopamine receptors within the left half of their brain (Stevens, 1992), as does the normal left amygdala (Bradbury et al., 1985). In fact, the left amygdala and left temporal lobe (Flor-Henry, 1969; Perez et al., 1985; Stevens, 1992) have long been thought to be a major component in the pathophysiology of psychosis and schizophrenia (Heath, 1954; Stevens, 1973; Torey and Peterson, 1974). For example, abnormal activity as well as size decrements have been noted in the left amygdala (Flor-Henry, 1969; Perez et al., 1985) as well as the left inferior temporal lobe in schizophrenic patients (see Stevens, 1992). Spiking has also been observed in the amygdala of psychotic individuals who are experiencing emotional and psychological stress (see Halgren, 1992). Thus there appears to be an abnormal concentration of dopamine in the left hemisphere and temporal lobe among psychotic individuals, which in some cases may be secondary to repeated traumas experienced during childhood.

Given that stress increases amygdaloid activity, induces kindling, and increases the production of opiates and dopamine, this may explain why psychotic disturbances can be exacerbated when patients are subject to emotionally arousing conditions. Moreover, given that stress and emotional arousal also alters serotonin activity, and given the role of this transmitter in sensory filtering, emotional inhibition, dream production, and hallucinatory states, the likelihood of severe psychological abnormalities is tremendously increased, particularly in those who may be predisposed to develop these disturbances because of adverse early environmental and childhood experiences.

SUMMARY

Under conditions of intense sexual or emotional arousal, the amygdala and hippocampus may become highly aroused and produce dissociative states, as well as hallucinations and abnormal memories. The hippocampus may also cease to participate in memory formation, as it tends to becomes deactivated under conditions of intense and repetitive emotional arousal. Under these conditions the amygdala may create an independent and isolated neural network due to a temporary disconnection involving the hippocampus. This network may be limited to associated emotional memories and/or personal reactions that subsequently come to be disconnected and dissociated.

Once the stressful conditions cease, these emotional and associated memories sometimes cannot be accessed via normal hippocampal retrieval circuits, as these memories were formed by the amygdala in the absence of hippocampal input. Because the hippocampus has been disconnected, later, when the hippocampus returns to full functioning, this information may appear to be forgotten or repressed. The hippocampus is unable to gain access to the neural circuits independently formed by the amygdala and other brain regions that maintain these particular memories. Thus these memories and associated behavioral/personality characteristics associated with these experiences are essentially disconnected, dissociated, and repressed.

In addition, it appears that alternate personalities may be formed during a dissociative state and/or as a consequence of severe and repetitive trauma. That is, under certain traumatic (or neurological) conditions, an aspect of the "ego" may fragment, break off, and thereafter variably act in a semi-independent or completely independent manner. This form of dissociation may be likened to a disconnection syndrome. However, these "broken off" or alternate personalities, in turn, may be "state dependent" and supported by isolated neural networks maintained by the amygdala in the absence of hippocampal participation. This alternate "personality" essentially splits off from the main personality, and, under certain conditions, can be reactivated and can function independently. However, because of the state-dependent nature of this partial "personality," its supporting neural network may again become isolated, such that the main personality essentially becomes amnesic regarding it. In part this is a consequence of hippocampal deactivation. When the hippocampus returns to normal, it can no longer gain access to these memories, such that both the memory and the associated, albeit disconnected personality will become isolated and inaccessible.

However, as argued in Chapters 3 and 15, disconnection, dissociative states, and repression are also a function of the differential effects of emotional stress and arousal on the right and left halves of the brain and their capacity to successfully share and transfer information. Among children, interhemispheric communication is incomplete because of the immaturity of the corpus callosum. In this regard, repression and dissociative states can be vertical (left versus right hemisphere and/or amygdala versus hippocampus) such that the left and right hemispheres and the amygdala and hippocampus become disconnected. Or, they may be horizontal (limbic versus the neocortex), such that personality attributes that are developed and the memories stored in the absence of hippocampal participation can no longer be retrieved or incorporated and made available to the language-dependent regions of the mind via normal hippocampal retrieval mechanisms.

Neuroanatomy of Psychosis

Depression, Mania, Hysteria, Obsessive-Compulsions, Hallucinations, Schizophrenia

The brain is susceptible to a variety of internal and external influences that can negatively impact and interfere with neurological, cognitive, and emotional functioning, disturbances that in turn can give rise, in some cases, to psychotic disorders. However, mild abnormalities and related deficits, such as those due to a small and circumscribed stroke, may go unnoticed by friends, family, teachers, coworkers, and even the patient, although aspects of cognition, memory, or social and emotional functioning may be altered.

Similarly, although gross congenital abnormalities may be identified prenatally, mild cerebral injuries of early onset and associated disruptions in emotional and cognitive functioning may not become fully manifest until much later in life, at which point they may be mistakenly assumed as having their onset at this later time period. This includes abnormalities referred to as depression and "schizophrenia" as well as a host of related disturbances involving intellect, cognition, and emotional functioning that may in fact be secondary to neurological abnormalities suffered congenitally (e.g., Akbarian et al., 1993, 1995; Benes, 1991, 1995; Lewis, 1995)—e.g., due to neural migration errors, viral infection, and subplate abnormalities.

FUNCTIONAL LOCALIZATION AND PSYCHOSIS

As detailed in previous chapters, a variety of anatomical as well as biochemical disturbances have been found in various populations of individuals suffering from the broad symptom complex that is associated with psychotic behavior and schizophrenia. For example, among those with schizophrenia neurological impairments include reduced interhemispheric transfer of visual, auditory, and somesthetic information (Beaumont and Dimond, 1973; Carr, 1980; Green and Kotenko, 1980); pathological abnormalities in the temporal lobes, amygdala, hippocampus, and hypothalamus (Akabarian et al., 1993; Berman et al., 1987; Bogerts et al., 1985; Brown et al., 1986; Crowe and Kuttner, 1991); reduced prefrontal cortical metabolic activity (Ariel et al., 1983; Berman et al., 1986; Ingvar and Franzen, 1974; Weinberger et al., 1986); increased lateral ventricular size (Berman et al., 1987; DeQuardo et al., 1994); medial frontal lobe dysfunction (Joseph, 1990d) (see Chapter 11); and disturbances involving the basal ganglia and the dopamine (Swerdlow and Koob, 1987) (see Chapter 9) and GABA and serotonin (5-HT) neurotransmitter systems.

Similarly, depression, mania, and obsessive-compulsive disorders have been associated with abnormalities involving various regions of the brain and neurotransmitter systems (Arranz et al., 1994; Delgado et al., 1994; Pearlson et al., 1995; Robinson et al., 1995; Rubin et al., 1995; Swann et al., 1994; Trimble, 1991; Umbricht et al., 1995). For example, among those with depression, the medial frontal, left frontal, and left temporal lobes, the adrenal medulla, and the norepinephrine (NE) and 5-HT neurotransmitter systems have been implicated.

Nevertheless, damage involving certain regions of the brain are more likely to give rise to specific psychiatric symptoms as compared to equal damage to a different portion of the brain. For example, mania is not uncommon, at least initially, following right frontal injuries (particularly in males), whereas left frontal injuries are more likely to produce negative symptoms such as blunted forms of schizophrenia and apathetic depression. Similarly, obsessive-compulsive disorders are frequently associated with caudate-frontal lobe abnormalities, whereas nonapathetic forms of psychotic depression tend to be associated with left temporal lobe dysfunction.

These disturbances and related neurological abnormalities are discussed to varying degrees in Chapters 3 through 16 and 18. The purpose here, therefore, is to integrate, summarize, and present this material in a single chapter. Specifically, the different neuroanatomical regions linked to specific psychiatric disorders are reviewed with an emphasis on functional localization and differences in symptomology. Neurochemical contributions to some forms of depression and psychosis will also be briefly reviewed.

DEPRESSION

Patients may psychologically react to a brain injury by becoming depressed or displaying depressive-like features because of a brain injury—e.g., apathy, reduced speech, psychomotor retardation, cognitive decline due to an expanding tumor in the medial frontal lobes. Although it could be argued that depression may follow injuries to most any region of the brain, disturbances involving the left frontal convexity, the medial frontal lobes, and the inferior temporal lobes, the left temporal lobe in particular, are the most commonly implicated.

FRONTAL LOBE DEPRESSION

Although depression may result from damage to either frontal lobe (e.g., Robinson et al., 1984; Sinyour et al., 1986), infarcts involving the left frontal convexity are more likely to induce depressive states (Robinson et al., 1984; Robinson and Szetela, 1981), whereas right frontal injuries are associated with manic states (Joseph, 1986a, 1988a, 1990d).

Patients with severe forms of Broca's expressive aphasia often become frustrated, tearful, sad, and depressed (Robinson and Benson, 1981). Possibly they may become depressed because they are aphasic. However, depressive-like symptoms may also result with left anterior damage sparing Broca's area such as when the frontal pole of either hemisphere is compromised (Robinson and Szetela, 1981; Robinson et al., 1984—e.g., Sinyour et al., 1986).

Similarly, psychiatric patients classified as depressed (who presumably show no obvious signs of neurological impairment) also demonstrate a left frontal impairment, e.g., reduced left frontal arousal (d'Elia and Perris, 1973, Perris, 1974). In fact, based on EEG and clinical observation, d'Elia and Perris have argued that the degree of reduction in left hemisphere functional activity is correlated with and proportional to the degree of depression. They have also noted that with recovery from depression the amplitude of the evoked response increases to normal right and left hemisphere levels (see also Tucker et al., 1981).

Patients with frontal lobe depression appear apathetic, indifferent, and/or irritable and easily confused, with reduced speech and motor output. In part, apathetic and depressive features may result from left frontal convexity and frontal pole damage due to a severance of fibers that link these areas to the cingulate gyrus and medial frontal lobe. In consequence, limbic motivational and emotional impulses can no longer be received in the frontal neocortex because of disconnection. The patient, therefore, becomes motorically hypoemotionally aroused (i.e., depressed), and demonstrates psychomotor retardation and thoughts and ideas no longer come to be infused or assigned emotional significance. In the extreme the motivational impetus to even engage in thought production is cut off and the patient appears apathetic and depressed with a paucity of speech and thought production.

Of course, lesions to the frontal lobe can also induce biochemical abnormalities, including, if restricted to the left frontal lobe, a localized reduction in norepinephrine (Robinson, 1979). As is well known, NE reductions are commonly associated with depression (see below).

In addition, lesions to the left frontal lobe may lead to depressive and apathetic states due to the loss of activational influences and/or as a consequence of excessive inhibitory influences exerted by the right frontal lobe. That is, the depression may be due to a disruption of the normal counterbalancing influences shared by the right and left frontal lobes.

In general, those with Broca's aphasia are genuinely depressed and upset about their loss of speech and cognitive impairment. However, common features of frontal lobe (left, bilateral, or medial) damage sparing Bruca's area include irritability, anger, apathy, indifference, hypoactivity, reduced speech output, confusion, disinterest, and emotional blunting coupled with a paucity of worrisome thoughts or depressive ideation.

Sex Differences

Females are far more likely to experience depression and depressive episodes than males (DSM IV). Although this sex difference is no doubt related to differential stresses and hormonal factors affecting women versus men, this may also be secondary to the fact that females are at greater risk for anterior cerebral artery dysfunction (e.g., embolism—Hier et al., 1994) and become depressed with even subtle injuries involving the left (and right) frontal cortices.

THE TEMPORAL LOBES AND DEPRESSION

Personality, emotional, mood, and sexual disturbances are a frequent complication of temporal lobe abnormalities. Such individuals may develop paranoid, hysterical, or depressive tendencies, deepening of mood, hyposexuality, and other characteristics suggestive of affective disorders (Bear et al., 1982; Gibbs, 1951; Gloor, 1986, 1992; Herman and Chambria, 1980; Strauss et al., 1982; Williams, 1956). Unfortunately, a po-

tential side effect of temporal lobe disturbances is the development of kindling and perhaps seizure activity. Be it kindling, subclinical seizure activity, or the development of temporal lobe epilepsy, as abnormal activity levels wax and wane, personality, emotion, and mood may be affected accordingly (e.g., Herman and Chambria, 1980; Strauss et al., 1982; Trimble, 1991; Williams, 1956).

For example, a depressive aura may precede and thus herald the coming of a seizure by hours or even days. Depression (lasting from hours to weeks) may also occur as an immediate sequela to the seizure (Trimble, 1991; Williams, 1956).

In large part the production of depression may be due to the rich interconnections maintained between the amygdala and temporal lobe. That is, abnormal temporal lobe activity may well affect the amygdala, and vice versa. In consequence, emotional and emotional-perceptual functioning can become abnormal to varying degrees and patients may appear not only depressed but perhaps psychotic.

As noted, frontal lobe depression may be characterized by apathy and even a paucity of depressive thoughts. By contrast, temporal lobe depression may result in language and thought becoming infected with depressive and psychotic ideation and related impulses. This is a consequence of the rich interconnections maintained between the amygdala and the primary and secondary auditory areas (including Wernicke's area in the left hemisphere), as well as the visual and somesthetic cortices. Hence, perceptual as well as thought and language formation may become infected with depressive and psychotic impulses and abnormally invested with emotional, paranoid, and even religious significance (Bear et al., 1979; Gibbs, 1951; Gloor, 1992; Gloor et al., 1982). Therefore, the patient may not only report being depressed but may appear labile, delusional, or paranoid or wracked by religious guilt and feelings of persecution.

Such patients may claim to be the most terrible person on Earth and express a depressive-related fear that they are in some way responsible for certain specified or unspecified ills that they perceive as afflicting the world. Sometimes they may become extremely depressed as well as exceedingly fearful of the terrible punishments they know they deserve for being so horribly bad.

In fact fear followed by depression is the most common emotional experience associated with abnormal temporal lobe activity (Gloor, 1992; Gloor et al., 1982; Halgren, 1992; Williams, 1956). Thus some patients may fear that they are responsible for what depresses them and/or what depresses others, and as such they may appear more paranoid and delusional than depressed. However, because abnormal temporal lobe activity may wax and wane, and as various transmitters such as norepinephrine and serotonin may also wax, wane, and rebound, patients may also display labile and manic-like abnormalities. That is, they may easily cry, becoming angry and accusatory, followed by behavior that is more or less elated, and/or they may become convinced that they must take some action to save the world.

For example, because religious ideation may be triggered by temporal lobe and amygdala hyperactivation (see Chapters 8 and 14), patients may come to believe that they have taken up the sins of the world (i.e., messiah complex) and that it is up to them to act at the behest of God. Because the amygdala is interconnected with Wernicke's area and the inferior parietal lobe, they may therefore "preach" and write out their psychotic beliefs. Moreover, because the amygdala maintains rich interconnections with the basal ganglia, as well as the frontal lobes and reticular formation, they may suddenly feel exceedingly energized and aroused and motorically attempt to act out their delusional thoughts and feelings in a hyperactivated fashion. A diagnosis of manic depression would therefore be likely.

Although in general left temporal lobe abnormalities are more likely to result in psychotic-affective disorders, depression and disorganized mood states are just as likely with right temporal lobe dysfunction (e.g., Flor-Henry, 1969, 1983; Joseph, 1988a; Offen et al., 1976; Schiff et al., 1982; Sherwin, 1981; Taylor, 1975; Weil, 1956) (see also Chapters 3 and 14). Moreover, in that the right and left temporal lobes are interconnected via the corpus callosum, as well as via the anterior commissure, it is likely that abnormalities in one temporal lobe may well affect the functional integrity of the other.

It is noteworthy, however, that patients with left temporal lobe damage are more likely than those with right-sided lesions to also complain of thought blocking and feelings of emptiness: "feelings don't reach me anymore" (Weil, 1956). Presumably this is a consequence of limbic disconnection. That is, the seizure foci (or lesion) acts to deconnect the limbic areas, such as the amygdala, from the temporal (or orbital frontal) lobe. In consequence, percepts and thoughts no longer come to be assigned emotional or motivational significance.

THE HYPOTHALAMUS AND DEPRESSION

A common feature of psychotic depression is loss of appetite, sleep disorders, and an inability to feel pleasure; or, conversely, an overwhelming sense of hopelessness and despair. In large part, disturbances such as these can also be attributed to amygdala/temporal lobe abnormalities as well as to the hypothalamus.

Although the amygdala is dominant in regard to all aspects of emotion, it is dependent on the hypothalamus and frequently acts at the behest of hypothalamic impulses and homeostatic needs (Joseph, 1992a). Conversely, the amygdala also exerts tremendous modulatory influences on the hypothalamus. Hence, abnormalities involving the amygdala can influence the hypothalamus, and vice versa. Moreover, both nuclei influence the sleep cycle, dream sleep, and even the capacity to experience pleasure when eating (see Chapters 5 and 6), and the amygdala as well as the hypothalamus is strongly influenced and dependent on NE and 5-HT, which are also directly implicated in the genesis of depression (see Chapters 5, 6, 16, and 18).

As detailed in Chapter 5, feelings of pleasure are associated with activation of a number of diverse limbic areas including the amygdala, cingulate, substantia nigra (a major source of dopamine), locus ceruleus (a major source of norepinephrine), raphe nucleus (serotonin), medial forebrain bundle, and orbital frontal lobes (Brady, 1960; Lilly, 1960; Olds and Forbes, 1981; Stein and Ray, 1959; Waraczynski and Stellar, 1987). The predominant locus for pleasurable feelings, however, is within the lateral hypothalamus (Olds, 1956; Olds and Forbes, 1981).

If the lateral region is destroyed, the experience of pleasure and emotional responsiveness is almost completely attenuated. For example, in primates, faces become blank and expressionless, whereas a unilateral lesion results in a marked neglect and indifference regarding all sensory events occurring on the contralateral side (Marshall and Teitelbaum, 1974). Animals will in fact cease to eat and will die.

In contrast, activation of the medial hypothalamus is associated with aversive and unpleasant sensations (Olds and Forbes, 1981). However, as the lateral and medial nuclei exert, in part, counterbalancing influences (Joseph, 1992a), abnormalities in the lateral region (and/or abnormalities associated with amygdala input) may not only result in a loss of pleasure, but create sensations of unpleasure, which, if prolonged, would probably lead to feelings of hopelessness and despair—i.e., severe depression.

THE SUPRACHIASMATIC NUCLEUS AND DEPRESSION

Hypothalamic depression is associated with disturbances involving the suprachiasmatic nucleus (SCN)—e.g., seasonal affective disorder (Morin, 1994). In humans and other species, the SCN is a direct recipient of retinal axons and receives indirect visual projections from the lateral geniculate nucleus of the thalamus (see Chapter 5). Presumably the visual system acts to synchronize the SCN to function in accordance with seasonal and day-to-day variations in the light/dark ratio, which in turn is directly associated with activation and arousal versus depressed activation and sleep. Presumably the SCN also acts to directly influence mood and arousal by adjusting those hormonal and neurochemical activities normally associated with activation and high (daytime) activity. If the SCN is deprived of or unable to effectively respond to light (or related neurotransmitters), individuals may become depressed.

For example, the hypothalamic-pituitary axis secretes melatonin in phase with the circadian rhythm, and phase-delayed rhythms in plasma melatonin secretion have been repeatedly noted in most (but not all) studies of individuals with seasonal affective disorder (reviewed in Wirz-Justice et al., 1993). This is significant, for melatonin is derived from tryptophan via serotonin, and low serotonin levels have been directly linked to depression (e.g., Van Praag, 1982). However, with light therapy, melatonin secretions return to normal and the depression may be relieved.

The presence or absence of light, however, may be only one factor among many affecting the SCN. For example, disturbances in temperature perception and the 5-HT system may deregulate SCN activity, as can the process of aging (Aronson et al., 1993). In consequence, rest versus active cycles also become abnormal, with reductions in arousal and activity—i.e., the patient becomes depressed.

In addition, depression and SCN abnormalities may be a consequence of emotional stress on the hypothalamus (Chauloff, 1993) as well as on the amygdala and the NE and 5-HT systems (see Chapters 16 and 18). For example, the hypothalamic-pituitary axis is tightly linked with and in fact mediates stress-induced alterations in serotonin (see Chauloff, 1993, for review) as well as norepinephrine (Swann et al., 1994), which has also been repeatedly implicated in the genesis of depression.

THE HYPOTHALAMUS-PITUITARY-ADRENAL AXIS

The hypothalamic, pituitary, adrenal system (HPA) is critically involved in the adaption to stressful changes in the external or internal environment, whereas hyperactivity of the adrenal gland is clearly associated with major depression. In fact, the adrenal glands have been shown to increase in size during episodes of major depression and to then revert to normal size during remission (Rubin et al., 1995). In part, this may be secondary to (internally or externally induced) stress and/or a disruption in the HPA feedback regulatory system.

For example, in response to fear, anger, anxiety, disappointment, and even hope, the hypothalamus begins to release corticotropin-releasing factor (CRF), which activates the adenohypophysis, which begins secreting ACTH, which stimulates the adrenal cortex, which secretes cortisol. These events in turn appear to be under the modulating influences of norepinephrine. That is, as stress increases, NE levels decrease, which triggers the activation of the HPA axis.

Normally, cortisol secretion is subject to the tonic influences of NE, whereas cortisol can indirectly reduce NE synthesis. Thus a feedback system is maintained via the interaction of these substances (in conjunction with ACTH). Moreover, cortisol and NE levels fluctuate in reverse, and thus maintain a reciprocal relationship with the circadian rhythm—i.e., in oppositional fashion they increase and then decrease throughout the day and evening.

Among certain subgroups suffering from depression, it appears that this entire feedback regulatory system and thus the HPA axis is disrupted (Caroll et al., 1976; Sachar et al., 1973). This results in the hypersecretion of ACTH and cortisol with a corresponding decrease in NE, which results in NE-induced depression. It was these findings that led to the development of the dexamethasone suppression test in the late 1960s—i.e., individuals suffering from depression show excess cortisol and an increased frequency of cortisol secretory episodes (Caroll et al., 1976; Sachar et al., 1973; Swann et al., 1994).

It is also noteworthy that dexamethasone nonsuppression rates are increased in mania—specifically "mixed manic" states that consist of lability, grandiosity, and lability superimposed over depression (see Swann et al., 1994). These "mixed manic" individuals also display elevated NE levels but respond poorly to lithium and show higher levels of cortisol during the depressed phase of their illness (Swann et al., 1994). Nevertheless, NE and 5-HT abnormalities do not exclusively influence the hypothalamus, but also exert widespread influences throughout the neuroaxis, especially the amygdala, which may well interact with the hypothalamus in the production of certain forms of depression.

DEPRESSION, SEROTONIN, AND NOREPINEPHRINE

As is now well known, major factors in the pathogenesis of depression include disturbances in the locus ceruleus (LC)-NE and 5-HT systems (Arranz et al., 1994; Delgado et al., 1994; Pandey et al., 1995), neurotransmitter systems that fuel the amygdala, inferior temporal lobe, and the frontal lobes and hypothalamus. For example, a disturbance in NE synthesis and metabolism, regardless of its cause, will disrupt limbic and cortical functioning and arousal and reduce neuroendocrine activity, and thus the organism's ability to mobilize its defenses in response to stress, adverse experience, or even self-defeating thoughts. Every component of social-affective expression will be altered, especially that related to the experience of pleasure and reward—a condition conducive to depression. Presumably this is also due to related abnormalities in the amygdala, inferior temporal lobe, etc.

Similarly, when 5-HT levels are reduced, individuals may become depressed. There is also an increased tendency to respond to nonrewarding situations (Soubrie, 1986) and to continue to respond regardless of punishment (Spoont, 1992). Affected individuals may fail to avoid or may even be drawn to abusive or potentially frightening or traumatic situations, and may seem helpless to alter their behavior, e.g., learned helplessness (see Charney et al., 1993; Krystal, 1990). Indeed, reduced 5-HT, including abnormalities in platelet serotonin receptors, has been repeatedly noted in the brains of those who have committed suicide and taken their lives violently (e.g., Brown et al., 1982; Cronwell and Henderson, 1995; Pandey et al., 1995). Presumably, in these cases, associated abnormalities in the hypothalamus (loss of pleasure) and amygdala (depression and violent suicide) are implicated.

Although it is difficult to classify a depressive episode or psychosis strictly in terms of an anatomical location or biochemical abnormality, it has been known for over two decades that there are at least two subgroups who differ in regard to depressive symptoms and NE and 5-HT activity levels (Goodwin and Guze, 1979; Van Dongen, 1981; Van Praag, 1982), as well as a third group with bipolar disorder who show elevated D2 dopamine receptors (Pearlson et al., 1995). For example, those with NE depression may behave in a more emotionally labile and expressive manner, whereas those with 5-HT depression may demonstrate a much more profound motor retardation coupled with social withdrawal and confusion due to sensory overload.

Moreover, both groups respond differently to pharmacological treatment. For example, some respond to best to imipramine, which inhibits NE reuptake (thus increasing NE levels in the synapse), whereas others react best to amitriptyline, which potentiates 5-HT (Goodwin et al., 1978; Van Praag, 1982) and inhibits NE reuptake. This suggests that among some individuals, NE and 5-HT may simultaneously contribute to a distinct form of depression.

It has also been argued that reduced 5-HT does not cause depression per se and is not linearly related to the level of depression (Delgado et al., 1994). Rather, 5-HT may be a predisposing factor in depression that in turn may be related to a postsynaptic deficit in 5-HT utilization or binding affinity (Arranz et al., 1994; Delgado et al., 1994). Presumably the structures most affected are the amygdala, inferior temporal and frontal lobes, and hypothalamus, which in turn give rise to a somewhat different symptom complex that may be broadly defined as depression.

As per the role of dopamine, it appears that depressed and bipolar patients who display elevations in D2 dopamine receptors are more likely to appear psychotic than to be suffering from a mood disorder (Pearlson et al., 1995). Hence, those with dopamine abnormalities tend to more closely resemble patients with schizophrenia.

MANIA
The Right Frontal and Right Temporal Lobes

As reviewed above, mania may be induced by damage or dysfunction involving the right frontal lobe and/or the amygdala/temporal lobe. With predominant amygdala (and temporal lobe) dysfunction, individuals may appear labile, delusional, and psychotic, and they may display what has been variably referred to as manic depression or bipolar affective disorder coupled with disorganized thinking, ideas of reference, sleep and appetite disorders, and inappropriate sexual activity, including, in some instances, hyper-religiousness and related delusions.

Mania in the absence of severe depression appears to be far more likely with right temporal, and in particular, right frontal lesions—especially in men (Joseph, 1986a, 1988a) (see Chapters 3 and 11). Given the dominant role of the frontal lobes, and the right frontal lobe in particular, in the control and regulation of arousal (see Chapter 11), it is perhaps not surprising that injuries involving this tissue may result in a loss of control over arousal. Moreover, as the right frontal lobe appears to exert bilateral influences on cerebral activation and inhibition, it is also this portion of the brain that when damaged is the most likely to produce activational disturbances, including mania.

In addition, since the right hemisphere is dominant in the perception and expression of facial, somesthetic, and auditory emotionality, damage to this half of the brain can result in a variety of affective and social-emotional abnormalities, including lability, florid manic excitement, pressured speech, bizarre confabulatory responding, childishness, irritability, euphoria, impulsivity, promiscuity, and abnormal sexual behavior (Bear, 1977, 1983; Bear and Fedio, 1977; Clark and Davison, 1987; M. Cohen and Niska, 1980; Cummings and Mendez, 1984; Erickson, 1945; Forrest, 1982; Gardner et al., 1983; Gruzelier and Manchanda, 1982; House et al., 1990; Jamieson and Wells, 1979; Jampala and Abrams, 1983; Joseph, 1986a, 1990d; Lishman, 1968; Offen et al., 1976; Rosenbaum and Berry, 1975; Spencer et al., 1983; Spreen et al., 1965; Starkstein et al., 1987; Stern and Dancy, 1942). Individuals so affected are likely to be viewed as suffering from mania.

For example, M. Cohen and Niska (1980) report an individual with a subarachnoid hemorrhage and right temporal hematoma who developed an irritable mood; shortened sleep time; loud, grandiose, tangential speech; flight of ideas; and lability and who engaged in the buying of expensive commodities. Similarly, Oppler (1950) documented an individual with a good premorbid history who began to deteriorate over many years' time. Eventually, the patient developed flight of ideas, emotional elation, increased activity, hypomanic behavior, lability, extreme fearfulness, distractibility, jocularity, and argumentativeness. The patient was also overly talkative and produced a great deal of tangential-circumstantial ideation with fears of persecution and delusions. Eventually a tumor was discovered (which weighed over 74 g) and removed from the right frontal-parietal area.

One patient, formerly very reserved, quiet, conservative, and dignified with more than 20 patents to his name, and who had been married to the same woman for over 25 years, began patronizing up to four different prostitutes a day and continued this activity for months. He left his job, began thinking up and attempting to act upon extravagant, grandiose schemes, and camped out at Disneyland and attempted to convince personnel there to finance his ideas for developing an amusement park on top of a mountain. At night he frequently had dreams in which either John F. or Robert Kennedy would appear and offer him advice—and he was a Republican!

Confabulation, hypersexuality, tangentiality, and manic-like behaviors seem to be more frequently associated with right frontal dysfunction (Joseph, 1986a, 1988a), particularly among men. Hence, when the right (or both) frontal lobes are severely damaged it is not uncommon for those afflicted to behave in an inappropriate, labile, and disinhibited fashion. Individuals may become hyperactive, distractible, hypersexual, tangential, and confabulatory and behave in a childish, impulsive, euphoric manner (e.g., Lishman, 1968). It is just as likely for such individuals to just as suddenly cease their antics and to become largely unresponsive and seemingly disinterested in their environment until provoked, sometimes by seemingly inconsequential and trivial stimuli.

Although men are probably more likely than women to develop mania with right frontal lesions, females may be similarly affected, particularly if both frontal lobes and the orbital areas have been damaged. In one case, for example, a woman of 46 was admitted to the hospital and observed to be careless about her person and room, incontinent of urine and feces, to sleep very little, and to act in a hypersexual manner. Her symptoms had developed several months earlier when she began accusing a neighbor of taking things she had misplaced and on other occasions of stripping in front of him. She began going about in just her nightdress, informing people she was descended from queens, was very rich, and that many men wanted to divorce their wives and marry her. During her hospitalization she was frequently quite loud, disoriented to time and place, and extremely tangential, jumping from subject to subject. Eventually a meningioma involving the orbital surface of her frontal lobes was discovered (Girgis, 1971).

As noted, although laughing and joking one moment, these same patients can quickly become apathetic and disinterested and/or irritated, angered, enraged, destructive, or conversely tearful with slight provocation. Presumably, patients with right frontal injury are more likely to produce not only manic-like behaviors but delusional and confabulatory speech due to disinhibition and the flooding of the speech areas with ideas that are normally filtered out (Joseph, 1986a, 1988a). Right frontal injuries are also more likely to mimic orbital injuries, including the loss of regulatory restraint over limbic and thus emotional as well as motivational functioning. Patients may not only talk too much, but eat and drink too much, as well as behave in a socially and emotionally inappropriate manner.

HYSTERIA
The Parietal Lobes and Hysteria

Parietal lobe injuries (particularly when secondary to tumor or seizure activity) can give rise to sensory misperceptions such as pain, swelling, twisting, burning, and so on (Davidson and Schick, 1935; Hernandez-Peon et al., 1963; Ruff, 1980; Wilkinson, 1973; York et al., 1979). That is, patients may experience sensory distortions that concern various body parts because of abnormal activation of those neurons subserving and maintaining the somesthetic body image (Joseph, 1988a).

Unfortunately, when such patients' symptoms are not considered from a neurological perspective, their complaints with regard to pain may be viewed as psychogenic. This is because the sensation of pain, stiffness, and engorgement is, indeed, entirely "in their head" and based on distorted neurological perceptual functioning. Physical examination will reveal little or

nothing wrong with the seemingly affected limb or organ, unless significant neocortical tissue damage is present. Thus such patients may be viewed as hysterical or hypochondriacal, particularly in that right hemisphere dysfunction also disrupts emotional functioning and as right cerebral lesions are the most likely to induce somesthetic distortions.

In this regard, it is noteworthy that individuals suspected of suffering from hysteria are two to four times more likely to experience pain and other distortions on the left side of the body (Axelrod et al., 1980; Galin et al., 1977; Ley, 1980; D. Stern, 1977), findings that in turn suggest that the source of the hysteria may be a damaged right hemisphere. At least one investigator, however, reporting on psychiatric patients has attributed hysteria to left cerebral damage (Flor-Henry, 1983).

Sex Differences, the Amygdala, and Hysteria

Although some patients with parietal lobe abnormalities may be misdiagnosed or viewed as hypochondriacal and hysterical, it is likely that hysteria is directly related to abnormal as well as normal functioning of the temporal lobe and amygdala, and indirectly to the hypothalamus and HPA axis.

As detailed in Chapters 5 and 7, the amygdala is responsive to tactile input, assists in the mediation of somesthetic sensation, and also directly influences the hypothalamus as well as the HPA axis. Under conditions of fear, anxiety, or chronic stress, these nuclei may therefore become abnormally activated and produce somesthetic, autonomic, gastrointestinal, and convulsive disorders that, when prolonged, may give rise to a variety of related and unrelated physical symptoms coupled with emotional disorganization. Afflicted individuals may therefore be diagnosed as "hysterical."

However, the amygdala is not only involved in all aspects of emotional expression, but attachment and the desire for close social-emotional relations, as well as sexuality, aggression, and the desire for group affiliation. In consequence, some "hysterical" disorders also arise in and afflict individuals involved in group activities—e.g., mass hysteria. In addition, because the limbic system (and probably the temporal and parietal lobes) are sexually differentiated, men and women sometimes manifest certain hysterical symptoms in a completely and sexually differentiated manner.

For example, partly because of sex differences in the amygdala, anterior commissure, and hypothalamus, females are more emotionally expressive and perceptive, and in this regard much more sensitive and far more likely than males to become upset and to experience fear (see Chapters 3, 5, 7). Because for much of their evolutionary history females foraged and gathered in groups, and as they are more socially emotionally inclined with a greater expressed need for social-emotional contact comfort, they are also more solicitous of nurturance and are more sensitive to the feelings and the fears of other group members—i.e., females. They are also more likely to affirm these feelings and to behave in a collective and cohesive, i.e., similar, manner. This also includes becoming "ill" if close friends are also suddenly and mysteriously incapacitated (e.g., Moss and McEvedy, 1966; Small et al., 1991).

In consequence, females are not only more emotionally sensitive, they are also more likely to respond to even subtle social-emotional nuances, and are therefore more subject to temporary emotional disorganization and far more likely than males to develop mood disorders and/or physical symptoms requiring medical attention, including certain forms of hysteria. Because women are so in tune and responsive to the feelings of other women, when they are in large groups, the tendency to develop hysterical disorders is sometimes heightened, particularly if they perceive that other women are upset, excited, or frightened (e.g., Moss and McEvedy, 1966; Small et al., 1991).

Sex Differences and Collective, Socially Sanctioned Hysteria

It is noteworthy that mass outbreaks of crying and screaming frenzies are typical of large groups of young women when in the presence of musicians, actors, and even politicians and dictators (e.g., Hitler). Consider, for example, the mass outbreaks of hysteria among young women and girls during the 1960s when the rock and roll group, the Beatles, made appearances in this or other countries. In such situations, many women and girls behave as if attempting to outdo one another in their attempts to demonstrate who is the most emotionally affected. In this regard, they reinforce and support their mutual hysteria.

However, it sometimes happens that a single woman can induce fear and anxiety in other females, rather than adoring, screaming, adulation. That is, if one female becomes sufficiently upset or stressed, even if there is no apparent cause, it sometimes happens that some (but not all) of the other women will also begin experiencing tremendous amounts of anxiety or fear such that they respond in a cohesive manner. However, in some cases, one female member of a group might not only become afraid, but she may respond as if physically sick, which in turn results in other female group members developing the same symptoms (Colligan and Murphy, 1979; Kerckhoff, 1982; Sirois, 1982). Of course, be it a rock and roll concert or mass hysterical epidemic in a factory or school recital, these girls and women are mutually supportive of one another and are thus able to mutually validate their illness, fear, anxiety, or excited devotion (Stahl, 1982).

Of course, men are not immune to emotional infection and can likewise become upset, or angry, or yell and scream in a like fashion when in large groups (e.g., a football stadium) or when congregating in mobs. However, whereas women are more likely to display physical symptoms (e.g., fainting, convulsing, crying, screaming), males are more likely to respond physically (e.g., charging, attacking, hitting, yelling, screaming). Consider, for example, during the 1930s the hysterical and irrational response of adult German men and, in fact, one-third of the German nation, to the hysterical rantings of the girlish Adolf Hitler: a mass hysteria (or psychosis) that resulted in World War II and over 40 million deaths. Indeed, throughout history men have commonly become caught up in mass (or mob) hysteria and have then proceeded to irrationally kill and maim, often indiscriminately.

Mass "Female" Hysteria

Outbreaks of "mass hysteria" and the development of "mass" physical (conversion) symptomology, be it in America (e.g., Montana Mills), Europe, Asia, or Africa, characteristically involve large groups of women (Colligan and Murphy, 1979; McGrath, 1982; Sirois, 1982) or young females (Moss and McEvedy, 1966; Small et al., 1991). Onset may be quite rapid or take place over the course of several days, with the "illness" being transmitted by sight or sound. Usually recovery is just as swift and no physical basis for the illness can be found.

For example, in the famous Montana Mills incident (Colligan and Murphy, 1979; Sirois, 1982), over the course of approximately a week over 50% of the female work force developed severe and disabling imaginary illness. However, male employees were generally not affected. This is because mass hysteria involving mass physical and associated emotional upset is generally a disorder involving large groups of females.

Indeed, it has been repeatedly reported that in cases of "mass hysteria" over 80–90% of those initially affected, and 80–90% of those who may be subsequently affected, are women (see Colligan and Murphy, 1979; Sirois, 1982). Although the work force may in fact be one-fifth to one-third male, men are by and large unaffected (see McGrath, 1982). Even though large numbers of (female) employees may be suddenly stricken with a variety of mys-

terious physical ailments, heterosexual males tend to be immune, or any symptoms they develop are quite mild.

Similarly, among school- and college-aged individuals, girls (Small et al., 1991) and young women (Shelokov et al., 1957) are predominately affected. Females show a higher rate of "illness" and more severe symptoms, and they are affected longer than males who are also much less likely to become emotionally upset even though their female classmates are quickly becoming incapacitated.

In numerous cases of "mass hysteria," although males may be largely immune, many of the women will begin crying, screaming, hyperventilating, fainting, and/or convulsing as if suffering from seizures, and many will later invent a host of physical complaints (Colligan and Murphy, 1979; McGrath, 1982; Sirois, 1982; Small et al., 1991). Typically, in these instances of mass hysteria, a single female may become "ill" for no apparent reason, and just as suddenly a second female, who may be close friends with, or standing next to and/or attending the first female, may also become "ill." Soon, more and more women or girls become "ill" and may hyperventilate, faint, fall to the ground, display tremors and convulsions, and/or become incapacitated to varying degrees for varying time periods.

Hence, "mass female hysteria" may be initially triggered by the reaction of a single woman who then literally emotionally infects her female coworkers and friends as the "disorder" spreads. Among some (but certainly not all) girls and women, hysteria can be contagious.

For example, in April of 1989, there was an outbreak of mass illness among student performers who were to participate in a recital in a high school in Santa Monica, California (Small et al., 1991). It was noted earlier in the day during rehearsal that two girls complained of faintness, nausea, and dizziness. However, once the recital began the performance was interrupted by a mystery illness involving headache, abdominal pain, nausea, and dizziness that spread swiftly among the female students, with 16 females initially fainting followed by over 247 students, mostly female, becoming ill and severely upset. Ambulances were called, students were rushed to the hospital, and no physical reason for their ailments could be found.

These hysteria epidemics have occurred in factories, hospitals, and schools, and in rural as well as suburban areas. Nor is mass female hysteria limited to certain cultural or age groups. Rather, it affects women regardless of age (though younger females tend to be affected), culture, religion, or nation of origin (Markush, 1973; Sirois, 1982). Even nuns have been affected (e.g., the mewing nuns; see Markush, 1973; Sirois, 1982). However, it has been reported that those of lower occupations are far more likely to be affected (Colligan and Murphy, 1979; McGrath, 1982), whereas among girls and young women, those who failed to complete high school or who have a history of recent grief or personal loss (death, divorce) tend to be at highest risk (see Small et al., 1991). In addition, risk of subsequent contamination has been shown to be positively correlated with friendship and the social hierarchy (Moss and McEvedy, 1966; Small et al., 1991). That is, close friends are more likely to "contaminate" each other (usually by simple observation), whereas younger girls are likely to imitate older girls and thus become similarly ill.

A variety of explanations have been proposed for mass female hysteria, including stress, anxiety, boredom, and the fact that some of the work settings where these outbreaks occur are noisy and prevent the women from socializing and talking to one another (see Colligan et al., 1982), which females tend to find stressful. Moreover, given the routine and the repetitive qualities of some work situations, it has been argued that the afflicted women have little or no opportunity to express their attachment needs or to attract attention. That is, they utilize hysteria as a means of attracting attention, to solicit nurturance, and to express their feelings.

Moreover, many of the females affected tend to have high rates of absenteeism for health problems prior to the outbreak, suggesting that in part these women are manifesting a hypochondriacal and hysterical coping pattern independent of these outbreaks. On the other hand, it is possible that the absenteeism and the hysteria in these women are manifestations of an increased susceptibility to internal or external stress. Hence, they become "ill" more frequently, and are therefore predisposed to developing "hysterical" disorders.

However, it has also been noted that women or girls who have been recently subject to stress, such as due to a divorce or death, are also at risk (e.g., Small et al., 1991). Hence, certain forms of stress, including, perhaps the stress associated with the excitement of a high school recital, can increase the risk of developing a hysterical reaction among women if other females are also observed to mysteriously become ill.

Of course, those afflicted (and their attorneys) will vigorously deny that their illness is imaginary. In fact, given the role of the limbic system in controlling emotional and autonomic nervous system functioning, if sufficiently stressed, such as by watching their friends and coworkers become ill, these women likely will subsequently become excessively emotionally aroused and will feel ill, and may even suffer convulsions and gastrointestinal and related limbic and autonomic abnormalities—in which case it may become almost impossible to determine if the disorder is "hysterical."

Consider, for example, the widely reported incident of possible mass hysteria that took place in the emergency room of General Hospital in Riverside, California (reviewed by Stone, 1995). Specifically, on the evening of February 19, 1994, a 31-year-old woman, Gloria R., was in the process of being resuscitated when a female nurse, S. K., noticed a chemical smell, similar to ammonia, coming from her blood. In addition, a female resident, J. G., noticed manila-colored particles floating in the blood. Suddenly the nurse, S. K., felt that her face was burning and she fell to the floor. She was placed on a gurney and taken to away. Suddenly the female resident, J. G., began feeling queasy and light-headed. She then slumped to the floor, shook and convulsed intermittently, and displayed apnea. Then a female respiratory therapist, M. W., also began displaying symptoms and "couldn't control the movement of" her "limbs." And then, a vocational nurse, S. B., felt a burning sensation and began retching. Eventually 23 of the 37 emergency room staff members were afflicted, and an intensive investigation was launched to determine the cause.

According to the California Department of Heath and Human Services, and two of its scientists, Drs. A. Osorio and K. Waller, those afflicted experienced "an outbreak of mass sociogenic illness, perhaps triggered by an odor," i.e., mass hysteria. It was noted that the two male paramedics who brought in the patient and who touched Gloria R.'s skin and blood did not become ill, and that women were predominantly and the most severely affected. They also noted that those who had skipped dinner and were working on an empty stomach (which would induce limbic arousal) were also more likely to be affected.

Nevertheless, the diagnosis of "mass hysteria" was not acceptable to those who became ill, or to their attorneys. Scientists at Livermore Laboratories were called onto the case. As often occurs when lawsuits are involved, completely different conclusions were reached and a completely different scenario involving a rather harmless substance, DMSO, was hypothesized as the main factor involved (see Stone, 1995 for details). That is, these female nurses and doctors became ill because of a sequence of chemical transformations involving the deceased patient's presumed excessive use of DMSO as a treatment for pain. The family of the patient (Gloria R.), however, denies that she ever used DMSO or had access to it.

As noted, some males were also affected to varying degrees in the above-mentioned cases. Hence, males are not immune from developing hysterical disorders. However, in contrast to women, who may develop such symptoms in isolation or in groups, often via the power of suggestion or through mild fear and social contagion, for heterosexual men the conditions must be generally quite horrendous and involve not imagined fears but death, destruction, and the killing of close friends and comrades such as in war (Grinker and Spiegel, 1945) (see Chapter 16).

That is, in contrast to some women who may develop any number of hysterical symptoms while working or studying in large female-dominated groups in otherwise safe environments where threats to their lives and health are for the most part imaginary, men are more likely to develop hysterical disorders when placed in horrible conditions where the threat to their life is not only real but is experienced daily, weekly, and monthly, and under conditions where they may see friends blown apart by gun fire or their airplanes blown to bits (Grinker and Spiegel, 1945). During World War II, in the "overseas Air Force it was a mathematical certainty that only a few men out of each squadron would finish a tour of duty" (Grinker and Spiegel, 1945:33). In this regard, in contrast to those females who may experience a single episode of terrible fear or tremendous excitement and then develop hysterical symptoms, when men develop hysteria involving physical symptoms, often the fear must be repeatedly experienced—as is common during war time.

"Fear itself is the most potent source of emotional stress in combat. Fear is cumulative, because the longer the individual stays in the battle, the more remote appears his chance of coming out alive or uninjured" (Grinker and Spiegel, 1945:33). Hence, under these repetitive traumatic conditions, some males may eventually convert their fears and feelings of stress, which are then manifested as physical symptoms. Presumably, it is the physical symptom that enables them to escape the horror and their terrible fears, without causing them to suffer overwhelming feelings of cowardice.

However, as detailed in Chapters 6, 7, 16, and 18, fear and emotional trauma can in fact damage the brain, which in turn can influence physical and perceptual functioning and induce psychotic and dissociative states as well as post-traumatic stress disorder. For example, it has been demonstrated that the hippocampus is damaged by repetitive stress (Uno et al., 1989), and that patients with combat-related post-traumatic stress disorder display a statistically significant, 8% reduction in the right hippocampus (Bremner et al., 1995). Similarly, abnormal early environmental influences not only affect the hippocampus and a variety of cerebral structures, but result in reductions in the size of the right amygdala (Diamond, 1985). As detailed in Chapter 16, limbic and amygdala hyperactivation and hippocampal dysfunction are directly associated with dissociative and psychotic states as well as the development of post-traumatic stress disorder.

Sex Differences, Hysteria, and the Limbic System

As detailed in Chapter 5, the amygdala and hypothalamus are the primary source for feelings of fear, and both nuclei are sexually differentiated. Hence, men and women respond differently to similar stimuli, and differentially develop hysterical and related mood disorders, which in turn reflects not only sex differences in the limbic system, but sex differences in self-concept and in reactions to stress and even the possibility of illness. As also noted in Chapter 5, the female limbic system may predispose women to becoming more easily stressed and emotionally upset, which in turn can disrupt the functioning of the HPA axis. Since females are more emotionally sensitive and expressive (see Chapters 3, 5, 7), they are also more likely to respond fearfully and thus develop emotional, physical, and/or "hysterical" disorders.

Indeed, as is well known, females are also far more likely to seek medical attention than males (see Chapters 3, 5, and 7), and this includes female "soldiers." For ex-

ample, female armed forces personnel generally have a much higher rate of "illness" that requires some sort of medical attention and thus relief from their duties (see Farrell, 1993)—e.g., almost a third of female enlisted personnel became "pregnant" or ill just prior to the 1993 Gulf War and thus could not ship out with their male counterparts (reviewed in Farrell, 1993).

By contrast, male soldiers and Air Force personnel (as well as men and boys in general) often try to downplay or hide their injuries or potentially disabling conditions. Moreover, in war time, many obviously injured male soldiers earnestly desire to rejoin their comrades and fighting units, in part because of a sense of loyalty and feeling of brotherhood, and, in some cases, a fear of appearing cowardly or weak; in others, this is due to a desire to be part of the continuing violent action. This is an innate male characteristic. As noted in Chapter 5, male chimpanzees enjoy and seek out violent encounters, which in turn is a manifestation of their enjoyment of aggression as well as their concern for and desire for status (Goodall, 1971, 1990)—functions that are clearly linked to sex differences in the functional integrity of the limbic system.

As is also detailed in Chapters 5 and 7, sex differences in the limbic system are reflected in those behaviors and activities that men and women are most likely to find pleasurable, aversive, or threatening to their self-concept as men or women. For example, an injured soldier may seek to rejoin his comrades not because he enjoys fighting, but because he fears being perceived as weak. Being viewed as weak or fragile is not something most women fear, whereas seeking nurturance and revealing, discussing, and sharing one's problems and troubles is often a source of considerable female pleasure (Glass, 1992; Tannen, 1990). Moreover, whereas males may be teased or tormented by other men regarding their perceived fears, illness, or disabilities, women respond with concern, nurturance, and sympathy.

Hence, it is more socially acceptable for women to become ill, to seek help, and to become upset, which (when coupled with limbic system sex differences) increases the likelihood that they may develop "imaginary" illness, especially if their close friends or associates also become "ill." By contrast, whereas a single instance of intense fear may induce hysteria and the development of physical disturbances in females, males tend to be intensely fearful of and reluctant to reveal similar problems, which in turn reduces the likelihood that they may develop imaginary illnesses. Of course, this male predisposition to "tough it out" probably also contributes to the development of real illnesses leading to death.

OBSESSIVE-COMPULSIVE DISORDERS
The Caudate and Frontal Lobes

Obsessive-compulsive disorders are characterized by the repetitive experience of unwanted and recurrent, perseverative thoughts (obsessions) and/or a compulsion to repetitively perform certain acts. However, some patients tend to suffer from compulsions, whereas others experience obsessions, with the majority experiencing both (DSM IV).

Obsessions not uncommonly tend to be religious or concerned with contamination coupled with cleaning and washing rituals. In some instances obsessive compulsions serve as a means of coping with unwanted or troubling impulses—i.e., by perseverating in thinking certain thoughts or engaging in certain actions, the individual is able to avoid focusing thoughts, actions, or impulses that he or she finds disturbing. Thus, the experience of intrusive and unpleasant (e.g., violent and/or sexual) images is not uncommon (Lipinski and Pope, 1994).

Obsessive-compulsive behavior is in part associated with disturbances of impulse control as well as memory, e.g., checking doors again and again because the patient can't remember if he or she really locked them. Hence, the frontal lobes are implicated in the genesis of this disor-

der (Flor-Henry et al., 1979), the orbital frontal lobes in particular (Baxter et al., 1988; Rappoport, 1991; Rauch et al., 1994).

As detailed in Chapters 9 and 11, compulsions involving simple as well as complex motor acts are frequently associated with frontal injuries. In addition, abnormal activation of the frontal lobes has been reported to elicit recurrent and intrusive ideational activity ("forced thinking"), as well as compulsive urges to perform aberrant actions, e.g., shouting (Penfield and Jasper, 1954; Ward, 1988). Perseverations of speech and motor functioning are also common.

The mixture of perseveration and compulsions, when subtle, may take on the appearance of obsessions. Obsessions are perseverative, often involving intrusive recurring thoughts, feelings, or impulses to perform certain actions (Goodwin and Guze, 1979).

Motorically obsessive compulsions may involve repetitive, stereotyped acts, including the perseverative manipulation and touching of objects (Goodwin and Guze, 1979)—abnormalities also associated with certain forms of frontal lobe damage (e.g., magnetic apraxia, utilization behavior—see Chapter 11).

It appears, however, that the frontal lobes are only part of an abnormal circuitry involved in this disorder, and that the actual locus may be within the basal ganglia, specifically the caudate nucleus (Baxter et al., 1992; Rappoport, 1991; Rauch et al., 1994; Robinson et al., 1995). As detailed in Chapter 9, the basal ganglia is intimately associated with the hypothalamus, amygdala, and orbital frontal lobes, nuclei that are also associated with sexual activity and the mediation of emotional functioning, with the orbital areas acting to gate and inhibit limbic system activity (Joseph, 1990d).

In that obsessive-compulsive disorders are also associated with hypermetabolism within the caudate nucleus (Baxter et al., 1992; Rappoport, 1991; Rauch et al., 1994)—as well as reduced caudate nucleus volume (Robinson et al., 1995) and reduced 5-HT activity (Goodman et al., 1990; Zohar and

Insel, 1987), which in turn results in disinhibition—this raises the possibility that the inhibitory circuit maintained by the orbital frontal lobes and basal ganglia, in some instances, may be deactivated or overwhelmed, thereby giving rise to perseverative disorders of thought and motor action—i.e., obsessive compulsion.

Given that it is via the frontal lobes that thoughts and ideas come to be integrated and infused with emotion, and/or inhibited and suppressed, this also suggests that caudate/orbital dysfunction may result in the inability to extinguish certain thoughts and actions, even in the absence of an underlying emotional problem (see Baxter et al. 1992; Rauch et al., 1994, for a different albeit related explanation). Therefore, patients may perseverate on a single thought or action, which in turn may interfere with their capability to engage in alternative modes of response.

Conversely, it is possible that "normally" this same frontal-inhibitory circuit may be employed so as to prevent such patients from engaging in certain behavioral acts that the "conscious" mind finds objectionable or "dirty" (see Chapters 11 and 15). Hence, they perseverate on a certain theme, thought, or action to the exclusion of others that prevents these unacceptable limbic impulse (or memories) from achieving conscious (neocortical) realization.

HALLUCINATIONS

Hallucinations may occur secondary to tumors or seizures involving the occipital, parietal, frontal, and temporal lobe, or arise secondary to drugs, toxic exposure, high fevers, general infections, exhaustion, starvation, extreme thirst, or partial or complete hearing loss including otosclerosis, and with partial or complete blindness such as that due to glaucoma (Bartlet, 1951; Flournoy, 1923; Lindal et al., 1994; Pesme, 1939; Rhein, 1913; Ross et al., 1975; Rozanski and Rosen, 1952; Semrad, 1938; Tarachow, 1941). Interestingly, when hallucinations are secondary to peripheral hearing loss, individuals frequently report hearing certain songs and melodies from their childhood—melodies that they had usually long forgotten. In addition, individuals suffering from cortical blindness, i.e., Anton's syndrome (Redlich and Dorsey, 1945), and deafness (Brown, 1972), as well as those recovering from Wernicke's aphasia, frequently experience hallucinations.

In general, hallucinations secondary to loss of visual or auditory input appears to be secondary to the interpretation of neural noise and the spontaneous activation of associated neural circuits. That is, with loss of input, various brain regions begin to extract or assign meaningful significance to random neural events, or to whatever input may be received. Thus we find that subjects will hallucinate when placed in sensory reduced environments or even when movement is restricted (Lilly, 1956, 1972; Lindsley, 1961; Shurley, 1960; Zuckerman and Cohen, 1964).

Conversely, hallucinations can occur because of increased levels of neural noise or spontaneous activation of various neural circuits. For example, if an area of the neocortex is abnormally activated, that area in turn may act to interpret its own neural activity and may even assign emotional or religious significance to whatever is perceived. However, the degree and type of interpretative activity depends on the type of processing performed in the region involved. For example, tumors or electrical stimulation of the occipital lobe produce simple hallucinations such as colors, stars, spots, balls of fire, and flashes of light. Tumors invading the parietal lobe may induce somesthetic hallucinations, including burning, engorgement, stiffness, or pain like that of an electric shock.

With superior temporal involvement the patient may experience crude noises, such as buzzing, roaring sounds, bells, and an occasional voice or sounds of music (Penfield and Jasper, 1954; Penfield and Perot, 1963), including clicking, ticking, humming, whispering, and ringing, most of which are localized as coming from opposite side of the room. Patients may complain that sounds seem louder and/or softer than normal, closer and/or more distant, strange or even unpleasant (Hecaen and Albert, 1978). There is often a repetitive quality that makes the experience even more disagreeable.

However, with anterior, inferior temporal lobe abnormalities, the hallucinations become increasingly complex, consisting of both auditory and visual features, including faces, people, objects, animals, etc. (Critchley, 1939; Penfield and Perot, 1963; Tarachow, 1941). Presumably, in part this is a consequence of the activation of specific neurons or neural assemblies that normally respond to specific environmental stimuli, such as faces—e.g., feature detector activation in the absence of appropriate external stimuli. As the inferior temporal lobe contains neurons that respond to a variety of complex stimuli, whereas tissues in the occipital lobe are more responsive to simple stimuli, hallucinations correspondingly increase in complexity as the disturbance expands from primary to association areas and as involvement moves toward the anterior temporal regions—which is one of the major interpretive regions of the neocortex (Gibbs, 1951; Gloor, 1990, 1992; Halgren, 1992; Penfield and Perot, 1963).

Presumably, the anterior-inferior temporal lobes and associated limbic nuclei give rise to the most complex forms of imagery because cells in these areas are specialized for the perception and recognition of specific forms, including faces and people. As noted in Chapters 3 and 5, it is the inferior temporal lobe, including the amygdala and hippocampus, which is also largely involved in the formation of dream images.

Indeed, it has frequently been reported that as compared to other cortical areas, the most complex and most forms of hallucination occur secondary to temporal lobe involvement (Critchley, 1939; Horowitz et al., 1968; Malh et al., 1964; Penfield and Perot, 1963; Tarachow, 1941) and that the hippocampus and amygdala (in conjunction with the temporal lobe) appear to be the responsible agents (Gloor, 1990, 1992; Gloor et al., 1982; Horowitz et al., 1968;

Halgren et al., 1978). For example, Bancaud and coworkers (1994), Halgren and associates (1978), and Horowitz and colleagues (1968) note that hippocampal stimulation was predominantly associated with fully formed and/or memory-like hallucinations, including feelings of familiarity, and secondarily dream-like hallucinations. However, stimulation limited to the neocortex had relatively little effect in this regard (Gloor et al., 1982). It appears, therefore, that limbic activation is necessary in order to bring to a conscious level percepts that are being processed in the temporal lobes.

Hallucinations and Hemispheric Laterality

In general, complex auditory verbal hallucinations seem to occur with right or left temporal destruction or stimulation (Hecaen and Albert, 1978; Penfield and Perot, 1963; Tarachow, 1941), although left temporal involvement is predominant. Left temporal lobe hallucination may involve single words, sentences, commands, advice, or distant conversations that can't quite be made out. According to Hecaen and Albert (1978), verbal hallucinations may precede the onset of an aphasic disorder, such as one due to a developing tumor or other destructive process. Patients may complain of hearing "distorted sentences," "incomprehensible words," etc.

By contrast, Penfield and Perot (1963) report that with electrical stimulation of the right superior temporal gyrus with tumors and seizure disorders involving the predominantly the right (versus the left) temporal region, patients may experience musical hallucinations. Frequently the same melody is heard over and over. In some instances patients have reported the sound of singing voices, and individual instruments may be heard (Hecaen and Albert, 1978). Similarly, complex visual and emotional hallucinations, such as typified by dream imagery or via LSD, are associated with the right temporal lobe (see Chapter 3).

As noted in Chapter 3, presumably it is the left hemisphere and temporal lobe/amygdala/hippocampal complex that provides the verbal monologue that is experienced during dream states and paradoxical sleep. Conversely, the right temporal lobe provides the visual and emotional hallucinatory mosaic that is commonly experienced during REM, and while under LSD, and presumably during related psychotic states. However, as noted above, the most complex hallucinations typically involve the anterior temporal lobes of either hemisphere—regions that are linked via the anterior commissure and that are therefore subject to abnormal influences that originate in either half of the brain.

SCHIZOPHRENIA

A variety of anatomical as well as biochemical disturbances have been found in various populations of individuals diagnosed as suffering from schizophrenia (Akbarian et al., 1993; Benes, 1995; Bruton et al., 1994; Buchanan et al., 1994; Crowe and Kuttner, 1991; Stanely et al., 1995). These include impaired interhemispheric transfer of visual, auditory, and somesthetic information (Beaumont and Dimond, 1973; Carr, 1980; Green and Kotenko, 1980), increased lateral ventricular size (Berman et al., 1987; DeQuardo et al., 1994), increased corpus callosum thickness (Rosenthal and Bigelow, 1972), pathological abnormalities in the temporal lobes, amygdala, hippocampus, and hypothalamus (Akbarian et al., 1993; Benes, 1995; Berman et al., 1987; Brown et al., 1986; Bogerts et al., 1985; Crowe and Kuttner, 1991), reduced prefrontal cortical metabolic activity (Ariel et al., 1983; Berman et al., 1986; Ingvar and Franzen, 1974; Weinberger et al., 1986), medial frontal lobe dysfunction (Joseph, 1990d) (see Chapter 11), dorsolateral frontal abnormalities (Akbarian et al., 1995; Stanley et al., 1995), and disturbances involving the basal ganglia and dopamine neurotransmitter system (Swerdlow and Koob, 1987) (see Chapter 9).

In part, the reason so many different brain areas are implicated is secondary to the wide range of symptoms that various researchers and physicians may diagnose as "schizophrenic" (e.g., Andreasen et al., 1995). Indeed, as is well known (at least among psychiatrists) a perusal of most any patient's diagnostic history will often reveal a variety of diagnoses that appear to change as frequently as the patient changes doctors (DSM I–IV notwithstanding).

Nevertheless, in the majority of studies that have addressed the issue of laterality, localization, and schizophrenia, left frontal and in particular left temporal lobe abnormalities have been reported (e.g., Abrams and Taylor, 1980; Flor-Henry, 1983; Morihisa et al., 1983; Morstyn et al., 1983; Trimble, 1991). Although bilateral temporal (or frontal) lobe abnormalities have been observed (Bruton et al., 1994) (see Chapters 11 and 14), unilateral dysfunction of the right half of the brain has not been demonstrated as significantly related to the manifestations of these disturbances. Thus, abnormalities predominantly involving the left hemisphere and the left frontal and temporal lobe are indicated in a sizable minority of patients diagnosed with schizophrenia, with the single exception that so called "catatonic schizophrenia" is predominantly (if not exclusively) associated with medial frontal lobe damage (see Chapters 9, 11) and related to dysfunction involving the head of the caudate and perhaps, indirectly, the amygdala (see Chapter 9).

The Frontal Lobes and Schizophrenia

Individuals classified as schizophrenic demonstrate EEG and other abnormalities suggestive of bilateral or left frontal dysfunction or hypoarousal (Akbarian et al., 1993, 1995; Ariel et al., 1983; Ingvar and Franzen, 1974a, b; Kolb and Whishaw, 1983; Levin, 1984; Stanley et al., 1995). Schizophrenic patients who are most likely to be suffering from frontal lobe dysfunction tend to display unusual mannerisms, catatonia, "emotional blunting," apathy, and/or "hebephrenia" (puerile, silly, childish, disinhibited, and jocular behavior), though negative symptoms predominate. Indeed, almost 50 years ago, Hillbom (1951) reported a strong association between individuals with head trauma and missile wounds to the frontal lobe who later developed schizophrenic-like symptoms, including catatonia and hebephrenia if the medial or right frontal lobes are also impacted.

As detailed in Chapters 9 and 11, if the medial frontal lobes are damaged or experimentally electrically stimulated, there sometimes results an inability to initiate a voluntary movement, and the "Will" to move or speak may be completely attenuated and abolished (Hassler, 1980; Laplane et al., 1977; Luria, 1980; Penfield and Jasper, 1954; Penfield and Welch, 1951). Similarly, lesions of the medial frontal lobes can cause subjects to simply sit or stand motionless, as if frozen, and display waxy flexibility, or "gegenhalten" or counterpull—i.e., involuntary resistance to movement of the extremities (Brutkowski, 1965; Hasslet, 1980; Laplane et al., 1977; Luria, 1980; Mishkin, 1964). Posturing is also noted, and patients might remain in odd and uncomfortable positions for exceedingly long time periods and make no effort to correct the situation (Freeman and Watts, 1942; Rose, 1950), as if they were dead and in the first stages of rigor mortise. These are all classic signs of catatonia.

By contrast, apathetic, blunted, and "negative" forms of schizophrenia coupled with "psychomotor retardation" reduced verbal and intellectual output and slowness of thought tend to be associated with the left frontal lobe dysfunction (Buchsbaum, 1990; Carpenter et al., 1993; Casanova et al., 1992; Crowe and Kuttner, 1991; Leven, 1994; Stanley et al., 1995; Weinberger, 1987). Nevertheless, because of the apathy and reduced speech and motor output, these patients may variably be diagnosed as depressed.

Lateral Frontal Psychosis and Neocortical Information Processing

A variety of studies that have implicated the lateral frontal lobes in the genesis of certain forms of schizophrenia (Akbarian et al., 1993, 1995; Buchsbaum, 1990; Carpenter et al., 1993; Casanova et al., 1992; Crowe and Kuttner, 1991; Leven, 1994; Stanley et al., 1995; Weinberger, 1987). As detailed in Chapter 11, the lateral frontal lobes receive converging sensory input from the primary, secondary, and association areas for audition, vision, and somethesis. In addition, whereas the orbital frontal lobes regulate subcortical and limbic arousal, the lateral frontal lobes, via massive interconnections with the dorsal medial thalamus and the posterior neocortex, are able to regulate neocortical information processing and arousal (Joseph, 1990d) (see Chapter 11).

However, in addition to frontal lobe hypoactivity and dysfunction and studies indicating a reduction in dendritic spines, synapses, and cortical neuropil, it has been reported that the thalamus is diminished in size. Specifically, a reduction in dorsomedial thalamic neurons has been reported (see Lewis, 1995).

As detailed in Chapter 11, disruptions of the integrity of the lateral frontal-thalamic-neocortical regulatory circuit can result in sensory overload, confusion, disorientation, and memory loss, therebycreating profound psychological and cognitive disturbances that may be manifested as psychosis, thought blocking, psychomotor retardation, etc. Conditions such as these, however, may also be induced by frontal lobe hypoactivity and/or disturbances in the frontal GABA neurotransmitter system (Akbarian et al., 1995; Benes, 1995; Lee and Tobin, 1995). GABA is an extremely potent inhibitor of neural activity. Reductions in GABA could result in neocortical disinhibition and thus sensory overload, confusion, and psychosis.

In this regard it is noteworthy that Akbarian and colleagues (1995) have presented evidence that suggests that the genesis for frontal lobe hypoactivity in schizophrenics is cellular abnormalities in the level of messenger RNA for glutamic acid decarboxylase (GAD). GAD is normally present in about 30% of lateral frontal neurons, particularly those which rely on GABA as a neurotransmitter.

Hence, as pointed out by Lee and Tobin (1995), decreases in frontal lobe GABA could well result in compensatory increases in GABA receptors in the caudate, temporal lobe, and cingulate, which have also been demonstrated in patients with schizophrenia. However, GABA overactivity in the caudate, and in particular the temporal lobe, is associated with positive symptoms, whereas negative symptoms are linked to the lateral frontal lobe, which suggests that these regional differences in GABA activity may not be necessarily linked.

Perhaps more importantly, lateral frontal GABA neurons receive significant input from DA axons. About 30% of frontal DA axons terminate on GABA dendrites, which also receive DA input from the striatum (see Chapters 6 and 8). In consequence, GABA abnormalities can induce DA abnormalities within the frontal lobe as well as within the basal ganglia.

THE CAUDATE AND PSYCHOSIS

It is likely, as in obsessive-compulsive disorders, that schizophrenic-like abnormalities associated with the frontal lobe may in fact be due to disruption of the basal ganglia-DA-medial/lateral frontal circuit. In this regard, as the extent of caudate-DA involvement increases, so too does the severity of schizophrenia. As is well known, many of the major antipsychotic drugs act on dopamine receptors within these nuclei.

In primates and humans, lesions, destruction, or shrinkage of the head of the caudate can result in sensory neglect, agitation, hyperactivity, and in some cases what appears to be a manic or "schizoaffective" psychosis (Aylward et al., 1994; Caplan et al., 1990; Castellanos et al., 1994; Chakos et al., 1994; Richfield et al., 1987), depending on the extent and laterality of the destruction.

Some patients will alternate between and/or display a variable mixture of hyperactivity and schizophrenic-like apathy, de-

pending, in part, on the laterality and extent of the lesion as well as associated biochemical alterations. For example, injuries or abnormalities restricted to the right caudate are more likely to result in a manic-like psychosis (Castellanos et al., 1994)—quite similar to what occurs after right frontal lobe injury (Joseph, 1986a, 1988a, 1990d), whereas left caudate lesions tend to mimic left and medial frontal dysfunction.

For example, with extensive left or bilateral caudate injuries it is not uncommon for patients to appear agitated and apathetic with decreased spontaneous activity and slowed and delayed, dysarthric (or stuttering) and emotionally flat speech, with some patients responding to questions only after a delay of 20–30 seconds (Caplan et al., 1990). This condition is particularly likely if the medial frontal lobes have been compromised as well.

Catatonic Schizophrenia

Perhaps because of its extensive interconnections with the medial frontal lobes and the supplementary motor areas (SMA)—a region that when destroyed can give rise to catatonia (Joseph, 1990d)—massive bilateral lesions to the anterior caudate and anterior putamen and surrounding tissue have been shown to produce catatonic or "frozen" states, where animals show a tendency to maintain a single posture or simply stand unmoving for weeks at a time (Denny-Brown, 1962).

Similarly, chemically lesions of the caudate can produce complete catatonia, posturing, and a cessation of all movement (Spiegel and Szekely, 1961). Conversely, because benzodiazepine receptors are located throughout the striatum, antipsychotic agents, because of their dopamine properties, can reduce this abnormal activation and reduce catatonic, schizophrenic, and related psychotic disturbances of thought and behavior.

Dopamine and the Caudate

A major function of the head of the caudate (particularly the caudate nucleus of the right hemisphere) appears to be inhibition. Similarly, the principle effect of striatal DA is inhibition (Ellison, 1994; Le Moal and Simon, 1991; Mercuri et al., 1985). In consequence, excessive and abnormal amounts of corpus striatal DA can result in increased levels of caudate-putamen activity such that limbic striatal and the reception of amygdaloid/emotional input into the striatum may be severely diminished and/or abnormally processed, thereby producing confusion and cognitive-emotional disorganization.

Similarly, individuals diagnosed as paranoid and schizophrenic with psychomotor retardation have been found to have elevated DA levels and increased DA receptor binding (Crow, 1980; Ellison, 1994; Ring et al., 1994). Therefore it appears that excessive corpus striatal activity and increases in striatal DA can induce massive inhibition and emotional blunting or distortion as well as cognitive abnormalities such that affected individuals appear psychotic (see Castellanos et al., 1994; Chakos et al., 1994; Ring et al., 1994).

Conversely, because benzodiazepine receptors are located throughout the striatum, antipsychotic agents, because of their dopamine properties, can reduce this abnormal activation. Presumably this is why dopamine blockers such as the phenothiazines are effective in reducing psychotic behavior—i.e., they reduce excessive DA binding and excessive striatal activity, which in turn restores the capability of the striatum (and related nuclei) to engage in cognitive-emotional integrative and related analytical activities.

TEMPORAL LOBES AND PSYCHOSIS

Psychotic disturbances, including schizophrenia, need not involve frontal lobe, basal ganglia, or DA dysfunction, as disorders of thought and mood may also appear following temporal lobe damage (see Chapter 4). For example, it has been repeatedly reported that a significant minority of individuals with temporal lobe epilepsy suffer from prolonged and/or repeated instances of major psychopathology (Bruton et al., 1994; Buchsbaum, 1990; Flor-Henry, 1969; Perez and Trimble, 1980; Slater and Beard, 1963; Taylor, 1975; Umbricht et al., 1995). These estimates range from 3% to 7% (see Trimble, 1991, 1992).

Similarly, it has been repeatedly demonstrated in histological studies that the temporal lobes of schizophrenics are characterized by excessive focal temporal lobe neuronal damage, including gliosis as well as evidence of neural migration errors (Akbarian et al., 1993; Bruton, 1988; Bruton et al., 1990; Stevens, 1982; Taylor, 1972). This includes volume reductions in the left anterior portions of the amygdala and hippocampus, parahippocampal gyrus, and superior temporal lobe and abnormalities in the orientation of and decreases in the number of hippocampal neurons, as demonstrated by postmortem and MRI studies (e.g., Trojanowski and Arnold, 1995; Wible et al., 1995).

Schizophrenia associated with the left temporal lobe, however, is markedly different from that manifested by left frontal, medial frontal, or basal ganglia lesions. For example, individuals with medial frontal lobe "schizophrenic" dysfunction are more likely to demonstrate thought blocking, severe apathy, catatonic symptoms, and mutism. Those with left frontal dysfunction may demonstrate poverty of speech, reduced motor activity, apathy, and depressive and blunted forms of schizophrenia. Those with basal ganglia involvement may present with a mixture of frontal and temporal lobe signs (see Chapters 4, 9, and 11).

However, in contrast with the "negative" symptoms associated with frontal lobe schizophrenia, temporal lobe schizophrenics are likely to present with "positive" symptoms, including the capacity to experience and express "warmth" (Crow, 1980; Crowe and Kuttner, 1991; Wible et al., 1995). Patients with schizophrenia who display "positive" symptoms tend to display temporal lobe but not frontal lobe abnormalities (Wible et al., 1995).

Emotional expression in temporal lobe schizophrenics, however, tends to be highly abnormal, inappropriate, incongruent, and/or exaggerated (Bear and Fedio, 1977; Crow, 1980; Crowe and Kuttner, 1991). It is also not uncommon for these individuals to demonstrate aphasic abnormalities in their thought and speech (Chaika, 1982; Clark et al., 1994; Flor-Henry, 1983; Hoffman, 1986; Hoffman et al., 1986; Perez and Trimble, 1980; Rutter, 1979; Slater and Beard, 1963) as well as to develop severe paranoid and related affective abnormalities, including auditory as well as visual hallucinations—particularly if the anterior-inferior temporal lobe and the amygdala and hippocampus are involved (Bogerts et al., 1985; Brown et al., 1986; Falkai and Bogerts, 1986), in which case memory may also be affected.

Temporal lobe schizophrenics with limbic involvement may also behave violently and aggressively (Crowe and Kuttner, 1991; Lewis et al., 1982; Taylor, 1969). They may also experience extreme feelings of religiousness (Bear, 1979; Trimble, 1991, 1992), fear, paranoia, and depression (Crowe and Kuttner, 1991; Perez and Trimble, 1980; Trimble, 1991, 1992) as well as auditory and visual hallucinations ranging from angry accusatory voices from disembodied spirits and dead relatives to frightening images of ghosts, demons, angels, and even God, as well as sensations of demonic and angelic possession (Bear, 1979; Daly, 1958; Gloor, 1986, 1992; Horowitz et al., 1968; MacLean, 1990; Mesulam, 1981; Penfield and Perot, 1963; Schenk and Bear, 1981; Slater and Beard, 1963; Subirana and Oller-Daurelia, 1953; Trimble, 1991; Weingarten et al., 1977; Williams, 1956).

THE AMYGDALA, ANTERIOR TEMPORAL LOBE, HALLUCINATIONS, AND PSYCHOSIS

Some authors have argued that, as with hallucinations, the nearer the seizure foci to the anterior temporal lobe, the greater the probability of significant psychiatric abnor-

mality (Gibbs, 1951; Gloor, 1992; Trimble, 1991). Presumably this is a function of hyperactivation of underlying limbic structures and thus abnormal attribution of emotional significance to different afferent streams of perceptual experience (Bear, 1979; Gibbs, 1951). Moreover, in that the amygdala contains multisensory neurons, once these cells are hyperactivated, patients may taste colors, see sound, feel music, as well as abnormally invest perceptual experience with emotional significance that in turn may induce fear, despair, or panic.

Complex visual hallucinations are far more likely with right temporal lobe and right hemisphere abnormalities, though visual and auditory hallucinations may occur with injuries involving either side of the brain (see Chapter 3). With complex auditory hallucinations, particularly when patients display disturbed comprehension, abnormalities of speech and thought, and schizophrenic psychosis, the left temporal lobe is generally implicated (Dauphinais et al., 1990; DeLisi et al., 1991; Flor-Henry, 1983; Perez et al., 1985; Rossi et al., 1990, 1991; Sherwin, 1981; Trimble, 1991), in particular the nuclei of the amygdala and hippocampus.

The Amygdala, DA, and Psychosis

In some instances of schizophrenia, the frontal lobe, basal ganglia, and DA transmitter systems are often affected (see Chapters 16 and 17). However, the amygdala is also part of the basal ganglia, and it is also significantly influenced by DA. Indeed, abnormal DA as well as NE and 5-HT have been reported in the amygdala of those diagnosed as psychotic (Spoont, 1993; Stevens, 1992).

However, a gross asymmetry in the postmortem levels of mesolimbic DA has been noted in the brains of those diagnosed as schizophrenic (reviewed in Le Moal and Simon, 1991). Specifically, an increase in DA has been noted in the central amygdala and in the striatum of the left hemisphere, whereas DA levels within the right cerebral hemisphere are similar to those of normals.

In fact, the left amygdala and left temporal lobe (Flor-Henry, 1969; Perez et al., 1985; Stevens, 1992) have long been thought to be a major component in the pathophysiology of psychosis and schizophrenia (Heath, 1954; Stevens, 1973). For example, abnormal activity as well as size decrements have been noted in the left amygdala (Flor-Henry, 1969; Perez et al., 1985) as well as the left inferior temporal lobe in schizophrenic patients (see Stevens, 1992). Spiking has also been observed in the amygdala of psychotic individuals who are experiencing emotional and psychological stress (see Halgren, 1992).

In this regard it has been suggested that stress, particularly when experienced prenatally, may be responsible, in some cases, for the development of DA abnormalities (see Le Moal and Simon, 1991), whereas adverse early environmental experiences and trauma may contribute to the development of kindling, spiking, and abnormal amygdala functioning (see Chapter 16).

Aphasic and Psychotic Speech

Given that Wernicke's receptive speech area is located within the left temporal lobe, afflicted individuals may experience auditory hallucinations and are more likely to demonstrate formal thought disorders and aphasic thinking, including the production of neologisms and syntactical speech errors, as well as emotional (or thought) blocking or blunting due to limbic system/temporal lobe disconnection (Perez and Trimble, 1980; Slater and Beard, 1963). As often occurs with receptive aphasia and left temporal lobe dysfunction, individuals suffering from schizophrenia may be unaware of their illness (Amador et al., 1994; Wilson et al., 1986).

Hence, certain subgroups of schizophrenics display significant abnormalities involving speech processing, such that the semantic, temporal-sequential, and lexical aspects of speech organization and comprehension are disturbed and deviantly constructed (Chaika, 1982; Clark et al., 1994; Flor-Henry, 1983; Hoffman, 1986;

Hoffman et al., 1986; Rutter, 1979). Indeed, significant similarities between schizophrenic discourse and aphasic abnormalities have been reported (Faber et al., 1983). Moreover, it has been reported that the severity of the thought disorder is directly related to abnormalities in the asymmetry of the planum temporale (Petty et al., 1995).

Conversely, because individuals with Wernicke's aphasia display unusual speech, loss of comprehension, a failure to comprehend that they no longer comprehend or "make sense" when speaking, and paranoia and/or euphoria, they are at risk for being misdiagnosed as psychotic or suffering from a formal thought disorder, i.e., "schizophrenia"

Hence, considerable evidence indicates that a significant relationship exists between abnormal left hemisphere and left temporal lobe functioning and schizophrenic language, thought, and behavior (see also Chapters 14 and 17).

Between-Seizure Psychoses

Estimates of psychosis and affective disorders among individuals suffering from temporal lobe epilepsy, be it a single episode or chronic psychiatric condition, have varied from 2% to 81% (Flor-Henry, 1983; Gibbs, 1951; Jensen and Larsen, 1979; Sherwin, 1977, 1981; Taylor, 1972; Trimble, 1991). Probably a more realistic figure is about 4.5–7% (Trimble, 1991).

Frequently these disturbances wax and wane, such that some patients experience islands of normality followed by islands of psychosis (e.g., Flor-Henry, 1983; Umbricht et al., 1995). Possibly these changes are a function in variations in subclinical seizure activity and kindling.

In some instances, the psychiatric disturbance may develop over days as a prelude to the actual onset of a seizure, presumably because of increasing levels of abnormal activity until the seizure is triggered. In other cases, the alterations in personality and emotionality may occur following the seizure and may persist for weeks or even months (Trimble, 1991; Umbricht et al., 1995).

In yet other instances, patients may act increasingly bizarre for weeks, experience a seizure, and then behave in a normal fashion for some time ("forced normalization," Flor-Henry, 1983), only to again begin acting increasingly bizarre (Umbricht et al., 1995). This has been referred to as an interictal (between-seizure) and postictal psychosis and is possibly secondary to a build-up of abnormal activity in the temporal lobes and limbic system (Bear, 1979; Flor-Henry, 1969, 1983) Hence, once the seizure occurs, the level of abnormal activity is decreased and the psychosis goes away only to gradually return as abnormal activity again builds up.

SOCIAL-EMOTIONAL AGNOSIA

Patients diagnosed with schizophrenia and children afflicted with autism often appear to suffer from abnormalities involving social-emotional intellectual perceptual and expressive functioning. That is, such individuals appear to have extreme difficulty determining and identifying the motivational and emotional significance of externally occurring events, to discern social-emotional nuances conveyed by others, or to select what behavior is appropriate given a specific social context. In consequence, in addition to their other afflictions, autistic children or schizophrenic adults may appear socially-emotionally blunted or agnosic to varying degrees.

Social-emotional agnosias are commonly observed following amygdala, and right cerebral lesions involving the temporal lobe in particular (Joseph, 1988a, 1992a). For example, with damage to the right temporal-occipital region, there can result a severe disturbance in the ability to recognize the faces of friends, loved ones, or pets (Braun et al., 1994; DeRenzi, 1986; DeRenzi et al., 1968; DeRenzi and Spinnler, 1966; Hanley et al., 1990; Hecaen and Angelergues, 1962; Landis et al., 1986; Levine, 1978; Whitely and Warrington, 1977) or to

discriminate and identify even facial affect (Braun et al., 1994)—prosopagnosia. In fact, with gradual deterioration and degeneration of the right inferior temporal lobe, patients may suffer progressive prosopagnosia (Evans et al., 1995). Some patients may in fact be unable to recognize their own face in the mirror.

In large part these neocortical deficits are due to limbic system disconnection and/or destruction of specific limbic nuclei such as the amygdala (Joseph, 1990a, 1992a) (see Chapter 5). For example, lesions that destroy amygdala fibers of passage to the right or left temporal lobe essentially disconnect the neocortex and amygdala such that they can no longer receive or transmit appropriate signals. In addition, lesions to only the right or left amygdala (in primates) may induce social-emotional agnosic states, but only in regard to stimuli and persons in the contralateral half of auditory, visual, and tactile space (Downer, 1961)—e.g., only the right or left hemisphere may become severely affected. Hence, in humans, given right cerebral dominance for social and emotional stimuli, a right-sided disconnection or destruction of the amygdala can exert profound influences regardless of which hemisphere is perceiving and responding to emotional stimuli.

Among primates and mammals, bilateral destruction of the amygdala significantly disturbs the ability to determine and identify the motivational and emotional significance of externally occurring events, to discern social-emotional nuances conveyed by others, or to select what behavior is appropriate given a specific social context (Bunnel, 1966; Fuller et al., 1957; Gloor, 1960; Kling and Brothers, 1992; Kluver and Bucy, 1939; Weiskrantz, 1956). Bilateral lesions lower responsiveness to aversive and social stimuli and reduce aggressiveness, fearfulness, competitiveness, dominance, and social interest (Rosvold et al., 1954). Indeed, this condition is so pervasive that subjects seem to have tremendous difficulty discerning the meaning or recognizing the significance of even common objects—a condition sometimes referred to as "psychic blindness" or the "Kluver-Bucy syndrome."

Thus, animals with bilateral amygdaloid destruction, although able to see and interact with their environment, may respond in an emotionally blunted manner, and seem unable to recognize what they see, feel, and experience. Things seem stripped of meaning (Brown and Schaffer, 1888; Gloor, 1960; Kluver and Bucy, 1939; Weiskrantz, 1956) such that even the ability to appropriately interact with loved ones is impaired—similar in some respects to what occurs in autism and with adults afflicted with schizophrenia.

For example, Terzian and Ore (1955) described a young man who, following bilateral removal of the amygdala, subsequently demonstrated an inability to recognize anyone, including close friends, relatives, and his mother. He ceased to respond in an emotional manner to his environment and seemed unable to recognize feelings expressed by others. In addition, he became extremely socially unresponsive such that he preferred to sit in isolation, well away from others.

Among primates who have undergone bilateral amygdaloid removal, once they are released from captivity and allowed to return to their social group, a social-emotional agnosia becomes readily apparent as they no longer respond to or seem able to appreciate or understand emotional or social nuances. Indeed, they appear to have little or no interest in social activity and persistently attempt to avoid contact with others (Dicks et al., 1969; Jonason and Enloe, 1972; Jonason et al., 1973; Kling and Brothers, 1992;). If approached they withdraw, and if followed they flee. Indeed, they behave as if they have no understanding of what is expected of them or what others intend or are attempting to convey, even when the behavior is quite friendly and concerned. Among adults with bilateral lesions, total isolation seems to be preferred.

In addition, they no longer display appropriate social or emotional behaviors, and if kept in captivity will fall in dominance in a group or competitive situation—even when formerly dominant (Bunnel, 1966; Dicks et al., 1969; Fuller et al., 1957; Jonason and Enloe, 1972; Jonason et al., 1973; Rosvold et al., 1954).

As might be expected, maternal behavior is severely affected. According to Kling (1972), mothers will behave as if their "infant were a strange object be mouthed, bitten, and tossed around as though it were a rubber ball."

Given the many similarities between amygdala destruction and the social-emotional agnosias demonstrated by certain subgroups of those classified as schizophrenic, as well as those diagnosed as autistic, it would thus appear that amygdala/temporal lobe dysfunction is implicated as a major source of psychopathology in these patient populations.

CONGENITAL AND EARLY ENVIRONMENTAL CONTRIBUTIONS TO PSYCHOSIS

The brain is susceptible to a variety of disruptive and damaging influences, including emotional trauma and stress, stroke, tumors, seizures, degenerative and biochemical abnormalities, drugs, alcohol, toxic exposure, and congenital trauma and related abnormalities.

However, the nature of the disruption, and the areas of the brain most affected, are determined not just by the causative agents but also by the timing. For example, prenatal disturbances of early onset (e.g., a viral infection) may selectively interfere with neural migration and thus cortical formation, whereas an identical disturbance experienced just prior to birth may have little or no effect on brain development.

As noted, be it a prenatal infection or a trauma, or a cerebral infarct suffered at age 60, in general it is only the more moderate to severe disturbances that are likely to come to the attention of the physician. However, as to mild abnormalities, such as a small and circumscribed stroke, there is the possibility that related deficits may go unnoticed by friends, family, and even the patient.

The same is true regarding prenatal damage, the effects of which may not become fully manifest until later in life; at which point the symptoms may be mistakenly assumed as having their onset at this later time period. Disturbances of prenatal origin but adult onset includes abnormalities referred to as schizophrenia as well as a host of related and unrelated disabilities involving intellect, cognition, and emotional functioning (e.g., Akbarian et al., 1993, 1995; Benes, 1991, 1995; Lewis, 1995).

Consider, for example, the broad symptom complex that is associated with schizophrenia. As detailed in previous chapters, a variety of anatomical as well as biochemical disturbances have been found in various populations of individuals diagnosed as suffering from this disorder. Although these abnormalities in some cases are due to infarcts, tumors, and degenerative disturbances experienced as an adult, they may just as likely be congenital and secondary to intrinsic and extrinsic prenatal influences that result in neuronal migration errors, abnormal synaptic interconnections, increased ventricular size, basal ganglia and cerebellar abnormalities, and so on.

As noted above, Akbarian and colleagues (1995) have demonstrated that the frontal lobes of individuals diagnosed as schizophrenic show a reduced expression of glutamic acid decarboxylase (GAD), which is a key enzyme in GABA synthesis. Presumably, reduced GAD expression is linked to or brought about by reduced frontal lobe activity (Akbarian et al., 1995). However, Akbarian and colleagues (1995) also argue that, since they failed to find any alterations in actual cell numbers in their studies, this condition may not be due to disturbances in neural migration per se.

On the other hand, Akbarian and coworkers (1993) have reported that a small population of nicotinamide-adenine dinucleotide phosphate-diaphorase neurons located in the white matter of the frontal and temporal lobes are also abnormal in schizophrenics, a disturbance that appears to be directly linked to congenital abnormalities in cerebral development. Coupling this finding with their 1995 data, Akbarian and colleagues (1995) suggest that these conditions (i.e., reduced GAD, altered neuronal distribution in the white matter) therefore may well be due to disturbances of neuronal migration or abnormalities involving the cortical subplate during early embryonic formation of the brain. Because the cortical subplate serves as the terminal junction for migrating neocortical neurons (see Chapter 18), abnormalities involving the subplate may in turn result in white matter abnormalities and displaced (subplate) neurons, as well as reduced GAD expression within all six layers of the neocortex of schizophrenics (see also Lewis, 1995).

In addition, reductions in layer II and III interneurons have been reported in the brains of schizophrenics (reviewed in Benes, 1995). An increase in synaptic interconnections between layers I and II has also been documented (Benes, 1995). Normally these initial interconnections (which serve to nourish and maintain the cell until its true synaptic counterpart is available) drop out as surrounding neurons and those which are still arriving begin to differentiate, mature, and grow axons and dendrites (see Chapter 18). Presumably, layer I neurons maintain these interconnections because of the failure of their normal counterparts to successfully complete their migration (thus creating white matter heterotopias).

Neurodevelopmental disturbances involving select aspects of neural migration or cortical formation would also explain the lack of gliosis. That is, these and other indications of preferential tissue or neurochemical loss in the absence of gliosis argues against a degenerative process of late onset—at least in these cases.

Hence, in some instances, individuals diagnosed as suffering from schizophrenia may well have suffered a prenatal insult to the brain—e.g., viral infection, malnutrition (see Chapter 18) that resulted in migration errors, and abnormal neuronal distribution and displacement, and thus neocortical as well as white matter aberrations.

Abuse and Cerebral Development

Cerebral development continues well past the first decade of life (see Chapter 18). Indeed, for the first few years of life, large regions of the neocortex are markedly immature, and this includes the frontal and temporal lobes, and in fact layers II and III of the neocortex—layers that contain local circuit neurons that interlink various neocortical areas. As neurons are still maturing, and as dendritic and axonal synaptic connections are still being formed, the brain of the child, like that of the neonate, is also at risk for developing abnormally.

Psychotic individuals often have children who are reared in a psychotic environment. Many children are also neglected and physically and sexually abused by seemingly "normal" parents (see Chapter 15). Babies may be shaken repeatedly, inducing axonal and neocortical shearing as well as coup and contra coupe lesions, and they may be battered, struck in the head, or held by the legs as parents or caretakers, siblings, etc., slam them against walls or furniture. Unless their lives are threatened, children who are abused do not always come to the attention of the authorities. Those who are emotionally abused rarely do.

Nevertheless, and as detailed in Chapters 7, 16, and 18, because different regions of the brain take years if not decades to completely mature (see Chapter 18), and because of the plastic capabilities of the immature cerebrum, severe emotional and environmental stress, be it neglect or physical abuse, can induce structural abnormalities in the brain, the temporal lobes in particular, as well as create and establish abnormal neural interconnections (see Chapter 18). Indeed, the brain of a young child is particularly susceptible to repetitive and severe emotional stress, as well as stroke or head injury, all of which can induce significant brain damage and functional/structural abnormalities (see Chapters 16, 18).

In addition, when these emotional stresses are coupled with the abnormal learning that characterizes abusive parent-child relationships and early rearing experiences, associated abnormal neural networks are created and abnormal emotional, cognitive, and behavioral functioning results (Joseph, 1992b). However, depending on the age of the child and even the type of stress or abuse (e.g., nutritional), different regions of the brain may be affected to different degrees.

Schizophrenia, Early-Onset Frontal Lobe Injury, and Symptom Formation

Although schizophrenia seems to develop in the late teens, early twenties, and beyond, as indicated above, the disturbance may well be congenital and/or due to early childhood emotional or physical trauma. However, if congenital or due to childhood trauma, why the long delay in the development of psychotic symptoms? Moreover, when considering, for example, the role of the frontal lobes in attention, arousal, inhibition, etc. or the temporal lobes in memory, it would seem that if these regions became abnormal early in life, or congenitally, related abnormalities would become immediately apparent.

Possibly, as has been demonstrated with primates, frontal dysfunction of early onset does not become fully manifest until later in life (e.g., Goldman-Rakic, 1974)—that is, when the frontal lobes are required to assume new functions. Similarly, the temporal lobe, the inferior regions in particular, continue to develop and mature well into late childhood and beyond (Blinkov and Glezer, 1968). On the other hand, it has also been demonstrated that some children suffering from autism (which has also been referred to as childhood schizophrenia) suffer from maturational delays involving the frontal cortex (Zibovicious et al., 1995). In these instances, related psychological abnormalities are apparent often within the first 2 years of life. Possibly, at least in cases of congenital or early childhood origin, what appears to be adult versus childhood onset is really a matter of degree. That is, those who are most severely affected demonstrate severe psychotic symptoms that are recognized at an earlier age, whereas those who are mildly to moderately abnormal are able to better manage and cope. Nevertheless, in this regard it is noteworthy that many adults diagnosed as schizophrenic were often viewed by schoolmates and teachers, if not by their parents, as odd, socially inept, or withdrawn, even as children. However, although "odd" (and probably psychotic), because these "preschizophrenic" children are still cared for by parents and spend much of their time alone in a structured environment (school), the full range of their pathology is not always evident.

However, once these individuals leave the structured environment provided by school (including college) and are expected to fend for themselves, it is not uncommon for their problems to finally come to the attention of the authorities, social workers, psychologists, and psychiatrists. This is particularly true of males (e.g., Szymanski et al., 1995), who are expected to become independent at an earlier age and who are far less likely, once they reach adulthood, to find someone to take care of them. By contrast, even psychotic females may be able to trade sexual favors so as to get married and/or have someone provide for them. Thus, a consistent finding is that males are hospitalized at an earlier age, have more frequent hospitalizations with longer stays, and suffer more frequent relapses (e.g., Szymanski et al., 1995). Of course, sex differences in cerebral functioning are probably also contributory. In this regard it is noteworthy that young males appear to be more vulnerable to early environmental trauma, particularly emotional neglect, and they are not as resilient and do not appear to recover as much or as

quickly as females after removal from these environments.

CONCLUSIONS

Although encased in a thick skull, covered by leathery tough meninges, and protected by a blood-brain barrier, the brain is subject to a myriad of environmental and intrinsic insults, beginning soon after conception, and continuing until death. Often it is only when the brain damage is moderate to severe, or involves speech and the motor system, that medical or psychological treatment may be sought.

The same is true regarding prenatal lesions. Moreover, unless moderate to severe, subsequent cognitive, intellectual, and emotional abnormalities due to congenital or early environmentally induced lesions may not become manifest until later in life. In fact, even after these related disturbances become manifest, the underlying neurological foundations may be completely overlooked.

The same is true regarding adults. It is not uncommon for patients to suffer from, for example, an expanding medial frontal tumor and to be diagnosed as depressed, but for the tumor to be overlooked. Or to suffer scrotal pain or electric shock–like sensations due to a parietal tumor, but to be viewed as hysterical. Nevertheless, be it congenital abnormalities, small infarcts, neoplasm, slowly progressive deteriorative or biochemical disorders, subclinical seizures, early environmental emotional trauma, or even "mild" head injuries, these insults are often the triggering event if not the root cause of many if not most psychotic and related cognitive and emotional disorders. The symptom complex, however, is dependent not only on the nature of the disturbance (e.g., tumor, neurotransmitter) but the location (e.g., frontal versus temporal) and the timing of the original intrinsic insult or environmental/emotional trauma.

SECTION VII

Neuroanatomy and Pathophysiology of Head Injury, Stroke, Neoplasm, and Abnormal Development

18

Neuroanatomy of Normal and Abnormal Cerebral Development:

Neuronal Migration Errors, Congenital Defects, Neural Plasticity, Environmental Influences, Trauma, Abuse, Psychosis

The purpose of this chapter is to detail those processes involved in both normal and abnormal neuroanatomical, neurological, and neuropsychological development, beginning soon after conception and continuing for the first 5 years of postnatal development. This is a time period when the developing brain is highly susceptible to a variety of disrupting influences, some of which are intrinsic, others of which are external and environmental, all of which can interfere with the maturational process and cerebral functional and neuroanatomical organization. However, the nature of the disruption, and the areas of the brain most affected, are determined not just by the causative agents but the timing of the interference or trauma.

For example, prenatal disturbances of early onset may selectively interfere with neural migration and thus cortical formation, whereas an identical disturbance experienced just prior to birth may have little or no effect on brain development.

However, the brain of the infant and child is also exceedingly immature and plastic, and its development can be significantly impacted by environmental insults such as neglect or repetitive head injuries as occur in child abuse.

Specific regions of the developing brain also require specific types of environmental input and stimulation in order for proper interconnections to be established and maintained (e.g., Casagrande and Joseph, 1978, 1980; Greenough and Chang, 1988). If denied that input early in postnatal development (e.g., visual or emotional stimulation), related cortical areas will be adversely affected.

Hence, during the prenatal period and for the first few years of life, the cerebrum is particularly vulnerable to a variety of damaging influences that may range from drug exposure, neglect, and malnourishment to physical abuse and blows to the head—all of which not only damage the brain but can result in the creation of abnormal neural circuitry and interconnections.

However, be it postnatal or prenatal influences, in general it is only the more severe disturbances that are likely to come to the attention of a physician. Mild to moderate abnormalities may not even be recognized, and repeated instances of trauma (such as in cases of abuse) may never be treated (e.g., Courtois, 1988; Leestma, 1995).

Nevertheless, although the initial deficit may be overlooked even when physicians are in attendance (such as at birth), it is not uncommon for individuals who are traumatized prenatally or during the first few years of life to later suffer from a variety of cognitive, intellectual, behavioral, and emotional problems that are a direct consequence of these cerebral irregularities and injuries. That is, cognitive, intellectual, and emotional abnormalities that are a direct consequence of an early cerebral trauma may not become fully manifest until later in life, because these functions are not required until later in development (e.g., Goldman-Rakic, 1971), at which point these disturbances may be mistakenly assumed as having their onset at this later time period. These include abnormalities referred to as "autism" and "schizophrenia" (e.g., Benes, 1991, 1995; Carpenter et al., 1991; Crow et al., 1989).

Given that it is not uncommon for pregnant mothers to abuse alcohol or drugs and/or to be beaten and traumatized, and that it is also not uncommon for infants to be shaken, beaten, and subjected to head injuries and physical and sexual abuse, it is likely that their developing brain will be negatively impacted. As these individuals grow older they may suffer from low intelligence, learning disabilities, and poor impulse control, as well as a host of disturbances involving motor, intellect, cognition, and emotional functioning.

PART I

ONTOGENY OF CEREBRAL AND CORTICAL DEVELOPMENT

OVERVIEW

Neocortical and brain development can be divided into nine stages:

1. The generation of the neuroectoderm from ectoderm, and the formation of the neural preplate
2. The splitting of the preplate and the formation of the neural plate
3. The rising up and the inward folding of the sides of the neural plate, thereby forming the neural tube
4. The generation of neuroepithelium and immature nerve cells, the neuroblasts
5. The flexure of the neural tube forming three then five vesicles
6a. The migration to and aggregation of neurons within specific brain regions, thereby forming nuclei
6b. The migration of the neuroblasts from the ventricular and subventricular zones so as to form limbic and striatal nuclei and specific neocortical layers and columns

7. Differentiation and growth of neurons, axons and dendrites, a process of organization and reorganization that can span an individual's lifetime
8. The formation of synaptic connections, a process that spans the individual's lifetime
9. The elimination of neurons, axons, and dendrites, a process that spans the individual's lifetime

Ontogeny of Cerebral and Cortical Development

Following fertilization, the rapidly multiplying cells form three principle germ layers: ectoderm, mesoderm, and endoderm. The mesoderm gives rise to the chordomesoderm and acts to stimulate the ectoderm to form the neuroectoderm and later the neuroepithelium. The neuroepithelium will give rise to those cells which are to form the central nervous system (CNS) (Arey, 1965; Cowan, 1979; Hamilton and Mossman, 1972; Joseph, 1982; Lund, 1978; Sidman and Rakic, 1973, 1982).

During the first few weeks of fetal development, the rapidly proliferating pre-CNS cells of the neuroectoderm give rise to a transient outer cortical tissue referred to as a "preplate." Although much of the preplate degenerates over the course of development, a long thin band of neural ectodermal tissue, i.e., the neural plate, is formed within the preplate, thereby splitting it into two layers.

The neural plate essentially begins as a thickening of the ectoderm that overlies the notochord (a derivative of the chordomesoderm). Beginning about the 22nd day of gestation, the outer edges of the neural plate begin to grow upward on each side of the midline, forming neural ridges with a deep central neural groove. The two ridges continue to grow upward and then meet in the midline, forming a neural tube—a process that is complete by about the 28th day of gestation, at which point the anterior neuropore and then the posterior neuropore have closed.

One of the most distinguishing features of the neural tube, however, is the eye-forming area, which is first discernible in its most anterior portion. The optic vesicles soon begin to extend outward so as to form the optic cup, which in turn will form the retina and then the eye.

At about the fourth week, the neonate is about 4 mm in length. After the neural tube is formed, the internal neural ectodermal tissue, referred to now as the neuroepithelium, continues to differentiate, and neuroblasts emerge and begin to migrate toward their terminal substrate (Bayer et al., 1995; Sidman and Rakic, 1973, 1982). As neuronal development and migration continues the most anterior half of the tube begins to swell in three different places. These three swellings constitute what will become the midbrain (where the optic and auditory lobes will form), the hindbrain (from which the pons and medulla will be fashioned), and the telencephalon/forebrain (from which the limbic system, striatum, and cerebral hemispheres will form). It is the caudal 50% of the neural tube that eventually becomes the spinal cord.

The formation of these three vesicles and their division into five vesicles (myelencephalon, metencephalon, mesencephalon, diencephalon, telencephalon) basically reflect the different stages of growth of the neuroepithelium (Bayer et al., 1995), which in turn heralds and thus predicts the lateral morphological differentiation and organization of the mature CNS.

Specifically, the organization of the neuroepithelium essentially reflects an intrinsic blueprint and organizational plan from which future neural populations will conform. That is, specific sites within the neuroepithelium give rise to discrete neuronal populations that will migrate and settle and thus form specific brain tissues (Bayer et al., 1995).

For example, according to Bayer and associates (1995:100), "each neuronal population in the central nervous system has three defining characteristics. (1) Each is generated in a specific site in the neuroepithelium. (2) Each one has a unique timetable of neurogenesis that is (3) linked to developmental patterns in other populations

with which it will have strong interconnections." These authors also argue that neurons begin migrating soon after neurogenesis is initiated in the third week of gestation and that neurons continue to develop and migrate after birth.

Because of this time-dependent relationship, insults to the brain can affect specific populations and thus specific regions of the developing brain while seemingly sparing other tissues. And yet, that is not always the case. For example, if the insult is experienced early during neurogenesis, the precursors to neurons in the germinal matrix are destroyed in vast quantities. "In cases where neuronal populations are produced over a short time span, a brief, but massive, insult may lead to the permanent decimation of the entire population" (Bayer et al., 1995). Nevertheless, the generation of one population of neurons and their migration can then trigger the generation of a second population of neurons. If the first wave is decimated, abnormalities in the second wave will also occur, although the initial trauma, infection, etc., has long passed (Bayer et al., 1995; Rakic, 1988).

CORTICAL FORMATION
Ontogeny and Evolution

As detailed in Chapter 1, the forebrain is largely (but not completely) a secondary elaboration and evolutionary outgrowth of the olfactory system, and in this regard, so too are layers II–VIa of the neocortex. That is, with the exception of layers I and VII, the basic pattern of olfactory "neural" organization is repeated and replicated throughout the telencephalon, which is quite olfactory- like in structure (e.g., Allman, 1990; Haberly, 1990; however, see Bayer et al., 1995).

However, the fetal olfactory bulb develops separately from the neural tube and has a semi-independent ectodermal origin, appearing as a posterior extension of mesodermally derived ectoderm, and growing in a posterior ventral direction and appearing as a cortical formation at about 6 weeks of embryonic development.

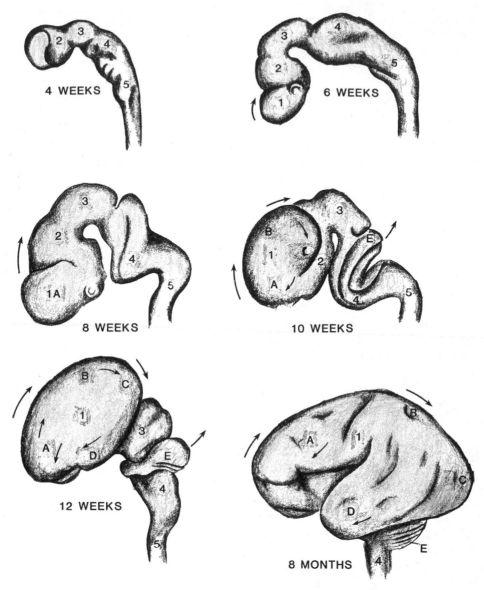

Figure 18.1   Sequences of embryonic and fetal development of the human brain. Not drawn to scale. Arrows suggest patterns and direction of growth. At approximately 4 weeks three swellings (vesicles) appear, followed by the formation of five vesicles. 5, Spinal cord; 4, medulla; 3, pons; 2, midbrain; 1, telencephalon. A, Frontal lobe; B, parietal lobe; C, occipital lobe; D, temporal lobe; E, cerebellum. (From Joseph R. The neuropsychology of development. J Clin Psychol 38:4–33, 1982.)

It is not known if the olfactory bulb or pre-olfactory ectoderm contributes neurons or neuroblasts to the developing forebrain. In this regard, ontogeny does not appear to reflect phylogeny.

There is, however, some evidence which suggests that the olfactory ectoderm exerts a tremendous influences on the growth and differentiation of the forebrain, even though telencephalic cells and nuclei are believed to originate in the neuroepithelium of the anterior-most portion of the neural tube. For example, the olfactory bulbs are invariably and severely affected even in mild prosencephalic disorders of forebrain development (see below).

The forebrain, however, is also an evolutionary derivative of visually sensitive nuclei from which portions of the midbrain, thalamus, and hippocampus are derived. Similarly, the fetal forebrain is derived from both midbrain-like reticular tissue (layers I and VII) and the anterior-most portion of the neural tube. Layers I and VII (also referred to as VIb) are organized in a fashion that is quite similar to the midbrain (Marin-Padilla, 1988a), and like portions of the midbrain these layers are heavily infiltrated by the norepinephrine (NE) and serotonin (5-HT) transmitter systems.

As will be detailed below, wave after wave of migrating neurons (neuroblasts) successively come to be sandwiched between the phylogenetically older two midbrain-like layers (I and VII) that in fact are extensions of the anterior-most portion of the neural tube and fetal midbrain. That is, initially, the most anterior portion of the neural tube is not forebrain but midbrain, the optic cup being its most distinguishing feature.

Indeed, even in the adult human brain there is a border area that separates and demarcates the most anterior portion of the midbrain from the forebrain. They are bordered by the lamina terminalis, which in turn originally forms the most rostral aspect of the embryonic neural tube.

The Limbic System, Striatum, Neocortex

Paralleling, in some respects, evolutionary development (see Chapter 1), the amygdala and the ventral floor of the telencephalon appears and expands prior to the appearance of the olfactory bulb (but not the pre-olfactory epithelium), followed by the ventral-most portion of the basal ganglia and the medially situated septal nuclei (Humphrey, 1972). This expansion is due to the continued proliferation of neocortical neuroepithelium, which also induces the lateral ventricles to balloon outward and reflects the rapid proliferation of neuroblasts and neurons that are acting to form various nuclei and the cortical layers of not just the neocortex but the limbic system (Sidman and Rakic, 1982). From the ventral floor and dorsal-lateral, dorsal-medial walls the primordium of the neopallium appears, which will give rise to the neocortex (Bayer et al., 1995; Marin-Padilla, 1988a).

However, just as the neocortex could be viewed as having two separate origins—i.e., midbrain and olfactory/limbic forebrain (see Chapter 1)—the layers of neurons that come to form the neocortex as well as the limbic cortex have two separate origins, the ventricular and subventricular zones of the neuroepithelium. That is, although the majority of neocortical neurons are generated in the ventricular zones, a significant minority are produced in the subventricular zone. By contrast, whereas the majority of amygdala, septal nuclei, and hippocampal neurons originate in the subventricular zones, these various tissues also receive neurons from the ventricular zones (Bayer et al., 1995; Nowakowski and Rakic, 1981; Sidman and Rakic, 1982).

This is not entirely surprising as both the amygdala and hippocampus contain phylogenetically old and more recently evolved neural tissue (see Chapter 1). For example, the lateral portions of the amygdala are quite cortical in appearance.

Hence, just as the amygdala and hippocampus contain both phylogenetically old and new neurons, so to does the neocortex, layers II–VIa being sandwiched between the midbrain-derived layers I and VII. Nevertheless, it is noteworthy that with the exception of hippocampal layer I, the other four layers of the hippocampal (e.g., entorhinal) cortex are substantially different from that of the neocortex (Bayer et al., 1995). By contrast, the lateral amygdala is quite similar to neocortex.

As is postulated in evolutionary theory (see Chapter 1), during fetal development the neocortex comes to be established via the development and cortical elaboration of the amygdala, hippocampus, and septal nuclei, followed by the striatum and then the anterior cingulate. Through their respective primordial contributions to different regions and layers of neocortex, the telencephalon expands and increasingly encapsulates the limbic system, forming the two cerebral hemispheres in the process.

Specifically, by the 5th week (35th day), there occurs a lateral outgrowth that forms the telencephalic vesicles. The telencephalic vesicles then begin to rise upward, forming paired lateral evaginations and cavities. From the medial portions of these outgrowths, the diencephalon comes to be fashioned. By the 7th to 8th week the choroid plexus begins to form and invaginates the medial portions of each lateral cavity (Bayer et al., 1995; Sidman and Rakic, 1982).

As noted, however, the neurons that will eventually compose the layers of the neocortex are predominantly generated in the germinal region of the ventricular surface. These immature neurons must then migrate outwards toward the surface (which becomes increasingly distant), thereby forming five layers of neocortex sand-

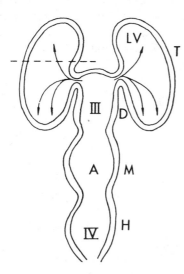

Figure 18.2 The outpouching of the telencephalon. T, Telencephalon; D, diencephalon; M, midbrain; H, hindbrain; IV, fourth ventricle; A, aqueduct; III, third ventricle; LV, lateral ventricle. (From Lund RD. Development and plasticity of the brain. Oxford University Press, New York, 1979.)

wiched between two phylogenetically old layers. Layers I and VII in fact appear to attract, nourish, stabilize, and initially maintain the cellular integrity of these arriving neurons until they are sufficiently mature and/or until their synaptic counterparts arrive. At this point, these neurons begin to lose their interconnections with these older layers and grow and establish dendritic and axonal interconnections so as to form patterns of connection that will be maintained through adulthood.

BRAINSTEM, CEREBELLUM, AND LIMBIC SYSTEM FETAL DEVELOPMENT

By the 32nd day of gestation, the remaining flexures begin to form, and by the 33rd day, the first stages of hemispheric construction begin, the two walls of which in turn induce an evagination. This marks the beginning of the development of the cerebral vesicles (Marin-Padilla, 1988a), which in turn contain the primordia of the forebrain, midbrain, and brainstem.

Brainstem

The pons and medulla of the brainstem constitute the middle portions of the neural tube, the caudal aspects of which form the spinal cord. From the 5th to 7th weeks of gestation, neurons begin to migrate toward and form the various nuclei of the brainstem (Bayer et al., 1995; Sidman and Rakic, 1982). This process continues, including the penetration and development of various fiber tracts, up until birth and for the first 6 months of life and beyond.

Midbrain

The midbrain constitutes the most anterior portion of the neural tube, and by the 5th through 7th weeks of gestation the neurons in the raphe nuclei followed by dopamine neurons in the substantia nigra begin to form (Bayer et al., 1995). From the 5th to 9th weeks neurons that will form the superior colliculus are generated, and from the 6th up to the 17th weeks of development, neurons of the inferior colliculus are formed (Bayer et al., 1995).

Cerebellum

It is around the 4th week that the pontine flexure begins to develop, which in turn will give rise to the cerebellum. Specifically, the lateral edges of the neural tube become quite wide at the level of the alar plates, forming the rhombencephalon around the 4th week. As the pontine region continues to flex it forms a transverse crease in the roof of the rhombencephalon, whereas the alar plates have become so thickened by the tremendous proliferation of neuroblasts that they form two lips. These are the rhombic lips from which will spring forth the cerebellum (Bayer et al., 1995; Sidman and Rakic, 1982).

Hypothalamus

Hypothalamic and preoptic nuclei are derived from the neuroepithelium of the floor of the third ventricle, immediately below the neuroepithelium, which generates thalamic neurons. Hypothalamic neurons are generated in the 5th to 7th weeks, whereas the preoptic neurons are formed from weeks 4 through 11 (Bayer et al., 1995).

Amygdala

From the ventral surface of the developing brain will arise the primordium of the olfactory striatum, the lateral-ventral portion giving rise to the amygdala, and the medial ventral region giving rise to the corpus and limbic striatum. The amygdala essentially forms the floor of the telencephalon, extending from the olfactory tubercle to the hippocampal region. Neurons that will form the amygdala begin to be generated at the end of the 4th week through the 7th week (Humphrey, 1972). However, different amygdala subnuclei form at different time periods, such that some neurons are not generated until the 16th week (Bayer et al., 1995).

As noted in Chapter 9, the striatum and amygdala maintain not only rich interconnections, but the dorsal-posterior tail of the amygdala merges with the caudate (i.e., forming the tail of the caudate). The cortico-medial portion of the amygdala gives rise to the limbic striatum. In addition, it is from the amygdaloid portion of the emerging forebrain that portions of the anterior-inferior and anterior and superior temporal lobe and overlying neocortical mantle will evolve and develop.

Limbic Striatum

Neurons that will form the limbic (ventral striatum), i.e., olfactory tubercle, ventral (and dorsal) globus pallidus, and substantia innominata, are generated and begin to migrate from weeks 4 through 7, with those forming the nucleus accumbens forming, migrating, and settling from weeks 10 through 24 (Bayer et al., 1995).

Corpus Striatum

Neurons that will form the caudate and putamen have an extended generation period. They are first formed early in the 7th week and continue to be generated and to migrate up until the 18th week of gestation (Bayer et al., 1995; Humphrey, 1972; Sidman and Rakic, 1982).

Septal Nuclei

Most neurons that will come to form the septal nuclei are generated within the neuroepithelium, which lines the medial wall of the lateral ventricle. These neurons are generated and begin to migrate from weeks 5 through 9 (Bayer et al., 1995).

Hippocampus

From the medial wall appears the primordium of the archipallium from which the archicortex and hippocampus will develop, followed by the septal nucleus, then the hippocampal commissure and the fornix, followed by the medial walls (Marin-Padilla, 1988a).

It is the anterior-dorsal portion of the hippocampus (and later the septal nucleus) that first begins to differentiate so as to form the medial walls (Sanides, 1969, 1970). However, with the evagination of the hemispheres, the hippocampus begins to be pushed downward and comes to extend in a caudal arc such that by the 16th week of gestation it has rotated and increasingly assumes the characteristic location and configuration as is seen in adults. However, during this process the hippocampus gives rise to neocortical tissue that will in turn form portions of the superior parietal, occipital, posterior-inferior, and middle temporal lobes.

THE TELENCEPHALON

The neocortical neuroepithelium gives birth to and generates neurons that will migrate to and assist in the formation of the limbic floor of the telencephalon (Bayer et al., 1995; Sidman and Rakic, 1973, 1982), creating the amygdala, piriform cortex, and hippocampus. The ventral floor and dorsal-lateral, dorsal-medial walls the primordium of the neopallium will give rise to the neocortex (Bayer et al., 1995; Marin-Padilla, 1988a; Sidman and Rakic, 1982). As noted, although both the amygdala and hippocampus contribute to the development of neocortex, large numbers of neocortical neurons are generated in the ventricular zone, whereas many of the neurons of the amygdala and hippocampus are first formed in the subventricular zone (Bayer et al., 1995; Nowakowski and Rakic, 1981).

Specifically, by the 5th week (35th day), there occurs a lateral outgrowth that forms the telencephalic vesicles. The telencephalic vesicles then begin to rise upward forming paired lateral evaginations and cavities. From the medial portions of these outgrowths, the diencephalon (thalamus, hypothalamus) comes to be fashioned. By the 7th to 8th weeks the choroid plexus begins to form and invaginates the medial portions of each lateral cavity.

ORIGIN AND ONTOGENESIS OF THE NEOCORTEX

Neurons and neuroglial tissue are derived from the neuroepithelium, which lines the neural tube. Basically, the neuroepithelium is made up of a pseudostratified layer of immature bipolar radial cells (Encha-Razavi, 1995).

Neurons destined for the neocortex first begin to proliferate as neuroblasts within two areas that line the walls of the lateral ventricles: the subventricular and the ventricular zones, which give birth to bipolar cells. The neurons that will eventually compose the layers of the neocortex are predominantly generated in the germinal region of the ventricular surface. These immature neurons must then migrate outwards toward the surface and layers I and VII (Sidman and Rakic, 1973, 1982).

Migration of the neuroblasts begins after they complete their last mitotic division. They form an association with the glial fibers, develop a leading process (what will become a dendrite) that is directed toward the pial surface of the developing brain, and begin to migrate. In some regions, migration is guided via chemical and trophic signals that are being secreted by their respective terminal substrates (Lund, 1978).

In addition, throughout wide regions of the developing neuroaxis, these preneurons are also assisted by radial glial cells that essentially provide a scaffolding via which successive waves of neurons may migrate. Radial glial fibers stretch from the ventricular lining all the way to the subpial glial limitans (Sidman and Rakic, 1973, 1982). It is likely that glia not only guide, but may nourish migrating neurons (Bayer et al., 1995). Hence, after the neuroblasts have completed their final mitotic division, they form an association with the glial fibers and begin to migrate.

However, not all migrating neurons follow a radial trajectory, e.g., pontine and medulla neurons. And, in certain regions of the neocortex, some neurons migrate laterally (Bayer et al., 1995). Nevertheless, as development ensues, the distance each neuroblast must traverse steadily increases such that it takes progressively longer to reach its final destination because of the expansion of the brain.

Neural Migration and Cortical Layers I and VII: The Preplate

The preplate of what will become the neocortex consists of layers I and VII, both of which develop quite early in cortical ontogenesis, and rapidly establish intimate interconnections.

As noted, layers I and VII appear to resemble midbrain reticular neural tissue. From a phylogenetic and evolutionary perspective these two cell layers in fact appear quite ancient, having characteristics (including a unique horizontal organization and abundance of fibers but scarcity of neurons) shared by all vertebrates except fish (Marin-Padilla, 1988a).

These two layers, however, appear to exert some type of trophic attractive influence on specific populations of ventricular and subventricular neuroblasts. In successive waves these pre-neurons migrate toward and crowd in between layers I and VII, thereby splitting them apart and forming the neocortical plate and cortical layers II through VIa in the process. That is, with the accumulation of neurons and the creation of layers II through VIa, layer I and VII are subsequently split apart, with all subsequent cortical layers forming between them.

Specifically, from the 7th week onward, neural migration occurs in wave after wave, accumulating at the juncture of the marginal and intermediate zones of the cortical mantle (layers I and VII), thus forming the cortical plate (Sidman and Rakic, 1973, 1982). However, with each successive wave the cortical plate becomes thicker and layers I and VII are pushed apart.

Because many of the migrating columns are guided by and utilize the same radial glial fiber scaffolding, they traverse the same path and leap-frog over previous waves that have already taken up residence, having arrived earlier. In this manner these successive waves of migrating neurons become oriented as columns as well as layers, thereby forming vertical radial cellular units (Sidman and Rakic, 1973, 1982) that later will respond to the same stimulus (Mountcastle, 1957).

Each successive wave of neurons, however, also attempts to locate close to the pial surface and thus next to layer I. Once this has been accomplished, these neurons lose their glia attachment and develop an apical dendrite, which forms a functional connection with layer I reticular neurons (Marin-Padilla, 1988a). Since layer I is located at the surface, all subsequently arriving neurons leap-frog over those which have already arrived. The next wave of newly arriving immature neurons then squeeze in close to layer I.

Hence, with the exception of the superficial marginal zone layer I and the deep subplate layer VII (Marin-Padilla, 1988a), neocortical ontogenesis follows an "inside-out" pattern of development, with the deepest layers forming first (VIa, V, IV), followed by layers of neuroblasts that successively leap-frog over them in order to reach layer I, thereby forming layers III and then II (Sidman and Rakic, 1973, 1982).

Specifically, around the 15th week of gestation, immature neurons and thalamic fibers (thalamic-cortical) begin to arrive just below layer VII, where they wait. Then they migrate and accumulate between layers VII and I, forming what will become layers VIa and V (which are initially formed as closed as possible to the pial surface). Soon thereafter these cells begin to differentiate into pyramidal neurons (Marin-Padilla, 1988a)—an event that may be triggered by thalamic axons. That is, the arrival of thalamic fibers may act to induce the differentiation of the neurons located in layers V and VIa.

This process is again repeated with the formation of layer IV followed by layers III and II. Finally, by the 40th week of gestation, the cortical plate has begun to display the columnar and horizontal stratification that is characteristic of adult human neocortex (Encha-Razavi, 1995; Sidman and Rakic, 1982).

Layer I and Neocortical Genesis

It thus appears that layer I plays a significant role in perhaps attracting migrating neurons in the first place. In part this may be due to the presence of nore-pinephrine, which maintains plasticity (see below), and layer I's mesencephalic and reticular organization, which may serve to activate and thus maintain functional viability. Indeed, layer I appears to maintain an intimate relationship with the reticular activating system, as well as the midbrain norepinephrine transmitter system, perhaps acting as its most distal component (Marin-Padilla, 1988a).

In this regard, layer I neurons may provide a low threshold background of excitation that acts to stimulate and maintain the functional and anatomical stability of newly arriving neurons, which in turn have not yet established their functional interconnections. If that is the case, this may also account for the inside-out pattern of subsequent neural migration and cortical formation.

Neural Migration and Cortical Layer VII

As subsequent layers are formed, the initial interconnections linking layers VII and I are progressively lost as these layers are pushed further away from one another. As this occurs and the successive layers are formed, the cells located in layer VII (also referred to as VIb) begin to undergo a process of transformation and become polymorphous in structure, some assuming the shape of curved sickles or inverted pyramids. They are then progressively (but not completely) replaced by the large pyramidal (mammalian) projecting neurons similar to those of layer V, thereby forming layer VIa.

However, like those pre-mammalian, reticular neurons of layer I, the cells of layer VII also play an important role in the establishment and maintenance of neuronal viability and future interconnections.

Immature neurons, led by their migratory (immature pre-dendritic) fibers, follow the radial glia fibers that lead them toward layers VII and I, where they take up temporary residence just below layer VII (in the

intermediate zone). Here they wait until some trophic signal initiates a final wave of migration, where they leap-frog over cells and layers already formed. Once they take up their new and final residence they lose their leading process and grow elaborate dendritic trees even as the next wave of young neurons climb over them to establish the next succeeding layer as close as possible to layer I.

By contrast, axonal fibers from subcortical nuclei, e.g., the thalamus, arrive in advance of the cortical neurons, often even before these target neurons are even generated in the ventricular zone (Rakic, 1977; Shatz et al., 1988; Sidman and Rakic, 1982). Because their neuronal targets have not yet arrived and/or have not yet developed dendrites, these thalamic axons will also accumulate just below layer VII in the intermediate zone (future white matter) where they also wait, sometimes for days or weeks.

Layer VII, being rich in NE and composed of midbrain reticular-like tissue (like layer I), apparently provides some type of activational input into these cells, thereby maintaining their functional viability. Some of these "waiting" neurons form synapses with the cells in area VII (Shatz et al., 1988). Hence, ("subplate") neurons in layer VII (like those in layer I) may act in some manner to presumably nourish and maintain the integrity of these "waiting" neurons until their counterparts are formed and they migrate to their respective cortical layers, where they then establish temporary interconnections with layer I.

That cells in layer VII serve some organizational role in neural migration, pathway establishment, and columnar organization is also reflected in the process of disorganization and cell death that then ensues. For example, once layers VI through II and their connections are formed, large numbers of cells in layer VII become increasingly polymorphous and/or are replaced and die. Moreover, cell in layer VII also appear to be subject to competitive influences with pyramidal neurons, which take up residence within VII, splitting it in half and thereby forming layer VI (also referred to as VIa).

In summary, neurons in layers I and VII exert an attractive influence on migrating neurons, acting to maintain cortical cells once they have arrived and while they "wait." In addition, layers I and VII act as a type of scaffold that makes possible the creation of short- and long-distance neural circuits and the creation of the cortical layers.

Once these cortical neurons have settled into their terminal substrates, their leading process differentiates into a dendrite tree whose branches slowly blossom, attracting axons even from far and distant neurons, which struggle and compete with one another in order to establish synaptic contact (Joseph, 1982).

Neural Migration: Axons and Dendrites

As neurons migrate they form and establish their axonal fibers, which trail behind them. These immature axons will later serve to form all subcortical and most corticocortical interconnections, including the corpus callosum.

Projection neurons (versus local circuit, interneurons) begin to extend their axons even before they have reached their final destination. This is especially true of the pyramidal neurons of the motor neocortex. Neurons and axons giving rise to the corticospinal tract, in fact, descend, cross-over at the level of the medulla (medullary pyramids), and descend into the spinal cord prior to birth. It is noteworthy that in the majority of humans, the axons from the left cerebral motor centers develop and cross-over the pyramidal decussation of the medulla, in advance of those from the right hemisphere (Kertesz and Geschwind, 1971; Rakic and Yakovlev, 1966), thus giving the left hemisphere a competitive advantage in the development of motor and right-hand dominance (Joseph, 1982).

However, in many regions of the developing brain, migrating neurons and their still-growing and still-maturing axons often transiently innervate various neurons with which they normally do not maintain contact (reviewed in Nowakowski, 1987). Presumably these transient connections serve to maintain mutual viability as they wait for those signals which will induce the formation of more permanent and "normal" neural circuits. Hence, during the course of maturation, these extra (transient) axonal collaterals are eliminated, in some cases because of cell death (Cowan et al., 1984; Oppenheim, 1981).

In general, two types of transient connections are made, convergent and divergent (reviewed in Nowakowski, 1987). Divergent transient interconnections are characterized by neurons (or populations of neurons) that exuberantly innervate more cells than is seen in the adult brain. However, via the retraction of the these axonal collaterals or via the shrinking of the axon's terminal arbors, these extra connections are eventually lost.

Convergent transient interconnections are characterized by a single target neuron (or population of neurons) that is simultaneously innervated by many different neurons. However, over the course of development, only one of these neurons maintains these connections, whereas axons from the others drop out and are eliminated. This "drop out" involves the elimination of extra and transient interconnections, the pruning of axons and axonal collaterals, a decrease in synaptic numbers and densities, and cell death.

Dendrites

The dendrite develops from the fiber that leads the migratory process. However, dendritic differentiation does not begin until these neurons have completed their migration, a development that then proceeds in correspondence with the ongoing process of axonal proliferation and growth in adjoining and distant neurons. Specifically, after the growth of the dendritic tree and its branches, dendritic spines begin to rapidly proliferate along the shaft of the dendrite. This is followed by a surge of spine formation and increased spine density over the length of the dendrite.

Dendrites that have not yet established synaptic contact or that have for some reason lost axonal input (such as occurs with trauma) exert some type of trophic influence that attracts and stimulates axonal growth. In some cases, the dendrite may actually grow and increase its dendritic tree (Coleman and Flood, 1988; Kolb, 1992), perhaps in search of or as part of the process of attracting new axons. Commonly, however, undamaged dendrites that lose their axonal synaptic counterparts, degenerate and will lose their terminal branches unless new synaptic contacts are established.

Neural Networks: Local and Long-Distance Projecting Neurons

Considered at its most generalized and basic level, the CNS is essentially composed of two types of neurons: long-distance projecting (extrinsic, Golgi type I, principle, pyramidal) and local circuit (Golgi type II, interneurons) cells that establish interconnections with adjacent neurons.

Local circuit neurons (interneurons) tend to contain high concentrations of GABA and possess axons that remain within the gray matter and that arborize in the vicinity of their own or adjacent dendritic fields (Miller, 1988). Within the cortex, there are three types of local circuit neurons (interneurons), all of which have dendrites that are smooth and lacking spines or are only sparsely spinous, i.e., stellate neurons, bipolar neurons, and bitufted neurons.

Local circuit interneurons tend to assume final cortical position beginning on the day of birth and continuing thereafter (Miller, 1988). Once migration has been completed the cell body begins to grow quite rapidly, and within a few weeks these neurons have reached their mature size (Miller, 1988). However, those located in the deeper cortical layers mature more rapidly than those in the superficial layers, which continue to develop and establish

synaptic connections for the first 5 years of life and beyond (Joseph, 1982; Marin-Padilla, 1988a, b).

The projecting neurons have axons that travel beneath the gray matter and that traverse long distances in order to make synaptic connections with spinal, brainstem, midbrain, thalamic, or limbic nuclei. Many projection neurons also interconnect different regions of the neocortex and form the corpus callosum.

There are basically two types of projecting neurons, both of which have dendrites that are covered with spines: pyramidal neurons and multipolar stellate neurons (reviewed in Miller, 1988). Pyramidal neurons are quite numerous and are generally found in layers II, V, and VI and comprise the majority of projecting neurons, whereas the multipolar stellate neurons are found in layer IV.

In contrast to local interneurons, projection neurons tend to migrate and assume their final position well before birth. Similarly, the axons of the projection neurons begin to extend and establish synapses in advance of local circuit neurons. However, it appears that many of these axons do not become functional until after the process of myelination has begun (e.g., Graftstein, 1963).

Fissures, Sulci, and Gyri

According to Sidman and Rakic (1973, 1982), by the 16th week of gestation most cortical neurons have arrived and taken their respective positions within the cortex. Horizontal stratification begins around 24 weeks and continues until about 28 weeks, at which time the first rudiment of six-layered neo/isocortex can be discerned (Norman and Ludwin, 1991).

The external surface of the hemispheres also becomes modified during this period, for with the exception of the major fissures (formed around the 16th–17th weeks of gestation) the brain remains quite smooth up to around the 24th week, though a few primary sulci can be discerned (e.g., hippocampal, Sylvian). It is after the 24th

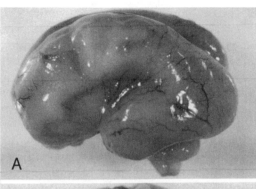

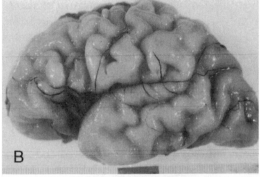

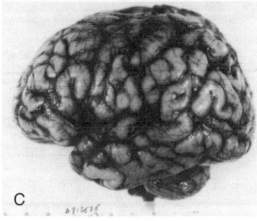

Figure 18.3   Brain development at (A) 20 weeks, (B) 30 weeks, and (C) at birth. (From Enach-Razavi. Fetal neuropathology. In Pediatric neuropathology. Williams & Wilkins, Baltimore, 1995.)

week that the secondary sulci begin to form. By the third trimester of gestation, the secondary and primary gyri begin to subdivide and third-order sulci appear.

Sulci formation is generally quite variable such that the final pattern is usually unique to each individual human brain, as well as to the right and left hemispheres.

The right hemisphere appears to be particularly unique in its variability, which may reflect its greater susceptibility to environmental influences due to its relatively longer period of immaturity (Joseph, 1982, 1988a, 1992b).

NEUROTRANSMITTERS AND BRAIN DEVELOPMENT

The many and varied neurotransmitter systems develop and mature in distinct and unique ontogenetic patterns. For example, the norepinephrine, dopamine, serotonin, and cholinergic systems mature early in postnatal and late in prenatal life (Johnston, 1988). Indeed, acetylcholine (AChE) activity in the substantia innominata (nucleus basalis) increases markedly from the 12th–22nd weeks of fetal development.

However, these neurotransmitters play roles that go far beyond information transmission but include growth promoting trophic influences on developing and adjacent neuronal tissue. For example, in the adult brain, loss of AChE activity can induce massive cortical atrophy and all the signs of Alzheimer's disease (see Chapter 9). Correspondingly, AChE activity may in turn promote cortical development. However, as AChE, as well as NE and 5-HT, innervates, for the most part, different layers and different neurons, to different and varying degrees, differences in maturation rate may not only influence but may determine differential rates of neuronal development in the various regions of the cortex (Johnston, 1988).

For example, NE has been shown electrophysiologically to inhibit or suppress irrelevant background activity while simultaneously enhancing evoked responses in both inhibitory and excitatory afferent neural circuits associated with the processing of relevant environmental input (Foote et al., 1983).

Hence, in the developing nervous system, NE may also act to suppress the establishment of irrelevant neural circuits, while simultaneously stabilizing and/or promoting the growth and formation of relevant synaptic neural networks (see Bear and Singer, 1986; Pettigrew and Kasamatsu, 1978).

Norepinephrine

NE axons arrive prior to migrating neocortical neurons and quickly infiltrate layers I and VII, and then proceed to innervate the cortex in a rostral-caudal direction. By birth, both cerebral hemispheres have been densely infiltrated by NE axons (Levitt and Moore, 1979), and NE levels are about 30–40% of adult levels (Johnston, 1988). As detailed in Chapter 10, NE innervation of the entire neocortex derives from the A6 catecholamine group in the locus ceruleus.

NE innervation of the different neocortical layers differs depending on brain areas. For example, in area 17, NE appears to be particularly dense in layers III, V, and VI. By contrast, 5-HT innervation is more extensive in layer IV (Foote and Morrison, 1984). However, in area 18, layer IV contains more NE. This also suggests that some counterbalancing or reciprocal influences are maintained by the NE and 5-HT systems throughout development (Foote and Morrison, 1984).

NE Influences on Neocortical Development

NE appears to play a significant role in cerebral cortical development, particularly in that NE axons arrive early, quickly infiltrate the various layers, and can therefore act to influence if not regulate the developmental process, including cell migration, neuronal and axonal maintenance, synaptic development, and neuronal differentiation (Parnavelas et al., 1988).

For example, NE innervation first occurs in layers I and VII. These layers also serve as the site where the first NE synapses are formed, and it is the subplate layer VII where the first non-NE synapses are formed. Moreover, those migrating cells which first differentiate are the first targets of NE axons.

Nevertheless, although the initial innervation of the neocortex by NE is paralleled by the capacity of these axons to engage in NE reuptake (Levitt and Moore, 1979), NE neurotransmitter levels, storage capacity, and rate of synthesis develop at a rather slow pace (Goldman-Rakic and Brown, 1982), a process in primates that continues well after age 3. Given that these NE synapses form when the capability of storing and synthesizing NE is not yet fully developed indicates that these axons are not concerned with NE transmission per se but, again, are probably more concerned with regulating the developmental process, including early synaptogenesis (Parnavelas et al., 1988). These are also functions of cortical layers I and VII.

As noted, thalamic axons "wait" in the intermediate zone for almost a week until the appropriate neuroblasts arrive and take position, at which point they are joined by the thalamic axons, which establish synaptic activity in these neurons. However, these thalamic axons actually wait and do not reach their termination site (layer IV) until NE synaptogenesis has occurred.

Apparently, NE stimulates cyclic AMP formation in the developing cortex, which in turn regulates protein kinase phosphorylation of intracellular proteins, as well as energy metabolism via glycogenolysis stimulation (reviewed in Johnston, 1988). Hence, NE is important in neural growth and neural metabolic activity.

In addition, NE plays an important role in neuronal plasticity and in guiding and inhibiting neural development, even in non-NE systems. For example, if deprived of sensory input (e.g., patterned vision) early in postnatal development, neurons and neuronal interconnections become abnormal in those pathways subserving the lost input. However, if NE is depleted in advance, subsequent deprivation has little or no influence on cortical structure (Kasamatsu and Pettigrew, 1976; Pettigrew and Kasamatsu, 1978), which presumably remains in an immature and plastic state.

By contrast, in the presence of normal environmental input, when the developing neocortex is deprived of NE, there results a profound disturbance in the pyramidal neuron morphology, coupled with reduction in cell densities (see Parnavelas et al., 1988).

NE and 5-HT

The cerebral cortex is widely innervated by 5-HT axons, which originate in the dorsal raphe nuclei. In primates, within 6 weeks of birth the adult pattern of innervation is evident (Goldman-Rakic and Brown, 1982). By 9 weeks, in primates, most regions of the neocortex contain adult levels of 5-HT (Goldman-Rakic and Brown, 1982).

As noted, 5-HT and NE appear to play counterbalancing roles in regard to numerous emotional and psychological functions (see Chapter 10). In this regard, some evidence suggests that in the absence of NE, 5-HT levels become unbalanced such that there is a massive increase in cortical 5-HT axons, which in turn results in increased 5-HT density and increased 5-HT innervation of target (Blue and Molliver, 1987) and possibly inappropriate nontarget sites. This suggests that under early environmental conditions (such as chronic emotional stress) that might result in NE depletion, 5-HT densities may be enhanced, resulting in altered and abnormal synaptic and cellular function and development (see below). Consider, for example, that a major role of 5-HT is sensory inhibition. Hence, nuclei that come to be abnormally innervated by 5-HT may also become abnormally inhibited.

Vascularization

The neuroectoderm of the embryonic neural tube is completely avascular and thus lacks an intrinsic blood supply (Marin-Padilla, 1988b). It is only over the course of subsequent differentiation that the CNS becomes progressively vascularized by extrinsic vessels that organize themselves around the developing brain. These vessels then perforate and penetrate the surface of the brain and begin to grow, and then later regress with the establishment of the intrinsic vasculature (Marin-Padilla, 1988b).

The process of vascularization follows a caudal-cephalad gradient that commences with the myelencephalon and then progressively encompasses the metencephalon, then the mesencephalon, followed by the diencephalon and then the telencephalon. Once established, the embryonic capillaries adapt to the structural and functional needs of the brain region innervated. Hence, the vasculature of the brain differs depending on region (Marin-Padilla, 1988b) (see Chapter 20 for further details).

Like the brain itself, the vascular system is also affected by trauma and environmental influences. For example, a dramatic and significant increase in capillaries is evident in animals reared in an enriched environment. Moreover, otherwise normally reared animals that are subsequently placed in an enriched (versus deprived) environment also display an increase in capillaries (Black et al., 1987, 1991).

Myelination

As neurons mature, some glial cells also undergo a process of differentiation and form oligodendrocytes, which in turn form myelin. However, prior to the onset of axonal myelination, there are marked proliferations in the number and density of glia (a process that has unfortunately been called "myelination gliosis," which suggests an abnormal process such as astrogliosis—which it is not). Characteristically, prior to the formation of myelin, large quantities of premyelin lipids accumulate in the glia cytoplasm. Blood flow to the white matter also is increased, as is glucose utilization because of the increased metabolic requirements of the growing of premyelinating cells. However, once the myelin sheath has formed and then encircles the axon, the glia are displaced and glial nuclear density decreases. In some brain regions, the process of myelination can continue up to age 30 and beyond (Bennes et al., 1994, Lecours, 1975; Yakovlev and Lecours, 1968).

The process of myelination can be severely disrupted even in the fetal brain because of such events as stress, direct trauma, or malnutrition—all of which can delay or even reverse the myelination process (e.g., Gilles, 1991; Rain and Path, 1991). However, the process of myelination begins and ends in different regions at different time periods (Lecours, 1975). Hence, a specific environmental trauma (e.g., malnutrition) may damage the myelination process of one brain region, whereas yet other tissues are relatively unaffected (Gilles, 1991).

This prolonged rate of maturation is also evident in the corpus callosum, whose axonal fibers originate from the immature neurons located in the superficial layers of the neocortex (Cummings, 1970; Hewitt, 1962; Rakic and Yakovlev, 1966, Ramaeker and Nkiokiktjien, 1991). Indeed, the pattern of callosal development appears to parallel neocortical developmental patterns (Cummings, 1970; Hewitt, 1962; Rakic and Yakovlev, 1966, Ramaeker and Nkiokiktjien, 1991) and the development of transmission capabilities (Grafstein, 1963), which is not surprising since synaptic interconnections are maintained and myelin is formed only when the cells are nearing functional maturity and become active.

It thus appears that it is not until after synaptic contacts have been formed that the process of myelination begins (see Grafstein, 1963), a process that takes well over 10 years to complete in the corpus callosum and over 30 years to complete in other brain tissues (Yakovlev and Lecours, 1967). Even so, just because an axon has begun to myelinate does not indicate that its cell body is completely functionally mature.

PART II

CONGENITAL MALFORMATIONS OF THE NERVOUS SYSTEM

Congenital malformations most commonly arise in the central nervous system, more so than any other bodily organ (Norman and Ludwin, 1991). It has been estimated that approximately 1% of newborns suffer from major defects and malforma-

tions of the CNS (Gilles, 1991). Lesions to the developing fetal brain can range in severity and diversity from complex malformations to a simple loss of neurons in a discrete region or layer of tissue. These conditions may be due to abnormalities in neuronal migration, repair defects, developmental arrests and delays, and interference due to drugs, malnutrition, toxic exposure, or direct injury, in which case glial scars may also form.

In the majority of cases asphyxia, e.g., hypoxia, anoxia, lowered pH, cyanosis, and increased carbon dioxide, is suspected (Gilles, 1991). About 20% of all congenital abnormalities are due to environmental activation of hereditary tendencies (reviewed in Norman and Ludwin, 1991). Twelve percent are due to toxins, maternal infections, drugs, and malnutrition, whereas from 1% to 6% are due to major chromosomal abnormalities.

In general, congenital disturbances can be considered of two types, primary and secondary. Primary malformations are due to an intrinsically abnormal developmental process that in turn is related to genetic or chromosomal abnormalities. These may be reflected in massive neuronal migration failure, and in the extreme, anencephaly and a complete failure of the forebrain and upper brainstem to develop such that the infant may only possess a spinal cord and caudal medulla.

Secondary malformations and defects are due to disruption of or interference with normal development (Encha-Razavi, 1995), e.g., infections, head injuries, toxins, perinatal cerebrovascular accidents. These injuries may be manifested as neural migration failures.

Malformation secondary to interference are usually characterized by a retardation in growth, such as in the inferior parietal lobule (which may be later reflected in the development of learning disabilities, e.g., dyslexia), a failure to acquire subsequent developmental components such as myelin (Gilles, 1991), or the premature arrest of migrating neurons (heterotopias) and/or their migration to inappropriate substrates.

For example, Galaburda and associates (1983) found islands of heterotopic neurons in the neocortex and inferior parietal region of the brain among individuals diagnosed as dyslexic.

In general, secondary malformations and developmental interference and delays are unique to the child, infant, fetal, and neonatal brain. However, disruptions suffered prior to 17 weeks of gestational age generally cannot be determined as resulting from primary or secondary factors. Injuries to the CNS after 17–18 weeks of gestational age are usually characterized by the presence of gliosis, macrophages, necrosis, and calcification (Norman and Ludwin, 1991).

In general, the more immature the organism, the more susceptible it is to injury and the more widespread and serious will be the consequences. However, the nature and distribution of the disturbances also depend on the timing, for not all regions of the CNS mature at the same time. Hence, abnormalities in development can at one extreme result in wide spread cell death and neural migration failure and thus anencephaly, whereas at the other extreme even functional abnormalities may not be readily detectable and the lesion may be confined to a specific area of the brain (Norman and Ludwin, 1991).

Disturbances of Neuronal Migration: Migration Failure and Heterotopias

Disturbances involving neuronal migration can lead to profound and grossly disabling abnormalities or subtle disturbances that may never be detected. This includes the development of structural lesions that may be typified by cortical cytoarchitectural disorganization coupled with abnormal laminar and columnar organization (Barth, 1987; Rakic, 1988). Many of those afflicted demonstrate focal or lateralized cortical abnormalities, including polymicrogyria, macrogyria, and heterotopias. However, in large part the nature of the disturbance depends on the timing of the insult or interference.

For example, disturbances early in fetal development may result in mass failures of neuronal migration from the germinal layer to the cortical mantle. These disturbances are manifested as anomalies in gyri, sulci, and fissure formation (reviewed in Norman and Ludwin, 1991) as well as anencephaly and microencephaly.

Disturbances later in the fetal period may instead result in abnormalities confined to the upper layers of the neocortex and/or specific brain regions, such as the cerebellum. Disturbances late in the third trimester of pregnancy or, given the prolonged maturation of the upper layers of the neocortex and hippocampus, even during the first year, could give rise to massive neuronal loss. Given that the upper layers are made up of interneurons (versus the projection neurons of the lower layers), deficits in multimodal informational analysis and cognitive functioning, including disturbances such as schizophrenia (Benes, 1995; Weinberger, 1987), would be a likely sequelae.

Neurodevelopmental Psychosis

In a number of studies, significant interneuron reductions in layers II and III have been reported in the brains of schizophrenics who are otherwise free of gross cerebral abnormalities (reviewed in Benes, 1995). These and other indications of preferential tissue loss in the absence of gliosis argue against a degenerative process of late onset—at least in these cases. Rather, onset may have occurred during a period extending from the third trimester to the first few years of life (reviewed in Benes, 1995).

It is noteworthy that an increase in synaptic interconnections between layers I and II has also been reported in the brains of schizophrenics (Benes, 1995). Normally these initial interconnections (which serve to nourish and maintain the cell until its true synaptic counterpart is available) drop out as surrounding neurons and those which are still arriving begin to differentiate, mature, and grow axons and dendrites. Presumably, in schizophrenics, layer I neurons maintain these interconnections with layer II and III interneurons because of the failure of their normal counterparts to successfully complete their migration (thus creating white matter heterotopias). This would also explain the lack of gliosis, i.e., neural migration failure. Moreover, given that other regions of the forebrain also continue to mature well beyond the first year of life, it would be expected that these tissues would show similar abnormalities.

As summed up by Benes (1995:105), "studies of schizophrenia have revealed preferential volume loss in the entorhinal cortex, hippocampus, amygdala, globus pallidus...prefrontal, anterior cingulate, [and] primary motor cortices" as well as reductions in neuronal "density in the entorhinal formation, hippocampal formation, medial dorsal thalamus, and nucleus accumbens."

In some cases, however, disorders of neural migration may result in the formation of anomalous and aberrant interconnections between tissues that normally do not directly communicate. Similarly, some neurons may migrate to the wrong region of the brain and establish abnormal interconnections and/or exert abnormal trophic influences on surrounding tissues. In some cases the migrating neurons lose their way, and some will form anomalous groupings in the white matter, in which case they may form heterotopic nests.

Heterotopias

Under some conditions the migration of a certain subset of neurons may be delayed such that, in consequence, their terminal and potential synaptic substrate comes to be occupied by other neurons and their axons and dendrites. If this occurs, the migration of the appropriate (albeit delayed) neurons may be cut short such that they form anomalous nuclei, beneath their terminal substrates, within the white matter. These displaced nerve cell masses are referred to as heterotopias.

Conversely, in some cases there may be an excessive production of neuroblasts.

However, because they are produced in excess, there is insufficient cortical space available. That is, all cortical space comes to be quickly occupied such that those neurons that were produced in excess cease their migration while they are still en route toward the neocortex (Roessman, 1995). Like their delayed counterparts, presumably these excess cells become isolated within white matter.

There are two types of heterotopia: laminar and nodular (Roessman, 1995). The laminar heterotopias are characterized by diffuse, bilateral, sometimes symmetrical gray matter deposits within the white matter. Nodular heterotopias tend to collect near the horns and walls of the lateral ventricles or they may be diffusely scattered throughout the white matter. The convolutions of the hemispheres, however, may appear normal (Norman and Ludwin, 1991). In fact, in some cases seemingly "normal" patients are subsequently discovered to have this condition but to have an otherwise normal-appearing neocortex (Roessman, 1995).

Heterotopias, Trophic Influences, and Neural Grafts

Although seemingly isolated, these displaced heterotopic neurons may form interconnections only with each other (such that they seemingly "talk" only to themselves). It is also likely that they exert abnormal trophic influences on adjacent axons that make up the white matter as they pass to and fro between divergent brain areas. Moreover, in many cases, these heterotopic nests may send forth axons and innervate nearby tissues and exert both indirect and direct influences.

Indeed, similar trophic influences and (in some cases) neural interconnections are formed by surgically implanted neural grafts. That is, it has been demonstrated that transplanted (grafted) fetal tissue not only survives but exerts trophic influences and, in some cases, forms interconnections with the host tissue (see Lescaudron and Stein, 1990; Sinden et al., 1992). Infant brains, however, are more likely to accept such grafts than adult brains, and they are also far more likely to establish interconnections with the grafted tissue.

For example, if somesthetic or visual cortex is implanted into the neonatal midbrain, visual fibers will sprout and innervate the superficial layers of the superior colliculus (which subserves visual functions), whereas the somesthetic fibers will innervate the intermediate layers that subserve movement and somesthesis (Marion and Lund, 1987; cited by Lescaudron and Stein, 1990). They thus form connections with functionally similar tissues. Moreover, in many instances following fetal transplants in adults' brains ravaged by various diseases (e.g., Parkinson's), partial behavioral recovery ensues.

For example, Lindvall and coworkers (1992) transplanted midbrain fetal dopamine neurons into the putamen in two patients with Parkinson's disease. According to these authors, both patients demonstrated a significant albeit gradual improvement in symptoms beginning 6–12 weeks after surgery, i.e., a reduction in rigidity and bradykinesia predominantly on the side of the body contralateral to the graft. Freed and colleagues (1990), Madrazo and associates (1990), and others have reported similar findings. Moreover, Lindvall and coworkers (1992:156) report that transplanted DA neurons not only survived, but that "fetal mesencephalic grafts can restore DA synthesis and storage in the denervated human striatum and that this restoration can lead to a significant and sustained improvement in motor function." However, recovery was nevertheless partial.

Although transplanted embryonic tissue may survive, the connections established between this and the host tissue do not resemble the original or normal cytoarchitectural organization (Lescaudron and Stein, 1990; Stein and Glasier, 1992) and instead take on a disorganized structural arrangement. Moreover, in some instances, wholly inappropriate tissues may be innervated. For example, Marion and Lund (cited Lescaudron and Stein, 1990) found that neonatal visual cortex implants into the pons innervated the substantia nigra and even sent axons into the spinal cord.

Therefore, like heterotopias, grafted tissue appears to exert some type of trophic influence on the damaged tissue (Lescaudron and Stein, 1990; Sinden et al., 1992), for even when synaptic connections fail to form, these grafts often exert a positive influence on functional recovery and act to reduce or diminish the degree of deficit or disability. In fact, even transplants of purified astrocytes and wound extracts are able to enhance behavioral recovery as much as neural tissue transplants. Hence, possibly the grafted tissue acts to release various hormones or neurotransmitters, which act to diffusely influence adjoining tissues. In fact, these influences appear to be exceedingly widespread and can even influence the functional activity in the hemisphere contralateral to the lesion (Lescaudron and Stein, 1990).

Heterotopias: Epilepsy and Psychosis

Unlike grafted and transplanted fetal tissue, which can exert positive influences on adjacent and distant brain structures, the trophic influences exerted by heterotopic neural nests appear to be abnormal.

For example, presumably because of these abnormal trophic influences and/or the abnormal activity they generate (thus influencing long-distance axons and adjacent neurons), these nests of ectopic neurons are associated with the development of seizures and epilepsy (Palmini et al., 1991; Spencer et al., 1984).

Similarly, Kovelman and Scheibel (1983) report that individuals diagnosed as schizophrenic display heterotopic neurons in the CA1 areas of the hippocampus. Even the offspring of schizophrenics demonstrate similar abnormalities (Cannon et al., 1994), which may well be due to being reared in a psychotic environment and/or due to an abnormal maternal internal environment during the fetal period—such as viral infection.

For example, Akbarian and associates (1993) have discovered populations of nicotinamide-adenine dinucleotide phosphate-diaphorase neurons abnormally located in the white matter of the frontal and temporal lobes of schizophrenics. Coupled with their 1995 data, Akbarian and colleagues (1995) suggest that these anomalies may be due to disturbances of neuronal migration (Benes, 1995; Weinberger, 1987) or abnormalities involving the cortical subplate during embryonic formation of the brain, perhaps secondary to maternal viral infection (Crow et al., 1989). Because the cortical subplate serves as the terminal junction for migrating neocortical neurons, abnormalities involving the subplate may in turn result in white matter abnormalities and displaced (subplate) neurons, as well as reduced GAD expression within all six layers of the neocortex of schizophrenics (see also Lewis, 1995).

Considering that it has been reported that these and related pathologies in schizophrenics is as severe in the late as well as the early stages of their illness, and as reviewed above, the findings indicating a selective loss of interneurons in layers II and III, argues in favor of a neurodevelopmental lesion or abnormal process of early onset and involving selective aspects of cortical formation and selective populations of migrating neurons (Akbarian et al., 1993, 1995; Benes, 1991, 1995; Carpenter et al., 1991; Crow et al., 1989; Weinberger, 1987), as well as their subsequent interconnections, neurotransmitter levels, etc.

Late-Onset Symptoms

Schizophrenia has classically been associated with onset in the late teens and early twenties. Not uncommonly, although many of these individuals appear to have their "nervous breakdown" at this early age, it turns out that some of these individuals were in fact always viewed as odd, socially inept, or withdrawn by schoolmates, teachers, siblings, and often parents. Also, in many cases, once these individuals leave the structured environment provided by

school (including college) and are expected to fend for themselves, their problems become fully manifest. This is particularly true of males (e.g., Szymanski et al., 1995). In this regard, it is likely that their underlying problem may well have been of early onset, but because of the routine, structure, and presumably at least semi-supportive structure of their early home environment (though in some cases the home environment is also psychotic), their symptoms were masked, or perhaps existed only as possibility until triggered by stress or some other environmental insult.

DEFECTS OF BRAINSTEM, SPINAL, AND FOREBRAIN DEVELOPMENT

Interference with neuronal maturation or the mass failure of dendrites and axons to form synaptic connections can result in cell death, which in turn will affect the formation of the forebrain, including gyri and sulci development. In general, the more severe forms of defective neuronal development and migration result in anencephaly, whereas less severe forms allow the forebrain to develop, although the convolutional pattern is exceedingly simplified, with only a few primary sulci and fissures present; conditions referred to as lissencephaly and agyria-pachygyria (Norman and Ludwin, 1991).

In addition, infection, ischemia, or anoxia may result in cell death and necrosis, which again would affect forebrain and gyral formation, conditions that also result in severe intellectual deficits and/or death. In cases where synaptic failure or necrosis is more subtle or restricted, behavioral, emotional, or intellectual abnormalities may also be more subtle.

Anencephaly

Anencephaly occurs in almost one-third of all major CNS malformations found at birth (Roessman, 1995). Commonly, no CNS tissue is formed or found above the foramen magnum, although the lower brainstem and spinal cord are usually intact. Because the deficit involves the midbrain and brainstem as well as the forebrain, anencephaly cannot be considered a true prosencephaly. Anencephaly is believed to be due to a malformation that has its onset by the 28th day of gestation, and is due to a defect in the closure of the neural tube (Roessman, 1995).

There are several subtypes of anencephaly, such as "cranioschisis," which is characterized by cranial openings and scalp defects, and "rachischisis," which refers to an opening in the spine that leads to destruction and exposure of the spinal cord. These conditions may also occur together.

Encephalocele

Encephalocele is characterized by brain tissue herniation outside the cranial vault and is analogous to spina bifida. However, in the vast majority of cases, the blow-out occurs in the occipital region (Roessman, 1995).

Spina Bifida

The most common (i.e., obvious) congenital malformations are spinal tube defects, the least severe of which is spinal bifida occulta. In spina bifida occulta, the underlying spinal cord is essentially normal, as instead this condition is due to a

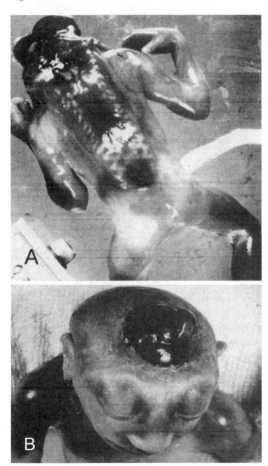

Figure 18.4   A, Anencephaly. B, Anencephaly with cranial defect. (From Norman MG, and Ludwin SK. Congenital malformations of the nervous system. In: Davis RL, Robertson DM, Eds. Textbook of neuropathology. Williams & Wilkins, Baltimore, 1991.)

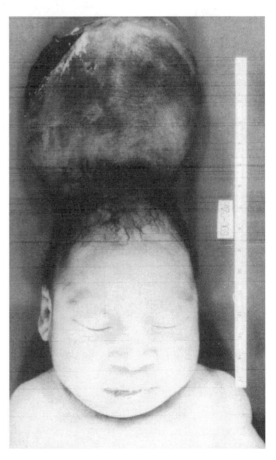

Figure 18.5   This large encephalocele constitutes most of the brain tissue, and the cranium is proportionately reduced in size. (From Roessmann V. Congenital malformation. In: Duckett S, Ed. Pediatric neuropathology, Williams & Wilkins, Baltimore, 1995.)

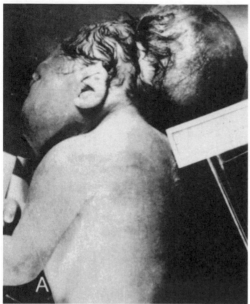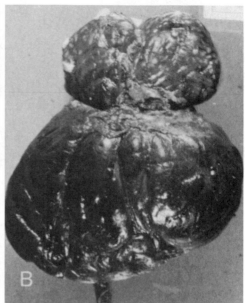

Figure 18.6 A, Occipital encephalocele that is covered by an intact scalp. B, The same encephalocele protrusion is viewed at the bottom, whereas the internally situated brain is at the top.

(From Norman MG, Ludwin SK. Congenital malformations of the nervous system. In: Davis RL, Robertson DM, Eds. Textbook of neuropathology. Williams & Wilkins, Baltimore, 1991.)

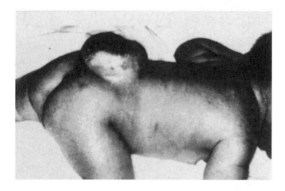

Figure 18.7 A lumbar spina bifida. (From Norman MG, Ludwin SK. Congenital malformations of the nervous system. In: Davis RL, Robertson DM, Eds. Textbook of neuropathology. Williams & Wilkins, Baltimore, 1991.)

failure of the bone to fuse, which in turn requires surgical intervention (Roessman, 1995).

In general, the majority of lesions occur in the lumbar-thoracic area and usually are due to dysraphias involving the ectoderm and mesoderm. Symptoms may range from mild gait abnormalities to profound sensory and motor deficits.

Forebrain Growth Failure

The forebrain (prosencephalon) first begins to develop around the early part of the 5th week of gestation, and then continues beyond birth. Abnormalities or a failure of the forebrain to develop characteristically results in a wide spectrum of malformations referred to as prosencephalies (Encha-Razavi, 1995). The mildest forms of this disorder are associated with an isolated absence of the olfactory bulbs. In fact, agenesis of the olfactory bulbs and olfactory tracts is common to all forms of prosencephalon malformation.

As noted, the most severe prosencephalies are characterized by anencephaly (in which case most of the brainstem also fails to develop) or profound microencephaly. Related or intermediate disturbances include agenesis of the corpus callosum and anomalies of the septal nuclei and septum pellucidum. These disorders are also associated with a failure of the telencephalic vesicles to form, which results in a single anterior ventricle (Roessman, 1995), i.e., holoprosencephaly.

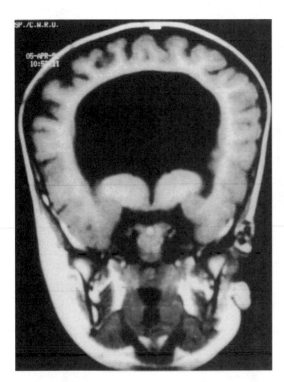

Figure 18.8   Holoprosencephaly. (From Roessmann V. Congenital malformation. In: Duckett S, Ed. Pediatric neuropathology, Williams & Wilkins, Baltimore, 1995.)

Figure 18.9   Cyclopia. There is one eye and the nose is missing. (From Norman MG, Ludwin SK. Congenital malformations of the nervous system. In: Davis RL, Robertson DM, Eds. Textbook of neuropathology. Williams & Wilkins, Baltimore, 1991.)

Prosencephalies are also commonly associated with facial abnormalities such as cyclopia or cleft lip. This is because the prechordal mesoderm, which induces forebrain and olfactory bulb development (via influences on the ectoderm), is also associated with facial development. Hence, abnormalities involving this tissue affect both the forebrain and the face.

LISSENCEPHALY, AGYRI, AND MICROPOLYGYRIA

Individuals suffering from severe congenital malformations of early onset tend to have small brains because of an arrest in development and neural migration defects (Roessman, 1995). These disturbances occur as early as the 7th week of gestation and may be related to cell multiplication failure, early cell death, and/or dendritic atrophy such as occurs with Down's syndrome (Norman and Ludwin, 1991).

Characteristic disturbances include "agyri," i.e., a defect that occurs around 11–15 weeks of gestation and is characterized by a smooth brain devoid of gyri (also referred to as lissencephaly). Micropolygyria is characterized by multiple smooth and/or small gyri, a condition that is due to developmental arrest at a somewhat later period.

MEGALENCEPHALY

Megalencephaly refers to a wide range of abnormalities that are characterized by increased brain size and excessive brain weight. This condition is usually associated with disorders of neuronal migration. Typically both the white matter and cortex are increased in thickness.

MICROENCEPHALY

Microencephaly refers to a smallness in brain size and is usually associated with disorders of cell migration, or with holoprosencephaly. Microencephaly may also be due to destructive, toxic, or degenerative lesions. Gyral patterns may range from the exceedingly simplified to hypergyral complexity.

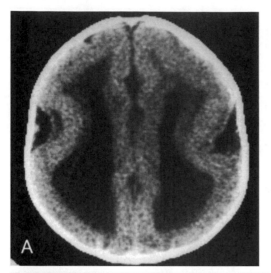

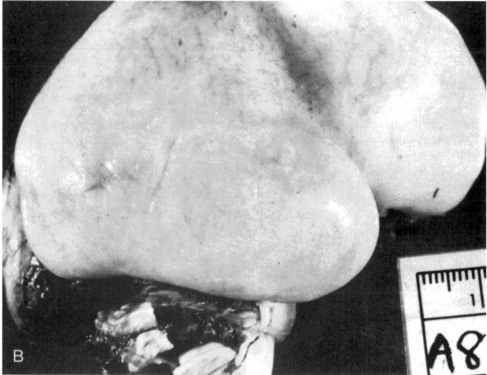

Figure 18.10 Lissencephaly. (From Norman MG, and Ludwin SK. Congenital malformations of the nervous system. (In: Davis RL, Robertson DM, Eds. Textbook of neuropathology. Williams & Wilkins, Baltimore, 1991.)

MICROCEPHALY

Microcephaly refers to smallness in head size. However, in determining if the size is abnormal, the most important variable may well be head proportion in relation to other body parts, including overall body length.

In neonates and infants (as well as young children), the cranium adjusts to the brain as it grows. Hence, there is a relationship between brain weight and volume and head circumference in neonates and infants. In consequence, congenital malformations of the brain can significantly affect the size, shape, and growth of the skull (Gilles, 1991).

For example, if the child suffers from microencephaly, the overall size of the brain and the skull will be dramatically reduced. If the malformations are predominantly unilateral, the skull may grow more slowly on that side. If the malformation or growth failure involves the frontal lobes, the frontal cranium will not properly develop and may take on a crumpled or sloped appearance. (Consider, for example, the flattened, slopping frontal cranium of Neanderthals; see Chapter 2). In addition, the skull may begin to thicken around the areas of malformation.

With lesions acquired later in development because of subsequent atrophy, the close fit between the brain and skull will be lost, and there may be empty spaces, including widened subarachnoids spaces, due to the loss of brain tissue (Gilles, 1991).

Hemorrhage

Hemorrhage is not uncommon in the fetal or neonatal brain (reviewed in Gilles,

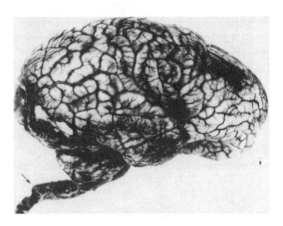

Figure 18.11   Microgyria, at 34 weeks of gestation. (From Enach-Razavi F. Fetal neuropathology. In: Duckett S, Ed. Pediatric neuropathology. Williams & Wilkins, Baltimore, 1995.)

1991). However, even hemorrhages at the same site and of the same size may have completely different etiologies even if they occur around the same time period. Unfortunately, only a few significant risk factors have been determined: trauma, intrapartum hypertension, and abnormal uterine environment (e.g., prior fetal waste, intrapartum

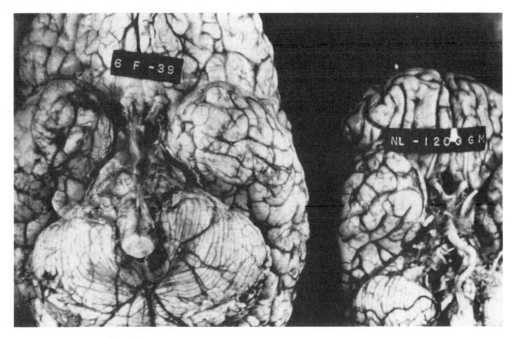

Figure 18.12   Megalencephaly (left) compared with a normal adult brain (right). (From Norman MG, and Ludwin SK. Congenital malformations of the nervous system. In: Davis RL, Robertson DM, Eds. Textbook of neuropathology. Williams & Wilkins, Baltimore, 1991.)

hypertension, maternal gonococcal infection). Hemorrhages in premature infants are not uncommon and are associated with low birth rate (Leestma, 1995).

Fetal and neonatal cerebral hemorrhage commonly occur in the leptomeninges, ventricular systems, the germinal matrix overlying the caudate nucleus and its tail, and in the molecular layers of the cerebrum and cerebellum, and in the neocortex of the temporal and parietal lobes (Gilles, 1991). In some cases the hemorrhage is secondary to capillary leakage. In consequence, the lateral, third, or fourth ventricles may be abnormally impacted, including the rhombic lip from which will spring the cerebellum.

Given that the germinal matrix of the ganglionic eminence is the embryological source for the development of the striatum, whereas the temporal lobe contains the amygdala, early-onset and even microscopic hemorrhages in these areas can induce profound or subtle disturbances in related emotional, cognitive, and motor functioning, and/or place the individual at risk for later developing such disturbances if later environmental insults are experienced.

As noted in Chapter 17, the caudate, amygdala/temporal lobe, cerebellum, and increased ventricle size have all been implicated in the pathogenesis of "schizophrenia." Indeed, whereas the amygdala is susceptible to kindling, the striatum (as well as the thalamus) appears to be a common fetal site of glial fibrillary scarring, acute neuronal necrosis, and neuronal mineralization. If this occurs, information transfer to the neocortex may be disrupted, whereas cognitive-motor activities may become and remain abnormal.

Hence, neonates can suffer brain injuries that may be quite subtle and that may be later manifested as learning, behavioral, intellectual, or emotional disturbances. In most cases, however, microscopic hemorrhages normally never come to the attention of a physician unless the infant or neonate dies from some other cause. By contrast, hemorrhages deep in the white matter, particularly those which lie close to the ventricular linings, are usually associated with profound cerebral abnormalities and usually result in death (Gilles, 1991).

Necrosis

Usually major cerebral trauma experienced during the early months of gestation result in neuronal death and/or gross cerebral malformations. During these early months, however, there is not the usual buildup of glial fibrils, macrophages, and hypertrophic astrocytes that characteristically follow cerebral injury.

It is only after the sixth fetal month that the brains response to injury begins to increasingly become similar to that of the adult cerebrum. Therefore, it is generally during the later stages of fetal development (e.g., late second trimester, early third trimester) that trauma may induce cellular and tissue necrosis. Moreover, because different regions mature and develop at different times, the location of the necrosis may indicate the age at which the injury was sustained. This is because mature neurons are more vulnerable than those which are immature to becoming necrotic.

In that the brainstem matures more rapidly than most forebrain nuclei, brainstem (including midbrain and thalamic) neurons, and to a lesser extent, the striatum, tend to show the earliest vulnerabilities. It is not until the end of gestation that cortical and white matter necrosis may be induced, in which case etiology can also be deduced.

For example, areas of necrosis that may develop in the white matter are often caused by hypoxia as well as hypotension or infarcts, in which case multiple lesions may be discovered, particularly in the frontal and the parietal-occipital-temporal junction (reviewed in Gilles, 1991). Lesions of this sort, even if quite mild, would place those infants at later risk for developing learning disabilities involving reading and writing (see Chapter 4).

Cysts

Cysts may develop at any time during gestation are usually located along the gyri and the distribution of the major arteries. Presumably these are a consequence of impaired blood flow and the development of massive necrosis, though in some cases they are due to developmental abnormalities and migration errors involving mantle formation or repair. In general, cysts are frequently found in the brains of children and infants with marked intellectual and neurological disturbances, though some children may be only mildly affected.

White Matter Tears

In contrast to the adult brain, the white matter of an infant's brain is highly susceptible to tearing, usually in a symmetrical and linear fashion, such as in response to head trauma (Gilles, 1991) and whiplash injuries, e.g., shaken baby syndrome (see below). In instances of white matter tears, wide-ranging areas of the neocortex, thalamus, and other subcortical structures can be completely disconnected because of the shearing of their axonal interconnections.

In consequence, neurons may die, abnormal connections may form, glia scarring may develop, and a host of abnormalities ranging from the mild to the profound may ensue. However, this is usually in those dramatic cases that come to a physician's attention, and patients with mild abnormalities may simply go through life suffering, in turn, a variety of mild (to severe) disturbances involving various aspects of behavior, intellect, emotional functioning, and cognition.

Hydrocephalus

Congenital hydrocephalus is always secondary to another disease or injury to the brain, including obstruction of the aqueduct of Sylvius, which prevents fluid that has formed in the third and lateral ventricular system from passing and flowing to the cerebral subarachnoid space, where it is absorbed. This condition is usually discovered within the first few months of life and may be due to congenital malformations,

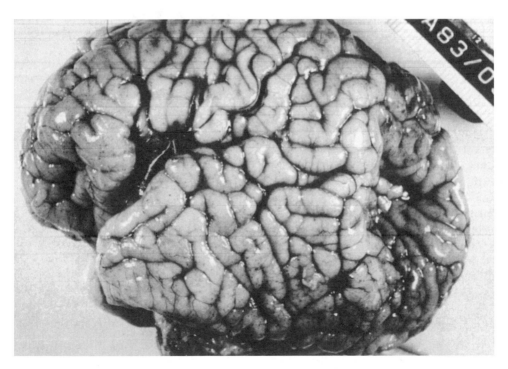

Figure 18.13   Microgyri in a distended brain that is swollen secondary to hydrocephalus. (From Norman MG, and Ludwin SK. Congenital malformations of the nervous system. In: Davis RL, Robertson DM, Eds. Textbook of neuropathology. Williams & Wilkins, Baltimore, 1991.)

cysts, tumors, space-occupying lesions, or failure of the foramina of Luschka and Magendie to develop (the Dandy-Walker malformation), which normally occurs during the 4th–5th months of gestation.

DANDY-WALKER SYNDROME

Migration errors involving the cerebellum include the "Dandy-Walker syndrome," which refers to malformation and hypoplasia of the vermis, as well as other hindbrain abnormalities involving the sinuses and ventricles. This condition is characteristically associated with hydrocephalus with variable degrees of ventricular dilation (Norman and Ludwin, 1991).

Hypoplasia is often found in trisomy 18 as well as with other malformations (Encha- Razavi, 1995).

In brief, the Dandy-Walker syndrome complex includes a wide spectrum of disorders that include hydrocephalus, posterior fossa cysts that are continuous with the fourth ventricle and cystic dilation of the fourth ventricle with absence of the foramina of Luschka and Magendie, and abnormal or partial agenesis of the cerebellar vermis.

ENVIRONMENTAL AGENTS AND NEURODEVELOPMENTAL ABNORMALITIES

Environmental agents are major factors in the development of congenital abnormalities. Environmental trauma may trigger a genetic susceptibility that results in congenital disturbances, or it may create discrete or widespread structural lesions.

For example, alcohol (e.g., fetal alcohol syndrome) can induce cerebral, cerebellar, and brainstem dysraphia and disorganization, microencephaly, microcephaly, and hydrocephaly. Similarly, rubella may induce microencephaly, although most viral infections induce destructive lesions, including leptomeningeal inflammation, mineralization, and necrosis.

The capacity of the brain to respond to infectious invaders is only slowly acquired and by different regions at different times. Hence, rubella is most destructive when acquired early during development and can induce germinolysis due to infection of the germinal matrix (Gilles, 1991).

Human Immunodeficiency Virus (HIV)

Neonates can become infected with the human immunodeficiency virus (HIV) via vertical transmission from the mother. In completed pregnancies, approximately 30–70% of infected women are found to have infected their newborns during the fetal period. HIV does not appear to significantly affect pregnancy, however (Kozolowski, 1995). Nevertheless, approximately 50% of children with congenital HIV infection die by 18 months of age (Kozolowski, 1995). Unfortunately, only a few neuropathological studies have been conducted on such fetuses and HIV-infected infants.

In general, head circumference of the newborn, as well as height and birth weight, does not appear to be significantly affected, and infected infants appear little different from "normal" infants. Nevertheless, a wide spectrum of abnormalities are associated with fetal HIV infection, including progressive encephalopathy, significant loss of neurons, abnormal microglia cell proliferation, nonspecific necrosis and hypoxic-ischemic necrosis, myelin abnormalities, vascular lesions, calcification of the basal ganglia, reductions in brain mass, and microencephaly (Encha-Razavi, 1995; Kozlowski, 1995).

Approximately 38% of HIV-infected children suffer from HIV encephalitis, and multinucleated giant cells are frequently found in the white matter and deeper layers of the neocortex. These giant cells are also believed to be replication cites for the virus (Kozlowski, 1995).

Maternal Malnutrition

As noted above, the neural tube is devoid of vasculature and thus a blood supply. Its main source of nutrition is amniotic fluid, including ependymal cell secretions. It is only around the 5th week, at the time the forebrain begins to be established, that the anterior choroidal artery penetrates the medial ventricular wall and begins to form the vascularized choroid plexus, which then begins secreting CSF.

The choroid plexus remains the primary source of nutrition for the developing brain until about the 12th to 16th weeks of gestation, at which point the cerebral vasculature becomes established and takes over the role of providing nutrition (reviewed in Duckett and Winick, 1995; Marin-Padilla, 1988b). Hence, there are some periods of overlapping nutritional sources during these various developmental stages.

Fetal nutrition is completely dependent on the mother—her diet is the fetal diet, and the fetus has no way of regulating or selecting intake. There is no protective system in the mother or fetus that controls or regulates the amounts and proportions of proteins, carbohydrates, and fats necessary for fetal growth (Duckett and Winick, 1995). In consequence, deficiencies in the mother's diet can affect fetal development, such as by slowing or retarding cellular division, growth, migration, and maturation—a secondary consequence of protein deficiency.

For example, in rats it has been shown that maternal malnourishment can result in a 20% reduction in the synapse/neuron ratio, as well as stunting of dendrites, decreases in cortical density, and abnormal cortical stratification (reviewed in Duckett and Winick, 1995). Myelin formation can also be reduced, and significant associations between maternal malnutrition and CNS malformations such as encephalopathy, hydrocephalus, and even spina bifida have been established.

Although authors have argued or presented evidence indicating otherwise, malnutrition is the primary killer of children and is one of the more important causes of congenital neurological and cognitive abnormalities (Duckett and Winick, 1995).

Maternal Drug Abuse

The ingestion of a wide variety of drugs and neurotoxicants is a common practice among "modern" pregnant women (Bates et al., 1995), and this includes both illicit and therapeutic substances. According to Bates and coworkers (1995), "as many as 90% of pregnant women take one drug and 60% of women use multiple drugs."

Of course, when a pregnant woman takes drugs, her fetus is also exposed to these substances. Indeed, although the placenta is impermeable to compounds whose molecular weight exceeds 1000 most medications have a molecular weight that is less than 500. Hence, the placenta does not act as a barrier to most drugs and neurotoxicants. In addition, the "blood-brain barrier" does not develop or mature until after the 28th week of gestation.

Not surprisingly, a small percentage of babies (3%) are born with gross congenital malformations that are secondary to the mother's drug use (Bates et al., 1995). The proportion of those with mild to moderate deficits is currently unknown.

Cocaine

Complications secondary to maternal cocaine abuse include premature labor, fetal death, and spontaneous abortion. Infants born to such mothers tend to suffer major disturbances, including small head size, cerebral infarcts, and abnormal EEGs in up to 40% of those exposed. Severe emotional disturbances, including postnatal irritability, have been reported in over 90% of those afflicted (Bates et al., 1995).

Heroin

Complications secondary to maternal heroin abuse include premature birth and fetal complications of pregnancy, small-for-gestational-age babies, sudden infant death, fetal addiction and withdrawal, and reduced head size. Abnormal sleep patterns and excessive wakefulness, irritability, hyperactivity, and seizures are also common (Bates et al., 1995).

Maternal Alcohol Abuse

Drinking as little as 30 ml of absolute alcohol per day is associated with the development of fetal alcohol syndrome (reviewed in Bates et al., 1995). Characteristic features include microcephaly and developmental retardation, hyperactivity, congenital heart defects, restrictions in mobility of hand and elbow joints with a tendency to dislocate the hip, and short palpable fissures.

Approximately 32% of infants born to heavy drinkers suffer gross congenital abnormalities, and up to 14% of infants born to moderate drinkers suffer similar disturbances. Of course, it is likely that factors such as poor maternal diet, smoking, frequent falls, etc., are contributory.

Maternal Trauma

It is not uncommon for pregnant women to fall and/or bump their swollen bellies and thus their fetus against sinks and other objects, sometimes repeatedly on a daily basis, e.g., as they wash or brush their teeth in the morning. In addition, women involved in abusive relationships usually continue to be struck, beaten, kicked, etc., while pregnant. Obviously, the fetus would be also affected by these traumas.

In fact, given that it has been demonstrated that newborn infants can recognize (e.g., prefer) the voice of their mother over that of other females, and demonstrate other "memories" and reactions to auditory stimuli experienced during the late fetal period, even nonphysical abuse (e.g., yelling, screaming) and associated maternal stress would likely exert negative influences on development and the functional integrity of the maturing nervous system (see also Chapter 16).

There are, however, no statistics as to the incidence of fetal injury and/or subsequent developmental disorders secondary to these physical (and emotional) traumas. On the other hand, it has been reported that approximately 7–8% of pregnancies are disrupted or complicated by maternal trauma (Leestma, 1995). That is, if sufficient trauma has been received, the result is usually fetal death.

Birth Trauma

It could well be argued that even under the best and most desirable of conditions, birth is a traumatic experience. According to Leestma (1995:255), "numerous studies of presumably normal births have revealed a remarkable incidence of sequelae in neonates, which include subarachnoid and retinal hemorrhages, cranial hematomas, and other injuries," e.g., skull fractures, intracranial and extracalvarial hemorrhages, brain contusions, brainstem and spinal cord injuries, and peripheral nerve damage. Skull fractures and temporal lobe injuries are also not uncommon with forceps delivery.

PART III

EARLY COGNITIVE AND INFANT BRAIN DEVELOPMENT

The brain continues to develop, grow, and mature after birth (Blinkov and Glezer, 1968; Conel, 1939; Farber and Njiokiktjien, 1993; Huttenlocher, 1990; Joseph, 1982; Lecours, 1975; Yakovlev and Lecours, 1968), a process that may in fact last a lifetime. Indeed, the brain is exceedingly plastic and is capable of undergoing tremendous functional reorganization over the course of just a few months, such as in response to injury or concentrated sensory stimulation (Feldman et al., 1992; Jenkins et al., 1990; Juliano et al., 1994; Ramachandran, 1993; Strauss et al., 1992; Weiller et al., 1993). Indeed, synaptic and dendritic changes have been repeatedly reported following simple learning trials (Barnes, 1979; Davies et al., 1989; Gustafsson and Wigstrom, 1988; Lynch, 1986; Lynch et al., 1990; W Singer, 1990; Stevens, 1989) and/or when subjects are repeatedly exposed to complex environmental stimuli (Diamond, 1985, 1991; Greenough and Chang, 1988; Rosenzweig, 1971; Rosenzweig et al., 1972).

Nevertheless, much of this growth occurs within the first 3–5 years, and the potential for neuronal plasticity and functional reorganization appears to be greatest during childhood as well. In large part this may be a function of immaturity and the fact that not all synaptic interconnections have been established, particularly in layers II and III of the neocortex.

Functional reorganization may also be maximal during infancy and childhood because of the excessive numbers of available axons, dendrites, and neurons. That is, during the course of development, neurons, dendrites and axons are produced in excess and in consequence may be pruned away depending on environmental demands, the remainder being depleted in massive amounts (Arey, 1965; Hamilton and Mossman, 1972; Huttenlocher, 1990; Joseph, 1982; Lund, 1978).

Because the nature of cerebral maturation and the growth and establishment of neural circuitry involves cell death and the elimination of excessive axons, dendrites, and neurons, these excess cells and their processes may be differentially recruited in cases of early injury and/or in response to changing environmental conditions and increased learning opportunities. Because the adult brain is not as well endowed with excess potential, the capacity for significant functional reorganization is comparably reduced. Indeed, depending on location, there are anywhere from 15% to 85% more neurons in the infant as compared to the 60- to 70-year-old adult brain (reviewed in Joseph, 1982). For example, neuronal density in layer III of the frontal cortex exceeds that of the adult brain by 55% by 5 months of age but only 10% above adult values by age 7 (reviewed in Huttenlocher, 1990).

Presumably, the rate of neocortical and dendritic drop-out is largely affected by environmental conditions and associated neural activity (i.e., the "use it or lose it" principle). For example, when animals are reared under conditions of social isolation and/or reduced sensory stimulation, neurons drop out and/or become smaller, and dendritic density is reduced (Casagrande and Joseph, 1978, 1980; Diamond, 1985, 1991; Greenough and Chang, 1988; Rosenzweig, 1971; Rosenzweig et al., 1972).

According to Greenough and Chang (1988:407), "it seems reasonable to assert that intrinsically generated overproduction followed by activity-dependent selective preservation or elimination of neuronal processes is a general feature of early CNS development in mammals."

Thus these "excessive" neurons, dendrites, etc., do not always simply die, such as occurs with programmed cell death. Rather, the nature of one's environment and the stresses involved in large part determine which neurons and neural pathways are developed, which die out, and which dendrites come to be pruned away.

REGIONAL DIFFERENCES IN NEOCORTICAL GROWTH AND MATURATION

Neuronal growth, maturation, and death occurs at different rates in different brain regions and within the different layers of the neocortex (Blinkov and Glezer, 1968; Conel, 1939; Huttenlocher 1990; Joseph, 1982; Lecours, 1975). Layers II and III of the neocortex, for example, remain functionally and anatomically immature, and do not fully develop their interconnections with distant and even adjacent neurons, for the first several years of life. Similarly, the hippocampus is only about 40% mature by birth, and although structurally it appears fully mature by 15 months of age (Kretschmann et al., 1986), myelination of the hippocampus and its key relay zones continues up and beyond the fifth and sixth decades of life (Benes et al., 1994).

The Corpus Callosum

This prolonged rate of maturation is also evident in the corpus callosum, whose axonal fibers originate from the immature neurons located in the superficial layers of the neocortex (Cummings, 1970; Hewitt,

1962). Indeed, the pattern of callosal development appears to parallel neocortical patterns (Cummings, 1970; Hewitt, 1962; Kretechmer, 1974), and age increments in interhemispheric and intrahemispheric information processing appear to parallel these maturational developments (Joseph and Gallagher, 1985; Joseph et al., 1984).

Specifically, the corpus callosum first makes it embryonic appearance during the third month of fetal development, arising as collaterals of the long association and projection fibers within layers II and III of the neocortex (Cummings, 1970). However, the splenium, body, and genu of the corpus callosum continue to develop and expand up to and beyond the first decade of life, and axonal myelination follows this same pattern of protracted development (Yakovlev and Lecours, 1967).

Because these axons originate in cell bodies located in these layers, maturation of these fiber processes and their synaptic connections reflects neocortical growth and functional readiness. In fact, the functional development of each individual neuron appears to initiate or trigger myelination (Grafstein, 1963), which in turn increases functional and transmission capacity by a magnitude of 400% (e.g., Rain and Path, 1991).

Correspondingly, the capacity to transfer tactile, visual, and cognitively complex information between the cerebral hemispheres parallels the development, maturation, and myelination of the corpus callosum (Deruelle and de Schonen, 1991; Finlayson, 1975; Galin et al., 1979a, b; Gallagher and Joseph, 1982; Joseph and Gallagher, 1985; Joseph et al., 1984; Kraft et al., 1980; Molfese et al., 1975; O'Leary, 1980; Ramaeker and Njiokiktjien, 1991; Salamy, 1978). As such, significant deficits are apparent up until age 7 (Joseph et al., 1984). However, transfer deficits are also a function of intrahemispheric immaturity.

The Lobes of the Brain

Because of intrahemispheric and interhemispheric neocortical immaturity, especially in the upper layers, those regions of the brain that are heavily dependent on multimodal input from other neocortical areas also remain exceedingly immature for years. For example, the inferior parietal lobule has reached only about 64% of adult size by age 4, and by age 7 is still 20% below adult average; in fact, it is the slowest portion of the neocortex to anatomically and functionally mature (Blinkov and Glezer, 1968; Conel, 1939; Joseph, 1982; Joseph and Gallagher, 1982; Joseph et al., 1984). This immaturity is also reflected in the slow pace of acquisition of functions associated with the inferior parietal lobe, such as reading and math ability (versus speaking, visual analysis).

In addition, the frontal, superior parietal, and inferior temporal lobes are only between 11% and 17% developmentally complete at birth, with the occipital lobe the most developmentally advanced, being 33.7% complete (Blinkov and Glezer, 1968). By the end of the first year all the neocortical regions are still less than 57% structurally complete, and by 2 years the level of development has reached approximately 75% of that of the adult brain. Even by age 4 most regions are at best 83% complete, and by age 7 the brain is still immature, although the superior parietal, temporal, and the frontal lobes have reached 90% of what is characteristic for the adult brain (Blinkov and Glezer, 1968).

The Right and Left Hemispheres

Differential rates of maturity are not just the characteristic of the different lobes and fiber tracts, but similar areas within the right and left hemispheres mature at different rates (Corballis and Morgan, 1978; Joseph, 1982). For example, the motor cortex of the left hemisphere develops in advance of the right, such that the axonal, corticospinal fibers from layers V and VIa descend to the brainstem, cross at the medullary pyramids, and establish synaptic contact with spinal motor neurons, thereby gaining a competitive advantage in motor processing and control that is re-

Table 18.1. Dimensions and Growth of Various Regions of Cerebral Cortex (in cm² and % of Size Compared to Adult)

Age	Occipital		Inferior Parietal		Superior Parietal	
	cm^2	%	cm^2	%	cm^2	%
6 Lunar months	—	—	2.43	3.1	—	—
Newborn	21.1	33.7	10.91	13.4	11.1	15.5
2 years	74.1	70.4	58.75	74.7	53.2	74.4
4 years	86.5	82.2	50.63	64.3	52.4	83.1
7 years	88.5	84.1	62.92	80.0	65.1	91.1
Adult	105.2	100	78.7	100	71.5	100
	Frontal		Inferior Temporal		Superior Temporal	
6 Lunar months	9.5	4.6	176.0	5.7	201.0	7.0
Newborn	23.3	11.1	532.4	17.3	632.2	22.0
2 years	151.3	72.8	2644.0	85.0	2322.6	80.0
4 years	165.7	79.7	2718.0	88.0	2560.2	88.0
7 years	193.6	93.1	2937.0	95.0	2705.3	93.0
Adult	207.8	100	3070.0	100	2907.0	100

Adapted from Blinkov, SM, and Glezer, II. (1968). The human brain in figures and tables. New York, Plenum

flected in right-hand dominance (Joseph, 1982).

Given the close association between nonemotional language and motor functioning (see Chapters 2 and 4), because the left hemisphere motor centers mature in advance of the right, this half of the brain also becomes dominant for expressive and receptive speech as well as handedness because of this competitive advantage (Joseph, 1982). Conversely, because the motor centers of the right hemisphere are slower to mature, early left cerebral injury, including left hemispherectomy, can result in language-motor competition for still vacant right cerebral synaptic space. Hence, for the first 5 or so years of life, the still immature right hemisphere can therefore acquire (to varying degrees) dominance for expressive and receptive speech following left cerebral injury (Joseph, 1986b; Novelly and Joseph, 1983).

By contrast, those areas of the neocortex which come to subsume limbic and related sensory receptive functions may begin to develop in the right hemisphere in advance of the left (Joseph, 1982), which would partly explain right hemisphere dominance and hierarchical representation of nonverbal visual-spatial, social, and related limbic and emotional functions. Limbic system nuclei develop earlier and mature at a

faster rate than neocortical tissues. In consequence, because of hierarchical representation, portions of the right hemisphere are larger than the left, the right hemisphere is heavier than the left, white matter (nonmotor axonal) interconnections with subcortical structures are more extensive (reviewed in Joseph, 1982, 1988a), and the formation of nonmotor gyri begins sooner and more rapidly (Chi et al., 1977).

Because portions of the right hemisphere begin to develop in advance of many nonmotor areas in the left, early and severe right cerebral damage can also result in axonal competition for vacant neocortical space in the left hemisphere. For example, because of early right cerebral injuries, the left hemisphere may come to subserve and become dominant in regard to emotional expression and perception (e.g., Joseph, 1986b). In fact, because of early right or left or bilateral hemisphere injury involving left cerebral motor centers and right cerebral sensory areas, there may result considerable functional reorganization such that the left hemisphere becomes dominant for emotion whereas the right hemisphere dominates language production and perception (Joseph, 1986b; Novelly and Joseph, 1983).

It also appears that although some nonmotor right cerebral tissues may begin to develop at an earlier age than is apparent in the left hemisphere, right cerebral maturation may also be more prolonged (Joseph, 1982). Because of its prolonged rate of maturation and development, the right hemisphere is not only subject to crowding effects and synaptic competition because of early left hemisphere lesion, but is more vulnerable to injury secondary to emotional trauma and social-emotional neglect (Joseph, 1982, 1992b), a function also of its role in emotional processing and expression and hierarchical limbic representation.

Disconnection, Neocortical Immaturity, and Cross-Modal Cognitions

Axons of different tissues mature and become myelinated at different rates (Lecours, 1975; Yakovlev and Lecours, 1967) with some areas taking over 50 years to myelinate (Benes et al., 1994). Similarly, maturation in neocortical layers II and III, the pathways that link different sensory and associational receiving areas, are also quite slow to develop (see Blinkov and Glezer, 1968; Joseph, 1982; Lecours, 1975), and different lobes mature at different rates with the inferior parietal lobule taking over a decade to fully develop, and other lobes reaching only about 90% of adult size by age 7.

In this regard, for the first 7 years of life, the first three in particular, large regions of the child's neocortex are in fact disconnected because of this slow rate of maturational development, e.g., different sensory association areas have not established sufficient interconnections until this later age. These regional and intrahemispheric (as well as interhemispheric) variations in substrate maturity are apparent not only neuroanatomically, but are reflected in the differential rates in the development of language, cognition, fine motor functioning, and multimodal sensory processing (Farber and Njiokiktjien, 1993).

For example, cognitive and neocortical sensory development parallels cortical development. Rudimentary and single modal processing is apparent at or soon after birth, and the brain becomes capable of more complex processing. That is, the primary sensory receiving areas mature before the association areas, and so on.

Hence, infants are capable of intramodal recognition of novel stimuli by 4–5 months of age (see Schacter and Moscovitch, 1984), and by 6 months of age they can make tactual to visual transfers so as to visually recognize objects they manipulate, though not vice versa (Gottfried et al., 1977). Nevertheless, infants younger than 6–9 months are not capable of cross-modal transfer and recognition (so as to distinguish a simple novel object), whereas those who are 12 months of age are (Mackay-Soroka et al., 1982; Rolfe and Day, 1981; Rose et al., 1981). However, even at 12 months information is processed at a very slow rate.

Not surprisingly, because of neocortical immaturity, sensory and perceptual functioning is initially quite rudimentary (see Milner, 1967; Salapatek and Banks, 1983)—that is, at the level of the neocortex (Joseph, 1982). For example, in the visual modality it is not until around 6 months of age that infants are able to recognize and differentiate between simple differences in objects, shapes, patterns, and details (Olson, 1976). The ability to recognize and differentiate between categories of perceptual stimuli based on form, size, color, and utility, i.e., toys, does not appear until around 9 months (Ruff, 1978).

Of course, information and stimuli are also processed subcortically by limbic system nuclei such as the amygdala, as well as via midbrain and brainstem nuclei. In this regard, although neocortically immature, a wide variety of functions and perceptual capabilities are present at birth—just as they are in creatures who never evolved a neocortex and in humans born without a forebrain (Joseph, 1982). Nevertheless, as limbic nuclei also continue to develop and mature, even those functions performed and expressed at birth correspondingly become more complex.

For example, although infants are able to identify faces in general by the first week of life they cannot differentiate between faces that vary by gender or age until after 5 months of age (Fagan, 1984). They also

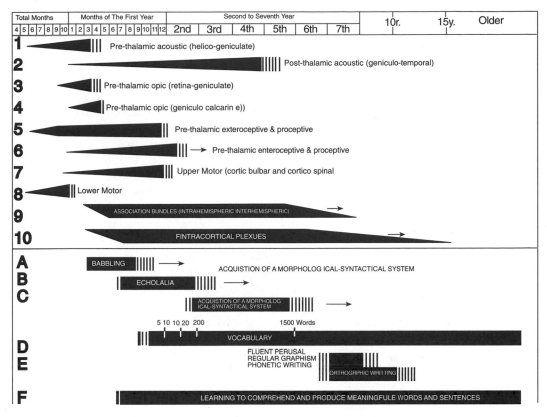

| Total Months | Months of The First Year | | Second to Seventh Year | | | | | 10r. | 15y. | Older |

Figure 18.14 Myelogenetic cycles of different regions of the brain. Reproduced with permission from Lecours A. In: Lenneberg E, Lenneberg E, Eds. Foundations of language development, Academic Press, New York, 1975.

lack the ability to differentiate between the faces of two same-sexed adults (other than close relatives, e.g., mother versus a female stranger) until about 7 months of age.

Given that infants (particularly infant females) are responsive to and enjoy looking at faces even soon after birth, responsiveness to faces indicates considerable maturity in the amygdala, which contains facial recognition neurons, including neurons that can recognize gender (Rolls, 1992). Presumably, increased facial recognition capability over the ensuing first year of life (and beyond) reflects continued maturation in the amygdala and the inferior temporal lobe, which also contains facial recognition neurons (see Chapter 14).

Cross-Modal Processing

Over the course of the first year of life considerable development and nerve cell drop-out as well as synaptic proliferation occurs throughout the neocortex, with primary and secondary receiving areas showing the greatest amounts of growth (Conel, 1939; Joseph, 1982). However, it is not until about 12 months of age that interconnections between different sensory modalities and the upper layers of the neocortex begin to mature, and it is at this age that infants are able to perform tasks involving cross-modal sensory integration (Mackay-Soroka et al., 1982; Rolfe and Day, 1981; Rose et al., 1981), tasks that require communication between different sensory association areas. Hence, by 12 months, infants can touch and handle objects that are hidden and then recognize them visually, or they can recognize them via touch if first presented in the visual modality, even when pictures are used (Rose et al., 1981).

These increased capacities are a function of increasing cortical development and the establishment of synaptic interconnections across wide regions of the infant cerebrum and throughout the neocortex, particularly the association areas. Hence, by 12 months of age visual and auditory and tactual association areas are able to interact and produce cross-modal perceptions and memories, as well as images, feelings, and visual ideas. Correspondingly, learning and memory capabilities also rapidly expand during the first few years.

Cross-modal, however, is not the same as multimodal. Cross-modal information analysis can take places within the visual, auditory, or tactile association areas (see Chapters 12 and 14), whereas multimodal processing is more greatly dependent on the inferior parietal lobule, inferior temporal lobe, and nonmotor frontal cortices. Because these cortical tissues, the inferior parietal lobule in particular, take many years to functionally develop and mature, multimodal processing, including those aspects of cognition dependent on this form of analysis (e.g., language, verbal memory, reading, writing) also take many years to develop and mature (Farber and Njiokik-tjien, 1993).

Neocortical Maturation: Language

The rate of language acquisition and verbal memory formation appears to coincide with the cycle of myelination of the Language Axis and its interlinking axonal pathways, the inferior, superior and arcuate fasciculi (Lecours, 1975). Neocortical immaturity (which is also reflected in the child's limited experience and knowledge base), particularly within the region of the inferior parietal lobe as well as in the hippocampal area (e.g., Benes et al., 1994), also accounts for why younger children are less likely to organize their verbal memories in regard to temporal sequences or related semantic categories (see Bjorklund, 1987) and why their verbal statements tend to be fragmented and incomplete (Ackerman, 1985).

Because of inferior parietal immaturity and the slow pace of development in those layers linking various tissues within the Language Axis, neocortical language acquisition and the capacity to generate multimodal concepts and classifications also occurs at a slow pace. For example, a 5-year-old child may have approximately 1500 words at his or her command, a 2-year-old child has a vocabulary of only 200 words (Freburg-Berry, 1969).

Moreover, neocortical immaturity creates temporary (developmental) disconnections syndromes that also affect language expression as well as comprehension. For example, similar to adults with injuries involving Wernicke's receptive speech area and the left inferior parietal lobe, children have difficulty communicating even when told they are not making themselves clear (Cashmore, 1990) and they have difficulty realizing or comprehending their lack of comprehension, i.e., anosognosia.

Hence, children often fail to understand that they do not understand (Flavell, 1981), and their ability to evaluate the quality of their own utterances and to respond to feedback is reduced. Even when they do not understand what is being said, they will not ask for clarification because they don't know they don't know and that they don't understand (Flavell, 1981).

Young children also have considerable difficulty coherently reporting and constructing a narrative of sequential events (Baker-Ward et al., 1993). Presumably this reflects immaturity within Wernicke's area and the inferior parietal lobule, which creates very discrete functional disconnections between these and other cortical tissues. Hence, children also have much difficulty providing verbal detail regarding their everyday experiences because of the enormous gaps in what is available to the Language Axis (e.g., Joseph et al., 1984).

It is presumably because of these early developmental neocortical disconnection syndromes, as well as the availability of alternative pathways, that children may seem inconsistent as they are also less likely to report the same information when questioned on separate occasions (Baker-Ward et al., 1993). In part, however, these inconsistencies may be due to differences in the nature or phrasing of the question, as well as the availability and utilization of alternative pathways to the Language Axis (see Chapter 16).

As the brain matures and new information processing strategies are learned and developed, the manner in which information is processed and stored is also altered, and more extensive interconnections and associations are formed (Joseph, 1982). Therefore, as children age they tend to utilize more features when expressing or encoding words (Ackerman, 1984, 1985; Ceci, 1980; Demster, 1981), a direct reflection not just of experience, but of neocortical maturation.

Neocortical Maturation: Memory

The immaturity of neocortical tissues and pathways limits not only language development and expression but the availability of verbal material that may be stored in memory. This is because those nuclei and pathways involved in language expression, perception, and acquisition also subserve verbal memory formation, whereas myelination of the hippocampus and surrounding neocortical tissue (which are involved in memory) occurs at a slow and protracted rate (Benes et al., 1994). Because of immaturity in these regions, children initially have difficulty learning or remembering certain words, and they have extreme difficulty constructing coherent or complex sentences.

As these tissues interact in regard to language production, verbal memory and verbal abilities tend to develop in parallel. For example, Waldvogel (1948) reported that when verbal "memories were plotted according to the age of their origin, it was found that there was an increment from year to year that took the form of an ogive and seemed to parallel the growth of language during childhood."

As the inferior parietal lobule takes the longest to mature of all neocortical structures (Blinkov and Glezer, 1968; Conel, 1939; Joseph, 1982) and as these associated interlinking pathways are slow to develop and are thus incomplete (e.g., myelination of the hippocampus), verbal memory and the capability to form multimodal verbal

classifications and to gain access to alternative or related neural networks is also deficient and incomplete.

Moreover, because of hippocampal immaturity and the partial disconnection syndromes that characterize cerebral functioning at this age, the ability of children to verbally encode and then verbally access their memories, including even those formed just hours or days (and even seconds) before, are not only simplified and incomplete, but fragmented, poorly organized, and filled with gaps (Ackerman, 1985; Joseph et al., 1984). Hence, young children, including those as old as age 5, require cues and reminders or very structured and repetitive interview situations so as to fill these gaps when asked about their everyday experiences (see Fivush, 1993; Pillemer and White, 1989). This is the case even well after the age of 5 as they do not possess the neurological sophistication or verbal skills to form complex verbal memories or to retrieve them. Hence, verbal "memories in narrative form do not exist before 5, 6, or 8 years" (White and Pillemer, 1979).

Overall, it therefore appears that the brain of the infant and child is characterized by islands of functional neocortical activity, and that a considerable portion of their neuroanatomy has not yet established complete or even partial synaptic interconnections. As such, a good deal of neocortical and cognitive functioning occurs in semi-isolation and is characterized by multiple disconnection syndromes (see Chapter 16).

PART IV

ABUSE AND EARLY ENVIRONMENTAL AND POSTNATAL DISTURBANCES OF NEUROANATOMICAL DEVELOPMENT, RECOVERY, AND NEURAL PLASTICITY

The nature of one's early learning and social-emotional and physical and visual environment can exert dramatic influences on brain structure, determining the size and thickness of the neocortex and nerve cells; the number and size of synaptic structures; the location, density, and number of vesicles in the presynaptic terminals; the shape of postsynaptic spines; and the perforations in postsynaptic density, as well as which neural pathways are strengthened and which neurons, axons, and dendrites will die and drop out (Casagrande and Joseph, 1978, 1980; Diamond, 1985; Denenberg, 1981; Greenough and Chang, 1988; Joseph, 1982; Rosenzweig, 1971; Rosenzweig et al., 1972). In the occipital cortex, for example, the dendritic fields are increased by about 20% among those reared in an enriched visual environment. In fact, "environmental conditions can affect synapse numbers on the order of thousands of synapses per neuron, or billions, perhaps trillions of synapses per brain" (Greenough and Chang, 1988:405).

Moreover, different regions of the brain may be differentially influenced, depending on the age of the brain and the nature of the experience. For example, when young animals were trained to use either the right or left paw for a reaching task that was rewarded by cookie crumbs, it was found that pyramidal cell apical dendritic growth and branching was more extensive in the trained versus nontrained hemisphere (reviewed in Greenough and Chang, 1988). Given that pyramidal neurons give rise to the corticospinal tract, presumably training involving just the right or left paw differentially acted and stimulated neuronal growth in those neurons.

Conversely, abnormal experience can also influence if not determine which neural pathways develop and cell density and size, as well as synaptic and dendritic development and vascular volume. Abnormal and deprived early experience during specific critical periods may even induce functional invasion and occupation by competing neuronal cell assemblies (Casagrande and Joseph, 1978, 1980).

Given that the vasculature, the vesicles, the synapses, the size and density of the dendritic tree, and the shape and number of spines are all influenced significantly by early experience, it is therefore not surprising that learning and memory (Joseph and Gallagher, 1980) and the conductive properties and ability to transmit and receive complex information are also significantly affected (Joseph and Casagrande, 1980; Joseph and Gallagher, 1982).

For example, if patterned visual input is prevented from reaching target cells in the lateral geniculate nucleus of the thalamus via the surgical suturing of the lids of one eye, target neurons become smaller, fewer in number, and functionally suppressed by adjacent cells receiving normal input (Casagrande and Joseph, 1978, 1980). If the sutures are reversed such that the formerly deprived eye is opened, the subject responds as if blind.

Similarly, in their classic experiments with visually deprived kittens, Wiesel and Hubel (1963, 1965) demonstrated that when they sutured shut a single eye during early development, cats became blind and/or displayed severe visual defects when that eye was later opened. Essentially, the deprived eye becomes functionally disconnected at the level of the thalamus (Casagrande and Joseph, 1980) and at the level of the visual cortex (LeVay et al., 1980) because of the competitive influences of normal neurons, which essentially take over vacant (or inactive) cortical space.

Nevertheless, these deprived neurons still retain some plastic capabilities. For example, if the normal eye is surgically removed and the deprived eye is opened and allowed visual experience, some vision returns to the formerly deprived eye and some functional neuronal recovery is observed (Casagrande and Joseph, 1980; Joseph and Casagrande, 1980). Visual recovery is due to the removal of competitive influences and functional sprouting in those axons and dendrites that are subsequently activated by the formerly deprived eye.

Although there may be some perceptual and neuronal recovery, this recovery is not complete or equivalent to what might normally be expected given normal rearing and perceptual experiences. This is because, although even the adult nervous system is capable of significant plastic changes and neural growth, certain neural assemblies require certain experiences at a specific time, in order for those circuits to stabilize or even form.

Greenough and coworkers (1987) describe this principle and process as "experience-expectant." That is, synapses are produced at a certain time period in expectation that experience will activate those which are most appropriate. Specifically, Greenough and associates (1987:540) argue that "since the normal environment reliably provides all species members with certain experiences, such as seeing contrasting borders, many mammalian species have evolved a neural mechanism that takes advantage of such experiences to shape developing sensory and motor systems. An important component... appears to be the intrinsically governed generation of an excess of synaptic connections among neurons, with experiential input subsequently determining which of them survive."

However, if environmental input is abnormal, abnormal interconnections may be formed in their place. Consider, for example, the classic studies of Hirsch and Spinelli (1970) on plasticity within the visual system. When kittens were reared in an environment that consisted of thick lines oriented only in a horizontal (or vertical) direction, they later appeared unable to see objects or lines placed in the opposite direction and would bump into objects and furniture that were oriented differently. Moreover, neurons in the visual cortex largely ceased to respond to stimuli to which it had not been exposed but instead would fire maximally to stimuli similar to those experienced during infancy.

In this regard, experience acts to organize neuronal interconnections and the establishment of not just synapses but vast neural networks subserving a variety of complex behaviors and perceptual activity.

Indeed, the brain is exceedingly plastic and is capable of undergoing tremendous functional reorganization not just over the course of evolution, but within a few years, months, or even weeks of a single individual's lifetime (Feldman et al., 1992; Jenkins et al., 1990; Juliono et al., 1994; Ramachandran, 1993; Strauss et al., 1992; Weiller et al., 1993). For example, a somatosensory map of the body is maintained within the parietal lobe, and in humans and primates, more cortical space is devoted to the hands than to the elbow, wrist, or forearm (see Chapter 12). However, if the forearm is repeatedly stimulated, correspondingly increases in forearm representation are noted in the parietal lobe. That is, an increased number of neurons is recruited so as to subserve and process this additional input.

However, if a single digit is amputated, the remaining fingers will increase their neocortical representation and will take over the temporarily vacated space (Jenkins et al., 1990; Juliano et al., 1994). As described by Jenkins and associates (1990:575), "representations of adjacent digits and palmar surfaces expanded topographically to occupy most or all of the cortical territories formerly representing the amputated digit. As a consequence, there were several-fold increases in the cortical magnification (cortical area/skin surface area) of adjacent digits." Presumably, the expansion of representation is due to the attractive influence unoccupied dendrites exert on adjacent neurons and their axons.

However, if the tissue subserving hand sensation is destroyed, because the hand is still intact, neurons surrounding the lesion become subject to innervation by those intact axons which have been disconnected. As described by Jenkins and coworkers (1990:579), "skin surfaces formerly represented in the cortical zone of the lesion came to be represented in entirety in the cortex surrounding the lesion." Presumably those axons which are still intact (though cut off from their terminal substrate because of tissue destruction) will search out and compete for dendrites that are already occupied, such that an axonal-dendritic battle for functional (living) space ensues.

By contrast, in cases where the entire hand is severed such that all "hand" and "finger" neurons are completely denervated, it appears that adjacent neuronal tissues that subserve completely different regions of the body will be attracted to and come to innervate these tissues. Therefore, if the hand is severed, as the "hand" area is adjacent to the "face" area, neurons that represent the face begin to expand and take over those sites no longer receiving "hand" input (Pons et al., 1991).

If the face is subsequently stimulated, patients may experience highly localized sensations that are attributed to a phantom hand and fingers (Ramachandran, 1993). That phantom hand and finger sensation is not completely overridden and extinguished, may well be due to the maintenance of an indelible memory of the body image, including the hand and fingers, by parietal neurons (see Chapter 12).

Abnormal Environmental Experience and Neural Plasticity

Social, emotional, and physical stimulation during early development constitutes a crucial aspect of the normal progression of experience-expectant neural development. As detailed in Chapters 7 and 16, if early experience is abnormal, repetitively traumatic, and/or insufficient or characterized by social, emotional, and physical neglect, the behavioral consequences can be likened to what occurs following bilateral amygdala destruction (Joseph, 1990a, 1992a, b). That is, following severe neglect or bilateral amygdalectomy, animals and humans cease to respond "normally" to social and emotional stimulation, withdraw when approached, display enhanced startle and fear responses, fail to accurately perceive social signals, and may neglect or physically abuse their infants.

Abnormalities in cellular development and function also occur within the amygdala, and probably the cingulate and septum secondary to social and emotional deprivation and neglect. For example, it has been demonstrated that the right amygdala is larger than the left in animals reared in enriched and normal environments, whereas these differences are insignificant in animals reared in a restricted setting (Diamond, 1985), suggesting that deprivation differentially reduces right amygdaloid growth. Similarly, it has been demonstrated that stress and emotional trauma during adulthood significantly affects and reduces the size of the right hippocampus—a condition that is undoubtedly paralleled by similar disturbances in other brain areas (see Chapter 16). These findings are significant because among humans, the amygdala and the right cerebral hemisphere have been shown to be dominant in regard to most aspects of social-emotional functioning (see Chapters 3 and 5), whereas hippocampal injury is associated with dissociative disturbances (see Chapter 16).

Given the immature state of not just the neocortex but various limbic nuclei, it may well be expected that a failure to receive proper stimulation may result in a significant reduction of those neurons, dendrites, and axons and their associated neural circuits, which subserve "normal" social-emotional intellectual development. Moreover, given that abnormal experience, including emotional and sexual trauma, also act to stimulate the brain, it is likely that abnormal experiences would result in the establishment of abnormal neural circuits (see also Chapter 16). That is, tissues that normally have few or no interconnections may subsequently come to be synaptically linked due to the abnormal activity and processing occurring in these regions.

Similarly, and as detailed in Chapter 6, learning and experience can induce synaptic growth, including long-term potentiation. Hence, children reared in abnormal and abusive environments not only learn from their experiences, but may act them out as they grow older (Joseph, 1992b), a consequence of the development of abnormal neural circuits that correspond to these abnormal experiences.

Thus, children who are neglected, emotionally or sexually abused, and/or raised in an impoverished environment, may correspondingly suffer a significant dropout of neurons, axons, dendrites, etc., throughout the cerebrum, as well as develop abnormal synaptic interconnections and associated neural networks (particularly within the limbic system), which later contribute to the development of abnormal behavior.

FALLS, HEAD INJURIES, AND ABUSE

Children and infants may be subject to multiple forms of abuse, which range from and include neglect, sexual molestation, beatings, and head injuries (e.g., Courtois, 1988; Kempe and Helfer, 1980). Indeed, it is not uncommon for children and even infants to be struck with fists and heavy objects, or to be thrown to the floor or across rooms, and even swung by the legs so that their head will strike walls and furniture. Of course those infants and children who suffer head injuries also suffer cerebral injuries, including infarcts and hemorrhage (Leestma, 1995).

Children and infants who are reared in homes where drug and alcohol abuse or domestic violence is common are at obvious risk for physical abuse. Moreover, children and infants who are abused are generally repeatedly abused (Courtois, 1988; Kempe and Helfer, 1980) (see Chapter 16) and thus suffer repeated cerebral and emotional traumas.

Abuse, however, does not occur exclusively in "abnormal" homes. Even a presumably "normal" mother or father may become frustrated by their crying infant and in consequence strike, violently shake, or even toss the infant against a wall or floor, inducing significant brain damage in the process—brain damage that may never come to the attention of a physician and that may not be obvious to the parents although years later they may notice that their child is intellectually "a little slow."

Head injuries secondary to nonabusive factors are also common in healthy as well as abnormal environments. It is common for children to fall and strike their head. Falls (or at least alleged falls), in fact, account for up to 50% of all emergency room visits (reviewed in Leestma, 1995).

Although it is not uncommon for children in urban settings to fall from buildings, falls down stairways are far more frequent. In more suburban and rural areas, falls are also common regardless of the quality of home life. These include head injuries secondary to infants toppling over while in walkers or when momentarily left on a table or bed (while being changed).

As stressed by Leestma (1995:250), however, "severe injuries, especially those involving major or multiple skull fractures and/or multiple truncal, or proximal extremity injuries attributed to a fall...demand further investigation and point toward a willful injury and child abuse."

The Battered Infant

Characteristically, the "classic" battered child is often less than 1 year of age; in the vast majority of cases the abuse occurs regularly, sometimes beginning the very day the infant is brought home from the hospital (e.g., Kempe and Helfer 1980). The range of abuse may include neglect, malnourishment, and sexual and physical trauma resulting in skull fractures and hemorrhages as well as cuts, lacerations, burns, bites, and tissue damage involving the viscera, genitals, and anus. It is not at all uncommon to find bite marks on the bodies of abused infants. Head injuries, however, account for 75% of those cases that come to the attention of authorities (Leestma, 1995), and up to 80% of all instances of fatal child abuse involves direct trauma to the head.

Unfortunately, even those infants who were initially spared abuse may be subject to repeated physical traumas once they reach the "terrible twos" and/or begin toilet training. Even those who pass through this difficult period unscathed may be increasingly subject to abuse, particularly of

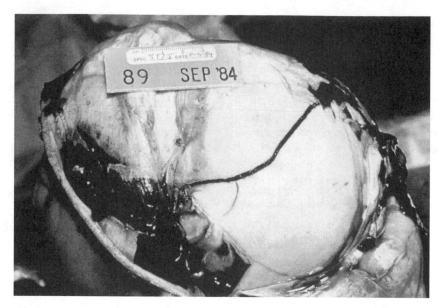

Figure 18.15   A lateral view of the skull with reflected scalp. Fatal child abuse victim who was either struck with a heavy object or was slung by the feet against a wall, floor, or furniture. Note widely spaced complex skull fracture with massive hemorrhage. (Courtesy of Dr. S. Teas, Office of the Medical Examiner, Cook County, Illinois.)

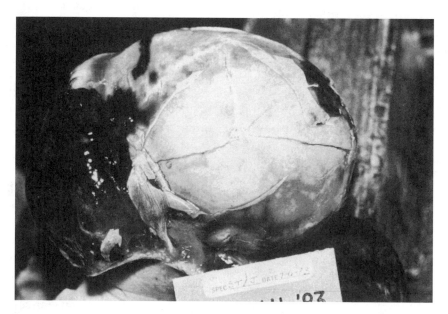

Figure 18.16   Lateral view of skull with reflected scalp. The child was 4 months old at death and had been abused since birth. Note depressed skull fracture. This child had received multiple head injuries. The skull fracture was probably caused by being swung by the legs against a wall or furniture. (Courtesy of Dr. S. Teas, Office of the Medical Examiner, Cook County, Illinois.)

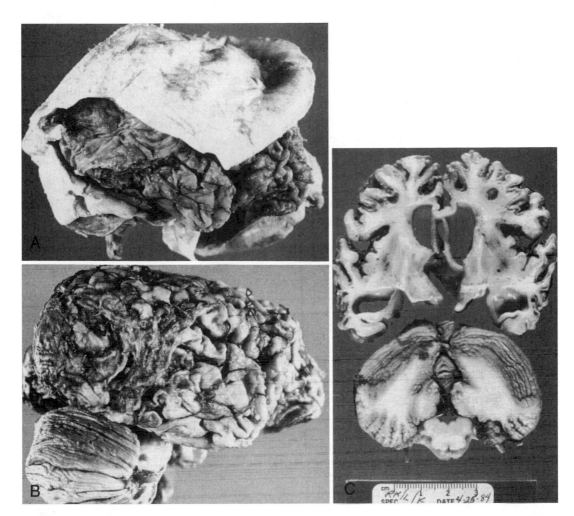

Figure 18.17 Multiple lesions secondary to repeated and chronic shaking and head injuries secondary to abuse. A, The dura is thickened with bilateral chronic subdural hematoma membranes, which are scarred down to the brain. B, The dura has been removed to reveal multiple and diffuse cerebral cortical necrosis. (Courtesy of Dr. Robert Kirschner, Office of the Medical Examiner, Cook County, Illinois.)

a sexual nature, as they approach puberty (e.g., Courtois, 1988; Kempe and Helfer, 1980). As detailed in Chapter 16, sexual abuse as well as emotional trauma can induce profound neurological abnormalities involving the hippocampus and temporal lobes.

The Shaken Baby

Not all infants who are abused and suffer significant brain damage receive head injuries. Instead, they may be shaken to death, or shaken sufficiently so as to induce axonal shearing injuries (see Chapter 19) as well as midline subdural hematomas. Direct blows to the head tend to produce lateral and basilar hematomas (Leestma, 1995). Unfortunately, not uncommonly, a crying, screaming infant may be shaken by a frustrated, immature, enraged caretaker, or even a "wonderful" and devoted mother (or father) who may not even realize the damage he or she is inducing in an attempt to quiet the baby.

Nevertheless, because of immature muscular development and coordination, and the disproportionate size of the infant's head relative to the body, the infant has little control over his or her head. In consequence, when infants are shaken the head will whip back and forth, which results in rapidly alternating decelerative and accelerative forces such that the brain is subject to shearing forces, axonal injury, subdural hemorrhages, edema, and in some cases, brainstem and cervical spinal injuries (see Chapter 19).

For example, when the infant is shaken, because of the various rotational forces and the swishing and swirling of the brain within the hard inner surface of the cranium, the upper layers of the neocortex may be sheared from the brain surface. In addition, long-distance axons as well as blood vessels may be stretched, torn, and severed.

Moreover, when the infant is shaken and the head is subject to oppositional and rapidly changing rotational forces, the cerebral hemispheres, brainstem, and different parts of the brain will actually move in somewhat different directions as well as at different speeds, thus inducing widespread stretching, straining, and snapping of axons running throughout the white matter, including the corpus callosum (Adams et al., 1981, 1989; Gennarelli, 1986; Zimmerman et al., 1978). Because these strains are not uniform but affect various parts of the cerebrum differently, such that one part of the brain is moving in one direction while another portion is sliding at a somewhat different angle, certain regions can literally snap and break loose.

Not surprisingly, rotational injuries are also associated with venous tears, arterial shearing stresses, and hemorrhagic lesions in the midline region and throughout the brain. If the vasculature is injured, the consequences may include hemorrhage, cerebral edema, and herniation.

In addition, rotational acceleration and deceleration injuries in infants secondary to being shaken can create extensive brainstem damage. For example, severe rotational forces sometimes cause tearing at the pontine-medullary junctions (where the brainstem meets the midbrain), and the brainstem may be partly torn loose and disconnected from the rest of the brain (Hardman, 1979; Jennett and Plum, 1972). Because the cerebral hemispheres sit atop the thin brainstem, there can result torque in the pontine-midbrain junction, thus disconnecting these lower from higher regions. When this occurs, consciousness is lost and the patient either dies or remains in a prolonged coma.

Although the shaken infant may suffer obvious cognitive and physical impairment, it is usually only when they are near death or severely injured that medical treatment is sought. Moreover, some parents may see their behavior as normal and "normally" rely on shaking to quiet their babies. This is because they are rewarded by the infant's ensuing silence and what appears to be sleep. Indeed, because infants may be rendered unconscious or stuporous when they are repeatedly shaken, they cease to cry, and the "relieved" parent may then put them to bed, where they may slip into a coma, though insofar as the parent is concerned, the baby is in fact "sleeping peacefully."

In some cases, however, because of the development of an expanding hemorrhage, the infant may eventually cease to breath and die. More commonly, however, he or she will later awaken (perhaps 24 or 48 hours later) and then interact in a lethargic or stuporous manner. Chronic lethargy and drowsiness are common symptoms in this population, disorders that may never completely resolve (Leestma, 1995). Even when the child may be shaken only once for a brief time period, the brain is subject to these rotational and shearing forces.

It is also not uncommon for crying infants to be smothered, which in turn results in oxygen deprivation and cell death. Some crying infants are in fact smothered to death. If there is no gross evidence of physical trauma the death may be erroneously attributed to "crib death" or "sudden infant death syndrome" (see Leestma, 1995). Those who do not die, however, like their shaken or battered counterparts, simply go through life with varying degrees of brain damage.

Repetitive Trauma and Aberrant Synaptic Recovery

When an infant is shaken, not uncommonly this results in cell death (necrosis), usually within the first 24 hours after injury—a process that may continue for up to a week. In addition, a few days after axons and nerve fibers have been torn, the nerve cell often dies, and the glial cellular reactions around these damaged, dying, and dead cells create microglia scars and cavities within the cortex and white matter that can also prevent functional recovery and reinnervation of normal neurons that have been disconnected. Often these scars can become epileptogenic and a source of abnormal activity.

Therefore, because of the immaturity of the cerebrum, infants who are shaken (and/or who receive direct blows to the head) not only suffer brain damage, but the maturational process itself is disrupted such that normal interconnections may not form, and abnormal pathways may develop in their place.

For example, following white matter or cortical injury, axons may sprout, grow, and reinnervate tissues that have lost their interconnections (Goldberger and Murray, 1985; Raisman, 1969), such as occurs with disconnection. This is not just a property of the damaged axon, however, for the denervated tissue is acting in some manner to attract these processes. For example, if an axon is severed, not only the severed axon but other axons may spout, including those which might not normally synapse or form connections with these regions. One synaptic reconnection has been established, growth influences cease and migrating axons may shrink, die, or become rerouted (Joseph, 1982).

Nevertheless, those axons which establish these new connections may be wholly inappropriate. That is, neurons and their axons that subserve completely different functions and that normally innervate completely different tissues and nuclei may be attracted to the denervated region. Abnormal interconnections may therefore be established.

Prenatal complications, or trauma experienced during infancy or the fetal period, is commonly associated with inappropriate neurons terminating in and establishing interconnections in inappropriate layers of the neocortex (e.g., Bruton, 1988; Akbarian et al., 1993; Cannon et al., 1994; Fish et al., 1992; Mednick and Cannon, 1991; Taylor, 1972). However, infants who are shaken or struck in the head not only may develop aberrant neural interconnections, but associated disturbances involving intellectual, emotional, and cognitive functioning. Affected children may seem slow, dull, overly aggressive, withdrawn, easily frightened and startled, etc. Unfortunately, as repeatedly stressed above, the etiology (e.g., child abuse, head injury, shaken baby) may never even be suspected or detected.

GLIA, NE, FUNCTIONAL RECOVERY AND REORGANIZATION

As detailed above, in many regions of the developing fetal brain, radial glia fibers act to guide migrating neurons to their terminal substrate. Similarly, when the brain has been injured, glia cells near the site of injury begin to secrete various peptides, proteins, and related neurotrophic substances (e.g., nerve growth factor) and laminin. These substances induce or stimulate neuronal repair as well as promote neuronal growth such as by attracting axonal terminals to newly vacated cortical areas. Indeed, glia cells continue to secret these substances for up to 15 days post injury (reviewed in Stein and Glasier, 1992).

Moreover, glia begin to secrete trophic substances, which also share a feedback relationship with NE terminals. That is, NE neurons respond to brain injuries by secreting NE, which in turn can influence the glia, some of which also have NE receptors (reviewed in Kolb, 1992). When these glia are stimulated they can release more trophic substances, which again act on NE neurons.

NE levels are differentially affected depending on the site of injury. For example, right cerebral injuries can induce widespread and bilateral alterations in NE levels, whereas left cerebral damage is characterized by local changes in NE (Robinson, 1979).

Anomalous cerebral "recovery" even at distant and inappropriate sites may ensue because of these widespread NE changes. Because some glia have NE receptors, glia far distant from the site of injury (Coleman and Flood, 1988) may be induced to secrete trophic factors that in turn induce axonal (Foerester, 1988) and dendritic growth (Coleman and Flood, 1988) and possible reorganization even at sites far from the original injury (Pons et al., 1991).

Thus injury to the brain can promote the establishment of abnormal as well as appropriate interconnections, as well as competition between adjacent tissues and those which have been disconnected in sites near and far from the injury.

As detailed above, following injury or denervation, axons from adjoining tissue often are stimulated to sprout terminals that may compete with severed and recovering axons for undamaged, albeit temporarily vacated, synaptic space. That is, since lesions not only destroy tissue but destroy fibers of passage, when those axons begin to regenerate, their original target may be occupied by a different axon, which may in fact subserve and transmit information quite different from that of the original neural circuit.

Misdirected axonal interconnections, even when terminating in completely unrelated areas subserving completely different functions, may in turn become functional. For example, when retinal axons were experimentally redirected to the medial (rather than the lateral) geniculate or the somesthetic areas of the thalamus, the auditory and somesthetic neocortex was found to subsequently respond to visual stimuli (reviewed in O'Leary, 1989).

Similarly, Dunnett and colleagues (1985) performed neonatal lesions of the basal nucleus (substantia innominata), destroying cholinergic neurons, which results in a loss of cortical cholinergic innervation. They then transplanted ventral forebrain tissue into the frontal cortex and found that neuropeptide Y-immunoreactive neurons innervated the (cholinergic) vacant cortical tissue, which was followed by partial behavioral recovery (see Lescaudron and Stein, 1990).

In some cases, depending on the timing and extent of the lesion, behavioral recovery may be characterized by functional substitution such that a pattern that differs from normal actions may ensue, or by restitution, such that the original behavior and pattern of action is recovered. However, as detailed above, sometimes wholly inappropriate and abnormal interconnections are established, which in turn results in perceptual limitations and a host of behavioral abnormalities.

Glia, Amino Acids, Repetitive Injuries, and Limitations on Recovery

Glia not only promote recovery but act as a limiting factor as well as a possible source of epileptic activity due to the presence of the buildup of a glial scar. Glial scarring may generate abnormal activity and essentially creates a barrier that prevents or blocks access to damaged tissue by disconnected and adjacent axons.

Similar disturbances and functional limitations may be induced by the release of various amino acids, substances that can also act as neurotransmitters and that assist in the regulation and control of neurotransmission. For example, following brain injury, large quantities of excitatory amino acids such as glutamate are released (reviewed in Stein and Glasier, 1992). However, when released in large amounts, glutamate can induce neuronal degeneration. For example, ischemia can produce a "glutamate cascade" that causes an increased and excessive flow of calcium ions into affected neurons, thereby producing cell death (see Stein and Glasier, 1992).

Given that children and infants who are abused are usually repeatedly abused and thus suffer multiple and repeated brain injuries, the presence of glia as well as amino acids such as glutamate—having been repeatedly generated, released, or activated because of previous and recent head injuries—can therefore act to interfere and disrupt subsequent attempts by the CNS to repair itself when subjected to repeated injury. Children who are repeatedly physically abused, therefore, not only suffer a high incidence of head trauma and brain damage, but because of the repetitive nature of the abuse and subsequent and repeated trophic alterations (as well as the incremental loss of brain tissue and available synapses and neurons), their ability to recover from subsequent injuries is also significantly diminished.

NOREPINEPHRINE, ABUSE, NEURAL RECOVERY, AND REORGANIZATION

As detailed above, during the course of neocortical development, wave after wave of migrating neurons sandwich themselves between the NE-rich, midbrain-reticular-like layers I and VII. Presumably, these layers act to attract as well as maintain the functional viability of migrating and "waiting" neurons until they establish their synaptic interconnections.

NE has been shown to be exceedingly important in maximizing neural growth and plasticity, not only during the early stages of development (Johnston, 1988; Parnavelas et al., 1988), but following traumatic experiences and brain injuries sustained as an adult, i.e., NE promotes functional and anatomical recovery and exerts significant influences on the differential rates of neuronal development in the various regions of the cortex (Johnston, 1988).

In the developing nervous system, NE also acts to suppress the establishment of irrelevant neural networks and pathways, while simultaneously stabilizing and/or promoting the growth and formation of relevant synaptic circuits (see Bear and Singer, 1986; Pettigrew and Kasamatsu, 1978). Similarly, NE has been shown to inhibit or suppress irrelevant background activity, while simultaneously enhancing evoked responses in both inhibitory and excitatory circuits associated with the processing of relevant environmental input (Foote et al., 1983).

In addition, NE plays an important role in neuronal plasticity (Kasamatsu and Pettigrew, 1976; Pettigrew and Kasamatsu, 1978), the guidance and inhibition of neural development (even in non-NE systems), and neuronal regeneration and recovery. For example, immediately following a brain injury, NE secretion is rapidly increased, which acts to promote plasticity, synaptic development, and thus functional recovery. If deprived of NE, functional recovery is retarded (see Parnavelas et al., 1988).

Similarly, in the presence of normal environmental input, when the developing neocortex is deprived of NE, there results a profound disturbance in neuron morphology, coupled with reduction in cell densities (Parnavelas et al., 1988). This is an exceedingly important finding, for as detailed in Chapter 16, adverse early environmental influences can induce significant alterations and reductions in NE, even when the trauma is mild and/or involves varying degrees of neglect (Higley et al., 1992; Kraemer et al., 1989; Rosenblum et al., 1994).

Consider, for example, that among its many diverse functions, NE appears to be directly involved in the experience of separation anxiety from the mother (Pankseep et al., 1988). Primates who are deprived of mothering or early socializing experience appear to suffer reductions in NE activity (Kraemer et al., 1989). However, the NE (as well as the 5-HT) neurotransmitter systems may be significantly and permanently altered by even mildly adverse early experiences (Higley et al., 1992; Rosenblum et al., 1994).

When primate mothers are repeatedly confronted by an unpredictable environment that affects their ability to forage, their ability to respond or attend to their infants is also affected. These infants in turn become less securely attached, more easily frightened and startled, and less social or independent, and they display significant NE-related abnormalities.

According to Rosenblum and associates (1994:226), "adverse conditions in infancy, whether the product of disturbed infant-mother relations and/or more general stressors in the environment, may lead to a state in which NE responses are exaggerated and 5-HT responses are blunted." Moreover, infants reared under these mildly stressful conditions display responses similar to those seen in humans suffering from post-traumatic stress disorder (Rosenblum et al., 1994).

Presumably, these stress-related disturbances involving the NE (and 5-HT) systems, particularly when experienced during infancy, abnormally influence neuronal and synaptic development, thereby producing a functional lesion as well as abnormal neural circuitry. Given that the limbic system (e.g., amygdala) is also affected by stress, emotional trauma, and alterations in NE and 5-HT, abnormalities in the development of associated limbic system circuitry may also be adversely impacted.

Stress, Emotional Abuse, and Limbic System Reorganization

In response to chronic stress or emotional upheavals, the limbic system begins to secrete massive amounts of cortisone and opiate peptides, while NE and 5-HT are simultaneously diminished or significantly altered (see Chapter 5), alterations that would exert profound influences on presynaptic and postsynaptic receptor sites in the amygdala, hypothalamus, and hippocampus. During the prenatal period and throughout early childhood, these alterations would result in abnormalities involving the circuitry of these limbic system neurons.

As detailed in Chapter 16, in some cases of childhood sexual and physical abuse, the amygdala and hippocampal temporal lobe complex may be injured following repeated instances of traumatic stress and/or via breath holding or temporary suffocation and/or as a consequence of high levels of opiate and NE activity. Given the important role of NE in the development of synaptic interconnections and neuronal maturity, disturbances of this system can therefore profoundly affect consequent cerebral organization and neural functional integrity.

As noted, stress increases NE turnover in the amygdala (Tanaka et al., 1982) and increases the production of opiates (Krystal, 1990). Opiate receptors appear to be located presynaptically on NE cells (Llorens et al., 1978), and opiate peptides appear to exert an inhibitory effect on the release of NE (Izquierdo and Graundenz, 1980). In the developing brain, particularly within the first 4 or 5 years, the repeated release of these substances coupled with fear-induced LTP and synaptic plasticity can in turn significantly and permanently alter the neuronal interconnections, neuronal drop-out rate, and thus the functioning of the wide areas of the cerebrum, especially within the amygdala and related limbic nuclei (Cain, 1992; Racine, 1978).

Normally, the release of NE retards kindling in the amygdala. However, with repetitive stress and high levels of arousal, opiates will inhibit NE release. This can lead to permanent structural and functional alterations within infant and adult amygdala neurons, affecting their neocortical interconnections, postsynaptic densities, and the size of the presynaptic terminals, as well as their capacity to process and transmit information (e.g., Cain, 1992; Racine, 1978).

In addition, when the amygdala is repeatedly stimulated by excitatory neurotransmitters, or via pharmacological agents including opiates, or through direct electrical activation (as would also occur under instances of abuse), the threshold for amygdala activation and information transmission may be reduced as well as become abnormal. Depending on the age of the brain, three major, albeit closely related, disturbances may variably result: neural migration errors, the development of abnormal interconnections, and the development of kindling activity.

Heightened activity within the immature amygdala or septal nuclei, for example, might induce the growth of additional dendrites, which in turn act to attract axons from alternative sites (e.g., Raisman, 1969). In consequence, abnormal neural networks and pathways may be formed between nuclei that "normally" do not directly interact. Conditions such as these could predispose the individual to behaving or reacting abnormally and to processing and even storing information abnormally.

With repeated instances of activation or heightened activity, an abnormal form of neuronal plasticity and a lowered threshold of responding also results. These changes are associated with increases in the size of the evoked potential amplitudes and can give rise to epileptiform afterdischarges, seizures, and convulsions, as well as induce kindling (subseizure activity) within the amygdala (Cain, 1992; Racine, 1978).

However, repetitive and traumatic stress may result in permanent alterations and lesions not just in the amygdala and septal nuclei, but within the hippocampus (Uno et al., 1989) and throughout the neocortex of the adult and infant brain (see Chapter 16).

Given the involvement of the amygdala, hippocampus, and temporal lobe in the development of psychotic and dissociative states (see Chapters 14 and 16), it therefore seems likely that these changes and reduced threshold for activation might well predispose an affected individual to chronically or repeatedly experience similar psychotic or dissociative states such as when emotionally stressed. Indeed, because of kindling, abnormal plasticity, and other changes, the amygdala, hippocampus, and temporal lobe may become abnormal and hypersensitive and potentially capable of inducing or maintaining long-term psy-

chotic episodes (see Halgren, 1992; Honer et al., 1994; Stevens, 1992). In children these episodes may be diagnosed as autism or childhood schizophrenia.

Alterations in neocortical interconnections, the pathways leading to and from the frontal-thalamic system, and the axonal interconnections with the hippocampus and amygdala might all be altered as a function of abnormal learning and early environmental stress. In children and adults, associated neural pathways would develop in accordance with these environmental stresses and associated stimuli. Perceptions as well as emotional reactions would be expressed in accordance with this abnormal circuitry.

Behaviorally and emotionally the consequences of developing these abnormal neural networks may include psychosis, chronic depression, schizophrenia, chronic dissociative disorders, social-emotional agnosia, obsessive-compulsiveness, hysteria, excessive emotionality, and so on, as well as a pronounced tendency, or predisposition, to form abnormal or abusive relationships or to seek traumatic experiences, as this is what the brain has become adapted to (Joseph, 1992b).

In this regard, brain structure and intellectual, emotional, and personality functioning may be irreversibly altered by repeated traumas suffered early as well as late in life (see Chapter 16). If these events occur during childhood when neural pathways are still being formed, the consequences of traumatic learning and memory formation may result in permanent alterations in brain structure, synaptic organization, and neurochemistry, which in turn would be expressed emotionally, behaviorally, and through speech, thought, and language.

Neocortical Forebrain Plasticity and Susceptibility to Permanent Injury

The forebrain owes its evolutionary origins in large part to the olfactory-limbic system, from which the amygdala, striatum, hippocampus, and portions of the septal nuclei are derived. The forebrain is also an evolutionary derivative of visually sensitive nuclei from which portions of the midbrain and thalamus are derived.

Similarly, over the course of ontogeny the fetal forebrain is derived from both midbrain-like reticular tissue (layers I and VII) and the anterior-most portion of the neural tube, as well as from the olfactory system (albeit to a much less extent). Hence, under conditions of even mild prosencephalic disorders, the olfactory bulbs are invariably affected.

The cellular composition and organization of layers I and VII, however, are also phylogenetically ancient and common in structure among all animals, except fish and other "lower" species who dwell on land or sea. Layers II through VI compose the neocortex.

Over the course of evolution, not only layers I and VII, but the midbrain, brainstem, and spinal cord became increasingly under the dominance and control of the telencephalon and the relatively recently acquired five layers of neocortex of the right and left cerebral hemispheres.

Unfortunately, as pointed out by Hughling Jackson over a hundred years ago (Jackson, 1884), brain structures that have been more recently acquired in many respects are also the most fragile and the least resistant to injury. When injury occurs, however, those more ancient brain structures that have become accustomed to forebrain control, also often cease to fully function or provide sensory data.

When neocortical structures are severely damaged, functions originally associated with the hindbrain (e.g., vision, audition, tactile sensation) may be lost even though these phylogenetically older nuclei are spared injury. Thus, if a human suffers a severe injury to the primary motor or visual neocortex, he or she may become paralyzed or (neocortically) blind, whereas if a more primitive mammal (such as an opossum or a rat) where to receive similar destructive lesions, they would quickly regain most of these lost functions.

However, in humans, lesions of early onset involving the forebrain, particularly if mild, may not be manifested until later in life. In part this is due to these "lost" functions being acquired later in life (Goldman, 1974), as initially and for the first few years, these functions may not be necessary and/or are subserved by the midbrain and brainstem, and thus are hierarchically acquired at a later age at which point the deficit becomes apparent. Hence, in many cases of late-onset disability or psychotic or emotional disturbance, the actual injury was probably sustained early in life and perhaps even during the fetal period of development.

Cerebral and Cranial Trauma:

Neuroanatomy and Pathophysiology of Mild, Moderate, and Severe Brain Injury

Traumatic brain injury may well be as common as stroke as a leading cause of neurological injury and death in America and Europe (Slagle, 1990; Teasdale, 1995), with well over a million individuals a years seeking medical attention in the United States alone. Almost half of such injuries are due to falls (especially in children), with assault, automobile accidents, and gunshot and stab wounds accounting for the bulk of the remainder. However, the actual yearly incidence of head injury may be much higher, particularly in regard to children and infants with abusive parents and because men with supposedly "mild" head injuries tend to downplay the significance of their injuries and are often reluctant to seek medical attention unless cognitive and physical functioning is significantly compromised (e.g., Kelly et al., 1991). Unfortunately, some physicians and emergency room personnel also downplay the significance of "mild" head injuries, particularly when the patient appears to have more immediate and life-threatening problems. In part this attitude is due to the erroneous assumption that "mild" injuries are in-themselves insignificant and/or that the brain is so well protected by the skull that only under conditions involving open head injuries and/or loss of consciousness should the possibility of an actual brain injury be entertained. As will be detailed below, this belief is in error.

THE MENINGES

The living brain is a soft and delicate tissue with a rather compact consistency. The brain is protected by several outercoatings (i.e., membranes) that act as a cushion between it and the hard inner shell of the cranium. The innermost membrane is a sheer sheet of translucent material that actually adheres to the brain surface. This is called the pia matter. Lying above the pia matter is yet another very thin, web-like, fibrous membrane referred to as the arachnoid. The space between the arachnoid and pia matter is called the subarachnoid space, through which circulates cerebrospinal fluid. Collectively the arachnoid and pia matter are called the leptomeninges (reviewed in Carpenter, 1991; Parent, 1995).

Sitting above the leptomeninges and partially adhering to the inside of the skull is a thick, tough, leathery-like membrane, the dura mater (i.e., "tough mother"). The dura mater is richly innervated by blood vessels, including the middle meningeal artery (which is sometimes subject to laceration following skull fractures, creating an epidural hemorrhage).

The dura not only acts as a hard shield protecting the brain from the skull but forms a number of compartments that partly encompass and support various portions of the cerebrum. One major compartment formed by the dura is the falx cerebri, which juts down between the cerebral hemispheres. The falx in turn gives rise (in a fan-like fashion) to the tentorium cerebelli, upon which sits the occipital lobes and below which is the cerebellum. With certain types of head traumas, movement of the brain against the falx and tentorium can cause contusions and shearing of cortical tissue.

THE SKULL

Completely encasing and supporting the brain is a thick bony covering that protects

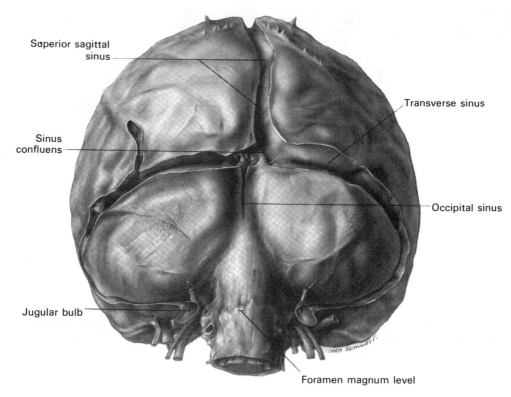

Figure 19.1   Postioror superior view of the dura surrounding the brain. (From Mettler FA. Neuroanatomy, 2nd ed. St. Louis: CV Mosby, 1948.)

it from damage, i.e., the skull. The cranium is composed of several bony sheets, e.g., the frontal, parietal, temporal, and occipital bones. The frontal bone constitutes the skeletal framework for the forehead (including the orbits for the eyes) and is joined to the parietal bones (the bulging topsides of the skull) by the coronal sutures. Composing the lower and lateral portion of the cranium, including parts of its floor, are the temporal bones, which also house the structures of the inner ear. The occipital bone makes up the lower posterior portion of the cranium.

Although seemingly smooth on the outside, the skull is not completely smooth and rounded on the inside as it consists of several small bony protrusions and cavities. These same protrusion can sometimes serve as a source of injury. For example, when the head is subject to accelerating-decelerating forces, the soft mass of the brain may be thrown against these ridges and be-

come torn and contused in consequence (Leestma, 1991). Nevertheless, the skull serves as the first line of defense against traumatic brain injury.

Skull Injuries

If struck by a blunt object, the skull is flattened with stress oscillating outwards laterally—like a rock hitting a pool of water. If sufficient pressure is applied the skull will fracture. Usually the break occurs at the site of impact.

Skull fractures are of three types: depressed, linear, or basilar. Basilar fracture are relatively uncommon, whereas approximately 75% are linear, the remainder being depressed. In general, head injuries are also considered as either closed or open if accompanied by scalp laceration and/or if the fracture extends into the sinuses or middle ear (Adams and Victor, 1994).

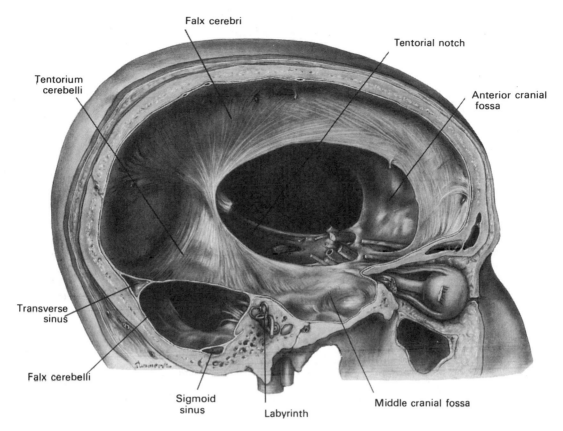

Falx cerebri

Tentorial notch

Tentorium
cerebelli

Anterior cranial
fossa

Transverse
sinus

Falx cerebelli

Sigmoid
sinus

Labyrinth

Middle cranial fossa

Figure 19.2 Parasagittal (lateral) view with brain removed depicting the dura and its compartments. (From Mettler FA. Neuroanatomy, 2nd ed. St. Louis: CV Mosby, 1948.)

BASILAR FRACTURES

Basilar fractures often extend into the base of the skull and are difficult to detect unless severe. However, existence of a basal skull fracture may be indicated by cranial nerve damage or hormonal-endocrine abnormalities (such as from damage to the pituitary). Fractures near the sella turcica (at the base of the skull) may tear the stalk of the pituitary such that in consequence diabetes, impotence, and reduced libido may result.

In some instances these fractures may extend in an anterior, posterior, or lateral direction. If they extend in an anterolateral direction, tearing of the olfactory, optic, oculomotor, trochlear, first and second branches of the trigeminal, and the facial and auditory nerves may occur, thus disrupting olfaction, vision, and eye move-

ments and/or causing unilateral facial paralysis and hearing loss. If the fractures extend laterally they may damage the mastoid bone and tympanic membrane of the inner ear, resulting in dizziness, disturbances involving equilibrium, and a loss of hearing.

Basilar fractures are sometimes associated with tearing of the dura as well as cerebrospinal fluid (CSF) leakage. Hence, a variety of related complications may occur, including infection.

DEPRESSED FRACTURES

Usually with depressed fractures, part of the skull will shatter into several fragments, which are driven downward toward the brain. If the dura is torn the brain is often lacerated as well. Moreover, if the dura has been torn the patient becomes

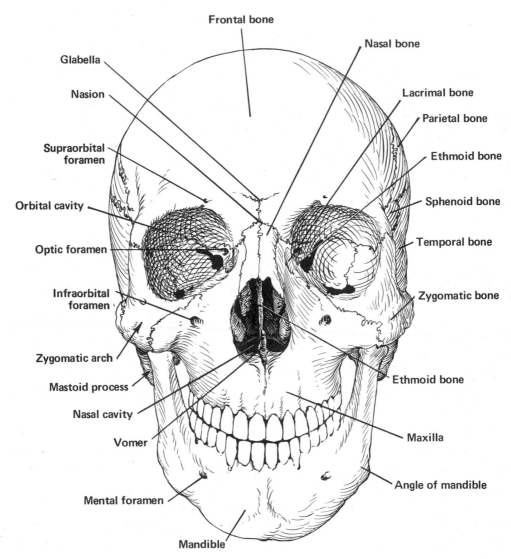

Figure 19.3 Anterior (frontal) view of human skull. (From Waxman SG, De Groot J. Correlative Neuroanatomy, 22nd ed. East Norwalk, CT: Appleton & Lange, 1994.)

vulnerable to infection, particularly in that pieces of hair or other debris may be driven into the cranial vault. This in turn will later give rise to a host of symptoms, including the possible development of meningitis (Jennett and Teasdale, 1981).

Frequently, but not always, the meningeal artery is torn and intracerebral or extracerebral hematomas or an epidural hematoma may develop. Laceration and contusions are usually found beneath the broken bone fragments, and subdural hematomas may develop on the contralat-

eral side (Bakay et al., 1980). If not accompanied by a laceration of the scalp, depressed fractures are described as closed.

In some cases, particularly if bone fragments have been driven into the brain and/or with the development of hematomas, patients develop focal neurological signs, depending on which part of the brain has been compromised.

Approximately 50% of those who suffer a depressed skull fracture do not lose consciousness (Bakay et al., 1980; Jennett and Teasdale, 1981), and in many instances the

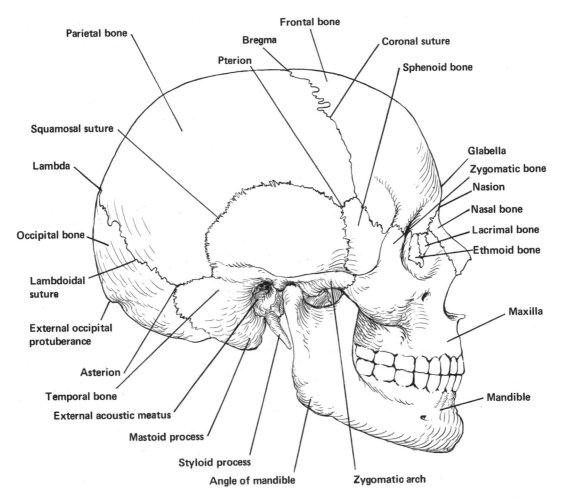

Figure 19.4 Lateral view of the skull. (From Waxman SG, De Groot J. Correlative Neuro- anatomy, 22nd ed. East Norwalk, CT: Appleton & Lange, 1994.)

dura is spared and there is no gross evidence of neurological compromise. This does not mean, however, that the brain has not been injured.

LINEAR FRACTURES

When the head is struck there usually results an inward deformation of the skull immediately beneath the site of impact, whereas the surrounding area is bent outward. In some instances the skull shatters (i.e., depressed fracture), whereas in the majority of cases it will crack. Linear fractures are of two types, longitudinal and transverse.

Like depressed fractures, patients may or may not lose consciousness. However, it has been reported that patients with linear fractures who retain consciousness are 400 times more likely to develop a mass lesion (e.g., hematoma) as compared to comatose patients, who are 20 times more likely to develop intracranial hemorrhage (Jennett and Teasdale, 1981).

The most common sites of linear fractures involve the temporal and parietal bones. Indeed, the temporal portion of the skull may fracture following trauma to any portion of the cranium.

Hearing Loss, Tinnitus, and Vertigo

Linear fractures involving the temporal-parietal bones may damage the auditory meatus, eustachian tube, and ear drum,

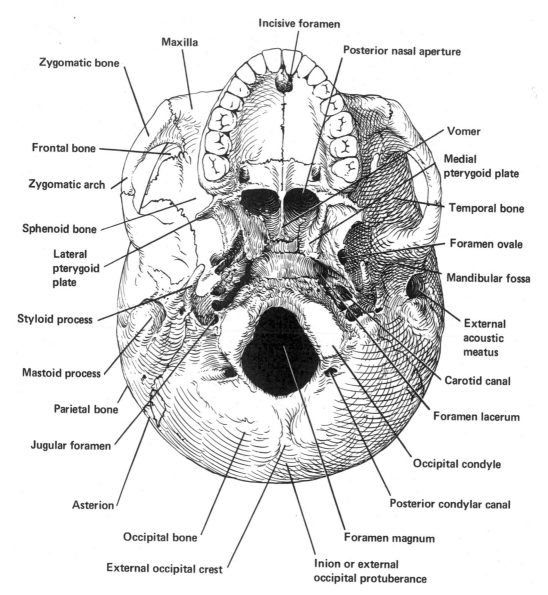

Figure 19.5 View from the base of the skull. (From Waxman SG, De Groot J. Correlative Neu- roanatomy, 22nd ed. East Norwalk, CT: Appleton & Lange, 1994.)

causing hearing loss, tinnitus, disorders of equilibrium, and vertigo.

Facial Paralysis

In some instances, longitudinal fractures may damage the cochlear nucleus and cause injuries to the 7th and 5th cranial nerves, which pass through this area before innervating the skin and muscles of the face. When these nerves are crushed or damaged, a unilateral facial paralysis and loss of sensation results. Transverse fractures can also cause stretching of the 7th and 8th nerves and may damage the vestibular and cochlear portions of the labyrinth. Hence, facial paralysis and hearing related abnormalities may also occur.

Anosmia

Anosmia (loss of the sense of smell) and an apparent loss of taste (loss of aromatic flavor perception) are frequent sequelae of

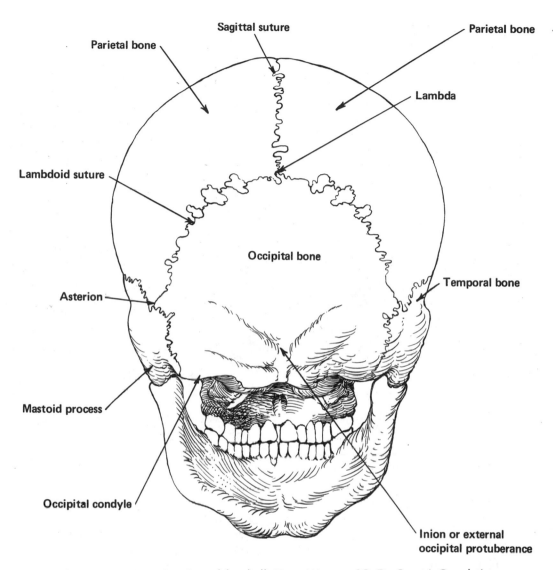

Figure 19.6   Posterior view of the skull. (From Waxman SG, De Groot J. Correlative Neuroanatomy, 22nd ed. East Norwalk, CT: Appleton & Lange, 1994.)

head injury, especially following injuries to the face and fractures involving the back of the head or frontal bone. Anosmia is due to damage to the olfactory nerve, usually in the vicinity of the cribriform plate.

The cribriform plate is a wafer-thin sheet of perforated bone through which the olfactory nerves pass on their journey from the nasal mucosa to the olfactory bulbs. Because this thin sheet of bone is perforated it is predisposed to fracture during head trauma regardless of where the patient was struck. This may cause the olfactory nerves to shear off, thus resulting in a permanent loss of smell—anosmia. Patients are unable to even detect markedly unpleasant odors. If odors can be detected, the olfactory nerve is intact.

If the shearing is unilateral the loss of smell will not be recognized by the patient. It is only with complete bilateral shearing that patients begin to complain, usually noting that they have suffered a loss of taste.

With damage to the olfactory nerve and cribriform plate, sometimes a laceration or rupture of the meninges results. If there is meningeal rupture, CSF will leak into the nose. Frequently the only symptom is what appears to be a continually "running nose."

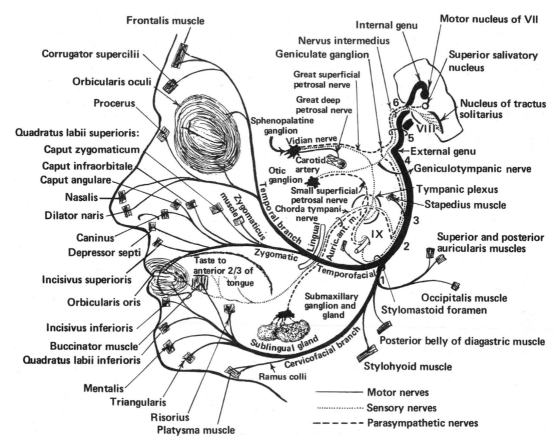

Figure 19.7 The distribution of the 7th (facial) nerve. (From Waxman SG, De Groot J. Correlative Neuroanatomy, 22nd ed. East Norwalk, CT: Appleton & Lange, 1994.)

In some instances CSF has gushed into the patient's nose when he or she has coughed or sneezed—well after the injury. Hence, if a patient has a runny nose, loss of smell, but no cold or allergy, and has had a head injury, a CSF fistula secondary to meningeal rupture and cribriform plate fracture should be considered. If this is suspected the patient should be referred immediately to a neurosurgeon. Sometimes a secondary consequence of rupture that goes untreated is bacterial infection, which may develop into meningitis.

Blindness

With some head injuries, the optic nerve may be stretched or torn and the eyes may pop out of the socket. Fractures near the sphenoid bone (which juts out beneath and below the frontal bone within the skull)

may result in laceration of the optic nerve. When this occurs the patient becomes immediately and permanently blind. The pupil becomes permanently dilated and is unreactive to light, although consensual reflexes are maintained.

HEMATOMAS

Following a head injury, with or without skull fracture, the arteries and veins running above, below, or through the meningeal membranes may be stretched, broken, pierced, or ruptured. This results in blood loss and the development of a blood clot, a hematoma.

There are various types of hematomas. Some form below the dura, i.e., **intradural hematoma,** whereas others develop between the skull and the meninges and are referred to as **extradural.** Both are due to

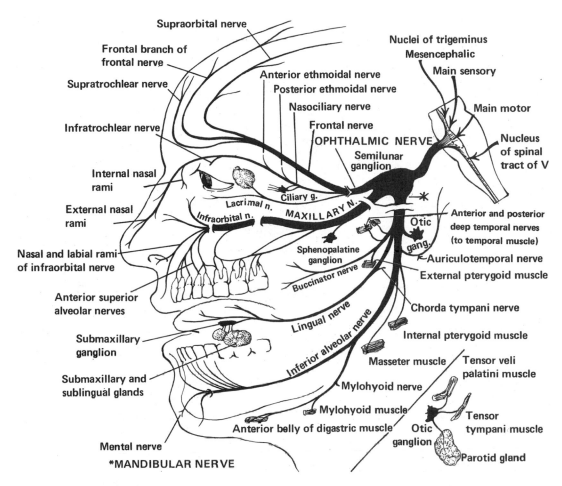

Figure 19.8 The distribution of the 5th (trigeminal) nerve. (From Waxman SG, De Groot J. Correlative Neuroanatomy, 22nd ed. East Norwalk, CT: Appleton & Lange, 1994.)

bleeding inside the skull and the formation of a clot. The clot in turn acts to compress the brain. Extradural hematomas are often secondary to a skull fracture and laceration of the meningeal arteries below the site of primary impact.

Intradural Hematomas

There are several subtypes of intradural hematoma, i.e., epidural, subdural, intracerebral, all of which may occur and develop immediately at the time of primary insult, or slowly, thus causing symptoms days, weeks, or even months after the original head injury. Although 30–55% of those with severe head injuries develop hematomas, they also occur with mild injuries.

Intracerebral hematomas are not very common. They are usually secondary to ruptured aneurysms, gunshot or stab wounds, and lacerations from depressed skull fractures.

Epidural hematomas are also not very common as they tend to occur in less than 10% of all patients with severe head injury. Most are secondary to fractures of the temporal-parietal area and laceration of the middle meningeal artery. They are frequently quite slow to develop, appearing hours or even days after the injury (Nikas, 1987a, b).

In most cases a patient will be struck and briefly lose consciousness. Upon waking they seem lucid, but then as the hematoma develops they begin to increasingly

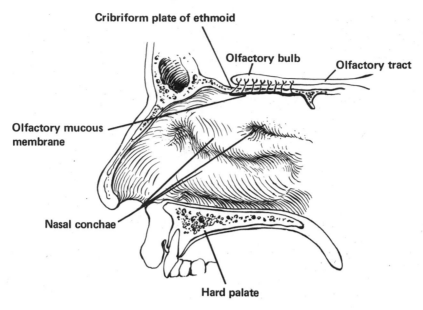

Figure 19.9 Lateral view of the nose and olfactory nerve. (From Waxman SG, De Groot J. Correlative Neuroanatomy, 22nd ed. East Norwalk, CT: Appleton & Lange, 1994.)

complain of headache, irritability, and confusion. As the hematoma increases in size and compresses and damages the brain, the patient's symptoms become more severe and consciousness may again be lost. If the hematoma is not evacuated the patient is likely to die (Gallagher and Browder, 1968).

SUBDURAL HEMATOMAS

Subdural hematomas are the most common of all and are associated with a very poor prognosis. These hematomas develop following damage to the various veins crisscrossing the subdural space. This results in a pool of blood developing over the surface of the entire brain. Patients may slowly or rapidly deteriorate and become markedly depressed or even stuporous as the hematoma enlarges and increasingly compresses the brain (Bucci et al., 1986). In slowly developing, undetected cases patients may continually complain of headache and develop alterations in personality as well as changes in level of consciousness. Eventually they may develop hemiparesis, language disorders, and other focal signs. Mortality has ranged from 30% to 70%—

even when removal of the hematoma is attempted (Nikas, 1987).

Unfortunately, computed tomography (CT) (Cooper et al., 1979), and in particular magnetic resonance imaging (MRI) (Snow et al., 1986), are not very useful in detecting the initial development of subdural hematomas, even when associated with hemorrhage (see Newton et al., 1992, for related discussion). Hence, although an initial CT scan may have been normal, if a patient subsequently deteriorates, the possibility of hematoma should be entertained.

CHRONIC HEMATOMAS

Acute and chronic subdural hematoma may be unilateral or bilateral with an onset latency period of days or weeks. There may be headache, drowsiness, confusion, and disturbances in consciousness, all of which progressively worsen. Focal or lateralizing signs are less prominent. In the chronic condition, the traumatic etiology may not be clear or may seem very minor (e.g., striking the head against branch of a tree), particularly among older individuals. Headaches, giddiness, and slowness in thinking may not develop until weeks after the initial injury and may give the initial impression of a vascular lesion, tumor, or drug intoxication.

HERNIATION

Hematomas are potentially life-threatening and can cause extensive brain injury (even when initially brain damage seemed minimal). This is due to the effects of compression as well as secondary increases in intracranial pressure (ICP).

The brain and skull are very tightly fitted and there is little room for expansion. Hence, when blood leaks into or around the brain, something has to give and make room. Because the skull is rigid and relatively inflexible, pressure is thus exerted on the soft tissue of the brain, which in turn becomes compressed. Because of the compression the brain will shift within the cranial vault, press up and/or beneath the various dural tentorial compartments, or even become forced down into the foramen magnum at the base of the skull, thus compressing the brainstem—which leads to death (see Fisher, 1995, for related discussion and opposing viewpoints). These latter conditions are referred to collectively as **herniations.** Not all herniations are secondary to hematoma, and many arise from tumors, stroke, or a combination of factors.

As noted the skull is subdivided into compartments via the dura. Because of its compartmental arrangement, pressure may be high in one region of the brain and less so in yet another. Nevertheless, force will be exerted from a high-pressure region into a low-pressure zone, causing displacement of the brain. Sometimes the brain is displaced over or under the dura, causing tearing and further compression effects. Nevertheless, depending on where the brain is displaced, various types of herniation and subsequent brain damage may result. These include subfalcial, tonsillar, and temporal lobe/tentorial herniations.

Subfalcial Herniation

Subfalcial herniation may occur secondary to frontal or parietal hematomas. With increased pressure in these areas the frontal-parietal areas tend to herniate medially under the falx cerebri, which separates the two hemispheres. That is, with subfalcial herniation the medial part of one frontal or parietal lobe is pushed under the falx, damaging not only these regions but the cingulate gyrus.

Tonsillar/Cerebellar Herniation

Pressure developing from a mass lesion tends to displace the brain and in particular the cerebellum in a downward direction. This also results in brainstem compression as the cerebellum is forced down in the foramen magnum. This lead to respiratory and vasomotor abnormalities, and eventually loss of consciousness. With increasing expansion there will be respiratory arrest and death. This form of herniation can also occur secondary to an expanding frontotemporal lobe hematoma.

Tentorial Herniation

A **tentorial/temporal lobe herniation** is due to swelling and pressure acting on the medial uncal region of the temporal lobe. As pressure develops the temporal lobe is forced into the tentorial opening within which sits the midbrain. Consequently the midbrain, diencephalon, and subthalamic regions are pressed against the opposite edge of the tentorial opening. Usually the oculomotor nerve becomes compressed, which causes ptosis, pupil dilation, and loss of eye movement.

As pressure continues to develop the midbrain and diencephalon are shoved down into the posterior fossa, compressing the brainstem (Adams and Victor, 1994). Because of interference with the reticular activating system, patients begin to experience rapid changes in conscious awareness; become stuporous; descend into somnolence, semicoma, and coma; and develop irregular respiration, hemiplegia (due to compression of descending motor tracts), and/or decerebrate rigidity (i.e., extension of the extremities), until finally respiration is arrested and the patient dies. In part this is a consequence of disconnection of the brainstem from upper cortical centers and can also occur with subfalcial or cerebellar herniation (however, see Fisher, 1995).

Even in less severe cases there may be infarction (i.e., stroke) within these regions secondary to compression of the blood vessels surrounding these various subcortical centers. In these instances death is also a likely sequela.

CONTUSIONS

Following a blow to the head the skull will bend inward and sometimes strike the brain, causing a contusion directly below the site of impact, i.e., a coup injury. This is because the brain, being more fluid, moves more slowly than the skull in response to force. However, except in cases of depressed skull fracture, contusions are not commonly found beneath the site of impact (Adams et al., 1980). Indeed, damage can be widespread or localized to the contralateral half of the brain.

Usually the force of the impact will force the brain to shift, and/or bounce against and strike the inner bony prominences of the opposite half of the skull (i.e., contre coup injury) as well as the falx cerebri and tentorium (Leestma, 1991). This results in contusions to the neocortical surface, particularly along the temporal and frontal lobes. Indeed, contusions and bruises are usually found in the orbital frontal regions and anterior and lateral temporal lobes bilaterally as well as along the convexity regardless of where the skull was struck (Adams et al., 1980). Frequently the medial portions of the brain (Leestma, 1991) including the corpus callosum may be traumatized because of rotational forces such that they bump and slide against the falx and/or tentorium. In contrast, contusions are rarely encountered along the occipital lobe, even when the back of the head has been struck.

In general, contusion may consist of multiple hemorrhages that develop immediately upon impact along the crests of the various cerebral gyri. Contusional hemor-

rhages may increase in size during the first few hours after injury, extending from the superficial layers of the neocortex into the white matter. Edema is usually a secondary consequence. Thus, widespread damage, varying in severity, is a likely consequence of head injury.

Necrosis

When the brain has been contused, severely compressed, or subjected to rotational shearing forces, axons are often severed and cells crushed. This results in cell death (necrosis) within the first 24 hours after injury—a process that may continue for up to a week. After a few months these cells are reabsorbed, leaving behind glia scars and small cavities in the cortex and white matter. These scars can in turn becomes a source of abnormal activity and may cause the development of seizures.

Axonal Transport Deficits

Trauma to the brain may interfere in axonal transport in otherwise normal neurons (see Burke et al., 1992). When this occurs there may be a buildup of neurotoxics, which in turn result in cell death throughout the neuroaxis. That is, if the postsynaptic neuron is injured, axonal transport to the damaged cell may become abnormal, such that in addition to a buildup of neurotoxicity, neurofibrillary tangles, amyloid plaques, and cell death result.

Loss of Consciousness

When consciousness is not lost following a head injury it is frequently erroneously assumed that the brain was not damaged (see Kelly et al., 1991 for related discussion). However, a patient may in fact suffer significant cognitive disturbances and extensive contusions and lacerations throughout the brain even with no loss of consciousness (Bakay et al., 1980; Teasdale and Mendelow, 1980). Moreover, there may be complete memory for the accident with no anterograde or retrograde amnesia.

Coup and Contre Coup Contusions

In some instances it has been postulated that because of the differential movement of the brain, both coup and contre coup lesions can result. Coup contusions are due to the slapping effect of the skull hitting the brain. Tissue is damaged directly below the point of impact. Generally the head is stationary at the time of the injury.

Contre coup contusions are usually associated with translational linear acceleration injuries and the free movement of the head. Translational (linear) acceleration of the brain occurs at the moment of impact and/or as there is a rapid linear acceleration in the motion of the head—such as when a person is thrown from a car. This causes the brain to bounce against and strike the rough bony protrusions on the opposite side of the skull.

Contre Cavitation

As pertaining to coup and contre coup lesions, rapid increases in acceleration sometimes cause the development of increased ICP at the trauma point, and decreased (contre coup) pressure at the opposite side of the cerebrum. In addition to the brain striking the opposite side of the skull, this decreased pressure at the contre coup site can cause the development of cavitation bubbles that, when they burst or collapse, cause local tissue damage, i.e., a contre coup lesion (Bakay et al., 1980). Contusions are often associated with this type of injury (Leestma, 1991; Unterharnscheidt and Sellier, 1966).

Differential Effects of Impact Site

The development of coup and/or contre coupe lesions also depends on what part of the skull was initially impacted. That is, impact to certain regions of the skull seem to commonly result in coup lesion, whereas other zones when struck give rise to coup and contre coup damage (Leestma, 1991). For example, impact at the occipital region often forces the brain to move in an anterior direction, causing a contre coup lesion

to the frontal lobe (Bakay et al., 1980; Lindgren, 1966). Indeed, although there is some evidence of considerable movement of the parietal-occipital region regardless of the impact site (see Bakay et al., 1980 for review), direct blows to the occiput seldom cause occipital damage.

In contrast, impact to the front of the skull almost always results in frontal lobe (coup) injuries. However, in about 50% of the cases, both coup and contre coup lesions result. Temporal impacts also create contre coup lesions. If a person is struck from the top of the head, contre coup contusions frequently develop in the corpus callosum and the orbital frontal lobes because of the downward force of the blow (Bakay et al., 1980).

Caveats

Not all investigators accept the notion of contra coupe damage, however. Indeed, the fact that the frontal and temporal regions are most severely affected regardless of impact site argues against the notion of contre coup injuries (Ommaya et al., 1971), as does the finding that when there is a skull fracture, contusions are maximal on the side of impact (Adams et al., 1980). Rather, it has been postulated that much of the damage that occurs secondary to impact or nonimpact (including coup and contre coup) injuries is due to the effects of rotational acceleration and shearing forces.

ROTATION AND SHEARING FORCES

In certain types of head injuries, such as those involving a great deal of force and movement, the brain will swirl, rotate, oscillate, and slosh around inside the skull in a rotary fashion (Pudenz and Shelden, 1946). This can occur at the moment of impact or if the head is suddenly subjected to rapid changes in motion (such as when the patient is airborne after being thrown through the windshield of their car versus a blow to the head when stationary). Indeed, rotational damage may occur even when the head has not been struck but can result from rapid skull movement alone, e.g., whiplash, severe and repeated shaking, or extreme motion of the head and neck (Adams et al., 1980; Uterharnschidt and Sellier, 1966). This is in contrast to linear acceleration injuries, which are not usually associated with this differential swirling movement of the brain.

Nevertheless, as the brain swirls about the neocortical surface becomes contused, sliced, and sheared by the bony protrusions (e.g., the sphenoid wing) within the cranium, and hemorrhages will develop over the surface of the cerebrum. The frontal-temporal regions are at particular risk for these type of injuries.

BRAINSTEM AND SUBCORTICAL LESIONS

However, rotational acceleration and deceleration injuries can create extensive subcortical and brainstem damage as well. For example, the cerebral hemispheres, the brainstem, and different parts of the brain will actually move in somewhat different directions as well as at different speeds, thus inducing widespread stretching, straining, and snapping of axons running throughout the white matter, including the corpus callosum (Adams et al., 1981, 1989; Gennarelli, 1986; Zimmerman et al., 1978). Indeed, because these strains are not uniform but affect various parts of the cerebrum differently, such that one part of the brain is moving in one direction whereas another portion is sliding at a somewhat different angle, certain regions can literally snap and break loose.

For example, when the brain is subjected to severe rotational forces, tearing sometimes occurs at the pontine-medullary junctions (where the brainstem meets the midbrain), such that in consequence the brainstem is partly torn loose and disconnected from the rest of the brain (Hardman, 1979; Jennett and Plum, 1972). Because the cerebral hemispheres sit atop the thin brainstem, there can result torque in the pontine-midbrain junction, thus disconnecting these lower regions from higher regions. When this occurs consciousness is lost and the patient either dies or remains in a prolonged coma.

In addition to contusions, rotational injuries are associated with venous tears, arterial shearing stresses, and hemorrhagic lesions in the midline region and throughout the brain. Nevertheless, the degree and type of damage that occurs is dependent on the speed and direction at which the head moves (Gennarelli, 1986). However, even presumably mild head injuries can induce brainstem injury (Montgomery et al., 1991).

DIFFUSE AXONAL INJURY

Independent of actual impact (Gennarelli et al., 1982), when the brain is subjected to severe rotational and acceleration/deceleration forces, there can result severe diffuse microscopic damage and profound shearing and stretching of axons throughout the brain and brainstem as well as focal lesions in the corpus callosum. Axons are literally torn in half because of the twisting strains placed on them. This condition is referred to as "diffuse axonal injury" (DAI).

When axons are severed or stretched, the capacity to transmit electrical-chemical impulses and thus transmit information is attenuated. If this stretching is severe, a permanent loss of functional capability can result, accompanied by an immediate and prolonged loss of consciousness. However, even in less severe head injuries subtle axonal injuries may result, including axonal swelling and impaired axonal transport and transmission (Povlishock, 1992).

Conditions that give rise to DAI are also associated with brainstem damage and decorticate and/or decerebrate posturing. When this occurs, many patients, if they survive, end up in a vegetative state. Most will die, however (Gennarelli et al., 1982a, b). If less severe there may be a temporary inability to function because of strains involving the axonal membrane (Adams et al., 1989; Gennarelli, 1986).

In general, DAI takes the form of axonal retraction balls in those who die within days, microglial scars in those who live days to weeks, and degeneration of the fiber tracts among those who survive for longer time periods, e.g., months (Adams et al., 1981, 1989). That is, a few days after an injury in which axons and nerve fibers have been torn, the nerve cell often dies and the glial cellular reactions around these damaged, dying, and dead cells create microglia scars. Often these scars become epileptogenic.

Moreover, because of this diffuse damage, various and widespread regions of the brain are no longer able to function or intercommunicate, i.e., disconnection. Because of this, patients with DAI who have lost consciousness yet continue to live tend not to regain consciousness but remain in a prolonged coma (Adams et al., 1981; Gennarelli et al., 1982a, b). Interestingly, this condition occurs more frequently among individuals who do not suffer skull fractures.

HYPOXIA AND BLOOD FLOW

The energy required to run the brain is produced via the oxidative metabolism of glucose. Indeed, the brain's need for these substances is quite substantial as it utilizes almost 20% of the body's oxygen supply (which the brain is unable to store) and consumes over 25% of total body glucose. The source for both is the vascular system.

Reduction in oxygen and glucose create disturbances involving neuronal functioning, including dizziness, confusion, seizures, and/or a temporary cessation of brain activity and coma—even if the deprivation is transient (Joseph, 1990f; Morgan and Gibson, 1991; Sokoloff, 1981). If prolonged the consequence is neuronal death. Reductions in the brain's oxygen supply are referred to as **hypoxia.**

In general, mild degrees of hypoxia are associated with dizziness, confusion, blurring of vision, or difficulty seeing in the dark, whereas moderate degrees produce loss of vision, nausea, coma, and amnesia (Joseph, 1990f; Morgan and Gibson, 1991;

Sokoloff, 1981). If hypoxia is prolonged there results severe permanent neuronal damage, coma, and death. This is a serious concern as anywhere from 50% to 70% of those with severe head trauma are hypoxic following injury (Klauber et al., 1985). However, as neocortical oxygen and glucose uptake reaches its peak between the ages of 5 and 6 (Sokoloff, 1981), with consumption levels approaching 50% of all body oxygen, children can often be more severely affected by hypoxia.

Characteristically, following head trauma there is an immediate and global reduction in cardiac output and blood flow for a few seconds, which in turn results in oxygen deprivation. This is followed by a transient hypertension and then a prolonged hypotension (Crockard et al., 1982). The overall consequences are reductions in arterial blood perfusion and thus hypoxia throughout the brain as well as decreased blood flow within the damaged tissue. Metabolism is disrupted and energy failure occurs. In addition, the brain's blood and oxygen supply can be reduced or diverted following chest injury, scalp laceration, or fractures involving the limbs.

Unfortunately a vicious deteriorative feedback cycle can be produced by these conditions, which further reduces the brain's supply of blood and oxygen. For example, hypoxia induces vasodilation and constriction of the blood vessels, a condition that results in **intracranial hypertension.** This creates an overall increase in **intracranial pressure.** When ICP increases, the brain and blood vessels become compressed (because of displacement pressure), which further reduces blood and oxygen flow within the brain and to the damaged tissue. This causes the development of ischemia and focal cerebral edema (Jennett and Teasdale, 1981; Miller, 1985), all of which in turn add to displacement pressures (herniation).

Respiratory Distress

With severe head injury many patients may also suffer from transient disturbances involving respiration and apnea (Miller, 1985), i.e., a decrease in respiratory rate. With decreased respiration there is decreased reoxygenation of the blood, which favors the development of hypotension and hypoxia (Staller, 1987).

With respiratory distress, hypotension, and hypoxia, there results pulmonary hypoperfusion and an increase in blood platelets in the capillaries. Unfortunately, the increased number of blood platelets actually act to obstruct the blood vessels, thus further reducing blood flow. This has been referred to as **neurogenic pulmonary edema.** Onset is often delayed and may occur 48 hours after injury. However, arterial hypotension may be an immediately sequela if there has been blood loss from a scalp laceration or other injury (Miller et al., 1981).

Vascular Trauma

Frequently, particularly with severe head injuries, the vascular system may be traumatized because of shearing and stretching of the blood vessels. Moreover, the blow itself or the action of rotational forces can cause vascular spasms of the carotid arteries as well as of the anterior and middle cerebral arteries (Bakay et al., 1980). Although vasospasm can temporarily increase cerebral blood flow, it also increases the likelihood of stroke and/or blockage of the small arteries because of the breaking off of tiny plaques and other debris lining the large vessel walls (see Chapter 20).

Myocardial Trauma

Head trauma can also result in secondary cardiovascular abnormalities including myocardial damage. For example, if the brainstem medulla has been compromised, there may be abnormal activation of the vagus nerve, which influences heart rate, the result being atrial fibrillations—a condition that in turn sometimes results in stroke (see Chapter 20).

Moreover, if the head-injured patient has suffered a chest injury the heart may be bruised and injured, i.e., myocardial contusion. In these instances the heart is unable to adequately contract. This leads to reduced output and hypotension—all of which can lead to respiratory distress and hypoxic conditions (Staller, 1987).

Chest Trauma

It is important to note, that even with a minor head injury, if there is an injury to the chest (i.e., a pneumothorax, hemothorax), hypoxia may be produced because of inadequate ventilation of air. For example, because of pulmonary edema or lung shunting, the ability of the lungs to transport oxygen to the blood is reduced (systemic hypoxia). This results in brain damage even if cerebral blood circulation is normal.

Summary

Hence, a variety of conditions associated with head injury can conspire to produce widespread hypoxia and disturbances involving cerebral metabolism. Neuronal death, independent of impact or rotational acceleration-deceleration forces, is the long-term consequence, even in mild cases. Hence, an individual with a minor head injury may suffer graver disturbances than an individual with a more serious injury if bones were broken or if the chest was injured.

COMA AND CONSCIOUSNESS
Reticular Damage and Immediate Cessation of Consciousness

In general, it is believed that the loss of consciousness following a brain injury is often because of abnormalities involving the reticular formation of the brainstem (see Chapter 10). Indeed, destruction of the brainstem reticular formation invariably induces states of prolonged unconsciousness and unresponsiveness, which are accompanied by a slow, synchronized EEG, and frequently irreversible coma, i.e., the remainder of the brain cannot be activated. However, damage need not be severe or extensive, for even small lesions in the brainstem, midbrain, and thalamus can induce a protracted loss of consciousness.

HEMATOMAS AND HEMORRHAGE

Frequently brainstem, midbrain, and thalamic damage is secondary to acceleration rotational forces and/or develops as a consequence of intracranial hemorrhage, ischemia, compression, and herniation. However, these effects of trauma may also be slow to develop (Teasdale, 1995). That is, a small artery or capillary may be damaged, resulting in a slow and gradual loss of blood (e.g., epidural hemorrhage), in which case patients will lose consciousness sometime after the injury. These conditions are also associated with temporal-parietal fractures and laceration of the middle meningeal artery and vein.

Nevertheless, in these later instances the patient may seem normal for some time after the accident. However, as blood continues to leak and the brain becomes increasingly compressed, within a few hours or even days the patient begins to complain of headache, drowsiness, and confusion. With increasing blood loss the patient may suddenly experience seizures, hemiparesis, and eventually loss of consciousness. If blood loss continues there is a great threat of temporal lobe herniation and eventual crushing of the midbrain. Hence, these disturbances are life-threatening and require surgical intervention and identification of the bleeding vessel (Teasdale, 1995).

Loss of consciousness sometime after the injury is not always due to the development of hemorrhage. In some instances pain and emotional upset give rise to a vasopressor syncopal attack. Thus, following the injury the patient may walk about the room, seemingly returning to normal, and then turn pale and fall unconscious to the ground. In contrast, those who develop a hemorrhage or other disturbance lose consciousness at the time of injury but regain it after a few moments or minutes, only to fall back into unconsciousness later.

Levels of Consciousness

In general, the ability to obey simple commands is often considered as indicative of the return of consciousness. However, patients may slide in and out of consciousness and/or display fluctuating levels of awareness. That is, the patient may appear confused, stuporous, or delirious.

Stupor

The patient is unresponsive and can only be aroused via intense, loud, or painful stimulation. They may open their eyes and in some cases make some type of verbal response only when vigorously stimulated.

Confusion

The patient has difficulty thinking coherently and cannot carry out two-step or complex commands. Their speech may consist of just a few words, they seem unaware of what is occurring around them, and they cannot grasp their immediate situation or the circumstances surrounding their condition. As such they are disoriented to time and place.

Delirium

This condition can be secondary to toxic disturbances as well as cerebral contusion. The patient is disoriented and shows fear and irritability, overreactively, and faulty perception of sensory stimuli, sometimes accompanied by visual hallucinations.

COMA

One need not actually suffer a severe head trauma in order to sustain significant damage to the brain and cognitive loss (reviewed in Miller, 1993). Nor is loss of consciousness at the time of injury a prerequisite. That is, in some cases patients may experience a severe injury to the skull (e.g., depressed fracture) but only suffer mild brain damage. In contrast, a patient may seem to have sustained a mild skull injury and later demonstrate moderate to severe cognitive loss (Miller, 1993).

Nevertheless, it does appear that when a patient has lost consciousness, the longer the period of unconsciousness, the more widespread and debilitating is the injury and the loss of cognitive capability, including memory functioning. However, as pertaining to mild head traumas, particularly in cases where consciousness was not lost, there simply have not been enough studies conducted by trained neuropsychologists so as to properly delineate the full potential ramifications of these injuries. Nevertheless, recent studies have indicated that significant cognitive and neuropsychological disturbances are associated with even mild injuries with no evidence of loss of consciousness (Leininger et al., 1990; Montgomery et al., 1991)

The Glasgow Coma Scale

Two general indicators of severity are length of unconsciousness and extent of memory loss, both retrograde and anterograde (i.e., post-traumatic amnesia) from the time of injury. One major advance in this regard was the development of the Glasgow Coma Scale (GCS) (Teasdale and Jennett, 1974). The GCS in effect provides standardized criteria for the assessment and description of head injury severity. Tremendous inter-rater reliability has been found in the use of this scale.

The GCS is composed of three major categories: **Eye Opening, Motor Response,** and **Verbal Response.** The patient's rating (or score) is based on his or her degree of ability to respond or react to various stimuli. If there is no response the patient receives a score of 1. Hence, the minimum score possible is 3.

Eye Opening

If the patient opens his eyes spontaneously he or she receives a score of 4. If the patient opens them only in response to speech and pain, he or she receives a score of 3. If only in reaction to pain the score is 2.

Motor Response

If the patient can adequately motorically respond to and obey commands his or her score is 6. If his or her reaction is more limited and he or she is capable of only responding to pain in a generalized, albeit localized manner, e.g., attempting to brush away the source, the score is 5. If the patient can only withdraw a limb in response to pain the score is 4. Responses limited to abnormal flexion (decorticate) are given a score of 3, whereas abnormal extensions (decerebrate) in response to pain receive a score of 2.

Verbal Response

If the patient is oriented to time, person, and place the score received is 5. Confused spontaneous verbalizations yield a score of 4. Inappropriate verbalizations (yelling, swearing) are given a rating of 3, and incomprehensible verbalizations limited, for example, to moaning are scored 2.

SEVERITY

Hence, a patient in a vegetative coma and completely unresponsive would receive a score of 3, whereas an individual who is fully responsive and alert would receive a score of 15. On the other hand, a patient who speaks inappropriate words, and opens his or her eyes and motorically withdraws only in response to painful stimuli will have a score of 9. In general, a GCS score of 8 or less is indicative of a severe head injury, 9–12 is considered moderate, and 13–15 is rated as mild.

Mortality

Mortality and GSC

Based on a review of a number of reports it appears that patients with a GCS score of 8 or less have a mortality rate of approximately 40% (35–50%), and that as the scores decrease the death rate increases. For example, those with a score between 3 and 5 have a mortality rate of approximately 60%, whereas individuals with a coma score of 6 or 7 who are between the ages of 30 and 50 have about a 50-50 chance of living or dying. (See Nikas, 1987a for a review of some of this literature.)

The mortality rate also is significantly high for individuals who remain in a coma for 6 hours or more (Jennett et al., 1977). Indeed, in one study, of 700 such cases who were still alive 6 months after their injury, half this group died over the course of the next 6 months, i.e., within 1 year of injury.

In addition, the longer the coma the worse are the prospects for rehabilitation and the greater is the long-term disability (Najenson et al., 1974). Similarly, among children intellectual deficits and academic difficulties becomes more severe as coma length increases (Brink et al., 1970; Dalby and Obrzut, 1991).

Mortality and Age

Mortality (as well as severity) is also a function of age. Supposedly children suffer a lower risk of mortality from head injury compared to adults (Berger et al., 1985; Luerssen et al., 1988). However, children who are physically abused and babies who are repeatedly shaken (thus inducing brain injury, i.e., "shaken baby syndrome") are not likely to come to the attention of a physician. Nevertheless, very young children and infants have a mortality rate that is higher than that of older children (Humphreys, 1983; Raimondi and Hirshauer, 1984).

In general, however, the mortality rate increases with increasing age. For example, those over 40 with severe injuries have a death rate of up to 70%, whereas the mortality rate for those below 40 is 23% (Nikas, 1987). Specifically, it has been estimated that for each year over age 35, the odds of death increase by 3.6% and decrease by the same amount for each year under 35 (Teasdale et al., 1982).

Mortality and Motor Functioning

The death rate is also inversely correlated with residual motor functioning. That is, the mortality rate increases as motor function deteriorates. Disturbances in motor function are often indicative of widespread cortical or brainstem damage. Prognosis is particularly poor among those who show posturing or flaccidity of the muscles (Nikas, 1987a).

Mortality: General Complications

Complications affecting recovery and leading to mortality most commonly include hypoxia, hypotension, seizures, infections (e.g., meningitis, abscess), and unrecognized mass lesions (Nikas, 1987; Teasdale, 1995). Indeed, many patients will talk and show some alertness after injury but die because of the later development of hypoxia and associated complications (Rose et al., 1977; Teasdale, 1995).

The majority of patients who die, however, when first seen are often in a state of shock with subnormal blood pressure, hypothermia, fast pulse, and pale-moist skin. If this state persists along with deep coma, widely dilated pupils, no eye movements, flaccid limbs, and irregular or rapid respiration, death usually follows. Hence, these are all grave prognostic signs.

At autopsy there is usually evidence of cerebral contusion, focal swelling, hemorrhage, and necrosis. A minority of such patients, however, don't die and instead persists in a state of profound disability.

Vegetative States

Some patients will remain in a coma for months or even years and never regain consciousness. Others, however, although having suffered profound and severe brain injuries, after a few weeks or months open their eyes and seemingly recover from their sleep-like coma, yet never truly regain consciousness. That is, recognizable mental and purposeful motor functioning never reappear. These patients are in a persistent vegetative state (Adams and Victor, 1994; Jennett and Plum, 1972). Often such individuals demonstrate an absence of cortical functioning, although the lesion may be in the brainstem. In the majority of cases the brain stem is actually well preserved, and it is the hemispheres and subcortical white matter that are most severely affected (Berrol, 1986).

According to Jennett and Plum (1972), after a few weeks these patients may open their eyes, first in response to pain, and then later spontaneously. They may grimace, flex the limbs, or blink to threat or to loud sounds and may make roving movements of the eyes. Sometimes they seemingly follow objects and may fixate on a physician or family member. This usually gives a family member or an inexperienced physician the erroneous impression that there is cognition.

Indeed, in one case a vegetative patient was misdiagnosed as oriented to person (although she could not communicate) simply because sometimes she would open her eyes and seemingly fixate (for 1–3 seconds) on people who made sounds or called her name. However, she also would fixate on empty space, and CT scan indicated extensive and severe damage throughout the brain.

Nevertheless, among those in a vegetative state there is wakefulness without awareness and the patient remains inattentive and never speaks or gives any consistent sign indicating cognition or awareness of inner need. The limbs are often extended in a posture called decerebrate rigidity (i.e., arms and legs extended with the feet and hand internally rotated). However, after a few weeks or months this may also wear off. Sometimes there are fragments of coordinated movements that look purposeful, such as scratching, or movement toward a noxious stimulus. Patients may swallow food and water and may chew and grind their teeth for long time periods. Grunting and groaning may also be provoked, though most of the time these patients are silent (see also Berrol, 1986). In this regard, however, their behavior is little different from that of an "anencephalic monster," i.e., a child born with only a brainstem.

It has been argued that the term "persistent" should not be applied until at least 1 year has elapsed since injury (Berrol, 1986). This is because some patients may actually improve from the vegetative state to one of severe disability—that is, those with traumatic versus hypoxic damage. However, the longer the patient is in a persistent vegetative state the less potential there is for any realistic change.

MEMORY LOSS

When the brain has been traumatized, the ability to process information or retain, store, and recall ongoing events may be seriously altered and disturbed—even following a mild head injury and/or when consciousness has not been lost (e.g., Yarnell and Lynch, 1970). That is, the patient may suffer from a brief or prolonged period of amnesia. If consciousness was merely altered (i.e., the patient is dazed), the amnesia may be only for the events that occurred at the moment of impact and for a few seconds thereafter.

Usually following a significant head injury there is a period of amnesia for events that occurred just before and just after the time of injury. However, because the brain may remain dysfunctional for some time even after the return of consciousness, memory functioning and new learning remain deficient for long time periods. Indeed, following the recovery of consciousness patients may be unable to recall anything that occurred for days, weeks, or even months after their injury. This condition has been referred to as **post-traumatic (anterograde) amnesia** (Russell, 1971) (see Chapter 6). Frequently the amnesia includes events that occurred well before the moment of impact as well, i.e., **retrograde amnesia.** Retrograde amnesia may extend for seconds, minutes, hours, days, months, or even years, depending on the severity of the injury (Blomert and Sisler, 1974). In part this is a consequence of damage to the hippocampus within the temporal lobe as well as frontal lobe injuries that disrupt attentional functioning.

It has been suggested that the length of both retrograde and anterograde amnesia may be used as indicants of the severity of the injury. The longer the memory lapse, the more severe the injury.

Retrograde Amnesia

Retrograde amnesia (RA) is measured by determining the last series of consecutive events recalled prior to impact. Patients may show islands of memory—hence the need to determine continuous memory.

RA is not necessarily a permanent condition, and slowly memory for various events may return; usually the older memories return before more recent experiences. This has been referred to as a shrinking retrograde amnesia. However, the shrinkage is not complete, as events occurring seconds or minutes before the trauma are permanently forgotten (Schacter and Crovitz, 1977).

Not all patients show a shrinking RA (Sisler and Penner, 1975). In fact, there is some possibility that although memory for some events well in the past may have returned, the return (or shrinkage) is not complete. Rather, only islands of events are recalled. As pointed out by Squire and coworkers (1975), "questions about the remote past tend to sample a greater time interval and tend to be more general than questions about the recent past" (p. 77). That is, recent events become telescoped such that lapses are more apparent. Hence, rather than shrinkage, it is probable that only islands of memory have returned.

With these caveats in mind, it is apparent that true retrograde amnesia is difficult to assess and probably should not be utilized as an indicant of brain injury severity except in the most general of terms.

Post-Traumatic/Anterograde Amnesia

Many individuals who suffer brief or extended periods of coma or unconsciousness experience protracted periods of disorientation and confusion upon recovery (Levin et al., 1982). Indeed, frequently there is a period of complete amnesia for continuously occurring events for some time after the return of consciousness (Russell, 1971). That is, although consciousness returns, since the brain is impaired information processing and thus memory remain faulty for a variable length of time, depending on the severity as well as the location of the injury. This has been referred to as post-traumatic amnesia (PTA).

Nevertheless, PTA is not because of an inability to register information, for immediate recall may be intact, whereas short- and long-term memory remain compromised (Yarnell and Lynch, 1970). Although patients are responsive and may interact somewhat appropriately with their environment, they may continue to have difficulty consolidating and transferring information from immediate to short- to longer-term memory.

PTA has long been considered a useful method for determining the severity of the brain injury (Dalby and Obrzut, 1991; Russell, 1971). According to Russell (1971), mild injuries are associated with PTA of less than 1 hour. PTA of 1–24 hours is associated with moderate injuries; PTA of 1–7 days suggests moderate-to-severe injuries, whereas PTA over 7 days is associated with severe injuries.

In attempting to calculate PTA it is important to bear in mind that the first appearance of normal memory does not indicate the end of the amnesic period. That is, although the memory is ostensibly normal, it may be temporary and followed by another period of PTA. Hence, PTA is determined by the return of continuous memory.

PTA seems to correlate well with severity of the trauma and degree of cognitive impairment during the first 3 months following injury (Klove and Cleeland, 1972)—at least for moderate and severe injuries. Beyond 3 months post injury, these correlations become increasingly weak except among those with profound injuries and long periods of PTA, in which case correlations remain constant for many years (Dailey, 1956; Wowern, 1966). In these instances, however, patients are so severely disabled that estimates of severity based on PTA are completely superfluous.

Some investigators have also argued that the duration of PTA is related to the development of subsequent impairments in memory functioning (Dalby and Obrzut, 1991; Russell, 1971). For example, individuals with coma lasting at least 1 day, a Glasgow Coma Scale of 8 or less, and PTA of 2 or more weeks continue to have significant memory problems for up to 1 year after injury (Dikmen et al., 1987). However, these relationships are largely dependent on how one measures memory and are a function of which brain regions may have been damaged.

There is some evidence, however, that PTA is somewhat correlated with the extent of RA. For example, among individuals with PTA exceeding 7 days, the RA is less than 30 minutes in over 50% of the patients (Russell, 1971). If PTA is less than 1 hour, the RA is often less than 1 minute (Bromert and Sisler, 1974).

Caveats: PTA and Long-Term Disorders

It seems that regardless of the nature and severity of the initial trauma, abnormalities involving memory are the most frequently reported disturbances following a head injury. Moreover, although memory functioning may improve, depending on which part of the brain has been compromised, it may remain abnormal and never completely recover even after 10–20 years have elapsed (Brooks, 1972; Schacter and Crovitz, 1977; Smith, 1974). This is not entirely surprising given the fact that the frontal and especially the temporal lobes are particularly vulnerable to the effects of hypoxia, herniation, hematomas, and contusions.

Because of this, PTA and the time period in which one is rendered unconscious should only be used as a general indicant of brain injury severity, for what appears to be a mild injury to one part of the brain may in fact yield more severe long-term consequences than what appears to be a more significant injury involving yet a different portion of the cerebrum. For example, a patient may suffer a brief loss of consciousness and a PTA of less than 20 minutes following a mild head injury involving the anterior-inferior temporal lobes, yet continue to suffer disturbances involving attention, motivational functioning, and verbal and/or visual memory for years. In this regard, one cannot with any sense of assurance make blanket statements that those with mild versus moderate head injuries (based on PTA estimates) will subsequently suffer less severe memory or cognitive impairment.

Moreover, when we consider the numerous intervening variables (e.g., hypoxia, a slowly developing hematoma, chest injury) that may add to the severity of what initially appears to be an inconsequential head injury, too heavy a reliance on PTA and length of coma in determining severity will certainly result in an erroneous diagnosis. Hence, rather than relying on PTA estimates, or the period in which the patient has lost consciousness, one must adequately examine the patient in question in order to arrive at a true measure of his or her capability and the possible extent and severity of the injury.

Thus, in general, PTA and time unconscious are probably only applicable in regard to those with moderate and severe injuries, and do not offer adequate estimates for large numbers of patients with what appears to be a mild injury.

PERSONALITY AND EMOTIONAL ALTERATIONS

Emotional and personality disturbances secondary to head injury are often more debilitating and disruptive than any residual cognitive or physical impairment (Slagle, 1990; see Miller, 1993 for a detailed discussion). Neurologically related emotional disturbances are also the most resistant to treatment. This is because the emotional and personality changes are frequently a direct consequence of the brain injury, and you cannot talk (or counsel) someone out of a damaged brain. That is, just as patients may become aphasic following an injury to Wernicke's area, they may become emotionally abnormal secondary to injuries involving those portions of their brain that govern personality and emotion.

Indeed, with severe injuries a considerable number of individuals may in fact develop what appears to be psychotic features, including schizophrenic-like psychosis, hysteria, euphoria, manic-excitement, and indifference (reviewed in Chapters 1–4) because of damage involving the limbic system, the frontal lobes, the temporal lobes, or the right parietal area. However, this does not mean to imply that such patients (and their families) should not be treated, as psychotherapy can be helpful (Miller, 1993).

A common sequela is depression and anxiety. Indeed, many patients seem anxious and depressed, stay depressed for long time periods, and tend to become socially withdrawn (Dikmen and Reitan, 1976; Fordyce et al., 1983; Miller, 1993; Slagle, 1990). Of course, in many instances disturbances such as depression are reactive and thus secondary to a realistic appraisal of the patient's injury. Nevertheless, marked alterations in personality and emotional functioning occurs even in those with little or no physical handicap (Bond, 1979; Miller, 1993; Slagle, 1990).

DELAYED DEVELOPMENT OF EMOTIONAL CHANGES

In some cases, emotional and personality disturbances may intensify over time (Fordyce et al., 1983; Oddy et al., 1985; Slagle, 1990), with the maximal amount of change noted during the first 3 months following injury (Brooks and Mckinlay, 1983). The deterioration is often unrelated to the severity of the initial injury or the degree of neuropsychological impairment (Fordyce et al., 1983). In these instances stress, self-image, loss of abilities, and the reactions of friends and family appear to be significant factors (Bergland and Thomas, 1991; Miller, 1993; Slagle, 1990).

Sometimes the delay in the development of emotional problems corresponds to the time period in which the patient has reached a high level of recovery (Brooks and McKinlay, 1983). Hence, this may be because of a greater or developing awareness of their disability as well as difficulties with social interaction and adjustment. Indeed, common complaints include feelings of worthlessness, uselessness, loneliness, and boredom (Bergland and Thomas, 1991; Lezak, 1987; Miller, 1993). Frequently, however, patients have a unrealistic appraisal of their condition (Tobias et al., 1982), and in fact will actively deny or refuse to acknowledge the extent of their deficits (Prigatano et al., 1990).

FAMILY STRESS

Alterations in personality also place severe strains on relatives and close friends—more so in fact than subsequent physical limitations (Jennett, 1975; Miller, 1993; Rosin, 1977; Thomsen, 1974). Often family members have much difficulty tolerating or accepting the personality changes, and frequently there is much denial, guilt, and disengagement, and sometimes rejection (Rosin, 1977; Tobias et al., 1982). There is frequently a perception that the head-injured family member is no longer the same person, particularly in that behaviorally they may seem drastically different (e.g., impulsive, amotivated). Because the patient seems different, he or she is easier to reject—the person being rejected is a stranger. Family members may treat the patient differently and spouses may in fact refuse to continue their marriages, depending on the nature of the change. For example, patients who develop sexual preoccupations are four times more likely to become separated or divorced (Dikmen and Reitan, 1976).

PREMORBID PERSONALITY

Patient who develop psychiatric abnormalities after head injury sometimes have premorbid indications of instability and other problems (Lishman, 1988; Slagle, 1990). This is not always the case, however

(Barth et al., 1983; Rimel et al., 1981). Personality changes may be completely opposite to what was observed premorbidly (Tobias et al., 1982)—particularly among those with frontal and temporal lobe injuries (Joseph, 1986a).

For example, one young man who was described as being very kind, courteous, law-abiding, and gentle, following a frontal head injury subsequently raped and savagely beat several women. Other examples of behavior change include mania and hysteria, even among formerly quite conservative and reserved individuals (Joseph, 1986a, 1988a). However, there have also been reports of improved personality functioning following injury (Ranseen, 1990). Indeed, in one case a patient whom I had evaluated in 1986 and who displayed a morose, angry, and sullen demeanor (which was similar to his premorbid personality) subsequently sustained a significant blow to the front of his head (coupled with loss of consciousness) and in consequence became a friendly, cheerful (albeit mildly impulsive) individual.

Children, however, commonly develop behavior patterns quite inconsistent with that observed premorbidly (Dalby and Obrzut, 1991). For example, an older child may begin to act like a younger child or an infant, such that he or she seems to have regressed. Common behavioral changes among children include enuresis, impulsiveness, hyperactivity, short attention span, destructive and aggressive behavior, and temper tantrums (Brink et al., 1970; Dalby and Obrzut, 1991).

Nevertheless, neurosis and even psychoses may develop in adults and children even after a mild head injury with little or no loss of consciousness (Bennett, 1969; Miller, 1993).

RECOVERY

Severe injuries to the brain are generally associated with profound disturbances involving all aspects of sensory, motor, cognitive, intellectual, and social-emotional functioning (Alfano et al., 1993; Miller,

1993; Teasdale, 1995). However, there is no general rule as to which functions may be the most seriously compromised, as this is determined by which parts of the brain have been most seriously damaged. Nevertheless, in comparing large groups of moderately and severely injured patients, certain generalizations may be made regarding recovery.

For example, sensory, motor, and language skills tend to show the greatest degree of recovery, whereas writing skills, math ability, memory, attention, intelligence, and social-emotional difficulties persist longer and tend to exert more devastating effects (Bergland and Thomas, 1991; Dikmen and Reitan, 1976; Jennett et al., 1981; Miller, 1993; Miller and Stern, 1965; Najenson et al., 1974, 1980; Tobias, 1982; Weddell et al., 1980). In general, the bulk of motor and speech recovery takes place within 3–6 months of injury (Bond, 1979; Brooks, 1975), and the recovery of speech parallels the recovery of motor functioning (Brink et al., 1970). Recovery of learning, memory, and constructional skills appears to proceed at a slower rate compared to recovery of other skills (Brooks and Aughton, 1979; Dalby and Obrzut, 1991), and attention and memory show the least recovery (Dikmen and Reitan, 1976).

INTELLECT

A number of studies have purported that even with severe injuries intellect may return to normal. These impressions are based on repeated administration of various measures over time and do not account for practice effects. Nor are premorbid IQs taken into account. Lastly, as noted with frontal injuries, although the patient may have an average or even superior IQ, this does not mean they can use their intelligence effectively.

EMPLOYMENT

Only approximately a third of all individuals with severe closed head injuries, even with extensive rehabilitation, return to gainful employment (Weddel et al., 1980). Those with personality and emotional problems tend to have the most difficulty (Miller, 1993; Prigatano et al., 1984).

Age

In general, the outlook for recovery is grim among individuals over 45 years of age and who suffered prolonged unconsciousness (Najenson et al., 1974). In fact, the rate of severe disability becomes greater in older individuals such that as age increases there is an exponential increase in the number of individuals who fail to make meaningful gains (Carlsson et al., 1968; Teasdale et al., 1982).

Pediatric patients with severe head injuries have a higher percentage of good outcome compared with adults, as well as a significantly lower mortality rate (Alberico et al., 1987). At 1 year post injury, the ratio was 2:1 in good outcome in comparing children with adults. However, children below the age of 4 seem to be more vulnerable than their older counterparts (Alberico et al., 1987). Indeed, some functions are more impaired after early versus late lesions (Schnieder, 1979), and in some cases the recovery of some functions occurs at the expense of those which originally were not compromised (Joseph, 1982, 1986b).

There is also some suggestion that long-term intellectual deficits are more pronounced among younger than older children (i.e., adolescents), even among those whose comas were briefer than those of the older groups (Brink et al., 1970). This is consistent with other findings suggesting that very young children are more susceptible than older individuals to the more debilitating effects of severe head injury, including a higher mortality rate and the development of epilepsy.

Nevertheless, in general children and young adults have better prospects for recovery and rehabilitation as compared to older individuals (Tobias et al., 1982), which indicates that the aged brain has a reduced potential for recovery.

RECOVERY OVER TIME

It has long been known that recovery is greatest over the course of the first 3 months after injury and that considerable recovery may continue during the first year (Dikmen and Reitan, 1976). Significant recovery also continues, albeit at a reduced rate, well beyond the first year for simple as well as complex abilities (Brink et al., 1970; Dikmen et al., 1983), even among those who suffered prolonged periods of unconsciousness (Najenson et al., 1980).

Severe Injuries and Recovery

Nevertheless, the longer the coma and the more severe the injury, the greater are the long-term disabilities and the worse are the prospects for rehabilitation among children and adults (Brink et al., 1970; Dalby and Obrzut, 1991; Najenson et al., 1974, 1980; Teasdale, 1995). Indeed, those with severe injuries often remain severely disabled, as only a small minority of patients may progress from severe to moderate disability, or from vegetative conditions to severe disability (Rosin, 1977; Teasdale, 1995). Moreover, the potential for severely head-injured individuals to return to a socially and vocationally functional life is severely limited (Tobias et al., 1982). Indeed, those with severe injuries seem to have "little potential for improvement over and beyond some very basic functional levels that they may reach fairly rapidly" (Dikmen et al., 1983:337). Some patients with severe injury may in fact deteriorate after having recovered to some degree (Brooks and Aughton, 1979; Miller, 1993; Montgomery et al., 1991).

EPILEPSY

When the brain has been damaged, neurons die and are replaced by glial scars. Sometimes these scars generate abnormal activity and/or disrupt neural activity in neighboring tissue, a phenomenon that has been described as "kindling." Kindling is the induction of epileptiform potentials in neighboring brain areas (Noebels and

Prince, 1978; Pedley, 1978; see also Chapter 16). Kindling can create an expanding lesion. Not all glial scars are epileptic, however. Nevertheless, certain regions of the brain are more susceptible to developing epileptic activity than others—the anterior and mesial temporal lobes, for example (see Chapter 14).

Patients at the most risk for immediately developing epilepsy are those with temporal lobe damage, focal neurological signs (e.g., hemiparesis), intracranial hematomas, depressed skull fractures, and penetrating head injuries, as well as those with closed head injuries with PTA exceeding 24 hours (Jennett and Teasdale, 1981). Anywhere from 25% to 50% of those with penetrating injuries develop epilepsy, and 10% will suffer an epileptic seizure and develop epilepsy within 1 week of the injury—the majority having their first attack within 24 hours (Jennett et al., 1974). Among those with nondepressed skull fracture the incidence of epilepsy is much lower, anywhere from 7% to 13% (Jennett et al., 1974; Najenson et al., 1980). In general, 20% of those with severe injuries develop epilepsy (Najenson et al., 1980), and children are more likely to develop epilepsy than adults (Jennett and Teasdale, 1981).

Post-traumatic seizures tend to decrease in frequency as the years pass, and 10–30% of patients may cease having seizures altogether, particularly if their first seizure occurred within a year after injury. Those who develop seizures 1 or more years following head trauma may in fact suffer increasing episodes unless their condition is properly controlled via medication.

Alcoholism has an adverse effect on these conditions, and seizures may be precipitated by a bout of hard drinking. Among women, seizures may occur more frequently around the time of menstruation.

CONCUSSION AND MILD HEAD INJURIES

Various definitions for what constitutes a "mild head" injury have been offered, many of which differ in regard to loss of consciousness, severity of injury, degree of memory loss, and so on (see Teasdale, 1995, for related discussion). In consequence, the ability to accumulate reliable data on associated disturbances and recovery is therefore compromised. Moreover, as noted, many view mild head injuries as inconsequential when in fact that may not be the case (Kelly et al., 1991; Leininger et al., 1990; Miller, 1993; Montgomery et al., 1991; Newton et al., 1992).

Moreover it is often erroneously assumed that consciousness must be lost in order for brain damage to occur. However, there is much evidence that indicates otherwise. In fact, as noted above, brain injury from shearing injuries (Adams et al., 1989), even in the absence of a head injury, has been documented (Gennarelli, 1986). In addition, what appears to be mild may be quite serious, even when CT scan and MRI fail to indicate brain damage (Newton et al., 1992). Indeed, some patients will die from a "mild" injury (Oppenheimer, 1968). On the other hand, the aftereffects may also be quite insignificant.

In general, the period that encompasses "mild" for many investigators includes PTA and a loss of consciousness for up to 1 hour. However, even a loss of consciousness for less than 20 minutes can be quite serious, regardless of the extent of PTA (Joseph, 1990e).

Concussion

Although an individual may suffer a concussion and memory loss without losing consciousness (Kelly et al., 1991; Yarnell and Lynch, 1970), concussion is characterized by a brief period of unconsciousness followed by an immediate return of consciousness. There are usually no focal neurological signs, and the loss of consciousness is usually because of mechanical forces such as a blow to the head. As the magnitude of the applied force is increased, the severity of the concussion increases.

Often immediately after an insult to the head the patient will drop motionless to the ground and there may be an arrest of respiration and an eventual fall in blood pressure (following a rise at impact)—death can occur from respiratory arrest. Vital signs usually return—even if the patient is unconscious—within a few seconds. Once respiration returns, the patient may begin to move about in a restless, random fashion and may speak, usually unintelligibly. They may become abusive and irritable, shout, and resist contact (Russell, 1971).

Mild and Classic Concussion

As described in detail by Gennarelli (1986), there are two broad categories of concussion as well as distinct subtypes. These include mild concussion, in which there is no loss of consciousness, and classic concussion, which involves a short period of coma.

MILD CONCUSSION

Mild concussion consists of several subtypes; these include focal concussion, which is due to involvement of a localized region of the cerebral cortex. For example, transient left-sided weakness may be due to right motor involvement, whereas the patient who sees "stars" may have received an occipital injury (Gennarelli, 1986). A diffuse concussion results in transient confusion and disorientation. In either instance these concussive injuries may or may not be accompanied by amnesia.

With greater impact, although the patient does not lose consciousness, he or she may be confused, disoriented, and continue to motorically interact with the environment (Kelly et al., 1991; Yarnell and Lynch, 1970). That is, such patients may walk around and even speak to others. However, although immediate recall is intact, they later become amnesic for the event as well as the events leading up to the injury, i.e., anterograde and retrograde amnesia. The anterograde and retrograde amnesia may extend forward and backward in time for up to 20 minutes, even without loss of consciousness (Yarnell and

Lynch, 1970). When this occurs it is evident that the brain has suffered a significant insult.

CLASSIC CONCUSSION

Classic concussion is accompanied by a brief loss of consciousness at the instant of injury and is almost always associated with post-traumatic amnesia. However, according to Gennarelli (1986), a patient with a classic concussion may remain unconscious for as long as 6 hours. If unconsciousness lasts longer than 6 hours, the patient has probably suffered diffuse axonal injuries such that large-scale regions of the cerebrum have become partially disconnected.

In general, a mild (classic) concussion is associated with amnesia of less than 1 half-hour, a moderate concussion with amnesia of 1–24 hours, and severe concussion with amnesia of 24 hours or more (Tubbs and Potter, 1970). Nevertheless, although the concussion may be considered mild, the accompanying brain damage may be minimal, moderate, or in a few rare cases severe.

In cases of classic concussion, immediately following the injury the patient develops apnea, hypertension, bradycardia, cardiac arrhythmias, and neurological disturbances such as decerebrate posturing and pupillary dilation. Thus the patient may seem temporarily paralyzed. Often, upon regaining consciousness the patient will vomit and complain of headache. Even in mild cases, there can result temporary alterations in the permeability of the blood-brain barrier and damage involving the neuronal mitochondria (Bakay et al., 1980).

Mild Head Injury

A mild head injury is usually indicated by a GCS score of 13–15, no CT or radiological abnormalities, and a loss of consciousness for less than 20 minutes (some authors, however include those with unconsciousness lasting up to 1 hour—e.g., Dikmen et al., 1986). This has also been referred to as cerebral concussion, and patients need not have lost consciousness at the time of injury (Dikmen et al., 1983).

Although described as "mild" these injuries can be quite serious and involve various degrees of permanent brain damage. For example, anywhere from 1% to 5% of those with mild head injuries may demonstrate focal neurological deficits (Coloban et al., 1986; Rimel et al., 1981). In addition, 2–3% of those with mild injuries deteriorate after initially appearing alert and responsive (Dacey et al., 1986; Fisher et al., 1981). Moreover, 3% of those with mild head injury will develop increased ICP and life-threatening hematomas, which must be removed (Dacey et al., 1986). Indeed, approximately 1 of 100 individuals with mild head injury dies (Luerssen et al., 1988).

Brainstem Abnormalities

It has been suggested that sheer strain secondary to mild acceleration-deceleration injuries causes tearing of axons, which is followed by degeneration of the neural tracts in the brainstem. Brainstem axonal degeneration has been experimentally demonstrated in monkeys with mild injuries induced by acceleration/deceleration (Jane et al., 1982), and among humans up to 40% of those with "mild" head injuries may demonstrate abnormal brainstem evoked potentials (Montgomery et al., 1991; Rowe and Carlson, 1980). Similarly, those who have died from other causes demonstrate microscopic neuronal damage, microglial scars, and fiber degeneration within the brainstem and cerebral hemispheres—even among those who received what were considered "trivial" injuries (Oppenheimer, 1968; Strich, 1969).

DAI and Contusions

Diffuse axonal injuries (DAIs) have been noted among those suffering brief periods of unconsciousness (Teasdale and Mendelow, 1980). Moreover, contusions within the anterior temporal regions and frontal lobes have been found among those with mild nonimpact concussions because of acceleration/deceleration injuries (Adams et al., 1981).

Neuropsychological Deficits

Individuals with mild head injuries have been repeatedly shown to suffer a variety of neuropsychological, psychosocial, and emotional impairments, even in the absence of gross or focal neurological deficits (Barth et al., 1983; Dikmen et al., 1986; Gronwall and Wrightson, 1974; Miller, 1993; Montgomery et al., 1991; Rimel et al., 1981). These include pervasive neurobehavioral impairments involving attention, memory, and rate of information processing (Levin et al., 1987; Montgomery et al., 1991). Patients may suffer word-finding and expressive speech difficulties, reduced reaction time, perceptual-spatial disorders, and abnormalities involving abstract reasoning abilities. Moreover, although generalized recovery is often noted within 3 months of injury, a memory disturbance often remains and may in fact persist for years (Lidvall et al., 1974; Jennett, 1978).

In some studies it has been reported that at 3 months post mild injury a significant number of individuals continue to demonstrate visual-spatial abnormalities, memory disorders, reduced ability to concentrate, and persistent headaches, and as many as 34% of those with mild injuries may be unemployed at this time (Barth et al., 1983; Rimel et al., 1981). Interestingly, these deficits are not in any manner correlated with length of unconscious or PTA, and in one study only 6 of over 400 patients were involved in litigation. Hence, purposeful malingering is not likely among the majority of such cases.

PREMORIBID INFLUENCES

It is noteworthy that recovery appears to be dependent on previous mental abilities (Dikmen and Reitan, 1976; Miller, 1993) such that the debilitating effects of mild head injury appear to affect individuals with minimal educational backgrounds more than they do those with initially higher levels of functioning and education. This may be because these individuals have "bad brains" to begin with and/or are less able to compensate for cognitive disturbances secondary to brain injury because of their more limited capabilities. That is, those with higher-level abilities have more to draw from and to fall back on. Of course, they also have more to lose.

POSTCONCUSSION SYNDROME

Following a mild (versus severe or moderate) trauma to the head, patients may not begin to complain of cognitive disturbances until days or even weeks have passed (Alves et al., 1986; Miller, 1993), even when consciousness has not been lost. Symptoms are often nonspecific and are hard to quantify or objectively document. In general, these are referred to as postconcussive disorders. Indeed, about 50% of those who suffer mild head injuries (Alves et al., 1986) are at risk for developing postconcussion symptoms (PCS). PCS is manifested somewhat differently for adults than for children.

Among adults, complaints regarding PCS may include persistent headaches, transient dizziness, nausea, impaired memory and attention, irritability, depression, anxiety, easy fatigability, nystagmus, and inner ear disturbances such as vertigo, tinnitus, and hearing loss (Alfano et al., 1993; Elia, 1972; Leininger et al., 1990; Levin et al., 1987; Lishman, 1973; Miller, 1993; Ommaya et al., 1968; Toglia et al., 1970). Headache, dizziness, memory problems, weakness, nausea, and tinnitus are often the most common complaints—even at 12 months following injury (Alves et al., 1986). Other symptoms may include hyperacoutism, photophobia, decreased judgment, loss of libido, and difficulty with self-restraint and inhibition.

In contrast, children often may become withdrawn, antisocial, and aggressive and develop enuresis as well as sleep disturbances (Dalby and Obrzut, 1991; Dillon et al., 1961). Moreover, it has also been reported that some individuals, children in particular, develop transient migraine-like attacks following even minimal injuries where consciousness was not necessarily lost (Haas and Lourie, 1988). These attacks include blurred, tunnel, and/or partial or complete loss of vision, paresthesias, dysphasia, confusion, agitation, headache, drowsiness, and vomiting—all of which come on like spells, particularly soon after the injury (Haas and Lourie, 1988). It has been suggested that this is because of traumatic spasm of the larger cerebral arteries.

In general, postconcussion symptoms vary in their degree and duration, although frequently they persist for long time periods, from months to years or sometimes indefinitely (Dikmen and Reitan, 1976; Merskey and Woodforde, 1972; Miller, 1993; Symonds, 1962). Nevertheless, even when PCS begins to wane it is sometimes followed by a long period of depression (Merskey and Woodorde, 1972; Miller, 1993).

Emotional Sequela

Many patients with PCS in fact develop a variety of emotional and personality difficulties that seem to be triggered by the experience of having been injured (Bennett, 1969; Miller, 1993). Patients may seem extremely anxious, fearful, depressed, and/or preoccupied with the details surrounding the accident (Bennett, 1969; Merskey and Woodforde, 1972; Miller, 1993), and they remain (albeit periodically and transiently) upset for long time periods. Others may appear intolerant of noise, emotional excitement, and crowds, and complain of tenseness, restlessness, inability to concentrate, and feelings of nervousness and fatigue. That is, they demonstrate the signs of posttraumatic stress disorder (Miller, 1993). Many patients with this syndrome also seem unable to tolerate the effects of alcohol. The emotional disturbances may persist for months or years, but usually lessen as time passes.

It has been argued, and frequently it is suspected, that PCS has in fact no objective basis and represents an exacerbation of premorbid personality characteristics or is motivated by a desire for financial compensation (Miller, 1961). Undoubtedly this is true in some cases. Nevertheless, frequently the majority of such individuals are not even involved in litigation (Barth et al., 1983; Merskey and Woodforde, 1972; Rimel et al., 1981). Moreover, sometimes the actual deficit is greater than the patient realizes (Prigatono et al., 1990; Waddell and Gronwall, 1984).

Cerebral Blood Flow

Neuronal damage involving the brainstem and cerebral hemispheres as well as significant decrease in cerebral blood flow have been demonstrated among patients with mild and trivial head injuries (Montgomery et al, 1991; Taylor and Bell, 1966). In fact, many of these same patients were without PCS complaints and demonstrated normal circulation during the first 12 hours after their injuries and then within a period of 3 days developed postconcussive symptoms and decreased cerebral blood circulation that lasted weeks and months. It has been suggested that this may be secondary to vasomotor abnormalities because of brainstem (medullary) impairment (Taylor and Bell, 1966).

Attention and Information Processing

It has also been shown that patients with PCS and mild head injury are unable to process information at a normal rate (Gronwall and Writhson, 1974) and tend to become overwhelmed. Frequently, however, patients do not complain about this until days or weeks later, as this does not become apparent to them until they return to work and thus discover that they are having problems. As such they may find that tasks that require attention to a number of details and that formerly were performed quite easily now seem difficult and beyond their capacity. Hence, the patient says that he or she cannot concentrate (Gronwall and Writhson, 1974; Miller, 1993).

Presumably, many of the postconcussion symptoms are secondary to shearing forces associated with the acceleration/deceleration nature of the injuries as well as torque and other forces exerted on the brainstem. Indeed, following whiplash, patients may develop PCS, including vestibular symptoms and dizziness (Ommaya et al., 1968; Toglia et al., 1970).

Whiplash and Blood Flow

In whiplash the head is like a ball at the end of a whip, and rapid or extreme rotations or extensions of the head may cause decreased blood flow by compressing the vertebral arteries that supply the brainstem, cerebellum, occipital lobe, and hippocampal region of the temporal lobe (Toole, 1984). This leads to symptomatic vestibular insufficiency and other brainstem responses (feelings of giddiness, lightheadedness, loss of balance), and can exert disrupting effects on memory and visual functioning. If the patient is suffering from atherosclerosis or abnormalities involving the cervical vertebrae (e.g., cervical osteoarthritis), the overall effects can become exacerbated even further (Toole, 1984).

Similarly, as is well known, the internal carotid artery is also very vulnerable to trauma, particularly in the neck, where it is exposed. Hence, rapid extensions and rotations, as well as karate-like blows, can tear and dissect the artery, resulting in obstruction.

In general, tall persons suffer whiplash more often than shorter individuals, and those riding in the front of an auto are 50% more likely than those in the rear of sustaining such an injury (Elia, 1972). Slow-speed crashes carry a greater chance of whiplash injury than high-speed crashes do (Elia, 1972), presumably because patients become tense in expectation of the impending accident.

As noted, acceleration/deceleration injuries are most likely to affect the brainstem and temporal and frontal lobes. In fact, there is some suggestion that frontal injuries are more likely to give rise to postconcussional disturbances (Rabavilas and

Scarpalezos, 1981). However, brainstem and temporal lobe abnormalities are obviously contributory.

Vertigo, Hearing Loss, and Tinnitus

Damage to the inner ear (which is situated within the temporal bone) is frequently associated with the development of vertigo, dizziness, and tinnitus. As noted, the temporal bone (which contains the auditory meatus) is often subject to injury, including fracture, regardless of where the head was initially struck. Hence, the presence of vertigo and tinnitus following a head injury suggests a concussion of the inner ear (Elia, 1972).

Some patients also complain of hearing loss or even unilateral deafness. In some cases this is secondary to an injury of the tympanic membrane, external auditory canal, or inner skin of the ear. This results in some bleeding and thus the potential development of dried blood clots (and cerumen) in the auditory canal, which subsequently plug the ear, reducing auditory acuity by impeding sound transmission. In addition, in cases of temporal bone fracture, CSF may escape into the ear, thus temporarily disrupting functioning.

On the other hand, some patients complain of increased sound sensitivity. It has been proposed that this is secondary to rotational acceleration forces that stretch and cause traction of the stapedius muscle, which is attached to the vestibule and stapes of the ear.

Diplopia and Photophobia

It has been proposed that visual problems such as diplopia occur secondary to traction of the eye muscles. However, if the oculomotor nerve is injured in any manner, this too may cause diplopia and may also contribute to the development of nystagmus (Elia, 1972). Nystagmus usually resolves over time.

Photophobia and excessive sensitivity to light may also be secondary to oculomotor traction. Interestingly, it has been demonstrated that many such patients with minor head injury are consequently more sensitive than they realize (Waddell and Gronwall, 1984), and some demonstrate a hypersensitivity to both sound and light when in fact this was not a complaint.

PCS and Severe Head Injury

It has been noted that those with mild injuries sometimes complain of PCS more than do those with severe injuries (Levin et al., 1987). This in turn has led to the un-

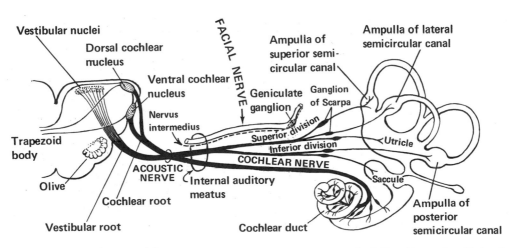

Figure 19.10 Distribution of the acoustic nerve as it projects from the ear to the brainstem. (From Waxman SG, De Groot J. Correlative Neuroanatomy, 22nd ed. East Norwalk, CT: Appleton & Lange, 1994.)

founded suspicion that these disturbances have no objective basis. However, in some cases, as patients recover and progress from severe to moderate or moderate to mild degrees of disability, they begin to make these same complaints (Miller, 1993). Moreover, there is some possibility that because of the greater degree of cognitive and personality disorganization among those with severe injuries, their ability to appreciate and complain of these disturbances is lessened (i.e., lack of insight, decreased motivation). In this regard, patients with severe brain injuries frequently do not become depressed or demonstrate depression until after there has been considerable recovery (Merskey and Woodforde, 1972).

PREMORBID CHARACTERISTICS

It has been noted that individuals who sustain head injuries sometimes have a history of emotional, impulse, and educational difficulties, as well as a history of previous head injuries (Haas et al., 1987; Joseph, 1990e; Miller, 1993). In fact it has been reported that as many as 50% of head-injured individuals have a history of poor premorbid academic performance, including learning disabilities, school drop-out, multiple failed subjects, and social difficulties (Fahy et al., 1967; Fuld and Isher, 1977; Haas et al., 1987; Miller, 1993).

The finding that many head- and even spinal-injured patients (Morris et al., 1986) have also suffered previous cerebral traumas has led some investigators to suggest that these individuals have a lifestyle that seems to predispose them to violence and injuries to the cranium (Tobias et al., 1982).

As suggested by Haas and colleagues (1987), this relationship may be because of such qualities as poor attention span, distractibility, limited frustration tolerance, poor judgment, impulsivity, difficulty anticipating consequences, and perceptual-motor abnormalities, all of which increases the likelihood of these individuals being involved in an auto accident. However, it is also possible that once someone has a head injury, because of subsequent decreases in overall functional efficiency, they are less likely to avoid situations where a second (or third) injury may occur.

Lifestyle and premorbid social-emotional stability are in fact important contributors to the possibility of suffering a head trauma. For example, chronic alcoholics are especially susceptible to cerebral injury (Bennett, 1969; Miller, 1993). It has also been reported that many head-injured patients have suffered a bout of depression or other emotional disturbances (e.g., fight with a girlfriend) immediately prior to their injury (Tobias et al., 1982).

Overall, the individuals most at risk for suffering a head injury at some point in time are young males (Teasdale, 1995), particularly those with a history of learning disability, alcohol abuse, social-emotional difficulties, and previous head trauma. Indeed, males are twice as likely to suffer head injuries as females (Teasdale, 1995) and are more likely than females to engage in "risky" and "macho" behaviors, which of course puts them at the greatest risk. Conversely, it is also males who are more likely to downplay the significance of their injury and who will display the most resistance to rehabilitation, and in particular, psychotherapy.

Stroke and Cerebral-Vascular Disease

When the brain is not adequately perfused with blood and is deprived of oxygen, glucose, and other nutrients, a variety of neurological, neuropsychiatric, and neuropsychological abnormalities may result depending on how long and which portions of the cerebrum are involved (Absher and Toole, 1995; Beckson and Cummings, 1991; Ginsberg, 1995; Starkstein and Robinson, 1992). As is well known, the complete deprivation of blood for longer than 3 minutes produces neuronal, glial, and vascular necrosis and irreversible brain damage, i.e., cerebral infarction (stroke) and ischemic necrosis. If blood flow is reduced for a shorter time period, transient neurological abnormalities may occur (Ginsberg, 1995).

For example, at rest the human brain consumes about 20% of the total oxygen consumed by the body and 70% of the total glucose (Morgan and Gibson, 1991). However, unlike other organs, the brain lacks the capacity to store oxygen or glucose and is therefore completely dependent on the blood supply. Thus disruption of the blood supply and/or a loss of oxygen and glucose for a little as 10–30 seconds can induce dizziness, confusion, and loss of consciousness (Morgan and Gibson, 1991). Similarly, if cerebral blood flow drops below 40% of what is normal, electrocortigraphic silence is produced, followed by anoxic depolarization and depletion of cellular energy metabolites coupled with massive increases in extracellular potassium ions, which is followed by neuronal death (Ginsberg, 1995). The patient has suffered a cerebral infarct, or stroke.

CEREBRAL INFARCT

Thus a cerebral infarct and the neurological symptoms that accompany it may develop quite abruptly and in a manner of seconds. However, the full symptomatic development of an infarct can be more prolonged, taking perhaps minutes, hours, or days, i.e., a stroke in evolution.

Specifically, cerebral infarction is characterized by lack of oxygen and nutrients and/or the impaired removal of metabolic products—conditions that result in the death of neurons, glia, and the vasculature. Moreover, when cells die their membranes burst, releasing lipids, fatty acids, and other substances that can produce systemic and local effects and that magnify damage in surrounding zones (Ginsberg, 1995; Welch and Levine, 1991).

Cerebral infarction or stroke is often secondary to cerebrovascular disease and vascular abnormalities involving the heart and/or blood vessels. This includes atherosclerosis, occulusion of the cerebral arteries by thrombus or embolus, or rupture of a vessel that causes hemorrhage. Other major risk factors for stroke include hypertension, diabetes mellitus, atrial fibrillation, transient ischemic attacks (TIAs), left ventricular hypertrophy, congestive heart failure, and coronary heart disease (Alter et al., 1994; Bornstein and Kelly, 1991; Toole, 1984; Wolf et al., 1983). Diseases or damage involving the heart, including myocardial infarction, are leading contributors to stroke.

FUNCTIONAL ANATOMY OF THE HEART AND ARTERIAL DISTRIBUTION

The heart is a four-chambered muscular organ about the size of a man's fist that lies predominantly to the left of the body's midline. The upper two chambers are referred to as the atria (or auricles), and the lower chambers are the ventricles. In general, blood flows from the atria to the ventricles (when the atria valves contract) and from the right and left ventricles into the pulmonary artery and aorta, respectively (when the ventricular valves contract).

OXYGENATION

From the pulmonary artery blood is shunted to the lungs where carbon dioxide is removed and oxygen replaced. This oxygenated blood is then transmitted to the left atrium via the pulmonary veins. From the left atrium, blood is transported to the left ventricle, which pumps the oxygenated blood through the aorta.

Arterial Cerebral Pathways

It is via the aorta that blood is transmitted directly to the brain. Hence, clots or other debris cast off from the heart commonly affect cerebral integrity. Specifically,

the aorta gives rise to the left common carotid and brachiocephalic artery. The brachiocephalic artery bifurcates into the right common carotid and right subclavian arteries. Each common carotid divides into the external and internal carotid arteries. The subclavian arteries give rise to the vertebral arteries.

Hence, the brain is actually nourished by two separate systems of vasculature, the carotid and the vertebral (reviewed in Carpenter, 1991; Parent, 1995). The carotid system supplies the frontal and parietal lobe, all but the inferior-posterior third of the temporal lobe, the hypothalamus, basal ganglia, and the eyes. The vertebral system nourishes the posterior temporal lobe, occipital lobe, upper part of the spinal cord, brainstem, midbrain, thalamus, cerebellum, and inner ear. Hence, abnormalities

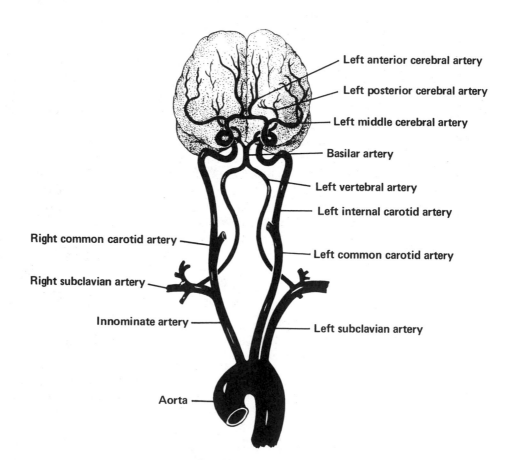

Figure 20.1. The arteries of the heart and brain. (From Waxman SG, De Groot J. Correlative Neu- roanatomy, 22nd ed. East Norwalk, CT: Appleton & Lange, 1994.)

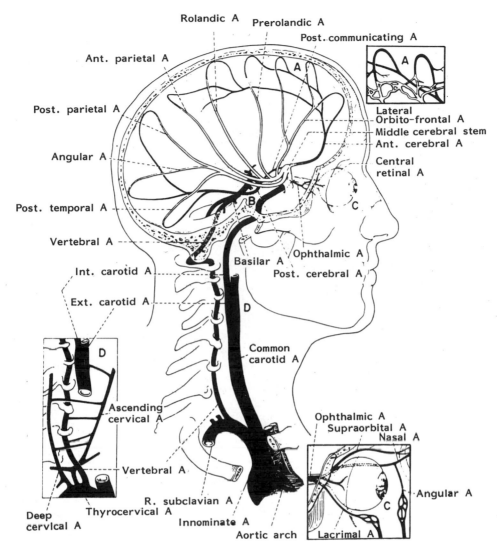

Figure 20.2.   The distribution of arteries from the heart to the right side of the brain. The arteries of the heart and brain. (From Waxman SG, De Groot J. Correlative Neuroanatomy, 22nd ed. East Norwalk, CT: Appleton & Lange, 1994.)

involving either of these circulatory systems can result in widespread albeit characteristic disturbances.

THE BLOOD SUPPLY OF THE BRAIN
The Carotid System

The internal carotid artery gives rise to four major arterial branches: the middle cerebral, anterior cerebral, opthalmic, and anterior choroidal arteries.

MIDDLE CEREBRAL ARTERY

The middle cerebral artery is a direct extension of the internal carotid and receives almost 80% of the blood passing through this arterial system. Being a major recipient of the carotid blood supply it is often the most commonly affected part of the vasculature when debris is cast off from the heart and vessel walls.

Through its branches this artery supplies the lateral part of the cerebral hemispheres, including the neocortex and white matter of

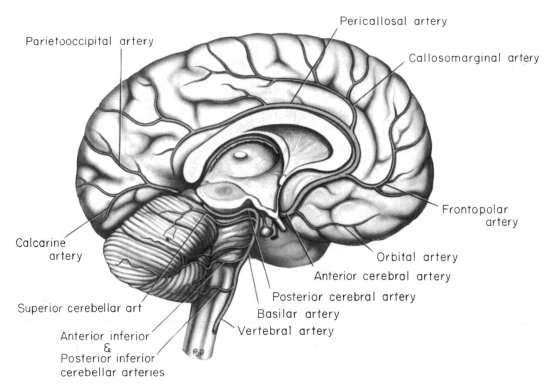

Figure 20.3. The principle arteries on the me-
dial surface of the cerebrum and cerebellum.
(From Carpenter M. Core Text of Neuroanatomy.
Baltimore: Williams & Wilkins, 1991.)

the lateral/inferior convexity of the frontal
lobes, the motor areas, the primary and association somesthetic areas, the inferior
parietal lobule, and the superior portion of
temporal lobe and insula. Its penetrating
branches supply the putamen, portions of
the caudate and globus pallidus, posterior
limb of the internal capsule, and corona radiata (Carpenter, 1991; Parent, 1995).

ANTERIOR CEREBRAL ARTERY

Through its branches the anterior cerebral
artery supplies the orbital frontal lobes, all
but the most posterior portion of the medial
walls of the cerebral hemispheres, and the
anterior four-fifths of the corpus callosum.
In this regard it might better be referred to
as the anterior medial- orbital artery.

OPTHALAMIC ARTERY

The opthalmic artery is the first large
branch given off by the internal carotid.
The opthalmic supplies and nourishes the
eye and the orbit as well as some skin on
the forehead. The opthalmic also gives off
branches that feed the posterior limb of the
internal capsule, the thalamus, the wall of
the third ventricle, and portions of the
optic chiasm.

ANTERIOR CHOROIDAL

The anterior choroidal artery arises from
the lateral side of the internal carotid and
supplies the medial portion of the globus
pallidus and Ammons horn, and parts of
the internal capsule.

In general, the terminal arterial branches
of the anterior, middle, and posterior cerebral arteries (which are part of the vertebral
system) interconnect with one another over
the surface of the brain such that there is
some overlap in distribution.

Vertebral System

Via its branches the vertebral system
supplies the brain stem, cerebellum, and
occipital and inferior temporal lobes.
Specifically, the vertebrals unite to form the

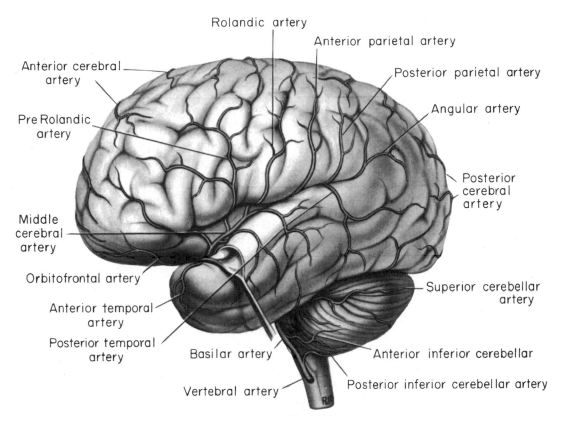

Figure 20.4.   The arteries of the lateral surface of the left cerebrum together with arteries of the cerebellum and the brain stem. (From Carpenter M. Core Text of Neuroanatomy. Baltimore: Williams & Wilkins, 1991.)

basilar artery, and the basilar artery gives rise to the cerebellar and posterior cerebral arteries (Carpenter, 1991; Parent, 1995).

VERTEBRAL ARTERIES

Each vertebral artery arises from the right and left subclavian and ascends to the medulla, where they unite to form the basilar artery at the caudal base of the pons. The vertebral arteries are the chief arteries of the medulla and supply the lower three-fourths of the pyramids, medial lemniscus, restiform body, and posterior-inferior cerebellum.

BASILAR ARTERY

The basilar artery supplies the pons and gives rise to the inferior and superior cerebellar arteries, which nourish the cerebellum. As the basilar artery ascends it bifurcates into the two posterior cerebral arteries.

POSTERIOR CEREBRAL ARTERY

The major trunks of the posterior cerebral artery irrigate the hippocampus, the medial portion of the temporal lobes, the occipital lobes including areas 17, 18, and 19, and the splenium of the corpus callosum. Via its deep penetrating branches it also feeds the thalamus, subthalamic nuclei, substantia nigra, midbrain, pineal body, and posterior hippocampus.

CEREBRAL-VASCULAR DISEASE

ARTERIES

By definition, an artery (with the exception of the pulmonary artery) is a vessel that carries oxygenated blood away from the heart to other parts of the body. Small arteries are referred to as arterioles. Capillaries are microscopic vessels that transport blood from the arterioles to the venules. Venules

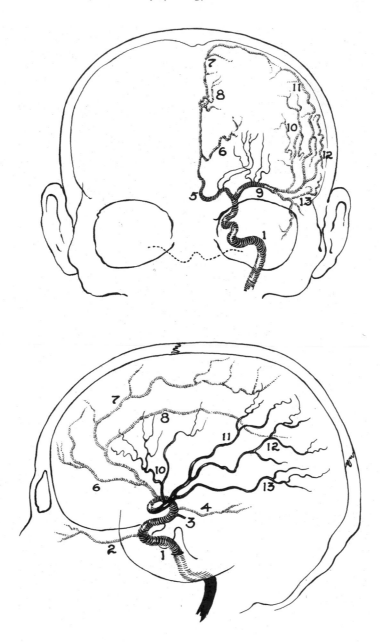

1. Internal carotid artery.
2. Ophthalmic artery.
3. Posterior communicating artery.
4. Anterior choroidal artery.
5. Anterior cerebral artery.
6. Frontopolar artery.
7. Callosomarginal artery.
8. Pericallosal artery.
9. Middle cerebral artery.
10. Ascending frontoparietal artery.
11. Posterior parietal artery.
12. Angular artery.
13. Posterior temporal artery.

Figure 20.5. The distribution of the internal carotid artery. (From Waxman SG, De Groot J. Correlative Neuroanatomy, 22nd ed. East Norwalk, CT: Appleton & Lange, 1994.)

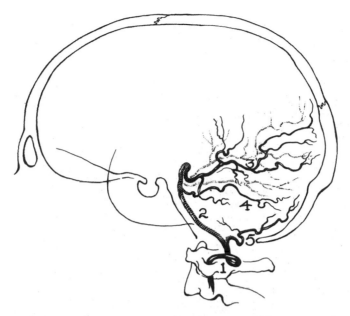

1. Vertebral artery.
2. Basilar artery.
3. Posterior cerebral artery.

4. Superior cerebellar artery.
5. Posterior inferior cerebellar artery.

Figure 20.6.   The distribution of the vertebral-basilar and posterior arteries. (From Waxman SG, De Groot J. Correlative Neuroanatomy, 22nd ed. East Norwalk, CT: Appleton & Lange, 1994: based on a figure from List, Burge, and Hodges. Intracranial angiography. Radiology, 45:1, 1945.)

carry blood to veins. In general, the capillary density is greater in the gray than the white matter, and the gray matter receives about 5 times as much blood and 7 times as much oxygen as the white matter.

Arteries consist of an outer coat of tough fibrous tissue, an inner lining of endothelium, and a smooth inner coat of muscle, which acts to constrict and dilate the vessel. Any alteration in the smoothness of this muscular coat can give rise to the development of blood clots and thrombi. Most rough spots are secondary to atherosclerosis. Atherosclerosis is a major risk factor for stroke (Bornstein and Kelly, 1991). If an artery is severed it has the capacity to regenerate.

Atherosclerosis

Atherosclerosis is a noninflammatory degenerative disease than can result in arterial abnormalities throughout the body. These include roughening of the blood vessel intima; elongation and stiffening of the artery, which may cause it to kink and buckle (which in turn reduces blood flow and pressure); reductions in the caliber or dissection or tearing of the lumen; obstruction or occlusion of the lumen with possible diversion of flow through collateral channels; focal dilations; and plaque formation (Absher and Toole, 1995).

Atherosclerosis exerts its effects indirectly on the brain by decreasing perfusion pressure and reducing flow through the tissues supplied by the various blood vessels. This is accomplished by promoting the development of thrombi and emboli (fibrin, platelets, or cholesterol crystals), which in turn increasingly occlude the arterial lumen. The metabolic demands of many areas of the brain can thus become partially deprived of oxygen and glucose.

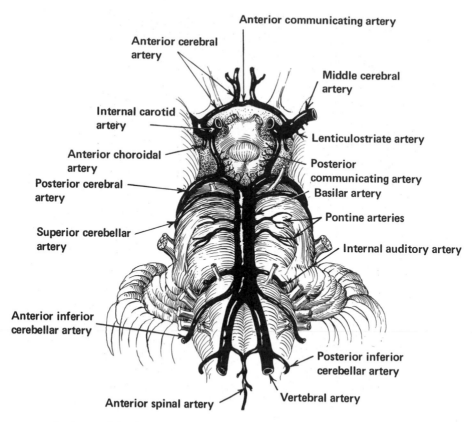

Figure 20.7.   The base of the brain. The circle of Willis and the principle arteries of the brain. (From Waxman SG, De Groot J. Correlative Neu- roanatomy, 22nd ed. East Norwalk, CT: Appleton & Lange, 1994.)

THROMBI

Clot Formation

Atherosclerotic development usually begins early in life, exerting its initial roughening effects at the bifurcation of the various blood vessels. Endothelial cells lining the blood vessel begin to die, ulceration occurs, and blood platelets, lipids, and cholesterol begin to be deposited in the arterial intima, thus forming a mound of tissue. Indeed, within a matter of seconds, blood platelets will begin to adhere to any portion of a vessel that is not perfectly smooth and contains rough spots (e.g., a patch-like accumulation of lipid or cholesterol). When this occurs, the adhering blood platelets will rupture, which in turn triggers the formation of thrombin and insoluble fibrin proteins—thromboplastin. These thrombin proteins act to create clots.

Specifically, these fibrin proteins resemble web-like threads that are tangled together. These tangles act like nets that act to trap more blood cells and other debris, which then clot together. Once started, the clot tends to grow as more and more platelets are enmeshed and rupture, thereby releasing more thromboplastin fibrin proteins. Furthermore, the plaques that narrow the lumen may serve as a nidus on which even more thrombi come to adhere and from which emboli may be dislodged. Hence, these elevated fibrous plaques (thrombi) invariably grow larger and may even become vascularized. A complicated lesion is thus formed.

OBSTRUCTIVE INFLUENCES

As the clot grows blood flow is reduced, which causes even more thromboplastin to

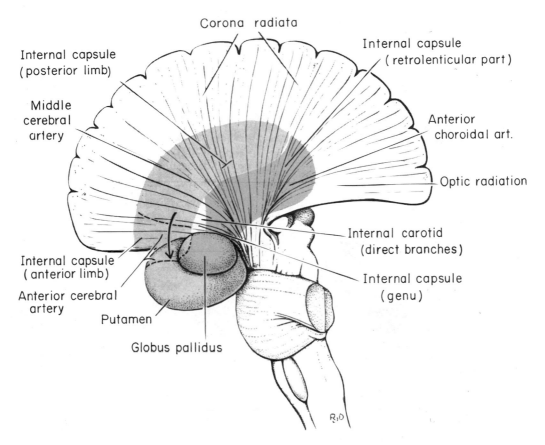

Figure 20.8. The arterial supply to the corpus striatum and thalamus. (From Carpenter M. Core Text of Neuroanatomy. Baltimore: Williams & Wilkins, 1991.)

accumulate, particularly in that these deposits create turbulence or eddy current, which further damage the smooth endothelium of the artery as well as reduce resilience.

As the clot grows and thrombi thicken, the lumen becomes progressively smaller, which results in arterial dilation. By reducing the resilience and the diameter of the large arteries, atherosclerosis can induce systolic hypertension. The driving force of the blood against vessel walls is therefore increased, which causes further roughening and more plaque formation. Atherosclerosis is thus an insidious process.

When the lumen of the artery is narrowed by about 50%, pressure in the proximal segment increases, whereas distal to the clot pressure begins to fall. With 70% reduction, flow is significantly decreased and neurological functioning may be altered.

Typically, however, patients remain asymptomatic until there is a sudden reduction in blood flow or thrombi or emboli are thrown off, creating a complete obstruction. However, if the core of the plaque becomes necrotic and calcifies, the surface may disintegrate, thus exposing the weakened underlying surface of the vessel. When this occurs the vessel may rupture and hemorrhage.

Obstruction of the artery due to atherosclerosis may produce disastrous neurological consequences or be associated with no symptoms whatsoever. The outcome depends on (1) the configuration of the arterial tree with which the patient was born, (2) the segment of artery involved, (3) the rapidity of obstruction development, (4) the presence or absence of collateral vessels, (5) associated diseases such as hypertension, and (6) triggering mechanisms, such as a sudden blow or turn of the head.

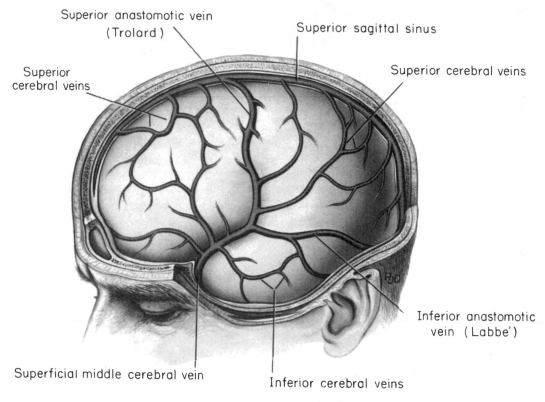

Figure 20.9.   The superficial cerebral veins. (From Carpenter M. Core Text of Neuroanatomy. Baltimore: Williams & Wilkins, 1991.)

Susceptible Arteries

There is a tendency for atheromatous plaques to form at branching and curves of cerebral arteries—most frequently the internal carotid artery and carotid sinus; the vertebral arteries and their junction, which forms from the basilar artery, the main bifurcation of the middle cerebral artery, and the posterior cerebral arteries as they wind around the midbrain; and in the anterior cerebrals as they curve over the corpus callosum (Adams and Victor, 1994; Toole, 1984).

Atherosclerosis seldom exists in either of the coronary arteries of the cerebral artery system without the other being involved as well—it occurs in both (Toole, 1984). When the heart is involved this can lead to myocardial infarction, and mural thrombi may be cast off that then flow directly to the lungs and brain. Atherosclerotic heart disease can also cause cardiac dysrhythmias or decreases in cardiac output—possibly to such a degree that cerebrovascular insufficiency results.

RISK FACTORS

Risk factors for cerebrovascular and carotid atherosclerosis include hypertension, increased age, diabetes mellitus, and cigarette smoking, age and hypertension being the most important (Bornstein and Kelly, 1991; Crouse et al., 1987; Wolf, 1985). In general, the onset of atherosclerotic disease is usually in childhood and remains silent, growing slowly for 20 or 30 years before becoming symptomatic. It appears to reach its peak incidence between the ages of 50 and 70, and men are twice as likely as women to suffer.

CEREBRAL ISCHEMIA

When emboli, thrombi, and/or atherosclerosis reduces perfusion of brain tissue below a critical level, ischemia occurs and the patient suffers a cerebral infarct. Cerebral infarction and ischemia are due to insufficiency of the oxygen and blood supply (Ginsberg, 1995).

Ischemia leads to hypoxia and impaired delivery of glucose as well as a failure to remove metabolites and waste such as lactic acid and carbon dioxide (Welch and Levine, 1991). As the infarction evolves, local edema results, which may cause an increase in tissue pressure and further reduce tissue perfusion by compressing capillaries, thereby enlarging the infarction. The severity and extent of the neurological deficit is increased even further.

There is usually reduced blood flow during ischemic attacks, which extends far beyond the infarct borders. This further reduces neuronal activity such that transient (as well as permanent) neurological deterioration occurs (Ginsberg, 1995; Welch and Levine, 1991). These transient blood flow reductions are due to metabolic changes, compression of the local vasculature, and temporary cessation of neural activity (due to disconnection). Initially this may make the functional effects of the stroke appear more pervasive and profound. Nevertheless blood flow is still sufficient in these border areas so as to keep this tissue alive (Torvik and Svinland, 1986). Hence, as the ischemia subsides and blood flow is increased to these depressed areas there occurs a considerable degree of what appears to be recovery.

The majority of infarcts are sharply delineated. Nevertheless, surrounding the area of ischemia is a concentric zone of hyperemia caused by a local loss of autoregulatory capacity and vasomotor paralysis (Ginsberg, 1995). That is, there is an increase in flow to the immediately surrounding area, which acts to divert blood from other areas of the brain as well as from the ischemic zone, thereby increasing ischemic damage to this area (Toole, 1984; Welch and Levine, 1991). This loss of autoregulation is most evident during the acute phase of cerebral infarction. As the infarction evolves there is local edema and

swelling. This causes an increase in tissue pressure and further reduced tissue perfusion by compressing capillaries. The infarction is enlarged even further.

The pathological effects of ischemia and reduced blood flow may be massive, if, for example, the internal carotid or middle cerebral artery is obstructed. Or, the effects may be minute if impaired circulation is limited to the smaller arteries and arterioles (Toole, 1984; Welch and Levine, 1991).

With moderate or massive infarction, even if quite focal, regions within the opposite hemisphere also exhibit similar but less extensive changes due perhaps to local reflexia, increased intracranial pressure, or spillage and spread of vasoactive chemicals that affect the entire brain (Toole, 1984). Moreover, infarction of one hemisphere may cause so much swelling that it leads to symptoms of increased intracranial pressure. When pressure increases, blood flow decreases. Blood flow tends toward normal in those who recover, but remains slow throughout the affected hemisphere in those who do poorly.

RISK FACTORS

Risk factors for ischemic stroke include older age, male sex, hypertension, TIAs, hypertensive heart disease, coronary heart disease, congestive heart failure, and diabetes mellitus (Alter et al., 1994; Bornstein and Kelly, 1991; Davis et al., 1987; Welch and Levine, 1991).

THROMBOSIS

When a thrombotic clot enlarges sufficiently the patient suffers a cerebral infarct, i.e., thrombosis. Thrombosis is due to a thrombus—a clot produced by coagulated blood (thromboplastic: a protein that forms clots). Thrombotic strokes often occur during periods of prolonged inactivity or sleep such that the symptoms are present upon arising in the morning. As the patient attempts to swing out of bed or a chair, they find that they are weak on one side, that they cannot speak, or that some other deficit has developed.

Thrombotic strokes developed gradually (Absher and Toole, 1995). Symptoms and signs usually progress in a stepwise fashion, sometimes taking 2–4 days to fully develop. Patients deteriorate, improve slightly, then deteriorate further—a process referred to as stroke in evolution. This stepwise evolution is due to edema and alterations in blood flow and metabolism as thrombotic particles continually form and break off. Stabilization with improvement begins in 7–14 days after stroke onset (Baxter, 1984; Meier and Strauman, 1991).

The size of the infarct in part is dependent on the presence and integrity of collateral vessels. If there are no collateral circulatory channels, occlusion exerts drastic effects and the infarct is quite large.

Warning signs include headache and/or previous transient neurological symptoms or a TIA. When TIAs precede a stroke, they almost always stamp the process as thrombotic (Adams and Victor, 1994; Toole, 1984).

The most common cause of thrombotic stroke is arteriosclerosis. Thrombosis follows clotting of the blood at a site where its flow is impeded by a sclerotic plaque, usually in the bifurcations of the carotids. Complete occlusion of the carotids is always the consequence of thrombosis (Toole, 1984).

EMBOLISM

In contrast to thrombosis, which is due to a localized buildup, emboli are free particles that become lodged within a vessel, causing occlusion. Like thrombosis, embolic strokes create an ischemic cerebral infarct. However, unlike thrombosis, embolic strokes are often followed later by hemorrhage (Chamorro et al., 1995). Both thrombotic and cerebral embolic strokes occur most commonly among older individuals. Among younger age groups, however, embolism is the most frequent culprit.

An embolism may be composed of thrombotic particles that have broken off from a clot, air bubbles, masses of bacteria, or blood clots. They may also consist of cholesterol crystals that are thrown off from

atheromatous plaques situated in the vertebral-basilar and carotid arteries; pieces of tumor cells from the lung, stomach, or kidney that subsequently take root in the brain after drifting in the bloodstream; and fat globules released into the bloodstream after trauma to the marrow of the long bones (Toole, 1984). Oral contraceptives among women can give rise to the development of embolism, and an association among those with mitral valve prolapse has been noted (Jackson, 1986; Toole. , 1984).

Cardiac Embolic Origins

Cerebral emboli may develop when there is plaque buildup on the cardiac or carotid valves and may also result from mitral stenosis, arterial or auricle fibrillation, aortic valve calcification, fragments of left atria or ventricular thrombus, mitral valve polyps, myocardial infarction, or vegetations of infective endocarditis. Heart disease, mitral stenosis, and atrial fibrillation are thus major risk factors for cerebral embolism. Chronic atrial fibrillation in particular carries a high risk of stroke (Bornstein and Kelly, 1991; Petersen and Godtfredsen, 1986; Wolf et al., 1983), the clot coming from a paralyzed atrium. When the embolus is infected, meningitis or abscess may develop. Indeed, it is estimated that approximately 90% of emboli from the heart end up in the brain. Thus embolic stroke is often a sign of systemic heart disease.

The Carotid and Left Middle Cerebral Arteries

In general, emboli expelled from the heart and the left ventricle are carried into the brachiocephalic artery and then the left common carotid. From the left carotid, emboli are transmitted and occlude the arteries supplying the left cerebral hemisphere. Hence, the left side of the brain is a common destination of emboli originating in the heart (Absher and Toole, 1995).

When emboli are repeatedly freed into the blood system they tend to lodge in the same artery, usually the middle cerebral.

This is because the middle cerebral artery is a direct extension of the internal carotid and receives almost 80% of the blood. Hence, most emboli end up within the middle cerebral system.

Focal Influences

Embolic cerebral strokes tend to produce focal effects such as expressive aphasia, monoplegia, or receptive aphasia (Absher and Toole, 1995; Beckson and Cummings, 1991; Starkstein and Robinson, 1992). Global effects are less common.

Vertebral Basilar Artery

Emboli entering the vertebral-basilar system produce deep coma and total paralysis. Indeed, a 1-mm speck lodged in an artery of the brainstem can cause catastrophic neurological consequences (Toole, 1984). Embolic infarction of the undersurface of the cerebellum (via the cerebellar arteries) may cause fatal brainstem compression. If it enters the posterior cerebral artery it may produce a unilateral or bilateral homonymous hemianopia and/or severe memory loss.

Onset

After the lodgment of an embolus, the vessel usually goes into spasm and thrombosis may occur. Embolic strokes characteristically begin suddenly, often while the patient is awake and active. There is usually no warning, and the effects of the infarction reach their peak almost immediately (Adams and Victor, 1994).

Cerebral embolization will be major if the embolus is large or temporary if the fragment is small. Smaller fragments are more likely to disintegrate. Not infrequently an extensive deficit from embolism reverses itself dramatically within a few hours or a day or two (Absher and Toole, 1995; Meier and Strauman, 1991). It is not uncommon for many to suffer a recurrence, frequently with severe damage. Early anticoagulant therapy should be emphasized as it is important in the prevention of em-

bolic stroke recurrence (Chomorro et al., 1995; Koller, 1982; Lodder and van der Lugt, 1983).

Embolic Hemorrhage

Approximately 65% of those with a cerebral embolic stroke develop hemorrhagic infarction. Conversely, less than 20% of those with thrombosis will develop hemorrhage (Ott et al., 1986). Similarly, approximately 50% of all hemorrhagic infarcts (HIs) are associated with embolic strokes (Fisher and Adams, 1951; Hart and Easton, 1986). Hence embolic strokes have a special propensity for hemorrhagic transformation, and hemorrhagic infarction nearly always indicates embolism.

Blood vessels affected by embolism characteristically hemorrhage within 12–48 hours (Cerebral Embolism Study Group, 1984; Hart and Easton, 1986). However, it may take several days to develop (Laureno et al., 1987). Hemorrhagic transformation occurs when the emboli disintegrate and/or migrate distally, thus allowing reperfusion of the damaged vessel (Fisher and Adams, 1951: Jorgensen and Torvic, 1969). That is, when a vessel is occluded, it is damaged and weakened, and both the brain and the involved portion of the vessel may become necrotic. When the weakened portion of a previously occluded vessel is subsequently reexposed to the full force of arterial pressure it ruptures and hemorrhages.

Frequently embolic hemorrhages are asymptomatic (Hakim et al., 1983; Ott et al., 1986) and benign since the tissue involved has already been damaged. However, if the hemorrhage is secondary to anticoagulant therapy the bleeding may be more profuse and cause significant neurological deterioration and even death (Hart and Easton, 1986; see Chamorro et al., 1995).

TRANSIENT ISCHEMIC ATTACKS

TIAs are due to a brief and temporary reduction in blood flow to a focal region within the brain. They are accompanied by transitory and focal neurological disturbances that seem to immediately and completely resolve with no apparent residuals.

The most frequent causes of TIAs are atherosclerosis and microembolization of platelet aggregates from ulcerated atherosclerotic plaques within the carotids or from the cardiac valves (Adams and Victor, 1994). These conditions act to reduce blood flow. Other causes include vascular spasm and aggregations of cholesterol crystals, fibrin, and/or blood platelets that temporarily occlude a vessel, or a combination of these factors.

TIAs are related to the microcirculation, where small irritants such as an embolus can lead to blockage. However they can involve any cerebral or cerebellar artery. They may last a few seconds or up to 12–24 hours. Most last 2–15 minutes (Toole, 1984). The symptoms are transient because the occluding embolus almost immediately disintegrates, which allows for the immediate reestablishment of blood flow and functional recovery.

TIAs that last less than 60 minutes are different from those of longer duration. The shorter ones are due to emboli that travel from one artery to another, whereas the longer ones are from emboli passed by the heart (Toole, 1984). Hence, artery-to-artery emboli tend to be smaller and more quickly disintegrate, whereas those from the heart tend to be larger and/or more numerous. Some patients may experience a single episode of TIA in a life time, whereas others may have as many as 20 attacks in a single day. Usually they are fewer than one or two a week (Toole, 1984)

Weakness or numbness of a single finger may be the only manifestation of a TIA. However, a person may suffer a TIA and not be conscious of it, or the manifestation may be a disturbance in memory, speech, and word finding, a brief confusion, etc. Usually with these kinds of disturbances the patients and their families ignore, fail to notice, or attribute little significance to these attacks as they are transient.

In some instances, the TIA may take the form of temporary blindness, like a shade being pulled over the eye. It is when the patient becomes aware that there is a loss of vision, or a loss of strength in the hand or foot that rapidly progresses to involve the entire extremity, that they become alarmed.

TIAs can occur within any region of the cerebral vasculature. Some investigators, however, have classified them as opthalmic, hemispheric, or vertebral-basilar.

HEMISPHERIC TIAS

In hemispheric attacks, ischemia occurs foremost in the distal territory of the middle cerebral artery, producing weakness or numbness in the opposite hand and arm—however, different combinations may occur: face and lips, hand and foot, fingers alone, etc. In ocular attacks, transient monocular blindness occurs and is described as a shade falling smoothly over the visual field until the eye is blind. The attack clears smoothly and uniformly.

VERTEBRAL-BASILAR TIAS

If the TIA occurs in the vertebral-basilar system, there may be dizziness, diplopia, dysarthria, bifacial numbness, and weakness or numbness involving part or parts of the body, or even both body halves (because of the involvement of diverse regions of the brainstem and thalamus supplied by the vertebral-basilar system). Other symptoms include, in order of frequency, headache, staggering, veering to one side, feeling of being cross-eyed, dark or blurred or tunnel vision, partial or complete blindness, pupillary changes, and paralysis of gaze (Toole, 1984). In basilar artery disease, each side of the body may be affected alternately. There is good evidence that TIAs may be abolished by anticoagulant drugs.

TIA and Stroke

Forty percent of individuals with TIAs subsequently suffer brain infarction, the majority of which occur within the first 6 months (Wolf et al., 1983). Stroke may occur after the first or second TIA or only after hundreds have occurred over a period of weeks or months. However, the greatest risk of stroke following TIA is within the first year and in particular the first 30 days, 20% occurring in the first month and 50% within a year (Bornstein and Kelly, 1991; Wolf et al., 1983).

When TIAs precede a stroke, however, they almost always stamp the process as thrombotic (atherosclerotic thrombosis). Nevertheless, it is noteworthy that although thrombotic strokes are said to occur predominantly at night, the majority of infarctions in general occur between 6 am and 6 pm (Van Der Windt and Gijn, 1988).

RISK FACTORS

Risk factors include cigarette smoking, previous history of stroke or TIA, ischemic heart disease, and diabetes (Bornstein and Kelly, 1991; Howard et al., 1987). However, as the number of risk factors increases, the prognosis for recovery following stroke declines (Meier and Strauman, 1991).

LACUNAR STROKES

The term *lacune* is associated with the development of small holes in the subcortical tissue, and lacunar strokes are often secondary to the occlusion of a deep single perforating artery (Cummings and Mahler, 1991; Fisher, 1969). The most common type is due to an occlusion via macrophage plaques. Approximately 20% of all infarcts are lacunar (Bamford et al., 1987; Kunitz et al., 1984). However, the incidence increases with age.

In lacunar strokes, there is usually no impairment of consciousness or higher cognitive functioning. However, if the lesion involves the internal capsule there may be motor disturbances, including paralysis or facial or arm weakness. If the lesion involves the thalamus, sensory sensation over various body parts may be attenuated. The incidence of fatality is very low. Nevertheless, over a third of affected patients may be seriously impaired as long as 1 year post stroke (Bamford et al., 1987;

Meier and Strauman, 1991). Usually, since the stroke is so tiny, computed tomography (CT) is negative.

Multi-Infarct Dementia

In some instances individuals will repeatedly suffer small strokes over a number of years. At first these strokes seem asymptomatic or have only mild consequences. However, with repeated strokes patients (or family members) may notice that their memory is not as good as it was, that they are having difficulty concentrating, that they are sometimes confused, and that they have increased word finding difficulty as well as other linguistic problems (Cummings and Mahler, 1991). Personality and emotional changes may be subtle at first as well.

Essentially, as the lesion expands because of repeated, albeit small strokes, the patient gradually deteriorates, until finally what was a series of minor deficits is suddenly a major alteration in personality functioning and intellectual capability. Frequently these individuals and family members erroneously describe the onset as sudden because they tended to ignore the stepwise progression of severity. However, careful questioning will reveal that the patient's symptoms had been progressing over time.

In addition, patients with multi-infarct dementia have been reported to have bilateral patchily reduced blood flow in the gray matter, particularly within the distribution of the middle cerebral artery (Cummings and Mahler, 1991; Yamaguchi et al., 1980). The reductions are related to the severity of dementia (Kitagawa et al., 1984; Meyer et al., 1988). In general the pattern of reduction is suggestive of widespread and diffuse cerebral ischemia or infarction, and the regions involved include the thalamus and the frontal and temporal lobes. Nevertheless, patients with multi-infarct dementia are frequently diagnosed as suffering from Alzheimer's disease (see Blumenthal and Premachandra, 1992).

HEART DISEASE, MYOCARDIAL INFARCTION, AND CARDIAC SURGERY
Ischemia and Heart Disease

The delivery of oxygen to the brain depends on the vascular, pulmonary, and cardiac systems, the flow characteristics of the blood, and the quality and quantity of the hemoglobin molecules and the binding of oxygen to those molecules (Grotta et al., 1986). Abnormalities involving red blood cell structure or concentration are frequent causes of cerebrovascular abnormalities (Grotta et al., 1986) and can create conditions that promote coronary heart disease, abnormal clotting, and thus reduced blood flow.

Coronary heart disease (even when asymptomatic) is a major underlying contributor to the development of regional decreases in cerebral blood flow, cerebral infarction, and ischemia (Di Pasquale et al., 1986). In many cases patients suffer cerebral deterioration even in the absence of stroke. It is not uncommon for patients to note that their memory or other cognitive disturbances began soon following a heart attack or major surgery.

However, in some cases, an otherwise normal heart may be injured secondary to pulmonary edema following hemorrhage. This in turn may lead to vasospasm and thus an additional cerebral infarction (Mayer et al., 1994b).

Myocardial Infarction

When an individual has a myocardial infarct, blood pressure, cerebral blood flow, and metabolism fall by as much as 50% (Toole, 1984). Cerebral-vascular insufficiency and cerebral infarction are thus a likely consequence. With cardiac arrest blood pressure and flow cease and widespread cerebral ischemic damage results (Welch and Levine, 1991).

Myocardial infarction is a major contributor to stroke. When myocardial infarction occurs, thrombi are cast off from the heart and supporting vessels into the blood and are carried to the cerebral vasculature, causing occlusion. In fact, following a heart attack embolic events may continue for up to a month, the patient suffering a series of small or major cerebral infarctions. This is because during the healing process mural thrombi, embolisms, and clots may continually form and break off. If major occlusions occur within the vertebral system the patient may die. Indeed, myocardial infarction is the leading cause of mortality among patients with stroke (Burke et al., 1982).

Global Ischemia

Global ischemia is due to a sudden and complete reduction in arterial pressure such as occurs during myocardial infarction. If it persists for more than a few minutes widespread damage in the cortex, hippocampus, basal ganglia, brainstem, and cerebellum results. Also areas at the junctions between the various vascular territories, i.e., watershed areas, are especially vulnerable.

VALVULAR INSUFFICIENCY

During a heart cycle, the atria and ventricles contract and relax successively; this is the heart beat. However, if any of these valves lose their ability to close, blood will leak backward, a condition referred to as valvular insufficiency. If the left atria-ventricular passageway were to become narrowed (mitral stenosis), blood flow is hindered and circulatory failure results, accompanied by massive cerebral ischemia and neuronal death. The death rate among stroke patients with coronary artery disease has been estimated to be over 50% (Di Pasquale et al., 1986).

FIBRILLATION

Atrial fibrillation and tachy-bradycardia are abnormalities of rhythm and conduction that deprive the brain of adequate perfusion. Moreover, thrombi can be thrown off, which in turn become a source of emboli and vascular occlusion (Barnett, 1984).

The clinical pictures may include syncope or severe diffuse hypoxia.

MITRAL VALVE PROLAPSE

Mitral valve prolapse is due to thickening, elongation, and altered collagen composition, as well as basic "wear and tear" of the mitral valve leaflets (Barnett, 1984; Cole et al., 1984; Jackson, 1986; Lucas and Edwards, 1982). This causes the mitral valves to balloon back (prolapse) into the left atrium during ventricular contraction. This disturbance can in turn disrupt blood flow to the brain and create ischemic conditions. However, thrombi also tend to form on the abnormal mitral valve and thus become thrown off into the blood supply (Barnett, 1984).

The opthalmic and posterior cerebral arteries seem to be particularly affected, as retinal ischemia as well as transient global amnesia seems to be a frequent consequence (Jackson, 1986; Wilson et al., 1977). However, the cerebral hemispheres are just as frequently affected (Barnett, 1984; Jackson, 1986). Nevertheless, the risk of stroke, at least among young individuals, is rather low. However, mitral valve prolapse is associated with the development of infective endocarditis.

Mitral valve prolapse has also been associated with the development of chronic anxiety, agoraphobia, and panic disorders (Klein and Gorman, 1984). The causal nature of this relationship, however, is unclear. Possibly individuals who are susceptible to anxiety disorders who also suffer from mitral valve prolapse respond to these symptoms with increased fear and anxiety. That is, mitral valve prolapse exacerbates an already disturbed emotional condition (Jackson, 1986).

VASOSPASM

Arteries may spasm when emboli pass and/or occlude an artery, if an aneurysm ruptures, or if a vessel is manipulated or traumatized in some manner (Toole, 1984) such as after a head injury. They have also been known to occur among individuals whose arteries tend to be hyperreactive, i.e., young women (Toole, 1984). Spasm may be diffuse or segmental along the artery.

When a spasm occurs there is a transient stoppage of the flow of blood. If prolonged the patient will suffer ischemic cerebral damage. Moreover, spasm sometimes causes embolic particles to be dislodged from the vessel wall, which then occludes smaller distal arteries. Cerebral infarction is therefore a likely consequence of spasm.

MURMURS

Murmurs (or bruits) are sometimes produced by changes in the direction of blood flow such as at the bifurcations of the arteries. Frequently they are a consequence of alterations in the wall of the blood vessel, including shrinkage of the luminal diameter. In general, murmurs become audible after the lumen has been reduced by about 50% (Toole, 1984), and with increasing stenosis the volume and pitch become louder and higher.

Murmurs are often indicative of atherosclerosis, and are most intense over the site of the lesion. However, they may be heard over vessels other than the affected artery. Bruits are graded from 1 to 6, 1 being the most mild and 6 being the loudest.

CARDIAC SURGERY

In the early 1980s we were witness to the catastrophic effects of artifical heart surgery and implantation on various human experimental subjects such as Dr. Barney Clark. All these unfortunate individuals suffered extremely serious cerebral damage due to repeated and massive strokes. Although not well publicized, the effects of heart transplantation (Hotson and Pedley, 1976; Montero and Martinez, 1986; Schober and Herman, 1973) and even open heart surgery can create similar disastrous consequences.

For example, a variety of studies have indicated that although motor and other initial impairments secondary to cerebrovascular disease generally improve after open heart surgery, significant deterioration in regard to personality, intellectual, visual-spatial, perceptual, memory, and attentional functioning often occurs (Aberg . and Kihlgren, 1974; Gilberstadt and Sako, 1967; Heller et al., 1974; Lee et al., 1969, 1971; Raymond et al., 1984; Shaw et al., 1987).

Open heart surgery can also cause and induce stroke (Smith et al., 1986). Frequently, the effects are diffuse rather than focal (Sotaneiemi et al., 1986).

Cardiac surgery may cause cerebral injury due to reduced blood blow and arterial pressure, macroembolization of fat, air, or other gasses, the dislodgement of debris from the valves, and the initiation of blood cell and platelet aggregation (see Shaw et al., 1987 for review). In fact, just the process of opening the chest cavity via sternal retraction can reduce oxygen intake if the brachial plexus is traumatized (Baisden et al., 1984).

In general, the cerebral effects of heart surgery appear to be age-dependent. Children seem to be more resilient as the long-term consequences on cerebral functioning are less drastic (Whitman et al., 1973).

Given the above, and the common knowledge that the majority of certain types of heart surgery (e.g., so-called "triple bypass") seem to have no proven benefit, it may be extremely wise to be cautious before undergoing these procedures. Certainly, to further clarify these risks, a series of rigorous neuropsychological studies needs to be conducted on such patients both pre and post surgery. Nevertheless, sometimes central nervous system (CNS) functioning is improved because of the correction of circulatory abnormalities and increases in blood flow (Sotaneiemi et al., 1986).

HEART TRANSPLANTS

A common supposed side effect of heart transplants was altered personality, the explanation being that patients would assume the personality of the donor. In reality, however, many of these patients experienced alterations in mental and personality functioning because they suffered strokes. Based on a review of the few studies conducted, it appears that heart transplantation carries very significant risks in regard to cerebral functional integrity. This is because many of these individuals have severe vascular disease that is not confined to the heart, such as atherosclerosis and the presence of mural and arterial thrombi. Hence, during the trauma of surgery, debris can be loosened and cast into the blood, and various vessels may go into spasm, creating serious complications involving the brain (Hotson and Pedley, 1976; Montero and Martinez, 1986; Schober and Herman, 1973).

Other primary and secondary consequences of heart transplanting on the brain include vascular lesions due to circulatory collapse, decreased cardiac output and chronic postoperative hypotension, thrombosis, embolism, hemorrhage, and hypoxic damage and opportunistic infection due to depressed immune status (Montero and Martinez, 1986). In this regard, it certainly seems that these individuals frequently trade an unhealthy heart for a damaged brain. The 5-year survival rate of heart transplantation is 50% (Montero and Martinez, 1986).

It is important to note, however, that there is a paucity of neuropsychological research on the effects of heart transplantation on cerebral functional integrity. Perhaps the immediate and long-term effects are not as dismal as this short review suggests.

HEMORRHAGE

Hemorrhage can occur anywhere throughout the brain and may be due to a number of causes, e.g., head injury, hypertension, rupture of an aneurysm or arterial venous malformation (AVM), weakening of a segment of the vasculature secondary to emboli or thrombus, or vessel wall necrosis due to occlusion and ischemia (Adams and Victor, 1994; Roos et al., 1995).

Hemorrhages are frequently classified in terms of gross anatomical location. These include, extradural, subdural, subarachnoid, intercerebral/cerebral, and cerebellar. Extradural and subdural hemorrhages are frequently secondary to head injury, whereas subarachnoid, cerebral, and cerebellar hemorrhages are often related to arterial abnormalities.

Bleeding from a hemorrhage may be minute and inconsequential, or profuse and extensive such that a large pool of blood rapidly develops. In some cases bleeding may occur at a very slow, albeit continuous pace such that the adverse effects are not detected for days.

Subarachnoid Hemorrhage

Subarachnoid hemorrhage results from any condition that causes blood to leak into the subarachnoid space. Massive subarachnoid hemorrhage is usually due to rupture of an intracranial aneurysm or bleeding from a cerebral angioma—in either case there is no warning and the onset is quite sudden and abrupt (Adams and Victor, 1994; Roos et al., 1995; Schievink et al., 1995). If the hemorrhage is severe it may lead to immediate coma and death, particularly if there is a buildup of over 20 ml of intraventricular blood (Roos et al., 1995). If moderate, the patient may pass into a semi-stuporous state and/or become confused and irritable. If minor patients may complain only of severe headache and possibly develop focal deficits after hours or over the course of the first few days or weeks following hemorrhage. Vasospasm and rebleeding are very common during the first two weeks.

Subarachnoid hemorrhage causes death in over 50% of the persons it affects, a third of whom will die immediately or within the first 24 hours (Schievink et al., 1995). Only approximately 25% who survive will make a good recovery (Hijdra et al., 1987).

Subarachnoid hemorrhage occurs most often among individuals above age 50. When it occurs among younger individuals it is often secondary to congenital vascular abnormalities, including angioma, ruptured aneurysms, or the rupture of an AVM on the brain surface (Adams and Victor,

1994; Brown et al., 1991; Schievink et al., 1995; Toole, 1984). Rupture or seepage into the ventricular system is not uncommon (Roos et al., 1995), and cerebrospinal fluid is bloody in 90% of cases. Anemic and hemorrhagic infarction may coexist in the same lesion.

Headache and vomiting are immediate common sequelae of hemorrhage (as well as cerebrovascular disease), and most patients will complain of severe backache and neck stiffness in the absence of demonstrable focal signs, which develop later (Gorelick et al., 1986; Portenoy et al., 1984). Onset is sudden and the intensity is described as severe or violent. These headaches are diffusely distributed over the cranium and/or are localized to the frontal-parietal region (Gorelick et al., 1986). Headaches are usually due the pressure effects of escaping blood, which distend, distort, or stretch pain-sensitive intracranial structures. Pain in or behind the eye is often associated with hemorrhage of posterior communicating-carotid aneurysms (Gorelick et al., 1986).

Cerebral/Intracerebral Hemorrhage

Cerebral hemorrhage is commonly caused by hypertension and associated ruptured aneurysm and degenerative changes in the vessel wall of penetrating arteries, which make hemorrhages susceptible to rupture (Kase, 1986; Schievink et al., 1995). Onset is always sudden.

The rupture may be brought on by mental excitement or physical effort, or may it occur during rest or sleep. Usually the patient complains of sudden headache and may vomit and become confused and dazed with progressive impairment of consciousness over several minutes or hours (except in the mildest of cases). However, the rupture may evolve gradually, taking hours or days to become fully developed, and there may be no warning signs (Mayer et al., 1994).

After a large hemorrhage the affected hemisphere becomes larger than the other because of swelling. As the pool of blood increases and begins to clot, surrounding tissues become compressed, the convolutions become flattened, and pressure may be exerted against the opposite half of the brain, causing damage in this region also. With large hemorrhages, coma and death may ensue because of compression of midline and vital brainstem nuclei (Adams and Victor, 1994; Mayer et al., 1994). If the patient survives and the clot is not surgically removed it is eventually absorbed and replaced by a glial scar. In these instances, however, patients are commonly incapacitated to varying degrees. Cerebral hemorrhages occur most often in the vicinity of the internal capsule, corona radiata, frontal lobe, pons, thalamus, and putamen (Adams and Victor, 1994; Kase, 1986).

The neurological deficit is never transitory (good functional recovery being attained by less than 40% of the survivors), and 30–75% die within 30 days (Adams and Victor, 1994; Fieschi et al., 1988; Portenoy et al., 1987). Most patients suffer persistent, permanent, and severe neurological abnormalities. Good clinical outcome is related to lower age, the size of hemorrhage, how long the patient was unconscious, high scores on the Glasgow Coma Scale, and postoperative neurological events (Meier, 1991; Portenoy et al., 1987; Tidswell et al., 1995; Toole, 1984).

In over 60% of the cases, intracerebral hemorrhage is related to hypertensive cerebrovascular disease (Mohr et al., 1978), which makes vessels susceptible to rupture. That is, hypertension can induce degenerative changes and may in fact induce the formation of microaneurysms, particularly in the subcortical and perforating arteries (Kase, 1986). However, not all cerebral hemorrhages are due to hypertension.

HYPERTENSION

Hypertension is a leading cause of vascular hypertrophy, arteriosclerosis, stroke, and hemorrhage. Conversely, approximately 40% of those with arteriosclerosis are hypertensive (Toole, 1984). However, in some respects hypertension and arteriosclerosis are mutually reinforcing in that

the development of thrombi can act to obstruct blood flow passage, thus increasing (at least proximal) pressure, whereas vessels damaged by increased pressure are likely to become infiltrated by thrombi.

Specifically, chronic hypertension acts to reduce arterial elasticity by stretching and thickening the walls of the blood vessels, including the capillaries (Garcia et al., 1981; Hart et al., 1980). It can also potentiate the development of thrombi via pressure-induced erosion or roughening of vessel walls. Hence, frequently the space within the vessel through which the blood flows will decrease in size. However, sometimes it is actually enlarged.

Chronic hypertension, although associated with increased blood pressure, actually can cause generalized blood flow reductions, particularly in regions served by the middle cerebral artery, i.e., the frontal-parietal and temporal lobes (Rodriguez, et al., 1987). Hence, one consequence of hypertension is an increased risk for ischemic infarcts (Alter et al., 1994; Bornstein and Kelly, 1991). That is, because the caliber of the lumen becomes fixed and rigid, the ability to dilate and thus compensate for alterations in blood pressure is lost. In consequence, if there were a decrease in blood pressure, this inability to compensate would result in a significant reduction in blood flow and thus cerebral ischemia.

If, on the other hand, blood pressure were to increase, this lack of flexibility would predispose the vessels to burst and hemorrhage (Toole, 1984). Hence, there is a strong correlation between the incidence of stroke and high blood pressure.

Hypertensive encephalopathy refers to an acute syndrome in which there is an absence of warning signs and a rapid development (from minutes to hours) of deficits. Patients often complain of headache, nausea, vomiting, visual disturbances, and confusion and they may develop convulsions, develop focal neurological signs, and/or lapse into a deepening coma or stuporous state.

Onset is often preceded by an extreme and rapid rise in blood pressure and severe hypertension. Autoregulatory responses may become abnormal such that widespread vasospasms occur accompanied by ischemia and/or the rupturing of arterioles and capillaries (Toole, 1984).

ANEURYSMS, AVMS, TUMORS, AMYLOID ANGIOPATHY

Hypertension is only one of many causes of hemorrhage (Alter et al., 1995; Brott et al., 1986; Hart and Easton, 1986; Laureno et al., 1987). Hemorrhage may occur secondary to drug use, anticoagulant therapy, medication, AVMs, aneurysms, vessel wall necrosis, brain tumors, and various types of arterial pathology, including small vascular malformations and cerebral amyloid angiopathy (Adams and Victor, 1994; Brown et al., 1991; Kase, 1986; Roos et al., 1995; Schievink et al., 1995; Toole, 1984).

Ischemia and Hemorrhagic Infarcts

Ischemia not only results in the death of brain cells but necrosis of the local vasculature, which is also deprived of metabolic support (Welch and Levine, 1991). These vessel wall ischemic structural alterations make them very susceptible to rupture. Indeed, over 40% of those with cerebral ischemic infarction will become hemorrhagic within 1–2 weeks (Hornig et al., 1986).

These ischemic-related HIs are not limited to a single vessel, however. Bleeding may be multifocal, particularly if the patient had suffered a large stroke.

In part, this is also due to the more extensive edema associated with large strokes. When swelling occurs, not only is brain tissue compressed but the endothelium of various small vessels is also crushed and damaged. When these vessels are compressed blood flow is prevented, which in turn makes these same vessels and their distal extensions more susceptible to rupture when blood flow is reestablished (Garcia et al., 1983). That is, with blockage of a vessel, the distal part of the vessel may become necrotic. Even with mild degrees of edema

there is compression and subsequent damage to various small vessels surrounding the lesion (Welch and Levine, 1991).

HIs are not usually associated with chronic hypertension (Hart and Easton, 1986). Frequently, however, they are secondary to the reestablishment of blood flow following occlusion. Because of this, when anticoagulants are employed to remove the clot, the necrotic vessels rupture and hemorrhage. Hence anticoagulants can increase the risk of secondary hemorrhagic infarction, particularly when used following large strokes and/or those accompanied by gross neurological disturbances (Cerebral Embolism Study Group, 1983; Hornig et al., 1986).

Aneurysm

Aneurysms (also called saccular or berry aneurysms) take the form of small, thin-walled blisters protruding from the various cerebral arteries (Brown et al., 1991). Aneurysms may be single or multiple, and are presumed to be due to developmental defects, e.g., a congenital weakness at the junction of two arteries. Often they are located at the bifurcations and branches of various arteries, particularly the internal carotid, the middle cerebral, or the junction of the anterior communicating and anterior cerebral arteries (Adams and Victor, 1994; Brown et al., 1991).

Symptoms secondary to aneurysm (due to rupture or compression) may occur at any age. Prior to rupture they are usually asymptomatic. However, as there is a tendency for them to enlarge over time, which in turn makes them more susceptible to rupture, with increasing age there is increasing risk. The peak incidence of rupture is between 40 and 55 (Adams and Victor, 1994).

Aneurysms may rupture because of sudden increases in blood pressure, while engaged in strenuous activity, during sexual intercourse, or while straining during a bowel movement (Adams and Victor, 1994). One patient I examined suffered a ruptured aneurysm when hyperventilating in his swimming pool so that he could remain submerged for a long time period.

Occasionally, if large and located near the base of the brain, aneurysms may compress the optic nerves, hypothalamus, or pituitary; and if within the cavernous sinus, compress the 3rd, 4th, 6th, or opthalmic division of the 5th nerve. Hence, a variety of visual, endocrine, and emotional alterations may herald the presence of an aneurysm prior to rupture (Brown et al., 1991).

With large aneurysms, when rupture occurs, blood under high pressure may be forced into the subarachnoid space, and the patient may be stricken with an excruciating generalized headache and/or almost immediately fall unconscious to the ground, or they may suffer a severe headache but remain relatively lucid (Adams and Victor, 1994; Brown et al., 1991; Jorensen et al., 1994; Schievink et al., 1995). If the hemorrhage is confined to the subarachnoid space there are few or no lateralizing signs and no warning symptoms. In some cases, however, patients may complain of headache, transitory unilateral weakness, numbness or tingling, or speech disturbance in the days and weeks preceding rupture—due to minor leakage of the aneurysm.

Often those who become unconscious following rupture develop decerebrate rigidity. This is usually due to compression effects (such as herniation) on the brainstem. Persistent deep coma is accompanied by irregular respiration, attacks of extensor rigidity, and finally respiratory arrest and circulatory collapse. In mild cases, consciousness, if lost, may be regained within minutes or hours. However, patients remain drowsy, and confused, and they complain of headache and neck stiffness for several days (Jorgenson et al., 1994). Unfortunately, in mild or severe cases there is a tendency for the hemorrhage to recur (Adams and Victor, 1994).

Cerebral Amyloid Angiopathy

Amyloid angiopathies are associated with the development of microaneurysms and the occlusion of arteries in the superficial layers of the cerebral cortex (Kase, 1986). Following amyloid occlusion, the arteries are often weakened, thus making them susceptible to rupture.

Amyloid angiopathies are often associated with recurrent hemorrhages over a period of months, which in turn may lead to the development of intracerebral hematomas. Sometimes a head trauma can trigger this latter form of hemorrhages.

Arteriovenous Malformations

AVMs consist of a tangle of dilated blood vessels and are sometimes referred to as angiomas. It is a developmental abnormality, and may become symptomatic at any age, but most commonly between the ages of 20 and 30.

Frequently AVMs form abnormal collateral channels between arteries and veins, thus bypassing the capillary system (Brown et al., 1991). When this occurs there may be an abnormal shunting of blood from the arteries to the veins. In consequence underlying brain tissue is not adequately irrigated and may become ischemic, depending on the size of the AVM.

AVMs vary in size and tend to be located in the posterior portion of the cerebral hemispheres, near the surface, as well as deep within the brainstem, thalamus, and basal ganglia. Frequently they are multiple and may be found in a variety of separate locations (Brown et al., 1991; Toole, 1984). They tend to be more common among males.

Like aneurysms, AVMs are present from birth and can grow larger and more complicated over time. It has been estimated that AVMs can increase in size by 2. 8% per year and can become 56% larger over the span of a 20-year time period (Mendelow et al., 1987). As they increase in size, the risk of them becoming symptomatic increases as there is a greater likelihood of collateral shunting.

AVMs are often a cause of intracerebral and subarachnoid hemorrhage (Brown et al., 1991; Drake, 1978). When hemorrhage occurs blood may enter the subarachnoid space, thus mimicking an aneurysm. However, although the first symptom is usually a hemorrhage, 30% of patients with this disorder may suffer a seizure, and 20% suffer headaches or focal neurological symptoms.

Small vascular malformations often become symptomatic during the 30s and 40s and occur more often among females (Kase, 1986). These often involve the subcortical white matter of the convexity.

Brain tumors can give rise to hemorrhage in a variety of ways, particularly if the tumor is malignant and is richly vascularized. That is, certain tumors have a tendency to become spontaneously necrotic. When this occurs, their supporting vasculature ruptures. However, frequently tumors are transmitted to the brain via the arterial system, whereas others, such as carcinomas, tend to invade the walls of blood vessels. In either instance, by adhering to or penetrating the walls of the blood vessels, tumors can make them more susceptible to rupture (see Chapter 20).

Drug-Induced Hemorrhages

Individuals with possible cerebrovascular abnormalities (such as aneurysm, AVM, or even tumor) and who abuse cocaine or amphetamines are at risk for suffering an intracranial HI (Golbe and Merkin, 1986; Lichtenfeld et al., 1984; Schwartz and Cohen, 1984; Wojak and Flamm, 1987). Presumably these drug-induced hemorrhages are due to transient increases in blood pressure and/or vasospasm, which in turn act to rupture abnormal vessels.

LOCALIZED HEMORRHAGIC SYMPTOMS

INTERNAL CAPSULE HEMORRHAGE

A patient who suffers an internal capsular hemorrhage is usually unconscious, pulse rate is slow, and there may be Cheyne-Stokes respiration. The head and eyes deviate to the side of the lesion (be-

cause of paralysis), and a divergent squint is common (Adams and Victor, 1994). The corneal reflex is often lost opposite to the lesion, and may be lost on both sides if coma is profound. Also there is paralysis of the contralateral side of the body and no response to pin prick on the paralyzed side. The limbs are extremely hypotonic such that when lifted by the physician they will fall inertly.

THALAMIC HEMORRHAGE

Thalamic hemorrhages also produce a hemiplegia or paresis due to compression of the adjacent internal capsule. Thalamic syndromes include severe sensory loss from both deep and cutaneous (contralateral) receptors, and transitory hemiparesis (Adams and Victor, 1994). Sensation may return to be replaced by pain and hyperatheia. Sensory deficits usually equal or outstrip the motor weakness, and an expressive aphasia may be present if the hemorrhage involves the left thalamus. The eyes may deviate downward, with palsies of vertical and lateral gaze, and inequality of pupils with absence of light reaction.

PONTINE HEMORRHAGE

Hemorrhage of the pontine brainstem is usually fatal. Deep coma ensues in a few minutes and there may be total paralysis, decerebrate rigidity, and pinpoint pupils that do not react to light. The head and eyes are turned toward the side of the hemorrhage if it is unilateral. Usually, however, even unilateral brainstem hemorrhages exert bilateral brainstem compression (Adams and Victor, 1994). Eyeballs are usually fixed.

CEREBELLAR HEMORRHAGE

Hemorrhage involving the cerebellum usually develops over a period of hours (Adams and Victor, 1994). However, there may be sudden onset with occipital headache, vomiting, inability to stand or walk, and loss of consciousness. There is a paresis of conjugate lateral gaze to the side of the hemorrhage and forced deviation contralaterally. Pupils are small and unequal but react to light. Also, there may be involuntary closure of one eye, as well as ocular bobbing.

FRONTAL, PARIETAL, TEMPORAL, AND OCCIPITAL HEMORRHAGE

These hemorrhagic conditions initially give rise to slow or rapidly developing focal syndromes, including headache (Jorgensen. et al., 1994). However, as the hemorrhage increases in size, syndromes becomes more global with impaired consciousness.

Hemorrhages and Strokes: Arterial Syndromes

INTERNAL CAROTID ARTERY SYNDROMES

Because the cerebral arteries arise from the internal carotid, hemorrhage or occlusion of this artery may be associated with extremely variable as well as widespread symptoms. As the artery becomes increasingly occluded an occasional patient may complain of hearing a disturbing noise (bruits)—a result of turbulence from stenosis of the carotid artery being relayed through the blood supply of the ear.

Stenosis of the internal carotid artery may cause massive infarction involving the anterior two-thirds of all of the cerebral hemisphere, including the basal ganglia, and can lead to death in a few days. Usually it produces a picture resembling middle cerebral artery occlusion. For example, if the left internal carotid is occluded the patient may become hemiplegic and globally aphasic. Because occlusion is accompanied by ischemia, the swelling of cerebral tissue may simulate an intracranial neoplasm. If there is massive edema, tentorial herniation and death may ensue.

Since this artery gives rise to the opthalmic artery, which nourishes the optic nerve and retina, carotid artery insufficiency may produce transient monocular blindness just prior to stroke onset. Unilateral blindness is the only feature distin-

guishing the carotid syndrome from that produced by obstruction of the middle cerebral artery.

Nevertheless, since the carotid also gives rise to the middle and anterior cerebral arteries the most distal parts of the vascular territories of these vessels will suffer as they are maximally subject to the influences of ischemia. That is, the most distal portion of any occluded artery is the most severely affected. These zones are also the most vulnerable to TIAs—giving rise to weakness or paresthesias of the arm, and if extensive, the face and tongue.

In some cases of internal carotid obstruction, the numerous branches of the external carotid (occipital, superficial temporal, and maxillary arteries) can serve as collateral blood supply channels. In these instances, although the arteries are occluded, symptoms are mild or nonexistent.

MIDDLE CEREBRAL ARTERY SYNDROMES

Through its branches the middle cerebral artery supplies the lateral part of the cerebral hemispheres, including the neocortex and white matter of the lateral/inferior frontal lobes (and motor areas), the superior and inferior parietal regions, and the superior portion of the temporal lobe (Carpenter, 1991; Parent, 1995). Its penetrating branches irrigate the putamen, caudate, globus pallidus, posterior limb of internal capsule, and corona radiata. Whether caused by trauma, embolus, or atherosclerosis, contralateral hemiplegia is the hallmark of infarction in the territory supplied by this artery (Absher and Toole, 1995; Adams and Victor, 1994).

The classic picture of total occlusion is contralateral hemiplegia, with hemiparesis involving the face and arm more than the leg. Sensory deficits may be severe or mild, with disorders of pain perception, touch, vibration, and position (Absher and Toole, 1995; Adams and Victor, 1994). This includes extinction of pin-prick sensation or touch, deficits in two-point discrimination, astereognosis, and perhaps dense sensory loss if the parietal lobe is involved. Visual

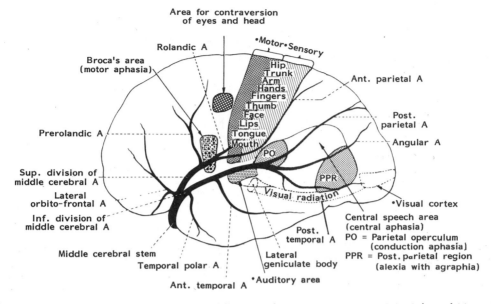

Figure 20.10. The distribution of the middle cerebral artery and its branches along the lateral surface of the left cerebral hemisphere. (From Adams RD, Victor M. Principles of Neurology. New York: McGraw-Hill, 1994.)

field defects consist of a homonymous hemianopsia or inferior quadrantanopia.

If the left hemisphere is involved, global aphasia is present. However, if the right hemisphere is affected, there may occur neglect, denial, and indifference, confabulation, and manic-like states, as well as disturbances involving visual-perceptual functioning and the ability to perceive and/or express musical and emotional nonverbal nuances.

It is important to note that only a branch of this artery may be obstructed. If the posterior branch of the middle cerebral artery is occluded, the parietal and inferior parietal lobes may be affected. If the left anterior branch is occluded, the patient may develop expressive aphasia.

ANTERIOR CEREBRAL ARTERY SYNDROMES

Through its branches the anterior cerebral artery supplies the anterior three-fourths of the medial surface of the hemispheres, including the medial-orbital frontal lobe, the anterior four-fifths of corpus collasum, and the head of the caudate nucleus and putamen (Carpenter, 1991; Parent,

1995). Hemorrhage is most often due to rupture of an aneurysm. With total occlusion or hemorrhage patients become obtunded and mute; they develop grasp, sucking, and snout reflexes, gait apraxia, gegenhalten, waxy flexibility, and catatonic-like postures; and they are incontinent (Beckson and Cummings, 1991; Starkstein and Robinson, 1992). (For a further discussion of orbital and medial symptoms, see Chapter 11).

VERTEBRAL-BASILAR ARTERY SYNDROMES

Because there may be collateral circulation from the vertebral-basilar system, obstruction of this artery on one side may not result in symptoms. However, if later there occurs obstruction of the opposite vertebral artery, consequences will be disastrous. Headache occurs in over 50% of those with ischemia, beginning as a pounding or throbbing behind the orbit or in the temporal region (Adams and Victor, 1994).

A wide variety of visual and vestibular symptoms also occurs when the vertebral-basilar system is compromised including diplopia, transient blindness or blurring, and visual hallucinations and illusions. Pa-

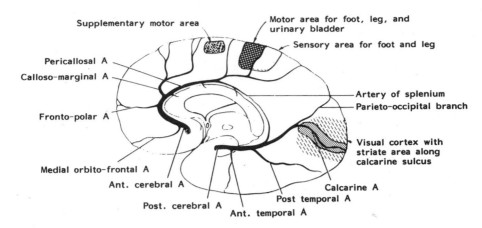

Figure 20.11. The distribution of the anterior cerebral artery along the medial aspect of the right hemisphere. (From Adams RD, Victor M. Principles of Neurology. New York: McGraw-Hill, 1994.)

tients will complain of severe pounding or throbbing headaches (usually localized behind the orbit of the eye or in the temporal region) and will experience episodes of dizziness and vertigo, with or without hearing loss. This is because the auditory and vestibular portions of the ear as well as the brainstem vestibular nuclei are supplied by the vertebral-basilar arterial system. In addition, other brainstem signs include numbness, diplopia, impaired vision in one or both fields, dysarthria, hiccups, and difficulty swallowing (Adams and Victor, 1994).

LATERAL MEDULLARY SYNDROME

The vertebral arteries are the chief arteries of the medulla and supply the lower three-fourths of the pyramids, the medial lemniscus, restiform body, and posterior-inferior cerebellum (Carpenter, 1991; Parent, 1995). However, rarely does an infarct involve the pyramids or medial lemniscus. Rather, two prominent sites of lesion are the medial and in particular the lateral medulla.

The classic lateral medullary syndrome is due to an infarction of a wedge-shaped area of the lateral medulla and inferior surface of the cerebellum. Onset is associated with severe vertigo, and vomiting may occur (Adams and Victor, 1994). Symptoms typically include contralateral impairment of pain and thermal sense; ipsilateral Horner's syndrome (miosis, ptosis, decreased sweating); ipsilateral paralysis of the soft palate, pharynx, and vocal cord (due to involvement of nucleus ambiguous); 9th and 10th nerve dysfunction (hoarseness, dysphagia, ipsilateral paralysis of the palate and vocal cords); loss of balance such that the patient falls to the side ipsilateral to the lesion; loss of taste sensation; hiccups; nystagmus; and nausea. There may also be dysphagia and pain or paresthesia—a sensation of hot water running over the face. There is some degree of cerebellar deficiency, with nystagmus, hypotonia, and incoordination on the side of the lesion (Adams and Victor, 1994).

MEDIAL MEDULLARY SYNDROME

This condition is a less common consequence of vertebral artery occlusion. Nevertheless, this causes contralateral hemiparesis, ipsilateral paralysis of the tongue due to 12th nerve involvement near the zone of infarction, and loss of position and vibratory perception with sparing of pain and temperature sensation. Vertical nystagmus implies a lesion at the pontomedullary junction, and a paralysis of gaze suggests a lesion above the medulla (in the pons or midbrain).

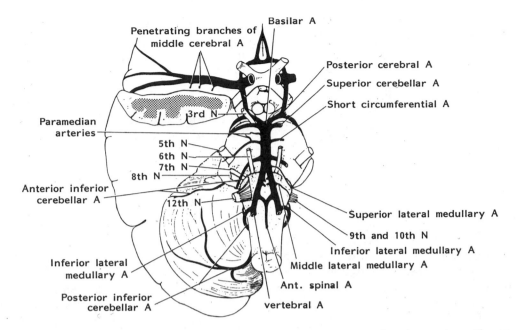

Figure 20.12.   The principle blood vessels of the brainstem vertebral-basilar system. (From Adams RD, Victor M. Principles of Neurology. New York: McGraw-Hill, 1994.)

VERTEBRAL ARTERY TRAUMA

The vertebral arteries are vulnerable to compressions or extremes in extension such as can occur during whiplash (Toole, 1984). Moreover, if the patient suffers from osteoarthritis of the cervical spine, the space through which the vertebrals pass may become narrowed, thus subjecting these arteries to pinching or other stresses with movement of the head. Hence, these patients may be subject to repeated instances of vertebral insufficiency.

BASILAR ARTERY SYNDROMES

The basilar artery supplies the pons (Carpenter, 1991; Parent, 1995). Infarcts characteristically involve the corticospinal and corticobulbar tracts, cerebellum, cerebellar peduncles, medial/lateral lemniscus, spinothalamic tracts, and 3rd through 8th nerves (Adams and Victor, 1994). Hence, the outstanding features of basilar artery infarct are a constellation of cranial nerve signs of both the sensory and motor varieties; cerebellar symptoms; bilateral pyramidal tract signs, including ataxia, dysarthria, and hemiplegia; and distur-

bances involving ocular movement and paralysis of gaze. If convergence is preserved the lesion is in the middle of the inferior pons. If there is no convergence the lesion is in the superior pons (medial longitudinal fasciculus).

Often patients become comatose because of ischemic compression effects on the reticular formation. Early signs of increasing occlusion or hemorrhage may occur in different combinations: somnolence, visual hallucinations, disorders of ocular movement, delirium, and Korsakoff's amnesic defects (Adams and Victor, 1994).

Medial pontine lesions result in ataxia of the ipsilateral limbs, a contralateral paresis of the face, arm, leg, and variable sensory loss. If Horner's syndrome is present or if cranial nerve 5, 7, or 8 is affected, the lesion is laterally placed. If paralysis of cranial nerve 3, 4, 6, or 7, or involvement of the pyramidal tract occurs, the lesion is probably medially placed.

In general, infarct of the medial pons results in (variably) vertigo, nausea, vomiting, nystagmus, ipsilateral ataxia, ipsilateral Horner's syndrome, paresis of conjugate gaze, contralateral loss of pain and temperature sensation over the face, arm, trunk and leg, and slurred speech.

LOCKED-IN SYNDROME

If occlusion spares the upper brainstem but involves the midpontine level, the "locked-in" syndrome may occur: the patient is alert, conscious of his surroundings, able to see and hear, but completely paralyzed and unable to communicate except through eye blinks. However, because the respiratory, vasomotor, and thermoregulatory centers and pathways are affected, there might be associated abnormalities. Unfortunately, because patients can only interact via eye blinks it is extremely difficult to ascertain how impaired the patient actually is.

POSTERIOR CEREBRAL ARTERY SYNDROMES

The posterior cerebral artery is a branch of the vertebral-basilar system and supplies via its own branches the inferomedial temporal and medial occipital regions, including areas 17, 18, and 19 and the posterior hippocampus (Carpenter, 1991; Parent, 1995). Via deep penetrating branches it also supplies the thalamus, subthalamic nuclei, substantia nigra, midbrain, and pineal body.

Thalamic infarct or hemorrhage includes severe sensory loss and possible receptive aphasia. However, sensation may return to be replaced by pain and hyperathesia. Midbrain infarcts include Weber syndrome (oculomotor palsy with contralateral hemiplegia) paralysis of vertical gaze, and stupor or coma (Adams and Victor, 1994).

Cortical syndromes include anomia, alexia apraxia, prosopagnosia (depending on which hemisphere is involved), and related temporal-occipital and parietal-occipital disturbances. Involvement of the optic radiations or infarction of the calcarine cortex causes visual field impairments, such as scotoma and homonymous hemianopias, particularly of the upper quadrants. With bilateral occlusion the patient will become cortically blind.

Ischemic lesions in the occipital area may cause variations in the nature and form of visual field defects experienced by the patient from day to day—occasionally leading to an erroneous diagnosis of hysteria. TIAs in this vicinity are also reflected by fleeting visual field defects of a hemianopic distribution.

Distal occlusion also causes medial temporal infarction, with hippocampal involvement and memory loss. Transient global amnesia is also sometimes a consequence of transient occlusion of the posterior cerebral arteries.

CEREBELLAR ARTERY SYNDROME

The cerebellar artery is also a branch of the vertebral-basilar system, and it supplies portions of the pons as well as the cerebellum (Carpenter, 1991; Parent, 1995). Deafness and tinnitus may occur with lateral inferior pontine lesions due to occlusion of the anterior-inferior cerebellar artery. Infraction in this area may also cause vertigo, nystagmus, ipsilateral ataxia of limbs, and a contralateral hemiparesis with no sensory defect. With mild lesions the patient may walk with an unsteady, wide-based gait. With more severe lesions patients may have extreme difficulty walking or even looking to the side of the lesion, although the pupils are normal and reactive to light.

DIFFERENTIAL DIAGNOSIS: CAROTID VERSUS VERTEBRAL SYSTEM

The development of deficits such as aphasia, agnosia, apraxia, constructional and manipulospatial deficits, emotional abnormalities, and/or delusions indicates a carotid circulatory disturbance. Dizziness, diplopia, ataxia, nystagmus, cranial nerve signs, internuclear ophthalmoplegia, dissociated sensory loss, and/or bilateral abnormalities are hallmarks of a brainstem lesion within the vertebral-basilar territory.

RECOVERY AND MORTALITY

Mortality rates for cerebrovascular disease have declined in the United States since the 1950s (Gillum, 1986; Meier and Strauman, 1991). Nevertheless, the initial death rate for individuals in the acute phase and up to 30 days after stroke is about 38%. However, 50% of those who survive this phase die over the course of the next 7 years (Dombovy et al., 1986).

The major determinants for short-term mortality are intraventicular hemorrhage, pulmonary edema, impaired consciousness, leg weakness, and increasing age (Chambers et al., 1987; Roos et al., 1995; Schievink et al., 1995), with level of consciousness following stroke being the single most important predictor of short-term survival (Chambers et al., 1987). The major determinants for long-term mortality are low activity level, advanced age, male sex, heart disease, and hypertension. However, those who suffer intraventricular HIs have a higher mortality rate than those with infarcts due to other causes (Chambers. et al., 1987; Roos et al., 1995; Schievink et al., 1995). As noted in Chapter 3, those with right hemisphere damage tend to have poorer outcomes as well as higher mortality rates. In particular right parietal lobule infarcts are associated with very poor outcomes (Valdimarsson et al., 1982).

Hyperglycemia and diabetes are also associated with poor neurological recovery, higher short-term mortality, and increasing risk for stroke in general. This is because diabetes and hyperglycemia both accentuate ischemic damage (Pulsinelli et al., 1983; Woo et al., 1988). Hyperglycemia also appears to have a negative effect on energy metabolism due to the generation of severe lactic acidosis (Rehncrona et al., 1980)—factors that act to retard neuronal recovery.

Some authors have argued that luxury perfusion and increased cerebral blood flow (CBF) within an infarcted cite are often indicative of a good prognosis, whereas low CBF is a bad prognosis (Olsen et al., 1981). Presumably increased flow acts to nourish damaged tissue. Other studies however, indicate that initial CBF levels are not predictive of clinical outcome (Burke et al., 1986). Apparently this is because once damage occurs during the initial period of ischemia, these cells cannot be salvaged (Heiss and Rosner, 1983). Hence, blood flow increases only when undamaged neurons return to a functionally active state (Burke et al., 1986) rather than acting to rejuvenate injured tissue. In fact, hyperperfusion may endanger neuronal recovery (Mies et al., 1983). On the other hand oxygen metabolism seems to correlate better with clinical status and functional recovery than does blood flow (Wise et al., 1983).

Recovery is often greatest during the first 30 days after stroke (Dombovy et al., 1986; Lind, 1982), but continues up to 6 months in some patients (Wade and Hewer, 1987). It has been estimated that although about 60% of stroke patients are able to achieve total independence in activities of daily living (Meier and Strauman, 1991; Wade and Hewer, 1987), only approximately 10–30% of initial survivors return to their jobs without gross or obvious disability (Beckson and Cummings, 1991). Depending on the nature of the stroke, about 40% demonstrate mild disability, 40% are severely disabled, and 10% require institutionalization (Stallones et al, 1972).

SUMMARY

Stroke is the third most common cause of death (after heart disease and cancer) in the United States and Europe. Thrombosis and embolism account for approximately 75% of all strokes, whereas about 20% are due to hemorrhage. Up to 70% of all major stroke victims are usually permanently and significantly disabled (Beckson and Cummings, 1991), with subarachnoid hemorrhages often resulting in sudden death (Schievink et al., 1995). Of those who survive, the 5-year accumulative risk of repeated stroke is about 42% in men and 24% in women (Baxter, 1984).

21

Cerebral Neoplasms

In the United states malignant brain tumors are the second most common cause of cancer-related death in individuals up to age 34, and the third most common cause in males up to age 54 (American Brain Tumor Association, 1992).

Brain tumors have a number of originating etiologies and may arise following head injury, infection, metabolic and other systematic diseases, and exposure to toxins and radiation. Some are due to tumors that developed in other parts of the body and have metastasized to the brain. Yet others are believed to have originated in embryonic cells left in the brain during development (Burger and Scheithauer, 1994). In some cases tumors are a consequence of embryological timing and migration errors. That is, if germ layers differentiate too rapidly or if cells migrate to the wrong location, they may exert neoplastic influences within their unnatural environment (see Chapter 18).

Indeed, among children, many tumors are congenital, developing from displaced embryonic cells, dysplasia of developing structures, due to the altered development of primitive cells that normally act as precursors to neurons and glia. Most of these "congenital" tumors tend to occur within the brainstem, cerebellum, and midline structures, including the third ventricle and cervical-medullary junction (Burger and Scheithauer, 1994; Robertson et al., 1994). Among older individuals, most tumors are due to dedifferentiation of adult elements.

TUMOR DEVELOPMENT: ONCOGENES AND DEFECTIVE DNA EXCISION REPAIR

Tumors are sometimes believed to be due to a loss of restraint in regard to cellular growth and differentiation such that a ma-

lignant progression ensues (Kim et al., 1994; Modrich, 1994; Sancar, 1994). In large part this is believed to be due to the loss or inactivation of tumor suppressor genes, and/or the abnormal presence of an enzyme referred to as telomerase (see below), and/or the loss of regulatory proteins (produced by oncogenes) that normally control cellular growth (Gilbert, 1983; Shapiro, 1986a, b; see Marx, 1994 for a recent review).

For example, the aberrant expression (or inhibition) of oncogenes is associated with cancer-causing abnormalities in chromosomal structure. If these oncogenes become abnormal or are inappropriately expressed, they in turn induce malignant cellular formation.

Presumably there are three groups of chromosomal abnormalities associated with altered oncogene expression, deletions, duplications, and abnormal reciprocal translocations. For example, depletions are associated with the removal of the constraints normally exerted on oncogenes, thus promoting malignant expression through the rapid and abnormal proliferation of cells as well as the possible loss of chromosomes during cell division (Rowley, 1983; Shapiro, 1986a, b; Unis, 1983). Deletions are associated with the development of neuroblastomas in children (Gilbert et al., 1984).

Duplications are associated with the transposition, amplification, and hyperactivation of oncogenes (Alitalo et al., 1983), such that numerical abnormalities in chromosomal expression result in addition to segregational errors involving the gain of chromosomes during cell division (Shapiro, 1986a, b). Again, this is presumed to be secondary to a loss of regulatory restraint. Hence, if a malignant recessive gene is present, its duplication may result in a malignant overrepresentation and thus the ex- pression of recessive malignant traits. The ability to suppress its expression has been overridden, and a malignant hybrid cell is developed (Modrich, 1994; Shapiro, 1986a, b).

As per oncogenes and translocation errors, it is believed that when two chromosomes abnormally exchange segments, there is a shift in the oncogene's position that results in its activation as well as the creation of abnormal fusion proteins. This again results in a series of replication and recombination errors that, under normal conditions, are corrected (Modrich, 1994; Sancar, 1994).

These translocation errors in turn produce base-pairing abnormalities within the DNA helix that violate the "Watson-Crick" pairing rules by creating unpaired bases within the helix. Normally, the "cellular mismatch repair system" recognizes these mismatches and eliminates them from the newly synthesized strands of DNA (Modrich, 1994). That is, they normally act to ensure genetic stability by removing the damaged nucleotide by replacing it with a normal base by inserting a complementary strand as a template, i.e., the duplex normally contains redundant information that can be reinserted when errors occur (Sancar, 1994). However, if these errors are not corrected, the cells mutate and begin to accrue additional mutations at rates hundreds of times that of normal cells (Modrich, 1994), thus resulting in uncontrolled proliferation.

TELOMERASE, TELOMERES, "IMMORTAL CELLS," AND TUMOR GROWTH

It has also been recently determined that a specific enzyme, telomerase, may be a crucial factor in unchecked tumor growth and proliferation (Kim et al., 1994). For example, mammalian cells (but not microorganisms) typically have a finite lifetime, and are able to replicate about 60 times before simply dying. Cell death, in part is believed to be due to the progressive loss of their DNA, i.e., the erosion of the chromosomal structure, the telomere.

Specifically, telomeres appear at the ends of chromosomes and consist of thousands of repetitions of the sequence TTAGGG and related proteins (reviewed in Kim et al., 1994). Telomeres presumably aid in the positioning, replication, and protection of chromosomes.

However, normally chromosomes will lose between 50 and 200 nucleotides per cell division of the telomeric sequence, which results in the net loss and shortening of the telomeres. Since the telomere supposedly acts as a "mitotic clock by which cells count their divisions" (see Kim et al., 1994), the net loss of these DNA fragments may eventually signal that replication has come to an end, which results in cell death.

Conversely it is believed that "immortal cells" and certain microorganisms continue to replicate because of the presence of a ribonucleoprotein "telomerase" that acts to synthesize and possibly replace these DNA fragments, the telomere. Because the telomere is resynthesized, the cell (and its DNA) can live on indefinitely. Similarly, human tumor cells have been recently discovered to also contain telomerase (Kim et al., 1994). Hence, like other "immortal cells" they too may grow and proliferate, presumably indefinitely, thus giving rise to abnormal cellular development and growth that in turn becomes cancerous. However, once a primary tumor begins to develop, there are a series of secondary events that contribute to their growth, invasive properties, and metastasis.

THE ESTABLISHMENT AND GROWTH OF A TUMOR
Vascularization and Necrosis

For a tumor to grow it must become vascularized. Indeed, it has been suspected that tumors may in fact secrete a substance that potentiates capillary development (Sherbet, 1982). Hence, the initial stages of tumor growth are sometimes associated with increased blood flow and local hyperemia.

Nevertheless, established tumors have a blood flow that is lower and more variable than surrounding normal brain tissue (Blasberg et al., 1981; Brooks et al., 1986). In fact, tumors have a lower oxygen utilization as well (Ito et al., 1982). Hence, tumors, or at least their central-most portions, often exist in a state of hypoxia, although the oxygen they receive seems adequate for their peripheral proliferation and metabolism. In consequence many tumors, particularly those which are rapidly growing, contain central regions of necrosis because of reduced reception of oxygen. As the tumor volume increases the surface area becomes increasing inadequate in regard to blood perfusion and the diffusion of nutrients and metabolic products, such that as the periphery proliferates, cells in the center become increasingly deprived of nutrients and undergo death and necrosis (Sherbet, 1982). Subsequently, underlying normal tissue is also destroyed.

Invasion and Dissemination

As tumors grow they increasingly encompass, surround, and destroy neighboring tissue. Invasion also sometimes occurs by the locomotive behavior of individual tumor cells that develop microfilaments. These microfilaments extend into the cytoplasm of surrounding tissue and then infiltrate them, dragging the tumor behind them.

When a tumor becomes vascularized its potential to disseminate is increased because of tumor material escaping into the blood stream. Capillaries offer little resistance to invading tumors, which destroy the endothelium of the vessel walls. Hence, these vessels may also become hemorrhagic.

Once tumor cells enter the blood stream they may be carried short or long distances, where they lodge and give rise to secondary growths. That is, like a thrombus they will attach to the vessel wall. Moreover, imperfections in the endothelial wall increases the likelihood of adhesion. However, unlike a thrombus or embolus, these tumors cells are able to penetrate the vessel walls so as to form attachments to new host material (Sherbet, 1982). Interestingly, anticoagulants, including aspirin, have been found to arrest or reduce the development of certain blood vessel borne tumor cell colonies (Sherbet, 1982). Tumor cells are also transported via the sympathetic system.

Metastasis

It has been estimated that anywhere from 10% to 33% of all cancers located in other body regions will metastasize to the brain (Adams and Victor, 1994; Bloom, 1979; Hirsh et al., 1979; Sherbet, 1982; Williams, 1979). These metastases may be single or multiple. In the majority of the cases the patient has lung cancer, whereas in other instances breast cancer is the culprit (Bloom, 1979; Hirsh et al., 1979; Williams, 1979). These metastases are usually passed via the blood or sympathetic system. However, surgery for tumor often results in the development of cerebral metastases, presumably due to the dislodgement of tumor fragments, which are then carried elsewhere.

Intracranial metastases (or carcinomas) are of three types: those which attach to the skull, those which metastasize to the meninges, and those which infiltrate the brain. Most form in the cerebral hemispheres (60%) and the cerebellum (30%) and seem to develop preferentially along the distribution of the middle cerebral artery (Bloom, 1979; Burger and Scheithauer, 1994). In more than 70% of reported cases the metastases are multiple and scattered throughout the brain, usually near the surface in the gray and subcortical white matter.

Malignancy

Malignancy refers the biological ability of a group of cells to become freed of hemostatic controls. When this occurs they progressively divide, disseminate, invade, and form distant metastases so as to eventually kill the host. However, in addition to its infiltrative and proliferative potential, a tumor malignancy is dependent on its anatomical location, the degree of elevated intracranial pressure it induces, the patient's age and neurological status, and its responsiveness to therapy (Salcman and Kaplan, 1986). In contrast, so-called "benign" tumors are slow-growing, generally encapsulated, and localized, and they have little metastatic potential.

Realistically, however, all tumors have the potential to kill the patient and thus all are potentially malignant (Salcman and Kaplan, 1986; Sherbet, 1982). For example, a so-called benign tumor that is located in the brainstem, or a low-grade astrocytoma located near vital centers, may quickly kill the patient. Moreover, a tumor that is recurrent and ineradicable, albeit "benign," is just as fatal as a "malignant" tumor because of the effects of pressure within limited cranial-cerebral space, even though they are not considered malignant, rarely metastasize, and may in fact be relatively small. Hence, the grading of a tumor or the determination of its malignant versus benign state must be done cautiously (Salcman and Kaplan, 1986).

Tumor Grading

Many types of tumors are graded in regard to their cellular architecture, their mitotic activity, and in particular their degree of differentiation (Burger and Scheithauer, 1994). Tumor grading reflects the degree of malignancy. Those which are least well differentiated are the most malignant. For example, a tumor that is three-fourths differentiated is considered a grade 1. If half the tumor is differentiated it is a grade 2. If only a quarter of the tumor contains differentiated cells, it is a grade 3. Grade 4 tumors are those where very little or no differentiation is seen. Some authors, however, have collapsed this point system to three grades.

Grade 1 tumors generally do not metastasize, whereas grade 3 and 4 tumors have the greatest degree of potential for forming metastases. Moreover, patients with grade 3 and 4 tumors have the highest and fastest mortality rate. For example, in one study, patients with astrocytomas graded 1–4 had average postoperative survival times of 74, 24, 12, and 7 months, respectively. (Sherbet, 1982).

Some tumors, however, may initially appear to be benign and receive a low grade, whereas as they grow they become progressively less differentiated and more malignant, and they increase their metastatic potential (Sherbet, 1982). For example, the daughter cells of transformed astrocytes, oligodendrocytes, microgliacytes, or ependymocytes tend to become variably dedifferentiated and thus more malignant as the tumor grows and cells divide and proliferate. Hence, what was a benign grade 1 astrocytoma is now a grade 4 glioblastoma multiforme.

AGE

There is a bimodal age distribution regarding the development and incidence of brain tumors as they tend to occur in childhood and then later with increasing risk during the 5th, 6th, and 7th decades of life, until a decline occurs during the 8th decade (Salcman and Kaplan, 1986).

Gliomas, pituitary adenomas, and meningiomas are the most frequently occurring types of brain tumor across all age groups (Salcman and Kaplan, 1986). However, among children (and individuals below age 20), ependymomas of the 4th ventricle, pinealomas, cerebellar astrocytomas, and medulloblastomas are the most common (Burger and Scheithauer, 1994). Meningiomas and gliomas are most frequent around 50 years.

Extrinsic and Intrinsic Tumors

Broadly, most cerebral tumors can be considered as either intracerebral (intrinsic) or extracerebral (extrinsic). Extrinsic tumors usually arise from the coverings of the brain and cranial nerves and, although benign, can exert compression on underlying structures. These include meningiomas, acoustic schwannomas, and pituitary adenomas.

Intrinsic tumors are often secondary to the dedifferentiation of glia, ependymal cells, and astrocytes. These include ependymomas, various grades of astrocytoma, and oligodendrogliomas, all of which (particularly gliomas) arise from adult elements. These tumors have the capability of becoming increasingly less well differentiated and more malignant. Frequently they involve the lateral portions of the hemispheres or the ventricles.

NEOPLASMS AND SYMPTOMS ASSOCIATED WITH TUMOR FORMATION

Initially, tumor development is associated with localized cerebral displacement, compression, edema, and regional swelling, particularly in the white matter. Because of this, although there are a variety of tumors (discussed below), the symptoms produced are more dependent on the tumor's rate and location of development rather than the type of tumor per se. Their influences are site-specific such that initial localized effects are the most prominent. Hence, the commonest modes of onset include progressive focal symptoms, e.g., focal epilepsy, monoplegia, hemiplegia, aphasia, cerebellar deficiency, and symptoms of increased intracranial pressure. Most patients also suffer from periodic bifrontal and bioccipital headaches that awaken the patient at night or are present upon waking. Vomiting, mental torpor, unsteady gait, sphincter incontinence, and papilledema are also frequent sequelae (Adams and Victor, 1994).

Nevertheless, as intracranial pressure increases, disturbances of consciousness, cognition, and neurological functioning become progressively more widespread. However, generalized symptoms occur late or not at all.

Even so, functional disruptions associated with tumor formation are usually slow to appear and develop, taking months or years. The sudden appearance of symptoms is generally indicative of vascular disease. However, in some cases a silent slow-growing tumor may hemorrhage, in which case the symptoms are of sudden onset.

Fast- Versus Slow-Growing Tumors

A tumor may appear, develop, and in fact grow for long time periods with hardly any symptoms. This is particularly true of slow-growing tumors that give the brain time to adjust. Rapidly growing tumors exert more drastic effects. In this regard, a tumor that takes 3 years to reach a size of 5 cm may exert little disruptive influences, whereas a rapidly growing tumor that reaches a size of 2 cm may exert profound consequences, even if localized to the same region. However, tumor location is also important, for even tiny tumors of the brainstem may cause death.

GENERALIZED NEOPLASTIC SYMPTOMS

INCREASED INTRACRANIAL PRESSURE

Because of limitations in cranial-cerebral space, when tumors (or any mass lesion) form, intracranial pressure is likely to rise. As intracranial pressure increases disturbances of consciousness, cognition, and neurological functioning become increasingly diffuse and progressively more widespread. This is because the skull is rigid and the development and enlargement of any type of neoplasm or space-occupying mass (such as a hematoma or hemorrhage) occurs at the expense of the brain, which becomes increasingly compressed. Hence, generalized symptoms begin to replace or at least override localized deficits as the tumor enlarges and spreads.

Moreover, a tumor growing in one part of the brain will not only displace underlying brain tissue, but can compress local blood vessels as well as decrease the amount of cerebrospinal fluid (CSF) in the ventricles and subarachnoid space (Burger and Scheithauer, 1994).

With increased displacement and edema, eventually CSF pressure begins to rise. Compression effects are thus exerted on midline structures as well as on contralateral brain tissue. If the brain becomes sufficiently swollen and displaced, herniation may result.

COGNITIVE ABNORMALITIES

As the tumor enlarges and intracranial pressure rises, often there is a slight bewilderment, slowness in comprehension, or loss of capacity to sustain continuous mental activity, although specific, localized signs of disease are diminished or lacking (Burger and Scheithauer, 1994). There may be a persistent lack of application to the tasks of the day, a slowing of thought processes and reaction time, undue irritability, emotional lability, inertia, faulty insight, forgetfulness, reduced range of mental activity, indifference to common social practices, etc. Also, there may be inordinate drowsiness, apathy, or stoicism that after a few weeks becomes more prominent (Adams and Victor, 1994).

With increased tumor development, when questioned a long pause precedes each reply and the patient may not even bother to reply, as if he or she did not hear the question. However, responses are more intelligent than one would expect. Much of the drowsiness, etc. and general restriction of the mental horizon is related to increased intracranial pressure and is unrelated to the site of the lesion.

HEADACHE

The most frequent initial nonlocalizing complaints include headaches, seizures, personality change, weakness, and confusion. Tumor headaches are an early symptom and are reported by approximately a third of all patients. Although not usually well localized they may be bifrontal or located along the vertex of the cranium. Frequently they are moderate to severe in intensity with a deep nonpulsatile quality, and are usually quite variable, occurring throughout the day as well as nocturnally (Adams and Victor, 1994). However, they tend to be worse in the morning when the patient first awakens, and may improve if the patient vomits. Tumor headaches are often due to local swelling of tissues and to distortion of blood vessels in and around the tumor.

Vomiting occurs in about one-third of all tumor patients and accompanies the headache. Vomiting is not related to the ingestion of food but is sometimes due to increased intracranial pressure or brainstem compression, and is found with tumors of the posterior fossa or with low brainstem gliomas.

SEIZURES

Generalized epileptiform convulsions are a common symptom of tumor and may precede other disturbances by many years. It is the first symptom in 30–50% of all patients, particularly those suffering from astrocytoma, meningioma, or glioblastoma. Indeed, the occurrence of a seizure for the first time during adult years and the existence of a localizing aura almost always suggest tumor. Glioblastoma, multiforme, oligodendroglioma, ependyma, metastatic carcinoma, and primary reticulum cell sarcoma are all associated with the development of seizures. Similarly, tumors of the frontal and temporal lobes frequently trigger seizure activity.

FOCAL SYNDROMES AND NEOPLASMS

FRONTAL LOBE TUMORS

Depending on which quadrant of the frontal lobes is involved symptoms associated with neoplastic growth can be quite variable. Regardless of location patients tend to complain of headache even during the early stages of tumor formation. In the later stages papilledema and frequent vomiting are characteristic—features that are usually due to increased intracranial pressure.

If the tumor is within the midline region, apathy, mutism, depressive and catatonic-like features with posturing, grasp reflexes, and gegenhalten may be prominent. Some patients are initially misdiagnosed as depressed.

Downward pressure on the olfactory nerve may lead to anosmia on the side of the lesion. Pressure on the corticospinal fibers may lead to contralateral weakness, most marked in face and tongue. If the tumor extends to the medial motor areas the lower extremities may become paralyzed.

Depressive-like symptoms, disturbances of speech output, and some degree of motor impairment are also characteristic of left frontal tumors. In contrast, right frontal tumors may initially give rise to manic and delusional symptoms (see Chapter 11).

In general, however, a progressive dementia becomes increasingly apparent regardless of which quadrant has been most compromised. Fifty percent of all patients with frontal tumors are likely to develop seizures.

Tumors of the motor areas are easy to localize because of focal convulsions in select body regions. Depending on the extensiveness of the tumor, weakness or paralysis may variably involve the fingers, hand, arm, tongue, face, and trunk. Motor weakness is the result of destruction of corticospinal tract fibers (Burger and Scheithauer, 1994).

TEMPORAL LOBE TUMORS

Tumors of the temporal lobe can cause receptive and fluent aphasia (if the left temporal region is involved), receptive amusia and agnosia for environmental and emotional sounds (if the right lobe is compromised), and visual field defects, i.e., a crossed upper quadrantic hemianopia with loss greatest in the ipsilateral field. These tumors also cause auditory hallucinations and tinnitus (see Chapter 7).

If the inferior regions are involved, there is an increased likelihood of seizures (in over 50% of all cases), emotional alterations, and disruptions of memory. Like frontal lobe tumors, personality changes may be the most prominent symptom.

PARIETAL LOBE TUMORS

With tumors involving the parietal lobes, sensory disturbances are most prominent. These tumors also may cause sensory Jacksonian fits, which consist of paresthesia, tingling, pain, or sensations of electric shock that correspond to the focus of damage. That is, a patient may feel spasms of shock-like sensations along his or her leg. Abnormalities involving the spatial and discriminative aspects of sensation may be prominent (e.g., astereognosis), especially of postural sensibility and tactile discrimination, while crude appreciation of pain, heat, and cold is intact. Parietal postcentral tumors are also associated with hypotonia and wasting of the various body parts (parietal wasting).

OCCIPITAL TUMORS

Tumors of the occipital lobe are not as common as frontal and temporal neoplasms or parietal tumors. However, visual abnormalities, including blindness if the tumor is situated medially, are common sequelae.

BRAINSTEM TUMORS

In 90% of all reported cases the initial manifestation of a brainstem tumor is a palsy of one or more cranial nerves, most often the 6th and 7th such that conjugate gaze, gaze deviation, and facial sensation are abnormal. In addition, patients develop hemiparesis, unilateral ataxia of gait, and paraparesis. If the tumor increasingly involves the medulla, patients may develop uncontrollable hiccups and disturbances of cardiac and respiratory rate (Adams and Victor, 1994). Death is likely as even very small tumors exert drastic effects (Packer et al., 1991). However, tumors in children arising in the brainstem-cervical junction are much more amenable to treatment and demonstrate good-quality, longer-term survival (Robertson et al., 1994).

CEREBELLAR TUMORS

The cerebellum is a common site of tumor, especially in childhood (Burger and Scheithauer, 1994). Symptoms are most marked on standing and walking, such that patients tend to fall backward or sometimes forward. Gait is ataxic, and nystagmus may be present, if the tumor is laterally placed, or absent if localized to the midline. Patients tend to walk with a wide-based gait and have much difficulty with balance. Many patients will list to the side of the tumor. Cerebellar tumor growth, via compression, also tends to affect cranial nerves 5–12.

Tumors of the Ventricles

EPENDYMOMAS

Ependymomas originate from the dedifferentiation of adult ependymal cells that line the ventricular walls (Burger and Scheithauer, 1994). The 4th ventricle is involved about 70% of the time. These tumors usually develop in young adults and especially children such that about 40% occur during the first year of life and 75% during the first decade.

Presumably these tumors are benign. However, as they can occlude the ventricles, CSF flow may be impeded and hydrocephalus and increased intracranial pressure may result. With increased pressure there is a danger of herniation and brainstem compression accompanied by paroxysmal headaches, vomiting, difficulty swallowing, paresthesias of the extremities, abdominal pain, and vertigo. Seizures occur in one-third of cases.

Among infants and young children, initial symptoms include headaches, lethargy, stupor, spastic weakness of the legs, and unsteadiness of gait. In some respects the symptoms are similar to medulloblastoma. The 5-year survival rate is about 50% (Bloom, 1979; Burger and Scheithauer, 1994).

Although these tumors form in the walls of the ventricle they can also grow into the brain. Moreover, they tend to form miniature ventricles, i.e., tubules of cells that are compacted in a radial manner like spokes in a wheel such that a central cavity is formed. Usually a small blood vessel forms the hub at the center.

MEDULLOBLASTOMA

Medulloblastoma is a rapidly growing embryonic neuroectodermal tumor that arises in the neuroepithelial roof of the 4th ventricle and posterior part of the cerebellar vermis (Allen et al., 1986; Burger and Scheithauer, 1994). Seedings of the tumor may be seen on the walls of the 3rd and lateral ven-

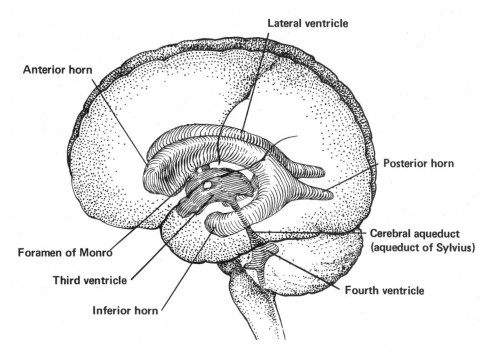

Figure 21.1. Schematic diagram of the ventricles. Lateral view. (From Waxman SG, De Groot J. Correlative Neuroanatomy, 22nd ed. East Norwalk, CT: Appleton & Lange, 1994.)

tricles, on the meningeal surfaces, and may metastasize into the spinal canal or supratentorial region (Allen and Epstein, 1982).

It occurs most frequently in children aged 4 to 8, with males outnumbering females 3:2. Symptoms associated with medulloblastomas, however, may not become manifest until the patient reaches his or her early 20s. Symptoms are usually present for 1–5 months before diagnosis.

Children with these tumors characteristically become listless, vomit repeatedly, and suffer early morning headache. Because of cerebellar involvement patients develop a stumbling gait and frequently fall. Examination will reveal papilledema (Burger and Scheithauer, 1994).

If the tumor grows unchecked there is danger of cerebellar herniation. As this develops patients may tend to hold their head tilted at an angle, the occiput being tilted back and away from the side of tumor.

The 5-year survival of patients who have been treated is about 50% (Allen et al., 1986; Berry et al., 1981). The risk of recurrence is limited to the first 3 years after treatment (Bloom, 1979; Burger and Scheithauer, 1994).

Tumors of the Pineal Gland: Pinealomas

The germinoma, which makes up more than half of all pineal tumors, is a firm, discrete mass that reaches up to 3–4 cm in diameter (Burger and Scheithauer, 1994). Although its effects on the pineal body are benign, it tends to compress the superior colliculi (thus causing visual and visual attentional disturbances) and sometimes the superior surface of the cerebellum and narrows the aqueduct of Sylvius. When these later regions are involved, gait and brainstem abnormalities may arise as well as hydrocephalus. Moreover, often these tumors expand into the 3rd ventricle and may compress the hypothalamus (Burger and Scheithauer, 1994). This can give rise to extreme changes in emotionality as well as exert profound disturbances on neuroendocrine and hormonal functioning.

Children, adolescents, and young adult males are most often affected, as these tumors rarely develop after age 30. Most characteristic localizing symptoms are an inability to look upward (Parinaud's syndrome) and slightly dilated pupils that react to accommodation but not to light.

MIDBRAIN TUMORS

Tumors of the midbrain usually cause internal hydrocephalus due to cerebral aqueduct obstruction, and may led to headache, papilledema, and vomiting. Ocular abnormalities are prominent, including weakness of upward conjugate deviation. Pupils are unequal and tend to be dilated, and reactions to light and convergence-accommodation may be lost. Corticospinal tracts are involved on both sides although asymmetrically such that the extremities become weak and possibly paralyzed.

PITUITARY TUMORS

Pituitary tumors (e.g., pituitary adenomas) typically produce a variety of symptoms as abnormalities of this gland frequently cause hypersecretion or hyposecretion and other endocrine changes. These include acromegaly, pituitary giantism, depression of sexual functioning, personality changes, and hypercortisolism. Overgrowth of bones may be evident in the skull, face, and mandible and the hands may become broad and spade-like. The patient (particularly if male) may also experience changes in the texture of the skin such that it becomes soft and pliable. There may also be a loss of hair over the body such that in the extreme, patients may no longer need to shave (Adams and Victor, 1994).

In that the optic nerves and chiasm pass directly beneath the pituitary, tumors in this vicinity often disrupt visual functioning via compression (Burger and Scheithauer, 1994). These visual changes may range from a loss of visual acuity to a progressive narrowing of the visual fields, i.e., tunnel vision. As these tumors expand they can also displace and compress the carotid arteries and thus cause ischemic conditions throughout the brain. Frequently these tumors spontaneously hemorrhage and in fact carry the highest risk for hemorrhage of all intracranial neoplasms.

Nevertheless, there are a variety of pituitary tumors, each of which gives rise to specific hypersecretory products (Burger and Scheithauer, 1994). These include prolactin-secreting adenomas, which in turn can cause amenorrhea and infertility; growth hormone-secreting adenomas, which give rise to giantism and acromegaly; adrenocorticotropic hormone-secreting adenomas, which are responsible for producing Cushing's syndrome; and oncocytomas, which are nonsecreting pituitary tumors (Salcman, 1986). Oncocytomas make their presence known via pituitary hypofunction.

The annual incidence of pituitary tumor is 1.0 per 100,000 individuals. However, there is an increased risk of developing a pituitary adenoma as patients grow older, and women are more at risk than men. For example, among women between the ages of 15 and 44, the incidence is 7.1 per 100,000 (Salcman, 1986). These tumors are also associated with "ice-cream cone" headaches in which the apex of pain points downward toward the center of the head.

Tumors of the Meninges: Meningioma

Meningiomas are tumors of the dura mater and/or arachnoid, and arise from abnormal arachnoidal cells of the arachnoid villae (Burger and Scheithauer, 1994). They account for 15% of all extracranial tumors. However, the risk for developing a meningioma increases with age. Individuals in their 70s have the highest incidence of all age groups. They also occur twice as often among women as men. Indeed, portions of these tumors actually contain estrogen and progesterone receptors (Salcman and Kaplan, 1986). Women with breast cancer appear to have the highest risk of all.

These tumors are very slow-growing and can become extremely immense before giving rise to symptoms. Although considered benign, they can kill and cause extensive damage via pressure effects. They also can form metastatic deposits that infiltrate and invade other organs such as the liver and lungs. Indeed, they may invade the sagittal sinus, envelop the carotid arteries, or even erode through the skull (Salcman and Kaplan, 1986).

Nevertheless, although originating within the meninges, via compression and infiltration they exert their predominant disruptive effects on the surface of the cerebral hemispheres, the base of the skull, and within regions where the dura forms tentorial compartments. They occur most often along the sylvian region, superior parasagittal surface of the frontal and parietal lobes, olfactory grooves, sphenoid bones, superior surface of the cerebellum, and spinal canal (Burger and Scheithauer, 1994).

Like many tumors their initial development may not cause any symptoms, particularly in that they are very slow-growing. Nevertheless, focal seizures are often an early sign. Patients may also develop focal neurological deficits such as motor weakness with frontal involvement, gait disturbances if the cerebellum is involved, or somesthetic abnormalities with parietal invasion.

PROGNOSIS AND RECOVERY

The survival of patients with meningeal tumors tends to be directly correlated with the extent of surgical resection. Unfortunately, incompletely removed meningiomas tend to regrow over a period of 5–10 years. In fact, they may reconstitute from what seems to be very tiny amounts of residual tumor. Nevertheless, 90% of those who have had the tumor surgically resected (and who survived the initial surgery) are alive after 5 years (Salcman and Kaplan, 1986).

Acoustic Neuromas and Schwannomas

Schwannomas and neuromas are tumors of the cranial nerves, usually the 8th, 5th,

9th, and 1st. They are thought to be benign and develop from the Schwann cells that cover the cranial nerves. Acoustic (8th nerve) neuromas (which are essentially a schwannoma) are the most common type (Burger and Scheithauer, 1994). These tumors usually have their onset at an early age. However, the highest incidence is during the 5th decade of life.

Acoustic neuroma occurs occasionally as part of neurofibromatosis, and may be bilateral and combined with multiple meningiomas. It originates in the junction between the nerve root and medulla, i.e., at the point where axons become enveloped in Schwann cells. Acoustic neuromas also attack the vestibular division of the 8th nerve just within the internal auditory canal. As it grows it increasingly occupies the angle between the cerebellum and pons, i.e., the posterior fossa, and may compress the 5th, 7th, and 9th or 10th nerves. With increased growth these tumors come to compress the pons and lateral medulla and obstruct CSF circulation (Burger and Scheithauer, 1994).

The earliest signs are loss of hearing, headache, disturbed sense of balance, unsteadiness of gait, and facial pain. Approximately a third of patients affected are troubled by vertigo associated with nausea, vomiting, and pressure in the ear.

ASTROCYTOMA, GLIOMA, GLIOBLASTOMA MULTIFORME, AND OLIGODENDROGLIOMA
Abnormal Glia Development

Glia (or nerve "glue") serves a variety of complex functions including providing nutrients to neurons, aiding in the synthesis and removal of neurotransmitters, maintaining synapses, and creating the myelin coat, which promotes neuronal transmission. In the developing organism, glia acts to direct embryonic and migrating neurons to their target cells.

Broadly, there are several types of glia, including Schwann cells, which are found in the peripheral nervous system, and microglia and macroglia, which are found in the spinal cord and throughout the cerebrum. Microglia serve, in part, as the brain's immune system, whereas macroglia aid in neuronal intercommunication. Macroglia subtypes include astrocytes and the oligodendrocytes that make up the myelin sheath. Unfortunately, glia, specifically unchecked glia growth and proliferation, are a primary source of brain tumors.

Astrocytomas

Astrocytomas are the most frequently occurring type of tumor and arise from the dedifferentiation of astrocytes that have lost their ability to limit their own replication (Burger and Scheithauer, 1994; Salcman and Kaplan, 1986). Although the abnormal cells themselves are not as numerous as what constitutes a higher-grade glioblastoma, these tumors tend to be evenly distributed and infiltrate a wide zone of tissue. Indeed, they frequently infiltrate and entrap normal neurons and glia (Burger, 1986; Burger and Scheithauer, 1994) and may occur anywhere in the brain or spinal cord.

Astrocytomas are slowly growing tumors that infiltrate and form large cavities in the regions occupied. Some are noncavitating and form in the white matter. Sometimes these tumors form microcysts (Kepes, 1987). If several tumors are adjacent these microcysts can coalesce to form larger cysts.

GRADING

These tumors are usually graded on a 4-point scale depending on their degree of differentiation, grade 1 being the most benign and the smallest. Typically grade 1 astrocytomas contain normal as well as abnormal astrocytes, and they tend to be well differentiated. They are apparently quite difficult to detect on gross inspection. Grade 2 represents an increased density of abnormal cells as well as reduced differentiation. Both grade 1 and grade 2 astrocytomas are often referred to as low-grade gliomas.

PREFERENTIAL REGIONS OF FORMATION

In general, astrocytomas occur most frequently along the frontal-temporal convexity (Burger and Scheithauer, 1994; Salcman and Kaplan, 1986)—areas that are most susceptible to contusion following head injury or acceleration-rotational traumas. In fact, grade 1 astrocytomas structurally and metabolically resemble a normal astrocytic reaction to an injury (Salcman and Kaplan, 1986). Indeed it appears that some individuals develop these tumors only after a head injury. Presumably the head injury acts as the environmental event that interacts with a genetic predisposition so as to trigger tumor formation. Nevertheless, they also tend to grow almost anywhere within the cerebrum, including the cerebellum, hypothalamus, pons, optic nerve, and chiasm.

Although considered relatively benign, as these tumors grow they can initially give rise to focal neurological deficits, may distort the lateral and third ventricle, displace the anterior and cerebral arteries, and give rise to seizures.

Fifty percent of patients present with initial focal or generalized seizure, and many of these individuals continue to have recurrent seizures throughout the illness. Headache and signs of increased intracranial pressure occur late. Age of onset is 20–60 years.

MORTALITY AND MALIGNANCY

These tumors are relatively dangerous and are associated with a high mortality rate, with 50% of patients dying over the course of the first 5 years (Salcman and Kaplan, 1986). In addition, although initially a grade 1 or 2, it appears that as patients age these initially slow-growing astrocytomas become less well differentiated and possibly progress to a more rapidly growing grade 3 or 4 tumor (Berger, 1986). Grade 3 and 4 astrocytomas are also referred to as gliomas. For example, the probability that an astrocytoma is malignant is less than 0.35, among individuals below age 45, whereas after age 60 it increases to 0.85. In this regard, as age increases, the likelihood of suffering a malignant astrocytoma (i.e., glioma) increases. Moreover, many malignant gliomas contain fields of benign tumor, which further suggests that astrocytomas progress from a slowly growing benign neoplasm to a rapidly growing malignant tumor (Shapiro and Shapiro, 1986).

Gliomas

Grade 3 lesions are very densely packed, are less well differentiated, contain necrotic tissue, and are very malignant. Sometimes they are accompanied by small areas of hemorrhage. These cells also tend to develop and divide much more rapidly than normal cells (Nelson et al., 1983). Patients who develop higher-graded gliomas are older than those with the lower-grade astrocytomas and typically become or have been symptomatic for a shorter time period before coming to the attention of a physician, i.e., on the average of 5.4 months versus 15.7 months (Burger, 1986). Of course, glioblastomas grow much more rapidly (Burger and Scheithauer, 1994).

They have their peak incidence between the ages of 40 and 60 and occur among males more frequently than females (Ransohoff et al., 1986). The 18-month survival of patients with glioblastoma is anywhere from 15% to 30% (Burger, 1986; Shapiro, 1986b). In part, the grim outlook for these patients is due to the development of local disease (Gaya, 1979; Nelson et al., 1986), as well as their potential for rapidly infiltrating wide regions of the cerebrum.

Glioblastoma Multiforme

Grade 4 astrocytomas are an extremely malignant form of neoplasm and are commonly referred to as glioblastoma multiforme. Some have argued, however, that grade 3 and 4 gliomas should be considered one and the same (Burger, 1986) and that grading should be based on a 3-point rather than a 4-point scale. Glioblastoma multiforme accounts for about 20% of all intracranial neoplasms, for about 55% of all glioma tumors, and for 90% of the gliomas of the adult cerebral hemisphere.

As the name implies, a glioblastoma multiforme is highly dedifferentiated, consisting of multiple cells types intermingled among others with various consistencies, cytoplasmic contents, colors, textures, and atypical nuclei (Shapiro et al., 1981). These tumors also contain large areas of necrosis (surrounded in a "picket fence" manner by astrocytic nuclei). In addition, they are heavily vascularized, and considerable abnormal capillary proliferation occurs in association with these tumors. Many of these capillaries also have vessel walls that demonstrate endothelial hyperplasia. Hence, there is a tendency for these tumors to spontaneously hemorrhage (Laurent et al., 1981).

These tumors also have a tendency to develop predominantly within the frontal-temporal regions (Burger, 1986; Burger and Scheithauer, 1994; Salcman and Kaplan, 1986). However, they are very rapidly growing and tend to be multifocal. If allowed to run their course, they will freely infiltrate the cerebrum, brainstem, spinal cord, and cerebellum, and will extend into the opposite cerebral hemisphere. About 50% are bilateral or occupy more than one lobe of a hemisphere (Burger and Scheithauer, 1994). Even when these tumors are unilateral, midline structures are often compressed, and the lateral ventricles are distorted and displaced contralaterally. In addition, they may attain enormous size before attracting medical attention.

These tumors are thus very dangerous and also very resistant to treatment due to their cellular, metabolic, vascular, and chromosomal diversity (Shapiro and Shapiro, 1986). Indeed, the central portion of the tumor (which contains areas of necrosis and cells that are hypoxic but viable) is highly resistant to radiation and chemotherapy (Salcman et al., 1982). This is because of their reduced blood flow, which in turn impedes the ability to treat them chemically. Moreover, because they are hypoxic they are more resistant to radiation than well-oxygenated cells (Nelson et al., 1986). Because the tumor is multicentric it is impossible to remove completely (Burger and Scheithauer, 1994).

In addition, even following resection and chemoradiation therapy, recurrence usually occurs within 1 year (Burger, 1986). The tumor will often revert back to its original size and distribution. Unfortunately, even among patients who survive, the long-term effects of treatment are associated with progressive cerebral atrophy and deterioration of cognitive capability (Burger, 1986; Ransohoff et al., 1986). Thus the outlook for patients with this form of tumor is extremely grim. Less than 20% of adults will survive 1 year. Among children, 50% will die within 18 months, and only 25% survive 5 years (Allen et al., 1986). Peak incidence, however, is during middle life.

Oligodendroglioma

These tumors are said to have a "fried egg" appearance and are not typically graded. They arise because of the dedifferentiation of oligodendrocytes—the cells that form myelin. Oligodendrogliomas occur most commonly in the deep white matter of the frontal and temporal lobes with little or no surrounding edema. The tumor is slow growing, and the first symptom in 50% of patients is a focal or generalized seizure. These have a tendency to calcify.

LYMPHOMAS, SARCOMAS, NEUROBLASTOMAS, AND CYSTS
Lymphomas

Lymphomas are tumors that are composed of cells that reside in the lymph nodes, i.e., lymphocytes, lymphoblasts, histocytes, and reticulum cells (Burger and Scheithauer, 1994). These tumors often develop preferentially within the adventitial spaces of the blood vessels (Kepes, 1987), and have also been referred to as reticulum cell sarcoma (Portlock, 1979).

They also develop within the meninges (Lister et al., 1979; Portlock, 1979) and are sometimes due to bone marrow disease (Young et al., 1979). However, lesions are often multifocal and may extend into the basal ganglia, thalamus, and corpus callosum (Portlock, 1979).

Lymphomas occur most commonly among individuals in their 50s. However, they have been known to affect infants as young as 16 days of age (Portlock, 1979). Symptoms develop quite rapidly and death ensues quite quickly, many dying in the first 13 months but a few living as long as 3 ½ years (Portlock, 1979).

Sarcomas

Sarcomas are malignant tumors composed of cells derived from mesenchymal tissues (fibroblasts, lipocytes, osteoblasts), and are rare. Many occur within the cranial cavity, or the temporal-frontal-basal regions during the 5th to 6th decades of life (Burger and Scheithauer, 1994).

Neuroblastomas

Neuroblastomas are tumors that occur predominantly among young children, infants, and even neonates (Burger and Scheithauer, 1994; Pochedly, 1976). They arise from sympathetic nervous tissue from any region of the body. Within the brain they originate in the sympathetic ganglia or the adrenal medulla (Burger and Scheithauer, 1994). Their effects on the brain are due primarily to compression, although they have a tendency to infiltrate the meninges and venous sinus. Usually, however, they infiltrate the cranium causing destruction and widening of the sutures, as well as the meninges, particularly the dura behind the orbit of the eyes and at the base of the brain (Pochedly, 1976). Their major effects are due to compression and increases in intracranial pressure.

Cysts

Cysts are pseudotumors that simulate intracranial neoplasm. They are usually found overlying the sylvian fissure or the cerebral convexity and may develop within the ventricular system. Sometimes, however, tumors will calcify and/or coalesce so as to form cysts.

Colloid cysts are congenital, but do not usually exert significant effects until adulthood, when they begin to compress and/or obstruct the 3rd ventricle, thus producing obstructive hydrocephalus.

UNILATERAL TUMORS AND BILATERAL DYSFUNCTION

In general, as a tumor grows it will exert pressure on underlying structures, and with sufficient development will not only compress but cause shifting of the brain such that subcortical and contralateral brain tissue becomes dysfunctional. In some cases, during the early stages of tumor growth patients may complain of symptoms of or demonstrate signs that seem to suggest bilateral lesions. If the tumor is too small to be visualized by computed tomography or magnetic resonance imaging, the bilateral nature of the patients deficits (which might only be detected by a competent neuropsychologist) may seem paradoxical.

As noted, neoplasms utilize less oxygen and have a lower blood flow as compared to normal tissue. However, possibly because of the effects of compression or bilateral arterial adjustments to localized blood flow reductions, cerebral tissue contralateral to the tumor also demonstrates lower-than-normal oxygen utilization and blood flow (Beaney et al., 1985). This in turn depresses the functioning of this otherwise normal tissue. However, with treatment of the tumor, blood flow returns to normal in this contralateral region (Brooks et al., 1986).

Hence, in some respects a tumor can exert both localized and seemingly paradoxical contralateral disruptive influences. Thus a patient with a tumor of the right motor strip may display not only unilateral left-sided weakness or paresis, but clouding of consciousness, confusion, depressive-like emotional changes, and expressive language difficulties (due to contralateral compression and hypoxia).

HERNIATION

With the formation of a neoplasmic growth, local tissue becomes depressed, the development of edema is incited, the brain begins to swell, and the volume of the intracranial contents is increased (Fisher, 1995). When brain tissue swells, midline structures shift and become compressed, and the lumen of the affected veins and capillaries collapse, as do the ventricles, sulci, and subarachnoid space of the swollen hemisphere. If intracranial pressure is raised sufficiently, the brain not only shifts, but the midbrain and brainstem may become compressed (Fisher, 1995), interfering with life-sustaining functions of respiration, blood pressure control, and temperature regulation.

The first compensation for increased pressure depends on the two rapidly mobile intracranial fluid pools, the CSF, and intravascular blood. If the hemispheres swelling exceeds the compensatory mechanisms, the only escape for the displaced hemisphere is herniation. The manner in which herniation occurs is dependent on the location of the tumor.

For example, frontal or parietal tumors tend to induce herniation medially under the falx cerebri, which separates the two hemispheres (subfalcial herniation). With subfalcial herniation the medial part of one frontal or parietal lobe is pushed under the falx, damaging not only these regions but the cingulate gyrus.

Temporal lobe tumors induce downward herniation over the edge of the tentorial notch, which surrounds the mesencephalon (transtentorial or temporal lobe herniation). That is, the medial portion of the temporal lobe is forced into the oval-shaped tentorial opening through which the midbrain and subthalamus pass, pushing these structures to the opposite side and thus exerting great pressure on them and their vessels. This results in a hemiparesis due to compression of the cerebral peduncle, which is ipsilateral to the lesion. In addition, the medial aspects of the temporal lobe including the amygdala and hippocampus become seriously compromised. Transtentorial herniation may be bilateral if there are multiple neoplasms.

Tumors that have invaded the cerebellum are likely to cause cerebellar herniation. This also results in brainstem compression as the cerebellum is forced down in the foramen magnum (however, see Fisher, 1995).

After subfalcial or transtentorial herniation, the diencephalon and mesencephalon become compressed and torqued and the brainstem becomes seriously damaged. Because of interference not only with vital centers but the reticular activating system, patients begin to experience rapid changes in conscious awareness, descending into somnolence, semicoma, coma, and then death (Mayer et al., 1994). Many patients, however, experience a transient phase of elevated consciousness, with excitement and delirium before gradually descending through lower levels of consciousness.

As transtentorial herniation increases, the uncus displaces the posterior cerebral artery and the 3rd nerve, which in turn causes the pupil to constrict. Eventually both 3rd nerves may cease to function, and both pupils become dilated and fixed, no longer responding to light.

Decerebrate rigidity is a postural syndrome of transtentorial herniation. The mouth is closed and the wrists, fingers, and toes are flexed, with the head, trunk, arms, and legs extended with hands and feet internally rotated. Breathing also becomes irregular and shallow. Appearance of this syndrome requires transection/compression of the mesencephalon and sparing of the vestibulospinal tract and dorsal/ventral roots as well as brainstem compression.

Tumors most likely to cause increased intracranial pressure without conspicuous focal or lateralizing signs are medulloblastoma, ependymoma of the 4th ventricle, glioblastoma of the cerebellum, and colloid cysts of the 3rd ventricle.

It is noteworthy that in a recent review Fisher (1995:83) argues that "temporal lobe herniation is not the means by which the midbrain sustains irreversible damage in acute cases, but rather that later displacement of the brain at the tentorium is the prime mover and herniation a harmless accompaniment [and that] bilateral brain stem compression in acute bilateral cases must be distinguished from herniation."

PROGNOSIS

Major prognostic indicators regarding so-called malignant neoplasms are the age at time of diagnosis, Karnofsky performance score (a measure of quality of life), the presence of seizures, tumor pathology, e.g., anaplastic astrocytoma versus glioblastoma multiforme, and even blood type (Burger and Scheithauer, 1994; Green et al., 1983; Shapiro, 1986b). Hence, young patients live longer than older individuals (the death rate for those under 45 being one-third that of those over 65), those with type B or O blood have a better prognosis than those with type A or A/B, and individuals who have developed seizures are more likely to survive than those who have not. In fact the longer they have had seizures, the longer the survival time (Shapiro, 1986b).

Some researchers have noted that preoperative tumor size is strongly related to survival (Androu et al., 1983), whereas oth-ers have not noted a strong relationship. Nevertheless, it has consistently been noted that the smaller the residual tumor following surgery, the longer the patient will live (Androu et al., 1983; Salcman, 1985; Wood et al., 1988).

Unfortunately, total resection is quite difficult in that it is not practical or ethical to remove a sufficient quantity of normal tissue so as to ensure total removal and thus total elimination. Hence, given the infiltrative and regenerative abilities of most tumors, recurrence is likely. Moreover, combined resection and chemotherapy is not always successful because of the protective influences of the blood-brain barrier (Blasberg and Groothuis, 1986; Shapiro and Shapiro, 1986). It is this same blood brain-barrier that protects the tumor from the antigens of the immune system—a functional result of the relative absence of vesicular transport across the endothelial cytoplasm between the tightly spaced capillary endothelial cells. Fortunately, as new methods of therapy are discovered and employed, and with early detection, prognosis for many patients should only improve.

References

Aberg T, Kihlgren M. Effect of open-heart surgery on intellectual function. Scand J Thorac Cardiovasc Surg 1987;15:1–63.

Abraham A, Mathai, KV. The effect of right temporal lesions on matching of smells. Neuropsychologia 1983;21:277–282.

Abrams R, Taylor M. Differential EEG patterns in affective disorder and schizophrenia. Arch Gen Psychiatry 1979;36:1355–1358.

Abrams R, Taylor MA. Psychopathology and the electroencephalogram. Biol Psychiatry 1980;15:871–878.

Absher JR, Toole JF. Neurobehavioral features of cerebrovascular disease. In: Fogel BS, Schiffer RB, eds. Neuropsychiatry. New York: Williams & Wilkins, 1996.

Ach N. Determining tendencies. In: Rapaport D, ed. Organization, and pathology of thought. New York: Columbia University Press, 1951.

Ackerly SS. Instinctive emotional and mental changes following pre-frontal lobe extirpation. Am J Psychiatry 1935;92:717–727.

Ackerman BP. Item specific and relational encoding effects in children's recall and recognition memory for words. J Exp Child Psychol 1984;37:426–450.

Ackerman BP. The effects of specific and categorical orienting in children's incidental and intentional memory for pictures and words. J Exp Child Psychol 1985;39:300–325.

Ackerman D. A natural history of the senses. New York: Vintage, 1990.

Achterberg J. Woman as healer. Boston: Shambhala, 1991.

Adam N. Effects of general anesthetics on memory functions in man. J Comp Physiol Psychol 1973;83:294–305.

Adams JH, Graham DI, Murray LS, Scott G. (1981). Diffuse axonal injury due to non-missile head injury in humans: an analysis of 45 cases. Ann Neurol 1981;12:557–563.

Adams JH, Graham DI, Scott G, et al. Brain damage in fatal non-missile head injury. J Clinical Pathol 1989;33:1132–1145.

Adams RD, Victor M. Principles of neurology. New York: McGraw-Hill, 1993.

Adey WR, Dunlop CW, Hendrix CE. Hippocampal slow waves: distribution and phase relations in the course of approach learning. Arch Neurol 1960;3:74–90.

Adey WR, Dunlop CW, Sunderland S. A survey of rhinencephalic interconnections with the brainstem. J Comp Neurol 1958;110:173–204.

Aggleton JP, Burton MJ, Passingham RE. (1980). Cortical and subcortical afferents to the amygdala of the rhesus monkey. Brain Res 1980;190:347–368.

Aggleton JP, Mishkin M. Visual recognition impairment following thalamic lesions in monkeys. Neuropsychologia 1983; 21:189–197.

Ahmad SS, Harvey JA. Long-term effect of spetal lesions and social experience on shock elicited fighting in rats. J Comp Physiol Psychol 1968;66:596–602.

Ainsworth MDS, Witig BA. Attachment and exploratory behavior of one-year-olds in a strange situation. NJ: Lawrence Erlbaum, 1969.

Akbarian S, Kim JJ, Potkin SG, et al. Gene expression for glutamic acid decarboxylase is reduced without loss of neurons in prefrontal cortex of schizophrenics. Arch Gen Psychiatry 1995;52:258–266.

Akbarian S, Vinuela A, Kim JJ, et al. Distorted distribution of nicotinamide-adenine dinucleotide phosphate-diaphroase neurons in temporal lobe of schizophrenics implies anomalous cortical development. Arch Gen Psychiatry 1993;50:178–187.

Akelaitis AJ. Studies on the corpus callosum. IV. Diagnostic dyspraxia in epileptics following partial and complete section of the corpus callosum. Am J Psychiatry 1945;101:594–599.

Alajoanine T. Aphasia and artistic realization. Brain 1948;71:229–241.

Alajoanine T. Verbal realization in aphasia. Brain 1956; 79:1–28.

Albe-Fessard D, Sanderson P. A demonstration of tonic inhibitory and faciliatory striatal actions on substantia nigra neurons. In: Carpenter MB, Jayaraman A. eds. The basal ganglia. New York: Plenum, 1987.

Alberico AM, Ward JD, Choi SC, Marmarou A, Young HF. Outcome after severe head injury. J Neurosurg 1987;67:648–656.

Albert ML. A simple test for visual neglect. Neurology 1973;23:658–664.

Albert ML. Alexia. In: Heilman KM, Valenstein E, eds. Clinical Neuropsychol, New York:Oxford University Press, 1993.

Albert ML, Bear D. Time to understand. A case study of word deafness with reference to the role of time in auditory comprehension. Brain 1974;97:383–394.

Albert ML, Reches A, Silverberg R. Associative visual agnosia without alexia. Neurology 1975;25:322–326.

Albert ML, Silverberg R, Reches A, Berman M. Cerebral dominance for consciousness. Arch Neurol 1976;33:453–454.

Albert ML, Sparks R, Helm N. Melodic intonation therapy for aphasia. Arch Neurol 1973;29:334–339.

Albert ML, Sparks R, von Strockert T, Sax D. A case of auditory agnosia. linguistic and nonlinguistic processing. Cortex 1972;8:427–443.

Albright TD, Desimone R, Gross CG. Columnar organization of directionally selective cells in visual area MT of the macaque. J Neurophysiol 1984;51:16–31.

Alexander GE, Crutcher MD. Preparation for movement: neural representations of intended direction in three motor areas of the monkey. J Neurophysiol 1990;64:133–150.

Alexander GE, Crutcher MD. Neural representations of the target (goal) of visually guided arm movements in three motor areas of the monkey. J Neurophysiol 1990;64:164–178.

Alexander GE, Crutcher MD, DeLong MR. Basal ganglia-thalamocortical circuits: parallel substrates for motor, oculomotor, prefrontal and limbic functions. Prog Brain Res 1990;85:119–146.

Alexander GE, DeLong MR, Crutcher MD. Do cortical and basal ganglionic motor areas use "motor programs" to control movement? Behav Brain Sciences 1992;15:656–665.

Alexander MP, Stuss DT, Benson DF. Capgras syndrome: a reduplicative phenomenon. Neurology 1979;29:130–131.

Alfano DP, Neilson PM, Fink MP. Long-term psychosocial adjustment following head or spinal cord injury. Neuropsychiatr Neuropsychol Behav Neurol 1993;6:117–125.

Alheid GF, Heimer L. New perspectives in basal forebrain organization of special relevance for neuropsychiatric disorders. Neuroscience 1988;27:1–39.

Alitalo K, Schwab M, Lin CC, et al. Homogenously staining chromosomal regions contain amplified copies of an abundantly expressed cellular oncogene in malignant neuroendocrine cells from a human. Proc Natl Acad Sci USA 1983;80:1701–1711.

Allen JC, Bloom J, Ertel I, et al. Brain tumors in children. Semin Oncol 1986;13:110–122.

Allen LS, Gorski RA. Sexual orientation and the size of the anterior commissure in the human brain. Proc Natl Acad Sci 1992;89:199–202.

Allen LS, Hines M, Shryne JE, Gorski RA. Two sexually dimorphic cell groups in the human brain. J Neurosci 1989;9:497–506.

Allman J. Evolution of neocortex. In: Jones ED, Peters A, eds. Cerebral Cortex. Vol 8. Comparative Structure and Evolution of Cerebral Cortex. New York: Plenum, 1990.

Allsworth-Jones P. The Szeletian and the stratigraphic succession in Central Europe and adjacent areas: main trends, recent results, and problems for solution. In: Mellars P, ed. The human revolution: behavioral and biological perspectives on the origins of modern humans, vol 2. Edinburgh: Edinburgh University Press, 1989.

Alpers N. Personality and emotional disorders associated with hypothalamic lesions. Association of Researchers in Nervous and Mental Disease 1940; 20:725–752.

Alter M, Friday G, Laid SM, et al. Hypertension and risk of stroke recurrence. Stroke 1994;25:1605–1610.

Alter-Reid K, Gibbs MS, Lachenmeyer, et al. Sexual abuse of children. Clin Psychol Rev 1986;6:249–266.

Alvarez G, Fuentes P. Recognition of facial expression in diverging socioeconomic levels. Brain Cogn 1994;25:235–239.

Alvarez W. Toward a theory of impact crisis. Eos 1986; 131:248–250.

Alves WM, Coloban ART, O'Leary TJ, Rimel RW, Jane JA. Understanding posttraumatic symptoms after minor head injury. Journal of Head Trauma Rehabilition 1986;1:1–12.

Amador ZF, Flaum M, Andreasen NC, et al. Awareness of illness in schizophrenia and schizoaffective and mood disorders. Arch Gen Psychiatry 1994; 51:826–836.

Amaral DG, Kurtz J. An analysis of the origins of the cholinergic and non-cholinergic septal projections to the hippocampal formation of the rat. J Comp Neurol 1985;24:37–59.

Amaral DG, Price JL. Amygdalo-cortical projections in the monkey (Macaca fascicularis). J Comp Neurol 1984;230:465–496.

Amaral DG, Price JL, Pitkanen A, Thomas S. Anatomical organization of the primate amygdaloid complex. In: Aggleton JP, ed. The Amygdala. New York: Wiley, 1992.

Anand BK, Dua S. Electrical stimulation in the limbic system of brain ("visceral brain") in the waking animal. Indian J Med Res 1956;44:107–119.

Anand BK, Dua S, China GS. Higher nervous control over food intake. Indian J Med Res 1958;46:277–287.

Anand BK, Malhortra CL, Singh V, Dua S. Cerebellar projections to limbic system. J Neurophysiol 1959; 22:451–457.

Andersen R. Differences in the course of learning as measured by various memory tasks after amygdalectomy in man. In: Hitchock ER, Laitinen L, Vaernet K, eds. Psychosurgery. Springfield: Thomas, 1972.

Andersen RA, Essics GK, Siegel RM. Encoding of spatial location by posterior parietal neurons. Science 1985;230:456–458.

Anderson-Gerfaud P. Aspects of behavior in the Middle Paleolithic: functional analysis of stone tools from Southwest France. In: Mellars P, ed. The human revolution: Behavioral and biological perspectives on the origins of modern humans. Vol 2. Edinburgh: Edinburgh University Press, 1989.

Andreasen NC, Arndt S, Alliger R, et al. Symptoms of schizophrenia. Arch Gen Psychiatry 1995;52:341–351.

Androu J, George AE, Wise A, et al. CT prognostic criteria of survival after malignant glioma sugery. Am J Neuroradiol 1983;4:488–490.

Andy OJ, Stephan H. Septal nuclei in the soricidae (insectivores). J Comp Neurol 1961;117:251–273.

Antrobus JS, Antrobus JS, Singer JL. Eye movements accompanying daydreaming, visual imagery, and thought suppression. J Abnorm Soc Psychol 1964; 69:341–344.

Antrobus JS, Dement W, Fisher C. Patterns of dreaming and dream recall. An EEG study. J Abnorm Soc Psychol 1964;69:341–344.

Applegate CD. 5,7,-dihydroxytryptamine-induced mouse killing and behavioral reversal with ventricular administation of serotonin in rats. Behav Neural Biol 1980;30:178–190.

Applewhite PB. The micrometazoa as a model system for studying the physiology of memory. Yale J Biol Med 1966;39:90–105.

Arensburg B, Schepartz LA, Tillier AM, Vandermersh B, Rak Y. A reappraisal of the anatomical basis for speech in Middle Palaeolithic hominids. Am J Phys Anthropol 1990;83:137–146.

Arey LB. Developmental anatomy. Philadelphia: WB Saunders, 1965.

Argyle M, Beit-Hallahmi B. The social psychology of religion. Boston:Routledge, 1975.

Ariel RN, Golden CJ, Berg RA, et al. Regional blood flow in schizophrenia. Arch Gen Psychiatry 1983; 40:258–263.

Ariens Kappers CU. The evolution of the nervous system. Bohn: Harlem, 1929.

Armstrong K. A history of god. New York: Ballantine, 1994.

Aronson BD, Bell-Pedersen D, Block GD, et al. Circadian rhythms. Brain Res Brain Res Rev 1993;18: 315–333.

Arranz B, Eriksson A, Mellerup E, et al. Brain 5-HT receptors in suicide victims. Biol Psychiatry 1994.

Arrigoni G, DeRenzi E. Constructional apraxia and hemispheric locus of lesion. Cortex 1964;1:170–197.

Arron W. Straight. New York:Holt, 1973.

Ashmead DH, Perlmutter M. Infant memory in everyday life. In: Perlmutter M, ed. Children's Memory. San Francisco: Josey-Bass, 1980.

Aslin RN. Visual and auditory development in infancy. In: Osofksy JD, ed. Handbook of Infant Development. New York: Wiley, 1987.

Astruc, J. Corticofugal connections of area 8 (frontal eye field) in Macaca mulatta. Brain Res 1972;33: 241–256.

Attkinson RC, Shiffrin RM. Human memory: a proposed system and its control processes. In: Spence KW, Spence JT, eds. The Psychology of Learning and Motivation. New York: Academic Press, 1968.

Atweh SF, Kuhar MJ. Autoradiographic localization of opiate receptors in rat brain. I. Brain Res 1977a; 129:1–12.

Atweh SF, Kuhar MJ. Autoradiographic localization of opiate receptors in rat brain. III. Brain Res 1977b; 134:273–405.

Auerbach SH, Allard T, Naeser M, et al. Pure word deafness. Brain 1982;105:271–300.

Auerbach S, Fornal C, Jacobs BL. Response of serotonin containing neurons in nucleus raphe magnus to morphine, noxious stimuli, and periaqueductal grey stimulation in freely moving cats. Exp Neurol 1985;88:609–628.

Axelrod S, Noonan M, Atanacio B. On the laterality of psychogenic somatic symptoms. J Nerv Ment Dis 1980;168:517–525.

Aylward EH, Roberts-Twille JV, Barta PE, et al. Basal ganglia volumes and white matter hyperintensities in patients with bipolar disorder. Am J Psychiatry 1994;151:687–693.

Azamita EC. The serotonin-producing neurons of the midbrain median and dorsal raphe nuclei. In: Iversen L, Iverson S, Snyder S, eds. Handbook of Psychopharmacology: Chemical Pathways in the Brain. New York: Plenum, 1978.

Azamitia EC, Gannon PJ. The primate serotonergic system. In: Fahn S, Marsden CD, Van Woert M, eds. Advances in Neurology: Myoclonus. New York: Raven Press, 1986.

"B." An introspective analysis of co-conscious life, by a personality (B) claiming to be co-conscious. J Abnorm Psychol 1908;3:311–334.

Bach-Y-Rita P. Brain plasticity as a basis for recovery of function in humans. Neuropsychologia 1990;28: 547–554.

Baer L, Rauch SL, Ballatine T, et al. Cingulotomy for intractable obsessive-compulsive disorder. Arch Gen Psychiatry 1995;52:384–392.

Bagshaw M, Benzies S. Multiple measures of the orienting reaction and their dissociation after amygdalectomy in monkeys. Exp Neurol 1968;27: 31–40.

Bailey P, Sweet WH. Effects on respiration, blood pressure and gastric motility of stimulation of orbital surface of frontal lobes. J Neurophysiol 1940; 3:276–281.

Baisden CE, Greenwald LV, Symbas PN. Occult rib fractures and brachial plexus injury following median sternotomy for open-heart operations. Ann Thorac Surg 1984;38:192–194.

Bakan P. Dreaming, REM sleep and the right hemisphere: a theoretical integration. J Altered States of Consciousness 1977;3:285–307.

Bakay L, Glasauer FE, Alker GJ. Head Injury. Boston: Little ,Brown & Co, 1980.

Baker-Ward L, Gordon BN, Ornstein PA, et al. Young children's long term retention of a pediatric examination. Child Development 1993;64:1519–1533.

Bakker RT. Dinosaur physiology and the origin of mammals. Evolution 1971;25:636–658.

Bakker RT. Tetrapod mass extinctions. In: Patterns of Evolution. Hallem A, ed. Amsterdam: Elsevier, 1977.

Baldwin M, Lewis SA, Bach SA. The effects of lysergic acid after cerebral ablation. Neurology 1959;9: 469–474.

Baleydier C, Maguiere F. The duality of the cingulate gyrus of the monkey. Brain 1980;103:525–554.

Balswick JO. Male inexpressiveness. In: Soloman K, Levy M, eds. Men in transition. New Yorkz: Plenum, 1982.

Balswick JO. The inexpressive male. Lexington: Lexington Books, 1988.

Bamford J, Sandercock P, Jones L, Warlow C. The natural history of lucunar infarction. Stroke 1987;18: 545–551.

Bancaud J, Brunet-Bourgin F, Chauvel P, Halgren E. Anatomical origin of deja vu and vivid memories in human temporal lobe epilepsy. Brain 1994;117: 71–90.

Ban HT, Sawyer CH. Autonomic and electroencephalographic responses to stimulation of the rabbit cerebellum. Anat Rec 1980;136:309.

Bandi HG. Art of the stone age. New York: Crown Publishers, 1961.

Banquet JP. Inter- and intrahemispheric relationships of the EEG activity during sleep in man. Electroencephalogr Clin Neurophysiol 1983;55:51–59.

Barbas H, Dubrovsky B. Excitatory and inhibitory interactions of extraocular and dorsal neck muscle afferents in the cat frontal cortex. Exp Neurol 1981; 74:51–66.

Barbas H, Mesulam MM. Organization of afferent input to subdivision of area 8 in the rhesus monkey. J Comp Neurol 1981;200:407–431.

Barbut D, Gazzaniga MS. Disturbances in conceptual space involving language and speech. Brain 1987; 110:1487–1496.

Bard P, Mountcastle VB. Some forebrain mechanisms involved in expression of rage with special reference to suppression of angry behavior. J Nerv Ment Dis 1948;27:362–404.

Barfield RJ, Chen JJ. Activation of estrous behavior in ovariectomized rats by intracranial implants of estradiol benzoate. Endocrinology 1977;101: 1716–1725.

Barnes C. Memory deficits associated with senescence: a neurophysiological and behavioral study in the rat. J Comp Physiol Psychol 1979;93:74–104.

Barnes CA, McNaughton BL. An age comparison of the rates of acquisition and forgetting of spatial information in relation to long-term enhancement of hippocampal synapses. Behav Neurosci 1985;99: 1040–1048.

Barnes GL. China, Korea, and Japan. The rise of civilization in East Asia. London: Thames & Hudson, 1993.

Barnett HJM. Cardiac causes of cerebral ischemia. In: Toole JF, ed. Cerebrovascular disorders. New York: Raven Press, 1984, pp. 174–177.

Baron JC, Petit-Tabou MC, Le Doze F, et al. Right frontal cortex hypometabolism in transient global amnesia. Brain 1994;117–545.

Baron RA. Environmentally-induced positive affect. J Appl Soc Psychol 1990;20:368–384.

Barr WB, Goldberg E, Wasserstein J, Novelly RA. Retrograde amnesia following unilateral temporal lobectomy. Neuropsychologia 1990;28:243–255.

Barris RW, Schuman, HR. Bilateral anterior cingulate gyrus lesions. Syndrome of the anterior cingulate gyri. Neurology 1953;3:44–52.

Barron, RW. Visual and phonological strategies in reading and spelling. In: Frith U, ed. Cognitive processes in spelling. London: Academic Press, 1980.

Bartlet JEA. A case of organized visual hallucination in an old man with cataracts, and their relation to the phenomena of the phantom limb. Brain 1951; 84:363–373.

Barth JT, Macciocchi SN, Giordani B, Rimel R, Jane JA, Boll TJ. Neuropsychological sequelae of minor head injury. Neurosurgery 1983;13:529–533.

Barth PG. Disorders of cerebral migration. Can J Neurol Sci 1987;14:1–16.

Bartlett FC. Remebering. Cambridge: Cambridge University Press, 1932.

Bartlett JC, Burleson G, Santrock JW. Emotional mood and memory in young children. J Exp Child Psychol 1982;34:59–76.

Bartoloemeo P, D'Erme P, Gainotti G. The relationship between visuo-spatial and representational neglect. Neurology 1994;44:1710–1714.

Bar-Yosef O. The date of the southwest Asian Neanderthals. In: Mellars P, Stringer CB, eds. The human revolution: behavioral and biological perspectives on the origins of modern humans. Vol 1. Edinburgh: Edinburgh University Press, 1989.

Bastiaanse R. Broca's aphasia: A syntactic and/or a morphological disorder? A case study. Brain Lang 1955;48:1–32.

Basso A, Taborelli A, Vignolo A. Dissociated disorders of speaking and writing in aphasia. J Neurol Neurosurg Psychiatry 1978;41:556–563.

Bates SR, Scalzo FM, Fowler KK. Drugs and neurotoxicants. In: Duckett S, ed. Pediatric neuropathology. Philadelphia: Williams & Wilkins, 1995.

Bathurst K, Kee DW. Finger-tapping interference as produced by concurrent verbal and nonverbal tasks. Brain Cogn 1994;24:123–136.

Batson DC, Ventis W. The religious experience. New York: Oxford University Press, 1982.

Batterman-Faunce JM, Goodman GS. Effects of context on the accuracy and suggestibility of child witnesses. In: Goodman GS, Bottoms BL. Child Victims, Child Witnesses. New York: Guilford, 1993.

Batuyev AS. The frontal lobes and the processes of synthesis in the brain. Brain Behav Evol 1969;2: 202–212.

Bauer PJ, Shore CM. Making a memorable event: effects of familiarity and organization on young children's recall of action sequences. Cognit Development 1987;2:327–338.

Bauman M, Kemper TL. Histoanatomic observations of the brain in early infantile autism. Neurology 1985;35:866–874.

Bayer SA, Altman J, Russo RJ, Zhang X. Embryology. In: Duckett S, ed. Pediatric Neuropathology. Philadelphia: Williams & Wilkins, 1995.

Bayley N. Behavioral correlates of mental growth. Am Psychol 1968;23:1–17.

Baylis GC, Rolls ET. Responses of neurons in the inferior temporal cortex in short term and serial recognition memory tasks. Exp Brain Res 1987;65:614–622.

Baxter D. Clinical syndromes associated with stroke. In: Brandstater ME, Bamajian JV, ed. Stroke Rehabilitation. Los Angeles: Williams & Wilkins, 1984.

Baxter LR, Schwartz JM, Bergman KS, et al. Caudate glucose metabolic rate changes with both drug and behavioral therapy for obsessive-compulsive disorder. Arch Gen Psychiatry 1992;49:681–689.

Baxter LR, Schwartz JM, Mazziotta JC, et al. Cerebral glucose metabolic rates in nondepressed patients with obsessive-compulsive disorder. Am J Psychiatry 1988;145:1560–1563.

Bayart F, Hayashi KT, Faull KF, Barchas JD, Levine S. Influence of maternal proximity on behavioral and physiological responses to separation in infant rhesus monkeys (Macaca mulatta). Behav Neurosci 1990;104:98–107.

Bayley N. Behavioral correlates of mental growth. Am Psychol 1968;23:1–17.

Beach, F.A. Hormonal modification of sexually dimorphic behavior. Psychoneuroendocrinology 1961; 3–23.

Beaney RP, Brooks DJ, Leenders KL, et al. (1985). Blood flow and oxygen utilisation in the contralateral cortex of patients with untreated intracranial tumors as studied by positron emission tomography with observations on the effect of decompressive surgery. J Neurol Neurosurg Psychiatry 1985; 48:310–319.

Bear D. The significance of behavioral change in temporal lobe epilepsy. McLean Hospital Journal 1977; 9:11–23.

Bear DM. Temporal lobe epilepsy: a sydnrome of sensory-limbic hyperconnexion. Cortex 1979;15: 357–384.

Bear DM. Hemispheric specialization and the neurology of emotion. Arch Neurol 1983;40:195–202.

Bear DM, Fedio P. Quantitative analysis of interictal behavior in temporal lobe epilepsy. Arch Neurol 1977;4:454–467.

Bear DM, Leven K, Blumer D, Chetam D, Ryder J. Interictal behavior in hospitalized temporal lobe epileptics: relationship to idiopathic psychiatric syndromes. J Neurol Neurosurg Psychiatry 1982; 45:481–488.

Bear MF, Singer W. Modulation of visual cortical plasticity by acetylcholine and noradrenaline. Nature 1986;320:172–176.

Beatty WW. Gonadal hormones and sex differences in nonreproductive behaviors. In: Gerall AA, et al., ed. Handbook of behavioral neurobiology. New York: Plenum, 1992.

Beaumont JG. Handedness and hemisphere function. In: Dimond SJ, Beaumont JG, eds. Hemispheric function in the human brain. New York: John Wiley & Sons, 1974.

Beaumont JG, Dimond SJ. Brain disconnection and schizophrenia. Br J Psychiatry 1973;23:661–662.

Beck AT, Rush AJ, Shaw BF, Emery G. Cognitive therapy of depression. New York: Guilford, 1979.

Beck EC, Dustman RE, Sakai M. Electrophysiological correlates of selective attention. In: Evans CR, Mulholland RB, eds. Attention in neurophysiology. New York: Appleton, 1969.

Beckers G, Zeki S. The consequences of inactivating areas V1 and V5 on visual motion perception. Brain 1995;118:49–60.

Beckson M, Cummings JL. Neuropsychiatric aspects of stroke. Int J Psychiatry Med 1991;21:1–5.

Bedichek R. The sense of smell. New York: Doubleday, 1960.

Beeman M. Semantic processing in the right hemisphere may contribute to drawing inferences from discourse. Brain Lang 1993;44:80–120.

Behrens MC. "The electrical response of the planarian photoreceptor. Comp Biochem Physiol 1961;5: 129–138.

Beier EG, Zautra AJ. Identification of vocal communication of emotions across cultures. J Consult Clin Psychol 1972;39:166–167.

Bekerian DA, Bowers JM. Eyewitness testimony: were we misled. J Experiment Psychol Learn Mem Cogn 1983;9:139–145.

Belfer-Cohen A, Hovers E. In the eye of the beholder: mousterian and Natufian burials in the levant. Curr Anthropol 1992;133:463–471.

Bell AP, Weinberg MS, Hammersmith SK. Sexual preference: its development in men and women. Bloomington: Indiana University Press, 1981.

Bell BD. Pantomime recognition impairment in aphasia. Brain Lang 1994;47:269–278.

Belsky J, Gilstrap B, Rovine M. The Pennsylvania infant and family development project. I. Child Dev 1984;55:692–705.

Benes FM. Evidence for neurodevelopmental disturbances in anterior cingulate cortex of post-mortem schizophrenic brain. Schizophr Bull 1991;5:187–188.

Benes FM. Is there a neuroanatomic basis for schizophrenia? An old question revisited. The Neuroscientist 1995;1:104–115.

Benes FM, McSparren J, Bird ED, et al. Deficits in small interneurons in prefrontal and anterior cingulate cortex of schizophrenic and schizoaffective patients. Arch Gen Psychiatry 1991;48:996–1001.

Benes FM, Turtle M, Khan Y, Farol P. Myelination of a key relay zone in the hippocampal formation occurs in the human brain during childhood, adolescence, and adulthood. Arch Gen Neurol 1994;51: 477–484.

Bennet HL. Perception and memory for events during adequate general anaestheesia for surgical operations. In: Pettinati HM, ed. Hypnosis and memory. New York: Guklford, 1988.

Bennet HL, Davis HS, Giannini JA. Post-hypnotic suggestions during general anesthesia and subsequent dissociated behavior. Society for Clinical and Experimental Hypnosis. Portland, 1981.

Bennett AE. Psychiatric and neurologic problems in head injury with medicolegal implications. Dis Nerv Syst 1969;30:314–318.

Benninger RJ. The role of serotonin and dopamine in learning to avoid aversive stimuli. In: Archer T, Nilsson L-G, eds. Aversion, avoidance and Anxiety. New Jersey: Erlbaum, 1989.

Ben-Itzhac S, Smith P, Bloom RA. Radiographic study of the humerus in Neanderthals and Homo sapiens sapiens. Am J Phys Anthropol 1988;77: 231–242.

Benson DF. Fluency in aphasis correlation with radioactive scan localization. Cortex 1967;3:373–394.

Benson DF. The third alexia. Arch Neurol 1977;34: 327–331.

Benson DF. Aphasia, alexia, agraphia. New York: Churchill-Livingston, 1979.

Benson D, Barton M. Disturbances in constructional ability. Cortex 1970;6:19–46.

Benson DF, Gardner H, Meadows JC. Reduplicative paramnesia. Neurology 1976;26:147–151.

Benson DF, Geschwind N. The alexias. In: Vinken PJ, Bruyn GW, eds. Handbook of clinical neruology. Vol. 4. Amsterdam: North Holland, 1969.

Benson DF, Sheremata WA, Bouchard R, et al. Conduction aphasia. Arch Neurol 1973;28:339–346.

Benson DF, Weir WF. Acalculia: acquired anarithmetria. Cortex 1972;8:465–472.

Benson GS. Male sexual function. In: Knobil E, et al, eds. The physiology of reproduction. New York: Raven Press, 1988.

Benton AL. Disorders of spatial disorientation. In: Vinken PJ, Bruyn GW, eds. Handbook of clinical neurology, vol. 3. New York: North Holland, 1969.

Benton A. Visuoperceptive, visuospatial and visuo-constructive disorders. In: Heilman KM, Valenstein E., eds. Clinical neuropsychology. Oxford: Oxford University Press, 1993:186–232.

Benton AL, Hannay J, Varney N. Arithmetic ability, finger-localization capacity and right-left discrimination in normal and defective children. Am J Orthopsychiatry 1975;21:756–766.

Benton AL, Hecaen H. Stereoscopic vision in patients with unilateral cerbral disease. Neurology 1970;20: 1084–1088.

Beppu H, Suyda M, Tanaka R. Analysis of cerebellar motor disorders by visually guided elbow tracking movements. Brain 1984;107:787–809.

Bergen M. Effect of severe trauma on a 4-year old child. Psychoanal Study Child 1958;13:407–429.

Berlucchi G, Rizolatti G. Binocularly driven neurons in visual cortex of split-chiasm cats. Science 1968; 159:308–310.

Berg WK, Berg KM. Psychophysiological development in infancy. State, sensory function, and attention. Psychol Bull 1978;47:103–170.

Berger MS, Pitts LH, Lovely M, et al. Outcome from severe head injury in children and adolescents. J Neurosurg 1985;62:194–199.

Bergland MM, Thomas KR. Psychosocial issues following severe head injury in adolescence. Rehabilitation Counseling Bulletin 1991;35:5–22.

Bergmann MS. In the shadow of moloch: the sacrifice of children and its impact on Western religions. New York: Columbia University Press; 1992.

Berkley KJ, Parmer R. Somatosensory cortical involvement in response to noxious stimulation in the cat. Exp Brain Res 1974;20:363–374.

Berlyne DE. Conflict, arousal, and curiosity. New York: McGraw-Hill, 1960.

Berman KF, Weinberger DR, Shelton RC, Zec RF. A relationship between anatomical and physiological brain pathology in schizophrenia. Am J Psychiatry 1987;144:1277–1282.

Berman KF, Zek RF, Weinberger DR. Physiological dysfunction of dorsolateral prefrontal cortex in schizophrenia. Arch Gen Psychiatry 1986;43: 126–143.

Berman P. Children's nurturance to younger children. Society for Research in Child Development Detroit. April 1983.

Berman P, Goodman W. Age and sex differences in children's responses to babies. Child Dev 1984;55: 1071–1077.

Berman RA. On the ability to relate events in narrative. Discourse Processes 1988;11:469–497.

Bernstein M. The search for Bridey Murphy. New York: Lancer, 1956.

Bernston GG, Potloicchio SJ, Miller NE. Evidence for higher functions of the cerebellum. Proc Natl Acad Sci U S A 1973;70:2497–2499.

Berrios GE. Musical hallucinations: a historical clinical study. Br J Psychiatry 1990;156:188–194.

Berrol S. Evolution and the persistent vegetative state. Head Trauma Rehab 1986;1:7–13.

Berry M, Jenkin R , Keen C, et al. Radiation therapy for medulloblastoma: a 21 year review. J Neurosurg 1981;55:43–51.

Bertini M. Violani C, Zoccolotti P, Antonelli A, DiStephano L. Performance on a unilateral tactile test during waking and upon awakenings from REM and NREM. In: Koella WP, ed. Sleep. Basel: Karger, 1983, pp. 122–155.

Bexton WA, Heron W, Scott TH. Effects of decreased variation in the sensory environment. Can J Psychol 1954;8:70–76.

Bhaskar PA. Scrotal pain with testicular jerking: an unusual manifestation of epilepsy. J Neurol Neurosurg Psychiatry 1987;50:1233–1234.

Bianchi L. The mechanism of the brain and the function of the frontal lobes. Edinburgh: Livingstone, 1922.

Bieber I. A psychoanalytic study. New York: Basic Books, 1962.

Bieber I, Dain HJ, Dince PR, et al. Homosexuality. New York: Basic Books, 1962.

Biemond A. The conduction of pain above the level of the thalamus. Arch Neurol Psychiatry 1956;75: 231–244.

Binder J, Marshall R, Lazar R, Benjamin J, Mohr JP. Distinct syndromes of hemineglect. Arch Neurol 1992;49:1187–1194.

Binet A. La suggestibilitie. Paris: Schleicher Freres, 1900.

Binford L. Bones: ancient men and modern myths. New York: Academic Press, 1981.

Binford SR. A structural comparison of disposal of the dead in the Mousterian and the Upper Paleolithic. Southwest J Anthropol 1968;24:139–54.

Binford SR. Interassemblage variability—the Mousterian and the 'functional' argument. In: Renfrew C, ed. The explanation of culture change. Models in prehistory. Pittsburgh: Pittsburgh University Press, 1973.

Binford SR. Rethinking the Middle/Upper Paleolithic transition. Curr Anthropol 1992;23:177–181.

Bird ED, Iversen LL. Huntington's chorea. Brain 1974; 97:457–467.

Bishop GH. Natural history of nerve impulses, Physiol Rev 1956;36:376–399.

Bisiach E, Berti A. Dyschiria. An attempt at its systemic explanation. In: Jeannerod M, ed. Neurophysiological and neuropsychological aspects of spatial neglect. Amsterdam: North-Holland, 1987.

Bisiach E, Bulgarelli C, Sterzi R, Vallar G. Line bisection and cognitive plasticity of unilateral neglect of space. Brain Cogn 1983;2:32–38.

Bisiach E, Geminiani G. Anosognosia related to hemiplegia and hemianopia. In: Prigatano GP, Schacter DL, eds. Awareness of deficit after brain injury. New York: Oxford University Press, 1991.

Bisiach E, Luzzatti C. Unilateral neglect of representational space. Cortex 1978;14:129–133.

Bisiach E, Luzzati C, Perani D. Unilateral neglect, representational schema and consciousness. Brain 1979; 102:609–618.

Bitter EE. Evolutionary biology. Connecticut, JAI Press, 1991.

Bjorklund DF. How age changes in knowledge base contribute to the development of children's memory: an interpretive review. Developmental Review 1987;7:93–130.

Black Elk, Neihardt JG. Black Elk speaks. Lincoln:University Nebraska Press, 1979.

Black FW, Bernard BA. Constructional apraxia as a function of lesion locus and size in patients with focal brain damage. Cortex 1984;20:111–120.

Black FW, Strub RL. Constructional apraxia in patients with discrete missile wounds of the brain. Cortex 1976;12:212–220.

Black JE, Sirevaag AM, Greenough WT. Complex experience promotes capillary formation in young rat visual cortex. Neurosci Lett 1987;83:351–355.

Black JE, Zelazny AM, Greenough WT. Capillary and mitochondrial support of neural plasticity in adult rat visual cortex. Exper Neurol 1991;111:204–209.

Blakemore C. The representation of three-dimensional space in the cat's striate cortex. J Physiol 1970;209: 155–178.

Blakemore C. Mechanics of the mind. New York: Cambridge Univesity Press, 1977.

Blakemore JEO. Age and sex differences in interactions with a human infant. Child Dev 1981;52: 386–388.

Blakemore JEO. Interaction with a baby by young adults. Sex Roles 1985;13:405–411.

Blakemore JEO. Children's nurturant interactions with their infant siblings. Sex Roles 1990;22:43–57.

Blakemore JEO, Baumgardner SR, Keniston AH. Male and female nurturing. Sex Roles 1988;18:449–459.

Blanchard RJ, Blanchard DC. Limbic lesions and reflexive fighting. J Comp Physiol Psychol 1968;66: 603–605.

Blaney PH. Affect and memory: a review. Psychol Bull 1986;99:229–246.

Blasberg RG, Groothuis DR. Chemotherapy of brain tumors. Semin Oncol 1986;13:70–82.

Blasberg RG, Groothuis DR, Molnar P. Application of quantitative autoradiographic measurements in experimental brain tumor models. Semin Neurol 1981;1:203–221.

Blier MJ, Blier-Wilson LA. Gender differences in self-rated emotional expressiveness. Sex Roles 1989;21: 287–299.

Blier R, Byne W, Siggelkow I. Cytoarchitectonic sexual dimorphisms of the medial preoptic and anterior hypothalamic area in guinea pig, rat, hamster, and mouse. J Comp Neurol 1982;212:118–130.

Blinkov SM, Glezer II. The human brain in figures and tables. New York: Plenum, 1968.

Bliss EL, Zwanziger J. Brain amines and emotional stress. J Psychiatr Res 1966;4:189–198.

Bliss EL, Ailion J, Zwanziger J. Metabolism of norepinephrine, serotonin, and dopamine in rat brain with stress. J Pharmac Exp Ther 1968;164:122–134.

Bliss EL. Multiple personalities: a report of 14 cases with implications for schizophrenia and hysteria. Arch Gen Psychiatry 1980;37:1388–1397.

Bloedel JR. Functional heterogeneity with structural homogeneity: how does the cerebellum operate. Behav Brain Sci 1992;15:666–678.

Blomert DM, Sisler GC. The measurement of retrograde post-traumatic amnesia. Can Psychiatr Assoc J 1974; 19:185–192.

Blonder LX, Burns AF, Bowers D, et al. Right hemisphere facial expressivity during natural conversation. Brain Cogn 1993;21:44–56.

Bloom FE. Advancing a neurodevelopmental origin for schizophrenia. Arch Gen Psychiatry 1993;50: 224–227.

Bloom HJM. Intracranial secondary carcinomas and disseminating gliomas: treatment and prognosis. In: Whitehouse JMA, Kay, HEM, eds. CNS complications of malignant disease. Baltimore: University Park Press, 1979.

Blue ME, Molliver ME. 6-Hydroxydopamine induces serotonergic sprouting in cerebral cortex of newborn rat. Dev Brain Res 1987;32:255–270.

Blugental DB, Blue J, Cortez V, Fleck K, Rodriguez A. Influences of witnessed affect on information processing in children. Child Dev 1992;63:774–786.

Blum JS, Chow KL, Pribram KH. A behavioural analysis of the organization of the parieto-temporopreoccipital cortex. J Comp Physiol 1950;93:53–199.

Blum MS. Alarm pheromones. In: Kerkut GA, Gilbert LI, eds. Comprehensive insect physiology, biochemistry, and pharmacology. New York: Pergammon Press, 1985.

Blumenthal HT, Premachandra BN. Cerebrovascular disease in dementia. In: Morley JE, et al., eds. Memory function and aging-related disorders. New York: Springer, 1992.

Blumer D, Benson DF. Personality changes with frontal and temporal lesions. In: Benson DF, Blumeer D, eds. Psychiatric aspects of neurologic disease. New York: Grune & Stratton, 1975.

Blumstein SE, Alexander MP, Ryalls JH, Katz W, Dworetzky B. On the nature of the foreign accent syndrome: a case study. Brain Lang 1987;31: 215–244.

Blumstein S, Cooper WE. Hemispheric processing of intonational contours. Cortex 1974;10:146–158.

Blumstein S, Goodglass H. Perception of stress as a semantic cue in aphasia. J Speech Hearing Research 1972;15:800–806.

Bobillier P, Sakai F, Sequin S, et al. The effect of sleep deprivation upon in vivo and in vitro incorporation of tritiated acids into brain proteins in the rat at three different age levels. J Neurochem 1974;22: 23–31.

Bock E. Nervous system specific proteins. J Neurochem 1978;30:7–14.

Bogen JE. The other side of the brain. Bull Los Angeles Neurol Soc 1969;34:135–162.

Bogen JE. The calosal syndrome. In: Heilman KM, Valenstein E, eds. Clinical neuropsychology. New York: Oxford University Press, 1979, pp. 308–358.

Bogen J, Bogen C. The other side of the brain: III. The corpus callosum and creativity. Bull Los Angeles Neurol Soc 1969;34:191–220.

Bogerts B, Meertz E, Schonfeldt-Bausch R. Basal ganglia and limbic system pathology in schizophrenia. Arch Gen Psychiatry 1985;42:784–791.

Bogousslavsky J, Ferrazzini M, Regli F, Assal G, Tanabe H, Delaloye-Bischof A. Manic delirium and frontal-like syndrome with paramedian infarction of the right thalamus. J Neurol, Neurosurg Psychiatry 1988;51:116–119.

Bohannon III, JN. Flashbulb memories for the space shuttle disaster: a tale of two theories. Cognition 1988;29:179–196.

Bohannon III, JN, Symons VL. Flashbulb memories: Confidence, consistency, and quantity. In: Winograd E, Neisser U, eds. Affect and accuracy in recall. Cambridge: Cambridge University Press, 1992.

Boller F, Cole M, Vrtunski PB, Patterson M, Kim Y. Paralinguistic aspects of auditory comprehension in aphasia. Brain Lang 1979;9:164–174.

Boller F, Grafman J. Acalculia: historical development and current significance. Brain Cogn 1983:2; 205–223.

Boller F, Green E. Comprehension in severe aphasics. Cortex 1972;8:382–390.

Bonanno GA, Singer JL. Repressive personality style: theoretical and methodological implications for health and pathology. In: Singer JL, ed. Repression and dissociation. Chicago: University of Chicago Press, 1990.

Bond MR. Assessment of the psychosocial outcome of severe head injury. Acta Neurochirurgica 1976;34: 57–70.

Bond MR. The states of recovery from severe head injury with special reference to late outcome. Int Rehabil Med 1979;1:155–159.

Bord H. The making of new Masculinities. Boston: Allen & Unwin, 1987.

Bornstein RA, Kelly MP. Risk factors for stroke and neuropsychological performance. In: Bornstein RA,

Brown GC, eds. Neurobehavioral aspects of cerebrovascular disease. New York: Oxford University Press, 1991.

Bornstein RA, Matarazzo JD. Wechsler VIQ versus PIQ differences in cerebral dysfunction: a review of the literature with emphasis on sex differences. J Clin Neuropsychol 1982;4:319–334.

Borod J. Interhemispheric and intrahemispheric control of emotion: a focus on unilateral brain damage. J Consult Clin Psychol 1992;60: 339–348.

Borod JC, Caron HS. Facedness and emotion related to lateral dominance, sex, and expression type. Neuropsychologia 1980;18:237–241.

Borod JC, Andelman F, Obler LK, et al. Right hemispheric specialization for the identification of emotional words and sentences: evidence from stroke patients. Neuropsychologia 1992;30:827–844.

Borod JC, Caron HS. Facedness and emotion related to lateral dominance, sex, and expression type. Neuropsychologia 1980;18:237–241.

Borson SM, Lampe T, Raskind MA, Veith RC. Autonomic nervous system dysfunction in Alzheimer's disease. In: Morley JE, et al, eds. Memory function and aging-related disorders. New York: Springer, 1992.

Botez MI, Attig E, Vezina J. Cerebellar atrophy in epileptic patients. Can J Neurol Sci 1988;15: 299–303.

Botez MI, Olivier M, Vezina J-L, Botez T, Kaufman B. Defective revisualization: dissociation between cognitive and imagistic thought case report and short review of the literature. Cortex 1985;21: 375–389.

Bouchard TJ Jr, McGee MG. Sex differences in human spatial ability. Soc Biol 1977;24:332–335.

Bower GH. Mood and memory. Am Psychol 1981;2: 129–148.

Bower GH, Mayer JD. In search of mood-dependent retrieval. In: Kuiken D, ed. Mood and memory. Newbury Park: Sage, 1991.

Bowers D, Blonder LX, Feinberg T, Heilman KM. Differential impact of right and left hemisphere lesions on facial emotion and object imagery. Brain 1991;114:2593–2609.

Bowers D, Coslett B, Bauer RM, Speedie LJ, Heilman K. Comprehension of emotional prosody following unilateral hemisphere lesions: processing defect versus distraction defect. Neuropsychologia 1987; 25:317–328.

Bowlby J. The influence of early environment in the development of neurosis and neurotic character. Int J Psychoanal 1940;21:154–178.

Bowlby J. Maternal care and mental health. Geneva: WHO, 1951.

Bowlby J. Separation anxiety. Int J Psychoanal 1960; 412:1–25.

Bowlby J. Attachment and loss. New York: Basic Books, 1982.

Braak H, Braak C. Neuronal types in the basolateral amygdaloid nuclei of man. Brain Res Bull 1983;11: 349–396.

Bracke-Tolkmitt R, Linden A, Canavan AGM, et al. The cerebellum contributes to mental skills. Behav Neurosci 1989;103:442–446.

Bradbury AJ, Costall B, Domeney AM, Naylor RJ. Laterality of dopamine function and neuroleptic action in the amygdala of the rat. Neuropharmacol 1985;24:1163–1170.

Bradford R. Nursing procedures and problems. In: Greenblatt M, Arnot R, Solomon HC, eds. Studies in lobotomy. New York: Grune & Stratton, 1950.

Bradshaw JL, Nettleton NC. Language lateralization to the dominant hemisphere: tool use, gesture and language in hominid evolution. Curr Psychol Rev 1982;2:171–192.

Bradshaw JL, Nettleton NC, Spher K. Braille reading and left and right hemispace. Neuropsychologia 1982;20:493–500.

Bradshaw JL, Taylor MJ, Patterson K, Nettleton N. Upright and inverted faces, and housefronts, in the two visual fields: a right and a left hemisphere contribution. J Clin Neuropsychol 1980;2:245–257.

Brady JV. Temporal and emotional effects related to intracranial electrical self-stimulation. In: Ramsey E. Doherty DO, eds. Electrical studies on the unanesthetized brain. New York: Hoeber, 1960.

Brady JV, Nauta WJH. Subcortical mechanisms in emotional behavior: affective changes following septal lesions in the rat. J Comp Physiol Psychol 1953;46:339–346.

Brain R. Visual disorientation with special references to lesions of the right cerebral hemisphere. Brain 1941;64:244–272.

Brain R, Walton JN. Brain's diseases of the nervous system. New York: Oxford University Press, 1969.

Brainerd C, Kingma J. On the independence of short-term memory and working memory in cognitive development. Cogn Psychol 1985;17:210–247.

Brainerd C, Ornstein PA. Children's memory for witnessed events. In: Doris J, ed. The suggestibility of children's recollections. Washington: American Psychological Association, 1991.

Brandon SGF. The judgment of the dead. New York: Scribners, 1967.

Bratgaard SO. RNA increase in ganglion cells of retina after stimulation by light. Acta Radiologica 1951; Supplements 96, 80.

Brauer G. The evolution of modern humans: a comparison of the African and non-African evidence. In: The human revolution: behavioral and biological perspectives on the origins of modern humans. Vol 1. Mellars P, Stringer CB, eds. Edinburgh: Edinburgh University Press, 1989.

Braun BG. Toward a theory of multiple personality and other dissociative phenomena. Psychiatr Clin North Am 1984;7:171–193.

Braun CMJ, Denault C, Cohen H, Rouleau I. Discrimination of facial identity and facial affect by temporal and frontal lobectomy patients. Brain Cogn 1994;117:71–90.

Brazier WAB. Stimulation of the hippocampus in man using implanted electrodes. In: Brazier MAB, ed. RNA and brain function, memory and learning. Berkeley: University of California, 1966.

Breasted JH. History of egypt. New York: Scribners, 1909.

Breitling D, Guenther W, Rondot P. Auditory perception of music measured by brain electrical activity mapping. Neuropsychologia 1987;25:765–774.

Bremner JD, Randall P, Scott TM, et al. MRI-based measurement of hippocampal volume in patients with combat-related posttraumatic stress disorder. Am J Psychiatry 1955;152:973–981.

Brend R. "Male female intonation patterns in American English. In: Thorne B, Henley N, eds. Language and sex. Massachussetts: Newbury House Publishers, 1975.

Breslow R, Kocsis J, Belkin B. Contribution of the depressive perspective to memory function in depression. Am J Psychiatry 1981;138:227–230.

Bretherton I, Fritz J, Zahn-Waxler C, Ridgeway D. Learning to talk about emotions. A functionalist perspective. Child Dev 1986;57:529–548.

Breuer J, Freud S. On the psychical mechanism of hysterical phenomena. In: Strachey J, ed. Standard Edition. London: Hogarth Press, 1893.

Breuer J, Freud S. Studies in hysteria. Standard Edition. 2. London: Hogarth Press, 1893–1895.

Brewer WF. Memory for randomly sampled autobiographical events. In: Neisser U, Winograd E, eds. Remembering reconsidered. New York: Cambridge University Press, 1988.

Brier J. Studying delayed memories of childhood sexual abuse. The Advisor 1992;5:17–18.

Brier J, Conte J. Self-reported amnesia for abuse in adults molested as children. J Traumatic Stress 1995.

Brindley GS, Smith GPC, Lewin W. Cortical blindness and the functions of the niongeniculate fibers of the optic tracts. J Neurol Neurosurg Psychiatry 1969;32:259–264.

Brink AS. Speculation on some advanced mammalian characteristics in the higher mammal-like reptiles. Palaeontology 1956;4:77–95.

Brink JD, Garrett AL, Hale WR, et al. Recovery of motor and intellectual function in children sustaining severe head injuries. Dev Med Child Neurol 1970;12:565–571.

Brinkman C. Lesions in supplementary motor area interfere with a monkey's performance on a bimanual coordination task. Neurosci Lett 1981;27: 267–270.

Brinkman C, Porter R. Supplementary motor area in the monkey: activity of neurons during performance of a learned motor task. J Neurophysiol 1979;42:681–709.

Brodal A. Neurological anatomy. New York: Oxford University Press, 1981.

Brody LR. Visual short-term cued recall memory in infancy. Child Dev 1981;52:242–250.

Brody L. Gender differences in emotional develoment. A review of theories and research. In: Steward A, Lyko M, eds. Gender and personality. Durham: Duke University Press, 1985.

Brodzinsky DM, et al. Sex differences in children's expression and control of fantasy and overt aggression. Child Dev 1979;50:372–379.

Broffman M. The lobotomized patient during the first year at home. In: Greenblatt M, Arnot R, Solomon HC, eds. Studies in lobotomy. New York: Grune & Stratton, 1950.

Bronson FH. Pheromonal influences on mammalian reproduction. In: Diamond MN, ed. Perspective in reproduction and sexual behavior. Bloomington: Indiana University Press, 1968.

Bronson FH, et al. Strange male pregnancy block in deer mice. Biology Reprod 1969;1:302–310;

Bronson FH, Whitten W. Oestrus-accelerating pheromone in mice. J Reprod Fertil 1968;15:131–140.

Brooks AS, Helgren DM, Cramer JS, et al. Dating and context of three middle stone age sites with bone points in the Upper Semliki Valley, Zaire. Science 1995;268:548–552.

Brooks DJ, Beaney RP, Thomas DGT. The role of positron emission tomography in the study of cerebral tumors. Semin Oncol 1986;13:83–93.

Brooks DN, Aughton ME. Cognitive recovery during the first year after severe blunt head injury. Int Rehabil Med 1979;1:166–172.

Brooks DN, Baddeley AD. What can amnesic patient's learn? Neuropsychologia 1976;14:111–122.

Brooks DN, McKinlay W. Personality and behavioural change after severe blunt head injury—a relative's view. J Neurol Neurosurg Psychiatry 1983;46:336–344.

Broom R. The mammal-like reptiles of South Africa and the origin of mammals. London: Witherby, 1932.

Brott T, Thalinger K, Hertzberg V. Hypertension as a risk factor for spontaneous intracerebral hemorrhage. Stroke 1986;17:1078–1083.

Broughton R. Human consciousness and sleep/waking rhythms: a review and some neuropsychological considerations. J Clin Neuropsychol 1982;4:193–218.

Broughton RJ. Narcolepsy. In: Thorpy MJ, ed. Handbook of sleep disorders. New York: Dekker, 1990.

Brown F, et al. Early Homo erectus skeleton from west of Lake Turkana, Kenya. Nature 1985;316:788–792.

Brown GC, Spicer KB, Malik G. Neurobehavioral correlates of arteriovenous malformations and cerebral aneurysms. In: Bornstein RA, Brown GC, eds. Neurobehavioral aspects of cerebrovascular disease. New York: Oxford University Press, 1991.

Brown GL, Goodwin FK, Bunney WE. Human aggression and suicide. In: Ho BT, ed. Serotonin in biological psychiatry. New York: Raven, 1982.

Brown JW. Aphasia, apraxia and agnosia. Springfield: CC Thomas, 1972.

Brown JW. Early prenatal development of the human precommissural septum. J Comp Neurol 1983;215:331–350.

Brown R, Colter N, Corsellis JAN, et al. Postmortem evidence of structural brain changes in schizophrenia. Arch Gen Psychiatry 1986;43:36–42.

Brown S, Schaefer AE. An investigation into the func-tions of the occipital and temporal lobe of the monkey's brain. Philadelphia Transactions of the Royal Society of Britain 1888;179:303–327.

Brownell HH, Potter HH, Bihrle AM. Inference deficits in right brain-damaged patients. Brain Lang 1986;27:310–321.

Broverman DM, et al. Roles of activation and inhibition in sex differences in cognitive abilities. Psychol Rev 1968;328–331.

Brown A. The development of memory: knowing about knowing, and knowing how to know. In: Reese HW, ed. Advances in child development and behavior. New York: Academic Press, 1975.

Brown R, Kulik J. Flashbulb memories. Cognition 1977;5:73–99.

Bruce CJ, Desimone R, Gross CG. Visual properties of neurons in a polysensory area in superior temporal sulcus of the macaque. J Neurophysiol 1982;46:369–384.

Bruce CJ, Desimone R, Gross CG. Both striate and superior colliculus contribute to visual properties of neruons in superior temporal polysensory area of Macaque monkey. J Neurophysiol 1986;58,1057–1076.

Bruce HM. Pheromones. Br Med Bull 1970;26:10–22.

Bruhn AR. Earliest childhood memories. New York: Praeger, 1990.

Bruner JS. On cognitive growth. In: Bruner JS, Oliver RR, Greenfield PM, eds. Studies in cognitive growth. New York: Wiley, 1966.

Bruner JS, Postman L. Emotional selection in perception and reaction. J Personality 1947a;16:69–77.

Bruner, JS, Postman L. Tension and tension release as organizing factors in perception. J Personality 1947b;15:300–308.

Brutkowski S. Functions of prefrontal cortex in animals. Physiol Rev 1965;45:721–746.

Brutkowski S, Mishkin M, Rosvold HE. Positive and inhibitory motor CRs in monkeys after ablation of orbital or dorsolateral surface of the frontal cortex. In: Guttman E, Hnik P, eds. Central and peripheral mechanisms of motor functions. Prague: Czechoslovak Academy of Science, 1963, pp. 133–141.

Bruton CJ. The Neuropathology of temporal lobe epilepsy. Oxford: Oxford University Press, 1988.

Bruton CJ, Crow TJ, Frith CD, et al. Schizophrenia and the brain. Psychol Med 1990;20:285–304.

Bruton CJ, Stevens JR, Frith CD. Epilepsy, psychosis, and schizophrenia. Neurology 1994;44:34–42.

Bryden MP. A model for the sequential organization of behaviour. Can J Psychol 1967;21:36–56.

Bryden MP, Allard F. Visual hemifield differences depend on typeface. Brain Lang 1976;3:191–200.

Bryden MP, Ley RG, Sugarman JH. A left-ear advantage for identifying the emotional quality of tonal sequences. Neuropsychologia 1982;20:83–87.

Bucci MN, Phillips TW, McGillicuddy JE. Delayed epidural hemorrhage in hypotensive multiple trauma patients. Neurosurgery 1986;19:65–68.

Buchanan RW, Strauss ME, Kirkpatrick B, et al. Neuropsychological impairments in deficit vs non-deficit forms of schizophrenia. Arch Gen Neurol 1994;51:804–811.

Buchsbaum MS. The frontal lobes, basal ganglia, and temporal lobe as sites for schizophrenia. Schizophr Bull 1990;156:216–227.

Buchsbaum MS, Nuechterlein KH, Haier RJ, et al. Glucose metabolism rate in normals and schizophrenics during the continuous performance test assessed by positron emission tomography, Br J Psychiatry 1990;156:216–227.

Buchsbaum MS, Neuchterlien KH, Manning RG, et al. Cerebral glycography with positron tomography. Br J Psychiatry 1990;156:216–227.

Buchtel H, Campari F, De Risio C, Rota R. Hemispheric differences in the discrimination reaction times to facial expression. Italian J Psychol 1978;, 5, 159–169.

Buchwald JS, Halas ES, Schramm S. Changes in cortical and subcortical unit activity during behavioral conditioning. Physiol Behav 1966;1:11–22.

Buck R. Nonverbal communication of affect in preschool children. J Personality Soc Psychol 1977; 35:225–236.

Buck R. The communication of emotion. New York: Guilford, 1984.

Buck R, Baron R, Barrette D. The temporal organization of spontaneous nonverbal expression. J Personality Soc Psychol 1982;42:506–517.

Buck R, Miller R, Caul W. Sex, personality and physiological variables in communication of affect via facial expressions. J Personality Soc Psychol 1974;30: 587–596.

Buck R, Savin V, Miller R, Caul W. Communication of affect through facial expressions in humans. J Personality Soc Psychol 1972;42:506–517.

Budge W. The Book of the Dead. New Jersey: Carol, 1994.

Bueresova O, Aleksanyan Z, Bures J. Electrophysiological analysis of retrieval of conditioned taste aversions in rats. Physiol Bohemoslov 1979;28: 525–539.

Bullard DE, Gillespie GY, Mahaley MS, Bigner D. Immunobiology of human gliomas. Semin Oncol 1986; 13:94–109.

Bullock M. Causal reasoning and developmental change over the preschool years. Hum Devel 1985; 28:169–191.

Bunnel BN. Amygdaloid lesions and social dominance in the hooded rat. Psychonomic Sci 1966;6:93–94.

Bures J, Buresova O. The use of Leao's spreading depression in the study of interhemispheric transfer of memory traces. J Comp Physiol Psychol 1960;59: 211–214.

Burger PC. Malignant astrocytic neoplasms. Semin Oncol 1986;13:16–25.

Burger PC, Scheithauer BW. Tumors of the central nervous system. Bethesda: Armed Forces Institute of Pathology, 1994.

Burke AM, Younkin D, Gordon J, et al. Changes in cerebral blood flow and recovery from acute stroke. Stroke 1986;17:173–178.

Burke PA, Callow AD, O'Donnel TF, et al. Prophylactic carotid endarterectomy for asymptomatic bruit. Arch Surg 1982;117:1222–1227.

Burke WJ, Chung HD, Marshall GL, et al. Defective axonal transport: a mechanism in the degeneration of neurons in Alzheimer's disease. In: Morely JE, et al., eds. Memory function and aging-related disorders. New York: Springer, 1992.

Burling R. Primate calls, human language, and nonverbal communication. Curr Anthropol 1993;34: 25–53.

Burnett AL, Diehl NA. The nervous system of hydra. J Exper Zoolol 1964;157:237–250.

Burnett SA, et al. Spatial visualization and sex differences in quantitative ability. Intelligence 1979;3: 345–354.

Burton H, Jones EG. The posterior thalamic region and its cortical projection in New World and Old World monkeys. J Comp Neurol 1976;168:249–302.

Burton H, Mitchell G, Brent D. Second somatic sensory area in the cerebral cortex of cats. J Comp Neurol 1982;210:109–135.

Burton LA, Levy J. Sex differences in the lateralized processing of facial emotion. Brain Cogn 1989;11: 210–228.

Burton R. The Language of smell. London: Routledge & Kegan Paul, 1976.

Bussey K. Children's lying and truthfulness: implications for children's testimony. In: Ceci SJ, Leichtman M, Putnick M, eds. Cognitive and social factors in preschoolers deception. New Jersey: Elrbaum, 1992.

Bussey K, Less K, Grimbeck EJ. Lies and secrets: implications for childrens' reporting of sexual abuse. In: Goodman GS, Bottoms BL, eds. Child victims, child witnesses. New York: Guilford Press, 1993.

Butler AB. The evolution of the dorsal thalamus of jawed vertebrates, including mammals. Brain Res Rev 1994;19:29–65.

Butter CM. Perseveration in extinction and in discrimination reversal tasks following selective frontal ablations in Macaca mulatta. Physiol Behav 1969;4: 163–171.

Butter CM, Mishkin M, Mirsky AF. Emotional response toward humans in monkeys with selective frontal lesions. Physiol Behav 1968;3:213–215.

Butter CM, Mishkin M, Rosvold HL. Conditioning and extinction of food rewarded responses after selective ablations of frontal cortex in rhesus monkey. Exp Neurol 1963;7:65–75.

Butter CM, Snyder DR. Alterations in aversive and aggressive behaviors following orbital frontal lesions in rhesus monkeys. Acta Neurobiol Exp 1972;32: 525–566.

Butter CM, Snyder DR, McDonald JA. Effects of orbital frontal lesions on aversive and aggressive behaviors in rhesus monkeys. J Comp Physiol Psychol 1970; 72:132–144.

Butters N, Barton M. Effect of parietal lobe damage on the performance of reversible operations in space. Neuropsychologia 1970;8:205–214.

Butters N, Cermak LS. Alcoholic Korsakoff's syndrome. New York: Academic Press, 1980.

Butters N, Miliotis P. Amnesic disorders. In: Heilman KM, Valenstein E, eds. Clinical neuropsychology. Oxford: Oxford University Press, 1985.

Butzer K. Geomorphology and sediment stratigraphy. In: Singer R, Wymer J, ed. The Middle Stone Age at Klasies River Mouth in South Africa. Chicago: University of Chicago Press, 1982.

Bygott JD. Antagonistic behaviour, dominance, and social structure in wild chimpanzees. In: Hamburg DA, McGown ER, eds. The Great Apes. Menlo Park: Benjamin/Cummings. 1979.

Byrne W, Parsons B. Human sexual orientation. Arch Gen Psychiatry 1993;50:228–239.

Cade WH. Insect mating and courtship behavior. In: Kerkut GA, Gilbert LI, eds. Comprehensive Insection Physiology. Biochemistry and Pharmacology. New York: Pergamon Press, 1985.

Cahill DW, Knippe H, Mosser J. Intracranial hemorrhage with amphetamine abuse. Neurology 1981; 31:1058–1059.

Cahill L, McGaugh, JL. Amygdaloid complex lesions differentially affect retention of tasks using appetitive and aversive reinforcement. Behav Neurosci 1990;104:532–543.

Caligiuri MP, Heindel WC, Lohr JB. Sensorimotor disinhibition in Parkinson's disease: effects of levodopa. Ann Neurol 1992;31:53–58.

Cain DP. Kindling and the amygdala. In: Aggleton JP, ed. The amygdala. New York: Wiley, 1992.

Cairns RB. The attachment behavior of mammals. Psychol Rev 1967;73:409–426.

Cajal, S. Ramon y. (1911). Histoligie du Systeme Nerveux de L'Homme et des Vertebres. Paris: Maloine, 1955.

Cajal S. Ramon y. Neuron theory or reticular theory (translated by Purkiss WU, Fox, CA). Madrid, 1954.

Calandre L, Oretega JF, Bermejo F. Anticoagulation and hemorrhagic infarction in cerebral embolism secondary to rheumatic heart disease. Arch Neurol 1984;41:1152–1154.

Calhoun JB. The study of wild animals under controlled conditions. Ann N Y Acad Sci 1950;51: 1113–1122.

Caligiuri MP, Heindel WC, Lohr JB. Sensorimotor disinhibition in Parkinson's disease: effects of levodopa. Ann Neurol 1992;31:53–58.

Calne DB. Is idiopathic parkinsonism the consequence of an event or a process. Neurology 1994; 44:5–10.

Calvanio R, Petrone PN, Levine DN. Left visual spatial neglect is both environment-centered and body-centered. Neurology 1987;37:1179–1183.

Calvo JM, Badillo S, Morales-Ramirez M, Palacios-Salas P. The role of the temporal lobe amygdala in ponto-geniculo-occipital activity and sleep organization in cats. Brain Res 1987;403:22–30.

Campbell A. Men, Women and Aggression. New York: Basic Books, 1993.

Campbell J. Historical Atlas of World Mythology. New York: Harper & Row, 1988.

Campbell R. Asymmetries in interpreting and expressing a posed facial expression. Cortex 1978;15: 327–342.

Campos JJ, Barrett KC, Lamb ME, Goldsmith HH, Sternberg C. Socio-emotional development. In: Handbook of Child Psychology. Vol. 2: Infancy and Development Psychobiology. Haith MM, Campos JJ, eds. New York: Wiley, 1983.

Camps M, Kelly PH, Palacios JM. Autoradiographic localization of dopamine D1 and D2 receptors in the brain of several mammalian species. J Neural Transm 1990;80:105–127.

Cancelliere AEB, Kertesz A. Lesion localization in acquired deficits of emotional expression and comprehension. Brain Cogn 1990;13:133–147.

Cann A, Ross D. Olfactory stimuli as context cues in human memory. Am J Psychol 1989;102:91–102.

Cannon TD, Sarnoff A, Mednick A, et al. Developmental brain abnormalities in the offspring of schizophrenic mothers. Arch Gen Psychiatry 1994;51: 955–962.

Cano RJ, Borucki MK. Revival and identification of bacterial spores in 25- to 40-million-year-old Dominican amber. Science 1955;268:1060–1064.

Caplan LR, Schmahmann JD, Kase CS, et al. Caudate infarcts. Arch Neurol 1990;47:133–143.

Caramazza A, Gordon J, Zurif EB, Deluca D. Right-hemispheric damage and verbal problem solving behavior. Brain Lang 1976;3:41–46.

Caramazza A, Zurif EB. Dissociation of algorithmic and heuristic process in language comprehension: evidence from aphasia. Brain Lang 1976;3:572–582.

Card AL, Jackson LA, Stollak GE, Ialongo NS. Gender role and person-perception accuracy. Sex Roles 1986;15:159–171.

Carlsen J, De Olmos J, Heimer L. Tracing of two-neuron pathways in the olfactory system by the aid of transneuronal degeneration: projection to the amygdaloid body and hippocampal formation. J Comp Neurol 1982;208:196–208.

Carlson EB, Rosser-Hogan R. Trauma experiences, posttraumatic stress, dissociation, and depression in Cambodian refugees. Am J Psychiatry 1991;148: 1548–1551.

Carlsson CA, von Essen C, Lofgren J. Factors affecting the clinical course of patients with severe head injuries. J Neurosurg 1968;29:242–251.

Carmon A, Bechtoldt HP. Dominance of the right cerebral hemisphere for stereopsis. Neuropsychologia 1969;7:29–39.

Carmon A, Benton AL. Tactile perception of direction and number in patients with unilateral cerebral disease. Neurology 1969;19:525–532.

Carmon A, Nachshon I. Effect of unilateral brain damage on perception of temporal order. Cortex 1971; 7:410–418.

Carmon A, Nachshon I. Ear asymmetry in perception of emotional non-verbal stimuli. Acta Psychologica 1973;37:351–357.

Caroll BJ. Limbic system adrenal cortex regulation in depression and schizophrenia. Psychosom Med 1976;38:106–121.

Caroll BJ, Curtis GC, Mendels J. Neuroendocrine regulation in depression. I & II. Arch Gen Psychiatry 1976;33:1039–1044, 1051–1058.

Carpenter MB. Human Neuroanatomy. Baltimore: Williams & Wilkins, 1991.

Carpenter WT, Buchanan RW, Breier A, et al. Psychopathology and the question of neurodevelopmental or neurodegenerative disorder. Schizophr Bull 1991;5:192–194.

Carpenter WT, Buchanan RW, Kirkpatrick B. Strong inference, theory testing, and the neuroanatomy of schizophrenia. Arch Gen Psychiatry 1993;50: 825–831.

Carr SA. Interhemispheric transfer of stereognostic information in chronic schizophrenics. Br J Psychiatry 1980;136:53–58.

Carrasco, D. Religions of MesoAmerica. San Francisco: Harper & Row, 1990.

Carroll J, Schaffer C, Spensley J, Abramowitz SI. Family experiences of self-mutilating patients. Am J Psychiatry 1980;137:852–853.

Cartwright RD, Bernick N, Borwitz G, Kling A. Effect of an erotic movie on sleep and dreams in young men. Arch Gen Psychiatry 1969;20:262–271.

Cartwright R, Romanek I. Nature and function of repetitive dreams in normal subjects. Psychiatry 1979;42:131–137.

Cartwright RD, Tipton LW, Wicklund J. Focusing on dreams. Arch Gen Psychiatry 1980;37:275–288.

Casagrande VA, Joseph R. Effects of monocular deprivation on geniculostriate connections in prosimian primates. Anat Rec 1978;190:359.

Casagrande VA, Joseph R. Morphological effects of monocular deprivation and recovery on the dorsal lateral geniculate nucleus in galago. J Comp Neurol 1980;194:413–426.

Case R, Kurlund M, Goldberg J. Operational efficiency and the growth of short-term memory span. J Exp Child Psychol 1982;30:352–371.

Casey, RJ. Children's emotional experience. Dev Psychol 1993;29:119–129.

Cashmore J. Problems and solutions in lawyer-child communication. Crim Law J 1990;15:193–202.

Casanova MF, Stevens JR, Kleinman JE. The neuropathology of schizophrenia. In: Lindenmayer JP, Kety SR, eds. New biological vistas for schizophrenia. New York: Mazel, Inc., 1992.

Cassiday KL, McNally RJ, Zeitlin SB. Cognitive processing of trauma cues in rape victims with post-traumatic stress disorder. Cognit Ther Res 1992;16: 283–295.

Castellanos FX, Giedd JN, Eckburg P, et al. Quantitative morphology of the caudate nucleus in attention deficit hyperactivity disorder. Am J Psychiatry 1994;151:1791–1786.

Catchpole CK. Vocal communication in birds. Baltimore: University Park Press, 1979.

Cavada C. Transcortical sensory pathways to the pre-frontal cortex with special attention to the olfactory and visual modalities. In: Reinoso-Suarez F, Ajmone-Marsan C, eds. Cortical integration. New York: Raven Press, 1984, pp. 317–328.

Cavada C, Goldman-Rakic PS. Posterior parietal cortex in the rhesus monkey. I & II. J Comp Neurol 1989;287:393–445.

Ceci SJ. A developmental study of multiple encoding and its relationship to age related changes in free recall. Child Dev 1980;51:892–895.

Ceci SJ, Bruck M. Suggestibility of the child witness: a historical review and synthesis. Psychol Bull 1993; 113:403–439.

Ceci SJ, Lea SEG, Ringstrom MD. Coding characteristics of normal and learning disabled 10-year-olds. J Exp Psychol [Hum Learn] 1980;6:285–297.

Ceci SJ, Ross DF, Toglia MP. Suggestibility of children's memory: psycholegal implications. J Exp Psychol Learn Mem Cogn 1987;116:38–49.

Celesia GG, Bushnell D, Toleikis C, Brigell MG. Cortical blindness and residual vision. Neurology 1991; 41:862–869.

Cendes F, Andermann F, Gloor P, et al. Relationship between atrophy of the amygdala and ictal fear in temporal lobe epilepsy. Brain 1994;117:739–736.

Cerebral Embolism Study Group. Immediate anticoagulation of embolic stroke. Stroke 1984;15: 779–789.

Cermak LS, Lewis R, Butters N, Goodglass H. Role of verbal mediation in performance by Korsakoff patients. Percept Mot Skills 1973;37:259–262.

Cermak LS, O'Connor U. The anterograde and retrograde retrieval ability of a patient with encephalitis. Neuropsychologia 1983;21:213–234.

Cespuglio R, Gomez ME, Faradji H, Jouvet M. Alterations in the sleep-waking cycle iduced by cooling of the locus coeruleus area. Electroencephalography and clinical neurophysiology 1982;54:570–578.

Chaika E. A unified explanation for the diverse structural deviations reported for adult schizophrenics with disrupted speech. J Commun Disord 1982;15: 167–189.

Chakos MH, Lieberman JA, Bilder RM, et al. Increase in caudate nuclei volumes of first-episode schizophrenic patients taking antipsychotic drugs. Am J Psychiatry 1994;151:1430–1436.

Chalouloff F. Physiopharmacological interactions between stress hormones and central serotonergic systems. Brain Res Rev 1993;18:1–32.

Chambers BR, Norris JW, Shurvell BL, Hachinski VC. Prognosis of acute stroke. Neurology 1987;37: 221–225.

Chamberlain AF. Some points in linguistic psychology. Am J Psychol 1903;5:116–119.

Chamorro A, Sacco RL, Cierceirsky BA, et al. Visual hemineglect and hemihallucinations in a patient with a subcortical infarction. Neurology 1990;40: 1463–1464.

Chamorro A, Vial N, Saiz A, et al. Early anticoagulation after large cerebral embolic infarction. Neurology 1995;45:861–865.

Chandler SHG, Chase MH, Nakamura Y. Intracellular analysis of synaptic mechanisms controlling trigeminal motorneuron activity during sleep and wakefulness. J Neurophysiol 1980;44:359–371.

Chaouloff F. Physiopharmacological interactions between stress hormones and central serotonergic systems. Brain Res Rev 1993;18:1–32.

Chapman J. The early symptoms of schizophrenia. Br J Psychiatry 1966;12:225–251.

Chapman LF, Markham CH, Rand RW, Crandall PH. Memory changes induced by stimulation of the hippocampus or amygdala in epilepsy patients implanted with electrodes. Trans Am Neurol Assoc 1967;92:50–56.

Chapman LF, Walter RD. Actions of lysergic acid dienthalamid on averaged human cortical evoked responses to light flash. Recent Adv Biol Psychiatry 1965;7:23–36.

Chapman LF, Walter RD, Ross W, et al. Altered electrical activity of human hippocampus and amygdala induced by LSD-25. Physiologist 1963;5:118.

Chapman PF, Kairiss EW, Keenan CL, Brown TH. Long-term synaptic potentiation in the amygdala. Synapse 1990;6:271–278.

Chapman WP. Studies of the periamygdaloid area in relation to human behavior. Proc Assoc Res Nerv Ment Dis 1958;36:258–277.

Chapman WP. Depth electrode studies in patients with temporal lobe epilepsy. In: Ramey ER, O'Doherty DS, eds. Electrical studies on the unanesthetized brain. New York: Hoeber, 1960, pp. 334–350.

Chapman WP, Livingston R, Livingston KE. The effects of lobotomy and of electrical stimulation of the orbital surface of the frontal lobe upon respiration and blood pressure in man. In: Greenblatt M, Arnot R, Solomon HC, eds. Studies in lobotomy. New York: Grune & Stratton, 1950.

Chapman WP, Schroeder HR, Geyer G, Brazier MAB, Fager C, Poppen TL, Soloman HC, Yakovlev PI. Physiological evidence concerning importance of the amygdaloid nuclear region in the integration of circulatory functioning and emotion and man. Science 1954;177:949–951.

Charney D, Deutch A, Krystal J, et al. Psychobiological mechanisms of posttraumatic stress disorder. Arch Gen Psychiatry 1993;50:294–305.

Chase MH, Morales FR. Postsynaptic modulation of spinal cord motorneuron membrane potential during sleep. In: McGinty DJ, et al, eds. Brain mechanisms of sleep. New York: Raven Press, 1985.

Chase PG. How different was Middle Palaeolithic subsistance? A zooarchaeological pespective on the Middle to Upper Paleolithic transition. In: The human revolution: behavioral and biological perspectives on the origins of modern humans, vol 1. Mellars P, Stringer CB, eds. Edinburgh: Edinburgh University Press, 1989.

Chase PG, Dibble HL. Middle Paleolithic symbolism: a review of current evidence and interpretations. J Anthropol Archaeol 1987;6:263–293.

Chase RA. Discussion. In: Darley FL, ed. Brain mechanisms underlying speech and language. New York: Grune & Stratton, 1967, pp. 136–139.

Chaurasis BD, Goswami HK. Functional asymmetry in the face. Acta Anatomica 1975;91:154–160.

Chen R, Forster FM. Cursive and gelastic epilepsy. Neurology 1973;23:1019–1029.

Cheney DL, Seyfarth RM. How monkeys see the world: inside the mind of another species. Chicago: Chicago University Press, 1990.

Chernigovskaya TV, Deglin VL. Brain functional asymmetry and neural organization of linguistic competence. Brain Lang 1986;29:141–153.

Chevalier-Skolnikoff S. Male-female, female-female, and male-male sexual behavior in stumptail monkey. Arch Sex Behav 1974;3:1974.

Chi CC. An experimental silver study of the ascending projections of the central gray substance and adjacent tegmentum in the rat with observation in the cat. J Comp Neurol 1970;139:259–273.

Chi JG, Dooling EC, Gilles FH. Gyral development of the human brain. Ann Neurol 1977;1:86–93.

Chi MTH. Age differences in memory span. J Exp Child Psychol 1977;23:266–281.

Chi MTH, Ceci SJ. Content knowledge: its role, representation, and restructuring in memory development. In: Reese HW, ed. Advances in child development and behavior. New York: Academic Press, 1987.

Chiera E. They wrote on clay. Chicago: University of Chicago Press, 1966.

Chomsky N. Syntactic structures. The Hague: Mouton, 1957.

Chomsky N. Language and the mind. New York: Harcourt, 1972.

Chow KL. A retrograde cell degeneration study of the cortical projection field of the pulvinar of the monkey. J Comp Neurol 1950;93:313–340.

Christianson, S-A. The relationship between induced emotional arousal and amnesia. Scand J Psychol 1984;25:147–160.

Christianson, S-A. Emotional stress and eyewitness memory: a critical review. Psychol Bull 1992;12: 284–309.

Christianson S-A, Fallman L. The role of age on reactivity and memory for emotional pictures. Scand J Psychol 1990;31:291–301.

Christianson S-A, Nilsson L-G. Functional amnesia as induced by a psychological trauma. Mem Cognit 1984;12:142–155.

Christianson S-A, Nilsson L-G. Hysterical amnesia. In: Archer T, Nilsson L-G, eds. Aversion, avoidance and anxiety: perspective on aversively motivated behavior. New Jersey: Erlbaum, 1989, pp. 289–310.

Christianson S-A, Nilsson L, Mjorndal T, et al. Psychological versus physiological determinants of emotional arousal and its relationship to laboratory induced amnesia. Scand J Psychol 1986;27: 300–310.

Christiaanson RE, Ochalek K. Editing misleading information from memory: evidence for the coexistence of original and postevent information. Mem Cognit 1983;11:467–475.

Christman SS. Target-related neologism formation in jargonaphasia. Brain Lang 1994;46:109–128.

Churchill SE, Trinkaus E. Neanderthal scapular glenoid morphology. Am J Phys Anthropol 1990; 83:147–160.

Churgani HT, Phelps ME, Mazziotta JC. Positron emission tomography study of human brain functional development. Ann Neurol 1987;22:487–497.

Cicone M, Wapner W, Gardner H. Sensitivity to emotional expressions and situations in organic patients. Cortex 1980;16:145–158.

Cimino CR, Verfaellie M, Bowers D, Heilman K. Autobiographical memory: Influences of right hemisphere damage on emotionality and specificity. Brain Cognit 1991;15:106–118.

Claparede E (1911). Recognition et moitie. Archives de Psychologie Geneve 11: 79–90. Reprinted in Rapaport D, ed. Organization and pathology of thought. New York: Columbia University Press, 1951.

Clark AF, Davison K. Mania following head injury. Br J Psychiatry 1987;150:841–844.

Clark A, Harvey P, Alpert M. Medication effects in referent communication in schizophrenic patients. Brain Lang 1994;46:392–401.

Clark CVH, Isaacson RL. Effects of bilateral hippocampal ablation on DRL performance. J Comp Physiol Psychol 1965;59:137–140.

Clark DM, Teasdale JD. Diurnal variation in clinical depression and accessibility to positive and negative experiences. J Abnorm Psychol 1983;91:87–95.

Clark JD, Harris JWK. Fire and its role in early hominid lifeways. African Archaeol Rev 1985;3:3–27.

Clark P, Dibble H. Middle Paleolithic symbolism. J Anthropol Archaeol 1987;6:263–296.

Clark GA, Lindly J. The case of continuity. In: Mellars P, Stringer C, eds. The human revolution. Edinburgh: Edinburgh University Press, 1989.

Clarke-Stewart A, Thompson W, Lepore S. Manipulating children's interpretations through interrogation. Paper presented at the meeting of the Society for Research in Development. Kansas City, MO, 1989.

Clarke-Stewart KA. And daddy makes three. The father's impact on the mother and young child. Child Devel 1978;44:466–478.

Clayton DE. A comparative study of the non-nervous elements in the nervous system of invertebrates. J Entomol Zool 1932;24:3–22.

Clemente CD, Chase MH, Knauss TK, Sauerland EK, Sterman MB. Inhibition of a monosynaptic reflex by electrical stimulation of the basal forebrain or the orbital gyrus in the cat. Experientia 1966;22: 844–845.

Clifford BR, Scott J. Individual and situational factors in eyewitness testimony. J Applied Psychol 1978; 63:352–359.

Clugnet MC, LeDoux JE. Synaptic plasticity in fear conditioning circuits: Induction of LTP in the lateral nucleus of the amygdala by stimulation of the medial geniculate body. J Neurosci 1990;10: 2818–2824.

Cockburn J. Task interruption in prospective memory: a frontal lobe function. Cortex 1995;31:87–97.

Cohen AH, Rossignol S, Gillner S, eds. Neural control of rhythmic movements in vertebrates. New York: Wiley, 1988.

Cohen DAD, Prud'Homme MJL, Kalasaka JF. Tactile activity in primate primary somatosensory cortex during active arm movements. J Neurophysiol 1994; 71:161–191.

Cohen DB. Remembering and forgetting dreaming. In: Kihkstrom JF, Evans FJ, eds. Functional disorders of memory. New Jersey:Erlbaum, 1979.

Cohen DB, Squire LR. Preserved learning and retention of pattern analyzing skills in amnesia. Science 1980;210:207–209.

Cohen DB, Wolfe G. Dream recall and repression. J Consult Clin Psychol 1973;41:349–355.

Cohen HD, Rosen RC, Goldstein I. Electroencephalographic laterality changes during human sexual orgasm. Arch Sex Behav 1976;5:189–200.

Cohen L, Dehaene S. Amnesia for arithmetic facts. Brain Lang 1994;47: 214–232.

Cohen MR, Niska RW. Localized right cerebral hemisphere dysfunction and recurrent mania. Am J Psychiatry 1980;137:847–848.

Cohen RM, Weingartner H, Smallberg SA, et al. Effort and cognition in depression. Arch Gen Psychiatry 1982;39:593–597.

Colbert EH. Evolution of Vertebrates. New York: John Wiley & Sons, 1980.

Cole CB, Loftus E. The memory of children. In: Ceci SJ, Toglia MP, Ross DF, eds. Children's eyewitness memory. New York: Springer–Verlag, 1987.

Cole J, Glees P. Effects of small lesions in sensory cortex in trained monkeys. J Neurophysiol 1954;17: 1–13.

Cole PM. Children's spontaneous control of facial expression. Child Devel 1986;57:1309–1321.

Cole WG, Chan D, Hickey AJ, et al. Collagen composition of normal and myxomatous human mitral valves. J Biochem 1984;219:451–460.

Coleman R. Male and female voice quality and its relationship to vowel format frequencies. J Speech Hear Res 1971;14:123–133.

Coleman P, Flood DG. Is dendritic proliferation of surviving neurons a compensatory response to loss of neighbors in the aging brain? In: S. Finger et al., eds. Brain injury and recovery. Controversial and theoretical issues. New York: Plenum, 1988.

Colgrove FW. Individual memories. Am J Psychol 1899;10:228–255.

Colligan M, Murphy LR. Mass psychogenic illness in organizations: an overview. J Occupat Psychol 1979;52:77–90.

Colligan MJ, Pennebaker JW, Murphy LR. Mass psychogenic illness. New Jersey: Erlbaum, 1982.

Collins WA. Learning of media content: a developmental study. Devel Psychol 1970;8:215–221.

Collins WA. Temporal integration and children's understanding of social information on television. Am J Orthopsychiatry 1978;48:198–204.

Collins WA, Wellman H, Keniston AH, Wesby SD. Age-related aspects of comprehension and inference from a televised dramatic narrative. Child Devel 1978;49:389–399.

Coloban ART, Dacey RG, Alves W, et al. Neurologic and neurosurgical implications of mild head injury. J Head Trauma Rehab 1986;1:13–21.

Como P, Joseph R, Fiducia D, Siegel J. Visually evoked potentials and after-discharge as a function of arousal and frontal lesion. Proc Soc Neuroscience 1979;5:202.

Conel JL. The postnatal development of the human cerebral cortex. Cambridge, MA: Harvard University Press, 1939.

Conte JR, Schuerman JR. Factors associated with an increased impact of child sexual abuse. Child Abuse Negl 1987;2:201–211.

Conwell Y, Henderson RE. The neuropsychiatry of suicide. In: Fogel BS, Schiffer RB, eds. Neuropsychiatry. Baltimore: Williams & Wilkins, 1996.

Cook DN, Frugh H, Mehr A, et al. Hemispheric cooperation in visuospatial rotations. Brain Cogn 1994; 25:240–249.

Cools AR, Van den Bos R, Ploeger G, Ellenbroek BA. Gating function of noradrenaline in the ventral striatum. In: Willner P, Scheel-Kruger J, eds. The mesolimbic dopamine system. New York: Wiley, 1991.

Coons PM, Milstein V. Psychosexual disturbances in multiple personality disorder. A review. Dissociation 1986;1:47–53.

Coons PM, Milstein V, Marley C. EEG studies of two multiple personalities. Arch Gen Psychiatry 1982; 39:823–825.

Cooper JR, Bloom F, Roth RH. The biochemical basis of neuropharmacology. New York: Oxford University Press, 1974.

Cooper RP, Aslin RN. Preference for infant-directed speech in the first month after birth. Child Devel 1990;61:1584–1595.

Cooper PR, Maravilla K, Moody S, et al. Serial computerized tomographic scanning and the prognosis of severe head injury. Neurosurgery 1979;5: 566–569.

Corballis MC. The lopsided ape. Boston: Harvard University Press, 1991.

Corballis MC, Beale IL. The ambivalent mind: the neuropsychology of left and right. Chicago: Nelson-Hall, 1983.

Corballis MC, Morgan MJ. On the biological basis of human laterality. I. Evidence for a maturational left-right gradient. Behav Brain Sciences 1978;1: 261–269.

Corina DP, Poizner H, Bellugi U, et al. Dissociation between linguistic and nonlinguistic gestural systems: a case for compositionality. Brain Lang 1992; 43:414–447.

Corina DP, Poizner H, Bellugi U, et al. Dissociation between linguistic and nonlinguistic gestural systems. Brain Lang 1992;43:414–447.

Corkin S. Tactual-guided maze learning in man. Neuropsychologia 1965;3:339–351.

Corkin S. Some relationships between global amnesias and the memory impairments in Alzheimer's disease. In: Corkin S, et al, eds. Alzheimer's disease. New York: Raven, 1982.

Corkin S, Milner B, Rasmussen T. Somotosensory thresholds: contrasting effects of post-central gyrus and posterior parietal-lobe excisions. Arch Neurol 1970;23:41–58.

Cornell LB. Infant's recognition memory, forgetting and savings. J Exp Child Psychol 1979;28:359–374.

Cornell LB. Distributed study facilitates infants' delayed recognition memory. Mem Cognit 1980;8: 539–542.

Cornell LB, Heth CD. Response versus place learning by human infants. J Exp Psychol [Hum Learn] 1979;5:188–196.

Cornell, LB, Heth CD. Spatial cognition: gathering strategies used by preschool children. J Exp Child Psychol 1983;35:93–110.

Cornford JM. Specialized resharpening techniques and evidence of handedness. In: Callow P, Cornford JM, eds. Excavations. Enald: Geo Books, 1986, pp. 337–362.

Constable G. The Neanderthals. New York: Time-Life Books, 1973.

Corteen RS, Wood B. Autonomic response to shock-associated words in an unattended channel. J Exp Psychol 1972;4:308–313.

Corter CM, Zucker KJ, Galligan RF. Patterns in the infant search for mother during brief separation. Devel Psychol 1980;16:62–69.

Cory TL, Ormiston DW, Simmel E, Dainoff M. Predicting the frequency of dream recall. J Abn Psychol 1975;84:261–266.

Courtois CA. Healing the incest wound. New York: Norton, 1988.

Cowan N. Activation, attention and short-term memory. Mem Cognit 1993;21:162–167.

Cowan WM. The development of the brain. Scientific American. 1979;241:112–133.

Cowan WM, Fawcett JW, O'Leary DD, Stanfield BB. Regressive events in neurogenesis. Science 1984; 225:1258–1265.

Cowen R. Hubble constant: controversy continues. Science News 1995;147: 198.

Cowey A. Why are there so many visual areas. In: Schmitt FO, et al, eds. The organization of the cerebral cortex. Cambridge: MIT Press, 1981.

Cowie RJ, Robinson DL. Subcortical contributions to head movements in Macaques. I. J Neurophysiol 1994;72:2648–2664.

Cowie RJ, Smith MK, Robinson DL. Subcortical contributions to head movements in Macques. I. J Neurophysiol 1994;72:2665–2885.

Craik KJW. The nature of explanation. Cambridge: Cambridge University Press, 1943.

Cremony JC. Life among the Apaches. San Francisco: A Roman & Co, 1868.

Critchly M. Neurogical aspects of visual and auditory hallucinations. Br Med J 1939;107:634–639.

Critchly M. The parietal lobes. New York: Hafner, 1953.

Critchly M. The neurology of psychotic speech. Br J Psychiatry 1964;40:353–364.

Critchly M. The problem of visual agnosia. J Neurol Sci 1964;1:274–290.

Crockard A, Iannotti F, Kang J. Posttraumatic edema in the gerbil. In: Grossman RG, Gildenberg PL, eds. Head injury: basic clincal aspects. New York: Raven Press, 1982.

Crockett D, Bilsker D, Hurwitz T, Kozak J. Clinical utility of three measures of frontal lobe dysfunction in neuropsychiatric samples. Int J Neurosci 1986;30:241–248.

Crompton AW, Jenkins FA. Mammals from reptiles. Ann Rev Earth Planet Sci 1973;1:131–155.

Crompton AW, Jenkins FA. Origin of mammals. In: Mesozoic mammals. Lillegraven JA, et al, eds. Berkeley: University of California Press, 1979.

Cronin-Galomb A. Subcortical transfer of cognitive information in subjects with complete forebrain commissurotomy. Cortex 1986;22:499–519.

Cropper EC, Eisenman JS, Azmitia EC. An immunocytochemical study of the serotonergic innervation of the thalamus of the rat. J Comp Neurol 1985;234: 38–50.

Crosby EC, DeJonge BD, Schneider RC. Evidence for some of the trends in the phylogenetic development of the vertebrate telencephalon. In: Hassler R, Stephan H, eds. Evolution of the forebrain. Stuttgart: Verlag, 1966, pp. 333–371.

Cross HA, Harlow HF. Prolonged and progressive effect of partial isolation on the behavior of macaque monkeys. J Exp Res Personality 1965;1:39–49.

Crossman AR, Sambrook MA, Mitchell IJ, et al. Basal ganglia mechanisms mediating experimental dyskinesia in the monkey. In: Carpenter MB, Jayaraman A, eds. The basal ganglia. New York: Plenum, 1987.

Crouse JR, Toole JF, Mckinney WM, et al. Risk factors for extracranial carotid artery atherosclerosis. Stroke 1987;18:990–996.

Crovitz HF, Harvey MT, McClanahan S. Hidden memory: a rapid method for the study of amnesia using perceptual learning. Cortex 1981;17:273–278.

Crow TJ. Molecular pathology of schizophrenia: more than one disease process. Br Med J 1980;28: 66–68.

Crow TJ, Ball J, Bloom SR, et al. Schizophrenia as an anomaly of development of cerebral asymmetry. Arch Gen Psychiatry 1989;46:1145–1150.

Crowder RG. The demise of short-term memory. Acta Psychologica 1982;50:291–323.

Crowder RG. Short-term memory: where do we sand? Mem Cognit 1993;21:142–145.

Crowe SF, Kuttner M. Differences between schizophrenia and the schizophrenia-like psychosis of temporal lobe epilepsy. Neuropsychiatr Neuropsychol Behav Neurol 1991;4:127–135.

Crowne DP. The frontal eye field and attention. Psychol Bull 1983;93:232–260.

Crutcher MD, Alexander GE. Movement-related neural activity selectively coding either direction of muscle pattern in three motor areas of the monkey. J Neurophysiol 1990;64:151–163

Crutcher MD, DeLong MR. Single cell studies of the primate putamen. Exp Brain Res 1984;53:233–287.

Cubelli R, Caselli M, Neri I. Pain endurance in unilateral cerebral lesions. Cortex 1984;20:369–375.

Cubelli R, Nichelli P, Bonito V, De Tanti A, Inzaghi M. Different patterns of dissociation in unilateral spatial neglect. Brain Cognit 1991;15:139–159.

Cummings JL, Mahler ME. Cerebrovascular dementia. In: Bornstein RA, Brown GC, eds. Neurobehavioral aspects of cerebrovascular disease. New York: Oxford University Press, 1991.

Cummings JL, Mendez MF. Secondary mania with focal cerebrovascular lesions. Am J Psychiatry 1984;41:1084–1087.

Cummings WJK. An anatomical review of the corpus callosum. Cortex 1970;6:1–18.

Currier RD, Little SC, Suess JF, Andy OJ. Sexual seizures. Arch Neurol 1971;25:260–264.

Curry FKW. A comparison of left-handed and right-handed subjects on verbal and non-verbal dichotic listening tasks. Cortex 1967;3:343–352.

Curtis EA, Jacobson S, Marcus EM. An introduction to the neurosciences. Philadelphia: Saunders, 1972.

Cushing H. Papers relating to the pituitary body, hypothalamus, and parasympathetic nervous system. Springfield: Thomas, 1932.

Cutting JE. Two left hemisphere mechanisms in speech perception. Percept Psychophys 1974;16: 601–612.

Dacey RG, Alves WM, Rimel RW, et al. Neurosurgical complications after apparently minor head injury. J Neurosurg 1986;65:203–210.

Dahlberg F. Woman the gatherer. New Haven: Yale University Press, 1981.

Dahlstrom A, Fuxe K. Experimentally induced changes in intraneuronal amine levels of bulbospinal neuron systems. Acta Physiol Scand 1965;64:1–38.

Dailey CA. Psychological findings five years after head injury. J Clin Psychol 1956;12:349–353.

Dalhberg F. Woman the gatherer. New York: Yale University Press, 1981.

Dalby PR, Obrzut JE. (1991). Epidemiologic characteristics and sequela of closed head-injured children and adolescents: a review. Devel Neuropsychol 1991;7:35–68.

Dalton P. The role of stimulus familiarity in context-dependent recognition. Mem Cognit 1993;21: 223–234.

Daly D. Ictal affect. Am J Psychiatry 1958;115:97–108.

Daly D, Moulder D. Gelastic epilepsy. Neurology 1957;7:26–36.

Daly M, Wilson M. Sex, evolution and behavior. New York: Duxbury Press, 1983.

Daniel PM, Whitteridge L. The representation of the visual field on the cerebral cortex in monkeys. J Physiol 1961;59:203–221.

D'Aquili EG, Newberg AB. Religious and mystical states. Zygon 1993;28:177–200.

Dart. The predatory implemental technique of Australopithecus. Am J Phys Anthropol 1949;7:1–38.

Darwin C. The origin of species by means of natural selection. London: Murray, 1859.

Darwin C. The origin of species and the descent of man. New York: Random House, 1871.

Darwin C. The expression of emotions in man and animals. (Reprinted 1965). Chicago: University of Chicago Press, 1872.

Dawson JL, Cheung YM, Lau RTS. Developmental effects of neonatal sex hormones on spatial ability and activity skills in the white rat. Biol Psychol 1975;3:213–229.

Dauphinais ID, DeLissi LE, Crow TJ, et al. Reduction in temporal lobe size in siblings with schizophrenia. Psychiatr Res 1990;35:137–147.

Davies SN, Lester RAJ, Reymann KG, Colingridge GL. Nature 1989;338:500–503.

Davidoff DA, Buitters N, Gerstman LJ, et al. Affective/motivational factors in the recall of prose passages by alcoholic Korsakoff patients. Alcohol 1984;1:63–69.

Davidoff J, De Bleser R. Impaired picture recognition with preserved object naming and reading. Brain Cognit 1994;24:1–23.

Davidson C, Schick W. Spontaneous pain and other subjective sensory disturbances. Arch Neurol Psychiatry 1935;34:1204–1237.

Davidson RJ. Affect, cognition, and hemispheric specialization. In: Izard CE, Kagan, Zajonc RB, eds. Emotions, cognition, and behavior. Cambridge: Cambridge University Press, 1984.

Davidson RJ, Fox NA. The relation between tonic EEG asymetry and ten-month-old infant emotional responses to separation. J Abn Psychol 1989;98: 127–131.

Davidson RJ, Mednick D, Moss E, Saron C, Schaffer CE. Ratings of emotion in faces are influenced by the visual field to which stimuli are presented. Brain Cognit 1987;6:403–411.

Davidson RJ, Schaffer CE, Saron C. Effects of lateralized presentation of faces on self-reports of emotion and EEG asymmetry in depressed and non-depressed subjects. Psychophysiology 1985; 22:353–364.

Davidson RM, Bender DB. Selectivity for relative motion in the monkey superior colliculus. J Neurophysiol 1991;65:1115–1130.

Davies G. Concluding comments. In: Doris J, ed. The suggestibility of chidren's recollections. Washington, DC: American Psychological Association, 1991.

Davies SN, Lester RAJ, Reymann KG, Colingridge GL. Nature 1989;338:500–503.

Davis M. The mammalian startle response. In: Eaton RC, ed. Neural mechanisms of startle behavior. New York: Plenum, 1984.

Davis M. The amygdala and conditioned fear. In: Ag-gleton JP, ed. The amygdala. New York: Wiley-Lis, 1992.

Davis PJ. Repression and the inaccessibility of emotional memories. In: Singer JL, ed. Repression and dissociation. Chicago: Chicago University Press, 1990.

Davis PG, McEwen BS, Pfaff DW. Localized behavioral effects of tritiated estradiol implants in the ventromedial hypothalamus of female rats. Endocrinology 1979;104:898–903

Davis PH, Dambrosia JM, Schoenberg BS, et al. Ann Neurol 1987;22:319–327.

Davison C, Kelman H. Pathological laughing and crying. Arch Neurol Psychiatry 1939;42:595–643.

Dawson JLM, Cheung YM, Lau RTS. Developmental effects of neonatal sex hormones on spatial and activity skills in the white rat. Biolol Psychiatry 1975;3:213–229.

Dax ED, Reitman F, Radley-Smith E. Prefrontal leucotomy. Digest Neurol Psychiatry 1948;16:533–534.

Day J. Right hemisphere language processing in normal right-handers. J Exp Psychol Hum Percept Perform 1977;3:518–528.

Day R, Cutting JE, Copeland P. Perception of linguistic and non-linguistic dimensions of dichotic stimuli. Status Report of Haskins Laboratories 1971;27:1–6.

de Beauvoir S. The second sex. New York: Bantam, 1952.

De Boysson-Bardies B, Bacri N, Sagart L, Poizat M. Timing in late babbling. J Child Lang 19180;8: 525–539.

DeCasper AJ, Spence MJ. Prenatal maternal speech influences newborn's perception of speech sounds. Infant Behav Devel 1986;9:133–150.

Decety J, Sjoholm H, Ryding E, et al. The cerebellum participates in mental activity. Brain Res 1990;535: 313–317.

Decker HS. The lure of nonmaterialism in materialistic Europe: investigations of dissociative phenomenon. In: Quen JM, ed. Split mind/split brains. New York: New York University Press, 1986.

Deffenbacher K. The influence of arousal on reliability of testimony. In: Clifford BR, Lloyd-Boistock S, eds. Evaluating witness evidence. New York: Wiley, 1983.

DeFrance JF, Marchand JF, Sikes RW, et al. Characterization of fimbria input to nucleus accumbens. J Neurophysiol 1985;54:1553–1567.

Deglin VL, Nikolaenko NN. Role of the dominant hemisphere in the regulation of emotional states. Hum Physiol 1975;1:394–402.

Degos J-D, da Fonseca N, Gray F, Cesaro P. Severe frontal syndrome associated with infarcts of the left anterior cingulate gyrus and the head of the right caudate nucleus. Brain 1993;116:1541–1548.

Deicken RF, Calabrese G, Merrin EL, et al. Basal ganglia phosphorous metabolism in chronic schizophrenia. Am J Psychiatry 1995;152:126–129.

DeKosky ST, Heilman KM, Bowers D, Valenstein E. Recognition and discrimination of emotional faces and pictures. Brain Lang 1980;9:206–214.

Delaney RC, Rosen AJ, Mattson RH, Novelly RA. Memory function in focal epilepsy: a comparison of non-surgical unilateral temporal lobe and frontal lobe samples. Cortex 1980;16:103–117.

Delgado J. Cerebral structures involved in transmission and elaboration of noxious stimulation. J Neurophysiol 1955;18:261–275.

Delego JMR, Anand BK. Increased food intake induced by electrical stimulation of the lateral hypothalamus. Am J Physiol 1953;172:162–168.

Delgado JMR, Livingston RB. Some respiratory, vascular, and thermal responses from stimulation of the orbital surface of frontal lobe. J Neurophysiol 1948;11:39–55.

Delgado PL, Price LH, Miller HL, et al. Serotonin and the neurobiology of depression. Arch Gen Psychiatry 1994;51:865–874.

d'Elia G, Perris C. Cerebral functional dominance and memory functioning. Acta Psychiatr Scand 1974; 255:143–157.

Delis DC, Robertson LC, Efron R. Hemispheric specialization of memory for visual hierarchical stimuli. Neuropsychologia 1986;24:410–433.

DeLisis LE, Hoff AL, Schwartz JE, et al. Brain morphology in first-episode schizophrenic-like psychotic patients. Biol Psychiatry 1991;29:159–175.

Deloche G, Serion X, Scius G, Segui J. Right hemisphere language processing: lateral difference with imageable and nonimageable ambiguous words. Brain Lang 1987;30:197–205.

DeLong MR. Anatomy of the basal ganglia. In: Fogel BS, Schiffer RB, eds. Neuropsychiatry. Baltimore: Williams & Wilkins, 1995.

Davies G. Concluding comments. In: Doris J, ed. The suggestibility of children's recollections. Washington, DC: American Psychological Association, 1991.

DeLong MR, Crutcher MD, Georgopoloulos AP. Primate globus pallidus and subthalamic nucleus. Functional organization. J Neurophysiol 1985;53:530–543.

DeLong MR, Georgopoloulos AP, Crutcher MD. Corticobasal ganglia relations and coding of motor performance. Exp Brain Res 1983;7:30–40.

Dement WC, Wolpert E. The relation of eye movements, bodily mobility and external stimuli to dream content. J Exp Psychol 1958;53:543–553.

Dempster FN. Memory span: sources of individual and developmental differences. Psychol Bull 1981; 89:63–100.

Dennell R. European prehistory. London: Academic Press, 1985.

Dennis M, Whitaker HA. Language acquisition following hemidecortication. Brain Lang 1976;3:494–533.

Denenberg VH. Hemispheric laterality in animals and the effects of early experience. Behav Brain Sci 1981;4:1–49.

Denenberg VH. Hemispheric laterality in animals and the effects of early experience. Behav Brain Sci 1981;4:1–49.

Denes G, Semenza C. Auditory modality-specific anomia. Evidence from a case of pure word deafness. Cortex 1975;11:401–411.

Denes G, Semenza C, Stoppa E, Lis A. Unilateral spatial neglect and recovery form hemiplegia. Brain 1982;105:543–552.

Dennis W. Causes of retardation among institutionalized children. Iran. J Gen Psychol 1960;96:47–59.

Dennis W. Causes of retardation among institutionalized children. J Gen Psychol 1975;96:47–59

Denny-Brown D. The nature of apraxia. J Nerv Ment Dis 1958;126:15–56.

Denny-Brown D. The cerebral control of movement. Liverpool: Liverpool University Press, 1966.

Denny-Brown D, Chambers RA. The parietal lobe and behavior. Proc Assoc Res Nerv Ment Dis 1958; 36:35–117.

Denny-Brown D, Meyer JS, Horenstein S. The significance of perceptual rivalry resulting from parietal lobe lesion. Brain 1952;75:433–471.

Dent HR, Stephenson GM. An experimental study of the effectiveness of different techniques of questioning child witnesses. Br J Soc Clin Psychol 1979; 18:41–51.

DeQuardo JR, Tandon R, Goldman R, et al. Ventricular enlargement, neuropsychological status, and premorbid function in schizophrenia. Biol Psychiatry 1994;36:323–340.

DeRenzi E. Disorder of space exploration and cognition. New York: Wiley, 1982.

DeRenzi E. Prosopagnosia in two patients with CT-scan evidence of damage confined to the right hemisphere. Neuropsychologia 1986;24:385–389.

DeRenzi E, Faglioni P. The comparative efficiency of intelligence and vigilance tests in detecting hemisphere cerebral damage. Cortex 1965;1:410–433.

DeRenzi E, Faglioni P, Scotti G, et al. Impairment of color sorting behavior after hemispheric damage. Cortex 1972;8:147–163.

DeRenzi E, Faglioni P, Spinnler H. The performance of patients with unilateral brain damage on face recognition tasks. Cortex 1968;4:17–34.

DeRenzi E, Pieczuro A, Vignolo LA. Oral apraxia and aphasia. Cortex 1966;2:56–73.

DeRenzi E, Scotti G. The influence of spatial disorders in impairing tactual discrimination of shapes. Cortex 1969;5:53–62.

DeRenzi E, Scotti G, Spinnler H. Perceptual and associative disorders of visual recognition: relationship to the side of the cerebral lesion. Neurology 1969; 19:634–642.

DeRenzi E, Spinnler H. Facial recognition in brain-damaged patients. An experimental approach. Neurology 1966;16:145–152.

DeRenzi E, Spinnler H. Impaired performance on color tasks in patients with hemispheric damage. Cortex 1967;3:194–216.

DeRenzi E, Vignolo LA. The token test. Brain 1962; 85:665–678.

DeRenzi E, Zambolini A, Crisi G. The pattern of neuropsychological impairment associated with left posterior cerebral artery infarcts. Brain 1987;110: 1099–1116.

de Ropp RS. Psychedelic drugs and religious experience. In: Lehmann AC, Myers JE, eds. Magic, witchcraft and religion. Mountain View, CA: Mayfield, 1993.

Deruelle C, de Schonen S. Hemispheric assymetries in visual pattern processing in infancy. Brain Cognit 1991;16:151–179.

Desimone R, Gross CG. Visual areas in the temporal cortex of the macaque. Brain Res 1979;178:363–380.

Desimone R, Schein SJ. Visual properties of neurons in area V4 of the Macaque: sensitivity to stimulus form. J Neurophysiol 1987;57:835–867.

Desmedt JE. Active touch exploration of extrapersonal space elicits specific electrogenesis in the right cerebral hemisphere of intact right-handed man. Proc Natl Acad Sci U S A 1977;74:4037–4040.

Desmo V. Children's understanding of affect terms. (Abstract). Doctorial Dissertation. Cambridge: Harvard University, 1975.

Desmone R, Gross CG. Visual areas in the temporal cortex of the macaque. Brain Res 1979;178:363–380.

DeUrso V, Denes G, Testa S, Semenza C. The role of the right hemisphere in processing negative sentences in context. Neuropsychologia 1986;24: 289–292.

De Vaus D, McAllister I. Gender differences in religion. Am Sociol Rev 1987;52:472–481.

Devinsky O, Putnam F, Grafman J, Bromfield E, Theodore WH. Dissociative states and epilepsy. Neurology 1989;39:835–840.

Devinksy O, Morrell MJ, Vogt BA. Contributions of anterior cingulate cortex to behavior. Brain 1995; 118:279–306.

Devore I. Mother-infant relations in free-ranging baboons. In: Rheingold HL, ed. Maternal behavior in mammals. New York: Wiley, 1963.

Devore I. Primate behavior. In: Tax S, ed. Horizons of anthropology. Chicago: Aldine, 1964.

Devore I. Male dominance and mating behavior in baboons. In: Beach FA, ed. Sex and behavior. New York: Wiley, 1977.

Dewson JH, Burlingame H, Kizer K, Dewson S, Kenney P, Pribram K. Hemispheric asymmetry of auditory function in monkeys. J Acoustical Soc Am 1975;58:Supplement 1.

Dewson JH, Pribram K, Lynch JC. Effects of ablations of temporal cortex upon speech sound discrimination in the monkey. Exp Neurol 1969;24:579–591.

Diamond A. Frontal lobe involvement in cognitive changes during the first year of life. In: Gibson KR, Petersen AC, eds. Brain maturation and cognitive development. New York: Aldine De Gruyter, 1991.

Diamond DM, Weinberger NM. Role of context in the expression of learning-induced plasticity of single neurons in auditory cortex. Behav Neurosci 1989; 103:471–494.

Diamond IT. The evolution of the tectal-pulvinar system in mammals: structural and behavioural studies of the visual system. Symp Zool Soc London 1973;33:205–233.

Diamond MC. Rat forebrain morphology: right-left; male-female; young-old; enriched-impoverished. In: Glick SD, ed. Cerebral lateralization in nonhuman primates. Orlando: Academic Press, 1985.

Diamond M. Environmental influences on the young brain. In: Brain maturation and cognitive development. Gibson KR, Petersen AC, eds. New York: Aldine De Gruyter, 1991.

Dicks D, Myers RE, Kling A. Uncus and amygdaloid lesions on social behavior in free ranging rhesus monkeys. Science 1969;160:69–71.

Dietz MA, Goetz CG, Stebbins GT. Evaluation of a modified inverted walking stick as a treatment for parkinsonian freezing episodes. Mov Disord 1990; 5:243–247.

Dikmen S, McLean A, Temkin N. Neuropsychological and psychosocial consequences of minor head injury. J Neurol Neurosurg Psychiatry 1986;49: 1227–1232.

Dikmen S, Reitan RM. Psychological deficits and recovery of functions after head injury. Trans Am Neurol Assoc 1976;101:72–79.

Dikmen S, Reitan RM, Temkin NR. Neuropsychological recovery in head injury. Arch Neurol 1983; 40:333–338.

Dikmen S, Temkin N, McLean A, Wyler A, Machamer J. Memory and head injury severity. J Neurol Neurosurg Psychiatry 1987;50:1613–1618.

Dillon H, Leopold RL. Children and the post-concussion syndrome. J Am Med Assoc 1961;14:175–186.

Dimond SJ. Depletion of attentional capacity after total commissurotomy in man. Brain 1876;99: 347–356.

Dimond SJ. Tactual and auditory vigilance in split-brain man. J Neurol Neurosurg Psychiatry 1979; 42:70–74.

Dimond SJ. Neuropsychology. London: Butterworths, 1980.

Dimond SJ, Beaumont JG. Experimental studies of hemisphere function in the human brain. In: Dimond SJ, Beaumont JG, eds. Hemisphere function in the human brain New York: John Wiley, 1974.

Dimond S, Farrington L. Emotional responses to films shown to the right or left hemisphere of the brain measured by heart rate. Acta Psychol 1977;41: 255–260.

Dimond S, Farrington L, Johnson P. Differing emotional responses from right and left hemispheres. Nature 1976;261:690–692.

Dimond SJ, Gazzaniga MS, Gibson AR. Cross field and within field integration of visual information. Neuropsychologia 1972;10:379–381.

Di Pasquale G, Andreoli A, Pinelli G, et al. Cerebral ischemia and asymptomatic coronary artery disease: a prospective study of 83 patients. Stroke 1986; 17:1098–1101.

Dixon NF. Subliminal perception: the nature of a controversy. New York: McGraw-Hill, 1971.

Dixon NF. Preconscious processing. New York: Wiley, 1981.

Dodds AG. Hemispheric differences in tactuo-spatial processing. Neuropsychologia 1978;16:247–254.

Dollard J, Miller NE. Personality and psychotherapy. New York: McGraw-Hill, 1950.

Dombovy ML, Sandok BA, Basford JB. Rehabilitation for stroke: a review. Stroke 1986;17:363–369.

Donaldson MA, Gardner R Jr. Diagnosis and treatment of traumatic stress among women after childhood incest. In: Figley CR, ed. Trauma and its wake. New York: Brunner/Mazel, 1985.

Donaldson SK, Westerman MA. Development of children's understanding of ambivalence and causal theories of emotions. Devel Psychol 1986;22:655–662.

Dong WK, Chudler EH, Sugiyama K, et al. Somatosensory, multisensory, and task-related neurons in cortical area 7b (PF) of unanesthetized monkeys. J Neurophysiol 1994;72:542–564.

Donkelaar HJT. Brainstem mechanisms of behavior: comparative aspects. In: Klemm WR, Vertes RP, eds. Brainstem mechanisms of behavior. New York: Wiley, 1990.

Dood DH, Bradshaw JM. Leading questions and memory: pragmatic constraints. J Verbal Learning and Verbal Behavior 1980;19:695–704.

Doty R, Reyes P, Gregor T. Presence of both odor identification and detection deficits in Alzheimer's disease. Brain Res Bull 1987;18:598.

Doty RL, et al. Sex differences in odor identification abilty: a cross cultural analysis. Neuropsychologia 1985;23:667–672.

Doty RW. Nongeniculate afferents to striate cortex in macaques. J Comp Neurol 1983;218:159–173.

Doty RW. Time and memory. In: McGaugh JL, Winberger NM, Lynch G, eds. Brain Organ Mem. New York: Oxford University Press, 1990.

Doty RW, Overman WH. Mnemonic role of forebrain commissures in macaques. In: S. Harnad, R. W. Doty, L. Goldstein, J. Jaynes, & G. Krauthamer, (Eds.), Lateralization in the nervous system. New York: Academic Press, 1977, pp. 75–88.

Douglas RJ. The hippocampus and behavior. Psychol Bull 1967;67:416–442.

Dow BM. Functional classes of cells and their laminar distribution in monkey visual cortex. J Neurophysiol 1974;37:927–946.

Dow RS. The evolution and anatomy of the cerebellum. Biol Rev Cambridge Philosophical Soc 1942; 17:179–220.

Dowling JE. Neurons and networks. Boston: Harvard University Press, 1992.

Downer J. Changes in visual gnostic functions and emotional behavior following unilateral temporal lobe damage in 'split-brain' monkey. Nature 1961; 191:50–51.

Drake CG. Cerebral arteriovenous malformation: considerations from an experience with surgical treatment of 166 cases. Clin Neurosurg 1978;26:145–208.

Drake ME. Epilepsy and multiple personality: clinical and EEG findings in 15 cases. Epilepsia 1986; 27:635.

Dreyfus-Brisac C. Sleep ontogenesis in human prematures after 32 weeks of age. Devel Psychobiol 1970; 3:91–121.

Dreifuss JJ, Murphy JT, Gloor P. Contrasting effects of two identified amygdaloid efferent pathways on single hypothalamic neurons. Neurophysiol 1968; 31:237–248.

Drewe EA. The effect of type and area of brain lesion on Wisconsin Card Sorting Task performance. Cortex 1974;10, 159–170.

Dreyfus-Brisac C, Monod N. The electroencephalogram of full-term newborns and premature infants. In: Drefus-Brisac C, ed. Handbook of electroencephalography and clinical neurophysiology. V. 6b. Amsterdam: Elsevier, 1975.

Duckett S, Winick M. Malnutrition. In: Duckett S, ed. Pediatric neuropathology. Philadelphia: Williams & Wilkins, 1955.

Dudycha GJ, Dudycha MM. Some factors and characteristics of childhood memories. Child Development 1933;4:265–278.

Dudycha GJ, Dudycha MM. Childhood memories. Psychol Bull 1941;38:668–682.

Dunnett SB, Toniolo G, Fine A, et al. Transplantation of embryonic ventral forebrain neurons to the neocortex of rats with lesions of the nucleus basalis magnocellularis—II. Neuroscience 1985;16:787–797.

Durant W. The life of Greece. New York: Simon & Shuster, 1939.

Duvall D. A new question of pheromones: aspects of possible chemical signaling and reception in the mammal-like reptiles. In: Maclean PE, Roth JJ, Roth EC, eds. The ecology and biology of mammal-like reptiles. Washington, DC: Smithsonian Institute Press, 1986, pp. 219–238.

Dwyer JW, Rinn WE. The role of the right hemisphere in contextual inference. Neuropsychologia 1981;19: 479–482.

Eadie BJ. Embraced by the light. California: Gold Leaf Press, 1992.

Eagly A, Steffen V. Gender and aggressive behavior. Psychol Bull 1986;100:309–330.

Earnest MP, Monroe PA, Yarnell PR. Cortical deafness. Neurology 1977;27:1172–1175.

Easterbrook JA. The effect of emotion on cue utilization and the organization of behavior. Psychol Rev 1959;66:183–201.

Eckenhoff MF, Rakic P. Synaptogenesis in the dentate gyrus of primates. Proc Soc Neurosci 1981;7:6.

Eckenhoff MF, Rakic P. Nature and fat of proliferative cells in the hippocampal dentate gyrus during the lifespan of the rhesus monkey. J Neurosci 1988; 8:2729–2747.

Eckenrode J, Laird M, Doris J. School performance and disciplinary problems among abused and neglected children. Devel Psychol 1993;29:53–62.

Eckenrode J, Larid M, Doris J. School performance and disciplinary problems among abused and neglected children. Devel Psychol 1993;29:53–62.

Edeline J-M, Massioui NN, Dutrieux G. Frequency-specific cellular changes in the auditory system during acquisition and reversal of discriminative conditioning. Psychobiology 1990;18:382–393.

Edeline J-M, Weinberger NM. Subcortical adaptive filtering in the auditory system: associative receptive field plasticity in the dorsal medial geniculate body. Behav Neurosci 1991;105:154–175.

Edelman GM. Neural darwinism. New York: Basic Books, 1987.

Edelsky C. Question intonation and sex role. Lang Sociol 1979;8:15–32.

Edwards SB. The deep cell layers of the superior colliculus. In: Hobson JA, Brazior MAB, eds. The reticular formation revisited. New York: Raven Press, 1980.

Efron R. The effect of handedness on the perception of simultaneity and temporal order. Brain 1963;86:261–284.

Egger MD, Flynn JP. Effect of electrical stimulation of the amygdala on hypothalamically elicited attach behavior. J Neurophysiol 1963;26:705–720.

Ehrhardt AA, Baker WW. Fetal androgens, human central nervous system differentiation and behavior sex differences. In: Freidman RC, et al, eds. Sex differences in behavior. New York: Wiley, 1974.

Ehrlichman HM, Antrobus JS, Wiener M. EEG assymetry and sleep mentation during REM and NREM. Brain Cognit 1985;4:477–485.

Ehrlichman H, Barrett J. Right hemisphere specialization for mental imagery: a review of the evidence. Brain Cognit 1983;2:55–76.

Ehrlichman H, Bastone L. Olfaction and emotion. In: Serby MJ, Chobor KL, eds. Science of Olfaction. New York: Springer-Verlag, 1992.

Eibl-Eibesfeldt I. Ethology. New York: Holt, 1990.

Eich E. The cue-dependent nature of state-dependent memory. Mem Cognit 1980;8:157–173.

Eich E. Theoretical issues in state-dependent memory. In: Roediger HL, Craik FIM, eds. Varies of memory and consciousness. Papers presented in honor of Endel Tulving. New Jersey: Erlbaum, 1987.

Eich E, Metcalfe J. Mood dependent memory for internal versus external events. J Exp Psychol Learn Mem Cognit 1989;15:443–455.

Eichenbaum H, Otto T, Cohen NJ. Two functional components of the hippocampal memory system. Behav Brain Sci 1994;17:449–518.

Eisele JA, Aram DM. Comprehension and imitation of syntax following early hemisphere damage. Brain Lang 1994;46:212–231.

Eisenberg N, Fabes RA, Miller PA, et al. Relation of sympathy and personal distress to prosocial behavior. J Person Soc Psychol 1989;57:55–66.

Ekman P. Universal and cultural differences in facial expressions of emotions. In: Cole J, ed. Nebraska Symposium on Motivation. Lincoln: University of Nebraska Press, 1972.

Ekman P. Facial expression and emotion. Am Psychol 1993;48:384–392.

Elia JC. The post concussion syndrome. Industrial Med 1972;41:23–31.

Elia I. The female animal. New York: Holt, 1988.

Elkins RL. Attentuation of x-ray induced taste aversion by the olfactory bulb or amygdaloid lesions. Physiol Behav 1980;24:515–521.

Ellen P, Wilson AS, Powell EW. Septal inhibition and timing behavior in the rat. J Comp Neurol 1964;10:120–132.

Ellis AW. Spelling and writing, and reading and speaking. In A. W., 1982

Ellis E, ed. Normality and pathology in cognitive functions. London: Academic Press, 1990.

Ellis E, Ashbrook PW. Resource allocation model of the effects of depressed mood states in memory. In: Fiedler K, Forgas J, eds. Affect, cognition, and social behavior. Toronto: Hogrefe, 1987.

Ellis HC, Ashbrook PW. The "state" of mood and memory research. In: Kuiken D, ed. Mood and memory. Newbury Park: Sage, 1991.

Ellis HD, Shepherd JW. Recognition of upright and inverted faces presented in the left and right visual fields. Cortex 1975;11:3–7.

Ellison G. Stimulant-induced psychosis, the dopamine theory of schizophrenia and the habenula. Brain Res Rev 1994;19:223–239.

Emde RN, Koenig KL. Neonatal smiling and rapid eye movement states. Am Acad Child Psychiatry 1969;8:57–67.

Emde RN, Metcalf DR. An electroencephalographic study of behavior and rapid eye movement states in the human newborn. J Nerv Ment Dis 1970;150:376–386.

Emson RH. The reactions of the sponge to applied stimuli. Comp Biochem Physiol 1966;18:805–827.

Encha-Razavi F. Fetal neuropathology. In: Duckett S, ed. Pediatric neuropathology. Philadelphia: Williams & Wilkins, 1995.

Endroczi E, Schreiberg G, Lissak K. The role of central activating and inhibitory structures in the control of pituitary-adrenocortical functions. Acta Physiol 1963;24:211–221.

Erdelyi MH. A new look at the new look: perceptual defense and vigilance. Psychol Rev 1974;81:1–25.

Erdelyi MH, Goldberg B. Let's not sweep repression under the rug. In: Kilhstrom JF, Evans FJ, eds. Functional disorders of memory. New Jersey: Erlbaum, 1979.

Erickson MH, Rossi EL. Experiencing hypnosis. New York: Irvington Press, 1981.

Erickson RP, Di Lorenzo PM, Woodburgy MA. Classification of taste responses in brain stem. J Neurophysiol 1994;71:2139–2147.

Erickson T. Erotomania (nymphomania) as an expression of cortical epileptiform discharge. Arch Neurol Psychiatry 1945;53:226–231.

Erwin J. Rhesus monkey vocal sounds. In: Bourne G, ed. The rhesus monkey. New York: Academic Press, 1975.

Eskandar EN, Optican LM, Richmond BJ. Role of inferior temporal neurons in visual memory II. Multiplying temporal waveforms related to vision and memory. J Neurophysiol 1992;68:1296–1306.

Eskandar EN, Richmond BJ, Optican LM. Role of inferior temporal neurons in visual memory I. Temporal encoding of information about visual images, recalled images, and behavioral context. J Neurophysiol 1992;68:1277–1295.

Eth S, Pynoos RS. Developmental perspective on psychic trauma in childhood. In: Figley CR, ed. Trauma and its wake. New York: Brunner/Mazel, 1985.

Ettlinger G. The description and interpretation of pictures in cases of brain lesion. J Ment Sci 1960;106: 1337–1346.

Evans C, Richardson PH. Improved recovery and reduced post-operative stay after therapeutic suggestions during general anaesthesia, Lancet 1988; 8609:491–493.

Evans FJ. Contextual forgetting: posthypnotic source amnesia. J Abn Psychol 1979;88:556–563.

Evans JJ, Heggs AJ, Hodges JR. Progressive prosopagnosia associated with right temporal lobe atrophy. Brain 1995;118:1–13.

Evarts EV. Activity of pyramidal tract neurons during postural fixation. J Neurophysiol 1969;32:375–385.

Evarts EV, Wise SP. Basal ganglia outputs and motor control. In: Evered D, O'Connor M, eds. Functions of the basal ganglia. London: Pitman, 1984, pp. 83–96.

Everitt BJ, Morris KA, O'Brien A, Robbins TW. The basolateral amygdala-ventral striatal system and conditioned place preference: further evidence of limbic striatal interactions underlying reward-related processes. Neuroscience 1991;42:1–18.

Eysenck MW. Anxiety, learning, and memory. J Res Personality 1979;13:363–385.

Faaborg-Anderson KC. Electromyographic investigation of intrinsic laryngeal muscles in humans. Acta Physiol Scand 1957;140:1–148.

Faber R, Abrams R, Taylor M, Kasprisin A, Morris C, Weisz R. Comparison of schizophrenic patients with formal thought disorder and neurologically impaired patients with aphasia. Am J Psychiatry 1983;140:1348–1351.

Fagan JF. Infant memory: history, current trends, relations to cognitive development. In: Moscovitch M, ed. Infant memory. New York: Plenum, 1984.

Fagot BI. Beyond the reinforcement principle. Dev Psychol 1985;21:1097–1104.

Fagot BI. Gender labeling and the development of sex-typed behaviors. Dev Psychol 1986;22:440–443.

Fahy TJ, Irving MH, Miller P. Severe head injuries. Lancet 1967;2:475–478.

Falk, D. Cerebral cortices of East African Early hominids. Science 1983a;221:1072–1074.

Falk D. The Taung endocast: a reply to Holloway. Am J Phys Anthropol 1983b;60:17–45.

Falk D. Brain evolution in Homo: the "radiator" theory. Behav Brain Sci 1990;13:333–381.

Falkai, Bogerts B. Cell loss in hippocampus of schizophrenics. Eur Arch Psychiatry Neurol Sci 1986; 236:154–161.

Fallon J. Histochemical characterization of dopaminergic, noradrenergic and serotonergic projections to the amgydala. In: Ben-Ari Y, ed. The amygdaloid complex. Amersterdam: Elsevier, 1981, pp. 175–184.

Fallon J, Ceiofi P. Distribution of monoamines within the amygdala. In: Aggleton JP, ed. The amygdala. New York: Wiley-Lis, 1992.

Farah MJ. The neural basis of imagery. Trends Neurosci 1989;12:395–399.

Farah MJ, Levine DN, Calvanio R. A case study of mental imagery deficit. Brain Cognit 1988;8: 147–164.

Farb P. Word play. New York: Bantam, 1975.

Farber D, Njiokiktjien C. Pediatric behavioural neurology, 4. Developing brain and cognition. Amsterdamn: Suyi, 1993.

Farrar MJ, Goodman GS. Developmental differences in the relation between scripts and episodes: do they exist? In: Fivush R, Hudson JA, eds. Knowing and remembering in young children. New York: Cambridge University Press, 1990.

Farrell W. The myth of male power. New York: Simon & Schuster, 1993.

Faull RLM, Villiger JW. Benzodiazepine receptors in the striatum of the human brain. In: Carpenter MB, Jayaraman A, eds. The basal ganglia. New York: Plenum, 1987.

Fedigan L. Primates and paradigms: sex roles and social bonds. Montreal: Elden Press, 1992.

Fedio P, Van Buren J. Memory deficits during electrical stimulation of the speech cortex in conscious man. Brain Lang 1974;1:29–42.

Feigley DA, Spear NE. Effect of age and punishment condition on long-term retention by the rat of active- and passive-avoidance learning. J Comp Physiol Psychol 1970;73:515–526.

Feldman S, Saphier D, Conforti N. Hypothalamic afferent connections mediating adrenocortical responses that follow hippocampal stimulation. Exp Neurol 1987;98:103–109.

Feldman HM, Holland AL, Kemp SS, Janosky JE. Language development after unilateral brain injury. Brain Lang 1992;42:89–102.

Feldman S, Saphier D, Conforti N. Hypothalamic afferent connections mediating adrenocortical responses that follow hippocampal stimulation. Exp Neurol 1987;98:103–109.

Feldman S. Neurophysiological mechanisms modifying afferent hypothalamic-hippocampal condition. Exp Neurol 1962;5:269–291.

Felleman DJ, Kaas JH. Receptive-field properties of neurons in middle temporal visual area (MT) of owl monkeys. J Neurophysiol 1974;52:488–513.

Fernald A. Prosody in speech to children: Prelinguistic and linguistic functions. In: Vasta, R, ed. Annals of Child Development 1991, pp 43–80.

Fernald A. Meaningful melodies in mother's speech to infants. In: Papousek H, Jurgens U, Papousek M, eds. Origins and development of nonverbal communication: evolutionary, comparative, and methodological aspects. Cambridge: Cambridge University Press, 1992.

Fernald A. Approval and disapproval: infant responsiveness to vocal affect in familiar and unfamiliar languages. Child Devel 1993;64:657–674.

Fernald A, Taescherner T, Dunn J, Papousek M, Boysson-Bardies B, Fukui I. A cross-language study of prosodic modifications in mothers' and fathers' speech to preverbal infants. J Child Lang 1989;16:477–501.

Fernandez A, Glenberg AM. Changing environmental context does not reliably affect memory. Mem Cognit 1985;13:333–345.

Ferro JM, Kertesz A, Black SE. Subcortical neglect. Neurology 1987;37:1487–1492.

Feshbach ND. Sex differences in empathy and social behavior in children. In: Eisenberg N, ed. The development of prosocial behavior. New York: Academic Press, 1982.

Feshbach ND, Roe K. Empathy in six and seven year olds. Child Devel 1968;39:133–145.

Fibiger HC, Phillips AG. Reward, motivation, cognition. In: Bloom FE, ed. Handbook of physiology, Vol. IV. Baltimore: American Physiological Society, 1986.

Fichtner CG, Kuhlman MJ, Hughes JR. Decreased episodic violence and increased control of dissociation in a carbamazepine-treated case of multiple personality. Biol Psychiatry 1990;27:1045–1052.

Fieschi C, Carolei A, Fiorelli M, et al. Changing prognosis of primary intracerebral hemorrhage: results of a clinical and computed tomographic follow up study of 104 patients. Stroke 1988;19:192–195.

Fiez JA, Petersen SE, Cheney MK, Raichle ME. Impaired non-motor learning and error detection associated with cerebellar damage. Brain 1992;115:155–178.

Findlay R, Ashton R, McFarland K. Hemispheric differences in image generation and use in the haptic modality. Brain Cognit 1994;25:67–78.

Finlay JM, Abercrombie ED. Stress induced sensitization of norepinephrine release in the medial prefrontal cortex. Proc Soc Neurosci 1991;17:151.

Finlayson MAJ. A behavioral manifestation of the development of interhemispheric transfer of learning in children. Cortex 1975;12:290–295.

Fischer RP, Carlson J, Perry JF. Postconcussive hospital observation of alert patients in a primary trauma center. J Trauma 1981;21:920–924.

Fischer RS, Alexander MP, D'Esposito MD, Otto R. Neuropsychological and neuroanatomical correlates of confabulation. J Clin Exp Neuropsychol 1995;17:20–28.

Fish B, Marcus J, Hans SK, et al. Infants at risk for schizophrenia. Arch Gen Psychiatry 1992;49:221–235.

Fisher CM. The arterial lesions underlying lacunes. Acta Neuropathol 1969;12:1–15.

Fisher CM. Transient global amnesia: precipitating activities and other observations. Arch Neurol 1982;39:605–608.

Fisher CM. Hunger and the temporal lobe. Neurology 1994;44:1577–1579.

Fisher CM. Brain herniation. Can J Neurol Sci 1995;22;83–91.

Fisher CM, Adams RD. Observations on brain embolism with special reference to the mechanisms of hemorrhage infarction. J Neuropathol Exp Neurol 1951;10:92–93.

Fivush R. Learning about school: the development of kindergartners' school scripts. Child Devel 1984;55:1697–1709.

Fivush R. Developmental perspectives on autobiographical recall. In: Goodman GS, Bottoms BL, eds. Child victims, child witnesses. New York: Guilford Press, 1993.

Fivush R, Hamond N. Autobiographical memory across the preschool years: toward reconceptualizing childhood amnesia. In: Fivsh R, Hudson J, eds. Knowing and remembering in young children. Cambridge: Cambridge University Press, 1990.

Fivush R, Gray JT, Fromhoff FA. Two-year-olds talk about the past. Cognit Devel 1987;2:393–409.

Fivush R, Hudson J, Nelson K. Children's long-term memory for a novel event. Merrill Palmer Quarterly 1984;30:303–316.

Flamsteed S. Crisis in the cosmos. Discover. March 1995, pp. 67–77.

Flavell JH. Cognitive monitoring. In Dickson, W. P. (Ed.). Children's Communicative Skills. New York: Academic Press, 1981.

Flor-Henry P. Psychosis and temporal lobe epilepsy: a controlled investigation. Epilepsia 1969;10:363–395.

Flor-Henry P. Cerebral basis of psychopathology. Boston: John Wright, 1983.

Flor-Henry P, Yeudall LT, Koles J, Howarth BG. Neuropsychological and power spectral EEG investigations of the obsessive-compulsive syndrome. Biol Psychiatry 1979;14:119–129.

Flournoy H. Hallucinations. Encephale 1923;2:566–572.

Flynn JP, Edwards SB, Bandler RJ. Changes in sensory and motor systems during centrally elicited attack. Behav Sci 1971;16:1–19.

Foa EB, Feske U, Murdock TB, et al. Procesing of threat-related information in rape victims. J Abn Psychol 1991;100:156–162.

Fodor J, Katz J. The structure of language. New Jersey: Prentice-Hall, 1964.

Foerester A. Return of function after optic tract lesions in adult rats. In: Flohr H, ed. Post lesion neural plasticity. New York: Springer-Verlag, 1988.

Foerster OO, Gagel O. Die Vorderseitenstrangdurschschneidun biem Menschen (English summary). Z. Neurology and Psychiatry 1932;138:1–92.

Foldi NS, Cicone M, Gardner H. Pragmatic aspects of communication in brain-damaged patients. In: Segalowitz SS, ed. Language functions and brain organization. New York: Academic Press.

Folstein SE, Brandt J, Folstein MF. Huntington's disease. In: Cummings E, ed. Subcortical dementia. New York: Springer, 1990.

Fontenot DJ. Visual field differences in the recognition

of verbal and nonverbal stimuli in man. J Comp Physiol Psychol 1973;85:564–569.

Fontenot DJ, Benton AL. Tactile perception of direction in relation to hemispheric locus of lesion. Neuropsychologia 1971;9:83–88.

Foote SL, Bloom FE, Aston-Jones G. Nucleus locus ceruleus. Physiol Rev 1983;63:844–914.

Foote SL, Morrison JH. Postnatal development of laminar innervation patterns by monoaminergic fibers in Macaca fascicularis primary visual cortex. J Neurosci 1984;4:2667–2680.

Ford CS, Beach FA. Patterns of sexual behavior. New York: Harper, 1951.

Fordyce DJ, Roueche JR, Prigatano GP. Enhanced emotional reactions in chronic head trauma patients. J Neurol Neurosurg Psychiatry 1983;46: 620–624.

Foreman N, Stevens R. Relationship between the superior colliculus and hippocampus. Neural and behavioral considerations. Behav Brain Sci 1987;10: 101–152.

Forgas JP, Bower GH, Krantz SE. The influence of mood on perception and social interactions. J Exp Soc Psychol 1984;20:497–513.

Forrest DV. Bipolar illness after right hemispherectomy. Arch Gen Psychiatry 1982;39:817–819.

Foulkes WD. Dream reports from different stages of sleep. J Abnorm Soc Psychol 1962;65:14–25.

Foulkes D. A grammar of dreams. New York: Basic Books, 1978.

Fox CL, Perez-Perez A. The diet of the Neanderthal child Gibraltar 2 (Devil's Tower) through the study of vestibular striation pattern. J Hum Evol 1993;24: 29–41.

Fox NA, Bell MA. Electrophysiological indices of frontal lobe development. In: Diamond A, ed. The development and neural bases of high cognitive functions. New York: New York Academy of Sciences, 1990.

Franco L, Sperry RW. Hemispheric lateralization for cognitive processing of geometry. Neuropsychologia 1977;15:107–111.

Frankel F. Hypnosis: trance as a coping mechanism. New York: Plenum, 1976.

Frank LA. The big splash. Avon: New York, 1990.

Frayer DW, Wolpoff MH, Thorne AG, Smith FH, Pope GO. Theories of modern human origins: the paleontological test. Am Anthropol 1993;95:14–50.

Frazier JG. The golden bough. New York: Macmillan, 1950.

Freburg Berry M. Language disorders in children. New York: Appleton, 1969.

Fredrikson R. Repressed memories: a journey to recovery from sexual abuse. New York: Simon & Shuster, 1992.

Freed CR, Breeze RE, Rosenberg NL, et al. Transplantation of human fetal dopamine cells for Parkinson's disease. Archives of Neurology 1990;47:505–512.

Freedman DG. Human infancy. An evolutionary perspective. Hillsdale, NJ: Erhlbaum, 1974.

Freedman DG. Sexual dimorphism and the status hi-

erarchy. In: Omark DR, Strayer FF, Freedman DG. Dominance relations. New York: Garland Press, 1980.

Freedman M. Parkinson's disease. In: Cummings JL, eds. Subcortical dementia. New York: Oxford, 1990.

Freedman SJ. Perceptual changes in sensory deprivation. J Nerv Ment Dis 1961;132:17–21.

Freedman WL. Science 1994;266:539–540.

Freeman FG, Traugott J. Hemispheric processing of emotional stimuli. Neuropsychiatr Neuropsychol Behav Neurol 1993;6:209–213.

Freeman W, Watts JW. Psychosurgery. Springfield, IL: Charles C. Thomas, 1942.

Freeman W, Watts JW. Prefrontal lobotomy. Am J Psychiatry 1943;99:798–806.

Freeman W, Williams J. Human sonar. J Nerv Ment Dis 1952;32:456–462.

Freeman W, Williams JM. Hallucinations in Braille. Arch Neurol Psychiatry 1953;70:630–634.

Freemon FR, Nevis AH. Temporal lobe sexual seizures. Neurology 1969;19:87–90.

French GM. A deficit associated with hypermotility in monkeys with lesions of the dorsolateral frontal granular cortex. J Comp Physiol Psychol 1959;52: 25–28.

French GM, Harlow HF. Locomotor reaction decrement in normal and brain damaged monkeys. J Comp Physiol Psychol 1955;48:496–501.

French JD. The frontal lobes and association. In: Warren JM, Akert K, eds. The frontal granular cortex and behavior. New York: McGraw-Hill, 1964, pp. 56–74.

French JD, Magoun HW. Effects of chronic lesions in central cephalic brain stem of monkeys. Arch Neurol Psychiatry 1952;68:591–604.

Freud S. On aphasia, 1991.

Freud S. The neuro-psychoses of defense. In: Strachey J, ed. The standard edition. Vol. 3. London: Hogarth Press, 1894.

Freud S. Further remarks on the neuro-psychoses of defense. In: Strachey J, ed. The standard edition. Vol. 3. London: Hogarth Press, 1896.

Freud S. Screen memories. In: Strachey J, ed. The standard edition. Vol. 3. London: Hogarth Press, 1899.

Freud S. The interpretation of dreams. The standard edition. Vol. 5. London: Hogarth Press, 1900.

Freud S. Notes upon a case of obsessional neurosis. In: Strachey J, ed. The standard edition. Vol. 10. London: Hogarth Press, 1909.

Freud S. Formulations regarding the two principles in mental functioning. The standard edition. Vol. 12. London: Hogarth Press, 1911.

Freud S. Three essays on the theory of sexuality. In: Strachey J, ed. The standard edition. Vol. 7. London: Hogarth Press, 1905.

Freud S. Repression. In: Strachey J, ed. The standard edition. Vol. 14. London: Hogarth Press, 1915a.

Freud S. The unconscious. In: Strachey J, ed. The standard edition. Vol. 14. London: Hogarth Press, 1915b.

Freud S. Introductory lectures on psychoanalysis. In: Strachey J, ed. The standard edition. Vol.17. London: Hogarth, 1916.

Freud S. The ego and the id. In: Strachey J, ed. The standard edition. Vol. 19. London: Hogarth Press, 1923.

Freud S. Inhibitions, symptoms and anxiety. In: Strachey J, ed. The Standard Edition. Vol. 20. London: Hogarth Press, 1926.

Freud S. Female sexuality. In: Strachey J, ed. The standard edition. Vol. 21. London: Hogarth Press, 1931.

Freud S. New introductory lectures in psychoanalysis. In: Strachey J, ed. The standard edition. Vol. 19. London: Hogarth Press, 1933.

Freud S. Moses and monotheism. In: Strachey J, ed. The standard edition. Vol. 23. London: Hogarth Press, 1937.

Friday N. My Mother My Self. New York: Dell, 1977.

Friday N. Women on top. New York: Pocket, 1991.

Friderici AD, Schoenle PW, Goodglass H. (1981). Mechanisms underlying writing and speech in aphasia. Brain Lang 1981;13:212–222.

Fried I, Mateer C, Ojemann G, Wohns R, Fedio P. Organization of visuospatial functions in human cortex. Brain 1982;105:349–371.

Fried I, Ojemann GA, Fetz EE. Language related potentials specific to human language cortex. Science 1981;212:353–356.

Friedman A, Pines AM. Increase in Arab women's perceived power in the second half of life. Sex Roles 1992;26:1–9.

Friedman RC, Richart RM, Vande-Wiele RL. Sex differences in behavior. New York: Wiley, 1974.

Frisk F, Milner B. The role of the left hippocampal region in the acquistion and retention of story content. Neuropsychologia 1990;28:349–359.

Frodi AM, Lamb ME. Sex differences n responsiveness to infants. Child Devel 1978;49:1182–1188.

Frodi AM, Lamb ME, Hwang CP, Frodi M. Father-mother-infant interaction in traditional and non-traditional Swedish families. Alternative Lifestyles 1982;4:6–13.

Fuchs D, Thelen MH. Children's expected interpersonal consequences of communicating their affective state and reported likelihood of expression. Child Devel 1988;59:1314–1322.

Fukuda M, Ono T, Nakamura K. Functional relation among inferiotemporal cortex, amygdala and lateral hypothalamus in monkey operant feeding behavior. J Neurophysiol 1987;57:1060–1077.

Fuld PA, Fisher P. Recovery of intellectual ability after closed head injury. Dev Med Child Neurol 1977; 19:495–502.

Fuller JL, Rosvold HE, Pribram KH. Effects on affective and cognitive behavior in the dog after lesions of the pyriform-amygdaloid-hippocampal complex. J Comp Physiol Psychol 1957;50:89–96.

Fulton JF. Grasping and groping in relation to the syndrome of the premotor area. Arch Neurol Psychiatry 1934;31:221–235.

Fulton JF, Jacobsen CF, Kennard MA. A note concerning the relation of the frontal lobes to posture and forced grasping in monkeys. Brain 1932;55:524–536.

Fulton JF, Ingraham FD. Emotional disturbances following experimental lesions of the base of the brain. J Physiol 1929;67:47–90.

Furlong MW. A randomized double-blind study of positive suggestions presented during anesthesia. In: Bonke B, Fitch W, Millar K, eds. Memory and awareness in anaesthesia. The Netherlands: Swets & Zeitlinger, 1990.

Fuster JM. The prefrontal cortex. Anatomy, physiology, and neuropsychology of the frontal lobes. New York: Raven Press, 1980.

Fuster JM. Neuropsychiatry of frontal lobe lesions. In: Fogel BS, Schiffer RB, eds. Neuropsychiatry. Baltimore: Williams & Wilkins, 1995.

Fuster JM, Bauer RH, Jervey JP. Cellular discharge in the dorsolateral prefrontal cortex of the monkey in cognitive tasks. Exp Neurol 1982;77:679–694.

Fuster JM, Jevey JP. Inferotemporal neurons distinguish and retain behaviouraly relevant features of visual stimuli, Science 1981;212:952–955.

Fuxe K. Distribution of monoamine nerve terminals in the central nervous system. Acta Physiol Scand 1965;64:39–85.

Gabriel M, Poremba AL, Ellison-Perrine C, Miller JD. Brainstem mediation of learning and memory. In: Klemm WR, Vertes RP, eds. Brainstem mechanisms of behavior. New York: Wiley, 1990.

Gaffan D. Amygdala and the memory of reward. In: Aggleton JP, ed. The amygdala. New York: Wiley, 1992.

Gaffan D, Harrison S. Amygdalectomy and disconnection in visual learning for auditory secondary reinforcement in monkeys. J Neurosci 1987;7: 2285–2292.

Gaffney GR, Tasai LY, Kuperman S, Minchin S. Cerebellar structure in autism. Am J Dis Child 1987; 141:1330–1332.

Gage DF, Safer MA. Hemisphere differences in the mood state dependent effect for recognition of emotional faces. J Exp Psychol Learn Mem Cognit 1985;11:752–763.

Gahwiler BH. The action of neuropeptides on the bioelectric activity of hippocampal neurons. In: Seifert W, ed. Neurobiology of the hippocampus. San Diego: Academic Press, 1983.

Gainotti G. Emotional behavior and hemispheric side of lesion. Cortex 1972;8:41–55.

Gainotti G, D'Erme P, Monteleone D, Silveri MC. Mechanisms of unilateral spatial neglect in relation to laterality of cerebral lesions. Brain 1986; 109:599–612.

Gainotti G, Lemmo M. Comprehension of symbolic gestures in aphasia. Brain Lang 1976;3:451–460.

Gainotti G, Messerlie P, Tissot R. Qualitative analysis of unilateral neglect in relation to laterality of cerebral lesion. J Neurol Neurosurg Psychiatry 1972; 35:545–550.

Galaburda AM, Sherman GF, Geschwind N. Developmental dyslexia. Proc Soc Neurosci 1983;9:940.

Galdikas M, Birute MF. Living with the great orange apes. National Geographic 1980;157.

Galin D. Implications for psychiatry of left and right cerebral specialization. Arch Gen Psychaitry 1974; 31:572–583.

Galin D, Diamond DR, Braff D. Lateralization of conversion symptoms: more frequent on the left. Am J Psychiatry 1977;134:578–580.

Galin D, Diamond R, Herron J. Development of crossed and uncrossed tactile localization on the fingers. Brain Lang 1979;4:588–590.

Galin D, Johnstone J, Nakell L, Herron J. Development of the capacity for tactile information transfer between hemipsheres in normal children. Science 1979;204:1330–1332.

Gallagher JB, Browder EF. Extradural hematoma. J Neurosurg 1968;8:434–437.

Gallagher RE, Joseph R. Non-linguistic knowledge, hemispheric laterality, and the conservaton of inequivalence. J Gen Psychol 1982;107:31–40.

Galperin CZ, Mari-Germaine P, Lamour M, et al. Functional analysis of the contents of maternal speech to infants of 5 and 13 months in four cultures: Argentina, France, Japan and the United States. Devel Psychol 1992;28:593–603.

Gamble C. The palaeolithic settlement of Europe. Cambridge: Cambridge University Press, 1986.

Gamier D. Fossil hominids from the early Upper Palaeolithic (Aurignacian) of France. In: The human revolution: behavioral and biological perspectives on the origins of modern humans. Vol 1. Edited by P. Mellars P, Stringer BC, eds. Edinburgh: Edinburgh University Press, 1989.

Gandevia SC, Burke D. Does the nervous system depend on kinesthetic information to control natural limb movements? Behav Brain Sci 1992;15:614–632.

Gans A. HIV/AIDS and safer sex survey of gay and bisexual men. Santa Clara County Health Dept. San Jose, California, 1993.

Garcha HS, Ettlinger G, MacCabel JJ. Unilateral removal of the second somatosensory projection cortex in the monkey. Brain 1982;105:787–810.

Garcia JH, Lowry SL, Briggs L, et. al. Brain capillaries expand and rupture in areas of ischemia and reperfusion. In: Reivich M, Hurtig HI, eds. Cerebrovascular diseases. New York: Raven Press, 1983.

Garcia JH, Ben-David E, Conger KA, et al. Arterial hypertension injured brain capillaries. Stroke 1981; 12:410–413.

Gardner A. Egypt of the pharoahs. New York: Oxford University Press, 1961.

Gardner EB, Boitana JJ, Manicino NS, et al. Environmental enrichement and deprivation. Physiol Behav 1975;14:321–327.

Gardner H. The shattered mind. New York: Vintage Books, 1975.

Gardner H, Brownell HH, Wapner W, Michelow D. Missing the point: the role of the right hemisphere in the processing of complex linguistic materials. In: Perceman E, ed. Cognitive processing in the right hemisphere. New York: Academic Press, 1983.

Gardner H, Ling PK, Flamm L, Silverman J. Comprehension and appreciation of humorus material following brain damage. Brain 1975;98:399–412.

Gargett BH. Grave shortcomings: the evidence for Neanderthal burial. Curr Anthropol 1989;30:157–90.

Gartman LM, Johnson NF. Massed versus disrupted repetition of homographs: a test of differential encoding hypothesis. J Verbal Learn Verbal Behav 1972;11:801–808.

Gartrell NK. Hormones and homosexuality. In: Homosexuality: social, psychological and biological issues. Weinrich PE, Gonsiorek JC, Foster DW, ed. Beverly Hills, Sage Publishers, 1982.

Gaspar P, Duyckaerts C, Alvarez C, et al. Alterations of dopaminergic and noradrenergic innervations in motor cortex in Parkinson's disease. Ann Neurol 1991;30:365–374.

Gasparrini WG, Satz P, Heilman KM, Coolidge FL. Hemispheric asymmetry of affective processing as determined by the MMPI. J Neurol Neurosurg Psychiatry 1978;41:470–473.

Gasquoine PG. Bilateral alien hand signs following destruction of the medial frontal cortices. Neuropsychiatr Neuropsychol Behav Neurol 1993;6:49–53.

Gates A, Bradhsaw JL. The role of the cerebral hemispheres in music. Brain Lang 1977;3:451–460.

Gaya H. Central nervous system infections in neoplastic disease. In: Whitehouse JMA, Kay HEM, eds. CNS complications of malignant disease. Baltimore: University Park Press, 1979.

Gaylin W. The male ego. New York: Viking, 1992.

Gazzaniga MS. The bisected brain. New York: Appleton, 1970.

Gazzaniga MS, Ladavas E. Disturbances in spatial attention following lesion or disconnection of the right parietal lobe. In: Jeannerod M, ed. Neurophysiological and neuropsychological aspects of spatial neglect. Amsterdam: North-Holland, 1987.

Gazzaniga MS, LeDoux JE. The integrated mind. New York: Plenum Press, 1978.

Gebhard JW. Hypnotic age regression: a review. Am J Clin Hypn 1961;3:139–168.

Geffen G, Bradshaw JL, Wallace G. Interhemispheric effects on reaction time to verbal and nonverbal visual stimuli. J Exp Psychol 1971;87:415–422.

Gennarelli TA. Mechanisms of pathophysiology of cerebral concussion. J Head Trauma Rehab 1986; 1:23–29.

Gennarelli TA, Speilman GM, Lanfitt TW, et al. Influence of the type of intracranial lesion on outcome from severe head injury. J Neurosurg 1982;56:26–32.

Gennarelli TA, Thibault LE, Adams H, et al. Diffuse axonal injury and traumatic coma in the primate. Ann Neurol 1982;12:564–574.

Gerendai I. Lateralization of neuroendocrine control. In: Geschwind N, Galaburda AM, eds. Cerebral dominance. Cambridge: Harvard University Press, 1984.

Gerfen CR. The neostriatal mosaic. Nature 1984; 311:461–464.

Gerfen CR. The neostriatal mosaic: compartmental organization of mesostriatal systems. In: Carpenter MB, Jayaraman A, eds. The basal ganglia. New York: Plenum, 1987.

Gerall AA, Moltz H, Ward IL. Handbook of behavioral neurobiology. New York: Plenum, 1992.

Gerstmann J. Syndrome of finger agnosia, disorientation for right and left, agraphia and acalculia. Arch Neurol Psychiatry 1930;44:398–408.

Gerstmann J. Problem of imperception of disease and of impaired body territories with organic lesions. Arch Neurol Psychiatry 1942;48:890–913.

Geschwind N. Disconnexion syndromes in animals and man. Brain 1965;88:585–644.

Geschwind N. The perverseness of the right hemisphere. Behav Brain Sci 1981;4:106–107.

Geschwind N, Galaburda AM. Cerebral lateralization. Arch Neurol 1985;42:428–654.

Geschwind N, Levitsky W. Human brain: leftright asymmetries in temporal speech regions. Science 1968;161:186–187.

Geschwind N, Quadfasel FA, Segarra JM. Isolation of the speech area. Neuropsychologia 1968;6:327–340.

Gibbs AF. Ictal and non-ictal psychiatric disorders in temporal lobe epilepsy. J Nerv Ment Dis 1951;113: 522–528.

Gibson KR. Myelination and behavioral development: a comparative perspective on questions of neoteny, altriciality, and intelligence. In: Gibson KR, Petersen AC, eds. Brain maturation and cognitive development. New York: Aldine, 1991, pp. 29–64.

Gies F, Gies J. Women in the middle ages. New York: Barnes & Noble, 1978.

Gilbert D. The young child's awareness of affect. Child Devel 1969;40, 629–640.

Gilbert F. Chromosomes, genese and cancer. J Natl Cancer Inst 1983;71:107–114.

Gilbert F, Feder M, Balaban G, et al. Human neuroblastomas and abnormalities of chromosomes 1 and 17. Cancer Res 1984;44:5444–5449.

Gilles FH. Perinatal neuropathology. In: Davis RL, Robertson DM, eds. Textbook of neuropathology. Baltimore: Williams & Wilkins, 1991.

Gilligan C. In a different voice. Cambridge: Harvard University Press, 1982.

Gillum RF. Cerebrovascular disease morbidity in the United States, 1970–1983. Stroke 1986;17:656–661.

Ginsberg MD. Neuroprotection in brain ischemia. The Neuroscientist 1995;1:95–103.

Girgis M. The orbital surface of the frontal lobe of the brain. Acta Psychiatr Scand (Supplement) 1971; 22:1–58.

Gladue BA, Beatty WW, Larson J, Staton RD. Sexual orientation and spatial ability in men and women. Psychobiol 1990;18:101–108.

Glass L. He says, she says. New York: G.P. Putnam's Sons, 1992.

Glees P, Cole J, Whitty CWM, Cairns H. The effects of lesions in the cingulate gyrus and adjacent areas in monkeys. J Neurol Neurosurg Psychiatry 1959;13: 178–190.

Gleick J. Chaos. New York: Penguin, 1987.

Gloning I, Gloning K, Hoff H. Neuropsychological symptoms and syndromes in lesions of the occipital lobe and the adjacent areas. Paris: Gauthier-Villars, 1968.

Gloor P. Electrophysiological studies on the connections of the amygdaloid nucleus of the cat. I & II. Electroencephalogr Clin Neurophysiol 1955;7: 223–262.

Gloor P. Amygdala. In: Field J, ed. Handbook of physiology. Washington DC: American Physiological Society, 1960, pp. 300–370.

Gloor P. Temporal lobe epilepsy. In: Elftheriou BE, ed. The neurobiology of the amygdala. New York: Plenum, 1972.

Gloor P. The role of the human limbic system in perception, memory, and affect. Lessons from temporal lobe epilepsy. In: Doane BK, Livingston KE, eds. The limbic system: functional organization and clinical disorder. New York, Raven Press, 1986.

Gloor P. Experiential phenomena of temporal lobe epilepsy. Brain 1990;113:1673–1694.

Gloor P. Role of the amygdala in temporal lobe epilepsy. In: Aggleton JP, ed. The amygdala. New York: Wiley-Liss, 1992.

Gloor P, Olivier A, Quesney LF, et al. The role of the limbic system in experimental phenomena of temporal lobe epilepsy. Ann Neurol 1982;12:129–144.

Gluksberg S, McCloskey M. Decisions about ignorance: knowing that you don't know. J Exp Psychol Hum Learn Mem 1981;7:311–325.

Gnepp J. Children's social sensitivity. Devel Psychol 1983;19: 805–814.

Gock CY, Ringer BB, Bobbie ER. To comfort and to challenge. Berkeley: University of California Press, 1967.

Godden DR, Baddeley AD. Context-dependent memory in two natural environments: on land and under water. Br J Psychol 1975;13:333–345.

Godden DR, Baddeley AD. When does context influence recognition memory? Br J Psychol 1980;18: 144–160.

Godschalk M, Lemon RN, Nijis HGT, Kuypers HGJM. Behavior of neurons in monkeys peri-arcuate and precentral cortex before and during visually guided arm and hand movements. Exp Brain Res 1981;44: 113–116.

Godwin M. Angels. New York: Simon & Schuster, 1990.

Goelet P, Kandel ER. Tracking the flow of learned information from membrane receptors to genome. Trends Neurosci 1986;9:492–499.

Golbe LI, Merkin MD. Cerebral infarction in a user of free-based cocaine ("crack"). Neurology 1986;36: 1602–1604

Gold E, Neisser U. Recollections of kindergarten. Q Newsl Lab Comp Hum Cognit 1980;2:77–809.

Gold JM, Hermann BP, Randolph C, et al. Schizophrenia and temporal lobe epilepsy. Arch Gen Psychiatry 1994;51:265–272.

Gold M, Adair JC, Jacobs DH, Heilman KM. Anosognosia for hemiplegia. Neurology 1994;44:1804–1808.

Gold PE. A proposed neurobiologic basis for regulating memory storage for significant events. In: Winograd E, Neisser U, eds. Affect and accuracy in recall. Cambridge, Cambridge University Press, 1992.

Goldberg G. Supplementary motor area structure and function: review and hypothesis. Behav Brain Sci 1985;8:567–616.

Goldberg G. From intent to action. Evolution and function of the premotor system of the frontal lobe. In: Perecman E, ed. The frontal lobes revisited. New York, IRBN Press, 1987.

Goldberg G, Bloom KK. The alien hand sign. Am J Phys Med Rehab 1990;69:228–38.

Goldberg G, Meyer NH, Toglia JU. Medial frontal cortex infarction and the alien hand sign. Arch Neurol 1981;38:683–686.

Goldberg H. The hazards of being male. Plainville: Nash, 1976.

Goldenberg G. The ability of patients with brain damage to generate mental visual imagery. Brain 1989;112:305–326.

Goldenberg G, Spatt J. Influence of size and site of cerebral lesions on spontaneous recovery of aphasia and on success of language therapy. Brain Lang 1994;47:684–698.

Goldman PS. Functional development of the prefrontal cortex in early life and the problem of neuronal plasticity. Exp Neurol 1971;32:366–387.

Goldman PS. An alternative to developmental plasticity. In: Stein DG, Rosen JJ, Butters N, eds. Plasticity and recovery of function in the central nervous system. New York: Academic Press, 1974.

Goldman PS, Rosvold HW, Mishkin M. Evidence for behavioral impairment following prefrontal lobectomy in the infant monkey. J Comp Physiol Psychol 1970;70:454–463.

Goldman-Rakic P. Cortical localization of working memory. In: McGaugh JL, Winberger NM, Lynch G, eds. Brain organization and memory. New York: Oxford University Press, 1990.

Goldman-Rakic PS, Brown RM. Postnatal development of monoamine content and synthesis in the cerebral cortex of rhesus monkey. Devel Brain Res 1982;4:339–349.

Goldstein K. The significance of the frontal lobes for mental performance. J Neurol Psychopathol 1936–37;17:27–40.

Goldstein K. After effects of brain injuries in war. New York: Grune & Stratton, 1942.

Goldstein K. The mental changes due to frontal lobe damage. J Psychol 1944;17:187–208.

Goldstein K. Language and language disturbances. New York: Grune & Stratton, 1948.

Goldstein L, Stoltzfus NW, Gardocki JF. Changes in interhemispheric amplitude relationships in the EEG during sleep. Physiol Behav 1972;8:811–815.

Goodall J. In the shadow of man. Boston: Houghton Mifflin, 1971.

Goodall J. The chimpanzees of the Gombe: patterns of behavior. Cambridge: Cambridge University Press, 1986.

Goodall J. Through a window. Boston: Houghton Mifflin Co, 1990.

Goodall J. Combe chimpanzee politics. In: Schubert G, Masers RD, eds. Primate politics. Carbondale. S. University Press, 1991.

Goodenough DR, Witkin HA, Lewis HB, et al. Repression interferences and field dependence as factors in dream forgetting. J Abn Psychol 1974;83:32–44.

Goodenough DR, Shapiro A, Holden M, Steinschriber R. Comparison of "dreamers" and "non-dreamers." J Nerv Ment Dis 1959;59:295–302.

Goodglass H, Berko J. Agrammatism and inflectional morphology in English. J Speech Hear Res 1960;3:257–267.

Goodglass H, Kaplan E. Boston diagnostic aphasia examination. New York: Lange, 1982.

Goodman GS. Commentary: on stress and accuracy in research on children's testimony. In: Doris J, ed. The suggestibility of children's recollections. Washington, DC: American Psychological Association, 1991.

Goodman GS, Aman C. Children's use of anatomically detailed dolls to recount and event. Child Devel 1990;61:1859–1871.

Goodman GS, Bottoms BL, Schwartz-Kenney B, Rudy L. Children's testimony about a stressful event: Improving children's reports. J Narrative Life History 1991;7:69–99.

Goodman GS, Clarke-Stewart A. Suggestibility in children's testimony: implications for sexual abuse investigations. In: Doris J, ed. The suggestibility of children's recollections. Washington DC: American Psychological Association, 1991.

Goodman GS, Rudy L, Bottoms B, Aman C. Childrens' concerns and memory: issues of ecological validity in the study of children's eyewitness testimony. In: Fivsh RJ, Hudson J, eds. Knowing and remembering in young children. New York: Cambridge University Press, 1990.

Goodman WK, Price LH, Delgado PL, et al. Specificity of serotonin reuptake inhibitors in the treatment of obsessive-compulsive disorder. Arch Gen Psychiatry 1990;47:577–585.

Goodwin DW, Guze SB. Psychiatric diagnosis. New York: Oxford University Press, 1979.

Goodwin DW, Powell B, Bremer D, et al. Alcohol and recall: state dependent effects in man. Science 1969;163:1358–1360.

Goodwin J, Simms M, Bergman R. Hysterical seizures: a sequel to incest. Am J Orthopsychiatry 1979;49:698–703.

Gordon KA. A study of memories. J Delinquency 1928;12:129–132.

Gordon T, Draper TW. Sex bias against male day care workers. Child Care Quarterly 1982;10:15–17.

Gordon HW. Hemispheric asymmetries in the perception of muscial chords. Cortex 1970;6:387–398.

Gordon HW, Bogen JE. Hemispheric lateralization of singing after intracarotid sodium amylobarbitone. J Neurol Neurosurg Psychiatry 1974;37:727–737.

Gordon HW, Freeman B, Lavie P. Shift in cognitive

asymmetries between wakings from REM and NREM sleep. Neuropsychologia 1982;20:99–103.

Gorelick PB, Hier DB, Caplan LR, Langenberg P. Headache in acute cerebrovascular disease. Neurology 1986;36:1445–1450.

Gorelick PB, Ross ED. The aprosodias: further functional-anatomical evidence for the organization of affective language in the right hemisphere. J Neurol Neurosurg Psychiatry 1987;50:553–560.

Gorski RA, Gordon JH, Shryne JE, Southam AM. Evidence for morphological sex differences within the medical preoptic area of the rat brain. Brain Res 1978;148:333–346

Goto S, Hirano A, Matsumoto S. Met-enkephalin immunoreactivity in the basal ganglia in Parkinson's disease and striatonigral degeneration. Neurology 1990;40:1051–1056.

Gottfried AW, Rose SA, Bridger WH. Cross-modal transfer in human infants. Child Devel 1977;48:118–123.

Gottlieb JP, MacAvoy MG, Bruce C. Neural responses related to smooth-pursuit eye movements and their correspondence with electrically elicited smooth eye movements in the primate frontal eye field. J Neurophysiol 1994;72:1634–1574.

Gowlett J. Ascent to civilization. New York: Alfred A. Knopf, 1984.

Goy RW, McEwen BS. Sexual differentiation of the brain. Cambridge: MIT Press, 1980.

Graeber RC. Telencephalic function in elasmobranchs. A behavioral perspective. In: Ebbesson SOE, ed. Comparative neurology of the telencephalon. New York: Plenum Press, 1980.

Graf P, Squire LR, Mandler G. The information that amnesic patients do not forget. J Exp Psychol Learn Mem Cognit 1984;10:164–178.

Graff-Radford NR, Cooper WE, Colsher PL, et al. An unlearned foreign "accent" in a patient with aphasia. Brain Lang 1986;28:86–94.

Graff-Radford P, Tranel D, Van Hoesen GW, Brandt J. Diencephalic amnesia. Brain 1990;113:1–25.

Grafman J, Vance SC, Weingartner H, et al. The effects of lateralized frontal lesions on mood regulation. Brain 1986;109:1127–1148.

Grafman J, Vance S, Weingartner H, Salazar AM, Amin D. The effects of lateralized frontal lesions on mood regulation. Brain 1986;109:1127–1148.

Graftstein B. Postnatal development of the transcallosal evoked response in the cerebral cortex of the cat. J Neurophysiol 1963;26:79–99.

Graham H. Caring. In: Finch J, Groves D, eds. A labour of love: women, work, & caring. London: Routledge, 1983.

Grastyan E, Lissak K, Madarasz I, Donhoffer H. Hippocampal electrical activity during the development of conditioned reflexes. Electroencephalogr Clin Neurophysiol 1959;11:409–430.

Graves R, Landis T, Goodglass H. Laterality and sex differences in visual recognition of emotional and nonemotional words. Neuropsychologia 1981;19:95–102.

Gray JA. Sodium amobarbital, the hippocampal theta rhythm and the partial reinforcement extinction effect. Psychol Rev 1970;77:465–480.

Gray JA. The psychology of fear and stress. New York: Oxford University Press, 1987.

Gray TS. The organization and possible function of amygdaloid corticotropin releasing factor pathways. In: DeSouza EB, Nemeroff CB, eds. Corticotropin releasing factor. Boca Raton: CRC Press, 1990.

Graybiel AM. Neuropeptides in the basal ganglia. In: Martin JB, Barchas JD, eds. Neuropeptides in neurologic and psychiatric disease. New York: Raven Press, 1986.

Greaves GB. Multiple personality 165 years after Mary Reynolds. J Nerv Ment Dis 1980;168:577–596.

Green GL, Lessel S. Acquired cerebral dyschromatopsia. Arch Ophthamol 1977;95:121–128.

Green JD, Adey WR. Electrophysiological studies of hippocampal connections and excitability. Electroencephalogr Clin Neurophysiol 1956;8:245–262.

Green JE, Arduini A. Hippocampal electrical activity in arousal. J Neurophysiol 1954;17:533–557.

Green P, Kontenko V. Superior speech comprehension in schizophrenics under monaural versus binaural listening conditions. J Abn Psychol 1980;89:339–408.

Green R. Gender identity in childhood and later sexual orientation. Am J Psychiatry 1985;142:339–341.

Green R. The "sissy boy syndrome" and the development of homosexuality. New Haven: Yale University Press, 1987.

Green SB, Byuar DB, Walker MD, et al. Comparisons of carmustine, procarbazine, and high-dose methylprednisolone as additons to surgery and radiotherapy for the treatment of malignant glioma. Cancer Treat Rep 1983;67:121–132.

Greenberg MS, Farah MJ. The laterality of dreaming. Brain Cognit 1986;5:307–321.

Greenblatt M. Studies in lobotomy. New York: Grune & Stratton, 1950.

Greenough W. Enduring effects of differential experience and training. In: Rosenzweig MR, Bennet EL, eds. Neural mechanisms of learning and memory. Cambridge: MIT Press, 1976.

Greenough W, Black JE, Wallace CW. Experience and brain development. Child Devel 1987;58:539–559.

Greenough W, Chang F-LF. Plasticity of synapse structure and pattern in the cerebral cortex. In: Peters A, Jones EG, eds. Cerbral cortex. Vol. 7. New York: Plenum, 1988.

Greenspan JD, Winfield JA. Reversible pain and tactile deficits associated with a cerebral tumor compressing the posterior insula and parietal operculum. Pain 1992;50:29–39.

Greenspan GS, Samuel SE. Self cutting after rape. Am J Psychiatry 1989;146:789–790.

Greenwood P, Wilson DH, Gazzaniga MS. Dream report following commissurotomy. Cortex 1977;13:311–316.

Greenblatt SH. Alexia without agraphia or hemianopia. Brain 1973;96:307–316.

Grillner S, Shik M. On the descending control of the lumbosacral spinal cords from the mescenphalic locomotor region. Acta Physiol Scand 1973;87:320–333.

Grine FE. Evolutionary history of the "robust" Australopithecines. New York: Aldine, 1988.

Grinker RR, Spiegel JP. Men under stress. New York: McGraw-Hill, 1945.

Groenewegen HJ, Berendse HW. Connections of the subthalamic nucleus with the ventral striatopallial parts of the basal ganglia in rats. J Comp Neurol 1990;294:607–522.

Groenewegen HJ, Berendse HW, Meredith GE, et al. Functional anatomy of the ventral, limbic system-innervated striatum. In: Willner P, Scheel-Kruger J, eds. The mesolimbic dopamine system. New York: Wiley, 1991.

Gronwall D. Performance changes during recovery from closed head injury. Proc Australian Assoc Neurol 1976;13:143–147.

Gronwall D, Wrightson P. Delayed recovery of intellectual function after minor head injury. Lancet 1974;2:607–609.

Gross CG, Bender DB, Gerstein GL. Activity of inferior temporal neurons in behaving monkeys. Neuropsychologia 1979;17:215–229.

Gross CG, Bender DB, Rocha-Miranda CE. Inferotemporal cortex. A single unit analysis. In: Schmitt FO, Worden FG, eds. The neurosciences: third study program. Cambridge: MIT Press, 1974.

Gross CG, Graziano MSA. Multiple representations of space in the brain. The Neuroscientist 1995;1:43–50.

Gross CG, Mishkin M. The neural basis of stimulus equivalence across retinal translation. In: Harnad S, et al, eds. Lateralization in the nervous system. New York: Academic Press, 1977.

Gross CG, Rocha-Miranda CE, Bender DB. Visual properties of neurons in inferotemporal cortex of the macaque. J Neurophysiol 1972;35:96–111.

Gross CG, Weiskrantz L. Some changes in behavior produced by lateral frontal lesions in the macaque. In: Warren JM, Akert K, eds. The frontal granular cortex and behavior. New York: McGraw-Hill, 1964, pp. 74–101.

Grotta JC, Manner C, Pettigrew C, Yatsu FM. Red blood cell disorders and stroke. Stroke 1986;17:811–817.

Grun R, Beaumont PB, Stringer CB. ESR dating evidence for the early modern humans at Border Cave in South Africa. Nature 1990;344:537–539.

Gruzelier J, Manchanda R. The syndrome of schizophrenia: relations between electrodermal response, lateral asymmetries and clinical rating. Br J Psychiatry 1982;141:488–495.

Guiard YG, et al. Left-hand advantage in right-handers for spatial constant error: preliminary evidence in a unimanual ballistic aimed movement. Neuropsychologia 1983;21:111–115.

Guilbaud G, Peschanski M, Gautron M, Binder D. Responses of neurons of the nucleus raphe magnus to noxious stimuli. Neurosci Lett 1980;17:149–154.

Guillemnault C, Mignot E, Grumet FC. Familial patterns of narcolepsy. Lancet 1989;2:1376–1379.

Guillary RW. A quantitative study of the mamillary bodies and their connexins. J Anat 1955;89:19–32.

Gunne LM, Lewander T. Monoamine in brain and adrenal glands of cats after electrically induced defense reactions. Acta Physiol Scand 1966;67:405–410.

Gustafsson B, Wigstrom H. Physiological mechanisms underlying long-term potentiation. Trend Neurosci 1988;11:156–162.

Gustafson GE, Harris KL. Women's responses to young infants' cries. Develop Psychol 1990;26:144–152.

Guttentag RE. Memory and aging. Implications for theories of memory development during childhood. Develop Rev 1985;5:56–82.

Gyllenstein L, Malmfors T, Norrlin ML. Growth alteration in the auditory cortex of visually deprived mice. J Comp Neurol 1966;126:463–470.

Haaland K, Harrington DL. The role of the right hemisphere in closed loop movements. Brain Cognit 1990; 16:104–122.

Haaland KY, Harrington DL. Limb-sequencing deficits after left but not right hemisphere damage. Brain Cognit 1994;24:104–122.

Haarmaann HJ, Kolk HHJ. On-line sensitivity to subject-verb agreement violations in Broca's aphasia. Brain Lang 1994;46:493–516.

Haas DC, Lourie H. Trauma-triggered migraine: an explanation for common neurological attacks after mild head injury. J Neurosurg 1988;68:181–188.

Haas JF, Cope DN, Hall K. Premorbid prevalence of poor academic performance in severe head injury. J Neurol Neurosurg Psychiatry 1987;50:52–56.

Haberly LB. Comparative aspects of olfactory cortex. In: Jones ED, Peters A, eds. Cerebral cortex. Vol 8. Comparative structure and evolution of cerebral cortex. New York: Plenum, 1990.

Habib M, Robichon F, Levrier O, et al. Diverging asymmetries of temporo-parietal cortical areas: a reappraisal of Geschwind/Galaburda theory. Brain Lang 1995;48:238–258.

Hagino N, Yamoaka S. A neuroendocrinological approach to the investigation of the septum. In: DeFrance JF, ed. The septal nuclei. New York: Plenum Press, 1976.

Haglund MM, Ojemann GA, Lettich E, et al. Dissociation of cortical and single unit activity in spoken and signed languages. Brain Lang 1993;44:19–27.

Hakan RL, Eyle C, Henriksen SJ. Neuropharmacology of the nucleus accumbens. Neuroscience 1994;63:85–93.

Hakim AM, Ryder-Cooke A, Melanson D. Sequential computerized tomographic appearance of strokes. Stroke 1983;14:893–897.

Halgren E. The amygdala contribution to emotion and memory. In: Ben-Ari Y, ed. The amygdaloid complex. Amsterdam: Elsevier, 1981.

Halgren E. Emotional neurophysiology of the amygdala within the context of human cognition. In: Aggleton JP, ed. The amygdala. New York: Wiley-Liss, 1992.

Halgren E, Babb TL, Crandall PH. Activity of human hippocampal formation and amygdala neurons during memory tests. Electroencephalogr Clin Neurophysiol 1978;45:585–601.

Halgren E, Walter RD, Cherlow DG, Crandal PH. Mental phenomena evoked by electrical stimulation of the human hippocampal formation and amygdala. Brain 1978;101:83–117.

Hall GS. Note on early memories. Pedagogical Seminary 1899;6:485–512.

Hall J. Gender effects in decoding nonverbal cues. Psychol Rev 1978;85:845–857.

Halperin Y, Nachshon I, Carmon A. Shift of ear superiority in dichotic listening to temporally patterned nonverbal stimuli. J Exp Psychol 1973;101:46–54.

Halstead WC. Brain and intelligence. Chicago: University of Chicago Press, 1947.

Hamburg DA. Aggressive behavior of chimpanzees and baboons in natural habitats. J Psychiatr Res 1971;8:385–398.

Hamilton CR, Vermeire BA. Complementary hemispheric specialization in monkeys. Science 1988; 242:1694–1696.

Hamilton M. Fish's schizophrenia. Bristol: John Wright & Sons, 1976.

Hamilton WJ, Mossman HW. Human embryology. Baltimore: Williams & Wilkins, 1976.

Hamrick MW, Inouye SE. Thumbs, tools and early humans. Science 1995;268:586–587.

Hanley JR, Pearson NA, Young AW. Impaired memory for new visual forms. Brain 1990;113:1131–1148.

Hannay HJ, Falgout JC, Leli DA, Katholi CR, Halsey JH, Wills EL. Focal right temporo-occipital blood flow changes associated with judgment of line orientation. Neuropsychologia 1987;25:755–763.

Hannay HJ, Varney NR, Benton AL. Visual localization in patients with unilateral brain disease. J Neurol Neurosurg Psychiatry 1976;39:307–313.

Hansen EW. The development of maternal and infant behavior in the rhesus monkey. Behaviour 1966; 27:107–149.

Harackiewicz JN, DePaulo BM. Accuracy of person perception. Person Soc Psychol Bull 1982;8: 247–256.

Hardman JM. The pathology of traumatic brain injuries. In: Thompson RA, Green JR, eds. Advances in neurology. Vol. 22. New York: Raven Press, 1979.

Harlow HF. The heterosexual affectional system in monkeys. Am Psychol 1962;17:1–9.

Harlow HF. Sexual behavior in the rhesus monkey. In: Beach F, ed. Sex and behavior. New York: Wiley, 1965.

Harlow H, Harlow MK. Effects of various mother-infant relationships on rhesus monkey behaviors. In: Foss BM, ed. Determinants of infant behavior. Methuen, 1965.

Harlow HF, Harlow MK. The effect of rearing conditions on behavior. Int J Psychiatry 1965;1:43–51.

Harmson E. Variations in sex-related cognitive abilities across the menstrual cycle. Brain Cognit 1990; 14:26–43.

Harold FB. A comparative analysis of Eurasian Palaeolithic burials. World Archaeol 1980;12:195–211.

Harold FB. Mousterian, Chatelperronian, and early Aurignacian in Western Europe: continuity or discontinuity? In: Mellars P, Stringer CB, eds. The human revolution: behavioral and biological perspectives on the origins of modern humans. Vol 1. Edinburgh: Edinburgh University Press, 1989.

Harris LJ. Sex differences in spatial ability. In: Kinsbourne M, ed. Asymmetrical function of the brain. Cambridge: Cambridge University Press, 1978.

Harris M. Why we became religious and the evolution of the spirit world. In: Lehmann AC, Myers JE, eds. Magic, witchcraft, and religion. Mountain View: Mayfield, 1993.

Harris PL. Examination and search by young infants. Br J Psychol 1971;45:71–77.

Harris PL. Perseverative errors in search by young infants. Child Devel 1973;44:28–33.

Harris PL, Olthof T, Terwogt MM. Children's knowledge of emotion. J Child Psychol Psychiatry 1981; 22:247–261.

Hart M, Heistad DD, Brody MJ. Effect of chronic hypertension and sympathetic denervation on wall-lumen ratio of cerebral vessels. Hypertension 1980; 2:419–423.

Hart RG, Easton JD. Hemorrhagic infarcts. Stroke 1986;17:586–589.

Harter S, Whitesell NR. Developmental changes in children's understanding of single, multiple, and blended emotion concepts. In: Saarni C, Harris P, eds. Children's understanding of emotion. New York: Cambridge University Press, 1989.

Hartmann JA, Wolz W, Roeltgen DP, Loverso FL. Denial of visual perception. Brain Cognit 1991;16: 29–40.

Hartshorne C. Born to sing. An interpretation and world survey of bird songs. London: Indiana University Press, 1973.

Haslam DR. Lateral dominance in the perception of size and pain. Q J Exp Psychol 1970;22:503–507.

Hassler R. Brain mechanisms of intention and attention with introductory remarks on other volitional processes. Prog Brain Res 1980;54:585–614.

Hatta T. The functional asymmetry of tactile pattern learning in normal subjects. Psychologia 1978;21: 83–89.

Halperin Y, Nachshon I, Carmon A. Shift of ear superiority in dichotic listening to temporally patterned nonverbal stimuli. J Exp Psychol 1973;101:46–54.

Hauser MD. Right hemisphere dominance for the production of facial expression in monkeys. Science 1993;261:475–477.

Hauser MD, Ybarra MS. The role of lip configuration in monkey vocalizations. Brain Lang 1994;46: 232–244.

Haviland JM, Lelwica M. The induced affect response:

10-week-old infants' responses to three emotional expressions. Devel Psychol 1987;23:97–104.

Hawkes J. The world of the past. Vol. 1. New York: Alfred A. Knopf, 1963.

Hay RL, Leakey MD. The fossil footprints of Laetoli. Sci Am 1982;2:47–87.

Hayden B. The cultural capacities of Neandertals: a review and re-evaluation. J Hum Evol 1993;24:113–146.

Hatta T. The functional asymmetry of tactile pattern learning in normal subjects. Psychologia 1978;21:83–89.

Hayes DS, Casey DM. Young children and television: the retention of emotional reactions. Child Devel 1992;63:1423–1436.

Hayes DS, Kelly SB. Sticking to syntax: the reflection of story grammar in children's and adult's recall of radio and television shows. Merrill-Palmer Q 1985;31:345–360.

Haynes KF, Birch MC. The role of other pheromones in the behavioral responses of insects. In: Kerkut GA, Gilbert LI, eds. Comprehensive insect physiology, biochemistry, and pharmacology. New York: Pergammon Press, 1985.

Head H, Holmes G. Sensory disturbances from cerebral lesions. Brain 1911;34:102–254.

Headley PM, Duggan AW, Griersmith BT. Selective reduction by noradrenaline and 5-hydroxytryptamine of nociceptive responses of cat dorsal horn neurons. Brain Res 1978;145:185–189.

Heath R. Studies in schizophrenia. Cambridge: Harvard University Press, 1954.

Heath RG. Pleasure response of human subjects to direct stimulation of the brain. In: Heath RG, ed. The role of pleasure in behavior. New York: Harper, 1964.

Heath RG. Physiological basis of emotional expression. Biol Psychiatry 1972;5:172–184.

Heath RG. Brain function in epilepsy: midbrain, medullary and cerebral interaction with the rostral forebrain. J Neurol Neurosurg Psychiatry 1976;39:1037–1051.

Heath RG. Modulation of emotion with a brain pacemaker. J Nerv Ment Dis 1977;165:300–317.

Heath RG, Franklin DE, Walker CF, Keating JW. Cerebellar vermal atrophy in psychiatric patients. Biol Psychiatry 1982;17:569–583.

Heath RG, Harper JW. Ascending projections of the cerebellar fastigial nucleus to the hippocampus, amygdala, and other temporal lobe sites. Exp Neurol 1974;45:268–287.

Heath RG, Monroe R, Mickle W. Stimulation of the amygdaloid nucleus in schizophrenic patients. Am J Psychiatry 1955;111:862–863.

Hebb DO. The organization of behavior. New York: Wiley, 1949.

Hebben N. The role of the frontal and temporal lobes in the phonetic organizations of speech stimuli: a multidimensional scaling analysis. Brain Lang 1986;19:342–357.

Hecaen H. Clinical symptomatology in right and left hemispheric lesions. In: Mountcastle VB, ed. Interhemispheric relations and cerebral dominance. Baltimore: Johns Hopkins Press, 1962.

Hecaen H. Mental changes associated with tumors of the frontal lobes. In: Warren JM, Akert K, eds. The frontal granular cortex and behavior. New York, McGraw-Hill, 1964.

Hecaen H, Albert ML. Human Neuropsychology. New York: John Wiley, 1978.

Hecaen H, Angelergues R. Agnosia for faces (prosopagnosia). Arch Neurol 1962;7:92–100

Hecaen H, Assal G. A comparison of construction deficits following right and left hemisphere lesions. Neuropsychologia 1970;8:289–304.

Hecaen H, De Ajuriaguerra J. Balint's syndrome (psychic paralysis of visual fixation) and its minor forms. Brain 1954;77:373–400.

Hecaen H, Kremin H. Neurolinguistic research on reading disorders from left hemisphere lesions. In: Whitkaer HA, Whitaker H, eds. Studies in neurolinguistics. New York: Academic Press, 1976.

Hecaen H, Penfield W, Bertand C, Malmo R. The syndromes of apractoognosia due to lesions of the minor cerebral hemisphere. Arch Neurol Psychiatry 1956;75:400–434.

Hefez A, Metz L, Lavie P. Long-term effects of extreme situational stress on sleep and dreaming. Am J Psychiatry 1987;144:345–347.

Heffner HE, Heffner RS. Temporal lobe lesions and perception of species-specific vocalizations in macaques. Science 1984;226:75–76.

Heilman KM. Ideational apraxia. Brain 1973;96:861–864.

Heilman KM. Reading and writing disorders caused by central nervous system defects. Geriatrics 1975;30:115–118.

Heilman KM. Neglect and related disorders. In: Heilman KM, Valenstein E, ed. Clinical neuropsychology. New York: Oxford University Press, 1993.

Heilman KM, Bowers D. Emotional disorders associated with hemispheric dysfunction. In: Fogel BS, Schiffer RB, eds. Neuropsychiatry. Baltimore: Williams & Wilkins, 1995.

Heilman KM, Bowers D, Speedie L, Coslett HB. Comprehension of affective and nonaffective prosody. Neurology 1984;34:917–921.

Heilman KM, Bowers D, Valenstein E. Emotional disorders associated with neurological diseases. In: Heilman K, Valenstein E, eds. Clinical Neuropsychology. Oxford: Oxford University Press, 1985.

Heilman KM, Bowers D, Valenstein E, Watson RT. Hemispace and hemispatial neglect. In: Jeannerod M, ed. Neurophysiological and neuropsychological aspects of spatial neglect. Amsterdam: North-Holland, 1987.

Heilman KM, Rothi L, Kertesz A. Localization of apraxia-producing lesions. In: Kertesz A, ed. Localization in neuropsychology. New York: Academic Press, 1983.

Heilman KM, Rothi LJ, Valenstein E. Two forms of ideomotor apraxia. Neurology 1982;32:342–346.

Heilman K, Scholes RJ. The nature of comprehension errors in Broca's conduction, and Wernicke's aphasia. Cortex 1976;12:258–265.

Heilman K, Scholes R, Watson RT. Auditory affective agnosia. J Neurol Neurosurg Psychiatry 1975;38:69–72.

Heilman KM, Schwartz HD, Watson RT. Hypoarousal in patients with a neglect syndrome and emotional indifference. Neurology 1978;28:229–232.

Heilman KM, Valenstein E. Frontal lobe neglect in man. Neurology 1972;22:660–664.

Heilman KM, Van Den Abell T. Right hemispheric dominance for mediating cerebral activation. Neuropsychologia 1979;17:315–321.

Heilman KM, Van Den Abell T. Right hemisphere dominance for attention. The mechanism underying hemispheric asymmetries of inattention (neglect). Neurology 1980;30:327–330.

Heilman KM, Watson RT. The neglect syndrome. In: Harnad S, et al, ed. Lateralization in the nervous system. New York: Academic Press, 1977.

Heilman KM, Watson RT, Valenstein E, et al. Localization of lesions in neglect. In Kertesz A, ed. Localization in neuropsychology. New York: Academic Press, 1983.

Heimer L, Alheid GF. Piecing together the puzzle of basal forebrain anatomy. In: Napier TC, et al, eds. The basal forebrain. New York: Plenum, 1991.

Heindel WC, Butters N, Salmon DP. Impaired learning of a motor skills in patients with Huntington's disease. Behav Neurosci 1988;102:141–147.

Heiss W-D, Rosner G. Functional recovery of cortical neurons as related to degree and duration of ischemia. Ann Neurol 1983;14:294–301.

Heit G, Smith ME, Halgren E. Neuronal activity in the human medial temporal lobe during recognition memory. Brain 1990;113:1093–1112.

Hellemans A. Dwarfs and dim galaxies mark limits of knowledge. Science 1995;268:366–367.

Heller SS, Frank KA, Kornfled DS, et al. Psychological outcome following open-heart surgery. Arch Int Med 1974;134:908–914.

Heller W, Levy J. Perception and expression of emotion in right-handers and left-handers. Neuropsychologia 1981;19:263–272.

Hellige JB, Webster R. Right hemisphere superiority for initial stages of letter processing. Neuropsychologia 1979;653–660.

Helm-Estabrooks N. Exploiting the right hemisphere for language rehabilitation: melodic intonation therapy. In: Perecman E, ed. Cognitive processing in the right hemisphere. New York: Academic Press, 1983.

Hendricks JC, Morrison AR, Mann GL. Different behaviors during paradoxical sleep without atonia depend on pontine lesion site. Brain Res 1982;239:81–105.

Henkel CK, Edwards SB. The superior colliculus control of pinna movements in the cat: possible anatomical connnections. J Comp Neurol 1978;182:763–776.

Henley N. Language and sex. Massachussetts: Newbury House Publishers, 1957.

Henri V, Henri C. Enquete sur les premiers souvenirs de L'enrance. L'Annee Psychologique 1897;3:184–198.

Henriksen SJ, Bloom FE, McCoy F, et al. B-endorphin induces nonconvulsive limbic seizures. Proc Nat Acad Sci 1978;75:5221–5225.

Herkenham M, Moon ES, Stuart J. Cell clusters in the nucleus accumbens of the rat, and the mosaic relationship of opiate receptors, and subcortical terminations. Neuroscience 1984;11:561–593.

Hermann BP, Chambria S. Interictal psychopathology in patients with ictal fear. Arch Neurol 1980;37:667–668.

Herman JL, Schatzow E. Recovery and verification of memories of childhood sexual trauma. Psychoanal Psychol 1987;4:1–14.

Hermelin B, O'Connor N. Functional asymmetry in reading of Braille. Neuropsychologia 1971;9:431–435.

Hernandez-Peon R, Chavez-Iberra G, Aguilar-Figuerua E. Somatic evoked potentials in one case of hysterical anesthesia. Electroencephalogr Clin Neurophysiol 1963;15:889–896.

Herrick CJ. The amphibian forebrain. J Comp Neurol 1925;39:400–489.

Herzog AG, Kemper TL. Amygdaloid changes in aging and dementia. Arch Neurol 1980;37:625–629.

Herzog AG, Van Hoesen GW. Temporal neocortical afferent connections in the amygdala in the rhesus monkey. Brain Res 1976;115:57–59.

Herzog M, Hopf S. Behavioral response to species-specific warning calls in infant squirrel monkeys reared in social isolation. Am J Primatol 1984;7:99–106.

Hess EH. Imprinting an effect of early experience. Science 1959;130:133–144.

Hewes GW. Primate communication and the gestural origin of language. Curr Anthropol 1973;14:5–24.

Hewitt W. The development of the corpus callosum. J Anat 1962;96:355–358.

Hicks RE. Asymmetry of bilateral transfer. Am J Psychol 1974;87:667–674.

Hicks RE. Intrahemispheric response competition between vocal and unimanual performance in normal adult human males. J Comp Physiol Psychol 1975;89:50–60.

Hier DB, Mondlock J, Caplan LR. Behavioral abnormalites after right hemisphere stroke. Neurology 1983;33:337–344.

Hier DB, Yoon WB, Mohr JP, et al. Gender and aphasia in the stroke data bank. Brain Lang 1994;47:155–167.

Higley JD, Suomi SJ, Linnoila M. A longitudinal assessment of CSF monoamine metabolite and plasma concentrations in young rhesus monkeys. Biol Psychiatry 1992;32:127–145.

Hijdra A, Braakman R, van Gijn J, Vermeulen M, van Crevel H. Aneurysmal subarachnoid hemorrhage. Stroke 1987;18:1061–1067.

Hilgard ER. Divided consciousness. New York: Wiley, 1977.

Hilgard ER, Hilgard JR. Hypnosis in the relief of pain. Los Altos: Kaufman, 1974.

Hilgard E, Marquis D. Conditioning and learning. New York: Appleton, 1940.

Hilgard JR. Imaginative involvement: some characteristics of the highly hypnotizable and nonhypnotizable. Int J Clin Exp Hypn 1974;22:238–256.

Hill KT. Anxiety in the evaluative context. In: Hartup WW, ed. Young child. Washington DC: NAEYC, 1972.

Hillbom E. Schizophrenia-like psychoses after brain trauma. Acta Psychiatr 1951;60:36–47

Hillbom E. After-effects of brain injuries. Acta Psychiatr Scand (Supplement) 1960;142.

Hines D. Recognition of verbs, abstract nouns and concrete nouns from the left and right visual half-fields. Neuropsychologia 1976;14:211–216.

Hirsch HVB, Spinelli DN. Visual experience modifies distribution of horizontally and vertically oriented receptive fields in cats. Science 1970;168:869–871.

Hirsch HUB, Spinelli DN. Modification of the distribution of receptive field orientation in cats by selective visual exposure during development. Exp Brain Res 1971;12:504–527.

Hirsh AR. Effect of an ambient odor on slot-machine usage in a Las Vegas Casino. Chemical Senses 1993;18:112.

Hirsh AR, Gomez R. Weight reduction through inhalation of odorants. J Neurol Orthoped Med Surg 1995;16:2–5.

Hirsh FR, Hansen HHM, Paulson OB, Vraa-Jensen J. Development of brain metastases in small-cell anaplastic carcinoma of the lung. In: Whitehouse JMA, Kay HEM, eds. CNS complications of malignant disease. Baltimore: University Park Press, 1979.

Hirst W, Johnson MK, Phelps EA, Volpe BT. More on recognition and recall in amnesics. J Exp Psychol Learn Mem Cognit 1988;14:758–762.

Hirst W, Volpe BT. Temporal order judgments with amnesia. Brain Cognit 1982;1:294–306.

Hisock M. Eye-movement asymmetry and hemisphere function. J Psychol 1977;97:49–52.

Hiscock M, Cohen DB. Visual imagery and dream recall. J Res Person 1973;7:179–188.

Hitchcock J, Davis M. Fear-potentiated startle using an auditory conditioned stimulus: effect of lesions on the amygdala. Physiol Behav 1987;10:403–408.

Hitchcock JM, Davis M. Efferent pathway of the amygdala involved in conditioned fear as measured with the fear-potentiated startle paradigm. Behav Neurosci 1991;105:826–842.

Hobhouse N. Prognosis in hemiplegia in middle life. Lancet 1936;1:327–328.

Hobson J. Reciprocal interaction model of sleep cycle control. Implications for PGO wave generation and dream amnesia. In: Olin D, McGough J, eds. Neurobiology of sleep and memory. New York: Academic Press, 1977.

Hobson JA, Lydic R, Baghdoyan HA. Evolving concepts of sleep cycle generation: From brain centers to neuronal populations. Behav Brain Sci 1986;9: 371–448.

Hocherman S, Yirmiya R. Neuronal activity in the medial geniculate nucleus and in the auditory cortex of the rhesus monkey reflects signal anticipation. Brain 1990;113:1707–1720.

Hodgson F. A case of double consciousness. Proc Soc Psychic Res 1892;7:22–257.

Hodoba D. Paradoxic sleep facilitation by interictal epileptic activity of right temporal origin. Biol Psychiatry 1986;21:1267–1278.

Hoebel BG, Monaco AP, Hernandez L, et al. Self-injection of amphetamine directly into the brain. Psychopharmacol 1983;81:158–163.

Hoff-Ginsberg E, Shatz M. Linguistic input and the child's acquisition of language. Child Dev 1982;17: 63–71.

Hoffman D, Beninger RJ. The effects of priozide on the establishment of conditioned reinforcement as a function of the amount of conditioning. Psychopharmacol 1985;87:454–460.

Hoffman ML. Sex differences in empathy and related behaviors. Psychol Bull 1977;84:712–720.

Hoffman RE. Verbal hallucinations and language production processes in schizophrenia. Behav Brain Sci 1986;9:503–548.

Hoffman R, Stopek S, Andreasen N. A discourse analysis comparing manic versus schizophrenic speech disorganization. Arch Gen Psychiatry 1986;43: 831–838.

Hofstede BTM, Kolk HHJ. The effects of task variation in the production of grammatical morphology in Broca's aphasia. Brain Lang 1994;46:278–328.

Hollander E, Liebowitz MR, DeCaria C, et al. Treatment of depersonalization with serotonin reuptake blockers. J Clin Psychopharmacol 1990;10:200–203.

Holloway RL. Revisiting the South African Taung australopithecine endocast. Am J Phys Anthropol 1981;56:43–58.

Holloway RL. Human paleontological evidence relevant to language behavior. Hum Neurobiol 1983; 2:105–114.

Holloway R. The poor brain of Homo sapiens neandertaliensis... see what you please. In Delson D, ed. Ancestors: the hard evidence. New York: Liss, 1985.

Holloway RL. The gyrification index to account for volumetric reorganization in the evolution of the human brain. J Hum Evol 1992;22:163–170.

Holloway RL, Anderson PJ, Defendini R, Harper C. Sexual dimorphism of the human corpus callosum from three independent samples: relative size of the corpus callosum. Am J Phys Anthropol 1993; 92:481–498.

Holmes G. The symptoms of acute cerebellar injuries due to gunshot injuries. Brain 1917;40:461–535.

Holmes G. The cerebellum in man. Brain 1939;62:1–30.

Hom J, Reitan R. Effects of lateralized cerebral damage on contalster and ipsilateral sensorimotor performance. J Clin Neuropsychol 1982;3:47–53.

Honer WG, Bassett AS, Smith GN, et al. Temporal lobe

abnormalities in multigenerational families with schizophrenia. Biol Psychiatry 1994;36:737–743.

Hopkins WD, Washburn DA, Rumbaugh DM. Processing of form stimuli presented unilaterally in humans, chimpanzee, and monkeys. Behav Neurosci 1990;104:577–582.

Hoppe KD. Split brains and psychoanalysis. Psychoanal Q 1977;46:220–244.

Hoppe KD, Bogen JE. Alexithymia in twelve commissurotomized patients. Psychother Psychosom 1977; 28:148–155.

Horak FE, Anderson ME. Influence of globus pallidus on arm movements in monkeys. I. J Neurophysiol 1984;52:290–304.

Hore J, Vilis T. Arm movement performance during reversible basal ganglia lesions in monkeys. Exp Brain Res 1980;39:217–228.

Horel JA. The neuroanatomy of amnesia. Brain 1978; 101:403–445.

Horel JA, Pytko-Joiner DE, Voytko M, Salsbury K. The performance of visual tasks while segments of the inferotemporal cortex are suppressed by cold. Behav Brain Res 1987;23:29–42.

Hornig CR, Dorndorf W, Agnoli AL. Hemorrhagic cerebral infarction—a prospective study. Stroke 1986; 17:179–185.

Hornykiesciz O. Brain neurotransmitter changes in Parkinson's disease. In: Marsden CD, Fahn S, eds. Movement disorders. London: Butterworth, 1982.

Horowitz MJ, Adams JE, Rutkin BB. Visual imagery on brain stimulation. Arch Gen Psychiatry 1968; 19:469–486.

Hosch HM, Cooper DS. Victimization as a determinant of eyewitness accuracy. J Applied Psychol 1982;67:649–652.

Hotson JR, Pedley TA. The neurological complications of cardiac transplantation. Brain 1976;99:673–694.

Hough MS. Narrative comprehension in adults with right and left hemisphere brain-damage: theme organization. Brain Lang 1990;38:253–277.

House A, Dennis M, Warlow C, et al. Mood disorders after stroke and their relation to lesion location. Brain 1990;113:1113–1129.

Howard G, Toole JF, Frye-Pierson J, Hinshelwood LC. Factors influencing the survival of 451 transient ischemic attack patients. Stroke 1987;18:552–557.

Howe ML, Courage ML. On resolving the enigma of infantile amnesia. Psychol Bull 1993;113:305–326.

Howe ML, Hunter MA. Adult age differences in storage-retrieval processes. Can J Psychol 1985;39: 130–150.

Howe ML, Hunter MA. Long-term memory in adulthood. Devel Rev 1986;6:334–364.

Howell A, Conway M. Mood and the suppression of positive and negative self-referent thoughts. Cognit Ther Res 1992;16:535–555.

Howells K. Adult sexual interest in children: conclusions relevant to theories of aetiology. In: Cook M, Howells K, eds. Adult sexual interest in children. New York: Academic Press, 1980.

Howes D, Boller F. Simple reaction times: evidence for focal impairment from lesions of the right hemisphere. Brain 1975;98:317–322.

Howes DH, Solomon RL. A note on McGinnies' "emotionality and perceptual defense." Psychol Rev 1950;57:229–234.

Hoyle F. The black cloud. London: Heinermann, 1957.

Hrbek V. Pathophysiologic interpretation of Gerstmann's syndrome. Neuropsychologia 1977;11: 377–388.

Hubel DH, Calvin OH, Rupert A, Galambos R. Attention units in the auditory cortex. Science 1959;129: 1279–1280.

Hubel DH, Wiesel TN. Receptive fields of single neruons in the cat's striate cortex. J Physiol 1959; 148:574–591.

Hubel DH, Wiesel TN. Receptive fields, binocular perception and functional architecture in the cat's visual cortex. J Physiol 1962;160:106–154.

Hubel DH, Wiesel TN. Receptive fields and functional architecture of monkey striate cortex. J Physiol 1968;195:215–243.

Hubel DH, Wiesel TN. The period of susceptibility to the physiological effects of unilateral eye closure in kittens. J Physiol 1970;206:419–436.

Hubel DH, Wiesel TN. Sequence regularity and geometry of orientation columns in monkey striate cortex. J Comp Neurol 1974;158:267–293.

Hubel DH, Wiesel TN. Brain mechanisms of vision. Sci Am 1979;241:150–163.

Hudson JA. Constructive processing in children's event memory. Devel Psychol 1990;26:180–187.

Hudson JA, Fivush R. As time goes by: sixth graders remember a kindergarten experience. Emory Cognition Project Report #13. Emory University, Atlanta, 1987.

Hughes M, Grieve R. On asking children bizarre questions. First Language 1980;1:149–160.

Humphrey ME, Zangwill OL. Cessation of dreaming after brain injury. J Neurol Neurosurg Psychiatry 1911;14:322–325.

Humphrey T. The development of the human amygdaloid complex. In: Elefterhiou BE, ed. The neurobiology of the amygdala. New York: Plenum, 1972.

Humphreyes RP. Outcome of severe head injury in children, In: Raimondi AJ, ed. Concepts in pediatric neurosurgery, Vol. 3. Basel: Karger, 1983.

Hupfer K, Jurgens U, Ploog D. The effect of superior temporal lesions on recognition of species-specific calls in the squirrel monkey. Exp Brain Res 1977; 30:75–87.

Huppert FA, Piercy M. Dissociation between learning and remembering in organic amnesia. Nature 1977;275:317–318.

Hurwitz LJ, Adams GF. Rehabilitation of hemiplegia: indices of assessment and prognosis. Br Med J 1972;1:94–98.

Huttenlocher J. The origins of language comprehension. In Solso RL, ed. Theories in cognitive psychology. New Jersey: Erlbaum, 1974.

Huttenlocher PR. Synaptic density in human frontal

cortex—developmental changes and effects of aging. Brain Res 1979;163:195–205.

Huttenlocher PR. Morphometric study of human cerebral cortex development. Neuropsychologia 1990; 28:517–527.

Hyden H. The differentiation of of brain cell protein, learning and memory. Biosystems 1977;8:213–218.

Hyman LH. The transition from the unicellular to the multicellular individual. Biol Symp 1942;8:27–42;

Hyvarinen J. The parietal cortex of monkey and man. Berlin: Spinger Verlag, 1982.

Hyvarinen J, Poranen A. Function of the parietal associative area 7 as revealed from cellular discharges in alert monkeys. Brain 1974;97:673–692.

Hyvarinen J, Shelepin Y. Distribution of visual and somatic fucntions in the parietal association area of the monkey. Brain Res 1979;169:561–564.

Ikeda A, Luders HO, Burgess RC, Shibasaki H. Movement-related potentials recorded from supplementary motor area and primary motor area. Brain 1992;115:1017–1043.

Ingram WR. Brainstem mechanisms and behavior. Electroencephalogr Clin Neurophysiol 1952;4: 395–406.

Ingvar DH, Franzen G. Abnormalities of cerebral blood flow distribution in patients with chronic schizophrenia. Acta Psychiatr Scand 1974;50: 425–462.

Insausti R, Amaral DG, Cowan WM. The enterorhinal cortex of the monkey: II. Cortical afferents. J Comp Neurol 1987;264:356–395.

Ironside R. Disorders of laughter due to brain lesions. Brain 1956;79:589–609.

Irvine WM, Leschine SB, Schloerb FP. Thermal history, chemical composition, and the relationship of comets to the origin of life. Nature 1980;283:748.

Isaacson RL. The limbic system. New York: Plenum, 1982.

Ito M, Lammsertsma AA, Wise RJS, et al. Measurement of regional cerebral blood flow and oxygen utilization in patients with cerebral tumours using positron emission tomography. Neuroradiol 1982; 23:63–74.

Iversen SD, Dunnett SB. Functional organization of striatum as studied with neural grafts. Neuropsychologia 1990;28:601–626.

Iversen SD, Mishkin M. Perseverative interference in monkeys following selective lesions of the inferior prefrontal convexity. Exp Brain Res 1970;11: 476–386.

Ivry RB, Keele SW. Timing functions of the cerebellum. J Cognit Neurosci 1989;1:136–152.

Iwai E, Yukie M. Amygdalofual and amygdalopetal connections with modality-specific cortical areas in macaques. J Comp Neurol 1987;261:362–387.

Iwamura Y, Tanaka M. Postcentral neurons in hand region of area 2. Brain Res 1978;150:662–666.

Izquierdo I, Graundenz M. Memory facilitation by naloxone is due to release of dopaminergic and beta-adrenergic system from tonic inhibition. Psychopharmacol 1980;67:265–268.

Jack RA, Rivers-Bulkeley NT, Rabin PL. Seconday mania as a presentation of progressive dialysis encephalopathy. J Nerv Ment Dis 1983;71:193–195.

Jackson AC. Neurologic disorders associated with mitral valve prolapse. Can J Neurol Sci 1986;13:15–20.

Jackson JH. Dreamy states. In: Taylor J, ed. Selected writings of John Hughlings Jackson. New York: Basic Books, 1958.

Jackson JH, Stewart. Epileptic attacks with a warning of a crude sensation of smell and with the intellectual aura (dreamy state) in a patient who had symptoms pointing to gross organic disease of the right temporo-sphenoidal lobe. In: Taylor J, ed. Selected writings of John Hughlings Jackson. New York: Basic Books, 1958.

Jacobesen E. Electrophysiology of mental activities. Am J Psychol 1932;44:677–694.

Jacobs BL, Aszmita EC. Structure and function of the brain serotonin system. Physiol Rev 1992;72: 165–245.

Jacobs KH. Evolution in the postcranial skeleton of late glacial and early postglacial European hominids. Zool Morphol Anthropol 1985;75:307–326.

Jacobson AL. Learning in flatworms and annelids. Psychol Bull 1963;60:74–94.

Jacobson M. Developmental neurobiology. New York: Plenum, 1978.

Jacombe DE, McLain LW, Fitzgerald R. Postural reflex gelastic seizures. Arch Neurol 1980;37:249–251.

James W. The varieties of religious experience. New York. New American Library.

James W. Psychology. New York: Harper & Row, 1961.

Jamieson RC, Wells CE. Manic psychosis in a patient with multiple metastatic brain tumors. J Clin Psychiatry 1979;40:280–282.

Jampala VC, Abrams R. Mania secondary to right and left hemisphere damage. Am J Psychiatry 1983;140: 1197–1199.

Jampel RS. Convergence, divergence, pupillary reactions and accomodation of the eyes from faradic stimulation of the macaque brain. J Comp Neurol 1960;115:371–397.

Jane JA, Riomel RW, Pobereskin LH, et al. Outcome and pathology of head injury. In: Granman RG, Gildenburg PL, eds. Head injury: basic and clinical aspects. New York: Raven Press, 1982

Janet P. The major symptoms of hysteria. New York: MacMillan, 1927.

Janro L. Korsakoff-like amnesic syndrome in penetrating brain injury. Acta Neurol Scand 1973;49:44–67.

Jarvik E. Basic structure and evolution of vertebrates. Vol. 2. New York: Academic Press, 1980.

Jasper HH, Rasmussen T. Studies of clinical and electrical responss to deep temporal stimulation in man with some considerations of functional anatomy. Res Publications Assoc Res Nerv Ment Dis 1958; 36:316–334.

Jaynes J. The origin of consciousness in the breakdown of the bicameral mind. Boston: Houghton Mifflin, 1976.

Jeannerod M. Neurophysiological and neuropsychological aspects of spatial neglect. New York: North-Holland, 1987.

Jeeves MA, Dixon NF. Hemispheric differences in response rates to visual stimuli. Psychonomic Sci 1970;20:249–251.

Jelinek A. The amudian in the context of the Mugharan tradition at the Tabun Cave (Mount Carmel), Israel. In: The human revolution: behavioral and biological perspectives on the origins of modern humans. Vol 2. Mellars P, ed. Edinburgh: Edinburgh University Press, 1989.

Jenkins WM, Merzenich MM, Recanzone G. Neocortical representational dynamics in adult primates. Neuropsychologia 1990;28:573–584.

Jennett B. Scale, scope and philosophy of the clinical problem. In: Porter R, Fitzsimons DW, eds. Outcome of severe damage to the central nervous system. (Ciba). Amsterdam: North Holland, 1975.

Jennett B. The problem of mild head injury. Practitioner 1978;221:77–82.

Jennett B, Plum F. Persistent vegetative states after brain damage. Lancet 1972;1:734–737.

Jennett B, Miller JD, Braakman R. Epilepsy after nonmissile depressed skull fracture. J Neurosurg 1974; 41:208–216.

Jennett B, Snoek J, Bond MR, Brooks N. Disability after severe head injury. J Neurol Neurosurg Psychiatry 1981;44:285–293.

Jennett B, Teasdale G. Aspects of coma after severe head injury. Lancet 1977;1:878–881.

Jennett B, Teasdale G. Management of head injuries. Philadelphia: FA Davis, 1981.

Jennett B, Teasdale G, Galbraith S, et al. Severe head injuries in three countries. J Neurol Neurosurg Psychiatry 1977;40:291–298.

Jennings VA, Lamour Y, Solis H, Fromm C. Somatosensory cortex activity related to position and force. J Neurophysiol 1983;49:1216–1229.

Jensen GD, Bobbit RA, Gordon BN. Sex differences in the development of independence in infant monkeys. Behaviour 1968;30:1–14.

Jensen I, Larsen JK. Psychoses in drug-resistant temporal lobe epilepsy. J Neurol Neurosurg Psychiatry 1979;42:948–955.

Jerison HJ. Evolution of the brain and intelligence. New York: Academic Press, 1973.

Jerison HJ. Fossil evidence on the evolution of neocortex. In: Jones ED, Peters A, eds. Cerebral cortex. Vol 8. Comparative structure and evolution of cerebral cortex. New York: Plenum, 1990.

Jerslid A. Memory for the pleasant as compared with the unpleasant. J Exp Psychol 1931;14:284–288.

Jetter W, Poser U, Freeman RB, Markowitsch JH. A verbal long term memory deficit in frontal damaged patients. Cortex 1986;22:229–242.

Jocic Z, Staton RD. Reduplication after right middle cerebral infarction. Brain Cognit 1993;23:222–230.

Johanson D, Shreeve J. Lucy's child. New York: Morrow, 1989.

Johanson DC, White TD. A systematic assessment of early African hominids. Science 1979;203:321–330.

Johanson D, Edey M. Lucy: the begnnings of humankind. New York: Simon & Schuster, 1981.

Johnson MK, Kim JK, Risse G. Do alcoholic Korsakof's syndrome patients acquire affective reactions? J Exp Psychol Learn Mem Cognit 1985;17: 210–223.

Johnson RN. Aggression in man and animals. Philadelphia: WB Saunders, 1972.

Jonason KR, Enloe LJ. Alterations in social behavior following septal and amygdaloid lesions in the rat. J Comp Physiol Psychol 1971;75:280–301.

Johnson TN, Rosvold HE, Mishkin M. Projections of behaviorally defined sectors of the prefrontal cortex to the basal ganglia, septum, and diencephalon of the monkey. Exp Neurol 1968;21:20–34.

Johnston MV. Biochemistry of neurotransmitters in cortical develoment. In: Peters A, Jones EG, eds. Cerebral cortex. Vol. 7. New York: Plenum, 1988.

Jolly A. The evolution of primate behavior. New York: MacMillan, 1972.

Jonason KR, Enloe LJ. Alterations in social behavior following septal and amygdaloid lesions in the rat. J Comp Physiol Psychol 1972;75:280–301.

Jonason KR, Enloe LJ, Contrucci J, Meyer PM. Effects of stimulation and successive septal and amygdaloid lesions on social behavior in the rat. J Comp Physiol Psychol 1973;54:625–628.

Jones B, Mishkin M. Limbic lesions and the problem of stimulus-reinforcement associations. Exp Neurol 1972;36:362–377.

Jones EG, Coulter JD, Hendry SHC. Intracortical connectivity of architectonic fields in the somatic sensory, motor, and parietal cortex of monkeys. J Comp Neurol 1978;181:291–348.

Jones EG, Powell TPS. An anatomical study of converging sensory pathways within the cerebral cortex of the monkey. Brain 1970;93:793–820.

Jones EG, Wise SP, Coulter JD. Differential thalamic relationships of sensory-motor and parietal cortical fields in monkeys. J Comp Neurol 1979;183: 830–853.

Jorgensen HS, Jespersen HF, Nakayama H, et al. Headache in stroke. Neurology 1994;44:1793–1797.

Jorgenson L, Torvic A. Ischaemic cerebrovascular disease in an autopsy series. J Neurol Sci 1969;9: 285–320.

Joseph JP, Lesevre N, Dreyfus-Bricsac C. Spatiotemporal organization of EEG in premature infants and full-term newborns. Electroencephalogr Clin Neurophysiol 1976;40:153–168.

Joseph R. Effects of rearing and sex on learning and competitive exploration. J Psychol 1979;101:37–43.

Joseph R. The neuropsychology of development: hemispheric laterality, Limbic Language, and the Origin of Thought. J Clin Psychol 1982;44:3–34.

Joseph R. Confabulation and delusional denial: frontal lobe and lateralized influences. J Clin Psychol 1986a; 42:845–860.

Joseph R. Reversal of cerebral dominance for language and emotion in a corpus callosotomy patient. J Neurol Neurosurg Psychiatry 1986b;49:628–634.

Joseph R. The right cerebral hemisphere: emotion, music, visual-spatial skills, body image, dreams, and awareness. J Clin Psychol 1988a;44:630–673.

Joseph R. Dual mental functioning in a split-brain patient. J Clin Psychol 1988b;44:770–779.

Joseph R. The limbic system. In: Neuropsychology, Neuropsychiatry, Behavioral Neurology. Puente AE, Reynolds CR, series eds. New York: Plenum Press, 1990a.

Joseph R. The left cerebral hemisphere: aphasia, alexia, agraphia, agnosia, apraxia, schizophrenia, language and thought. In: Neuropsychology, neuropsychiatry, behavioral neurology. Puente AE, Reynolds CR, series eds. New York: Plenum, 1990b.

Joseph R. The parietal lobes. In: Neuropsychology, neuropsychiatry, behavioral neurology. Puente AE, Reynolds CR, series eds. New York: Plenum, 1990c.

Joseph R. The frontal lobes. In: Neuropsychology, neuropsychiatry, behavioral neurology. Puente AE, Reynolds CR, series eds. New York: Plenum, 1990d.

Joseph R. The right cerebral hemisphere. In: Neuropsychology, neuropsychiatry, behavioral neurology. Puente AE, Reynolds CR, series eds. New York: Plenum, 1990e.

Joseph R. The temporal lobe. In: Neuropsychology, neuropsychiatry, behavioral neurology. Puente AE, Reynolds CR. New York: Plenum, 1990f.

Joseph R. The limbic system: emotion, laterality, and unconscious mind. Psychoanal Rev 1992a;79: 405–456.

Joseph R. The right brain and the unconscious. New York: Plenum, 1992b.

Joseph R. The naked neuron: evolution and the languages of the body and brain. New York: Plenum, 1993.

Joseph R. The limbic system and the foundations of emotional experience. In: Ramachandran VS, ed. Encyclopedia of human behavior. San Diego: Academic Press, 1994.

Joseph R, Casagrande VA. Visual field defects and morphological changes resulting from monocular deprivation in primates. Proc Soc Neuroscience 1978;4:2021.

Joseph R, Casagrande VA. Visual deficits and recovery following monocular lid closure in a prosimian primate. Behav Brain Res 1980;1:165–186.

Joseph R, Forrest N, Fiducia N, Como P, Siegel J. Electrophysiological and behavioral correlates of arousal. Physiol Psychol 1981;9:90–95.

Joseph R, Gallagher RE. Gender and early environmental influences on activity, overresponsiveness, and exploration. Devel Psychobiol 1980;13:527–544.

Joseph R, Gallagher RE. Interhemispheric transfer and the completion of reversible operations in non-conserving children. J Clin Psychol 1985;41: 796–800.

Joseph R, Gallagher RE, Holloway J, Kahn J. Two brains, one child: interhemispheric transfer and confabulation in children aged 4, 7, 10. Cortex, 1984;20:317–331.

Joseph R, Hess S, Birecree E. Effects of sex hormone manipulations and exploration on sex differences in learning. Behav Biol 1978;24:364–377.

Jouvet M. Neurophysiology of the states of sleep. Physiol Rev 1967;47:117–177.

Joynt RJ. Verney's concept of the osmoreceptor. Arch Neurol 1966;14:331–334.

Juliano SL, Eslin DE, Tommerdahl M. Developmental regulation of plasticity in cat somatosensory cortex. Journal of Neurophysiology 1994;72:1706–1716.

Jung CG. Collected papers on analytical psychology. London: Bailliere, Tidall and Cox. London, 1922.

Jung CG. On the nature of dreams. (Translated by RFC Hull.) The collected works of CG Jung. Princeton: Princeton University Press, 1945, pp. 473–507.

Jung C. Experimental researches. Collected works II. New Jersey: Princeton University Press, 1954.

Jung CG. Man and his symbols. New York: Dell, 1964.

Jung CG. Psychology and religion. Vol 11. New Jersey: Princeton University Press, 1966.

Jung CG. Experimental researches. Vol. 2. New Jersey: Princeton University Press. 1973.

Jung CG. The structure and dynamics of the psyche. New Jersey: Princeton University Press, 1973.

Jung R. Neuronal integration in the visual cortex and its significance for visual information. In: Rosenblith WA, ed. Sensory communication. Cambridge: MIT Press, 1961.

Jurgens U. Vocal communication in primates. In: Kesner RP, Olton DS, eds. Neurobiology of comparative cognition. New Jersey, Erlbaum, 1990.

Jurgens U, Kirzinger A, von Cramon D. The effects of deep-reaching lesions in the cortical face area on phonation. A combined case report and experimental monkey study. Cortex 1982;18:125–140.

Jurgens U, Muller-Preuss P. Convergent projections of different limbic vocalization areas in the squirrel monkey. Exp Brain Res 1977;29:75–83.

Kaada BR. Somato-motor, autonomic and electrocortical responses to electrical stimulation of rhienncephalon and other structures in primates, cat, and dog. Acta Physiol Scand 1951;24:1–170.

Kaada BR. Cingulate, posterior orbital, anterior insular and temporal pole cortex. In: Field J, Magoun HW, Hall VA, eds. Handbook of physiology. Vol. II. Washington DC: American Physiological Society, 1972:1345–1372.

Kaada BR, Jansen J, Andersen P. Stimulation of hippocampus and medial cortical areas in unanesthetized cats. Neurology 1953;3:844–857.

Kaas JH, Krubitzer LA. Neuroanatomy of visual pathways and their retinotopic organization. In: Dreher B, Robinson SR, eds. New York: Macmillan, 1991.

Kaas JH, Nelson RJ, Sur M, Merzenich. Multiple representations of the body in the post central somatosensory cortex of primates. In: Woolsey CN, ed. Cortical sensory organization (1). New Jersey: Humana Press, 1981.

Kaminer H, Lavie P. Sleep and dreaming in holocaust survivors. J Nerv Ment Dis 1991;179:664–669.

Kamiya J. Behavioral, subjective and physiological aspects of drowsiness and sleep. In: Fiske DW,

Maddi SR, eds. Function of varied experience. Homewood, IL: Dorsey Press, 1961, pp. 145–174.

Kaplan JA, Brownell HH, Jacobs JR, Gardner H. The effects of right hemisphere damage on the pragmatic interpretation of conversational remarks. Brain Lang 1990;38:315–333.

Kapp BC, Whalen PJ, Supple WF, Pascoe JP. Amygdala contributions to conditioned arousal and sensory information processing. In: Aggleton JP, ed. The amygdala. New York: Wiley-Lis, 1992.

Kapp BS, Supple WF, Whalen PJ. Effects of electrical stimulation of the amygdaloid central nucleus on neocortical arousal in the rabbit. Behav Neurosci 1994;108:81–93.

Kapur N, Coughlan AK. Confabulation and frontal lobe dysfunction. J Neurol Neurosurg Psychiatry 1980;43:461–463.

Kapur N, Friston KJ, Young A, et al. Activation of human hippocampal formation during memory for faces: a PET study. Cortex 1995;31: 99–108.

Kapura N, Ellison D, Smith MP, Mclellan DL. Focal retrograde amnesia following bilateral temporal lobe pathology. Brain 1992;115:73–85.

Karli P, Vergnes M. Interspecific aggressive behavior and its manipulation by brain ablation and stimulation. In: Garattini S, Sig EB, eds. Aggressive behavior. Amsterdam, Excerpta Medica, 1969.

Kasamatsu T, Pettigrew JD. Depletion of brain catecholamines. Science 1976;194:206–208.

Kase CS. Intracerebral hemorrhage: non-hypertensive causes. Stroke 1986;17:590–595.

Katz J, Halstead W. Protein organization and mental functions. Compr Physiol 1950;20:1–38.

Kauer JA, Malenka RC, Nicoll RA. Nature 1988;334: 249–252.

Kavey NB, Whyte J, Resor SR, Gidro-Frank S. Somnambulism in adults. Neurology 1990;40:749–752.

Kawaguchi S. Regeneration of cerebellofugal projections in kittens. In: Florh H, ed. Post-lesion neural plasticity. New York: Springer-Verlag, 1988.

Kawamura S, Sprague JM, Niimi K. Corticofugal projections from the visual cortices to the thalamus and superior colliculus. J Comp Neurol 1974;158: 339–362.

Kawano K, Sasaki M. Response properties of neurons in posterior parietal cortex of monkey during visual-vestibular stimulation. J Neurophysiol 1984; 51:352–360.

Kawano K, Sasakii M, Yamashita M. Response properties of neurons in posterior parietal cortex of monkey during visual-vestibular stimulation. J Neurophysiol 1984;51:340–351.

Kay J, Ellis A. A cognitive neuropsychological case study of anomia. Brain 1987;110:613–629.

Kazui S, Naritomi H, Sawada T, Inoue N. Subcortical auditory agnosia. Brain Lang 1990;38:476–487.

Kebeck G, Lohaus A. Effect of emotional arousal on free recall of complex material. Percept Motor Skills 1986;63:461–462.

Keller EL, Edelman JA. Use of interrupted saccade paradigm to study spatial and temporal dynamics

of saccadic burst cells in superior colliculus in monkey. J Neurophysiol 1994;72:2754–2770.

Kelly AE, Domesick VB, Nauta WJH. The amygdalostriatal projection in the rat. Neuroscience 1982;7: 615–630.

Kelly F. My captivity among the Sioux Indians. New York: Corinth Books, 1962.

Kelly JP, Nichols JS, Filley CM, et al. Concussion in sports. J Am Med Assoc 1991;266:2867–2869.

Kelsey JE, Arnold SR. Lesions of the dorsomedial amygdala, but not the nucleus accumbens, reduce the aversiveness of morphine withdrawal in rats. Behav Neurosci 1994;108:1119–1127.

Kemp JM, Powell TPS. The cortico-striate projection in monkey. Brain 1970;93:525–546.

Kempe CH, Helfer RE. The battered child. Chicago: Chicago University Press, 1980.

Kemper T. Toward a sociology of emotions. Am Sociol 1978;13:30–41.

Kennard MA. Alterations in response to visual stimuli following lesions of frontal lobes in monkeys. Arch Neurol Psychiatry 1939;41:1153–1165.

Kennard MA. Focal autonomic representation in the cortex and its relation to sham rage. J Neurophysiol Exp Neurol 1945;4:295–304.

Kennard MA. Effects of bilateral ablation of cingulate area on behavior in cats. J Neurophysiol 1955;18: 159–169.

Kennard MA, Spencer S, Fountain G. Hyperactivity in monkeys following lesions of the frontal lobes. J Neurophysiol 1941;4:512–522.

Kepes JJ. Astrocytomas: old and newly reconized variants, their spectrum of morphology and antigen expression. Can J Neurol Science 1987;14: 109–121.

Kerr NH, Foulkes D. Reported absence of visual dream imagery in a normally sighted subject with Turner's syndrome. J Ment Imag 1978;2:247–264.

Kerr NH, Foulkes D. Right hemisphere mediation of dream visualization: a case study. Cortex 1981;17: 603–611.

Kertesz A. Localization of lesions in Wernicke's aphasia. In: Kertesz A, ed. Localization in neuropsychology. New York: Academic Press, 1983a.

Kertesz A. Right-hemisphere lesions in constructional apraxia and visuospatial deficit. In: Kertesz A, ed. Localization in neuropsychology. New York: Academic Press, 1983b.

Kertesz A, Ferro JM, Shewan CM. Apraxia and aphasia: the functional anatomical basis for their dissociation. Neurology 1984;30:40–47.

Kertesz A, Geschwind N. Patterns of pyramidal decussation and their relationship to handedness. Arch Neurol 1971;24:326–332.

Kertesz A, Lesk D, McCabe P. Isotope localization of infarcts in aphasia. Arch Neurol 1977;34:590–601.

Kesner RP. Learning and memory in rats with an emphasis on the role of the amygdala. In: Aggleton JP, ed. The amygdala. New York: Wiley-Liss, 1992.

Kesner RB, Andrus RG. Amygdala stimulation disrupts the magnitude of reinforcement contribution to long-term memory. Physiol Psychol 1982;10: 55–59.

Kesner RP, DiMattia BV. Neurobiology of an attribute model of memory. In: Epstein AN, Morrison AR, eds. Progr Psychobiol Physiol Psychol. San Diego: Academic Press, 1987.

Kester DB, Saykin AJ, Sperling MR, et al. Acute effect of anterior temporal lobectomy on musical processing. Neuropsychologia 1991;29:703–708.

Kihlstrom JF, Harackiewiz JM. The earliest recollection: a new survey. J Personality 1982;50:134–148.

Kihlstrom JF, Schacter DL, Cork RC, et al. Implicit and explicit memory following surgical anesthesia. Psychol Sci 1990;1:303–306.

Kilpatrick DG, Veronen LJ, Best CL. Factors predicting psychological distress among rape victims. In: Figely CR, ed. Trauma and its wake. New York: Brunner/Mazel, 1985.

Kim JJ, Fanselow MS. Modality-specific retrograde amnesia of fear. Science 1992;256, 675–677.

Kim NW, Mieczslaw A, Piatyszek A, et al. Specific association of human telomerase activity with immortal cells and cancer. Science 1994;266: 2011–2014.

Kim Y, Morrow L, Passafiume D, Boller F. Visuoperceptual and visuomotor abilities and locus of lesion. Neuropsychologia 1984;2:177–185.

Kimbel WH, White TD. A reconstruction of the adult cranium of Australopithecus afarensis. Am J Phys Anthropol 1980;52:244.

Kimura D. Cerebral dominance and the perception of verbal stimuli. Can J Psychol 1961;15:156–171.

Kimura D. Right temporal lobe damage: perception of unfamiliar stimuli after damage. Arch Neurol 1963;18:264–271

Kimura D. Left-right differences in the perception of melodies. Q J Psychol 1964;16:355–358.

Kimura D. Dual functional asymmetry of the brain in visual perception. Neuropsychologia 1966;4: 275–285.

Kimura D. Spatial localization in left and right visual fields. Can J Psychol 1969;23:445–448.

Kimura D. Manual activity during speaking. Neuropsychologia 1973;11:45–55.

Kimura D. The neural basis of language qua gesture. In: Whitaker H, Whitaker HA, eds. Studies in neurolinguistics. Vol. 2. New York: Academic Press, 1976.

Kimura D. Acquisition of a motor skill after left-hemisphere damage. Brain 1977;100:527–542.

Kimura D. Neuromotor mechanisms in the evolution of human communication. In: Steklis HD, Raleigh MJ, eds. Neurobiology of social communication in primates. New York: Academic Press, 1979.

Kimura D. Left-hemisphere control of oral and brachial movement and their relation to communication. Philosoph Trans Royal Soc London 1982; 298:135–149.

Kimura D. Neuromotor mechanisms in human communication. New York: Oxford University Press, 1993.

Kimura D, Archibald Y. Motor functions of the left hemisphere. Brain 1974;97:337–350.

Kimura D, Folb S. Neural processing of backward speech sounds. Science 1968;161:395–396.

Kimura M. The putamen neuron. In: Carpenter MB, Jayaraman A, eds. The basal ganglia. New York: Plenum, 1987.

Kimura M, Rajkowski J, Evarts E. Tonically discharging putamen neurons exhibit set-dependent responses. Proc Nat Acad Sci 1984;81:4998–5001.

King FA. Effects of septal and amygdaloid lesions on emotional behavior and conditioned avoidance responses in the rat. Science 1958;128:655–656.

King FA, Meyer PM. Effect of amygdaloid lesions upon septal hyperemotionality in the rat. Science 1958;128:655–656.

King FL, Kimura D. Left ear superiority in dichtoic perception of vocal nonverbal sounds. Can J Psychol 1972;26:111–116.

King MA, Yuille JC. Suggestability and the child witness. In: Ceci SJ, Toglia MP, Ross DF. Child eyewitness memory. New York: Springer-Verlag, 1987.

Kinsbourne M, Cook J. Generalized and lateralized effect of concurrent verbalization on a unimanual skill. Q J Exp Psychol 1971;23:341–345.

Kinsbourne M, Warrington EK. A variety of reading disabilities associated with right hemisphere lesions. J Neurol Neurosurg Psychiatry 1962;25: 339–344.

Kinsbourne M, Warrington EK. Jargon aphasia. Neuropsychologia 1963;1:27–37.

Kinsbourne M, Warrington EK. Disorders of spelling. J Neurol Neurosurg Psychiatry 1964;27:224–228.

Kish SJ, Shannak K, Hornykiewicz O. Uneven pattern of dopamine loss in the striatum of patients with idopathic Parkinson's disease. New Engl J Med 1988;318:876–880.

Kitagawa Y, Meyer JS, Rogers RL, et al. Ct-CBF correlations of cognitive deficits in multi-infract dementia. Stroke 1984;15:1000–1008.

Klauber MR, Marshall LF, Toole BM, et al. Cause of decline in head-injury mortality rate in San Diego County, California. J Neurosurg 1985;62:528–531.

Klein DF, Gorman JM. Panic disorders and mitral valve prolapse. J Clin Psychiatr Monogr 1984;2:14–17.

Klein R, Armitage R. Rhythms in human performance: 1 1/2 hour oscillations in cognitive style. Science 1979;204:1326–1328.

Kleinsmith L, Kaplan S. The interaction of arousal and recall interval in nonsense syllable paired associate learning. J Exp Psychol 1964;67:124–126.

Klemm WR. Behavioral inhibition. In: Klemm WR, Vertes RP, eds. Brainstem mechanisms of behavior. New York: Wiley, 1990.

Klemm WR. The behavioral readiness response. In: Klemm WR, Vertes RP, eds. Brainstem mechanisms of behavior. New York: Wiley, 1990.

Kling A. Effects of amygdalectomy on social-affective behavior in non-human primates. In: Eleftherious BE, ed. The neurobiology of the amygdala. New York, Plenum, 1972, pp. 127–170.

Kling AS, Brothers LA. The amygdala and social behavior. In: Aggleton JP, ed. The amygdala. New York: Wiley-Liss, 1992.

Kling AS, Lloyd RL, Perryman KM. Slow wave changes in amygdala to visual, auditory and social stimuli following lesions of the inferior temporal cortex in squirrel monkey. Behav Neural Biol 1987; 47:54–72.

Kling A, Mass R. Alterations of social behavior with neural lesions in nonhuman primates. In: Holloway B, ed. Primate aggression, territoriality, and zenophobia. New York: Academic Press, 1974, pp. 361–386.

Kling A, Steklis HD. A neural substrate for affiliative behavior in nonhuman primates. Brain Behav Evol 1976;13:216–238.

Klinger J, Gloor P. The connections of the amygdala and of the anterior temporal cortex in the human brain. J Comp Neurol 1960;115:333–352.

Klockgether T, Schwarz M, Turski L, et al. Neurotransmitters in the basal ganglia and motor thalamus. In: Carpenter MB, Jayaraman A, eds. The basal ganglia. New York: Plenum, 1987.

Kluft RP. Childhood antecedents of multiple personality. Washington DC: American Psychiatric Press, 1985.

Kluft RP. Treatment of multiple personality disorder: a study of 33 cases. Psychiatr Clin North Am 1984; 7:9–29.

Kluver H, Bucy PC. Preliminary analysis of functions of the temporal lobes in monkeys. Arch Neurol Psychiatry 1939;542:979–1000.

Knapp ME. Problems in rehabilitation of the hemiplegic patient. J Am Med Assoc 1959;169:224–229.

Knight RT, Hillyard SA, Woods DL, Neville HJ. The effects of frontal cortex lesions on event-related potentials during auditory selective attention. Electroencephalogr Clin Neurophysiol 1981;52:571–582.

Knobler RL. Demyelinating and dysmyelinating diseases. In: Duckett S. ed. Pediatric neuropathology. Philadelphia: Williams & Wilkins, 1995.

Knox C, Kimura D. Cerebral processing of nonverbal sounds in boys and girls. Neuropsychologia 1970; 8:227–237.

Knuezle H, Akert K. Efferent connections of cortical area 8. J Comp Neurol 1977;173:147–164.

Koenigsknecht RA, Friedman R. Syntax development in boys and girls. Child Devel 1976;47:1109–1115.

Kolb B. Mechanisms underlying recovery from cortical injury. In: Rose FD, Johnson DA, eds. Recovery from brain damage. New York: Plenum, 1992.

Kolb B, Milner B. Performance on complex arm and facial movement after focal brain lesions. Neuropsychologia 1981;19:491–503.

Kolb B, Nonneman AJ, Singh RJ. Double dissociation of spatial impairments and perseveration following selective prefrontal lesions in rats. J Comp Physiol Psychol 1974;87:772–780.

Kolb B, Whishaw W. IQ effects of brain lesions and atropine on hippocampal and neocortical EEG in the rat. Exp Neurol 1977;56:1–22.

Kolb B, Whishaw IQ. Performance of schizophrenic patients on tests sensitive to left or right frontal, temporal, or parietal function in neurological patients. J Nerv Ment Dis 1983;171:435–443.

Kolb B, Wilson B, Taylor L. Developmental changes in the recognition and comprehension of facial expression: implications for frontal lobe function. Brain Cognit 1992;20:74–84.

Kolb LC. Comments on post-traumatic stress disorder and dissociation. In: Quen JM, ed. Split mind/split brains. New York: New York University Press, 1986.

Koller RI. Recurrent embolic cerebral infarction and anticoagulation. Neurology 1982;32:283–285.

Koller WC. Handbook of Parkinson's disease. New York: Marcel-Dekker, 1987.

Konner M. Universals of behavioral development in relation to brain myelination. In: Gibson KR, Petersen AC, eds. Brain maturation and cognitive development. New York: Aldine De Gruyter, 1991.

Koob GF, Swerdlow NR, Vaccarino F, et al. Functional output of the basal forebrain. In: Napier TC, et al, ed. The basal forebrain. New York: Plenum, 1991.

Kornhuber HM. Motor functions of cerebellum and basal ganglia. Kybernetic 1971;8:157–162.

Koulack D, Goodenough DR. Dream recall and dream failure: an arousal-retrieval model. Psychol Bull 1976;83:975–984.

Kovelman JA, Scheibel AB. A neuroanatomical correlate of schizophrenia. Proc Soc Neurosci 1983; 9:850.

Kozlowski PB. Pediatric human immunodeficiency virus (HIV) infection. In: Duckett S, ed. Pediatric neuropathology. Philadelphia: Williams & Wilkins, 1995.

Kraemer GW, Ebert MH, Schmidt DE, et al. A longitudinal study of the effects of differential social rearing conditions on cerebrospinal fluid norepinephrine and biogenic amine metabolites in rhesus monkey. Neuropsychopharmacology 1989; 2:175–189.

Kraft RH, Mitchell OR, Languis ML, Wheatley GH. Hemispheric asymmetries during six-to-eight year-olds' performance on Piagetian conservation and reading tasks. Neuropsychologia 1980;18: 637–644.

Kramer GW. A psychobiological theory of attachment. Behav Brain Sci 1992;15:493–541.

Kramer D. Elimination of verbal cues in judgment of emotion from voice. J Abn Soc Psychol 1964;68: 390–396.

Kramer M, Schoen LS, Kinney L. Nightmares in Vietnam veterans. J Am Acad Psychoanal 1987;15: 67–81.

Kramer M, Buckout C, Eugenio E. Weapon focus, arousal, and eyewitness memory: attention must be paid. Law Hum Behav 1990;14:167–184.

Kramer HC. Laughing spells in patients after lobotomy. J Nerv Ment Dis 1954;140:517–522.

Kramer M, Whitman RM, Baldridge BJ, Lansky LM. Patterns of dreaming: the interrelationship of the dreams of a night. J Nerv Ment Dis 1964;139: 4526–439.

Kramer S. History begins at Sumer. Philadelphia: The University of Pennsylvania Press, 1981.

Krames L, et al. A pheromone associated with social dominance among male rats. Psychonomic Sci 1969;16:11–12.

Kretschmann H-J, Kammradt G, Krauthausen B, et al. Growth of the hippocampal formation in man. Bibliotheca Anatomica 1986;28:27–52.

Kretschmer H. Callosal tumors. In: Vinken PJ, Bruyn GW, eds. Handbook of clinical neurology. Vol. 17. New York: Elsevier, 1974.

Krettek JE, Price JL. A direct input from the amygdala to the thalamus and the cerebral cortex. Brain Res 1974;67:169–174.

Krettek JE, Price JL. Projections from the amygdaloid complex. J Comp Neurol 1977;172:687–752.

Kriendler A, Steriade M. EEG patterns of arousal and sleep induced by stimulating various amygdaloid levels in the cat. Arch Ital Biol 1964;102: 576–586.

Kripke DF, Sonnenschein D. A 90 minute daydream cycle. Sleep Res 1973;2:187–188.

Kritchevsky MD, Squire LR. Transient global amnesia. Neurology 1989;39:213–218.

Kritchevsky MD, Squire LR, Zouzounis JA. Transient global amensia. Neurology 1988;38:213–219.

Krystal H. Integration and self-healing. Hillsdale, NJ: The Analytic Press, 1988.

Krystal JH. Animal models for post traumatic stress disorder. In: Giller Jr EI, ed. Biological assessment and treatment of postraumatic stress disorder. Washington: American Psychiatric Press, 1990.

Kuehn L. Looking down a gun barrel. Percept Motor Skills 19174;39:1159–1164.

Kuhar M. Cholinergic neurons: septal-hippocampal relationships. In: Isaacson RL, Pribram KH, eds. The hippocampus. New York: Plenum, 1975, pp. 269–283.

Kuhar MJ. Opioid peptides and receptors in the rat brainstem. In: Hobson JA, Brazier MAB, eds. The reticular formation revisited. New York: Raven Press, 1980.

Kuhn TS. The structure of scientific revolutions. Chicago: University of Chicago Press, 1962.

Kummer H. Social organization of Hamadryas baboons. Chicago: University of Chicago Press, 1968.

Kummer H. Primate societies. Chicago: Aldine, 1971.

Kunitz SC, Gross CR, Heyman A, et al. The pilot stroke data bank. Stroke 1984;15:740–746.

Kunzle H. Thalamic projections from the precentral motor cortex in Macaca fascicularis. Brain Res 1976;105:253–280.

Kurcharski D, Hall WG. New routes to early memories. Science 1987;238:786–788.

Kurten B. The cave bear story. New York: Columbia University Press, 1976.

Kurthen M, Helmstaedter C, Linke DB, et al. Interhemispheric dissociation of expressive and receptive language functions in patients with complex-partial seizures: an amobarbital study. Brain Lang 1992;43:694–712.

Kurtz KH, Siegel A. Conditioned fear and magnitude of startle response. A replication and extension. J Comp Physiol Psychol 1966;62:8–14.

Kuypers HGJM. Corticobulbar connexions to the pons and lower brain stem in man. Brain 1958;81: 364–388.

Kuypers HGJM, Catsman-Berrevoets CE. Frontal corticosubcortical projections and their cells of origin. In: Reinoso-Suarez F, Ajmone-Marsan C, eds. Cortical integration. New York: Raven Press, 1984:171–194.

Kuypers HGJM, Szwarcbart MK, Mishkin M, Rosvold HE. Occipitotemporal cortico-cortical connections in the Rhesus monkey. Experimental Neurology 1965;11:245–262.

Lackner JL, Teuber H-L. Alterations in auditory fusion thresholds after cerebral injury in man. Neuropsychologia 1973;11:409–415.

Lakotos I, Musgrave A. Criticism and the growth of scientific knowledge. Cambridge, UK: Cambridge University Press, 1970.

Lalande S, Braun CMJ, Charlebois N, Whitaker HA. Effects of right and left hemisphere cerebrovascular lesions on discrimination of prosodic and semantic aspects of affect in sentences. Brain Lang 1992;42:165–186.

Lamotte RH, Acuna C. Defects in accuracy of reaching after removal of posterior parietal cortex in monkeys. Brain Res 1978;139:309–326.

Lamotte RH, Mountcastle VB. Disorders in somesthesis following lesions of the parietal lobe. J Neurophysiol 1979;42:400–419.

Landis T, Assal G, Perrot E. Opposite cerebral hemisphere superiorities for visual associative processing of emotional facial expressions and objects. Nature 1979;278:739–740.

Landis T, Cummings JL, Christen L, et al. Are unilateral right posterior cerebral lesions sufficient to cause prospagnosia? Clinical and radiological findings in six additional patients. Cortex 1986; 22:243–252.

Landis T, Graves R, Goodglass H. Aphasic reading and writing: possible evidence for right hemisphere participation. Cortex 1982;18:105–112.

Lang PJ, Bradley MM, Cuthbert BN. Emotion, attention, and the startle reflex. Psychol Rev 1990; 97:377–395.

Langevin R. Sexual trends: understanding and treating sexual anomalies in men. Hillsdale, NJ: Erlbaum, 1990.

Langmeier J, Matejcek Z. Psychological deprivation in childhood. New York: John Wiley, 1975.

Langworthy OR, Richter CP. Increased activity produced by frontal lobe lesions in cats. Am J Physiol 1939;126:158–161.

Lansdell H. Extent of temporal lobe albations on two lateralized deficits. Physiol Behav 1968;3:271–273.

Lansdell H. Relation of extent of temporal removal to

closure and visuomotor factors. Percept Motor Skills 1970;31:491–498.

Laplane D, Degos JD, Baulac M, Gray F. Bilateral infarction of the anterior cingulate gyri and of the fornices. J Neurol Sci 1981;51:289–300.

Laplane, D, Degos JD, Baulac M, Gray F. Bilateral infarction of the anterior cingulate gyri and of the fornices. J Neurol Sci 1981;51:289–300.

Laplane D, Talairach J, Meininger V, et al. Clinical consequences of cortisectomies involving the supplementary motor area in man. J Neurol Sci 1977; 34:301–314.

Larson CR, Yajima Y, Ko P. Modification in activity of medullary respiratory-related neurons for vocalizing and swallowing. J Neurophysiol 1994;71: 2294–2304.

Latto R, Cowery A. Fixation changes after frontal eyefield lesions in monkeys. Brain Res 1971a;30:25–36.

Latto R, Cowery A. Visual field defect after frontal eye field lesions in monkeys. Brain Res 1971b;30:1–24.

Laureno R, Shields RW Jr, Narayan T. The diagnosis and management of cerebral emoblism and haemorrhagic infarction with sequential computerized cranial tomography. Brain 1987;110:93–105.

Laurent J, Bruce D, Schut L. Hemorrhagic brain tumors in pediatric patients. Child Brain 1981;8: 263–270.

Lauterbach EC, Spears TE, Prewett MJ, et al. Neuropsychiatric disorders, myoclonus, and dystonia in calcification of basal ganglia pathways. Biol Psychiatry 1994;35:345–351.

Lavi P, Kripke DF. Ultradian rhythms: the 90-minute clock inside us. Psychol Today 1975;8:54–56.

Lavond DG, Logan CG, Sohn JH, et al. Lesions of the cerebellar interpositus nucleus abolish both nictitating membrain and eyelid EMG conditioned response. Brain Res 1990;514:238–248.

Lawrence RD. In praise of wolves. New York: Holt, 1986.

Lawson KR. Spatial and temporal congruity and auditory-visual integration in infants. Dev Psychol 1980;16:185–192.

Lazerwitz B. Some factors associated with variations in church attendance. Social Forces 1961;39:301–309.

Leakey MD. Footprints in the ashes of time. Natl Geogr 1979;155:446–457.

Leakey RE. Hominids of Africa. Am Sci 1976;64: 174–178.

Leakey R, Lewin R. Origins. New York: Dutton, 1977.

Leakey REF, Walker A. New Australopithecus boisea specimens from East and West Lake Turkana, Kenya. Am J Phys Anthropol 1988;76:1–24.

Lebedev MA, Denton JM, Nelson RJ. Vibration-entrained and premovement activity in monkey primary somatosensory cortex. J Neurophysiol 1994; 72:1654–1670.

Le Beau J. Anterior cingulotomy in man. J Neurosurg 1954;11:268–276.

Lebrun Y. Anosognosia in aphasics. Cortex 1987;23: 251–263.

Lecours AR. Myelogenetic correlates of the develop-

ment of speech and language. In: Lenneberg E, Lenneberg E, eds. Foundations of language development. New York: Academic Press, 1975.

Lederer W. The fear of women. New York: Harcourt-Brace, 1968.

LeDoux JE. Emotion and the amygdala. In: Aggleton JP, ed. The amygdala. New York: Wiley-Liss, 1992.

Lee DE. Reduced inhibitory capacity in prefrontal cortex of schizophrenics. Arch Gen Psychiatry 1995; 52:267–268.

Lee GP, Loring DW, Dahl JL, Meador KJ. Hemispheric specialization for emotional expression. Neuropsychiatry Neuropsychol Behav Neurol 1993;6: 143–148.

Lee GP, Loring DW, Meader KJ, Brooks BB. Hemispheric specialization for emotional expression: a reexamination of results from intracarotid administration of sodium amobarbital. Brain Cogn 1990; 12:267–280.

Lee WH, Brady MP, Rowe JM, et al. Effects of extracorporeal circulation on behaviour, personality, and brain function. Ann Surg 1971;173:1031–1023.

Lee WH, Miller W, Rowe JM, et al. Effects of extracorporeal circulation on personality and cerebration. Ann Thorac Surg 1969;7:562–569.

Leestma J. Forensic neuropathology. In: Duckett S, ed. Pediatric neuropathology. Philadelphia: Williams & Wilkins, 1995.

Leestman JE. Neuropathology and pathophysiology of trauma and toxicity. In: Doerr HO, Carlin AS, eds. Forensic neuropsychology. New York: Guilford, 1991.

Legatt AD, Rubin MJ, Kaplan LR, et al. Global aphasia without hemiparesis. Neurology 1987;37:201–205.

LeGros Clark WE. The antecedents of man. Chicago: Quadrangle Books, 1962.

Lehmann AC, Myers JE. Magic, witchcraft, and religion. Mountain View: Mayfield, 1993.

Lehman RAW. Manual preference in prosimians, monkeys and apes. In: Ward JP, Hopkins WD, eds. Primate laterality. New York: Springer-Verlag, 1993.

Leicester J, Sidman M, Stoddard LT, Mohr JP. Some determinants of visual neglect. J Neurol Neurosurg Psychiatry 1969;32:580–587.

Leichnetz GR, Astruc J. The efferent projections of the medical prefrontal cortex in the squirrel monkey (Saimiri sciureus). Brain Res 1976;109:455–472.

Leichnietz G, Smith DJ, Spencer RF. Cortical projections to the paramedian tegmental and basila pons in the monkey. J Comp Neurol 1984;228: 388–408.

Leiner HC, Leiner AL, Dow RS. Coes the cerebellum contribute to mental skills. Behav Neurosci 1986; 100:443–454.

Leininger BE, Gramling SE, Farrell AD, et al. Neuropsychological deficits in symptomatic minor head injury patients after concussion and mild concussion. J Neurol Neurosurg Psychiatry 1990; 53:293–296.

Leinonen L, Hyvarinen J, Nyman, et al. Functional properties of neurons in the lateral part of associative area 7 of awake monkeys. Exp Brain Res 1979; 34:299–320.

LeMay M. Morphological cerebral asymmetries of modern man, fossil man, and nonhuman primate. Ann N Y Acad Sci 1976;280:349–366.

Le Moal M, Simon H. Mesocorticalimbic dopaminergic network: functional and regulatory roles. Physiol Rev 1991;71:155–235.

Lenneberg E. Biological foundations of language. New York: John Wiley, 1967.

Lentz TL. Primitive nervous systems. New Haven: Yale University Press, 1968.

Leopold WE. Speech development of a binlingual child, vol 2. Evanston, IL: Northwestern University Press, 1947.

Leroi-Gourhan A. Treasure of prehistoric art. New York: HN Abrams, 1964.

Leroi-Gourhan A. The archaeology of Lascauz Cave. Sci Am 1982;24:104–112.

Lescaudron L, Stein DG. Functional recovery following transplants of embryonic brain tissue in rats with lesions of visual, frontal and motor cortex. Neuropsychologia 1990;28:585–599.

Lesse H, Heath RG, Micle WA, et al. Rhinencephalic activity during thought. J Nerv Ment Dis 1955;112: 433–440.

Lesser RP, Luders, H, Morris HH, et al. Electrical stimulation of Wernicke's area interferes with comprehension. Neurology 1986;36:658–663.

Lesser RP, Lueders, H, Dinner DS, et al. The location of speech and writing function in the frontal language area. Brain 1984;107:275–291.

LeVay SA. A difference in hypothalamic structure between heterosexual and homosexual men. Science 1991;253:1034–1037.

Lever J. Sex differences and the games children play. Soc Probl 1976;23:478–487.

Levin HS. The acalculias. In: Heilman KM, Valenstein E, eds. Clinical neuropsychology. New York: Oxford University Press, 1993.

Levin HS, Benton AL, Grossman RG. Neurobehavioral consequences of closed head injury. New York: Oxford University Press, 1982.

Levin HS, High WM, Goethe KE, et al. The neurobehavioral rating scale: assessment of the behavioural sequelae of head injury by the clinician. J Neurol Neurosurg Psychiatry 1987;50:183–193.

Levin HS, Mattis, S, Ruff RM, et al. Neurobehavioral outcome following minor head injury. J Neurosurg 1987;66:234–243.

Levin S. Frontal lobe dysfunction in schizophrenia—I and II. J Psychiatr Res 1984;18:27–55, 57–72.

Levine DN. Prosopagnosia and visual object agnosia: a behavioral study. Brain Lang 1978;5:341–365.

Levine DN, Sweet E. Localization of lesions in Broca's motor aphasia. In: Kertesz A, ed. Localization in neuropsychology. New York: Academic Press, 1983:185–207.

Levine SC, Levy J. Perceptual asymmetry for faces across the life span. Brain Cogn 1986;5:291–306.

Levitt M, Levitt J. Sensory hind-limb representaiton in cortex of cat after spinal tractotomies. Exp Neurol 1968;22:276–302.

Levitt P, Moore RY. Noradrenaline neuron innervation of the neocortex of the rat. Brain Res 1978;139: 219–231.

Levitt P, Moore RY. Origin and organization of brainstem catecholamine innervation in the rat. J Comp Neurol 1979;186:505–528.

Levitt P, Moore RY. Organization of brainstem noradrenalin hyperinnervation following neonatal 6-hydroxzydopamine treatment in rat. Anat Embryol 1980;158:122–150.

Levy-Agresti J, Sperry RW. Differential perceptual capacities in major and minor hemispheres. Proc U S Natl Acad Sci 1968;61:1151.

Levy J. Psychological implications of bilateral asymmetry. In: Dimond S, Beaumont JG, eds. Hemisphere function in the human brain. London: Paul Elek, 1974:121–183.

Levy J. Language, cognition and the right hemisphere. Am Psychol 1983;38:538–541.

Levy J, Heller W. Gender differences in human neuropsychological function. In: Gerall AA, et al, eds. Sexual differentiation. New York: Plenum, 1992.

Levy J, Trevarthen C. Metacontrol of hemispheric function in human split-brain patients. J Exp Psychol Hum Percept Perform 1976;2:299–312.

Levy J, Trevarthen C, Sperry RW. Perception of bilateral chimeric figures following hemispheric deconnection. Brain 1972;95:61–78.

Lewicki P. Nonconscious social information processing. New York: Academic Press, 1986.

Lewis DA, Neural circuitry of the prefrontal cortex in schizophrenia. Arch Gen Psychiatry 1995;52: 269–272.

Lewis DO, Pincus JH, Shanok SS, Glaser GH. Psychomotor epilepsy and violence in a group of incarcerated adolescent boys. Am J Psychiatry 1982; 139:882–887.

Lewis H. Freud and modern psychology. New York: Plenum, 1983.

Lewis J, Hoover HD. Sex differences on standardized academic achievement tests. Montreal: American Educational Research Association, 1983.

Lewis N, Reinhold M, eds. Roman civilization source books: the republic; the empire. New York: Harper & Row, 1966.

Lewis WC, Wolfman RN, King M. The development of the language of emotions: II. Intentionality in the experience of affect. J Genet Psychol 1979;120: 303–316.

Ley RG. An archival examination of an asymmetry of hysterical conversion symptoms. J Clin Neuropsychol 1980;2:1–9.

Ley RG, Bryden MP. Hemispheric differences in processing emotions and faces. Brain Lang 1979;7: 127–138.

Lezak MD. Relatinships between personality disorders, social disturbances and physical disability following traumatic brain injury. J Head Trauma Rehab 1987;2:57–69.

Lhermitte F. "Utilization behaviour" and its relation to lesions of the frontal lobes. Brain 1983;106:237–255.

Libet B, Electrical stimulation of cortex in humans and conscious sensory aspects. In: Iggo A, ed. Handbook of sensory physiology. New York: Springer, 1973:744–790.

Lidvall H, Linderoth B, Norlin B. Causes of the post concussional syndrome. Acta Neurol Scand 1974; 56:1–87.

Lieberman A, Benson DF. Control of emotional expression in pseudobulbar palsy. Arch Neurol 1977; 34:717–719.

Lieberman DE, Shea JJ. Behavioral differences between archaic and modern humans in the Levantine Mousterian. Am Anthropol 1994;96:300–332.

Lieberman P. The biology and evolution of language. Cambridge, MA: Harvard University Press, 1984.

Lieberman P. Uniquely human. Boston: Harvard University Press, 1991.

Lieberman P. On Neanderthal speech and Neanderthal extinction. Curr Anthropol 1992;33:409–410.

Lieberman P, Crelin ES. On the speech of Neanderthal man. Lingustic Inquiry 1971;2:203–222.

Lieberman P, Laitman JT, Reidenberg JS, Gannon PJ. The anatomy, physiology, acoustics, and perception of speech: essential elements in analysis of the evolution of human speech. J Hum Evol 1992;23: 447–467.

Lilly JC. Mental effects of reduction of ordinary levels of physical stimuli on intact, healthy persons. Psychiatr Res Rep 1956;5:1–9.

Lilly JC. Learning motivated by subcortical stimulation. In: Ramsey E, O'Doherty E, eds. Electrical studies on the unanethetized brain. New York: Hoeber, 1960.

Lilly JC. The center of the cyclone. New York: Julian Press, 1972.

Lin C-S, Weller RE, Kaas JH. Cortical connections of striate cortex in the owl monkey. J Comp Neurol 1982;211:165–176.

Lin L-D, Murray GM, Sessle BJ. Functional properties of single neurons in the primate primary somatosensory Cortex. I. J Neurophysiol 1994;71: 2377–2390.

Lin L-D, Sessle BJ. Functional properties of single neurons in the primate face primary somatosensory Cortex. III. J Neurophysiol 1994;71:2401–2420.

Lind K. A synthesis of studies on storke rehabilitation. J Chron Dis 1982;35:133–149.

Lindal E, Stefansson JG, Stefansson SB. The qualitative difference of visions and visual hallucinations. Comp Psychiatry 1994;35:405–408.

Lindly JM, Clark GA. Symbolism and modern human origins. Curr Anthropol 1990;31:233–261.

Lindberg M. An interactive approach to assessing the suggestibility and testimony of eyewitnesses. In: Doris J, ed. The suggestibility of children's recollections. Washington, DC: American Psychological Association, 1991.

Lindgren SO. Experimental studies on mechanical effects in head injury. Acta Scand 1966;132:1–87.

Lindsey LL. Gender roles: a sociological perspective. Englewood Cliffs, NJ: Prentice Hall, 1990.

Lindsley D. Common factors in sensory deprvation. In: Solomon P, ed. Sensory deprivation. Cambridge, MA: Harvard University Press, 1961.

Lindvall O, Widner H, Rehncrona S, et al. Transplantation of fetal dopamine neurons in Parkinson's disease. Ann Neurol 1992;31:155–165.

Lineberry CG, Siegel J. EEG synchronization, behavioral inhibition, and mesencephalic unit effect produced by stimulation of orbital cortex basal forebrain and caudate nucleus. Brain Res 1971;34: 143–161.

Lings M. Muhammad, his life based on the earliest sources. New York: Ballintine Books, 1983.

Lipinski JF, Pope HG. Do flashbacks represent obsessional imagery? Comp Psychiatry 1994;35:245–247.

Lishman WA. Brain damage in relation to psychiatry disability after head injury. Br J Psychiatry 1968; 114:373–410.

Lishman WA. The psychiatric sequelae of head injury: a review. Psychol Med 1973;3:304–322.

Lishman WA. The psychiatric sequela of head injuries. J Ir Med Assoc 1978;71:306–314.

Lishman WA. Physiogenesis and psychogenesis in the "post-concussional syndrome." Br J Psychiatry 1988;153:460–469.

Lisk WG. Neural localization for androgen activation of copulatory behavior in the male rat. Endocrinology 1967;80:754–780.

Lisk WG. Diencephalic placement of estrodiol and sexual receptivity in the female rat. Am J Physiol 1971;203:493–500.

List CF, Dowman CE, Bagheiv R. Posterior hypothalamic hermatomas and gangliomas causing precious puberty. Neurology 1958;8:164–174.

Lister TA, Sutcliffe SBJ, Brearley RL, Cullen MH. Patterns of CNS involvement in malignant lymphoma. In: Whitehouse JMA, Kay EM, eds. CNS complications of malignant disease. Baltimore: University Park Press, 1979.

Litchfield PJ, Rubin DM, Feldman RS. Subarachnoid hemorrhage precipitated by cocaine snorting. Arch Neurol 1984;41:223–224.

Livesley WJ, Bromley DB. Person perception in childhood and adolescence. New York: Wiley, 1973.

Livingston KE, Escobar A. Anatomical bias of the limbic system concept. Arch Neurol 1971;24:26–32.

Livingston M, Hubel D. Segregation of form, color, movement, and depth. Science 1988;240:740–749.

Livingston R, Chapman W, Livingston K. Stimulaton of orbital surface of man prior to lobotomy. Research Publication: The frontal lobes, (Chapter 16). Washington, DC: U.S. Government Printing Office, 1948:421–433.

Llorens, C, Martres MP, Baudry M, Schwartz JC. Hypersensitivity to anoradrenaline in cortex after chronic morphine. Nature 1978;274:603–605.

Lloyd GG, Lishman WA. Effect of depression on the speed of recall of pleasant and unpleasant experiences. Psychol Med 1975;5:173–180.

Lodder J, van der Lugt PJM. Evaluation of the risk of immediate anti-coagulant treatment in patients with embolic strokes of cardiac origin. Stroke 1983; 14:42–46.

Loewenstein G. Psychogenic amnesia and psychogenic fugue: In: Goldfinger TA, ed. American Psychiatric Press review of psychiatry. Washington, DC: American Psychiatric Press, 1991.

Loftus EF, Burns T. Mental shock can produce retrograde amnesia. Memory Cognition 1982;10: 318–323.

Loiseau P, Cohandon F, Cohandon S. Gelastic epilepsy, a review and report of five cases. Epilepsia 1971; 12:313–320.

Lomas J, Kimura D. Intrahemispheric interactions between speaking and sequential manual activity. Neuropsychologia 1976;14:23–33.

Lombardo JP, Lavine LO. Sex-role stereotyping and patterns of self-disclosure. Sex Roles 1981;7: 403–411.

London P, Cooper LM. Norms of hypnotic susceptibility in children. Dev Psychol 1969;1:113–124.

Lorente de No R. Analysis of the activity of the chains of internuncial neurons. J Neurophysiol 1938;1: 207–244.

Lovejoy CO. Evolution of human walking. Sci Am 1988;11:118–125.

Lovelock JE. Gaia. A new look at life on Earth. New York: Oxford University Press, 1979.

Lucas RV, Edwards JE. The floppy mitral valve. Curr Prob Cardiol 19982;7:1–48.

Luerssen TG, Klauber MR, Marshall LF. Outcome from head injury related to patient's age. J Neurosurg 1988;68:409–416.

Lund RD. Development and plasticity of the brain. New York: Oxford University Press, 1978.

Luria A. The working brain. New York: Basic Books, 1973.

Luria A. Higher cortical functions in man. New York: Basic Books,1980.

Lutz C. Emotional words and emotional development [Abstract of doctoral dissertation]. Cambridge, MA. Harvard University, 1980.

Lynch G. Synapses, circuits, and the beginnings of memory. Cambridge, MA: MIT Press, 1986.

Lynch G, Larson J, Muller D, Granger R. Neural networks and networks of neurons. In: McGaugh JL, Winberger NM, Lynch G, eds. Brain organization and memory. New York: Oxford University Press, 1990.

Lynch JC. The functional oranization of posterior parietal association cortex. Behav Brain Sci 1980;3: 485–499.

Lynch JC, Mountcastle VB, Talbot WH, Yin TCT. Parietal lobe mechanisms for directed visual attention. J Neurophysiol 1977;40:362–389.

Machne X, Segundo J. Unitary reponses to afferent volleys in amygdaloid complex. J Neurophysiol 1956;19:232–240.

Mack J. Abduction: human encounters with aliens. New York: Scribners, 1994.

Mack JL, Boller F. Associative visual agnosia and its related deficits. Neuropsychologia 1977;15: 345–349.

Mackay-Soroka S, Trehub SE, Bull DH, Corter CM. Effects of encoding and retrieval conditions on infant's recognition memory. Child Dev 1982;53: 815–818.

MacKinon D, Squire LR. Autobiographical memory in amnesia. Psychobiology 1989;17:247–256.

MacLean PD. Psychosomatic disease and the "visceral brain." Recent developments bearing on the Papex theory of emotion. Psychosom Med 1949;11: 338–353.

MacLean P. The hypothalamus and emotional behavior. In: Haymaker W, ed. The hypothalamus. Springfield, IL: Charles C Thomas, 1969.

MacLean P. New findings on brain function and socio-sexual behavior. In: Zubin J, ed. Contemporary sexual behavior. Baltimore: John Hopkins Press, 1973.

MacLean P. The evolution of the triune brain. New York: Plenum, 1990.

MacNeilage PF. Implications of primate functional asymmetries for the evolution of cerebral hemispheric specialization. In: Ward JP, Hopkins WD, eds. Primate laterality. New York: Springer-Verlag, 1993.

MacNeilage PF, Studdert-Kennedy MG, Lindblom B. Primate handedness reconsidered. Behav Brain Sci 1987;11:748–758.

Madrazo I, Franco-Bourland R, Ostroysky-Solis F, et al. Fetal homotransplants to the striatum in parkinsonian subjects. Arch Neurol 1990;47:1281–1285.

Maglio VJ. Evolution of African mammals. Cambridge, MA: Harvard University Press, 1978.

Magnuson DJ, Gray TS. Central nucleus of amygdala and bed nucleus of stria terminalis projections to serotonin or tyrosine hydroxylase immunoreactive cells in the dorsal and median raphe nucleus in the rat. Soc Neurosci Abstr 1990;16:121.

Magnusson A, Stefansson JG. Prevalence of seasonal affective disorder in Iceland. Arch Gen Psychiatry 1993;50:941–946.

Maher LM, Rothi LJ G, Heilman KM. Lack of error awareness in an aphasic patient with relatively preserved auditory comprehension. Brain Lang 1994;46:402–418.

Mahoney AM, Sainsbury RS. Hemispehric asymmetry in the perception of emotional sounds. Brain Cogn 1987;6:216–233.

Mahowald MW, Schenck CH. Status dissociatus— a perspective on states of being. Sleep 1991;14: 69–79.

Maier SF. Learned helplness, depression, analesia, and endogenous opiates. Psychopharmacol Bull 1983; 19:531–536.

Mair WGP, Warrigton EK, Weiskrantz L. Memory disorder in Korsakoff's psychosis. Brain 1979;102: 749–783.

Malh GF, Rothenberg A, Delgado JMR, Hamlin H. Psychological response in the human to intracerebral electrical stimulation. Psychosom Med 1964; 26:337–368.

Malinowski B. Magic, science and religion. New York: Doubleday, 1954.

Maltman I. Orienting reflexes and classical conditioning in humans. In: Kimmell HD, van Olst EH, Orlebke JF, eds. The orienting reflex. Hillsdale, NJ: Erlbaum :323–353.

Mandler G. Mind and emotion. New York: Wiley, 1975.

Mandler JM. A code in the node: the use of story schema in retrieval. Discourse Processes 1978; 1:14–35.

Mandler JM. Representation and recall in infancy. In: Moscovitch M, ed. Infant memory. New York: Plenum, 1974.

Mann DMA. The neuropathology of the amygdala in ageing and in dementia. In: Aggleton JP, ed. The amygdala. New York: Wiley, 1992.

Mannhaupt HR. Processing of abstract and concrete nouns in lateralized memory-search tasks. Psychol Res 1983;45:91–105.

Manning A. An introduction to animal behavior. Menlo Park: Addison-Wesley, 1972.

Manoach DS, Sandson TA, Wentrub S. The developmental social-emotional processing disorder is associated with right hemisphere abnormalities. Neuropsychiatry Neuropsychol Behav Neurol 1995;8:99–105.

Maple TL, Hoff MP. Gorilla behavior. New York: Van Nostrand, 1982.

Marcel AG, Patterson KE. Word recogniton and production. In: Requin J, ed. Attention and performance. Hillsdale, NJ: Erlbaum, 1978.

Marchman VA, Miller R, Bates EA. Babble and first words in children with focal brain injury. Appl Psycholing 1991;12:1–22.

Marcie P, Hecaen H. Agraphia. In: Heilman KM, Valenstein E, eds. Clinical neuropsychology. New York: Oxford University Press, 1979.

Margulis L. Origin of eukaryotic cells. New Haven: Yale University Press, 1970.

Marin-Padilla M. Early ontogenesis of the human cerebral cortex. In: Peters A, Jones EG, eds. Cerebral cortex, vol 7. New York: Plenum, 1988a.

Marin-Padilla M. Embryonic vascularization of the mammalian cerebral cortex. In: Peters A, Jones EG, eds. Cerebral cortex, vol 7. New York: Plenum, 1988b.

Marion DW, Lund RD. Connections of neocortex xenografts in the rat brainstem. Neuroscience Abstracts 1987;141–149.

Mark VH, Ervin FR, Sweet WH. Deep temporal lobe stimulation in man. In: Eleftheriou BE, ed. The neurobiology of the amygdala. New York: Plenum Press, 1972.

Marks AE. The middle and upper palaeolithic of the Near East and the Nile Valley: the problem of cultural transformations. In: Mellars P, ed. The human revolution: behavioral and biological perspectives on the origins of modern humans, vol 2. Edinburgh: Edinburgh University Press, 1989.

Marquardsen J. The natural history of acute cerebrovascular disease. Acta Neurol Scand Suppl 1969;45:1–133.

Marrazzi AS, Hart ER. Relationship of hallucinogens to adrenergic cerebral neurohumors. Science 1955; 121:365–367.

Marsden CD, Obesco JA. Brain. The functions of the basal ganglia and the paradox of stereotaxic surgery in Parkinson's disease. Brain 1994;117:877–897.

Marshach A. Early hominid symbol and evolution of human capacity. In: Mellars P, ed. The emergence of modern humans. New York: Cornell University Press, 1990.

Marshall JC, Newcombe F. Syntactic and semantic errors in paralexia. Neuropsychologia 1966;4: 169–176.

Marshall JF, Teitelbaum P. Further analysis of sensory inattention following lateral hypothalamic damage in rats. J Comp Physiol Psychol 1974;86:375–395.

Marslen-Wilson WD, Teuber HL. Memory for remote events in anteriograde amnesia. Neuropsychologia 1975;13:353–364.

Mark VH, Sweet WH, Erwin FR. The effect of amygdalectomy on violent behavior in patients with temporal lobe epilepsy. In: Hitchcock ER, Laitinen L, Vaernet K, eds. Psychosurgery. Springfield, IL: Charles C Thomas, 1972.

Martin JP Fits of laughter (sham mirth) in organic cerebral disease. Brain 1950;73:453–464.

Martin MK, Voorhies B. Female of the species. New York: Columbia University Press, 1975.

Martinez-Millan L, Hollander H. Cortico-cortical projections from striate cortex of the squirrel monkey. Brain Res 1975;83:405–417.

Marx J. Oncogenes reach a milestone. Science 1994; 266:1942–1944.

Marzi CA. Transfer of visual information after unilateral input to the brain. Brain Cogn 1986;5:163–173.

Marzi IA, Berlucchi G. Right visual field superiority for accuracy of recognition of famous faces in normals. Neuropsychologia 1977;15:751–756.

Maslowski-Cobuzzi RJ, Napier T. Neuroscience 1994; 62:1103–1120.

Masino T. Brainstem control of orienting movements. Brain Behavior Evol 1992;40:98–111.

Masters JC, Barden RC, Ford ME. Affective states, expressive behavior, and learning in children. J Personal Soc Psychol 1979;37:380–390.

Mateer CA. Motor and perceptual functions of the left hemisphere and their interaction. In: Segalowitz SJ, ed. Language functions and brain organization. New York: Academic Press, 1983.

Mateer C, Kimura D. Impairment of nonverbal oral movements in aphasia. Brain Lang 1976;4:262–276.

Mathew RJ, Jack RA, West WS. Regional cerebral blood flow in a patient with multiple personality. Am J Psychiatry 1985;142:504–505.

Matthews WS, Barbas G, Ferrari M. Emotional concomitants of childhood epilepsy. Epilepsia 1982; 23:671–681.

Mattingly IG, Studdert-Kennedy M. Modularity and

the motor theory of speech perception. Hillsdale, NJ: Erlbaum, 1991.

Maunsell JHR, Van Essen DC. Functional properties of neurons in middle temporal visual area of the macaque. J Neurophysiol 1983;49:1127–1165.

Mayer MS, Mankin RW. Neurobiology of pheramone perception. In: Kerkut GA, Gilbert LI, eds. Comprehensive inspection physiology, biochemistry and pharmacology. New York: Pergamon Press, 1985.

Mayer SA, Sacco RL, Shi T, Mohr JP. Neurologic deterioration in noncomatose patients with supratentorial intracerebral hemorrhage. Neurology 1994a;44: 1379–1384.

Mayer SA, Fink ME, Hommma S, et al. Cardiac injury associated with neurogenic pulmonary edema following subarachnoid hemorrahage. Neurology 1994b;44:815–820.

Maynard Smith J, Haigh J. The hitch-hiking effect of a favorable gene. Gen Res 1974;23:23–35.

Maziotta JC, Phelps ME, Carson RE, Kull DE. Tomographic mapping of human cerebral metabolism: auditory stimulation. Neurology 1982;32:921–937.

Mazur A, Lamb TA. Testosterone, status and mood in human males. Horm Behav 1980;14:236–246.

McAndrews MP, Milner B. The frontal cortex and memory for temporal order. Neuropsychologia 1991;29:849–859.

McCarley RW, Ito K. Intracellular evidence linking medial pontine reticular formation neurons to PGO generation. Brain Res 1983;280:343–348.

McClain EG. The Pythagorian Plato. Denver: Great Eastern Book, 1978.

McClary RA. A response specificity in the behavioral effect of limbic system lesions in the cat. J Comp Physiol Psychol 1961;54:605–613.

McClary RA. Response-modulating functions of the limbic sytem. In: Stellar E, Sprague JM, eds. Progress in physiological psychology. New York: Academic Press, 1966.

McClintock B. The significance of responses to the genome to challenge. Science 1984;226:792–801.

McCloskey M, Wible CG, Cohen NJ. Is there a special flashbulb memory mechanism? J Exp Psychol Gen 1988;114:375–380.

McConnel JV, et al. The effects of regeneration upon retention of a conditioned response in the planarian. J Comp Physiol Psychol 1959;52:1–5.

McConnel JV, et al. The effects of ingestion of conditioned planaria on the response level of naive planaria. Worm Runnders Digest 1961;3:41–47.

McCown T. Mugharet es-Skhul: description and excavation. In: Garrod DAE, Bate D, eds. The stone age of Mount Carmel. Oxford: Clarendon Press, 1937.

McDonald AJ. Cell types and intrinsic connections of the amygdala. In: Aggleton JP, ed. The amygdala. New York: Wiley, 1992.

McDonald, S, Tate RL, Rigby J. Error types in ideomotor apraxia: a qualitate analysis. Brain Cogn 1994; 25:250–270.

McFarland HR, Fortin D. Amusia due to right temporal-parietal infarct. Arch Neurol 1982;39:725–727.

McFie J, Thompson JA. Picture arrangement: a measure of frontal lobe function? Br J Psychiatry 1972; 121:547–552.

McFie J, Zangwill OL. Visual-constructive disabilities associated with lesions of the left cerebral hemisphere. Brain 1960;83:243–260.

McGaugh JL. Involvement of hormonal and neuromodulatory systems in the regulation of memory storage. Annu Rev Neurosci 1989;12:255–287.

McGaugh JL, Gold PE. Hormonal modulation of memory. In: Brush RB, Levine S, eds. Psychoendocrinology. New York: Academic Press, 1992.

McGaugh JL, Introini-Collison IB, Decker M. Interactions of hormones and neurotransmitters in the modulation of memory storage. In: Morely JE, et al, eds. Memory function and aging-related disorders. New York: Springer, 1992.

McGeer PL, McGeer EG. Integration of motor functioning in the basal ganglia. In: Carpenter MB, Jayaraman A, eds. The basal ganglia. New York: Plenum, 1987.

McGinnies E. Emotionality and perceptual defense. Psychol Rev 1949;56:244–251.

McGlone J. Sex differences in human brain asymmetry. Brain Behav Sci 1980;3:215–263.

McGlone J. Sex differences in human laterality. Brain Behav Sci 1982;4:3–33.

McGraw MB. The neuromuscular maturation of the human infant. New York: Hafner, 1969.

McGrew WC. An ethological analysis of children's behavior. New York: Academic Press, 1972.

McGrew WC. Thumbs, tools and early humans. Science 1995;268:586.

McGrew WC, Marchant LF. Chimpanzees, tools, and terminates: hand preference or handedness. Curr Anthropol 1992;33:114–119.

McGrew WC, McGrew PL. Group formation in preschool children. Proceedings of the Third International Congress on Primatology 1970; 3:71–78.

McGuigan FJ. Imagery and thinking: covert functioning of the motor system. In: Schwartz GF, Shapiro D, eds. Consciousness and self regulation, vol 2. New York: Plenum, 1978.

McGuiness D. Sex differences in the organization of perception and cognition. In: Loyd B, Archer J, eds. Exploring sex differences. New York. Academic Press, 1976.

McLoughlin DM, Lucey JV, Dinan TG. Cerebral serotonergic hyperresponsivity in late-onset Alzheimer's disease. Am J Psychiatry 1994;151: 1701–1703.

McNabb AW, Caroll WM, Mastaglia FL. "Alien hand" and loss of bimanual coordination after dominant anterior cerebral artery territory infaction. J Neurol Neurosurg Psychiatry 1988;51:218–222.

McNally RJ, Kaspi SP, Riemann BC, Zietlin SB. Selective processing of threat cues in post-traumatic stress disorder. J Abnormal Psychol 1990;99: 398–402.

McNally RJ, Luedke DL, Besyner JK, et al. Sensitivity

to stress-relevant stimuli in post-traumatic stress disorder. J Anxiety Disord 1987;1:105–116.

McNeill TH, Brown SA, Shoulson I, et al. Regression of striatal dendrites in Parkinson's disease. In: Carpenter MB, Jayaraman A, eds. The basal ganglia. New York: Plenum, 1987.

Meador KJ, Loring DW, Bowers D, Heilman KM. Remote memory and neglect syndrome. Neurology 1987;37:522–526.

Meador KJ, Thompson JL, Loring DW, et al. Behavioral state-specific changes in human hippocampal theta activity. Neurology 1991;41:869–872.

Meadows JC. (1974a). Disturbed perception of colours associated with localized cerebral lesions. Brain 1974a;97:615–632.

Meadows JC. (1974b). The anatomical basis of prosopagnosia. J Neurol Neurosurg Psychiatry 1974b;37:489–501.

Mech LD. The wolf. Garden City: Natural History Press, 1970.

Meck WH, Church RM, Olton DS. Hippocampus, time, and memory. Behav Neurosci 1984;98:3–22.

Mecklenbrauker S, Hager W. Effects of mood on memory. Psychol Res 1984;46:355–376.

Mednick SA, Cannon TD. Fetal development, birth and the syndromes of adult schizophrenia. In: Mednic SA et al, eds. Fetal neural development and adult schizophrenia. New York: Cambridge University Press, 1991.

Mehler WR. Subcortical afferent connections of the amygdala in the monkey. J Comp Neurol 1980; 190:733–762.

Mehta Z, Newcombe F, Damasio H. A left hemisphere contribution to visiospatial processing. Cortex 1987;23:447–461.

Mellars P. Major issues in the emergence of modern humans. Curr Anthropol 1989;30:349–385.

Melson GF, Fogel A. Young children's interest in unfamiliar infants. Child Dev 1982;53:693–700.

Meltzer H. Individual differences in forgetting pleasant and unplesant experiences. J Educ Psychol 1930;21:399–409.

Meltzer H. Sex differences in forgetting pleasant and unpleasant experiences. J Abnormal Psychol 1931; 25:450–464.

Meltzoff AN. Towards a development cognitive science. Ann N Y Acad Sci 1990;608:1–37.

Meltzoff AN, Borton R. Intermodal matching by human neonates. Nature 1979;282:403–404.

Meltzoff AN, Moore MK. Immitation of facila and manusal gestures by human neonates. Science 1977;198:75–78.

Meltzoff AN, Moore MK. (1983b). Newborn infants immitate adult facial gestures. Child Dev 1983;54: 702–709.

Mendelow AD, Erfurth A, Grossart K, Macpherson P. Do cerebral arterovenous malformations increase in size? J Neurol Neurosurg Psychiatry 1987;50: 980–987.

Mendez MF, Geehand GR. J Neurol Neurosurg Psychiatry 1980;51:1–9.

Mennemeier M, Wertman E, Heilman KM. Neglect of near peripersonal space. Brain 1992;115:37–50.

Mercuri N, Bernardi G, Calabresi P, et al. Dopamine decreases cell excitability in rat striatal neurons by pre- and post-synaptic mechanisms. Brain Res 1985;358:110–121.

Meredith GE, Blank B, Groenewegen HJ. The distribution and compartmental organization of the cholinergic neurons in nucleus accumbens of the rat. Neuroscience 1989;31:327–345.

Merriam AE, Gardner EB. Corpus callosum function in schizophrenia: a neuropsychological assessment of interhemispheric information processing. Neuropsychologia 1987;25:185–193.

Merskey H, Watson GD. The lateralization of pain. Pain 1979;7:271–280.

Merskey H, Woodforde JM. Psychiatrric sequelae of minor head injury. Brain 1972;95:521–528.

Mertin P. The memory of young children for eyewitness events. Aust J Soc Iss 1989;24:23–32.

Merzenich MM, Brugge JF. Representation of the cochlear partition of the superior temporal plan of the macaque monkey. Brain Res 1973;50:275–296.

Mesulam MM. Dissociative states with abnormal temporal lobe EEG: multiple personality and the illusion of possession. Arch Gen Psychiatry 1981; 38:176–181.

Mesulam MM. Large scale neurocognitive networks and distributed processes for attention, language, and memory. Ann Neurol 1990;28:597–613.

Mesulam MM, Mufson EJ. Insula of the old world monkey. III. Efferent cortical output and comments on function. J Comp Neurol 1982;212:38–52.

Mesulam M-M, Mufson EJ, Levey AI, Wainer BH. Cholinergic innervation of the cortex by the basal forebrain. J Comp Neurol 1983;214:170–197.

Mesulam M-M, Van Hoesen GW, Pandya DN, Geschwind N. Limbic and sensory connections of the inferior parietal lobule in the rhesus monkey. Brain Res 1977;136:393–414.

Mettler FA, Spindler J, Mettler CC, Combs JD. Disturbances in gastrointestinal funcion after localized ablations of cerebral cortex. Arch Surg 1936;32: 618–620.

Meudell PR, Mayes AR. The Claparede phenomenon. Curr Psychol Res 1981;1:75–88.

Meyer-Bahlburg HFL. Psychoendocrine research on sexual orientation: current status and future options. Prog Brain Res 1984;13:505–544.

Meyer C, MacElhaney M, Martin W, MacGraw CP. Stereotaxic cingulotomy with results of acute stimulation and serial psychogoical testing. In: Laitinen LV, Livingston KE, eds. Surgical approaches to psychiatry. Lancaster: Medical Publishing, 1973:38–57.

Meyer DR, Ruth RA, Lavond DG. The septal social cohesiveness effect. Physiol Behav 1978;21:1027–1029.

Meyer JS, Ishikawa Y, Hata T, Karacan I. Cerebral blood flow in normal and abnormal sleep and dreaming. Brain Cogn 1987;6:266–294.

Meyer JS, Rogers RL, Jusdd BW, et al. Cognition and

cerebral blood flow fluctuate together in multi-infract dementia. Stroke 1988;19:163–169.

Meyer JS, Ishikawa Y, Hata T, Karacan I. Cerebral blood flow in normal and abnormal sleep and dreaming. Brain Cogn 1987;6:266–294.

Meyer V, Yates A. Intellectual changes following temporal lobecotmy for psychomotor epilepsy. J Neurol Neurosurg Psychiatry 1955;18:44–52.

Miall RC, Weir DJ, Stein JF. Visuo-motor tracking during reversible inactivation of the cerebelluym. Exp Brain Res 1987;65:455–454.

Michael RP, Kaverne EB. Pheromones in the communication of sexual status in primates. In: Ver der Kloot W, Walcott C, Dane B, eds. Readings in behavior. New York: Holt, Rhinehart & Winston, 1974.

Mies G, Auer LM, Ebbardt G, et al. Flow and neuronal density in tissue surrounding chronic infarction. Stroke 1983;14:22–27.

Millar K. Assessment of memory for anesthesia. In: Hindmarch I, Jones JG, Moss E, eds. Aspects of recovery from anaesthesia. New York: Wiley, 1987.

Miller BL, Cummings JL, McIntyre H, et al. Hypersexuality or altered sexual preferences following brain injury. J Neurol Neurosurg Psychiatry 1986;49:867–873.

Miller E. The long term consequence of head injury. Br J Soc Clin Psychol 1979;18:87–98.

Miller GA, Galanter E, Pribram KH. Plans and the structure of behavior. New York: Holt, 1960.

Miller H. Accident neurosis. BMJ 1961;1:919–925, 992–998.

Miller H, Stern G. The long term prognosis of severe head injuries. Lancet 1965;1:225–227.

Miller JD. Head injury and brain ischaemia—implication for therapy. Br J Anesthesiol 1985;57:120–129.

Miller JD, Butterworth JF, Gudeman SK, et al. Further experience in the management of severe head injury. J Neurosurg 1981;54:289–299.

Miller JW, Petersen RC, Metter EJ, et al. Transient global amnesia: clinical characteristics and prognosis. Neurology 1987;37:733–737.

Miller L. Neuropsychological conceptus of somatoform disorders. Int J Psychiatry Med 1984;14:31–46.

Miller L. Inner natures. New York: St. Martins Press, 1990.

Miller L. Freud's brain. New York: Guilford, 1991.

Miller L. Psychotherapy with the brain injured patient. New York: Norton, 1993.

Miller MW. Development of projection and local circuit neurons in neocortex. In: Peters A, Jones EG, eds. Cerebral cortex, vol 7. New York: Plenum, 1988.

Miller NE. Learnable drives and reward. In: Stevens SS, ed. Handbook of experimental psychology. New York: Wiley, 1951.

Miller WC, DeLong MR. Altered tonic activity of neurons in the globus pallidus and subthalamic nucleus in the primate MPTP model of parkinsonism. In: Carpenter MB, Jayaraman A, eds. The basal ganglia. New York: Plenum, 1987.

Mills L, Rollman GB. Hemispheric asymmetry for auditory perception of temporal order. Neuropsychologia 1980;18:41–47.

Milner B. Psychological defects produced by temporal lobe excisions. Research Publication of the Association for Research in Nervous and Mental Disease 1958;36:244–257.

Milner B. Laterality effect in audition. In: Mountcastle V, ed. Interhemispheric relations and cerebral dominance. Baltimore: John Hopkins University Press, 1962.

Milner B. Some effects of frontal lobectomy in man. In: Warren JM, Akert K, eds. The frontal granular cortext and behavior. New York: McGraw-Hill, 1964:313–334.

Milner B. Amnesia following operation on the temporal lobes. In: Whitty CWM, Zangwill OL, eds. Amnesia. London: Butterworths, 1966.

Milner B. Discussion. In: Millikan CH, Darley FL, eds. Brain mechanisms underlying speech and language. New York: Grune & Stratton, 1967.

Milner B. Visual recognition and recall after right temporal lobe excision in man. Neuropsychologia 1968;6:191–209.

Milner B. Memory and the medial temporal regions of the brain. In: Pribram K, Broadbent DE, eds. Biology of memory. New York: Academic Press, 1970.

Milner B. Interhemispheric differences in the location of psychological processes in man. Br Med Bull 1971;27:272–277.

Milner B. Hemispheric specialization: scope and limits. In: Schmitt FE, Worden FG, eds. The neurosciences. Third study program. Cambridge, MA: MIT Press, 1974.

Milner B, Corkin S, Teuber HL. Further analysis of the hippocampal amnesic syndrome. Neuropsychologia 1968;6:215–234.

Milner B, Teuber HL. Alteration of perception and memory in man. In: Weiskrantz L, ed. Analysis of behavioral changes. New York: Harper & Row, 1968.

Milner E. Human neural and behavioral development. Springfield, IL: Charles C Thomas, 1967.

Mink JW, Thach WT. Basal ganglia motor control. I, II, III. J Neurophysiol 1991;65:273–351.

Mintz M, Myslobodsky MS. Two types of hemisphere imbalance in hemi-Parkinsonian codes by brain electrical activity and electodermal activity. In: Myslobodsky MS, ed. Hemisyndromes: psychobiology, neurology, psychiatry. New York: Academic Press, 1983:213–238.

Mishkin M. Perseveration of central sets after frontal lesions in monkeys. In: Warren JM, Akert K, eds. The frontal granular cortex and behavior. New York: McGraw-Hill, 1964:219–241.

Mishkin M. Cortical visual areas and their interaction. In: Karczman AG, Eccles JC, eds. Brain and human behavior. Berlin: Spinger-Verlag, 1972.

Mishkin M. Memory in monkeys severely impaired by combined but not by separate removal of amygdala and hippocampus. Nature 1978;273:297–299.

Mishkin M. A memory system in the monkey. Philos Trans R Soc Lond 1982;298:85–92.

Mishkin M, Malamut B, Backevalier J. Memories and habits: two neural systems. In: Lynch G, McGaugh JL, Weinberger MN, eds. Neurobiology of learning and memory. New York: Guilford, 1984.

Mishkin M, Pribram KH. Analysis of the effects of frontal lesion in monkeys, II. Variation of delayed response. J Comp Physiol Psychol 1956;49:36–40.

Mitchell G. Paternalistic behavior in primates. Psychol Bull 1968;71:339–417.

Mitchell G. Behavioral sex differences in nonhuman primates. New York: Van Nostrand, 1979.

Mitchell GD. Attachment differences in male and female infant monkeys. Child Dev 1968;39:611–620.

Miyashita Y, Chang H. Neuronal correlate of pictorial short-term memory in the primate temporal cortex. Nature 1988;331:66–70.

Miyashita Y, Rolls ET, Cahusac PMB, et al. Activity of hippocampal formation neurons in the monkey related to conditional response tasks. J Neurophysiol 1989;61:669–678.

Miyata M, Sasaki K. HRP studies on thalamocortical neurons related to the cerebellocerebral projection in monkey. Brain Res 1983;274:213–224.

Modrich P. Mismatch repair, genetic stability, and cancer. Science 1994;266:1959–1960.

Mogenson G. Septal-hypothalamic relationships. In: DeFrance JF, ed. The septal nuclei. New York: Plenum Press, 1976.

Mogenson GJ. Limbic-motor integration. Progr Psychobiol Physiol Psychol 1987;12:117–170.

Mogenson GJ, Yang CR. The contribution of basal forebrain to limbic integration and the mediation of motivation to action. In: Napier TC et al, eds. The basal forebrain. New York: Plenum, 1991.

Molfese D, Freeman RB, Palermo DS. The ontogeny of brain lateralization for speech and nonspeech stimuli. Brain Lang 1975;2:356–368.

Mollica RF. Mood disorders: epidemiology. In: Kaplan HI, Sadock BJ, eds. Comprehensive textbook of psychiatry. Baltimore, Williams & Wilkins, 1989.

Molnar S. Tooth wear and culture. A survey of tooth functions among some prehistoric populations. Curr Anthropol 1972;85:31–49.

Molloy R, Brownell HH, Gardner H. Discourse comprehension by right hemisphere stroke patients. In: Joanette Y, Brownell HH, eds. Discourse ability and brain damage: theoretical and empirical perspectives. New York: Springer-Verlag, 1990.

Mohr JP, Caplan LR, Melski JW, et al. The Harvard cooperative resgtistry. Neurology 1978;28:754–762.

Monaco G, Gaier D. Developmental level and children's responses to the explosion of the space shuttle Challenger. Early Child Res Q 1987;2: 83–95.

Moncrieff RW. Odours. London: Heinemann, 1970.

Money J, Ehrhard AA. Man & woman, boy & girl. Baltimore: Johns Hopkins University Press, 1972.

Money J, Hosta G. Laughing seizures and sexual precocity. J Hopk Med J 1967;120:326–330.

Monrad-Krohn GH. Dysprosody or altered "melody of language." Brain 1947;70:405–415.

Monrad-Krohn GH. The third element of speech: prosidy in the neuropsychiatric clinic. J Ment Sci 1957;103:326–331.

Monrad-Krohn G. Prosody and its disorders. In: Halpern L, ed. Problems of dynamic neurology. Jerusalem: Jerusalem Post Press, 1963:237–291.

Monroe B, Rechtschaffen A, Foulkes, D, Jensen J. Discriminability of REM and NREM reports. Personality Soc Psychol 1965;2:456–460.

Monroe RA. Ultimate journey. New York: Doubleday, 1994.

Montero CG, Martinez AJ. Neuropathology of heart transplantation: 23 cases. Neurology 1986;36: 1149–1154.

Montgomery EA, Fenton GW, McClelland RJ, et al. The psychobiology of minor head injury. Psychol Med 1991;21:375–384.

Mood DW, Johnson JE, Shantz CU. Social comprehension and affect mathcing in young children. Merrill-Palmer Q 1978;24:63–66.

Moody R. Life after life. Georgia: Mockingbird Books, 1977.

Moore EH. A note on the recall of pleasant vs unpleasant experiences. Psychol Rev 1935;42:214–215.

Moore T. Language and intelligence. Hum Dev 1967;10:88–106.

Morath M. Inborn vocalizations of human baby and communicative value for the mother. Exp Brain Res 1979;2:236–244.

Morell V. The earliest art becomes older—and more common. Science 1995;267:1908–1909.

Moreno CR, Borod JC, Welkowitz J, Alpert M. Lateralization for the expression and perception of facial emotion as a function of age. Neuropsychologia 1990;28:199–209.

Morgan B, Gibson KR. Nutritional and environmental interactions in brain development. In: Gibson KR, Petersen AC, eds. Brain maturation and cognitive development. New York: Aldine, 1991: 91–106.

Morgan MJ, Corballis MC. On the biological basis of human laterality, II: The mechanisms of inheritance. Behav Brain Sci 1978;1:270–277.

Mori A. Signals found in the grooming interactions of wild Japanese monkeys. Primates 1975;16:107–140.

Mori U, Kawai M. Social relations and behavior of gelada baboons. Contemporary primatology. Basel: Karger, 1975.

Morihisa A, Duffy FH, Wyatt RJ. Brain electrical activity mapping (BEAM) in schizophrenic patients. Arch Gen Psychiatry 1983;40:719–728.

Morin LP. Circadian rhythms and the suprachiasmatic nucleus. Brain Res Rev 1994;67:102–127.

Moritz C, Dowling TE, Brown WM. Evolution of animal mitochondrial DNA: relevance for population biology and systematics. Annu Rev Ecol System 1987;18:269–292.

Morris ET. Fragrance. New York: Scribner's, 1986.

Morris J, Roth E, Davidoff G. Mild closed head injury

and cognitive deficits in spinal-cord-injured patients. J Head Trauma Rehabil 1986;1:31–42.

Morrison AR, Pompeiana O. An analysis of the supraspinal influences acting on motoneurons during sleep in the unrestrained cat. Arch Ital Biol 1965;103:497–516.

Morrison JH, Foote SL. Noradrenergic and serotonergic innervation of cortical, thalamic, and tectal visual structures in Old and New World monkeys. J Comp Neurol 1986;243:117–138.

Morrison JH, Hof PR, Campbell MJ, et al. Cellular pathology in Alzheimer's disease. In: Rapoport SR, et al, eds. Imaging, cerebral topography and Alzheimer's disease. New York: Springer, 1990.

Morstyn R, Duffy FH, McCarley R. Altered topography of EEG spectral content in schizophrenia. Electroenecephalogr Clin Neurophysiol 1983;56: 263–271.

Moscovitch M. Language and the cerebral hemispheres. In: Pliner P, et al, eds. Communication and affect: language and thought. New York: Academic Press, 1973.

Moscovitch M. Recovered consciousness: a hypothesis concerning modularity and episodic memory. J Clin Exp Neuropsychol 1995;17:276–290.

Moss HA. Early sex differences in mother-infant interaction. In: Friedman RD, Richart RM, Van de Wiele R, eds. Sex differences in behavior. New York: Wiley, 1974.

Moss PD, McEvedy CP. An epidemic of overbreathing among school girls. Br Med J 1966;2:1295–1300.

Motter BC, Mountcastle VB. The functional properties of the light sensitivity neurons of the posterior parietal cortex studies in waking monkey. J Neurosci 1981;1:3–26.

Motomura N, Yamadori A, Mori E, et al. Unilateral spatial neglect due to hemorrhage in the thalamic region. Acta Neurol Scand 1986;74:190–194.

Mountcastle VB. Modalities and topographic properties of single neurons of cat's sensory cortex. J Neurophysiol 1957;20:408–434.

Mountcastle VB. The world around us: neural command functions for selective attention. Neurosci Res Prog Bull 1976;14:1–47.

Mountcastle VB, Lynch JC, Georgopoulos A, et al. Posterior parietal assocation cortex of the monkey. J Neurophysiol 1975;38:871–908.

Mountcastle VB, Motter BC, Andersen RA. Some further observations on the functional properties of neurons in the parietal lobe of the waking monkey. Brain Behav Sci 1980;3:520–529.

Mountcastle VB, Powell TPS. Central nervous mechanisms subserving postion sense and kinesthesis. Bull Johns Hopkins Hosp 1959;105:173–200.

Moyer KE. The physiology of aggression and the implications for aggression control. In: Singer JL, ed. The control of aggression and violence. New York: Academic Press, 1971.

Moyer KE. Sex differences in aggression. In: Friedman RC, Richart RM, Van de Wiele RL, eds. Sex differences in behavior. New York: Wiley, 1974.

Mueser K, Butler R. Auditory hallucinations in combat-related chronic postraumatic stress disorder. Am J Psychiatry 1987;144:299–302.

Mullan S, Penfield W. Epilepsy and visual hallucinations. Arch Neurol Psychiatry 1959;81:269–281.

Murakami JW, Courchesne E, Press GA, et al. Reduced cerebellar hemisphere size and its relation to vermal hypoplasia in autism. Arch Neurol 1989;46:689–694.

Murdock GP, Provost C. Factors in the division of labor by sex. Ethnology 1981;12:203–235.

Murray EA. Medial temporal lobe structures contributing to recognition memory: the amygdaloid complex versus rhinal cortex. In: Aggleton JP, ed. The amygdala. New York: Wiley-Liss, 1992.

Murray EA, Gaffan D. Removal of the amygdala plus subjacent cortex disrupts the retention of both intramodal and crossmodal associative memories in monkeys. Behav Neurosci 1994;108:494–500.

Murray EA, Mishkin M. Amygdalectomy impairs cross modal associations in monkeys. Science 1985; 228:604–606.

Murray FS, Hagan BC. Pain threshold and tolerance of hands and feet. J Comp Physiol Psychol 1973; 84:639–643.

Murri L, Arena R, Siciliano G, et al. Dream recall in patients with focal cerebral lesions. Arch Neurol 1984;41:183–185.

Murrow RW, Schweiger GD, Kepes JJ, Koller WC. Parkinsonism due to a basal ganglia lucar state. Neurology 1990;40:897–900.

Myers FWH. Human personality and its survival after death. London: Freen & Co., 1903.

Myers RE. Interhemispheric communication through the corpus callosum: limitations under conditions of conflict. J Comp Physiol Psychol 1959;52:6–9.

Myers RE. Transmission of visual information within and between the hemispheres: a behavioral study. In: Mountcastle VB, ed. Interhemispheric relations and cerebral dominance. Baltimore: Johns Hopkins Press, 1962.

Myers RE, Swett C, Miller M. Loss of social group affinity following prefrontal lesions in free-ranging macaques. Brain Res 1973;64:257–269.

Myers RF. Comparative neurology of vocalization and speech: proof of a dichotomy. Ann N Y Acad Sci 1976;280:755–757.

Myles-Worsely W, Cromer C, Dodd D. Children's preschool script construction: reliance on general knowledge after memory fades. Dev Psychol 1986; 22:22–30.

Nadeau SE, Roletgen DP, Sevush S, et al. Apraxia due to a pathologically documented thalamic infarction. Neurology 1994;44:2133–2137.

Nadel L. The hippocampus and space revisited. Hippocampus 1991;1:221–229.

Nadler RD, Braggio JT. Sex and species differences in captive reared juvenile chimpanzees and orangutans. J Hum Evol 1974;5:24–33.

Najenson T, Groswasser Z, Mendelson, L, Hackett P. Rehabilitation outcome of brain damaged patients after severe head injury. Int Rehabil Med 1980; 2:17–22.

Najenson T, Mendelson L, Schechter I, et al. Rehabilitation after severe head injury. ScandJ Rehabil Med 1974;6:5–14.

Nakamura K, Ono T. Lateral hypothalamus neuron involvement in integration of natural and artifical rewards and cue signals. J Neurophysiol 1986;55: 163–181.

Nakamura K, Matsumoto K, Mikami, A, Kubota K. Visual response properties of single neurons in the temporal pole of behaving monkeys. J Neurophysiol 1994;71:1206–1240.

Nannis ED, Cowan PA Emotional understanding: a matter of age, dimension, and point of view. J Applied Dev Psychol 1987;8:289–304.

Napier B, Clube V. A theory of terrestrial catastrophism. Nature 1979;282:455.

Narabayashi H. The place of amygdalectomy in the treatment of aggressive behavior with epilepsy. In: Mosely TP, ed. Current controversies in neurosurgery. Toronto: WB Saunders, 1970.

Narabayashi H. Two groups of extrapyramidal involuntary movements. In: Carpenter MB, Jayaraman A, eds. The basal ganglia. New York: Plenum, 1987.

Nash M. What, if anything, is regressed about hypnotic age regression? A review of the empirical literature. Psychol Bull 1987;102:42–52.

Nash M, Lynn S. Child abuse and hypnotic ability. Imagin Cogn Personal 1985;5:211–217.

Nash SC, Feldman SS. Sex-role and sex-related attributions. In: Lamb ME, Brown AL, eds. Advances in developmental psychology. Hillsdale, NJ: Erlbaum, 1981.

Natale M, Hantas M. Effects of temporary mood states on selective memory about the self. J Personal Soc Psychol 1982;42:927–934.

Nathanson M, Bergman PS, Gordon GG. Denial of illness. Arch Neurol Psychiatry 1952;68:380–387.

Nauta WJH. An experimental study of the fornix in the rat. J Comp Neurol 1956;104:247–272.

Nauta WJH. Hippocampal projections and related neural pathways to the midbrain in cat. Brain 1958; 81:319–340.

Nauta WJH. Some efferent connections of the prefrontal cortex in the monkey. In: Warren JM, Akert K, eds. The frontal granular cortex and behavior. New York: McGraw-Hill, 1964:397–407.

Nauta WJH. The problem of the frontal lobe: a reinterpretation. J Psychiatr Res 1971;8:167–187.

Neary H. The pallium of anuran amphibians. In: Jones ED, Peters A, eds. Cerebral cortex, vol 8: Comparative structure and evolution of cerebral cortex. New York: Plenum, 1990.

Nebes RB. Handedness and the perception of whole-part relationship. Cortex 1971;7:350–356.

Neihardt JG, Black Elk. Black Elk speaks. Lincoln: University of Nebraska Press, 1979.

Neilsen JM. Agnosia, apraixa, aphasia, their value in cerebral localization. New York: Hoeber, 1946.

Neisser U. Cognitive psychology. New York: Appleton Crofts, 1967.

Neisser U, Harsch N. Phantom flashbulbs: false recollections of hearing the news about Challenger. In: Winograd E, Neisser U, eds. Affect and accuracy in recall. Cambridge, UK: Cambride University Press, 1992.

Neithammer C. Daughters of the earth. New York: Collier, 1977.

Nelson JP, McCarley RW, Hobson JA. REM sleep burst neurons, PGO waves, and eye movement information. J Neurophysiol 1983;50:784–797.

Nelson KE. Infants short term progress toward one component of object permanence. Merrill-Palmer Q 1974;20:3–8.

Nelson K. The transition from infant to child memory. In: Moscovitch M, ed. Infant memory. New York: Plenum, 1984.

Nelson K, Gruendel J. Childrens' scripts. In: Nelson K, ed. Event knowledge: structure and function in development. Hillsdale, NJ: Erlbaum, 1986.

Nelson K, Ross G. The generalities and specifics of long-term memory in infants and young children. In: Perlmutter M, ed. Children's memory. San Francisco: Josey-Bass, 1980.

Nelson DF, Urtasum RC, Saunders WM, et al. Recent and current investigations of radiation therapy of malignant gliomas. Semin Oncol 1986;13:46–55.

Nemiah J. Dissociative amnesia. In: Kilhstrom JF, Evans FJ, eds. Functional disorders of memory. Hillsdale, NJ: Erlbaum, 1979.

Newcombe F, Russell WR. Dissociated visual perceptual and spatial deficits in focal lesions of the right hemisphere. J Neurol Neurosurg Psychiatry 1969; 32:73–81.

Newman DB, Cruce WLR. The organization of the reptilian brainstem reticular formation. J Morphol 1982;173:325–349.

Newman JD, Wollberg Z. Multiple coding of species-specific vocalizations in the auditory cortex of squirrel monkey. Brain Res 1973;54:287–304.

Newton MR, Greenwood RJ, Britton KE, et al. A study comparing SPECT with CT and MRI after closed head injury. J Neurol Neurosurg Psychiatry 1992; 55:92–94.

Nichols I, Hunt J, McV. A case of partial bilateral frontal lobectomy. Am J Psychiatry 1940;96:1063–1087.

Niehoff DL, Kuhar MJ. Benzodiapine receptors: localization in the rat amygdala. J Neurosci 1983;3: 2091–2097.

Nielsen JM. Unilateral cerebral dominance as related to mind blindness: minimal lesions capable of causing visual agnosia for objects. Arch Neurol Psychiatry 1937;38:108–135.

Nieuwehuhs R, Meek J. The telencephalon of actinopterygian fishes. In: Jones ED, Peters A, eds. Cerebral cortex, vol 8: Comparative structure and evolution of cerebral cortex. New York: Plenum, 1990a.

Nieuwehuhs R, Meek J. The telencephalon of Sarcopterygian fishes. In: Jones ED, Peters A, eds. Cerebral cortex, vol 8: Comparative structure and evolution of cerebral cortex. New York: Plenum, 1990b.

Nikas DL. Prognostic indicators in patients with severe head injury. Crit Care Nurs Q 1987a;10:25–34.

Nikas DL. Critical aspects of head trauma. Crit Care Nurs Q 1987b;10:19–44.

Nisbett RE, Wilson TD. Telling more than we can know. Psychol Rev 1977;84:231–259.

Nishijo H, Ono T, Nishino H. Topographic distribution of modality-specific amygdalar neurons in alert monkey. J Neurosci 1988;8:3556–3569.

Nitecki MH, Nitecki DV, eds. Origins of anatomical modern humans. New York: Plenum, 1994.

Njiokiktjien C. Pediatric behavioural neurology. Amsterdam: Suyi, 1988.

Njemanze PC. Cerebral lateralization in linguistic and nonlinguistic perception. Brain Lang 1991;41: 367–380.

Noebels JL, Prince DA. Development of focal seizures in cerebral cortex: role of axons terminal bursting. J Neurophysiol 1978;41:1267–1281.

Norgren, R. Gustatory afferents to ventral forebrain. Brain Res 1974;81:285.

Norman MG, Ludwin SK. Congenital malformations of the nervous system. In: Davis RL, Robertson DM, eds. Textbook of neuropathology. Baltimore: Williams & Wilkins, 1991.

Nottebohm F. From bird song to neurogenesis. Sci Am 1989;260:74–79.

Novelly RA, Joseph R. Complex partial epilepsy of early development: gender specific effects on IQ with right hemisphere speech. 15th Annual Epilepsy International Symposium, Florida, September, 1983.

Novoa OP, Ardila A. Linguistic abilities in patients with prefrontal damage. Brain Lang 1987;30: 206–225.

Nowalowski RS. Basic concepts in CNS development. Child Dev 1987;58:568–595.

Nowalowski RS, Rakic P. The site of origin and route and rate of migration of neurons to the hippocampal region of the rhesus monkey. J Comp Neurol 1981;196:129–154.

Noyes R, Kletti R. Depersonalization in response to life threatening danger. Comp Psychiatry 1977; 18:375–384.

O'Callaghan G, D'Arcy H. Use of props in questioning preschool witnesses. Aust J Psychol 1989;41: 187–195.

Oddy M, Coughlan T, Tyerman, A, Jenkins D. Social adjustement after closed head injury. J Neurol Neurosurg Psychaitry 1985;48:564–568.

Offen ML, Davidoff RA, Troost BT, Richey, ET. Dacrystic epilepsy. J Neurol Neurosurg Psychiatry 1976 39:829–834.

Ogden JA. The "neglected" left hemisphere and its contributions to visuo-spatial neglect. In: Jeannerod M, ed. Neurophysiological and neuropsychological aspects of spatial neglect. Amsterdam: North-Holland, 1987.

Ojemann GA. Brain organization for language from the perspective of electrical stimulation mapping. Behav Brain Sci 1983;32:921–937.

Ojemann GA, Blick KI, Ward AA. Improvement and disturbance of short-term verbal memory with human ventrolateral stimulation. Brain 1971;94: 225–240.

Ojemann GA, Fedio P. Effect of stimulation of the human thalamus, parietal, and temporal white matter on short-term memory. J Neurosurg 1968; 29:51–59.

Ojemann GA, Fedio P, van Buren J. Anomia from pulvinar and subcortical parietal stimulation. Brain 1968;91:99–116.

Ojemann GA, Whitaker HA. Language localization and variability. Brain Lang 1978;6:239–260.

Okasha A, Saad A, Khalil A, et al. Phenomenology of obsessive-compulsive disorder: a transcultural study. Comp Psychiatry 1994;35:191–197.

Oke A, Keller R, Mefford I, Adams RN. Lateralization of norepinephrine in human thalamus. Science 1978;200:1411–1413.

O'Keefe DL. Stolen lightning. New York: Vintage Books, 1982.

O'Keefe J. Place units in the hippocampus of the freely moving rat. Exp Neurol 1976;51:78–109.

O'Keefe J, Bouma H. Complex sensory properties of certain amygdala units in the freely moving cat. Exp Neurol 1969;23:384–398.

O'Keefe J, Bouma H. Complex sensory properties of certain amygdala units in the freely moving cat. Exp Neurol 1969;23:384–398.

Olds JA. A preliminary mapping of electrical reinforcing effects in rat brain. J Comp Physiol Psychol 1956;49:281–285.

Olds JA, Milner P. Positive reinforcement produced by electrical stimulation of septal areas and other regions of the rat brain. J Comp Physiol Psychol 1954;47:419–427.

Olds ME, Forbes JL. The central basis of motivation: intracranial self-stimulation studies. Annu Rev Psychol 1981;32:523–574.

O'Leary DD. Do cortical areas emerge from a protocortex? Trends Neurosci 1989;12:400–406.

O'Leary DS. A developmental study of interhemispheric transfer in children aged 5 to 10. Child Dev 1980;51:743–750.

Olmstead CE, Best PJ, Mays LW. Neural activity in the dorsal hippocampus during paradoxical sleep, slow wave sleep and waking. Brain Res 1973; 60:381–391.

Olsen TS, Larsen, B, Kriver EE, et al. Focal cerebral hyperemia in acute stroke. Stroke 1981;2:598–606.

Olson EC. Vertebrate paleozoology. New York: Wiley, 1970.

Olson GM. An information processing analysis of visual memory and habituation in infants. In: Tighe TJ, Leaton RN, eds. Infant social cognition. Hillsdale, NJ: Erlbaum, 1976.

Olson GM. Infant recognition memory for briefly presented visual stimuli. Infant Behav Dev 1979;2: 123–134.

Olton DS, Branch M, Best PJ. Spatial correlates of hippocampal unit activity. Exp Neurol 1978;58:397–409.

Olton D, Markowska, A, Voytko ML, et al. Basal forebrain cholinergic system. In Napier TC et al, eds. The basal forebrain. New York: Plenum, 1991.

Ommaya AK, Fass F, Yarnell P. Whiplash injury and brain damage: an experimental study. JAMA 1968;204:285–289.

Ommaya AK, Grubb RL, Naumann RA. Coup and contrecoup injury. J Neurosurg 1971;35:503–516.

O'Neil JM. Gender and sex role conflicts in men's lives. In: Soloman K, Levy M, eds. Men in transition. New York: Plenum, 1982.

Ono T, Nishino H, Sasaki K, et al. Role of the lateral hypothalamus and the amygdala in feeding behavior. Brain Res Bull 1980;5:143–149.

Oppenheim RW. Neuronal cell death and some related regressive phenomon during neurogenesis. In: Cowan WM, ed. Studies in developmental neurobiology. New York: Oxford University Press, 1981.

Oppenheimer DR. Microscopic brain lesions after trauma. J Neurol Neurosurg Psychiatry 1968;31:299–306.

Oppler W. Manic psychosis in a case of parasagital meningioma. Arch Neurol Psychiatry 1950;47:417–430.

Orem J. Neural basis of behavioral and state dependent control of breathing. In: Lydic R, Beibuyck JF, eds. Clinical physiology of sleep. Bethesda, MD: American Physiological Society, 1988.

Orgel LE. The origin of life on earth. Sci Am 1994;271:76–83.

Orgogozo JM, Larsen B. Activation of the suplementary motor area during voluntary movement in man suggest it works as a supramodal motor area. Science 1979;206:847–850.

Orne MT. The nature of hypnosis: artifact and essence. J Abnorm Soc Psychol 1959;58:277–299.

Ornstein PA, Gordon BN, Larus DM. Children's memory for a personally experienced event: implications for testimony. Appl Cogn Psychol 1992;6:49–60.

Ornstein R. The psychology of consciousness. San Francisco: WH Freeman, 1972.

Orton JH. An experimental effect of light on the sponge. Nature 1924;113:924–925.

Oscar-Berman M. The effects of dorsolateral-frontal and ventrolateral orbitofrontal lesions on spatial-discrimination learning and delayed response in two modalities. Neuropsychologia 1975;13:237–246.

Ostrove JM, Simpson T, Gardner H. Beyond scripts: a note on the capacity of right hemisphere-damaged pateints to process social and emotional content. Brain Cogn 1990;12:144–154.

Ott BR, Zamani A, Kleefield J, Funkenstein HH. The clinical spectrum of hemorrhagic infarction. Stroke 1986;17:630–637.

Otto MW, Yeo RA, Dougher MJ. Right hemisphere involvement in depression: toward a neuropsychological theory of negative affective experiences. Biol Psychiatry 1987;22:1201–1215.

Oviatt SL. The emerging ability to comprehend language. Child Dev 1980;51:97–106.

Packard MG, Hirsh R, White NM. Differential effects of fornix and caudate nucleus lesions on two radial maze tasks: evidence for multiple memory systems. J Neurosci 1989;9:1465–1472.

Packard MG, White NM. Dissociation of hippocampus and caudate nucleus memory systems by post-training intracerebral injection of dopamine agonists. Behav Neurosci 1991;105:295–306.

Packer RJ, Allen JC, Goldwein JW, et al. Hyperfractionated radiotherapy for children with brainstem glioma. Ann Neurol 1991;27:167–173.

Palmini A, Andermann F, Olivier A, et al. Focal neuronal migration disorders and intractable partial epilepsy. Ann Neurol 1991;30:741–749.

Pandey GN, Pandey SC, Dwivedi Y, et al. Platelet serotonin-2A receptors: a potential biological marker for suicidal behavior. Am J Psychiatry 1995;152:850–855.

Pandya DN, Kuypers HGJM. Corticocortical connections in the rhesus monkey. Brain Res 1969;13:13–36.

Pandya DN, Vignolo LA. Corticocortial connections in the rhesus monkey. Brain Res 1969;13:13–16.

Pandya DN, Van Hoesen GW, Domeskick VB. A cinguloamygdaloid projection in the rhesus monkey. Brain Res 1973;61:369–373.

Pandya DN, Vignolo L. Intra and inter-hemispheric projections of the precentral, premotor, and arcuate areas of the rhesus monkey. Brain Res 1971;26:217–233.

Panksepp J, Normansell L, Herman B, et al. Neural and neurochemical control of the separation distress call. In: Newman JD, ed. The physiological control of mammalian behavior. New York: Plenum, 1988.

Pantev C, Hoke M, Lutkenhorner B, Lehnertz K. Tonotopic organization of the auditory cortex. Science 1989;246:486–488.

Panula P, Revuelta AV, Cheney DL, et al. An immunohistochemical study on the location of GABAergic neurons in rat septum. J Comp Neurol 1984;222:69–80.

Papcun G, Krashen S, Terbeek D, et al. Is the left hemisphere specialized for speech, language and-or something else. J Acoust Soc Am 1974;55:319–327.

Papez JW. Comparative neurology. New York: Hafner, 1967.

Pardo JV, Fox PT, Reichle ME. Localization of a human system for sustained attention by positron emission tomography. Nature 1991;349:61–64.

Parent A. Human neuroanatomy. Baltimore: Williams & Wilkins, 1995.

Parent A, Hazrati L-N. Functional anatomy of the basal ganglia. I, II. Brain Res Rev 1995;20:91–154.

Parham K, Willott JF. Effects of inferior colliculus lesions on the acoustic startle response. Behav Neurosci 1990;104:831–840.

Paris S. Integration and inference in children's comprehension and memory. In: Restle F; et al. eds. Cognitive theory. Hillsdale, NJ: Erlbaum, 1975.

Parkin AJ. Memory and amnesia. Oxford: Blackwell, 1987.

Parlow SE, Kinsbourne M. Asymmetric transfer of training between the hands. Implications for interhemispheric communication in the normal brain. Brain Cogn 1989;11:98–113.

Parmelee AH, Wenner WH, Akiyama Y, et al. Sleep states in premature infants. Dev Med Child Neurol 1967;14:70–77.

Parnavelas JG, Papadopoulos GC, Cavanagh ME. Changes in neurotransmitters during development. In: Peters A, Jones EG, eds. Cerebral cortex, vol 7. New York: Plenum, 1988.

Parrinder G. Sex in the world's religions. New York: Oxford University Press, 1980.

Parson ER. Post-traumatic self disorders (PTsfD): theoretical and practical considerations in psychotherapy of Vietnam War Veterans. In: Wilson JP, Harel Z, Kahana B, eds. Human adaptation to extreme stress. New York: Plenum, 1988.

Partridge M. Pre-frontal leucotomy. A survey of 300 cases personally followed over 1½–3 years. Oxford: Blackwell, 1950.

Pascual-Leone A, Valls-Sole JM, Brasil-Neto JP, et al. Akinesia in Parkinson's disease. Neurology 1994; 44:892–898.

Passingham RE. Broca's area and the origins of human vocal skills. Philos Trans R Soc London Biol 1981; 292:167–175.

Passinghmam RE. The frontal lobes and voluntary action. Oxford: Oxford University Press, 1993.

Passingham RE. The human primate. San Franciso: WH Freeman, 1982.

Patterson K, Bradshaw JL. Differential hemispheric mediation of nonverbal visual stimuli. J Exp Psychol Hum Percept Perform 1975;1:246–252.

Paul GS. Predatory dinosaurs. New York: Simon & Schuster, 1988.

Paulson PE, Robinson TE. Relationship between circadian changes in spontaneous motor activity and dorsal versus ventral striatal dopamine neurotransmission assessed with on-line microdialysis. Behav Neurosci 1994;108:624–635.

Pearlson GD, Robinson RG. Sunction lesions of the frontal cortex in the rat induce asymmetrical behavior and catecholaminergic responses. Brain Res 1981;218:233–242.

Pearlson GD, Wong DF, Tune LE, et al. In vivo D2 dopamine receptor density in psychotic and nonpsychotic patients with bipolar disorder. Arch Gen Psychiatry 1995;52:471–477.

Pechtel C, McAvoy T, Levitt M, et al. The cingulate and behavior. J Nerv Mental Dis 1958;126: 148–152.

Pedley TA. The pathophysiology of focal epilepsy: neurophysiological considerations. Ann Neurol 1978;3:2–9.

Pedley TA, Guilleminault C. Episodic nocturnal wandering responsivity to anticonvulsant drug therapy. Ann Neurol 1977;2:30–35.

Pena-Casanova J, Roig-Rovira T. Optic aphasia, optic apraxia, and loss of dreaming. Brain Lang 1985; 26:63–71.

Penfield W. Memory mechanisms. Arch Neurol Psychiatry 1952;67:178–191.

Penfield W. The permanent records of the stream of consciousness. Acta Psychologica 1954;11:47–69.

Penfield W, Evans J. Functional defects produced by cerebral lobectomies. Publication of the Association for Research in Nervous and Mental Disease 1934;13:352–377.

Penfield W, Jasper H. Epilepsy and the functional anatomy of the human brain. Boston: Little-Brown, 1954.

Penfield W, Mathieson G. Memory: autopsy findings and comments on the role of the hippocampus in experimental recall. Arch Neurol 1974;31:145–154.

Penfield W, Milner B. Memory deficit produced by bilateral lesions in the hippocampal zone. Arch Neurol Psychiatry 1958;79:475–497.

Penfield W, Perot P. The brains record of auditory and visual experience. Brain 1963;86:595–695.

Penfield W, Rasmussen T. The cerebral cortex of man. New York: Macmillan, 1950.

Penfield W, Roberts L. Speech and brain mechanisms. Princeton, NJ: Princeton University Press, 1959.

Penfield W, Welch K. Supplementary motor area of cerebral cortex. Clinical and experimental study. Arch Neurol Psychiatry 1951;66:289–317.

Perelman A. Gender and sex-role influences on emotional intimacy in marriage. Montreal: American Psychological Association, 1980.

Perez MM, Trimble MR. Epileptic psychosis—diagnositic comparison with process schizophrenia. Br J Psychiatry 1980;137:245–249.

Perez NM, Trimble MR, Murray NMF, Reider I. Epileptic psychosis. Br J Psychiatry 1985;146: 155–163.

Perris C. Averaged evoked responses (AER) in patients with affective disorders. Acta Psychiatr Scand 1974;255:1–107.

Perris EE, Myers NA, Clifton RK. Long term memory for a single infancy experience. Child Dev 1990; 61:1796–1807.

Perryman KM, Kling AS, Lloyd RL. Differential effects of inferior temporal cortex lesions upon visual and auditory-evoked potentials in the amygdala of the squirrel monkey. Behav Neural Biol 1987;47:73–79.

Pesme P. Auditory hallucinations in a deaf person. Rev Neuropathol Opthalmol 1939;17:280–291.

Peters DP. The impact of naturally occuring stress on children's memory. In: Ceci SJ, Toglia MP, Ross DF, eds. Children eyewitness memory. New York: Springer-Verlag, 1987.

Peters DP. The influence of stress and arousal on the child witness. In: Doris J, ed. The suggestibility of chidren's recollections. Washington, DC: American Psychological Association, 1991.

Peters M. Cerebral asymmetry for speech and the

asymmetry in path lengths for the right and left recurrent nerves. Brain Lang 1992;43:349–352.

Petersen AC, Crokett L. Factors influencing sex differences in spatial ability during adolescene. In: Sex differences in spatial ability across the lifespan [Abstract]. Symposium conducted at the 1993 APA convention, Los Angeles, 1985.

Petersen P, Godtfedsen J. Embolic complications in paroxysmal atrial fibrillation. Stroke 1986;17:622–629.

Petersen SE, Fox PR, Posner MI, et al. Positron emission tomography studies of the cortical anatomy of single-word processing. Nature 1988;331:585–589.

Petty RG, Barta PE, Pearlson GD, et al. Reversal of asymmetry of the planum temporale in schizophrenia. Am J Psychiatry 1995;152:715–721.

Petrides M. Deficits on conditional associative learning tasks after frontal- and temporal-lobe lesions in man. Neuropsychologia 1985;23:601–614.

Petrides M, Milner N. Deficits on subject-ordered tasks after frontal temporal lobe lesions in man. Neuropsychologia 1982;20:249–262.

Petrie A. Personality and the frontal lobes. New York: Blakiston, 1952.

Petsche H, Gogolack G, Van Zwieten PA Rhythmicity of septal cell discharges at various levels of reticular excitation. Electroencephalogr Clin Neurophysiol 1965;19:25–33.

Pettigrew JD, Kasamatsu T. Local perfusion of noradrenaline maintains visual plasticity. Nature 1978;271:761–763.

Pfieffer J. The emergence of humankind. New York: Harper & Row, 1985.

Phillips RG, LeDoux JE. Differential contributions of amygdala and hippocampus to cued and contextual fear conditioning. Behav Neurosci 1992;106:274–285.

Piaget J. The origins of intelligence in children. New York: Norton, 1952.

Piaget J. Play, dreams and imitation in childhood. New York: Norton, 1962.

Piaget J. The child and reality. New York: Viking Press, 1974.

Piaget J. The moral judgement of children. New York: Penguin, 1965.

Piaget J, Inhelder B. Memory and intelligence. London, Routledge & Kegan Paul, 1973.

Piazza DM. The influence of sex and handedness in the hemispheric specialization of verbal and non-verbal tasks. Neuropsychologia 1980;18:163–176.

Piercy M, Hecaen H, Ajuriaguerra J. Constructional apraxia associated with unilateral cerebral lesions—left and right sided cases compared. Brain 1960;83:225–242.

Pietro MJS, Rigdrodsky MS. Patterns of oral-verbal perseveration in adult aphasics. Brain Lang 1986;29:1–17.

Pigarev IN. Neurons of visual cortex respond to visceral stimulation during slow wave sleep. Neuroscience 1994;62:1237–1243.

Pillemer DB. Preschool children's memories of personal circumstances: the fire alarm study. In: Winograd E, Neisser U, eds. Affect and accuracy in recall. Cambridge, UK: Cambride University Press, 1992.

Pillemer D, White SH. Childhood events recalled by children. In: Neisser U, ed. Memory observed. New York: Freeman, 1989.

Pilleri G, Poeck K. Sham rage-like behavior in a case of traumatic decerebration. Confina Neurol 1965;25:156–166.

Pipe M, Gee S, Wilson C. Cues, props, and context: do they facilitate children's event report. In: Goodman GS, Bottoms BL, eds. Child victims, child witnesses. New York: Guilford Press, 1993.

Pitman RK. Self-mutiliation in combat-related PTSD. Am J Psychiatry 1990;147:123–124.

Playford ED, Jenkins LH, Passingham RE, et al. Impaired mesial frontal and putamen activation in Parkinson's disease. Ann Neurol 1992;32:151–161.

Pochedly C. Neuroblastomas in the head and central nervous system. In: Pochedly C, ed. Neuroblastoma. Acton, MA: Publishing Sciences Group, 1976.

Poeck K, Kerschensteiner M, Hartje, W, Orgass B. Impairment in visual recognition of geometric figures in patients with circumscribed retrorolandic brain lesions. Neuropsychologia 1973;11:311–319.

Pohl P. Central auditory processing. Ear advantages for acoustic stimuli in baboons. Brain Lang 1983;20:44–53.

Poizner H, Klima ES, Bellugi U. What the hands reveal about the brain. Cambridge, MA: MIT Press, 1987.

Poletti CE, Sujatanond M. Evidence for a second hippocampal efferent pathway to hypothalamus and basal forebrain comparable to fornix system: a unit study in the monkey. J Neurophysiol 1980;44:514–531.

Pollak M. Homosexual rituals and safer sex. J Homosexual 1993;25:307–317.

Polster MR. Drug-induced amnesia: implications for cognitive neuropsychological investigations of memory. Psychol Bull 1993;114:477–493.

Pompeiano O. Reticular formation. In: Iggo A, ed. Handbook of sensory physiology, vol II. Berlin: Springer, 1973.

Pons T, Garraghty PE, Ommaya AK, et al. Massive cortical reorganization after sensory deafferentation in adult macaques. Science 1991;252:1857–1860.

Poplawsky A, Isaacson RL. The GM1 ganglioside hastens the reduction of hyperemotionality after septal lesions. Behav Neural Biol 1987;48:150–158.

Popper KR, Eccles JC. The self and its brain. New York: Springer, 1977.

Porrino LJ, Crane AM, Goldman-Rakic PS. Direct and indirect pathways from the amygdala top the frontal lobe in rhesus monkeys. J Comp Neurol 1981;205:30–48.

Portenoy RK, Abissi CJ, Lipton RB. Headache in cerebrovascular disease. Stroke 1984;15:1009–1012.

Portenoy RK, Lipton RB, Berger AR, et al. Intracerebral haemorrhage: a model for the prediction of outcome. J Neurol Neurosurg Psychiatry 1987;50: 976–979.

Porter R. The Kugelberg lecture. Brain mechanisms of voluntary motor commands—a review. Electroencephalogr Clin Neurophysiol 1990;76:282–293.

Porteus SD, Peters HN. Psychosurgery and test validity. J Abnorm Soc Psychol 1947;42:473–488.

Portlock CS. Lymphomatous involvement of the central nervous system. In: Whitehouse JMA, Kay HEM, eds. CNS complications of malignant disease. Baltimore: University Park Press, 1979.

Potts R. Home bases and early hominids. Am Sci 1984;72:338–347.

Potwin EB. Study of early memories. Psychol Rev 1901;8:596–601.

Povlishock JT. Traumatically induced axonal injury. Brain Pathol 1992;2:1–12.

Powell EW The cingulate bridge between allocortex, isocortex, and thalamus. Anat Rec 1978;190: 783–794.

Powell EW, Akagi K, Hatton JB. Subcortical projections of the cingulate gyrus in cat. J Hirnforsch 1974;15:269–278.

Prado-Alcala RA. Is cholinergic activity of the caudate nucleus involved in memory? Life Sci 1985;37: 2135–2142.

Pragay EB, Mirskey AF, Nakamura RK. Attention-related activity in the frontal association cortex during a go/no-go discrimination task. Exp Neurol 1987;96:841–500.

Pratt MW, Golding G, Hunter WJ. Does morality have a gender. Merrill-Palmer Q 1984;30:321–348.

Premack D, Premack AJ. The mind of an ape. New York: Dutton, 1983.

Previc FH. Functional specialization in the lower and upper visual field in humans. Behav Brain Sci 1990;13:519–575.

Pribram KH, Ahumada A, Hartog J, Ross LA. A progress report on the neurological processes disturbed by frontal lesions in primates. In: Warrent JM, Akert K, eds. The frontal granular cortex and behavior. New York: McGraw-Hill, 1964:28–55.

Pribram KH, Chow KL, Semmes J. Limit and organization of the cortical projections from the medial thalamic nucleus in monkey. J Comp Neurol 1953; 98:433–448.

Prideaux T. Cro-Magnon. New York: Time-Life, 1973.

Prigatano GP, Altman JM, O'Brien KP. Behavioral limitations that traumatic brain-injured patients tend to underestimate. Clin Neuropsychol 1990;4: 163–176.

Prigatano, G, Fordyce DJ, Zeiner H, et al. Neuropsychological rehabilitation after closed head injury in young adults. J Neurol Neurosurg Psychiatry 1984;47:505–513.

Prince M. Some of the revelations of hypnotism. Boston Med Surg J 1890;122:463–467.

Prince M. The unconscious. New York: Macmillan, 1924.

Prud'homme JJL, Kalaska JF. Proprioceptive activity in primate somatosensory cortex during active arm reaching movements. J Neurophysiol 1994; 72:2280–2299.

Prytulak LS. Natural language mediation. Cogn Psychol 1971;2:1–56.

Pudenz RH, Shelden CH. The lucite calvarium—a method for direct observation of the brain. II. J Neurosurg 1946;3:487–505.

Pulsinelli WA, Levy DE, Sigsbee B, et al. Increased damage after ischemic stroke in patients with hyperglycemia with or without established diabetes mellitus. Am J Med 1983;74:540–544.

Purpura DP. Electrophysiological analysis of psychotogenic drug action. I, II. Arch Neurol Psychiatry 1956;40:122–143.

Putnam FW. Dissociation as a response to extreme trauma. In: Kluft RP, ed. The childhood antecedents of multiple personality. Washington, DC: American Psychiatric Press, 1985.

Putnam FW. The scientific investigation of multiple personality disorder. In: Quen JM, ed. Split mind/split brains. New York: New York University Press, 1986.

Putnam FW, Guroff JJ, Silberman EK, et al. The clinical phenomenology of multiple personality disorder. Review of 100 recent cases. J Clin Psychiatry 1986;47:285–293.

Pynoos RS, Eth S. Children traumatized by witnessing acts of personal violence: homicide, rape, or suicide. In: Eth S, Pynoos RS, eds. Post-traumatic stress disorder in children. Washington, DC: American Psychiatric Press, 1985.

Quiroga JC. The brain of two mammal-like reptiles. J Hirnforsch 1979;20:341–350.

Quiroga JC. The brain of the mammal-like reptile. J Hirnforsch 1980;21:299–336.

Rabavilas AD, Scarpalezos S. The post-traumatic syndrome. Some considerations related to psychiatric prevention. Bibl Psychiatr 1981;160:73–77.

Raimondi AJ, Hirschauer J. Head injury in the infant and toddler. Childs Brain 1984;11:344–351.

Rainbow TC, Parsons B, McEwen BA. Sex differences in rat brain oestrogen and progestin receptors. Nature 1982;300:648–649.

Raine A, Buchsbaum MS, Stanely J, et al. Selective reductions in prefrontal glucose metabolism in murders. Biol Psychiatry 1994;29:14–25.

Raine CL, Path FRC. Oligodendrocytes and central nervous system myelin. In: Davis RL, Robertson DM, eds. Textbook of neuropathology. Baltimore: Williams & Wilkins, 1991.

Raisman G. Neuronal plasticity in the septal nuclei of the adult rat. Brain Res 1969;14:25–48.

Raisman G, Field PM. Sexual dimorphism in the preoptic area of the rat. Science 1971;173:731–733.

Raisman G, Field PM. Sexual dimorphism in the neuropil of the preoptic area of the rat and its dependence on neonatal androgen. Brain Res 1973; 54:1–29.

Rajput AH. Prevalence of dementia in Parkinson's disease. In: Huber SJ, Cummings JL, eds. Parkinson's disease. New York: Oxford, 1992.

Rakic P. Neurons in the rhesus monkey visual cortex. Systematic relationship between time of origin and eventual disposition. Science 1974;183:425–427.

Rakic P. Specification of cerebral cortical areas. Science 1988;241:170–176.

Rakic P. Defects of neuronal migration and the pathogenesis of cortical malformations. Progr Brain Res 1988;73:15–37.

Rakic P, Nowakowski RS. The time of origin of neurons in the hippocampal region of the rhesus monkey. J Comp Neurol 1981;96:99–128.

Rakic P, Yakovlev PI. Development of the corpus callosum and cavum septum in man. J Comp Neurol 1966;132:45–72.

Raleigh MJ, Brammer GL, McGuire MT. Male dominance, serotonergic systems, and the behavioral and physiological effects of drugs in vervet monkeys (Cerocopthecus aethips sabaeus). In: Miczek KA, ed. Ethnopharmacology: primate models of neuropsychiatric disorders. New York: Liss, 1983.

Ramachandran VS. Behavioral and magnetoencephalographic correlates of plasticity in the adult human brain. Proc Natl Acad Sci U S A 1993;90: 10413–10420.

Ramaeker G, Njiokiktjien C. Pediatric behavioural neurology 3. The child's corpus callosum. Amsterdam: Suyi, 1991.

Randolf M, Semmes J. Behavioral consequences of selective subtotal ablations in the posterior gyrus of Macaca mulatta. Brain Res 1974;70:55–70.

Ranseen JD. Positive personality change following traumatic brain injury. Cogn Rehabil 1990;8:8–12.

Ransohoff J, Kelly P, Laws E. The role of intracranial surgery for the treatment of malignant gliomas. Semin Oncol 1986;13:27–37.

Ransom TW, Powell TE. Early social development of feral baboons. In: Poirer FE, ed. Primate socialization. New York: Random House, 1972.

Ranson SW, Kabat H, Magoun HW. Autonomic responses to electrical stimulation of the hypothalamus, preoptic region and septum. Arch Neurol Psychiatry 1935;33:467–477.

Rao SM, Binder JR, Hammeke TA, et al. Somatotopic mapping of the human primary motor cortex with functional magnetic resonance imaging. Neurology 1995;45:919–924.

Rapcsak SZ, Cimino CR, Heilman KM. Altitudinal neglect. Neurology 1988;38:277–281.

Rapcsak SZ, Ochipa C, Beeson PM, Rubens AB. Praxis and the right hemisphere. Brain Cogn 1993;23: 181–202.

Rapoport JL. Topography of Alzhiemer's disease. In Rapoport SR et al, eds. Imaging, cerebral topography and Alzheimer's disease. New York: Springer, 1990.

Rapoport JL. Recent advances in obsessive-compulsive disorder. Neuropsychopharmacology 1991;5: 1–10.

Rasheed BM, Manchanda SK, Anand BK. Effects of stimulation of paelocerebellum on certain vegetative functions in the cat. Brain Res 1970;20:293–308.

Ratcliff DA, Davies-Jones GAB. Defective visual localization in focal brain wounds. Brain 1972;95:49–60.

Ratner HH, Smith BS, Dion SE. Development of memory for events. J Exp Child Psychol 1986;41: 411–428.

Rauch SL, Jenike MA, Alpert NM, et al. Regional cerebral blood flow measured during symptom provocation in obsessive-compulsive disorder. Arch Gen Psychiatry 1994;51:62–70.

Raup DM. Extinction. New York: Norton, 1991.

Rawlings M. Beyond death's door. London: Sheldon Press, 1978.

Rawlins JNP. Associations across time: the hippocampus as a temporary memory store. Behav Brain Sci 1985;8:479–496.

Raymond M, Conklin C, Schaeffer J, et al. Coping with transient intellectual dysfunction after coronary bypass surgery. Heart Lung 1984;13:531–539.

Rebek J. Synthetic self-replicating molecules. Sci Am 1994;271:48–57.

Redding FK. Modification of sensory cortical evoked potentials by hippocampal stimulation. Electroencephalogr Clin Neurophysiol 1967;22:74–83.

Redlich FC, Dorsey JE. Denial of blindness by patients with cerebral disease. Arch Neurol Psychiatry 1945;53:407—417.

Redmond DE, Hunag YH. Current concepts 2. New evidence for a locus coeruleus-norepinephrine connection with anxiety. Life Sci 1979;25: 2149–2162.

Reed G. Everyday anomalies of recall and recognition. In: Kihlstrom Jf, Evans FJ, eds. Functional disorders of memory. Hillsdale, NJ: Erlbaum, 1979.

Reese HH. The relation of music to diseases of the brain. Occupat Ther Rehabil 1948;27:12–18.

Rehncrona S, Rsoen I, Siesjo BK. Excessive culluar acidosis: an important mechanism of neuronal damage in the brain? Acta Physiol Scand 1980;110: 435–437.

Rehak A, Kaplan JA, Gardner H. Sensitivity to conversational deviance in right-hemisphere-damaged patients. Brain Lang 1992;42:203–217.

Reid GC, et al. Influence of ancient solar proton events on the evolution of life. Nature 1976;259:177, 1976.

Reiner A, Albin RL, Anderson KD, et al. Differential loss of striatal projection neurons in Huntington's disease. Proc Natl Acad Sci 1988;85:5733–5737.

Reinhold M. A case of pure auditory agnosia. Brain 1950;73:203–223.

Reinisch JM. Fetal hormones, the brain, and human sex differences. Arch Sex Behav 1974;3:51–90.

Reinisch JM, Sanders SA. Prenatal hormonal contributions to sex differences in human cognitive and personality development. In Gerall AA, Moltz H, Ward IL, eds. Handbook of behavioral neurobiology. New York: Plenum, 1992.

Reisberg R, Heuer H. Remembering emotional events.

In: Winograd E, Neisser U, eds. Affect and accuracy in recall. Cambridge, UK: Cambridge University Press, 1992.

Reitman F. Orbital cortex syndrome following leucotomy. Am J Psychiatry 1946;103:238–241.

Reitman F. Observations of personality changes after leucotomy. J Nerv Ment Dis 1947;105:582–589.

Remillard GM, et al. Sexual ictal manifestations predominate in women with temporal lobe epilepsy: a finding suggesting sexual dimorphism in the human brain. Neurology 1983;33:323–330.

Represa A, Temblay E, Ben-Ari Y. Transient increase of NMDA-Binding sites in human hippocampus during development. Neurosci Lett 1989;99:61–66.

Reuter-Lorenz P, Davidson RJ. Differential contributions of the two cerebral hemispheres to the perception of happy and sad faces. Neuropsychologia 1981;19:609–613.

Reynolds GP. Increased concentrations and lateral asymmetry of amygdala dopamine in schizophrenia. Nature 1983;305:527–529.

Rhein JHW. Hallucinations of hearing and diseases of the ear. N Y Med J 1913;97:1236–1238.

Rholes WS, Riskind JH, Lane JW. Emotional states and memory biases. Effects of cognitive priming and mood. J Personal Soc Psychol 1987;52:91–99.

Rhue JW, Lynn SJ, Henry S, et al. Child abuse, imagination, and hypnotizability. Imagin Cogn Personal 1991;10:53–63.

Richards G. Freed hands or enslaved feet. J Hum Evol 1986;15:143–150.

Riches IP, Wilson FA, Brown MW. The effects of visual stimualtion and memory on neurons of hippocampal formation and the neighboring parahippocampal gyrus and inferior temporal cortex of the primate. J Neurosci 1991;11:1763–1770.

Richfield EK, Twyman R, Berent S. Neurological syndrome following bilateral damage to the head of the caudate nuclei. Ann Neurol 1987;22:768–771.

Richmond BJ, Optican LM, Podel M, Spitzer H. Temporal encoding of two-dimensional patterns by single units in primate inferior temporal cortex. J Neurophysiol 1987;57:132–162.

Richmond BJ, Wurtz RH, Sato T. Visual responses of inferior temporal neurons in awake rhesus monkey. J Neurophysiol 1983;50:1415–1432.

Ridgeway D, Waters E, Kuczaj SA. Acquisition of emotion-descriptive language. Dev Psychol 1985;21:901–908.

Ridley RM, Ettlinger G. Impaired tactile learning and retention after removals of the second somatic sensory projection cortex (SII) in the monkey. Brain Res 1976;109:656–660.

Rightmire GP. Homo sapiens in Sub-Saharan Africa. In: Smith FH, Spencer F, eds. The origins of modern humans: a world survey of the fossil evidence. New York: Alan R. Liss, 1984.

Rightmire GP. Africa and the origins of modern humans. In: Singer R, Lundy J, eds. Variation, culture and the evolution of of African populations. Johannesburg: Witwatersrand University Press, 1986.

Rightmire GP. The evolution of Homo erectus. New York: Cambridge University Press, 1990.

Rimel RW, Giordani B, Barth JT, et al. Disability caused by minor head injury. Neurosurgery 1981;9:221–228.

Rinck M, Glowalla U, Schneider K. Mood-congruent and mood-incongruent learning. Memory Cogn 1992;20:29–39.

Ring HA, Trimble MR, Costa DC, et al. Striatal dopamine receptor binding in epileptic psychoses. Biol Psychiatry 1994;35:375–380.

Ring K. Life at death. New York: Coward, McCann & Geoghegan, 1980.

Rinkel M, Greenblatt M, Coon GP, Solomon HC. Relations of the frontal lobe to autonomic nervous system in man. In: Greenblatt M, Arnot M, Solomon HC, eds. Studies in lobotomy. New York: Grune & Stratton, 1950.

Rinkel M, Solomon HC, Rosen D, Levine J. Lobotomy and urinary bladder. In: Greenblatt M, Arnot M, Solomon HC, eds. Studies in lobotomy. New York: Grune & Stratton, 1950.

Ritaccio AL, Hickling EJ, Ramani V. The role of dominant premotor cortex and grapheme to phoneme transformation in reading epilepsy. Arch Neurol 1992;49:933–939.

Ritchie JM. Pathophysiology of conduction in demyelinated nerve fibers. In: Morell P, ed. Myelin. 2nd ed. New York: Plenum, 1984.

Risse GL, Gazzaniga MS. Well-kept secrets of the right hemisphere: a carotid amytal study of restricted memory transfer. Neurology 1979;28:950–953.

Rizzo M, Hurtig R. Looking but not seeing. Neurology 1987;37:1642–1648.

Rizzo M, Robin DA. Simultagnosia. Neurology 1990;40:447–455.

Rizzolatti G, Scandolara C, Matelli, M, Gentillucci M. Afferent properties of periacuate neurons in macaque monkeys. 1, 2. Behav Brain Res 1981a;2:125–163.

Rizzolatti G, Scandolara C, Gentillucci, M, Camarda R.. Response properties and behavioral modulation of "mouth" neurons on the post-arcuate cortex (area 6) in macaque monkeys. Brain Res 1981b;225:421–424.

Rizzolatti G, Umilta C, Berlucchi G. Opposite superiorities of the right and left cerebral hemispheres in discriminative reaction time to physiognomical and alphabetical material. Brain 1971;94:431–442.

Roberts MH. 5-Hydroxytryptamine and antinociception. Neuropharmacology 1984;23:1529–1536.

Robertson PL, Allen JC, Abott IR, et al. Cervicomedullary tumors in children. Neurology 1994;44:1798–1803.

Robbins TW, Everitt BJ. Functions of dopamine in the dorsal and ventral striatum. Semin Neurosci 1992;4:119–127.

Robinson BW. Vocalizations evoked from forebrain in Macaca mulatta. Physiol Behav 1967;2:345–352.

Robinson BW. Anatomical and physiological contrasts between human and other primate vocalizations.

In: Washburn SL, Dolhinow P, eds. Perspect Hum Evol 1972;2:438–443.

Robinson BW, Mishkin M. Alimentary responses evoked from forebrain structures in Macaca mulatta. Science 1968;136:260–261.

Robinson D, Wu H, Munne RA, et al. Reduced caudate nucleus volume in obessive-compulsive disorder. Arch Gen Psychiatry 1995;52:393–398.

Robinson DL. Electrophysiological analysis of interhemispheric relations in the second somatosensory cortex of the cat. Exp Brain Res 1973;18:131–144.

Robinson DL, Goldberg ME, Stanton GB. Parietal association cortex in the primate: sensory mechanisms and behavioral modulation. J Neurophysiol 1978;41:910–932.

Robinson RG. Differential behavioral effects of right versus left hemispheric cerebral infarctions. Science 1979;205:707–710.

Robinson RG, Benson DF. Depression in aphasic patients: frequency, severity, and clinical-pathological correlations. Brain Lang 1981;14:282–291.

Robinson RG, Coyle JT. The differential effect of right vesus left hemisphere cerebral infarction on catecholamines and behavior in the rat. Brain Res 1980;188:63–78.

Robinson RG, Kubos KL, Starr LB, et al. Mood disorders in stroke patients. Brain 1984;107:81–93.

Robinson RR, Szetela B. Mood change following left hemisphere brain injury. Ann Neurol 1981;9:447–453.

Rochas-Miranda CE, Bender DB, Gross CG, Mishkin M. Visual activation of neruons in the inferiotemporal cortex depends on striate cortex and forebrain commissures. J Neurophysiol 1975;38:475–491.

Rodriguez G, Arvigo F, Marenco S, et al. Regional cerebral blood flow in essential hypertension. Stroke 1987;18:13–20.

Roediger H. Implicit memory: retention without rembering. Am Psychol 1990;45:1043–1056.

Roeltgen DP, Sevush S, Heilman KM. Pure Gerstmann's syndrome from a focal lesion. Arch Neurol 1983;40:46–47.

Roessmann U. Congenital malformations. In: Duckett S, ed. Pediatric neuropathology. Philadelphia: Williams & Wilkins, 1995.

Rogart RB, Ritchie JM. Pathophysiology of conduction in demyelinated nerve fibers. In: Morell P, ed. Myelin. New York: Plenum Press, 1977:144–288.

Roland PE. Astereognosis. Arch Neurol 1976;33:543–550.

Roland PE, Skinhoj E, Lassen NA, Larsen B. Different cortical areas in man in organiztion of voluntary movements in extrapersonal space. J Neurophysiol 1980;43:137–150.

Roland PE, Skinhoj E, Lassen NA. Focal activation of human cerebral cortex during auditory discrimination. J Neurophysiol 1981;45:1139–1150.

Rolfe SA, Day RH. Effects of similarity and dissimilarity between familiarization and test objects on recognition memory in infants following unimodal and bimodal familiarization. Child Dev 1981;52:1308–1312.

Rolland A. Middle Palaeolithic socioeconomic formations in Western Eurasia: an exploratory survey. In: Mellars P. The human revolution: behavioral and biological perspectives on the origins of modern humans, vol 2. Edinburgh: Edinburgh University Press, 1989.

Rolls ET. Neurons in the cortex of the temporal lobe and in the amygdala of the monkey with responses selective for faces. Hum Neurobiol 1984;3:209–222.

Rolls ET. Connections, functions, and dysfunctions of limbic structures, the prefrontal cortex and hypothalamus. In: Swash M, Kennard C, eds. The scientific basis of clinical neurology. London: Livingstone, 1985:201–213.

Rolls ET. Information representation, processing and storage in the brain: analysis at the single neuron level. In Changeuz J-P, Konishi M, eds. The neural and molecular bases of learning. New York: Wiley, 1987:503–540.

Rolls ET. Functions of neural networks in the hippocampus and neocortex in memory. In: Byrne JH, Berry WO, eds. Neural models of plasticity: theoretical and empirical approaches. New York: Academic Press, 1988.

Rolls ET. Functions of neural networks in the hippocampus and of backprojections in the cerebral cortex in memory. In: McGaugh JL, Winberger NM, Lynch G, eds. Brain organization and memory. New York: Oxford University Press, 1990a.

Rolls ET. Principles underlying the representation and storage of information in neuronal networks in the primate hippocampus and cerebral cortex. In: Zornetzer SF, Davis JL, Lau C, eds. An introduction to neural and electronic networks. San Diego: Academic Press, 1990b.

Rolls ET. Neurophysiology and functions of the primate amygdala. In: Aggleton JP, ed. The amygdala. New York: Wiley-Liss, 1992.

Rolls ET, Burton MJ, Mora F. Hypothalamic neuronal response associated with the sight of food. Brain Res 1976;111:53–66.

Rolls ET, Perret, D, Thorpe SJ, et al. Responses of neurons in area 7 of the parietal cortex to objects of different significance. Brain Res 1979;169:194–198.

Rolls ET, Williams GV. Neuronal activity in the ventral striatum of the primate. In: Carpenter MB, Jayaraman A, eds. The basal ganglia. New York: Plenum, 1987.

Romer AS. Vertebrate paleontology. Chicago: University of Chicago Press, 1966.

Roos Y, Hasan D, Vermeulen M. Outcome in patients with large intraventicular haemorrhages. J Neurol Neurosurg Psychiatry 1995;58:622–624.

Roozendaal B, Cools AR. Influence of the noradrenergic state of the nucleus accumbens in basolateral amygdala mediated changes in neophobia of rats. Behav Neurosci 1994;108:1107–1118.

Rose AS. Postoperative behavior. In: Greenblatt M, Arnot M, Solomon HC, eds. Studies in lobotomy. New York: Grune & Stratton, 1950.

Rose JD. Brainstem influences on sexual behavior. In: Klemm WR, Vertes RP, eds. Brainstem mechanisms of behavior. New York: Wiley, 1990.

Rose J, Valtonen S, Jennett B. Avoidable factors contributing to death after head injury. Br Med J 1977; 2:615–618.

Rose SA. Development changes in hemispheric specialization for tatual processing in very young children: evidence from cross-modal transfer. Dev Psychol 1984;20:568–574.

Rose SA, Gottfield AW, Bridger WH. Cross-modal transfer and information processing by the sense of touch in infancy. Dev Psychol 1981;17:90–98.

Rosen AD, Gur RC, Sussman N, et al. Hemispheric asymmetry in control of heart rate. Proc Soc Neurosci 1982;8:917.

Rosen JB, Hitchcock JM, Sananes CB, et al. A direct projection from the central nucleus of the amygdala to the acoustic startle pathway: anterograde and retrograde tracing studies. Behav Neurosci 1991;105:817–825.

Rosenbaum AH, Berry MJ. Positive therapeutic response to lithium in hypomania secondary to organic brain syndrome. Am J Psychiatry 1975;132:1072–1073.

Rosenbaum AH, Weaver W. Dissociated state. J Nerv Ment Dis 1980;168:597–603.

Rosenberg KR. The functional significance of Neanderthal public length. Curr Anthropol 1988;29:595–606.

Rosenblum LA. Sex differences in mother-infant attachment in monkeys. In: Friedman RD, Richart RM, Vande Wiele R, eds. Sex differences in behavior. New York: Wiley, 1974.

Rosenblum LA, Coplan JD, Friedman T, et al. Adverse early experiences affect noradrenergic and serotonergic functioning in adult primates. Biol Psychiatry 1994;35:221–227.

Rosenkilde CE. Functional heterogenety of the prefrontal cortex in the monkey: a review. Behav Neural Biol 1979;25:301–345.

Rosenthal PA, Rosenthal S. Suicidal behavior by preschool children. Am J Psychiatry 1984;141:520–525.

Rosenzweig MR. Effects of environment on development of brain and behavior. In: Tolbach E, Aronson LR, Shaw E, eds. The biopsychology of development. New York: Academic Press, 1971.

Rosenzweig MR, et al. Chemical and antomical plasticity of the brain. In: Gaito J, ed. Macromolecules and behavior. New York: Appleton, 1972.

Rosin AJ. Reactions of families of brain-injured patients who remain in a vegetative state. Scand J Rehabil Med 1977;9:1–5.

Ross ED. The aprosodias: functional-anatomic organization of the affective components of language in the right hemisphere. Arch Neurol 1981;38:561–589.

Ross ED, Jossman PB, Bell B, et al. Musical hallucinations in deafness. JAMA 1975;231:620–622.

Ross ED, Mesulam MM. Dominant language functions of the right hemisphere? Prosody and emotional gesturing. Arch Neurol 1979;36:144–148.

Ross RR, McKay HB. Self-mutilation. Lexington, MA: Lexington Books, 1979.

Rossi A. Gender and parenthood. In: Rossi A, ed. Gender and the life curse. Aldine: Hawthorne, 1985.

Rossi A, Stratta P, D'Albenzio, et al. Reduced temporal lobe areas in schizophrenia. Biol Psychiatry 1990; 27:61–68.

Rossi A, Stratt P, di Michele V, et al. Temporal lobe structure by magnetic resonance in bipolar affective disorders and schizophrenia. J Affect Disord 1991;21:19–22.

Rossi GF, Brodal A. Corticofugal fibers to the brain stem reticular formation. J Anat 1956;90:42–62.

Rossi GF, Rosadini G. Experimental analysis of cerebral dominance in man. In: Millikan CH, Darley FL, eds. Brain mechanisms underling speech and language. New York: Grune & Stratton, 1967.

Rosvold HE, Mirsky AF, Pribram KH. Influences of amygdalectomy on social behavior in monkeys. J Comp Physiol Psychol 1954;47:173–178.

Roth D, Rehm LP. Relationships among self-monitoring processes, memory and depression. Cogn Ther Res 1980;4:149–159.

Roth M. Disorders of the body image caused by lesions of the right parietal lobe. Brain 1949; 72:89–111.

Roth N. Unusual types of anosognosia and their relation to the body-image. J Nerv Ment Dis 1944; 100:35–43.

Rothi LJG, Mack L, Heilman KM. Pantomime agnosia. J Neurol Neurosurg Psychiatry 1986;49:451–454.

Rothi LJG, Ochipa C, Heilman KM. A cognitive neuropsychological modeal of limb praxis. Cogn Neuropsychol 1991;8:443–458.

Rothwell JC, Thompson PD, Day BL, et al. Motor cortex stimulation in intact man. Brain 1987;110:1173–1190.

Routtenberg A. The two arousal hypothesis: reticular formation and limbic system. Psychol Rev 1968; 75:51–80.

Rovee-Collier C, Borza MA, Adler SA, Bolle K. Infants eyewitness testimony: effects of postevent information on a prior memory representation. Memory Cogn 1993;21:267–279.

Rovee-Collier C, Hayne H. Reactivation of infant memory: implications for cognitive development. Adv Child Dev Behav 1987;20:185–238.

Rovee-Collier CK, Shyi G. A functional and and cognitive analysis of infant long-term retention. In: Howe ML, Brainer CJ, Reyna VF, eds. Development of long-term retention. New York: Springer-Verlag, 1992.

Rovee-Collier CK, Sullivan MW. Organization of infant memory. J Exp Psychol Hum Learn Mem 1980;6:798–807.

Rowe MJ, Carlson C. Brainstem auditory evoked potentials in post-concussion dizziness. Arch Neurol 1980;37:679–683.

Rowell TE, et al. The social development of baboons in their first three months. J Zool 1968;155: 461–483.

Rowell TE. On the significance of the concept of the harem when applied to animals. In: Schubert G, Masers RD. Primate politics. Carbondale: University Press, 1991.

Rowley JD. Human oncogene locations and chromosome aberrations. Nature 1983;301:290–291.

Royce GJ. Recent research on the centromedian and parafascicular nuclei. In: Carpenter MB, Jayaraman A, eds. The basal ganglia. New York: Plenum, 1987.

Rozanski J, Rosen H. Musical hallucinosis in otosclerosis. Cinfina Neurol 1952;12:49–54.

Rozin P. Psychobiological approach to human memory. In: Rosenzweig MR, Bennet EL, eds. Natural mechanisms of learning and memory. Cambridge, MA: MIT Press, 1976.

Rubens AB. Agnosia. In: Heilman KM, Valenstein E, eds. Clinical neuropsychology. New York: Oxford Univesity Press, 1993.

Rubens AB, Benson DF. Associative visual agnosia. Arch Neurol 1971;24:305–316.

Rubens AB, Garrett MF. Anosognosia of linguistic deficits in patients with neurological deficits. In: Prigatano GP, Schacter DL, eds. Awareness of deficit after brain injury. New York: Oxford University Press, 1991.

Rubenstein R, Newman R. The living out the "future" experiences under hypnosis. Science 1954;119: 472–473.

Rubin DC, Kozin M. Vivid memories. Cognition 1984; 16:81–95.

Rubin L. Intimate strangers. New York: Harper & Row, 1983.

Rubin RT, Phillips JJ, Sadow Tf, McCracken JT. Adrenal gland volume in major depression. Arch Gen Psychiatry 1995;52:213–218.

Ruch TC, Shenkin HA. The relationship of area 13 on the orbital surface of the frontal lobes to hyperactivity and hyperphagia. J Neurophysiol 1943;6: 349–360.

Rudy JW, Morledge P. Ontogeny of contextual fear conditioning in rats. Behav Neurosci 1994;108: 227–234.

Rudy L, Goodman GS. Effects of participation on children's reports: implications for children's testimony. Dev Psychol 1991;27:527–538.

Ruff HA. Infant recognition of the invariant form of objects. Child Dev 1978;49:293–306.

Ruff RL. Orgasmic epilepsy. Neurology 1980;30: 1252–1253.

Rumsey JM, Andreason, P, Zametkin AJ, et al. Failure to activate the left temporoparietal cortex in dyslexia. Arch Neurol 1992;49:527–534.

Russchen FT. Amygdalopetal projection in the cat. J Comp Neurol 1982;207:157–176.

Russchen FT, Smeets WJAJ, Lohman AHM. On the

basal ganglia of a reptile. In: Carpenter MB, Jayaraman A, eds. The basal ganglia. New York: Plenum, 1987.

Russell DEH. The secret trauma: incest in the lives of girls and women. New York: Basic Books, 1986.

Russell JA, Bullock M. On the dimensions preschoolers use to interpret facial expressions of emotion. Dev Psychol 1986;22:97–102.

Russell PN, Beekuis ME. Organization of memory. J Abnormal Psychol 1976;42:927–934.

Russell WR. Transient disturbances following gunshot wounds of the head. Brain 1945;68:79–97.

Russell WR. The traumatic amnesias. London: Oxford Univesity Press, 1971.

Russell WR, Whitty WM. Studies in traumatic epilepsy. J Neurol Neurosurg Psychiatry 1955;18:79–96.

Russo M, Vignolo L. Visual figure-ground discrimination in patients with unilateral cerbral disease. Cortex 1967;3:111–127.

Rutter D. The reconstruction of schizphrenic speech. Br J Psychiatry 1979;134:356–359.

Rylander G. Personality changes after operation on the frontal lobes. A clinical study of 32 cases. Acta Psychiatr Neurol (suppl) 1939;20:3–327.

Rylander G. Personality analysis before and after frontal lobotomy. Research Publication of the Association of Nervous and Mental Disease 1948;27: 691–700.

Saarni C. An observational study of chidlren's attempts to monitor their expressive behavior. Child Dev 1984;55:1504–1513.

Sabom MB. Recollections on death. New York: Harper & Row, 1982.

Sachar EJ, Hellman L, Roffwarg H, et al. Disrupted 24-hour patterns of cortisol secretion in psychotic depression. Arch Gen Psychiatry 1973;28:19–24.

Sacher GA, Staffeldt EF. Relation of gestation time to brain weight for placental mammals: implications for the theory of vertebrate growth. Am Naturalist 1974;108:593–615.

Sackheim HA, Gur RC. Lateral asymmetry in intensity of emotional expression. Neuropsychologia 1978; 16:473–481.

Sackheim HA, Gur RC, Saucy MC. Emotions are expressed more intensely on the left side of the face. Science 1978;202:424–435.

Sackheim HA, Weinman AL, Gur RC, et al. Pathological laughing and crying: functional brain asymmetry in the expression of positive and negative emotions. Arch Neurol 1982;39:210–218.

Sackett GP. Monkeys reared in isolation with pictures of visual input. Evidence for an innate releasing mechanism. Science 1966;154:1468–1473.

Sackett GP. Sex differences in rhesus monkeys following varied rearing experiences. In: Friedman RD et al, eds. Sex differences in behavior. New York: Wiley, 1974.

Safer M. Sex and hemisphere differences in access to codes for processing emotional expression and faces. J Exp Psychol Gen 1981;110:86–100.

Safer M, Leventhal H. Ear differences in evaluating

emotional tones and verbal content. J Exp Psychol Hum Percept Perform 1977;3:75–82.

Saghir GB, Robins E. Male and female homosexuality. Baltimore: Williams & Wilkins, 1973.

Sakaguchi T, Nakamura S. The mode of projections of single locus coeruleus neurons to the cerebral cortex in rats. Neuroscience 1987;20:221–230.

Sakata H, Iwamura Y. Cortical procssing of tactile information in the first somatosensory and parietal association areas in the monkey. In: Gordon G, ed. Active touch. Oxford: Pergamon, 1978.

Sakata H, Shibutani H, Kawano K. Spatial properties of visual fixation neurons in posterior parietal association cortex of the monkey. J Neurophysiol 1980;43:1654–1672.

Salama AA. Multiple personality. A case study. Can J Psychiatry 1980;25:569–572.

Salamy A. Commissural transmission: maturational changes in humans. Science 1978;200:1409–1411.

Salapatek P, Banks MS. Infant visual perception. In: Haith M, Campos J, eds. Infancy and developmental psychobiology. New York: Wiley, 1983.

Salcman M. Tumors of the pituitary gland. In: Moossa AR, Robson MC, Schimpff SC, eds. Oncology. Los Angeles: Williams & Wilkins, 1986.

Salcman M, Kaplan RS. Intracranial tumors in adults. In: Moossa AR, Robson MC, Schimpff SC, eds. Oncology. Los Angeles: Williams & Wilkins, 1986.

Salcman M, Kaplan RS, Ducker TB, et al. Effects of age and reoperation on survival in the combined modality treatment of malignant astrocytoma. Neurosurgery 1982;10:454–463.

Salovey P, Singer JA. Mood congruency effect in recall of childhood versus recent memories. In: Kuiken D, ed. Mood and memory. Newbury Park: Sage, 1991.

Sallanon M, Janin M, Buda, C, Jouvet M. Serotoninergic mechanisms and sleep rebound. Brain Res 1983;268:95–104.

Salzinger, S, Feldman RS, Hammar M, Rosario M. The effects of physical abuse on children's social relationships. Child Dev 1993;64:169–187.

Sameroff AJ, Cavanagh PJ. Learning in infancy. In: Osofsky JD, ed. Handbook of infant development. New York: Wiley, 1979.

Samson S, Zatorre RJ. Melodic and harmonic discrimination following unilateral cerebral excision. Brain Cogn 1988;7:348–360.

Samson S, Zatorre RJ. Learning and retention of melodic and verbal information after unilateral temporal lobectomy. Neuropsychologia 1992;30:815–826.

Samuels JA, Benson DF. Some aspects of language comprehension in anterior aphasia. Brain Lang 1979;8:275–286.

Sancar A. Mechanisms of DNA excision repair. Science 1994;266:1954–1956.

Sanchez-Longo LP, Forster FM. Clinical significance of impairment of sound localization. Neurology 1958;8:119–125.

Sanders B, et al. The sex differences on one test of spatial visualization. Child Development. 1982;53:106–110.

Sanders G, Ross-Field L. Sexual orientation, cognitive abilities, and cerebral asymmetry, a review and a hypothesis tested. Ital J Zool 1986;20:459–470.

Sandifer PH. Anosognosia and disorders of body scheme. Brain 1946;69:122–137.

Sanes JN, Dimitrov B, Hallett M. Motor learning in patients with cerebellar dysfunction. Brain 1990;113:103–120.

Sandson J, Albert ML. Perseveration in behavioral neurology. Neurology 1987;37:1736–1741.

Sanguet J, Benton AL, Hecaen H. Disturbances of the body schema in relation to language impairment and hemispheric locus of lesion. J Neurol Neurosurg Psychiatry 1971;34:496–501.

Sanides F. The cyto-myeloarchitecture of the human frontal lobe and its relation to phylogenetic differentiation of the cerebral cortex. J Himforsch 1964;6:269–282.

Sanides F. Comparative architectonics of the neocortex of mammals and their evolutionary interpretation. Ann N Y Acad Sci 1969;167:404–423.

Sanides F. Functional architecture of motor and sensory cortices in primates in light of a new concept of neocortex evolution. In: Noback C, Montagna W, eds. The primate brain. New York: Appleton, 1970.

Sanides F. Representation in cerebral cortex. In: Bourne GH, ed. The structure and function of nervous tissue, vol 5. New York: Academic Press, 1972:329–453.

Sapir E. Culture, language and personality. Berkeley: University of California Press, 1966.

Sapiro V. Women in American society. Mountain View: Mayfield, 1990.

Sarich V. Primate systematics. In: Napier JR, Napier PH, eds. Old Word Monkeys. New York: Academic Press, 1970.

Sarnat HB, Netsky MG. Evolution of the nervous system. New York: Oxford University Press, 1981.

Sarter M, Markowitsch EK. The amygdala's role in human mnemonic processing. Cortex 1985;21:7–24.

Sartre, J-P. Being and nothingness. New York: Philosophical Library, 1956.

Sasaki K. Cerebro-cerebellar interconnections in cats and monkeys. In: Massion J, Sasaki K, eds. Cerebro-cerebellar interactions. Amsterdam: Elsevier, 1979.

Sattel JW. Men, inexpressiveness and power. In: Richardson L, Taylor V, eds. Feminist frontiers II. New York: Random House, 1989.

Sauerland EK, Nakamura Y, Clemente CD. The role of the lower brain stem in cortically induced inhibition of somatic reflexes in the cat. Brain Res 1967;6:164–180.

Savage GE. The fish telencephalon and its relation to learning. In: Ebbesson SOE, ed. Comparative neurology of the telencephalon. New York: Plenum Press, 1980.

Sawa M, Delgado JMR. Amygdala unitary activity in

the unrestrained cat. Electroencephalogr Clin Neurophysiol 1963;15:637–650.

Sawa M, Ueki Y, Arita M, Harada T. Preliminary report on the amygdalectomy on the psychotic patients with interpretation or oral-emotional manifestations in schizphrenics. Folia Psychiatr Neurol Japon 1954;7:309–329.

Sawyer CH, Hilliard J, Ban T. Autonomic and EEG responses to cerebellar stimulation in rabbits. Am J Physiol 1961;200:405–412.

Saywitz KI, Goodman GS, Nicholas E, Moan SF. Children's memory of a physical examination involving genital touch: implications for reports of child sexual abuse. J Consult Clin Psychol 1991;59: 682–691.

Schachtel E. On memroy and childhood amnesia. Psychiatry 1947;10:1–26.

Schacter DL, Crovitz HF. Memory function after closed head injury: a review of the quantitative research. Cortex 1977;13:150–176.

Schacter DL, Harbluk JL, McLachlan DR. Retrieval without recollection: an experimental analysis of source amnesia. J Verbal Learn Verbal Behav 1984; 23:593–611.

Schacter DL, Moscovitch M. Infants, amnesics, and dissociable memory systems. In: Moscovitch M, ed. Infant memory. New York: Plenum, 1984.

Schacter DL, Wang PL, Tulving E, Freedman M. Functional retrograde amnesia: a quantitative case study. Neuropsychologia 1982;20:523–532.

Schaffer HR. The onset of fear of strangers and the incongruity hypothesis. J Child Psychol Psychiatry 1966;7:95–106.

Schaller GB. The mountain gorilla. Chicago: University of Chicago Press, 1963.

Schanfald D, Pearlman C, Greenberg R. The capacity of stroke patients to report dreams. Cortex 1985; 21:237–247.

Scharwz DWF, Tomlinson RWW. Spectral response patterns of auditory cortex neurons to harmonic complex tones in alert monkey (Macaca mulatta). J Neurophysiol 1990;64:282–298.

Schatzberg A, Westfall M, Blumeti A, Birk C. Effeminacy. Arch Sex Behav 1975;4:31–41.

Scheibel A. Anatomical and physiological substates of arousal. In: Hobson JA, Brazier MAB, eds. The reticular formation revisited. New York: Raven Press, 1980.

Scheibel AB. Some structural and developmental correlates of human speech. In: Gibson KR, Petersen AC, eds. Brain maturation and cognitive development. New York: De Gruyter, 1991.

Scheibel ME, Scheibel AB. Patterns of organization in specific and nonspecific thalamic fields. In: Papura DP, Yahr MD, eds. The thalamus. New York: Columbia University Press, 1966:13–46.

Schenck CH, Mahowald MW. Motor dyscontrol in narcolepsy. Ann Neurol 1992;32:3–10.

Schenk L, Bear D. Multiple personality and related dissociative phenomenon in patients with temporal lobe epilepsy. Am J Psychiatry 1981;138:1311–1316.

Schenkenberg PH, Dustman RE, Beck EC. Changes in evoked responses related to age, hemisphere and sex. Electroencephalogr Clin Neurophysiol 1971; 30:163.

Schievink WI, Wijdicks, E, Parisi JE, et al. Sudden death from aneurysmal subarachnoid hemorrage. Neurology 1995;45:871–874.

Schiff BB, Lamon M. Inducing emotion by unilateral contraction of facial muscles: a new look at hemispheric specialization and the experience of emotion. Neuropsychologia 1989;27:923–935.

Schiff HB, Sabin TD, Geller A, et al. Lithium in aggressive behaivor. Am J Psychiatry 1982;139:1346–1348.

Schilder P. The image and appearance of the human body. London: Routledge, Kegan Paul, 1935.

Schilder P. On the development of thoughts. In: Rappaport D, ed. Organization and pathology of thought. New York: Columbia University Press, 1951.

Schiller P, Stryker M. Single unit recording and stimulation in superior colliculus of the alert rhesus monkey. J Neurophysiol 1972;35:915–924.

Schneider GE. Two visual systems. Science 1969;163: 895–902.

Schneider GE. Is it really better to have your brain lesion early? A revision of the "Kennard principle." Neuropsychologia 1979;17:557–583.

Schneider JS. Basal ganglia role in behavior: importance of sensory gating and its relevance to psychiatry. Biol Psychiatry 1984;19:1693–1710.

Schneider JS, Lidsky TI. Processing of somatosensory information in striatum of behaving cats. J Neurophysiol 1981;45:841–851.

Schneider W. Varieties of working memory as seen in biology and in connectionist/control architectures. Memory Cogn 1993;21:184–192.

Schnider A, Benson DF, Alexander DN. Schnider-Klaus A. Non-verbal environmental sound recognition after unilateral hemispheric stroke. Brain 1994;117:281–287.

Schober R, Herman MM. Neuropathology in cardiac transplantation: survey of 31 cases. Lancet 1973; 1:962–967.

Schopf JW. The evolution of the earliest cells. Sci Am 1978;239:84–100.

Schopf JW. The oldest fossils and what they mean. In: Schopf JW, ed. Major events in the history of life. New York: Bartlett, 1992.

Schrandt NJ, Tranel D, Damasio H. The effect of focal cerebral lesion on skin conductance responses to "signal" stimuli. Neurology 1989;39:223.

Schreiber F. Sybil. New York: Warner, 1974.

Schriner H, Kling A. Behavioral changes following rhinencephalic injury in the cat. J Neurophysiol 1953;16:643–659.

Schreiner L, Kling A. Rhinencephalon and behavior. Am J Physiol 1956;184:486–490.

Schubert G, Masters RD. Primate politics. Carbondale, IL: Southern Illinois University Press, 1991.

Schutze I, Knuepfer MM, Eismann A, et al. Sensory input to single neurons in the amygdala of the cat. Exp Neurol 1987;97:499–515.

Schwarcz A, et al. ESR dates for the hominid burial site of Qafzeh. J Hum Evol 1988;17:733–737.

Schwartz GE. Psychobiology of repression and health: a systems approach. In: Singer JL, ed. Repression and dissociation. Chicago: University of Chicago Press, 1990.

Schwartz GE, Davidson RJ, Maer F. Right hemisphere lateralization for emotion in the human brain: interaction with cognition. Science 1975;190:286–288.

Schwartz KA, Cohen JA. Subaranoid hemorrhage precipitated by cocaine snorting. Arch Neurol 1984; 41:705.

Schwarz DWF, Deecke L, Fredrickson JM. Cortical projection of group 1 muscle afferents to areas 2, 3a, and the vestibular fields in the rhesus monkey. Exp Brain Res 1973;17:516–526.

Schwarz DWF, Tomlinson RWW. Spectral response patterns of auditory cortex neurons to harmonic complex tones in alert monkey. J Neurophysiol 1990;64:282–298.

Scoville WB, Milner B. Loss of recent memory after bilateral hippocampal lesions. J Neurol Neurosurg Psychiatry 1957;20:11–21.

Scrivner E, Safer SA. Eyewitnesses show hypermnesia for details about a violent event. J Appl Psychol 1988;73:371–377.

Sears RR. Sex and behavior. In: Beach FA, ed. Sex and behavior. New York: Wiley, 1965.

Segal M. The action of serotonin in the rat hippocampus. J Physiol 1980;303:375–390.

Segalowitz SJ, Plantery P. Music draws attention to the left and speech draws attention to the right. Brain Cogn 1985;4:1–6.

Segawa M, Nomura Y, Hikosaka O, et al. Roles of the basal ganglia and related structure in symptoms of dystonia. In: Carpenter MB, Jayaraman A, eds. The basal ganglia. New York: Plenum, 1987.

Segraves MA, Goldberg ME. Functional properties of corticotectual neurons in the monkey's frontal eye field. J Neurophysiol 1987;58:1387–1419.

Sejnowski TJ, Tesauro G. Building network learning algorithms from Hebbian synapses. In: McGaugh JL, Winberger NM, Lynch G, eds. Brain organization and memory. New York: Oxford University Press, 1990.

Seldon NRW, Everitt BJ, Jarrard LE, Robbins TW. Complementary roles for the amygdala and hippocampus in aversive conditioning to explicit and contextual cues. Neuroscience 1991;42: 335–350.

Selemon LD, Goldman-Rakic PS, Tamminga CA. Prefrontal cortex. Am J Psychiatry 1995;152:5.

Selnes OA. The corpus callosum: some anatomical and functional considerations with special reference to language. Brain Lang 1974;1:111–139.

Seltzer B, Pandya DN. Afferent cortical connections and architectonics of the superior temporal sulcus and surround cortex in the rhesus monkey. Brain Res 1978;149:1–24.

Semmes J. A non-tactual factor in astereognosis. Neuropsychologia 1965;3:295–315.

Semmes J. Somesthetic effects of damage to the central nervous system. In: Iggo A, ed. Handboook of sensory physiology. Berlin: Spinger, 1973.

Semmes J, Turner B. Effects of cortical lesions on somatosensory tasks. J Invest Dermatol 1977;69: 181–189.

Semmes J, Weinstein S, Ghent L, Teuber HL. Somatosensory changes after penetrating head wounds in man. Cambridge, MA: Harvard University Press, 1960.

Semrad EV. Study of the auditory apparatus in patients experiencing auditory hallucinations. Am J Psychiatry 1938;95:53–63.

Serafetinides EA. The significance of the temporal lobes and of hemisphere dominance in the production of the LSD-25 symptomology in man. Neuropsychologia 1965;3:69–79.

Sereno MI, Dale AM, Reppas JB, et al. Borders of multiple visual areas in humans revealed by functional magnetic resonance imaging. Science 1995;268: 889–892.

Sergent J. The neuropsychology of visual image generation. Brain Cogn 1990;13:98–129.

Sergent J, Ohta S, Macdonald B. Functional neuroanatomy of face and object processing. Brain 1992;115:15–36.

Sethi PK, Raio ST. Gelastic, quiritarian, and cursive epilepsy. J Neurol Neurosurg Psychiatry 1976; 39:823–828.

Seyfarth RM, Cheney DL, Marler P. Monkey responses to three different alarm calls. Evidence of predator classification and semantic communication. Science 1980;210:801–803.

Seymour SE, Reuter-Lorenz PA, Gazzaniga MS. The disconnection syndrome. Brain 1994;117:105–115.

Shagass CM, Roemer RA, Straumanis JJ, Amadeo M. Evoked potential evidence of lateralized hemispheric dysfunction in psychosis. In: Gruzelier J, Flor-Henry P, eds. Hemisphere asymmetries of function in psychopathology. New York: Elsevier, 1979.

Shankweiler D. Performance of brain-damaged patients on two tests of sound localization. J Comp Physiol Psychol 1961;54:375–381.

Shankweiler D. Effects of temporal lobe damage on the perception of dichotically presented melodies. J Comp Physiol Psychol 1966;62:115–122.

Shankweiler D, Studdert-Kennedy M. Lateral differences in perception of dichotically presented synthetic consonant-vowel syllables and steady-state vowels. J Acoust Soc Am 1966;39:1256A.

Shankweiler D, Studdert-Kennedy M. Identification of consonants and vowels presented to left and right ears. Q J Exp Psychol 1967;19:59–63.

Shantz CU. Social cognition. In: Flavell JH, Markman EM, eds. Cognitive development. New York: Wiley, 1983.

Shapiro BE, Alexander MP, Gardner H, Mercer S. Mechanisms of confabulations. Neurology 1981; 31:1070–1076.

Shapiro BE, Danly M. The role of the right hemisphere in the control of speech prosody in propositional and affective contexts. Brain Lang 1985;1: 111–139.

Shapiro JR. Biology of gliomas: heterogeniety, oncogenes, growth factors. Semin Oncol 1986a;13:4–15.

Shapiro JR, Yung W-KA, Shapiro WR. Isolation, karyotype and clonal growth of heterogenous subpopulations of human malignant gliomas. Cancer Res 1981;41:2349–2359.

Shapiro WR. Therapy of adult malignant brain tumors. Semin Oncol 1986;13:38–45.

Shapiro WR, Shapiro JR. Principles of brain tumor chemotherapy. Semin Oncol 1986;13:56–69.

Shelton PA, Bowers D, Duara R, Heilman KM. Apperceptive visual agnosia: a case study. Brain Cogn 1994;25:1–23.

Shatz CJ, Chun JJM, Luskin MB. The role of the subplate in the development of the mammalian telencephalon. In: Peters A, Jones EG, eds. Cerbral cortex, vol 7. New York: Plenum, 1988.

Shaw PJ, Bates D, Cartlidge NEF, et al. Neurologic and neuropsychological morbidity following major surgery. Stroke 1987;18:700–707.

Shea JJ. A functional study of the lithic industries associated with hominid fossils in the Kebara and Qafzeh caves, Israel. In: Mrllars P, Stringer CB, eds. The human revolution: behavioral and biological perspectives on the origins of modern humans, vol 1. Edinburgh: Edinburgh University Press, 1989.

Shealy C, Peel J. Studies of amygdaloid nucleus of cat. J Neurophysiol 1957;20:125–139.

Sheingold K, Tenney YJ. Memory for a salient childhood event. In: Neisser U, ed. Memory observed. San Francisco: Freeman, 1982.

Shelton PA, Bowers D, Duara R, Heilman KM. Apperceptive visual agnosia: a case study. Brain Cogn 1994;25:1–23.

Shelton PA, Bowers D, Heilman KM. Peripersonal and vertical neglect. Brain 1990;113:191–205.

Shennum W, Begental D. The develoment of control over affective expression in nonverbal behavior. In: Feldman RS, ed. Development of nonverbal behavior in children. New York: Springer-Verlag, 1982.

Sherbet GV. The biology of tumour malignancy. New York: Academic Press, 1982.

Sherwin I. Clinical and EEG aspects of temporal lobe epilepsy with behavior disorder. McLean Hospital J 1977;June:40–50.

Sherwin I. Psychosis associated with epilepsy. J Neurol Neurosurg Psychiatry 1981;44:83–85.

Sherwin I, Peron-Magnana P, Bancard J, et al. Prevalence of psychosis in epilepsy as a fucntion of the laterality of the epileptogenic lesion. Arch Neurol 1982;39:621–625.

Shibasaki H, Sadato N, Lyshkow H, et al. Both primary motor cortex and supplementary motor area play an important role in complex finger movement. Brain 1993;116:1298–1387.

Shields WM, Shields LM. Forcible rape. An evolutionary perspective. Ethnol Sociobiol 1983;4:115–136.

Shiffrin RN, Schneider W. Controlled and automatic

human information processing: II. Perceptual learning, automatic attending, and a general theory. Psychol Rev 1977;84:129–190.

Shore WH. Mysteries of life and the universe. New Jersey: Harcourt Brace, 1994.

Shores TJ, Seib TB, Levine S, et al. Inescapable versus escapable shock modulates long-term potentiation in the rat hippocampus. Science 1989;244:224–226.

Shuping JR, Rollinson R, Toole JF. Transient global amnesia. Ann Neurol 1980;7:281–285.

Shurley J. Profound experimental sensory isolation. Am J Psychiatry 1960;117:539–545.

Sidman RL, Rakic P. Neural migration with special reference to developing human brain. A review. Brain Res 1973;62:1–87.

Sidman RL, Rakic P. Development of the human nervous system. In: Haymaker W, Adams RD, eds. Histology and histopathology of the central nervous system. Springfield, IL: Charles C Thomas, 1982.

Siegel A, Edinger H. Organization of the hippocampal-septal axis. In: DeFrance JF, ed. The septal nuclei. New York: Plenum Press, 1976.

Siegel A, Fukushima T, Meibach R, Burke L, et al. The origin of the afferent supply to the mediodorsal thalamic nucleus: enhanced HRP transport by selective lesions. Brain Res 1977;135:11–23.

Siegel A, Skog D. Effects of electrical stimulation of the septum upon attack behavior elicited from the hypothalamus in the cat. Brain Res 1970;23: 371–380.

Siegel JM, Tomaszewski KS, Nienhus R. Behavioral states in the chronic medullary and midpontine cat. Electroencephalogr Clin Neurophysiol 1986; 63:274–288.

Siegel J, Wang RY. Electroencephalographic, behavioral, and single unit activity produced by stimulation of forebrain inhibitory structures in cats. Exp Neurol 1974;42:28–50.

Silberman EK, Weingartner H. Hemispheric lateralization of functions related to emotion. Brain Cogn 1986;5:322–353.

Silverberg R, Bentin S, Gaziel T, et al. Shift of visual field preference for English words in native Hebrew speakers. Brain Lang 1979;8:184–190.

Silveri MC, Leggio MG, Molinari M. The cerebellum contributes to linguistic production. Neurology 1994;44:2047–2050.

Simmons T, Smith FH. Human population relationships in the Late Pleistocence. Curr Anthropol 1991;32:623–627.

Sinden JD, Marsden KM, Hodges H. Neural transplantation and recovery of function. In: Rose FD, Johnson DA, eds. Recovery from brain damage. New York: Plenum, 1992.

Singer B, Wymer J. The Middle Stone Age at Klasies River Mouth in South Africa. Chicago: University of Chicago Press, 1982.

Singer J. Repression and dissociation. Chicago: University of Chicago Press, 1990.

Singer JA, Salovey P. Mood and memory: evaluating

the network theory of affect. Clin Psychol Rev 1988;8:211–251.

Singer W. Ontogenetic self-organization and learning. In: McGaugh JL, Winberger NM, Lynch G, eds. Brain organization and memory. New York: Oxford University Press, 1990.

Sinyour D, Jacques P, Kaloupek DG, et al. Poststroke depression and lesion location. Brain 1986;109:537–546.

Sirois F. Perspectives on epidemic hysteria. In: Colligan M, Pennebaker JW, Murphy LR, eds. Mass psychogenic illness. Hillsdale, NJ: Erlbaum, 1982.

Sisler G, Penner H. Amnesia following severe head injury. Can Psychiatr Assoc J 1975;20:333–336.

Sitchin Z. The lost realms. New York: Avon, 1990.

Skelton RR, McHenry HM. Evolutionary relationships among early hominids. J Hum Evol 1992;23:303–349.

Skinner JE, Lindsley DB. Enhancement of visual and auditory evoked potentials during blockage of the non-specific thalamo-cortical system. Electroencephalogr Clin Neurophysiol 1971;31:1–6.

Skinner JE, Yingling CD. Central gating mechanisms that regulate event related potentials and behavior. In: Desmedt J, ed. Attention, voluntary contraction, and event related cerebral potentials. Basel, Switzerland: S. Karger, 1977:30–69.

Skinner RD, Garcia-Rill E. Brainstem modulation of rhythmic functions and behaviors. In: Klemm WR, Vertes RP, eds. Brainstem mechanisms of behavior. New York: Wiley, 1990.

Slagle DA. Psychiatric disorders following closed head injury. Int J Psychiatry Med 1990;20:1–35.

Slater E, Beard AW. The schizophrenia-like psychoses of epilepsy. Br J Psychiatry 1963;109:95–112.

Slater E, Beard AW, Glithero E. The schizophrenia-like psychosis of epilepsy, I, II, V. Br J Psychiatry 1963;109:95–112, 130–133, 143–150.

Slobin D. Psycholinguistics. New Jersey: Scott Foresman, 1971.

Small GW, Propper MW, Randolph ET, Eth S. Mass hysteria among student performers. Am J Psychiatry 1991;148:1200–1205.

Smart N. The religious experience of mankind. Chicago: University of Chicago Press, 1969.

Smeets WJAJ. The telencephalon of cartilaginous fishes. In: Jones ED, Peters A, eds. Cerebral cortex, vol 8: Comparative structure and evolution of cerebral cortex. New York: Plenum, 1990.

Smirnov YA. On the evidence for Neanderthal burial. Curr Anthropol 1989;30:324.

Smith A. Intellectual functions in patients with lateralized frontal tumors. J Neurol Neurosurg Psychiatry 1966a;29:52–59.

Smith A. Speech and other functions after left (dominant) hemispherectomy. J Neurol Neurosurg Psychiatry 1966b;29:467–471.

Smith A, Burklund CW. Dominant hemispherectomy. Science 1966;153:1280–1282.

Smith BS, Ratner HH, Hobart CJ. The role of cuing and organization in children's memory for events. J Exp Child Psychol 1987;44:1–24.

Smith E. Influence of site of impact on cognitive impairment persisting long after severe closed head injury. J Neurol Neurosurg Psychiatry 1974;37:719–726.

Smith FH. Upper pleistocene hominid evolution in south-central Europe: a review of the evidence and analysis of trends. Curr Anthropol 1982;23:667–703.

Smith FH. The role of continuity in modern human origins. In Brauer G, Smith FH, eds. Continuity or replacement. Rotterdam: Balkema, 1992.

Smith OA, DeVito JL, Astley CA. Neurons controlling cardiovascular responses to emotion are located in lateral hypothalamus-perifornical region. Am J Physiol 1990;259:943–954.

Smith PH, Arehart DM, Haaf RA, deSaintVictor CM. Expectancies and memory for spatiotemporal events in 5 month old infants. J Exp Child Psychol 1989;5:136–150.

Smith PLC, Treasure, T, Newman SP, Joseph P, et al. Cerebral consequences of cardiopulmonary bypass. Lancet 1986;1:823–825.

Smith SM. Environmental context effects on memory. In: Davies GM, Thomson DM, eds. Memory in context: context in memory. New York: Wiley, 1988.

Smith SM, Glenberg A, Bjork RA. Environmental context and human memory. Memory Cogn 1978;6:342–353.

Smith WK. The results of ablation of the cingular region of the cerebral cortex. Fed Proc 1944;3:42–55.

Snider RS, Maiti M. Septal after-discharge and their modification by the cerebellum. Exp Neurol 1975;49:529–539.

Snow RB, Zimmerman RD, Gandy SE, et al. Comparison of magnetic resonance imaging and computed tomograph in the evaluation of head injury. J Neurosurg 1986;18:45–52.

Snyder M, Diamond IT. The organization and function of the visual cortex in the tree shrew. Brain Behav Evol 1968;1:244–288.

Snyder M, White P. Moods and memories: elation, depression and the remembering of the events of one's life. J Personal 1982;42:221–238.

Sokoloff L. Circulation and energy metabolism in the brain. In: Siegel GJ et al, eds. Basic neurochemistry. New York: John Wiley & Sons, 1981:471–495.

Solecki R. Shanidar: the first flower people. New York: Knopf, 1971.

Soloman D, Ali F. Age trends in the perception of verbal reinforcement. Dev Psychol 1972;7:238–243.

Soloman DH, Barohn RJ, Bazan C, Grissom J. The thalamic ataxia syndrome. Neurology 1994;44:810–814.

Sotaniemi KA, Mononen H, Hokkanen TE. Long-term cerebral outcome after open-heart surgery. A five year neuropsychological follow-up study. Stroke 1986;17:410–416.

Soubrie P. Reconciling the role of central serotonin neurons human and animal behavior. Behav Brain Sci 1986;9:319–364.

Southard EE. Shell-shock and other neuropsychiatric problems. Boston: Hoeber, 1919.

Spear NE. Experimental analysis of infantile amnesia. In: Kilstrom JF, Evans FJ, eds. Functional disorders of memory. Hillsdale, NJ: Erlbaum, 1979.

Spellacy F. Lateral preference in the identification of patterned stimuli. J Acoust Soc Am 1970;47: 574–578.

Spencer DD, Spenser SS, Mattson RH, Williamson PD. Intracerebral masses in patients with intractable partial epilepsy. Neurology 1984;34:432–436.

Spencer SS, Spencer DD, Williamson PD, Mattson RH. Sexual automatisms in complex partial seizures. Neurology 1983;33:527–533.

Sperry R. Brain bisection and the neurology of consciousness. In: Eccles JC, ed. Brain and conscious experience. New York: Springer Verlag, 1966: 298–313.

Sperry RW. Hemisphere dissconnection and unity in conscious awareness. Am Psychol 1968;23:723–733.

Sperry R. Lateral specialization in the surgically separated hemispheres. In: Schmitt FO, Worlden FG, eds. The neurosciences: third study program. Cambridge, MA: MIT Press, 1974:1–12.

Sperry R. Some effects of disconnecting the cerebral hemispheres. Science 1982;217:1223–1226.

Sperry RW, Zaidel E, Zaidel D. Self recognition and social awareness in the deconnected minor hemisphere. Neuropsychologia 1979;17:153–166.

Spiegel D. Dissociative disorders. Maryland: Sidran Press, 1993.

Spiegel EA, Szekely EG. Prolonged stimulation of the head of the caudate nucleus. Arch Neurol 1961; 4:67–77.

Spiegler BJ, Mishkin M. Evidence for the sequential participation of inferior temporal cortex and amygdala in the acquisition of stimulus-reward associations. Behav Brain Res 1981;3:303–317.

Spinelli DN. OCCAM: a computer model for a content of addressable memory in the central nervous system. In: Pribram KH, Broadbent DE, eds. Biology of memory. New York: Academic Press, 1970.

Spinnler H, Vignolo LA. Impaired recognition of meaningful sounds in aphasia. Cortex 1966;2: 337–348.

Spitz RA. Hospitalism: an inquiry into the genesis of psychiatric conditions in early childhood. Psychoanal Study Child 1945;1:53–74.

Spitz RA, Wolf KM. The smiling response: a contribution to the ontogenesis of social relations. Genet Psychol Monogr 1946;34:57–125.

Spoont MR. Modulatory role of serotonin in neural information processing: implications for human psychopathology. Psychol Bull 1992;112:330–350.

Spreen O, Benton AL, Fincham RW. Auditory agnosia without aphasia. Arch Neurol 1965;13:84–92.

Squire LR. Two forms of human amnesia: an analysis of forgetting. J Neurosci 1981;5:241–273.

Squire LR. Comparisons between forms of amnesia. J Exp Psychol Learn Mem Cogn 1982;8:560–573.

Squire L. Memory and brain. New York: Oxford University Press, 1987.

Squire LR. Memory and the hippocampus: a synthesis from findings with rats, monkeys, and humans. Psychol Rev 1992;99:195–231.

Squire LR, Cohen NJ, Nadel L. The medial temporal region and memory consolidation. In: Weingartner H, Parker E, eds. Memory consolidation. New York: Plenum, 1984.

Squire LR, Haist F, Shimamura AP. The neurology of memory. J Neurosci 1989;9:828–839.

Squire LR, Slater PC, Chace PM. Retrograde amnesia. Science 1975;187:77–79.

Stacy M, Jankovic J. Clinical and neurobiological aspects of Parkinson's disease. In: Huber SJ, Cummings JL, eds. Parkinson's disease. New York: Oxford, 1992.

Stagner R. The reintegration of pleasant and unpleasant experiences. Am J Psychol 1931;43:463–468.

Stahl SM. Illness as an emergent norm or doing what comes naturally. In: Colligan M, Pennebaker JW, Murphy LR, eds. Mass psychogenic illness. Hillsdale, NJ: Erlbaum, 1982.

Staller AG. Systemic effects of severe head trauma. Crit Care Nurs Q 1987;10:58–68.

Stallones RA, Dyken ML, Fang HCH, et al. Epidemiology for stroke facilities planning. Stroke 1972;3: 360–371.

Stamm JS, Rosen SC. The locus and crucial time of implication of prefrontal cortex in the delayed response task. In: Pribram KH, Luria AR, eds. Psychophysiology of the frontal lobes. New York: Academic Press, 1973:139–153.

Stanley JA, Williamson PC, Drost DJ, et al. An in vivo study of the prefrontal cortex of schizophrenic patients at different stages of illness via phosphorus magnetic resonance spectroscopy. Arch Gen Psychiatry 1995;52:399–406.

Stanley SM, Yang X. A double mass extinction at the end of the Paleozoic era. Science 1994;266: 1340–1345.

Starkstein SE, Fedoroff JP, Price TR, et al. Neuropsychological deficits in patients with anosognosia. Neuropsychiatry Neuropsychol Behav Neurol 1993;6:43–48.

Starkstein SE, Fedoroff JP, Price TR, et al. Neuropsychological and neuroradiologic correlates of emotional prosody comprehension. Neurology 1994;44: 515–522.

Starkstein SE, Pearlson GE, Boston J, Robinson RG. Mania after brain injury. Arch Neurol 1987; 44:1069–1073.

Starkstein SE, Robinson RG. Neuropsychiatric aspects of cerebral vascular disorders. In: Yudofsi SC, Hales RE, eds. Textbook of neuropsychiatry. Washington DC: American Psychiatric Press, 1992.

Starr A, Phillips L. Verbal and motor memory in the amenstic syndrome. Neuropsychologia 1970; 8:75–88.

Starr MS, Summerhayes M. Role of the ventromedial nucleus of the thalamus in motor behavior. Neuroscience 1983;10:1157–1169.

Steklis HD, Kling A. Neurobiology of affiliative behavior in nonhuman primates. In: Reite M, Fields

T, eds. The psychobiology of attachment and separation. Orlando: 1985:93–134.

Stein DG, Glasier MM. An overview of research on recovery from brain injury. In: Rose FD, Johnson DA, eds. Recovery from brain damage. New York: Plenum, 1992.

Stein JF. The representation of egocentric space in the posterior parietal cortex. Behav Brain Sci 1992;15:691–700.

Stein JF, Glickstein M. Role of the cerebellum in visual guidance of movement. Physiol Rev 1992;72:967–1017.

Stein L, Ray OS. Self-regulation of brain stimulating current intensity in the rat. Science 1959;130:570–572.

Steklis HD, Kling A. Neurobiology of affiliative behavior in non-human primates. In: Reite M, Field T, eds. The psychobiology of attachment and separation. New York: Academic Press, 1985:93–134.

Stepien I, Stamm JS. Impairments on locomotor tasks involving spatial opposition between cue and reward in frontally ablated monkeys. Acta Neurobiol Exp 1970;30:1–12.

Steriade M. Development of evoked responses and self-sustained activity within amygdalo-hippocampal circuits. Electroencephalogr Clin Neurophysiol 1964;16:221–231.

Steriade M, Datta S, Pare D, et al. Neuronal activities in brainstem cholinergic nuclei related to tonic activation processes in thalamocortical sytems. J Neurosci 1990;10:2541–2559.

Steriade M, McCarley W. Brainstem control of wakefulness and sleep. New York: Plenum, 1990.

Stern DB. Handedness and the lateral distribution of conversion reactions. J Nerv Ment Dis 1977;164:122–130.

Stern G. The effects of lesions in the substantia nigra. Brain 1966;84:449–478.

Stern K, Dancy T. Glioma of the diencephalon in a manic patient. Am J Psychiatry 1942;98:716.

Stern LD. A review of theories of amnesia. Memory Cogn 1981;9:247–262.

Stern MM. Anxiety, trauma, and shock. Psychoanal Q 1951;20:179–203.

Sternberg K, Lamb ME, Greenbaum C, Cicchetti D, et al. Effects of domestic violence on children's behavior problems and depression. Dev Psych 1993;291:44–52.

Stevens CF. Strengthening the synapses. Nature 1989;338:460–461.

Stevens JR. Psychosis and the temporal lobe. In: Smith DB, et al, eds. Neurobehavioral problems in epilepsy. New York: Raven, 1991.

Stevens JR. Abnormal reinnervation as a basis for schizophrenia. Arch Gen Psychiatry 1992;49:238–243.

Stevens JR, Bigelow L, Denney D, et al. Telemetred EEG in schizophrenia. J Neurol Neurosurg Psychiatry 1979;36:251–262.

Steward O. Topographic organization of the projections from the entorhinal area to the hippocampal formation of the rat. J Comp Neurol 1976;167:285–314.

Stewart J, Skavarenina A, Pottier J. Effects of neonatal androgen on open field behavior and maze learning in the prepubescent and adult rat. Physiol Behav 1975;14:291–295.

Stone R. Analysis of a toxic death. Discover 1995;16:66–87.

Stoof JC, Drukarch B, DE Boer P, et al. Regulation of the activity of striatal cholinergic neurons by dopamine. Neuroscience 1992;47:755–770.

Stoneking M, Cann RL. African origin of human mitochondrial DNA. In: Mellars P, Stringer CB, eds. The human revolution: behavioral and biological perspectives on the origins of modern humans, vol 1. Edinburgh: Edinburgh University Press, 1989.

Strauss E, Moscovitch M. Perception of facial expressions. Brain Lang 1981;13:308–332.

Strauss E, Risser A, Jones MW. Fear responses in patients with epilepsy. Arch Neurol 1982;39:626–630.

Strauss E, Wada J, Goldwater B. Sex differences in interhemispheric reorganization of speech. Neuropsychologia 1992;30:353–359.

Strauss PR, Wilson SH. The eukaryotic nucleus. New Jersey: Telford Press, 1990.

Strayer FF, Strayer J. An ethological analysis of social agonism and dominance relations among preschool chidlren. Child Dev 1976;47:980–989.

Strayer J. A naturalistic study of emphatic behaviors and their relation to affective states and perspective-taking skills in preschool children. Child Dev 1980;51:815–822.

Stayer J, Schroeder M. Children's helping strategies. In: Eisenberg N, ed. Empathy and related emotional responses. San Francisco: Jossey-Bass, 1989.

Strich SJ. Shearing of nerve fibers as a cause of brain damage due to head injury. Lancet 1961;2:443–448.

Stringer CB. Documenting the origin of modern humans. In: Trinkaus T, ed. The emergence of modern humans. Cambridge, UK: Cambridge University Press, 1988.

Stringer CB. Paleoanthropology. The dates of Eden. Nature 1988;331:565–566.

Stringer CB. Replacement, continuity, and the origin of Homo sapiens. In: Brauer G, Smith FH, eds. Continuity or replacement. Rotterdam: Balkema, 1992.

Strom-Olsen J. Discussion on prefrontal leucotomy with reference to indication and results. Proc R Soc Med 1946;39:443–444.

Strub RL, Geschwind N. Localization in Gerstmann syndrome. In: Kertesz A, ed. Localization in neuropsychology. New York: Academic Press, 1983.

Strub RL, Black FW. The mental status examination in neurology. Philadelphia: FA Davis, 1993.

Studdert-Kennedy M, Shankweiler D. Hemispheric specialization for speech perception. J Acoust Soc Am 1970;48:579–594.

Strum SC. Almost human. New York: Random House, 1987.

Stuss DT, Alexander MP, Lieberman A, Levine H. An

extraordinary form of confabulation. Neurology 1978;28:1166–1172.

Stuss DT, Benson DF. The frontal lobes. New York: Raven, 1984.

Suberi M, McKeever WF. Differential right hemispehric memory storage of emotional and non-emotional faces. Neuropsychologia 1977;5:757–768.

Subirana A, Oller-Daurelia L. The seizures with a feeling of paradisiacal happiness as the onset of certain temporal symptomatic epilepsies. Congr Neurol Int Lisbonne 1953;4:246–250.

Summit RC. The child sexual abuse accommodation syndrome. Child Abuse Negl 1983;7:177–193.

Suomi SJ. Social development of rhesus monkeys reared in an enriched laboratory environment [Abstract]. Proceedings of the 20th International Congress of Psychology. Tokyo: Japan Science Press, 1972.

Sur M, Nelson RJ, Kaas JH. Representations of the body surface in cortical areas 3b and 1 of squirrel monkeys. J Comp Neurol 1982;211:177–192.

Suskind P. Perfume. New York: Knopf, 1987.

Susman RL. Thumbs, tools and early humans. Science 1995;268:589.

Swaab DF, Fliers E. A sexually dimorphic nucleus in the human brain. Science 1985;228:1112–1114.

Swaab DF, Hoffman MA. Sexual differentiation of the human hypothalamus: ontogeny of the sexually dimorphic nucleus of the preoptic area. Dev Brain Res 1988;44:314–318.

Swaab DF, Hoffman MA. An enlarged suprachiasmatic nucleus in homosexual men. Brain Res 1990;537:141–148.

Swann AC, Stokes PE, Secunda SK, et al. Depressive mania vs agitated depression. Biogenic amine and hypothalamic-pituitary-adrenocortical function. Biol Psychiatry 1994;35:803–813.

Swanson LW, Cowan WM. The connections of the septal region in the cat. J Comp Neurol 1979;186:621–656.

Sweeney JE, Lamour Y, Bassant MH. Arousal-dependent properties of medial septal neurons in the unanesthetized rat. Neuroscience 1992;48:353–362.

Sweet JJ, Newman P, Bell B. Signifance of depression in clinical neuropsychological management. Clin Psychol Rev 1992;12:21–45.

Sweet WH. Intracranial aneurysm simulating neoplasm. Syndrome of the corpus callosum. Arch Neurol Psychiatry 1945;45:86–103.

Sweet WH, Ervin F, Mark VH. The relationship of violent behavior in focal cerebral disease. In: Garattini S, Sigg E, eds. Aggressive behavior. New York: Wiley, 1969.

Swenson LW, Mogenson GJ, Simerly RB, Wu M. Anatomical and electrophysiological evidence for a projection from the medial preoptic area to the mescencephalic and subthalamic locomotor regions in the rat. Brain Res 1987;405:108–122.

Swerdlow NR, Koob GF. Dopamine, schizophrenia, mania, and depression. Behav Brain Sci 1987;10:197–245.

Swimme B, Berry T. The universe story. San Francisco: Harper San Francisco, 1992.

Swisher LP, Dudley JG, Doehring DG. Influence of contralateral noise on auditory intensity discrimination. J Acoust Soc Am 1969;45:1532–1536.

Symons D. The evolution of human sexuality. New York: Oxford University Press, 1979.

Symonds C. Concussion and its sequela. Lancet 1962;1:1–5.

Szentagothai J. Growth of the nervous system. In: Wolsteholme GEW, O'Connor M, eds. Growth of the nervous system. Boston: Little Brown, 1969.

Szymanksi S, Lieberman JA, Alvir JM, et al. Gender differences in onset of illness, treatment response, course, and biological indexes in first-episode schizophrenic patients. Am J Psychiatry 1995;152:698–703.

Takeuchi Y, McLean JH, Hopkins DA. Reciprocal connections between the amygdala and parabachial nuclei. Brain Res 1982;239:583–588.

Talland GA. Confabulation in the Wernicke-Korsakoff syndrome. J Nerv Ment Dis 1961;132:361–381.

Tamaki Y. Sex pheromones. In: Kerkut GA, Gilbert LI, eds. Comprehensive insect physiology, biochemistry, and pharmacology. New York: Pergammon Press, 1985.

Tanabe T, Yarita H, Lino M, et al. An olfactory projection area in orbitofrontal cortex of the monkey. J Neurophysiol 1975;38:1269–1283.

Tanaka M, Kohno Y, Nakagawa R, et al. Naloxone enhances stress-induced increases in noradrenaline turnover in specific brain regions in rats. Life Sci 1982;30:1663–1669.

Tanaka Y, Kamo TM, Yoshida M, Yamadori A. "So-called" cortical deafness. Brain 1991;114:2385–2401.

Tanaka Y, Yamadori A, Mori E. Pure word deafness following bilateral lesions. Brain 1987;110:381–403.

Tanji J, Kurata K. Comparison of movement-related neurons in two cortical motor areas of primates. J Neurophysiol 1982;40:644–653.

Tanji J, Tanguchi K, Saga T. Supplementary motor area: neuronal response to motor instructions. J Neurophysiol 1980;43:60–68.

Tannen D. You just don't understand. New York: Ballantine, 1990.

Tarachow S. The clinical value of hallucinations in localizing brain tumors. Am J Psychiatry 1941;99:1434–1442.

Taylor AR, Bell TK. Slowing of cerebral circulation after concussional head injury. Lancet 1966;2:178–180.

Taylor DA, Harris PL. Knowledge of the link between emotion and memory among normal and maladjusted boys. Dev Psychol 1983;19:832–838.

Taylor DC. Aggression and epilepsy. J Psychosomat Res 1969;13:229–236.

Taylor DC. Mental state and temporal lobe epilepsy. Epilepsia 1972;13:727–765.

Taylor DC. Factors influencing the occurrence of

schizophrenia-like psychosis in patients with temporal lobe epilepsy. Psychol Med 1975;5:429–254.

Taylor GH, Heilman K. Left-hemispheric motor dominance in right handers. Cortex 1980;16:587–603.

Taylor LB. Psychological assessment of neurosurgical patients. In: Rasmussen T, Marino R Jr, eds. Functional neurosurgery. New York: Raven Press, 1979.

Taylor MA. The role of the cerebellum in the pathogenesis of schizophrenia. Neuropsychiatry Neuropsychol Behav Neurol 1991;4:251–280.

Taylor PJ, Kopelman MD. Amnesia for criminal offenses. Psychol Med 1984;14:581–588.

Teasdale JD, Fogarty SJ. Differential effects of induced mood on retrieval of pleasant and unpleasant events from episodic memory. J Abnormal Psychol 1979;88:248–257.

Teasdale JD, Taylor R, Fogarty SJ. Effects of induced elation-depression on the accessibility of happy and unhappy experiences. Behav Res Ther 1980; 18:339–346.

Teasdale GM. Head injury. J Neurol Neurosurg Psychiatry 1995;58:526–539.

Teasdale G, Mendelow D. Pathophysiology of head injuries. In: Brooks N, ed. Closed head injury. New York: Oxford University Press, 1984.

Teasdale G, Skene A, Spiegelhater D, Murry L. Age, severity, and outcome of head injury. In: Grossman RG, Gidenberg PL, eds. Head injury: basic and clinical aspects. New York: Raven Press, 1982.

Teitelbaum P. Disturbances in feeding and drinking behavior after hypothalamic lesions. In: Jones MR, ed. Nebraska symposium on motivation. Lincoln: University of Nebraska Press, 1961.

Teitelbaum P, Epstein AN. The lateral hypothalamic syndrome. Psychol Rev 1962;69:74–90.

Tellegen A, Horn JM, Legrand RG. Psychonom Sci 1969;14:104.

Templeton AR. Human origins and analysis of mitochondrial DNA sequences. Science 1992;255:737.

Terr L. What happens to early memories of trauma? J Am Acad Child Adolesc Psychiatry 1988;27: 96–104.

Terr L. Too scared to cry. New York: Harper & Row, 1990.

Terzian H. Behavioural and EEG effect of intracarotid sodium amytal injections. Acta Neurochirurg 1964; 12:230–239.

Terzian H, Ore GD. Syndrome of Kluver and Bucy in man by bilateral removal of temporal lobes. Neurology 1955;5:373–380.

Teuber HL. The riddle of frontal lobe function in man. In: Warren JM, Akert K, eds. The frontal granular cortex and behavior. New York: McGraw-Hill, 1964:410–477.

Teuber HL. Disorders of memory following penetrating missile wounds of the brain. Neurology 1968; 18:287–288.

Teuber HL, Battersfy WS, Bender MB. Visual field defects after penetrating missile wounds of the brain. Cambridge, MA: Harvard University Press, 1960.

Teuber H-L, Weinstein S. Ability to discover hidden figures after cerebral lesions. Arch Neurol Psychiatry 1956;76:369–379.

Thach WT. Correlation of neural discharge with pattern and force of muscular activity. J Neurophysiol 1978;41:654–676.

Thigpen CH, Checkley HM. The three faces of Eve. New York: McGraw Hill, 1957.

Thomas DG, Campos JJ, Shucard DW, et al. Semantic comprehension in infancy. Child Dev 1981;52: 798–803.

Thompson CI, Neely JE. Retrograde amnesia—effects of periodicity and degree of training. Physiol Behav 1970;5:783–786.

Thompson CJ, Meyers NA. Inferences and recall at ages four and seven. Child Dev 1985;56:1134–1144.

Thompson RF. Neuronal substrates of simple associative learning: classical conditioning. Trends Neurosci 1986;6:270–283.

Thompson T. Visual reinforcement in Siamese fighting fish. Science 1963;141:55–57.

Thomsen I. The patient with severe head injury and his family. Scand J Rehabil Med 1974;6:180–183.

Tidswell, P, Dias PS, Sagar H. Cognitive outcome after aneurysm rupture. Neurology 1995;45:875–882.

Tigges J, Tigges M, Anschell S, et al. Areal and laminar distribution of neurons interconnecting the central visual cortical areas, 17, 18, 19 and MT. J Comp Neurol 1981;202:539–560.

Tigges J, Walker LC, Tigges M. Subcortical projections to the occipital and parietal lobes of the chimpanzee brain. J Comp Neurol 1983;220:106–115.

Tilney F. The brain from ape to man. New York: PB Hoeber, 1928.

Tobias PV. The brain in hominid evolution. New York: Columbia University Press, 1971.

Tobias JS, Puria KB, Sheridan J. Rehabilitation of the severely brain-injured. Scand J Rehabil Med 1982; 14:83–88.

Todd C, Permultter M. Reality recalled by preschool children. In: Permulter M, ed. New directions for child development no. 19: children's memory. New York: Cambridge University Press, 1980.

Todd JW, Kesner RP. Effects of posttraining injections of cholinergic agonist and antagonists into the amygala on retention of passive avoidance in rats. J Comp Physiol Psychol 1978;22:958–968.

Toglia JU, Rosenberg PE, Ronis ML. Post traumatic dizziness. Arch Otolaryngol 1970;92:7–13.

Tohgi H, Saitoh K, Takahashi S, et al. Agraphia and acalculia after a left prefrontal (F1, F2) infarction. J Neurol Neurosurg Psychiatry 1995;58:629–632.

Tompkins P, Bird C. The secret life of plants. New York: Avon, 1992.

Toole JF. Cerebrovascular disease. New York: Raven Press, 1984.

Torrey EF, Peterson MR. Schizophrenia and the limbic system. Lancet 1974;2:942–946.

Torvik A, Svinland A. Is there a transitional zone between brain infarcts and the surrounding brain? A histological study. Acta Neurol Scand 1986;74: 365–370.

Toth N. Archeological evidence for preferential right-handedness in Lower and Middle Pleistocene, and its possible implications. J Hum Evol 1985;14:607–614.

Tovee MJ, Rolls ET, Azzopardi P. Translation invariance in the responses to faces of single neurons in the temporal visual cortical areas of the alert macaque. J Neurophysiol 1994;72:1049–1060.

Tow PM. Personality changes following frontal leucotomy. New York: Oxford University Press, 1955.

Tow PM, Whitty CWM. Personality changes after operations of the cingulate gyrus in man. J Neurol Neurosurg Psychiatry 1953;16:186–193.

Tramo MJ, Baynes K, Volpe BT. Impaired syntactive comprehension and production in Broca's aphasia: CT lesion localization and recovery patterns. Neurology 1988;38:95–98.

Travis AM. Neurological deficiencies following supplementary motor area lesions in macaca mulatta. Brain 1955;78:174–198.

Trimble MR. The psychoses of epilepsy. New York: Raven Press, 1991.

Trimble MR. The schizophrenia-like psychosis of epilepsy. Neuropsychiatry Neuropsychol Behav Neurol 1992;5:103–107.

Trinkaus E. Western Asia. In: Smith FH, Spencer F, eds. the origins of modern humans: a world survey of the fossil evidence. New York: Alan R. Liss, 1984.

Trinkaus E. The Neanderthals and modern human origins. Annu Rev Anthropol 1986;15:193–211.

Trinkaus E. On neanderthal public morphology and gestation length. Curr Anthropol 1987;27:91.

Tripp CA. The homosexual matrix. New York: New American Library, 1987.

Trojana L, Grossi D. A critical review of mental imagery defects. Brain Cogn 1994;24:213–243.

Trojanowski JQ, Arnold SE. In pursuit of the molecular neuropathology of schizophrenia. Arch Gen Psychiatry 1995;52:274–276.

Truby HM, Lind J. Cry sounds of the new born infant. Acta Paediatr Scand 1965;163:1–57.

Truelle J-L, Le Gall D, Joseph P-A, et al. Movement disturbances following frontal lobe lesions: qualitative analysis of gesture and motor programming. Neuropsychiatry Neuropsychol Behav Neurol 1995;8:14–19.

Tsunoda T. Functional differences between right- and left-cerebral hemispheres detected by the key-tapping method. Brain Lang 1975;2:152–170.

Tubbs ON, Potter JM. Early post concussion headache. Lancet 1970;2:128–129.

Tucker D. Lateral brain function, emotion, and conceptualization. Psychol Bull 1981;89:19–46.

Tucker DM, Frederick SL. Emotion and brain lateralization. In: Wagner HL, Manstead ASR, eds. Handbook of social psychophysiology. New York: Wiley, 1989.

Tucker DM. Lateral brain, function, emotion, and conceptualization. Psychol Bull 1981;89:19–46.

Tucker DM, Stenslie CE, Roth RS, Shearer SL. Right frontal lobe activation and right hemisphere performance: decrement during a depressed mood. Arch Gen Psychiatry 1981;38:169–174.

Tucker DM, Watson RT, Heilman KM. Affective discrimination and evocation in patients with right parietal disease. Neurology 1977;27:947–950.

Tulving E. Elements of episodic memory. Oxford: Oxford University Press, 1983.

Tulving E, Schacter DL. Priming and human memory systems. Science 1990;247:301–306.

Tulving E, Schacter DL, Stark HA. Priming effects in word-fragment completion are independent of recognition memory. J Exp Psychol Learn Mem Cogn 1982;8:336–342.

Tumarkin A. Evolution of the auditory conducting apparatus in terrestrial vertebrates. In: de Reuck AVS, Knight J, eds. Hearing mechanisms in vertebrates (Ciba Foundation Symposium). Boston: Little Brown, 1968.

Turner BH. The cortical sequence and terminal distribution of sensory related afferents to the amygdaloid complex of the rat and monkey. In: Ben Ari Y, ed. The amygdaloid complex. Amsterdam: Elsevier, 1981.

Turner BH, Mishkin M, Knapp M. Organization of the amygdalopetal projections from modality-specific cortical association areas in the monkey. J Comp Neurol 1980;191:515–543.

Tyler LK, Ostrin RK, Cooke M, Moss HE. Automatic access of lexical information in Broca's aphasics. Brain Lang 1995;48:131–162.

Uhl RG, Kuhar BR, Snyder SH. Enkephalin containing pathways: amygdaloid efferents in the stria terminalis. Brain Res 1978;149:223–228.

Ulinksi PS. The cerebral cortex of reptiles. In: Jones ED, Peters A, eds. Cerebral cortex, vol 8: Comparative structure and evolution of cerebral cortex. New York: Plenum, 1990.

Umbricht D, Degreef G, Barr WB, et al. Postictal and chronic psychoses in patients with temporal lobe epilepsy. Am J Psychiatry 1995;152:224–231.

Umilta C. Domain-specific forms of neglect. J Clin Exp Neuropsychol 1995;17:209–219.

Underwood BJ. Attributes of memory. Psychol Rev 1969;76:559–573.

Underwager R, Wakefield H. The real world of child interrogations. Springfield, IL: Charles C Thomas, 1990.

Ungerleider LG, Mishkin M. Two cortical visual systems. In: Ingle DG, Goodale MA, Mansfield RJW, eds. Cambridge, MA: MIT Press, 1982.

Uno H, Tarara R, Else J, et al. Hippocampal damage associated with prolonged and fatal stress in primates. J Neurosci 1989;9:1705–1711.

Unterharnscheidt F, Sellier K. Mechanisms and pathomorphology of closed head injury. In: Caveness WF, Walker AE, eds. Head injury. Philadelphia: JB Lippincott, 1966.

Updyke BV. The patterns of projection of cortical areas 17, 18, 19, onto the laminae of the dorsal lateral geniculate nucleus of the cat. J Comp Neurol 1975;163:377–396.

Ursin H, Kaada BR. Functional localization within the amygdaloid complex in the cat. Electroencephalogr Clin Neurophysiol 1960;12:1–20.

Vallada H, et al. Thermoluminescence dating of Mousterian "Proto-Cro-Magnon" remains from Israel and the origin of modern man. Nature 1988;331:614–616.

Valdimaersson E, Bervall U, Samuelson K. Prognostic significance of cerebral computed tomograph results in supratentorial infarction. Acta Neurol Scand 1982;65:133–145.

van Bergeijk WA. Evolution of the sense of hearing in vertebrates. Am Zool 1966;6:371–377.

Van Buren JM, Fedio P. Functional representation on the medial aspect of the frontal lobes in man. J Neurosurg 1976;44:275–289.

Van Den Aardweg GJM. Parents of homosexuals—not guilty? Am J Psychother 1980;38:181–189.

van der Kolk BA. Psychological trauma. Washington, DC: American Psychiatric Press, 1987.

van der Kolk B, Greenberg M, Boyd H, Krystal J. Inescapable shock, neurotransmitters, and addiction to trauma. Biol Psychiatry 1985;20:314–325.

van der Kolk BA, Perry JC, Herman JL. Childhood origins of self-destructive behavior. Am J Psychiatry 1991;148:1665–1671.

Van Der Windt C, Van Gijn J. Cerebral infarction does not occur typically at night. J Neurol Neurosurg Psychiatry 1988;51:109–111.

Vanderwolf CH, Leung L-WS. Hippocampal rhythmical slow activity. In: Seifert W, ed. Neurobiology of the hippocampus. San Diego, Academic Press, 1983.

van Dongen HR, Catsman-Berrevoets CE, van Mourik M. The syndrome of cerebellar mutism and subsequent dysarthria. Neurology 1994;44:2040–2046.

Van Dongen PAM. The human locus coeruleus in neurology and psychiatry. Progr Neurobiol 1981;17:97–139.

Van Hoesen GW. The differential distribution, diversity and sprouting of cortical projections to the amygdala in the rhesus monkey. In: Ben Ari Y, ed. The amygdaloid complex. Amsterdam: Elsevier, 1981.

Van Hoesen GW. Functional neuroanatomy of the frontal and limbic systems. In: Fogel BS, Schiffer RB, eds. Neuropsychiatry. Baltimore: Williams & Wilkins, 1995.

Van Hoesen GW, Pandya DN. Some connections of the entorhinal (area 28) and perirhinal (area 35) cortices of the rhesus monkey, III. Efferent connections. Brain Res 1975;95:48–67.

Van Hoesen G, Pandya DN, Butters N. Some connections of the entorhinal (area 28) and perirhinal (area 35) cortices of the rhesus monkey. Brain Res 1975;95:25–38.

van Kammen WB, Christiansen C, van Kammen DP, Reynolds CF. Sleep and the prisoner of war experience—40 years later. In: Giller EL Jr, ed. Biological assessment and treatment of posttraumatic stress disorder. Washington, DC: American Psychiatric Press, 1990.

Van Lawick-Goodall J. The behavior of free-living chimpanzees in the Gombe Stream Reserve. Animal Behav Monogr 1968;1:Part III.

Van Praag HM. The significance of biological factors in the diagnosis of depression. I, II. Comp Psychiatry 1982;23:124–148.

Van Putten T, Emory WH. Traumatic neuroses in Vietnam returnees. Arch Gen Psychiatry 1973;29:695–698.

Van Strien JW, Morpurgo M. Opposite hemispheric activations as a result of emotionally threatening and non-threatening words. Neuropsychologia 1992;30:845–848.

Varney NR. Linguistic correlates of pantomime recognition in aphasic patients. J Neurol Neurosurg Psychiatry 1978;41:564–568.

Vellutino FR, Scanlon DM. Free recall of concrete and abstract words in poor and normal readers. J Exp Child Psychol 1985;39:363–380.

Velten E. A laboratory task for induction of mood states. Behav Res Ther 1968;6:473–482.

Vertes RP. An analysis of ascending brain stem systems involved in hippocampal synchronizaton and desynchronization. J Neurophysiol 1981;46:1140–1159.

Vertes RP. Brainstem control of events of REM sleep. Progr Neurobiol 1984;22:241–288.

Vertes RP. Fundamental of brainstem anatomy: a behavioral perspective. In: Klemm WR, Vertes RP, eds. Brainstem mechanisms of behavior. New York: Wiley, 1990.

Vertes RP. Brainstem mechanisms of slow-wave sleep and REM sleep. In: Klemm WR, Vertes RP, eds. Brainstem mechanisms of behavior. New York: Wiley, 1990.

Victor M, Agamanolis J. Amnesia due to lesions confined to the hippocampus: a clinical-pathological study. J Cogn Neurosci 1990;2:246–257.

Victor, M, Adams RD, Collins GH. The Wernicke-Korsakoff syndrome and related neurological disorders due to alcoholism and malnutrition. Philadelphia: FA Davis, 1989.

Vignolo LA. Modality-specific disorders of written language. In: Kertesz A, ed. Localization in neuropsychology. New York: Academic Press, 1983.

Vikki J. Amnesic syndromes after surgery of anterior communicating artery aneurysms. Cortex 1985;21:431–444.

Vikki J. Perseveration in memory for figures after frontal lobe lesion. Neuropsychologia 1989;27:1101–1104.

Vitek JL, Ashe J, DeLong MR, Alexander GE. Physiological properties and somatotopic organization of the primate motor thalamus. J Neurophysiol 1994;71:1498–1513.

Vochteloo JD, Koolhaas JM. Medial amygdala lesions in male rats reduce aggressive behavior. Physiol Behav 1987;41:99–102.

Vogt BA, Rosene DL, Pandya DN. Thalamic and cortical afferents differentiate anterior from posterior cingulate cortex in the monkey. Science 1979;204: 205–212.

Volman TP. Early prehistory of Southern Africa. In: Klein RG, ed. Southern African prehistory and palaeoenvironments. Rotterdam: Balkema, 1984.

von Cramon DY, Hebel N, Schuri U. A contribution to the anatomical basis of thalamic amnesia. Brain 1985;108:993–1008.

Vonderache AR. Changes in the hypothalamus in organic disease. J Nerv Ment Dis 1940;20:689–712.

Vonsattel J-P, Myers RH, Stevens TJ, et al. Huntington's disease: neuropathological grading. In: Carpenter MB, Jayaraman A, eds. The basal ganglia. New York: Plenum, 1987.

Vygotsky LS. Thought and language. Cambridge, MA: MIT Press, 1962.

Wada J, Clarke R, Hamm A. Cerebral hemispheric asymmetry in humans. Cortical speech zones in 100 adults and 100 infant brains. Arch Neurol 1975;32:239–246.

Waddel PA, Gronwall DMA. Sensitivity to light and sound following minor head injury. Acta Neurol Scand 1984;69:270–278.

Wade DT, Hewer RL. Functional abilities after stroke: measurement, natural history and prognosis. J Neurol Neurosurg Psychiatry 1987;50:177–182.

Wadfogel S. The frequency and affective characer of childhood memories. Psychol Monogr 1948;62:291.

Wagenaar B. My memory: a study of autobiographical memory over six years. Cogn Psychol 1986;18: 225–252.

Wagman IH, Krieger HP, Papetheodorou CA, Bender MB. Eye movements elicited by surface and depth electrode stimulation of the frontal lobe of Macaca mulatta. J Comp Neurol 1961;117:179–188.

Walker AE. The medial thalamic nucleus. A comparative anatomical, physiological, and clinical study. J Comp Neurol 1940;73:87–115.

Walker EL. Action decrement and its relation to learning. Psychol Rev 1958;65:129–142.

Walker LJ, de Vries B, Trevethan SD. Moral stages and moral orientations in real life and hypothetical dilemmas. Child Dev 1987;58:842–858.

Walker SF. The possible role of asymmetric laryngeal innvervation in language lateralization. Brain Lang 1994;46:482–489.

Wall JT, Symonds LL, Kaas JH. Cortical and subcortical projections of the middle temporal area (MT) and adjacent cortex in galagos. J Comp Neurol 1982;211:193–214.

Wall PD, Davis GD. Three cerebral cortical systems affecting autonomic function. J Neurophysiol 1951; 14:507–517.

Wallace AR. Natural selection and tropical nature. London: Macmillan, 1895.

Wallesch C-W, Horn A. Long-term effects of cerebellar pathology on cognitive functions. Brain Cogn 1990;14:19–25.

Walsh BW, Rosen PM. Self-mutilation. New York: Guilford, 1988.

Wang J, Aigner T, Mishkin M. Effects of neostriatal lesions on visual habit formation in rhesus monkey. Proc Soc Neurosci 1990;16:617.

Wang L, Goodglass H. Pantomime, praxis, and aphasia. Brain Lang 1992;42:402–418.

Wapner W, Hamby S, Gardner H. The role of the right hemisphere in the apprehension of complex linguistic materials. Brain Lang 1981;14:15–33.

Waraczynski M, Stellar JR. Reward saturation in medial forebrain bundle self-stimulation. Physiol Behav 1987;41:585–593.

Ward CD. Transient feelings of compulsion caused by hemispheric lesions: three cases. J Neurol Neurosurg Psychiatry 1988;51:266–268.

Ward JP, Hopkins WD. Primate laterality. New York: Springer-Verlag, 1993.

Wardlaw KA, Kroll NE. Autonomic responses to shock-associated words in a nonattended message: a failure to replicate. J Exp Psychol Hum Percept Perform 1976;2:357–360.

Warren AR, Swartwood JN. Developmental issues in flashbulb memory research: children recall the Challenger event. In: Winograd EE, Neisser U, eds. Affect and accuracy in recall. Cambridge, UK: Cambridge University Press, 1992.

Warren S, Hamalainen HA, Gardner E. Coding of the spatial period of gratings rolled across receptive fields of somatosensory neruons in awake monkeys. J Neurophysiol 1986a;56:623–639.

Warren S, Hamalainen HA, Gardner E. Objective classification of motion and direction sensitive neruons in primary somatosensory cortex of awake monkeys. J Neurophysiol 1986b;56: 598–622.

Warrington EK. Constructional apraxia. In: Vinken P, Bruyn G, eds. Handbook of clinical neurology. Amsterdam: North-Holland, 1969.

Warrington EK, James M, Kinsbourne M. Drawing disability in relation to laterality of lesion. Brain 1966;89:53–92.

Warrington EK, James M, Maciejewski C. The WAIS as a lateralizing and localizing instrument: a case study of 656 patients with unilateral cerebral lesions. Neuropsychologia 1986;24:223–239.

Warrington EK, Rabin P. Visual span of apprehension in patients with unilateral cerebral lesions. Q J Exp Psychol 1971;23:423–431.

Warrington EK, Shallice T. The selective impairment of auditory verbal short-term memory. Brain 1969; 92:885–896.

Warrington EK, Weiskrantz L. The amensic syndrome: consolidation or retrieval? Nature 1970;228:628–630.

Warrington EK, Weiskrantz L. Conditioning in amnesic patients. Neuropsychologia 1979;17:187–194.

Was-Hockert O, Lind J, Vuorenkoski V, et al. The infant cry. A spectrographic and auditory analysis. Clin Dev Med 1968;29:33–73.

Wasman M, Flynn JP. Directed attack elicited from the hypothalamus. Arch Neurol 1962;6:220–227.

Watanabe E. Neuronal events correlated with long-term adaption of the horizontal vestibuloocular reflex in the primate flocculus. Brain Res 1984;297: 169–174.

Waters E, Matas L, Stroufe LA. Infant's reactions to an approaching stranger: description, validation and functional significance of wariness. Child Dev 1975;46:348–356.

Watson RT, Fleet S, Gonzalez-Rothi L, Heilman KM. Apraxia and the supplementary motor area. Arch Neurol 1986;43:787–792.

Watson RT, Valenstein E, Heilman KM. Thalamic neglect: the possible role of the medial thalamus and nucleus reticularis in behavior. Arch Neurol 1981; 38:501–506.

Watts JW, Fulton JF. Intussuseption—the relation of the cerebral cortex to intestinal motility in the monkey. N Engl J Med 1934;210:883–890.

Webb WB, Kersey J. Recall of dreams and the probability of stage 1-REM sleep. Percept Motor Skills 1967;24:627–630.

Weber EH. Weber on sensory asymmetry (by JD Mollon). In: Kinsbourne M, ed. Asymmetrical function of the brain. Cambridge, UK: Cambridge University Press, 1834/1978.

Wechsler AF. The effect of organic brain disease on recall of emotionally charged versus neutral narrative texts. Neurology 1973;23:130–135.

Warren AR, Swartwood JN. Developmental issues in flashbulb memory research. In: Winograd E, Neisser U, eds. Affect and accuracy in recall: the problem of flashbulb memories. New York: Cambridge University Press, 1992.

Weddell RA. Recognition memory for emotional facial expressions in patients with focal cerebral lesions. Brain Cogn 1989;11:1–17.

Weddell R, Oddy M, Jenkins D. Social adjustment after rehabilitation. Psychol Med 1980;10:257–263.

Weil AA. Ictal depression and anxiety in temporal lobe disorders. Am J Psychiatry 1956;113:149–157.

Weiller C, Ramsay SC, Wise JS, et al. Individual patterns of functional reorganization in the human cerebral cortex after capsular infarction. Ann Neurol 1993;33:181–189.

Weinberger DA. The construct validity of the repressive coping style. In: Singer JL, ed. Repression and dissociation. Chicago: University of Chicago Press, 1990.

Weinberger DR. Implications of normal brain development for the pathogenesis of schizophrenia. Arch Gen Psychiatry 1987;44:660–669.

Weinberger DR, Berman KF, Zek RF. Physiological dysfunction of dorsolarteral prefrontal cortex in schizophrenia. Arch Gen Psychiatry 1986;114: 114–125.

Weinberger DR, Klienman J, Luchins D, et al. Cerebellar pathology in schizophrenia. Am J Psychiatry 1980;137:359–361.

Weinberger DR, Torrey E, Wyatt RJ. Cerebellar atrophy in chronic schizophrenics. Lancet 1979;1:718–791.

Weingartner H, Miller H, Murphy DL. Mood-state dependent retrieval of verbal associations. J Abnormal Psychol 1977;86:276–284.

Weingarten SM, Cherlow DG, Holmgren E. The rela-

tionship of hallucinations to depth structures of the temporal lobe. Acta Neurochirugica 1977;24: 199–216.

Weingartner H. Verbal learning in patients with temporal lobe lesions. J Verb Learn Verb Behav 1968; 7:520–526.

Weiller C, Ramsay SC, Wise RJS, et al. Individual patterns of functional reorganization in the human cerebral cortex after capsular infarction. Ann Neurol 1993;33:181–189.

Weinrich M, Wise SP, Mauritz KH. A neurophysiological study of the premotor cortex in rhesus monkey. Brain 1984;107:385–414.

Weinstein EA, Kahn RL. The syndrome of anosognosia. Arch Neurol Psychiatry 1950;64:772–791.

Weinstein EA, Kahn RL. Non-aphasic misnaming (paraphasia) in organic brain disease. Arch Neurol Psychiatry 1952;67:72–78.

Weinstein EA, Sersen EA. Tactual sensitivity as a function of handedness and laterality. J Comp Physiol Psychol 1961;54:665–669.

Weinstein EA, Lyerly OG, Cole M, Ozer MS. Meaning in jargon aphasia. Cortex 1966;2:165–187.

Weinstein S. Functional cerebral hemispheric asymmetry. In: Kinsbourne M, ed. Asymmetrical function of the brain. Cambridge: Cambridge University Press, 1978.

Weintraub SW, Mesulam MM, Kramer L. Disturbances of prosody: a right hemisphere contribution to language. Arch Neurol 1981;38:742–744.

Weiskrantz L. Behavioral changes associated with ablation of the amygdaloid complex in monkeys. J Comp Physiol Psychol 1956;49:381–391.

Weiskrantz L. Contour discrimination in a young monkey with striate cortex ablation. Neuropsychologia 1963;1:145–164.

Weiskrantz L. Behavioral changes associated with ablation of the amygdaloid complex in monkeys. J Comp Physiol Psychol 1956;49:381–391.

Weiskrantz L. A comparison of hippocampal pathology in man and other animals. In: Elliot K, Whelan J, eds. Functions of the sept-hippocampal system. Amerstadam: Elsevier, 1978.

Weiskrantz L. Issues and theory in the study of the amnesic syndrome. In Weinberger NM, et al., eds. Memory systems of the brain. New York: Guilford, 1985.

Weiskrantz L. Blindsight: a case study and implications. Oxford: Clarendon Press, 1986.

Weiskrantz L. Neuroanatomy of memory and amnesia: a case for multiple memory systems. Hum Neurobiol 1987;6:93–105.

Weiskrantz L, Cowey A. Striate cortex lesions and visual acuity in rhesus monkey. J Comp Physiol Psychol 1963;56:225–231.

Weiskrantz L, Mishkin M. Effect of temporal and frontal cortical lesions on auditory functions in monkey. Brain 1958;81:233–275.

Weiskrantz L, Warrington EK, Sanders MD, Marshall J. Visual capacity in the hemianopic field following a restricted occipital ablation. Brain 1974;97:709–728.

Weinstein EA, Kahn RL. The syndrome of anosognosia. Arch Neurol Psychiatry 1950;64:772–791.

Weinstein EA, Kahn RL. Non-aphasic misnaming (paraphasia) in organic brain disease. Arch Neurol Psychiatry 1952;67:72–78.

Weinstein EA, Sersen EA. Tactual sensitivity as a function of handedness and laterality. J Comp Physiol Psychol 1961;54:665–669.

Weinstein S. Functional cerebral hemispheric asymmetry. In: Kinsbourne M, ed. Asymmetrical function of the brain. New York: Cambridge University Press, 1978:17–48.

Weintraub S, Mesulam M-M, Kramer L. Disturbances in prosody: a right hemisphere contribution to language. Arch Neurol 1981;38:742–744.

Welch KMA, Levine SR. Focal brain ischemia and stroke. In: Bornstein RA, Brown GC, eds. Neurobehavioral aspects of cerebrovascular disease. New York: Oxford University Press, 1991.

Welch K, Stuteville P. Experimental production of unilateral neglect in monjey. Brain 1958;81:341–347.

Weller RE. Two cortical visual systems in old world and new world primates. In: Hicks TP, Benedek G, eds. Vision within extra-geniculo-striate systems. Amsterdam: Elsevier, 1988.

Werker JF, Tees RC. Cross language speech perception? Infant Behav Dev 1984;7:49–63.

Werner JS, Siqueland ER. Visual recognition in the preterm infant. Infant Behav Dev 1978;1:79–84.

Wertheim N. The amusias. In: Vinkin PJ, Bruyn GW, eds. Handbook of clincial neurology, vol 4. Amsterdam: North-Holland, 1969:195–206.

Whalen R. Differentiation of the neural mechanisms which control gonadrotropin secretion and sexual behavior. In: Diamond M, ed. Reproduction and sexual behavior. Indiana University Press, 1980.

Whallon R. Elements of cultural change in the Later Palaeolithic. In: Mellars P, Stringer CB, eds. The human revolution: behavioral and biological perspectives on the origins of modern humans, vol 1. Edinburgh: Edinburgh University Press, 1989.

Wheatley MD. The hypothalamus and affective behavior. Arch Neurol Psychiatry 1944;52:296–316.

Weller RE. Two cortical visual systems in old and new world primates. In: Hicks TP, Benedek G, eds. Vision within extrageniculo-striate systems, vol 75. Amsterdam: Elsevier, 1988.

Wheldall K, Poborca B. Conservation without conversation? An alternative, non-verbal paradigm for assessing conservation of liquid quantity. Br J Psychol 1980;71:117–134.

White R. Visual thinking in the ice age. Sci Am 1989;261:92–99.

White R. Rethinking the Middle/Upper Paleolithic transition. Curr Anthropol 1982;23:169–192.

White SH, Pillemer DB. Childhood amnesia and the development of a socially accessible memory system. In: Kihlstrom JF, Evans FJ, eds. Functional disorders of memory. Hillsdale, NJ: Erlbaum, 1979.

White SR, Neuman RS. Facilitation of spinal motorneuron excitability by 5-hydroxytryptamine and noradrenaline. Brain Res 1980;185:1–9.

Whitehead R. Right hemisphere processing superiority during sustained visual attention. J Cogn Neurosci 1991;4:329–334.

Whitehouse PJ. Imagery and verbal encoding in left and right hemisphere damage patients. Brain Lang 1981;14:315–332.

Whiteley AM, Warrington EK. Prosopagnosia: a clinical, psychological and anatomical study of three patients. J Neurol Neurosurg Psychiatry 1977;40:395–403.

Whitlock DG, Nauta WJH. Subcortical projections from the temporal neocortex in Macaca mulatta. J Comp Neurol 1956;106:183–212.

Whitsel BL, Perrucelli LM, Werner G. Symmetry and connectivity in the map of the body surface in somatosensory area II of primates. J Neurophysiol 1969;32:170–183.

Whitsel BL, Roppolo JR, Werner G. Cortical information processing of stimulus motion on primate skin. J Neurophysiol 1972;35:691–717.

Whitty CW, Lewin W. Vivid day dreaming—an unusual form of confusion following anterior cingulectomy. Brain 1957;80:72–76.

Whitty CW, Lewin W. A Korsakoff syndrome in the post cingulectomy confusional state. Brain 1960;83:648–653.

Whorf B. Language, thought and reality. New York: Wiley, 1956.

Wichman T, Bergman H, DeLong MR. The primate subthalamic nucleus. I–III. J Neurophysiol 72:494–530.

Wible CG, Shenton ME, Hokama H, et al. Prefrontal cortex and schizophrenia. Arch Gen Psychiatry 1995;52:279–288.

Wickler W. The sexual code. Garden City: Anchor, 1973.

Wiener SG, Bayart F, Faull KF, Levine S. Behavioral and physiological response to maternal separation in squirrel monkeys (Saimiri sciureus). Behav Neurosci 1990;104:108–115.

Wiesel TN, Hubel DH. Effects of visual deprivation on morphology and physiology of cells in the cats' lateral geniculate body. J Neurophysiol 1963;26:978–993.

Wiesel TN, Hubel DH. Extent of recovery from effects of visual deprivation in kittens. J Neurophysiol 1965;28:1060–1072.

Wilbur CB. Multiple personality and child abuse. Psychiatr Clin North Am 1984;7:3–8.

Wilcott RC. Skeletal and autonomic inhibition from low frequency electrical stimulation of the cat's brain. Neuropsychologia 1974;12:487–495.

Wilcott RC. Electrical stimulation in the prefrontal cortex and delayed response in the cat. Neuropsychologia 1977;15:115–121.

Wilcox J, Briones D, Suess L. Auditory hallucinations, postraumatic stress disorder, and ethnicity. Comp Psychiatry 1991;32:320–323.

Wilkins AJ, Shallice T, McCarthy R. Frontal lesions and sustained attention. Neuropsychologia 1987; 25:359–365.

Wilkinson DA, Carlen PL. Chronic organic brain syndromes associated with alcoholism. In: Isreal Y, et al, eds. Res Adv Alcohol Drug Probl New York: Plenum, 1982.

Wilkinson HA. Epileptic pain. Neurology 1973;23: 518–520.

Wilkinson J. Context in children's event memory. In: Gruneberg MM, Morris PE, Sykes RN, eds. Practical aspects of memory. New York: Wiley, 1988.

Willatts W. Development of problem-solving strategies in infancy. In: Bjorklund DF, ed. Children's strategies: contemporary views of cognitive development. Hillsdale, NJ: Erlbaum, 1990.

Williams CL, Barnett AM, Meck WH. Organizational effects of early gonadal secretions on sexual differentiation in spatial memory. Behav Neurosci 1990; 104:84–97.

Williams CJ. The prevention of CNS metastases in small-cell carcinoma of the bronchus. In: Whitehouse JMA, Kay HEM, eds. CNS complications of malignant disease. Baltimore: University Park Press, 1979.

Williams D. The structure of emotions reflected in epileptic experiences. Brain 1956;79:29–67.

Williams GC. Adaptation and natural selection. Princeton: Princeton University Press, 1966.

Williams JMG, Broadbent K. Autobiographical memory in suicide attempters. J Abnormal Psychol 1986;95:144–149.

Williams LM. Adult memories of child abuse: preliminary findings from a longitudinal study. Advisor 1992;5:19–21.

Williamsen JA, Johnson HJ, Eriksen CW. Some characteristics of posthypnotic amnesia. J Abnormal Psychol 1965;70:123–131.

Wilmot M, Brierley R. Cognitive characteristics and homosexuality. Arch Sex Behav 1984;13:311–319.

Wilson EO. Chemical systems. In: Sebeok TA, ed. Animal communication. Bloomington, Indiana: Indiana University Press. 1962.

Wilson EO. The insect societies. Boston: Harvard University Press, 1971.

Wilson EO. Sociobiology. Cambridge, MA: Harvard University Press, 1980.

Wilson EO. The diversity of life. New York: Norton, 1992.

Wilson EO, Regnier FE. The evolution of the alarm-defense system in formicine ants. Am Natural 1971; 105:279–289.

Wilson I. Reincarnation? The claims investigated. New York: Penguin, 1982.

Wilson I. The after death experience. New York: Morrow, 1987.

Wilson JA. The culture of ancient Egypt. Chicago: University of Chicago Press, 1951.

Wilson JC, Pipe ME. The effects of cues on young children's recall of recal events. N Z J Psychol 1989; 18:65–70.

Wilson LA, Keeling PWN, Malcolm AD, et al. Visual complications of mitral leaflet prolapse. Br Med J 1977; 2:86–88.

Wilson MA, McNaughton BL. Dynamics of the hippocampal ensemble for space. Science 1993;261: 1055–1058.

Wilson MA, Molliver ME. The organization of serotonergic projections to cerebral cortex in primates: regional distribution of axon terminals. Neuroscience 1991a;44:537–553.

Wilson MA, Molliver ME. The organization of serotonergic projections to cerebral cortex in primates. Retrograde transport studies. Neuroscience 1991b; 44:555–570.

Wingartner H, Cohen RM, Murphy DL, et al. Cognitive processes in depression. Arch Gen Psychiatry 1981;38:42–47.

Winocur G. The amnesic syndrome: a deficit in cue utilization. In: Cermak LS, ed. Human memory and amnesia. Hillsdale, NJ: Erlbaum, 1982.

Winograd E, Killinger W. Relating age at encoding in early childhood to adult recall. J Exp Psychol Gen 1983;112:412–422.

Winson J. Brain and psyche: the biology of the unconscious. New York: Anchor Press, 1985.

Winson J. Behaviorally dependent neuronal gating in the hippocampus. In: Isaacson RL, Pribram K, eds. The hippocampus. New York: Plenum, 1986.

Winter P, Handley P, Ploog D, Schott D. Ontogeny of squirrel monkey calls under normal conditions and under acoustic isolation. Behaviour 1973;47: 230–239.

Wirz-Justice A, Graw P, Krauchi K, et al. Light therapy in seasonal affective disorder is independent of time of day or circadian phase. Arch Gen Psychiatry 1993;50:929–937.

Wise RJS, Bernardi S, Frackowiack RSJ, et al. Serial observations on the pathophysiology of acute stroke. Brain 1983;106:197–222.

Wiskrantz ML, Mihailovic LJ, Gross CG. Effects of stimulation of frontal cortex and hippocampus on behavior in monkeys. Brain 1962;85:487–504.

Wittling W. Psychophysiological correlates of human brain asymmetry: blood pressure changes during lateralized presentations of an emotionally laden film. Neuropsychologia 1990;28:457–470.

Wittling W, Pfluger M. Neuroendocrine hemisphere asymmetries: salivary cortisol secretion during lateralized viewing of emotion-related and neutral films. Brain Cogn 1990;14:243–265.

Wojak JC, Flamm ES. Intracranial hemorrhage and cocaine use. Stroke 1987;18:712–715.

Wolf PA. Risk factors for stroke. Neurol Clin 1985;16: 359–360.

Wolf PA, Kannel WB, McGee DL, et al. Duration of atrial fibrillation and imminence of stroke: the Framingham Study. Stroke 1983;14:664–667.

Wolf PA, Kannel WB, Verter J. Current status of risk factors for stroke. Neurol Clin 1983;1:317–343.

Wolf SM. Diufficulties in right-left discrimination in a normal population. Arch Neurol 1973;29:128–129.

Wolfenstein M. How is mourning possible? Psychoanal Study Child 1966;21:93–123.

Wollberg Z, Newman VD. Auditory cortex of squirrel monkey. Science 1972;175:212–214.

Wolpoff MH. Paleoanthropology. New York: Knopf, 1980.

Wolpoff MH. Multiregional evolution: the fossil alternative to Eden. In: Mellars P, Stringer CB, eds. The human revolution: behavioral and biological perspectives on the origins of modern humans, vol 1. Edinburgh: Edinburgh University Press, 1989.

Woo, E, Chan YW, Yu YL, Huang CY. Admission glucose level in relation to mortality and morbidity outcome in 252 stroke patients. Stroke 1988;19: 185–191.

Wood JR, Green SB, Shapiro WR. The prognostic importance of tumor size in malignant gliomas. J Clin Oncol 1988;6:338–343.

Woolf NJ, Butcher LL. Cholinergic projections to the basolateral amygdala. Brain Res Bull 1982;8: 751–763.

Wooley CL. Ur of the Chaldees. New York: Norton, 1965.

Woolsey CN. Organization of somatic sensory and motor area of the cerebral cortex. In: Harlow HF, Woolsey CN, eds. Biological and biochemical bases of behavior. Madison, WI: University of Wisconsin Press, 1958:63–81.

Woolsey CN, Fairman D. Contralteral, ipsilateral, and bilateral representation of cutaneous receptors in somatic areas 1 and II of the cerebral cortex. Surgery 1946;19:684–702.

Wolpert EA, Trosman H. Studies in psychophysiology of dreams. I. Experimental evocation of sequential dream episodes. Arch Neurol 1958;79:603–606.

Worner KH. History of music. New York: Free Press, 1973.

Wowern, Von F. Post traumatic amnesia and confusion as an index of severity in head injury. Acta Neurol Scand 1966;42:373–378.

Wright J, Kunkel D, Pinon M, Huston A. Children's affective and cognitive reactions to televised coverage of the space shuttle disaster. Paper presented at the Biennial Meeting of the Society for Research in Child Development, Baltimore, 1987.

Wurtz RH, Goldberg ME, Robinson DL. Behavioral modulation of visual response in the monkey. In: Sprague JM, Epstein AN, eds. Progress in psychobiology and physiological psychology. New York: Academic Press, 1980.

Wurtz RH, Mohler CW. Enhancement of visual response in monkey striate cortex and frontal eye fields. J Neurophysiol 1976;39:766–772.

Wyke M. The effects of lesions in the performance of an arm-hand precision task. Neuropsychologia 1968;6:125–134.

Yakovlev PI, Lecours A. The myelogenetic cycles of regional maturation of the brain. In: Minkowski A, ed. Regional development of the brain in early life. London: Blackwell, 1967:404–491.

Yakovlev PI, Rakic P. Patterns of decussation of bulbar pyramids and distribution of pyramidal tracks on two sides of the spinal cord. Trans Am Neurol Assoc 1966;91:366–367.

Yamadori A, Osumi U, Mashuara, S, Okuto M. Preservation of singing in Broca's aphasia. J Neurol Neurosurg Psychiatry 1977;40:221–224.

Yamaguchi, F, Meyer JS, Yamamoto M, et al. Non-invasive regional cerebral blood flow measurements in dementia. Arch Neurol 1980;37:114–119.

Yamour BJ, Sridhakan MR, Rice JR, Floers WG. Electrocardiographic changes in cerebrovascular hemorrhage. Am Heart J 1980;99:294–300.

Yarcorzynski GK, Davis L. Modifications of perceptual responses with unilateral lesions of the frontal lobes. Trans Am Neurol Assoc 1942;68:122–130.

Yarnell PR, Lynch S. Retrograde memory immediately after concussion. Lancet 1970;1:863–864.

Yates AJ. Hypnotic age regression. Psychol Bull 1961;88:429–440.

Yates LJ, Nasby W. Dissociation, affect, and network models of memory. J Trauma Stress 1993;6:305–326.

Yellen JE, Brooks AS, Cornelissen E, et al. A middle stone age worked bone industry from Katanda, Upper Semliki Valley, Zaire. Science 1995;268: 553–556.

Yim CY, Mogenson GJ. Response of nucleus accumbens neurons to amygdala stimulation and its modification by dopamine. Brain Res 1982;239:401–415.

Yim CY, Mogenson GJ. Response of ventral pallidal neurons to amygdala stimulation and its modification by dopamine projections to nucleus accumbens. J Neurophysiol 1983;50:148–161.

Yin TCT, Mountcastle VB. Visual input to the visomotor mechanisms of the monkey's parietal lobe. Science 1977;197:1381–1383.

Yingling CD, Skinner JE. Gating of thalamic input to cerebral cortex by nucleus reticularis thalami. In: Desmedt J, ed. Attention, voluntary contraction and event related cerebral potentials. Basel, Switzerland: S. Karger, 1977:70–96.

Young AB. Huntington's disease: lessons from and for molecular neuroscience. Neuroscientist 1995; 1:51–55.

Young AW, Aggleton JP, Hellawell DJ, et al. Face processing impairments after amygdalotomy. Brain 1995;118:15–24.

Young AW, Bion PJ. Absence of any develomental trend in right hemisphere superiority in face recognition. Cortex 1980;17:97–106.

Young AW, Hellawell DJ, Welch J. Neglect and visual recognition. Brain 1992;115:51–71.

Young J, Miller R. Incidence of malignant tumors in U.S. children. J Pediatr 1975;86:254–258.

Young RC, Howser DM, Anderson T, et al. CNS infiltration: a complication of diffuse lymphomas. In: Whitehouse JMA, Kay HEM, eds. CNS complications of malignant disease. Baltimore: University Park Press, 1979.

Young WS, Alehid GF, Heimer L. The ventral pallidal projection to the mediodorsal thalamus. J Neurosci 1984;4:1626–1638.

York GK, Gabor AJ, Dreyfus PM. Paroxysmal genital pain: an unusual manifesation of epilepsy. Neurology 1979;29:516–519.

Yuille JC, Cutshall JL. A case study of eyewitness memory of a crime. J Appl Psychol 1986;71:291–301.

Yunis JJ. The chromsomal basis of human neoplasia. Science 1983;221:227–236.

Zaborszky L, Cullinan WE, Braun A. Afferents to basal forebrain projection neurons. In: Napier TC et al, eds. The basal forebrain. New York: Plenum, 1991.

Zahn-Waxler C, Friedman SL, Cummings EM. Children's emotions and behaviors in response to infant cries. Child Dev 1983;54:1522–1528.

Zaidel E. Unilateral auditory language comprehension on the token test following cerebral commissurotomy and hemispherectomy. Neuropsychologia 1977;15:1–13.

Zaidel E. Language in the right hemisphere, convergent perspectives. Am Psychol 1983;38:542–546.

Zamrini EY, Meador KJ, Loring DW, et al. Unilateral cerebral inactivation produces differential left/right heart rate responses. Neurology 1990;40:1408–1411.

Zanni RR, Offerman JT. Eyewitness testimony. Percept Motor Skills 1978;46:163–166.

Zaragoza MS. Memory, suggestibilty, and eyewitness testimony in children and adults. In: Cecci SJ, Toglia MP, Ross DF, eds. Children's eyewitness memory. New York: Springer-Verlag, 1987.

Zaragoza MS. Preschool children's susceptibility to memory impairment. In: Doris J, ed. The suggestibility of chidren's recollections. Washington, DC: American Psychological Association, 1991.

Zajonc RB. Feeling and thinking: preferences need no inferences. Am Psychol 1980;35:151–175.

Zeki SM. Functional organization of a visual area in the posterior bank of the superior temporal sulcus of the rhesus monkey. J Physiol 1974;236:549–573.

Zeki SM. Functional specialisation in the visual cortex of the rhesus monkey. Nature 1978a; 274:423–428.

Zeki SM. The cortical projections of foveal striate cortex in the rhesus monkey. J Physiol 1978b;277:227–244.

Zemlan FP Influence of p-choloramphetamine and p-chlorophenylalanine on female mating behavior. Ann N Y Acad Sci 1978;305:621–626.

Zhang SP, Davis PJ, Bandler R, Carrive P. Brainstem integration of vocalization: role of the midbrain periaqueductal gray. J Neurophysiol 1994;72:1337–1356.

Zilbovicius M, Garreau B, Samson Y, et al. Delayed maturation of the frontal cortex in childhood autism. Am J Psychiatry 1995;152:248–252.

Zihlman AL, et al. Pygmy chimpanzee as a possible prototype for the common andestor of humans, chimpanzees and gorillas. Nature 1978;275:744–746.

Zilman AL. Women as shapers of the human adaptation. In: Dahlberg F, ed. Woman the gatherer. New York: Yale University Press, 1981.

Zimmerman RA, Bilianiuk LT, Gennarelli TA. Computerized tomography of shearing injuries of the cerebral white matter. Radiology 1978;127:393–396.

Zlotnick C, Begin A, Shea MT, et al. The relationship between characteristics of sexual abuse and dissociative experiences. Comprehens Psychiatry 1994;35:465–470.

Zohar J, Insel TR. Obsessive-compulsive disorder. Biol Psychiatry 1987;22:667–687.

Zola-Morgan S, Amaral DG, Squire R. Human amnesia and the medial temporal region. J Neurosci 1986;6:2950–2967.

Zola-Morgan S, Squire LR. Preserved learning in monkeys with medial temporal lesions: sparing of motor and cognitive skills. J Neurosci 1984;4:1072–1085.

Zola-Morgan S, Squire LR. Medial temporal lesions in monkeys impair memory on a variety of tasks sensitive to human amnesia. Behav Neurosci 1985;99:22–34.

Zola-Morgan S, Squire LR. Memory impairment in monkeys following lesions of the hippocampus. Behav Neurosci 1986;100:155–160.

Zola-Morgan S, Squire LR, Alvarez-Royo P, Clower R. Independence of memory functions and emotional behavior: separate contributions of the hippocampal formation and the amygdala. Hippocampus 1991;1:207–220.

Zola-Morgan S, Squire R, Amaral DG. Lesions of the hippocampal formation but not lesions of the fornix of the mammillary nuclei produce long-lasting memory impairment in monkeys. J Neurosci 1989;9:898–913.

Zuckerman M, Cohen N. Sources of reports of visual and auditory sensations in perceptual-isolation experiments. Psychol Bull 1964;62:1034–1956.

Zurif EB. Auditory lateralization: prosodic and syntactic factors. Brain Lang 1974;1:391–404.

Zufif EB, Caramazza A, Myerson R. Grammatical judgments on agrammatic aphasics. Neuropsychologia 1972;10:405–417.

Zurif EB, Carson G. Dyslexia in relation to cerebral dominance and temporal analysis. Neuropsychologia 1970;8:239–244.

Zurif EB, Carson G. Dyslexia in relation to cerebral dominance and temporal analysis. Neuropsychologia 1970;8:239–244.

Zybrozyna AW. The anatomical basis of patterns of autonomic and behavior responses affected via the amygdala. In: Bargmann W, Schade JP, eds. Progress in Brain Research 13. Amsterdam: Elsevier, 1963.

Index

Note: Page numbers in *italics* refer to illustrations; page numbers preceded by P refer to plate page numbers.

Abducens nerve, *377*, 380
Abraham, 295–296, 299–300
Acalculia, 144, 147, 463–464
Acetylcholine
  vs dopamine, 343
  in Parkinson's disease, 339
Acoustic nerve, 17, *705*
Acoustic neuroma, 745–746
Ad Extirpanda, 289
Adversive conditioning, hippocampal lesion and, 587
Afterlife, belief in, 268. See also *Religious experience*
Aggression, amygdala and, 183, 189–190
Aging, memory loss and, 231–232
Agnosia, 138–139, 144
  auditory, 496–497
  finger, 144–145, 459–461
  number, 463
  reading and, 480
  social-emotional, 184–185, *188*, 618–619
  visual, 145–146, 479–480
Agraphia, 156–158, 453
  alexia and, 154–155, 157
  aphasia and, 158
  apraxic, 157–158
  Exner's writing area and, 408–409, *408*
  pure, 157
  spatial, 158
Agyria, 647
Alcohol, congenital malformations and, 654
Alexia, 154–156
  agraphia and, 154–155, 157
  frontal, 155
  global, 155
  literal, 155
  number, 463
  partial, 157
  for sentences, 155
  spatial, 155–156
  verbal, 155
Alexic agraphia, 154–155, 157
Alien abduction, amygdaloid hyperactivation and, 583
Alien hand syndrome, 401, 403–404
Alzheimer's disease
  axonal transport in, 345
  globus pallidus and, 344
  limbic striatum and, 342–346
  memory loss and, 231–232, 342–344
  neuronal death and, 345

  olfactory disorders in, 164
  substantia innominata in, 344
Amber, microbes from, 309–310
American Sign Language, 60–61
Amir, Yigal, 292
Amnesia, 206–208. See also *Memory; Memory loss; Repression*
  anterograde, 209–210, 695–696
  bilateral hippocampal ablation and, 218–219
  dissociative, 571–572
  fear and, 548–549
  functional, 539–541, 572–573
  hippocampal, 577, 585–587
  hysterical, 572–573
  of Korsakoff's syndrome, 223–224, 426
  learning and, 208–209
  post-traumatic, 206–207, 209–210, 226, 230–231, 554–555, 694–696
  recognition memory and, 209, 539
  vs repression, 209, 523–530, 536
  retrograde, 210–211, 695
  for sexual abuse, 534, 539–541
  for sexual assault, 539–541, 584
  shrinking, 211
  source, 208–209, 224, 539
  temporary, 207–208
  terror and, 548–549
  verbal, 207–209, 230, 539, 549–550, 554–555
Amoeba, 4
Amphetamines, rewarding effects of, 327–328
Amphibians
  hearing in, 22–23, *23*
  nervous system of, 19–21
Amusia, 85
Amygdala, *14*, 178–193
  ablation of, 183–185, 187–189, *188*
  aggression and, 183, 189–190
  Alzheimer's disease and, 344
  anatomy of, *179*, *180*, *336*, *348*, P8
  anxiety and, 228–229
  attachment behavior and, 260–261, 264–265
  attention and, 181–182
  axonal pathways of, *123*
  basal ganglia control by, 20
  burial practices and, 48
  catatonia and, 333
  cross-modal association formation and, 227
  dopamine neurons of, 343, 592–593
  dreaming and, 201–203, 301–302

Amygdala, *continued*
  embryology of, 179
  emotional expression and, 188–189
  emotional intelligence and, 178–193. See also *Emotional intelligence*
  emotional learning and, 227
  emotional memory and, 226–227, 533–534, 581
  emotional-melodic speech and, 81–82, 185, 188–189
  environmental surveillance of, 200, 205, 226
  evolution of, 14–15, *14*, 20
  facial expression and, 182
  fear and, 182–183, 227–228, 333–335, 580–581, 585–587
  fear-induced long-term synaptic potentiation in, 585–587
  fetal, 629–630, 631
  in forebrain arousal, 365
  freezing behavior and, 228
  frontal lobe seizures and, 187
  glioblastoma multiforme of, 183
  of Gymnophiona, 9
  hallucinations and, 200–201
  hemispheric connections of, *82*, 240–241
  heterosexuality and, 191
  hippocampus interactions with, 193–194, 197–205
  homosexuality and, 191–192
  hyperactivation of
    alien abduction and, 582–583
    dreaming and, 201–203
    religious experience and, 201
    sexual abuse and, 581–583
    sexual behavior and, 580–583
  hypothalamus interaction with, 181
  hysteria and, 603–604
  kindling in, 592
  language connections of, *82, 185,* 240–241
  lateral, *179,* 181
  laterality of, 199–200, 241
  learning and, 198–199, 226–227, 231
  lesions of
    aggression and, 183
    docility and, 183–184
    emotional perception and, 142
    emotional-melodic speech and, 185, 240
    hallucinations and, 200–201
    hypersexuality and, 187–189, *188*
    rage and, 183
    social-emotional agnosia and, 184–185, *188,* 618–619
    social-emotional deprivation and, 264–265
    vocalization and, 240
  limbic language and, 239–242
  long-term potentiation in, 216, 585–587
  in mammalian social behavior, 27–28
  maturation of, 659, 664
  medial, *179,* 179–181
  memory and, 197–200, 226–227, 230–232. See also *Amnesia; Memory; Memory loss; Repression*
  mesolimbic DA system and, 343, 592–593
  movement and, 333–335, 340, 348, 353
  neocortical interactions with, 226
  opiate secretion of, 587–588
  overview of, 232–234
  in PGO activity, 372–373
  pleasure and, 190–191
  polymodal response to, 226–227
  psychosis and, 340, 616–618
  rage and, 182–183
  religious experience and, 201, 273–277, 279–280
  seizure origin in, 185–187
  septal nuclei interactions with, 260–261
  sex differences in, 170, 188–189
  sexual behavior and, 187, 191–192, 533, 580–583
  social-emotional deprivation and, 264–265
  social and emotional development and, 226–228, 231, 239–242
  social-emotional intelligence and, 226–227, 231–232, 239–242. See also *Social-emotional intelligence*
  startle reaction and, 227–228, 240
  temporal lobe seizures and, 185–187
  thalamus interactions with, 232
Amyloid angiopathy, 728
Anarithmetria, parietal lobe lesions and, 463
Anarthria, 359
Anaximander, 308
Anencephaly, 644–645, *645*
Anesthesia, unconscious learning and, 208
Aneurysm, 727–728
Angiogenesis, in tumor development, 737–738
Angular gyrus
  axonal pathways of, *123*
  evolution of, 124, 254–255
Animal spirits, 301
Anomia, 134–135, 144, 453
Anosmia, 165
  linear fracture and, 680–682
Anosognosia, 140
Anterograde amnesia, 209–210, 695–696. See also *Amnesia*
Anticipation, amygdala and, 182–183, 228–229
Anton's syndrome, 479, 610
Anxiety, amygdala and, 227–229
Apathy, 137
  caudate lesions and, 330–332
  frontal lobe lesions and, 401, 402, 431–432
Apes
  evolution of, 32
  limbic system in, 60, 237
  right hemisphere in, 70
Aphasia, 59–61
  agraphia and, 158
  Broca's (expressive), 76, 134–137, 253–254, 409–410
    depression and, 137, 432–433
  conduction, 133–134
  emotion and, 140–141
  fluent, 139–140
  global, 141
  parietal lobe lesions and, 457
  psychosis and, 516–517, 617

schizophrenia and, 141
sex differences in, 69
temporal lobe seizures and, 516–517
transcortical, 141–143
Wernicke's (receptive), 60, 76–78, 139–141, 499–500
Aphonia, 359
Appetite, frontal lobe lesions and, 428
Apraxia, 52, 144, 147–148
aphasia and, 457
constructional, 458
dressing, 457
frontal lobe lesions and, 433
ideational, 456
ideomotor, 456
parietal lobe lesions and, 451, 454–458
unilateral, 456–457
Apraxic agraphia, 157–158
Archetypes, 269
Arnhem Land, rituals of, 299
Arousal, 365
frontal lobes and, 395, 438–439, 551
hippocampus and, 194–197
norepinephrine in, 367
right hemisphere and, 93–94, 438–439
serotonin in, 367
thalamus and, 551–553
Art, of Upper Paleolithic, 55, 55, 56
Arteriovenous malformation, 728–729
Artery (arteries), 707–709, 708, 709, 710, 711, 713
Arthropod, neural organization of, 9
Asahara, Shoko, 292
Astereognosis, 446–447
Astrocytoma, 746–747
Ataxia, optic, 481
Atherosclerosis, 713–716
Atonia, muscle, 376
during sleep, 374
Atrial fibrillation, ischemia and, 722–723
Attachment behavior, 257–261
septal nuclei—amygdala interactions and, 260–261
Attention, 205. See also Neglect
after mild head injury, 704
amygdala and, 181–182
auditory cortex and, 493–495
frontal eye fields and, 407, 408
frontal lobe lesions and, 416, 421–423, 438
hippocampus and, 197
orbital frontal lobes and, 551
parietal lobes and, 464–470
perseverative, 416
right cerebral hemisphere and, 550–551
shifting of, 421–423, 438
verbal, 438
visual
inferior temporal lobe and, 505–506
parietal lobes and, 450, 451
in simultanagnosia, 481
Auditory association areas, 497–500, 498
Auditory closure, 489–490

Auditory cortex, 255, 483–484
amygdala connection with, 498, 501
attention and, 493–495
bird sound reception by, 493
language acquisition and, 490–492
lesions of, 495–497
musical sensitivity of, 491
neuroanatomical organization of, 486, 489
neuronal fine tuning of, 490–492
signal transmission to, 363, 484–486, 485
spatial localization and, 493–495
sustained activity of, 486–488
Auditory stimuli, temporal sequencing of, 59
Auditory system, 361–363
evolution of, 17, 22–23
signal transmission pathway of, 363, 484–486, 485
Auditory tectum, evolution of, 24
Aura
depressive, 597
of temporal lobe seizure, 512, 514
Aurignacian technology, 57
Australopithecus, 32–34, 33
Authority, memory and, 558–559
Automatisms, of temporal lobe seizure, 512
Autonomic nervous system, orbital mediation of, 414–415
Aversion
hippocampus and, 196–197
hypothalamus and, 172
septal nuclei and, 258–259
Axon, post-traumatic growth of, 669
Axonal transport
in Alzheimer's disease, 345
trauma and, 686
Aztec, religious practice of, 290

Babbling, in language development, 129–130
Baboon, right hemisphere in, 70
Babylonians, mathematics of, 462
Balint's syndrome, 481
Basal ganglia, 323–328, 336. See also Caudate; Corpus striatum; Putamen
Basilar artery, 711, 711, 713
embolism of, 719
obstruction of, 731–734
Basilar fracture, 677
Battered infant. See also Sexual abuse
brain development and, 665–667, 666
Big bang theory, 313
Bipedalism, evolution of, 34, 34, 384
Bird
brain of, 19
songs of, 88
as soul symbol, 275
Birth trauma, congenital malformations and, 654
Black Elk, 285
Blake, William, 304
Blind sight, 478–479, 566
Blindness
cortical, 478–479, 566
linear fracture and, 682

Blood flow
  hypoxia and, 689–690
  mild head injury and, 704–705
Body image
  pain response and, 447
  representation of, 441–443, 444, 446, 448–449, 466.
    See also *Parietal lobe(s)*
  right hemisphere lesions and, 94–96, 528–529
  three-dimensional analysis of, 448–449
Bone tools, 54–57, *57*
Bosch, Hieronymus, *272*
Bourne, Ansel, 572–573
Brachial nerves, evolution of, 17
Brainstem, 9–10, 354–355, *368*
  anatomy of, 354–355, *355, 356, 381,* P7, P9, P10
  auditory perception and, 361–363
  crying and, 359–360
  depression and, 370–371
  developmental defects of, 644–652, *646*
  dopamine production of, 366
  dorsal view of, *355*
  dreaming and, 371–374
  evolution of, 9–10, 355–356
  fetal, 630
  vs forebrain, 16–17
  functions of, 354
  inferior colliculus of, 361–362
  lateral view of, *381*
  laughter and, 359–360
  lesions of
    coma and, 364–365
    lethargy and, 364–365
    in mild head injury, 702
    traumatic, 688
  medial view of, *12*
  midbrain of, 10, 358–362, *362.* See also *Midbrain*
  multimodal neurons of, 358
  neoplasms of, 742
  norepinephrine production of, 366–367
  oscillating rhythms of, 371–374
  PGO waves of, 372–373
  reflexive motor action and, 356–357
  REM-off neurons of, 371–372
  REM-on neurons of, 371–372
  reticular formation of, 363–365
  rhythm sensitivity of, 13
  serotonin production of, 367–369
  sex and, 357–358
  sleep and, 371–376. See also *Sleep*
  superior colliculus of, 360–361
  vasculature of, *733*
  ventral view of, *356*
  vocalization and, 358–360
Breath holding, sexual abuse and, 583
Broca's (expressive) aphasia, 76, 134, 135–137, 254,
    409–410
  depression and, 137, 432–433
  sex differences in, 136–137
Broca's area, 409–410
  axonal pathways of, *123*
  evolution of, 27, 121, 253

  lesions of, 59–61, 409–413
Burial, 268
  in Middle Paleolithic, 269–270, *270*
  in Upper Paleolithic, *271*

Calendars, 462
  Mayan, 276
Callosal apraxia, left parietal lobe lesion and, 456–457
Cano, Raul, 309–310
Capgras syndrome, 99, 407
Cardiac surgery, 723–724
Carotid artery (arteries), 709–710
  embolism of, 718–719
  internal, *712, 730*
  stenosis of, 730
Cartilaginous fish, 13–16, *13, 15*
Cataplexy, 376
Catatonia, 328
  amygdala and, 333
  caudate lesions and, 332, 333
  evolutionary significance of, 334–335
  frontal lobe lesions and, 333, 399–400, 402
  frontal-caudate-amygdala neural network and, 333
Catatonic schizophrenia, 614
Category sorting task, frontal lobe lesions and, 437
Caudate, *324, 326,* 328–332, *348*
  functional differentiation of, 329–330, *329*
  lesions of, 330–332
  motor functions and, 328–329, 346
  obsessive-compulsive disorders and, 341, 609–610
  patches and matrix of, 328
  psychosis and, 339–340, 614–615
  schizophrenia and, 614–615
Cave drawings, of Upper Paleolithic, *149,* 273
Cerebellar artery, *713*
Cerebellar artery syndrome, 734
Cerebellar cortex, 385
Cerebellar gait, 386
Cerebellar hemorrhage, 729–730
Cerebellar herniation, 685
Cerebellar peduncles, 387–388
Cerebellum, 383–385, *383, 384*
  cytoarchitecture of, 385
  emotion and, 387–389
  evolution of, 383–385
  fetal, 631
  hemorrhage of, 729–730
  limbic system and, 388
  motor learning and, 385–386
  movement disorders and, 386–387
  neoplasms of, 742–743
  psychosis and, 388
  speech disorders and, 386–387
  structure of, 385
  vision and, 387
Cerebral amyloid angiopathy, 728
Cerebral artery (arteries)
  anterior, 710, *732*
    aneurysm of, 731
  middle, 709–710, *731*
    embolism of, 718–719

stenosis of, 730–731
posterior, 711, *711, 713*
obstruction of, 734
Cerebral development, 625–674. See also *Fetal brain development; Neonatal brain development*
Cerebral hemisphere(s)
left, 118–158, *240*
  constructional skills and, 90–91
  dreaming and, 110, 542–543
  egocentric speech and, 130–133
  emotion and, 527
  emotional intelligence and, 123–124
  evolution of, 253
  facial recognition and, 540–541
  imagery and, 110
  inferior view of, P4
  language and, 77, 118–120, 241–242, *242*, 526. See also *Language*
  lateral view of, *76, 119, 156*, P3
  learning by, 106–107
  lesions of, 93, 110, 133–148, *145*, 154–158
    acalculia and, 147
    agraphia and, 156–158
    alexia and, 154–156
    alexic agraphia and, 157
    anomia and, 134–135
    anosognosia and, 140
    apathetic states and, 137
    aphasia and, 76, 133–141, 158, 409–410
    apraxia and, 147–148
    apraxic agraphia and, 157–158
    Broca's aphasia and, 76, 134–137, 409–410
    conduction aphasia and, 133–134
    confabulation and, 126–127
    constructional deficit and, 90–91
    finger agnosia and, 144–145
    frontal alexia and, 155
    gap filling and, 126–127
    global aphasia and, 141
    inattention and, 93
    literal alexia and, 155
    memory loss and, 106
    pantomime recognition and, 148
    psychosis and, 137–138
    pure agraphia and, 157
    right-left disorientation and, 146, 464
    schizophrenia and, 137–138, 141
    sentence alexia and, 155
    simultanagnosia and, 146, 481
    spatial agraphia and, 158
    spatial alexia and, 155–156
    transcortical aphasia and, 141–143
    verbal alexia and, 155
    visual agnosia and, 145–146, 479–480
    Wernicke's aphasia and, 60, 76–78, 139–141, 449–500
    word blindness and, 154–155
    word deafness and, 138–139
  limbic system for, *162*
  maturation of, 656–658
  medial surface of, *195*

memory and, 106
music and, 86–87
oscillation of, 108–109
overview of, 75–78, 118
positive emotions and, 99, 530–532
sex differences in, 136–137, 158
specialization of, 526
superior view of, P4
temporal sequencing and, 120
verbal thought and, 127–133
right, 78–83, *241, 255*
  arousal and, 93–94
  bilateral representation of, 94, 96
  body image and, 94–95
  in children, 113–114
  constructional skills and, 88–91
  dreaming and, 109–111, 202, 304–305, 542–543
  emotion and, 78–83, 95–96
  emotional-melodic speech and, 78–83, 95–96, *185*, 252–253, 410–413, *411*
  environmental sound recognition by, 85–86
  facial-emotional recognition and, 97–99
  inferior view of, P4
  lateral view of, *76, 81*, P3, P5, P6
  learning by, 106–107
  lesions of, 80, 89–90, 92–93, *92*
    body image disturbance and, 94–96, 528–529
    confabulation and, 83–84, 412–413, 468–469
    constructional deficit and, 90–91
    delusions and, 84, 98–99, 467–468
    disconnection and, 568–570
    drawing deficit and, 90
    dreaming and, 109, 202
    emotional disturbance and, 95–96, 99–101, 554
    fugue states and, 573–574
    gap filling and, 84, 126–127, 468–469
    hallucinations and, 109–110
    hysteria and, 97
    inattention and, 92–94, *92, 93*
    language and, 79–81, 410–413, 550–551
    memory loss and, 89–90, 93, 106, 550–551
    norepinephrine levels and, 100
    pain and, 96–97, 529–530
    personality disturbances and, 99–101
    prosopagnosia and, 97–99, 480–481
    reading and, 91–92
    visual-spatial neglect and, 92–93, *92, 93*
  limbic language and, 79–81. See also *Limbic language*
  math and, 91
  maturation of, 656–658
  medial view of, *334, 400, 401*, P5
  memory and, 106–116. See also *Memory*
  mental functioning and, 101–102
  music and, 84–85
  in nonhuman primates, 70
  oscillation of, 108–109
  reading and, 81, 91–92
  religious experience and, 304–305
  sex differences in, 82–83, 89

Cerebral hemisphere(s), *continued*
    sexual behavior and, 529–530
    singing and, 84–85
    social-emotional discrimination and, 527–528
    spatial perceptual skills and, 88–91
    spatial reasoning and, 525–526
    specialization of, 527–528
    superior surface of, P4
    tactual sensation and, 94, 528–530
    visual-spatial neglect and, 92–93, *92*
    transfer between, 114–115. See also *Split-brain*
        egocentric speech and, 132–133
Cerebral hemorrhage, 725–726
    drug-induced, 729
    fetal, 649–650
    loss of consciousness and, 691
Cerebral infarction, 707
Cerebral ischemia, 716–717
Cerebrovascular disease, 711, 713–716
    mortality from, 734–735
Challenger spacecraft explosion, memory for, 585
Chatelperronian technology, 57
Chemicals, receptors for, 6
Chest, lesions of, 690
Children
    abuse of. See also *Sexual abuse*
        brain development and, 664–674, *666, 667*
    amnesia in, 536
    competing memories in, 557–558
    confabulation in, 559–560
    corpus callosum immaturity in, 113, 115–116,
        524–526, 554
    egocentric speech in, 130–133, 469–470
    emotional memory in, 113–114, 531, 532
    face recognition in, 531–532
    false memory in, 556–557
    genital examination in, 530
    memory distortion in, 559
    peer pressure on, 558–559
    repression in, 536
    right hemisphere in, 113–114
    spatial reasoning in, 525
    split-brain in, 115–116
    suggestibility in, 557–559, 561–562
    traumatic memory loss in, 541, 549
    verbal memory in, 522–523, 531
    verbal reports of, 558–559
Chimpanzee, brain of, *41, 42*
Chorea, Huntington's, 352–353
Choroidal artery, 710
Chromosome abnormalities, in tumor development,
    736–737
Cingulate gyrus, 242–246, *243*
    anatomy of, *243, 245*
    anterior, 243–244, *245*, 405–406
        emotional free will and, 244–246
        evolution of, 25–28, *26, 83*
        language connections of, *82*
        lesions of, 406
            maternal behavior and, 246
            stuttering and, 245

        in mammalian social behavior, 27–28
        maternal behavior and, 83, 244, 246
        seizures of, 245
        sex differences in, 82–83, 249–251
        in vocalization, 27, 82, 245, 246
    hemispheric connections of, *82, 243*
    posterior, 242–243
Cingulotomy, 243–244
Circadian rhythms
    limbic system and, 174–175
    of norepinephrine, 367
Circle of Willis, *714*
Circumcision, 294
Circumlocutory speech, 411–412
Classical conditioning, 219
Cocaine
    congenital malformations and, 653
    rewarding effects of, 327–328
Cochlea, 363
    auditory transmission from, 484–486, *485*
Cochlear nerve, 377–378
Co-conscious personalities, 574–576
Codex Magliabechiano, *291*
Coelenterates, nerve net of, 8, *8*
Cognition
    in infant, 128–129
    sex differences in, 66–69
    tumor development and, 740–741
Colliculus
    inferior, 24, 361–362
    superior, 24, 360–361
Colloid cyst, 749
Colors, impaired recognition of, 481–482
Coma, 691–694
    Glasgow Coma Scale for, 692–693
    mortality and, 692–693
    reticular activating system and, 364–365
    severity of, 692
Combat, emotional stress of, 607–608
Commissure, anterior, sex differences in, 188, 249
Commissurotomy, functional, 114–115
Communication. See also *Language; Speech; Vocalization*
    evolution of, 24, 25
Complexity, evolution and, 310–311
Compulsive utilization, 403, 427
Concussion, 700–701
Conduction aphasia, 133–134
Confabulation, 126–127
    in children, 559–560
    in cortical blindness, 479
    frontal eye fields and, 407–408
    frontal lobe lesions and, 412–413
    memory, 525, 560–561
    parietal lobe lesions and, 468–469
    right hemisphere lesions and, 83–84, 412–413,
        468–469
Conflict
    in disconnection syndromes, 102–105, 568–569
    psychic, 115, 553–555
    of split-brain, 103–105
Confusion, 691

Congenital malformations, 640–654
    birth trauma and, 654
    cocaine and, 653
    cystic, 650–651
    environmental agents in, 652–654
    forebrain growth failure in, 646–649, *647, 648,
        649*
    hemorrhage-associated, 649–650
    heroin and, 653
    human immunodeficiency virus in, 652
    maternal alcohol abuse and, 654
    maternal drug abuse in, 653
    maternal malnutrition in, 652–653
    maternal trauma and, 654
    neuronal heterotopia in, 641, 642–644
    neuronal migration failure in, 641
    primary, 640
    in psychosis, 641–642
    secondary, 640–641, 651–652
    trauma-associated, 649–652, 654
Consciousness, 564–565
    after corpus callosotomy, 101–102
    contusion and, 686
    language and, 492–493, 564–565
    levels of, 691
    loss of, 690–694
    multiple personality and, 574–576
Consonants, 489
Constructional apraxia, 458
Constructional skills
    left hemisphere lesions and, 90–91
    right hemisphere and, 88–89
    right hemisphere lesions and, 90–91
Contact comfort, septal nuclei and, 259–266
Contextual cues, in memory recall, 534, 535, 538–539,
    541–545, 549–550
Contusion, 686–687
    contre coup, 687
    in mild head injury, 702
Corpus callosotomy, 525. See also *Disconnection syn-
    dromes; Split-brain*
    consciousness after, 101–102
    dreaming after, 109
Corpus callosum, P6
    in children, 113–114, 115–116, 524–526, 554
    fetal, 640
    lesions of, apraxia and, 456–457
    maturation of, 524–526, 655–656
    repression and, 524–526
    sex differences in, 68, 136–137
    unilateral memory storage and, 534–536
Corpus striatum, *14,* 323, 324, 328–332, *336.* See also
    Caudate; Putamen
    amygdala control of, 20
    cortical projections of, *331*
    evolution of, 14, 20, 22, 324–327
    fetal, 631
    functional differentiation of, 329–330, *329*
    limbic overlap with, *332*
    motor functions and, 20–21, 324–332, 328–329, *329,*
        331

obsessive-compulsive disorders and, 340–342
Parkinson's disease and, 20, 336–340
patches and matrix of, 328
psychosis and, 339–342. See also *Psychosis;*
    Schizophrenia
of reptiles, 22
of sharks, 14
social-emotional behavior and, 20–21, 22, 24,
    324–327
Cortex. See also *Neocortex*
    evolution of, 17
    of reptiles, 21–22
Cortical blindness, 478–479
Corticospinal pyramidal tract, 396–400
Corticostriatal projection system, *331*
Corticotropin-releasing factor, amygdala secretion of,
    588
Cortisol, in depression, 177, 600
Cortisone
    hippocampal effects of, 229
    stress-related secretion of, 588
Cosmology, 87
Counting, 461
    disorders of, 463–464
Cowering reaction, 229, 580–581
Cranial nerve(s), *377*
    I, 379
    II, 380–382
    III, 382
    IV, 382
    V, 380
    VI, 380
    VII, 379–380
    VIII, 377–379
    IX, 377
    X, 376
    XI, 376
    XII, 376
Cribriform plate, fracture of, 681
Criminal behavior
    amnesia for, 541
    in males, 189, 190
    witnesses to, 547, 560–561
Cro-Magnon
    cave paintings of, 271
    hunting behavior of, 46
    Neanderthal replacement by, 65–66
    religious experience of, 277–278
    skull of, *33, 37*
    symbol use of, *153*
Cross, symbol of, 49, 274, *276, 277*
Cry, of infant, 26, 235–237
Crying, 359–360
    temporal lobe seizures and, 187, 512
Cuneiform characters, 152, *153*
Cyclopia, 647, *647*
Cyclostomes, central nervous system of, 13–16,
    13, 15
Cyst(s), 749
    colloid, 749
    fetal, 650–651

Dandy-Walker syndrome, 652
Daydreams, 108–109, 202, 305
Deafness, 378
    neocortical, 495–497
    word, 496
Death, 317–318
    feigning of, 334
    limbic system at, 286
    merciful, 308
Decimal system, 461
Delirium, 691
Delusion, right hemisphere lesions and, 84, 98–99,
    467–468
Delusional denial, 84, 467–468
Delusional playmates
    vs multiple personalities, 578
    parietal lobe lesions and, 469–470
Dementia, multi-infarct, 721
Dendrites, in neural migration, 635–636
Denial. See also *Repression*
    delusional, right hemisphere lesions and, 84,
        467–468
Dentate gyrus, *384*, 384–385
    cognition and, 387
    movement and, 385–386
Deoxyribonucleic acid (DNA)
    origin of, 310
    in tumor development, 736–737
Depersonalization, emotional stress and, 548
Depression, 137, 596
    after temporal lobe seizure, 513
    Broca's aphasia and, 137
    cortisol in, 177
    dopamine and, 601
    EEG in, 433
    frontal lobe, 431–434, 596–597
    hypothalamic-pituitary-adrenal axis in, 175, 600
    hypothalamus and, 598–599
    left hemisphere lesions and, 137, 530
    memory and, 522
    norepinephrine in, 175, 177, 370–371, 600–601
    seasonal occurrence of, 175
    serotonin in, 370–371, 591, 600–601
    sex differences in, 137, 189, 597
    suprachiasmatic nucleus in, 175, 599
    temporal lobe, 597–598
    treatment of, 371, 601
Diffuse axonal injury, 688–689
    in mild head injury, 702
Dinosaurs, 31
Diplopia, mild head injury and, 705
Disconnection syndromes, 253–254, 554–555, 565–570
    apraxia in, 456–457
    in child, 658
    conflict in, 102–105, 568–569
    language and, 566–568, *567*
    parietal lobe lesions and, 468–469
    right hemisphere lesions and, 568–570
    trauma and, 524
Disinhibition
    frontal lobe lesions and, 415, 422, 429–430

hippocampal lesions and, 197
Dissociation, 566–568, *567*
    DSM-IV definition of, 570
    emotional trauma and, 571
    fear and, 576–577
    hippocampus and, 577
    predisposition to, 580
    right hemisphere lesions and, 568–570
    sex abuse and, 577–579
Dissociative amnesia, 571–572
Distractibility
    frontal lobe lesions and, 421–423
    hippocampus and, 196
Dizziness, 378–379
Docility, amygdala ablation and, 183–184
Dog
    brain of, 42
    social-emotional deprivation in, 263–264
Dopamine
    vs acetylcholine, 343
    depression and, 601
    GABA interactions with, 351–352
    language and, 61
    memory and, 342–344
    Parkinson's disease and, 337–339, 345–346,
        351–352
    production of, 366
    psychosis and, 339–340, 592–593, 614–618
    schizophrenia and, 616–617
Dopaminergic system
    mesolimbic, 323–324, 366
        memory and, 342–343
        in pleasure circuit, 327–328
    nigrostriatal, 323, 366
        in Parkinson's disease, 337–338
Dorsal medial thalamus, 221–224
    lesions of, 223, 224
Dorsal raphe nuclei, *368*
    serotonin secretion by, 368–369, 589–590
Dostoevsky, Feodor, 296–297
Drawing, hemisphere lesions and, 90–91
Dream(s), 48
    amygdala and, 301–302
    hippocampus and, 302
    limbic system and, 302–304
    patterns of, 111–112
    recall of, 110, 202–203, 511
    recovery of, 542–543
    recurrent, 112–113
    religious experience and, 301–305
    soulful, 301–302
    temporal lobe and, 304–305
Dreaming
    after corpus callosotomy, 109
    amygdala and, 201–203, 301–302
    backwards, 111
    emotional trauma and, 112–113
    hemispheric oscillation and, 108
    hippocampus and, 201–203
    in infancy, 203
    interhemispheric communication in, 107–108

left hemisphere and, 110, 542–543
norepinephrine in, 367
patterns of, 111–112
PGO waves and, 372–373
right hemisphere and, 107–111, 202, 304–305,
    542–543
serotonin in, 368
stimulants to, 110–111
temporal lobes and, 510–511
Dressing apraxia, 457
Drugs, congenital malformations and, 653
Duplication, genetic, adaptive, 311–312
Dynein, of microtubules, 4
Dysarthria, 359
Dysnomia, 134–235
Dysosmia, 379

Eadie, Betty J., 285
Ear. See also *Auditory cortex*
    auditory transmission from, 484–486, *485*
    evolution of, 23–24
Ectoderm, 626
Ego splitting
    emotional trauma and, 570–571
    fear and, 576–577
    hippocampal stimulation and, 577
    sexual abuse and, 582
Egocentric speech, 130–133
    external components of, 132–133
    internalization of, 131–132
    parietal lobes and, 469–470
    self-explanation and, 132–133
Egyptians, symbol use by, 150–154, *151, 152*
Electroencephalogram, in depression, 433
Embolism, 718–719
Emotion. See also *Limbic system*
    amygdala, 182–183
    aphasia and, 140–141
    aversive, 196–197
    cerebellum and, 387–389
    in children, 113–114, 168
    contextual cues and, 538–539
    disinhibition of, 197, 415, 428–430
    dreaming and, 112–113
    facial expression of, 97–99
    in females, 188–189
    frontal lobe lesions and, 415
    frontal lobe seizures and, 187
    hippocampus and, 194, 196–197, 664
    hypothalamus and, 172–174, 177–178
    in infant, 168
    inflectional communication of, 79–80
    integration of, 554
    in Klüver-Bucy syndrome, 184–185
    left hemisphere and, 99, 527, 530–532
    in males, 189
    memory and, 226–227, 521–522
    memory loss and, 546–549, 583–585
    music and, 86
    negative, 99
    paleocerebellum and, 387–388

parietal lobes and, 451–452
    positive, 99, 530–532
    right hemisphere and, 95–101, 527–528
    somesthesis and, 168–169
    temporal lobe seizures and, 185–187, 513
Emotional attachment, 257–261
Emotional free will, 244–246
Emotional ideas, 183, 229
Emotional incontinence, right hemisphere injury and,
    100–101
Emotional intelligence, 161–205. See also *Limbic lan-
    guage; Social-emotional intelligence*
    aggression and, 182–183, 189–190
    amygdala and, 178–193. See also *Amygdala*
    amygdala lesions and, 184–185
    arousal and, 194–197
    attention and, 181–182, 197
    aversion and, 172, 196–197
    circadian rhythms and, 174–175
    docility and, 183–184
    fear and, 182–183. See also *Fear*
    frontal lobe seizures and, 187
    hippocampus and, 193–220. See also *Hippocampus*
    hypothalamus and, 169–178. See also
        Hypothalamus
    hypothalamus–pituitary–adrenal axis and,
        175–178
    id of, 177–178
    laughter and, 172–173
    memory and, 198–200
    origins of, 162–164
    pheromones and, 164–169
    pleasure and, 171–172, 178, 200
    punishment and, 196–197
    rage and, 173–174, 182–183
    sexual orientation and, 190–192
    sexuality and, 187–189
    somesthesis and, 168–169
    temporal lobe seizures and, 185–187
Emotional learning, amygdala and, 227
Emotional trauma, 570–571. See also *Sexual abuse*
    addiction to, 591
    attentional capacity and, 584
    cortisone secretion and, 588
    ego splitting and, 570–571
    flashbacks and, 587
    fugue states and, 572–573
    memory loss and, 546–549, 583–585
    multiple personalities and, 576, 577–579
    neural network formation and, 588–589
    norepinephrine and, 591–593
    opiate secretion and, 587–588
    recognition memory and, 584–585
    serotonin reduction and, 589–590
Emotionality, function of, 162–163
Emotional-melodic speech, 251–253, 410–413, *411*. See
    also *Limbic language*
    amygdala and, 81–82, 185, 188–189
    anterior cingulate and, 82, 245
    frontal lobe lesions and, 410–411
    maternal, 248

Emotional-melodic speech, *continued*
    right hemisphere and, 78–83, 95–96, 252–253,
        410–413
    sex differences in, 82–83
Emotional-melodic speech area, *185*
Encephalocele, 645, *645, 646*
Encoding specificity principle, in memory recall, 544
Endocast
    language areas on, 63–65
    of modern human, *65*
    of Neanderthal, 63, *64*
Endoderm, 626
Enkephalins, in religious experience, 280
Environmental sounds
    evolutionary impact of, 24
    right hemisphere recognition of, 85–86
    temporal lobe recognition of, 494–495, 500–501
Environmental surveillance, amygdala and, 200, 205,
    226
Ependymoma, 743
Epilepsy. See *Seizures; Temporal lobe seizures*
Estrus cycle, 35
Eusthenopterons, nervous system of, 19–21
Eve hypothesis, 36
Evolutionary metamorphosis, 315–316
Excessive daytime sleepiness, 375
Exner's writing area, 156–157, 408–409, *408*
Explanation, 127–128
    by children, 130–133
Extra-terrestrial life, 314–315
Eye movement, 382
    abducens nerve in, 380
    disturbances of, 382–383
    parietal lobes and, 449–450
    in simultanagnosia, 481
    vestibular system in, 379
Eye opening, on Glasgow Coma Scale, 692

Faces
    famous, 97
        memory for, 540–541
    neuronal response to, 49
    recognition of
        in children, 531–532
        in infant, 658–659
        inferior temporal lobe and, 505
        left hemisphere and, 97
        occipital lobe lesions and, 480–481
        right hemisphere and, 97–99
Facial expression, 98, 527–528
    amygdala and, 182
    right hemisphere and, 97–99
Facial nerve (VII), *377, 379–380, 682*
Facial paralysis, linear fracture and, 680
Fear. See also *Frozen states*
    amnesia and, 548–549
    amygdala and, 182–183, 227–228, 333–335,
        580–581, 585–587
    catatonia and, 333
    depression after, 598
    ego splitting and, 576–577

hippocampus and, 586–587
life-threatening, 333–335
long-term synaptic potentiation and, 585–587
out-of-body experience and, 282–283, 576–577
religious experience and, 286–287, 300–301
slow-wave theta loss and, 229–230, 586
during temporal lobe seizure, 512
temporal lobe seizures and, 186
Fecal incontinence, 428
Feeling. See *Emotion; Emotional-melodic speech; Limbic
    language*
Fetal brain development, 625–654
    amygdala in, 631
    brainstem in, 630
    cerebellum in, 631
    corpus striatum in, 631
    cortical formation in, 627–630
    dendrites in, 635–636
    fissures in, 636–637
    hippocampus in, 631–632
    hypothalamus in, 631
    limbic system in, 629–630, 631
    malformations of, 640–654. See also *Congenital
        malformations*
    midbrain in, 630–631
    myelination in, 639–640
    neocortex in, 632–637
    neural networks in, 636
    neuronal migration in, 633–636, 641
    neurotransmitters and, 637–640
    norepinephrine in, 637–639
    ontogeny of, 626–627
    overview of, 626
    septal nuclei in, 631
    serotonin in, 639
    stages of, 626
    striatum in, 629–630
    sulci in, 636–637, *637*
    telencephalon in, 632
    vascularization in, 639
Finger agnosia, 144–145
    parietal lobe lesions and, 459–461
Fish
    brain of, *19*
    jawless and cartilaginous, 13–16, *13, 15*
    lobed-finned, 17–18
    lung, 18–19
Fissures, fetal development of, 636–637
Flashbacks, emotional trauma and, 571, 587
Flat worms, nervous system of, 7–9, *8*
Fluent aphasia, 139–140
Forebrain
    vs brainstem, 16–17
    developmental defects of, 646–649
    evolution of, 16–17
    fetal, 627–629, 646–649
    lesions of, 21
    plasticity of, 673–674
    of reptiles, 21
Form recognition, inferior temporal lobe and, 504–505
Free will, 244–246

Freezing behavior. See *Frozen states*
Frog, brain of, *19*
Frontal alexia, 155
Frontal apraxia, 456–457
Frontal eye fields, 406–407
    anticipatory activity of, 407
    lesions of, 407–408
Frontal hemorrhage, 730
Frontal lobe(s), 44–45, 393–440, *394*
    anterior, 44–45
    anterior cingulate of, *394*, 405–406. See also *Cingulate gyrus, anterior*
    arousal and, 93–94, 365, 395, 438–439
    attention and, 421–423
    Broca's speech area of, 409–410. See also *Broca's area*
    corticospinal pyramidal tract of, 396–400
    evolution of, 38–45, 57–58, 406
    Exner's writing area of, 156–157, 408–409, *408*
    frontal eye fields of, 406–408
    functional overview of, 44–45, 393, 395, 426–427, 439–440
    in goal-oriented behavior, 45
    inferior convexity of
        attention and, 421–423
        persevervation and, 416–419
    lateral convexity of
        attention and, 421–423
        neocortical regulation and, 419–421, *420*
        perceptual processing by, 421
        reticular thalamus and, 420–421
        schizophrenia and, 613–614
    lesions of, 45, *428, 429, 430*
        agraphia and, 408–409
        alien hand syndrome and, 401, 403–404
        apathy and, 401, 431–432
        aphasia and, 409–411
        appetite changes and, 428–429
        attentional dysfunction and, 93–94, 416, 422–423, 438
        behavioral alterations and, 427–432
        catatonia and, 333, 399–400, 402
        circumlocutory speech and, 411–412
        compulsive utilization and, 403
        confabulation and, 412–413
        depression and, 431–434, 596–597
        disinhibition and, 415, 422, 429–430
        emotional disinhibition and, 415, 422, 429–430
        emotional unresponsiveness and, 415
        fecal incontinence and, 428
        forced grasping and, 403
        frozen states and, 333
        gegenhalten and, 401–402, 403
        groping and, 403
        impulsiveness and, 429–430
        intellectual alterations and, 434, 436–437
        intelligence quotient and, 437–438
        internal utilization and, 404
        laughter and, 430–431
        mania and, 435–436
        memory retrieval and, 223–224

        mutism and, 400
        neglect and, 407–408
        neuropsychiatric disorders and, 432–436
        obsessive-compulsive disorders and, 434–435, 609–610
        paralysis and, 397–398
        Parkinson's disease and, 339
        perseveration and, 45, 341–342, 412, 416–419, *416, 417, 418, 419, 434*
        personality disorders and, 427–428
        posturing and, 401–402
        schizophrenia and, 435, 612–614, 621–622
        sexuality and, 431
        speech dysfunction and, 409–411
        tangential speech and, 411–412
        urinary incontinence and, 428
        visual scanning deficit and, 407–408
        waxy flexibility and, 401–402
    maturation of, 656, 657
    medial, *334*, 400–401
        alien hand and, 403–404
        catatonia and, 333, 402
        compulsive utilization and, 403
        evolution of, 406
        forced grasping and, 403
        functional anatomy of, 400–401
        gegenhalten and, 401–402, 403
        groping and, 403
        internal utilization and, 404
        lesions of, 401–404
        waxy flexibility and, 401–402
    memory and, 220–221, 223–224, 423–426
    Middle-to-Upper Paleolithic transition and, 42–43, 45–47
    motor homunculus of, *95, 397, 398*
    motor regions of, 395–396
    of Neanderthal, 38–43, *38*
    neoplasms of, 741–742
    orbital, 44, 413–419
        autonomic influences of, 414–415
        lesions of
            attentional disorders and, 416
            emotional disinhibition and, 415
            emotional unresponsiveness and, 415
            perseveration and, 416–419, *416, 417, 418, 419*
        limbic arousal and, 413–414
        repression and, 551
    posterior convexity of, 406–410
        eye movement and, 406–407
        lesions of, 407–408
    premotor neocortex of, 398–399
    primary motor area of, 396–398, *398*
    regions of, 393
    religious experience and, 286–291
    reticular thalamus connection with, 420–421
    right
        arousal and, 93–94, 438–439
        emotional-melodic speech and, 410–411
        lesions of, 411–413
        mania and, 601–603

Frontal lobe(s), *continued*
    repression and, 550–553
    supplementary motor area of, *334*, 345–346,
      399–400
      lesions of, 399–402
      Parkinson's disease and, 338, 339
      will and, 402
    ventral view of, *414*
Frontal lobe personality, 427–428
Frontal lobe seizures, sexual behavior and, 187
Frontal lobotomy, *428, 429, 430*
Frontal pole, lesions of, apathetic states and, 137
Frozen states, 334
    amygdala and, 228
    caudate injury and, 332
    evolutionary significance of, 334–335
    frontal lobe injury and, 333
    hippocampus and, 229
    life-threatening fear and, 333–335
    sexual abuse and, 582, 585
Fugue states
    predisposition to, 580
    right hemisphere lesions and, 572–574
Future, foreseeing of, 305

Gait, cerebellar, 386
Galaxies, 313
Gamma-aminobutyric acid (GABA)
    in movement disorders, 351–352
    in schizophrenia, 614, 620
Ganglia, neural
    evolution of, 7–9
    of prevertebrate, 9
Ganglion cell (protoneuron), 8
Gap filling, right hemisphere lesions and, 84, 126–127,
    468–469
Gathering
    language evolution and, 61, 66–67
    tool making for, 58–59
Gegenhalten, frontal lobe lesions and, 401–402, 403
Geniculate nucleus, 11
Genital examination, verbal report of, 530
Geometric images, prehistory of, 461
Geometric shapes, neuronal response to, 49, 274–275,
    277
Geometry, 87–88
    evolution of, 461–463
Gerstmann's syndrome, 145, 157, 459–464
Gesture, reading and, 457
Glasgow Coma Scale, 692–693
Glenohumeral joint, of Neanderthal, 54
Glia
    post-traumatic response of, 669–670
    tumors of, 746–747
Glial scarring
    epilepsy and, 699–700
    fetal, 651
    post-traumatic, 670–671
Glioblastoma multiforme, 747–748
    of amygdaloid nucleus, 183
Glioma, 747

Gliosis, temporal lobe, 540
Global alexia, 155
Global aphasia, 141
Global auditory agnosia, 138
Globus pallidus, *348*
    Alzheimer's disease and, 344
    dorsal, 323, 342
    lateral, 346–348
    lesions of, 328, 347
    medial, 346–348
    memory and, 343
    movement and, 346–348
    Parkinson's disease and, 337
    ventral, 323, 324, 342, 344
Glossopharyngeal nerve, *377, 377*
Glutamic acid decarboxylase, in schizophrenia,
    613–614, 620
Goal-oriented behavior, frontal lobe in, 45
God. See *Religious experience*
Goddess Nut, *277*
Golden rectangle, 87
Grammar. See also *Language*
    auditory closure and, 489–490
    inferior parietal lobe and, 453–454
    reality and, 492–493
    temporal-sequential activity and, 59
    universal, 490
Grasping
    frontal lobe lesions and, 403
    parietal lobes and, 450–451
Gravity, parietal lobe lesions and, 466
"Great Lion Hunt," *150*
Groping, frontal lobe lesions and, 403
Group behavior, limbic nuclei in, 27–28
Guiding hand, 308–310
Gymnophiona, olfactory system of, 9
Gyrus (gyri), fetal, 636–637

Hallucinations, 610–612, 616
    after death, 307–308
    amygdala-hippocampal interactions and, 200–201
    hemispheric laterality and, 611–612
    hypnogogic, 375
    in infancy, 203
    limbic system and, 280, 283–284
    LSD and, 510
    musical, 85, 495, 611
    neural noise and, 509–510
    occipital lobe lesions and, 510
    olfactory, 512
    post-traumatic stress disorder and, 583
    recollection of, 511
    right hemisphere lesions and, 109–110
    striate cortex lesions and, 476
    temporal lobe lesions and, 495, 502–503, 509–510,
      543, 610–611
    temporal lobe seizures and, 512
    verbal, 495
    visual association area lesions and, 478
Hand
    gender specialization of, 35

infant's use of, 143–144
neocortical representation of, 34, 122, 663–664
parietal lobe control of, 50–52, *50*, 459, *460*
social-emotional communication with, 34
Handedness, 50
evolution of, 121–122
language and, 50, 68–69, 120–121, 143
Hand-manipulation cells, 446
Hate, limbic system and, 266–267
Head injury, 675–706. See also *Skull fracture*
amnesia and, 206–207, 209–210, 226, 230–231,
554–555, 694–696
attention after, 704
axonal transport deficits and, 686
brain development and, 665–667, *666*
brainstem lesions and, 688
cerebellar herniation and, 685
cerebral blood flow after, 704–705
chest trauma and, 690
cognitive disturbances and, 686
coma and, 690–694
confusion and, 691
contre cavitation and, 687
contusion and, 686–687
delirium and, 691
diffuse axonal injury and, 688–689
diplopia and, 705
emotional disturbances and, 696–698, 703–704
employment and, 698–699
epilepsy and, 699–700
family stress and, 697
Glasgow Coma Scale in, 692–693
hearing loss and, 705
hematoma and, 682–685, 691
hemorrhage and, 691
herniation and, 685–686
hypoxia and, 689–690
intellect after, 698
memory and, 694–696
mild, 701–706
attention after, 704
blood flow after, 704–705
diplopia and, 705
hearing loss and, 705
photophobia and, 705
mortality and, 692–693
myocardial trauma and, 690
personality disturbances and, 696–698
photophobia and, 705
recovery from, 698–699
respiratory distress and, 689–690
reticular damage and, 690–694
rotation forces and, 687–688
shearing forces and, 687–688
stupor and, 691
subcortical lesions and, 688
subfalcial herniation and, 685
tentorial herniation and, 685–686
tinnitus and, 705
tonsillar herniation and, 685
vascular trauma and, 690

vegetative states and, 693–694
vertigo and, 705
Headache, in tumor development, 741
Hearing. See also *Auditory cortex*
evolution of, 22–24, *23*
Hearing loss
linear fracture and, 679–680
mild head injury and, 705
Heart
anatomy of, 707–709, *708*
lesions of, 690, 722
Heart rate, stress and, 228
Heart transplantation, 724
Helplessness, learned, 601
Hematoma, 682–685
chronic, 684–685
epidural, 683
intracerebral, 683
intradural, 683–685
loss of consciousness and, 691
subdural, 684
Hemianopsia, homonymous, 477–478
Hemorrhage, 724–726
cerebellar, 729–730
cerebral, 725–726
drug-induced, 729
fetal, 649–650
loss of consciousness and, 691
embolic, 719
fetal, 649–650
frontal, 730
internal capsule, 729
intracerebral, 725–726
occipital, 730
parietal, 730
pontine, 729
subarachnoid, 725
temporal, 730
thalamic, 729
Hemorrhagic infarction, ischemia-related, 727
Herniation
head injury and, 685–686
neoplastic growth and, 749–751
Heroin, congenital malformations and, 653
Heterosexuality. See also *Sexual behavior*
amygdala and, 191
Heterotopia, neuronal, 641, 642–644
Hindbrain, 354
Hippocampal amnesia, 577, 585–588
Hippocampus, 193–220, *194, 216, 218*
amygdala interactions with, 193–194, 197–205
anatomy of, *194, 216, 218*
arousal and, 194–197
attention and, 197
aversive emotion and, 196–197
dissociation and, 577
dream recall and, 202–203
dreaming and, 201–203, 302
emotion and, 194, 196–197, 664
fear and, 229–231, 586–587
fetal, 631–632

Hippocampus, *continued*
  in forebrain arousal, 365
  of Gymnophiona, 9
  hyperactivation of
    fear and, 586–587
    memory loss and, 225–226
  in infant, 203–205
  inferior temporal lobe connections with,
    507–508
  laterality of, 199–200
  learning and, 197–198, 217–218, 225–226
  lesions of
    disinhibition and, 197
    hallucinations and, 200–201
    memory loss and, 218–220
  long-term synaptic potentiation in, 215–216,
    229–230, 586
  maturation of, 538
  memory and, 197–200, 215–220, 534, 538. See also
    Amnesia; Memory; Memory loss; Repression
  mesolimbic DA system and, 343
  neocortical interactions with, 194, 196, 225
  overview of, 232–234
  parietal lobe interconnections with, 449–450
  Parkinson's disease and, 338–339
  recognition memory and, 219–220
  reticular activating system interaction with, 230
  septal interactions with, 193
  of sharks, 14
  slow-wave theta activity of, 196, 202, 224, 225
    loss of, 229–230, 586
  steering function of, 215–216
  thalamus interactions with, 196, 232
  visual-spatial activity of, 10
Homo erectus, 34–35
  endocast of, *39*
  skull of, *33*
Homo habilis, 32–34
  skull of, *33*
Homo sapiens sapiens, 32
  early modern, 43–44
  endocast of, *39, 65*
  frontal lobe in, 45–47
  hunting behavior of, 46, 47
  language of, 69–70
  mortuary practices of, 47–49, 270
  multiregional evolution of, 35, 316
  replacement evolution of, 35–36
  skull of, *47*
Homonymous hemianopsia, 477–478
Homosexuality. See also *Sexual behavior*
  amygdala and, 191–192
  etiology of, 170–171
Homunculus
  motor, *95, 397, 398*
  sensory, *442*
Horus, *278*
Human immunodeficiency virus, in congenital malformations, 652
Hunger
  hypothalamus and, 171

in infant, 204–205
Hunting behavior
  in hominid evolution, 46
  vs language, 61–63
  of Neanderthal, 46–47, 62
  silence and, 61–62
Huntington's chorea, 352–353
Hydra (fresh water flatworms), nervous system of, 7
Hydrocephalus, congenital, 651–652, *651*
Hyperresponsiveness, hippocampus and, 196
Hypertension, 726–727
Hypnogogic hallucinations, 375
Hypnosis, multiple personality and, 579–580
Hypoglossal nerve, 376
Hypothalamic-pituitary-adrenal axis, 175–177, *176*
  depression and, 175, 177, 599, 600
Hypothalamus, *168, 169–178, 169, 176*
  amygdala interaction with, 181
  aversion and, 172
  circadian rhythms and, 174–175
  depression and, 175, 177, 598–599
  fetal, 631
  in forebrain arousal, 365
  homeostatic function of, 177–178
  hunger and, 171
  lateral nuclei of, 171
  lateralization of, 177
  lesions of
    emotional incontinence and, 172–174
    laughter and, 173
    rage and, 173–174
  pleasure and, 171–172, 177–178
  psychic manifestations of, 177–178
  religious experience and, 279
  reward and, 171–172
  sex differences in, 169–171
  sexual behavior and, 279
  thirst and, 171
  ventromedial nuclei of, 171
  vocalization and, 237–238
Hypoxia
  blood flow and, 689–690
  sexual abuse and, 582
Hysteria, 603–609
  amygdala and, 603–604
  limbic system and, 603–604, 608–609
  mass, 605–608
  pain and, 448
  parietal lobes and, 603
  right hemisphere lesions and, 97
  sex differences in, 603–605, 607, 608–609
  socially sanctioned, 604–605
Hysterical amnesia, 572–573

Ichthyostega, central nervous system of, 19–21
Id, 177–178, 203–204
Ideational apraxia, left parietal lobe lesion and, 456
Ideomotor apraxia, left parietal lobe lesion and, 456
Imagery
  left hemisphere and, 110
  PGO waves and, 372–373

Imaginary friends
   vs multiple personalities, 578
   parietal lobes and, 469–470
Immortal cells, in tumor development, 737
Impulsiveness, frontal lobe lesions and, 429–430
Inattention. See *Neglect*
Infant. See also *Maternal behavior; Neonatal brain development*
   awareness of, 178
   contact seeking of, 260–261, 262
   emotion in, 168
   hand use of, 459
   hunger in, 204–205
   memory in, 227, 522–523, 544
   primary process in, 204–205
   separation cry of, 26
   social-emotional deprivation in, 261–266
   social-emotional discriminations of, 261, 527
Inflection. See *Emotional-melodic speech*
Inhibition, internal, septal nuclei and, 258–259
Intelligence, emotional, 161–205. See also *Emotional intelligence*
   frontal lobe lesions and, 434, 436–438
   social-emotional and, 235–267. See also *Social-emotional intelligence*
Intelligence Quotient (IQ), frontal lobe lesions and, 437–438
Interictal psychosis, 517
Internal capsule hemorrhage, 729
Internal utilization, frontal lobe lesions and, 404
Intonation. See *Emotional-melodic speech*
Intracranial pressure, in tumor development, 740
Irrational mind, 161–162. See also *Limbic system*
Ischemia, global, 722–724
Islam, 294–295
Isolated memories, 215, 536–539, 540
Isolated neural networks, 215
Isolation, limbic hyperactivation and, 297–298

Jacob's ladder, *303*
Jawless and cartilaginous fish, nervous system of, 13–16, *13, 15*
Jesus Christ, 297, *298*
John F. Kennedy, death of, memory for, 585

Kepler, Johannes, 87
Kinesin, of microtubules, 4
Klüver-Bucy syndrome, 184–185, 404, 509
Korsakoff's syndrome, 223–224, 426

Lamprey, third eye of, 11–12
Language. See also *Egocentric speech; Emotional-melodic speech; Limbic language; Speech; Vocalization*
   acquisition of, 128–133, 490–492, 660–661
   bird sound reception and, 493
   consciousness and, 492–493, 564–565
   consonants of, 489
   deep structure of, 489–490
   disconnection syndromes and, 566–568, *567*
   disorders of, 133–138. See also specific disorders
   dopamine and, 61
   endocasts and, 63–65
   evolution of, 59, 121–122
      in females, 59, 61, 66–67
   gathering and, 58–59, 61
   handedness and, 50, 68–69, 120–121, 143
   of Homo sapiens sapiens, 69–70
   vs hunting, 61–63
   inferior parietal lobe and, 124, 452–454
   innateness of, 490
   lateralization of, 60–61, 70–71, 77, 118–120, 241–242, *242*, 526
   left temporal lobe and, 487–488
   middle temporal lobe and, 502
   multiple personalities and, 576
   of Neanderthal, 59, 63
   in nonhuman primates, 60–61, 237
   parietal lobe and, 468–368
   phonemes of, 487–489, 491
   reality and, 492–493
   right hemisphere lesions and, 79–81, 410–413, 550–551
   sex differences in, 66–69
   split-brain and, 576
   structure of, 488–489
   temporal lobe epilepsy and, 515–516
   temporal-sequential activity and, 68–69, 143, 255–256
   vowels of, 489
Language axis, 241–242, *242*
Lateral line system
   evolution of, 22–23
   of lobed-finned fish, 18
Lateral medullary syndromes, 377, 732
Lateralization, 105
   of language, 60–61, 70–71, 77, 118–120, 241–242, *242*, 526
   of memory, 105–116, 534–536
   of spatial perceptual skills, 62
Laughter, 359–360
   frontal lobe lesions and, 430–431
   hypothalamic lesions and, 172–173
   temporal lobe seizures and, 187, 512
Learned helplessness, 601
Learning, 219–220
   amnesia for, 208–209
   amygdala and, 198–199, 226–227, 231
   anesthesia and, 208
   cerebellum and, 385–386
   emotional, 219, 227
   hippocampus and, 197–198, 217, 218, 225–226
   long-term potentiation and, 217
   neural networks and, 211–213
   unconscious, 208
   unilateral, 106–107
Left hemisphere neglect, 93, 466–467, *467*
Left-sided apraxia, 456–457
Legs, evolution of, 17
Lethargy, reticular activating system and, 364–365
Life, origins of, 309–310, 316–317
Life review, 281
Light therapy, in depression, 599

Limbic language, 235–266
    amygdala and, 239–242
    cingulate gyrus and, 242–246
    evolution of, 26–27, 238–239, 246–247, 254–255
    in females, 248–249
    hierarchical organization of, 237–239
    of infant, 128–129, 235–237, 246–251
    inferior parietal lobe and, 123–124, 242, 254–255
    maternal, 246–251
    music and, 86
    neocortical processing of, 251–255
    of nonhuman primates, 60, 235, 237
    overview of, 256–257
    periaqueductal gray and, 239
    right hemisphere and, 79–81, 251–253
    segmentation of, 59
    sex differences in, 82–83, 248–249
Limbic love, 166–167, 257–261
Limbic striatum, 161, 323
    Alzheimer's disease and, 342–345
    fetal development of, 631
    mesolimbic DA system and, 342–343
    in pleasure circuit, 327–328
Limbic system, 9, 26, 161–205
    aggression and, 189–190
    amygdala of, 178–193. See also Amygdala
    anatomy of, 195, 236, 238, 244, P8
    aversion and, 172, 258–259
    aversive emotion and, 196–197
    cerebellum and, 388
    circadian rhythm generation and, 174–175
    at death, 286
    depression and, 174–177
    dreams and, 302–304
    emotional expression and, 187–189, 251–253
    emotional intelligence and, 161–205. See also Emo-
        tional intelligence
    evolution of, 9, 27–28
    female, 248–251
    fetal, 629–630
    hallucinations and, 280, 283–284
    hate and, 266–267
    hippocampus of, 193–220. See also Hippocampus
    in homosexuality, 170–171
    hunger and, 171
    hyperactivation of
        birth of god and, 298–300
        inheritance of, 298
        religious experience and, 283–284, 296–298
    hypothalamus of, 169–178, 169. See also
        Hypothalamus
    hypothalamus-pituitary-adrenal axis in, 175–177,
        176
    hysteria and, 608–609
    in infant, 235–237
    laughter and, 172–173
    of left hemisphere, 162
    love and, 266–267
    maternal behavior and, 249–251
    murder and, 286–291
    of Neanderthal, 48
        olfactory, 9, 162–164, 164
        orbital mediation of, 413–414
        pheromonal communication and, 162–169, 164
        pleasure and, 171–172
        in primates, 60, 237
        rage and, 173–174
        relationships and, 266–267
        religious experience and, 269, 273–280, 283–284,
            296–298
        reward and, 171–172
        schizophrenia and, 616
        seasonal affective disorder and, 174–175
        septal nuclei of, 257–261, 258
        sex differences in, 188–189, 248–251
        sexuality and, 289–290, 292–294
        social-emotional attachment and, 257–261
        social-emotional intelligence and, 235–267. See
            also Social-emotional intelligence
        social-emotional deprivation and, 261–266
        species differences in, 163
        stress effects on, 672–673
        testosterone and, 190–191
        thirst and, 171
        traumatic memory and, 524
Linear fractures, 679–682
Lissencephaly, 647, 648
Literal alexia, 155
Lizard, brain of, 21
Lobed-finned fish, 17–18
Local-circuit neurons, 636
Locked-in syndrome, 734
Locus ceruleus
    depression and, 600–601
    lesions of, 367
    norepinephrine production of, 366–367
Long-distance projecting neurons, 636
Long-term synaptic potentiation
    amygdala and, 216
    fear and, 585–587
    hippocampus and, 215–216
    learning and, 217
    neural networks and, 213–214
Lot, 296
Love, limbic system and, 266–267
Lung fish
    central nervous system of, 18–19
    evolution of, 310
Lymphoma, 748–749
D-Lysergic acid diethylamide (LSD), 368
    hallucinations and, 201, 510

Magnetic groping, frontal lobe lesions and, 403
Magnocellular dorsal medial thalamic nucleus,
    413–414
Malignancy, 738–739. See also Neoplasm(s)
Malleus Maleficarum, 289
Malnutrition, congenital malformations and, 652–653
Mammals, evolution of, 24–30
Mammillary glands, evolution of, 247
Mania, 601–603
    caudate lesions and, 330–332

frontal lobe lesions and, 435–436
right hemisphere lesions and, 100–101
Maternal behavior, 26, 246–251
amygdala lesions and, 185, 264–265, 619
anterior cingulate gyrus and, 25–26
anterior cingulate gyrus lesions and, 246
in males, 251
in nonhuman primates, 249–251
Math
evolution of, 461–463
left hemisphere lesions and, 147
music and, 87–88
right hemisphere and, 91
Medial forebrain bundle, self-stimulation and, 171–172
Medial medullary syndromes, 732
Median raphe nuclei, *368*
serotonin secretion by, 368–369, 589–590
Medulla, 354–355. See also *Brainstem*
cranial nerves of, 376–377, *377*
infarction of, 732
Medullary syndromes, 732
Medulloblastoma, 743–744
Megalencephaly, 647, *649*
Melatonin, depression and, 599
Melodic-intonation therapy, 84–85
Melodic-intonational axis, 500–501, *500*, 527. See also *Emotional-melodic speech*
Membrane, of single-celled organism, 3
Memory. See also *Amnesia; Memory loss; Repression*
aging and, 231–232
Alzheimer's disease and, 344
amygdala and, 197–200, 216, 226–227, 230–232
anesthesia and, 208
anterior hippocampus and, 217
authority and, 558–559
for childhood events, 113, 521
competing, 557–558
contextual cues and, 534, 535, 543–546, 549–550
corpus callosotomy and, 526
earliest, 113, 521
emotional, 219, 521–522, 537
amygdala and, 226–227, 533–534, 581
in children, 531, 532
contextual cues and, 538–539
earliest, 521
right temporal lobe and, 532
emotional stress and, 546–549
false, 556–559
in females, 189
frontal lobes and, 220–221
frontal-thalamic damage and, 223–224
gaps in, 207–208, 561–562
confabulation and, 525
globus pallidus and, 343
hippocampus and, 197–200, 215–220, 534, 538
in infant, 227, 522–523, 544
inferior temporal lobe and, 507–509
inhibition of, 229. See also *Repression*
instability of, 215
isolation of, 215, 536–539, 540

in Korsakoff's syndrome, 223–224
lateralization of, 105–116, 534–536
long-term, 217–218
long-term synaptic potentiation and, 213–216, 217
loss of. See *Amnesia; Memory loss*
maturation of, 661
mesolimbic DA system and, 342–343
middle temporal lobe and, 504
multicontextual, 544
neocortical maturation and, 536–538
neural networks in, 211–213, 537
norepinephrine and, 343–344
olfaction and, 9, 163–164
opiate secretion and, 587–588
overview of, 232–234
parietal lobe and, 468
peer pressure and, 558–559
permanence of, 215
posterior hippocampus and, 218
prenatal, 522
recognition, 539
amnesia and, 209, 539
emotional trauma and, 584–585
hippocampus and, 219–220
recovery of, 549–550
right hemisphere lesions and, 93
serotonin and, 344
short-term, 217–218
state-dependent, 588
synaptic overlay hypothesis of, 537
thalamus and, 220, 221, 221–225, 222
traumatic, 226, 540–541, 545–546, 588–589. See also *Post-traumatic stress disorder*
contextual cues and, 534, 540–541
limbic system and, 524
recall of, 545–546
triggers of, 542. See also *Contextual cues*
unilateral storage of, 106–107, 534–536
verbal
for childhood events, 521, 522–523
in children, 522–523, 531
distortion in, 559–560
vs emotional memory storage, 532–533
left temporal lobe and, 532
visual, 559
of witnesses, 547, 560–561
Memory loss, 231. See also *Amnesia; Repression*
amygdala dysfunction and, 231–232
context and, 544
emotional arousal and, 571, 583–585
frontal lobe lesions and, 413, 423–426
functional, 539–541
hippocampal lesions and, 218–219
hippocampus and, 225–226
memory recovery after, 211, 231, 538–539, 541–542, 543–546
right hemisphere lesions and, 89–90, 93, 106, 550–551
sexual assault and, 584
temporal lobe lesions and, 508–509, 532

Meninges, 675
    tumors of, 745–746
Meningioma, 745–746
Menstrual cycle, pheromones and, 167
Mesoderm, 626
Metastasis, in tumor development, 738
Microcephaly, 648–649
Microencephaly, 647
Microgyria, *649, 651*
Micropolygyria, 647
Microtubules
    in multicellular organism origin, 5
    of single-celled organisms, 4–5
    of sponges, 6
Midbrain, 354, 358–362, *362*
    auditory functions of, 361–363
    cranial nerves of, 380–382
    dopamine production of, 366
    dorsal view of, *11, 362*
    evolution of, 9–10, 23–24, 360–361
    fetal, 630–631
    inferior colliculus of, 361–362
    medial view of, *12*
    neoplasm of, 744
    of reptiles, 21
    superior colliculus of, 361
Middle ear, evolution of, 23–24, *23*
Middle Paleolithic
    religious experience of, 269–273
    tool technology of, 54–57
Middle-to-Upper Paleolithic transition
    frontal lobe and, 42–43, 45–47
    gathering and, 62–63
    hunting and, 62–63
    inferior parietal lobe and, 42–43, 53–54
    language and, 62–63
    tool technology of, 54–57
Mitral valve prolapse, ischemia and, 723
Monkeys
    limbic vocalization in, 60, 237, 248
    right hemisphere of, 70
    social-emotional deprivation in, 262, *263*
Monologue. See *Egocentric speech*
Montana Mills incident, 605
Mood. See also *Emotional-melodic speech; Limbic language*
    limbic nuclei in, 26–27
    of sounds, 237
Morphemes, 488
Mortuary practices
    of Homo sapiens, 270
    of Neanderthals, 47–49, 269–270, *270*
Moses, 297
Motivation, parietal lobes and, 450–451
Motor circuit, 323–354, *324, 325, 326, 336*
    in Alzheimer's disease, 342–345
    amygdala in, 333–335, 340, 348, 353
    in catatonia, 333–335
    caudate in, 346
    cerebellum in, 385–386
    corpus striatum and, 20–22, 24, 324–332, *329, 331*

dopamine in, 351–352
    evolution of, 22
    feedback loops in, 345–348
    GABA in, 351–352
    globus pallidus in, 346–348
    in Huntington's chorea, 352–353
    motor thalamus in, 348–349, *350, 351*
    obsessive-compulsive disorders, 340–342
    in Parkinson's disease, 335–339. See also *Parkinson's disease*
    in psychosis, 339–340
    putamen in, 346–347
    split-brain and, 104–105
    subthalamic nucleus in, 349–351
    supplementary motors areas in, *334,* 339–401, 345–346
Motor engram, 454, 456
Motor homunculus, *95*
Motor response, on Glasgow Coma Scale, 692
Motor thalamus, 348–349, *350, 351*
Movement. See also *Apraxia; Motor circuit*
    anterior cingulate gyrus and, 405–406
    ballistic, 347
    brainstem and, 356–357
    cerebellar disturbances of, 386–387
    cerebellum and, 385–386
    corpus striatum and, 324–332, *329, 331*
    Exner's writing area and, 408–409, *408*
    GABA and, 352–353
    globus pallidus and, 346–348
    hand-manipulation cells and, 446
    in Huntington's chorea, 352–353
    midbrain response to, 360–361
    motor thalamus and, 348–349, *350, 351*
    parietal lobes and, 445, 446, 450–451
    perseverative, frontal lobe lesions and, 417–419, *417, 418, 419*
    planning for, parietal lobes and, 450–451
    premotor neocortex and, 398–399
    primary motor area and, 396–397
    putamen and, 346–347
    sensory guidance of, 454–455
    subthalamic nucleus and, 349–351
    supplementary motor area and, 345–346, 399–400
    timing of, 386
Muhammad, 294–295
Multicellular organisms, origin of, 5
Multi-infarct dementia, 721
Multiple personality, 574–576
    brain injury and, 578–579
    hypnosis and, 579–580
    language and, 576
    sexual abuse and, 577–579
Multipolar stellate neurons, 636
Multiregional replacement theory, of Homo sapiens evolution, 36
Multiregional theory, of Homo sapiens evolution, 35, 316
Murder
    limbic system and, 286–291
    religious, 290–292

Murmurs, 723
Muscle atonia, 376
    during sleep, 374
Music
    emotion and, 86
    environmental sounds and, 86
    hallucinations of, 495
    left hemisphere and, 86–87
    limbic language and, 86
    math and, 87–88
    neuronal experience of, 491
    right hemisphere and, 84–85
Musical hallucinations, 495
Mutation, adaptive, 311–312
Mutism, 253–254, 400
Myelination, cerebral development and, 639–640
Myocardial infarction, 722
Mystical state, 279–280. See also *Religious experience*

Naming, inferior parietal lobe and, 126
Narcolepsy, 374–375
Narrative, vs reconstruction, 523
Neanderthals
    angular gyrus in, 54
    brain development in, 40–42
    endocast of, *39, 64*
    environmental influences on, 42
    frontal lobe of, 38–43, *38, 45–47*
    hunting behavior of, 46, 62
    language of, 59, 63
    limbic system of, 48
    mobility of, 46–47
    mortuary practices of, 47–49, 269–270, *270*
    planning behavior of, 45–46
    religious rituals of, 48
    replacement of, 38–43, 65–66
    ritual symbolism of, 4
    skull of, *33, 37, 40, 44*
    speech of, 63
    temporal lobe of, 49
    tool making of, 54–57
    Wernicke's area of, 63
Near death experience, 280–282. See also *Religious experience*
    dark tunnel of, 271–272
    limbic hyperactivation and, 283–284
Necrosis
    fetal, 650
    traumatic, 686
    tumor, 738
Neglect
    frontal lobe lesions and, 93–94
    left hemisphere lesions and, 93, 466–467, *467*
    parietal lobe lesions and, 464–470
    right hemisphere lesions and, 92–94, *92, 93*
Neocerebellum, 384–386, *384*
    cognition and, 387
Neocortex, 28–29, 161
amygdala interactions with, 226
cellular organization of, 395–396
convolutions of, 28

crowding effect in, 70–71
evolution of, 16–17, 28–30, *29*
fetal, 629–630, 632–637
    neural migration in, 635–636
gyri of, 28
hippocampus interactions with, 194, 195–196, 225
layer I of, 632–634
layer VII of, 632, 634–635
layers of, 28
lobes of, 28–29
Neonatal brain development, 654–674, *659*
    abuse and, 664–674
    corpus callosum in, 655–656
    cross-modal processing and, 658–660
    disturbances of, 661–665
    environmental influences on, 661–665
    hemispheres in, 656–658
    language acquisition and, 660–661
    limbic system in, 672–673
    lobes in, 656, 657
    memory and, 661
    neural plasticity and, 664–665, 673–674
    norepinephrine and, 669–674
    traumatic injury during, 665–674, *666, 667*
        recovery after, 669–674
Neoplasm(s), 311, 736–751
    astrocytic, 746–747
    bilateral, 749
    brainstem, 742
    cerebellar, 742–743
    cognitive changes with, 740–741
    dissemination of, 738
    extrinsic, 739
    frontal lobe, 741–742
    glial, 747–748
    grading of, 739
    growth rate of, 740
    headache and, 741
    herniation and, 749–751
    intracranial pressure increase and, 740
    intrinsic, 739–740
    invasion of, 738
    lymphatic, 748–749
    meningeal, 745–746
    metastasis from, 738
    midbrain, 744
    necrosis and, 737–738
    neuroblastoma, 749
    occipital lobe, 742
    oligodendroglial, 748
    oncogenes and, 736–737
    parietal lobe, 742
    pineal gland, 744
    pituitary, 744–745
    prognosis for, 751
    sarcomatous, 749
    seizures and, 741
    telomerase and, 736–737
    temporal lobe, 742
    unilateral, 749
    vascularization and, 737–738

Neoplasm(s), *continued*
    ventricle, 743–744
    vomiting and, 741
Nerve cords, 8, 9
Nerve net, 8–9, *8*
Neural networks
    brain injury and, 578–579
    emotional trauma and, 588–589
    isolated, 215
    long-term potentiation and, 213–214
    in memory, 211–213, 537
    parallel, 214–215
    selective activation of, 223
    sequential, 214–215
    trauma-related development of, 588–589
Neural noise, hallucinations and, 509–510
Neural plasticity, 661–663
    abnormal environmental experience and, 664–665
    norepinephrine in, 671–674
    in visual system, 663
Neural plate, 626–627
Neural tube, 626–627
Neuroblast, migration of, 629
Neuroblastoma, 749
Neuroepithelium, 627
Neuroma, acoustic, 745–746
Neuron(s)
    evolution of, 7–9
    experience-expectant development of, 663
    god, 305–308
    local-circuit, 636
    long-distance, 636
    migration of, 633–636
    nasal, 6
    photosensitive, 10, 12–13
    post-traumatic, 669
    pyramidal, 636
    vs single-celled organisms, 4–5
    stellate, multipolar, 636
    transplantation of, 642–643
Neuronal grafts, 642–643
Neuronal heterotopia, 641, 642–644
Neurotransmitters, 10
    cerebral development and, 637–640
    long-term potentiation and, 214
    microtubule transport of, 4
    rhythmic release of, 13
    of sponges, 6
Nigrostriatal system, *326*
Nipples, evolution of, 247
Norepinephrine
    in arousal, 367
    brainstem production of, 366–367
    cerebral development and, 637–639
    depression and, 175, 177, 370–371, 600–601
    emotional trauma and, 591–593
    memory and, 343–344
    pain and, 369
    in Parkinson's disease, 338
    post-traumatic secretion of, 669–670
    right hemisphere lesions and, 100

    serotonin interaction with, 369–371
    stress and, 370
Nucleus accumbens, 342
    in pleasure circuit, 327–328
Nucleus basalis, 323, 324

Object localization, parietal lobes and, 451
Object recognition, parietal lobes and, 466
Obsessive-compulsive disorders, 609–610
    basal ganglia and, 340–342
    cingulotomy in, 243–244
    frontal lobe lesions and, 434–435
Occipital hemorrhage, 730
Occipital lobe(s), 471–482
    binocular input to, 476
    cells of, 472–474
    laterality of, 478, 481–482
    lesions of, 477–482
        blindness and, 478–479
        color recognition and, 481–482
        hallucinations and, 476, 478, 510, 610
        homonymous hemianopsia and, 477–478
        pantomime recognition and, 457
        prosopagnosia and, 480–481
        quadrantanopsia and, 477–478
        simultanagnosia and, 481
        visual agnosia and, 479–480
    maturation of, 656, 657
    neoplasms of, 742
    stimulation of, hallucinations and, 610
    striate cortex of, 471, 474–476, 477
        columnar organization of, 472, *475*
        complex cells of, 472–474
        feature detectors of, 472–474, *475*
        hypercomplex cells of, 474
        nonvisual interconnections of, 475–476
    visual association areas of, 471, 476–478
    visual information pathway of, 471–472, *473*
Oculomotor nerve, *377, 382*
Olfaction, *164*
    disorders of, 379
    evolution of, 165–166
    hallucinations of, 512
    memory and, 9, 163–164
    pheromonal communication and, 162–164
    sex differences in, 166
Olfactory bulb, 9, *12*
    fetal, 627–629
Olfactory hallucinations, of temporal lobe seizure, 512
Olfactory lobe, 6
Olfactory nerve, 379, *684*
Oligodendroglioma, 748
Oncogenes, 311, 736–737
Ophthalmic artery, 710
Opiates
    at death, 308
    in frozen state, 335
    limbic nuclei secretion of, 280
    memory and, 587–588
    norepinephrine release and, 591–592
    receptors for, in Parkinson's disease, 337

rewarding effects of, 327–328
secretion of, 280
    stress and, 585–588
withdrawal from, 327–328
Optic ataxia, 481
Optic nerve, 380–382
Optic tectum, 15–16
Oral utilization, frontal lobe lesions and, 404
Out-of-body experience, 280–282. See also *Religious experience*
    fear and, 282–283, 576–577
    limbic hyperactivation and, 283–284
    temporal lobe seizure and, 284–286, 513–514
Oval window, evolution of, 23–24, *23*
Ovulation, pheromones and, 167
Ovum, meiosis of, 5
Oxygen
    neuronal evolution and, 3
    single-cell diversification and, 4
Ozone, single-cell diversification and, 4

Pain
    anterior cingulate gyrus and, 243–244
    hysteria and, 448
    norepinephrine and, 369
    parietal lobe lesions and, 529–530
    response to, parietal lobes and, 447
    right hemisphere lesions and, 96–97, 529–530
    serotonin and, 369
    transmission of, 114
Paleocerebellum, emotion and, 387–388
Paleolithic cultural transition, 42–43
Paleoneurology, 42–43
Pallidum, ventral, 323, 324, 342, 344
Palsy, pseudobulbar, 359
Pantomime recognition, left hemisphere lesions and, 148, 457–458
Parallel neural networks, 214–215
Paralysis
    facial, 680
    motor area lesion and, 397–398
    sleep, 374, 375
    supplementary motor area lesion and, 399–400
Paramecium, 4
Paramnesia, reduplicative, frontal eye fields and, 407
Paranoia, dopamine in, 339–340
Parietal eye, 11–12, *12*
Parietal hemorrhage, 730
Parietal lobe(s), 441–470
    attention and, 464–470
    body image representation and, 44, 443, *444*, 446, 448–449
    command functions of, 450–451
    delusional playmates and, 469–470
    egocentric speech and, 469–470
    evolution of, 254
    expansion of, 40
    eye movements and, 449–450
    functional laterality of, 443
    grasping and, 450–451
    hand control of, 450–451, 459, *460*

hand-manipulation cells of, 446
hemorrhage of, 730
hippocampus interconnections with, 449–450
inferior, 49–54, 452–454, *453*
    anatomy of, 452
    evolution of, 57–58, 121, 124, 149
    in females, 68–69
    language and, 69, 124, 453–454
    lesions of, 52, 453–454
    Middle-Upper Paleolithic transition and, 53–54
    multimodal processing of, 124–125, *124*, 452–453, *453*
    naming and, 126
    of Neanderthal, 54
    reading and, 125
    sensory areas of, 124–125, *124*
    sex differences in, 68–69
    temporal-sequential activity and, 52, 123–124, 453–454
    tool making and, 52–53
    train of associations and, 125–126
    writing and, 53, *453*
language and, 69, 124, 453, 468–469
laterality of, 455
left, 53, 443
    lesions of, 451–452
        apraxia and, 451, 455–458
        constructional apraxia and, 458
        neglect and, 466–467, *467*
        pantomime recognition and, 457–458
    movement functions of, 455
    seizures of, 448
    lesions of, 441, 451–452
        acalculia and, 463, 464
        agnosia for numbers and, 463
        agraphia and, 453
        agraphia for numbers and, 463–464
        alexia for numbers and, 463–464
        anarithmetria and, 463
        aphasia and, 457
        apraxia and, 451, 454–458, 455–458
        astereognosis and, 446–447
        confabulation and, 468–469
        constructional apraxia and, 458
        delusional denial and, 467–468
        disconnection and, 468–469
        dressing apraxia and, 53, 457
        egocentric speech and, 469–470
        emotion and, 451–452
        finger agnosia and, 459–461
        gap filling and, 468–469
        Gerstmann's syndrome and, 459–464
        hysteria and, 603
        ideational apraxia and, 456
        ideomotor apraxia and, 456
        left-sided apraxia and, 456–457
        neglect and, 450, 464–467, *467*
        object localization and, 451, 466
        pain and, 448, 529–530
        pain tolerance and, 447
        pantomime recognition and, 457–458

Parietal lobe(s), *continued*
    position sense and, 446–447
    reading and, 457
    right-left disorientation and, 464
    somesthetic agnosias and, 443–445
    spatial acalculia and, 464
    tactile discrimination and, 446–447
    temporal sequencing and, 454
    visual-spatial function and, 451
  maturation of, 656, 657
  motivation and, 450–451
  in motor functioning, 50–52, 446, 454–455
  multimodal information processing of, 124–125,
    124, 448, 452–453, *453*
  neoplasms of, 742
  pain areas of, 447
  right, 53, 443
    lesions of, 451–452
      constructional apraxia and, 458
      delusional denial and, 467–468
      dressing apraxia and, 53, 457
      emotional functioning and, 451–452
      neglect and, 464–466, *467*
      somesthesis and, 446–447
      visual object localization and, 451
      visual-spatial function and, 451
    movement functions of, 455
    seizures of, 448
  seizures of, 448
  sensory homunculus of, *442*
  sex differences in, 68–69
  somesthetic association area of, 445–451
    lesions of, 446–447
  somesthetic receiving area of, 441–443, *442*, *444*,
    445
    lesions of, 443–445
  spatial body position and, 446, 448–449
  stereognosis and, 446–447
  superior-posterior, 448–451
  supramarginal gyrus of, 447
  temporal lobe interaction with, 52
  three-dimensional analysis of, 448–449
  topography of, 441
  visual association areas of, 476–477
  visual attention and, 450
  visual receptive fields of, 449
  visual-spatial areas of, 449–450
Parietal wasting, 454
Parkinson's disease, 328, 335–340, 352
  acetylcholine in, 339
  basal ganglia and, 336–340
  dopamine in, 337–339, 345–346, 352
  frontal lobe and, 339
  hippocampus and, 338–339
  neuronal transplantation in, 642–643
  putamen in, 337, 348
  striatal imbalance and, 337–339
  supplementary motor areas and, 338–339
Peer pressure, memory and, 558–559
Periaqueductal gray, 358–359
  vocalization and, 239, 358–360

Perseveration
  frontal lobe lesions and, 341–342, 412, 416–419,
    416, 417, 418, 419, 434
  in Middle Paleolithic, 45
  of obsessive-compulsive disorder, 340–342
Personal identity, 564
  loss of, 572–574
Personality. See also *Emotion; Multiple personality*
  frontal lobe lesions and, 427–428
  right hemisphere lesions and, 99–101
PGO waves, 372–373
Pheromones, 6, 164–169
  composition of, 165
  evolution of, 165–166
  mating behavior and, 165, 167
  receptors for, 6, 166
  sex and, 166–168
Phonemes, 487–489, 491–492
Phonetics, 488–489
Photophobia, mild head injury and, 705
Photosensitive cells, 7, 10
  rhythm sensitivity of, 12–13
Photosensitivity, 7, 10
Picture Arrangement, frontal lobe lesions and, 438
Picture Completion, frontal lobe lesions and, 437–438
Picture writing, of Upper Paleolithic, *70*
Pinealoma, 744
Pithecanthropus, endocranial cast of, *39*
Pituitary gland, neoplasm of, 744–745
Pituitary/pineal eye, 11–12, *12*
Play, of females, 250
Playmates, delusional, 469–470, 578
Pleasure, 203–204
  aggression as, 189–190
  hypothalamus and, 171–172, 177–178
Pleasure circuit, 327–328
  norepinephrine in, 367
Pleasure principle, 178
Pons, 12, 354. See also *Brainstem*
  cranial nerves of, 377–379, *377*
  dopamine production of, 366–367
Pontine, 729–730
Pontine flexure, fetal, 631
Postcentral gyrus, 441–443. See also *Parietal lobe(s)*
Postconcussion syndrome, 703–706
Post-traumatic amnesia, 209–210
Post-traumatic stress disorder, 229, 370, 571–572
  flashbacks in, 587
  hallucinations and, 583
  hippocampus in, 608
  startle response in, 588–589
Posturing, frontal lobe lesions and, 401–402
Predmost, endocranial cast of, *39*
Prefrontal lobe, lesions of, 45
Premotor neocortex, 398–399
Preoptic area, sexual dimorphism in, 170
Preplate, 626
Prevertebrates, neural organization of, 9
Primary motor area, 396–398
Primary process, in infant, 204–205
Primates

evolution of, 31–32
language in, 60–61
limbic vocalizations in, 60, 237
Prosencephaly, 646–649, *647*
Prosody. See also *Emotional-melodic speech*
emotional, 251–253
frontal lobe lesions and, 410–411
Prosopagnosia
right hemisphere lesions and, 97–99, 480–481
temporal lobe lesions and, 506–507
Protochordates, nervous system of, 3–10, *8*
Pseudobulbar palsy, 359
Pseudotumor, 749
Psychic blindness, 184–185, 509
Psychosis, 328, 339–340, 595–622. See also
Schizophrenia
abuse and, 591–592, 621
amygdala and, 340, 616–618
aphasia and, 516–517
basal ganglia and, 339–340
between-seizure, 617–618
caudate injury and, 330
cerebellum and, 387–389
child abuse and, 591–592, 621
congenital, 619–622
dopamine in, 592–593, 614–618
environment in, 619–622
functional localization and, 595–596
heterotopia in, 643–644
left hemisphere lesions and, 137–138
neurodevelopmental, 641–642
norepinephrine and, 591–593
prenatal insults in, 619–621
speech in, 617
stress and, 369–370
temporal lobe epilepsy and, 514–518
temporal lobe lesions and, 340, 615–616
Pulse rate, recognition memory and, 584–585
Punishment, hippocampus and, 196–197
Pure word blindness, 154
Pure word deafness, 138–139
Putamen, *326*, 328–332, *329*, *348*
functional differentiation of, 329–330, *329*
motor functions and, 328–329, 346–347
Parkinson's disease and, 337, 348
patches and matrix of, 328
topographical input to, 330
Pyramidal neurons, 636
Pythagoras, 87–88

Quadrantanopsia, 477–478
Quetzacoatl, *276*
Quiescence, septal nuclei and, 259

Rage
amygdala and, 182–183
hypothalamus and, 173–174
septal nuclei and, 259
sham, 389
Rain dance, 268
Ramapithecus, 32

Raphe nuclei, *368*
serotonin secretion by, 368–369, 589–590
Rational mind, 161–162
Ravel, Maurice, 85
Reading
abnormalities of, 91–92, 154–156
agnosia and, 480
evolution of, 148–150
inferior parietal lobe and, 125, 453
left parietal lobe lesions and, 457
right hemisphere lesions and, 91–92
sex differences in, 68, 158
Reasoning, evolution of, 16–17
Receptive aphasia, 139–140, 499–500
Recognition memory, 539
amnesia and, 209, 539
emotional trauma and, 584–585
hippocampus and, 219–220
Relationships, limbic system and, 266–267
Religious experience, 268–319
amygdala and, 201, 273–277, 279–280
animal spirits and, 301
astral projection and, 284–286
bird symbol and, 275, *275*
cross symbol and, 274–275, *276, 277*
day dreams and, 305
death and, 286, 308, 317–318
dreams and, 301–305, *303, 304*
evolutionary significance of, 286, 305–309, 312–316
exo-biological cerebral organization and, 314–315
fear and, 282–283, 286–287, 300–301
frontal lobe and, 286–291
geometric shapes and, 274–275, *276, 277*
hallucinations and, 283–284, 297–298
hypothalamus and, 278–279
infinite aggress and, 312–314
isolation and, 297–298
life's origins and, 309–310
limbic hyperactivation and, 283–284, 296–300
limbic system and, 278–280, 286–291
of Middle Paleolithic, 269–270, *270, 271*
murder and, 286–292, *290, 291*
of Neanderthals, 48
near death experience and, 280–282
neocortical origins of, 314–315
neuronal origins of, 305–308
otherworldly experience and, 284
out-of-body experience and, 280–282, 284
right hemisphere and, 304–305
sexual experience and, 286–296
soul dreams and, 301–302
temporal lobe and, 273–277, 279–280
temporal lobe hyperactivation and, 304–305
temporal lobe lesions and, 598
temporal lobe seizure and, 284–286, 296–297,
513–514
of Upper Paleolithic, 270–273, *271, 272, 273*, 277–278
wondjina figures and, *274*
REM sleep behavior disorder, 375
Remembering. See also *Memory*
evolution of, 16–17

Replacement theory, of Homo sapiens evolution, 35–36
Representational neglect, 93
Repression, 224, 229, 521–563. See also *Amnesia; Memory; Memory loss*
  vs amnesia, 209, 523–530, 536
  amygdala and, 533–534
  in children, 536
  coping style of, 555–556
  corpus callosum immaturity and, 524–526, 536, 554–555
  disconnection and, 554–555
  dream recall after, 542–543
  emotional arousal and, 546–549, 552–553, 571
  experts on, 562
  Freud's views on, 523, 571, 580
  frontal lobes and, 550–553
  hemispheric specialization and, 526–532
  individual differences in, 552, 555–556
  infant memory and, 522–530
  memory isolation and, 536–539
  psychic cleavage of, 523–530
  psychic conflicts and, 553–555
  sex differences in, 556
  sexual abuse and, 533–534, 538, 581–582
  synaptic overlay hypothesis of, 537
  thalamus and, 551–553
  triggered recall after, 538–539, 541–542, 543–546
  unilateral memory storage and, 534–536
Reptiles
  nervous system of, 21–22
  skeletal structure of, *25*
Repto-mammals
  nervous system of, 24–25
  skeletal structure of, *25*
  social behavior of, 27
Respiration, amygdala and, 228
Reticular activating system, 355, 363–365
  hippocampus interaction with, 230
  REM sleep and, 373
Reticular formation, evolution of, 19
Reticular thalamus, 420–421
Retina, signal transmission from, 11–12
Retrograde amnesia, 210–211. See also *Amnesia*
  post-traumatic, 695
  shrinking, 211
Reward
  hypothalamus and, 171–172
  putamen and, 347
Rhinencephalon, 9. See also *Amygdala; Hippocampus*
Rhythmicity
  limbic system and, 174–175
  of music, 86–87
  neuronal, 12–13
  of norepinephrine, 367
  of salamander motion, *18*
Ribonucleic acid, origin of, 310
Right-left disorientation
  left hemisphere lesions and, 146
  parietal lobe lesions and, 464
Rigidity. See *Catatonia; Frozen states*

Rotation forces, in head injury, 687–688
Running
  of Neanderthal, 54
  during temporal lobe seizure, 512

Salamanders, rhythmic motions of, *18*
Sapir, Edward, 492
Sarcoma, 749
Sarcopterygian fish, nervous system of, 17–18
Satan, amygdaloid hyperactivation and, 583
Schizophrenia, 612–614
  amygdala and, 340
  aphasia and, 141
  catatonic, 612, 614
  child abuse and, 621
  dopamine in, 339–340, 592–593, 616–617
  emotional expression in, 616
  frontal lobe lesions and, 340, 435, 612–614, 621–622
  heterotopia in, 643–644
  language in, 515–516
  late-onset symptoms of, 644
  left hemisphere lesions and, 137–138, 141
  limbic system and, 616
  neurodevelopmental failure in, 641–642
  prenatal insults in, 619–621
  septal nuclei in, 264
  sex differences in, 621–622
  speech in, 616, 617
  temporal lobe epilepsy and, 514–518
  temporal lobe lesions and, 340, 615–616
Schwannoma, 745–746
Seasonal affective disorder, 174–175
Seizures
  amygdala origin of, 185–187
  anterior cingulate origin of, 245
  glial scarring and, 699–700
  heterotopia in, 643–644
  temporal lobe, 511–513. See also *Temporal lobe seizures*
  in tumor development, 741
Self-concept, 564
Self-explanation, of egocentric speech, 132–133
Self-stimulation, medial forebrain bundle and, 171–172
Sensorimotor cortex, 454
Separation cry, 26, 247–248
Septal nuclei, 257–261, *258*
  amygdala interactions with, 260–261
  aversion and, 258–259
  contact comfort and, 259–266
  fetal, 631
  hippocampus interactions with, 193
  internal inhibition and, 258–259
  lesions of, 259–260
    aggression and, 259
    contact seeking and, 259–260
    social-emotional deprivation and, 264–265
  quiescence and, 259
  rage and, 259
  socialization and, 259–260
Sequential neural networks, 214–215
Serotonin, 589–591

brainstem production of, 367–369, 589–590
cerebral development and, 639
depression and, 370–371, 591, 600–601
in dreaming, 368
fantasy life and, 579–580
function of, 589
hypothalamic regulation of, 175
memory and, 344
norepinephrine interaction with, 343–344, 369–371
pain and, 369–370
reduction in, 343–344, 369–370, 589–591
sexual behavior and, 357, 370
sources of, 589–590
stress and, 369–370, 589–590
Sewing needle, of Upper Paleolithic, 57
Sex differences
in aggression, 189–190
in amygdala, 170, 188–189
in anterior cingulate gyrus, 82–83, 249–251
in anterior commissure, 188, 249
in aphasia, 69
in cognition, 66–69
in corpus callosum, 68, 136–137
in depression, 137, 189, 597
in emotional expression, 82–83, 188–189
in expressive aphasia, 136–137
in hypothalamus, 169–171
in hysteria, 603–605, 607, 608–609
in language, 66–69
in left cerebral hemisphere, 136–137, 158
in left parietal lobe, 68–69
in limbic language, 82–83, 248–249
in limbic system, 188–189, 248–251
in olfaction, 166
in reading, 68, 158
in repression, 556
in right cerebral hemisphere, 82–83, 89
in schizophrenia, 621–622
in spatial perceptual skills, 89, 192
in stria terminalis, 179–180, 188
in tool making, 58–59
in writing, 68
Sexagesimal system, 462
Sexual abuse. See also Emotional trauma
amnesia for, 539–541, 545–546
amygdala hyperactivation and, 533–534, 581–582
breath holding and, 583
dissociation and, 577–579
ego splitting and, 582
frozen states in, 582
hypoxia in, 582
multiple personality and, 577–579
recall of, 541, 545–546
repression of, 533–534, 581–582
trance state and, 579
Sexual assault, amnesia for, 539–541, 584
Sexual behavior
amygdala ablation and, 187–189, 188
amygdala and, 191–192, 533
amygdala hyperactivation and, 580–583
frontal lobe lesions and, 431
frontal lobe seizures and, 187
hypothalamus and, 279
limbic system and, 289–290, 292–294
pheromones and, 165, 167
right cerebral hemisphere and, 529–530
serotonin and, 370
temporal lobe and, 280
temporal lobe seizures and, 187
Sexual posturing, brainstem lesions and, 357–358
Sexual receptivity, 35
Shaken baby syndrome, brain development and, 667–669, 667
Shaman, of Upper Paleolithic, 273, 273
Sharks
nervous system of, 13–16, 15
optic tectum of, 15–16
striatum of, 14
Shearing forces, in head injury, 687–688
Shrinking retrograde amnesia, 211. See also Amnesia
Simultanagnosia, 146, 481
Singing, 84–85
amygdala and, 185
Single-celled organisms
evolution of, 3–4
movement of, 4
vs neuron, 4–5
semipermeable membrane of, 3
Skull, 675–676, 679, 680, 681
Skull fracture, 676–682
basilar, 677
depressed, 677–679
linear, 679–682
anosmia and, 680–682
auditory system and, 679–680
blindness and, 682
facial paralysis and, 680
Sleep, 371–374
daytime, 374–375
depression and, 598
disorders of, 374–376
motor inhibition during, 374
NREM (nonrapid eye movement), 371–372, 543
dreams of, 108
slow-wave activity of, 373–374
thalamus and, 374
walking during, 375
paradoxical, 108
PGO waves and, 372–373
REM (rapid eye movement), 371–372, 542–543
cataplexy and, 376
disorders of, 375
dreams of, 108
in infancy, 203
muscle atonia of, 374
reticular activating system and, 373
right cerebral hemisphere in, 202
sleep spindles and, 374
Sleep paralysis, 374, 375
Sleep spindles, 374
Sleep walking, 375
Social-emotional agnosia, 184–185, 618–619

Social-emotional behavior
    in blind and deaf child, *327*
    limbic nuclei in, 27–28
    striatum and, 20–21, 22, 24, 325–327
Social-emotional communication. See *Emotional-melodic speech; Limbic language*
Social-emotional deprivation, 261–266
    abuse and, 266
    amygdala and, 264–265
    in dogs, 262–263
    in humans, 262–266
    in monkeys, 262, *263*, 264, *265*
    septal nuclei and, 264–265
    temporary, 265–266
Social-emotional intelligence, 235–267. See also *Emotional intelligence; Limbic language*
    amygdala and, 239–242. See also *Amygdala*
    amygdaloid-septal interactions and, 260–261
    cingulate gyrus and, 242–246
    contact comfort and, 259–266
    female superiority in, 248–249
    free will and, 244–246
    hierarchical vocal organization and, 237–239
    infant vocalization and, 235–237
    maternal, 246–251
    paternal, 251
    periaqueductal gray and, 239
    prosodic language and, 251–253
    septal nuclei and, 257–260
    social-emotional deprivation and, 262–264, *263*, *265*
Socialization, septal nuclei and, 259–260
Sociopathology, in males, 189
Somesthetic association area, 445–451
Somesthetic receiving areas, primary, 441–443, *442*, *444, 445*. See also *Parietal lobe(s)*
Somnambulism, 375
Songs
    of birds, 88
    learning of, 84–85
Soul, 300–301. See also *Religious experience*
    lost, 301
    of Middle Paleolithic, 269–273
    Neanderthal belief in, 48
    transmigration of, 268
    wanderings of, 301–302
Sounds. See also *Auditory cortex*
    bird, 493
    environmental
        evolutionary impact of, 24
        right hemisphere recognition of, 85–86
        temporal lobe recognition of, 494–495, 500–501
    localization of, 493–495
    midbrain response to, 361–362
    neural duration of, 486–488
    production of
        limbic nuclei in, 26–27
        repto-mammalian, 25
Source amnesia, 208–209, 224
Spatial perceptual skills
    right hemisphere and, 62, 88–91

sex differences in, 89, 192
Spatial reasoning, right hemisphere and, 525–526
Speech. See also *Egocentric speech; Emotional-melodic speech; Language; Limbic language; Vocalization*
    ataxic, 386–387
    brainstem and, 358–360
    of Neanderthals, 63
    in Parkinson's disease, 337
    psychotic, 617
    in schizophrenia, 616, 617
    social, 131
    tangential, 411–412
Speech arrest, 83
Speech release. See *Confabulation*
Sphenoid bone, fracture of, 682
Spina bifida, 645–646, *646*
Spinal accessory nerve, 376
Spinal cord, P12
Spinal tube, developmental defects of, 645–646
Spirituality. See *Religious experience*
Split-brain
    in children, 115–116
    conflicts of, 102–105, 568–569
    consciousness and, 101–102
    dreaming and, 109
    information transfer in, 526
    language and, 576
    medial view of, *103*
    mental functioning of, 101–105, *102*
    motor functioning and, 104–105
Sponges, 6
Stapedius muscle, 380
Stapes, evolution of, 23–24, *23*
Stars, 313
Startle response, 370
    amygdala and, 227–228, 240
    in post-traumatic stress disorder, 588–589
Stereognosis, somesthetic association area and, 446–447
Stickiness behavior, septal nuclei and, 260
Stimulus anchor, train of thought and, 125–126
Stress. See also *Emotional trauma; Sexual abuse; Trauma*
    combat, 370, 607–608
    cortisone secretion and, 588
    depersonalization and, 548
    hallucinations and, 583
    head injury and, 697
    heart rate and, 228
    hypothalamus-pituitary-adrenal axis response to, 175, 177
    limbic system effects of, 672–673
    memory loss and, 546–549, 583–585
    norepinephrine and, 338, 370
    opiate secretion and, 585–588
    psychosis and, 369–370
    serotonin and, 338, 369–370, 589–590
    startle response and, 588–589
Stria terminalis, 179–180
    sex differences in, 179–180, 188
Striate cortex, 474–476, *475, 477*
    anatomy of, 471

columnar organization of, 472, *475*
complex cells of, 472–474
feature detectors of, 472–474, *475*
hypercomplex cells of, 474
lesions of, 476, 477–478
nonvisual interconnections of, 475–476
Stroke, 707–735
  amyloid angiopathy and, 728
  aneurysm and, 727–728
  arterial syndromes and, 729–734
  arteriovenous malformations and, 728–729
  atherosclerosis and, 713–716
  drug-induced hemorrhages and, 729
  embolism and, 718–719
  heart disease and, 722–724
  hemorrhage and, 724–726, 729–734
  hypertension and, 726–727
  lacunar, 721
  thrombosis and, 717–718
  transient ischemic attacks and, 719–721
Stromatolites, 3
Stupor, 691
Stuttering, 68
  cingulate-frontal lesions in, 245
Subfalcial herniation, 685
Substantia innominata, 323, 324, 342
  in Alzheimer's disease, 344
Substantia nigra, 323
Subthalamic nucleus, *336*, 349–351
  lesions of, 352
Suggestibility
  anesthesia-induced, 208
  in children, 556–559
  memory gaps and, 561–562
Suicide, serotonin and, 370, 591
Sulcus (sulci), fetal, 636–637
Sumerians
  mathematics of, 462
  symbol use by, 150–154, *153*
Supplementary motor areas, *334*, 345–346, 399–400
  Parkinson's disease and, 338–339
Suprachiasmatic nucleus
  circadian rhythms and, 174–175
  depression and, 175, 599
Supralaryngeal airway, evolution of, 63
Supramarginal gyrus, pain response and, 447
Surrender reaction, 334–335
Survival of the fittest, 37–38
Symbol(s)
  of bird, 275, *275*
  Cro-Magnon use of, *153*
  of cross, 49, 274, *276*, 277
  Egyptian use of, 150–154, *151*, *152*
  Neanderthal use of, 4
  Sumerian use of, 150–154, *153*
Synaptic overlay hypothesis, of repression, 537

Tactile discrimination
  right cerebral hemisphere and, 528–530
  somesthetic association area and, 446–447
Tangential speech, 411–412

Tantra, 293
Tegmentum, midbrain, 323
Telencephalic vesicles, 630, 632
Telencephalon, 14
  fetal, 632
Teleosts, nervous system of, 17
Telomerase, in tumor development, 737
Telomeres, in tumor development, 737
Temporal hemorrhage, 730
Temporal lobe(s), 483–518
  auditory association areas of, 497–500, *499*
    lesions of, 499–500
  auditory closure and, 489–490
  auditory cortex of, 483–484
    amygdala connection with, *498*, 501
    bird sound reception by, 493
    language acquisition and, 490–492
    lesions of, 495–497
    neuroanatomical organization of, 486, 489
    neuronal fine tuning of, 490–492
    phonemic analysis of, 486–489, 491–492
    spatial sound localization and, 493–495
    sustained activity of, 486–488
    tonotopic organization of, 486, 489, 491–493
  auditory transmission to, 363, 484–486, *485*
  blood flow in, in multiple personality dissociative
    disorder, 579
  cochlea transmission to, 363, 484–486, *485*
  dreaming and, 304–305, 510–511
  face recognition and, 505, 506–507
  form recognition and, 504–505
  gliosis of, 540
  hemorrhage of, 730
  hippocampus connections with, 507–508
  inferior
    memory and, 507–509
    visual discrimination and, 504–505
  laterality of, 487–488, 495, 500–501, 506–507
  left
    language and, 484, 487–488
    lesions of, 60, 142
    verbal memory and, 532
  lesions of, 495–497
    amnesia and, 540
    aphasia and, 499–500
    auditory agnosia and, 496–497
    deafness and, 495–497
    depression and, 597–598
    discrimination deficits and, 506–507
    emotional perception and, 142
    facial recognition and, 506–507
    fugue states and, 573–574
    hallucinations and, 495, 502–503, 509–510, 543,
      610–611
    mania and, 601–603
    melodic-intonational dysfunction and, 500–501
    memory loss and, 508–509, 532
    prosopagnosia and, 506–507
    psychosis and, 615–616
    receptive aphasia and, 499–500
    social-emotional agnosia and, 618–619

Temporal lobe(s), *continued*
visual discrimination deficits and, 506–507
word deafness and, 496
word recall and, 502
word recognition deficit and, 504
maturation of, 656, 657
medial, 501–503
melodic-intonational axis and, 500–501, *500*
memory and, 504, 507–509
middle, 501–504
auditory, 502–503
visual, 503–504, *503*
neoplasms of, 742
parietal lobe interaction with, 52
psychosis and, 340, 615–616
religious experience and, 273–277, 280
right
emotional memory and, 532
lesions of, 89–90
amnesia and, 540
emotional perception and, 142
mania and, 601–603
social-emotional agnosia and, 618–619
topography of, 483
tumors of, 750
visual attention and, 504–506
visual cortex of, 503–504, *503*
Wernicke's area of, 497, *498, 499*
Temporal lobe seizures, 511–513
aphasia and, 516–517
auras of, 512
automatisms of, 512
behavior during, 512–513
between-seizure psychosis in, 617–618
depressive aura in, 597
emotion and, 185–187
emotional disturbances during, 513
interictal behavior and, 517
out-of-body experience and, 284–286
psychiatric disorders and, 514–515
psychosis and, 514–518
psychosis between, 517
religious experience and, 296–297, 513–514
schizophrenia and, 514–518
sexual behavior and, 187
speech processing and, 515–516
Temporal-sequential activity
grammar and, 59, 454
language and, 59, 130, 143, 255–256
left hemisphere dominance for, 120–121
parietal lobe and, 50–52, 58–59, 123–124, 255–256,
453–454
Tentorial herniation, 685–686
Terror, amnesia after, 548–549
Testosterone
in limbic system development, 170, 190–191
in sex-related visual-spatial function, 89
Thalamic hemorrhage, 729
Thalamus, *14*, 220–221, *326, 336, 424, 425,* P7, P8
amygdala interactions with, 232
anatomy of, *220, 221, 222*

dorsal medial, 221–224, *221,* 423–424
dorsal view of, *11*
evolution of, 10–13, 21–22
in forebrain arousal, 365
geniculate nucleus of, *11*
hippocampus interactions with, 196, 232
lesions of
amnesia and, 222–224, 232
memory retrieval and, 223–224
medial view of, *12*
memory and, 221–225, 423–426
motor, 348–349, *350, 351*
repression and, 551–553
of reptiles, 21–22
sleep and, 374
Therapsids
nervous system of, 24–25
skeletal structure of, *25*
social behavior of, 27
Thinking
evolution of, 16–17
verbal, 127–128. See also *Language*
Third eye, 11–12
Thirst, hypothalamus and, 171
Thought, 127–128
Thrombus, 714–718
Throwing, of Neanderthal, 54
Thumb, of primates, 34, *34*
Time-gap experiences, 207–208
Tinnitus, 378
linear fracture and, 679–680
mild head injury and, 705
Tongue, evolution of, 13
Tonsillar herniation, 685
Tool making
Aurignacian, 57
Chatelperronian, 57
evolution of, 54–57
for gathering, 58–59
handedness and, 52–53, 122
inferior parietal lobe and, 52–53
sex differences in, 58–59
Train of associations, inferior parietal lobe and,
125–126
Trance state, sexual abuse and, 579
Transcortical aphasia, 141–143
Transient ischemic attacks, 719–721
hemispheric, 720
stroke and, 720–721
vertebral-basilar, 720
Translocation, genetic, 312
Transplantation, neuronal, 642–643
Trauma, 675–706. See also *Emotional trauma; Sexual
abuse*
amnesia after, 206–207, 230–231
amygdala and, 227–228
brainstem, 688
chest, 690
diffuse axonal injury and, 688–689
emotional sequela of, 696–698, 703–704
fetal, 651

hematoma and, 682–685
herniation and, 685–686
maternal, congenital malformations and, 654
memory and, 226, 694–696
myocardial, 690
neonatal, 664–674, 666, 667
    recovery after, 669–674
personality disturbances after, 696–698
recovery after, 669–674, 698–699
respiratory distress and, 689–690
rotation forces in, 687–688
shearing forces in, 687–688
skull, 675–682. See also *Skull fracture*
to meninges, 675
vascular, 690
Trigeminal nerve (V), 380, *683*
Trochlear nerve, *377*, 382
Tumor. See *Neoplasm(s)*

Unconscious, 564–565, 571
    traumatic memories of, 588–589
Unconscious knowledge, 208–209
Unconsciousness, anesthesia-induced, 208
Unilateral apraxia, left parietal lobe lesion and, 456–457
Universal grammar, 490
Universe, 312–314
Upper Paleolithic. See also *Middle-to-Upper Paleolithic transition*
    cave paintings of, *149*, 273
    language evolution during, 69–70
    mortuary practices of, *271*
    picture writing of, *70*
    religious experience of, 277–278
    shaman of, *273*, 273
    Venus figurines of, *272*
    wondjina figures of, *274*
Urinary incontinence, frontal lobe lesions and, 428

Vagus nerve, 376, *377*
Valvular insufficiency, ischemia and, 722
Vascular system, lesions of, 690
Vascularization
    in cerebral development, 639
    in tumor development, 737–738
Vasospasm, 723
Vegetative states, 693–694
Ventricles, *743*, P11
    neoplasms of, 743–744
Venus figurines, *272*
Verbal alexia, 155
Verbal amnesia, 207–209, 230, 539, 554–555. See also *Amnesia*
Verbal hallucinations, 495
Verbal response, on Glasgow Coma Scale, 692
Verbal thought, 127–128
Vertebral artery, 710–711, *713*
    obstruction of, 731–732
    trauma to, 733
Vertigo, 378–379
    linear fracture and, 679–680

mild head injury and, 705
Vestibular nerve, 378–379
Vestibular nucleus, 23, 363
Vestibular system, 378–379
Virgin birth, 294
Vision, 380–382, 471–482. See also *Eye movement; Occipital lobe(s)*
    cerebellum and, 387
    frontal eye fields and, 406–408
    parietal lobes and, 449
Visual agnosia, 145–146, 479–480
Visual analysis, precortical, 471–474
Visual cortex, 472–476, *473*, *477*. See also *Striate cortex*
    evolution of, 471
    lesions of, 477–479
Visual discrimination defect, temporal lobe lesions and, 506–507
Visual scanning, frontal eye fields and, 407–408
Visual system
    of primates, 32
    of thalamus, 11–12
Visual tectum, 11–12, 24
Visual-spatial mapping, hippocampus in, 10
Visual-spatial neglect, right hemisphere lesions and, 92–93, *92*
Vocalization. See also *Egocentric speech; Emotional-melodic speech; Language; Limbic language*
    amygdala and, 27–28, 185, 239–242
    anterior cingulate gyrus and, 27, 28, 245, 405
    brainstem and, 358–360
    emotional, 26–27, 527–528
    hypothalamus and, 237–238
    infant, 129–130, 235–237
    limbic nuclei in, 26–27
    mood-specific, 26–27, 244
    mother-infant, 246–248
    periaqueductal gray and, 239, 358–360
    primate recognition of, 60
    repto-mammalian, 25
    right hemisphere and, 527–528
Vomiting, in tumor development, 741
Vowels, 489

Wallace, A.R., 308, 310
Warning calls, 237
Waxy flexibility, frontal lobe lesions and, 401
Wernicke's (receptive) aphasia, 60, 76–78, 139–141, 499–500
Wernicke's area, 77, 497, *498*, 499
    axonal pathways of, *123*
    of Neanderthal, 63
Whiplash, 704–705
Whitman, Charles, 183
Whorf, Benjamin, 492
Will
    free, 244–246
    frontal lobe lesions and, 402
Witches, 289–290
Witnesses, memory of, 547, 560–561
Wondjina figures, 274

Word, context-dependent memory for, 544
Word blindness, 154–155
Word completion, in amnesiacs, 208
Word deafness, 138–139
Word finding, difficulty in, 134–135
Writing
    evolution of, 92, 150–154, *151, 152, 153*

hemisphere injury and, 90–91
    inferior parietal lobe and, 53, 453
    sex differences in, 68

Yin and Yang, 293

Zero, 462